Pre-Algebra

Solutions Manual

New York, New York
Columbus, Ohio
Chicago, Illinois
Peoria, Illinois
Woodland Hills, California

The McGraw-Hill Companies

Send all inquiries to:
Glencoe/McGraw-Hill
8787 Orion Place
Columbus, OH 43240-4027

ISBN: 0-07-827793-0

Pre-Algebra
Solutions Manual

1 2 3 4 5 6 7 8 9 10 045 07 06 05 04 03 02

CONTENTS

Chapter 1 The Tools of Algebra

Page 5 Getting Started

1. $\begin{array}{r} 6.6 \\ +\ 8.2 \\ \hline 14.8 \end{array}$

2. $\begin{array}{r} \overset{1}{4}.7 \\ +\ 8.5 \\ \hline 13.2 \end{array}$

3. $\begin{array}{r} 5.4 \\ -\ 2.3 \\ \hline 3.1 \end{array}$

4. $\begin{array}{r} \overset{7}{\not 8}.{}^{1}6 \\ -\ 4.9 \\ \hline 3.7 \end{array}$

5. $\begin{array}{r} 2.65 \\ +\ 0.3 \end{array} \rightarrow \begin{array}{r} 2.65 \\ +\ 0.30 \\ \hline 2.95 \end{array}$

6. $\begin{array}{r} 1.08 \\ +\ 1.2 \end{array} \rightarrow \begin{array}{r} 1.08 \\ +\ 1.20 \\ \hline 2.28 \end{array}$

7. $\begin{array}{r} 4.25 \\ -\ 0.7 \end{array} \rightarrow \begin{array}{r} \overset{3}{\not 4}.{}^{1}25 \\ -\ 0.70 \\ \hline 3.55 \end{array}$

8. $\begin{array}{r} 4.3 \\ -\ 2.89 \end{array} \rightarrow \begin{array}{r} \overset{3}{\not 4}.\overset{12}{\not 3}{}^{1}0 \\ -\ 2.8\ 9 \\ \hline 1.4\ 1 \end{array}$

9. $\begin{array}{r} \overset{8}{\not 9}.\overset{9}{\not 0}{}^{1}6 \\ -\ 1.1\ 8 \\ \hline 7.8\ 8 \end{array}$

10. Sample answer:
$\begin{array}{r} 1800 \\ +\ 285 \end{array} \rightarrow \begin{array}{r} 2000 \\ +\ 300 \\ \hline 2300 \end{array}$

11. Sample answer:
$\begin{array}{r} 328 \\ +\ 879 \end{array} \rightarrow \begin{array}{r} 300 \\ +\ 900 \\ \hline 1200 \end{array}$

12. Sample answer:
$\begin{array}{r} 22{,}431 \\ -\ 13{,}183 \end{array} \rightarrow \begin{array}{r} 20{,}000 \\ -\ 10{,}000 \\ \hline 10{,}000 \end{array}$

13. Sample answer:
$\begin{array}{r} 659 \\ -\ 536 \end{array} \rightarrow \begin{array}{r} 660 \\ -\ 540 \\ \hline 120 \end{array}$

14. Sample answer:
$68 \times 12 \rightarrow 70 \times 10 = 700$

15. Sample answer:
$189 \times 89 \rightarrow 200 \times 100 = 20{,}000$

16. Sample answer:
$3845 \div 82 \rightarrow 4000 \div 80 = 50$

17. Sample answer:
$21{,}789 \div 97 \rightarrow 22{,}000 \div 100 = 220$

18. Sample answer:
$\$1951 \div 49 \rightarrow \$2000 \div 50 = \$40$

19. Sample answer:
$\begin{array}{r} 8.8 \\ +\ 5.3 \end{array} \rightarrow \begin{array}{r} 9 \\ +\ 5 \\ \hline 14 \end{array}$

20. Sample answer:
$\begin{array}{r} 47.20 \\ +\ 9.75 \end{array} \rightarrow \begin{array}{r} 47 \\ +\ 10 \\ \hline 57 \end{array}$

21. Sample answer:
$\begin{array}{r} \$7.34 \\ -\ 2.16 \end{array} \rightarrow \begin{array}{r} \$7.00 \\ -\ 2.00 \\ \hline \$5.00 \text{ or } \$5 \end{array}$

22. Sample answer:
$\begin{array}{r} 83.60 \\ -\ 75.32 \end{array} \rightarrow \begin{array}{r} 84 \\ -\ 75 \\ \hline 9 \end{array}$

23. Sample answer:
$4.2 \times 29.3 \rightarrow 4 \times 30 = 120$

24. Sample answer:
$18.8(5.3) \rightarrow 20 \times 5 = 100$

25. Sample answer:
$7.8 \div 2.3 \rightarrow 8 \div 2 = 4$

26. Sample answer:
$54 \div 9.1 \rightarrow 54 \div 9 = 6$

27. Sample answer:
$21.3 \div 1.7 \rightarrow 20 \div 2 = 10$

1-1 Using a Problem-Solving Plan

Page 6 Why is it helpful to use a problem-solving plan to solve problems?

a. 0.34 0.55 0.76 0.97 1.18
(+0.21 +0.21 +0.21 +0.21)

The cost increases by \$0.21 for each additional ounce.

b. Add \$0.21 to \$1.18.

c. Sample answer: Extend the pattern.

Page 9 Check for Understanding

1. When an exact answer is not needed.
2. Sample answer: 0, 4, 8, 12, ...
3. Find the earliest departure time after 1:30 P.M. First look for a pattern in the departure times.
   Departures:
   8:45 A.M. 9:33 A.M. 10:21 A.M. 11:09 A.M.
   (+48 min +48 min +48 min)
   Each consecutive departure time is 48 min later.
   Next, extend the pattern.
   11:09 A.M. 11:57 A.M. 12:45 P.M. 1:33 P.M.
   (+48 min +48 min +48 min)
   The earliest a passenger can catch a ferry after 1:30 P.M. is 1:33 P.M.
4. 10 20 30 40 50 ?
   (+10 +10 +10 +10 +10)
   Assuming the pattern continues, the next term is $50 + 10$ or 60.
5. 37 33 29 25 21 ?
   (−4 −4 −4 −4 −4)
   Assuming the pattern continues, the next term is $21 - 4$ or 17.
6. 12 17 22 27 32 ?
   (+5 +5 +5 +5 +5)
   Assuming the pattern continues, the next term is $32 + 5$ or 37.
7. 3 12 48 192 768 ?
   (×4 ×4 ×4 ×4 ×4)
   Assuming the pattern continues, the next term is 768×4 or 3072.

$$\begin{array}{r} \$419.25 \\ 12\overline{)5031.00} \\ -48 \\ 23 \\ -12 \\ 111 \\ -108 \\ 30 \\ -24 \\ 60 \\ -60 \\ 0 \end{array}$$

8. (see division above)

Pages 9–10 Practice and Apply

9.

Age	20	25	30	35	40	45
Heart Rate (beats/min)	174	170	166	162	158	154

−4 −4 −4 −4 −4

Each consecutive heart rate decreases by 4 beats per minute. Extend the pattern backwards. The heart rate a 15-year-old should maintain while exercising at this intensity is 178 beats per min because 174 + 4 = 178.

10. (154 − 4) − 4 = 146 beats per min.

11. 2, 5, 8, 11, 14, ? (+3, +3, +3, +3, +3)

Assuming the pattern continues, the next term is 14 + 3 or 17.

12. 4, 8, 12, 16, 20, ? (+4, +4, +4, +4, +4)

Assuming the pattern continues, the next term is 20 + 4 or 24.

13. 0, 5, 10, 15, 20, ? (+5, +5, +5, +5, +5)

Assuming the pattern continues, the next term is 20 + 5 or 25.

14. 2, 6, 18, 54, 162, ? (×3, ×3, ×3, ×3, ×3)

Assuming the pattern continues, the next term is 162 × 3 or 486.

15. 54, 50, 46, 42, 38, ? (−4, −4, −4, −4, −4)

Assuming the pattern continues, the next term is 38 − 4 or 34.

16. 67, 61, 55, 49, 43, ? (−6, −6, −6, −6, −6)

Assuming the pattern continues, the next term is 43 − 6 or 37.

17. 2, 5, 9, 14, 20, ? (+3, +4, +5, +6, +7)

Assuming the pattern continues, the next term is 20 + 7 or 27.

18. 3, 5, 9, 15, 23, ? (+2, +4, +6, +8, +10)

Assuming the pattern continues, the next term is 23 + 10 or 33.

19. In the pattern, the shaded triangles move clockwise.

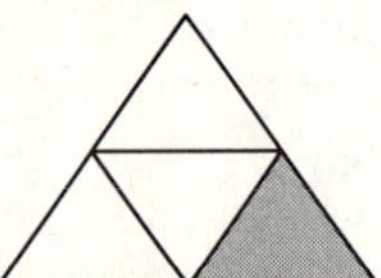

20. In the pattern, the circle is halved, then quartered, then divided into eighths.
Assuming the pattern continues, the circle will be divided into sixteenths.

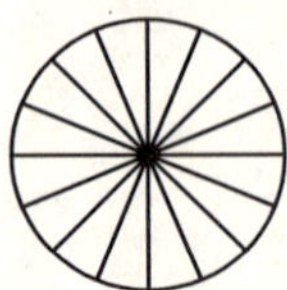

21. Since \$68 + \$15 + \$20 + \$16 = \$119,
Ryan does not have enough money for the ski trip.

22. By making a list and guessing and checking, you will find that 2 quarters, 1 dime, and 5 pennies make change for 65 cents but not for a quarter.

23. Sample answer:
The question uses the word *about*, so an estimate is all that is needed.
280 ÷ 55 → 300 ÷ 60 = 5
The car took about 5 h.

24. 100,000 × 800 = 80,000,000 jelly beans

25. Sample answer:
The question uses the word *about*, so an estimate is all that is needed.
2200 + 4700 + 12,500 + 0 + 900 + 400 + 100 + 1000 = 21,800
About 21,800 transplants were performed.

26.

Time (min)	People Notified
0	1
2	3
4	9
6	27
8	81
10	243

= 364 people

27a. There are more even products. Since any even number multiplied by any number is even, and only an odd number multiplied by an odd number is odd, there are more even products in the table. There are about 3 times as many evens as odd.

27b. Yes; in the addition table, there is only one more even number than odd.

28. A problem-solving plan helps you to be organized in solving a math problem. Answers should include the following.

- It is important to perform each step. This enables you to understand the problem and its solution.

- By estimating the answer, you will know if your answer is reasonable.

29. B; In the pattern, the shape is rotated 90° clockwise. Assuming the pattern continues, the shape will have the position in answer choice B.

30. C; Divide each of the answer choices by 8000. Only answering choice C gives you a reasonable number of working days for the year, namely 300.

31. 2.8 rounds to 3 because .8 is closer to 1 than to 0.

32. 5.2 rounds to 5 because .2 is closer to 0 than to 1.

33. 35.4 rounds to 35 because .4 is closer to 0 than to 1.

34. 49.6 rounds to 50 because .6 is closer to 1 than to 0.

35. 109.3 rounds to 109 because .3 is closer to 0 than to 1.

36. 999.9 rounds to 1000 because .9 is closer to 1 than to 0.

Page 11 Reading Mathematics

1. Sample answer: the product of 5 and 8; 2 plus 4; 16 divided by 2; the difference of 8 and 6; 2 times 5; 2 less than 5

2. c; 3 less than 9

3. e; 3 divided by 9

4. d; 9 multiplied by 3

5. a; the sum of 3 and 9

6. b; the quotient of 9 and 3

7. Sample answer: 5 plus 1; the sum of 5 and 1

8. Sample answer: the sum of 8 and 6; 6 more than 8

9. Sample answer: 9 times 5; the product of 9 and 5

10. Sample answer: 2 multiplied by 4; 2 times 4

11. Sample answer: 12 divided by 3; the quotient of 12 and 3

12. Sample answer: 20 divided by 4; the quotient of 20 and 4

13. Sample answer: 8 decreased by 7; 7 less than 8

14. Sample answer: 11 minus 5; the difference of 11 and 5

1-2 Numbers and Expressions

Page 12 Why do we need to agree on an order of operations?

a. For $1 + 2 \times 5$, multiplication then addition; for $8 - 4 \div 2$, division then subtraction; for $10 \div 5 + 14 \times 2$, division, multiplication, then addition.

b. No; $(1 + 2) \times 5 = 15$, $(8 - 4) \div 2 = 2$, $[(10 \div 5) + 14] \times 2 = 32$

c. $12 - 3 \times 2 = 6$; $16 \div 4 - 2 = 2$; $18 + 6 - 8 \div 2 \times 3 = 12$

d. Sample answer:
Multiplication and division first in the order they appear and then addition and subtraction in the order they appear.

Page 14 Check for Understanding

1. Sample answer: $(8 - 3) \cdot 2$
Simplify the expression inside the grouping symbols. Then multiply.

2. No; the value of $2 \times 4 + 3$ is 11 and the value of $2 \times (4 + 3)$ is 14.

3. Emily; she followed the order of operations and divided first.

4. Multiplication; $3 \cdot 6 - 4 = 18 - 4$
$= 14$

5. Division; $32 - 24 \div 2 = 32 - 12$
$= 20$

6. Multiplication; $5(8) + 7 = 40 + 7$
$= 47$

7. Subtraction; $6(15-4) = 6(11)$
$= 66$

8. Subtraction or addition;
$\frac{10 - 4}{1 + 2} = (10 - 4) \div (1 + 2)$
$= 6 \div 3$
$= 2$

9. Multiplication;
$11 + 56 \div (2 \cdot 7) = 11 + 56 \div (14)$
$= 11 + 4$
$= 15$

10. *Phrase* the quotient of fifteen and five
Key Word quotient
Expression $15 \div 5$

11. *Phrase* the difference of twelve and nine
Key Word difference
Expression $12 - 9$

12. 3 CDs for \$13 each and 2 cassette tapes for \$9 each
3×13 + 2×9
$(3 \times 13) + (2 \times 9) = 39 + 18$
$= 57$
The total cost is \$57.

Pages 14–16 Practice and Apply

13. $2 \cdot 6 - 8 = 12 - 8$
$= 4$

14. $12 - 3 \times 3 = 12 - 9$
$= 3$

15. $12 \div 3 + 21 = 4 + 21$
$= 25$

16. $9 + 18 \div 3 = 9 + 6$
$= 15$

17. $8 + 5(6) = 8 + 30$
$= 38$

18. $4(7) - 11 = 28 - 11$
$= 17$

19. $\frac{15+9}{32-20} = (15 + 9) \div (32 - 20)$
$= 24 \div 12$
$= 2$

20. $\frac{45-18}{9 \div 3} = (45 - 18) \div (9 \div 3)$
$= 27 \div 3$
$= 9$

21. $11(6 - 1) = 11(5)$
$= 55$

22. $(9 - 7) \cdot 13 = (2) \cdot 13$
$= 26$

23. $56 \div (7 \cdot 2) \times 6 = 56 \div (14) \times 6$
$= 4 \times 6$
$= 24$

24. $75 \div (7 + 8) - 3 = 75 \div (15) - 3$
$= 5 - 3$
$= 2$

25. $2[5(11 - 3)] - 16 = 2[5(8)] - 16$
$= 2(40) - 16$
$= 80 - 16$
$= 64$

26. $5[4 + (12 - 4) \div 2] = 5[4 + (8) \div 2]$
$= 5(4 + 4)$
$= 5(8)$
$= 40$

27. $9[(22 - 17) + 5(1 + 2)] = 9[5 + 5(3)]$
$= 9(5 + 15)$
$= 9(20)$
$= 180$

28. $10[9(2 + 4) - 6 \cdot 2] = 10[9(6) - 12]$
$= 10(54 - 12)$
$= 10(42)$
$= 420$

29. $6 + (4 \cdot 11) = 6 + 44$
$= 50$

30. $60 \div (2 + 10) = 60 \div 12$
$= 5$

31. *Phrase* six minus three
Key Word minus
Expression $6 - 3$

32. *Phrase* seven increased by two
Key Word increased
Expression $7 + 2$

33. *Phrase* nine multiplied by five
Key Word multiplied
Expression 9×5

34. *Phrase* eleven more than fifteen
Key Word more than
Expression $15 + 11$

35. *Phrase* twenty-four divided by six
Key Word divided
Expression $24 \div 6$

36. *Phrase* four less than eighteen
Key Word less than
Expression $18 - 4$

37. *Phrase* 3 notebooks at $6 each
Key Word at
Expression 3×6

38. *Phrase* the total amount of CDs if Erika has 4 and Roberto has 5
Key Word and
Expression $4 + 5$

39. 4 bags of soil at $2 | and | 2 bags of fertilizer at $13
4×2 | $+$ | 2×13

Expression: $(4 \times 2) + (2 \times 13)$

40. $(4 \times 2) + (2 \times 13) = 8 + 26$
$= 34$
The total cost is $34.

41. 3 suitcases at 57 pounds each | and | 2 sports bags at 12 pounds each
3×57 | $+$ | 2×12

Expression: $(3 \times 57) + (2 \times 12)$

42. $(3 \times 57) + (2 \times 12) = 171 + 24$
$= 195$
195 pounds is less than 200 pounds.
So, Miko's luggage is within the limit.

43. $61 - (15 + 3) = 43$
$61 - 18 = 43$
$43 = 43$

44. $12 \times 3 \div (1 + 2) = 12$
$12 \times 3 \div 3 = 12$
$36 \div 3 = 12$
$12 = 12$

45. $56 \div (2 + 6) - 4 = 3$
$56 \div 8 - 4 = 3$
$7 - 4 = 3$

46. $(5 + 2) \cdot (9 - 3) = 42$
$7 \cdot 6 = 42$
$42 = 42$

47. 50 first-place votes $= 50 \times 25$
7 second-place votes $= 7 \times 24$
4 fourth-place votes $= 4 \times 2$
3 tenth-place votes $= 3 \times 16$
$(50 \times 25) + (7 \times 24) + (4 \times 22) + (3 \times 16)$

48. $(50 \times 25) + (7 \times 24) + (4 \times 22) + (3 \times 16) =$
$1250 + 168 + 88 + 48 = 1554$

49. 0-07-825200-8

50. $[(10 \times 0) + (9 \times 0) + (8 \times 7) + (7 \times 8) + (6 \times 2) + (5 \times 5) + (4 \times 2) + (3 \times 0) + (2 \times 0) + (1 \times 8)] \div 11 = 165 \div 11 = 15$
Yes; the total of the expression is 165, and 165 can be divided by 11 with no remainder.

51. Sample answer: $111 - (1 + 1 + 1) \times (11 + 1)$
$111 - (1 + 1 + 1) \times (11 + 1) = 111 - 3 \times 12$
$= 111 - 36$
$= 75$

52. We need to agree on an order of operations so that each expression has one unique value. Answers should include the following.

- When evaluating a numerical expression, simplify any expressions inside grouping symbols. Then do all multiplication and/or division in order from left to right. Then do all addition and/or subtraction in order from left to right.
- When the order of operations is not followed, an incorrect value for the expression may result.

53. C;
$$\begin{aligned}(9 \times 3) - 63 \div 7 &= 27 - 63 \div 7\\ &= 27 - 9\\ &= 18\end{aligned}$$

54. B; the *quotient of ten and two* means to start with the 10 and divide by 2, so the expression is $\frac{10}{2}$.

Page 16 Maintain Your Skills

55. 2 4 8 16 32 ?
×2 ×2 ×2 ×2 ×2

Assuming the pattern continues, the next term is 32×2 or 64.

56. 45 42 39 36 33 ?
−3 −3 −3 −3 −3

Assuming the pattern continues, the next term is $33 - 3$ or 30.

57. 1 3 6 10 15 21 ?
+2 +3 +4 +5 +6 +7

Assuming the pattern continues, the next term is $21 + 7$ or 28.

58. 15 18 22 25 29 ?
+3 +4 +3 +4 +3

Assuming the pattern continues, the next term is $29 + 3$ or 32.

59. 100 125 150 175 ?
+25 +25 +25 +25

The pattern is to add 25. From 8 to 16 packages is another 4 steps. So, if Mrs. Lewis sells 16 packages, her bonus will be $\$175 + (4 \times \$25) = \$275$.

60. 31 million because 102 million − 71 million = 31 million

61. Sample answer: about 26 compact cars
$65{,}000 \div 2450 \approx 65{,}000 \div 2500 = 26$

62. $\begin{array}{r} \scriptstyle 1 \\ 18 \\ +\ 34 \\ \hline 52 \end{array}$

63. $\begin{array}{r} 85 \\ +\ 41 \\ \hline 126 \end{array}$

64. $\begin{array}{r} 342 \\ +\ 50 \\ \hline 392 \end{array}$

65. $\begin{array}{r} \scriptstyle 1 \\ 535 \\ +\ 28 \\ \hline 563 \end{array}$

1-3 Variables and Expressions

Page 17 How are variables used to show relationships?

a. $\$5 \cdot 10 = \$50$

b. The amount earned is five times the number of hours.

c. The amount of money earned is \$5 for each hour (h) worked or 5 h.

Page 19 Check for Understanding

1. Sample answer: $7n$ and $3x - 1$; $2 + 3$ and 3×8. $7n$ and $3x - 1$ are algebraic expressions because they contain the sums and/or products of variables *and* numbers. $2 + 3$ and 3×8 contain only numbers.

2. A variable is a place holder for a number.

3. Sample answer: $4 \times c \times d$
$4cd$ means $4 \times c \times d$

4. $$\begin{aligned}b + 6 &= 12 + 6\\ &= 18\end{aligned}$$

5. $$\begin{aligned}18 - 3c &= 18 - 3(4)\\ &= 18 - 12\\ &= 6\end{aligned}$$

6. $$\begin{aligned}\tfrac{2b}{8} &= 2b \div 8\\ &= 2(12) \div 8\\ &= 24 \div 8\\ &= 3\end{aligned}$$

7. $$\begin{aligned}5a - (b - c) &= 5(5) - (12 - 4)\\ &= 5(5) - (8)\\ &= 25 - 8\\ &= 17\end{aligned}$$

8. Let s represent the amount Kira saved.

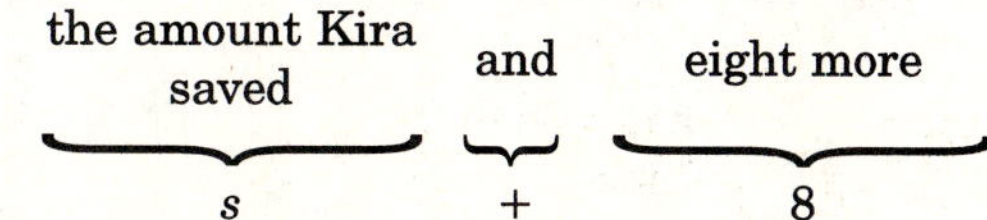

The expression is $s + 8$.

9. Let g represent goals the Pirates scored.

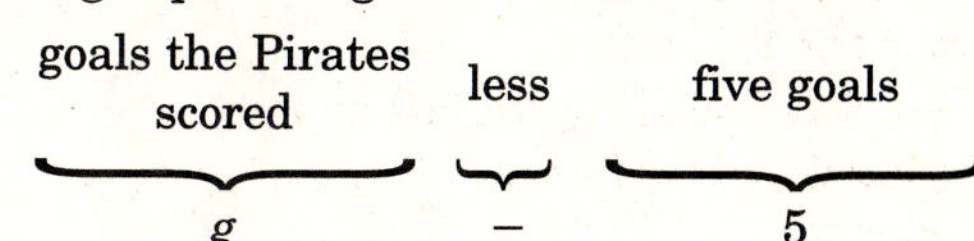

The expression is $g - 5$.

10. Let k represent the number.

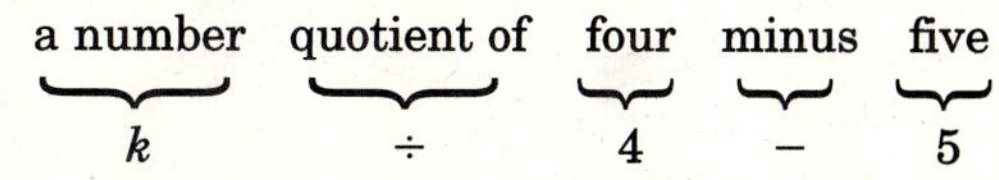

The expression is $k \div 4 - 5$.

11. Let n represent the number.

seven	increased by	a number	quotient of	eight
7	+	n	÷	8

The expression is $7 + n \div 8$.

12a. Three times as much on Earth
Let w = an object's weight on Mercury.
The expression is $3w$.

12b. $3w = 3(25)$
$= 75$ lbs

Pages 20–21 Practice and Apply

13. $z + 2 = 9 + 2$
$= 11$

14. $5 + x = 5 + 7$
$= 12$

15. $2 + 4z = 2 + 4(9)$
$= 2 + 36$
$= 38$

16. $15 - 2x = 15 - 2(7)$
$= 15 - 14$
$= 1$

17. $\frac{6y}{z} = 6y \div z$
$= 6(3) \div 9$
$= 18 \div 9$
$= 2$

18. $\frac{9x}{y} = 9x \div y$
$= 9(7) \div 3$
$= 63 \div 3$
$= 21$

19. $\frac{xy}{3} + 2 = xy \div 3 + 2$
$= 7 \cdot 3 \div 3 + 2$
$= 21 \div 3 + 2$
$= 7 + 2$
$= 9$

20. $10 - \frac{xz}{9} = 10 - (xz \div 9)$
$= 10 - (7 \cdot 9 \div 9)$
$= 10 - (7)$
$= 3$

21. $4z - 3y = 4(9) - 3(3)$
$= 36 - 9$
$= 27$

22. $3x - 2y = 3(7) - 2(3)$
$= 21 - 6$
$= 15$

23. $2x + 3z + 5y = 2(7) + 3(9) + 5(3)$
$= 14 + 27 + 15$
$= 56$

24. $5z - 3x - 2y = 5(9) - 3(7) - 2(3)$
$= 45 - 21 - 6$
$= 24 - 6$
$= 18$

25. $7z - (y + x) = 7(9) - (3 + 7)$
$= 7(9) - (10)$
$= 63 - 10$
$= 53$

26. $(8y + 5) - 2z = (8 \cdot 3 + 5) - 2(9)$
$= (24 + 5) - 2(9)$
$= 29 - 2(9)$
$= 29 - 18$
$= 11$

27. $3y + (7z - 4x) = 3 \cdot 3 + (7 \cdot 9 - 4 \cdot 7)$
$= 3 \cdot 3 + (63 - 28)$
$= 3 \cdot 3 + 35$
$= 9 + 35$
$= 44$

28. $6x - (z - 2y) + 15 = 6(7) - [9 - 2(3)] + 15$
$= 6(7) - (9 - 6) + 15$
$= 6(7) - (3) + 15$
$= 42 - 3 + 15$
$= 54$

29. $2x + (4z - 13) - 5 = 2(7) + (4 \cdot 9 - 13) - 5$
$= 2(7) + (36 - 13) - 5$
$= 2(7) + (23) - 5$
$= 14 + 23 - 5$
$= 32$

30. $9 - 3y) + 4z - 5 = (9 - 3 \cdot 3) + 4 \cdot 9 - 5$
$= (9 - 9) + 4 \cdot 9 - 5$
$= 0 + 36 - 5$
$= 31$

31. $C \div 4 + 37 = 136 \div 4 + 37$
$= 34 + 37$
$= 71°F$
The temperature is about 71°F.

32. $C \div 4 + 37 = 100 \div 4 + 37$
$= 25 + 37$
$= 62°F$
The temperature is about 62°F.

33. Let s represent Mark's salary.

Mark's salary	plus	a \$200 bonus
s	+	200

The expression is $s + \$200$.

34. Let c represent the number of cakes baked.

the number of cakes baked	(and)	three more
c	+	3

The expression is $c + 3$.

35. Let h represent the mountain's height.

the mountain's height	shorter	six feet
h	−	6

The expression is $h - 6$.

36. Let t represent Sara's time.

Sara's time	faster than	two seconds
t	+	2

The expression is $t + 2$.

37. Let q represent the number.

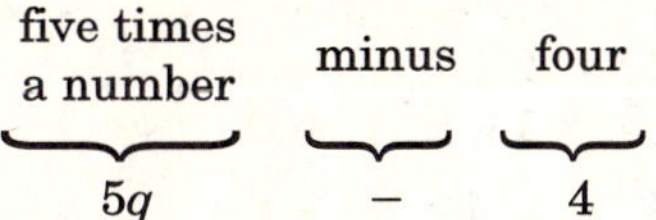

The expression is $5q - 4$.

38. Let n represent the number.

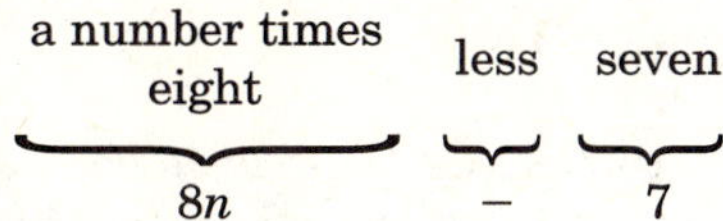

The expression is $8n - 7$.

39. Let n represent the number.

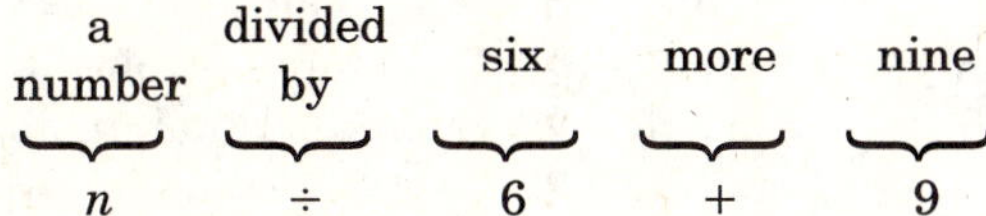

The expression is $n \div 6 + 9$.

40. Let b represent the number .

eight	quotient	twice a number
8	$\div$	$2b$

The expression is $8 \div 2b$.

41. Let w represent the number.

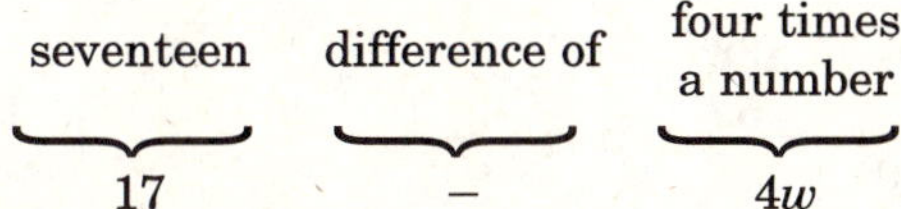

The expression is $17 - 4w$.

42. Let n represent the number.

three times		twenty-five	product	a number	
3	(	25	$\times$	n	)

The expression is $3(25 \times n)$.

43. $\frac{10mn}{3p - 3} = (10 \times 6 \times 3) \div (3 \times 7 - 3)$
$= (60 \times 3) \div (21 - 3)$
$= 180 \div 18$
$= 10$

44. $\frac{3(4a - 3b)}{b - 4} = [3(4 \times 6 - 3 \times 7)] \div (7 - 4)$
$= [3(24 - 21)] \div (3)$
$= 3 \times 3 \div 3$
$= 9 \div 3$
$= 3$

45. If x represents the *age now*, then the expression representing the *age in three years* would be $x + 3$.

46. If n represents the *number of items*, then the expression representing the *total cost* would be $5n$.

47. If $\$p$ represents the *regular price*, then the expression representing the *sale price* is $\$p - 4$.

48. Let c represent the number of premium channels.

\$32.50	and an additional	\$4.95 per premium channel
32.50	$+$	$4.95c$

The expression is $32.50 + 4.95c$.

49.

the selling price	is	the cost	plus	the markup	minus	the discount
s	$=$	c	$+$	m	$-$	d

The expression is $s = c + m - d$.

50. $s = c + m - d$
$= 25 + 20 - 6$
$= 45 - 6$
$= 39$
The selling price is \$39.

51. $6t \stackrel{?}{=} t + 5 \stackrel{?}{=} 2t + 4$
Use guess and check.
$6(1) = 1 + 5 = 2(1) + 4$
$6 = 6 = 6$
$t = 1$

52. Expressions show relationships, and the variables in the relationships are placeholders for numbers. Answers should include the following.
- Variables are placeholders that are represented by letters.
- For example, $x + y = 2$. This shows that the sum of two numbers is equal to 2.

53. D; With $c + 5 = 18$, c equals 13 because $13 + 5 = 18$.

54. B; Let n represent the number.

twice a number	less	four
$2n$	$-$	4

The expression is $2n - 4$.

Page 21 Maintain Your Skills

55. $3 + (6 \times 2) - 8 = 3 + (12) - 8$
$= 15 - 8$
$= 7$

56. $5(16 - 5 \times 3) = 5(16 - 15)$
$= 5(1)$
$= 5$

57. $36 \div (9 \cdot 2) + 7 = 36 \div 18 + 7$
$= 2 + 7$
$= 9$

58. Sample answer:
The question asks *about* how many pounds of pasta were sold, so we can estimate using rounding.
$310 + 120 + 70 + 50 + 50 + 40 + 20 = 660$
So, about 660 million pounds of pasta were sold.

59. $\begin{array}{r} \overset{4}{\cancel{5}}{}^{1}3 \\ -\ 17 \\ \hline 36 \end{array}$

60. $\begin{array}{r} \overset{8}{\cancel{9}}{}^{1}7 \\ -\ 28 \\ \hline 69 \end{array}$

61. $\begin{array}{r} 104 \\ -\ 82 \\ \hline 22 \end{array}$

62. $\begin{array}{r} 1\overset{4}{\cancel{5}}{}^{1}2 \\ -\ 123 \\ \hline 29 \end{array}$

Page 21 Practice Quiz 1

1. $4 \underset{+1}{\smile} 5 \underset{+2}{\smile} 7 \underset{+3}{\smile} 10 \underset{+4}{\smile} ?$

   Assuming the pattern continues, the next term is 10 + 4 or 14.

2. $28 \div 4 \times 2 = 7 \times 2$
   $= 14$

3. $7(3 + 10) - 2 \cdot 6 = 7(13) - 2 \cdot 6$
   $= 91 - 12$
   $= 79$

4. $3[6(12 - 3)] - 17 = 3[6(9)] - 17$
   $= 3(54) - 17$
   $= 162 - 17$
   $= 145$

5. $7x - 3y = 7(4) - 3(2)$
   $= 28 - 6$
   $= 22$

Page 22 Spreadsheet Investigation (Follow-Up of Lesson 1-3)

1. B1 = 6
   2*B1 = 2 × 6
   B2 = 12
   B2 + 6 = 12 + 6
   B3 = 18
   B3/2 = 18 ÷ 2
   B4 = 9
   B4 − B1 = 9 − 6
   = 3

2. B1 = 8
   2*B1 = 2 × 8
   B2 = 16
   B2 + 6 = 16 + 6
   B3 = 22
   B3/2 = 22 ÷ 2
   B4 = 11
   B4 − B1 = 11 − 8
   = 3

3. B1 = 25
   2*B1 = 2 × 25
   B2 = 50
   B2 + 6 = 50 + 6
   B3 = 56
   B3/2 = 56 ÷ 2
   B4 = 28
   B4 − B1 = 28 − 25
   = 3

4. B1 = 100
   2*B1 = 2 × 100
   B2 = 200
   B2 + 6 = 200 + 6
   B3 = 206
   B3/2 = 206 ÷ 2
   B4 = 103
   B4 − B1 = 103 – 100
   = 3

5. B1 = 1500
   2*B1 = 2 × 1500
   B2 = 3000
   B2 + 6 = 3000 + 6
   B3 = 3006
   B3/2 = 3006 ÷ 2
   B4 = 1503
   B4 − B1 = 1503 − 1500
   = 3

6. 3, 3

7. Let x represent the number. When you double the number, the result is $2x$. Adding 6, you get $2x + 6$. Dividing by 2, $x + 3$ results. Subtracting the original number, you get 3. Thus, regardless of what x represents, the result is always 3.

8. Sample answer: Think of a number. Double it, add 4, divide by 2, and subtract the original number. The result is always 2.

1-4 Properties

Page 23 How are real-life situations commutative?

a. 52 + 238

b. 238 + 52

c.
$$\begin{array}{r} \overset{1}{5}2 \\ +\ 238 \\ \hline 290 \end{array} \qquad \begin{array}{r} 2\overset{1}{3}8 \\ +\ 52 \\ \hline 290 \end{array}$$

   The values are the same.

d. No; the results are the same.

Page 26 Check for Understanding

1. Sample answer: $3 \cdot 4 = 4 \cdot 3$. The order in which the numbers are multiplied does not change the product.

2. The Commutative Property states that the order in which numbers are added or multiplied does not change the sum or product. The Associative Property states that the way numbers are grouped when added or multiplied does not change the sum or product.

3. Kimberly, because the Associative Property only holds true if all numbers are added or all numbers are multiplied, not a combination of the two.

4. The order of the numbers changed. This is the Commutative Property of Addition.

5. Zero is added to a number. This is the Additive Identity Property.

6. The order of the numbers changed. This is the Commutative Property of Multiplication.

7. $13 + 8 + 7 = 8 + (13 + 7)$
   $= 8 + 20$
   $= 28$

8. $6 \cdot 9 \cdot 5 = 9(6 \cdot 5)$
   $= 9(30)$
   $= 270$

9. $8 + 11 + 22 + 4 = (8 + 22) + (11 + 4)$
$= 30 + 15$
$= 45$

10. False; a counterexample is $4 \div 2 \neq 2 \div 4$.

11. $6 + (n + 7) = n + (7 + 6)$
$= n + 13$

12. $(3 \cdot w) \cdot 9 = (3 \cdot 9)w$
$= 27w$

13. $\$26 + \$12 + \$4 = \$12 + (\$26 + \$4)$
$= \$12 + \$30$
$= \$42$

Since the order in which the costs are added does not matter, the Commutative Property of Addition holds true and makes the addition easier.

Pages 26–27 Practice and Apply

14. The order of the numbers changed. This is the Commutative Property of Multiplication.

15. A number is multiplied by one. This is the Multiplicative Identity Property.

16. A number is multiplied by zero. This is the Multiplicative Property of Zero.

17. The order of the numbers changed. This is the Commutative Property of Multiplication.

18. The order of the numbers changed. This is the Commutative Property of Addition.

19. The grouping of the numbers changed. This is the Associative Property of Addition.

20. A number or variable is multiplied by one. This is the Multiplicative Identity Property.

21. Zero is added to a number or variable. This is the Additive Identity Property.

22. The grouping of the numbers or variables changed. This is the Associative Property of Addition.

23. The grouping of the numbers or variables changed. This is the Associative Property of Multiplication.

24. The grouping of the numbers or variables changed. This is the Associative Property of Multiplication.

25. The order of the numbers or variables changed. This is the Commutative Property of Addition.

26. $11 + 8 + 19 = 8 + (19 + 11)$
$= 8 + 30$
$= 38$

27. $17 + 5 + 33 = 5 + (17 + 33)$
$= 5 + 50$
$= 55$

28. $15 \cdot 0 \cdot 2 = (15 \cdot 0) \cdot 2$
$= 0 \cdot 2$
$= 0$

29. $5 + 18 + 15 + 2 = (5 + 15) + (18 + 2)$
$= 20 + 20$
$= 40$

30. $2 \cdot 7 \cdot 30 = (2 \cdot 30) \cdot 7$
$= 60 \cdot 7$
$= 420$

31. $11 \cdot 9 \cdot 10 = (9 \cdot 10) \cdot 11$
$= 90 \cdot 11$
$= 990$

32. $23 + 3 + 17 + 7 = (23 + 7) + (17 + 3)$
$= 30 + 20$
$= 50$

33. $125 \cdot 4 \cdot 0 = 125 \cdot (4 \cdot 0)$
$= 125 \cdot 0$
$= 0$

34. $16 + 57 + 94 + 33 = (16 + 94) + (57 + 33)$
$= 110 + 90$
$= 200$

35. False; because $(100 \div 10) \div 2 = 10 \div 2 = 5$ and $100 \div (10 \div 2) = 100 \div 5 = 20$, so $(100 \div 10) \div 2 \neq 100 \div (10 \div 2)$.

36. False, because $2 + 0 = 2$ and 2 is not greater than 2.

37. False, because $9 - 3 = 6$ and $3 - 9 = -6$, so $9 - 3 \neq 3 - 9$.

38. No, because the acid can be poured into the water, but the water cannot be poured into the acid.

39. $(m + 8) + 4 = m + (8 + 4)$
$= m + 12$

40. $(17 + p) + 9 = p + (17 + 9)$
$= p + 26$

41. $15 + (12 + a) = a + (15 + 12)$
$= a + 27$

42. $21 + (k + 16) = k + (21 + 16)$
$= k + 37$

43. $6 \cdot (y \cdot 2) = (6 \cdot 2) \cdot y$
$= 12y$

44. $7 \cdot (d \cdot 4) = (7 \cdot 4) \cdot d$
$= 28d$

45. $(6 \cdot c) \cdot 8 = (6 \cdot 8) \cdot c$
$= 48c$

46. $(3 \cdot w) \cdot 5 = (3 \cdot 5) \cdot w$
$= 15w$

47. $25s(3) = (25)(3)s$
$= 75s$

48. The set of whole numbers is not closed under subtraction and division because $2 - 3 = -1$ and -1 is not a whole number. Also, $1 \div 2 = 0.5$ and 0.5 is not a whole number.

49. There are many real-life situations in which the order in which things are completed does not matter, for example, reading the sports page and then the comics, or reading the comics and then the sports page. No matter the order, both parts of the newspaper will be read. When washing clothes, you would add the detergent and then wash the clothes, not wash the clothes and then add the detergent. Order matters.

50. A; the order of the numbers has changed. This is the Commutative Property of Addition.

51. B; the Associative Property of Multiplication allows us to regroup numbers.
$(7 \cdot m) \cdot 8 = 7 \cdot (m \cdot 8)$

Page 27 Maintain Your Skills

52. $a + c - b = 6 + 5 - 4$
$= 11 - 4$
$= 7$

53. $8a - 3b = 8 \cdot 6 - 3 \cdot 4$
$= 48 - 12$
$= 36$

54. $4a - (b + c) = 4 \cdot 6 - (4 + 5)$
$= 4 \cdot 6 - (9)$
$= 24 - 9$
$= 15$

55. *Phrase* the difference of w and 12
Key Word difference
Expression $w - 12$

56. $7 - 2 \times 3 = 7 - 6$
$= 1$

57. $21 \div 3 \times 5 = 7 \times 5$
$= 35$

58. $4 \cdot (8 + 9) + 6 = 4 \cdot (17) + 6$
$= 68 + 6$
$= 74$

59. 0, 1, 3, 6, 10, ? (+1, +2, +3, +4, +5)

Assuming the pattern continues, the next term is 10 + 5 or 15, and the next term after that is 15 + 6 or 21.

60. $48 \times 5 = 240$

61. $37 \times 8 = 296$

62. 16×12: 32, 16, 192

63. 25×42: 50, 100, 1050

64. 106×13: 318, 106, 1378

65. 127×59: 1143, 635, 7493

1-5 Variables and Equations

Page 28 How is solving an open sentence similar to evaluating an expression?

a. *Words* Emilio is seven years older than Rebecca.
Variable Let x represent Rebecca's age.
Expression $x + 7$

b. $x + 7$ and 19

c. Find the number that added to 7 is 19.
$12 + 7 = 19$

Page 30 Check for Understanding

1. Sample answer: $b + 7 = 12$ and $8 - h = 3$.
If $b = 5$, then $5 + 7 = 12$, which is true.
If $h = 5$, then $8 - 5 = 3$, which is true.

2. To solve an equation, find a value for the variable that makes a true statement.

3.

Value for h	$h + 15 = 21$	True or False?
5	$5 + 15 \stackrel{?}{=} 21$	false
6	$6 + 15 \stackrel{?}{=} 21$	true ✓
7	$7 + 15 \stackrel{?}{=} 21$	false

Therefore, the solution of $h + 15 = 21$ is 6.

4.

Value for m	$13 - m = 4$	True or False?
7	$13 - 7 \stackrel{?}{=} 4$	false
8	$13 - 8 \stackrel{?}{=} 4$	false
9	$13 - 9 \stackrel{?}{=} 4$	true ✓

Therefore, the solution of $13 - m = 4$ is 9.

5. $a + 8 = 13$
$5 + 8 = 13$
$a = 5$

6. $12 - d = 9$
$12 - 3 = 9$
$d = 3$

7. $3x = 18$
$3 \cdot 6 = 18$
$x = 6$

8. $4 = \frac{36}{t}$
$4 = \frac{36}{9}$
$t = 9$

9. If $a = b$, then $b = a$. This is the Symmetric Property of Equality.

10. If $a = b$ and $b = c$, then $a = c$. This is the Transitive Property of Equality.

11. Let n = the number.

$\underbrace{\text{A number increased by 8}}_{n + 8}\ \underbrace{\text{is}}_{=}\ \underbrace{23}_{23}$

$n + 8 = 23$
$15 + 8 = 23$
$n = 15$

12. Let d = the number.

$\underbrace{\text{Twenty-five}}_{25}\ \underbrace{\text{is}}_{=}\ \underbrace{\text{10 less than a number}}_{d - 10}$

$25 = d - 10$
$25 = 35 - 10$
$d = 35$

13. C;

Value for k	$6 = \frac{48}{k}$	True or False
6	$6 \stackrel{?}{=} \frac{48}{6}$	false
7	$6 \stackrel{?}{=} \frac{48}{7}$	false
8	$6 \stackrel{?}{=} \frac{48}{8}$	true ✓
12	$6 \stackrel{?}{=} \frac{48}{12}$	false

Therefore, the solution of $6 = \frac{48}{k}$ is 8.

Pages 31–32 Practice and Apply

14.

Value for c	$c + 12 = 30$	True or False?
8	$8 + 12 \stackrel{?}{=} 30$	false
16	$16 + 12 \stackrel{?}{=} 30$	false
18	$18 + 12 \stackrel{?}{=} 30$	true ✓

Therefore, the solution of $c + 12 = 30$ is 18.

15.

Value for g	$g + 17 = 28$	True or False?
9	$9 + 17 \stackrel{?}{=} 28$	false
11	$11 + 17 \stackrel{?}{=} 28$	true ✓
13	$13 + 17 \stackrel{?}{=} 28$	false

Therefore, the solution of $g + 17 = 28$ is 11.

16.

Value for m	$23 - m = 14$	True or False?
7	$23 - 7 \stackrel{?}{=} 14$	false
9	$23 - 9 \stackrel{?}{=} 14$	true ✓
11	$23 - 11 \stackrel{?}{=} 14$	false

Therefore, the solution of $23 - m = 14$ is 9.

17.

Value for k	$18 - k = 6$	True or False?
8	$18 - 8 \stackrel{?}{=} 6$	false
10	$18 - 10 \stackrel{?}{=} 6$	false
12	$18 - 12 \stackrel{?}{=} 6$	true ✓

Therefore, the solution of $18 - k = 6$ is 12.

18.

Value for k	$14k = 42$	True or False?
2	$14(2) \stackrel{?}{=} 42$	false
3	$14(3) \stackrel{?}{=} 42$	true ✓
4	$14(4) \stackrel{?}{=} 42$	false

Therefore, the solution of $14k = 42$ is 3.

19.

Value for n	$75 = 15n$	True or False?
3	$75 \stackrel{?}{=} 15(3)$	false
4	$75 \stackrel{?}{=} 15(4)$	false
5	$75 \stackrel{?}{=} 15(5)$	true ✓

Therefore, the solution of $75 = 15n$ is 5.

20.

Value for z	$\frac{51}{z} = 3$	True or False?
15	$\frac{51}{15} \stackrel{?}{=} 3$	false
16	$\frac{51}{16} \stackrel{?}{=} 3$	false
17	$\frac{51}{17} \stackrel{?}{=} 3$	true ✓

Therefore, the solution to $\frac{51}{z} = 3$ is 17.

21.

Value for p	$\frac{60}{p} = 4$	True or False?
15	$\frac{60}{15} \stackrel{?}{=} 4$	true ✓
16	$\frac{60}{16} \stackrel{?}{=} 4$	false
17	$\frac{60}{17} \stackrel{?}{=} 4$	false

Therefore, the solution to $\frac{60}{p} = 4$ is 15.

22.

Value for n	$3n + 13 = 25$	True or False?
2	$3(2) + 13 \stackrel{?}{=} 25$	false
3	$3(3) + 13 \stackrel{?}{=} 25$	false
4	$3(4) + 13 \stackrel{?}{=} 25$	true ✓

Therefore, the solution to $3n + 13 = 25$ is 4.

23.

Value for w	$7 = 4w - 29$	True or False?
8	$7 \stackrel{?}{=} 4(8) - 29$	false
9	$7 \stackrel{?}{=} 4(9) - 29$	true ✓
10	$7 \stackrel{?}{=} 4(10) - 29$	false

Therefore, the solution is $7 = 4w - 29$ is 9.

24. Sometimes; an equation is an open sentence when it contains a variable.

25. Always; the definition of an open sentence states that it contains a variable.

26. $d + 7 = 12$
$5 + 7 = 12$
$d = 5$

27. $19 = 4 + y$
$19 = 4 + 15$
$y = 15$

28. $8 + j = 27$
$8 + 19 = 27$
$j = 19$

29. $22 + b = 22$
$22 + 0 = 22$
$b = 0$

30. $20 - p = 11$
$20 - 9 = 11$
$p = 9$

31. $15 - m = 0$
$15 - 15 = 0$
$m = 15$

32. $16 = x - 7$
$16 = 23 - 7$
$x = 23$

33. $12 = y - 5$
$12 = 17 - 5$
$y = 17$

34. $7s = 49$
$7 \cdot 7 = 49$
$s = 7$

35. $8c = 88$
$8 \cdot 11 = 88$
$c = 11$

36. $63 = 9h$
$63 = 9 \cdot 7$
$h = 7$

37. $72 = 8w$
$72 = 8 \cdot 9$
$w = 9$

38. $\frac{30}{r} = 3$
$\frac{30}{10} = 3$
$r = 10$

39. $\frac{24}{y} = 8$
$\frac{24}{3} = 8$
$y = 3$

40. $12 = \frac{36}{p}$
$12 = \frac{36}{3}$
$p = 3$

41. $14 = \frac{56}{d}$
$14 = \frac{56}{4}$
$d = 4$

42. Let k = the number.

$$\underbrace{\text{The sum of 7 and a number}}_{7 + k} \quad \underbrace{\text{is}}_{=} \quad \underbrace{23.}_{23}$$

$7 + k = 23$
$7 + 16 = 23$
$k = 16$

43. Let h = the number.

$$\underbrace{\text{A number minus 10}}_{h - 10} \quad \underbrace{\text{is}}_{=} \quad \underbrace{27.}_{27}$$

$h - 10 = 27$
$37 - 10 = 27$
$h = 37$

44. Let a = the number.

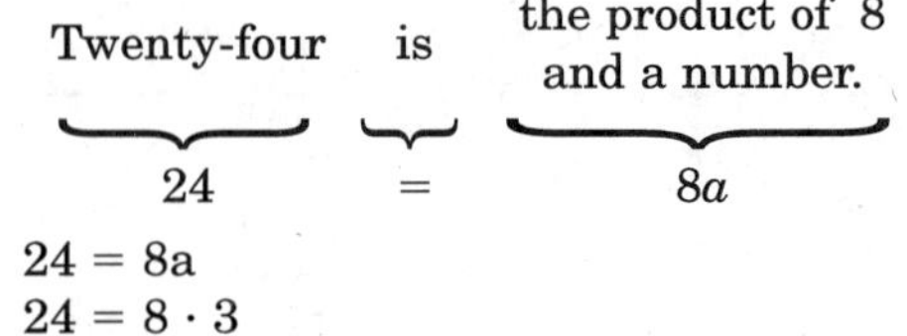

$24 = 8a$
$24 = 8 \cdot 3$
$a = 3$

45. Let w = the number.

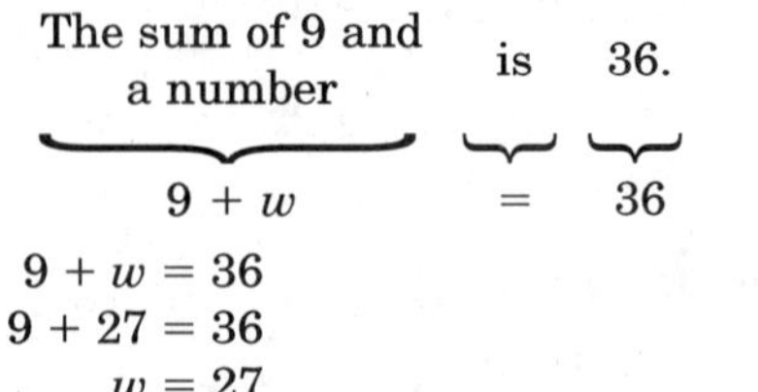

$9 + w = 36$
$9 + 27 = 36$
$w = 27$

46. Let z = the number.

$$\underbrace{\text{The difference of a number and 12}}_{z - 12} \quad \underbrace{\text{is}}_{=} \quad \underbrace{54.}_{54}$$

$z - 12 = 54$
$66 - 12 = 54$
$z = 66$

47. Let x = the number.

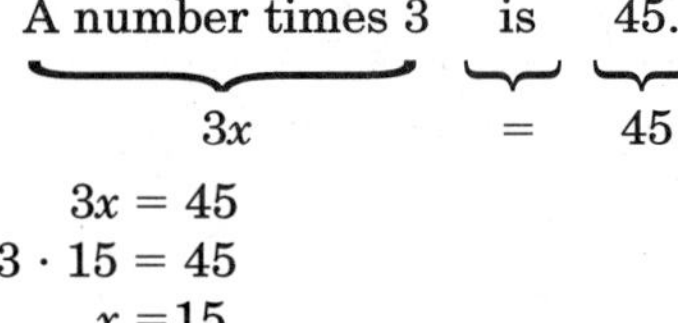

$3x = 45$
$3 \cdot 15 = 45$
$x = 15$

48. Let c = the cost of each ticket.
$3c = 24$

49. $3c = 24$
$3 \cdot 8 = 24$
$c = 8$
The cost of each ticket was $8.

50. If $a = b$ and $b = c$, then $a = c$. This is the Transitive Property of Equality.

51. If $a = b$, then $b = a$. This is the Symmetric Property of Equality.

52. If $a = b$, then $b = a$. This is the Symmetric Property of Equality.

53. If $a = b$, then $b = a$. This is the Symmetric Property of Equality.

54. Let h = the increase in height.
$65 + h = 68$

55. $65 + h = 68$
$65 + 3 = 68$
$h = 3$
Sean grew 3 inches.

56. Sample answer:

$4a = 21$

$a = \frac{21}{4}$; not a whole number

$7z = 3$

$z = \frac{3}{7}$; not a whole number

$14 = 5g$

$g = \frac{14}{5}$; not a whole number

57. Sample answer: Once the variable(s) are replaced in the open sentence, the order of operations is used to find the value of the expression. Answers should include the following.

- To evaluate an expression, replace the variable(s) with the given values, and then find the value of the expression.
- To solve an open sentence, find the value of the variable that makes the sentence true.

58. D;

Value for m	$9m = 54$	True or False?
4	$9 \cdot 4 \stackrel{?}{=} 54$	false
7	$9 \cdot 7 \stackrel{?}{=} 54$	false
5	$9 \cdot 5 \stackrel{?}{=} 54$	false
6	$9 \cdot 6 \stackrel{?}{=} 54$	true ✓

Therefore, the solution of $9m = 54$ is 6.

59. B;

Value for n	$2n - 5 = 19$	True or False?
11	$2 \cdot 11 - 5 \stackrel{?}{=} 19$	false
12	$2 \cdot 12 - 5 \stackrel{?}{=} 19$	true ✓
13	$2 \cdot 13 - 5 \stackrel{?}{=} 19$	false
14	$2 \cdot 14 - 5 \stackrel{?}{=} 19$	false

Therefore, the solution of $2n - 5 = 19$ is 12.

60a. See students' work.

60b. For equations like those in the table, an equation with one variable has one unique solution; an equation with more than one variable has more than one solution.

Page 32 Maintain Your Skills

61. $16 + (7 + d) = (16 + 7) + d$
$= 23 + d$

62. $(4 \cdot p) \cdot 6 = (4 \cdot 6) \cdot p$
$= 24p$

63. Let n represent the number.

$\underbrace{\text{ten}}_{10}\quad \underbrace{\text{decreased by}}_{-}\quad \underbrace{\text{a number}}_{n}$

The expression is $10 - n$.

64. Let n represent the number.

$\underbrace{\text{three times a number}}_{3n}\quad \underbrace{\text{and}}_{+}\quad \underbrace{\text{four}}_{4}$

The expression is $3n + 4$.

65. $3 \cdot 7 - 2(1 + 4) = 3 \cdot 7 - 2(5)$
$= 21 - 10$
$= 11$

66. $3[(17 - 7) - 2(3)] = 3[(10) - 2(3)]$
$= 3(10 - 6)$
$= 3(4)$
$= 12$

67. 67 62 57 52 47 ? (each step −5)

Assuming the pattern continues, the next term is $47 - 5$ or 42.

68. $4x; x = 3$
$4(3) = 12$

69. $3m; m = 6$
$3(6) = 18$

70. $2d; d = 8$
$2(8) = 16$

71. $5c; c = 10$
$5(10) = 50$

72. $8a; a = 9$
$8(9) = 72$

73. $6y; y = 15$
$6(15) = 90$

Page 32 Practice Quiz 2

1. A number is multiplied by one. This is the Multiplicative Identity Property.

2. The order of the numbers changed. This is the Commutative Property of Addition.

3. $8 \cdot (h \cdot 3) = (8 \cdot 3) \cdot h$
$= 24h$

4.

Value for w	$2w - 6 = 14$	True or False?
8	$2(8) - 6 \stackrel{?}{=} 14$	false
10	$2(10) - 6 \stackrel{?}{=} 14$	true ✓
12	$2(12) - 6 \stackrel{?}{=} 14$	false

Therefore, the solution to $2w - 6 = 14$ is 10.

5. $72 = 9x$
$72 = 9 \cdot 8$
$x = 8$

1-6 Ordered Pairs and Relations

Page 33 How are ordered pairs used to graph real-life data?

a. Sample answer: 3 over and 2 up; this move would block Hiroshi.

b. (3, 2)

c. In (5, 1), you go over 5 and up 1. In (1, 5), you go over 1 and up 5.

d. See students' work. It depends on where the third X is placed.

e. See students' work.

Page 36 Check Understanding

1. Sample answer: (3, 5); the x-coordinate is 3 and the y-coordinate is 5.
2. as a set of ordered pairs, in a table, in a graph
3. The domain of a relation is the set of x-coordinates. The range is the set of y-coordinates.
4. Start at the origin. Move 5 units right and 3 units up.

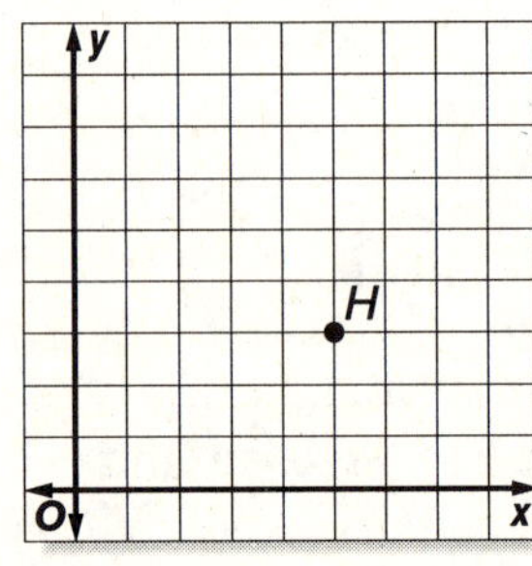

5. Start at the origin. Move 6 units right. Since the y-coordinate is 0, do not move up or down.

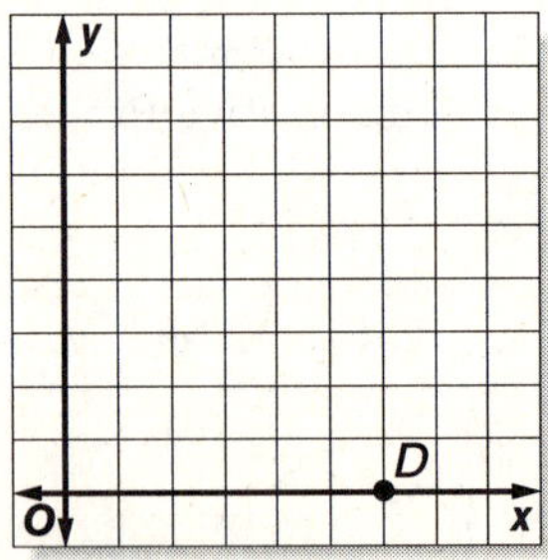

6. Start at the origin, move 2 units right and 3 units up. The ordered pair for point Q is (2, 3).
7. Start at the origin, move 6 units right and 5 units up. The ordered pair for point P is (6, 5).
8.

x	y
2	5
0	2
5	5

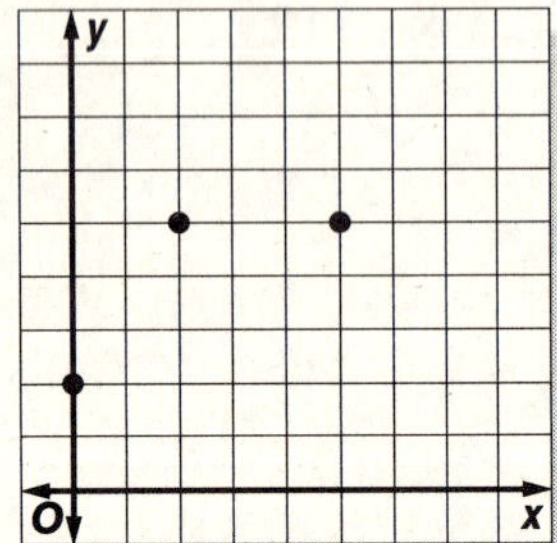

The domain is the set of x-coordinates: {2, 0, 5}
The range is the set of y-coordinates: {5, 2}

9.

x	y
1	6
6	4
0	2
3	1

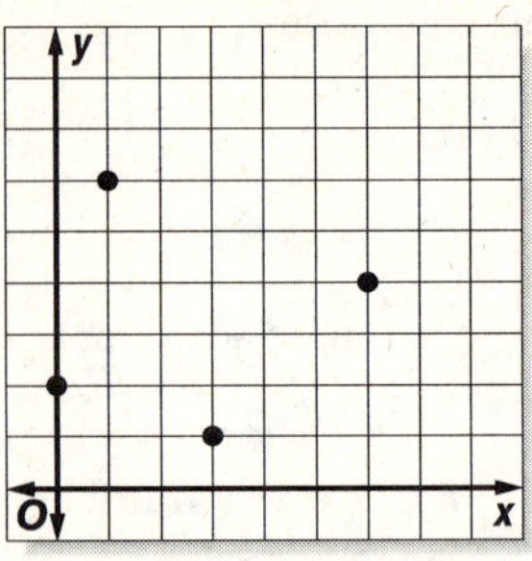

The domain is the set of x-coordinates: {1, 6, 0, 3}
The range is the set of y-coordinates: {6, 4, 2, 1}

10. y is always 4 times as much as x.

x	y
2	8
4	16
5	20

11.

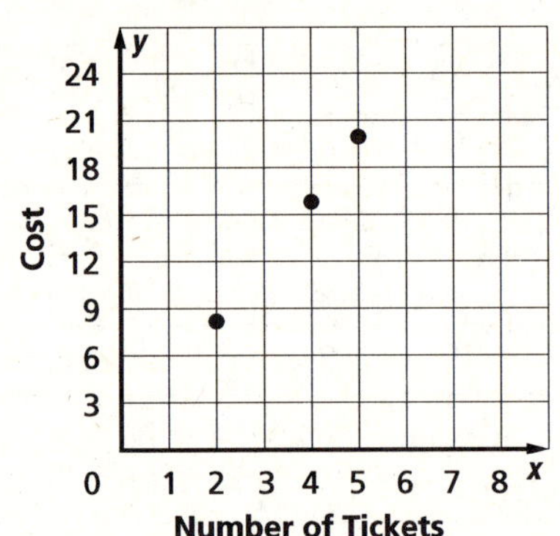

Pages 36–38 Practice and Apply

12. Start at the origin. Move 3 units right and 3 units up.

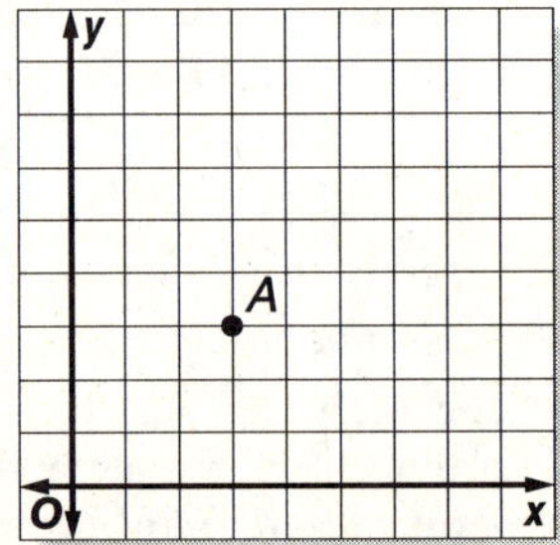

13. Start at the origin. Move 1 unit right and 8 units up.

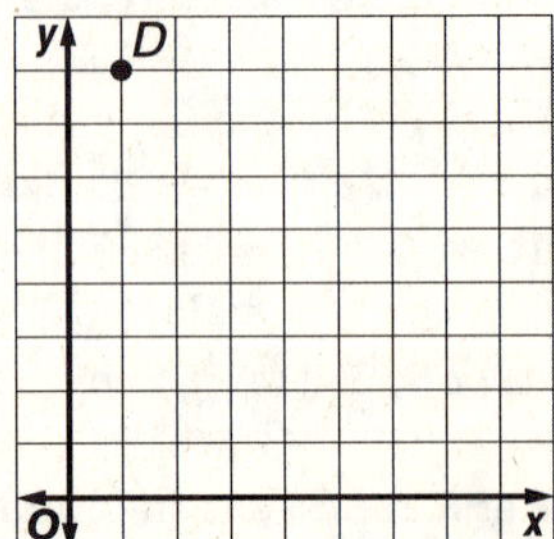

14. Start at the origin. Move 2 units right and 7 units up.

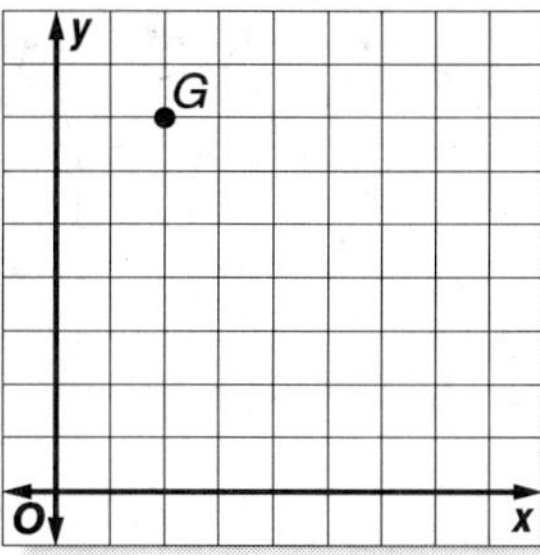

15. Start at the origin. Move 7 units right and 2 units up.

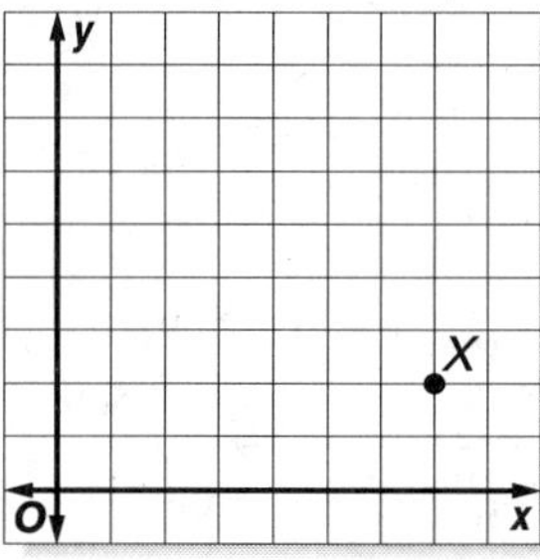

16. Start at the origin. Move 6 units up. Since the x-coordinate is 0, do not move right or left.

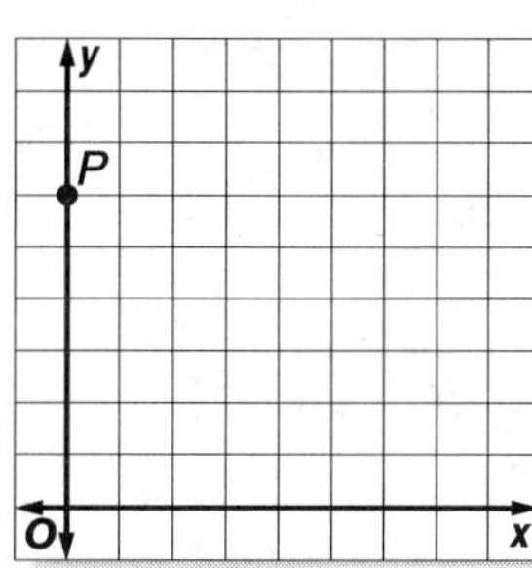

17. Start at the origin. Move 4 units right. Since the y-coordinate is 0, do not move up or down.

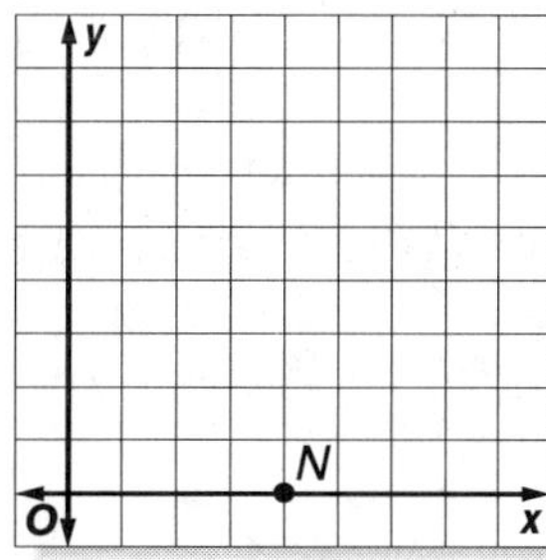

18. Start at the origin, and move 1 unit right and 7 units up. The ordered pair for point C is (1, 7).

19. Start at the origin, and move 7 units right and 3 units up. The ordered pair for point J is (7, 3).

20. Start at the origin, and move 4 units up. The ordered pair for point N is (0, 4).

21. Start at the origin, and move 6 units right and 6 units up. The ordered pair for point T is (6, 6).

22. Start at the origin, and move 2 units right and 1 unit up. The ordered pair for point Y is (2, 1).

23. Start at the origin, and move 3 units right and 4 units up. The ordered pair for point B is (3, 4).

24. (0, 0), the point at which the axes intersect is called the origin.

25. Points of the form (x, 0) lie on the x-axis; points of the form (0, y) lie on the y-axis.

26. 3×12 ft/sec = 36 ft
5×12 ft/sec = 60 ft
7×12 ft/sec = 84 ft

27. The ordered pairs are (3, 36), (5, 60), and (7, 84).

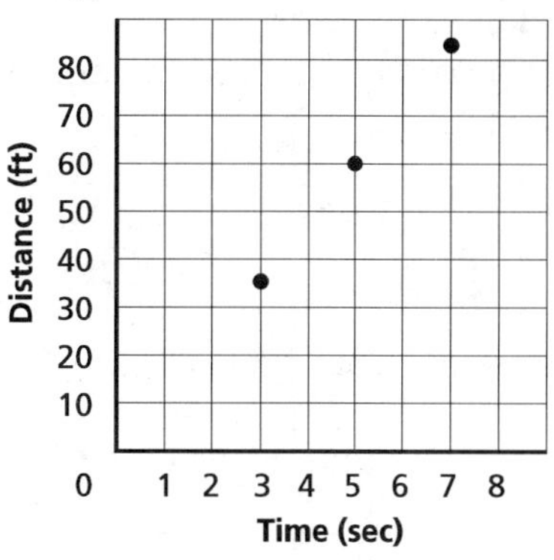

28. In each ordered pair, the first number is the bounce number and the second number is the height. (0, 100), (1, 50), (2, 25), (3, 13), (4, 6)

29.

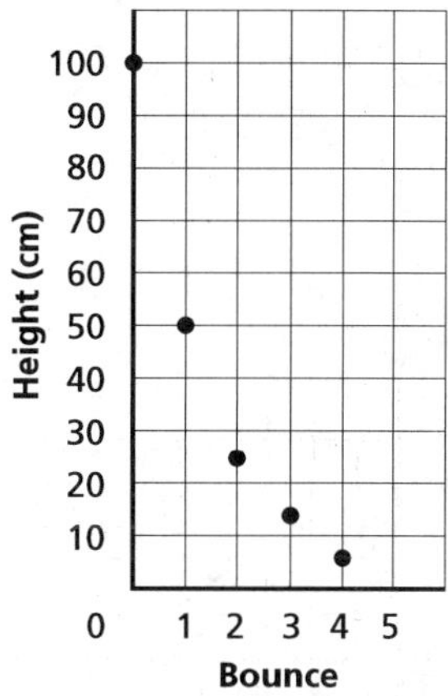

30. The height of each successive bounce is about half of the previous. Since the 4th bounce is 6 cm, the 5th bounce will be about $6 \div 2$ or 3 cm.

31.

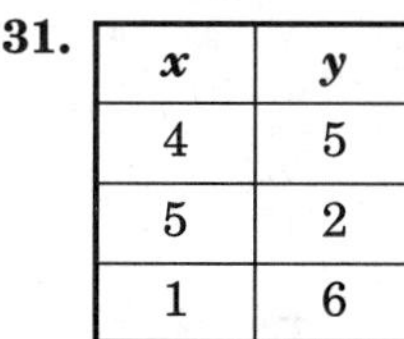

x	y
4	5
5	2
1	6

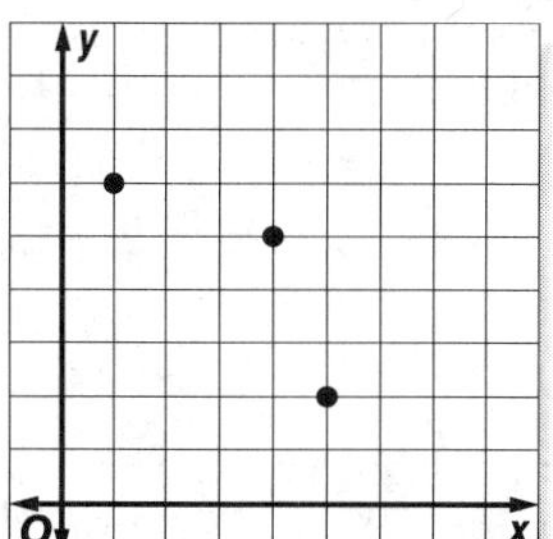

domain = {4, 5, 1}; range = {5, 2, 6}

32.

x	y
6	8
2	9
0	1

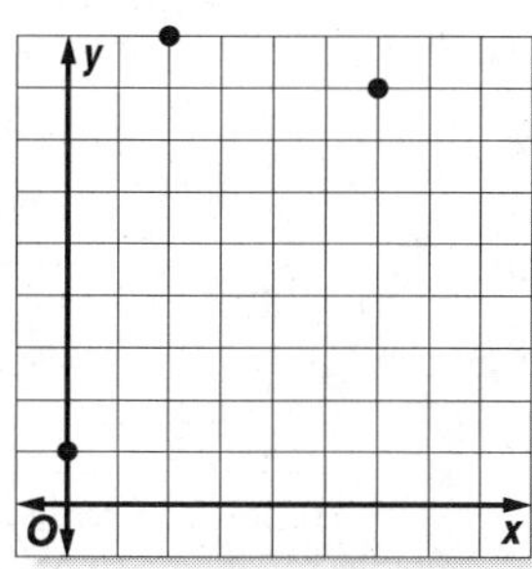

domain = {6, 2, 0}; range = {8, 9, 1}

33.

x	y
7	0
3	2
4	4
5	1

domain = {7, 3, 4, 5}; range = {0, 2, 4, 1}

34.

x	y
2	4
1	3
5	6
1	1

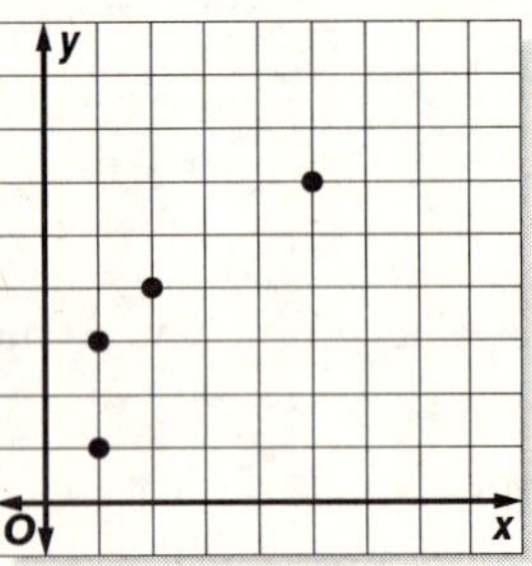

domain = {2, 1, 5}; range = {4, 3, 6, 1}

35.

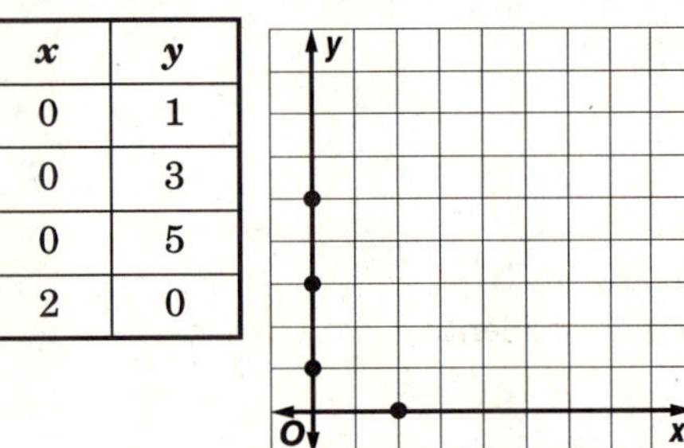

x	y
0	1
0	3
0	5
2	0

domain = {0, 2}; range = {1, 3, 5, 0}

36.

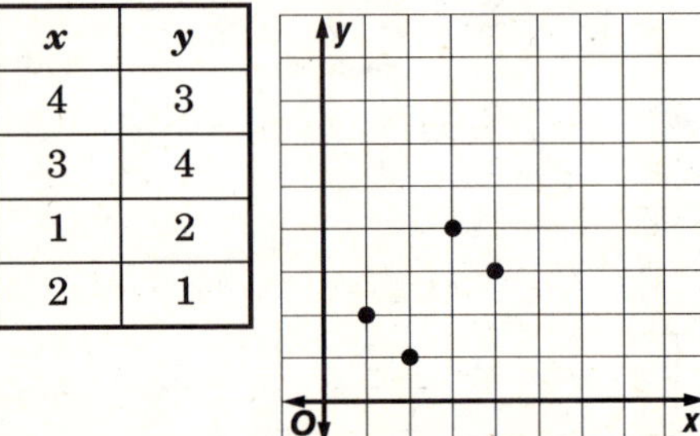

x	y
4	3
3	4
1	2
2	1

domain = {4, 3, 1, 2}; range = {3, 4, 2, 1}

37. In each ordered pair, the first number is the height and the second number is the pressure. (0, 14.7), (1, 10.2), (2, 6.4), (3, 4.3), (4, 2.7), (5, 1.6)

38.

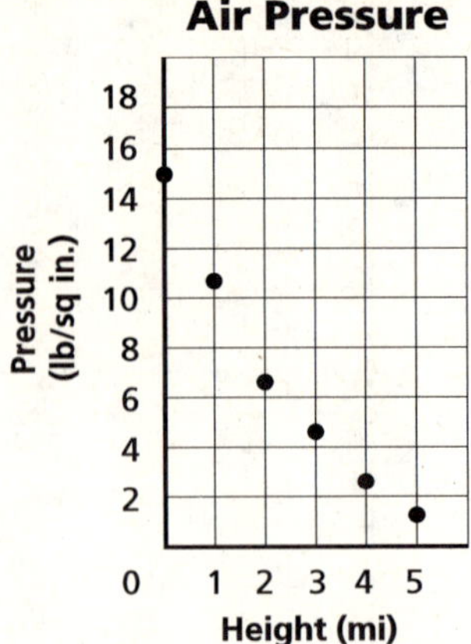

39. domain = {0, 1, 2, 3, 4, 5};
range = {14.7, 10.2, 6.4, 4.3, 2.7, 1.6}

40.

x	y
0	100
1	95
2	90
3	85
4	80
5	75

41. {(0, 100), (1, 95), (2, 90), (3, 85), (4, 80), (5, 75)}

42.

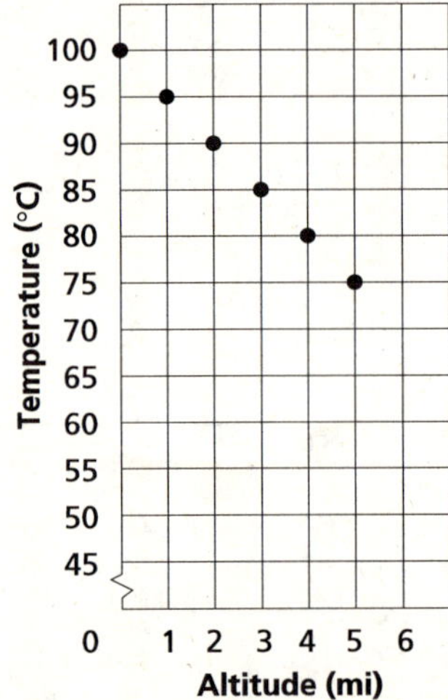

43. $7200 \div 5280 \approx 1.4$
Albuquerque is about 1.4 miles above sea level.
$1.4 \times 5 = 7$
The boiling point is at about 7° below 100°.
$100 - 7 = 93$°C
Water boils at about 93°C.
$4490 \div 5280 \approx 0.85$
Alpine is about 0.85 miles above sea level.
$0.85 \times 5 \approx 4$
The boiling point is at about 4° below 100°.
$100 - 4 = 96$°C
Water boils at about 96°C.

44. The points of the form $(4, y)$ all lie along the vertical line where $x = 4$.

45. Sample answer: Ordered pairs can be used to graph real-life data by expressing the data as ordered pairs and then graphing the ordered pairs. Answers should include the following.

- The x- and y-coordinate of an ordered pair specifies the point on the graph.
- Ordered pairs can be used to graph longitude and latitude lines.

46a. The x- and y-coordinates form the sequence 1, 2, 3 … . The next ordered pair is (9, 10).

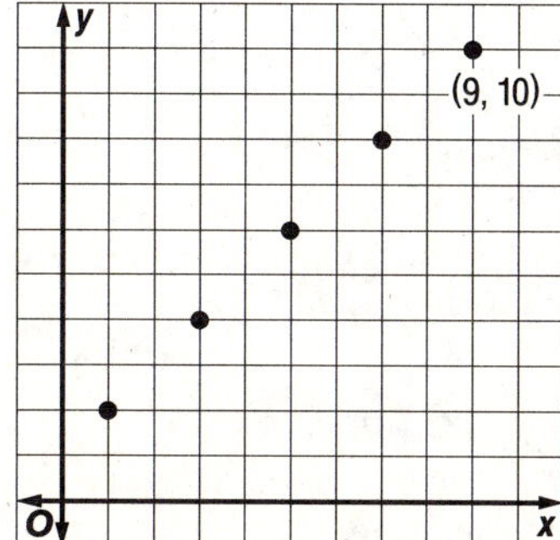

46b. The x-coordinates decrease by 2 as the y-coordinates increase by 2, so the sum of the coordinates is always 10. The next ordered pair is (8, 2).

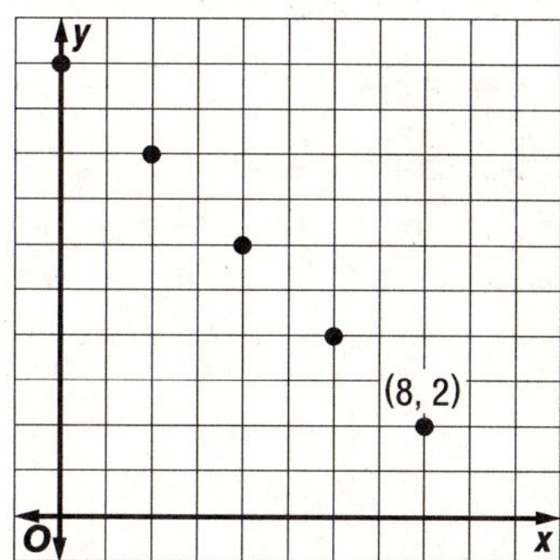

47. D; The domain is all the x-coordinate values, so the correct answer is {1, 2, 5, 6, 7}.

48. D; For all points on the graph the x-coordinate varies and the y-coordinate stays the same.

49a.

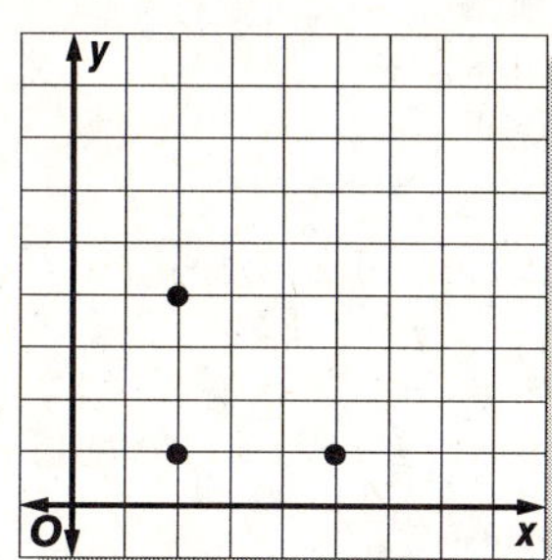

49b. a triangle

49c. (2, 1), (2, 4), (5, 1)

↓ ↓ ↓

(4, 2), (4, 8), (10, 2)

49d. a triangle

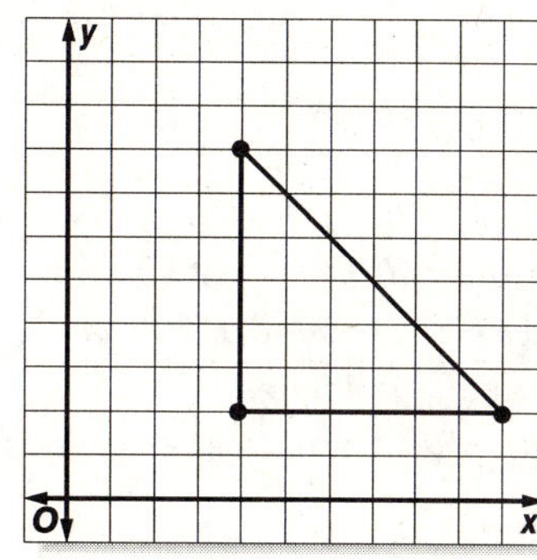

49e. The figures have the same shape but not the same size. The new triangle is twice as large.

Page 38 Maintain Your Skills

50. $a + 6 = 17$
$11 + 6 = 17$
$a = 11$

51. $7t = 42$
$7 \cdot 6 = 42$
$t = 6$

52. $\frac{54}{n} = 6$
$\frac{54}{9} = 6$
$n = 9$

53. A number is multiplied by 1. This is the Multiplicative Identity Property.

54. $ca - cb = (3)(5) - (3)(1)$
$= 15 - 3$
$= 12$

55. $5a - 6c = 5 \cdot 5 - 6 \cdot 3$
$= 25 - 18$
$= 7$

56. *Phrase* fifteen less than twenty-one
Key Word less than
Expression $21 - 15$

57. *Phrase* the product of ten and thirty
Key Word product
Expression $10 \cdot 30$

58. $2\overline{)74}$ = 37
-6
14
-14
0

59. $8\overline{)96}$ = 12
-8
16
-16
0

60. $3\overline{)102}$ = 34
-9
12
-12
0

61. $4\overline{)112}$ = 28
-8
32
-32
0

62. $16\overline{)80}$ = 5
-80
0

63. $13\overline{)91}$ = 7
-91
0

64. $22\overline{)132}$ = 6
$- 132$
0

65. $17\overline{)153}$ = 9
$- 153$
0

Page 39 Algebra Activity (Preview of Lesson 1-7)

1. Yes; as height increases, arm span increases.
2. about 60 inches; about 72 inches
3. A person's arm span is about equal to his/her height.
4. $y = x$
5. See students' work; generally, as height increases, shoe length increases.

1-7 Scatter Plots

Page 40 How can scatter plots help spot trends?

a. Sample answer: Consumers are buying fewer movies on videocassette.

b. Look for a pattern.

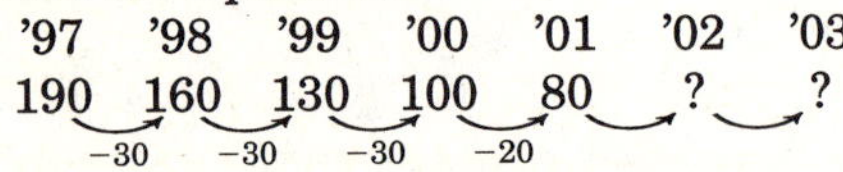

Assume a decrease of about 20 for future years.
2003 is 2 steps down from 2000.
$80 - (2 \cdot 20) = 40$
About 40 movies will be sold in 2003.

Page 42 Check for Understanding

1. Sample answer: A scatter plot can be used to make predictions, draw conclusions, and to spot trends.
2. Sample answer:

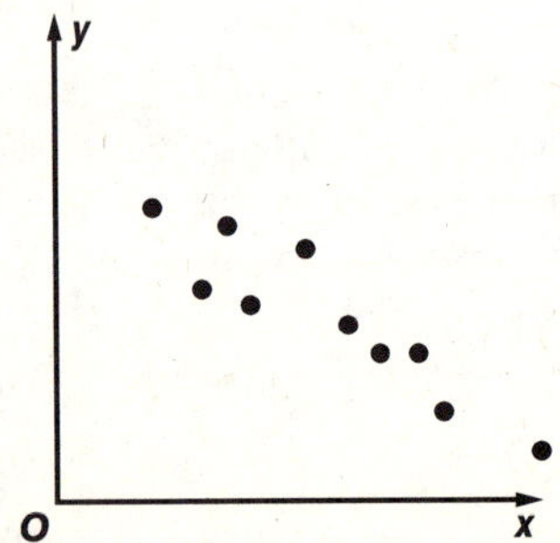

The downward slant of the pattern represents a negative relationship.

3. A scatter plot can show a positive relationship, a negative relationship, or no relationship.
4. Positive; as the number of hours worked increases, so do the earnings.
5. No; hair color is not related to height.
6.

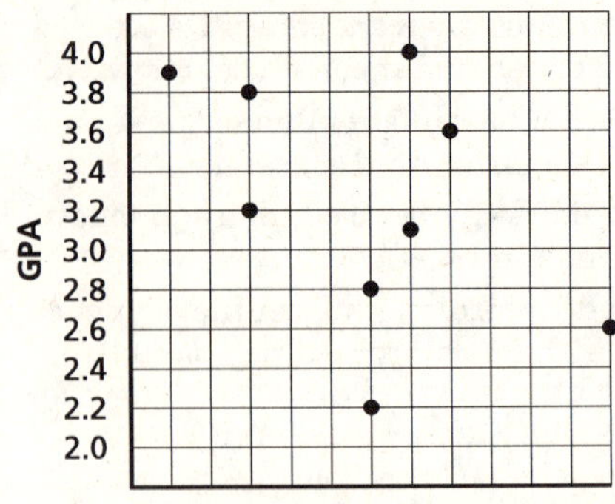

7. Since the points appear to be random, there is no relationship.

Pages 43–44 Practice and Apply

8. As a household size increases, the amount of the water bill increases; positive.
9. The number of songs on a CD usually does not affect the cost of the CD; no.
10. As the size of a car's engine increases, the miles per gallon decrease; negative.
11. As speed increases, distance traveled increases; positive.
12. As the outside temperature increases, the amount of the heating bill decreases; negative.
13. The size of a television screen and the number of channels it receives are not related; no.
14. There is no obvious pattern, so the graph shows no relationship.
15. The number decreases.
16. The number increases.
17.

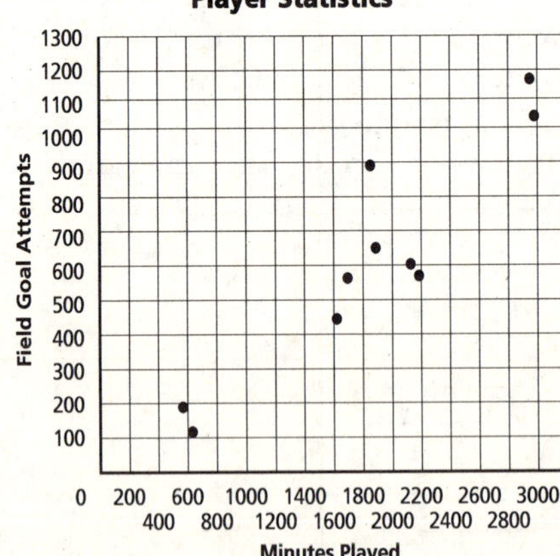

18. Yes; positive; as the number of minutes increases, the number of field goal attempts increases.
19. about 800
20. See students' work.
21. Sample answer: Yes; as more emphasis is placed on standardized tests, students will become more comfortable taking the tests; and the scores will increase.

22. By studying scatter plots and identifying the type of relationship the scatter plot shows, you can determine trends. Answers should include the following.
   - A positive relationship means that as one set of data increases, the other set also increases. A negative relationship means that as one set of data increases, the other set decreases. No relationship means that one set of data is not related to the other set of data.
   - The amount of money spent and the amount of money saved would be an example of a negative relationship. The number of hours spent studying and test scores would represent a positive relationship. The number of days in a month and the number of rainy days is an example of no relationship.

23. C; A student who studies for 60 minutes can expect a score between 80 and 90.

24. B; A student who studies for 90 minutes can expect a score around 95.

Page 44 Maintain Your Skills

25. Start at the origin. Move 3 units right and 2 units up.

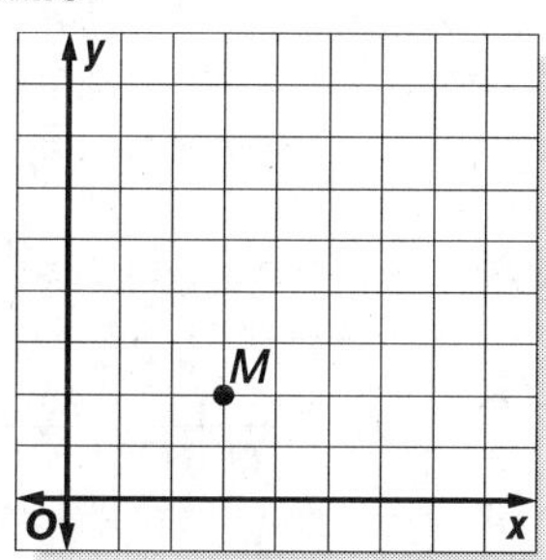

26. Start at the origin. Move 5 units right. Since the y-coordinate is 0, do not move up or down.

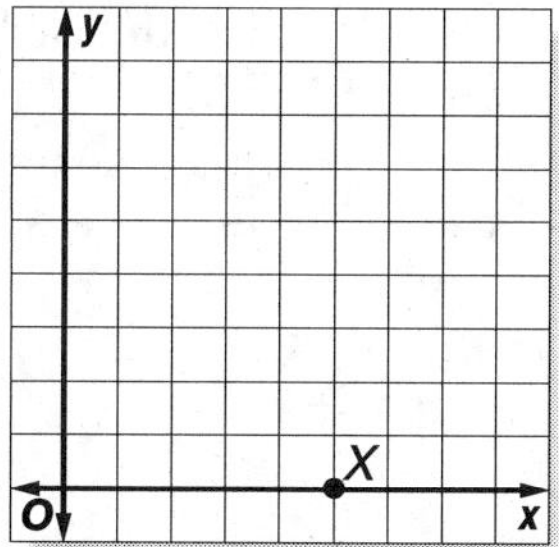

27. Start at the origin. Move 2 units up. Since the x-coordinate is 0, do not move right or left.

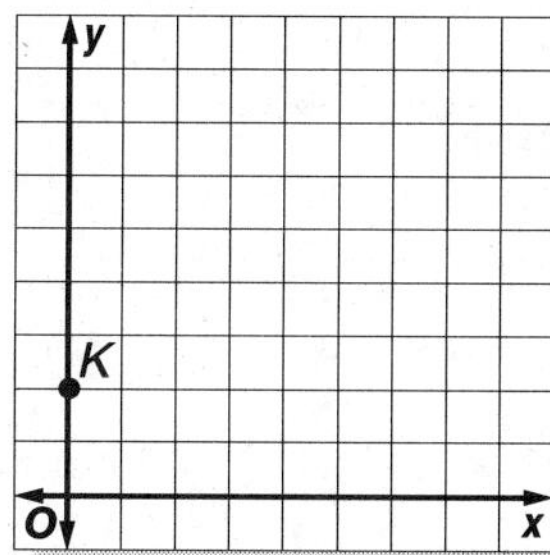

28. Start at the origin. Move 2 units right and 3 units up. The ordered pair for point D is (2, 3).

29. Start at the origin. Move 4 units up. Do not move right or left. The ordered pair for point W is (0, 4).

30. Start at the origin. Move 3 units right. Do not move up or down. The ordered pair for point J is (3, 0).

31. domain = {0, 4, 2, 6}
(consists of all the x-coordinates);
range = {9, 8, 3, 1}
(consists of all the y-coordinates)

32. $3c = 81$
$3 \cdot 27 = 81$
$c = 27$

33. $15 - x = 8$
$15 - 7 = 8$
$x = 7$

34. $8 = \frac{32}{m}$
$8 = \frac{32}{4}$
$m = 4$

35. $15 + (b + 3) = b + (15 + 3)$
$= b + 18$

36. $(2m + 3y) - m = (2 \cdot 8 + 3 \cdot 6) - 8$
$= (16 + 18) - 8$
$= (34) - 8$
$= 26$

37. $3m + (y - 2) + 3 = 3 \cdot 8 + (6 - 2) + 3$
$= 3 \cdot 8 + (4) + 3$
$= 24 + 4 + 3$
$= 31$

Page 46 Graphing Calculator Investigation (Follow-Up of Lesson 1-7)

1. The x-coordinate represents average number of nests and the y-coordinate represents extinction time for each bird.
2. Sample answer: The points seem to be slanting up from left to right. One point seems much higher than the others.
3. Sample answer: Generally, as the number of nests increases, the time it takes to become extinct increases also. There is a positive relationship.
4. Sample answer: Large birds tend to have longer extinction times.
5. 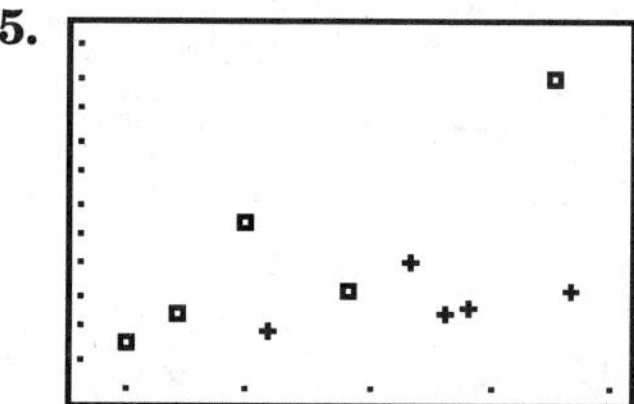

The plot shows larger birds tend to have longer extinction times.

6. 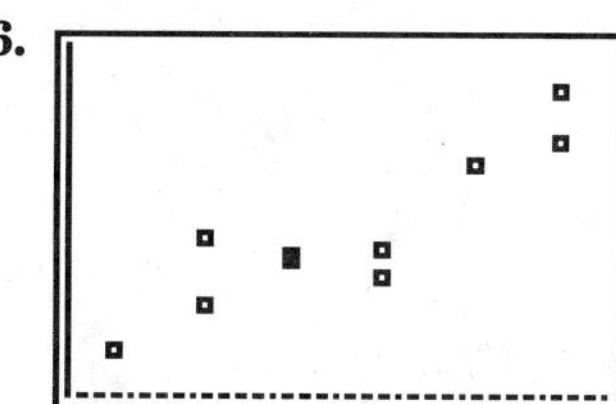

As x increases, y increases. There is a positive relationship.

7. 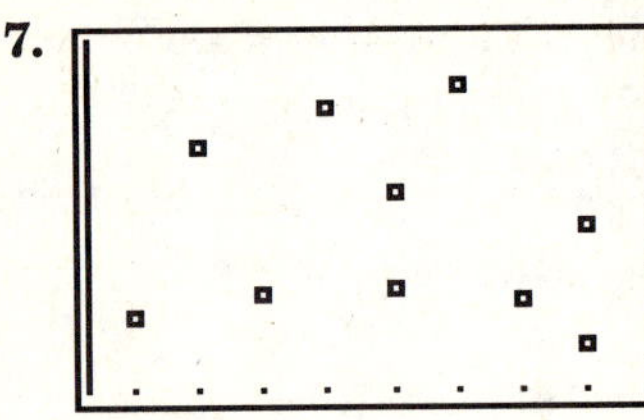

No relationship

8. 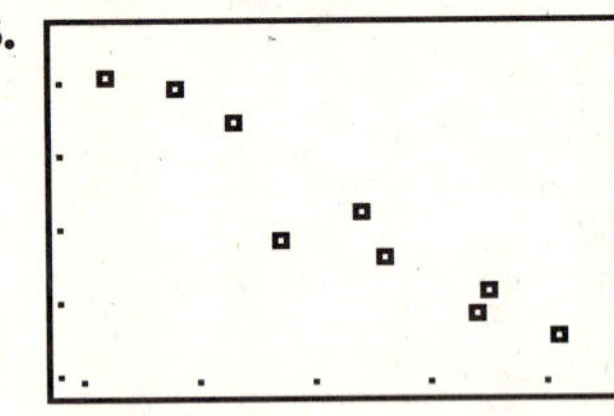

As x increases, y decreases. There is a negative relationship.

9. Sample data: year versus 100-meter dash times; year versus school enrollment.

Chapter 1 Study Guide and Review

Page 47 Vocabulary and Concept Check

1. d; algebraic expression
2. b; evaluate
3. e; range
4. a; numerical expression
5. c; domain

Pages 47–50 Lesson-by-Lesson Review

6. 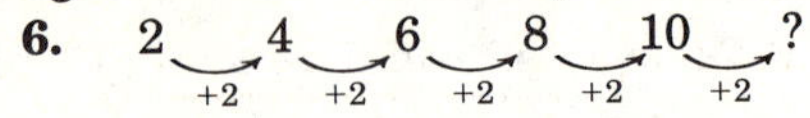

Assuming the pattern continues, the next term is 10 + 2 or 12.

7. 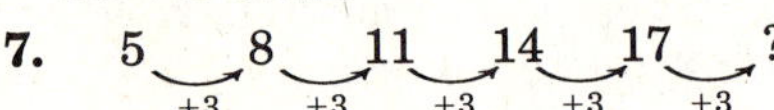

Assuming the pattern continues, the next term is 17 + 3 or 20.

8. 2 6 18 54 162 ?
×3 ×3 ×3 ×3 ×3

Assuming the pattern continues, the next term is 162 × 3 or 486.

9. 1 2 4 7 11 16 ?
+1 +2 +3 +4 +5 +6

Assuming the pattern continues, the next term is 16 + 6 or 22.

10. Each pound costs \$4.38. A 7-pound ham would cost 7 × \$4.38 or \$30.66.

11. $7 + 3 \cdot 5 = 7 + 15$
$= 22$

12. $36 \div 9 - 3 = 4 - 3$
$= 1$

13. $5 \cdot (7 - 2) - 9 = 5 \cdot (5) - 9$
$= 25 - 9$
$= 16$

14. $\frac{2(17 + 4)}{3} = 2(17 + 4) \div 3$
$= 2(21) \div 3$
$= 42 \div 3$
$= 14$

15. $18 \div (7 - 4) + 6 = 18 \div (3) + 6$
$= 6 + 6$
$= 12$

16. $4[9 + (1 \cdot 16) - 8] = 4[9 + 16 - 8]$
$=4(17)$
$=68$

17. $y + 6 = 8 + 6$
$= 14$

18. $17 - 2x = 17 - 2 \cdot 3$
$= 17 - 6$
$= 11$

19. $z - 3 + y = 5 - 3 + 8$
$= 2 + 8$
$= 10$

20. $6x - 2z + 7 = 6 \cdot 3 - 2 \cdot 5 + 7$
$= 18 - 10 + 7$
$= 8 + 7$
$= 15$

21. $\frac{6y}{x} + 9 = (6 \cdot 8) \div 3 + 9$
$= 48 \div 3 + 9$
$= 16 + 9$
$= 25$

22. $9x - (y + z) = 9 \cdot 3 - (8 + 5)$
$= 9 \cdot 3 - 13$
$= 27 - 13$
$= 14$

23. The order of the numbers changed. This is the Commutative Property of Addition.

24. Zero is added to a number. This is the Additive Identity Property.

25. A number is multiplied by zero. This is the Multiplicative Property of Zero.

26. The grouping of the numbers or variables changed. This is the Associative Property of Multiplication.

27. $n + 3 =13$
$10 + 3 = 13$
$n = 10$

28. $9 = k - 6$
$9 = 15 - 6$
$k = 15$

29. $24 = 7 + g$
$24 = 7 + 17$
$g = 17$

30. $6x = 48$
$6 \cdot 8 = 48$
$x = 8$

31. $54 = 9h$
$54 = 9 \cdot 6$
$h = 6$

32. $\frac{56}{a} = 14$

$\frac{56}{4} = 14$

$a = 4$

33.

x	y
2	3
6	1
7	5

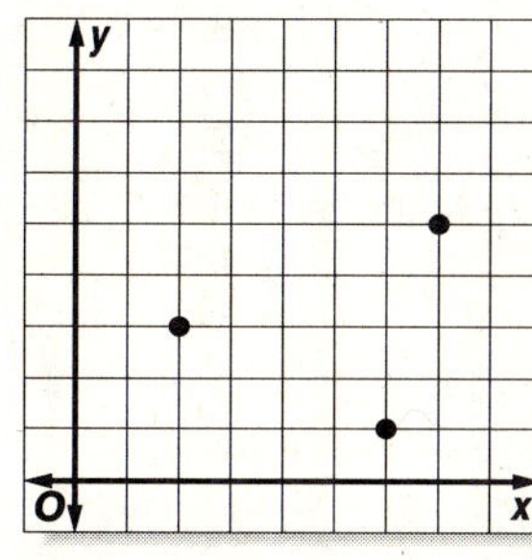

domain: {2, 6, 7}; range: {3, 1, 5}

34.

x	y
0	2
1	7
5	2
6	5

domain: {0, 1, 5, 6}; range: {2, 7, 5}

35. Positive; as the height increases, the circumference increases.

36. By looking at the pattern in the scatter plot, we can predict that a tree of height 245 ft would have a circumference of about 90 to 95 feet.

Chapter 1 Practice Test

Page 51

1. Explore, Plan, Solve, Examine
2. Simplify the expressions inside grouping symbols. Do all multiplication and/or division in order from left to right. Do all addition and/or subtraction in order from left to right.
3. $24 - 8 \div 2 \cdot 3 = 24 - 4 \cdot 3$
   $= 24 - 12$
   $= 12$
4. $16 \div 4 + 3(9 - 7) = 16 \div 4 + 3(2)$
   $= 4 + 3(2)$
   $= 4 + 6$
   $= 10$
5. $3[18 - 5(7 - 5 + 1)] = 3[18 - 5(3)]$
   $= 3(18 - 15)$
   $= 3(3)$
   $= 9$
6. *Phrase* three less than fifteen
   Key Word less than
   Expression $15 - 3$
7. *Phrase* twelve increased by seven
   Key Word increased by
   Expression $12 + 7$
8. *Phrase* the quotient of twelve and six
   Key Word quotient
   Expression $12 \div 6$
9. $4a - 3c = 4 \cdot 7 - 3 \cdot 5$
   $= 28 - 15$
   $= 13$
10. $42 \div [a(c - b)] = 42 \div [7(5 - 3)]$
    $= 42 \div (7 \cdot 2)$
    $= 42 \div 14$
    $= 3$
11. $5c + (a + 2b - 8) = 5 \cdot 5 + (7 + 2(3) - 8)$
    $= 5 \cdot 5 + (7 + 6 - 8)$
    $= 5 \cdot 5 + (5)$
    $= 25 + 5$
    $= 30$
12. The grouping of the numbers changed. This is the Associative Property of Multiplication.
13. $9 + (p + 3) = p + (9 + 3)$
    $= p + 12$
14. $6 \cdot (7 \cdot k) = (6 \cdot 7) \cdot k$
    $= 42k$
15. $4m = 20$
    $4 \cdot 5 = 20$
    $m = 5$
16. $16 - a = 9$
    $16 - 7 = 9$
    $a = 7$
17. Start at the origin. Move 2 units right and 5 units up.

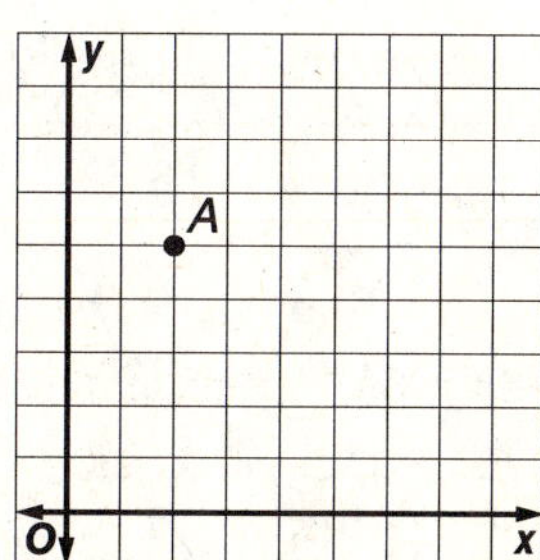

18. Start at the origin and move 3 units right. Do not move up or down. The ordered pair for point C is (3, 0).
19. Start at the origin. Move 1 unit right and 3 units up. The ordered pair for point D is (1, 3).
20.

x	y
8	5
4	3
2	2
6	1

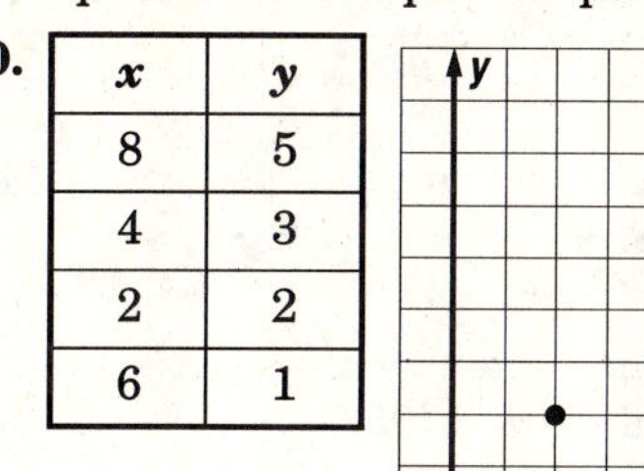

domain: {8, 4, 2, 6}; range: {5, 3, 2, 1}

21. Positive; as the outside temperature increases, an air conditioning bill increases.
22. No; height is not related to the number of siblings.

23. $3 \underset{+2}{\frown} 5 \underset{+4}{\frown} 9 \underset{+6}{\frown} 15 \underset{+8}{\frown} ?$

The pattern is to add consecutive even numbers starting with 2, so the next three terms in the list are 23, 33, and 45 because $15 + 8 = 23$, $23 + 10 = 33$, $33 + 12 = 45$.

24. Renting a car for \$350 a week was the better deal because $79 + (6 \cdot 49) = \$373$, which is \$23 more than \$350.

25. B; Let x = the price per loaf
6 = the number of loaves
$\$12$ = the total cost
The expression is $6x = 12$.

Chapter 1 Standardized Test Practice

Pages 52–53

1. D; $4 \underset{\times 3}{\frown} 12 \underset{\times 3}{\frown} 36 \underset{\times 3}{\frown} 108 \underset{\times 3}{\frown} ?$

Assuming the pattern continues, the next term is 108×3 or 324, and the next term after that is 324×3 or 972.

2. A; $2(15 - 3 \cdot 4) = 2(15 - 12)$
$= 2(3)$
$= 6$

3. C; 4 books with 200 pages = 4(200) and 2 books with 300 pages = 2(300).
The expression is 4(200) + 2(300).

4. D; 1 ounce at \$0.34 = 0.34 and 4 ounces at \$0.21 = 0.21(4). The expression is 0.34 + 0.21(4).

5. C; The grouping of the numbers changed. This is the Associative Property of Multiplication.

6. B;

Value for x	$17 - 2x = 9$	True or False?
2	$17 - 2(2) \stackrel{?}{=} 9$	false
4	$17 - 2(4) \stackrel{?}{=} 9$	true ✓
6	$17 - 2(6) \stackrel{?}{=} 9$	false
8	$17 - 2(8) \stackrel{?}{=} 9$	false

7. D;

a number	more than	nine	is	15
n	$+$	9	$=$	15

$n + 9 = 9 + n$, so the sentence is:
Nine more than a number is 15.

8. B; Start at the origin. Move 5 units right and 3 units up. The ordered pair for point P is (5, 3).

9. C

10. B; Negative because as the values of x increase, the values of y decrease.

11. The fewest number is represented by the shortest bar, which is Thursday.

12. about 80 because $130 - 50 = 80$

13. 16 because
$8 \times 6 = 48$ feet
$48 \div 3 = 16$ yards

14. $5 + 4 \times 6 \div 3 = 5 + 24 \div 3$
$= 5 + 8$
$= 13$

15. $x(xy + 3) = 5(5 \cdot 2 + 3)$
$= 5(10 + 3)$
$= 5(13)$
$= 65$

16.

14	is	twice the value of x	less	12
14	$=$	$2x$	$-$	12

The equation is $14 = 2x - 12$.

17a. {(2, 13), (3, 20), (5, 35), (7, 53), (9, 72)}

17b. **Running Distance and Time**

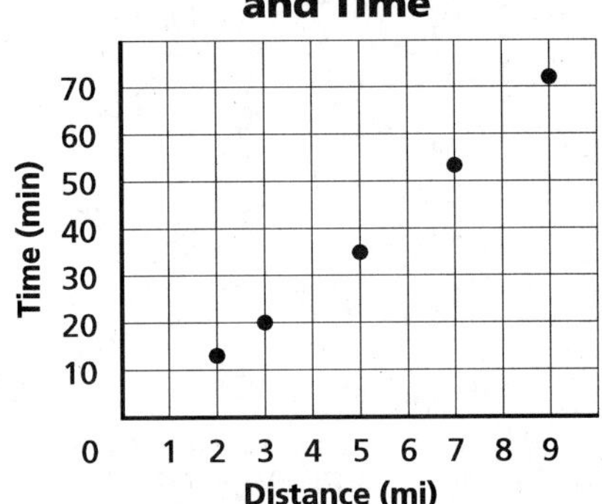

17c. By looking at the pattern in the graph, we can predict that it will take about 25–30 minutes for Kenneth to run 4 miles.

17d. By looking at the pattern in the graph, we can predict that in 60 minutes Kenneth will run about 8 miles.

18a.

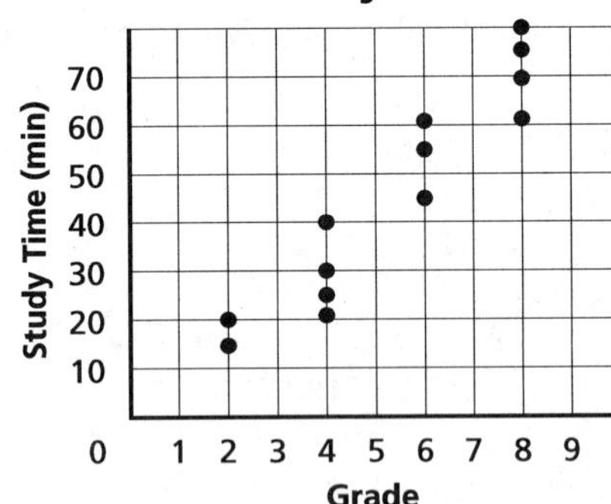

18b. (8, 80)

18c. Yes, there is a positive relationship between grade and time spent on homework. As the grade level increases, the time spent on homework increases.

Chapter 2 Integers

Page 55 Getting Started

1. $a + b + c = 4 + 10 + 8$
   $= 22$
2. $bc - ab = 10 \cdot 8 - 4 \cdot 10$
   $= 80 - 40$
   $= 40$
3. $b + ac = 10 + 4 \cdot 8$
   $= 10 + 32$
   $= 42$
4. $4c + 3b = 4 \cdot 8 + 3 \cdot 10$
   $= 32 + 30$
   $= 62$
5. $2b - (a + c) = 2 \cdot 10 - (4 + 8)$
   $= 20 - 12$
   $= 8$
6. $2c - b + a = 2 \cdot 8 - 10 + 4$
   $= 16 - 10 + 4$
   $= 10$
7. 34 28 22 16 10 ?
   −6 −6 −6 −6 −6

   Assuming the pattern continues, the next term is $10 - 6$ or 4.
8. 120 105 90 75 ?
   −15 −15 −15 −15

   Assuming the pattern continues, the next term is $75 - 15$ or 60.

9. *T* 10. *L* 11. *V* 12. *U*
13. *Q* 14. *R*

2-1 Integers and Absolute Value

Page 56 How are integers used to model real-world situations?

a. 7 in. below normal
b. Jackson, MS
c. +5

Page 59 Check for Understanding

1. Draw a number line. Draw a dot at −4.
2. Sample answer: $3 > -2$, $|-5| < |-8|$
3. The absolute value of a number is its distance from 0 on a number line.
4. −8;
   −8 −7 −6 −5 −4 −3 −2 −1 0
5. +15;
   8 9 10 11 12 13 14 15 16
6. −3 −2 −1 0 1 2 3 4 5 6
7. $-4 < 2$; $2 > -4$
8. $-6 > -10$; $-10 < -6$
9. $-18 < -8$
10. $0 > -3$
11. $9 > -9$
12. $\{-52, -6, -2, 0, 5, 28, 115\}$
13. $|-10| = 10$
14. $|10| - |-4| = 10 - 4$
    $= 6$
15. $|16| + |-5| = 16 + 5$
    $= 21$
16. $9 + |a| = 9 + |-8|$
    $= 9 + 8$
    $= 17$
17. $|a| - b = |-8| - 5$
    $= 8 - 5$
    $= 3$
18. $2|a| = 2|-8|$
    $= 2 \cdot 8$
    $= 16$
19. −54, −52, −45, −37, −36, −34, −27, −27, −2

Pages 59–61 Practice and Apply

20. −100;
    −100 −50 0 50 100
21. −6;
    −8 −7 −6 −5 −4 −3 −2 −1 0
22. +250;
    0 100 200 300
23. +9;
    5 6 7 8 9 10 11 12 13
24. +12;
    0 2 4 6 8 10 12 14 16
25. −5;
    −8 −7 −6 −5 −4 −3 −2 −1 0
26. −4 −3 −2 −1 0 1 2 3 4
27. −3 −2 −1 0 1 2 3 4 5
28. −8 −7 −6 −5 −4 −3 −2 −1 0
29. −8 −6 −4 −2 0 2 4 6 8
30. $3 > 2$; $2 < 3$
31. $-5 > -10$; $-10 < -5$
32. $55 < 65$; $65 > 55$
33. $248 < 425$; $425 > 248$
34. $-2 < 23$; $23 > -2$
35. $212 > 32$; $32 < 212$
36. $-6 < -2$
37. $-10 > -13$
38. $0 > -9$
39. $14 > 0$
40. $-18 < 8$
41. $5 > -23$
42. $|9| = |-9|$, since both equal +9.

43. $|-20| > |-4|$, since $20 > 4$.

44. $\{-8, 0, 5\}$

45. $\{-15, -4, -2, -1\}$

46. $\{-46, -3, 0, 5, 9, 24\}$

47. $\{-60, -57, 38, 98, 188\}$

48. $|-15| = 15$

49. $|46| = 46$

50. $-|20| = -20$

51. $-|5| = -5$

52. $|0| = 0$

53. $|7| = 7$

54. $|-5| + |4| = 5 + 4$
$= 9$

55. $|0| + |-2| = 0 + 2$
$= 2$

56. $|15| - |-1| = 15 - 1$
$= 14$

57. $|0 + 9| = |9|$
$= 9$

58. $-|-24| = -24$

59. $-||-6| + |14|| = -|6 + 14|$
$= -|20|$
$= -20$

60. $14 + |b| = 14 + |3|$
$= 14 + 3$
$= 17$

61. $|c| - a = |-4| - 0$
$= 4 - 0$
$= 4$

62. $a + b + |c| = 0 + 3 + |-4|$
$= 0 + 3 + 4$
$= 7$

63. $ab + |-40| = 0 \cdot 3 + |-40|$
$= 0 + 40$
$= 40$

64. $|c| - b = |-4| - 3$
$= 4 - 3$
$= 1$

65. $|ab| + b = |0 \cdot 3| + 3$
$= |0| + 3$
$= 0 + 3$
$= 3$

66. −8685 ft

67.
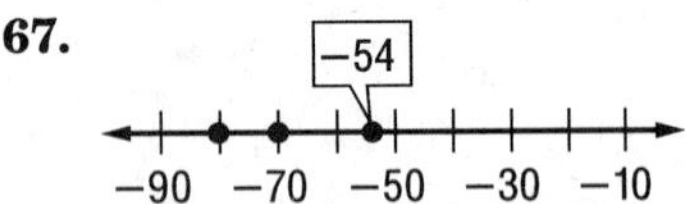

68. $-80 < -54$

69. $-54 > -70$

70. −54°, −70°, −80°

71. −4 and 3 are 7 units apart, since $|-4 - 3| = |-7| = 7$.

72. Sometimes; if X and Y are both negative the statement will not be true.

73. Sometimes; if A and B are both positive, both negative, or one is 0, it is true. If one number is negative and the other is positive, it is false.

74. Integers describe real-world situations involving measures above and below zero. Answers should include the following.
- Rainfall above normal is described using positive integers, below normal by negative integers.
- Some examples are: accounting uses + for deposits and − for withdrawals; gains are represented by +, losses by −; in golf, scores above par are +, below par are − ; temperatures above zero are +, below zero are −.

75. B; The absolute value of −2 is the distance between −2 and 0 on a number line, and a thermometer is a number line with integer degrees.

76. D

Page 61 Maintain Your Skills

77. Positive; as height increases, so does arm length.

78. None; birth month and weight are not related.

79.

x	y
3	2
3	4
2	1
2	4

{(3, 2), (3, 4), (2, 1), (2, 4)}

80.

x	y
1	4
6	2
9	6
1	9

{(1, 4), (6, 2), (9, 6), (1, 9)}

81. Commutative Property of Multiplication

82. Multiplicative Property of Zero

83. Commutative Property of Multiplication

84. $18 + 29 + 46 = 93$

85. $232 + 156 = 388$

86. $451 + 629 + 1027 = 2107$

87. $36 - 19 = 17$

88. $479 - 281 = 198$

89. $2011 - 962 = 1049$

Page 63 Algebra Activity (Preview of Lesson 2-2)

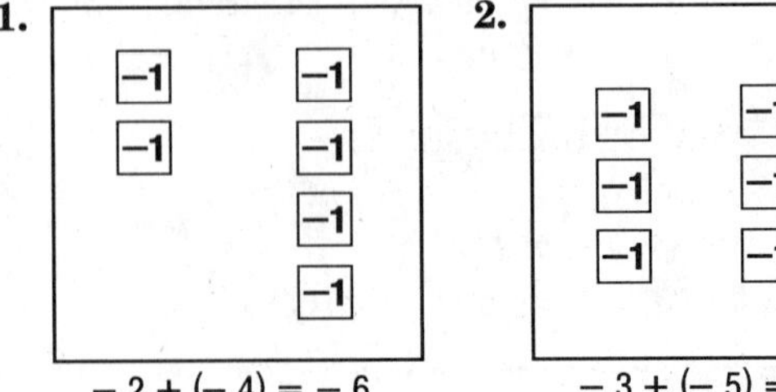

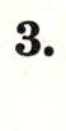

3.

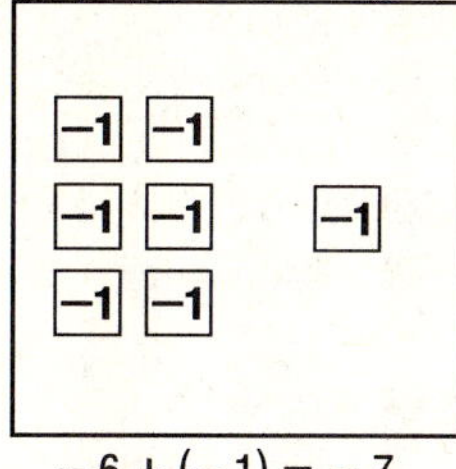

$-6 + (-1) = -7$

4.

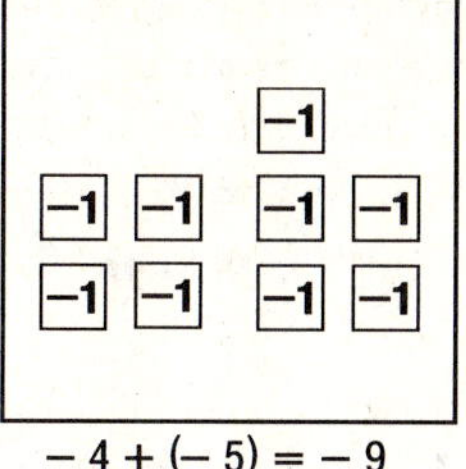

$-4 + (-5) = -9$

5.

$-4 + 2 = -2$

6.

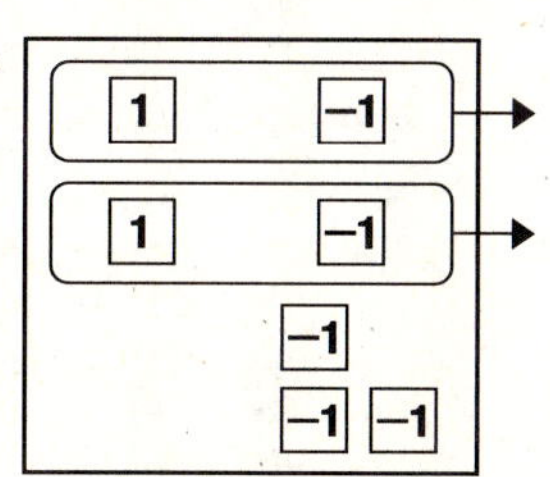

$2 + (-5) = -3$

7.

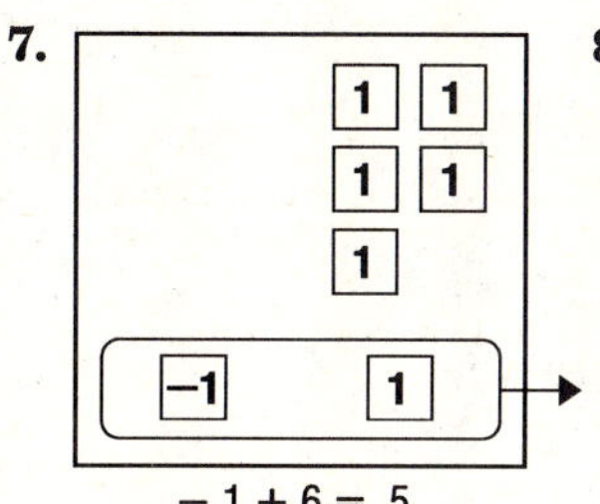

$-1 + 6 = 5$

8.

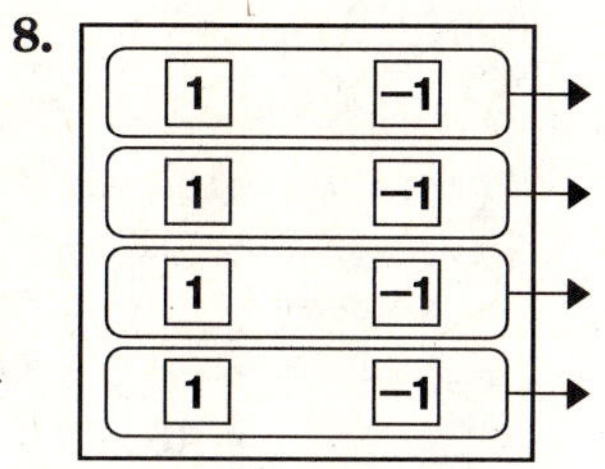

$4 + (-4) = 0$

9. The addends are both +, or one is + and one − with the + addend having the greater absolute value.

10. The addends are both −, or one is + and one − with the − addend having the greater absolute value.

11. They are additive inverses.

12. Yes; sample answers: $-1 + 0 = -1$; $0 + (-2) = -2$

13. Yes; sample answers: $-1 + 3 = 2$ and $3 + (-1) = 2$; $-2 + (-1) = -3$ and $-1 + (-2) = -3$

14. Yes; sample answers: $[3 + (-2)] + (-1) = 3 + [(-2) + (-1)]$ and $[1 + (-2)] + 2 = 1 + [-2 + 2]$

2-2 Adding Integers

Page 64 How can a number line help you add integers?

a. -7 **b.** $-5 + (-2) = -7$

Page 67 Check for Understanding

1a. Negative; both addends are negative.

1b. Positive; $|12| > |-2|$

1c. Negative; $|-11| > |9|$

1d. Positive; both addends are positive.

2. Sample answer: -1 and 1

3. $-2 + (-4) = -6$ **4.** $-10 + (-5) = -15$

5. $7 + (-2) = 5$ **6.** $11 + (-3) = 8$

7. $8 + (-5) = 3$ **8.** $9 + (-12) = -3$

9. $8 + (-6) + 2 = 8 + 2 + (-6)$
$= (8 + 2) + (-6)$
$= 10 + (-6)$
$= 4$

10. $-6 + 5 + (-10) = -6 + (-10) + 5$
$= [-6 + (-10)] + 5$
$= -16 + 5$
$= -11$

11. yards gained + yards lost = change in yardage
$4 + (-5) = -1$

Pages 67–68 Practice and Apply

12. $-4 + (-1) = -5$ **13.** $-5 + (-2) = -7$

14. $-4 + (-6) = -10$ **15.** $-3 + (-8) = -11$

16. $-7 + (-8) = -15$ **17.** $-12 + (-4) = -16$

18. $-9 + (-14) = -23$ **19.** $-15 + (-6) = -21$

20. $-11 + (-15) = -26$ **21.** $-23 + (-43) = -66$

22. $8 + (-5) = 3$ **23.** $6 + (-4) = 2$

24. $3 + (-7) = -4$ **25.** $4 + (-6) = -2$

26. $-15 + 6 = -9$ **27.** $-5 + 11 = 6$

28. $18 + (-32) = -14$ **29.** $-45 + 19 = -26$

30. -14, since $14 + (-14) = 0$

31. 21, since $-21 + 21 = 0$

32. $6 + (-9) + 9 = 6 + [(-9) + 9]$
$= 6 + 0$
$= 6$

33. $7 + (-13) + 4 = 7 + 4 + (-13)$
$= (7 + 4) + (-13)$
$= 11 + (-13)$
$= -2$

34. $-9 + 16 + (-10) = -9 + (-10) + 16$
$= [-9 + (-10)] + 16$
$= -19 + 16$
$= -3$

35. $-12 + 18 + (-12) = -12 + (-12) + 18$
$= [-12 + (-12)] + 18$
$= -24 + 18$
$= -6$

36. $14 + (-9) + 6 = 14 + 6 + (-9)$
$= (14 + 6) + (-9)$
$= 20 + (-9)$
$= 11$

37. $28 + (-35) + 4 = 28 + 4 + (-35)$
$= (28 + 4) + (-35)$
$= 32 + (-35)$
$= -3$

38. $-41 + 25 + (-10) = -41 + (-10) + 25$
$= [-41 + (-10)] + 25$
$= -51 + 25$
$= -26$

39. $-18 + 35 + (-17) = -18 + (-17) + 35$
$= [-18 + (-17)] + 35$
$= -35 + 35$
$= 0$

40. $50 + (-25) + (-32) = 50 + [(-25) + (-32)]$
$= 50 + (-57)$
$= -7$

The balance was $-\$7$.

41. $-4 + 2 + 0 + (-3) = -4 + (-3) + 2 + 0$
$= [-4 + (-3)] + (2 + 0)$
$= -7 + 2$
$= -5$

The finial score was -5, or 5 under par.

42. $|18 + (-13)| = |5|$
$= 5$

43. $|-27 + 19| = |-8|$
$= 8$

44. $|-25 + (-12)| = |-37|$
$= 37$

45. $|-28 + (-12)| = |-40|$
$= 40$

46. Dallas: 1,006,877 + 181,703 = 1,188,580;
Honolulu: 365,272 + 6385 = 371,657;
Jackson: 196,637 − 12,381 = 184,256;
Philadelphia: 1,585,577 − 68,027 = 1,517,550

47. 181,703 + 6385 + (−12, 381) + (−68,022)
= (181,703 + 6385) + [(−12,381) + (−68,027)]
= 188,088 + (−80,408)
= 107,680
The change in population is +107,680.

48. False; sample counterexample:
If $n = -2$, then $-(-2)$ is positive.

49. To add integers on a number line, start at 0. Move right to show positive integers and left to show negative integers. Answers should include the following.

- Sample answer:

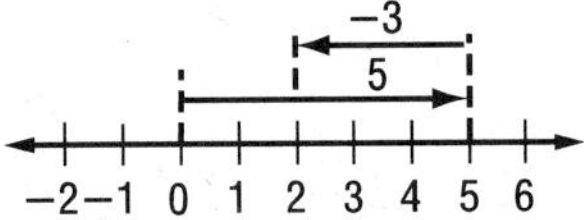

- Sample answer:

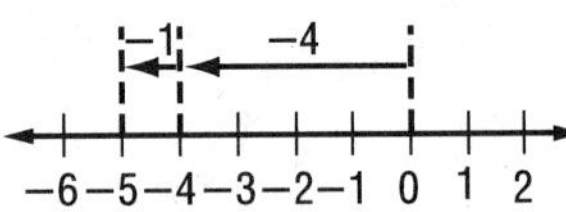

50. C; $-32 + 20 = -12$

51. D; $-|-2 + 8| = -|6|$
$= -6$

Page 68 Maintain Your Skills

52. −219°C

53. {−12, −9, −8, 0, 3, 14}

54. {−242, −158, −24, 35, 99}

55. no relationship

56. Negative, since mitten sales would decrease as the temperature increases.

57. $18 - n = 12$
$-n = 12 + (-18)$
$-n = -6$
$n = 6$

58. $25 = 16 + x$
$25 + (-16) = x$
$9 = x$

59. $\frac{x}{2} = 10$
$x = 10 \cdot 2$
$x = 20$

60. $7a = 49$
$a = \frac{49}{7}$
$a = 7$

61. $a + 19 = 6 + 19$
$= 25$

62. $2b - 6 = 2 \cdot 10 - 6$
$= 20 - 6$
$= 14$

63. $ab - ac = 6 \cdot 10 - 6 \cdot 3$
$= 60 - 18$
$= 42$

64. $3a - (b + c) = 3 \cdot 6 - (10 + 3)$
$= 18 - 13$
$= 5$

65. $5b + 5c = 5 \cdot 10 + 5 \cdot 3$
$= 50 + 15$
$= 65$

66. $\frac{6b}{c} = \frac{6 \cdot 10}{3}$
$= \frac{60}{3}$
$= 20$

Page 69 Reading Mathematics

1. Sample answer: simplify—make simpler; coordinate—put in the same order
2. Answers should include two of the following: divisor, quotient, integer, algebraic expression.
3. Sample answer: simplify—also means to simplify a fraction; factor—also means to find the factors of a number.

2-3 Subtracting Integers

Page 70 How are addition and subtraction of integers related?

a. $6 - 8$ is -2.

b. You move left on a number line to indicate subtracting a positive integer.

c. $6 + (-8) = -2$

Page 72 Check for Understanding

1. Sample answer: positive integer 5 and its additive inverse; -5; negative integer -9 and its additive inverse, 9
2. José; Reiko added -2 instead of its additive inverse, 2.
3. $8 - 11 = 8 + (-11)$
   $= -3$
4. $-9 - 3 = -9 + (-3)$
   $= -12$
5. $5 - (-4) = 5 + 4$
   $= 9$
6. $7 - (-10) = 7 + 10$
   $= 17$
7. $-6 - (-4) = -6 + 4$
   $= -2$
8. $-2 - (-8) = -2 + 8$
   $= 6$
9. $x - (-10) = 10 - (-10)$
   $= 10 + 10$
   $= 20$
10. $y - x = -4 - 10$
    $= -4 + (-10)$
    $= -14$
11. $x + y - z = 10 + (-4) - (-15)$
    $= 10 + (-4) + 15$
    $= 10 + 15 + (-4)$
    $= 25 + (-4)$
    $= 21$
12. To find the range, subtract the lowest temperature from the highest temperature.
    $105 - (-50) = 105 + 50$
    $= 155$
    The range for Vermont is 155°.
13. Explore: You need to find the range for a state in order to compare it to Vermont's range.
    Plan: To find the range, subtract the lowest temperature from the highest temperature for a state. Then compare the range to Vermont's, which is 155°.
    Solve: Utah: $117 - (-69) = 186°$
    Virginia: $110 - (-30) = 140°$
    Washington: $118 - (-48) = 166°$
    West Virginia: $112 - (-37) = 149°$
    Wisconsin: $114 - (-54) = 168°$
    Wyoming: $114 - (-66) = 180°$
    States with a greater range than Vermont's are Utah, Washington, Wisconsin, and Wyoming.
    Examine: All states have a positive difference in temperatures. Therefore, those differences greater than Vermont's have a greater range.

Pages 73–74 Practice and Apply

14. $3 - 8 = 3 + (-8)$
    $= -5$
15. $4 - 5 = 4 + (-5)$
    $= -1$
16. $2 - 9 = 2 + (-9)$
    $= -7$
17. $9 - 12 = 9 + (-12)$
    $= -3$
18. $-3 - 1 = -3 + (-1)$
    $= -4$
19. $-5 - 4 = -5 + (-4)$
    $= -9$
20. $-6 - 7 = -6 + (-7)$
    $= -13$
21. $-4 - 8 = -4 + (-8)$
    $= -12$
22. $6 - (-8) = 6 + 8$
    $= 14$
23. $4 - (-6) = 4 + 6$
    $= 10$
24. $7 - (-4) = 7 + 4$
    $= 11$
25. $9 - (-3) = 9 + 3$
    $= 12$
26. $-9 - (-7) = -9 + 7$
    $= -2$
27. $-7 - (-10) = -7 + 10$
    $= 3$
28. $-11 - (-12) = -11 + 12$
    $= 1$
29. $-16 - (-7) = -16 + 7$
    $= -9$
30. $10 - 24 = 10 + (-24)$
    $= -14$
31. $45 - 59 = 45 + (-59)$
    $= -14$
32. $-27 - 14 = -27 + (-14)$
    $= -41$
33. $-16 - 12 = -16 + (-12)$
    $= -28$
34. $48 - (-50) = 48 + 50$
    $= 98$
35. $125 - (-114) = 125 + 114$
    $= 239$
36. $-320 - (-106) = -320 + 106$
    $= -214$
37. $-2200 - (-3500) = -2200 + 3500$
    $= 1300$
38. $16 - (-2) = 16 + 2$
    $= 18$
    The difference is 18°F.

39. $14{,}494 - (-282) = 14{,}494 + 282$
$= 14{,}776$
The difference is 14,776 ft.

40. $y - 10 = 8 - 10$
$= 8 + (-10)$
$= -2$

41. $12 - z = 12 - (-12)$
$= 12 + 12$
$= 24$

42. $3 - x = 3 - (-3)$
$= 3 + 3$
$= 6$

43. $z - 24 = -12 - 24$
$= -12 + (-24)$
$= -36$

44. $x - y = -3 - 8$
$= -3 + (-8)$
$= -11$

45. $z - x = -12 - (-3)$
$= -12 + 3$
$= -9$

46. $y - z = 8 - (-12)$
$= 8 + 12$
$= 20$

47. $z - y = -12 - 8$
$= -12 + (-8)$
$= -20$

48. $x + y - z = -3 + 8 - (-12)$
$= -3 + 8 + 12$
$= 5 + 12$
$= 17$

49. $z - y + x = -12 - 8 + (-3)$
$= -12 + (-8) + (-3)$
$= -20 + (-3)$
$= -23$

50. $x - y - z = -3 - 8 - (-12)$
$= -3 + (-8) + 12$
$= -11 + 12$
$= 1$

51. $z - y - x = -12 - 8 - (-3)$
$= -12 + (-8) + 3$
$= -20 + 3$
$= -17$

52. Airedale Terrier: 2950 − 2891 = +59;
Beagle: 49,080 − 53,322 = −4242;
Shar-Pei: 6845 − 8614 = −1769;
Chow Chow: 4342 − 6241 = −1899;
Labrador Retriever: 154,897 − 157,936 = −3039;
Pug: 21,555 − 21,487 = +68

53. Use the answers from Exercise 52:
59 + (−4242) + (−1769) + (−1899) + (−3039) + 68 = −10,822

54a. $P = I - E$
$= 19{,}592 - 20{,}345$
$= -\$753$ profit

54b. Expenses are greater than income.

55a. False; $3 - 4 \neq 4 - 3$

55b. False; $(5 - 2) - 1 \neq 5 - (2 - 1)$

56. Addition and subtraction of integers are related because a subtraction problem can be rewritten as an addition problem. Answers should include the following.

- 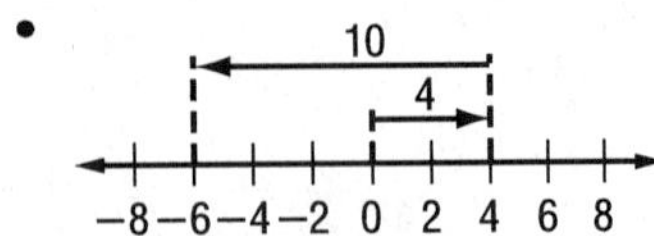

- 4 − 10 is rewritten as the addition expression 4 + (−10).

57. A; 13 ‿ 8 ‿ 3 ‿ −2 ‿ ? (each step −5)

Assuming the pattern continues, the next term is −2 − 5 or −7.

58. B; Let x = the number.
$x - 5 = -4$
$x = -4 + 5$
$x = 1$

Page 74 Maintain Your Skills

59. $-1300 - 1150 = -1300 + (-1150)$
$= -2450$
The submarine is 2450 m below sea level.

60. $|b| - |a| = |-4| - |2|$
$= 4 - 2$
$= 2$

61. $x + 9 = 12$
$3 + 9 = 12$
$x = 3$

62. $18 = w - 2$
$18 = 20 - 2$
$w = 20$

63. $5a = 35$
$5 \cdot 7 = 35$
$a = 7$

64. $\frac{64}{b} = 8$
$\frac{64}{8} = 8$
$b = 8$

65. $\frac{x}{5}$ **66.** $t + 9$ **67.** $\frac{86}{b}$ **68.** $s - 8$

69. $2x(5 + 8) - 6 = 2 \times 13 - 6$
$= 26 - 6$
$= 20$

70. $96 \div (6 \times 8) \div 2 = 96 \div 48 \div 2$
$= (96 \div 48) \div 2$
$= 2 \div 2$
$= 1$

71. $5 \cdot 15 = 75$ **72.** $8 \cdot 12 = 96$

73. $3 \cdot 5 \cdot 8 = (3 \cdot 5) \cdot 8$
$= 15 \cdot 8$
$= 120$

74. $2 \cdot 7 \cdot 5 \cdot 9 = (2 \cdot 7) \cdot 5 \cdot 9$
$= 14 \cdot 5 \cdot 9$
$= (14 \cdot 5) \cdot 9$
$= 70 \cdot 9$
$= 630$

Page 74 Practice Quiz 1

1. $-80, -70, -69$ **2.** $-5 + (-15) = -20$

3. $-5 + 11 = 6$

4. $-6 + 9 + (-8) = -6 + (-8) + 9$
$= -14 + 9$
$= -5$

5. $16 - 23 = 16 + (-23)$
$= -7$

6. $-15 - 8 = -15 + (-8)$
$= -23$

7. $25 - (-7) = 25 + 7$
$= 32$

8. $x - y = 5 - (-2)$
$= 5 + 2$
$= 7$

9. $z - 6 = -3 - 6$
$= -3 + (-6)$
$= -9$

10. $x - y - z = 5 - (-2) - (-3)$
$= 5 + 2 + 3$
$= 7 + 3$
$= 10$

2-4 Multiplying Integers

Page 75 How are the signs of factors and products related?

a. $4(-7)$

b. $4(-7) = -28$

Page 77 Check for Understanding

1. $3(-5) = -15$

2a. negative, since the factors have different signs

2b. negative, since the factors have different signs

2c. positive, since the factors have the same signs

2d. positive, since the factors have the same signs

2e. positive, since $-2(9)$ results in a negative product which has the same sign as (-3)

2f. negative, since $-7(-5)$ results in a positive product which has a different sign from (-11)

3. Sample answer: $(-4)(9)(2)$

4. $-3 \cdot 8 = -24$ **5.** $5(-8) = -40$

6. $4 \cdot 30 = 120$ **7.** $-7(-4) = 28$

8. $-4(2)(-6) = [-4(2)](-6)\}$
$= -8(-6)$
$= 48$

9. $-5(-9)(-12) = [-5(-9)]\,(-12)$
$= 45(-12)$
$= -540$

10. $-4 \cdot 3x = (-4 \cdot 3)x$
$= -12x$

11. $7(-3y) = [7 \cdot (-3)]y$
$= -21y$

12. $-8a(-3b) = (-8)(a)(-3)(b)$
$= [-8(-3)](a \cdot b)$
$= 24ab$

13. $-6h = -6(-20)$
$= 120$

14. $-4st = -4(-9) \cdot 3$
$= [-4(-9)] \cdot 3$
$= 36 \cdot 3$
$= 108$

15. A; $-100 \cdot 5 = -500$

Pages 78–79 Practice and Apply

16. $-3 \cdot 4 = -12$ **17.** $-7 \cdot 6 = -42$

18. $4(-8) = -32$ **19.** $9 \cdot (-8) = -72$

20. $-12 \cdot 3 = -36$ **21.** $14(-5) = -70$

22. $6 \cdot 19 = 114$ **23.** $4(32) = 128$

24. $-8(-11) = 88$ **25.** $-15(-3) = 45$

26. $-5(-4)(6) = [-5(-4)] \cdot 6$
$= 20 \cdot 6$
$= 120$

27. $5(-13)(-2) = [5(-13)](-2)$
$= -65 \cdot (-2)$
$= 130$

28. $-7(-8)(-3) = [-7(-8)](-3)$
$= 56(-3)$
$= -168$

29. $-11(-4)(-7) = [-11(-4)](-7)$
$= 44(-7)$
$= -308$

30. $-12(-9)(6) = [-12(-9)] \cdot 6$
$= 108 \cdot 6$
$= 648$

31. $-6(-8)(11) = [-6(-8)] \cdot 11$
$= 48 \cdot 11$
$= 528$

32. $2(-8)(-9)(10) = [2(-8)](-9)(10)$
$= -16(-9)(10)$
$= [-16(-9)] \cdot 10$
$= 144 \cdot 10$
$= 1440$

33. $4(-7)(-4)(-12) = [4(-7)](-4)(-12)$
$= -28(-4)(-12)$
$= [-28(-4)](-12)$
$= 112(-12)$
$= -1344$

34. $-1500 \cdot 24 = -36{,}000$ ft or about -6.8 mi

35. $44 - 10 \cdot 10 = 44 - 100$
$= -56°\text{F}$

36. $-5 \cdot 7x = (-5 \cdot 7)x$
$= -35x$

37. $-8 \cdot 12y = (-8 \cdot 12)y$
$= -96y$

38. $6(-8a) = [6(-8)]a$
$= -48a$

39. $5(-11b) = [5(-11)]b$
$= -55b$

40. $-7s(-8t) = (-7)(s)(-8)(t)$
$= [(-7)(-8)](s \cdot t)$
$= 56st$

41. $-12m(-9n) = (-12)(m)(-9)(n)$
$= [(-12)(-9)](mn)$
$= 108\ mn$

42. $2ab(3)(-7) = [2(3)(-7)](ab)$
$= [6(-7)](ab)$
$= -42ab$

43. $3x(5y)(-9) = (3)(x)(5)(y)(-9)$
$= [3 \cdot 5(-9)](xy)$
$= [15(-9)](xy)$
$= -135xy$

44. $-4(-p)(-q) = -4(-1)p(-1)q$
$= [-4(-1)(-1)](pq)$
$= [4(-1)](pq)$
$= -4pq$

45. $-8(-11b)(-c) = -8(-11)(b)(-1)(c)$
$= [-8(-11)(-1)](bc)$
$= [88(-1)](bc)$
$= -88bc$

46. $9(-2c)(3d) = 9(-2)(c)(3)(d)$
$= [9(-2)(3)]cd$
$= [-18(3)](cd)$
$= -54cd$

47. $-6j(3)(5k) = -6(j)(3)(5)(k)$
$= [-6(3)(5)](jk)$
$= [-18(5)](jk)$
$= -90jk$

48. $-7n = -7(-4)$
$= 28$

49. $9s = 9(-11)$
$= -99$

50. $ab = (9)(8)$
$= 72$

51. $-2xy = -2(-8)(5)$
$= 16(5)$
$= 80$

52. $-16cd = -16(4)(-5)$
$= -64(-5)$
$= 320$

53. $18gh = 18(-3)(4)$
$= -54(4)$
$= -216$

54. $-17 \cdot 6 = -102$ ft

55. $350 - 17 \cdot 6 = 350 - 102$
$= 248$ ft

56. If there is an even number of negative numbers being multiplied, the product will be positive. If there is an odd number of negative numbers being multiplied, the product will be negative.

57a. True; $3(-5) = -5(3)$

57b. True; $-2(3 \cdot 5) = (-2 \cdot 3)(5)$

58. When you multiply two integers with the same sign, the product is positive; when you multiply two integers with different signs, the product is negative. Answers should include the following.

- −4 −4
  −8 −7 −6 −5 −4 −3 −2 −1 0
- If the positive integer represents the number of groups made up of the negative integer, the total will be negative.
- $-3 \cdot 2 = -6$
  $-3 \cdot 1 = -3$
  $-3 \cdot 0 = 0$
  $-3 \cdot (-1) = 3$
  $-3 \cdot (-2) = 6$
  $-3 \cdot (-3) = 9$
  As the second factor decreases by 1, the product increases by 3.

59. B

60. D;
$y = -3x$
$y = -3(-2)$ $y = -3(-1)$
$y = 6$ $y = 3$
Choice D begins with 6, 3.

Page 79 Maintain Your Skills

61. $a - c = -2 - 14$
$= -2 + (-14)$
$= -16$

62. $b - a = -6 - (-2)$
$= -6 + 2$
$= -4$

63. $a - b = -2 - (-6)$
$= -2 + 6$
$= 4$

64. $a + b + c = -2 + (-6) + 14$
$= -8 + 14$
$= 6$

65. $b - a + c = -6 - (-2) + 14$
$= -6 + 2 + 14$
$= -4 + 14$
$= 10$

66. $a - b - c = -2 - (-6) - 14$
$= -2 + 6 + (-14)$
$= 4 + (-14)$
$= -10$

67. $104 - (-22) = 104 + 22$
$= 126°\text{F}$

68. $-10 + 8 + 4 = -10 + (8 + 4)$
$= -10 + 12$
$= 2$

69. $-4 + (-3) + (-7) = [-4 + (-3)] + (-7)$
$= -7 + (-7)$
$= -14$

70. $9 + (-14) + 2 = 9 + 2 + (-14)$
$= 11 + (-14)$
$= -3$

71. (6, 2) 72. (3, 4) 73. (1, 5) 74. (8, 0)

75. (5, 5) 76. (0, 3)

77. $3 \cdot 8 \cdot 20 = (3 \cdot 20) \cdot 8$
$= 60 \cdot 8$
$= 480$

78. $8 + 98 + 102 = 8 + (98 + 102)$
$= 8 + 200$
$= 208$

79. $5 \cdot 11 \cdot 10 = (5 \cdot 11) \cdot 10$
$= 55 \cdot 10$
$= 550$

80. $8\overline{)40}$ = 5; -40; 0

81. $15\overline{)90}$ = 6; -90; 0

82. $3\overline{)45}$ = 15; -3; 15; -15; 0

83. $7\overline{)105}$ = 15; -7; 35; -35; 0

84. $6\overline{)240}$ = 40; -24; 00; 0; 0

85. $24\overline{)96}$ = 4; -96; 0

2-5 Dividing Integers

Page 80 How is dividing integers related to multiplying integers?

a. There are 3 groups.

b. $-12 \div (-4) = 3$

c. $-4 \cdot 3 = -12$

d. $-10 \div (-2) = 5$

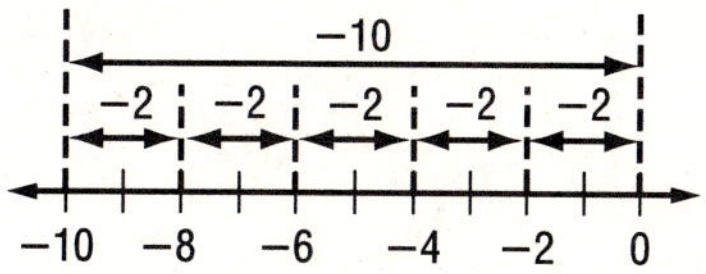

Page 83 Check for Understanding

1. Sample answer: $-16 \div 4 = -4$

2. Find the sum of the numbers. Divide by the number in the set.

3. $88 \div 8 = 11$

4. $-20 \div (-5) = 4$

5. $-18 \div 6 = -3$

6. $\frac{-36}{-4} = -36 \div -4$
$= 9$

7. $\frac{70}{-7} = 70 \div -7$
$= -10$

8. $\frac{-81}{9} = -81 \div 9$
$= -9$

9. $x \div 4 = -52 \div 4$
$= -13$

10. $\frac{s}{t} = \frac{-45}{5}$
$= -45 \div 5$
$= -9$

11. $\frac{-2 + 0 + 5 + (-1) + (-4) + 2 + 0}{7} = \frac{0}{7}$
$= 0$

Pages 83–84 Practice and Apply

12. $54 \div 9 = 6$

13. $45 \div 5 = 9$

14. $-27 \div (-3) = 9$

15. $-64 \div (-8) = 8$

16. $-72 \div (-9) = 8$

17. $-60 \div (-6) = 10$

18. $-77 \div 7 = -11$

19. $-300 \div 6 = -50$

20. $480 \div (-12) = -40$

21. $\frac{132}{-12} = 132 \div -12$
$= -11$

22. $\frac{175}{-25} = 175 \div -25$
$= -7$

23. $\frac{143}{-13} = 143 \div -13$
$= -11$

24. $-91 \div (-7) = 13$

25. $-76 \div (-4) = 19$

26. $\frac{x}{-5} = \frac{85}{-5}$
$= 85 \div -5$
$= -17$

27. $\frac{108}{m} = \frac{108}{-9}$
$= 108 \div -9$
$= -12$

28. $\frac{c}{d} = \frac{-63}{-7}$
$= -63 \div -7$
$= 9$

29. $\frac{s}{t} = \frac{52}{-4}$
$= 52 \div -4$
$= -13$

30. $xy \div (-3) = (9)(-7) \div (-3)$
$= -63 \div (-3)$
$= 21$

31. $ab \div 6 = (-12)(-8) \div 6$
$= 96 \div 6$
$= 16$

32. $\frac{4 + (-8) + 9 + (-3) + (-7) + 10 + 2}{7} = \frac{7}{7}$
$= 1$

33. $\frac{46 + 52 + 49 + 53 + 45}{5} = \frac{245}{5}$
$= 49$

The average points per game was 49.

34. $d = \left|65 - \frac{h + l}{2}\right| = \left|65 - \frac{81 + 65}{2}\right|$
$= \left|65 - \frac{146}{2}\right|$
$= |65 - 73|$
$= |-8|$
$= 8$ degree days

35. $d = \left|65 - \frac{h + l}{2}\right| = \left|65 - \frac{8 + 0}{2}\right|$
$= \left|65 - \frac{8}{2}\right|$
$= |65 - 4|$
$= |61|$
$= 61$ degree days

36. See students' work.
37. Sample answer: $x = -144$; $y = 12$; $z = -12$
38. addition, subtraction, and multiplication
39. When the signs of the integers are the same, both a product and a quotient are positive; when the signs are different, the product and quotient are negative. Answers should include the following.
- Sample answer: $4 \cdot (-6) = -24$ and $-24 \div 4 = -6$; $-3 \cdot 2 = -6$ and $-6 \div (-3) = 2$
- Sample answers: same sign: $-30 \div (-5) = 6$, $30 \div 5 = 6$;
different signs: $-24 \div 8 = -3$, $24 \div (-8) = -3$

40. C; $-10 \div 2 = -5°$

41. B; $\frac{8 + 7 + 8 + 9 + x}{5} = 8$
$\frac{32 + x}{5} = 8$
$32 + x = 8 \cdot 5$
$32 + x = 40$
$32 + 8 = 40$
$x = 8$

Page 84 Maintain Your Skills

42. $-8 - (-25) = -8 + 25$
$= 17$

43. $75 - 114 = 75 + (-114)$
$= -39$

44. $2ab \cdot (-2) = (2)(ab)(-2)$
$= [(2)(-2)](ab)$
$= -4ab$

45. $(-10c)(5d) = (-10)(c)(5)(d)$
$= [(-10)(5)](cd)$
$= -50cd$

46. 5, 4, 2, −1, ?, ? (differences −1, −2, −3, −4, −5)

Assuming the pattern continues, the next two terms are $-1 - 4 = -5$ and $-5 - 5 = -10$.

47. Start at the origin, and move 1 unit right and 5 units up. The ordered pair for point B is (1, 5).
48. Start at the origin, and move 6 units right and 2 units up. The ordered pair for point F is (6, 2).
49. Start at the origin, and move 4 units right and 5 units up. The ordered pair for point D is (4, 5).
50. Start at the origin and move 3 units up. The ordered pair for A is (0, 3).

Page 84 Practice Quiz 2

1. $-12 \cdot 7 = -84$
2. $-6(-15) = 90$
3. $-3(-7)(-6) = [-3(-7)](-6)$
$= 21(-6)$
$= -126$
4. $3(-8)(-5) = [3(-8)](-5)$
$= -24(-5)$
$= 120$
5. $-124 \div 4 = -31$
6. $-90 \div (-6) = 15$
7. $125 \div (-5) = -25$
8. $-126 \div (-9) = 14$
9. $4x(-5y) = (4)(x)(-5)(y)$
$= [4(-5)](xy)$
$= -20xy$
10. $-9a = -9(-6)$
$= 54$

2-6 The Coordinate System

Page 85 How is a coordinate system used to locate places on Earth?

a. The latitude of New Orleans is 30°N.
b. The longitude of New Orleans is 90°W.
c. 30° north of the equator; 90° west of the prime meridian

Page 87 Check for Understanding

1. Sample answer: (3, 6) represents a point 3 units to the right and 6 units up from the origin. (6, 3) represents a point 6 units to the right and 3 units up from the origin.
2. Sample answer: (0, 0), (4, 0)

3. Keisha; a point in Quadrant I has two positive coordinates. Interchanging the coordinates will still result in two positive coordinates, and the point will be in Quadrant I.
4. The x-coordinate is -4.
   The y-coordinate is 5.
   The ordered pair is $(-4, 5)$.
5. The x-coordinate is 1.
   The y-coordinate is 3.
   The ordered pair is $(1, 3)$.
6. The x-coordinate is -3.
   The y-coordinate is -2.
   The ordered pair is $(-3, -2)$.
7. The x-coordinate is 5.
   The y-coordinate is -4.
   The ordered pair is $(5, -4)$.
8. Start at the origin. Move 3 units to the right. Then move 4 units down and draw a dot. Point J is in Quadrant IV.
9. Start at the origin. Move 2 units to the left. Then move 2 units up and draw a dot. Point K is in Quadrant II.
10. Start at the origin. Since the x-coordinate is 0, the point lies on the y-axis. Then move 4 units up and draw a dot. Point L is not in any quadrant.
11. Start at the origin. Move 1 unit to the left. Then move 2 units down and draw a dot. Point M is in Quadrant III.

Exercises 8–11

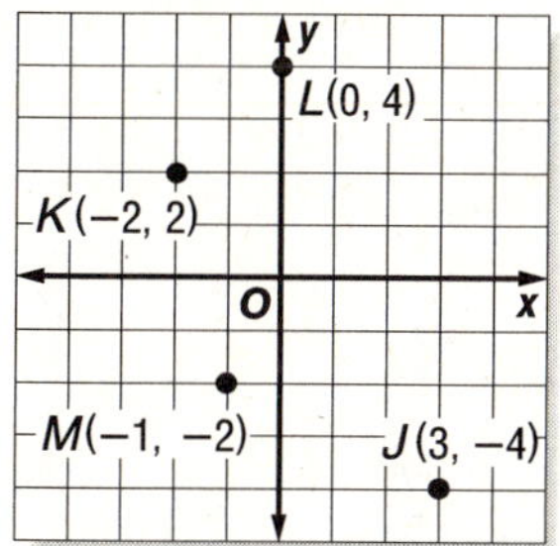

12. Sample answer:

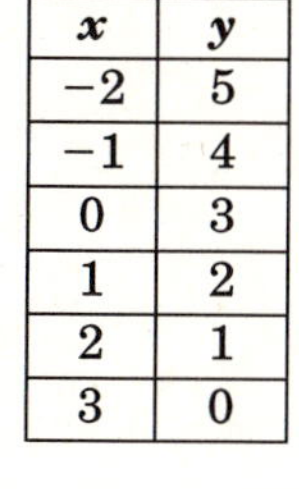

x	y
−2	5
−1	4
0	3
1	2
2	1
3	0

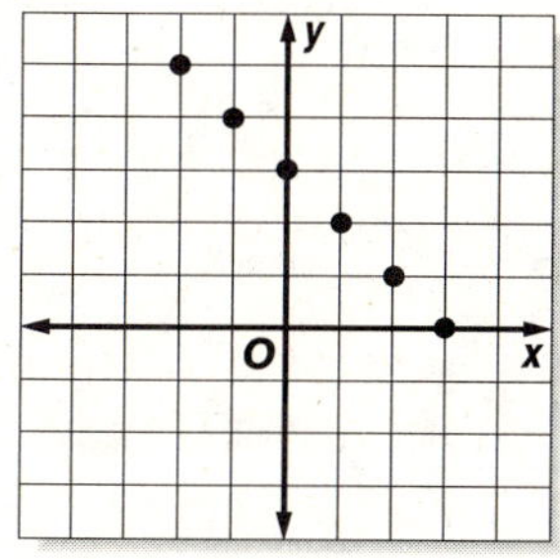

The points are along a diagonal line that crosses the y-axis at $y = 3$ and the x-axis at $x = 3$.

Pages 88–89 Practice and Apply

13. The x-coordinate is -2.
    The y-coordinate is 4.
    The ordered pair is $(-2, 4)$.
14. The x-coordinate is -4.
    The y-coordinate is 2.
    The ordered pair is $(-4, 2)$.
15. The x-coordinate is 4.
    The y-coordinate is -2.
    The ordered pair is $(4, -2)$.
16. The x-coordinate is 1.
    The y-coordinate is -4.
    The ordered pair is $(1, -4)$.
17. The x-coordinate is 2.
    The y-coordinate is 2.
    The ordered pair is $(2, 2)$.
18. The x-coordinate is 4.
    The y-coordinate is 3.
    The ordered pair is $(4, 3)$.
19. The x-coordinate is 0.
    The y-coordinate is -2.
    The ordered pair is $(0, -2)$.
20. The x-coordinate is 3.
    The y-coordinate is 0.
    The ordered pair is $(3, 0)$.
21. The x-coordinate is -3.
    The y-coordinate is -5.
    The ordered pair is $(-3, -5)$.
22. The x-coordinate is -2.
    The y-coordinate is -2.
    The ordered pair is $(-2, -2)$.
23. I
24. II
25. IV
26. III
27. IV
28. II
29. none
30. none
31. none
32. IV
33. none
34. none

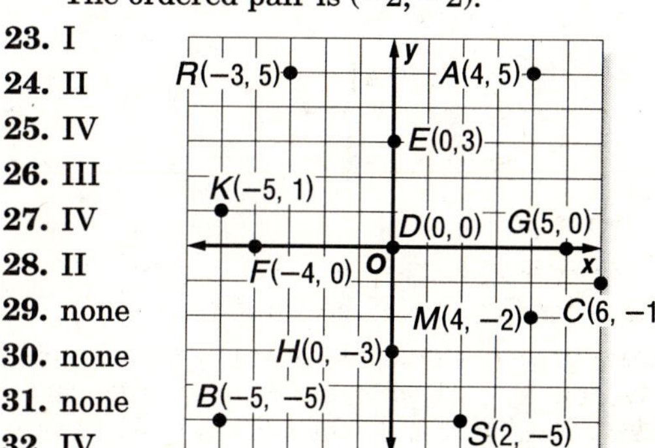

35. Sample answer:

x	y
1	4
2	3
3	2
4	1
5	0
6	−1

The points are along a line slanting down to the right, crossing the y-axis at 5 and the x-axis at 5.

36. Sample answer:

x	y
−2	0
−1	−1
0	−2
1	−3
2	−4
3	−5

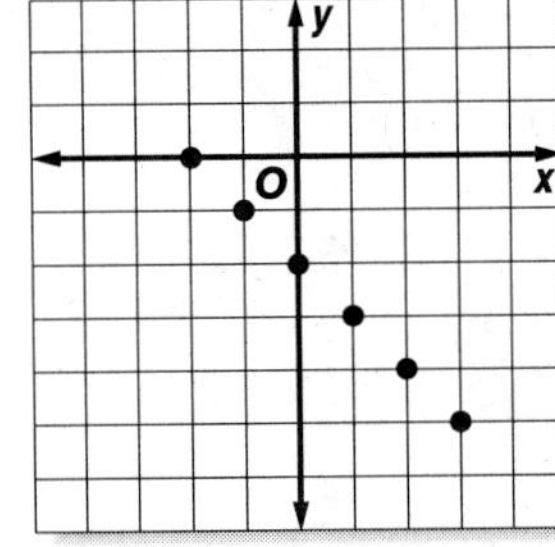

The points are along a line slanting down, crossing the x-axis at -2 and the y-axis at -2.

37. Sample answer:

x	y
−2	−4
−1	−2
0	0
1	2
2	4
3	6

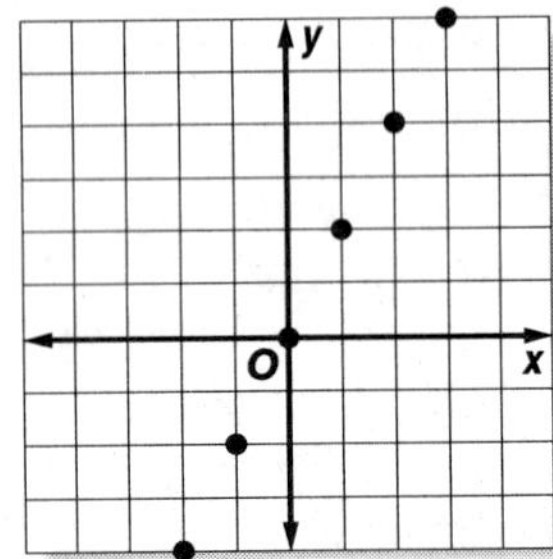

The points are along a line slanting up, through the origin.

38. Sample answer:

x	y
−2	4
−1	2
0	0
1	−2
2	−4
3	−6

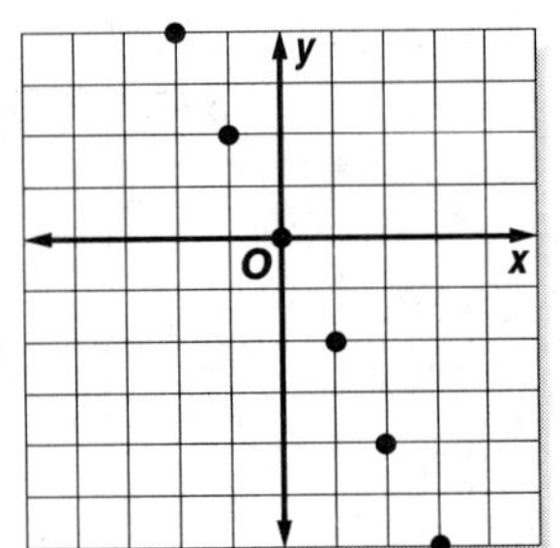

The points are along a line slanting down, through the origin.

39. Sample answer:

x	y
−3	−1
−2	0
−1	1
0	2
1	3
2	4

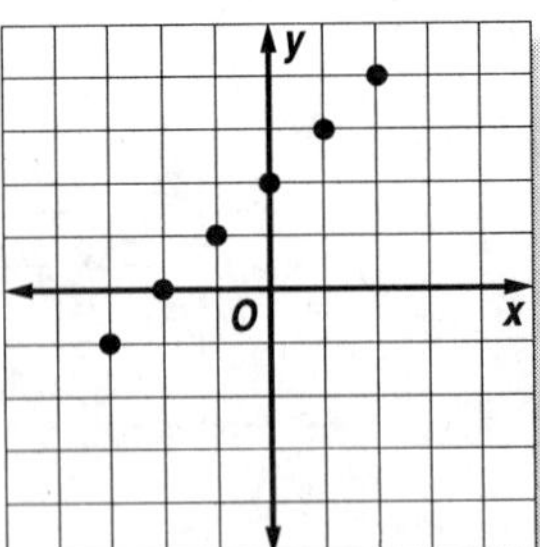

The points are along a line slanting up, crossing the y-axis at 2 and the x-axis at -2.

40. Sample answer:

x	y
−2	−3
−1	−2
0	−1
1	0
2	1
3	2

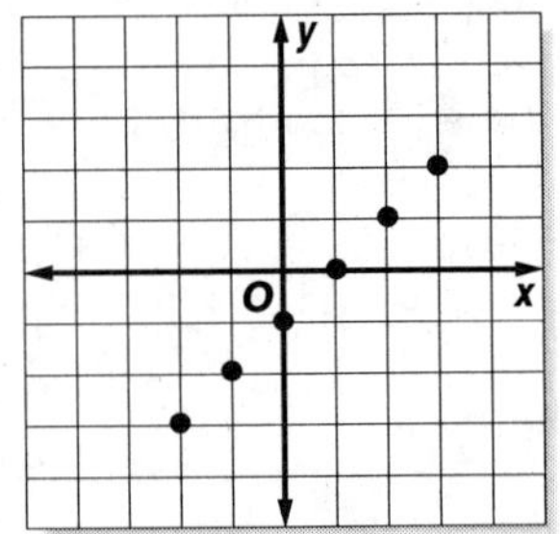

The points are along a line slanting up, crossing the y-axis at -1 and the x-axis at 1.

41. 5-point star

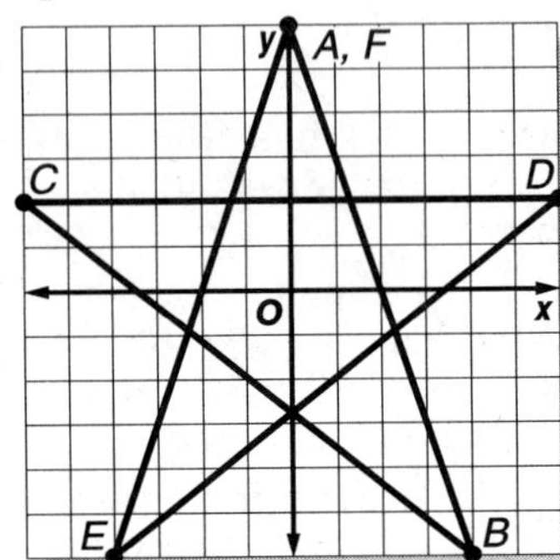

42. Kite

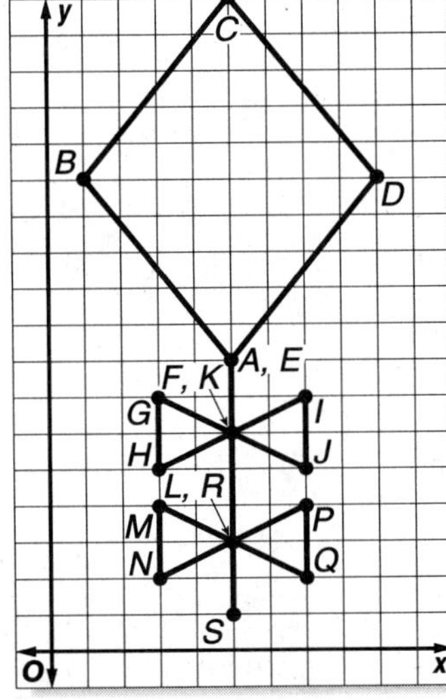

43. Sample answer:

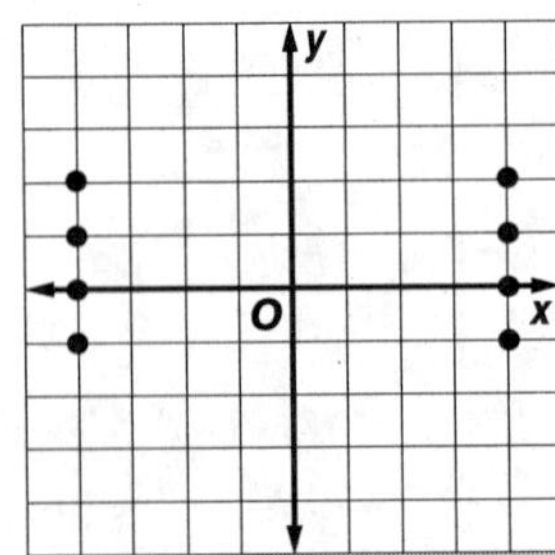

The graph can include any integer pairs where $x > 3$ or $x < -3$.

44.

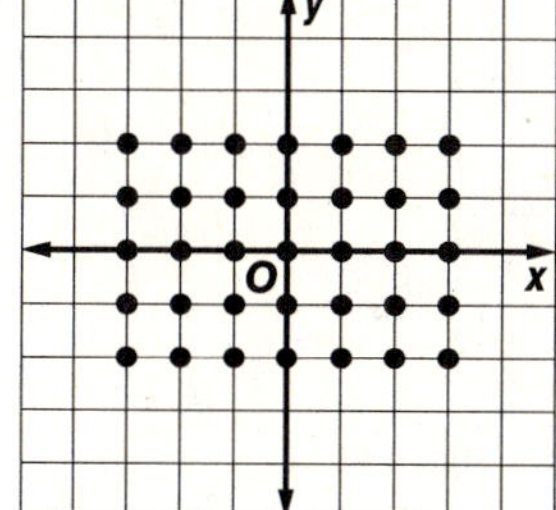

45.

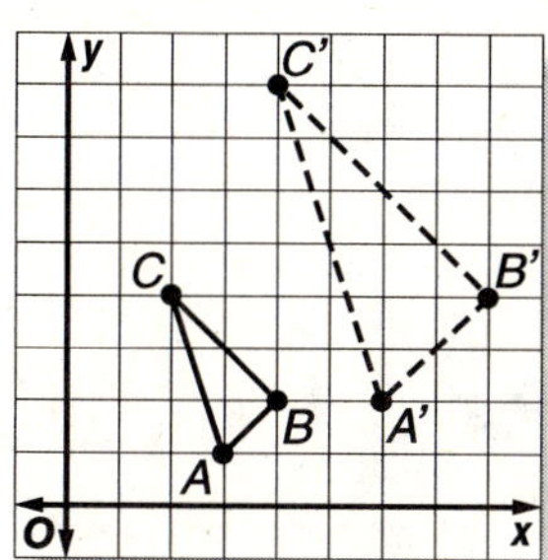

The new triangle is twice the size of the original triangle and is moved to the right and up.

46.

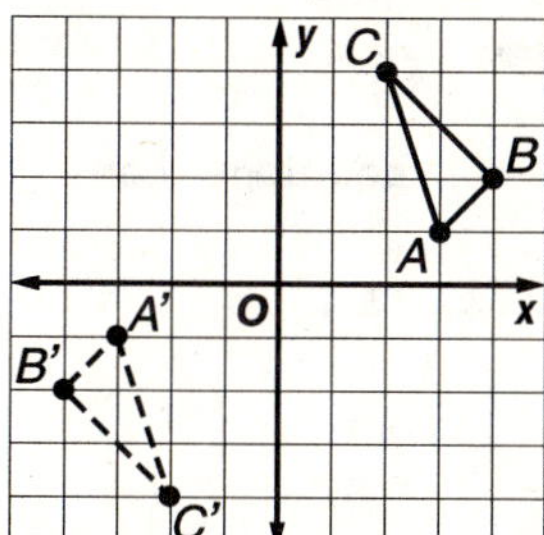

The new triangle is reflected over the x- and y-axes; it is the same size as the original triangle.

47.

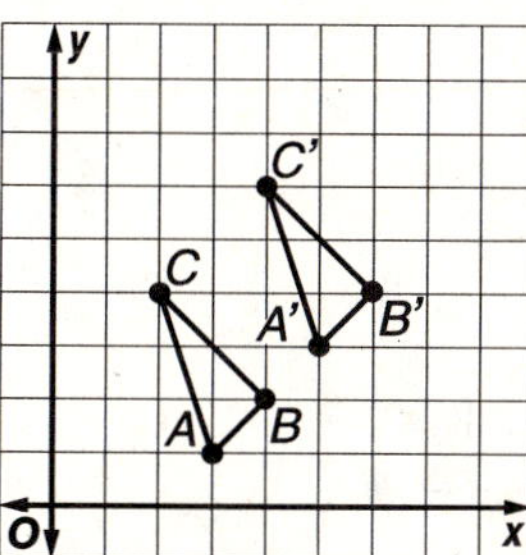

The new triangle is translated right 2 units and up 2 units; it is the same size as the original triangle.

48.

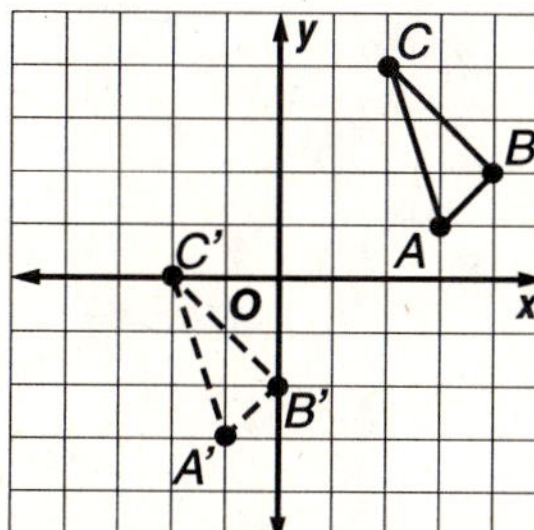

The new triangle is translated left 4 and down 4; it is the same size as the original triangle.

49. See students' work.

50a. I **50b.** III **50c.** II

51. Sample answer:

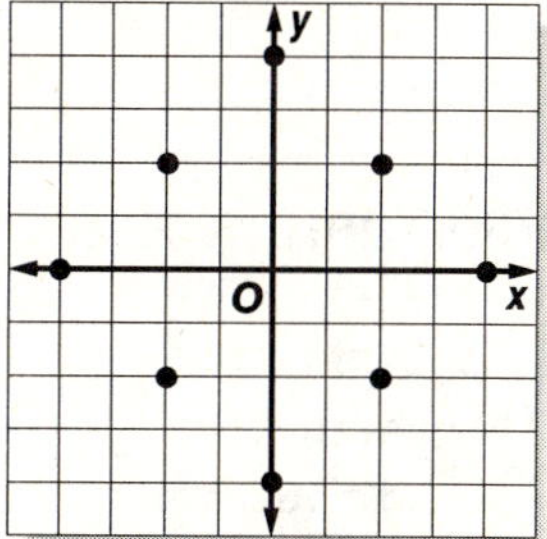

The points lie outside a rhombus defined by (0, 4), (4, 0), (0, −4), and (−4, 0).

52. Latitude and longitude lines are similar to a coordinate system. Answers should include the following.

- Coordinates describe a location by telling how far and in what direction the location is from a given starting point.
- In the latitude and longitude system, longitude is similar to the x-coordinate of a point in a coordinate plane. Latitude is similar to the y-coordinate. The point where the prime meridian and the equator meet is similar to the origin in the coordinate system.

53. D **54.** B

Page 89 Maintain Your Skills

55. $-24 \div 8 = -3$ **56.** $105 \div (-5) = -21$

57. $-400 \div (-50) = 8$

58. $-5fg = -5(-9)(-6)$
$= [-5(-9)](-6)$
$= 45(-6)$
$= -270$

59. $2gh = 2(-6)(8)$
$= [2(-6)](8)$
$= -12(8)$
$= -96$

60. $-10fh = -10(-9)(8)$
$= [-10(-9)](8)$
$= 90(8)$
$= 720$

61. $8 - (-5) = 8 + 5$
$= 13°F$

62. $(a + 8) + 6 = a + (8 + 6)$
$= a + 14$

63. $4(6h) = (4 \cdot 6)h$
$= 24h$

64. $(n \cdot 7) \cdot 8 = n \cdot (7 \cdot 8)$
$= n \cdot 56$
$= 56n$

65. $(b \cdot 9) \cdot 5 = b \cdot (9 \cdot 5)$
$= b \cdot 45$
$= 45b$

66. $(16 + 3y) + y = 16 + (3y + y)$
$= 16 + 4y$
$= 4y + 16$

67. $0(4z) = (0 \cdot 4)z$
$= 0$

Chapter 2 Study Guide and Review

Page 90 Vocabulary and Concept Check

1. negative number
2. quadrants
3. coordinate
4. additive inverses
5. integers
6. absolute value
7. inequality

Pages 90–92 Lesson-by-Lesson Review

8. $8 > -8$
9. $-3 = -3$
10. $-2 < 0$
11. $-12 > -21$
12. $|-32| = 32$
13. $|25| = 25$
14. $-|15| = -(15) = -15$
15. $|-8| + |-14| = 8 + 14$
$= 22$
16. $-6 + (-3) = -9$
17. $-4 + (-1) = -5$
18. $-2 + 7 = 5$
19. $4 + (-8) = -4$
20. $6 + (-9) + (-8) = 6 + [(-9) + (-8)]$
$= 6 + (-17)$
$= -11$
21. $4 + (-7) + (-3) + (-4) = 4 + [(-7) + (-3) + (-4)]$
$= 4 + [-10 + (-4)]$
$= 4 + (-14)$
$= -10$
22. $4 - 9 = 4 + (-9)$
$= -5$
23. $-3 - 5 = -3 + (-5)$
$= -8$
24. $7 - (-2) = 7 + 2$
$= 9$
25. $-1 - (-6) = -1 + 6$
$= 5$
26. $-7 - 8 = -7 + (-8)$
$= -15$
27. $6 - 10 = 6 + (-10)$
$= -4$
28. $-3 - (-7) = -3 + 7$
$= 4$
29. $6 - (-3) = 6 + 3$
$= 9$
30. $-9(5) = -45$
31. $11(-6) = -66$
32. $-4(-7) = 28$
33. $-3(-16) = 48$
34. $-2a(4b) = (-2)(a)(4)(b)$
$= [(-2)(4)](ab)$
$= -8ab$
35. $-14 \div (-2) = 7$
36. $-52 \div (-4) = 13$
37. $-36 \div 9 = -4$
38. $88 \div (-4) = -22$
39. $\frac{-3 + (-6) + 9 + (-3) + 13}{5} = \frac{10}{5}$
$= 2$
40. I
41. III
42. II
43. none

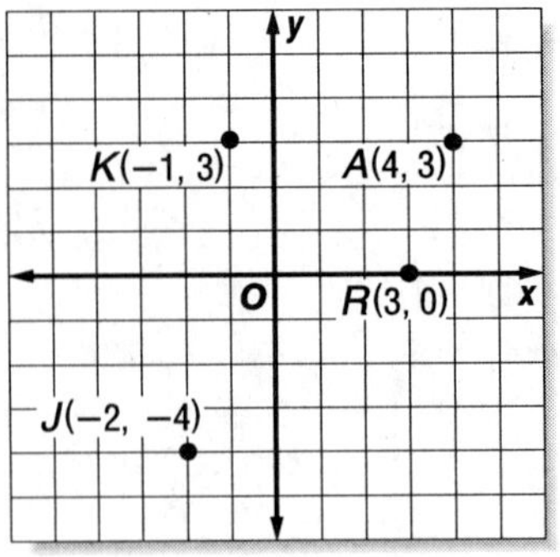

Chapter 2 Practice Test

Page 93

1. Find the difference of their absolute values. Give the result the sign of the integer having the larger absolute value.
2. Add the additive inverse of the integer being subtracted.
3. 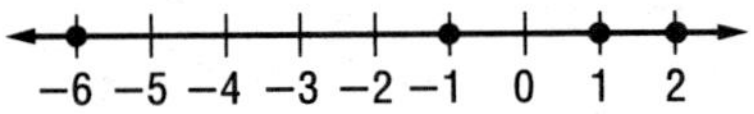

4. $-5 < 2;\ 2 > -5$
5. $12 > -15;\ -15 < 12$
6. $-5 < -3$
7. $-5 > -14$
8. $4 < |-7|$
9. $-4 + (-8) = -12$
10. $-9 + 15 = 6$
11. $12 + (-15) = -3$
12. $14 + (-7) + -11 = 14 + [(-7) + (-11)]$
$= 14 + (-18)$
$= -4$
13. $4 - 13 = 4 + (-13)$
$= -9$
14. $8 - (-6) = 8 + 6$
$= 14$
15. $-6 - (-10) = -6 + 10$
$= 4$
16. $-14 - (-7) = -14 + 7$
$= -7$
17. $6(-8) = -48$
18. $-9(8) = -72$
19. $-7(-5) = 35$
20. $2(-4)(11) = -8(11)$
$= -88$
21. $54 \div (-9) = -6$
22. $-64 \div (-4) = 16$
23. $-250 \div 25 = -10$
24. $-144 \div (-6) = 24$

25. $ab - c = (-5)(3) - (-10)$
$= -15 + 10$
$= -5$

26. $c \div a = -10 \div (-5)$
$= 2$

27. $4c + |a| = 4(-10) + |-5|$
$= -40 + 5$
$= -35$

28. II
29. IV
30. III

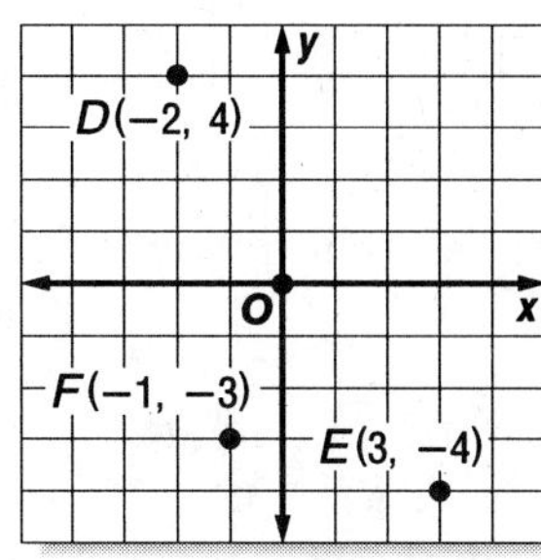

31. $\frac{-12 + 3 + (-7) + 0 + (-4) + 1 + (-2)}{7} = \frac{-21}{7}$
$= -3$

The average low temperature was −3°F.

32. $-7 + 3 \cdot (-4) = -7 + (-12)$
$= -19$

Total yards at the end of four plays was −19 yd.

33. A; On a coordinate plane, west would be to the left and north would be up. So, 2 miles west would be represented by 2 units left, and 4 miles north would be represented by 4 units up.

Chapter 2 Standardized Test Practice

Pages 94–95

1. C; Each hour the number of cells is 3 times the number from the previous hour. After 4 hours, there will be $3 \cdot 27 = 81$. After 5 hours, there will be $3 \cdot 81 = 243$ cells.
2. B; Since your sister has 3 more CDs than you, you have 3 less than your sister. $y = s - 3$
3. C; $|-8| = 8$, the largest of the numbers
4. B; $4 - 5 = 4 + (-5)$
$= -1$ foot
5. B; $-5 + 2 = -3$
6. D; $x = 7 - (-3)$
$= 7 + 3$
$= 10$
7. D; $-3t + 7 = -3(-5) + 7$
$= 15 + 7$
$= 22$
8. D; $\frac{b - 13}{a} = \frac{5 - 13}{-2}$
$= \frac{5 + (-13)}{-2}$
$= \frac{-8}{-2}$
$= -8 \div -2$
$= 4$
9. B; The point X is two units left and 5 units up from the origin.
10. C; Point U is 5 units left and 2 units down from the origin.
11. The height of the bars shows the number of girls and boys in each grade. The bars showing the greatest difference between the number of boys and girls are grade 6.
12. Let x = the number.
$x - 9 = 15$
$x = 15 + 9$
$x = 24$
13. $7 + (-11) = -4$ yards
14. $3 - (-15) = 3 + 15$
$= 18$°F
15. $y = x - 7$
$5 = x - 7$
$5 + 7 = x$
$12 = x$
16. $\frac{2 + (-7) + (-12) + 9}{4} = \frac{-8}{4}$
$= -2$°F
17.

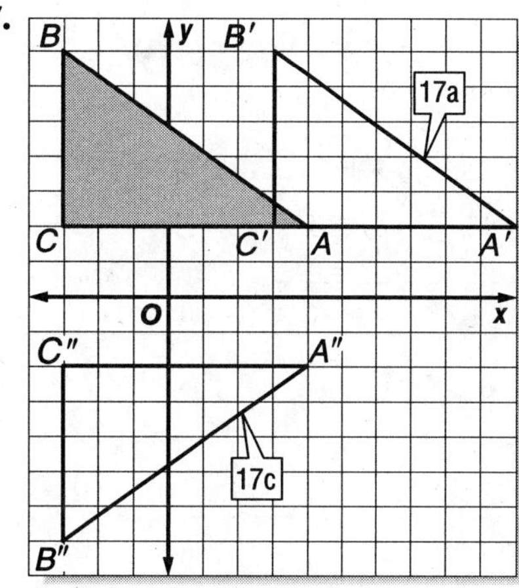

17a. Yes, the new figure is the same size and shape as the original triangle.
The original triangle could be slid (translated) 6 units to the right to create the new figure, A′B′C′.

17b. If −6 were added to each x-coordinate, the original triangle could be slid 6 units to the left (or slid −6 units) to create the new figure.

17c. Yes, the new figure is the same size and shape as the original triangle. The original triangle could be flipped (reflected) across the x-axis to create the new figure, A″B″C″.

17d. If you multiplied the x-coordinate of each original coordinate pair by −1, the original triangle would be flipped (reflected) over the y-axis.

Chapter 3 Equations

Page 97 Getting Started

1. $2(-3) = -6$
2. $-4(3) = -12$
3. $-5(-2) = 10$
4. $-4(6) = -24$
5. $5 - 7 = 5 + (-7)$
6. $6 - 10 = 6 + (-10)$
7. $-5 - 9 = -5 + (-9)$
8. $11 - 10 = 11 + (-10)$
9. $6 + (-9) = -3$
10. $-8 + 4 = -4$
11. $4 + (-4) = 0$
12. $7 + (-10) = -3$
13. $5 + 2n$
14. $n - 15$
15. $n - 3$
16. $n \div 10$

3-1 The Distributive Property

Page 98 How are rectangles related to the Distributive Property?

a.

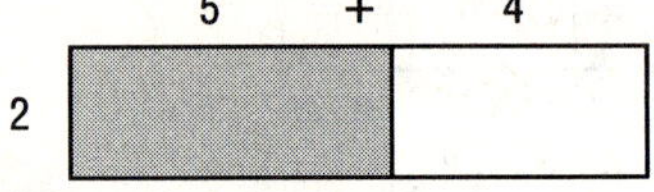

$2(5 + 4) = 2 \cdot 9$
$= 18$

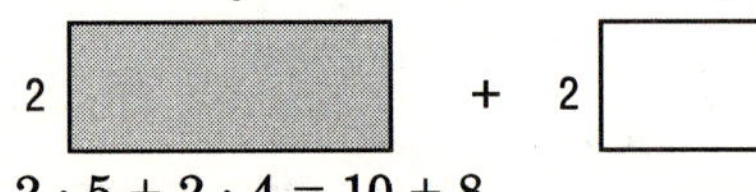

$2 \cdot 5 + 2 \cdot 4 = 10 + 8$
$= 18$

b.

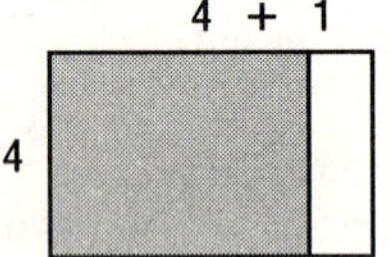

$4(4 + 1) = 4 \cdot 5$
$= 20$

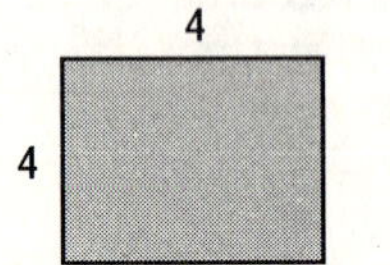

$4 \cdot 4 + 4 \cdot 1 = 16 + 4$
$= 20$

c.

2 2
3

$3(2 + 2) = 12$

2 2
3 + 3

$3 \cdot 2 + 3 \cdot 2 = 12$

d. The areas are equal.

Page 100 Check for Understanding

1. Sample answer: $2(3 + 4) = 2 \cdot 3 + 2 \cdot 4$
2. Catelyn; each number inside the parentheses should be multiplied by 3.
3. $5(7 + 8) = 5 \cdot 7 + 5 \cdot 8$
   $= 35 + 40$
   $= 75$
4. $2(9 + 1) = 2 \cdot 9 + 2 \cdot 1$
   $= 18 + 2$
   $= 20$
5. $(2 + 4)6 = 2 \cdot 6 + 4 \cdot 6$
   $= 12 + 24$
   $= 36$
6. $4(x + 3) = 4x + 4 \cdot 3$
   $= 4x + 12$
7. $(n + 2)3 = n \cdot 3 + 2 \cdot 3$
   $= 3n + 6$
8. $8(y - 2) = 8[y + (-2)]$
   $= 8y + 8 \cdot (-2)$
   $= 8y + (-16)$
   $= 8y - 16$
9. $-6(x - 5) = -6[x + (-5)]$
   $= -6x + (-6)(-5)$
   $= -6x + 30$
10. \$6.25(4 + 5); \$6.25(4) + \$6.25(5)
11. \$6.25(4 + 5) = \$6.25(4) + \$6.25(5)
    = \$25 + \$31.25
    = \$56.25
    The total wages are \$56.25.

Pages 101–102 Practice and Apply

12. $2(6 + 1) = 2 \cdot 6 + 2 \cdot 1$
    $= 12 + 2$
    $= 14$
13. $5(7 + 3) = 5 \cdot 7 + 5 \cdot 3$
    $= 35 + 15$
    $= 50$

14. $(4 + 6)9 = 4 \cdot 9 + 6 \cdot 9$
$= 36 + 54$
$= 90$

15. $(4 + 3)3 = 4 \cdot 3 + 3 \cdot 3$
$= 12 + 9$
$= 21$

16. $(9 + 2)4 = 9 \cdot 4 + 2 \cdot 4$
$= 36 + 8$
$= 44$

17. $(8 + 8)2 = 8 \cdot 2 + 8 \cdot 2$
$= 16 + 16$
$= 32$

18. $7(3 - 2) = 7[3 + (-2)]$
$= 7 \cdot 3 + 7 \cdot (-2)$
$= 21 + (-14)$
$= 7$

19. $6(8 - 5) = 6[8 + (-5)]$
$= 6 \cdot 8 + 6 \cdot (-5)$
$= 48 + (-30)$
$= 18$

20. $-5(8 - 4) = -5[8 + (-4)]$
$= -5 \cdot 8 + (-5)(-4)$
$= -40 + 20$
$= -20$

21. $-3(9 - 2) = -3[9 + (-2)]$
$= -3 \cdot 9 + (-3)(-2)$
$= -27 + 6$
$= -21$

22. $(8 - 4)(-2) = [8 + (-4)](-2)$
$= 8(-2) + (-4)(-2)$
$= -16 + 8$
$= -8$

23. $(10 - 3)(-5) = [10 + (-3)](-5)$
$= 10(-5) + (-3)(-5)$
$= -50 + 15$
$= -35$

24. 4(\$7 + \$3), 4(\$7) + 4(\$3);
4(\$7) + 4(\$3) = \$28 + \$12
= \$40
The cost is \$40.

25. 12(\$15 + \$10 + \$8), 12(\$15) +12(\$10) + 12(\$8);
12(\$15) + 12(\$10) + 12(\$8) = \$180 + \$120 + \$96
= \$396
The cost is \$396.00.

26. $2(x + 3) = 2x + 2 \cdot 3$
$= 2x + 6$

27. $5(y + 6) = 5y + 5 \cdot 6$
$= 5y + 30$

28. $3(n + 1) = 3n + 3 \cdot 1$
$= 3n + 3$

29. $7(y + 8) = 7y + 7 \cdot 8$
$= 7y + 56$

30. $(x + 3)4 = x \cdot 4 + 3 \cdot 4$
$= 4x + 12$

31. $(y + 2)10 = y \cdot 10 + 2 \cdot 10$
$= 10y + 20$

32. $(3 + y)6 = 3 \cdot 6 + y \cdot 6$
$= 18 + 6y$

33. $(2 + x)5 = 2 \cdot 5 + x \cdot 5$
$= 10 + 5x$

34. $3(x - 2) = 3[x + (-2)]$
$= 3x + 3(-2)$
$= 3x + (-6)$
$= 3x - 6$

35. $9(m - 2) = 9[m + (-2)]$
$= 9m + 9(-2)$
$= 9m + (-18)$
$= 9m - 18$

36. $8(z - 3) = 8[z + (-3)]$
$= 8z + 8(-3)$
$= 8z + (-24)$
$= 8z - 24$

37. $15(s - 3) = 15[s + (-3)]$
$= 15s + 15(-3)$
$= 15s + (-45)$
$= 15s - 45$

38. $(r - 5)6 = [r + (-5)]6$
$= r \cdot 6 + (-5) \cdot 6$
$= 6r + (-30)$
$= 6r - 30$

39. $(x - 3)12 = [x + (-3)]12$
$= x \cdot 12 + (-3)(12)$
$= 12x + (-36)$
$= 12x - 36$

40. $(t - 4)5 = [t + (-4)]5$
$= t \cdot 5 + (-4)(5)$
$= 5t + (-20)$
$= 5t - 20$

41. $(w - 10)2 = [w + (-10)]2$
$= w \cdot 2 + (-10) \cdot 2$
$= 2w + (-20)$
$= 2w - 20$

42. $-2(z + 4) = -2z + (-2) \cdot 4$
$= -2z + (-8)$
$= -2z - 8$

43. $-5(a + 10) = -5a + (-5) \cdot 10$
$= -5a + (-50)$
$= -5a - 50$

44. $-2(x - 7) = -2[x + (-7)]$
$= -2x + (-2)(-7)$
$= -2x + 14$

45. $-5(w - 8) = -5[w + (-8)]$
$= -5w + (-5)(-8)$
$= -5w + 40$

46. $(y - 4)(-2) = [y + (-4)](-2)$
$= y(-2) + (-4)(-2)$
$= -2y + 8$

47. $(a - 6)(-5) = [a + (-6)](-5)$
$= a(-5) + (-6)(-5)$
$= -5a + 30$

48. $2(x + y) = 2 \cdot x + 2 \cdot y$
$= 2x + 2y$

49. $3(a + b) = 3 \cdot a + 3 \cdot b$
$= 3a + 3b$

50. 2(\$59.20) + 2(\$38.55) = \$118.40 + \$77.10
= \$195.50

The total spent was \$195.50.

51. 5(\$59.20 + \$38.55) = 5(\$59.20) + 5(\$38.55)
= \$296 + \$192.75
= \$488.75

The total spent was \$488.75.

52. You can compute the area of two rectangles with the same width in two ways. You can put them together and multiply to find the total area or find each area separately and then add to find the total. The Distributive Property states that the expressions $a(b + c)$, and $ab + ac$ are equivalent. Answers should include the following.

- 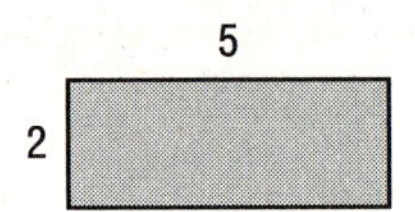

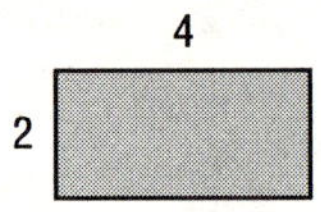

- $2(5 + 4) = 18;\ 2 \cdot 5 + 2 \cdot 4 = 18$

53. No; $3 + (4 \cdot 5) = 23$, $(3 + 4)(3 + 5) = 56$

54. D; $p(t + s)$

55. C; $5(a + b) = 5(b + a)$, using the Commutative Property of Addition.

56. $7 \cdot 14 = 7(10 + 4)$
$= 7 \cdot 10 + 7 \cdot 4$
$= 70 + 28$
$= 98$

57. $8 \cdot 23 = 8(20 + 3)$
$= 8 \cdot 20 + 8 \cdot 3$
$= 160 + 24$
$= 184$

58. $9 \cdot 32 = 9(30 + 2)$
$= 9 \cdot 30 + 9 \cdot 2$
$= 270 + 18$
$= 288$

59. $16 \cdot 11 = 16(10 + 1)$
$= 16 \cdot 10 + 16 \cdot 1$
$= 160 + 16$
$= 176$

60. $14 \cdot 12 = 14(10 + 2)$
$= 14 \cdot 10 + 14 \cdot 2$
$= 140 + 28$
$= 168$

61. $9 \cdot 103 = 9(100 + 3)$
$= 9 \cdot 100 + 9 \cdot 3$
$= 900 + 27$
$= 927$

62. $11 \cdot 102 = 11(100 + 2)$
$= 11 \cdot 100 + 11 \cdot 2$
$= 1100 + 22$
$= 1122$

63. $12 \cdot 1004 = 12(1000 + 4)$
$= 12 \cdot 1000 + 12 \cdot 4$
$= 12{,}000 + 48$
$= 12{,}048$

Page 102 Maintain Your Skills

64a.

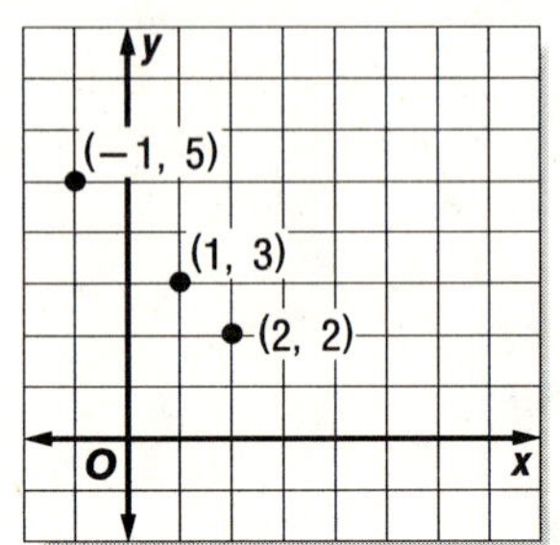

64b. The points are in a straight line.

65. $\frac{-4y}{x} = \frac{-4(-3)}{2}$
$= \frac{12}{2}$
$= 12 \div 2$
$= 6$

66.

Value of n	$2n + 3 = 9$	True or False?
1	$2(1) + 3 \stackrel{?}{=} 9$	false
2	$2(2) + 3 \stackrel{?}{=} 9$	false
3	$2(3) + 3 \stackrel{?}{=} 9$	true ✓
4	$2(4) + 3 \stackrel{?}{=} 9$	false
5	$2(5) + 3 \stackrel{?}{=} 9$	false

The solution of $2n + 3 = 9$ is 3.

67.

Value of n	$3n - 4 = 8$	True or False?
1	$3(1) - 4 \stackrel{?}{=} 8$	false
2	$3(2) - 4 \stackrel{?}{=} 8$	false
3	$3(3) - 4 \stackrel{?}{=} 8$	false
4	$3(4) - 4 \stackrel{?}{=} 8$	true ✓
5	$3(5) - 4 \stackrel{?}{=} 8$	false

The solution of $3n - 4 = 8$ is 4.

68.

Value of x	$4x - 9 = -5$	True or False?
1	$4(1) - 9 \stackrel{?}{=} -5$	true ✓
2	$4(2) - 9 \stackrel{?}{=} -5$	false
3	$4(3) - 9 \stackrel{?}{=} -5$	false
4	$4(4) - 9 \stackrel{?}{=} -5$	false
5	$4(5) - 9 \stackrel{?}{=} -5$	false

The solution of $4x - 9 = -5$ is 1.

69. 5 9 13 17 ? ? ?
(+4 +4 +4 +4 +4 +4)

Assuming the pattern continues, the next three terms are 17 + 4, or 21; 21 + 4, or 25; and 25 + 4, or 29.

70. 20 22 26 32 ? ? ?
(+2 +4 +6 +8 +10 +12)

Assuming the pattern continues, the next three terms are 32 + 8, or 40; 40 + 10, or 50; and 50 + 12, or 62.

71. 5 10 20 40 ? ? ?
(+5 +10 +20 +40 +80 +160)

Assuming the pattern continues, the next three terms are 40 + 40, or 80; 80 + 80, or 160; and 160 + 160, or 320.

72. $5 - 3 = 5 + (-3)$

73. $-8 - 4 = -8 + (-4)$

74. $10 - 14 = 10 + (-14)$

75. $3 - 9 = 3 + (-9)$

76. $-2 - (-5) = -2 + 5$

77. $-7 - 10 = -7 + (-10)$

3-2 Simplifying Algebraic Expressions

Page 103 How can you use algebra tiles to simplify an algebraic expression?

a.

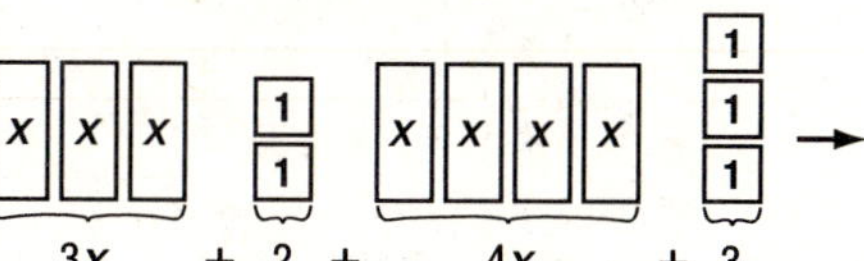

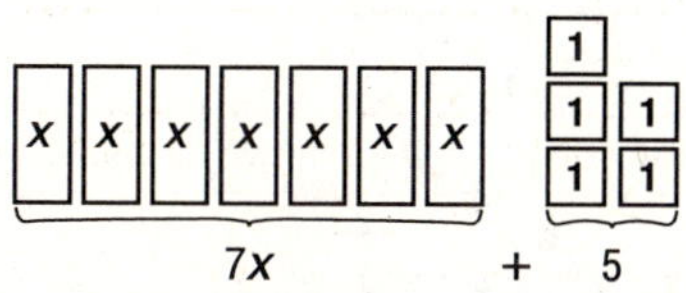

b.

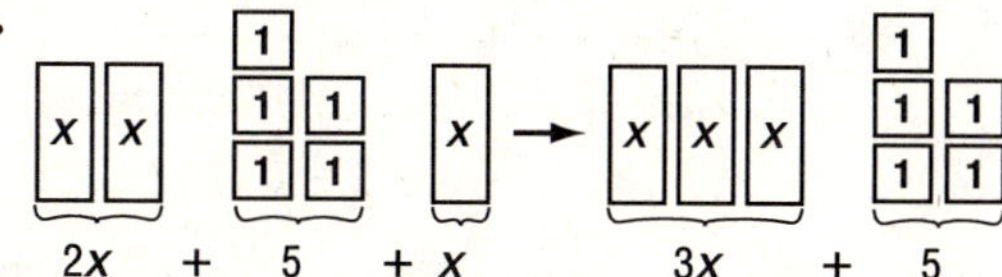

c.

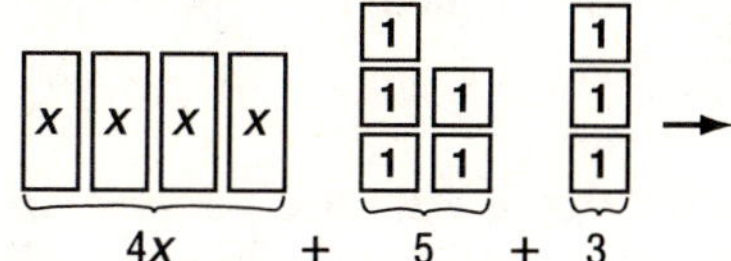

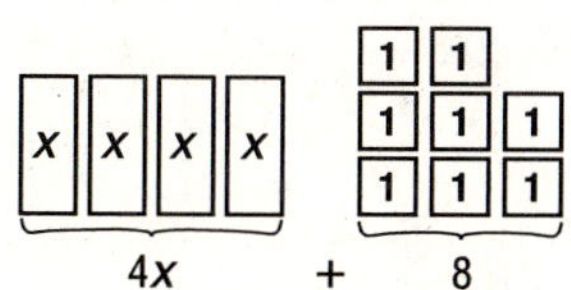

d.

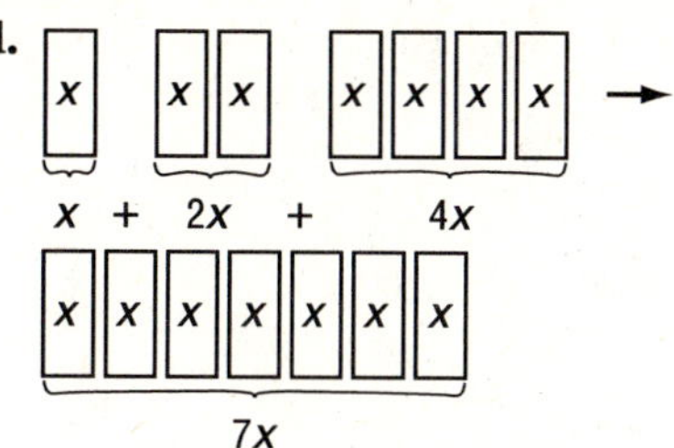

Page 105 Check for Understanding

1. terms that contain the same variable or are constants

2. Sample answer: $4x + 5y - 1$

3. Koko; $5x + x = 6x$, not $5x$.

4. terms: $4x$, 3, $5x$, y; like terms: $4x$, $5x$; coefficients: 4, 5, 1; constant: 3

5. $2m + (-1n) + 6m$; terms: $2m$, $-1n$, $6m$; like terms: $2m$, $6m$; coefficients 2, -1, 6; constant: none

6. $4y + (-2x) + (-7)$; terms: $4y - 2x$, -7; like terms: none; coefficients: 4, -2; constant: -7

7. $6a + 2a = (6 + 2)a$
$= 8a$

8. $x + 9x + 3 = 1x + 9x + 3$
$= (1 + 9)x + 3$
$= 10x + 3$

9. $6c + 4 + c + 8 = 6c + 1c + 4 + 8$
$= (6 + 1)c + 12$
$= 7c + 12$

10. $7m - 2m = 7m + (-2)m$
$= [7 + (-2)]m$
$= 5m$

11. $9y + 8 - 8 = 9y + 8 + (-8)$
$= 9y + 0$
$= 9y$

12. $2x - 5 - 4x + 8 = 2x + (-5) + (-4x) + 8$
$= 2x + (-4x) + (-5) + 8$
$= [2 + (-4)]x + 3$
$= -2x + 3$

13. $5 - 3(y + 7) = 5 + (-3)(y + 7)$
$= 5 + (-3y) + (-3)(7)$
$= 5 + (-3y) + (-21)$
$= (-3y) + 5 + (-21)$
$= -3y + (-16)$
$= -3y - 16$

14. $3x + 2y + 4y = 3x + (2 + 4)y$
$= 3x + 6y$

15. $x + 3(x + 4y) = 1x + 3x + 3(4y)$
$= (1 + 3)x + 12y$
$= 4x + 12y$

16. Let x = amount of money you saved.
Let $x + 20$ = amount your friend saved.
total amount = your money + your friend's
$= x + (x + 20)$
$= 1x + 1x + 20$
$= (1 + 1)x + 20$
$= 2x + 20$

Pages 106–107 Practice and Apply

17. terms: 3, $7x$, $3x$, x; like terms: $7x$, $3x$, x; coefficients: 7, 3, 1; constant: 3
18. terms: y, $3y$, $8y$, 2; like terms: y, $3y$, $8y$; coefficients: 1, 3, 8; constant: 2
19. $2a + 5c + (-1a) + 6a$; terms: $2a$, $5c$, $-1a$, $6a$; like terms: $2a$, $-1a$, $6a$; coefficients: 2, 5, -1, 6; constant: none
20. $5c + (-2d) + 3d + (-1d)$; terms: $5c$, $-2d$, $3d$, $-1d$; like terms: $-2d$, $3d$, $-1d$; coefficients: 5, -2, 3, -1; constant: none
21. $6m + (-2n) + 7$; terms: $6m$, $-2n$, 7; like terms: none; coefficients: 6, -2; constant: 7
22. $7x + (-3y) + 3z + (-2)$; terms: $7x$, $-3y$, $3z$, -2; like terms: none; coefficients: 7, -3, 3; constant: -2
23. $2x + 5x = (2 + 5)x$
$= 7x$
24. $7b + 2b = (7 + 2)b$
$= 9b$
25. $y = 10y = 1y + 10y$
$= (1 + 10)y$
$= 11y$
26. $5y + y = 5y + 1y$
$= (5 + 1)y$
$= 6y$
27. $2a + 3 + 5a = 2a + 5a + 3$
$= (2 + 5)a + 3$
$= 7a + 3$
28. $4 + 2m + m = 4 + 2m + 1m$
$= 4 + (2 + 1)m$
$= 4 + 3m$
29. $2y + 8 + 5y + 1 = 2y + 5y + 8 + 1$
$= (2 + 5)y + 9$
$= 7y + 9$
30. $8x + 5 + 7 + 2x = 8x + 2x + 5 + 7$
$= (8 + 2)x + 12$
$= 10x + 12$
31. $5x - 3x = 5x + (-3x)$
$= [5 + (-3)]x$
$= 2x$
32. $10b - 2b = 10b + (-2b)$
$= [10 + (-2)]b$
$= 8b$
33. $4y - 5y = 4y + (-5y)$
$= [4 + (-5)]y$
$= (-1)y$
$= -y$
34. $r - 3r = 1r + (-3r)$
$= [1 + (-3)]r$
$= -2r$
35. $8 + x - 5x = 8 + 1x + (-5x)$
$= 8 + [1 + (-5)]x$
$= 8 + (-4x)$
$= (-4x) + 8$
$= -4x + 8$
36. $6x + 4 - 7x = 6x + (-7x) + 4$
$= [6 + (-7)]x + 4$
$= -1x + 4$
$= -x + 4$ or $4 - x$
37. $8y - 7 + 7 = 8y + (-7) + 7$
$= 8y + 0$
$= 8y$
38. $9x + 2 - 2 = 9x + 2 + (-2)$
$= 9x + 0$
$= 9x$
39. $2x + 3 - 3x + 9 = 2x + (-3x) + 3 + 9$
$= [2 + (-3)]x + 12$
$= -1x + 12$
$= -x + 12$ or $12 - x$
40. $5t - 3 - t + 2 = 5t + (-1t) + (-3) + 2$
$= [5 + (-1)]t + (-1)$
$= 4t + (-1)$
$= 4t - 1$
41. $3(b + 2) + 2b = 3b + 3 \cdot 2 + 2b$
$= 3b + 2b + 3 \cdot 2$
$= (3 + 2)b + 6$
$= 5b + 6$
42. $5(x + 3) + 8x = 5x + 5 \cdot 3 + 8x$
$= 5x + 8x + 5 \cdot 3$
$= (5 + 8)x + 15$
$= 13x + 15$
43. $-3(a + 2) - a = -3a + (-3) \cdot 2 + (-1a)$
$= -3a + (-1a) + (-3)2$
$= [-3 + (-1)]a + (-6)$
$= -4a + (-6)$
$= -4a - 6$
44. $-2(x + 3) + 2x = -2x + (-2) \cdot 3 + 2x$
$= -2x + 2x + (-2) \cdot 3$
$= (-2 + 2)x + (-6)$
$= 0x + (-6)$
$= 0 + (-6)$
$= -6$

45. $4x - 4(2 + x) = 4x + (-4)(2 + x)$
$= 4x + (-4) \cdot 2 + (-4x)$
$= 4x + (-4x) + (-4) \cdot 2$
$= [4 + (-4)]x + (-8)$
$= 0x + (-8)$
$= 0 + (-8)$
$= -8$

46. $8a - 2(a - 7) = 8a + (-2)[a + (-7)]$
$= 8a + (-2a) + (-2)(-7)$
$= [8 + (-2)]a + 14$
$= 6a + 14$

47. $6m + 2n + 10m = 6m + 10m + 2n$
$= (6 + 10)m + 2n$
$= 16m + 2n$

48. $-2y + x + 3y = -2y + 3y + x$
$= (-2 + 3)y + x$
$= 1y + x$
$= y + x$ or $x + y$

49. $c + 2(d - 5c) = 1c + 2[d + (-5c)]$
$= 1c + 2d + 2(-5c)$
$= 1c + 2d + (-10c)$
$= 1c + (-10c) + 2d$
$= [1 + (-10)]c + 2d$
$= -9c + 2d$

50. $5x + 45 + 3 = 5x + 48$

51. $3s + 50 + 30 = 3s + 80$

52. $y + (y - 5) = 1y + 1y - 5$
$= (1 + 1)y - 5$
$= 2y - 5$

53. $d + (2d) + (2d - 2) = 1d + 2d + 2d - 2$
$= (1 + 2)d + 2d - 2$
$= 3d + 2d - 2$
$= (3 + 2)d - 2$
$= 5d - 2$

54. $3x + 5x + 4x = (3 + 5)x + 4x$
$= 8x + 4x$
$= (8 + 4)x$
$= 12x$

55. $x + (2x + 1) + x + (2x + 1)$
$= 1x + 2x + 1 + 1x + 2x + 1$
$= (1 + 2)x + 1 + (1 + 2)x + 1$
$= 3x + 1 + 3x + 1$
$= 3x + 3x + 1 + 1$
$= (3 + 3)x + 2$
$= 6x + 2$

56. You can use algebra tiles to simplify an algebraic expression by grouping the tiles with the same size and shape together. Answers should include the following.

- 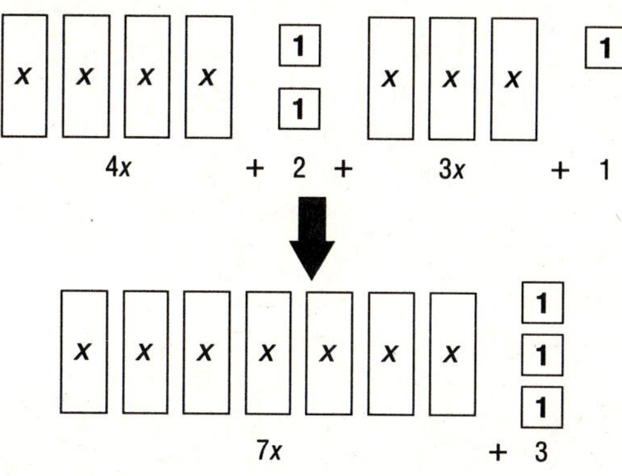

- *Like terms* are terms that have the same variable(s) or are constants.
- You use the Commutative Property when you change the order of 2 and $3x$. You use the Distributive Property when you write $4x + 3x$ as $7x$.

57a. Distributive

57b. Commutative

57c. Substitution

57d. Distributive

58. C; each of the other expressions is equal to $-6x + 12$.

59. C; $m + (m - 10) = 1m + 1m - 10$
$= 2m - 10$

Page 107 Maintain Your Skills

60. $3(a + 5) = 3a + 3 \cdot 5$
$= 3a + 15$

61. $-2(y + 8) = -2y + (-2) \cdot 8$
$= -2y + (-16)$
$= -2y - 16$

62. $-3(x - 1) = -3[x + (-1)]$
$= -3x + (-3)(-1)$
$= -3x + 3$

63. Start at the origin. Move 5 units to the left. Then move 4 units down and draw a dot. Point $P(-5, -6)$ is in Quadrant III.

64.

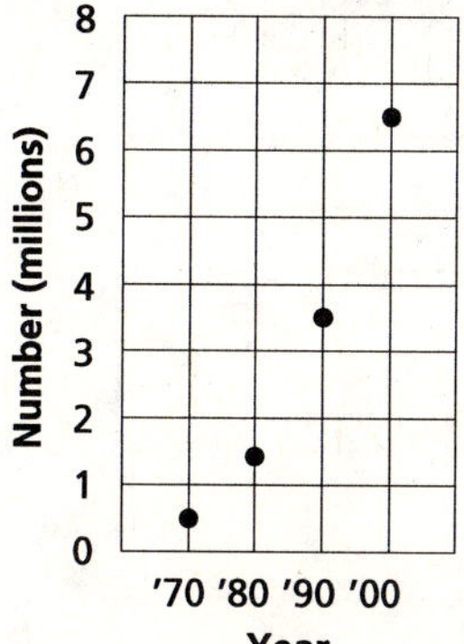

65. $2 + 3 \cdot 5 = 2 + 15$
$= 17$

66. $8 \div 2 \cdot 4 = (8 \div 2) \cdot 4$
$= 4 \cdot 4$
$= 16$

67. $10 - 2 \cdot 4 = 10 - 8$
$= 2$

68. $-5 + 4 = -1$

69. $-8 + (-3) = -11$

70. $10 + (-1) = 9$

71. $4 + (-9) = -5$

72. $11 + (-7) = 4$

73. $-4 + (-9) = -13$

Page 107 Practice Quiz 1

1. $6(x + 2) = 6x + 6 \cdot 2$
$= 6x + 12$

2. $5(x - 7) = 5[x + (-7)]$
$= 5x + 5 \cdot (-7)$
$= 5x + (-35)$
$= 5x - 35$

3. $6y - 4 + y = 6y + (-4) + 1y$
$= 6y + 1y + (-4)$
$= (6 + 1)y + (-4)$
$= 7y + (-4)$
$= 7y - 4$

4. $2a + 4(a - 9) = 2a + 4[a + (-9)]$
$= 2a + 4a + 4(-9)$
$= (2 + 4)a + (-36)$
$= 6a + (-36)$
$= 6a - 36$

5. $m + (m + 15) = 1m + 1m + 15$
$= (1 + 1)m + 15$
$= 2m + 15$

Pages 108–109 Algebra Activity (Preview of Lessons 3-3 and 3-4)

1.

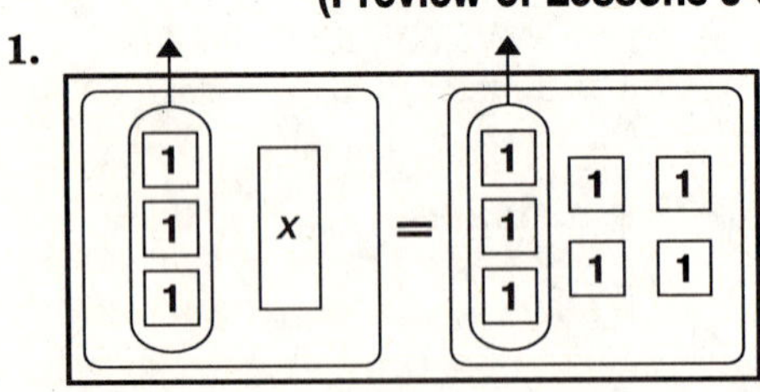

$3 + x = 7; x = 4$

2.

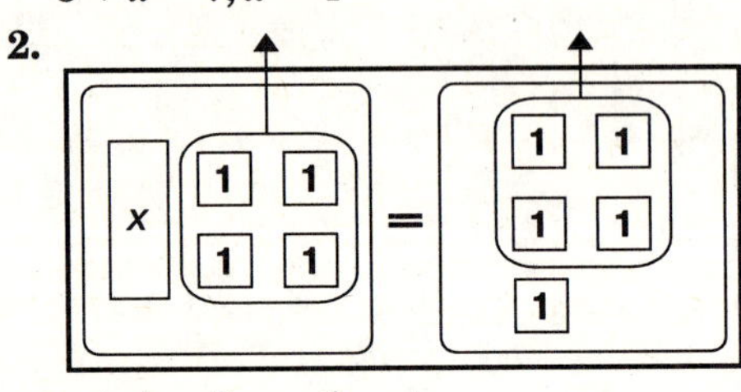

$x + 4 = 5; x = 1$

3.

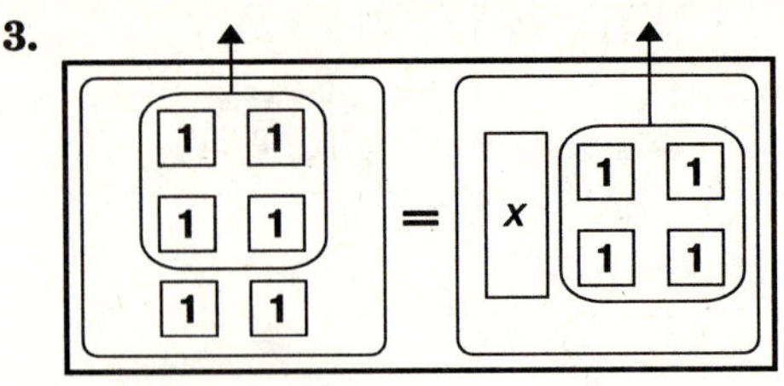

$6 = x + 4; x = 2$

4.

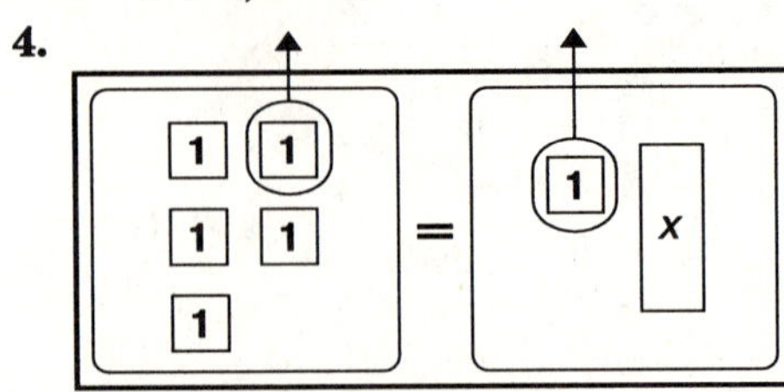

$5 = 1 + x; x = 4$

5.

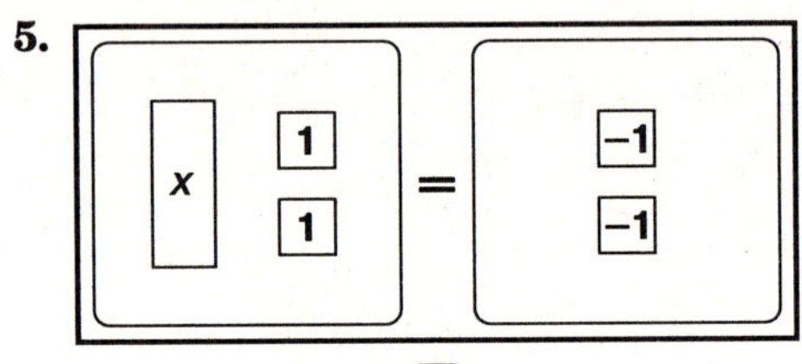

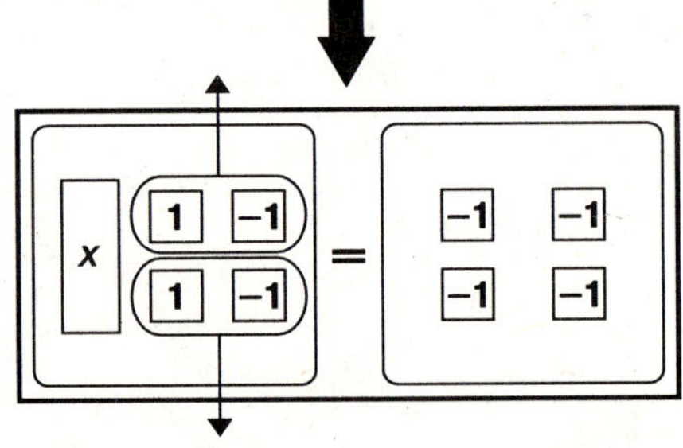

$x + 2 = -2; x = -4$

6.

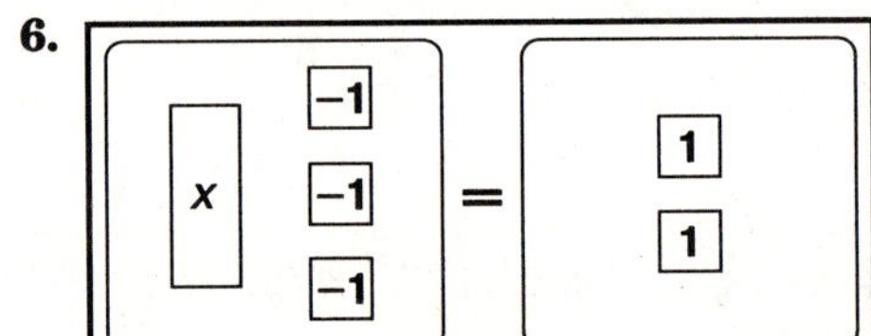

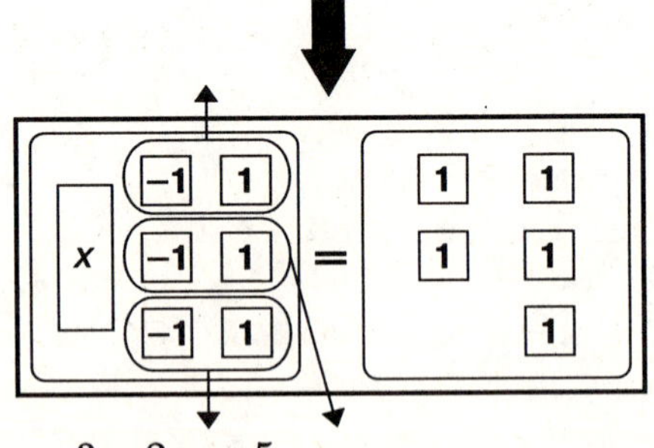

$x - 3 = 2; x = 5$

7. 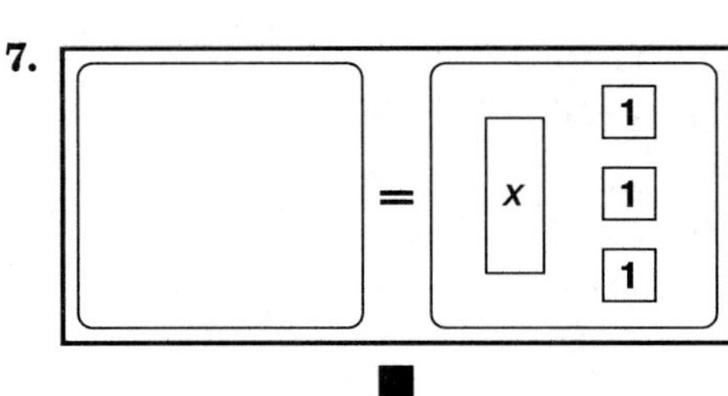

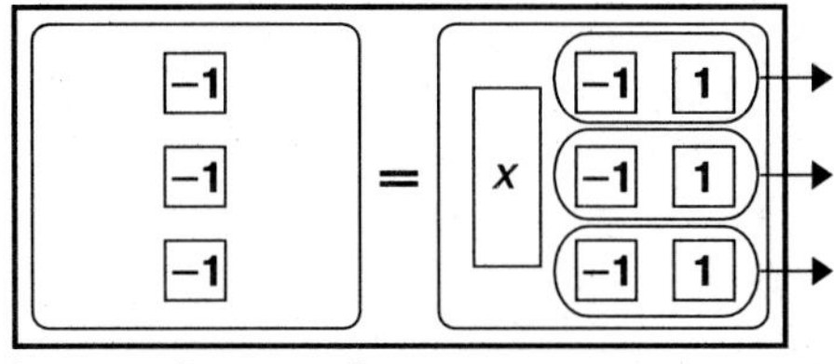

$0 = x + 3; x = -3$

8. 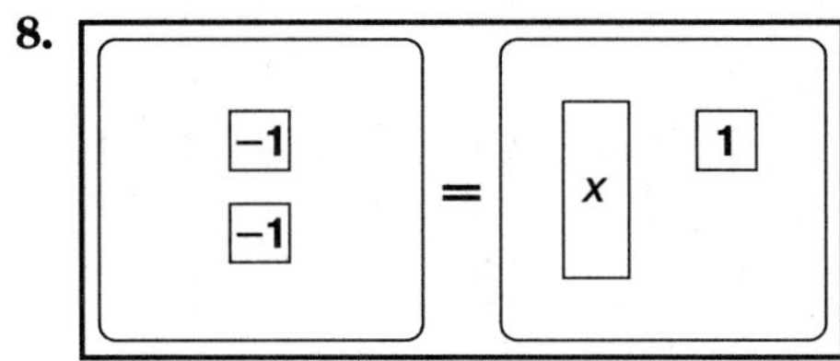

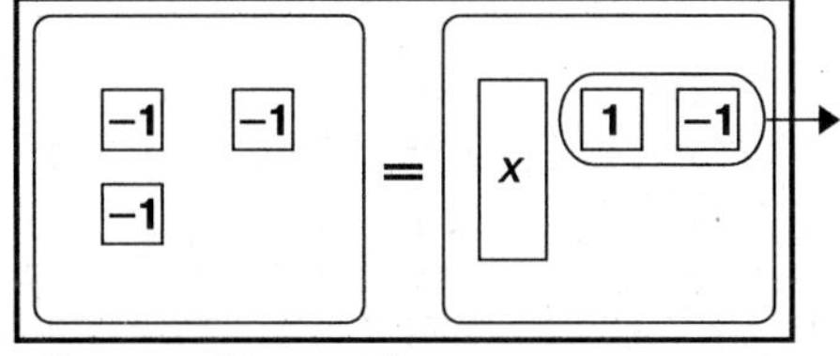

$-2 = x + 1; x = -3$

9. 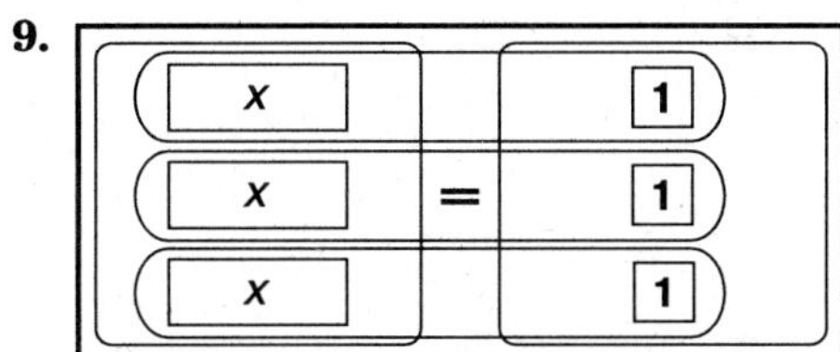

$3x = 3; x = 1$

10. 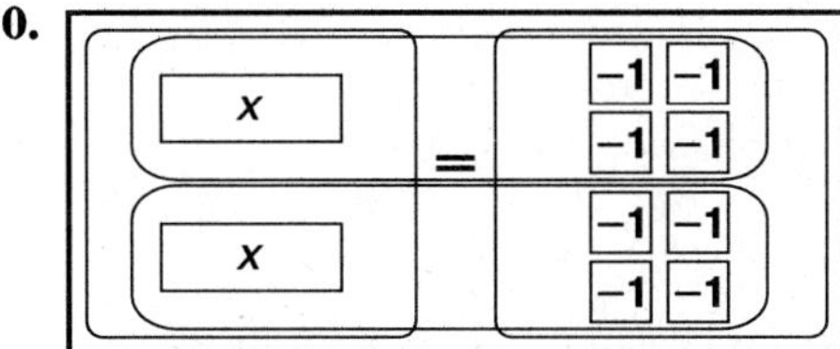

$2x = -8; x = -4$

11. 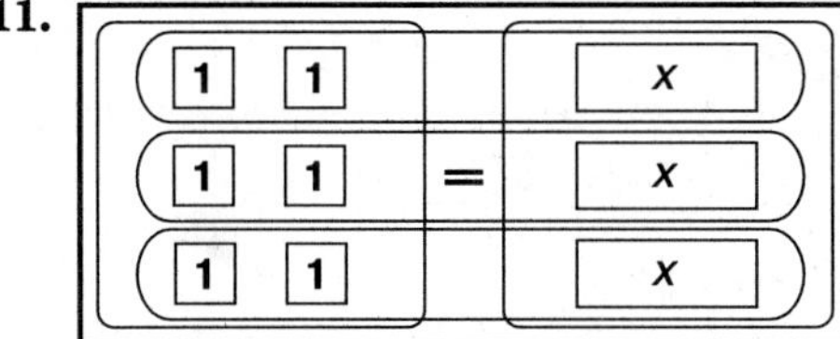

$6 = 3x; x = 2$

12. 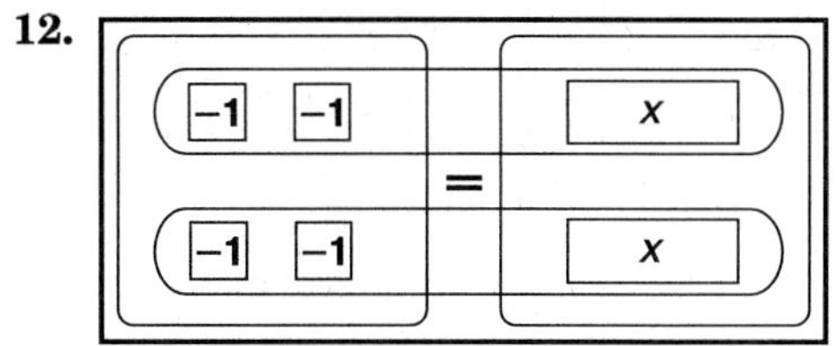

$-4 = 2x; x = -2$

3-3 Solving Equations by Adding or Subtracting

Page 110 How is solving an equation similar to keeping a scale in balance?

a. Remove 4 blocks from each side. There are 3 blocks remaining on the right, so there must be 3 blocks in the bag.

b. Removing the same number of blocks from each side keeps the scale in balance.

Page 113 Check for Understanding

1. Addition Property of Equality

2. Sample answer: equivalent—$x + 3 = 5, x = 2$; not equivalent—$x + 5 = 2, x + 1 = 5$

3.
$x + 14 = 25$
$x + 14 - 14 = 25 - 14$
$x + 0 = 11$
$x = 11$
Check: $x + 14 = 25$
$11 + 14 \stackrel{?}{=} 25$
$25 = 25$ ✓
The solution is 11.

4.
$w + 4 = -10$
$w + 4 - 4 = -10 - 4$
$w + 0 = -14$
$w = -14$
Check: $w + 4 = -10$
$-14 + 4 \stackrel{?}{=} -10$
$-10 = -10$ ✓
The solution is -14.

5. $16 = y + 20$
$y + 20 = 16$
$y + 20 - 20 = 16 - 20$
$y + 0 = -4$
$y = -4$
Check: $16 = y + 20$
$16 \stackrel{?}{=} -4 + 20$
$16 = 16$ ✓
The solution is -4.

6. $n - 8 = 5$
$n + (-8) = 5$
$n + (-8) + 8 = 5 + 8$
$n + 0 = 13$
$n = 13$
Check: $n - 8 = 5$
$13 - 8 \stackrel{?}{=} 5$
$5 = 5$ ✓
The solution is 13.

7. $k - 25 = 30$
$k + (-25) = 30$
$k + (-25) + 25 = 30 + 25$
$k + 0 = 55$
$k = 55$
Check: $k - 25 = 30$
$55 - 25 \stackrel{?}{=} 30$
$30 = 30$ ✓
The solution is 55.

8. $r - 4 = -18$
$r + (-4) = -18$
$r + (-4) + 4 = -18 + 4$
$r + 0 = -14$
$r = -14$
Check: $r - 4 = -18$
$-14 - 4 \stackrel{?}{=} -18$
$-18 = -18$ ✓
The solution is -14.

9. $-8 = x - 6$
$x - 6 = -8$
$x + (-6) = -8$
$x + (-6) + 6 = -8 + 6$
$x + 0 = -2$
$x = -2$

−3 −2 −1 0 1 2 3 4

10. $y - 3 = -1$
$y + (-3) = -1$
$y + (-3) + 3 = -1 + 3$
$y + 0 = 2$
$y = 2$

−3 −2 −1 0 1 2 3 4

11. C; $x - 10 = -5$
$x + (-10) = -5$
$x + (-10) + 10 = -5 + 10$
$x + 0 = 5$
$x = 5$

Pages 113–114 Practice and Apply

Exercises 12–32 For checks, see students' work.

12. $y + 7 = 21$
$y + 7 - 7 = 21 - 7$
$y = 14$
The solution is 14.

13. $x + 5 = 18$
$x + 5 - 5 = 18 - 5$
$x = 13$
The solution is 13.

14. $m + 10 = -2$
$m + 10 - 10 = -2 - 10$
$m = -12$
The solution is -12.

15. $x + 5 = -3$
$x + 5 - 5 = -3 - 5$
$x = -8$
The solution is -8.

16. $a + 10 = -4$
$a + 10 - 10 = -4 - 10$
$a = -14$
The solution is -14.

17. $t + 6 = -9$
$t + 6 - 6 = -9 - 6$
$t = -15$
The solution is -15.

18. $y + 8 = 3$
$y + 8 - 8 = 3 - 8$
$y = -5$
The solution is -5.

19. $9 = 10 + b$
$9 = b + 10$
$9 - 10 = b + 10 - 10$
$-1 = b$
$b = -1$
The solution is -1

20. $k - 6 = 13$
$k + (-6) = 13$
$k + (-6) + 6 = 13 + 6$
$k = 19$
The solution is 19.

21. $r - 5 = 10$
$r + (-5) = 10$
$r + (-5) + 5 = 10 + 5$
$r = 15$
The solution is 15

22. $8 = r - 5$
$8 = r + (-5)$
$8 + 5 = r + (-5) + 5$
$13 = r$
$r = 13$
The solution is 13.

23. $19 = g - 5$
$19 = g + (-5)$
$19 + 5 = g + (-5) + 5$
$24 = g$
$g = 24$
The solution is 24.

24. $x - 6 = -2$

$x + (-6) = -2$

$x + (-6) + 6 = -2 + 6$

$x = 4$

The solution is 4.

25. $y - 49 = -13$

$y + (-49) = -13$

$y + (-49) + 49 = -13 + 49$

$y = 36$

The solution is 36.

26. $-15 = x - 16$

$-15 = x + (-16)$

$-15 + 16 = x + (-16) + 16$

$1 = x$

$x = 1$

The solution is 1.

27. $-8 = t - 4$

$-8 = t + (-4)$

$-8 + 4 = t + (-4) + 4$

$-4 = t$

$t = -4$

The solution is -4.

28. $23 + y = 14$

$y + 23 = 14$

$y + 23 - 23 = 14 - 23$

$y = -9$

The solution is -9.

29. $59 = s + 90$

$59 - 90 = s + 90 - 90$

$59 + (-90) = s$

$-31 = s$

$s = -31$

The solution is -31.

30. $x - 27 = -63$

$x + (-27) = -63$

$x + (-27) + 27 = -63 + 27$

$x = -36$

The solution is -36.

31. $84 = r - 34$

$84 = r + (-34)$

$84 + 34 = r + (-34) + 34$

$118 = r$

$r = 118$

The solution is 118.

32. $y - 95 = -18$

$y + (-95) = -18$

$y + (-95) + 95 = -18 + 95$

$y = 77$

The solution is 77.

33. $n + 9 = -2$

$n + 9 - 9 = -2 - 9$

$n = -11$

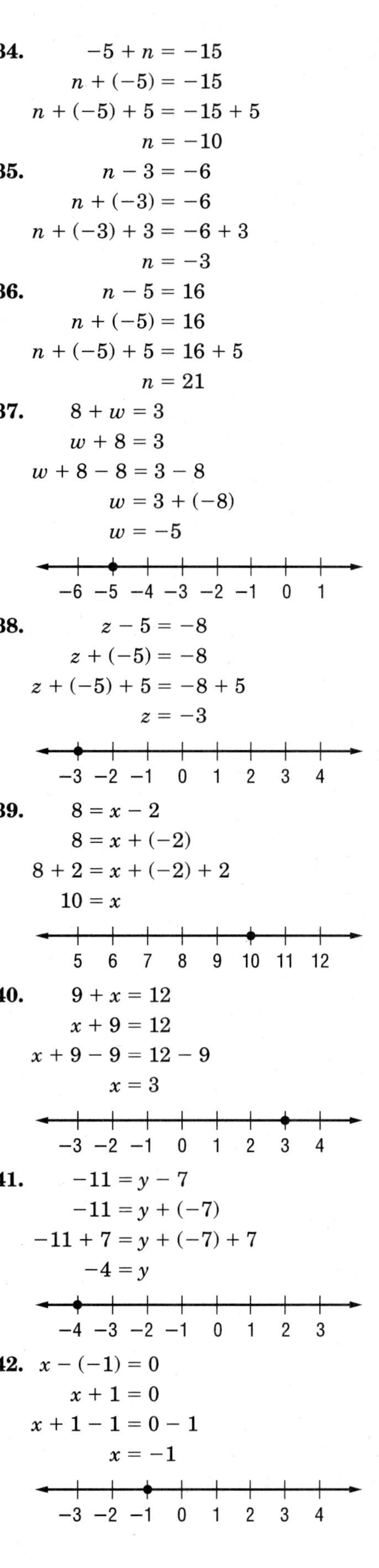

34. $-5 + n = -15$

$n + (-5) = -15$

$n + (-5) + 5 = -15 + 5$

$n = -10$

35. $n - 3 = -6$

$n + (-3) = -6$

$n + (-3) + 3 = -6 + 3$

$n = -3$

36. $n - 5 = 16$

$n + (-5) = 16$

$n + (-5) + 5 = 16 + 5$

$n = 21$

37. $8 + w = 3$

$w + 8 = 3$

$w + 8 - 8 = 3 - 8$

$w = 3 + (-8)$

$w = -5$

38. $z - 5 = -8$

$z + (-5) = -8$

$z + (-5) + 5 = -8 + 5$

$z = -3$

39. $8 = x - 2$

$8 = x + (-2)$

$8 + 2 = x + (-2) + 2$

$10 = x$

40. $9 + x = 12$

$x + 9 = 12$

$x + 9 - 9 = 12 - 9$

$x = 3$

41. $-11 = y - 7$

$-11 = y + (-7)$

$-11 + 7 = y + (-7) + 7$

$-4 = y$

42. $x - (-1) = 0$

$x + 1 = 0$

$x + 1 - 1 = 0 - 1$

$x = -1$

43. Let x = number of electoral votes in Texas.

$12 = x - 20$
$12 = x + (-20)$
$12 + 20 = x + (-20) + 20$
$32 = x$
$x = 32$

There were 32 electoral votes in Texas.

44. Let x = the record high temperature.

$x - (-5) = 109$
$x + 5 = 109$
$x + 5 - 5 = 109 - 5$
$x = 104$

The record high temperature was 104°F.

45. See students' work.

46. Let x = New York's population.

$27 = x + 10$, $13 = x - 4$

47. $27 = x + 10$ / $27 - 10 = x + 10 - 10$ / $17 = x$

$13 = x - 4$ / $13 = x + (-4)$ / $13 + 4 = x + (-4) + 4$ / $17 = x$

The population of New York is 17 million.

48. Sample answer: $x - 2 = -7$, $x + 2 = -3$

49. When you solve an equation, you perform the same operation on each side so that the two sides remain equal. Answers should include the following.

- In an equation, both sides are equal. In a balance scale, the weight of the items on both sides are equal.
- The Addition and Subtraction Properties of Equality allow you to add or subtract the same number from each side of an equation. The two sides of the equation remain equal.

50. B; $x + 4 = -2$

$x + 4 - 4 = -2 - 4$
$x = -6$
$-3x - 2 = -3(-6) - 2$
$= 18 - 2$
$= 16$

The answer is 16.

51. C; Let x = the number.

$x - 5 \cdot 7 = 3$
$x - 35 = 3$
$x + (-35) = 3$
$x + (-35) + 35 = 3 + 35$
$x = 38$

Page 114 Maintain Your Skills

52. $-2(x + 5) = -2x + (-2) \cdot 5$
$= -2x + (-10)$
$= -2x - 10$

53. $(t + 4)3 = t \cdot 3 + 4 \cdot 3$
$= 3t + 12$

54. $-4(x - 2) = -4x + (-4)(-2)$
$= -4x + 8$

55. $6z - 3 - 10z + 7 = 6z + (-3) + (-10z) + 7$
$= 6z + (-10z) + (-3) + 7$
$= [6 + (-10)]z + 4$
$= -4z + 4$

56. $2(x + 6) + 4x = 2x + 2 \cdot 6 + 4x$
$= 2x + 12 + 4x$
$= 2x + 4x + 12$
$= (2 + 4)x + 12$
$= 6x + 12$

57. $3 - 4(m + 1) = 3 - 4m + (-4) \cdot 1$
$= 3 + (-4m) + (-4)$
$= -4m + 3 + (-4)$
$= -4m + (-1)$
$= -4m - 1$

58. Commutative Property of Addition

59. Additive Inverse

60. Associative Property of Multiplication

61. $9a + 4b = 9(8) + 4(3)$
$= 72 + 12$
$= 84$

62. $-100 \div 10 = -10$

63. $50 \div (-2) = -25$

64. $-49 \div (-7) = 7$

65. $\frac{72}{-8} = -9$

66. $\frac{-18}{-6} = 3$

67. $\frac{-36}{9} = -4$

3-4 Solving Equations by Multiplying or Dividing

Page 115 How are equations used to find the U.S. value of foreign currency?

a. $d = 72$

b. Divide by 9.

Page 117 Check for Understanding

1. Multiplication Property of Equality

2. Divide each side by -5; $y = 9$.

3. Sample answer: $-5x = -20$

4. $4x = 24$

$\frac{4x}{4} = \frac{24}{4}$
$1x = 6$
$x = 6$

Check: $4x = 24$
$4(6) \stackrel{?}{=} 24$
$24 = 24$ ✓

The solution is 6.

5. $-2a = 10$

$\frac{-2a}{-2} = \frac{10}{-2}$

$1a = -5$

$a = -5$

Check: $-2a = 10$

$-2(-5) \stackrel{?}{=} 10$

$10 = 10$ ✓

The solution is -5.

6. $-42 = -7t$

$-7t = -42$

$\frac{-7t}{-7} = \frac{-42}{-7}$

$1t = 6$

$t = 6$

Check: $-42 = -7t$

$-42 \stackrel{?}{=} -7(6)$

$-42 = -42$ ✓

The solution is 6.

7. $\frac{k}{3} = 9$

$\frac{k}{3} \cdot 3 = 9 \cdot 3$

$k = 27$

Check: $\frac{k}{3} = 9$

$\frac{27}{3} \stackrel{?}{=} 9$

$9 = 9$ ✓

The solution is 27.

8. $\frac{y}{5} = -8$

$\frac{y}{5} \cdot 5 = -8 \cdot 5$

$y = -40$

Check: $\frac{y}{5} = -8$

$\frac{-40}{5} \stackrel{?}{=} -8$

$-8 = -8$ ✓

The solution is -40.

9. $-11 = \frac{n}{-6}$

$\frac{n}{-6} = -11$

$\frac{n}{-6} \cdot (-6) = -11(-6)$

$n = 66$

Check: $-11 = \frac{n}{-6}$

$-11 \stackrel{?}{=} \frac{66}{-6}$

$-11 = -11$ ✓

The solution is 66.

10. Let x = the number of toys.

$80x = 4000$

$\frac{80x}{80} = \frac{4000}{80}$

$x = 50$

50 toys can be made.

Pages 118–119 Practice and Apply

Exercises 11–34 For checks, see students' work.

11. $3t = 21$

$\frac{3t}{3} = \frac{21}{3}$

$t = 7$

The solution is 7.

12. $8x = 72$

$\frac{8x}{8} = \frac{72}{8}$

$x = 9$

The solution is 9.

13. $-32 = 4y$

$4y = -32$

$\frac{4y}{4} = \frac{-32}{4}$

$y = -8$

The solution is -8.

14. $5n = -95$

$\frac{5n}{5} = \frac{-95}{5}$

$n = -19$

The solution is -19.

15. $-56 = -7p$

$-7p = -56$

$\frac{-7p}{-7} = \frac{-56}{-7}$

$p = 8$

The solution is 8.

16. $-8j = -64$

$\frac{-8j}{-8} = \frac{-64}{-8}$

$j = 8$

The solution is 8.

17. $\frac{h}{4} = 6$

$\frac{h}{4} \cdot 4 = 6 \cdot 4$

$h = 24$

The solution is 24.

18. $\frac{c}{9} = 4$

$\frac{c}{9} \cdot 9 = 4 \cdot 9$

$c = 36$

The solution is 36.

19. $\frac{g}{2} = -7$

$\frac{g}{-2}(-2) = -7(-2)$

$g = 14$

The solution is 14.

20. $-42 = \frac{x}{-2}$

$\frac{x}{-2} = -42$

$\frac{x}{-2}(-2) = -42(-2)$

$x = 84$

The solution is 84.

21. $11 = \frac{b}{-3}$

$\frac{b}{-3} = 11$

$\frac{b}{-3}(-3) = 11(-3)$

$b = -33$

The solution is −33.

22. $\frac{h}{-7} = 20$

$\frac{h}{-7} \cdot (-7) = 20(-7)$

$h = -140$

The solution is −140.

23. $45 = 5x$

$5x = 45$

$\frac{5x}{5} = \frac{45}{5}$

$x = 9$

The solution is 9.

24. $3u = 51$

$\frac{3u}{3} = \frac{51}{3}$

$u = 17$

The solution is 17.

25. $86 = -2v$

$-2v = 86$

$\frac{-2v}{-2} = \frac{86}{-2}$

$v = -43$

The solution is −43.

26. $-8a = 144$

$\frac{-8a}{-8} = \frac{144}{-8}$

$a = -18$

The solution is −18.

27. $\frac{m}{45} = -3$

$\frac{m}{45} \cdot 45 = -3 \cdot 45$

$m = -135$

The solution is −135.

28. $\frac{d}{3} = -3$

$\frac{d}{3} \cdot 3 = -3 \cdot 3$

$d = -9$

The solution is −9.

29. $\frac{f}{-13} = -10$

$\frac{f}{-13}(-13) = -10 \cdot (-13)$

$f = 130$

The solution is 130.

30. $\frac{v}{-11} = -132$

$\frac{v}{-11}(-11) = -132(-11)$

$v = 1452$

The solution is 1452.

31. $-116 = -4w$

$-4w = -116$

$\frac{-4w}{-4} = \frac{-116}{-4}$

$w = 29$

The solution is 29.

32. $-68 = -4m$

$-4m = -68$

$\frac{-4m}{-4} = \frac{-68}{-4}$

$m = 17$

The solution is 17.

33. $-21 = \frac{k}{8}$

$\frac{k}{8} = -21$

$\frac{k}{8} \cdot 8 = -21 \cdot 8$

$k = -168$

The solution is −168.

34. $-56 = \frac{t}{9}$

$\frac{t}{9} = -56$

$\frac{t}{9} \cdot 9 = -56 \cdot 9$

$t = -504$

The solution is −504.

35. $6x = -42$

$\frac{6x}{6} = \frac{-42}{6}$

$x = -7$

The solution is −7.

36. $-7x = -35$

$\frac{-7x}{-7} = \frac{-35}{-7}$

$x = 5$

The solution is 5.

37. $\frac{x}{-4} = 8$

$\frac{x}{-4}(-4) = 8(-4)$

$x = -32$

The solution is −32.

38. $\frac{x}{-5} = -2$

$\frac{x}{-5}(-5) = -2(-5)$

$x = 10$

The solution is 10.

39. $48 = -6x$

$-6x = 48$

$\frac{-6x}{-6} = \frac{48}{-6}$

$x = -8$

−9 −8 −7 −6 −5 −4 −3 −2

40. $-32t = 64$

$\frac{-32t}{-32} = \frac{64}{-32}$

$t = -2$

−3 −2 −1 0 1 2 3 4

41. $-6r = -18$

$\frac{-6r}{-6} = \frac{-18}{-6}$

$r = 3$

−3 −2 −1 0 1 2 3 4

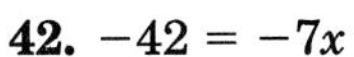

42. $-42 = -7x$

$-7x = -42$

$\frac{-7x}{-7} = \frac{-42}{-7}$

$x = 6$

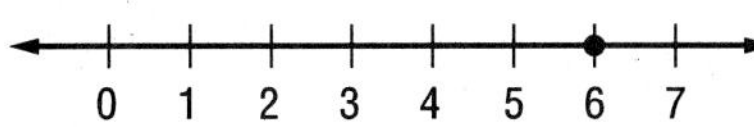

43. $\frac{n}{12} = 3$

$\frac{n}{12} \cdot 12 = 3 \cdot 12$

$n = 36$

30 31 32 33 34 35 36 37

44. $\frac{y}{-4} = -1$

$\frac{y}{-4}(-4) = -1(-4)$

$y = 4$

−3 −2 −1 0 1 2 3 4

45. Let x = the size of the largest U.S. ranch.

$5x = 12{,}000$

$\frac{5x}{5} = \frac{12{,}000}{5}$

$x = 2400$

The largest U.S. ranch is 2400 mi^2.

46. Let x = the number of cows.

$40x = 720$

$\frac{40x}{40} = \frac{720}{40}$

$x = 18$

18 cows can graze on the land.

47. Let p = the number of painters.

$6p = 24$

$\frac{6p}{6} = \frac{24}{6}$

$p = 4$

It will take 4 painters to paint the house in 6 days.

48a. $12f = 132$

$\frac{12f}{12} = \frac{132}{12}$

$f = 11$ *ft*

48b. $3y = 15$

$\frac{3y}{3} = \frac{15}{3}$

$y = 5$ yd

48c. $5280m = 10{,}560$

$\frac{5280m}{5280} = \frac{10{,}560}{5280}$

$m = 2$ mi

49a. True; one pyramid balances two cubes, so this is the same as adding one cube to each side.

49b. True; one pyramid and one cube balance three cubes, which balance one cylinder.

49c. False; one cyinder and one pyramid balance five cubes.

50. Sample answer: You can use the exchange rate to write and solve an equation involving foreign currency. Answers should include the following.

- The equation is $4d = 12$, where d is the number of U.S. dollars. The bus trip costs \$3 U.S. dollars.
- The equation is $16d = 3040$, where d is the number of U.S. dollars. The hotel room costs \$190 U.S. dollars.

51. B; $\frac{mx}{n} = p$

$\frac{mx}{n} \cdot n = p \cdot n$

$mx = pn$

$\frac{mx}{m} = \frac{pn}{m}$

$x = \frac{pn}{m}$

52. C; $\frac{x}{-6} = 24$

$\frac{x}{-6}(-6) = 24(-6)$

$x = -144$

Page 119 Maintain Your Skills

53. $3 + y = 16$

$y + 3 = 16$

$y + 3 - 3 = 16 - 3$

$y = 13$

54. $29 = n + 4$

$29 - 4 = n + 4 - 4$

$25 = n$

55. $k - 12 = -40$

$k + (-12) = -40$

$k + (-12) + 12 = -40 + 12$

$k = -28$

56. $4x + 7x = (4 + 7)x$

$= 11x$

57. $2y + 6 + 5y = 2y + 5y + 6$

$= (2 + 5)y + 6$

$= 7y + 6$

58. $3 - 2(y + 4) = 3 + (-2)(y + 4)$

$= 3 + (-2y) + (-2)4$

$= 3 + (-2y) + (-8)$

$= -2y + (-8) + 3$

$= -2y + (-5)$

$= -2y - 5$

59. $3ab = 3(-6)(2)$

$= -18(2)$

$= -36$

60. $-14 < -4$

61. $8 - (-2) = 8 + 2$

$= 10$

62. $-5 - 5 = -5 + (-5)$

$= -10$

63. $-10 - (-8) = -10 + 8$

$= -2$

64. $-18 - 4 = -18 + (-4)$
$= -22$

65. $-3 - (-5) = -3 + 5$
$= 2$

66. $-45 - (-9) = -45 + 9$
$= -36$

67. $-24 - (-5) = -24 + 5$
$= -19$

68. $-15 - (-15) = -15 + 15$
$= 0$

69. $-8 - 19 = -8 + (-19)$
$= -27$

3-5 Solving Two-Step Equations

Page 120 How can algebra tiles show the properties of equality?

a. Subtraction Property of Equality
b. Division Property of Equality
c. The solution of $2x + 1 = 9$ is 4.

Page 122 Check for Understanding

1. You undo the operations in reverse order.

2. Sample answer: $\frac{x}{2} - 3 = 5$

3.
$2x - 7 = 9$
$2x - 7 + 7 = 9 + 7$
$2x = 16$
$\frac{2x}{2} = \frac{16}{2}$
$x = 8$
Check: $2x - 7 = 9$
$2(8) - 7 \stackrel{?}{=} 9$
$16 - 7 \stackrel{?}{=} 9$
$9 = 9$ ✓
The solution is 8.

4.
$3t + 5 = 2$
$3t + 5 - 5 = 2 - 5$
$3t = -3$
$\frac{3t}{3} = \frac{-3}{3}$
$t = -1$
Check: $3t + 5 = 2$
$3(-1) + 5 \stackrel{?}{=} 2$
$-3 + 5 \stackrel{?}{=} 2$
$2 = 2$ ✓
The solution is -1.

5.
$-16 = 6a - 4$
$-16 + 4 = 6a - 4 + 4$
$-12 = 6a$
$\frac{-12}{6} = \frac{6a}{6}$
$-2 = a$
Check: $-16 = 6a - 4$
$-16 \stackrel{?}{=} 6(-2) - 4$
$-16 \stackrel{?}{=} -12 - 4$
$-16 = -16$ ✓
The solution is -2.

6.
$\frac{y}{3} + 2 = 10$
$\frac{y}{3} + 2 - 2 = 10 - 2$
$\frac{y}{3} = 8$
$\frac{y}{3} \cdot 3 = 8 \cdot 3$
$y = 24$
Check: $\frac{y}{3} + 2 = 10$
$\frac{24}{3} + 2 \stackrel{?}{=} 10$
$8 + 2 \stackrel{?}{=} 10$
$10 = 10$ ✓
The solution is 24.

7.
$1 + \frac{k}{4} = -9$
$\frac{k}{4} + 1 = -9$
$\frac{k}{4} + 1 - 1 = -9 - 1$
$\frac{k}{4} = -10$
$\frac{k}{4} \cdot 4 = -10 \cdot 4$
$k = -40$
Check: $1 + \frac{k}{4} = -9$
$1 + \frac{-40}{4} \stackrel{?}{=} -9$
$1 + -10 \stackrel{?}{=} -9$
$-9 = -9$ ✓
The solution is -40.

8.
$8 = \frac{n}{-7} - 5$
$8 + 5 = \frac{n}{-7} - 5 + 5$
$13 = \frac{n}{-7}$
$13(-7) = \frac{n}{-7}(-7)$
$-91 = n$
Check: $8 = \frac{n}{-7} - 5$
$8 \stackrel{?}{=} \frac{-91}{-7} - 5$
$8 \stackrel{?}{=} 13 - 5$
$8 = 8$ ✓
The solution is -91.

9.
$3 - c = 7$
$3 - 1c = 7$
$3 + (-1c) = 7$
$-3 + 3 + (-1c) = -3 + 7$
$-1c = 4$
$\frac{-1c}{-1} = \frac{4}{-1}$
$c = -4$
Check: $3 - c = 7$
$3 - (-4) \stackrel{?}{=} 7$
$3 + 4 \stackrel{?}{=} 7$
$7 = 7$ ✓
The solution is -4.

10. $2a - 8a = 24$
$$-6a = 24$$
$$\frac{-6a}{-6} = \frac{24}{-6}$$
$$a = -4$$
Check: $2a - 8a = 24$
$$2(-4) - 8(-4) \stackrel{?}{=} 24$$
$$-8 + 32 \stackrel{?}{=} 24$$
$$24 = 24 \checkmark$$
The solution is -4.

11. $8y - 9y + 6 = -4$
$$-1y + 6 = -4$$
$$-1y + 6 - 6 = -4 - 6$$
$$-1y = -10$$
$$\frac{-1y}{-1} = \frac{-10}{-1}$$
$$y = 10$$
Check: $8y - 9y + 6 = -4$
$$8(10) - 9(10) + 6 \stackrel{?}{=} -4$$
$$80 - 90 + 6 \stackrel{?}{=} -4$$
$$-10 + 6 \stackrel{?}{=} -4$$
$$-4 = -4 \checkmark$$
The solution is 10.

12. $8 + 4d = 36$
$$8 - 8 + 4d = 36 - 8$$
$$4d = 28$$
$$\frac{4d}{4} = \frac{28}{4}$$
$$d = 7$$
She will take the tablets for 7 days.

Pages 123–124 Practice and Apply

Exercises 13–30 For checks, see students' work.

13. $3x + 1 = 7$
$$3x + 1 - 1 = 7 - 1$$
$$3x = 6$$
$$\frac{3x}{3} = \frac{6}{3}$$
$$x = 2$$
The solution is 2.

14. $5x - 4 = 11$
$$5x - 4 + 4 = 11 + 4$$
$$5x = 15$$
$$\frac{5x}{5} = \frac{15}{5}$$
$$x = 3$$
The solution is 3.

15. $4h + 6 = 22$
$$4h + 6 - 6 = 22 - 6$$
$$4h = 16$$
$$\frac{4h}{4} = \frac{16}{4}$$
$$h = 4$$
The solution is 4.

16. $8n + 3 = -5$
$$8n + 3 - 3 = -5 - 3$$
$$8n = -8$$
$$\frac{8n}{8} = \frac{-8}{8}$$
$$n = -1$$
The solution is -1.

17. $37 = 4d + 5$
$$37 - 5 = 4d + 5 - 5$$
$$32 = 4d$$
$$\frac{32}{4} = \frac{4d}{4}$$
$$8 = d$$
The solution is 8.

18. $9 = 15 + 2p$
$$9 - 15 = 15 - 15 + 2p$$
$$-6 = 2p$$
$$\frac{-6}{2} = \frac{2p}{2}$$
$$-3 = p$$
The solution is -3.

19. $2n - 5 = 21$
$$2n - 5 + 5 = 21 + 5$$
$$2n = 26$$
$$\frac{2n}{2} = \frac{26}{2}$$
$$n = 13$$
The solution is 13.

20. $3j - 9 = 12$
$$3j - 9 + 9 = 12 + 9$$
$$3j = 21$$
$$\frac{3j}{3} = \frac{21}{3}$$
$$j = 7$$
The solution is 7.

21. $-1 = 2r - 7$
$$-1 + 7 = 2r - 7 + 7$$
$$6 = 2r$$
$$\frac{6}{2} = \frac{2r}{2}$$
$$3 = r$$
The solution is 3.

22. $12 = 5k - 8$
$$12 + 8 = 5k - 8 + 8$$
$$20 = 5k$$
$$\frac{20}{5} = \frac{5k}{5}$$
$$4 = k$$
The solution is 4.

23. $10 = 6 + \frac{y}{7}$
$$10 - 6 = 6 - 6 + \frac{y}{7}$$
$$4 = \frac{y}{7}$$
$$4 \cdot 7 = \frac{y}{7} \cdot 7$$
$$28 = y$$
The solution is 28.

24. $14 = 6 + \frac{n}{5}$

$14 - 6 = 6 - 6 + \frac{n}{5}$

$8 = \frac{n}{5}$

$8 \cdot 5 = \frac{n}{5} \cdot 5$

$40 = n$

The solution is 40.

25. $3 + \frac{t}{2} = 35$

$3 - 3 + \frac{t}{2} = 35 - 3$

$\frac{t}{2} = 32$

$\frac{t}{2} \cdot 2 = 32 \cdot 2$

$t = 64$

The solution is 64.

26. $13 + \frac{p}{3} = -4$

$13 - 13 + \frac{p}{3} = -4 - 13$

$\frac{p}{3} = -17$

$\frac{p}{3} \cdot 3 = -17 \cdot 3$

$p = -51$

The solution is -51.

27. $\frac{k}{5} - 10 = 3$

$\frac{k}{5} - 10 + 10 = 3 + 10$

$\frac{k}{5} = 13$

$\frac{k}{5} \cdot 5 = 13 \cdot 5$

$k = 65$

The solution is 65.

28. $\frac{w}{8} - 4 = -7$

$\frac{w}{8} - 4 + 4 = -7 + 4$

$\frac{w}{8} = -3$

$\frac{w}{8} \cdot 8 = -3 \cdot 8$

$w = -24$

The solution is -24.

29. $8 = \frac{c}{-3} + 15$

$8 - 15 = \frac{c}{-3} + 15 - 15$

$-7 = \frac{c}{-3}$

$-7(-3) = \frac{c}{-3}(-3)$

$21 = c$

The solution is 21.

30. $\frac{b}{-4} + 8 = -42$

$\frac{b}{-4} + 8 - 8 = -42 - 8$

$\frac{b}{-4} = -50$

$\frac{b}{-4}(-4) = -50(-4)$

$b = 200$

The solution is 200.

31. $2n + 5 = 27$

$2n + 5 - 5 = 27 - 5$

$2n = 22$

$\frac{2n}{2} = \frac{22}{2}$

$n = 11$

32. $4n - 3 = -7$

$4n - 3 + 3 = -7 + 3$

$4n = -4$

$\frac{4n}{4} = \frac{-4}{4}$

$n = -1$

33. $\frac{n}{2} - 10 = 5$

$\frac{n}{2} - 10 + 10 = 5 + 10$

$\frac{n}{2} = 15$

$\frac{n}{2} \cdot 2 = 15 \cdot 2$

$n = 30$

34. $\frac{n}{6} + 6 = -3$

$\frac{n}{6} + 6 - 6 = -3 - 6$

$\frac{n}{6} = -9$

$\frac{n}{6} \cdot 6 = -9 \cdot 6$

$n = -54$

Exercises 35–46 For checks, see students' work.

35. $8 - t = -25$

$8 + (-1t) = -25$

$8 - 8 + (-1t) = -25 - 8$

$-1t = -33$

$\frac{-1t}{-1} = \frac{-33}{-1}$

$t = 33$

The solution is 33.

36. $3 - y = 13$

$3 + (-1y) = 13$

$3 - 3 + (-1y) = 13 - 3$

$-1y = 10$

$\frac{-1y}{-1} = \frac{10}{-1}$

$y = -10$

The solution is -10.

37. $-5 - b = 8$

$-5 + (-1b) = 8$

$-5 + 5 + (-1b) = 8 + 5$

$-1b = 13$

$\frac{-1b}{-1} = \frac{13}{-1}$

$b = -13$

The solution is -13.

38. $10 = -9 - x$

$10 = -9 + (-1x)$

$10 + 9 = -9 + 9 + (-1x)$

$19 = -1x$

$\frac{19}{-1} = \frac{-1x}{-1}$

$-19 = x$

The solution is -19.

39. $2w - 4w = -10$

$$\begin{aligned} -2w &= -10 \\ \frac{-2w}{-2} &= \frac{-10}{-2} \\ w &= 5 \end{aligned}$$

The solution is 5.

40. $3x - 5x = 22$

$$\begin{aligned} -2x &= 22 \\ \frac{-2x}{-2} &= \frac{22}{-2} \\ x &= -11 \end{aligned}$$

The solution is −11.

41. $x + 4x + 6 = 31$

$$\begin{aligned} 5x + 6 &= 31 \\ 5x + 6 - 6 &= 31 - 6 \\ 5x &= 25 \\ \frac{5x}{5} &= \frac{25}{5} \\ x &= 5 \end{aligned}$$

The solution is 5.

42. $5r + 3r - 6 = 10$

$$\begin{aligned} 8r - 6 &= 10 \\ 8r - 6 + 6 &= 10 + 6 \\ 8r &= 16 \\ \frac{8r}{8} &= \frac{16}{8} \\ \mathrm{r} &= 2 \end{aligned}$$

The solution is 2.

43.

$$\begin{aligned} 1 - 3y + y &= 5 \\ 1 + (-3y) + 1y &= 5 \\ 1 + (-2y) &= 5 \\ 1 - 1 + (-2y) &= 5 - 1 \\ -2y &= 4 \\ \frac{-2y}{-2} &= \frac{4}{-2} \\ y &= -2 \end{aligned}$$

The solution is −2.

44.

$$\begin{aligned} 16 &= w - 2w + 9 \\ 16 &= -1w + 9 \\ 16 - 9 &= -1w + 9 - 9 \\ 7 &= -1w \\ \frac{7}{-1} &= \frac{-1w}{-1} \\ -7 &= w \end{aligned}$$

The solution is −7.

45.

$$\begin{aligned} 23 &= 4t - 7 - t \\ 23 &= 4t + (-7) + (-1t) \\ 23 &= 4t + (-1t) + (-7) \\ 23 &= 3t + (-7) \\ 23 + 7 &= 3t + (-7) + 7 \\ 30 &= 3t \\ \frac{30}{3} &= \frac{3t}{3} \\ 10 &= t \end{aligned}$$

The solution is 10.

46.

$$\begin{aligned} -4 &= -a + 8 - 2a \\ -4 &= -1a + 8 + (-2a) \\ -4 &= -1a + (-2a) + 8 \\ -4 &= -3a + 8 \\ -4 - 8 &= -3a + 8 - 8 \\ -12 &= -3a \\ \frac{-12}{-3} &= \frac{-3a}{-3} \\ 4 &= a \end{aligned}$$

The solution is 4.

47.

$$\begin{aligned} 1600 &= 320t + 640 \\ 1600 - 640 &= 320t + 640 - 640 \\ 960 &= 320t \\ \frac{960}{320} &= \frac{320t}{320} \\ 3 &= t \end{aligned}$$

It will take 3 hours to fill the pool.

48.

$$\begin{aligned} 25m + 75 &= 500 \\ 25m + 75 - 75 &= 500 - 75 \\ 25m &= 425 \\ \frac{25m}{25} &= \frac{425}{25} \\ m &= 17 \end{aligned}$$

You can use the card 17 minutes.

49.

$$\begin{aligned} 6200 &= 50b - 350 \\ 6200 + 350 &= 50b - 350 + 350 \\ 6550 &= 50b \\ \frac{6550}{50} &= \frac{50b}{50} \\ 131 &= b \end{aligned}$$

He sold 131 bikes.

50. Sample answer: You can add or remove tiles from each side of a mat. This models the Addition and Subtraction Properties of Equality. Also, separating tiles into groups models the Division Property of Equality.

•

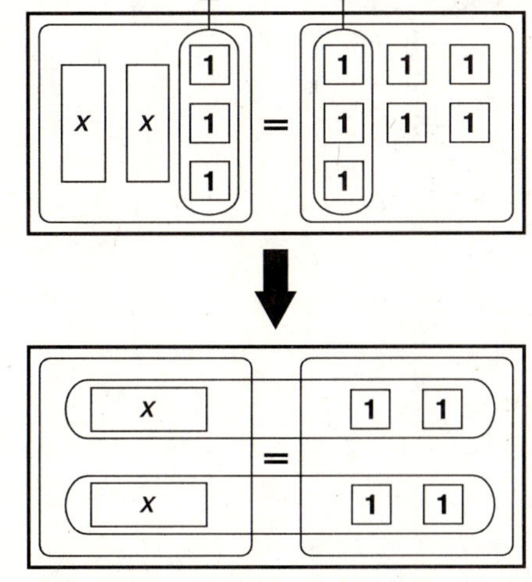

- Subtraction Property of Equality, Division Property of Equality

51. $5x - 2 = 8$

52. The graph shows that the cost increases \$0.50 every hour and that the flat rate is \$1.00. The expression for the total charge is $0.50x + 1.00$.
$0.50(7) + 1.00 = 3.50 + 1.00 = 4.50$
The total charge is \$4.50.

53. C; The table shows an increase of \$32 each month. The cost for 6 months is \$205, so the cost for 10 months is $205 + 4 \cdot 32 = 205 + 128 = \$333$.

Page 124 Maintain Your Skills

Exercises 54–59 For checks, see students' work.

54. $5y = 60$

$\frac{5y}{5} = \frac{60}{5}$

$y = 12$

The solution is 12.

55. $14 = -2n$

$\frac{14}{-2} = \frac{-2n}{-2}$

$-7 = n$

The solution is -7.

56. $\frac{x}{3} = -9$

$\frac{x}{3} \cdot 3 = -9 \cdot 3$

$x = -27$

The solution is -27.

57. $x - 4 = -6$

$x - 4 + 4 = -6 + 4$

$x = -2$

The solution is -2.

58. $-13 = y + 5$

$-13 - 5 = y + 5 - 5$

$-18 = y$

The solution is -18.

59. $18 = 20 + x$

$18 - 20 = 20 - 20 + x$

$-2 = x$

The solution is -2.

60. $4(x + 1) = 4x + 4 \cdot 1$

$= 4x + 4$

61. $-5(y + 3) = -5y + (-5) \cdot 3$

$= -5y + (-15)$

$= -5y - 15$

62. $3(k - 10) = 3[k + (-10)]$

$= 3k + 3 \cdot (-10)$

$= 3k + (-30)$

$= 3k - 30$

63. $-9(y - 4) = -9[y + (-4)]$

$= -9y + (-9)(-4)$

$= -9y + 36$

64. $7(a - 2) = 7[a + (-2)]$

$= 7a + 7(-2)$

$= 7a + (-14)$

$= 7a - 14$

65. $-8(r - 5) = -8[r + (-5)]$

$= -8r + (-8)(-5)$

$= -8r + 40$

66. Start at the origin, and move 3 units to the left and 1 unit up. The ordered pair for point T is $(-3, 1)$.

67. Start at the origin, and move 2 units to the right and 3 unit down. The ordered pair for point C is $(2, -3)$.

68. Start at the origin, and move 4 units to the right. Do not move up or down. The ordered pair for point R is $(4, 0)$.

69. Start at the origin, and move 3 units to the left and 4 unit down. The ordered pair for point P is $(-3, -4)$.

70. $2x - 6$

71. $x \div 5$

72. $2x - 8$

73. $2x + 10$

74. $2x + 7x + 4 = 9x + 4$

Page 125 Reading Mathematics

1a. Lucas

1b. 5 inches

1c. $x + 5$

1d. $x + (x + 5)$ or $2x + 5$

1e. $2x + 5 = 137$

2a. students

2b. 5 times

2c. $5x$

2d. $x + 5x$ or $6x$

2e. $5x + x = 132$ or $6x = 132$

3-6 Writing Two-Step Equations

Page 126 How are equations used to solve real-world problems?

a. $4n + 99$

b. $4n + 99 = 299$; 50

c. It has two operations, multiplication and addition.

Page 128 Check for Understanding

1. is, equals, is equal to

2. Sample answer: Three more than twice a number is 15.

3. Ben; *Three less than* means that three is subtracted *from* a number.

4. $4n + 3 = 23$

$4n + 3 - 3 = 23 - 3$

$4n = 20$

$\frac{4n}{4} = \frac{20}{4}$

$n = 5$

The number is 5.

5. $2n - 4 = -2$
$$2n - 4 + 4 = -2 + 4$$
$$2n = 2$$
$$\frac{2n}{2} = \frac{2}{2}$$
$$n = 1$$
The number is 1.

6. Let x = the number of hours.
$$17 + 3x = 32$$
$$17 - 17 + 3x = 32 - 17$$
$$3x = 15$$
$$\frac{3x}{3} = \frac{15}{3}$$
$$x = 5$$
It will take 5 hours.

7. Let x = Cole's age and $x + 5$ = Lawana's age.
$$x + (x + 5) = 37$$
$$x + x + 5 = 37$$
$$2x + 5 = 37$$
$$2x + 5 - 5 = 37 - 5$$
$$2x = 32$$
$$\frac{2x}{2} = \frac{32}{2}$$
$$x = 16$$
Lawana is 16 + 5, or 21 years old.

Pages 129–130 Practice and Apply

8. $2n + 7 = 17$
$$2n + 7 - 7 = 17 - 7$$
$$2n = 10$$
$$\frac{2n}{2} = \frac{10}{2}$$
$$n = 5$$
The number is 5.

9. $3n + 20 = -4$
$$3n + 20 - 20 = -4 - 20$$
$$3n = -24$$
$$\frac{3n}{3} = \frac{-24}{3}$$
$$n = -8$$
The number is −8.

10. $3n - 4 = 20$
$$3n - 4 + 4 = 20 + 4$$
$$3n = 24$$
$$\frac{3n}{3} = \frac{24}{3}$$
$$n = 8$$
The number is 8.

11. $10n - 8 = 82$
$$10n - 8 + 8 = 82 + 8$$
$$10n = 90$$
$$\frac{10n}{10} = \frac{90}{10}$$
$$n = 9$$
The number is 9.

12. $\frac{n}{-2} + 10 = 3$
$$\frac{n}{-2} + 10 - 10 = 3 - 10$$
$$\frac{n}{-2} = -7$$
$$\frac{n}{-2}(-2) = -7(-2)$$
$$n = 14$$
The number is 14.

13. $\frac{n}{-4} - 8 = -42$
$$\frac{n}{-4} - 8 + 8 = -42 + 8$$
$$\frac{n}{-4} = -34$$
$$\frac{n}{-4}(-4) = -34(-4)$$
$$n = 136$$
The number is 136.

14. $2n - 9 = 17$
$$2n - 9 + 9 = 17 + 9$$
$$2n = 26$$
$$\frac{2n}{2} = \frac{26}{2}$$
$$n = 13$$
The number is 13.

15. $3n - 8 = -2$
$$3n - 8 + 8 = -2 + 8$$
$$3n = 6$$
$$\frac{3n}{3} = \frac{6}{3}$$
$$n = 2$$
The number is 2.

16. $5 - 3n = -4$
$$5 + (-3n) = -4$$
$$5 - 5 + (-3n) = -4 - 5$$
$$-3n = -9$$
$$\frac{-3n}{-3} = \frac{-9}{-3}$$
$$n = 3$$
The number is 3.

17. $17 - 2n = 5$
$$17 + (-2n) = 5$$
$$17 - 17 + (-2n) = 5 - 17$$
$$-2n = -12$$
$$\frac{-2n}{-2} = \frac{-12}{-2}$$
$$n = 6$$
The number is 6.

18. $3n + 2n + 1 = -4$
$$5n + 1 = -4$$
$$5n + 1 - 1 = -4 - 1$$
$$5n = -5$$
$$\frac{5n}{5} = \frac{-5}{5}$$
$$n = -1$$
The number is −1.

19. $4n + 3n + 5 = 47$
$$7n + 5 = 47$$
$$7n + 5 - 5 = 47 - 5$$
$$7n = 42$$
$$\frac{7n}{7} = \frac{42}{7}$$
$$n = 6$$
The number is 6.

20. Let x = the cost of a bag of birdseed.
$$3x + 18 = 45$$
$$3x + 18 - 18 = 45 - 18$$
$$3x = 27$$
$$\frac{3x}{3} = \frac{27}{3}$$
$$x = 9$$
A bag of birdseed costs $9.

21. Let x = the number of hours.
$$8 - 5x = -7$$
$$8 + (-5x) = -7$$
$$8 - 8 + (-5x) = -7 - 8$$
$$-5x = -15$$
$$\frac{-5x}{-5} = \frac{-15}{-5}$$
$$x = 3$$
It will take 3 hours.

22. Let x = the amount you spend for lunch.
$$x + x + 3 = 15$$
$$2x + 3 = 15$$
$$2x + 3 - 3 = 15 - 3$$
$$2x = 12$$
$$\frac{2x}{2} = \frac{12}{2}$$
$$x = 6$$
You spent $6 for lunch.

23. Let x = the expected senior citizen population of Florida in 2020.
$$x + x + 2 = 12$$
$$2x + 2 = 12$$
$$2x + 2 - 2 = 12 - 2$$
$$2x = 10$$
$$\frac{2x}{2} = \frac{10}{2}$$
$$x = 5$$
The expected population is 5 million people.

24. Let x = New York's Native-American population. Then $x + 22{,}000$ = North Carolina's Native-American population, and $x + 187{,}000$ = Oklahoma's Native-American population.
$$x + (x + 22{,}000) + (x + 187{,}000) = 437{,}000$$
$$x + x + x + 22{,}000 + 187{,}000 = 437{,}000$$
$$3x + 209{,}000 = 437{,}000$$
$$3x + 209{,}000 - 209{,}000 = 437{,}000 - 209{,}000$$
$$3x = 228{,}000$$
$$\frac{3x}{3} = \frac{228{,}000}{3}$$
$$x = 76{,}000$$
Therefore, $x + 22{,}000 = 76{,}000 + 22{,}000 = 98{,}000$, and $x + 187{,}000 = 76{,}000 + 187{,}000 = 263{,}000$. The populations are as follows:

New York: 76,000

North Carolina: 98,000

Oklahoma: 263,000

25. Sample answer: By 2020, Texas is expected to have 10 thousand more people age 85 or older than New York will have. Together, they are expected to have 846 thousand people age 85 or older. Find the expected number of people age 85 or older in New York by 2020.

26. Let x = an even number and $x + 2$ = the next consecutive even number.
$$2x + 2 = 50$$
$$2x + 2 - 2 = 50 - 2$$
$$2x = 48$$
$$\frac{2x}{2} = \frac{48}{2}$$
$$x = 24$$
Then $x + 2 = 24 + 2 = 26$ is the next consecutive number. The two numbers are 24 and 26.

27. Sample answer: Two-step equations can be used when you start with a certain amount and increase or decrease at a certain rate. Answers should include the following.

- You've been running 15 minutes each day as part of a fitness program. You plan to increase your time by 5 minutes each week. After how many weeks do you plan to run 30 minutes each day? ($5w + 15 = 30$, 3 weeks)
- You are three years older than your sister is. Together the sum of your ages is 21. How old is your sister? ($2x + 3 = 21$, 9 years old)

28. C; $5n - 3$

29. D; Let x = the height of the Transamerica Pyramid. Then $x - 74$ = the height of the Bank of America building.
$$x + (x - 74) = 1632$$
$$2x - 74 = 1632$$
$$2x - 74 + 74 = 1632 + 74$$
$$2x = 1706$$
$$\frac{2x}{2} = \frac{1706}{2}$$
$$x = 853 \text{ ft}$$

Page 130 Maintain Your Skills

30. $6 - 2x = 10$
$$6 + (-2x) = 10$$
$$6 - 6 + (-2x) = 10 - 6$$
$$-2x = 4$$
$$\frac{-2x}{-2} = \frac{4}{-2}$$
$$x = -2$$

31. $-4x = -16$
$$\frac{-4x}{-4} = \frac{-16}{-4}$$
$$x = 4$$

32. $y - 7 = -3$
$y - 7 + 7 = -3 + 7$
$y = 4$

33. $|x| - 7 = |-12| - 7$
$= 12 - 7$
$= 5$

34. $|x| + |y| = |-12| + |4|$
$= 12 + 4$
$= 16$

35. $|z| - |x| = |-1| - |-12|$
$= 1 - 12$
$= -11$

36. Associative Property of Addition

Exercises 37–42 For checks, see students' work.

37. $2x = -8$
$\frac{2x}{2} = \frac{-8}{2}$
$x = -4$
The solution is −4.

38. $24 = 6y$
$\frac{24}{6} = \frac{6y}{6}$
$4 = y$
The solution is 4.

39. $5w = -25$
$\frac{5w}{5} = \frac{-25}{5}$
$w = -5$
The solution is −5.

40. $15s = 75$
$\frac{15s}{15} = \frac{75}{15}$
$s = 5$
The solution is 5.

41. $108 = 18x$
$\frac{108}{18} = \frac{18x}{18}$
$6 = x$
The solution is 6.

42. $25z = 175$
$\frac{25z}{25} = \frac{175}{25}$
$z = 7$
The solution is 7.

Page 130 Practice Quiz 2

1. $4h = -52$
$\frac{4h}{4} = \frac{-52}{4}$
$h = -13$

2. $\frac{x}{-3} = 4$
$\frac{x}{-3}(-3) = 4(-3)$
$x = -12$

3. $y - 5 = -23$
$y - 5 + 5 = -23 + 5$
$y = -18$

4. $2v - 11 = -5$
$2v - 11 + 11 = -5 + 11$
$2v = 6$
$\frac{2v}{2} = \frac{6}{2}$
$v = 3$

5. Let n = the number.
$3n + 20 = 32$
$3n + 20 - 20 = 32 - 20$
$3n = 12$
$\frac{3n}{3} = \frac{12}{3}$
$n = 4$
The number is 4.

3-7 Using Formulas

Page 131 Why are formulas important in math and science?

a. $65t$

b. You cannot show all of the possible relationships in a table.

c. Write an equation that relates speed, time, and distance.

Page 133 Check for Understanding

1. $d = rt$

2. Perimeter is the measure of the distance around a rectangle; area is a measure of the surface the perimeter encloses.

3. Sample answer:

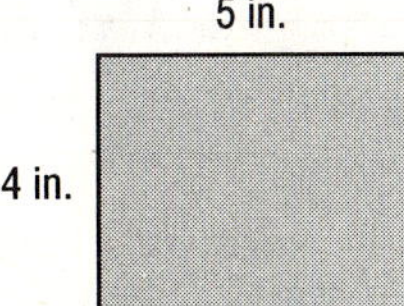

4. $P = 2(\ell + w)$
$P = 2(8 + 3)$
$P = 2(11)$
$P = 22$
The perimeter is 22 ft.
$A = \ell w$
$A = 8 \cdot 3$
$A = 24$
The area is 24 ft^2.

5. $P = 2(\ell + w)$

$P = 2(15 + 2)$

$P = 2(17)$

$P = 34$

The perimeter is 34 km.

$A = \ell w$

$A = 15 \cdot 2$

$A = 30$

The area is 30 km^2.

6. $P = 2(\ell + w)$

$P = 2(15 + 6)$

$P = 2(21)$

$P = 42$

The perimeter is 42 ft.

$A = \ell w$

$A = 15 \cdot 6$

$A = 90$

The area is 90 ft^2.

7. $P = 2(\ell + w)$

$P = 2\ell + 2w$

$32 = 2(12) + 2w$

$32 = 24 + 2w$

$32 - 24 = 24 - 24 + 2w$

$8 = 2w$

$4 = w$

The width is 4 in.

8. $A = \ell w$

$96 = \ell \cdot 8$

$\frac{96}{8} = \frac{\ell \cdot 8}{8}$

$12 = \ell$

The length is 12 m.

9. $d = rt$

$5200 = 650t$

$\frac{5200}{650} = \frac{650t}{650}$

$8 = t$

It will take 8 hours.

Pages 134–136 Practice and Apply

10. $d = rt$

$d = 55 \cdot 3$

$d = 165$

The distance traveled is 165 mi.

11. $d = rt$

$60 = r \cdot 4$

$\frac{60}{4} = \frac{r \cdot 4}{4}$

$15 = r$

The rate is 15 mph.

12. $P = 2(\ell + w)$

$P = 2(3 + 2)$

$P = 2(5)$

$P = 10$

The perimeter is 10 mi.

$A = \ell w$

$A = 3 \cdot 2$

$A = 6$

The area is 6 mi^2.

13. $P = 2(\ell + w)$

$P = 2(18 + 9)$

$P = 2(27)$

$P = 54$

The perimeter is 54 cm.

$A = \ell w$

$A = 18 \cdot 9$

$A = 162$

The area is 162 cm^2.

14. $P = 2(\ell + w)$

$P = 2(12 + 5)$

$P = 2(17)$

$P = 34$

The perimeter is 34 ft.

$A = \ell w$

$A = 12 \cdot 5$

$A = 60$

The area is 60 ft^2.

15. $P = 2(\ell + w)$

$P = 2(50 + 18)$

$P = 2(68)$

$P = 136$

The perimeter is 136 in.

$A = \ell w$

$A = 50 \cdot 18$

$A = 900$

The area is 900 in^2.

16. $P = 2(\ell + w)$

$P = 2(17 + 6)$

$P = 2(23)$

$P = 46$

The perimeter is 46 m.

$A = \ell w$

$A = 17 \cdot 6$

$A = 102$

The area is 102 mi^2.

17. $P = 2(\ell + w)$

$P = 2(12 + 12)$

$P = 2(24)$

$P = 48$

The perimeter is 48 m.

$A = \ell w$

$A = 12 \cdot 12$

$A = 144$

The area is 144 m^2.

18. $P = 2(\ell + w)$
$P = 2(38 + 10)$
$P = 2(48)$
$P = 96$
The perimeter is 96 m.
$A = \ell w$
$A = 38 \cdot 10$
$A = 380$
The area is 380 m^2.

19. $P = 2(\ell + w)$
$P = 2(5 + 5)$
$P = 2(10)$
$P = 20$
The perimeter is 20 m.
$A = \ell w$
$A = 5 \cdot 5$
$A = 25$
The area is 25 m^2.

20. $A = \ell w$
$270 = \ell \cdot 15$
$\frac{270}{15} = \frac{\ell \cdot 15}{15}$
$18 = \ell$
The length is 18 cm.

21. $A = \ell w$
$176 = 16w$
$\frac{176}{16} = \frac{16w}{16}$
$11 = w$
The width is 11 yd.

22. $P = 2(\ell + w)$
$P = 2\ell + 2w$
$70 = 2\ell + 2(11)$
$70 = 2\ell + 22$
$70 - 22 = 2\ell + 22 - 22$
$48 = 2\ell$
$24 = \ell$
The length is 24 m.

23. $P = 2(\ell + w)$
$P = 2\ell + 2w$
$24 = 2 \cdot 7 + 2w$
$24 = 14 + 2w$
$24 - 14 = 14 - 14 + 2w$
$10 = 2w$
$5 = w$
The width is 5 m.

24. $A = \ell w$
$154 = 14w$
$\frac{154}{14} = \frac{14w}{14}$
$11 = w$
The width is 11 in.

25. $A = \ell w$
$468 = \ell \cdot 12$
$\frac{468}{12} = \frac{\ell \cdot 12}{12}$
$39 = \ell$
The length is 39 ft.

26. $P = 2(\ell + w)$
$P = 2\ell + 2w$
$46 = 2\ell + 2 \cdot 5$
$46 = 2\ell + 10$
$46 - 10 = 2\ell + 10 - 10$
$36 = 2\ell$
$18 = \ell$
The length is 18 cm.

27. $A = \ell w$
$323 = 17w$
$\frac{323}{17} = \frac{17w}{17}$
$19 = w$
The width is 19 yd.

28. Since fencing is measured around the outside of the garden, the formula for perimeter is used.
$P = 2(\ell + w)$
$P = 2(45 + 18)$
$P = 2(63)$
$P = 126$
126 ft of fencing is needed.

29. $P = 2(\ell + w)$
$P = 2(120 + 75)$
$P = 2(195)$
$P = 390$
The perimeter is 390 yd.
$A = \ell w$
$A = 120 \cdot 75$
$A = 9000$
The area is 9000 yd^2.

30. $s = \ell - d$

31. $d = 2r$

32. Runner A:
$r = \frac{n}{t} = \frac{20}{5} = 4$
Runner B:
$r = \frac{n}{t} = \frac{30}{10} = 3$
Runner A has the greater stride rate.

33. The lawn is inside a rectangle 80 ft long and 75 ft wide.
$A = \ell w$
$A = 80 \cdot 75$
$A = 6000$
The areas of the house and the driveway must be subtracted from the area of the outer rectangle.
Area of house:
$A = \ell w$
$A = 50 \cdot 28$
$A = 1400$

Area of driveway:

$A = \ell w$

$A = 20 \cdot 15$

$A = 300$

Area of lawn:

$6000 - 1400 - 300 = 4300 \text{ ft}^2$.

34. $4300 \div 2500 = 1.72$

Since you cannot buy 1.72 bags, you must buy 2 bags of fertilizer.

35. 92 hours 33 minutes 8 seconds is converted to hours:

$92 + \frac{33}{60} + \frac{8}{3600} \approx 92.55$ h

$$\begin{aligned} d &= rt \\ 2178 &= r \cdot 92.55 \\ \frac{2178}{92.55} &= \frac{r \cdot 92.55}{92.55} \\ 23.5 &\approx r \end{aligned}$$

The rate is about 23.5 mph.

36. 1999; He traveled a greater distance in less time.

37. $P = 2(\ell + w)$ $\quad A = \ell w$

$14 = 2(\ell + w)$ $\quad 12 = \ell w$

$7 = \ell + w$

The length and width must have a sum of 7 and a product of 12. Use 3 and 4.

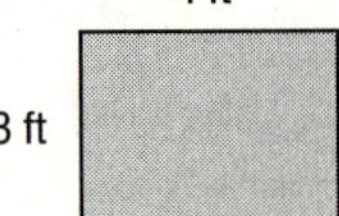

38. $P = 2(\ell + w)$ $\quad A = \ell w$

$16 = 2(\ell + w)$ $\quad 12 = \ell w$

$8 = \ell + w$

The length and width must have a sum of 8 and a product of 12. Use 6 and 2.

39. $P = 2(\ell + w)$ $\quad A = \ell w$

$16 = 2(\ell + w)$ $\quad 16 = \ell w$

$8 = \ell + w$

The length and width must have a sum of 8 and a product of 16. Use 4 and 4.

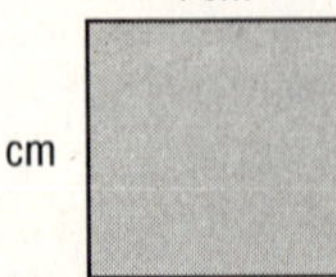

40. $P = 2(\ell + w)$ $\quad A = \ell w$

$12 = 2(\ell + w)$ $\quad 8 = \ell w$

$6 = \ell + w$

The length and width must have a sum of 6 and a product of 8. Use 2 and 4.

4 in.

2 in.

41. Sometimes; a 3-inch by 4-inch rectangle has a perimeter of 14 inches and an area of 12 square inches; a 6-inch by 8-inch rectangle has a perimeter of 28 inches and an area of 48 square inches.

42. Both planes travel d, the same distance. The plane traveling 600 mph travels for time t, so

$d = 600t$.

The plane traveling 500 mph travels for time $t + 0.5$, so

$d = 500(t + 0.5)$.

Next, we can put the two right-hand parts of the equations into a new equation, with only the variable t, and solve for t.

$$\begin{aligned} 600t &= 500(t + 0.5) \\ 600t &= 500t + 500(0.5) \\ 600t - 500t &= 500t - 500t + 250 \\ 100t &= 250 \\ \frac{100t}{100} &= \frac{250}{100} \\ t &= 2.5 \end{aligned}$$

The second plane will catch up with the first plane in 2.5 hours.

43. Formulas are important in math and science because they summarize the relationships among quantities. Answers should include the following.

- Sample answer: The formula to find acceleration is $a = \frac{v_f - v_i}{t}$ where v_f is the final velocity and v_i is the initial velocity.
- You can find the acceleration of an automobile with this formula.

44. B; Since d stays the same, but is divided by a larger value of t, $\frac{d}{t}$ will decrease.

45. C; $A = \ell w$

$16 = \ell w$

Since the figures are squares, the length and width are equal. So $\ell = 4$ and $w = 4$.

The perimeter is the distance around the figure. There are 12 sides, and each side has a length of 4, so the perimeter is $12 \cdot 4 = 48$ units.

Page 136 Maintain Your Skills

46.

$$\begin{aligned} 5n + 8 &= 78 \\ 5n + 8 - 8 &= 78 - 8 \\ 5n &= 70 \\ \frac{5n}{5} &= \frac{70}{5} \\ n &= 14 \end{aligned}$$

The number is 14.

Exercises 47–49 For checks, see students' work.

47.

$$\begin{aligned} -5x + 8 &= 53 \\ -5x + 8 - 8 &= 53 - 8 \\ -5x &= 45 \\ \frac{-5x}{-5} &= \frac{45}{-5} \\ x &= -9 \end{aligned}$$

The solution is -9.

48. $4y = -24$

$\frac{4y}{4} = \frac{-24}{4}$

$y = -6$

The solution is -6.

49. $m + 5 = -3$

$m + 5 - 5 = -3 - 5$

$m = -8$

The solution is -8.

50. $3y + 5 - 2y = 3y + 5 + (-2y)$

$= 3y + (-2y) + 5$

$= 1y + 5$

$= y + 5$

51. $9 + x - 5x = 9 + 1x + (-5x)$

$= 9 + (-4x)$

$= -4x + 9$

52. $3(r + 2) + 6r = 3r + 3 \cdot 2 + 6r$

$= 3r + 6 + 6r$

$= 3r + 6r + 6$

$= 9r + 6$

53a. {(1870, 14), (1881, 600), (1910, 1000), (2000, 1500)}

53b. domain: {1870, 1881, 1910, 2000}, range: {14, 600, 1000, 1500}

Page 137 Spreadsheet Investigation (Follow-Up of Lesson 3-7)

1. 0, -22

2. No; area cannot be zero or a negative number.

3. B1

4. 10 ft by 20 ft; 12.5 ft by 25 ft; 15 ft by 30 ft

5. 25 ft by 50 ft; the opposite wall is twice the attached wall.

Chapter 3 Study Guide and Review

Page 138 Vocabulary and Concept Check

1. like terms

2. perimeter

3. Multiplication Property of Equality

4. equivalent

5. Distributive

6. coefficient

7. coordinate

8. area

9. constant

10. inverse operations

Pages 138–140 Lesson-by-Lesson Review

11. $3(h + 6) = 3h + 3 \cdot 6$

$= 3h + 18$

12. $7(x + 2) = 7x + 7 \cdot 2$

$= 7x + 14$

13. $-5(k + 1) = -5k + (-5) \cdot 1$

$= -5k - 5$

14. $-2(a + 8) = -2a + (-2) \cdot 8$

$= -2a - 16$

15. $(t - 5)9 = [t + (-5)]9$

$= t \cdot 9 + (-5) \cdot 9$

$= 9t - 45$

16. $(x - 3)7 = [x + (-3)]7$

$= x \cdot 7 + (-3) \cdot 7$

$= 7x - 21$

17. $-2(b - 4) = -2[b + (-4)]$

$= -2b + (-2)(-4)$

$= -2b + 8$

18. $-6(y - 3) = -6[y + (-3)]$

$= -6y + (-6)(-3)$

$= -6y + 18$

19. $4a + 5a = (4 + 5)a$

$= 9a$

20. $3x + 7 + x = 3x + 7 + 1x$

$= 3x + 1x + 7$

$= (3 + 1)x + 7$

$= 4x + 7$

21. $8(n - 1) - 10n = 8[n + (-1)] + (-10n)$

$= 8n + 8(-1) + (-10n)$

$= 8n + (-10n) + 8(-1)$

$= [8 + (-10)]n + (-8)$

$= -2n - 8$

22. $6w + 2(w + 9) = 6w + 2w + 2 \cdot 9$

$= (6 + 2)w + 18$

$= 8w + 18$

23. $t + 5 = 8$

$t + 5 - 5 = 8 - 5$

$t = 3$

The solution is 3.

24. $12 = x + 4$

$12 - 4 = x + 4 - 4$

$8 = x$

The solution is 8.

25. $k - 1 = 4$

$k - 1 + 1 = 4 + 1$

$k = 5$

The solution is 5.

26. $-7 = n - 6$

$-7 + 6 = n - 6 + 6$

$-1 = n$

The solution is -1.

27. $6n = 48$

$\frac{6n}{6} = \frac{48}{6}$

$n = 8$

The solution is 8.

28. $-3x = 30$

$\frac{-3x}{-3} = \frac{30}{-3}$

$x = -10$

The solution is -10.

29. $\frac{t}{2} = -9$

$\frac{t}{2} \cdot 2 = -9 \cdot 2$

$t = -18$

The solution is -18.

30. $\frac{r}{-5} = -2$

$\frac{r}{-5}(-5) = -2(-5)$

$r = 10$

The solution is 10.

31. $6 + 2y = 8$

$6 - 6 + 2y = 8 - 6$

$2y = 2$

$y = 1$

The solution is 1.

32. $3n - 5 = -17$

$3n - 5 + 5 = -17 + 5$

$3n = -12$

$n = -4$

The solution is -4.

33. $\frac{t}{3} + 4 = 2$

$\frac{t}{3} + 4 - 4 = 2 - 4$

$\frac{t}{3} = -2$

$\frac{t}{3} \cdot 3 = -2 \cdot 3$

$t = -6$

The solution is -6.

34. $\frac{c}{9} - 3 = 2$

$\frac{c}{9} - 3 + 3 = 2 + 3$

$\frac{c}{9} = 5$

$\frac{c}{9} \cdot 9 = 5 \cdot 9$

$c = 45$

The solution is 45.

35. $2n + 3 = 53$

$2n + 3 - 3 = 53 - 3$

$2n = 50$

$n = 25$

36. $4x - 16 = 52$

$4x - 16 + 16 = 52 + 16$

$4x = 68$

$\frac{4x}{4} = \frac{68}{4}$

$x = 17$

37. $P = 2(\ell + w)$

$P = 2(9 + 8)$

$P = 2(17)$

$P = 34$

The perimeter is 34 ft.

$A = \ell w$

$A = 9 \cdot 8$

$A = 72$

The area is 72 ft^2.

38. $P = 2(\ell + w)$

$P = 2(15 + 5)$

$P = 2(20)$

$P = 40$

The perimeter is 40 m.

$A = \ell w$

$A = 15 \cdot 5$

$A = 75$

The area is 75 m^2.

Chapter 3 Practice Test

Page 141

1. Sample answer: like terms—$2x$ and $3x$, not like terms—$5y$ and 2.

2. Sample answer: To multiply a number by a sum, multiply the number outside the parentheses by each number inside the parentheses.

3. Perimeter is the distance around a figure; area is a measure of the surface the perimeter encloses.

4. $9x + 5 - x + 3 = 9x + 5 + (-1x) + 3$

$= 9x + (-1x) + 5 + 3$

$= [9 + (-1)]x + (5 + 3)$

$= 8x + 8$

5. $-3(a - 8) = -3[a + (-8)]$

$= -3a + (-3)(-8)$

$= -3a + 24$

6. $10(y + 3) - 4y = 10y + 10 \cdot 3 - 4y$

$= 10y + 30 + (-4y)$

$= 10y + (-4y) + 30$

$= [10 + (-4)]y + 30$

$= 6y + 30$

7. $19 = f + 5$

$19 - 5 = f + 5 - 5$

$14 = f$

Check: $19 = f + 5$

$19 \stackrel{?}{=} 14 + 5$

$19 = 19$ ✓

The solution is 14.

8. $-15 + z = 3$

$-15 + 15 + z = 3 + 15$

$z = 18$

Check: $-15 + z = 3$

$-15 + 18 \stackrel{?}{=} 3$

$3 = 3$ ✓

The solution is 18.

9. $x - 7 = 16$
$x - 7 + 7 = 16 + 7$
$x = 23$
Check: $x - 7 = 16$
$23 - 7 \stackrel{?}{=} 16$
$16 = 16$ ✓
The solution is 23.

10. $g - 9 = -10$
$g - 9 + 9 = -10 + 9$
$g = -1$
Check: $g - 9 = -10$
$-1 - 9 \stackrel{?}{=} -10$
$-10 = -10$ ✓
The solution is -1.

11. $-8y = 72$
$\frac{-8y}{-8} = \frac{72}{-8}$
$y = -9$
Check: $-8y = 72$
$-8(-9) \stackrel{?}{=} 72$
$72 = 72$ ✓
The solution is -9.

12. $\frac{n}{-30} = -6$
$\frac{n}{-30}(-30) = -6(-30)$
$n = 180$
Check: $\frac{n}{-30} = -6$
$\frac{180}{-30} \stackrel{?}{=} -6$
$-6 = -6$ ✓
The solution is 180.

13. $25 = 2d - 9$
$25 + 9 = 2d - 9 + 9$
$34 = 2d$
$\frac{34}{2} = \frac{2d}{2}$
$17 = d$
Check: $25 = 2d - 9$
$25 \stackrel{?}{=} 2(17) - 9$
$25 \stackrel{?}{=} 34 - 9$
$25 = 25$ ✓
The solution is 17.

14. $4w - 18 = -34$
$4w - 18 + 18 = -34 + 18$
$4w = -16$
$\frac{4w}{4} = \frac{-16}{4}$
$w = -4$
Check: $4w - 18 = -34$
$4(-4) - 18 \stackrel{?}{=} -34$
$-16 - 18 \stackrel{?}{=} -34$
$-34 = -34$ ✓
The solution is -4.

15. $6v + 10 = -62$
$6v + 10 - 10 = -62 - 10$
$6v = -72$
$\frac{6v}{6} = \frac{-72}{6}$
$v = -12$
Check: $6v + 10 = -62$
$6(-12) + 10 \stackrel{?}{=} -62$
$-72 + 10 \stackrel{?}{=} -62$
$-62 = -62$ ✓
The solution is -12.

16. $-7 = \frac{d}{-5} + 1$
$-7 - 1 = \frac{d}{-5} + 1 - 1$
$-8 = \frac{d}{-5}$
$-8(-5) = \frac{d}{-5}(-5)$
$40 = d$
Check: $-7 = \frac{d}{-5} + 1$
$-7 \stackrel{?}{=} \frac{40}{-5} + 1$
$-7 \stackrel{?}{=} -8 + 1$
$-7 = -7$ ✓
The solution is 40.

17. $7 - x = 18$
$7 + (-1x) = 18$
$7 - 7 + (-1x) = 18 - 7$
$-1x = 11$
$\frac{-1x}{-1} = \frac{11}{-1}$
$x = -11$
Check: $7 - x = 18$
$7 - (-11) \stackrel{?}{=} 18$
$7 + 11 \stackrel{?}{=} 18$
$18 = 18$ ✓
The solution is -11.

18. $b - 7b + 6 = -30$
$b + (-7b) + 6 = -30$
$-6b + 6 = -30$
$-6b + 6 - 6 = -30 - 6$
$-6b = -36$
$\frac{-6b}{-6} = \frac{-36}{-6}$
$b = 6$
Check: $b - 7b + 6 = -30$
$6 - 7(6) + 6 \stackrel{?}{=} -30$
$6 - 42 + 6 \stackrel{?}{=} -30$
$6 + (-42) + 6 \stackrel{?}{=} -30$
$6 + 6 + (-42) \stackrel{?}{=} -30$
$12 + (-42) \stackrel{?}{=} -30$
$-30 = -30$ ✓
The solution is 6.

19. $\frac{n}{8} - 17 = -15$

$\frac{n}{8} - 17 + 17 = -15 + 17$

$\frac{n}{8} = 2$

$\frac{n}{8} \cdot 8 = 2 \cdot 8$

$n = 16$

20. $3n - 5 = 25$

$3n - 5 + 5 = 25 + 5$

$3n = 30$

$\frac{3n}{3} = \frac{30}{3}$

$n = 10$

21. $P = 2(\ell + w)$

$P = 2(48 + 20)$

$P = 2(68)$

$P = 136$

The perimeter is 136 m.

$A = \ell w$

$A = 48 \cdot 20$

$A = 960$

The area is 960 m^2.

22. $P = 2(\ell + w)$

$P = 2(100 + 75)$

$P = 2(175)$

$P = 350$

The perimeter is 350 yd.

$A = \ell w$

$A = 100 \cdot 75$

$A = 7500$

The area is 7500 yd^2.

23a. $15(2 + 4)$, $15 \cdot 2 + 15 \cdot 4$

23b. $15(2 + 4) = 15(6)$

$= \$90$

24. Let x = Todd's height,
and $x - 5$ = his brother's height.

$x + (x - 5) = 139$

$(x + x) - 5 = 139$

$2x - 5 = 139$

$2x - 5 + 5 = 139 + 5$

$2x = 144$

$x = 72$

Todd's height is 72 in.

25. D; (16 yards) · (cost for 1 yard) + installation charge = total cost

$16 \cdot x + 60 = 300$

$16x + 60 = 300$

Chapter 3 Standardized Test Practice

Pages 142–143

1. C; $159x = 20$

$\frac{1.59x}{1.59} = \frac{20}{1.59}$

$x \approx 12.6$

She can buy about 12 gallons.

2. D; $3 \times 8 - 6 \div 2 = (3 \times 8) - 6 \div 2$

$= 24 - 6 \div 2$

$= 24 - (6 \div 2)$

$= 24 - 3$

$= 21$

3. B; $5 \cdot w \cdot 8 = 5 \cdot 8 \cdot w$

4. B; N

5. A; $-5 + 7 = 2°F$

6. B;

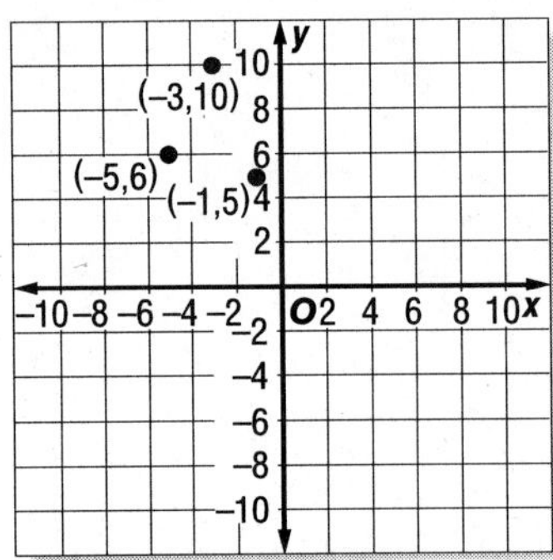

The graphs are located in Quadrant II.

7. C; $5 \times 3 + 5 \times 12 = 5 \times (3 + 12)$, using the Distributive Property.

8. B; $x - 4(x + 3) = x + (-4)(x + 3)$

$= 1x + (-4x) + (-4) \cdot 3$

$= -3x + (-12)$

$= -3x - 12$

9. A; $y - (-4) = 6 - 8$

$y + 4 = 6 + (-8)$

$y + 4 = -2$

$y + 4 - 4 = -2 - 4$

$y = -6$

10. C; Total money = bonus for each car + salary

$T = 100n + 400$

11. C; $P = S - E$

$4000 = S - 1850$

$4000 + 1850 = S - 1850 + 1850$

$5850 = S$

12. There is a \$0.50 increase for each additional person. The charge for 8 people will be $4 \times 0.50 =$ \$2 more than the charge for 4 people.

Charge = \$3 + \$2

= \$5

13. $-2(-8 + 5) = (-2)(-8 + 5)$

$= (-2)(-8) + (-2) \cdot 5$

$= 16 + (-10)$

$= 6$

14. $10 \div 2 = 5$

15.

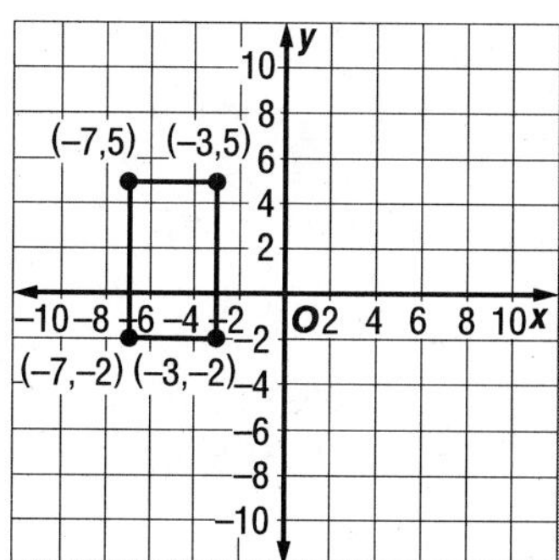

The fourth vertex is at $(-7, 5)$, so the y-coordinate is 5.

16.

$$x - 4 = -2$$
$$x - 4 + 4 = -2 + 4$$
$$x = 2$$

17. distance = (miles per gallon) × (number of gallons)

$$d = 24 \cdot 15$$
$$d = 360 \text{ mi}$$

18. total spent = charge for each box + handling charge

$$26 = 3x + 2$$
$$26 - 2 = 3x + 2 - 2$$
$$24 = 3x$$
$$8 = x$$

You can order 8 boxes.

19. The total area of the new room will be $360 + 180 = 540 \text{ ft}^2$.

$$A = \ell w$$
$$540 = \ell \cdot 18$$
$$\frac{540}{18} = \frac{\ell \cdot 18}{18}$$
$$30 = \ell$$

The length of the new room will be 30 ft.

20a.

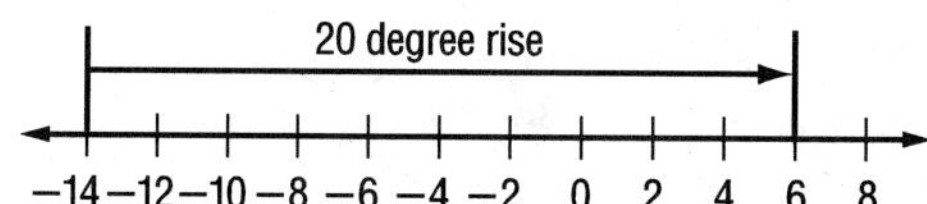

20b. $6 - (-14) = 6 + 14$

$$= 20°\text{F}$$

20c. The increase from the low temperature to 0° is the absolute value of the low temperature. The increase from 0° to the high temperature is the absolute value of the high temperature. The total increase can be found by adding the absolute values of the two temperatures, $|-14| + |6| = 20$.

21a. $f + 2g = p$

21b. If Soledad and Josh are both correct, then $f = 12$ and $p = 63$.

$$f + 2g = p$$
$$12 + 2g = 63$$
$$12 - 12 + 2g = 63 - 12$$
$$2g = 51$$
$$\frac{2g}{2} = \frac{51}{2}$$
$$g = 25.5$$

A team can't score half a basket, so they cannot both be correct.

Chapter 4 Factors and Fractions

Page 147 Getting Started

1. $2(x + 1) = 2 \cdot x + 2 \cdot 1$
$= 2x + 2$
2. $3(n - 1) = 3 \cdot n - 3 \cdot 1$
$= 3n - 3$
3. $-2(k + 8) = -2 \cdot k - 2 \cdot 8$
$= -2k - 16$
4. $-4(x - 5) = -4 \cdot x - 4(-5)$
$= -4x + 20$
5. $6(2c + 4) = 6(2c) + 6 \cdot 4$
$= 12c + 24$
6. $5(-3s + t) = 5(-3s) + 5 \cdot t$
$= -15s + 5t$
7. $7(a + b) = 7 \cdot a + 7 \cdot b$
$= 7a + 7b$
8. $9(b - 2c) = 9 \cdot b - 9(2c)$
$= 9b - 18c$
9. $x + 12 = 2 + 12$
$= 14$
10. $z + (-5) = -1 + (-5)$
$= -6$
11. $4y + 8 = 4(5) + 8$
$= 20 + 8$
$= 28$
12. $10 + 3z = 10 + 3 \cdot (-1)$
$= 10 - 3$
$= 7$
13. $(2 + y)9 = (2 + 5)9$
$= (7)9$
$= 63$
14. $6(x - 4) = 6(2 - 4)$
$= 6(-2)$
$= -12$
15. $3xy = 3 \cdot 2 \cdot 5$
$= 30$
16. $2z + y = 2 \cdot (-1) + 5$
$= -2 + 5$
$= 3$

17.
$$\begin{array}{r} 4.5 \\ \underline{\times\ 10} \\ 00 \\ \underline{450} \\ 45.0 \end{array}$$

18.
$$\begin{array}{r} 3.26 \\ \underline{\times\ 100} \\ 000 \\ 000 \\ \underline{32600} \\ 326.00 \end{array}$$

19.
$$\begin{array}{r} 0.1 \\ \underline{\times\ 780} \\ 00 \\ 080 \\ \underline{0700} \\ 78.0 \end{array}$$

20.
$$\begin{array}{r} 15 \\ \underline{\times\ 0.01} \\ 15 \\ 000 \\ \underline{0000} \\ 0.15 \end{array}$$

21.
$$\begin{array}{r} 3.9 \\ \underline{\times\ 0.1} \\ 39 \\ \underline{000} \\ 0.39 \end{array}$$

22.
$$\begin{array}{r} 63.2 \\ \underline{\times\ 0.1} \\ 632 \\ \underline{0000} \\ 6.32 \end{array}$$

23.
$$\begin{array}{r} 0.01 \\ \underline{\times\ 0.5} \\ 005 \\ \underline{0000} \\ 0.005 \end{array}$$

24.
$$\begin{array}{r} 301.8 \\ \underline{\times\ 0.001} \\ 3018 \\ 00000 \\ 000000 \\ \underline{0000000} \\ 0.3018 \end{array}$$

4-1 Factors and Monomials

Page 148 How are side lengths of rectangles related to factors?

a. Students should draw rectangles with dimensions 1×36, 2×18, 3×12, and 6×6.
b. No, if the length were 5, there is no whole number width that would give an area of 36.
c. 1 and 36, 2 and 18, 3 and 12, 4 and 9, 6 and 6; they are the same.
d. 4 rectangles; find factor pairs whose product is 64; 1×64, 2×32, 4×16, 8×8.

Page 150–151 Check for Understanding

1. Use the rules for divisibility to determine whether 18,450 is divisible by both 2 and 3. If it is, then the number is also divisible by 6 and there is no remainder.
2. No; 132 is divisible by 3, but 125 is not.

3a. Sample answer: 102
3b. Sample answer: 1035
3c. Sample answer: 343

4.

Number	Divisible?	Reason
2	no	The ones digit is 1, and 1 is not divisible by 2.
3	yes	The sum of the digits is 5 + 1 or 6, and 6 is divisible by 3.
5	no	The ones digit is 1, not 0 or 5.
6	no	51 is not divisible by 2 and 3.
10	no	The ones digit is 1, not 0.

So, 51 is divisible by 3.

5. **Number Divisible? Reason**

Number	Divisible?	Reason
2	yes	The ones digit is 6, and 6 is divisible by 2.
3	no	The sum of the digits is 1 + 4 + 6 or 11, and 11 is not divisible by 3.
5	no	The ones digit is 6, not 0 or 5.
6	no	146 is not divisible by 2 and 3.
10	no	The ones digit is 6, not 0.

So, 146 is divisible by 2.

6. **Number Divisible? Reason**

Number	Divisible?	Reason
2	yes	The ones digit is 6, and 6 is divisible by 2.
3	yes	The sum of the digits is 8 + 7 + 6 or 21, and 21 is divisible by 3.
5	no	The ones digit is 6, not 0 or 5.
6	yes	876 is divisible by 2 and 3.
10	no	The ones digit is 6, not 0.

So, 876 is divisible by 2, 3, and 6.

7. **Number Divisible? Reason**

Number	Divisible?	Reason
2	yes	The ones digit is 0, and 0 is divisible by 2.
3	no	The sum of the digits is 3 + 0 + 5 + 0 or 8, and 8 is not divisible by 3.
5	yes	The ones digit is 0 or 5.
6	no	3050 is not divisible by 2 and 3.
10	yes	The ones digit is 0.

So, 3050 is divisible by 2, 5, and 10.

8. Use the divisibility rules to determine whether 203 is divisible by 2, 3, 5, and so on. Then use division to find other factors of 203.

Number	203 Divisible by Number?	Factor Pairs
1	yes	1 · 203
2	no	—
3	no	—
4	no	—
5	no	—
6	no	—
7	yes	7 · 29
8	no	—
9	no	—

So, the factors of 203 are 1, 7, 29, and 203.

9. Use the divisibility rules to determine whether 80 is divisible by 2, 3, 5, and so on. Then use division to find other factors of 80.

Number	80 Divisible by Number?	Factor Pairs
1	yes	1 · 80
2	yes	2 · 40
3	no	—
4	yes	4 · 20
5	yes	5 · 16
6	no	—
7	no	—
8	yes	8 · 10
9	no	—

So, the factors of 80 are 1, 2, 4, 5, 8, 10, 16, 20, 40, and 80.

10. Use the divisibility rules to determine whether 115 is divisible by 2, 3, 5, and so on. Then use division to find other factors of 115.

Number	115 Divisible by Number?	Factor Pairs
1	yes	1 · 115
2	no	—
3	no	—
4	no	—
5	yes	5 · 23
6	no	—
7	no	—
8	no	—
9	no	—

So, the factors of 115 are 1, 5, 23, and 115.

11. Yes; a number
12. No; one term is subtracted from another term.
13. No; two terms are added—$5(x + y) = 5x + 5y$.
14. Yes; the product of numbers and a variable
15. **Year Yes/No Reason**

Year	Yes/No	Reason
2000	yes	The number ends in "00."
2004	yes	The last two digits are divisible by 4.
2015	no	The last two digits are not divisible by 4, and the number does not end in "00."
2018	no	The last two digits are not divisible by 4, and the number does not end in "00."
2022	no	The last two digits are not divisible by 4, and the number does not end in "00."
2032	yes	The last two digits are divisible by 4.

The years 2000, 2004, and 2032 are leap years.

Pages 151–152 Practice and Apply

16. 2—no, the ones digit is 9, and 9 is not divisible by 2.
3—yes, 3 + 9 = 12, and 12 is divisible by 3.
5—no, the ones digit is not 0 or 5.
6—no, 39 is not divisible by 2 and 3.
10—no, the ones digit is not 0.
So, 39 is divisible by 3.

17. 2—no, the ones digit is 5, and 5 is not divisible by 2.
3—yes, 1 + 3 + 5 = 9, and 9 is divisible by 3.
5—yes, the ones digit is 0 or 5.
6—no, 135 is not divisible by 2 and 3.
10—no, the ones digit is not 0.
So, 135 is divisible by 3 and 5.

18. 2—yes, the ones digit is 2, and 2 is divisible by 2.
3—no, 8 + 2 = 10, and 10 is not divisible by 3.
5—no, the ones digit is not 0 or 5.
6—no, 82 is not divisible by 2 and 3.
10—no, the ones digit is not 0.
So, 82 is divisible by 2.

19. 2—yes, the ones digit is 0, and 0 is divisible by 2.
3—yes, 1 + 2 + 0 = 3, and 3 is divisible by 3.
5—yes, the ones digit is 0 or 5.
6—yes, 120 is divisible by both 2 and 3.
10—yes, the ones digit is 0.
So, 120 is divisible by 2, 3, 5, 6, and 10.

20. 2—yes, the ones digit is 0, and 0 is divisible by 2.
3—no, 2 + 5 + 0 = 7, and 7 is not divisible by 3.
5—yes, the ones digit is 0 or 5.
6—no, 250 is not divisible by 2 and 3.
10—yes, the ones digit is 0.
So, 250 is divisible by 2, 5, and 10.

21. 2—yes, the ones digit is 8, and 8 is divisible by 2.
3—no, 1 + 1 + 8 = 10, and 10 is not divisible by 3.
5—no, the ones digit is not 0 or 5.
6—no, 118 is not divisible by 2 and 3.
10—no, the ones digit is not 0.
So, 118 is divisible by 2.

22. 2—yes, the ones digit is 8, and 8 is divisible by 2.
3—yes, 3 + 7 + 8 = 18, and 18 is divisible by 3.
5—no, the ones digit is not 0 or 5.
6—yes, 378 is divisible by 2 and 3.
10—no, the ones digit is not 0.
So, 378 is divisible by 2, 3, and 6.

23. 2—no, the ones digit is 5, and 5 is not divisible by 2.
3—no, 9 + 5 + 5 = 19, and 19 is not divisible by 3.
5—yes, the ones digit is 0 or 5.
6—no, 955 is not divisible by 2 and 3.
10—no, the ones digit is not 0.
So, 955 is divisible by 5.

24. 2—yes, the ones digit is 0, and 0 is divisible by 2.
3—yes, 5 + 0 + 1 + 0 = 6, and 6 is divisible by 3.
5—yes, the ones digit is 0 or 5.
6—yes, 5010 is divisible by 2 and 3.
10—yes, the ones digit is 0.
So, 5010 is divisible by 2, 3, 5, 6, and 10.

25. 2—yes, the ones digit is 4, and 4 is divisible by 2.
3—yes, 6 + 8 + 4 = 18, and 18 is divisible by 3.
5—no, the ones digit is not 0 or 5.
6—yes, 684 is divisible by 2 and 3.
10—no, the ones digit is not 0.
So, 684 is divisible by 2, 3, and 6.

26. 2—no, the ones digit is 3, and 3 is not divisible by 2.
3—no, 1 + 0 + 5 + 2 + 3 = 11, and 11 is not divisible by 3.
5—no, the ones digit is not 0 or 5.
6—no, 10,523 is not divisible by 2 and 3.
10—no, the ones digit is not 0.
So, 10,523 is divisible by none of the numbers.

27. 2—yes, the ones digit is 0, and 0 is divisible by 2.
3—no, 2 + 4 + 6 + 4 + 0 = 16, and 16 is not divisible by 3.
5—yes, the ones digit is 0.
6—no, 24,640 is not divisible by 2 and 3.
10—yes, the ones digit is 0.
So, 24,640 is divisible by 2, 5, and 10.

28. Use the divisibility rules to determine whether 75 is divisible by 2, 3, 5, and so on. Then use division to find other factors of 75.

Number	75 Divisible by Number?	Factor Pairs
1	yes	$1 \cdot 75$
2	no	—
3	yes	$3 \cdot 25$
4	no	—
5	yes	$5 \cdot 15$
6	no	—
7	no	—
8	no	—
9	no	—

So, the factors of 75 are 1, 3, 5, 15, 25, and 75.

29. Use the divisibility rules to determine whether 114 is divisible by 2, 3, 5, and so on. Then use division to find other factors of 114.

Number	114 Divisible by Number?	Factor Pairs
1	yes	1 · 114
2	yes	2 · 57
3	yes	3 · 38
4	no	—
5	no	—
6	yes	6 · 19
7	no	—
8	no	—
9	no	—

So, the factors of 114 are 1, 2, 3, 6, 19, 38, 57, and 114.

30. Use the divisibility rules to determine whether 57 is divisible by 2, 3, 5, and so on. Then use division to find other factors of 57.

Number	57 Divisible by Number?	Factor Pairs
1	yes	1 · 57
2	no	—
3	yes	3 · 19
4	no	—
5	no	—
6	no	—
7	no	—
8	no	—
9	no	—

So, the factors of 57 are 1, 3, 19, and 57.

31. Use the divisibility rules to determine whether 65 is divisible by 2, 3, 5, and so on. Then use division to find other factors of 65.

Number	65 Divisible by Number?	Factor Pairs
1	yes	1 · 65
2	no	—
3	no	—
4	no	—
5	yes	5 · 13
6	no	—
7	no	—
8	no	—
9	no	—

So, the factors of 65 are 1, 5, 13, and 65.

32. Use the divisibility rules to determine whether 90 is divisible by 2, 3, 5, and so on. Then use division to find other factors of 90.

Number	90 Divisible by Number?	Factor Pairs
1	yes	1 · 90
2	yes	2 · 45
3	yes	3 · 30
4	no	—
5	yes	5 · 18
6	yes	6 · 15
7	no	—
8	no	—
9	yes	9 · 10

So, the factors of 90 are 1, 2, 3, 5, 6, 9, 10, 15, 18, 30, 45, and 90.

33. Use the divisibility rules to determine whether 124 is divisible by 2, 3, 5, and so on. Then use division to find other factors of 124.

Number	124 Divisible by Number?	Factor Pairs
1	yes	1 · 124
2	yes	2 · 62
3	no	—
4	yes	4 · 31
5	no	—
6	no	—
7	no	—
8	no	—
9	no	—

So, the factors of 124 are 1, 2, 4, 31, 62 and 124.

34. Use the divisibility rules to determine whether 102 is divisible by 2, 3, 5, and so on. Then use division to find other factors of 102.

Number	102 Divisible by Number?	Factor Pairs
1	yes	1 · 102
2	yes	2 · 51
3	yes	3 · 34
4	no	—
5	no	—
6	yes	6 · 17
7	no	—
8	no	—
9	no	—

So, the factors of 102 are 1, 2, 3, 6, 17, 34, 51, and 102.

35. Use the divisibility rules to determine whether 135 is divisible by 2, 3, 5, and so on. Then use division to find other factors of 135.

Number	135 Divisible by Number?	Factor Pairs
1	yes	$1 \cdot 135$
2	no	—
3	yes	$3 \cdot 45$
4	no	—
5	yes	$5 \cdot 27$
6	no	—
7	no	—
8	no	—
9	yes	$9 \cdot 15$

So, the factors of 135 are 1, 3, 5, 9, 15, 27, 45, and 135.

36. Yes; a variable
37. Yes; a number
38. No; two terms are added.
39. No; one term is subtracted from another term.
40. No; two terms are added.
41. No; two terms are added.
42. No; two terms are added.
43. No; one term is subtracted from another term.
44. Yes; the product of a number and a variable
45. Yes; the product of a number and a variable
46. Yes; the product of a number and variables
47. Yes; the product of numbers and variables
48. No; 72 is not divisible by 7.
49. 6 ways; 1×72, 2×36, 3×24, 4×18, 6×12, 8×9
50. 1912, a 6-by-8 arrangement; 1959, a 7-by-7 arrangement
51. Alternating rows of a flag contain 6 stars and 5 stars, respectively. Fifty is not divisible by a number that would make the arrangement of stars in an appropriate-sized rectangle.
52. Sometimes; if a number is divisible by 2 and by 3, then it is also divisible by 6. If a number is divisible by 3, but not divisible by 2, then it is not divisible by 6.
53. Never; a number that has 10 as a factor is divisible by $2 \cdot 5$, so it is always divisible by 5.
54. Always; a number that has a factor of 10 is divisible by $2 \cdot 5$. Since such a number is divisible by 2, it is always an even number.

55a. $144 ÷ $6 = 24 cases

55b. $144 ÷ $4 = 36 bags

55c. Sample answer: 12 cases, 18 bags; 14 cases, 15 bags; 16 cases, 12 bags

56a. 997

56b. 101

57. The side lengths or dimensions of a rectangle are factors of the number that is the area of the rectangle. Answers should include the following.
    - A rectangle with dimensions and area labeled; for example, a 4×5 rectangle would have length 5 units, width 4 units, and area 20 square units.
    - Factors are numbers that are multiplied to form a product. The dimensions of a rectangle are factor pairs of the area since they are multiplied to form the area.
58. B; $4 + 4 + 4 = 12$, 12 is divisible by 3.
59. C; One term is subtracted from another term.

Page 152 Maintain Your Skills

60. $P = 2(\ell + w)$, $P = 2(4.9 + 3.5)$, $P = 2(8.4)$, $P = 16.8$ m

    $A = \ell w$, $A = 4.9 \cdot 3.5$, $A = 17.15\text{ m}^2$

61. $P = 2(\ell + w)$, $P = 2(12 + 5)$, $P = 2(17)$, $P = 34$ in.

    $A = \ell w$, $A = 12 \cdot 5$, $A = 60\text{ in}^2$

62. Let n = the number.

$$2n + 8 = -16$$
$$2n + 8 - 8 = -16 - 8$$
$$2n = -24$$
$$n = -12$$

63. Let n = the number.

$$5n - 2 = 3$$
$$5n - 2 + 2 = 3 + 2$$
$$5n = 5$$
$$n = 1$$

64.

$$2x - 1 = 9$$
$$2x - 1 + 1 = 9 + 1$$
$$2x = 10$$
$$\frac{2x}{2} = \frac{10}{2}$$
$$x = 5$$

Check: $2x - 1 = 9$

$$2(5) - 1 \stackrel{?}{=} 9$$
$$10 - 1 \stackrel{?}{=} 9$$
$$9 = 9 \checkmark$$

65.

$$14 = 8 + 3n$$
$$14 - 8 = 8 - 8 + 3n$$
$$6 = 3n$$
$$\frac{6}{3} = \frac{3n}{3}$$
$$2 = n$$

Check: $14 = 8 + 3n$

$$14 \stackrel{?}{=} 8 + 3(2)$$
$$14 \stackrel{?}{=} 8 + 6$$
$$14 = 14 \checkmark$$

66. $7 + \frac{k}{5} = -1$

$7 - 7 + \frac{k}{5} = -1 - 7$

$\frac{k}{5} = -8$

$\left(\frac{k}{5}\right)5 = -8 \cdot 5$

$k = -40$

Check: $7 + \frac{k}{5} = -1$

$7 + \frac{(-40)}{5} \stackrel{?}{=} -1$

$7 - 7 + \frac{(-40)}{5} \stackrel{?}{=} -1 - 7$

$\frac{(-40)}{5} \stackrel{?}{=} -8$

$5 \cdot \frac{(-40)}{5} \stackrel{?}{=} -8 \cdot 5$

$-40 = -40$ ✓

67. $4 \cdot 4 \cdot 4 = (4 \cdot 4) \cdot 4$
$= (16) \cdot 4$
$= 64$

68. $10 \cdot 10 \cdot 10 \cdot 10 = (10 \cdot 10) \cdot 10 \cdot 10$
$= (100 \cdot 10) \cdot 10$
$= 1000 \cdot 10$
$= 10{,}000$

69. $(-3)(-3)(-3) = [(-3)(-3)](-3)$
$= 9(-3)$
$= -27$

70. $(-2)(-2)(-2)(-2) = [(-2)(-2)](-2)(-2)$
$= [(4)(-2)](-2)$
$= (-8)(-2)$
$= 16$

71. $8 \cdot 8 \cdot 6 \cdot 6 = (8 \cdot 8)(6)(6)$
$= [64(6)](6)$
$= 384(6)$
$= 2304$

72. $(2)(2)(-5)(-5)(-5) = [(2)(2)](-5)(-5)(-5)$
$= [4(-5)](-5)(-5)$
$= [-20(-5)](-5)$
$= 100(-5)$
$= -500$

4-2 Powers and Exponents

Page 153 Why are exponents important in comparing computer data?

a. $2 \times 2 \times 2 \times 2$; 4 factors

b. 7 factors

c. 10 factors

Page 155 Check for Understanding

1. Sample answer: 2^5, x^5
2. 6 cubed means $6 \cdot 6 \cdot 6$, so 6 is used as a factor 3 times.
3. When n is even, $1^n = (-1)^n = 1$. When n is odd, $1^n = 1$ and $(-1)^n = -1$.
4. $n \cdot n \cdot n = n^3$
5. $7 \cdot 7 = 7^2$
6. $3 \cdot 3 \cdot x \cdot x \cdot x \cdot x = (3 \cdot 3) \cdot (x \cdot x \cdot x \cdot x)$
$= 3^2x^4$
7. $2695 = 2000 + 600 + 90 + 5$
$= (2 \times 1000) + (6 \times 100) + (9 \times 10) + (5 \times 1)$
$= (2 \times 10^3) + (6 \times 10^2) + (9 \times 10^1) + (5 \times 10^0)$
8. $2^4 = 2 \cdot 2 \cdot 2 \cdot 2$
$= 16$
9. $x^3 - 3 = (-2)^3 - 3$
$= [(-2)(-2)(-2)] - 3$
$= -8 - 3$
$= -11$
10. $5(y - 1)^2 = 5(4 - 1)^2$
$= 5(3)^2$
$= 5(9)$
$= 45$
11. $169 = 13 \cdot 13$
$= 13^2$

Pages 155–157 Practice and Apply

12. $4 \cdot 4 \cdot 4 \cdot 4 \cdot 4 \cdot 4 = 4^6$
13. $6 = 6^1$
14. $(-5)(-5)(-5) = (-5)^3$
15. $(-8)(-8)(-8)(-8) = (-8)^4$
16. $k \cdot k = k^2$
17. $(-t)(-t)(-t) = (-t)^3$
18. $(r \cdot r)(r \cdot r) = r \cdot r \cdot r \cdot r$
$= r^4$
19. $m \cdot m \cdot m \cdot m = m^4$
20. $a \cdot a \cdot b \cdot b \cdot b \cdot b = (a \cdot a) \cdot (b \cdot b \cdot b \cdot b)$
$= a^2b^4$
21. $2 \cdot x \cdot x \cdot y \cdot y = 2 \cdot (x \cdot x) \cdot (y \cdot y)$
$= 2x^2y^2$
22. $7 \cdot 7 \cdot 7 \cdot n \cdot n \cdot n \cdot n = (7 \cdot 7 \cdot 7) \cdot (n \cdot n \cdot n \cdot n)$
$= 7^3n^4$
23. $9 \cdot (p + 1) \cdot (p + 1) = 9(p + 1)^2$
24. $452 = 400 + 50 + 2$
$= (4 \times 100) + (5 \times 10) + (2 \times 1)$
$= (4 \times 10^2) + (5 \times 10^1) + (2 \times 10^0)$
25. $803 = 800 + 0 + 3$
$= (8 \times 100) + (0 \times 10) + (3 \times 1)$
$= (8 \times 10^2) + (0 \times 10^1) + (3 \times 10^0)$
26. $6994 = 6000 + 900 + 90 + 4$
$= (6 \times 1000) + (9 \times 100) + (9 \times 10) + (4 \times 1)$
$= (6 \times 10^3) + (9 \times 10^2) + (9 \times 10^1) + (4 \times 10^0)$

27. $23{,}781 = 20{,}000 + 3000 + 700 + 80 + 1$
$= (2 \times 10{,}000) + (3 \times 1000) + (7 \times 100) + (8 \times 10) + (1 \times 1)$
$= (2 \times 10^4) + (3 \times 10^3) + (7 \times 10^2) + (8 \times 10^1) + (1 \times 10^0)$

28. $7^2 = 7 \cdot 7$
$= 49$

29. $10^3 = 10 \cdot 10 \cdot 10$
$= 1000$

30. $(-9)^3 = (-9)(-9)(-9)$
$= -729$

31. $(-2)^5 = (-2)(-2)(-2)(-2)(-2)$
$= -32$

32. $b^4 = 4^4$
$= 4 \cdot 4 \cdot 4 \cdot 4$
$= 256$

33. $c^4 = (-3)^4$
$= (-3)(-3)(-3)(-3)$
$= 81$

34. $5a^4 = 5(2)^4$
$= 5 \cdot 2 \cdot 2 \cdot 2 \cdot 2$
$= 80$

35. $ac^3 = 2(-3)^3$
$= 2(-3)(-3)(-3)$
$= -54$

36. $b^0 - 10 = (4)^0 - 10$
$= 1 - 10$
$= -9$

37. $c^2 + a^2 = (-3)^2 + (2)^2$
$= (-3)(-3) + (2)(2)$
$= 9 + 4$
$= 13$

38. $3a + b^3 = 3(2) + (4)^3$
$= 3(2) + (4)(4)(4)$
$= 6 + 64$
$= 70$

39. $a^2 + 3a - 1 = 2^2 + 3(2) - 1$
$= 2 \cdot 2 + 3 \cdot 2 - 1$
$= 4 + 6 - 1$
$= 9$

40. $b^2 - 2b + 6 = 4^2 - 2 \cdot 4 + 6$
$= 4 \cdot 4 - 2 \cdot 4 + 6$
$= 16 - 8 + 6$
$= 14$

41. $3(b - 1)^4 = 3(4 - 1)^4$
$= 3(3)^4$
$= 3(3 \cdot 3 \cdot 3 \cdot 3)$
$= 243$

42. $2(3c + 7)^2 = 2[3(-3) + 7]^2$
$= 2(-9 + 7)^2$
$= 2(-2)^2$
$= 2(-2)(-2)$
$= 8$

43. $81 = 9 \cdot 9 = 9^2$ or $81 = 3 \cdot 3 \cdot 3 \cdot 3 = 3^4$; $64 = 8 \cdot 8 = 8^2$ or $64 = 4 \cdot 4 \cdot 4 = 4^3$ or $64 = 2 \cdot 2 \cdot 2 \cdot 2 \cdot 2 \cdot 2 = 2^6$

44. $7^3 \cdot x^2 = 7 \cdot 7 \cdot 7 \cdot x \cdot x$

45. $(-8)^3$; $(-8)(-8)(-8)$; -512

46. $96^0, 96, 96^5, 96^{10}$; As the exponents increase, the additional factors of 96, and therefore the values of the expressions, increase. Since $96^0 = 1$, it has the least value.

47. Always; the product of two negative numbers is always positive.

48. $121 = 11 \cdot 11 = 11^2$; $100 = 10 \cdot 10 = 10^2$

49. $2^1, 2^2, 2^3, 2^4, 2^5$

50. $2^{13} = 2 \cdot 2 \cdot 2 \cdot 2 \cdot 2 \cdot 2 \cdot 2 \cdot 2 \cdot 2 \cdot 2 \cdot 2 \cdot 2 \cdot 2$
$= 8192$ noodles

51. After 10 folds, the noodles are $5(2^{10}) = 5(1024)$ or 5120 feet long, which is slightly less than a mile. So, after 11 folds, the length of the noodles will be greater than a mile.

52. Since $3^7 = 3 \cdot 3 \cdot 3 \cdot 3 \cdot 3 \cdot 3 \cdot 3$ or 2187 and $7^3 = 7 \cdot 7 \cdot 7$ or 343, $3^7 > 7^3$.

53. Since $2^4 = 2 \cdot 2 \cdot 2 \cdot 2$ or 16 and $4^2 = 4 \cdot 4$ or 16, $2^4 = 4^2$.

54. Since $6^3 = 6 \cdot 6 \cdot 6$ or 216 and $4^4 = 4 \cdot 4 \cdot 4 \cdot 4$ or 256, $6^3 < 4^4$.

55. The area of each side is $A = \ell w$, with $\ell = 3$ and $w = 3$.
$(3 \cdot 3) + (3 \cdot 3) + (3 \cdot 3) + (3 \cdot 3) + (3 \cdot 3) + (3 \cdot 3) = 6 \cdot 3^2$ cm^2

56. Volume is length · width · height.
$3 \cdot 3 \cdot 3 = 3^3$ cm^3.

57. If the sides are doubled, the length and width are each 6 cm.
Surface area $= (6 \cdot 6) + (6 \cdot 6) + (6 \cdot 6) + (6 \cdot 6) + (6 \cdot 6) + (6 \cdot 6)$
$= 6 \cdot 6^2$ or 216 cm^2
Volume $= 6 \cdot 6 \cdot 6$
$= 6^3$ or 216 cm^3
Since the original surface area was $6 \cdot 3^2$ or 54 cm^2, the surface area is multiplied by 4 when the lengths are doubled. Since the original volume was 3^3 or 27 cm^3, the volume is multiplied by 8 when the lengths are doubled.

58a. The perimeter is doubled, and the area is four times the area of the original square.

58b. original area: n^2, new area: $(3n)^2$ or $9n^2$

58c. original volume: n^3, new volume: $(3n)^3$ or $27n^3$

59. As the capacity of computer memory increases, the factors of 2 in the number of megabytes increases. Answers should include the following.
- Computer data are measured in small units that are based on factors of 2.
- In describing the amount of memory in modern computers, it would be impractical to list all the factors of 2. Using exponents is a more efficient way to describe and compare computer data.

60. C; $10{,}000{,}000 = 10 \cdot 10 \cdot 10 \cdot 10 \cdot 10 \cdot 10 \cdot 10$
$= 10^7$

61. B; $256 = 2 \cdot 2 \cdot 2 \cdot 2 \cdot 2 \cdot 2 \cdot 2 \cdot 2$
$= 2^8$

Page 157 Maintain Your Skills

62. 2—yes, the ones digit is 8, and 8 is divisible by 2.
3—no, $1 + 2 + 8 = 11$, and 11 is not divisible by 3.
5—no, the ones digit is not 0 or 5.
6—no, 128 is not divisible by 2 and 3.
10—no, the ones digit is not 0.
So, 128 is divisible by 2.

63. 2—yes, the ones digit is 0, and 0 is divisible by 2.
3—no, $3 + 7 + 0 = 10$, and 10 is not divisible by 3.
5—yes, the ones digit is 0 or 5.
6—no, 370 is not divisible by 2 and 3.
10—yes, the ones digit is 0.
So, 370 is divisible by 2, 5, and 10.

64. 2—no, the ones digit is 5, and 5 is not divisible by 2.
3—yes, $9 + 4 + 5 = 18$, and 18 is divisible by 3.
5—yes, the ones digit is 0 or 5.
6—no, 945 is not divisible by 2 and 3.
10—no, the ones digit is not 0.
So, 945 is divisible by 3 and 5.

65. $d = rt$
$300 = r \cdot 2$
$\frac{300}{2} = \frac{r \cdot 2}{2}$
$150 = r$
The speed of the tornado is 150 mph.

66. $2x + 1 = 7$
$2x + 1 - 1 = 7 - 1$
$\frac{2x}{2} = \frac{6}{2}$
$x = 3$
Check: $2x + 1 = 7$
$2(3) + 1 \stackrel{?}{=} 7$
$6 + 1 \stackrel{?}{=} 7$
$7 = 7$ ✓

67. $16 = 5k - 4$
$16 + 4 = 5k - 4 + 4$
$\frac{20}{5} = \frac{5k}{5}$
$4 = k$
Check: $16 = 5k - 4$
$16 \stackrel{?}{=} 5(4) - 4$
$16 \stackrel{?}{=} 20 - 4$
$16 = 16$ ✓

68. $\frac{n}{3} + 8 = 6$
$\frac{n}{3} + 8 - 8 = 6 - 8$
$\frac{n}{3} = -2$
$3\left(\frac{n}{3}\right) = 3(-2)$
$n = -6$
Check: $\frac{n}{3} + 8 = 6$
$\frac{-6}{3} + 8 \stackrel{?}{=} 6$
$\frac{-6}{3} + 8 - 8 \stackrel{?}{=} 6 - 8$
$\frac{-6}{3} \stackrel{?}{=} -2$
$-2 = -2$ ✓

69. $4(y + 2) - y = 4y + 8 - y$
$= 3y + 8$

70. 1, 11

71. 1, 5

72. 1, 3, 9

73. 1, 2, 4, 8, 16

74. 1, 19

75. 1, 5, 7, 35

Page 158 Algebra Activity

1. $1011_2 = \underbrace{(1 \times 2^3)}_{8} + \underbrace{(0 \times 2^2)}_{0} + \underbrace{(1 \times 2^1)}_{2} + \underbrace{(1 \times 2^0)}_{1}$
$= 8 + 0 + 2 + 1$
$= 11$

2. The greatest factor of 2 that is less than 6 is 4. Place a 1 in that place value.

		1		
16	8	4	2	1

Subtract $6 - 4 = 2$. Place a 1 in that place value.

		1	1	
16	8	4	2	1

There are no factors of 2 left, so place a 0 in any unfilled spaces.

		1	1	0
16	8	4	2	1

So, $6 = 110_2$.

3. The greatest factor of 2 that is less than 9 is 8. Place a 1 in that place value.

	1			
16	8	4	2	1

Subtract $9 - 8 = 1$. Place a 1 in that place value.

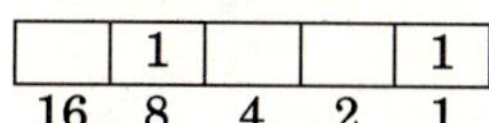

	1			1
16	8	4	2	1

There are no factors of 2 left, so place a 0 in any unfilled spaces.

	1	0	0	1
16	8	4	2	1

So, $9 = 1001_2$.

4. The greatest factor of 2 that is less than 15 is 8. Place a 1 in that place value.

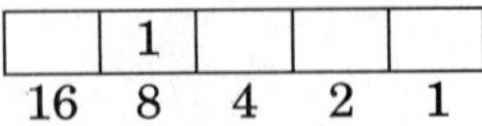

	1			
16	8	4	2	1

Subtract $15 - 8 = 7$. Now find the greatest factor of 2 that is less than 7. Place a 1 in the place value of 4.

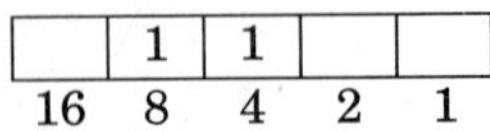

	1	1		
16	8	4	2	1

Subtract $7 - 4 = 3$. Now find the greatest factor of 2 that is less than 3. Place a 1 in the place value of 2.

	1	1	1	
16	8	4	2	1

Subtract $3 - 2 = 1$. Place a 1 in that place value.

	1	1	1	1
16	8	4	2	1

So, $15 = 1111_2$.

5. The greatest factor of 2 that is less than 21 is 16. Place a 1 in that place value.

1				
16	8	4	2	1

Subtract $21 - 16 = 5$. Now find the greatest factor of 2 that is less than 5. Place a 1 in the place value of 4.

1		1		
16	8	4	2	1

Subtract $5 - 4 = 1$. Place a 1 in that place value.

1		1		1
16	8	4	2	1

There are no factors of 2 left, so place a 0 in any unfilled spaces.

1	0	1	0	1
16	8	4	2	1

So, $21 = 10101_2$.

6. Find the greatest factor of 5 that is less than 179. Place a 1 in the place value of 125.

	1			
625	125	25	5	1

Subtract $179 - 125 = 54$. Now find the greatest factor of 5 that is less than 54. Place a 2 in the place value of 25, since the largest value is $2 \cdot 25$ or 50.

	1	2		
625	125	25	5	1

Subtract $54 - 50 = 4$. The greatest factor of 5 that is less than 4 is 1. Place a 4 in the place value of 1, since the largest value is $1 \cdot 4$ or 4.

	1	2		4
625	125	25	5	1

There are no factors of 5 left, so place a 0 in any unfilled spaces.

	1	2	0	4
625	125	25	5	1

So, $179 = 1204_5$.

7. Sample answer: Write 314 in base 4.

1	0	3	2	2
256	64	16	4	1

$314 = 10322_4$

8. Sample answer: Write 5123 in base 8.

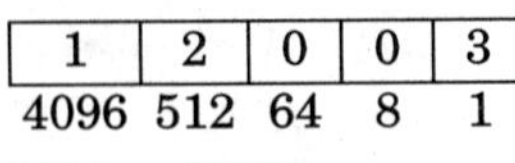

1	2	0	0	3
4096	512	64	8	1

$5123 = 12003_8$

4-3 Prime Factorization

Page 159 How can models be used to determine whether numbers are prime?

a. See students' work.

b. 4, 6, 8, 9

c. 2, 3, 5, 7

d. They all have a width of 1 because no other pair of factors can be found.

Pages 161–162 Check for Understanding

1. A prime number has exactly two factors: 1 and itself. A composite number has more than two factors.

2. Sample answer:

```
  12
  /\
 2 · 6
     /\
    2 · 3
```

3. Francisca; 4 is not prime.

4. prime; $7 = 1 \times 7$ is the only factorization.

5. prime; $23 = 1 \times 23$ is the only factorization.

6. composite; $15 = 1 \times 15$

$15 = 3 \times 5$

7.

```
   18
   /\
  2 · 9
  /   /\
 2 · 3 · 3
```

The prime factorization of 18 is $2 \cdot 3 \cdot 3$ or $2 \cdot 3^2$.

8.

```
  39
  /\
 3 · 13
```

The prime factorization of 39 is $3 \cdot 13$.

9.

```
   50
   /\
  2 · 25
  /   /\
 2 · 5 · 5
```

The prime factorization of 50 is $2 \cdot 5 \cdot 5$ or $2 \cdot 5^2$.

10. $4c^2 = 2 \cdot 2 \cdot c \cdot c$

11. $5a^2b = 5 \cdot a \cdot a \cdot b$

12. $-70xyz = -1 \cdot 2 \cdot 5 \cdot 7 \cdot x \cdot y \cdot z$

13. 3 and 5, 5 and 7, 11 and 13, 17 and 19, 29 and 31, 41 and 43

Pages 162–163 Practice and Apply

14. composite; $21 = 1 \times 21$
$21 = 3 \times 7$

15. composite; $33 = 1 \times 33$
$33 = 3 \times 11$

16. prime; $23 = 1 \times 23$ is the only factorization.

17. composite; $70 = 1 \times 70$
$70 = 2 \times 35$
$70 = 5 \times 14$
$70 = 7 \times 10$

18. prime; $17 = 1 \times 17$ is the only factorization.

19. composite; $51 = 1 \times 51$
$51 = 3 \times 17$

20. prime; $43 = 1 \times 43$ is the only factorization.

21. prime; $31 = 1 \times 31$ is the only factorization.

22. 26
/\
2 · 13

The prime factorization of 26 is $2 \cdot 13$.

23. 81
/\
9 · 9
/\ /\
3 · 3 · 3 · 3

The prime factorization of 81 is $3 \cdot 3 \cdot 3 \cdot 3$ or 3^4.

24. 66
/\
11 · 6
/ /\
11 · 2 · 3

The prime factorization of 66 is $2 \cdot 3 \cdot 11$.

25. 63
/\
7 · 9
/ /\
7 · 3 · 3

The prime factorization of 63 is $3 \cdot 3 \cdot 7$ or $3^2 \cdot 7$.

26. 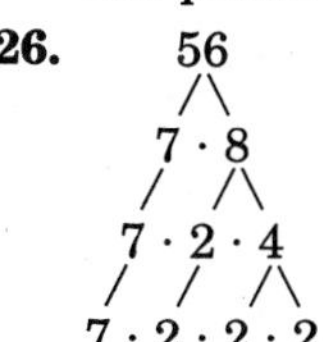

The prime factorization of 56 is $2 \cdot 2 \cdot 2 \cdot 7$ or $2^3 \cdot 7$.

27. 100
/\
4 · 25
/\ /\
2 · 2 · 5 · 5

The prime factorization of 100 is $2 \cdot 2 \cdot 5 \cdot 5$ or $2^2 \cdot 5^2$.

28. 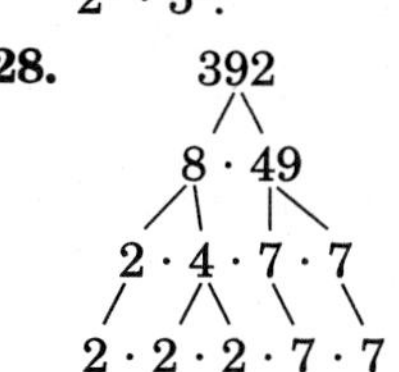

The prime factorization of 392 is $2 \cdot 2 \cdot 2 \cdot 7 \cdot 7$ or $2^3 \cdot 7^2$.

29. 110
/\
11 · 10
/ /\
11 · 2 · 5

The prime factorization of 110 is $2 \cdot 5 \cdot 11$.

30. $14w = 2 \cdot 7 \cdot w$

31. $9t^2 = 3 \cdot 3 \cdot t \cdot t$

32. $-7c^2 = -1 \cdot 7 \cdot c \cdot c$

33. $-25z^3 = -1 \cdot 5 \cdot 5 \cdot z \cdot z \cdot z$

34. $20st = 2 \cdot 2 \cdot 5 \cdot s \cdot t$

35. $-38mnp = -1 \cdot 2 \cdot 19 \cdot m \cdot n \cdot p$

36. $28x^2y = 2 \cdot 2 \cdot 7 \cdot x \cdot x \cdot y$

37. $21gh^3 = 3 \cdot 7 \cdot g \cdot h \cdot h \cdot h$

38. $13q^2r^2 = 13 \cdot q \cdot q \cdot r \cdot r$

39. $64n^3 = 2 \cdot 2 \cdot 2 \cdot 2 \cdot 2 \cdot 2 \cdot n \cdot n \cdot n$

40. $-75ab^2 = -1 \cdot 3 \cdot 5 \cdot 5 \cdot a \cdot b \cdot b$

41. $-120r^2st^3 = -1 \cdot 2 \cdot 2 \cdot 2 \cdot 3 \cdot 5 \cdot r \cdot r \cdot s \cdot t \cdot t \cdot t$

42. prime; $n^2 - n + 41 = 3^2 - 3 + 41$
$= 9 - 3 + 41$
$= 47$

47 is prime since $47 = 1 \times 47$ is the only factorization.

43. Sample answer: $-25x = -1 \cdot 5 \cdot 5 \cdot x$

44a. prime; $2^5 - 1 = 32 - 1$ or 31, and 31 has only factors of 1 and 31.

44b. $2^6 - 1 = 3 \cdot 3 \cdot 7$
$= 3^2 \cdot 7$

44c. prime; $2^7 - 1 = 128 - 1$ or 127, and 127 has only factors of 1 and 127.

44d. $2^8 - 1 = 3 \cdot 5 \cdot 17$

45. The number of rectangles that can be modeled to represent a number indicate whether the number is prime or composite. Answers should include the following.

- If a number is prime, then only one rectangle can be drawn to represent the number. If a number is composite, then more than one rectangle can be drawn to represent the number.
- If a model has a length or width of 1, then the number may be prime or composite. If a model does not have a length or width of 1, then the number must be composite.

46. $12 = 2^2 \cdot 3$, $60 = 2^2 \cdot 3 \cdot 5$, $84 = 2^2 \cdot 3 \cdot 7$, $132 = 2^2 \cdot 3 \cdot 11$, $180 = 2^2 \cdot 3^2 \cdot 5$;
Sample answer: A number is divisible by 12 if the number is divisible by 3 and 4.

47. C;

Input	Output
1	$1 + 1^2 = 2$
2	$2 + 2^2 = 2 + 4$ or 6
4	$4 + 4^2 = 4 + 16$ or 20

48. D; $70 = 2 \cdot 5 \cdot 7$

Page 163 Maintain Your Skills

49. $(-5) \cdot (-5) \cdot (-5) \cdot h \cdot h \cdot k = (-5)^3h^2k$

50. Yes; the product of numbers and variables

51. Yes; a number

52. No; one term is subtracted from another term.

53. No; $3(1 + 3r) = 3 + 9r$, two terms are added.

54. $\frac{n}{8} = -4$

$8\left(\frac{n}{8}\right) = 8(-4)$

$n = -32$

55. $2x = -18$

$\frac{2x}{2} = \frac{-18}{2}$

$x = -9$

56. $30 = 6n$

$\frac{30}{6} = \frac{6n}{6}$

$5 = n$

57. $-7 = \frac{y}{4}$

$4(-7) = \left(\frac{y}{4}\right)4$

$-28 = y$

58. $2(n + 4) = 2 \cdot n + 2 \cdot 4$

$= 2n + 8$

59. $5(x - 7) = 5 \cdot x + 5 \cdot (-7)$

$= 5x - 35$

60. $-3(t + 4) = -3 \cdot t + (-3)4$

$= -3t + (-12)$

$= -3t - 12$

61. $(a + 6)10 = a \cdot 10 + 6 \cdot 10$

$= 10a + 60$

62. $(b - 3)(-2) = (-2)(b) - (3)(-2)$

$= -2b - (-6)$

$= -2b + 6$

63. $8(9 - y) = 8 \cdot 9 - 8 \cdot y$

$= 72 - 8y$

Page 163 Practice Quiz 1

1. 2—no, the ones digit is 5, and 5 is not divisible by 2.
3—yes, $1 + 0 + 5 = 6$, and 6 is divisible by 3.
5—yes, the ones digit is 0 or 5.
6—no, 105 is not divisible by 2 and 3.
10—no, the ones digit is not 0.
So, 105 is divisible by 3 and 5.

2. 2—yes, the ones digit is 0, and 0 is divisible by 2.
3—yes, $2 + 7 + 0 = 9$, and 9 is divisible by 3.
5—yes, the ones digit is 0 or 5.
6—yes, 270 is divisible by 2 and 3.
10—yes, the ones digit is 0.
So, 270 is divisible by 2, 3, 5, 6, and 10.

3. 2—no, the ones digit is 1, and 1 is not divisible by 2.
3—no, $5 + 1 + 1 = 7$, and 7 is not divisible by 3.
5—no, the ones digit is not 0 or 5.
6—no, 511 is not divisible by 2 and 3.
10—no, the ones digit is not 0.
So, 511 is divisible by none of the numbers.

4. 2—yes, the ones digit is 8, and 8 is divisible by 2.
3—yes, $1 + 3 + 6 + 8 = 18$, and 18 is divisible by 3.
5—no, the ones digit is not 0 or 5.
6—yes, 1368 is divisible by 2 and 3.
10—no, the ones digit is not 0.
So, 1368 is divisible by 2, 3, and 6.

5. $b^2 - 4ac = 5^2 - [4(-1)(3)]$

$= 25 - (-12)$

$= 25 + 12$

$= 37$

6a. $2^0, 2^1, 2^2$

6b. $2^7 = 2 \cdot 2 \cdot 2 \cdot 2 \cdot 2 \cdot 2 \cdot 2$

$= 128$ cents

7. $77x = 7 \cdot 11 \cdot x$

8. $18st = 2 \cdot 3 \cdot 3 \cdot s \cdot t$

9. $-23n^3 = -1 \cdot 23 \cdot n \cdot n \cdot n$

10. $30cd^2 = 2 \cdot 3 \cdot 5 \cdot c \cdot d \cdot d$

4-4 Greatest Common Factor (GCF)

Page 164 How can a diagram be used to find the greatest common factor?

a. 2, 2

b. $2 \cdot 2 = 4$

c. yes

d.

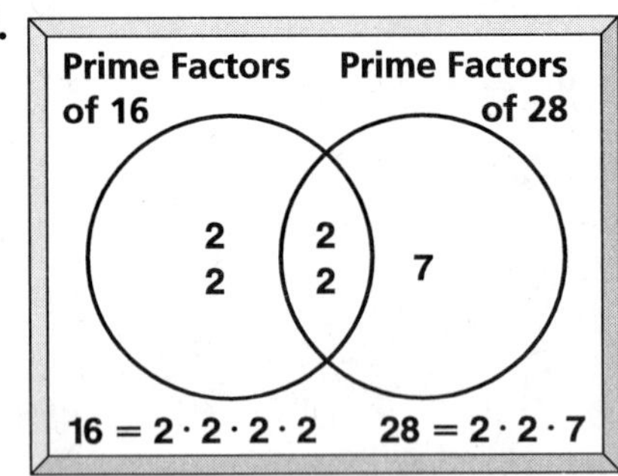

The common factors are 2, 2.

Pages 166–167 Check for Understanding

1. Sample answer: Find the prime factorization of each number. Multiply the factors that are common to both.

2. Sample answer: 12 and 24

3. Jack; the common prime factors of the expressions are 2 and 11, so the GCF is $2 \cdot 11$ or 22.

4. **Method 1** List the factors.
factors of 6: 1, ②, 3, 6
factors of 8: 1, ②, 4, 8
Method 2 Use prime factorization.
$6 = 2 \cdot 3$
$8 = 2 \cdot 2 \cdot 2$
The GCF of 6 and 8 is 2.

5. **Method 1** List the factors.
factors of 21: 1, ③, 7, 21
factors of 45: 1, ③, 5, 9, 15, 45
Method 2 Use prime factorization.
$21 = 3 \cdot 7$
$45 = 3 \cdot 3 \cdot 5$
The GCF of 21 and 45 is 3.

6. **Method 1** List the factors.
factors of 16: 1, 2, 4, ⑧, 16
factors of 56: 1, 2, 4, 7, ⑧, 14, 28, 56
Method 2 Use prime factorization.
$16 = 2 \cdot 2 \cdot 2 \cdot 2$
$56 = 2 \cdot 2 \cdot 2 \cdot 7$
The GCF of 16 and 56 is 8.

7. **Method 1** List the factors.
factors of 28: 1, 2, 4, 7, ⑭, 28
factors of 42: 1, 2, 3, 6, 7, ⑭, 21, 42
Method 2 Use prime factorization.
$28 = 2 \cdot 2 \cdot 7$
$42 = 2 \cdot 3 \cdot 7$
The GCF of 28 and 42 is 14.

8. **Method 1** List the factors.
factors of 7: ①, 7
factors of 30: ①, 2, 3, 5, 6, 10, 15, 30
Method 2 Use prime factorization.
$7 = 7$
$30 = 2 \cdot 3 \cdot 5$
The GCF of 7 and 30 is 1.

9. $108 = 2 \cdot 2 \cdot 3 \cdot 3 \cdot 3$
$144 = 2 \cdot 2 \cdot 2 \cdot 2 \cdot 3 \cdot 3$
The GCF of 108 and 144 is $2 \cdot 2 \cdot 3 \cdot 3$ or 36.

10. $12 = 2 \cdot 2 \cdot 3$
$24 = 2 \cdot 2 \cdot 2 \cdot 3$
$36 = 2 \cdot 2 \cdot 3 \cdot 3$
The GCF of 12, 24, and 36 is $2 \cdot 2 \cdot 3$ or 12.

11. $14n = 2 \cdot 7 \cdot n$
$42n^2 = 2 \cdot 3 \cdot 7 \cdot n \cdot n$
The GCF of $14n$ and $42n^2$ is $2 \cdot 7 \cdot n$ or $14n$.

12. $36a^3b = 2 \cdot 2 \cdot 3 \cdot 3 \cdot a \cdot a \cdot a \cdot b$
$56ab^2 = 2 \cdot 2 \cdot 2 \cdot 7 \cdot a \cdot b \cdot b$
The GCF of $36a^3b$ and $56ab^2$ is $2 \cdot 2 \cdot a \cdot b$ or $4ab$.

13. $3n = 3 \cdot n$
$9 = 3 \cdot 3$
The GCF is 3.
$3n + 9 = 3(n) + 3(3)$
$= 3(n + 3)$
So, $3n + 9 = 3(n + 3)$.

14. $t^2 = t \cdot t$
$4t = 2 \cdot 2 \cdot t$
The GCF is t.
$t^2 + 4t = t(t) + t(4)$
$= t(t + 4)$
So, $t^2 + 4t = t(t + 4)$.

15. $15 = 3 \cdot 5$
$20x = 2 \cdot 2 \cdot 5 \cdot x$
The GCF is 5.
$15 + 20x = 5(3) + 5(4x)$
$= 5(3 + 4x)$
So, $15 + 20x = 5(3 + 4x)$.

16. Find the GCF of 36 and 120.
$36 = 2 \cdot 2 \cdot 3 \cdot 3$
$120 = 2 \cdot 2 \cdot 2 \cdot 3 \cdot 5$
The GCF is $2 \cdot 2 \cdot 3$ or 12.
The greatest number of students in each row is 12.

Pages 167–168 Practice and Apply

17. $12 = 2 \cdot 2 \cdot 3$
$8 = 2 \cdot 2 \cdot 2$
The GCF of 12 and 8 is $2 \cdot 2$ or 4.

18. $3 = 3$
$9 = 3 \cdot 3$
The GCF of 3 and 9 is 3.

19. $24 = 2 \cdot 2 \cdot 2 \cdot 3$
$40 = 2 \cdot 2 \cdot 2 \cdot 5$
The GCF of 24 and 40 is $2 \cdot 2 \cdot 2$ or 8.

20. $21 = 3 \cdot 7$
$14 = 2 \cdot 7$
The GCF of 21 and 14 is 7.

21. $20 = 2 \cdot 2 \cdot 5$
$30 = 2 \cdot 3 \cdot 5$
The GCF of 20 and 30 is $2 \cdot 5$ or 10.

22. $12 = 2 \cdot 2 \cdot 3$
$18 = 2 \cdot 3 \cdot 3$
The GCF of 12 and 18 is $2 \cdot 3$ or 6.

23. $18 = 2 \cdot 3 \cdot 3$
$45 = 3 \cdot 3 \cdot 5$
The GCF of 18 and 45 is $3 \cdot 3$ or 9.

24. $22 = 2 \cdot 11$
$21 = 3 \cdot 7$
There are no common factors. The GCF of 22 and 21 is 1.

25. $16 = 2 \cdot 2 \cdot 2 \cdot 2$
$40 = 2 \cdot 2 \cdot 2 \cdot 5$
The GCF of 16 and 40 is $2 \cdot 2 \cdot 2$ or 8.

26. $42 = 2 \cdot 3 \cdot 7$
$56 = 2 \cdot 2 \cdot 2 \cdot 7$
The GCF of 42 and 56 is $2 \cdot 7$ or 14.

27. $30 = 2 \cdot 3 \cdot 5$
$35 = 5 \cdot 7$
The GCF of 30 and 35 is 5.

28. $12 = 2 \cdot 2 \cdot 3$
$60 = 2 \cdot 2 \cdot 3 \cdot 5$
The GCF of 12 and 60 is $2 \cdot 2 \cdot 3$ or 12.

29. $116 = 2 \cdot 2 \cdot 29$
$100 = 2 \cdot 2 \cdot 5 \cdot 5$
The GCF of 116 and 100 is $2 \cdot 2$ or 4.

30. $135 = 3 \cdot 3 \cdot 3 \cdot 5$
$315 = 3 \cdot 3 \cdot 5 \cdot 7$
The GCF of 135 and 315 is $3 \cdot 3 \cdot 5$ or 45.

31. $9 = 3 \cdot 3$
$15 = 3 \cdot 5$
$24 = 2 \cdot 2 \cdot 2 \cdot 3$
The GCF of 9, 15, and 24 is 3.

32. $20 = 2 \cdot 2$
$21 = 3 \cdot 7$
$25 = 5 \cdot 5$
There are no common factors. The GCF of 20, 21, and 25 is 1.

33. $20 = 2 \cdot 2 \cdot 5$
$28 = 2 \cdot 2 \cdot 7$
$36 = 2 \cdot 2 \cdot 3 \cdot 3$
The GCF of 20, 28, and 36 is $2 \cdot 2$ or 4.

34. $66 = 2 \cdot 3 \cdot 11$
$90 = 2 \cdot 3 \cdot 3 \cdot 5$
$150 = 2 \cdot 3 \cdot 5 \cdot 5$
The GCF of 66, 90, and 150 is $2 \cdot 3$ or 6.

35. $12x = 2 \cdot 2 \cdot 3 \cdot x$
$40x^2 = 2 \cdot 2 \cdot 2 \cdot 5 \cdot x \cdot x$
The GCF of $12x$ and $40x^2$ is $2 \cdot 2 \cdot x$ or $4x$.

36. $18 = 2 \cdot 3 \cdot 3$
$45mn = 3 \cdot 3 \cdot 5 \cdot m \cdot n$
The GCF of 18 and $45mn$ is $3 \cdot 3$ or 9.

37. $4st = 2 \cdot 2 \cdot s \cdot t$
$10s = 2 \cdot 5 \cdot s$
The GCF of $4st$ and $10s$ is $2 \cdot s$ or $2s$.

38. $5ab = 5 \cdot a \cdot b$
$6b^2 = 2 \cdot 3 \cdot b \cdot b$
The GCF of $5ab$ and $6b^2$ is b.

39. $14b = 2 \cdot 7 \cdot b$
$56b^2 = 2 \cdot 2 \cdot 2 \cdot 7 \cdot b \cdot b$
The GCF of 14b and 56b[2] is $2 \cdot 7 \cdot b$ or $14b$.

40. $30a^3b^2 = 2 \cdot 3 \cdot 5 \cdot a \cdot a \cdot a \cdot b \cdot b$
$24a^2b = 2 \cdot 2 \cdot 2 \cdot 3 \cdot a \cdot a \cdot b$
The GCF of $30a^3b^2$ and $24a^2b$ is $2 \cdot 3 \cdot a \cdot a \cdot b$ or $6a^2b$.

41. $32mn^2 = 2 \cdot 2 \cdot 2 \cdot 2 \cdot 2 \cdot m \cdot n \cdot n$
$16n = 2 \cdot 2 \cdot 2 \cdot 2 \cdot n$
$12n^3 = 2 \cdot 2 \cdot 3 \cdot n \cdot n \cdot n$
The GCF of $32mn^2$, $16n$, and $12n^3$ is $2 \cdot 2 \cdot n$ or $4n$.

42. $15v^2 = 3 \cdot 5 \cdot v \cdot v$
$70vw = 2 \cdot 5 \cdot 7 \cdot v \cdot w$
$36w^2 = 2 \cdot 2 \cdot 3 \cdot 3 \cdot w \cdot w$
There are no common factors. The GCF of $15v^2$, $70vw$, and $36w^2$ is 1.

43. Sample answer: $2x$, $6x^2$
$2x = 2 \cdot x$
$6x^2 = 2 \cdot 3 \cdot x \cdot x$
The GCF of $2x$ and $6x^2$ is $2 \cdot x$ or $2x$.

44. $2x = 2 \cdot x$
$8 = 2 \cdot 2 \cdot 2$
The GCF is 2.
$2x + 8 = 2(x) + 2(4)$
$= 2(x + 4)$
So, $2x + 8 = 2(x + 4)$.

45. $3r = 3 \cdot r$
$12 = 2 \cdot 2 \cdot 3$
The GCF is 3.
$3r + 12 = 3(r) + 3(4)$
$= 3(r + 4)$
So, $3r + 12 = 3(r + 4)$.

46. $8 = 2 \cdot 2 \cdot 2$
$32a = 2 \cdot 2 \cdot 2 \cdot 2 \cdot 2 \cdot a$
The GCF is $2 \cdot 2 \cdot 2$ or 8.
$8 + 32a = 8(1) + 8(4a)$
$= 8(1 + 4a)$
So, $8 + 32a = 8(1 + 4a)$.

47. $6 = 2 \cdot 3$
$3y = 3 \cdot y$
The GCF is 3.
$6 + 3y = 3(2) + 3(y)$
$= 3(2 + y)$
So, $6 + 3y = 3(2 + y)$.

48. $9 = 3 \cdot 3$
$3t = 3 \cdot t$
The GCF is 3.
$9 + 3t = 3(3) + 3(t)$
$= 3(3 + t)$
So, $9 + 3t = 3(3 + t)$.

49. $14 = 2 \cdot 7$
$21c = 3 \cdot 7 \cdot c$
The GCF is 7.
$14 + 21c = 7(2) + 7(3c)$
$= 7(2 + 3c)$
So, $14 + 21c = 7(2 + 3c)$.

50. $k^2 = k \cdot k$
$5k = 5 \cdot k$
The GCF is k.
$k^2 + 5k = k(k) + 5(k)$
$= k(k + 5)$
So, $k^2 + 5k = k(k + 5)$.

51. $4y = 2 \cdot 2 \cdot y$
$-16 = -1 \cdot 2 \cdot 2 \cdot 2 \cdot 2$
The GCF is $2 \cdot 2$ or 4.
$4y - 16 = 4(y) - 4(4)$
$= 4(y - 4)$
So, $4y - 16 = 4(y - 4)$.

52. $5n = 5 \cdot n$
$-10 = -1 \cdot 2 \cdot 5$
The GCF is 5.
$5n - 10 = 5(n) - 5(2)$
$= 5(n - 2)$
So, $5n - 10 = 5(n - 2)$.

53a. The GCF is 7; Sample answer:

7	14	21	28	35
↓	↓	↓	↓	↓
7(1)	7(2)	7(3)	7(4)	7(5)

The terms increase by a factor of 7.

53b. $7(6) = 42$, $7(7) = 49$
The next two terms are 42 and 49.

54. Since 16 divides evenly into 48 and 12 divides evenly into 72, the shelves can be made with no leftover material.

Area of plywood:	Area of each shelf:
$A = \ell w$	$A = \ell w$
$A = 48 \cdot 72$	$A = 12 \cdot 16$
$A = 3456 \text{ in}^2$	$A = 192 \text{ in}^2$

$3456 \div 192 = 18$ shelves

55a. Find the GCF of 30 and 24.
$30 = 2 \cdot 3 \cdot 5$
$24 = 2 \cdot 2 \cdot 2 \cdot 3$
The GCF is $2 \cdot 3$ or 6.
She can use 6-in. square tiles.

55b.

Area of surface	Area of a tile:
$A = \ell w$	$A = \ell w$
$A = 30 \cdot 24$	$A = 6 \cdot 6$
$A = 720 \text{ in}^2$	$A = 36 \text{ in}^2$

$720 \div 36 = 20$ tiles

56a. $b - a = 110 - 86$
$= 24$

56b. 24 is not a factor of either 110 or 86.
$a - 24 = 86 - 24$
$= 62$

56c. Subtract the two smallest numbers from the preceding calculation.
$62 - 24 = 38$; 38 is not a factor of 110 or 86.
$38 - 24 = 14$; 14 is not a factor of 110 or 86.
$24 - 14 = 10$; 10 is not a factor of 86.
$14 - 10 = 4$; 4 is not a factor of 110 or 86.
$10 - 4 = 6$; 6 is not a factor of 110 or 86.
$6 - 4 = 2$; 2 is a factor of both 110 and 86, so the GCF is 2.

57. Yes; the GCF of 2 and 8 is 2.

58. A Venn diagram can be used to display the prime factorizations of two or more numbers. Answers should include the following.

- Each circle in a Venn diagram represents a number. The prime factors of the number are listed inside its circle. The overlapping part(s) of the circles contain the common prime factors.
- The overlapping area in the Venn diagram is used to find the greatest common factor. Multiply the numbers found in this area.

59. D; $6y = 2 \cdot 3 \cdot y$
$21 = 3 \cdot 7$
The GCF is 3.
$6y + 21 = 3(2y) + 3(7)$
$= 3(2y + 7)$

60. C; $42x^2y = 2 \cdot 3 \cdot 7 \cdot x \cdot x \cdot y$
$38xy^2 = 2 \cdot 19 \cdot x \cdot y \cdot y$
The GCF is $2xy$.

61. Yes; $7 = 1 \cdot 7$
$8 = 2 \cdot 2 \cdot 2$
The GCF is 1.

62. Yes; $13 = 1 \cdot 13$
$11 = 1 \cdot 11$
The GCF is 1.

63. No; $27 = 3 \cdot 3 \cdot 3$
$18 = 2 \cdot 3 \cdot 3$
The GCF is $3 \cdot 3$ or 9.

64. No; $20 = 2 \cdot 2 \cdot 5$
$25 = 5 \cdot 5$
The GCF is 5.

65. Yes; $22 = 2 \cdot 11$
$23 = 1 \cdot 23$
The GCF is 1.

66. No; $8 = 2 \cdot 2 \cdot 2$
$12 = 2 \cdot 2 \cdot 3$
The GCF is $2 \cdot 2$ or 4.

Page 168 Maintain Your Skills

67. $9n = 3 \cdot 3 \cdot n$

68. $15x^2 = 3 \cdot 5 \cdot x \cdot x$

69. $-5jk = -1 \cdot 5 \cdot j \cdot k$

70. $22ab^3 = 2 \cdot 11 \cdot a \cdot b \cdot b \cdot b$

71. $7x^2 + y^3 = 7(-2)^2 + 4^3$
$= 7(-2)(-2) + (4)(4)(4)$
$= 28 + 64$
$= 92$

72. $69 \div 23 = 3$

73. $48 \div (-8) = -6$

74. $-24 \div (-12) = 2$

75. $-50 \div 5 = -10$

76. 1 ft = 12 in.

77. 1 yd = 36 in.

78. 1 lb = 16 oz

79. 1 day = 24 h

80. 1 m = 100 cm

81. 1 kg = 1000 g

4-5 Simplifying Algebraic Fractions

Page 169 How are simplified fractions used in representing measurements?

a. Yes; the same portion of each circle is shaded.
b. the third figure
c. $\frac{1}{4}$; the shaded part is not divided into smaller parts.

Page 171 Check for Understanding

1. The GCF of the numerator and denominator is 1.
2. Sample answer: $\frac{2}{3}$ and $\frac{x}{4}$ are in simplest form; $\frac{2}{4}$ and $\frac{3n}{n^2}$ are not in simplest form.
3. $2 = \boxed{2}$
$14 = \boxed{2} \cdot 7$
The GCF of 2 and 14 is 2.
$\frac{2}{14} = \frac{2 \div 2}{14 \div 2} = \frac{1}{7}$
4. $9 = \boxed{3} \cdot 3$
$15 = \boxed{3} \cdot 5$
The GCF of 9 and 15 is 3.
$\frac{9}{15} = \frac{9 \div 3}{15 \div 3} = \frac{3}{5}$
5. 5 and 11 have no common factors. $\frac{5}{11}$ is simplified.
6. $25 = \boxed{5} \cdot 5$
$40 = 2 \cdot 2 \cdot 2 \cdot \boxed{5}$
The GCF of 25 and 40 is 5.
$\frac{25}{40} = \frac{25 \div 5}{40 \div 5} = \frac{5}{8}$
7. $64 = \boxed{2} \cdot \boxed{2} \cdot 2 \cdot 2 \cdot 2 \cdot 2$
$68 = \boxed{2} \cdot \boxed{2} \cdot 17$
The GCF of 64 and 68 is $2 \cdot 2$ or 4.
$\frac{64}{68} = \frac{64 \div 4}{68 \div 4} = \frac{16}{17}$
8. $\frac{x}{x^3} = \frac{\cancel{x}^1}{\cancel{x}_1 \cdot x \cdot x} = \frac{1}{x^2}$
9. $\frac{8a^2}{16a} = \frac{\cancel{2}^1 \cdot \cancel{2}^1 \cdot \cancel{2}^1 \cdot \cancel{a}^1 \cdot a}{\cancel{2}_1 \cdot \cancel{2}_1 \cdot \cancel{2}_1 \cdot 2 \cdot \cancel{a}_1} = \frac{a}{2}$
10. $\frac{12c}{15d} = \frac{2 \cdot 2 \cdot \cancel{3}^1 \cdot c}{\cancel{3}_1 \cdot 5 \cdot d} = \frac{4c}{5d}$
11. 24 and 5k have no common factors. $\frac{24}{5k}$ is simplified.
12. There are 36 inches in 1 yard. Write the fraction $\frac{9}{36}$ in simplest form.
$\frac{9}{36} = \frac{\cancel{3}^1 \cdot \cancel{3}^1}{2 \cdot 2 \cdot \cancel{3}_1 \cdot \cancel{3}_1} = \frac{1}{4}$
So, 9 inches is $\frac{1}{4}$ of a yard.
13. B; $\frac{25mn}{65n} = \frac{\cancel{5}^1 \cdot 5 \cdot m \cdot \cancel{n}^1}{\cancel{5}_1 \cdot 13 \cdot \cancel{n}_1} = \frac{5m}{13}$

Page 172–173 Practice and Apply

14. $\frac{3}{18} = \frac{\cancel{3}^1}{2 \cdot \cancel{3}_1 \cdot 3} = \frac{1}{6}$
15. $\frac{10}{12} = \frac{\cancel{2}^1 \cdot 5}{\cancel{2}_1 \cdot 2 \cdot 3} = \frac{5}{6}$
16. $\frac{15}{21} = \frac{\cancel{3}^1 \cdot 5}{\cancel{3}_1 \cdot 7} = \frac{5}{7}$
17. $\frac{8}{36} = \frac{\cancel{2}^1 \cdot \cancel{2}^1 \cdot 2}{\cancel{2}_1 \cdot \cancel{2}_1 \cdot 3 \cdot 3} = \frac{2}{9}$
18. 17 and 20 have no common factors. $\frac{17}{20}$ is simplified.
19. $\frac{18}{4} = \frac{\cancel{2}^1 \cdot 3 \cdot 3}{\cancel{2}_1 \cdot 2 \cdot 11} = \frac{9}{22}$
20. $\frac{16}{64} = \frac{\cancel{2}^1 \cdot \cancel{2}^1 \cdot \cancel{2}^1 \cdot \cancel{2}^1}{\cancel{2}_1 \cdot \cancel{2}_1 \cdot \cancel{2}_1 \cdot \cancel{2}_1 \cdot 2 \cdot 2} = \frac{1}{4}$
21. 30 and 37 have no common factors. $\frac{30}{37}$ is simplified.
22. $\frac{34}{38} = \frac{\cancel{2}^1 \cdot 17}{\cancel{2}_1 \cdot 19} = \frac{17}{19}$
23. $\frac{17}{51} = \frac{\cancel{17}^1}{3 \cdot \cancel{17}_1} = \frac{1}{3}$
24. $\frac{51}{60} = \frac{\cancel{3}^1 \cdot 17}{2 \cdot 2 \cdot \cancel{3}_1 \cdot 5} = \frac{17}{20}$
25. $\frac{25}{60} = \frac{\cancel{5}^1 \cdot 5}{2 \cdot 2 \cdot 3 \cdot \cancel{5}_1} = \frac{5}{12}$
26. $\frac{36}{96} = \frac{\cancel{2}^1 \cdot \cancel{2}^1 \cdot \cancel{3}^1 \cdot 3}{\cancel{2}_1 \cdot \cancel{2}_1 \cdot 2 \cdot 2 \cdot 2 \cdot \cancel{3}_1} = \frac{3}{8}$

27. $\frac{133}{140} = \frac{\overset{1}{\cancel{7}} \cdot 19}{2 \cdot 2 \cdot 5 \cdot \underset{1}{\cancel{7}}}$

$= \frac{19}{20}$

28. $\frac{765}{2023} = \frac{3 \cdot 3 \cdot 5 \cdot \overset{1}{\cancel{17}}}{7 \cdot \underset{1}{\cancel{17}} \cdot 17}$

$= \frac{45}{119}$

29. There are 12 inches in 1 foot, so 46 ft is 46 · 12 or 552 in.

$\frac{18 \text{ in}}{46 \text{ ft}} = \frac{18}{552} = \frac{\overset{1}{\cancel{2}} \cdot \overset{1}{\cancel{3}} \cdot 3}{\underset{1}{\cancel{2}} \cdot 2 \cdot 2 \cdot \underset{1}{\cancel{3}} \cdot 23}$

$= \frac{3}{92}$

So, the model is $\frac{3}{92}$ of the actual airplane.

30. $\frac{a}{a^4} = \frac{\overset{1}{\cancel{a}}}{\underset{1}{\cancel{a}} \cdot a \cdot a \cdot a}$

$= \frac{1}{a^3}$

31. $\frac{y^3}{y} = \frac{\overset{1}{\cancel{y}} \cdot y \cdot y}{\underset{1}{\cancel{y}}}$

$= \frac{y^2}{1}$ or y^2

32. $\frac{12m}{15m} = \frac{2 \cdot 2 \cdot \overset{1}{\cancel{3}} \cdot \overset{1}{\cancel{m}}}{\underset{1}{\cancel{3}} \cdot 5 \cdot \underset{1}{\cancel{m}}}$

$= \frac{4}{5}$

33. $\frac{40d}{42d} = \frac{\overset{1}{\cancel{2}} \cdot 2 \cdot 2 \cdot 5 \cdot \overset{1}{\cancel{d}}}{\underset{1}{\cancel{2}} \cdot 3 \cdot 7 \cdot \underset{1}{\cancel{d}}}$

$= \frac{20}{21}$

34. $4k$ and $19m$ have no common factors. So, $\frac{4k}{19m}$ is simplified.

35. $\frac{8t}{64t^2} = \frac{\overset{1}{\cancel{2}} \cdot \overset{1}{\cancel{2}} \cdot \overset{1}{\cancel{2}} \cdot \overset{1}{\cancel{t}}}{\underset{1}{\cancel{2}} \cdot \underset{1}{\cancel{2}} \cdot \underset{1}{\cancel{2}} \cdot 2 \cdot 2 \cdot 2 \cdot \underset{1}{\cancel{t}} \cdot t}$

$= \frac{1}{8t}$

36. $\frac{16n}{18n^2p} = \frac{\overset{1}{\cancel{2}} \cdot 2 \cdot 2 \cdot 2 \cdot \overset{1}{\cancel{n}}}{\underset{1}{\cancel{2}} \cdot 3 \cdot 3 \cdot \underset{1}{\cancel{n}} \cdot n \cdot p}$

$= \frac{8}{9np}$

37. $\frac{28z^3}{16z} = \frac{\overset{1}{\cancel{2}} \cdot \overset{1}{\cancel{2}} \cdot 7 \cdot \overset{1}{\cancel{z}} \cdot z \cdot z}{\underset{1}{\cancel{2}} \cdot \underset{1}{\cancel{2}} \cdot 2 \cdot 2 \cdot \underset{1}{\cancel{z}}}$

$= \frac{7z^2}{4}$

38. $\frac{6r}{15rs} = \frac{2 \cdot \overset{1}{\cancel{3}} \cdot \overset{1}{\cancel{r}}}{\underset{1}{\cancel{3}} \cdot 5 \cdot \underset{1}{\cancel{r}} \cdot s}$

$= \frac{2}{5s}$

39. $12cd$ and $19e$ have no common factors. So, $\frac{12cd}{19e}$ is simplified.

40. $\frac{30x^2}{51xy} = \frac{2 \cdot \overset{1}{\cancel{3}} \cdot 5 \cdot \overset{1}{\cancel{x}} \cdot x}{\underset{1}{\cancel{3}} \cdot 17 \cdot \underset{1}{\cancel{x}} \cdot y}$

$= \frac{10x}{17y}$

41. $\frac{17g^2h}{51g} = \frac{\overset{1}{\cancel{17}} \cdot \overset{1}{\cancel{g}} \cdot g \cdot h}{3 \cdot \underset{1}{\cancel{17}} \cdot \underset{1}{\cancel{g}}}$

$= \frac{gh}{3}$

42. There are 24 hours in 1 day.

$\frac{15}{24} = \frac{\overset{1}{\cancel{3}} \cdot 5}{2 \cdot 2 \cdot 2 \cdot \underset{1}{\cancel{3}}}$

$= \frac{5}{8}$

So, 15 hours is $\frac{5}{8}$ of a day.

43. There are 100 centimeters in a meter.

$\frac{96}{100} = \frac{\overset{1}{\cancel{2}} \cdot \overset{1}{\cancel{2}} \cdot 2 \cdot 2 \cdot 2 \cdot 3}{\underset{1}{\cancel{2}} \cdot \underset{1}{\cancel{2}} \cdot 5 \cdot 5}$

$= \frac{24}{25}$

So, 96 centimeters is $\frac{24}{25}$ of a meter.

44. There are 16 ounces in a pound.

$\frac{12 \text{ oz}}{1 \text{ lb}} = \frac{12 \text{ oz}}{16 \text{ oz}} = \frac{\overset{1}{\cancel{2}} \cdot \overset{1}{\cancel{2}} \cdot 3}{\underset{1}{\cancel{2}} \cdot \underset{1}{\cancel{2}} \cdot 2 \cdot 2}$

$= \frac{3}{4}$

So, 12 ounces is $\frac{3}{4}$ of a pound.

45a. Yes; $\frac{E}{A} = \frac{330}{440} = \frac{\overset{1}{\cancel{2}} \cdot 3 \cdot \overset{1}{\cancel{5}} \cdot \overset{1}{\cancel{11}}}{\underset{1}{\cancel{2}} \cdot 2 \cdot 2 \cdot \underset{1}{\cancel{5}} \cdot \underset{1}{\cancel{11}}}$

$= \frac{3}{4}$

45b. No; $\frac{D}{F} = \frac{294}{349}$ cannot be simplified.

45c. Yes; $\frac{\text{first C}}{\text{last C}} = \frac{264}{528} = \frac{\overset{1}{\cancel{2}} \cdot \overset{1}{\cancel{2}} \cdot \overset{1}{\cancel{2}} \cdot \overset{1}{\cancel{3}} \cdot \overset{1}{\cancel{11}}}{\underset{1}{\cancel{2}} \cdot \underset{1}{\cancel{2}} \cdot \underset{1}{\cancel{2}} \cdot 2 \cdot \underset{1}{\cancel{3}} \cdot \underset{1}{\cancel{11}}}$

$= \frac{1}{2}$

46. elephant, $\frac{450}{9000} = \frac{\overset{1}{\cancel{2}} \cdot \overset{1}{\cancel{3}} \cdot \overset{1}{\cancel{3}} \cdot \overset{1}{\cancel{5}} \cdot \overset{1}{\cancel{5}}}{\underset{1}{\cancel{2}} \cdot 2 \cdot 2 \cdot \underset{1}{\cancel{3}} \cdot \underset{1}{\cancel{3}} \cdot \underset{1}{\cancel{5}} \cdot \underset{1}{\cancel{5}} \cdot 5}$

$= \frac{1}{20}$;

hummingbird, $\frac{2}{3}$;

polar bear, $\frac{25}{1500} = \frac{\overset{1}{\cancel{5}} \cdot \overset{1}{\cancel{5}}}{2 \cdot 2 \cdot 3 \cdot \underset{1}{\cancel{5}} \cdot \underset{1}{\cancel{5}} \cdot 5}$

$= \frac{1}{60}$;

tiger, $\frac{20}{500} = \frac{\overset{1}{\cancel{2}} \cdot \overset{1}{\cancel{2}} \cdot \overset{1}{\cancel{5}}}{\underset{1}{\cancel{2}} \cdot \underset{1}{\cancel{2}} \cdot \underset{1}{\cancel{5}} \cdot 5 \cdot 5}$

$= \frac{1}{25}$

47. $\frac{62}{100} = \frac{\overset{1}{\cancel{2}} \cdot 31}{\underset{1}{\cancel{2}} \cdot 2 \cdot 5 \cdot 5}$

$= \frac{31}{50}$

48. U.S. students save $\frac{21}{100}$ of a dollar.

49. $\frac{\text{U.S. students}}{\text{French students}} = \frac{21}{30}$

$= \frac{\overset{1}{\cancel{3}} \cdot 7}{2 \cdot \underset{1}{\cancel{3}} \cdot 5}$

$= \frac{7}{10}$

50. No, $23 \neq 2 \cdot 3$ and $53 \neq 5 \cdot 3$.

51. Fractions represent parts of a whole. So, measurements that contain parts of units can be represented using fractions. Answers should include the following.

- Measurements can be given as parts of a whole because smaller units make up larger units. For example, inches make up feet.

- Twelve inches equals 1 foot. So, 3 inches equals $\frac{3}{12}$ or $\frac{1}{4}$ foot.

52. B; $\frac{18}{60} = \frac{\overset{1}{\cancel{2}} \cdot \overset{1}{\cancel{3}} \cdot 3}{\underset{1}{\cancel{2}} \cdot 2 \cdot \underset{1}{\cancel{3}} \cdot 5}$
$= \frac{3}{10}$

53. A; $\frac{15ab}{25b^2} = \frac{3 \cdot \overset{1}{\cancel{5}} \cdot a \cdot \overset{1}{\cancel{b}}}{5 \cdot \underset{1}{\cancel{5}} \cdot \underset{1}{\cancel{b}} \cdot b}$
$= \frac{3a}{5b}$

Page 173 Maintain Your Skills

54. $9 = 3 \cdot 3$
$15 = 3 \cdot 5$
The GCF is 3.

55. $4 = 2 \cdot 2$
$12 = 2 \cdot 2 \cdot 3$
$10 = 2 \cdot 5$
The GCF is 2.

56. $40x^2 = 2 \cdot 2 \cdot 2 \cdot 5 \cdot x \cdot x$
$16x = 2 \cdot 2 \cdot 2 \cdot 2 \cdot x$
The GCF is $2 \cdot 2 \cdot 2 \cdot x$ or $8x$.

57. $25a = 5 \cdot 5 \cdot a$
$30b = 2 \cdot 3 \cdot 5 \cdot b$
The GCF is 5.

58. prime; $13 = 1 \times 13$ is the only factorization.

59. composite; $34 = 1 \times 34$
$34 = 2 \times 17$

60. composite; $99 = 1 \times 99$
$99 = 3 \times 33$
$99 = 9 \times 11$

61. prime; $79 = 1 \times 79$ is the only factorization.

62. $t - 18 = 24$
$t - 18 + 18 = 24 + 18$
$t = 42$
Check: $t - 18 = 24$
$42 - 18 \stackrel{?}{=} 24$
$24 = 24$ ✓
The solution is 42.

63. $30 = 3 + y$
$30 - 3 = 3 - 3 + y$
$27 = y$
Check: $30 = 3 + y$
$30 \stackrel{?}{=} 3 + 27$
$30 = 30$ ✓
The solution is 27.

64. $-7 = x + 11$
$-7 - 11 = x + 11 - 11$
$-18 = x$
Check: $-7 = x + 11$
$-7 \stackrel{?}{=} -18 + 11$
$-7 = -7$ ✓
The solution is -18.

65. $6 \cdot 7 \cdot k^3 = (6 \cdot 7)(k^3)$

66. $s \cdot t^2 \cdot s \cdot t = (s \cdot s)(t^2 \cdot t)$

67. $3 \cdot x^4 \cdot (-5) \cdot x^2 = (3 \cdot -5)(x^4 \cdot x^2)$

68. $5 \cdot n^3 \cdot p \cdot 2 \cdot n \cdot p = (5 \cdot 2)(n^3 \cdot n)(p \cdot p)$

page 174 Reading Mathematics

1. Sample answer: four times a squared
2. Sample answer: ten times x the quantity to the fifth power
3. Sample answer: five divided by n cubed
4. Sample answer: four divided by r the quantity squared
5. Sample answer: the sum of m and n the quantity cubed
6. Sample answer: a minus b the quantity to the fourth power
7. Sample answer: a minus b to the fourth power
8. Sample answer: a divided by b to the fourth power
9. Sample answer: four times c squared the quantity cubed
10. Sample answer: eight divided by c squared the quantity cubed
11. No; $4(ab)^5 = 4a^5 \cdot b^5$
12. Yes; $(2x)^3 = 2^3x^3$ or $8x^3$
13. Yes; $(mn)^4 = m^4 \cdot n^4$
14. No; cd^3 is not the same as $(cd)^3$. $cd^3 = c \cdot d \cdot d \cdot d$ and $(cd)^3 = cd \cdot cd \cdot cd$.
15. No; $\left(\frac{x}{y}\right)^5 = \frac{x^5}{y^5}$
16. Yes; $\left(\frac{n}{r}\right)^2 = \frac{n^2}{r^2}$

4-6 Multiplying and Dividing Monomials

Page 175 How are powers of monomials useful in comparing earthquake magnitudes?

a. The exponents of the factors are added to get the exponent of each product.

b. Sample answer: Add the exponents to determine the exponent of the product when you multiply powers with the same base.

Page 177 Check for Understanding

1. Neither; the factors have different bases.
2. $4^8 \cdot 4^6 = 4^{8+6} = 4^{14}$.
$4^4 \cdot 4^{10} = 4^{4+10} = 4^{14}$.
Yes; they both equal 4^{14}.
3. Sample answer: $5 \cdot 5^2 = 5^3$
4. $9^3 \cdot 9^2 = 9^{3+2}$
$= 9^5$
5. $a \cdot a^5 = a^{1+5}$
$= a^6$
6. $n^4 \cdot n^4 = n^{4+4}$
$= n^8$

7. $-3x^2(4x^3) = (-3 \cdot 4)(x^2 \cdot x^3)$
$= (-12)(x^{2+3})$
$= -12x^5$

8. $\frac{3^8}{3^5} = 3^{8-5}$
$= 3^3$

9. $\frac{10^5}{10^3} = 10^{5-3}$
$= 10^2$

10. $\frac{x^3}{x} = x^{3-1}$
$= x^2$

11. $\frac{a^{10}}{a^6} = a^{10-6}$
$= a^4$

12. $\frac{10^8}{10^5} = 10^{8-5}$
$= 10^3$ or 1000 times greater

Pages 178–179 Practice and Apply

13. $3^3 \cdot 3^2 = 3^{3+2}$
$= 3^5$

14. $6 \cdot 6^7 = 6^1 \cdot 6^7$
$= 6^{1+7}$
$= 6^8$

15. $d^4 \cdot d^6 = d^{4+6}$
$= d^{10}$

16. $10^4 \cdot 10^3 = 10^{4+3}$
$= 10^7$

17. $n^8 \cdot n = n^8 \cdot n^1$
$= n^{8+1}$
$= n^9$

18. $t^2 \cdot t^4 = t^{2+4}$
$= t^6$

19. $9^4 \cdot 9^5 = 9^{4+5}$
$= 9^9$

20. $a^6 \cdot a^6 = a^{6+6}$
$= a^{12}$

21. $2y \cdot 9y^4 = (2 \cdot 9)(y^1 \cdot y^4)$
$= 18(y^{1+4})$
$= 18y^5$

22. $(5r^3)(4r^4) = (5 \cdot 4)(r^3 \cdot r^4)$
$= 20(r^{3+4})$
$= 20r^7$

23. $ab^5 \cdot 8a^2b^5 = (1 \cdot 8)(a^1 \cdot a^2)(b^5 \cdot b^5)$
$= 8(a^{1+2})(b^{5+5})$
$= 8a^3b^{10}$

24. $10x^3y \cdot (-2xy^2) = [10 \cdot (-2)](x^3 \cdot x^1)(y^1 \cdot y^2)$
$= -20(x^{3+1})(y^{1+2})$
$= -20x^4y^3$

25. $\frac{5^5}{5^2} = 5^{5-2}$
$= 5^3$

26. $\frac{8^4}{8^3} = 8^{4-3}$
$= 8^1$ or 8

27. $b^6 \div b^3 = \frac{b^6}{b^3}$
$= b^{6-3}$
$= b^3$

28. $10^{10} \div 10^2 = \frac{10^{10}}{10^2}$
$= 10^{10-2}$
$= 10^8$

29. $\frac{m^{20}}{m^8} = m^{20-8}$
$= m^{12}$

30. $\frac{a^8}{a^8} = a^{8-8}$
$= a^0$ or 1

31. $\frac{(-2)^6}{(-2)^5} = (-2)^{6-5}$
$= (-2)^1$ or -2

32. $\frac{(-x)^5}{(-x)} = \frac{(-x)^5}{(-x)^1}$
$= (-x)^{5-1}$
$= (-x)^4$

33. $\frac{n^3(n^5)}{n^2} = \frac{n^{3+5}}{n^2}$
$= \frac{n^8}{n^2}$
$= n^{8-2}$
$= n^6$

34. $\frac{s^7}{s \cdot s^2} = \frac{s^7}{s^{1+2}}$
$= \frac{s^7}{s^3}$
$= s^{7-3}$
$= s^4$

35. $\left(\frac{k^3}{k}\right)\left(\frac{m^2}{m}\right) = (k^{3-1})(m^{2-1})$
$= k^2m$

36. $\left(\frac{15}{5}\right)\left(\frac{n^9}{n}\right) = 3(n^{9-1})$
$= 3n^8$

37. $9^4 \cdot 9^3 = 9^{4+3}$
$= 9^7$

38. $\frac{k^5}{k^2} = k^{5-2}$
$= k^3$

39. $7^3 \cdot 7^5 \cdot 7 = (7^3 \cdot 7^5) \cdot 7$
$= (7^{3+5}) \cdot 7^1$
$= 7^8 \cdot 7^1$
$= 7^{8+1}$
$= 7^9$

40. $a^4 \cdot a^6 \div a^2 = \frac{a^4 \cdot a^6}{a^2}$
$= \frac{a^{4+6}}{a^2}$
$= \frac{a^{10}}{a^2}$
$= a^{10-2}$
$= a^8$

41. $\frac{10^7}{10^5} = 10^{7-5}$
$= 10^2$ or 100 times

42. $\frac{10^9}{10^3} = 10^{9-3}$
$= 10^6$ or 1,000,000 times

43. Let x represent the pH level of cola.
$\frac{10^7}{10^x} = 10^4$
$10^{7-x} = 10^4$
Since the bases are equal, the exponents must be equal.

$7 - x = 4$

$7 - 7 - x = 4 - 7$

$-x = -3$

$\frac{-1x}{-1} = \frac{-3}{-1}$

$x = 3$

The pH value of cola is 3.

44a. Since the bacteria reproduce every 15 minutes, after 30 minutes, they have gone through 2 cycles of reproduction.
$100 \cdot 2^2 = 100 \cdot 4$ or 400 bacteria

44b. Since the bacteria reproduce every 15 minutes, they go through 4 cycles in 1 hour. In 3 hours, they have gone through $3 \cdot 4$ or 12 cycles.
$\frac{2^{12}}{2^4} = 2^{12-4}$
$= 2^8$
$= 256$ times more

45. $\frac{4}{2} = 2$ times

46. $\frac{2^6}{2^3} = 2^{6-3}$
$= 2^3$ or 8 times

47. 8; $(4^{\bullet})(4^3) = 4^{11}$
$4^{\bullet} = \frac{4^{11}}{4^3}$
$4^{\bullet} = 4^{11-3}$
$4^{\bullet} = 4^8$

48. 16; $\frac{t^{\bullet}}{t^2} = t^{14}$
$t^{\bullet} = t^{14} \cdot t^2$
$t^{\bullet} = t^{14+2}$
$t^{\bullet} = t^{16}$

49. 5; $\frac{13^5}{13^{\bullet}} = 1$
$13^5 = 13^{\bullet}$

50. Sample answer: By the law of exponents, $\frac{a^n}{a^n} = a^{n-n}$ or a^0 for $a \neq 0$. Since $\frac{a^n}{a^n} = 1$, then $a^0 = 1$. So, any number raised to the zero power must equal 1. Or, $a^0 \cdot a^m = a^{0+m}$ or a^m. Therefore, by the Identity Property of Multiplication, a^0 must equal 1.

51. Each level on the Richter scale is 10 times greater than the previous level. So, powers of 10 can be used to compare earthquake magnitudes. Answers should include the following.

- On the Richter scale, each whole-number increase represents a 10-fold increase in the magnitude of seismic waves.
- An earthquake of magnitude 7 is 10^5 times greater than an earthquake of magnitude 2 because $10^7 \div 10^2 = 10^{7-2}$ or 10^5.

52. A; $(7xy)(x^{14}z) = (7 \cdot 1)(x^1 \cdot x^{14})(y)(z)$
$= 7(x^{1+14})yz$
$= 7x^{15}yz$

53. B; $a^5 \div a = \frac{a^5}{a^1}$
$= a^{5-1}$
$= a^4$

Page 179 Maintain Your Skills

54. $\frac{12}{40} = \frac{\overset{1}{\cancel{2}} \cdot \overset{1}{\cancel{2}} \cdot 3}{\underset{1}{\cancel{2}} \cdot \underset{1}{\cancel{2}} \cdot 2 \cdot 5}$
$= \frac{3}{10}$

55. 20 and 53 have no common factors, so $\frac{20}{53}$ is simplified.

56. $\frac{8n^2}{32n} = \frac{\overset{1}{\cancel{2}} \cdot \overset{1}{\cancel{2}} \cdot \overset{1}{\cancel{2}} \cdot \overset{1}{\cancel{n}} \cdot n}{\underset{1}{\cancel{2}} \cdot \underset{1}{\cancel{2}} \cdot \underset{1}{\cancel{2}} \cdot 2 \cdot 2 \cdot \underset{1}{\cancel{n}}}$
$= \frac{n}{4}$

57. $\frac{6x^3}{4x^2y} = \frac{\overset{1}{\cancel{2}} \cdot 3 \cdot \overset{1}{\cancel{x}} \cdot \overset{1}{\cancel{x}} \cdot x}{\underset{1}{\cancel{2}} \cdot 2 \cdot \underset{1}{\cancel{x}} \cdot \underset{1}{\cancel{x}} \cdot y}$
$= \frac{3x}{2y}$

58. $36 = \boxed{2} \cdot \boxed{2} \cdot 3 \cdot 3$
$4 = \boxed{2} \cdot \boxed{2}$
The GCF of 36 and 4 is $2 \cdot 2$ or 4.

59. $18 = \boxed{2} \cdot 3 \cdot 3$
$28 = \boxed{2} \cdot 2 \cdot 7$
The GCF of 18 and 28 is 2.

60. $42 = \boxed{2} \cdot \boxed{3} \cdot 7$
$54 = \boxed{2} \cdot \boxed{3} \cdot 3 \cdot 3$
The GCF of 42 and 54 is $2 \cdot 3$ or 6.

61. $9a = 3 \cdot 3 \cdot \boxed{a}$
$10a^3 = 2 \cdot 5 \cdot \boxed{a} \cdot a \cdot a$
The GCF of $9a$ and $10a^3$ is a.

62. $|a| - |b| \cdot |c| = |-16| - |2| \cdot |3|$
$= 16 - 2 \cdot 3$
$= 16 - 6$
$= 10$

63. Positive; as the high temperature increases, the amount of electricity that is used also increases.

64. $\frac{1}{x} = \frac{1}{10}$

65. $\frac{y}{50} = \frac{-5}{50} = \frac{(-1) \cdot \overset{1}{\cancel{5}}}{2 \cdot 5 \cdot \underset{1}{\cancel{5}}}$
$= \frac{-1}{10}$

66. $\frac{z}{100} = \frac{4}{100} = \frac{\overset{1}{\cancel{2}} \cdot \overset{1}{\cancel{2}}}{\underset{1}{\cancel{2}} \cdot \underset{1}{\cancel{2}} \cdot 5 \cdot 5}$
$= \frac{1}{25}$

67. $\frac{1}{zy} = \frac{1}{4 \cdot (-5)}$
$= \frac{1}{20}$

68. $\frac{1}{x \cdot x} = \frac{1}{10 \cdot 10}$
$= \frac{1}{100}$

69. $\frac{1}{(z)(z)(z)} = \frac{1}{(4)(4)(4)}$
$= \frac{1}{64}$

Page 180 Algebra Activity (Follow-Up of Lesson 4-6)

1. Sample answer:

Number of Half-Lives	Number of Pennies That Remain
1	27
2	14
3	6
4	3
5	1

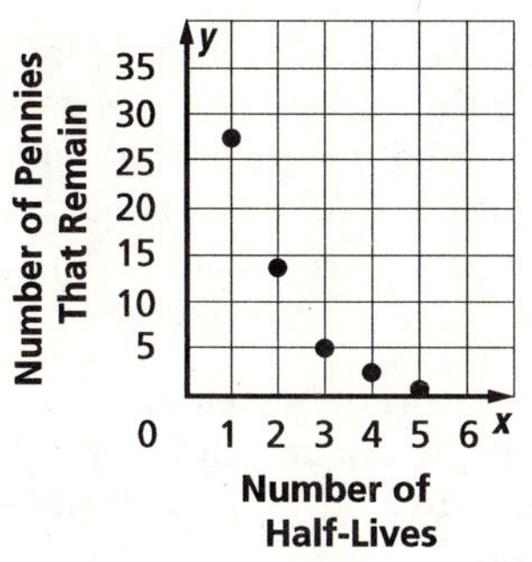

2. Points descend from left to right in a curve.
3. $50\left(\frac{1}{2}\right)\left(\frac{1}{2}\right)\left(\frac{1}{2}\right) = 50\left(\frac{1}{2}\right)^3$
$= 6.25$ or about 6 pennies; see students' work.
4. $1000\left(\frac{1}{2}\right)\left(\frac{1}{2}\right)\left(\frac{1}{2}\right) = 1000\left(\frac{1}{2}\right)^3$
$= 125$ pennies

4-7 Negative Exponents

Page 181 How do negative exponents represent repeated division?

a. The exponents decrease by 1; 1, $\frac{1}{2}$
b. Each number is divided by 2.
c. See students' work.
d. The pattern in powers of 3 shows that each number is divided by 3 as the exponents decrease by 1: $3^2 = 9$, $3^1 = 3$, $3^0 = 1$. Therefore, $3^{-1} = \frac{1}{3}$.

Page 183 Check for Understanding

1. To get each successive power, divide the previous power by 3. Therefore, $3^0 = 3 \div 3$ or 1.
2. 8^3, 8^0, 8^{-8}; Sample answer: 8^3 is the greatest because its exponent is positive. 8^0 is next because it equals 1. The expression 8^{-8} is a fraction less than 1.
3. $5^{-2} = \frac{1}{5^2}$
4. $(-7)^{-1} = \frac{1}{(-7)^1}$ or $\frac{1}{-7}$
5. $t^{-6} = \frac{1}{t^6}$
6. $n^{-2} = \frac{1}{n^2}$
7. $\frac{1}{3^4} = 3^{-4}$
8. $\frac{1}{9^2} = 9^{-2}$
9. $\frac{1}{49} = \frac{1}{7 \cdot 7} = \frac{1}{7^2} = 7^{-2}$
10. $\frac{1}{8} = \frac{1}{2 \cdot 2 \cdot 2} = \frac{1}{2^3} = 2^{-3}$
11. $a^{-5} = 2^{-5} = \frac{1}{2^5} = \frac{1}{32}$
12. $(ab)^{-2} = [(2)(-3)]^{-2} = (-6)^{-2} = \frac{1}{(-6)^2} = \frac{1}{36}$
13. $0.001 = \frac{1}{1000} = \frac{1}{10^3} = 10^{-3}$

Pages 183–185 Practice and Apply

14. $4^{-1} = \frac{1}{4^1}$ or $\frac{1}{4}$
15. $5^{-3} = \frac{1}{5^3}$
16. $(-6)^{-2} = \frac{1}{(-6)^2}$
17. $(-3)^{-3} = \frac{1}{(-3)^3}$
18. $3^{-5} = \frac{1}{3^5}$
19. $10^{-4} = \frac{1}{10^4}$
20. $p^{-1} = \frac{1}{p^1}$ or $\frac{1}{p}$
21. $a^{-10} = \frac{1}{a^{10}}$
22. $d^{-3} = \frac{1}{d^3}$
23. $q^{-4} = \frac{1}{q^4}$
24. $2s^{-5} = 2\left(\frac{1}{s^5}\right)$
25. $\frac{1}{x^{-2}} = x^2$
26. $10^{-6} = \frac{1}{10^6} = 0.000001$
27. $5^{-4} = \frac{1}{5^4} = 0.0016$
28. $\frac{1}{9^4} = 9^{-4}$
29. $\frac{1}{5^5} = 5^{-5}$
30. $\frac{1}{8^3} = 8^{-3}$
31. $\frac{1}{13^2} = 13^{-2}$
32. $\frac{1}{100} = \frac{1}{10 \cdot 10} = \frac{1}{10^2} = 10^{-2}$
33. $\frac{1}{81} = \frac{1}{9 \cdot 9} = \frac{1}{9^2} = 9^{-2}$
34. $\frac{1}{27} = \frac{1}{3 \cdot 3 \cdot 3} = \frac{1}{3^3} = 3^{-3}$
35. $\frac{1}{16} = \frac{1}{2 \cdot 2 \cdot 2 \cdot 2} = \frac{1}{2^4} = 2^{-4}$ or $\frac{1}{16} = \frac{1}{4 \cdot 4} = \frac{1}{4^2} = 4^{-2}$
36. $0.1 = \frac{1}{10^1} = 10^{-1}$

37. $0.01 = \frac{1}{100}$ or $0.01 = \frac{1}{100}$

$= \frac{1}{10 \cdot 10}$ $= 100^{-1}$

$= \frac{1}{10^2}$

$= 10^{-2}$

38. $0.0001 = \frac{1}{10{,}000}$ or $0.0001 = \frac{1}{10{,}000}$

$= \frac{1}{10 \cdot 10 \cdot 10 \cdot 10}$ $= \frac{1}{100 \cdot 100}$

$= \frac{1}{10^4}$ $= \frac{1}{100^2}$

$= 10^{-4}$ $= 100^{-2}$

39. $0.00001 = \frac{1}{100{,}000}$

$= \frac{1}{10 \cdot 10 \cdot 10 \cdot 10 \cdot 10}$

$= \frac{1}{10^5}$

$= 10^{-5}$

40. $x^{-4} = 3^{-4}$

$= \frac{1}{3^4}$

$= \frac{1}{81}$

41. $w^{-7} = (-2)^{-7}$

$= \frac{1}{(-2)^7}$

$= -\frac{1}{128}$

42. $8^w = 8^{-2}$

$= \frac{1}{8^2}$

$= \frac{1}{64}$

43. $(xy)^{-6} = [(3) \cdot (-1)]^{-6}$

$= (-3)^{-6}$

$= \frac{1}{(-3)^6}$

$= \frac{1}{729}$

44a. $\frac{1}{1{,}000{,}000{,}000}\text{ m} = \frac{1}{10 \cdot 10 \cdot 10 \cdot 10 \cdot 10 \cdot 10 \cdot 10 \cdot 10 \cdot 10}$

$= \frac{1}{10^9}$

$= 10^{-9}\text{ m}$

44b. $10 \cdot \frac{1}{1{,}000{,}000{,}000}\text{ m}$

$= \frac{\cancel{10}^{\,1}}{\cancel{10}_{\,1} \cdot 10 \cdot 10 \cdot 10 \cdot 10 \cdot 10 \cdot 10 \cdot 10 \cdot 10}$

$= \frac{1}{10 \cdot 10 \cdot 10 \cdot 10 \cdot 10 \cdot 10 \cdot 10 \cdot 10}$

$= \frac{1}{10^8}$

$= 10^{-8}\text{ m}$

45. $\frac{2^3}{2^{-4}} = 2^{3 - (-4)}$

$= 2^7$

$= 128$ times

46. penicillin, since 10^{-18} is larger than 10^{-23};

$\frac{10^{-18}}{10^{-23}} = 10^{-18 - (-23)}$

$= 10^5$ times greater

47. $x^{-2} \cdot x^{-3} = x^{-2 + (-3)}$

$= x^{-5}$

48. $r^{-5} \cdot r^9 = r^{-5 + 9}$

$= r^4$

49. $\frac{x^4}{x^7} = x^{4 - 7}$

$= x^{-3}$

50. $\frac{y^6}{y^{-10}} = y^{6 - (-10)}$

$= y^{16}$

51. $\frac{a^4b^{-4}}{ab^{-2}} = (a^{4 - 1})(b^{-4 - (-2)})$

$= a^3b^{-2}$ or $\frac{a^3}{b^2}$

52. $\frac{36s^3t^5}{12s^6t^{-3}} = \left(\frac{36}{12}\right)(s^{3 - 6})(t^{5 - (-3)})$

$= 3s^{-3}t^8$ or $\frac{3t^8}{s^3}$

53. yes; $(x^3)^{-2} = \frac{1}{(x^3)^2}$

$= \frac{1}{x^3 \cdot x^3}$ or $\frac{1}{x^6}$

$(x^{-2})^3 = (x^{-2})(x^{-2})(x^{-2})$

$= x^{-6}$ or $\frac{1}{x^6}$

54. Exponents are used to represent repeated factors in a multiplication. If the exponent is a negative number, then the repeated factors act as repeated divisors in the denominator. Answers should include the following.

- $4^{-2} = \frac{1}{4^2} = \frac{1}{4 \cdot 4}$ or $\frac{1}{16}$
- In 2^{-3}, the negative sign indicates that the factor 2 is repeated three times in the denominator, or 1 is divided by 2 three times. So, $2^{-3} = \frac{1}{2 \cdot 2 \cdot 2}$. If the exponent were -4, then 1 would be divided by 2 four times, and so on. So, as the value of n increases, the value of $\frac{1}{2^n}$ decreases.

55. C; $15^{-5} = \frac{1}{15^5}$

56. B; $1\text{ mm}^2 = \frac{1}{100\text{ cm}^2}$

$= 10^{-2}\text{ cm}^2$

57. $0.9 = (9 \times 10^{-1})$

58. $0.24 = 0.2 + 0.04$

$= (2 \times 10^{-1}) + (4 \times 10^{-2})$

59. $0.173 = 0.1 + 0.07 + 0.003$

$= (1 \times 10^{-1}) + (7 \times 10^{-2}) + (3 \times 10^{-3})$

60. $0.5875 = 0.5 + 0.08 + 0.007 + 0.0005$

$= (5 \times 10^{-1}) + (8 \times 10^{-2}) + (7 \times 10^{-3}) + (5 \times 10^{-4})$

Page 185 Maintain Your Skills

61. $3^6 \cdot 3 = 3^6 \cdot 3^1$

$= 3^{6 + 1}$

$= 3^7$

62. $x^2 \cdot x^4 = x^{2 + 4}$

$= x^6$

63. $\frac{5^5}{5^2} = 5^{5 - 2}$

$= 5^3$

64. $\frac{16n^3}{8n} = \frac{\cancel{2}^1 \cdot \cancel{2}^1 \cdot \cancel{2}^1 \cdot 2 \cdot \cancel{n}^1 \cdot n \cdot n}{\cancel{2}_1 \cdot \cancel{2}_1 \cdot \cancel{2}_1 \cdot \cancel{n}_1} = 2n^2$

65. $8(y + 6) = 8 \cdot y + 8 \cdot 6$

$= 8y + 48$

66. $(9 + k)(-2) = 9 \cdot (-2) + k(-2)$

$= -18 - 2k$

67. $(n - 3)5 = n \cdot 5 + (-3) \cdot 5$

$= 5n - 15$

68. Start at the origin. Move 1 unit right and 4 units down. The ordered pair that names point P is (1, −4).

69.
$$\begin{array}{r} 100 \\ \times\ 7.2 \\ \hline 200 \\ 7000 \\ \hline 720.0 \end{array}$$
So, 7.2 × 100 = 720.

70.
$$\begin{array}{r} 1000 \\ \times\ 1.6 \\ \hline 6000 \\ 10000 \\ \hline 1600.0 \end{array}$$
So, 1.6 × 1000 = 1600.

71.
$$\begin{array}{r} 4.05 \\ \times\ 10 \\ \hline 000 \\ 4050 \\ \hline 40.50 \end{array}$$
So, 4.05 × 10 = 40.5.

72.
$$\begin{array}{r} 3.8 \\ \times\ 0.1 \\ \hline 38 \\ 000 \\ \hline .038 \end{array}$$
So, 3.8 × 0.01 = 0.038.

73.
$$\begin{array}{r} 0.0001 \\ \times\ 5.0 \\ \hline 00000 \\ 000050 \\ \hline 0.00050 \end{array}$$
So, 5.0 × 0.0001 = 0.0005.

74.
$$\begin{array}{r} 9.24 \\ \times\ 0.1 \\ \hline 924 \\ 0000 \\ \hline 0.924 \end{array}$$
So, 9.24 × 0.1 = 0.924.

Page 185 Practice Quiz 2

1. $15 = 3 \cdot \boxed{5}$
$20 = 2 \cdot 2 \cdot \boxed{5}$
The GCF of 15 and 20 is 5.

2. $24 = \boxed{2} \cdot 2 \cdot 2 \cdot \boxed{3}$
$30 = \boxed{2} \cdot \boxed{3} \cdot 5$
The GCF of 24 and 30 is 2 · 3 or 6.

3. $2ab = \boxed{2} \cdot \boxed{a} \cdot b$
$6a^2 = \boxed{2} \cdot 3 \cdot \boxed{a} \cdot a$
The GCF of $2ab$ and $6a^2$ is $2 \cdot a$ or $2a$.

4. $\frac{2}{44} = \frac{\overset{1}{\cancel{2}}}{\underset{1}{\cancel{2}} \cdot \underset{1}{\cancel{2}} \cdot 11}$
$= \frac{1}{22}$
You were absent $\frac{1}{22}$ of the days.

5. $\frac{6}{15} = \frac{2 \cdot \overset{1}{\cancel{3}}}{\underset{1}{\cancel{3}} \cdot 5}$
$= \frac{2}{5}$
You responded to $\frac{2}{5}$ of the messages.

6. $4^2 \cdot 4^4 = 4^{2+4}$
$= 4^6$

7. $(n^4)(-2n^3) = [1 \cdot (-2)](n^{4+3})$
$= -2n^7$

8. $\frac{q^9}{q^4} = q^{9-4}$
$= q^5$

9. $b^{-6} = \frac{1}{b^6}$

10. $x^{-5} = (-2)^5$
$= \frac{1}{(-2)^5}$
$= -\frac{1}{32}$

4-8 Scientific Notation

Page 186 Why is scientific notation an important tool in comparing real-world data?

a. Since 1000 mm = 1 m, 5000 meters = 5000 · 1000 or 5,000,000 mm.

b. Since 1 micron = 0.001 mm, 0.5 micron = 0.5 · (0.001) or 0.0005 mm.

Page 188 Check for Understanding

1. Sample answer: Numbers that are greater than 1 can be expressed as the product of a factor and a positive power of 10. So, these numbers are written in scientific notation using positive exponents. Numbers between 0 and 1 cannot be expressed as the product of a factor and a whole number power of 10, so they are written in scientific notation using negative exponents.

2. Sample answer: 312 is in standard form. First, move the decimal point in 312 two places to the left, 3.12 × 100. Next, write 100 using a positive exponent, 3.12×10^2.

3. $3.08 \times 10^{-4} = 3.08 \times 0.0001$
$= 0.000308$

4. $1.4 \times 10^2 = 1.4 \times 100$
$= 140$

5. $8.495 \times 10^5 = 8.495 \times 100{,}000$
$= 849{,}500$

6. $80{,}000{,}000 = 8.0 \times 10{,}000{,}000$
$= 8.0 \times 10^7$

7. $697{,}000 = 6.97 \times 100{,}000$
$= 6.97 \times 10^5$

8. $0.059 = 5.9 \times 0.01$
$= 5.9 \times 10^{-2}$

9. $0.001 = 1.0 \times 0.001$
$= 1.0 \times 10^{-3}$

10. $d = rt$

$1.55 \times 10^8 = (3.0 \times 10^5)t$

$\frac{1.55 \times 10^8}{3.0 \times 10^5} = \frac{(3.0 \times 10^5)t}{3.0 \times 10^5}$

$0.517 \times 10^3 \approx t$

$t \approx 517$

It takes about 517 seconds.

11. First, order the numbers according to their exponents. Then, order the numbers with the same exponent by comparing the factors.

$6.790 \times 10^3 < \begin{matrix} 1.208 \times 10^4 \leftarrow \text{Venus} \\ 1.276 \times 10^4 \leftarrow \text{Earth} \end{matrix}$

$1.208 \times 10^4 < 1.276 \times 10^4$

The order from least to greatest is Mars, Venus, Earth.

Pages 189–190 Practice and Apply

12. $4.24 \times 10^2 = 4.24 \times 100 = 424$

13. $5.72 \times 10^4 = 5.72 \times 10{,}000 = 57{,}200$

14. $3.347 \times 10^{-1} = 3.347 \times 0.1 = 0.3347$

15. $5.689 \times 10^{-3} = 5.689 \times 0.001 = 0.005689$

16. $6.1 \times 10^4 = 6.1 \times 10{,}000 = 61{,}000$

17. $9.01 \times 10^{-2} = 9.01 \times 0.01 = 0.0901$

18. $1.399 \times 10^5 = 1.399 \times 100{,}000 = 139{,}900$

19. $2.505 \times 10^3 = 2.505 \times 1000 = 2505$

20. $1.5 \times 10^{-4} = 1.5 \times 0.0001 = 0.00015$

21. $2{,}000{,}000 = 20 \times 1{,}000{,}000 = 20 \times 10^6$

22. $499{,}000 = 4.99 \times 100{,}000 = 4.99 \times 10^5$

23. $0.006 = 6.0 \times 0.001 = 6.0 \times 10^{-3}$

24. $0.0125 = 1.25 \times 0.01 = 1.25 \times 10^{-2}$

25. $50{,}000{,}000 = 5.0 \times 10{,}000{,}000 = 5.0 \times 10^7$

26. $39{,}560 = 3.956 \times 10{,}000 = 3.956 \times 10^4$

27. $5{,}894{,}000 = 5.894 \times 1{,}000{,}000 = 5.894 \times 10^6$

28. $0.000078 = 7.8 \times 0.00001 = 7.8 \times 10^{-5}$

29. $0.000425 = 4.25 \times 0.0001 = 4.25 \times 10^{-4}$

30. $0.00031 = 3.1 \times 0.0001 = 3.1 \times 10^{-4}$

31. 6.25 million $= 6{,}250{,}000 = 6.25 \times 1{,}000{,}000 = 6.25 \times 10^6$

32. 3 billion $= 3{,}000{,}000{,}000 = 3.0 \times 1{,}000{,}000{,}000 = 3.0 \times 10^9$

33. $0.000000753 = 7.53 \times 0.0000001 = 7.53 \times 10^{-7}$

34. $d = rt$

$1.03 \times 10^8 = (3.0 \times 10^5)t$

$\frac{1.03 \times 10^8}{3.0 \times 10^5} = \frac{(3.0 \times 10^5)t}{(3.0 \times 10^5)}$

$0.343 \times 10^3 \approx t$

$343 \text{ s} \approx t$

35. 2.3×10^5 because $2.3 > 1.7$.

36. 1.8×10^3 because $10^3 > 10^{-1}$.

37. 5000 because $5.2 \times 10^2 = 520$ and $520 < 5000$.

38. 1.6×10^{-1} because $0.012 = 1.2 \times 10^{-2}$ and $10^{-2} < 10^{-1}$.

39. $5.44 \times 10^6 < \begin{matrix} 3.18 \times 10^7 \leftarrow \text{Atlantic} \\ 2.89 \times 10^7 \leftarrow \text{Indian} \\ 6.40 \times 10^7 \leftarrow \text{Pacific} \end{matrix}$

$2.89 \times 10^7 < 3.18 \times 10^7 < 6.40 \times 10^7$

Arctic, Indian, Atlantic, Pacific

40. $10^{15} > 10^9 > 10^3 > 10^{-9} > 10^{-12} > 10^{-18}$

petameter, gigameter, kilometer, nanometer, picometer, attometer

41. $6100 = 6.1 \times 10^3$

$0.0061 = 6.1 \times 10^{-3}$

$6.1 \times 10^{-5} < 6.1 \times 10^{-3} < 6.1 \times 10^{-2} < 6.1 \times 10^3 < 6.1 \times 10^4$

42. $(6 \times 10^0) = 6 \times 1 = 6$

$(4 \times 10^{-3}) = 4 \times 0.001 = 0.004$

$(3 \times 10^{-5}) = 3 \times 0.00001 = 0.00003$

$6 + 0.004 + 0.00003 = 6.00403$

43. $(4 \times 10^4) = 4 \times 10{,}000 = 40{,}000$

$(8 \times 10^3) = 8 \times 1000 = 8000$

$(3 \times 10^2) = 3 \times 100 = 300$

$(9 \times 10^1) = 9 \times 10 = 90$

$(6 \times 10^0) = 6 \times 1 = 6$

$40{,}000 + 8000 + 300 + 90 + 6 = 48{,}396$

44. $\frac{20{,}000}{0.01} = \frac{2 \times 10^4}{1 \times 10^{-2}}$

$= 2 \times 10^6$; 2,000,000

45. $\frac{(420{,}000)(0.015)}{0.025} = \frac{(4.2 \times 10^5)(1.5 \times 10^{-2})}{2.5 \times 10^{-2}}$

$= \frac{6.3 \times 10^3}{2.5 \times 10^{-2}}$

$= 2.52 \times 10^5$; 252,000

46. $\frac{(0.078)(8.5)}{0.16(250{,}000)} = \frac{(7.8 \times 10^{-2})(8.5 \times 10^0)}{(1.6 \times 10^{-1})(2.5 \times 10^5)}$

$= \frac{66.3 \times 10^{-2}}{4.0 \times 10^4}$

$= 16.575 \times 10^{-6}$

$= 1.6575 \times 10^{-5}$; 0.000016575

47. $4.0 \times 10^4 > 3.0 \times 10^4 > 2.0 \times 10^4 > 4.0 \times 10^3 > 2.0 \times 10^3$

Bezymianny; Santa Maria; Agung; Mount St. Helens tied with Helka 1947; Helka, 1970; Ngauruhoe

48. $\frac{4.0 \times 10^4}{2.0 \times 10^4} = 2$ times

49. $3.14 \times 10^0 = 3.14 \times 1$
$= 3.14$

50. Scientific notation is a shorthand way of writing very large or very small numbers. Answers should include the following.

- Examples of data that can be written in scientific notation, such as the surface area of planets and the diameter of atoms.
- Scientific notation is useful because you can compare quantities by simply looking at the exponent of the power of 10, rather than counting decimal places.

51. B; $4.7 \times 10^3 < 5.7 \times 10^4 < 6.9 \times 10^4 < 4.4 \times 10^5$
Lake Victoria is third in the list.

52. D; $5.80 \times 10^{-4} = 5.80 \times 0.0001$
$= 0.00058$

Page 190 Maintain Your Skills

53. $t^{-4} = 3^{-4}$
$= \frac{1}{3^4}$
$= \frac{1}{81}$

54. $s^{-5} = (-2)^{-5}$
$= \frac{1}{(-2)^5}$
$= -\frac{1}{32}$

55. $7^s = 7^{-2}$
$= \frac{1}{7^2}$
$= \frac{1}{49}$

56. $4^4 \cdot 4^7 = 4^{4+7}$
$= 4^{11}$

57. $3a^2 \cdot 5a^2 = (3 \cdot 5)(a^2 \cdot a^2)$
$= 15(a^{2+2})$
$= 15a^4$

58. $c^5 \div c^2 = \frac{c^5}{c^2}$
$= c^{5-2}$
$= c^3$

59. c = cost of books
\$2.50 = shipping
total cost of order = $c + \$2.50$

Chapter 4 Study Guide and Review

Page 191 Vocabulary and Concept Check

1. true
2. true
3. true
4. false; product
5. true
6. true
7. true
8. false; 1

Pages 191–194 Lesson-by-Lesson Review

9. 2—no, the ones digit is 1, and 1 is not divisible by 2.
3—yes, 1 + 1 + 1 = 3, and 3 is divisible by 3.
5—no, the ones digit is not 0 or 5.
6—no, 111 is not divisible by 2 and 3.
10—no, the ones digit is not 0.
So, 111 is divisible by 3.

10. 2—no, the ones digit is 5, and 5 is not divisible by 2.
3—yes, 4 + 0 + 5 = 9, and 9 is divisible by 3.
5—yes, the ones digit is 0 or 5.
6—no, 405 is not divisible by 2 and 3.
10—no, the ones digit is not 0.
So, 405 is divisible by 3 and 5.

11. 2—no, the ones digit is 5, and 5 is not divisible by 2.
3—no, 6 + 3 + 5 = 14, and 14 is not divisible by 3.
5—yes, the ones digit is 0 or 5.
6—no, 635 is not divisible by 2 and 3.
10—no, the ones digit is not 0.
So, 635 is divisible by 5.

12. 2—no, the ones digit is 3, and 3 is not divisible by 2.
3—no, 8 + 6 + 3 = 17, and 17 is not divisible by 3.
5—no, the ones digit is not 0 or 5.
6—no, 863 is not divisible by 2 and 3.
10—no, the ones digit is not 0.
So, 863 is divisible by none of these numbers.

13. 2—yes, the ones digit is 2, and 2 is divisible by 2.
3—yes, 5 + 8 + 2 = 15, and 15 is divisible by 3.
5—no, the ones digit is not 0 or 5.
6—yes, 582 is divisible by 2 and 3.
10—no, the ones digit is not 0.
So, 582 is divisible by 2, 3, and 6.

14. 2—yes, the ones digit is 4, and 4 is divisible by 2.
3—yes, 2 + 1 + 2 + 4 = 9, and 9 is divisible by 3.
5—no, the ones digit is not 0 or 5.
6—yes, 2124 is divisible by 2 and 3.
10—no, the ones digit is not 0.
So, 2124 is divisible by 2, 3, and 6.

15. 2—yes, the ones digit is 0, and 0 is divisible by 2.
3—no, 7 + 0 + 0 = 7, and 7 is not divisible by 3.
5—yes, the ones digit is 0 or 5.
6—no, 700 is not divisible by 2 and 3.
10—yes, the ones digit is 0.
So, 700 is divisible by 2, 5, and 10.

16. 2—yes, the ones digit is 0, and 0 is divisible by 2.
3—yes, 4 + 2 + 0 + 0 = 6, and 6 is divisible by 3.
5—yes, the ones digit is 0 or 5.
6—yes, 4200 is divisible by 2 and 3.
10—yes, the ones digit is 0.
So, 4200 is divisible by 2, 3, 5, 6, and 10.

17. $3^3 = 3 \cdot 3 \cdot 3$
$= 27$

18. $10^4 = 10 \cdot 10 \cdot 10 \cdot 10$
$= 10{,}000$

19. $(-5)^2 = (-5)(-5)$
$= 25$

20. $y^3 = 4^3$
$= 4 \cdot 4 \cdot 4$
$= 64$

21. $10x^2 = 10(-3)^2$
$= 10(-3)(-3)$
$= 90$

22. $xy^3 = (-3)(4)^3$
$= (-3)(4)(4)(4)$
$= -192$

23. $7y^0z^4 = (7)(4)^0(-2)^4$
$= (7)(1)(-2)(-2)(-2)(-2)$
$= 112$

24. $2(3z + 4)^5 = 2[3(-2) + 4]^5$
$= 2(-6 + 4)^5$
$= 2(-2)^5$
$= 2 \cdot (-2)(-2)(-2)(-2)(-2)$
$= -64$

25.
45
/\
$3 \cdot 15$
/ /\
$3 \cdot 3 \cdot 5$

The prime factorization of 45 is $3 \cdot 3 \cdot 5$ or $3^2 \cdot 5$.

26. $55 = 5 \cdot 11$
/\
$5 \cdot 11$

The prime factorization of 55 is $5 \cdot 11$.

27. $68 = 2^2 \cdot 17$
/\
$4 \cdot 17$
/\ \
$2 \cdot 2 \cdot 17$

The prime factorization of 68 is $2 \cdot 2 \cdot 17$ or $2^2 \cdot 17$.

28.
200
/\
$4 \cdot 50$
/| /\
$2 \cdot 2 \cdot 2 \cdot 25$
/ / / /\
$2 \cdot 2 \cdot 2 \cdot 5 \cdot 5$

The prime factoization of 200 is $2 \cdot 2 \cdot 2 \cdot 5 \cdot 5$ or $2^3 \cdot 5^2$.

29. $49k = 7 \cdot 7 \cdot k$

30. $-15n^2 = -1 \cdot 3 \cdot 5 \cdot n \cdot n$

31. $26p^3 = 2 \cdot 13 \cdot p \cdot p \cdot p$

32. $10a^2b = 2 \cdot 5 \cdot a \cdot a \cdot b$

33. $6 = \boxed{2} \cdot \boxed{3}$
$48 = \boxed{2} \cdot 2 \cdot 2 \cdot 2 \cdot \boxed{3}$

The GCF of 6 and 48 is $2 \cdot 3$ or 6.

34. $16 = \boxed{2} \cdot \boxed{2} \cdot \boxed{2} \cdot 2$
$24 = \boxed{2} \cdot \boxed{2} \cdot \boxed{2} \cdot 3$

The GCF of 16 and 24 is $2 \cdot 2 \cdot 2$ or 8.

35. $4n = 2 \cdot 2 \cdot \boxed{n}$
$5n^2 = 5 \cdot \boxed{n} \cdot n$

The GCF of $4n$ and $5n^2$ is n.

36. $20c^3d = \boxed{2} \cdot \boxed{2} \cdot 5 \cdot \boxed{c} \cdot c \cdot c \cdot \boxed{d}$
$12cd = \boxed{2} \cdot \boxed{2} \cdot 3 \cdot \boxed{c} \cdot \boxed{d}$

The GCF of $20c^3d$ and $12cd$ is $2 \cdot 2 \cdot c \cdot d$ or $4cd$.

37. $2t = \boxed{2} \cdot t$
$20 = \boxed{2} \cdot 2 \cdot 5$

The GCF is 2.

$2t + 20 = 2(t) + 2(10)$
$= 2(t + 10)$

So, $2t + 20 = 2(t + 10)$.

38. $3x = \boxed{3} \cdot x$
$24 = 2 \cdot 2 \cdot 2 \cdot \boxed{3}$

The GCF is 3.

$3x + 24 = 3(x) + 3(8)$
$= 3(x + 8)$

So, $3x + 24 = 3(x + 8)$.

39. $30 = \boxed{2} \cdot 3 \cdot 5$
$4n = \boxed{2} \cdot 2 \cdot n$

The GCF is 2.

$30 + 4n = 2(15) + 2(2n)$
$= 2(15 + 2n)$

So, $30 + 4n = 2(15 + 2n)$.

40. $\frac{6}{21} = \frac{2 \cdot \cancel{3}^1}{\cancel{3}_1 \cdot 7}$
$= \frac{2}{7}$

41. $\frac{24}{40} = \frac{\cancel{2}^1 \cdot \cancel{2}^1 \cdot \cancel{2}^1 \cdot 3}{\cancel{2}_1 \cdot \cancel{2}_1 \cdot \cancel{2}_1 \cdot 5}$
$= \frac{3}{5}$

42. $\frac{15}{16} = \frac{3 \cdot 5}{2 \cdot 2 \cdot 2 \cdot 2}$; 15 and 16 have no common factors. $\frac{15}{16}$ is simplified.

43. $\frac{30}{51} = \frac{2 \cdot \cancel{3}^1 \cdot 5}{\cancel{3}_1 \cdot 17}$
$= \frac{10}{17}$

44. $\frac{st}{t^4} = \frac{s \cdot \cancel{t}^1}{\cancel{t}_1 \cdot t \cdot t \cdot t}$
$= \frac{s}{t^3}$

45. $\frac{23x}{32y} = \frac{23 \cdot x}{2 \cdot 2 \cdot 2 \cdot 2 \cdot 2 \cdot y}$; $23x$ and $32y$ have no common factors. $\frac{23x}{32y}$ is simplified.

46. $\frac{9mn}{18n^2} = \frac{\cancel{3}^1 \cdot \cancel{3}^1 \cdot m \cdot \cancel{n}^1}{2 \cdot \cancel{3}_1 \cdot \cancel{3}_1 \cdot \cancel{n}_1 \cdot n}$
$= \frac{m}{2n}$

47. $\frac{15ac^2}{24ab} = \frac{\cancel{3}^1 \cdot 5 \cdot \cancel{a}^1 \cdot c \cdot c}{2 \cdot 2 \cdot 2 \cdot \cancel{3}_1 \cdot \cancel{a}_1 \cdot b}$
$= \frac{5c^2}{8b}$

48. $8^4 \cdot 8^5 = 8^{4+5}$
$= 8^9$

49. $c \cdot c^3 = c^{1+3}$
$= c^4$

50. $\frac{3^7}{3^2} = 3^{7-2}$
$= 3^5$

51. $\frac{r^{11}}{r^9} = r^{11-9}$
$= r^2$

52. $7x \cdot 2x^6 = (7 \cdot 2)(x^1 \cdot x^6)$
$= 14(x^{1+6})$
$= 14x^7$

53. $7^{-2} = \frac{1}{7^2}$

54. $10^{-1} = \frac{1}{10^1}$ or $\frac{1}{10}$

55. $b^{-4} = \frac{1}{b^4}$

56. $t^{-8} = \frac{1}{t^8}$

57. $(-4)^{-3} = \frac{1}{(-4)^3}$

58. $6.1 \times 10^2 = 6.1 \times 100$
$= 610$

59. $2.9 \times 10^{-3} = 2.9 \times 0.001$
$= 0.0029$

60. $1.85 \cdot 10^{-2} = 1.85 \times 0.01$
$= 0.0185$

61. $7.045 \times 10^4 = 7.045 \times 10{,}000$
$= 70{,}450$

62. $1200 = 1.2 \times 1000$
$= 1.2 \times 10^3$

63. $0.008 = 8.0 \times 0.001$
$= 8.0 \times 10^{-3}$

64. $0.000319 = 3.19 \times 0.0001$
$= 3.19 \times 10^{-4}$

65. $45{,}710{,}000 = 4.571 \times 10{,}000{,}000$
$= 4.571 \times 10^7$

Chapter 4 Practice Test

Page 195

1. A number is divisible by: 2 if the ones digit is even; 3 if the sum of the digits is divisible by 3; 5 if the ones digit is 0 or 5; 6 if the number is divisible by 2 and 3; 10 if the ones digit is zero.
2. A prime number has exactly two factors, 1 and the number itself. A composite number has more than two factors.
3. Sample answer: $\frac{x}{2}$
4. yes; product of a number and variables
5. no; sum of two terms
6. $3 \cdot 3 \cdot 3 \cdot 3 = 3^4$
7. $-2 \cdot -2 \cdot -2 \cdot a \cdot a \cdot a \cdot a = (-2)^3 a^4$
8. $12r^2 = 2 \cdot 2 \cdot 3 \cdot r \cdot r$
9. $50xy^2 = 2 \cdot 5 \cdot 5 \cdot x \cdot y \cdot y$
10. $7 = 7$
$21p = 3 \cdot 7 \cdot p$
The GCF is 7.
$7 + 21p = 7(1) + 7(3p)$
$= 7(1 + 3p)$
So, $7 + 21p = 7(1 + 3p)$.

11. $70 = 2 \cdot 5 \cdot 7$
$28 = 2 \cdot 2 \cdot 7$
The GCF of 70 and 28 is $2 \cdot 7$ or 14.

12. $36 = 2 \cdot 2 \cdot 3 \cdot 3$
$90 = 2 \cdot 3 \cdot 3 \cdot 5$
$180 = 2 \cdot 2 \cdot 3 \cdot 3 \cdot 5$
The GCF of 36, 90, and 180 is $2 \cdot 3 \cdot 3$ or 18.

13. $12a^3b = 2 \cdot 2 \cdot 3 \cdot a \cdot a \cdot a \cdot b$
$40ab^4 = 2 \cdot 2 \cdot 2 \cdot 5 \cdot a \cdot b \cdot b \cdot b \cdot b$
The GCF of $12a^3b$ and $40ab^4$ is $2 \cdot 2 \cdot a \cdot b$ or $4ab$.

14. $\frac{57}{95} = \frac{3 \cdot \cancel{19}^1}{5 \cdot \cancel{19}_1}$
$= \frac{3}{5}$

15. $\frac{240}{360} = \frac{\cancel{2} \cdot \cancel{2} \cdot \cancel{2} \cdot 2 \cdot \cancel{3} \cdot \cancel{5}}{\cancel{2} \cdot \cancel{2} \cdot \cancel{2} \cdot \cancel{3} \cdot 3 \cdot \cancel{5}}$
$= \frac{2}{3}$

16. $\frac{56m^3n}{32mn} = \frac{\cancel{2} \cdot \cancel{2} \cdot \cancel{2} \cdot 7 \cdot \cancel{m} \cdot m \cdot m \cdot \cancel{n}}{\cancel{2} \cdot \cancel{2} \cdot \cancel{2} \cdot 2 \cdot 2 \cdot \cancel{m} \cdot \cancel{n}}$
$= \frac{7m^2}{4}$

17. $5^3 \cdot 5^6 = 5^{3+6}$
$= 5^9$

18. $(4x^7)(-6x^3) = (4 \cdot -6)(x^7 \cdot x^3)$
$= -24(x^{7+3})$
$= -24x^{10}$

19. $w^9 \div w^5 = \frac{w^9}{w^5}$
$= w^{9-5}$
$= w^4$

20. $4^{-2} = \frac{1}{4^2}$

21. $t^{-6} = \frac{1}{t^6}$

22. $(yz)^{-3} = \frac{1}{(yz)^3}$

23. $9.0 \times 10^{-2} = 9.0 \times 0.01$
$= 0.09$

24. $5.206 \times 10^{-3} = 5.206 \times 0.001$
$= 0.005206$

25. $3.71 \times 10^4 = 3.71 \times 10{,}000$
$= 37{,}100$

26. $345{,}000 = 3.45 \times 100{,}000$
$= 3.45 \times 10^5$

27. $1{,}680{,}000 = 1.680 \times 1{,}000{,}000$
$= 1.680 \times 10^6$

28. $0.00072 = 7.2 \times 0.0001$
$= 7.2 \times 10^{-4}$

29. $\frac{12}{16} = \frac{\cancel{2} \cdot \cancel{2} \cdot 3}{\cancel{2} \cdot \cancel{2} \cdot 2 \cdot 2}$
$= \frac{3}{4}$

30. A; 93 million = 93,000,000
$= 9.3 \times 10{,}000{,}000$
$= 9.3 \times 10^7$

Chapter 4 Standardized Test Practice

Pages 196–197

1. D; Let x = Ling's computer games,
   $x - 7$ = Andy's computer games, and
   $2(x - 7)$ = Carlos' computer game.
2. D; subtraction is not commutative.
3. B; $\frac{-2+0+4+5+4}{5} = \frac{11}{5} = 2.2°F$
4. A; start at the origin and move about 13 units left and 7 units up. The coordinates are about (−13, 7).
5. B; $3n - 6 = -39$
   $3n - 6 + 6 = -39 + 6$
   $3n = -33$
   $\frac{3n}{3} = \frac{-33}{3}$
   $n = -11$
6. D; n = the number of items
   $\$3n$ = cost for shipping n items
   \$5 = handling charge
   $3n + 5$ = the cost for shipping and handling of n items
7. C; $A = \ell w$
   $18 = 6w$
   $\frac{18}{6} = \frac{6w}{6}$
   $3 = w$
   The width is 3 units.
8. D; 84
   4 · 21
   2 · 2 · 3 · 7
   The prime factorization of 84 is 2 · 2 · 3 · 7 or $2^2 \cdot 3 \cdot 7$.
9. C; 28 = 2 · 2 · 7
   42 = 2 · 3 · 7
   The GCF of 28 and 42 is 2 · 7 or 14.
10. C; $3^{-3} = \frac{1}{3^3}$
    $= \frac{1}{27}$
11. B; 17,400,000 = 1.74 × 10,000,000
    $= 1.74 \times 10^7$
12. Total membership cost
    = (initial fee) + (monthly membership fee) × (number of months after first month)
    = 60 + (28) × (8)
    = 60 + 224
    = \$284
13. $-3 - 9 = -12$ or −12 degrees
14. Let x = the number of hours he can skate.
    $2 + x(1.50) = \$8.00$
    $2 - 2 + 1.50x = 8 - 2$
    $1.50x = 6$
    $\frac{1.50x}{1.50} = \frac{6}{1.50}$
    $x = 4$ hours
15. Let x = the number of friends who got 2 gel pens.
    $6 \cdot 1 + 4 \cdot 3 + x \cdot 2 = 30$
    $6 + 12 + 2x = 30$
    $18 + 2x = 30$
    $18 - 18 + 2x = 30 - 18$
    $2x = 12$
    $\frac{2x}{2} = \frac{12}{2}$
    $x = 6$
    So, 6 friends got 2 gel pens.
16. The length and width need to be increased by 1 foot on each side for the part that will hang down.
    length = 8 + 1 + 1 or 10
    width = 2 + 1 + 1 or 4
    $A = \ell w$
    $A = 10 \cdot 4$
    $A = 40$
    They need 40 square feet of covering.
17. Since 15 is the least 2-digit number divisible by both 3 and 5, you can find the smallest multiple of 15 to give you a 3-digit number.
    15 · 1 = 15, 15 · 2 = 30, 15 · 3 = 45, 15 · 4 = 60, 15 · 5 = 75, 15 · 6 = 90, 15 · 7 = 105.
    So, 105 is the least 3-digit number that is divisible by 3 and 5.
18. $10 \cdot 10 \cdot 10 \cdot 10 = 10^4$
19. 18 = 2 · 3 · 3
    44 = 2 · 2 · 11
    12 = 2 · 2 · 3
    The GCF of 18, 44, and 12 is 2.
20. $\frac{1}{5 \cdot 5 \cdot 5} = \frac{1}{5^3}$
    $= 5^{-3}$
21. 200 billion = 200,000,000,000
    = 2 × 100,000,000,000
    $= 2 \times 10^{11}$

22a. CDBargains $S = 4 + n$; WebShopper $S = 6 + 3n$; EverythingStore $S = 2.5 + 1.5n$

22b. Let $n = 2$. CDBargains $S = 4 + 2$ or \$6; WebShopper $S = 6 + 3(2)$ or \$12; EverythingStore $S = 2.5 + 1.5(2)$ or \$5.50; EverythingStore has the least shipping cost.

22c. Let $n = 10$. CDBargains $S = 4 + 10$ or \$14; WebShopper $S = 6 + 3(10)$ or \$36; EverythingStore $S = 2.5 + 1.5(10)$ or \$17.50; CDBargains has the least shipping cost.

22d. $4 + n = 2.5 + 1.5n$
$4 - 2.5 + n = 2.5 - 2.5 + 1.5n$
$1.5 + n = 1.5n$
$1.5 + n - n = 1.5n - n$
$1.5 = 0.5n$
$\frac{1.5}{0.5} = \frac{0.5n}{0.5}$
$3 = n$
The charges are the same for 3 CDs.

Chapter 5 Rational Numbers

Page 199 Getting Started

1. $\begin{array}{r} 0.6 \\ 5\overline{)3.0} \\ \underline{-30} \\ 0 \end{array}$

   $3 \div 5 = 0.6$

2. $\begin{array}{r} 0.12 \\ 8\overline{)1.00} \\ \underline{-8} \\ 20 \\ \underline{-16} \\ 4 \end{array}$

   $-1 \div 8 \approx -0.1$

3. $2 \cdot 17 = 34$
4. $-12 \cdot 3 = -36$
5. $\begin{array}{r} 0.22 \\ 9\overline{)2.00} \\ \underline{-18} \\ 20 \\ \underline{-18} \\ 2 \end{array}$

   $-2 \div (-9) \approx 0.2$

6. $-4(-6) = 24$
7. $5(-15) = -75$
8. $\begin{array}{r} 0.26 \\ 15\overline{)4.00} \\ \underline{-3\ 0} \\ 1\ 00 \\ \underline{-90} \\ 10 \end{array}$

   $4 \div (-15) \approx -0.3$

9. $\begin{array}{r} 1.71 \\ 14\overline{)24.00} \\ \underline{-14} \\ 100 \\ \underline{-98} \\ 20 \\ \underline{-14} \\ 6 \end{array}$

   $-24 \div 14 = -1.7$

10. $\frac{5}{40} = \frac{1 \times \cancel{5}^{1}}{8 \times \cancel{5}_{1}}$

    $= \frac{1}{8}$

11. $\frac{12}{20} = \frac{3 \times \cancel{4}^{1}}{5 \times \cancel{4}_{1}}$

    $= \frac{3}{5}$

12. $\frac{14}{39}$ is simplified.
13. $\frac{36}{50} = \frac{18 \times \cancel{2}^{1}}{25 \times \cancel{2}_{1}}$

    $= \frac{18}{25}$

14. $4 + (-9) = -5$
15. $-10 + 16 = 6$
16. $20 - 12 = 8$
17. $19 - 32 = 19 + (-32)$

    $= -13$

18. $7 + (-5) = 2$
19. $26 - 11 = 15$
20. $(-3) + (-8) = -11$
21. $-1 - (-10) = -1 + 10$

    $= 9$

5-1 Writing Fractions as Decimals

Page 200 How were fractions used to determine the size of the first coins?

a. $0.50

b. Quarter, $0.25; dime, $0.10; nickel, $0.05.

c. $\frac{1}{20}, \frac{1}{10}, \frac{1}{4}$

Page 202 Check for Understanding

1. Sample answer: write the fractions as decimals and then compare.
2. $0.5 = 0.50$ and $0.\overline{5} = 0.55...$; $0.\overline{5}$ is greater because in the hundredths place, $5 > 0$.
3. Sample answer: $0.\overline{14}$
4. $\begin{array}{r} 0.875 \\ 8\overline{)7.000} \\ \underline{-6\ 4} \\ 60 \\ \underline{-56} \\ 40 \\ \underline{-40} \\ 0 \end{array}$

   0.875 is a terminating decimal.

5. $2\frac{2}{25} = 2 + \frac{2}{25}$

   $\begin{array}{r} 0.08 \\ 25\overline{)2.00} \\ \underline{-2\ 00} \\ 0 \end{array}$

   $2 + 0.08 = 2.08$

   $2\frac{2}{25} = 2.08$

6. $-\frac{5}{9} \rightarrow \begin{array}{r} 0.555... \\ 9\overline{)5.000...} \end{array}$

   So, $-\frac{5}{9} = -0.\overline{5}$.

7. $\frac{4}{15} \rightarrow \begin{array}{r} 0.266... \\ 15\overline{)4.000...} \end{array}$

   So, $\frac{4}{15} = 0.2\overline{6}$.

8. $\frac{9}{10}$ ● 0.90

   0.90 ● 0.90

   $0.90 = 0.90$

9. $0.3 \bullet \frac{1}{3}$

$0.3 \bullet 0.\overline{3}$

$0.3 < 0.\overline{3}$

10. $\frac{1}{4} \bullet \frac{1}{3}$

$0.25 \bullet 0.\overline{3}$

$0.25 < 0.\overline{3}$

11. $-\frac{3}{4} \bullet -\frac{7}{8}$

$-0.75 \bullet -0.875$

$-0.75 > -0.875$

12. pickups: $\frac{1}{5} = 0.2$

SUVs: 0.17

On a number line, 0.17 is to the left of 0.2. Since $0.17 < 0.2$, $0.17 < \frac{1}{5}$.

So, more pickups are sold.

Pages 203–204 Practice and Apply

13. $5\overline{)1.0}$ = 0.2

$-1\,0$

0

0.2 is a terminating decimal.

14. $20\overline{)3.00}$ = 0.15

$-2\,0$

$1\,00$

$-1\,00$

0

0.15 is a terminating decimal.

15. $25\overline{)8.00}$ = 0.32

$-7\,5$

50

-50

0

0.32 is a terminating decimal.

16. $-8\overline{)5.000}$ = 0.625

$-4\,8$

20

-16

40

-40

0

-0.625 is a terminating decimal.

17. $7\frac{3}{10} = 7 + \frac{3}{10}$

$= 7 + 0.3$

$= 7.3$

18. $1\frac{1}{2} = 1 + \frac{1}{2}$

$= 1 + 0.5$

$= 1.5$

19. $5\frac{1}{8} = 5 + \frac{1}{8}$

$8\overline{)1.000}$ = 0.125

-8

20

-16

40

-40

0

$5 + 0.125 = 5.125$

$5\frac{1}{8} = 5.125$

20. $-3\frac{3}{4} = -\left(3 + \frac{3}{4}\right)$

$-(3 + 0.75) = -3.75$

$-3\frac{3}{4} = -3.75$

21. $\frac{1}{9} \rightarrow 9\overline{)1.000...}$ = 0.111...

So, $\frac{1}{9} = 0.\overline{1}$.

22. $-\frac{2}{9} \rightarrow -9\overline{)2.000...}$ = 0.22...

So, $-\frac{2}{9} = -0.\overline{2}$.

23. $-\frac{5}{11} \rightarrow -11\overline{)5.0000...}$ = 0.4545...

So, $-\frac{5}{11} = -0.\overline{45}$.

24. $\frac{4}{11} \rightarrow 11\overline{)4.0000...}$ = 0.3636...

So, $\frac{4}{11} = 0.\overline{36}$.

25. $\frac{1}{6} \rightarrow 6\overline{)1.000...}$ = 0.166...

So, $\frac{1}{6} = 0.1\overline{6}$.

26. $15\overline{)7.000...}$ = 0.466...

So, $\frac{7}{15} = 0.4\overline{6}$.

27. $16\overline{)5.0000}$ = 0.3125

$-4\,8$

20

-16

40

-32

80

-80

0

0.3125 is a terminating decimal.

28. $\frac{7}{12} \rightarrow 12\overline{)7.0000...}$ = 0.5833...

So, $\frac{7}{12} = 0.58\overline{3}$.

29. $\frac{5}{6} \rightarrow 6\overline{)5.000...}$ = 0.833...

So, $\frac{5}{6} = 0.8\overline{3} \approx 0.83$.

30. 17 out of 20 $= \frac{17}{20}$

$$\begin{array}{r} 0.85 \\ 20\overline{)17.00} \\ \underline{-16\;0} \\ 1\;00 \\ \underline{-1\;00} \\ 0 \end{array}$$

17 out of 20 $= 0.85$

31. $\frac{7}{8} \rightarrow$
$$\begin{array}{r} 0.875 \\ 8\overline{)7.000} \\ \underline{-6\;4} \\ 60 \\ \underline{-56} \\ 40 \\ \underline{-40} \\ 0 \end{array}$$

$\frac{7}{8} = 0.875$

$\frac{7}{9} \rightarrow$
$$\begin{array}{r} 0.77 \\ 9\overline{)7.00} \\ \underline{-6\;3} \\ 70 \\ \underline{-63} \\ 7 \end{array}$$

$\frac{7}{9} = 0.\overline{7}$

Since $0.\overline{7} < 0.8 < 0.875$,

$\frac{7}{9} < 0.8 < \frac{7}{8}$.

Therefore, the correct order is $\frac{7}{9}$, 0.8, $\frac{7}{8}$.

32. $0.3 \bigcirc \frac{1}{4}$

$0.3 \bigcirc 0.25$

$0.3 > 0.25$

33. $\frac{5}{8} \bigcirc 0.65$

$0.625 \bigcirc 0.65$

$0.625 < 0.65$

34. $\frac{2}{5} \bigcirc 0.4$

$0.4 \bigcirc 0.4$

$0.4 = 0.4$

35. $\frac{1}{3} \bigcirc \frac{1}{2}$

$0.\overline{3} \bigcirc 0.5$

$0.\overline{3} < 0.5$

36. $\frac{1}{5} \bigcirc 0.\overline{5}$

$0.2 \bigcirc 0.\overline{5}$

$0.2 < 0.\overline{5}$

37. $1\frac{1}{20} \bigcirc 1.01$

$1.05 \bigcirc 1.01$

$1.05 > 1.01$

38. $\frac{7}{8} \bigcirc \frac{8}{9}$

$0.875 \bigcirc 0.\overline{8}$

$0.875 < 0.\overline{8}$

39. $3\frac{4}{9} \bigcirc 3.\overline{4}$

$3.\overline{4} \bigcirc 3.\overline{4}$

$3.\overline{4} = 3.\overline{4}$

40. $6.18 \bigcirc 6\frac{1}{5}$

$6.18 \bigcirc 6.2$

$6.18 < 6.2$

41. $-0.75 \bigcirc -\frac{7}{9}$

$-0.75 \bigcirc -0.\overline{7}$

$-0.75 > -0.\overline{7}$

42. $0.3\overline{4} \bigcirc \frac{34}{99}$

$0.3\overline{4} \bigcirc 0.\overline{34}$

$0.3\overline{4} > 0.\overline{34}$

43. $-2\frac{1}{12} \bigcirc -2.09$

$-2.08\overline{3} \bigcirc -2.09$

$-2.08\overline{3} > -2.09$

44. $\frac{11}{15} = 0.7\overline{3}$

$\frac{3}{4} = 0.75$

Since $0.7\overline{3} < 0.75$, $\frac{11}{15} < \frac{3}{4}$.

$\frac{11}{15}$ would be to the left of $\frac{3}{4}$.

45. Sample answer: 0.7 and $0.\overline{7}$; $\frac{1}{6} = 0.1\overline{6}$ and $\frac{8}{9} = 0.\overline{8}$; 0.7 and $0.\overline{7}$ are both greater than $0.1\overline{6}$ and less than $0.\overline{8}$.

46. $\frac{1}{4} = 0.25$

$0.28 > 0.25$

Therefore, more than $\frac{1}{4}$ of students chose math.

47. $\frac{1}{7} \approx 0.14$

$0.14 > 0.13$

Therefore, this is greater than those who chose English in the survey.

48a. $2 = 2, 3 = 3, 4 = 2^2, 5 = 5, 6 = 2 \cdot 3, 8 = 2^3,$
$9 = 3^2, 10 = 2 \cdot 5, 12 = 2^2 \cdot 3, 15 = 3 \cdot 5,$
$20 = 2^2 \cdot 5$

48b. $\frac{1}{2} = 0.5, \frac{1}{3} = 0.\overline{3}, \frac{1}{4} = 0.25, \frac{1}{5} = 0.2, \frac{1}{6} = 0.1\overline{6},$
$\frac{1}{8} = 0.125, \frac{1}{9} = 0.\overline{1}, \frac{1}{10} = 0.1, \frac{1}{12} = 0.08\overline{3},$
$\frac{1}{15} = 0.0\overline{6}, \frac{1}{20} = 0.05$

48c. Sample answer: Fractions whose denominators have only 2 or 5 as prime factors are terminating decimals. Fractions whose denominators have 3 as a prime factor are repeating decimals.

49. All coins were made with a fraction of silver that was contained in a silver dollar. Answers should include the following.

- A quarter had one-fourth the amount of silver as a silver dollar, a dime had one-tenth the amount, and a nickel had one-twentieth the amount.
- It is easier to perform arithmetic operations using decimals rather than using fractions.

50. B;
$$\begin{array}{r} 0.01 \\ 100\overline{)1.00} \\ \underline{-1\;00} \\ 0 \end{array}$$

$\frac{1}{100} = 0.01$

51. D; 19 out of 30 squares are shaded.

$$\frac{19}{30} = 30\overline{)19.000}$$

0.633
−18 0
1 00
−90
100
−90
10

$\frac{19}{30} = 0.6\overline{3}$

Page 204 Maintain Your Skills

52. $854{,}000{,}000 = 8.54 \times 100{,}000{,}000$
$= 8.54 \times 10^8$

53. $0.077 = 7.7 \times 0.01$
$= 7.7 \times 10^{-2}$

54. $0.00016 = 1.6 \times 0.0001$
$= 1.6 \times 10^{-4}$

55. $925{,}000 = 9.25 \times 100{,}000$
$= 9.25 \times 10^5$

56. $10^{-5} = \frac{1}{10^5}$

57. $(-2)^{-7} = \frac{1}{(-2)^7}$

58. $x^{-4} = \frac{1}{x^4}$

59. $y^{-3} = \frac{1}{y^3}$

60. $(a \cdot a \cdot a)(a \cdot a) = a^3 \cdot a^2$
$= a^{3+2}$
$= a^5$

61. $16\overline{)464}$

29
−32
144
−144
0

The car gets 29 miles per gallon.

62. $4n = 32$
$\frac{4n}{4} = \frac{32}{4}$
$n = 8$
The solution is 8.

63. $-64 = 2t$
$\frac{-64}{2} = \frac{2t}{2}$
$-32 = t$
The solution is −32.

64. $\frac{9}{5} = -9$
$(5)\frac{a}{5} = -9(5)$
$a = -45$
The solution is −45.

65. $-8 = \frac{x}{-7}$
$(-7)-8 = \frac{x}{-7}(-7)$
$56 = x$
The solution is 56.

66. $\frac{4}{30} = \frac{2 \times 2}{15 \times 2}$
$= \frac{2}{15}$

67. $\frac{5}{65} = \frac{1 \times \cancel{5}^1}{13 \times \cancel{5}_1}$
$= \frac{1}{13}$

68. $\frac{36}{60} = \frac{3 \times \cancel{12}^1}{5 \times \cancel{12}_1}$
$= \frac{3}{5}$

69. $\frac{12}{18} = \frac{2 \times \cancel{6}^1}{3 \times \cancel{6}_1}$
$= \frac{2}{3}$

70. $\frac{21}{24} = \frac{3 \times \cancel{7}^1}{4 \times \cancel{7}_1}$
$= \frac{3}{4}$

71. $\frac{16}{28} = \frac{4 \times \cancel{4}^1}{7 \times \cancel{4}_1}$
$= \frac{4}{7}$

72. $\frac{32}{48} = \frac{2 \times \cancel{16}^1}{3 \times \cancel{16}_1}$
$= \frac{2}{3}$

73. $\frac{125}{1000} = \frac{1 \times \cancel{125}^1}{8 \times \cancel{125}_1}$
$= \frac{1}{8}$

5-2 Rational Numbers

Page 205 How are rational numbers related to other sets of numbers?

a. yes; yes; yes

b. The set of rational numbers include natural numbers, whole numbers, and integers.

c. Yes; any natural number can be written as a fraction with a denominator of 1. No; $\frac{8}{9}$ is rational but it is not a natural number.

Page 207 Check for Understanding

1. any number that can be written as a fraction

2. Sample answer: 1.2424424444... is not rational because it is nonrepeating and nonterminating.

3. $-2\frac{1}{3} = -\frac{7}{3}$

4. $10 = \frac{10}{1}$

5. $0.8 = \frac{8}{10}$
$= \frac{4}{5}$

6. $6.35 = 6\frac{35}{100}$
$= 6\frac{7}{20}$

7. $N = 0.777...$
$10N = 10(0.777...)$
$10N = 7.777...$
$\underline{-(N = 0.777...)}$
$\frac{9N}{9} = \frac{7}{9}$
$N = \frac{7}{9}$

Therefore, $-0.\overline{7} = -\frac{7}{9}$.

8. $N = 0.4545...$
$100N = 100(0.4545...)$
$100N = 45.4545...$
$\underline{-(N = 0.4545...)}$
$\frac{99N}{99} = \frac{45}{99}$
$N = \frac{45}{99}$
$\frac{45}{99} = \frac{5 \times \cancel{9}^{1}}{11 \times \cancel{9}_{1}}$
$= \frac{5}{11}$

Therefore, $0.\overline{45} = \frac{5}{11}$.

9. -5 is an integer and a rational number.

10. Because $6.05 = 6\frac{5}{100} = \frac{605}{100}$, it is a rational number. It is neither a whole number nor an integer.

11. $0.000039 = \frac{39}{1{,}000{,}000}$

Pages 208-209 Practice and Apply

12. $5\frac{2}{3} = \frac{17}{3}$

13. $-1\frac{4}{7} = -\frac{11}{7}$

14. $-21 = -\frac{21}{1}$

15. $60 = \frac{60}{1}$

16. $0.4 = \frac{4}{10}$
$= \frac{2}{5}$

17. $0.09 = \frac{9}{100}$

18. $5.22 = 5\frac{22}{100}$
$= 5\frac{11}{50}$

19. $1.68 = 1\frac{68}{100}$
$= 1\frac{17}{25}$

20. $0.625 = \frac{625}{1000}$
$= \frac{5}{8}$

21. $8.004 = 8\frac{4}{1000}$
$= 8\frac{1}{250}$

22. $N = 0.222...$
$10N = 10(0.222...)$
$10N = 2.222....$
$\underline{-(N = 0.222...)}$
$\frac{9N}{9} = \frac{2}{9}$
$N = \frac{2}{9}$

Therefore, $0.\overline{2} = \frac{2}{9}$.

23. $N = 0.333...$
$10N = 10(0.333...)$
$10N = 3.333...$
$\underline{-(N = 0.333...)}$
$\frac{9N}{9} = \frac{3}{9}$
$N = \frac{3}{9}$
$N = \frac{1}{3}$

Therefore, $-0.333... = -\frac{1}{3}$.

24. $N = 4.555...$
$10N = 10(4.555...)$
$10N = 45.555...$
$\underline{-(N = 4.555)}$
$\frac{9N}{9} = \frac{41}{9}$
$N = \frac{41}{9}$
$N = 4\frac{5}{9}$

Therefore, $4.\overline{5} = 4\frac{5}{9}$.

25. $N = 5.666...$
$10N = 10(5.666...)$
$10N = 56.666...$
$\underline{-(N = 5.666...)}$
$\frac{9N}{9} = \frac{51}{9}$
$N = \frac{51}{9}$
$N = 5\frac{6}{9} = 5\frac{2}{3}$

Therefore, $5.\overline{6} = 5\frac{2}{3}$.

26. $N = 0.3232...$
$100N = 100(0.3232...)$
$100N = 32.3232...$
$\underline{-(N = 0.3232...)}$
$\frac{99N}{99} = \frac{32}{99}$
$N = \frac{32}{99}$

Therefore, $0.\overline{32} = \frac{32}{99}$.

27. $N = 2.2525...$
$100N = 100(2.2525...)$
$100N = 225.2525...$
$\underline{-(N = 2.2525...)}$
$\frac{99N}{99} = \frac{223}{99}$
$N = \frac{223}{99}$
$N = 2\frac{25}{99}$

Therefore, $2.\overline{25} = 2\frac{25}{99}$.

28. $0.028 = \frac{28}{1000}$
$= \frac{7}{250}$.

29. $0.06 = \frac{6}{100}$
$= \frac{3}{50}$

30. $0.095 = \frac{95}{1000}$
$= \frac{19}{200}$

31. $0.295 = \frac{295}{1000}$
$= \frac{59}{200}$

32. $0.07 = \frac{7}{100}$

33. $0.16 = \frac{16}{100}$
$= \frac{4}{25}$

34. 4 is a natural number, a whole number, an integer, and a rational number.

35. −7 is an integer and a rational number.

36. Because $-2\frac{5}{8} = -\frac{21}{8}$, it is a rational number. It is neither a whole number nor an integer.

37. Because $\frac{6}{3} = 2$, it is a natural number, a whole number, an integer, and a rational number.

38. Because $15.8 = 15\frac{8}{10} = 15\frac{4}{5} = \frac{79}{5}$, it is a rational number only.

39. $N = 9.0202...$
$100N = 100(9.0202...)$
$100N = 902.0202...$
$-(N = 9.0202...)$
$\frac{99N}{99} = \frac{893}{99}$
$N = \frac{893}{99}$

Because $9.0202020... = \frac{893}{99}$, it is a rational number only.

40. 1.2345... is a nonterminating, nonrepeating decimal. So, it is not a rational number.

41. 30.151151115... is a nonterminating, nonrepeating decimal. So, it is not a rational number.

42. $0.125 = \frac{125}{1000}$
$= \frac{1}{8}$

43. $200.19 = 200\frac{19}{100}$

44. Always; every integer can be written as a fraction with 1 as the denominator; $5 = \frac{5}{1}$, so 5 is a rational number.

45. Sometimes; $\frac{1}{2}$ and 2 are both rational numbers, but only 2 is an integer.

46. Never; every whole number can be written as a fraction with 1 as the denominator. 3 is a whole number and $3 = \frac{3}{1}$, so it is a rational number.

47. $0.0008 = \frac{8}{10{,}000}$
$= \frac{1}{1250}$ inch

48. $\frac{22}{7} = 7\overline{)22.000000}$ = 3.142857

$$\begin{array}{r} 3.142857 \\ 7\overline{)22.000000} \\ -21 \\ 10 \\ -7 \\ 30 \\ -28 \\ 20 \\ -14 \\ 60 \\ -56 \\ 40 \\ -35 \\ 50 \\ -49 \\ 10 \end{array}$$

Since $\frac{22}{7} \approx 3.142857$ and $3.142857 > 3.141592$, the

estimate is greater than the actual value.

49. $\frac{3}{8} = 8\overline{)3.000}$

$$\begin{array}{r} 0.375 \\ 8\overline{)3.000} \\ -2\,4 \\ 60 \\ -56 \\ 40 \\ -40 \\ 0 \end{array}$$

Since $2\frac{3}{8} = 2.375$ and $2.375 > 2.37$, the peg will fit.

50. Let $N = 0.999...$ and let $10N = 9.999...$
$10N = 9.999...$
$-(N = 0.999...)$
$9N = 9$
$\frac{9N}{9} = \frac{9}{9}$
$N = 1$

51. The set of rational numbers includes the set of natural numbers, whole numbers, and integers. In the same way, the set of natural numbers is part of the set of whole numbers and the set of whole numbers is part of the set of integers. Answers should include the following.
- The number 5 belongs to the set of natural numbers, whole number integers, and rational numbers.
- The number $\frac{1}{2}$ belongs only to the set of rational numbers.

52. A; There are 4 integers, 4 whole numbers, and 4 natural numbers.

53. C; $0.56 = \frac{56}{100}$
$= \frac{14}{25}$

Page 209 Maintain Your Skills

54. $5\overline{)2.0}$ = 0.4
$-2\,0$
0

$\frac{2}{5} = 0.4$

55. $-7\frac{4}{5} = -\left(7 + \frac{4}{5}\right)$
$= -(7 + 0.8)$
$= -7.8$

56. $20\overline{)13.00}$ = 0.65
$-12\,0$
$1\,00$
$-1\,00$
0

$-\frac{13}{20} = -0.65$

57. $2\frac{5}{9} = 2 + \frac{5}{9}$

$9\overline{)5.00}$ = 0.55
$-4\,5$
50
-45
5

$2 + 0.\overline{5} = 2.\overline{5}$

$2\frac{5}{9} = 2.\overline{5}$

58. $2 \times 10^3 = 2 \times 1{,}000$
$= 2{,}000$

59. $3.05 \times 10^6 = 3.05 \times 1{,}000{,}000$
$= 3{,}050{,}000$

60. $7.4 \times 10^{-4} = 7.4 \times 0.0001$
$= 0.00074$

61. $1.681 \times 10^{-2} = 1.681 \times 0.01$
$= 0.01681$

62. $\frac{12n^2}{3an} = \frac{\cancel{3} \cdot 4 \cdot n \cdot \cancel{n}}{\cancel{3} \cdot a \cdot \cancel{n}}$
$= \frac{4n}{a}$

63. $(4 \times 10^2) + (8 \times 10^1) + (3 \times 10^0)$
$= (4 \times 100) + (8 \times 10) + (3 \times 1)$
$= 400 + 80 + 3$
$= 483$

64. $P = 2(l + w)$ $\quad A = l \cdot w$
$P = 2(16 + 7)$ $\quad A = 7 \cdot 16$
$P = 2(23)$ $\quad A = 112$ in^2
$P = 46$ in.

65. $P = 2(l + w)$ $\quad A = l \cdot w$
$P = 2(9 + 3)$ $\quad A = 9 \cdot 3$
$P = 2(12)$ $\quad A = 27$ cm^2
$P = 24$ cm

66. $3(5 + 9) = 3 \cdot 5 + 3 \cdot 9$

67. $(8 + 1)2 = 8 \cdot 2 + 1 \cdot 2$

68. $6(b - 5) = 6b - 6 \cdot 5$
$= 6b - 30$

69. $(x + 4)7 = x \cdot 7 + 4 \cdot 7$
$= 7x + 28$

70. $1\frac{2}{3} \cdot 4\frac{1}{8} \approx 2 \cdot 4 = 8$

71. $-5\frac{2}{5} \cdot 3\frac{4}{5} \approx -5 \cdot 4 = -20$

72. $2\frac{1}{4} \cdot 2\frac{1}{9} \approx 2 \cdot 2 = 4$

73. $6\frac{7}{8} \cdot 1\frac{9}{10} \approx 7 \cdot 2 = 14$

74. $9\frac{1}{8} \cdot \left(-4\frac{3}{4}\right) \approx 9 \cdot (-5) = -45$

75. $15\frac{5}{7} \cdot 2\frac{1}{3} \approx 16 \cdot 2 = 32$

5-3 Multiplying Rational Numbers

Page 210 How is multiplying fractions related to areas of rectangles?

a. $\frac{6}{12}$ or $\frac{1}{2}$

b. $\frac{1}{6}$

c. $\frac{3}{20}$

d. $\frac{3}{12}$ or $\frac{1}{4}$

e. The numerator of the product equals the product of the numerators of the factors. The denominator of the product equals the product of the denominators of the factors.

Page 212 Check for Understanding

1. Sample answer: $\frac{1}{2}, \frac{1}{3}$

2. Terrence; Marie incorrectly divided the 18 in the numerator of the second fraction.

3. $\frac{1}{4} \cdot \frac{3}{5} = \frac{1 \cdot 3}{4 \cdot 5}$
$= \frac{3}{20}$

4. $\frac{1}{2}\left(-\frac{5}{6}\right) = \frac{1}{2} \cdot \frac{-5}{6}$
$= \frac{1 \cdot (-5)}{2 \cdot 6}$
$= -\frac{5}{12}$

5. $-\frac{2}{3}\left(-\frac{5}{6}\right) = \frac{-2}{3} \cdot \frac{-5}{6}$
$= \frac{\overset{-1}{\cancel{-2}}(-5)}{3 \cdot \underset{3}{\cancel{6}}}$
$= \frac{5}{9}$

6. $7\left(\frac{8}{21}\right) = \frac{\overset{1}{\cancel{7}}}{1} \cdot \frac{8}{\underset{3}{\cancel{21}}}$
$= \frac{1 \cdot 8}{1 \cdot 3}$
$= \frac{8}{3}$ or $2\frac{2}{3}$

7. $3\frac{1}{4} \cdot \frac{2}{11} = \frac{13}{\underset{2}{\cancel{4}}} \cdot \frac{\overset{1}{\cancel{2}}}{11}$
$= \frac{13 \cdot 1}{2 \cdot 11}$
$= \frac{13}{22}$

8. $-5\frac{1}{3} \cdot 3\frac{3}{8} = -\frac{\overset{2}{\cancel{16}}}{\underset{1}{\cancel{3}}} \cdot \frac{\overset{9}{\cancel{27}}}{\underset{1}{\cancel{8}}}$
$= \frac{-2 \cdot 9}{1 \cdot 1}$
$= -\frac{18}{1}$ or -18

9. $\frac{2}{x}\cdot\frac{3x}{7}=\frac{2}{\underset{1}{\cancel{x}}}\cdot\frac{3\cdot\overset{1}{\cancel{x}}}{7}$
$=\frac{6}{7}$

10. $\frac{a}{b}\cdot\frac{5b}{c}=\frac{a}{\underset{1}{\cancel{b}}}\cdot\frac{5\cdot\overset{1}{\cancel{b}}}{c}$
$=\frac{5a}{c}$

11. $\frac{4t}{9r}\cdot\frac{18r}{t^2}=\frac{4\overset{1}{\cancel{t}}}{\underset{1}{\cancel{9}}\underset{1}{\cancel{r}}}\cdot\frac{\overset{2}{\cancel{18}}\overset{1}{\cancel{r}}}{\underset{1}{\cancel{t}}\cdot t}$
$=\frac{8}{t}$

12. $d=65\,\frac{\text{mi}}{\cancel{h}}\cdot\frac{7}{2}\cancel{h}$ or $227\frac{1}{2}$ mi

Pages 213–214 Practice and Apply

13. $\frac{6}{7}\cdot\frac{2}{7}=\frac{6\cdot 2}{7\cdot 7}$
$=\frac{12}{49}$

14. $\frac{4}{9}\cdot\frac{2}{3}=\frac{4\cdot 2}{9\cdot 3}$
$=\frac{8}{27}$

15. $\frac{1}{5}\left(-\frac{1}{8}\right)=\frac{1}{5}\cdot\frac{-1}{8}$
$=\frac{1\cdot(-1)}{5\cdot 8}$
$=-\frac{1}{40}$

16. $-\frac{3}{4}\cdot\frac{3}{5}=\frac{-3}{4}\cdot\frac{3}{5}$
$=\frac{-3\cdot 3}{4\cdot 5}$
$=\frac{-9}{20}$

17. $\frac{5}{9}\cdot\frac{8}{25}=\frac{\overset{1}{\cancel{5}}}{9}\cdot\frac{8}{\underset{5}{\cancel{25}}}$
$=\frac{1\cdot 8}{9\cdot 5}$
$=\frac{8}{45}$

18. $-\frac{1}{2}\left(-\frac{2}{7}\right)=\frac{-1}{\underset{1}{\cancel{2}}}\cdot\frac{\overset{1}{\cancel{-2}}}{7}$
$=\frac{-1(-1)}{1\cdot 7}$
$=\frac{1}{7}$

19. $\frac{2}{5}\cdot\frac{5}{6}=\frac{\overset{1}{\cancel{2}}}{\underset{1}{\cancel{5}}}\cdot\frac{\overset{1}{\cancel{5}}}{\underset{3}{\cancel{6}}}$
$=\frac{1\cdot 1}{1\cdot 3}$
$=\frac{1}{3}$

20. $\frac{8}{9}\cdot\frac{27}{28}=\frac{\overset{2}{\cancel{8}}}{\underset{1}{\cancel{9}}}\cdot\frac{\overset{3}{\cancel{27}}}{\underset{7}{\cancel{28}}}$
$=\frac{2\cdot 3}{1\cdot 7}$
$=\frac{6}{7}$

21. $\frac{3}{4}\left(-\frac{1}{3}\right)=\frac{\overset{1}{\cancel{3}}}{4}\cdot\frac{-1}{\underset{1}{\cancel{3}}}$
$=\frac{1(-1)}{4\cdot 1}$
$=-\frac{1}{4}$

22. $-\frac{7}{8}\cdot\frac{2}{5}=\frac{-7}{\underset{4}{\cancel{8}}}\cdot\frac{\overset{1}{\cancel{2}}}{5}$
$=\frac{-7\cdot 1}{4\cdot 5}$
$=-\frac{7}{20}$

23. $\frac{3}{5}\cdot\frac{15}{24}=\frac{\overset{1}{\cancel{3}}}{\underset{1}{\cancel{5}}}\cdot\frac{\overset{3}{\cancel{15}}}{\underset{8}{\cancel{24}}}$
$=\frac{1\cdot 3}{1\cdot 8}$
$=\frac{3}{8}$

24. $\frac{3}{32}\cdot\frac{24}{39}=\frac{\overset{1}{\cancel{3}}}{\underset{4}{\cancel{32}}}\cdot\frac{\overset{3}{\cancel{24}}}{\underset{13}{\cancel{39}}}$
$=\frac{1\cdot 3}{4\cdot 13}$
$=\frac{3}{52}$

25. $2\cdot\frac{7}{12}=\frac{\overset{1}{\cancel{2}}}{1}\cdot\frac{7}{\underset{6}{\cancel{12}}}$
$=\frac{1\cdot 7}{1\cdot 6}$
$=\frac{7}{6}$ or $1\frac{1}{6}$

26. $\frac{6}{15}(-3)=\frac{6}{\underset{5}{\cancel{15}}}\cdot\frac{\overset{1}{\cancel{-3}}}{1}$
$=\frac{6(-1)}{5\cdot 1}$
$=-\frac{6}{5}$ or $-1\frac{1}{5}$

27. $6\frac{2}{3}\cdot\frac{1}{2}=\frac{\overset{10}{\cancel{20}}}{3}\cdot\frac{1}{\underset{1}{\cancel{2}}}$
$=\frac{10\cdot 1}{3\cdot 1}$
$=\frac{10}{3}$ or $3\frac{1}{3}$

28. $\frac{5}{12}\cdot 3\frac{1}{9}=\frac{5}{\underset{3}{\cancel{12}}}\cdot\frac{\overset{7}{\cancel{28}}}{9}$
$=\frac{5\cdot 7}{3\cdot 9}$
$=\frac{35}{27}$ or $1\frac{8}{27}$

29. $2\frac{2}{6}\cdot 6\frac{2}{7}=\frac{\overset{2}{\cancel{14}}}{\underset{3}{\cancel{6}}}\cdot\frac{\overset{22}{\cancel{44}}}{\underset{1}{\cancel{7}}}$
$=\frac{2\cdot 22}{3\cdot 1}$
$=\frac{44}{3}$ or $14\frac{2}{3}$

30. $3\frac{1}{3}\cdot 2\frac{5}{8}=\frac{\overset{5}{\cancel{10}}}{\underset{1}{\cancel{3}}}\cdot\frac{\overset{7}{\cancel{21}}}{\underset{4}{\cancel{8}}}$
$=\frac{5\cdot 7}{1\cdot 4}$
$=\frac{35}{4}$ or $8\frac{3}{4}$

31. $-6\frac{2}{3}\left(-1\frac{1}{2}\right)=\frac{\overset{10}{\cancel{-20}}}{\underset{1}{\cancel{3}}}\cdot\frac{\overset{1}{\cancel{-3}}}{\underset{1}{\cancel{2}}}$
$=\frac{-10(-1)}{1\cdot 1}$
$=\frac{10}{1}$ or 10

32. $1\frac{3}{7}\left(-9\frac{4}{5}\right)=\frac{\overset{2}{\cancel{10}}}{\underset{1}{\cancel{7}}}\cdot\frac{\overset{7}{\cancel{-49}}}{\underset{1}{\cancel{5}}}$
$=\frac{2(-7)}{1\cdot 1}$
$=-\frac{14}{1}$ or -14

33. $-1\frac{1}{4}\cdot 3\frac{5}{9}=\frac{-5}{\underset{1}{\cancel{4}}}\cdot\frac{\overset{8}{\cancel{32}}}{9}$
$=\frac{-5\cdot 8}{1\cdot 9}$
$=\frac{-40}{9}$ or $-4\frac{4}{9}$

34. $\frac{5}{\underset{1}{\cancel{6}}}\,\cancel{\text{mile}}\cdot\frac{\overset{880}{\cancel{5280}}\text{ feet}}{1\,\cancel{\text{mile}}}=\frac{5\cdot 880\text{ ft}}{1\cdot 1}$
$=4400$ ft

35. $\frac{3}{\underset{1}{\cancel{8}}}\ \cancel{\text{pound}} \cdot \frac{\overset{2}{\cancel{16}}\text{ ounces}}{1\ \cancel{\text{pound}}} = \frac{3 \cdot 2\text{ ounces}}{1 \cdot 1}$

$= 6\text{ ounces}$

36. $\frac{2}{\underset{1}{\cancel{3}}}\ \cancel{\text{hour}} \cdot \frac{\overset{20}{\cancel{60}}\text{ minutes}}{1\ \cancel{\text{hour}}} = \frac{2 \cdot 20\text{ minutes}}{1 \cdot 1}$

$= 40\text{ minutes}$

37. $\frac{3}{\underset{1}{\cancel{4}}}\ \cancel{\text{yard}} \cdot \frac{\overset{9}{\cancel{36}}\text{ inches}}{1\ \cancel{\text{yard}}} = \frac{3 \cdot 9\text{ inches}}{1 \cdot 1}$

$= 27\text{ inches}$

38. $\frac{4a}{5} \cdot \frac{3}{a} = \frac{4\overset{1}{\cancel{a}}}{5} \cdot \frac{3}{\underset{1}{\cancel{a}}}$

$= \frac{4 \cdot 3}{5 \cdot 1}$

$= \frac{12}{5}$ or $2\frac{2}{5}$

39. $\frac{3x}{y} \cdot \frac{9y}{x} = \frac{3 \cdot \overset{1}{\cancel{x}}}{\underset{1}{\cancel{y}}} \cdot \frac{9 \cdot \overset{1}{\cancel{y}}}{\underset{1}{\cancel{x}}}$

$= \frac{3 \cdot 9}{1 \cdot 1}$

$= \frac{27}{1}$ or 27

40. $\frac{12}{jk} \cdot \frac{3k}{4} = \frac{\overset{3}{\cancel{12}}}{j \cdot \underset{1}{\cancel{k}}} \cdot \frac{3 \cdot \overset{1}{\cancel{k}}}{\underset{1}{\cancel{4}}}$

$= \frac{3 \cdot 3}{j \cdot 1}$

$= \frac{9}{j}$

41. $\frac{8}{c} \cdot \frac{c^2}{11} = \frac{8}{\underset{1}{\cancel{c}}} \cdot \frac{\overset{1}{\cancel{c}} \cdot c}{11}$

$= \frac{8c}{11}$

42. $\frac{n}{18} \cdot \frac{6}{n^4} = \frac{\overset{1}{\cancel{n}}}{\underset{3}{\cancel{18}}} \cdot \frac{\overset{1}{\cancel{6}}}{\underset{1}{\cancel{n}} \cdot n \cdot n \cdot n}$

$= \frac{1 \cdot 1}{3 \cdot n^3}$

$= \frac{1}{3n^3}$

43. $\frac{x}{2z} \cdot \frac{2z^3}{3} = \frac{x}{\underset{1}{\cancel{2}}\underset{1}{\cancel{z}}} \cdot \frac{\overset{1}{\cancel{2}} \cdot \overset{1}{\cancel{z}} \cdot z \cdot z}{3}$

$= \frac{x \cdot z^2}{1 \cdot 3}$

$= \frac{xz^2}{3}$

44. $\left(-\frac{1}{2}\right)^2 = -\frac{1}{2} \cdot \left(-\frac{1}{2}\right)$

$= \frac{-1}{2} \cdot \frac{-1}{2}$

$= \frac{-1(-1)}{2 \cdot 2}$

$= \frac{1}{4}$

45. $\left(\frac{3}{4} \cdot \left(-\frac{4}{5}\right)\right)^2 = \left(\frac{3}{\underset{1}{\cancel{4}}} \cdot \frac{-\overset{1}{\cancel{4}}}{5}\right)^2$

$= \left(\frac{3(-1)}{1 \cdot 5}\right)^2$

$= \left(-\frac{3}{5}\right)^2$

$= -\frac{3}{5}\left(-\frac{3}{5}\right)$

$= \frac{-3(-3)}{5 \cdot 5}$

$= \frac{9}{25}$

46. $\frac{5}{8} \cdot \frac{2}{5} = \frac{\overset{1}{\cancel{5}}}{\underset{4}{\cancel{8}}} \cdot \frac{\overset{1}{\cancel{2}}}{\underset{1}{\cancel{5}}}$

$= \frac{1 \cdot 1}{4 \cdot 1}$

$= \frac{1}{4}$

$\frac{1}{4}$ of the eighth-grade boys talk about school at home.

47. $\frac{5}{9} \cdot \frac{33}{100} = \frac{\overset{1}{\cancel{5}}}{\underset{3}{\cancel{9}}} \cdot \frac{\overset{11}{\cancel{33}}}{\underset{20}{\cancel{100}}}$

$= \frac{1 \cdot 11}{3 \cdot 20}$

$= \frac{11}{60}$

$\frac{11}{60}$ of the twelfth-grade girls talk about school at home.

48. $\frac{2}{3} \cdot 7\frac{1}{2} = \frac{\overset{1}{\cancel{2}}}{\underset{1}{\cancel{3}}} \cdot \frac{\overset{5}{\cancel{15}}}{\underset{1}{\cancel{2}}}$

$= \frac{1 \cdot 5}{1 \cdot 1}$

$= \frac{5}{1}$ or 5

Jamal needs 5 bags of fertilizer.

49. 5 inches $\cdot \frac{2.54\text{ cm}}{1\text{ in.}} = \frac{5}{1}\ \cancel{\text{in.}} \cdot \frac{2.54\text{ cm}}{1\ \cancel{\text{in.}}}$

$= \frac{5 \cdot 2.54\text{ cm}}{1 \cdot 1}$

$= 12.7\text{ cm}$

50. 10 kilometers $\cdot \frac{0.62\text{ mi}}{1\text{ km}} = \frac{10}{1}\ \cancel{\text{km}} \cdot \frac{0.62\text{ mi}}{1\ \cancel{\text{km}}}$

$= \frac{10 \cdot 062\text{ mi}}{1 \cdot 1}$

$= 6.2\text{ mi}$

51. 26.3 centimeters $\cdot \frac{0.39\text{ in.}}{1\text{ cm}} = \frac{26.3}{1}\ \cancel{\text{cm}} \cdot \frac{0.39\ \cancel{\text{in.}}}{1\ \cancel{\text{cm}}}$

$= \frac{26.3 \cdot 0.39\text{ in.}}{1 \cdot 1}$

$= 10.257\text{ in.}$

52. $8\frac{2}{3}\text{ ft}^2 \cdot \frac{0.09\text{ m}^2}{1\text{ ft}^2} = \frac{26}{\underset{1}{\cancel{3}}}\ \cancel{\text{ft}^2} \cdot \frac{\overset{0.03}{\cancel{0.09}}\text{ m}^2}{1\ \cancel{\text{ft}^2}}$

$= \frac{26 \cdot 0.03\text{ m}^2}{1 \cdot 1}$

$= 0.78\text{ m}^2$

53a. Sample answer: $\frac{8}{4} \times \frac{3}{5}$

53b. Sample answer: $\frac{3}{4} \times \frac{5}{6}$

54. Fractions can be used to represent parts of rectangles. The product of the fractions equals a portion of the rectangle's area. Answers should include the following.

- A rectangle that is divided into rows and columns; one fraction represents the fraction of columns that are shaded and another fraction represents the fraction of rows that are shaded.
- The product of the fractions is the overlapping shaded area.

55. A; $\frac{8}{15} \cdot \frac{3}{8} = \frac{\overset{1}{\cancel{8}}}{\underset{5}{\cancel{15}}} \cdot \frac{\overset{1}{\cancel{3}}}{\underset{1}{\cancel{8}}}$

$= \frac{1}{5}$

$\frac{1}{5}$ is between 0 and 1.

56. C; 52 feet $\cdot \frac{1 \text{ yard}}{3 \text{ feet}} = \frac{52 \text{ ft}}{1} \cdot \frac{1 \text{ yd}}{3 \text{ ft}}$
$= \frac{52}{3}$ yd
$= 17\frac{1}{3}$ yd
= 17 yards 1 foot

57. $5(2)^{-3} = \frac{5}{1} \cdot \frac{1}{2^3}$
$= \frac{5}{2^3}$
$= \frac{5}{8}$

58. $7 \cdot 3^{-4} = \frac{7}{1} \cdot \frac{1}{3^4}$
$= \frac{7}{3^4}$
$= \frac{7}{81}$

59. $13(2)^{-1}(-4)^{-2} = \frac{13}{1} \cdot \frac{1}{2} \cdot \frac{1}{(-4)^2}$
$= \frac{13}{1} \cdot \frac{1}{2} \cdot \frac{1}{16}$
$= \frac{13}{32}$

Page 214 Maintain Your Skills

60. $0.18 = \frac{18}{100}$
$= \frac{9}{50}$

61. $-0.2 = -\frac{2}{10}$
$= -\frac{1}{5}$

62. $3.04 = 3\frac{4}{100}$
$= 3\frac{1}{25}$

63.
$N = 0.777...$
$10N = 10(0.777...)$
$10N = 7.777...$
$-(N = 0.777...)$
$\frac{9N}{9} = \frac{7}{9}$
$N = \frac{7}{9}$

64.
$$\begin{array}{r} 0.85 \\ 20\overline{)17.00} \\ \underline{-16\,0} \\ 1\,00 \\ \underline{-1\,00} \\ 0 \end{array}$$

65.
$$\begin{array}{r} 0.166... \\ 6\overline{)1.000...} \end{array}$$
So, $\frac{1}{6} = 0.1\overline{6}$.

66. $2\frac{2}{11} = 2 + \frac{2}{11}$
$$\begin{array}{r} 0.1818... \\ 11\overline{)2.0000...} \end{array}$$
$= 2 + 0.\overline{18}$
$= 2.\overline{18}$

67. $-4\frac{7}{8} = -\left(4 + \frac{7}{8}\right)$
$$\begin{array}{r} 0.875 \\ 8\overline{)7.000} \\ \underline{-6\,4} \\ 60 \\ \underline{-56} \\ 40 \\ \underline{-40} \\ 0 \end{array}$$
$= -(4 + 0.875) = -4.875$

68. $x^2 \cdot x^4 = x^{2+4}$
$= x^6$

69. $8n$: (2) · (2) · (2) · (n)
$16n$: (2) · (2) · (2) · 2 · (n)
The GCF is $2 \cdot 2 \cdot 2 \cdot n$ or $8n$.

70. $5ab$: $5 \cdot a \cdot$ (b)
$8b$: $2 \cdot 2 \cdot 2 \cdot$ (b)
The GCF is b.

71. $12t$: (2) · 2 · 3 · (t)
$10t$: (2) · 5 · (t)
The GCF is $2 \cdot t$ or $2t$.

72. $2rs$: 2 · (r) · (s)
$3rs$: 3 · (r) · (s)
The GCF is $r \cdot s$ or rs.

73. $9k$: (3) · (3) · k
27: (3) · (3) · 3
The GCF is $3 \cdot 3$ or 9.

74. $4p^2$: (2) · 2 · (p) · p
$6p$: (2) · 3 · (p)
The GCF is $2 \cdot p$ or $2p$.

5-4 Dividing Rational Numbers

Page 215 How is dividing by a fraction related to multiplying?

a. 6; $2 \times 3 = 6$
b. 8; $4 \times 2 = 8$
c. 12; $3 \times 4 = 12$
d. Dividing by a fraction with 1 in the numerator is the same as multiplying by the number that is in the denominator.

Pages 217–218 Check for Understanding

1. Dividing by a fraction is the same as multiplying by its reciprocal.
2. Sample answer: $14 \div \frac{5}{7}$
3. $\frac{4}{5}\left(\frac{5}{4}\right) = 1$
   The multiplicative inverse of $\frac{4}{5}$ is $\frac{5}{4}$.
4. $-16\left(-\frac{1}{16}\right) = 1$
   The multiplicative inverse of -16 is $-\frac{1}{16}$.

5. $3\frac{1}{8} = \frac{25}{8}$

$\frac{25}{8}\left(\frac{8}{25}\right) = 1$

The multiplicative inverse of $3\frac{1}{8}$ is $\frac{8}{25}$.

6. $\frac{1}{2} \div \frac{6}{7} = \frac{1}{2} \cdot \frac{7}{6}$

$= \frac{7}{12}$

7. $-\frac{2}{3} \div \left(-\frac{5}{6}\right) = -\frac{2}{3} \cdot -\frac{6}{5}$

$= -\frac{2}{\cancel{3}_1} \cdot -\frac{\cancel{6}^2}{5}$

$= \frac{4}{5}$

8. $\frac{7}{9} \div \frac{2}{3} = \frac{7}{9} \cdot \frac{3}{2}$

$= \frac{7}{\cancel{9}_3} \cdot \frac{\cancel{3}^1}{2}$

$= \frac{7}{6}$ or $1\frac{1}{6}$

9. $7\frac{1}{3} \div 5 = \frac{22}{3} \div \frac{5}{1}$

$= \frac{22}{3} \cdot \frac{1}{5}$

$= \frac{22}{15}$ or $1\frac{7}{15}$

10. $-\frac{8}{9} \div 3\frac{1}{5} = -\frac{8}{9} \div \frac{16}{5}$

$= \frac{8}{9} \cdot \frac{5}{16}$

$= -\frac{\cancel{8}^1}{9} \cdot \frac{5}{\cancel{16}_2}$

$= -\frac{5}{18}$

11. $2\frac{1}{6} \div \left(-1\frac{1}{5}\right) = \frac{13}{6} \div \frac{-6}{5}$

$= \frac{13}{6} \cdot \frac{-5}{6}$

$= -\frac{65}{36}$ or $-1\frac{29}{36}$

12. $\frac{14}{n} \div \frac{1}{n} = \frac{14}{n} \cdot \frac{n}{1}$

$= \frac{14}{\cancel{n}_1} \cdot \frac{\cancel{n}^1}{1}$

$= \frac{14}{1}$ or 14

13. $\frac{ab}{4} \div \frac{b}{6} = \frac{ab}{4} \cdot \frac{6}{b}$

$= \frac{a\cancel{b}^1}{\cancel{4}_2} \cdot \frac{\cancel{6}^3}{\cancel{b}_1}$

$= \frac{3a}{2}$

14. $\frac{x^2}{5} \div \frac{ax}{2} = \frac{x^2}{5} \cdot \frac{2}{ax}$

$= \frac{x \cdot \cancel{x}}{5} \cdot \frac{2}{a\cancel{x}}$

$= \frac{2x}{5a}$

15. $16 \div 2\frac{8}{12} = \frac{16}{1} \div 2\frac{2}{3}$

$= \frac{16}{1} \div \frac{8}{3}$

$= \frac{16}{1} \cdot \frac{3}{8}$

$= \frac{\cancel{16}^2}{1} \cdot \frac{3}{\cancel{8}_1}$

$= \frac{6}{1}$ or 6 boards

Pages 218–219 Practice and Apply

16. $\frac{6}{11}\left(\frac{11}{6}\right) = 1$

The multiplicative inverse of $\frac{6}{11}$ is $\frac{11}{6}$.

17. $-\frac{1}{5}\left(-\frac{5}{1}\right) = 1$

The multiplicative inverse of $-\frac{1}{5}$ is $-\frac{5}{1}$ or -5.

18. $-7\left(-\frac{1}{7}\right) = 1$

The multiplicative inverse of -7 is $-\frac{1}{7}$.

19. $24\left(\frac{1}{24}\right) = 1$

The multiplicative inverse of 24 is $\frac{1}{24}$.

20. $5\frac{1}{4} = \frac{21}{4}$

$\frac{21}{4}\left(\frac{4}{21}\right) = 1$

The multiplicative inverse of $5\frac{1}{4}$ is $\frac{4}{21}$.

21. $-3\frac{2}{9} = -\frac{29}{9}$

$-\frac{29}{9}\left(-\frac{9}{29}\right) = 1$

The multiplicative inverse of $-3\frac{2}{9}$ is $-\frac{9}{29}$.

22. $\frac{1}{4} \div \frac{3}{5} = \frac{1}{4} \cdot \frac{5}{3}$

$= \frac{5}{12}$

23. $\frac{2}{9} \div \frac{1}{4} = \frac{2}{9} \cdot \frac{4}{1}$

$= \frac{8}{9}$

24. $-\frac{1}{2} \div \frac{5}{6} = -\frac{1}{2} \cdot \frac{6}{5}$

$= \frac{-1}{\cancel{2}_1} \cdot \frac{\cancel{6}^3}{5}$

$= -\frac{3}{5}$

25. $\frac{6}{11} \div \left(-\frac{4}{5}\right) = \frac{6}{11} \cdot -\frac{5}{4}$

$= \frac{\cancel{6}^3}{11} \cdot -\frac{5}{\cancel{4}_2}$

$= -\frac{15}{22}$

26. $\frac{8}{9} \div \frac{4}{3} = \frac{8}{9} \cdot \frac{3}{4}$

$= \frac{\cancel{8}^2}{\cancel{9}_3} \cdot \frac{\cancel{3}^1}{\cancel{4}_1}$

$= \frac{2}{3}$

27. $\frac{7}{8} \div \frac{14}{15} = \frac{7}{8} \cdot \frac{15}{14}$

$= \frac{\cancel{7}^1}{8} \cdot \frac{15}{\cancel{14}_2}$

$= \frac{15}{16}$

28. $\frac{3}{4} \div \frac{3}{4} = \frac{3}{4} \cdot \frac{4}{3}$

$= \frac{\cancel{3}^1}{\cancel{4}_1} \cdot \frac{\cancel{4}^1}{\cancel{3}_1}$

$= \frac{1}{1}$ or 1

29. $\frac{2}{9} \div \left(-\frac{2}{9}\right) = \frac{2}{9} \cdot \frac{-9}{2}$

$= \frac{\cancel{2}^1}{\cancel{9}_1} \cdot \frac{-\cancel{9}^1}{\cancel{2}_1}$

$= \frac{-1}{1}$ or -1

30. $\frac{3}{5} \div \frac{5}{9} = \frac{3}{5} \cdot \frac{9}{5}$

$= \frac{27}{25}$ or $1\frac{2}{25}$

31. $\frac{3}{10} \div \frac{1}{5} = \frac{3}{10} \cdot \frac{5}{1}$

$= \frac{3}{\cancel{10}_2} \cdot \frac{\cancel{5}^1}{1}$

$= \frac{3}{2}$ or $1\frac{1}{2}$

32. $12 \div \frac{4}{9} = \frac{12}{1} \cdot \frac{9}{4}$

$= \frac{\cancel{12}^3}{1} \cdot \frac{9}{\cancel{4}_1}$

$= \frac{27}{1}$ or 27

33. $-8 \div \frac{4}{5} = \frac{-8}{1} \cdot \frac{5}{4}$

$= \frac{\cancel{-8}^2}{1} \cdot \frac{5}{\cancel{4}_1}$

$= \frac{-10}{1}$ or -10

34. $-\frac{5}{8} \div (-4) = \frac{-5}{8} \cdot \frac{-1}{4}$

$= \frac{5}{32}$

35. $6\frac{2}{3} \div 5 = \frac{20}{3} \cdot \frac{1}{5}$

$= \frac{\cancel{20}^4}{3} \cdot \frac{1}{\cancel{5}_1}$

$= \frac{4}{3}$ or $1\frac{1}{3}$

36. $-1\frac{1}{9} \div \frac{2}{3} = \frac{-10}{9} \cdot \frac{3}{2}$

$= \frac{\cancel{-10}^5}{\cancel{9}_3} \cdot \frac{\cancel{3}^1}{\cancel{2}_1}$

$= -\frac{5}{3}$ or $-1\frac{2}{3}$

37. $-\frac{2}{3} \div \left(-\frac{1}{3}\right) = \frac{-2}{3} \cdot \frac{-3}{1}$

$= \frac{-2}{\cancel{3}_1} \cdot \frac{\cancel{-3}^1}{1}$

$= \frac{2}{1}$ or 2

38. $3\frac{3}{10} \div 1\frac{5}{6} = \frac{33}{10} \div \frac{11}{6}$

$= \frac{33}{10} \cdot \frac{6}{11}$

$= \frac{\cancel{33}^3}{\cancel{10}_5} \cdot \frac{\cancel{6}^3}{\cancel{11}_1}$

$= \frac{9}{5}$ or $1\frac{4}{5}$

39. $7\frac{1}{2} \div \left(-1\frac{1}{5}\right) = \frac{15}{2} \div \frac{-6}{5}$

$= \frac{15}{2} \cdot \frac{-5}{6}$

$= \frac{\cancel{15}^5}{2} \cdot \frac{-5}{\cancel{6}_2}$

$= \frac{-25}{4}$ or $-6\frac{1}{4}$

40. $\frac{a}{7} \div \frac{a}{42} = \frac{a}{7} \cdot \frac{42}{a}$

$= \frac{\cancel{a}^1}{\cancel{7}_1} \cdot \frac{\cancel{42}^6}{\cancel{a}_1}$

$= \frac{6}{1}$ or 6

41. $\frac{10}{3x} \div \frac{5}{2x} = \frac{10}{3x} \cdot \frac{2x}{5}$

$= \frac{\cancel{10}^2}{3\cancel{x}_1} \cdot \frac{2\cancel{x}^1}{\cancel{5}_1}$

$= \frac{4}{3}$ or $1\frac{1}{3}$

42. $\frac{c}{8} \div \frac{cd}{5} = \frac{c}{8} \cdot \frac{5}{cd}$

$= \frac{\cancel{c}^1}{8} \cdot \frac{5}{\cancel{c}_1 d}$

$= \frac{5}{8d}$

43. $\frac{5s}{t} \div \frac{6rs}{t} = \frac{5s}{t} \cdot \frac{t}{6rs}$

$= \frac{5\cancel{s}^1}{\cancel{t}_1} \cdot \frac{\cancel{t}^1}{6r\cancel{s}_1}$

$= \frac{5}{6r}$

44. $\frac{k^3}{9} \div \frac{k}{24} = \frac{k^3}{9} \cdot \frac{24}{k}$

$= \frac{k \cdot k \cdot \cancel{k}^1}{\cancel{9}_3} \cdot \frac{\cancel{24}^8}{\cancel{k}_1}$

$= \frac{8k^2}{3}$

45. $\frac{2s}{t^2} \div \frac{st^3}{8} = \frac{2s}{t^2} \cdot \frac{8}{st^3}$

$= \frac{2\cancel{s}^1}{t \cdot t} \cdot \frac{8}{\cancel{s}_1 \cdot t \cdot t \cdot t}$

$= \frac{16}{t^5}$

46. $2\frac{3}{4} \div \frac{1}{4} = \frac{11}{4} \cdot \frac{4}{1}$

$= \frac{11}{\cancel{4}_1} \cdot \frac{\cancel{4}^1}{1}$

$= \frac{11}{1}$ or 11 hamburgers

47. 9 inch $= \frac{9}{36} = \frac{1}{4}$ yard

$1\frac{1}{2} \div \frac{1}{4} = \frac{3}{2} \cdot \frac{4}{1}$

$= \frac{3}{\cancel{2}_1} \cdot \frac{\cancel{4}^2}{1}$

$= \frac{6}{1}$ or 6 ribbons

48. $m \div n = \frac{-8}{9} \div \frac{7}{18}$

$= \frac{-8}{9} \cdot \frac{18}{7}$

$= \frac{-8}{\cancel{9}_1} \cdot \frac{\cancel{18}^2}{7}$

$= \frac{-16}{7}$ or $-2\frac{2}{7}$

49. $r^2 \div s^2 = \left(\frac{-3}{4}\right)^2 \div \left(1\frac{1}{3}\right)^2$

$= \left(\frac{-3}{4} \cdot \frac{-3}{4}\right) \div \left(\frac{4}{3} \cdot \frac{4}{3}\right)^2$

$= \frac{9}{16} \div \frac{16}{9}$

$= \frac{9}{16} \cdot \frac{9}{16}$

$= \frac{81}{256}$

50. $280 \div 80 = 3.5$

miles $\div \frac{\text{miles}}{\text{hour}} = \cancel{\text{miles}} \cdot \frac{\text{hour}}{\cancel{\text{miles}}}$

$=$ hour

The result is expressed as hour(s). The answer is 3.5 hours.

51. $12 \div 1\frac{1}{2} = 12 \div \frac{3}{2}$
$= \frac{12}{1} \cdot \frac{2}{3}$
$= \frac{\overset{4}{\cancel{12}}}{1} \cdot \frac{2}{\underset{1}{\cancel{3}}}$
$= 8$

$\text{cans} \div \frac{\text{cans}}{\text{day}} = \cancel{\text{can}} \cdot \frac{\text{day}}{\cancel{\text{cans}}}$
$= \text{day}$

The result is expressed as day(s). The answer is 8 days.

52a. $\frac{3}{4} \div \frac{1}{2} = \frac{3}{4} \cdot \frac{2}{1}$
$= \frac{3}{\underset{2}{\cancel{4}}} \cdot \frac{\overset{1}{\cancel{2}}}{1}$
$= \frac{3}{2} \text{ or } 1\frac{1}{2}$

$\frac{3}{4} \div \frac{1}{4} = \frac{3}{4} \cdot \frac{4}{1}$
$= \frac{3}{\underset{1}{\cancel{4}}} \cdot \frac{\overset{1}{\cancel{4}}}{1}$
$= 3$

$\frac{3}{4} \div \frac{1}{8} = \frac{3}{4} \cdot \frac{8}{1}$
$= \frac{3}{\underset{1}{\cancel{4}}} \cdot \frac{\overset{2}{\cancel{8}}}{1}$
$= 6$

$\frac{3}{4} \div \frac{1}{12} = \frac{3}{4} \cdot \frac{12}{1}$
$= \frac{3}{\underset{1}{\cancel{4}}} \cdot \frac{\overset{3}{\cancel{12}}}{1}$
$= 9$

52b. The quotient increases.

52c. The quotient decreases.

53. Dividing by a fraction is the same as multiplying by its reciprocal. Answers should include the following.

- For example, a model of two circles, each divided into four sections represents $2 \div \frac{1}{4}$. Since there are 8 sections, $2 \div \frac{1}{4} = 8$.
- Division of fractions and multiplication of fractions are inverse operations. So, $2 \div \frac{1}{4}$ equals $2 \cdot 4$ or 8.

54. D; $11.25 = 11\frac{1}{4}$
$11\frac{1}{4} \div 2\frac{1}{4} = \frac{45}{4} \div \frac{9}{4}$
$= \frac{45}{4} \cdot \frac{4}{9}$
$= \frac{\overset{5}{\cancel{45}}}{\underset{1}{\cancel{4}}} \cdot \frac{\overset{1}{\cancel{4}}}{\underset{1}{\cancel{9}}}$
$= \frac{5}{1}$ or \$5/hour

55. C; $\frac{3}{10} \div 1\frac{4}{5} = \frac{3}{10} \div \frac{9}{5}$
$= \frac{\overset{1}{\cancel{3}}}{\underset{2}{\cancel{10}}} \cdot \frac{\overset{1}{\cancel{5}}}{\underset{3}{\cancel{9}}}$
$= \frac{1}{6}$

Page 219 Maintain Your Skills

56. $\frac{3}{5} \cdot \frac{1}{3} = \frac{\overset{1}{\cancel{3}}}{5} \cdot \frac{1}{\underset{1}{\cancel{3}}}$
$= \frac{1}{5}$

57. $\frac{2}{9} \cdot \frac{15}{16} = \frac{\overset{1}{\cancel{2}}}{\underset{3}{\cancel{9}}} \cdot \frac{\overset{5}{\cancel{15}}}{\underset{8}{\cancel{16}}}$
$= \frac{5}{24}$

58. $2\frac{4}{5} \cdot \frac{3}{8} = \frac{14}{5} \cdot \frac{3}{8}$
$= \frac{\overset{7}{\cancel{14}}}{5} \cdot \frac{3}{\underset{4}{\cancel{8}}}$
$= \frac{21}{20} \text{ or } 1\frac{1}{20}$

59. $-\frac{5}{12} \cdot 1\frac{1}{7} = -\frac{5}{12} \cdot \frac{8}{7}$
$= \frac{-5}{\underset{3}{\cancel{12}}} \cdot \frac{\overset{2}{\cancel{8}}}{7}$
$= -\frac{10}{21}$

60. 16 is a natural number, a whole number, an integer, and a rational number.

61. Because $-2.8888... = -2\frac{8}{9} = -\frac{26}{9}$, it is a rational number only.

62. Because $0.\overline{9}$ is a repeating decimal, it is a rational number only.

63. 5.121221222... is a non-terminating, nonrepeating decimal. So, it is not a rational number.

64. $150 = 15 \cdot 10$
$= 3 \cdot 5 \cdot 2 \cdot 5$
$= 2 \cdot 3 \cdot 5^2$

65.
$$3x - 5 = 16$$
$$3x - 5 + 5 = 16 + 5$$
$$\frac{3x}{3} = \frac{21}{3}$$
$$x = 7$$

66. $\frac{9}{4} = 2\frac{1}{4}$

67. $\frac{8}{7} = 1\frac{1}{7}$

68. $\frac{17}{2} = 8\frac{1}{2}$

69. $\frac{25}{4} = 6\frac{1}{4}$

70. $\frac{24}{5} = 4\frac{4}{5}$

71. $\frac{22}{6} = \frac{11 \times 2}{3 \times 2}$
$= \frac{11}{3}$
$= 3\frac{2}{3}$

72. $\frac{15}{6} = \frac{5 \times 3}{2 \times 3}$
$= \frac{5}{2}$
$= 2\frac{1}{2}$

73. $\frac{30}{18} = \frac{5 \times 6}{3 \times 6}$
$= \frac{5}{3}$
$= 1\frac{2}{3}$

74. $\frac{18}{15} = \frac{6 \times 3}{5 \times 3}$
$= \frac{6}{5}$
$= 1\frac{1}{5}$

5-5 Adding and Subtracting Like Fractions

Page 220 Why are fractions important when taking measurements?

a. $\frac{4}{8}$ or $\frac{1}{2}$ in.

b. $\frac{7}{8}$ in.

c. $\frac{8}{8}$ or 1 in.

d. $\frac{3}{8}$ in.

Page 222 Check for Understanding

1.

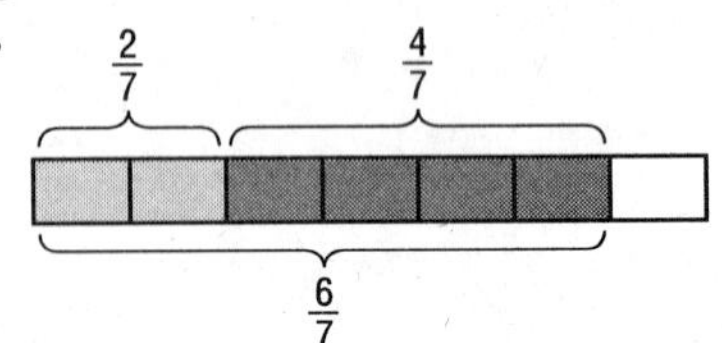

2. Sample answer: $\frac{19}{25} - \frac{1}{25}$

3. Kayla; Ethan incorrectly left out the negative sign on the first term.

4. $\frac{1}{7} + \frac{5}{7} = \frac{1+5}{7}$
$= \frac{6}{7}$

5. $\frac{11}{14} - \frac{3}{14} = \frac{11-3}{14}$
$= \frac{8}{14}$ or $\frac{4}{7}$

6. $\frac{3}{10} + \frac{3}{10} = \frac{3+3}{10}$
$= \frac{6}{10}$ or $\frac{3}{5}$

7. $-\frac{1}{8} - \frac{5}{8} = \frac{-1-5}{8}$
$= \frac{-6}{8}$ or $-\frac{3}{4}$

8. $-2\frac{4}{5} + \left(-\frac{2}{5}\right) = \frac{-14}{5} + \frac{-2}{5}$
$= \frac{-14+(-2)}{5}$
$= \frac{-16}{5}$ or $-3\frac{1}{5}$

9. $7\frac{1}{8} - \left(-1\frac{3}{8}\right) = \frac{57}{8} - \frac{-11}{8}$
$= \frac{57+11}{8}$
$= \frac{68}{8}$
$= \frac{17}{2}$ or $8\frac{1}{2}$

10. $x + y = 2\frac{4}{9} + 8\frac{7}{9}$
$= (2 + 8) + \left(\frac{4}{9} + \frac{7}{9}\right)$
$= 10 + \frac{4+7}{9}$
$= 10 + \frac{11}{9}$
$= 10 + 1\frac{2}{9}$
$= 11\frac{2}{9}$

11. $\frac{6r}{11} + \frac{2r}{11} = \frac{6r+2r}{11}$
$= \frac{8r}{11}$

12. $\frac{19}{a} - \frac{12}{a} = \frac{19-12}{a}$
$= \frac{7}{a}$

13. $\frac{5}{3x} - \frac{6}{3x} = \frac{5-6}{3x}$
$= -\frac{1}{3x}$

14. $63\frac{7}{8} - 62\frac{1}{8} = \frac{511}{8} - \frac{497}{8}$
$= \frac{511-497}{8}$
$= \frac{14}{8}$
$= \frac{7}{4}$ or $1\frac{3}{4}$ in.

Hoai grew $1\frac{3}{4}$ in.

Pages 222–224 Practice and Apply

15. $\frac{2}{5} + \frac{1}{5} = \frac{2+1}{5}$
$= \frac{3}{5}$

16. $\frac{10}{11} - \frac{8}{11} = \frac{10-8}{11}$
$= \frac{2}{11}$

17. $\frac{17}{18} - \frac{5}{18} = \frac{17-5}{18}$
$= \frac{12}{18}$ or $\frac{2}{3}$

18. $\frac{3}{10} + \frac{7}{10} = \frac{3+7}{10}$
$= \frac{10}{10}$ or 1

19. $\frac{1}{12} + \left(-\frac{7}{12}\right) = \frac{1+(-7)}{12}$
$= \frac{-6}{12}$ or $-\frac{1}{2}$

20. $\frac{9}{20} - \left(-\frac{7}{20}\right) = \frac{9-(-7)}{20}$
$= \frac{16}{20}$ or $\frac{4}{5}$

21. $-\frac{3}{4} + \left(-\frac{3}{4}\right) = \frac{-3+(-3)}{4}$
$= \frac{-6}{4}$
$= -\frac{3}{2}$ or $-1\frac{1}{2}$

22. $-\frac{13}{16} + \left(-\frac{9}{16}\right) = \frac{-13+(-9)}{16}$
$= \frac{-22}{16}$
$= -\frac{11}{8}$ or $-1\frac{3}{8}$

23. $-\frac{7}{9} + \frac{5}{9} = \frac{-7+5}{9}$
$= \frac{-2}{9}$

24. $-\frac{17}{20} + \frac{9}{20} = \frac{-17+9}{20}$
$= \frac{-8}{20}$ or $-\frac{2}{5}$

25. $7\frac{2}{5} + 4\frac{2}{5} = (7 + 4) + \left(\frac{2}{5} + \frac{2}{5}\right)$
$= 11 + \frac{2+2}{5}$
$= 11\frac{4}{5}$

26. $-4\frac{5}{8} - \frac{3}{8} = \frac{-37}{8} - \frac{3}{8}$
$= \frac{-37-3}{8}$
$= \frac{-40}{8}$ or -5

27. $5\frac{7}{9} - \left(-3\frac{5}{9}\right) = \frac{52}{9} - \left(\frac{-32}{9}\right)$
$= \frac{52 - (-32)}{9}$
$= \frac{84}{9}$
$= \frac{28}{3} \text{ or } 9\frac{1}{3}$

28. $2\frac{5}{12} + \left(-2\frac{7}{12}\right) = \frac{29}{12} + \frac{-31}{12}$
$= \frac{29 + (-31)}{12}$
$= \frac{-2}{12} \text{ or } -\frac{1}{6}$

29. $2\frac{3}{8} - 1\frac{5}{8} = \frac{19}{8} - \frac{13}{8}$
$= \frac{19 - 13}{8}$
$= \frac{6}{8} \text{ or } \frac{3}{4}$

30. $8\frac{9}{10} - 6\frac{1}{10} = \frac{89}{10} - \frac{61}{10}$
$= \frac{89 - 61}{10}$
$= \frac{28}{10}$
$= \frac{14}{5} \text{ or } 2\frac{4}{5}$

31. $7\frac{4}{7} - 2\frac{5}{7} = \frac{53}{7} - \frac{19}{7}$
$= \frac{53 - 19}{7}$
$= \frac{34}{7} \text{ or } 4\frac{6}{7}$

32. $-8\frac{6}{11} - \left(-2\frac{5}{11}\right) = \frac{-94}{11} - \frac{-27}{11}$
$= \frac{-94 - (-27)}{11}$
$= \frac{-67}{11} \text{ or } -6\frac{1}{11}$

33. $12\frac{7}{8} - 7\frac{3}{8} + 2\frac{5}{8} = \frac{103}{8} - \frac{59}{8} + \frac{21}{8}$
$= \frac{103 - 59}{8} + \frac{21}{8}$
$= \frac{44}{8} + \frac{21}{8}$
$= \frac{44 + 21}{8}$
$= \frac{65}{8} \text{ or } 8\frac{1}{8}$

34. $5\frac{5}{6} + 3\frac{5}{6} - 2\frac{1}{6} = (5 + 3) + \left(\frac{5}{6} + \frac{5}{6}\right) - 2\frac{1}{6}$
$= 8\frac{10}{6} - 2\frac{1}{6}$
$= \frac{58}{6} - \frac{13}{6}$
$= \frac{58 - 13}{6}$
$= \frac{45}{6}$
$= \frac{15}{2} \text{ or } 7\frac{1}{2}$

35. $x + y = \frac{8}{15} + 2\frac{1}{15}$
$= (0 + 2) + \left(\frac{8}{15} + \frac{1}{15}\right)$
$= 2 + \frac{9}{15}$
$= 2 + \frac{3}{5}$
$= 2\frac{3}{5}$

36. $z + y = \frac{11}{15} + 2\frac{1}{15}$
$= (0 + 2) + \left(\frac{11}{15} + \frac{1}{15}\right)$
$= 2 + \frac{12}{15}$
$= 2 + \frac{4}{5}$
$= 2\frac{4}{5}$

37. $z - x = \frac{11}{15} - \frac{8}{15}$
$= \frac{11-8}{15}$
$= \frac{3}{15} \text{ or } \frac{1}{5}$

38. $y - x = 2\frac{1}{15} - \frac{8}{15}$
$= \frac{31}{15} - \frac{8}{15}$
$= \frac{23}{15} \text{ or } 1\frac{8}{15}$

39. $\frac{x}{8} = \frac{4x}{8} = \frac{x + 4x}{8}$
$= \frac{5x}{8}$

40. $\frac{3r}{10} + \frac{3r}{10} = \frac{3r + 3r}{10}$
$= \frac{6r}{10}$
$= \frac{3r}{5}$

41. $\frac{12}{m} - \frac{9}{m} = \frac{12 - 9}{m}$
$= \frac{3}{m}$

42. $\frac{10a}{3b} - \frac{7a}{3b} = \frac{10a - 7a}{3b}$
$= \frac{3a}{3b}$
$= \frac{a}{b}$

43. $5\frac{4}{7}c - 3\frac{1}{7}c = \frac{39}{7}c - \frac{22}{7}c$
$= \frac{39c - 22c}{7}$
$= \frac{-17c}{7} \text{ or } 2\frac{3}{7}c$

44. $-2\frac{1}{6}y + 8\frac{5}{6}y = -\frac{13}{6}y + \frac{53}{6}y$
$= \frac{-13y + 53y}{6}$
$= \frac{40y}{6}$
$= \frac{20y}{3} \text{ or } 6\frac{2}{3}y$

45. $5 \text{ feet} \cdot \frac{12 \text{ inches}}{1 \text{ foot}} = 5 \cancel{\text{feet}} \cdot \frac{12 \text{ in.}}{1 \cancel{\text{ft}}}$
$= 60 \text{ inches}$

$60 - 54\frac{5}{8} = 59\frac{8}{8} - 54\frac{5}{8}$
$= (59 - 54) + \left(\frac{8}{8} - \frac{5}{8}\right)$
$= 5 + \frac{3}{8}$
$= 5\frac{3}{8} \text{ inches}$

46. $2\frac{5}{8} + \frac{7}{8} = (2 + 0) + \left(\frac{5}{8} + \frac{7}{8}\right)$
$= 2 + \frac{12}{8}$
$= 2 + \frac{3}{2}$
$= 2 + 1\frac{1}{2}$
$= 3\frac{1}{2} \text{ yd}$

47. 9 inches $\cdot \frac{1 \text{ ft}}{12 \text{ in.}} = \frac{9 \cancel{\text{in.}}}{1} \cdot \frac{1 \text{ ft}}{12 \cancel{\text{in.}}}$

$= \frac{9}{12}$ ft

$= \frac{3}{4}$ ft

Added perimeter will be:

$2\left(2\frac{3}{4}\right) + 2\left(3\frac{3}{4}\right) = 2\left(\frac{11}{4}\right) + 2\left(\frac{15}{4}\right)$

$= \frac{22}{4} + \frac{30}{4}$

$= \frac{52}{4}$

$= 13$ feet

New perimeter is 25 + 13 = 38 feet.

48a. $A = \frac{1}{4}, B = \frac{1}{4}, C = \frac{1}{8}, D = \frac{1}{16}, E = \frac{1}{8}, F = \frac{1}{16},$

$G = \frac{1}{8}$

48b. $A + B = \frac{1}{4} + \frac{1}{4}$

$= \frac{1+1}{4}$

$= \frac{2}{4}$ or $\frac{1}{2}$

48c. $F + D = \frac{1}{16} + \frac{1}{16}$

$= \frac{1+1}{16}$

$= \frac{2}{16}$ or $\frac{1}{8}$

48d. $C + E = \frac{1}{8} + \frac{1}{8}$

$= \frac{1+1}{8}$

$= \frac{2}{8}$ or $\frac{1}{4}$

48e. $E + G = \frac{1}{8} + \frac{1}{8}$

$= \frac{1+1}{8}$

$= \frac{2}{8}$ or $\frac{1}{4}$

Pieces A and B equal the sum of E and G.

49. When you use a ruler or a tape measure, measurements are usually a fraction of an inch. Answers should include the following.

- The marks on a ruler represent $\frac{1}{16}$ of an inch, $\frac{1}{8}$ of an inch, $\frac{1}{4}$ of an inch, and $\frac{1}{2}$ of an inch.
- Fractional measures are used in sewing and construction.

50. D; $\frac{13}{20} - \frac{7}{20} = \frac{13-7}{20}$

$= \frac{6}{20}$ or $\frac{3}{10}$

51. A; $1\frac{9}{16} + \frac{15}{16} = (1 + 0) + \left(\frac{9}{16} + \frac{15}{16}\right)$

$= 1 + \frac{24}{16}$

$= 1 + \frac{3}{2}$

$= 1 + 1\frac{1}{2}$

$= 2\frac{1}{2}$

Page 224 Maintain Your Skills

52. $\frac{1}{6} \div \frac{3}{4} = \frac{1}{6} \cdot \frac{4}{3}$

$= \frac{1}{\cancel{6}_3} \cdot \frac{\cancel{4}^2}{3}$

$= \frac{2}{9}$

53. $-\frac{5}{8} \div \frac{1}{3} = -\frac{5}{8} \cdot \frac{3}{1}$

$= \frac{-15}{8}$ or $-1\frac{7}{8}$

54. $\frac{2}{5} \div 1\frac{1}{2} = \frac{2}{5} \div \frac{3}{2}$

$= \frac{2}{5} \cdot \frac{2}{3}$

$= \frac{4}{15}$

55. $\frac{2}{5} \cdot \frac{3}{4} = \frac{\cancel{2}^1}{5} \cdot \frac{3}{\cancel{4}_2}$

$= \frac{3}{10}$

56. $\frac{1}{6} \cdot \left(-\frac{8}{9}\right) = \frac{1}{\cancel{6}_3} \cdot \frac{-\cancel{8}^4}{9}$

$= -\frac{4}{27}$

57. $\frac{4}{7} \cdot 2\frac{1}{3} = \frac{4}{7} \cdot \frac{7}{3}$

$= \frac{4}{\cancel{7}_1} \cdot \frac{\cancel{7}^1}{3}$

$= \frac{4}{3}$ or $1\frac{1}{3}$

58. $4y^2 \cdot 8y^5 = 4 \cdot 8 \cdot y^{2+5}$

$= 32y^7$

59. $P = 2(l + w)$ $\quad A = l \cdot w$

$P = 2(15 + 6)$ $\quad A = 15 \cdot 6$

$P = 2(21)$ $\quad A = 90 \text{ cm}^2$

$P = 42$ cm

60. $60 = 6 \cdot 10$

$= 2 \cdot 3 \cdot 2 \cdot 5$

$= 2^2 \cdot 3 \cdot 5$

61. $175 = 7 \cdot 25$

$= 7 \cdot 5 \cdot 5$

$= 5^2 \cdot 7$

62. $112 = 2 \cdot 56$

$= 2 \cdot 7 \cdot 8$

$= 2 \cdot 2 \cdot 2 \cdot 2 \cdot 7$

$= 2^4 \cdot 7$

63. $12n = 2 \cdot 6 \cdot n$

$= 2 \cdot 2 \cdot 3 \cdot n$

$= 2^2 \cdot 3 \cdot n$

64. $24s^2 = 8 \cdot 3 \cdot s^2$

$= 2 \cdot 2 \cdot 2 \cdot 3 \cdot s^2$

$= 2^3 \cdot 3 \cdot s^2$

65. $42a^2b = 6 \cdot 7 \cdot a^2 \cdot b$

$= 2 \cdot 3 \cdot 7 \cdot a^2 \cdot b$

Page 224 Practice Quiz 1

1. $25\overline{)4.00}$ = 0.16

$-2\,5$

$1\,50$

$-1\,50$

0

2. $9\overline{)2.00...}$ with quotient $0.22...$

So, $-\frac{2}{9} = -0.\overline{2}$.

3. $3\frac{1}{8} = 3 + \frac{1}{8}$
$= 3 + 0.125$
$= 3.125$

4. $-6.75 = -6\frac{75}{100}$
$= -6\frac{3}{4}$

5. $0.12 = \frac{12}{100}$
$= \frac{3}{25}$

6. $N = 0.5555...$
$10N = 10(0.5555...)$
$10N = 5.555...$
$-(N = 0.555...)$
$\frac{9N}{9} = \frac{5}{9}$
$N = \frac{5}{9}$

7. $\frac{5}{18} \cdot \frac{4}{15} = \frac{\overset{1}{\cancel{5}}}{\underset{9}{\cancel{18}}} \cdot \frac{\overset{2}{\cancel{4}}}{\underset{3}{\cancel{15}}}$
$= \frac{2}{27}$

8. $\frac{7}{8} \div \left(-\frac{1}{4}\right) = \frac{7}{8} \cdot -\frac{4}{1}$
$= \frac{7}{\underset{2}{\cancel{8}}} \cdot -\frac{\overset{1}{\cancel{4}}}{1}$
$= -\frac{7}{2}$ or $-3\frac{1}{2}$

9. $\frac{11}{12} - \frac{6}{12} = \frac{11-6}{12}$
$= \frac{5}{12}$

10. $6\frac{3}{5}a + 2\frac{4}{5}a = \left[(6+2) + \left(\frac{3}{5} + \frac{4}{5}\right)\right]a$
$= \left(8 + \frac{7}{5}\right)a$
$= \left(8 + 1\frac{2}{5}\right)a$
$= 9\frac{2}{5}a$

Page 225 Reading Mathematics

1. Sample answer: a factor is found by dividing a number; a multiple is found by multiplying a number.

2a. A factotum is a person that has many diverse activities or responsibilities. A factotum produces results, just as factors produce products.

2b. Multicultural means many diverse cultures, just as a multiple is one of many numbers that can be formed.

2c. Multimedia means using several media, or multiple media.

3. Sample answer: words with fact- generally mean make or do, and words with multi- mean many.

5-6 Least Common Multiple

Page 226 How can you use prime factors to find the least common multiple?

a. 2004, 2008, 2012

b. 2006, 2012, 2018

c. 2012

Page 228 Check for Understanding

1. The LCM involves the common multiples of a set of whole numbers; the LCD is the LCM of the denominators of two or more fractions.

2. Sample answer: $\frac{1}{7}$ and $\frac{1}{5}$

3. $6 = 2 \cdot 3 = 2 \cdot 3$
$8 = 2 \cdot 2 \cdot 2 = 2^3$
$\text{LCM} = 2^3 \cdot 3$
$= 24$
The LCM of 6 and 8 is 24.

4. $7 = 7 = 7$
$9 = 3 \cdot 3 = 3^2$
$\text{LCM} = 7 \cdot 3^2$
$= 63$
The LCM of 7 and 9 is 63.

5. $10 = 2 \cdot 5$
$14 = 2 \cdot 7$
$\text{LCM} = 2 \cdot 5 \cdot 7$
$= 70$
The LCM of 10 and 14 is 70.

6. $12 = 2 \cdot 2 \cdot 3 = 2^2 \cdot 3$
$30 = 2 \cdot 3 \cdot 5 = 2 \cdot 3 \cdot 5$
$\text{LCM} = 2^2 \cdot 3 \cdot 5$
$= 60$
The LCM of 12 and 30 is 60.

7. $16 = 2 \cdot 2 \cdot 2 \cdot 2 = 2^4$
$24 = 2 \cdot 2 \cdot 2 \cdot 3 = 2^3 \cdot 3$
$\text{LCM} = 2^4 \cdot 3$
$= 48$
The LCM of 16 and 24 is 48.

8. $36ab = 2 \cdot 2 \cdot 3 \cdot 3 \cdot a \cdot b = 2^2 + 3^2 \cdot a \cdot b$
$4b = 2 \cdot 2 \cdot b = 2^2 \cdot b$
$\text{LCM} = 2^2 \cdot 3^2 \cdot a \cdot b$
$= 36ab$
The LCM of $36ab$ and $4b$ is $36ab$.

9. $2 = 2 = 2$
$8 = 2 \cdot 2 \cdot 2 = 2^3$
$\text{LCM} = 2^3$
$= 8$
The LCD of $\frac{1}{2}$ and $\frac{3}{8}$ is 8.

10. $3 = 3$
$10 = 2 \cdot 5$
$\text{LCM} = 2 \cdot 3 \cdot 5$
$= 30$
The LCD of $\frac{2}{3}$ and $\frac{7}{10}$ is 30.

11. $25x = 5 \cdot 5 \cdot x = 5^2 \cdot x$
$20x = 2 \cdot 2 \cdot 5 \cdot x = 2^2 \cdot 5 \cdot x$
$\text{LCM} = 2^2 \cdot 5^2 \cdot x$
$= 100x$
The LCD of $\frac{2}{25x}$ and $\frac{13}{20x}$ is $100x$.

12. $\frac{1}{4} = \frac{1 \cdot 2^2}{2^4} = \frac{4}{16}$
$\frac{3}{16} = \frac{3}{2^4} = \frac{3}{16}$
Since $\frac{4}{16} > \frac{3}{16}$, then $\frac{1}{4} > \frac{3}{16}$.

13. $\frac{10}{45} = \frac{10}{3^2 \cdot 5} = \frac{10}{45}$
$\frac{2}{9} = \frac{2 \cdot 5}{3^2 \cdot 5} = \frac{10}{45}$
Since $\frac{10}{45} = \frac{10}{45}$, then $\frac{10}{45} = \frac{2}{9}$.

14. $\frac{5}{7} = \frac{5 \cdot 3^2}{3^2 \cdot 7} = \frac{45}{63}$
$\frac{7}{9} = \frac{7 \cdot 7}{3^2 \cdot 7} = \frac{49}{63}$
Since $\frac{45}{63} < \frac{49}{63}$, then $\frac{5}{7} < \frac{7}{9}$.

15. $52 = 2 \cdot 2 \cdot 13 = 2^2 \cdot 13$
$20 = 2 \cdot 2 \cdot 5 = 2^2 \cdot 5$
$\text{LCM} = 2^2 \cdot 5 \cdot 13$
$= 260$
front gear:
$260 \div 52 = 5$ revolutions
back gear:
$260 \div 20 = 13$ revolutions

Pages 229–230 Practice and Apply

16. $4 = 2 \cdot 2 = 2^2$
$10 = 2 \cdot 5 = 2 \cdot 5$
$\text{LCM} = 2^2 \cdot 5$
$= 20$
The LCM of 4 and 10 is 20.

17. $20 = 2 \cdot 2 \cdot 5 = 2^2 \cdot 5$
$12 = 2 \cdot 2 \cdot 3 = 2^2 \cdot 3$
$\text{LCM} = 2^2 \cdot 3 \cdot 5$
$= 60$
The LCM of 20 and 12 is 60.

18. $2 = 2 = 2$
$9 = 3 \cdot 3 = 3^2$
$\text{LCM} = 2 \cdot 3^2$
$= 18$
The LCM of 2 and 9 is 18.

19. $16 = 2 \cdot 2 \cdot 2 \cdot 2 = 2^4$
$3 = 3 = 3$
$\text{LCM} = 2^4 \cdot 3$
$= 48$
The LCM of 16 and 3 is 48.

20. $15 = 3 \cdot 5 = 3 \cdot 5$
$75 = 3 \cdot 5 \cdot 5 = 3 \cdot 5^2$
$\text{LCM} = 3 \cdot 5^2$
$= 75$
The LCM of 15 and 75 is 75.

21. $21 = 3 \cdot 7 = 3 \cdot 7$
$28 = 2 \cdot 2 \cdot 7 = 2^2 \cdot 7$
$\text{LCM} = 2^2 \cdot 3 \cdot 7$
$= 84$
The LCM of 21 and 28 is 84.

22. $14 = 2 \cdot 7 = 2 \cdot 7$
$28 = 2 \cdot 2 \cdot 7 = 2^2 \cdot 7$
$\text{LCM} = 2^2 \cdot 7$
$= 28$
The LCM of 14 and 28 is 28.

23. $20 = 2 \cdot 2 \cdot 5 = 2^2 \cdot 5$
$50 = 2 \cdot 5 \cdot 5 = 2 \cdot 5^2$
$\text{LCM} = 2^2 \cdot 5^2$
$= 100$
The LCM of 20 and 50 is 100.

24. $18 = 2 \cdot 3 \cdot 3 = 2 \cdot 3^2$
$32 = 2 \cdot 2 \cdot 2 \cdot 2 \cdot 2 = 2^5$
$\text{LCM} = 2^5 \cdot 3^2$
$= 288$
The LCM of 18 and 32 is 288.

25. $24 = 2 \cdot 2 \cdot 2 \cdot 3 = 2^3 \cdot 3$
$32 = 2 \cdot 2 \cdot 2 \cdot 2 \cdot 2 = 2^5$
$\text{LCM} = 2^5 \cdot 3$
$= 96$
The LCM of 24 and 32 is 96.

26. $10 = 2 \cdot 5 = 2 \cdot 5$
$20 = 2 \cdot 2 \cdot 5 = 2^2 \cdot 5$
$40 = 2 \cdot 2 \cdot 2 \cdot 5 = 2^3 \cdot 5$
$\text{LCM} = 2^3 \cdot 5$
$= 40$
The LCM of 10, 20, and 40 is 40.

27. $7 = 7 = 7$
$21 = 3 \cdot 7 = 3 \cdot 7$
$84 = 2 \cdot 2 \cdot 3 \cdot 7 = 2^2 \cdot 3 \cdot 7$
$\text{LCM} = 2^2 \cdot 3 \cdot 7$
$= 84$
The LCM of 7, 21, and 84 is 84.

28. $9 = 3 \cdot 3 = 3^2$
$12 = 2 \cdot 2 \cdot 3 = 2^2 \cdot 3$
$15 = 3 \cdot 5 = 3 \cdot 5$
$\text{LCM} = 2^2 \cdot 3^2 \cdot 5$
$= 180$
The LCM of 9, 12, and 15 is 180.

29. $45 = 3 \cdot 3 \cdot 5 = 3^2 \cdot 5$
$30 = 2 \cdot 3 \cdot 5 = 2 \cdot 3 \cdot 5$
$35 = 5 \cdot 7 = 5 \cdot 7$
$\text{LCM} = 2 \cdot 3^2 \cdot 5 \cdot 7$
$= 630$
The LCM of 45, 30, and 35 is 630.

30. $20c = 2 \cdot 2 \cdot 5 \cdot c = 2^2 \cdot 5 \cdot c$
$12c = 2 \cdot 2 \cdot 3 \cdot c = 2^2 \cdot 3 \cdot c$
$\text{LCM} = 2^2 \cdot 3 \cdot 5 \cdot c$
$= 60c$
The LCM of $20c$ and $12c$ is $60c$.

31. $16a^2 = 2 \cdot 2 \cdot 2 \cdot 2 \cdot a^2 = 2^4 \cdot a^2$
$14ab = 2 \cdot 7 \cdot a \cdot b = 2 \cdot 7 \cdot a \cdot b$
$\text{LCM} = 2^4 \cdot 7 \cdot a^2 \cdot b$
$= 112a^2b$
The LCM of $16a^2$ and $14ab$ is $112a^2b$.

32. $7x = 7 \cdot x = 7 \cdot x$
$12x = 2 \cdot 2 \cdot 3 \cdot x = 2^2 \cdot 3 \cdot x$
$\text{LCM} = 2^2 \cdot 3 \cdot 7 \cdot x$
$= 84x$
The LCM of $7x$ and $12x$ is $84x$.

33. $75n^2 = 3 \cdot 5 \cdot 5 \cdot n^2 = 3 \cdot 5^2 \cdot n^2$
$25n^4 = 5 \cdot 5 \cdot n^4 = 5^2 \cdot n^4$
$\text{LCM} = 3 \cdot 5^2 \cdot n^4$
$= 75n^4$
The LCM of $75n^2$ and $25n^4$ is $75n^4$.

34. $4 = 2 \cdot 2 = 2^2$
$8 = 2 \cdot 2 \cdot 2 = 2^3$
$\text{LCM} = 2^3$
$= 8$
The LCD of $\frac{1}{4}$ and $\frac{7}{8}$ is 8.

35. $15 = 3 \cdot 5$
$3 = 3$
$\text{LCM} = 3 \cdot 5$
$= 15$
The LCD of $\frac{8}{15}$ and $\frac{1}{3}$ is 15.

36. $5 = 5$
$2 = 2$
$\text{LCM} = 2 \cdot 5$
$= 10$
The LCD of $\frac{4}{5}$ and $\frac{1}{2}$ is 10.

37. $5 = 5$
$7 = 7$
$\text{LCM} = 5 \cdot 7$
$= 35$
The LCD of $\frac{2}{5}$ and $\frac{6}{7}$ is 35.

38. $9 = 3 \cdot 3 = 3^2$
$12 = 2 \cdot 2 \cdot 3 = 2^2 \cdot 3$
$\text{LCM} = 2^2 \cdot 3^2$
$= 36$
The LCD of $\frac{4}{9}$ and $\frac{5}{12}$ is 36.

39. $8 = 2 \cdot 2 \cdot 2 = 2^3$
$6 = 2 \cdot 3 = 2 \cdot 3$
$\text{LCM} = 2^3 \cdot 3$
$= 24$
The LCD of $\frac{3}{8}$ and $\frac{5}{6}$ is 24.

40. $3t = 3 \cdot t$
$5t^2 = 5 \cdot t^2$
$\text{LCM} = 3 \cdot 5 \cdot t^2$
$= 15t^2$
The LCD of $3t$ and $5t^2$ is $15t^2$.

41. $8cd = 2 \cdot 2 \cdot 2 \cdot c \cdot d = 2^3 \cdot c \cdot d$
$16c^2 = 2 \cdot 2 \cdot 2 \cdot 2 \cdot c^2 = 2^4 \cdot c^2$
$\text{LCM} = 2^4 \cdot c^2 \cdot d$
$= 16c^2d$
The LCD of $\frac{7}{8cd}$ and $\frac{5}{16c^2}$ is $16c^2d$.

42. $12 = 2 \cdot 2 \cdot 3 = 2^2 \cdot 3$
$30 = 2 \cdot 3 \cdot 5 = 2 \cdot 3 \cdot 5$
$84 = 2 \cdot 2 \cdot 3 \cdot 7 = 2^2 \cdot 3 \cdot 7$
$\text{LCM} = 2^2 \cdot 3 \cdot 5 \cdot 7$
$= 420$
The planets align every 420 years.

43. $30 = 2 \cdot 3 \cdot 5 = 2 \cdot 3 \cdot 5$
$20 = 2 \cdot 2 \cdot 5 = 2^2 \cdot 5$
$\text{LCM} = 2^2 \cdot 3 \cdot 5$
$= 60$
They will be together in 60 seconds.

44. $\frac{1}{2} = \frac{1 \cdot 2 \cdot 3}{2^2 \cdot 3} = \frac{6}{12}$
$\frac{5}{12} = \frac{5}{2^2 \cdot 3} = \frac{5}{12}$
Since $\frac{6}{12} > \frac{5}{12}$, then $\frac{1}{2} > \frac{5}{12}$.

45. $\frac{7}{9} = \frac{7 \cdot 2}{2 \cdot 3^2} = \frac{14}{18}$
$\frac{5}{6} = \frac{5 \cdot 3}{2 \cdot 3^2} = \frac{15}{18}$
Since $\frac{14}{18} < \frac{15}{18}$, then $\frac{7}{9} < \frac{5}{6}$.

46. $\frac{3}{5} = \frac{3 \cdot 7}{5 \cdot 7} = \frac{21}{35}$
$\frac{4}{7} = \frac{4 \cdot 5}{5 \cdot 7} = \frac{20}{35}$
Since $\frac{21}{35} > \frac{20}{35}$, then $\frac{3}{5} > \frac{4}{7}$.

47. $\frac{21}{100} = \frac{21}{2^2 \cdot 5^2} = \frac{21}{100}$
$\frac{1}{5} = \frac{1 \cdot 2^2 \cdot 5}{2^2 \cdot 5^2} = \frac{20}{100}$
Since $\frac{21}{100} > \frac{20}{100}$, then $\frac{21}{100} > \frac{1}{5}$.

48. $\frac{17}{34} = \frac{17}{2 \cdot 17} = \frac{17}{34}$
$\frac{1}{2} = \frac{1 \cdot 17}{2 \cdot 17} = \frac{17}{34}$
Since $\frac{17}{34} = \frac{17}{34}$, then $\frac{17}{34} = \frac{1}{2}$.

49. $\frac{12}{17} = \frac{12 \cdot 3}{3 \cdot 17} = \frac{36}{51}$
$\frac{36}{51} = \frac{36}{3 \cdot 17} = \frac{36}{51}$
Since $\frac{36}{51} = \frac{36}{51}$, then $\frac{12}{17} = \frac{36}{51}$.

50. $\frac{8}{9} = \frac{8 \cdot 7}{3^2 \cdot 7} = \frac{56}{63}$
$\frac{19}{21} = \frac{19 \cdot 3}{3^2 \cdot 7} = \frac{57}{63}$
Since $\frac{56}{63} < \frac{57}{63}$, then $\frac{8}{9} < \frac{19}{21}$.

51. $\frac{-9}{11} = \frac{-9 \cdot 2 \cdot 3}{2 \cdot 3 \cdot 11} = \frac{-54}{66}$
$\frac{-5}{6} = \frac{-5 \cdot 11}{2 \cdot 3 \cdot 11} = \frac{-55}{66}$
Since $\frac{-54}{66} > \frac{-55}{66}$, then $\frac{-9}{11} > \frac{-5}{6}$.

52. $\frac{-14}{15} = \frac{-14 \cdot 2}{2 \cdot 3 \cdot 5} = \frac{-28}{30}$

$\frac{-9}{10} = \frac{-9 \cdot 3}{2 \cdot 3 \cdot 5} = \frac{-27}{30}$

Since $\frac{-28}{30} < \frac{-27}{30}$, then $\frac{-14}{15} < \frac{-9}{10}$.

53. $\frac{7}{39} = \frac{7 \cdot 3}{3^2 \cdot 13} = \frac{21}{117}$

$\frac{5}{9} = \frac{5 \cdot 13}{3^2 \cdot 13} = \frac{65}{117}$

Since $\frac{21}{117} < \frac{65}{117}$, then $\frac{7}{39} < \frac{5}{9}$.

There are more amphibians.

54. $\frac{11}{20} = \frac{11 \cdot 5}{2^2 \cdot 5^2} = \frac{55}{100}$

$\frac{14}{25} = \frac{14 \cdot 2^2}{2^2 \cdot 5^2} = \frac{56}{100}$

Since $\frac{55}{100} < \frac{56}{100}$, then $\frac{11}{20} < \frac{14}{25}$.

More people in Anchorage have cell phones.

55. $36 = 2 \cdot 2 \cdot 3 \cdot 3$

$2 \cdot 2 \cdot 3 = 12$

$2 \cdot 3 \cdot 3 = 18$

The LCM of 12 and 18 is 36.

56. Sometimes; sample answer: the LCM of 2, 4, and 8, is 8, which is one of the numbers. The LCM of 2, 3, and 4, is 12, which is not one of the numbers.

57. Never; sample answer: 5 and 6 do not contain any factors in common, and the LCM of 5 and 6 is 30.

58. Always; sample answer: the GCF of 6 and 12 is 6 and the LCM is 12. The GCF of 3 and 5 is 1 and the LCM is 15. In each case, LCM > GCF.

59a. If two numbers are relatively prime, then their LCM is the product of the two numbers. For example, the LCM of 4 and 5 is $2^2 \cdot 5$ or 20; the LCM of 6 and 25 is $2 \cdot 3 \cdot 5^2$ or 150.

59b. Always; the LCM contains all of the factors of both numbers. Therefore, it must contain any common factors.

60. The LCM of two or more numbers involves the product of prime factors. Answers should include the following.

- The LCM of two numbers is the least number into which both of the numbers will divide evenly.
- Write the prime factorization of each number in exponential form. Multiply each factor the greatest number of times that it appears in either factorization. If the same factors appear in more than one number, multiply the greatest power of the factor that appears.

61. B; $12a^2b = 2 \cdot 2 \cdot 3 \cdot a^2 \cdot b = 2^2 \cdot 3 \cdot a^2 \cdot b$

$9ac = 3 \cdot 3 \cdot a \cdot c = 3^2 \cdot a \cdot c$

$\text{LCM} = 2^2 \cdot 3^2 \cdot a^2 \cdot b \cdot c$

$= 36a^2\,bc$

62. D; $\frac{7}{8} = \frac{7 \cdot 2}{2^4} = \frac{14}{16}$

$\frac{3}{4} = \frac{3 \cdot 2^2}{2^4} = \frac{12}{16}$

$\frac{5}{8} = \frac{5 \cdot 2}{2^4} = \frac{10}{16}$

$\frac{7}{16} = \frac{7}{2^4} = \frac{7}{16}$

$\frac{13}{16} = \frac{13}{2^4} = \frac{13}{16}$

The next smaller size is $\frac{13}{16}$.

Page 230 Maintain Your Skills

63. $\frac{7}{8} - \frac{3}{8} = \frac{7-3}{8}$

$= \frac{4}{8}$ or $\frac{1}{2}$

64. $3\frac{9}{11} - \frac{5}{11} = \frac{42}{11} - \frac{5}{11}$

$= \frac{42-5}{11}$

$= \frac{37}{11}$ or $3\frac{4}{11}$

65. $\frac{13}{14} + \frac{3}{14} = \frac{13+3}{14}$

$= \frac{16}{14}$

$= \frac{8}{7}$ or $1\frac{1}{7}$

66. $2\frac{5}{6} + 4\frac{1}{6} = (2 + 4) + \left(\frac{5}{6} + \frac{1}{6}\right)$

$= 6 + \frac{6}{6}$

$= 6 + 1$

$= 7$

67. $\frac{3}{n} \div \frac{1}{n} = \frac{3}{n} \cdot \frac{n}{1}$

$= \frac{3}{\cancel{n}_1} \cdot \frac{\cancel{n}^1}{1}$

$= \frac{3}{1}$ or 3

68. $\frac{x}{8} \div \frac{x}{6} = \frac{x}{8} \cdot \frac{6}{x}$

$= \frac{\cancel{x}^1}{\cancel{8}_4} \cdot \frac{\cancel{6}^3}{\cancel{x}_1}$

$= \frac{3}{4}$

69. $\frac{ac}{5} \div \frac{c}{d} = \frac{ac}{5} \cdot \frac{d}{c}$

$= \frac{a\cancel{c}^1}{5} \cdot \frac{d}{\cancel{c}_1}$

$= \frac{ad}{5}$

70. $\frac{6k}{7m} \div \frac{3}{14m} = \frac{6k}{7m} \cdot \frac{14m}{3}$

$= \frac{\cancel{6}^2k}{\cancel{7}_1\cancel{m}_1} \cdot \frac{\cancel{14}^2\cancel{m}^1}{\cancel{3}_1}$

$= \frac{4k}{1}$ or $4k$

71. $7 + 2n = 11$

$7 + 2n - 7 = 11 - 7$

$\frac{2n}{2} = \frac{4}{2}$

$n = 2$

72. $9 = x - 4$

$9 + 4 = x - 4 + 4$

$13 = x$

73. $\frac{a}{-2} = 10$

$\frac{a}{-2}(-2) = 10(-2)$

$a = -20$

74. $-5c = -105$

$\frac{-5c}{-5} = \frac{-105}{-5}$

$c = 21$

75. $\frac{3}{8} + \frac{3}{4} \approx 0 + 1 = 1$

76. $\frac{9}{10} + \frac{14}{15} \approx 1 + 1 = 2$

77. $\frac{4}{7} + 2\frac{1}{5} \approx 1 + 2 = 3$

78. $5\frac{7}{8} + \frac{2}{3} \approx 6 + 1 = 7$

79. $8\frac{3}{11} + 7\frac{2}{9} \approx 8 + 7 = 15$

80. $20\frac{5}{16} + 6\frac{1}{9} \approx 20 + 6 = 26$

Page 231 Algebra Activity (Follow-Up of Lesson 5-6)

1. The first number selected must be even to guarantee that the second player can circle a number; for example, if the first number were 97, there is no factor or multiple on the hundreds chart.
2. Sample answer: finding patterns in the chart

5-7 Adding and Subtracting Unlike Fractions

Page 232 Can the LCM be used to add and subtract fractions with different denominators?

a. 6

b. $\frac{5}{6}$

c. 3; 2

d. a figure with 12 parts; $\frac{7}{12}$

Pages 234–235 Check for Understanding

1. Find the least common denominator.
2. Sample answer: a piece of string $15\frac{3}{4}$ inches long is cut so that it is $2\frac{1}{8}$ inches shorter. How long is the piece of string after it is cut?
3. José; he finds a common denominator by multiplying the denominators. Daniel incorrectly adds the numerators and the denominators of unlike fractions.
4. $\frac{1}{10} + \frac{1}{3} = \frac{1}{10} \cdot \frac{3}{3} + \frac{1}{3} \cdot \frac{10}{10}$
$= \frac{3}{30} + \frac{10}{30}$
$= \frac{13}{30}$
5. $-\frac{1}{6} + \frac{7}{18} = -\frac{1}{6} \cdot \frac{3}{3} + \frac{7}{18}$
$= -\frac{3}{18} + \frac{7}{18}$
$= \frac{4}{18}$ or $\frac{2}{9}$
6. $\frac{1}{4} - \frac{2}{3} = \frac{1}{4} \cdot \frac{3}{3} - \frac{2}{3} \cdot \frac{4}{4}$
$= \frac{3}{12} - \frac{8}{12}$
$= -\frac{5}{12}$
7. $-\frac{7}{10} - \frac{2}{15} = -\frac{7}{10} \cdot \frac{3}{3} - \frac{2}{15} \cdot \frac{2}{2}$
$= -\frac{21}{30} - \frac{4}{30}$
$= -\frac{25}{30}$ or $-\frac{5}{6}$
8. $6\frac{4}{5} + \left(-1\frac{3}{4}\right) = \frac{34}{5} + \left(-\frac{7}{4}\right)$
$= \frac{34}{5} \cdot \frac{4}{4} + \left(-\frac{7}{4}\right) \cdot \frac{5}{5}$
$= \frac{136}{20} + \left(-\frac{35}{20}\right)$
$= \frac{101}{20}$
$= 5\frac{1}{20}$
9. $-9\frac{3}{4} - \left(-5\frac{1}{2}\right) = \frac{-39}{4} - \frac{-11}{2}$
$= \frac{-39}{4} - \left(\frac{-11}{2}\right) \cdot \frac{2}{2}$
$= \frac{-39}{4} - \left(\frac{-22}{4}\right)$
$= \frac{-17}{4}$
$= -4\frac{1}{4}$
10. $1\frac{5}{8} + 3\frac{1}{2} = \frac{13}{8} + \frac{7}{2}$
$= \frac{13}{8} + \frac{7}{2} \cdot \frac{4}{4}$
$= \frac{13}{8} + \frac{28}{8}$
$= \frac{41}{8}$
$= 5\frac{1}{8}$
Jessica needs $5\frac{1}{8}$ yd. of fabric.

Pages 235–236 Practice and Apply

11. $\frac{3}{5} + \frac{3}{10} = \frac{3}{5} \cdot \frac{2}{2} + \frac{3}{10}$
$= \frac{6}{10} + \frac{3}{10}$
$= \frac{9}{10}$
12. $\frac{9}{26} + \frac{3}{13} = \frac{9}{26} + \frac{3}{13} \cdot \frac{2}{2}$
$= \frac{9}{26} + \frac{6}{26}$
$= \frac{15}{26}$
13. $\frac{3}{7} + \left(-\frac{1}{4}\right) = \frac{3}{7} \cdot \frac{4}{4} + \left(\frac{-1}{4}\right) \cdot \frac{7}{7}$
$= \frac{12}{28} + \frac{-7}{28}$
$= \frac{5}{28}$
14. $-\frac{5}{8} + \left(-\frac{1}{3}\right) = -\frac{5}{8} \cdot \frac{3}{3} + \left(-\frac{1}{3}\right) \cdot \frac{8}{8}$
$= \frac{-15}{24} + \frac{-8}{24}$
$= \frac{-23}{24}$ or $-\frac{23}{24}$
15. $\frac{7}{8} - \left(-\frac{3}{16}\right) = \frac{7}{8} \cdot \frac{2}{2} - \left(-\frac{3}{16}\right)$
$= \frac{14}{16} - \left(-\frac{3}{16}\right)$
$= \frac{17}{16}$
$= 1\frac{1}{16}$
16. $-\frac{2}{5} - \frac{7}{8} = -\frac{2}{5} \cdot \frac{8}{8} - \frac{7}{8} \cdot \frac{5}{5}$
$= \frac{-16}{40} - \frac{35}{40}$
$= -\frac{51}{40}$
$= -1\frac{11}{40}$

17. $\frac{3}{4} - \frac{5}{8} = \frac{3}{4} \cdot \frac{2}{2} - \frac{5}{8}$
$= \frac{6}{8} - \frac{5}{8}$
$= \frac{1}{8}$

18. $\frac{5}{7} + \left(-\frac{10}{21}\right) = \frac{5}{7} \cdot \frac{3}{3} + \left(-\frac{10}{21}\right)$
$= \frac{15}{21} + \left(-\frac{10}{21}\right)$
$= \frac{5}{21}$

19. $-\frac{1}{2} + \frac{3}{8} = -\frac{1}{2} \cdot \frac{4}{4} + \frac{3}{8}$
$= \frac{-4}{8} + \frac{3}{8}$
$= -\frac{1}{8}$

20. $-\frac{2}{3} + \frac{7}{12} = -\frac{2}{3} \cdot \frac{4}{4} + \frac{7}{12}$
$= \frac{-8}{12} + \frac{7}{12}$
$= -\frac{1}{12}$ or $-\frac{1}{12}$

21. $1\frac{2}{5} - \frac{1}{3} = \frac{7}{5} - \frac{1}{3}$
$= \frac{7}{5} \cdot \frac{3}{3} - \frac{1}{3} \cdot \frac{5}{5}$
$= \frac{21}{15} - \frac{5}{15}$
$= \frac{16}{15}$
$= 1\frac{1}{15}$

22. $\frac{7}{8} + 4\frac{1}{24} = \frac{7}{8} + \frac{97}{24}$
$= \frac{7}{8} \cdot \frac{3}{3} + \frac{97}{24}$
$= \frac{21}{24} + \frac{97}{24}$
$= \frac{118}{24}$
$= 4\frac{22}{24}$ or $4\frac{11}{12}$

23. $-6\frac{2}{3} - \frac{8}{9} = -\frac{20}{3} - \frac{8}{9}$
$= -\frac{20}{3} \cdot \frac{3}{3} - \frac{8}{9}$
$= -\frac{60}{9} - \frac{8}{9}$
$= \frac{-68}{9}$
$= -7\frac{5}{9}$

24. $2\frac{16}{30} - \frac{7}{15} = \frac{76}{30} - \frac{7}{15}$
$= \frac{76}{30} - \frac{7}{15} \cdot \frac{2}{2}$
$= \frac{76}{30} - \frac{14}{30}$
$= \frac{62}{30}$
$= \frac{31}{15}$ or $2\frac{1}{15}$

25. $-4\frac{1}{6} + \left(-7\frac{11}{18}\right) = -\frac{25}{6} + \left(-\frac{137}{18}\right)$
$= -\frac{25}{6} \cdot \frac{3}{3} + \left(-\frac{137}{18}\right)$
$= \frac{-75}{18} + \frac{-137}{18}$
$= \frac{-212}{18}$
$= -11\frac{14}{18}$ or $-11\frac{7}{9}$

26. $3\frac{1}{2} - \left(-7\frac{1}{3}\right) = \frac{7}{2} - \left(-\frac{22}{3}\right)$
$= \frac{7}{2} \cdot \frac{3}{3} - \left(-\frac{22}{3}\right) \cdot \frac{2}{2}$
$= \frac{21}{6} - \left(-\frac{44}{6}\right)$
$= \frac{65}{6}$
$= 10\frac{5}{6}$

27. $-19\frac{3}{8} - \left(-4\frac{3}{4}\right) = \frac{-155}{8} - \left(-\frac{19}{4}\right)$
$= \frac{-155}{8} - \left(-\frac{19}{4}\right) \cdot \frac{2}{2}$
$= \frac{-155}{8} - \left(-\frac{38}{8}\right)$
$= \frac{-117}{8}$
$= -14\frac{5}{8}$

28. $-3\frac{2}{5} - \left(-2\frac{4}{7}\right) = -\frac{17}{5} - \left(-\frac{18}{7}\right)$
$= -\frac{17}{5} \cdot \frac{7}{7} - \left(-\frac{18}{7}\right) \cdot \frac{5}{5}$
$= \frac{-119}{35} - \left(-\frac{90}{35}\right)$
$= -\frac{29}{35}$

29. $x - y = 4\frac{7}{18} - 1\frac{1}{12}$
$= \frac{79}{18} - \frac{13}{12}$
$= \frac{79}{18} \cdot \frac{2}{2} - \frac{13}{12} \cdot \frac{3}{3}$
$= \frac{158}{36} - \frac{39}{36}$
$= \frac{119}{36}$
$= 3\frac{11}{36}$

30. $\frac{64}{143} - \frac{21}{208} = a$
$\frac{64}{143} \cdot \frac{16}{16} - \frac{21}{208} \cdot \frac{11}{11} = a$
$\frac{1024}{2288} - \frac{231}{2288} = a$
$\frac{793}{2288} = a$
$\frac{61}{176} = a$

31. $62\frac{1}{2} - 56\frac{9}{10} = \frac{125}{2} - \frac{569}{10}$
$= \frac{125}{2} \cdot \frac{5}{5} - \frac{569}{10}$
$= \frac{625}{10} - \frac{569}{10}$
$= \frac{56}{10}$
$= \frac{28}{5}$ or $5\frac{3}{5}$

32. $\frac{4}{25} + \frac{27}{50} = \frac{4}{25} \cdot \frac{2}{2} + \frac{27}{50}$
$= \frac{8}{50} + \frac{27}{50}$
$= \frac{35}{50}$ or $\frac{7}{10}$

33. $10 - \left(\frac{1}{2} + \frac{3}{4}\right) = 10 - \left(\frac{1}{2} \cdot \frac{2}{2} + \frac{3}{4}\right)$
$= 10 - \left(\frac{2}{4} + \frac{3}{4}\right)$
$= \frac{10}{1} - \frac{5}{4}$
$= \frac{10}{1} \cdot \frac{4}{4} - \frac{5}{4}$
$= \frac{40}{4} - \frac{5}{4}$
$= \frac{35}{4}$ or $8\frac{3}{4}$ in.

34. $1 - \left(\frac{1}{3} + \frac{2}{5}\right) = 1 - \left(\frac{1}{3} \cdot \frac{5}{5} + \frac{2}{5} \cdot \frac{3}{3}\right)$
$= 1 - \left(\frac{5}{15} + \frac{6}{15}\right)$
$= \frac{1}{1} - \frac{11}{15}$
$= \frac{1}{1} \cdot \frac{15}{15} - \frac{11}{15}$
$= \frac{15}{15} - \frac{11}{15}$
$= \frac{4}{15}$

Makayla got $\frac{4}{15}$ of the votes.

35. Sample answer: Fill the $\frac{1}{2}$-cup. From the $\frac{1}{2}$-cup, fill the $\frac{1}{3}$-cup. $\frac{1}{6}$ cup will be left in the $\frac{1}{2}$-cup because $\frac{1}{2} - \frac{1}{3} = \frac{1}{6}$.

36. Rational numbers are closed under addition, subtraction, and multiplication because the sum, difference, or product of two rational numbers is a rational number. They are not closed under division because a rational number divided by zero is undefined.

37. Find the LCM of the denominators. Then rename the fractions as like fractions with the LCM as the denominators. Answers should include the following.

- For example, the LCM of 4 and 6 is 12. So, $\frac{1}{4} + \frac{5}{6} = \frac{3}{12} + \frac{10}{12} = \frac{13}{12}$ or $1\frac{1}{12}$.
- Writing the prime factorization of the denominators is the first step in finding the LCM of the denominators, which is the LCD. Then the fractions can be added or subtracted.

38. A; $11\frac{3}{8} + 6\frac{7}{9} \approx 11 + 7 = 18$

39. B; $9\frac{1}{2} - \frac{6}{15} = \frac{19}{2} - \frac{6}{15}$
$= \frac{19}{2} \cdot \frac{15}{15} - \frac{6}{15} \cdot \frac{2}{2}$
$= \frac{285}{30} - \frac{12}{30}$
$= \frac{273}{30}$
$= 9\frac{3}{30}$ or $9\frac{1}{10}$

40. $\frac{7}{12} = \frac{4}{12} + \frac{3}{12}$
$= \frac{1}{3} + \frac{1}{4}$

41. $\frac{3}{5} = \frac{6}{10}$
$= \frac{5}{10} + \frac{1}{10}$
$= \frac{1}{2} + \frac{1}{10}$

42. $\frac{2}{9} = \frac{4}{18}$
$= \frac{1}{18} + \frac{3}{18}$
$= \frac{1}{18} + \frac{1}{6}$

Page 236 Maintain Your Skills

43. $9 = 3^2$
$12 = 2^2 \cdot 3$
$\text{LCM} = 2^2 \cdot 3^2$
$= 36$

The LCD of $\frac{4}{9}$ and $\frac{7}{12}$ is 36.

44. $15t = 3 \cdot 5 \cdot t$
$5t = 5 \cdot t$
$\text{LCM} = 3 \cdot 5 \cdot t$
$= 15t$

The LCD of $\frac{3}{15t}$ and $\frac{2}{5t}$ is $15t$.

45. $3n = 3 \cdot n$
$6n^3 = 2 \cdot 3 \cdot n^3$
$\text{LCM} = 2 \cdot 3 \cdot n^3$
$= 6n^3$

The LCD of $\frac{1}{3n}$ and $\frac{7}{6n^3}$ is $6n^3$.

46. $\frac{4}{7} + \frac{6}{7} = \frac{10}{7}$
$= 1\frac{3}{7}$

47. $2\frac{3}{4} + 6\frac{3}{4} = (2 + 6) + \left(\frac{3}{4} + \frac{3}{4}\right)$
$= 8 + \frac{6}{4}$
$= 8 + \frac{3}{2}$
$= 8 + 1\frac{1}{2}$
$= 9\frac{1}{2}$

48. $\frac{7}{8} - \frac{5}{8} = \frac{7 - 5}{8}$
$= \frac{2}{8}$ or $\frac{1}{4}$

49. $3\frac{2}{5} - \frac{3}{5} = \frac{17}{5} - \frac{3}{5}$
$= \frac{17 - 3}{5}$
$= \frac{14}{5}$ or $2\frac{4}{5}$

50. $4\frac{1}{6} + 5\frac{5}{6} = (4 + 5) + \left(\frac{1}{6} + \frac{5}{6}\right)$
$= 9 + \frac{1 + 5}{6}$
$= 9 + \frac{6}{6}$
$= 9 + 1$ or 10

51. $8 - 6\frac{1}{5} = \frac{8}{1} - \frac{31}{5}$
$= \frac{8}{1} \cdot \frac{5}{5} - \frac{31}{5}$
$= \frac{40}{5} - \frac{31}{5}$
$= \frac{9}{5}$ or $1\frac{4}{5}$

52. $124 = 4 \cdot 31$
$= 2 \cdot 2 \cdot 31$
$= 2^2 \cdot 31$

53. $24 + (-12) + 15 = 12 + 15$
$= 27$

54. $(-2) + 5 + (-3) = 3 + (-3)$
$= 0$

55. $4 + (-9) + (-9) + 5 = -5 + (-9) + 5$
$= -14 + 5$
$= -9$

56. $-10 + (-9) + (-11) + (-8)$
$= -19 + (-11) + (-8)$
$= -30 + (-8)$
$= -38$

Page 237 Algebra Activity (Preview of Lesson 5-8)

1. See students' work.
2. See students' work.
3. See students' work.

5-8 Measures of Central Tendency

Page 238 How are measures of central tendency used in the real world?

a. 11

b. 12

c. 13.0

d. Sample answer: 11, because it occurs more often than any other number.

Page 241 Check for Understanding

1. Mean; when an extreme value is added to the other data, it can raise or lower the sum, and therefore the mean.
2. Sample answer: 4, 7, 9, 10, 10
3. mean $= \frac{4 + 5 + 7 + 3 + 9 + 11 + 23 + 37}{8}$

   $= \frac{99}{8} \approx 12.4$

   The mean is 12.4.

   To find the median, order the numbers from least to greatest.

   3, 4, 5, $\underbrace{7, 9}$, 11, 23, 37

   $\frac{7 + 9}{2} = 8$

   The median is 8.

   There is no mode because each number in the set occurs only once.
4. mean $= \frac{7.2 + 3.6 + 9.0 + 5.2 + 7.2 + 6.5 + 3.6}{7}$

   $= \frac{42.3}{7} \approx 6.0$

   The mean is 6.0.

   To find the median, order the numbers from least to greatest.

   3.6, 3.6, 5.2, $\underline{6.5}$, 7.2, 7.2, 9.0

   The median is 6.5.

   3.6 and 7.2 are the modes as they both occur twice.
5. mean $= \frac{1(2) + 2(3) + 3(3) + 4(4) + 6(3) + 7}{16}$

   ≈ 3.6

   The mean is 3.6.

   There are 16 numbers. So the mean of the 2 middle numbers is the median.

   $\frac{3 + 4}{2} = 3.5$

   The median is 3.5.

   You can see from the graph that 4 occurs most often in the data set. The mode is 4.
6. mean

   $= \frac{34 + 26 + 37 + 35 + 42 + 25 + 25 + 28 + 13}{9}$

   $= \frac{265}{9} \approx 29.4$

   median: 13, 25, 25, 26, $\underbrace{28}$, 34, 35, 37, 42

   28

   mode: 25
7. Sample answer: 13 could be an extreme value because it is 12 less than the next value. It lowers the mean by 2.1.
8. Sample answer: median, because half the data are less than the median and half are greater than the median.
9. Find the sum of the first 5 quiz scores x.

   $86 = \frac{x}{5}$

   $(86)5 = \left(\frac{x}{5}\right)^5$

   $430 = x$

   Find the sixth score y.

   mean $= \frac{\text{sum of first 5 scores} + \text{6th score}}{6}$

   $88 = \frac{430 + y}{6}$

   $528 = 430 + y$

   $98 = y$

   The lowest score Brad can receive is 98.

Pages 241–242 Practice and Apply

10. mean $= \frac{41 + 37 + 43 + 43 + 36}{5}$

    $= \frac{200}{5} = 40$

    The mean is 40.

    To find the median, order the numbers from least to greatest.

    36, 37, $\underline{41}$, 43, 43

    The median is 41.

    Since 43 occurs twice, it is the mode.
11. mean $= \frac{2 + 8 + 16 + 21 + 3 + 8 + 9 + 7 + 6}{9}$

    $= \frac{80}{9} \approx 8.9$

    The mean is 8.9.

    To find the median, order the numbers from least to greatest.

    2, 3, 6, 7, $\underline{8}$, 8, 9, 16, 21

    The median is 8.

    Since 8 occurs twice, it is the mode.
12. mean $= \frac{14 + 6 + 8 + 10 + 9 + 5 + 7 + 13}{8}$

    $= \frac{72}{8} = 9$

    The mean is 9.

    To find the median, order the numbers from least to greatest.

    5, 6, 7, $\underbrace{8, 9}$, 10, 13, 14

    $\frac{8 + 9}{2} = 8.5$

    The median is 8.5.

    There is no mode because each number in the set occurs only once.

13. mean $= \frac{7.5 + 7.1 + 7.4 + 7.6 + 7.4 + 9.0 + 7.9 + 7.1}{8}$

$= \frac{61}{8} \approx 7.6$

To find the median, order the numbers from least to greatest.

7.1, 7.1, 7.4, $\underbrace{7.4, 7.5}$, 7.6, 7.9, 9.0

$\frac{7.4 + 7.5}{2} \approx 7.5$

The median is 7.5.

7.1 and 7.4 are the modes as they both occur twice.

14. mean $= \frac{16(3) + 17 + 18(3) + 19(2) + 20(3) + 21}{13}$

$= \frac{238}{13} \approx 18.3$

The mean is 18.3.

There are 13 numbers. So the 7th number, or 18, is the median.

You can see from the graph that 16, 18, and 20 occur most often in the data set. The modes are 16, 18, and 20.

15. mean $= \frac{4.1(4) + 4.2(4) + 4.3(2) + 4.4 + 4.8}{12}$

≈ 4.3

The mean is 4.3.

There are 12 numbers. So, the mean of the 2 middle numbers is the median.

$\frac{4.2 + 4.2}{2} = 4.2$

The median is 4.2.

You can see from the graph that 4.1 and 4.2 occur most often in the data set. The modes are 4.1 and 4.2.

16. Mean; the height of the tallest basketball player increases the mean height of the team.

17. The data value 60 appears to be an extreme value.

mean with extreme value	mean without extreme value
$= \frac{1365}{15}$	$\frac{1305}{14}$
$= 91$	≈ 93.2

The median, 95, or the modes, 95 and 97 are better descriptors because most students scored higher than the mean, which is 91.

18. Sample answer: The mean is greater than the median because of the high salaries of the top baseball players. These are the extreme values that affect the mean.

19. Sample answer: The median home price would be useful because it is not affected by the cost of the very expensive homes. The cost of half the homes in the county would be greater than the median cost and half would be less.

20. Measures of central tendency are used to describe or represent real-world data. Answers should include the following.

- Test scores, heights of students, and number of hours worked per week can be described using the mean, median, or mode.
- Newspaper articles may discuss the average cost of living, the average family size, or the average income for people in a certain age bracket.

21. C;

mean without 18	mean with 18
$= \frac{142}{8}$	$= \frac{160}{9}$
$= 17.75$	≈ 17.78

The mean increases.

22. A; mean; the extreme value increases the mean tip amount.

Page 242 Maintain Your Skills

23. $9\frac{2}{3} + \frac{1}{6} = \frac{29}{3} + \frac{1}{6}$
$= \frac{29}{3} \cdot \frac{2}{2} + \frac{1}{6}$
$= \frac{58}{6} + \frac{1}{6}$
$= \frac{59}{6}$ or $9\frac{5}{6}$

24. $\frac{7}{8} - \frac{3}{10} = \frac{7}{8} \cdot \frac{5}{5} - \frac{3}{10} \cdot \frac{4}{4}$
$= \frac{35}{40} - \frac{12}{40}$
$= \frac{23}{40}$

25. $-2\frac{3}{4} - 1\frac{1}{8} = -\frac{11}{4} - \frac{9}{8}$
$= \frac{-11}{4} \cdot \frac{2}{2} - \frac{9}{8}$
$= \frac{-22}{8} - \frac{9}{8}$
$= \frac{-31}{8}$ or $-3\frac{7}{8}$

26. $\frac{1}{2} = \frac{1 \cdot 6}{2^2 \cdot 3} = \frac{6}{12}$
$\frac{5}{12} = \frac{5}{2^2 \cdot 3} = \frac{5}{12}$
Since $\frac{6}{12} > \frac{5}{12}$, then $\frac{1}{2} > \frac{5}{12}$.

27. $\frac{16}{50} = \frac{16 \cdot 3}{2 \cdot 3 \cdot 5^2} = \frac{48}{150}$
$\frac{9}{30} = \frac{9 \cdot 5}{2 \cdot 3 \cdot 5^2} = \frac{45}{150}$
Since $\frac{48}{150} > \frac{45}{150}$, then $\frac{16}{50} > \frac{9}{30}$.

28. $\frac{4}{5} = \frac{4 \cdot 12}{2^2 \cdot 3 \cdot 5} = \frac{48}{60}$
$\frac{48}{60} = \frac{48}{2^2 \cdot 3 \cdot 5} = \frac{48}{60}$
Since $\frac{48}{60} = \frac{48}{60}$, then $\frac{4}{5} = \frac{48}{60}$.

29. $y - 5 = 13$
$y - 5 + 5 = 13 + 5$
$y = 18$
Check: $y - 5 = 13$
$18 - 5 \stackrel{?}{=} 13$
$13 = 13$ ✓

30. $10 = 14 + n$
$10 - 14 = 14 + n - 14$
$-4 = n$
Check: $10 = 4 + n$
$10 \stackrel{?}{=} 14 + (-4)$
$10 = 10$ ✓

31. $-4w = 20$

$\frac{-4w}{-4} = \frac{20}{-4}$

$w = -5$

Check: $-4w = 20$

$-4(-5) \stackrel{?}{=} 20$

$20 = 20$ ✓

32.
$$\begin{array}{r} 8.53 \\ 3\overline{)25.6} \\ -24 \\ 1\,6 \\ -1\,5 \\ 10 \\ -9 \\ 1 \end{array}$$

$25.6 \div 3 \approx 8.5$

33.
$$\begin{array}{r} 7.89 \\ 4.7\overline{)370.00} \\ -329 \\ 41\,0 \\ -37\,6 \\ 4\,40 \\ -4\,23 \\ 17 \end{array}$$

$37 \div 4.7 \approx 7.9$

34.
$$\begin{array}{r} 2.72 \\ 11.2\overline{)30.5\,00} \\ -22\,4 \\ 8\,1\,0 \\ -7\,8\,4 \\ 2\,60 \\ -2\,24 \\ 36 \end{array}$$

$30.5 \div 11.2 \approx 2.7$

35.
$$\begin{array}{r} 3. \\ 15.6\overline{)46.8} \\ -46\,8 \\ 0 \end{array}$$

$46.8 \div 15.6 = 3$

Page 243 Graphing Calculator Investigation (Follow Up of Lesson 5-8)

1. 5.85; 6.05
2. −20.57; −21
3. 210.22; 190
4. 5.5; 1.8
5. The median is a term when the number of terms is odd.

6a. Median; 68 in the data set increases the mean so that it does not best represent the data.

6b. 1.13; 1.8; There is a significant difference between the pair of means. The second mean is 4.37 less than the first mean. The median remains the same.

6c. Median; an error can change the mean significantly because it affects the sum of the data. However, the median could remain the same.

5-9 Solving Equations with Rational Numbers

Page 244 How are reciprocals used in solving problems involving music?

a. $\frac{5}{3}n = 440$; $5n = 1320$

b. Divide each side by 5.

c. Multiply each side by $\frac{3}{5}$.

d. 264 vibrations per second

Page 246 Check for Understanding

1. Subtraction Property of Equality
2. Sample answer: $\frac{1}{6}x = 2$
3. Ling; dividing 0.3 by 3 does not isolate the variable on one side.
4. $y + 3.5 = 14.9$

   $y + 3.5 - 3.5 = 14.9 - 3.5$

   $y = 11.4$

   Check: $y + 3.5 = 14.9$

   $11.4 + 3.5 \stackrel{?}{=} 14.9$

   $14.9 = 14.9$ ✓
5. $b - 5 = 13.7$

   $b - 5 + 5 = 13.7 + 5$

   $b = 18.7$

   Check: $b - 5 = 13.7$

   $18.7 - 5 \stackrel{?}{=} 13.7$

   $13.7 = 13.7$ ✓
6. $\frac{3}{2} = w + \frac{3}{5}$

   $\frac{3}{2} - \frac{3}{5} = w + \frac{3}{5} - \frac{3}{5}$

   $\frac{3}{2} - \frac{3}{5} = w$

   $\frac{15}{10} - \frac{6}{10}$ or $\frac{9}{10} = w$

   Check: $\frac{3}{2} = w + \frac{3}{5}$

   $\frac{3}{2} \stackrel{?}{=} \frac{9}{10} + \frac{3}{5}$

   $\frac{3}{2} \stackrel{?}{=} \frac{9}{10} + \frac{6}{10}$

   $\frac{3}{2} = \frac{15}{10}$ or $\frac{3}{2}$ ✓
7. $c - \frac{3}{5} = \frac{5}{6}$

   $c - \frac{3}{5} + \frac{3}{5} = \frac{5}{6} + \frac{3}{5}$

   $c = \frac{5}{6} + \frac{3}{5}$

   $c = \frac{25}{30} + \frac{18}{30}$

   $c = \frac{43}{50}$ or $1\frac{13}{30}$

   Check: $c - \frac{3}{5} = \frac{5}{6}$

   $\frac{43}{30} - \frac{3}{5} \stackrel{?}{=} \frac{5}{6}$

   $\frac{43}{30} - \frac{18}{30} \stackrel{?}{=} \frac{5}{6}$

   $\frac{25}{30}$ or $\frac{5}{6} = \frac{5}{6}$ ✓

8. $x + \frac{5}{8} = 7\frac{1}{2}$

$x + \frac{5}{8} - \frac{5}{8} = 7\frac{1}{2} - \frac{5}{8}$

$x = \frac{15}{2} - \frac{5}{8}$

$x = \frac{60}{8} - \frac{5}{8}$

$x = \frac{55}{8}$ or $6\frac{7}{8}$

Check: $x + \frac{5}{8} = 7\frac{1}{2}$

$\frac{55}{8} + \frac{5}{8} \stackrel{?}{=} 7\frac{1}{2}$

$\frac{60}{8} \stackrel{?}{=} 7\frac{1}{2}$

$\frac{15}{2} \stackrel{?}{=} 7\frac{1}{2}$

$7\frac{1}{2} = 7\frac{1}{2}$ ✓

9. $4\frac{1}{6} = r + 6\frac{1}{4}$

$4\frac{1}{6} - 6\frac{1}{4} = r + 6\frac{1}{4} - 6\frac{1}{4}$

$\frac{25}{6} - \frac{25}{4} = r$

$\frac{50}{12} - \frac{75}{12} = r$

$\frac{-25}{12}$ or $-2\frac{1}{12} = r$

Check: $4\frac{1}{6} = r + 6\frac{1}{4}$

$4\frac{1}{6} \stackrel{?}{=} -2\frac{1}{12} + 6\frac{1}{4}$

$4\frac{1}{6} \stackrel{?}{=} \frac{-25}{12} + \frac{25}{4}$

$4\frac{1}{6} \stackrel{?}{=} \frac{-25}{12} + \frac{75}{12}$

$4\frac{1}{6} \stackrel{?}{=} \frac{50}{12}$

$4\frac{1}{6} = \frac{25}{6}$ or $4\frac{1}{6}$ ✓

10. $3.5a = 7$

$\frac{3.5a}{3.5} = \frac{7}{3.5}$

$a = 2$

Check: $3.5a = 7$

$3.5(2) \stackrel{?}{=} 7$

$7 = 7$ ✓

11. $\frac{-1}{6}s = 15$

$-6\left(\frac{-1}{6}s\right) = (-6)(15)$

$s = -90$

Check: $\frac{-1}{6}s = 15$

$\frac{-1}{6}(-90) \stackrel{?}{=} 15$

$15 = 15$ ✓

12. $9 = \frac{3}{4}g$

$\frac{4}{3}(9) = \frac{4}{3}\left(\frac{3}{4}g\right)$

$\frac{36}{3} = g$

$12 = g$

Check: $9 = \frac{3}{4}g$

$9 \stackrel{?}{=} \frac{3}{4}(12)$

$9 = 9$ ✓

13. $28.79 = x - 0.36$

$28.79 + 0.36 = x - 0.36 + 0.36$

$29.15 = x$

The barometric pressure was 29.15 inches.

Pages 246–247 Practice and Apply

Exercises 14–40 For checks, see students' work.

14. $y + 7.2 = 21.9$

$y + 7.2 - 7.2 = 21.9 - 7.2$

$y = 14.7$

15. $4.7 = a + 7.1$

$4.7 - 7.1 = a + 7.1 - 7.1$

$-2.4 = a$

16. $x - 5.3 = 8.1$

$x - 5.3 + 5.3 = 8.1 + 5.3$

$x = 13.4$

17. $n - 4.72 = 7.52$

$n - 4.72 + 4.72 = 7.52 + 4.72$

$n = 12.24$

18. $t + 3.17 = -3.17$

$t + 3.17 - 3.17 = -3.17 - 3.17$

$t = -6.34$

19. $a - 2.7 = 3.2$

$a - 2.7 + 2.7 = 3.2 + 2.7$

$a = 6.2$

20. $\frac{2}{3} = \frac{1}{8} + b$

$\frac{2}{3} - \frac{1}{8} = \frac{1}{8} + b - \frac{1}{8}$

$\frac{16}{24} - \frac{3}{24} = b$

$\frac{13}{24} = b$

21. $m + \frac{7}{12} = -\frac{5}{18}$

$m + \frac{7}{12} - \frac{7}{12} = -\frac{5}{18} - \frac{7}{12}$

$m = -\frac{10}{36} - \frac{21}{36}$

$m = \frac{-31}{36}$

22. $g + \frac{2}{3} = 2$

$g + \frac{2}{3} - \frac{2}{3} = 2 - \frac{2}{3}$

$g = 1\frac{1}{3}$

23. $7 = \frac{2}{9} + k$

$7 - \frac{2}{9} = \frac{2}{9} + k - \frac{2}{9}$

$\frac{63}{9} - \frac{2}{9} = k$

$\frac{61}{9}$ or $6\frac{7}{9} = k$

24. $n - \frac{3}{8} = \frac{1}{6}$

$n - \frac{3}{8} + \frac{3}{8} = \frac{1}{6} + \frac{3}{8}$

$n = \frac{4}{24} + \frac{9}{24}$

$n = \frac{13}{24}$

25. $x - \frac{2}{5} = -\frac{8}{15}$

$x - \frac{2}{5} + \frac{2}{5} = \frac{-8}{15} + \frac{2}{5}$

$x = \frac{-8}{15} + \frac{6}{15}$

$x = \frac{-2}{15}$ or $-\frac{2}{15}$

26. $7\frac{1}{3} = c - \frac{4}{5}$

$7\frac{1}{3} + \frac{4}{5} = c - \frac{4}{5} + \frac{4}{5}$

$\frac{22}{3} + \frac{4}{5} = c$

$\frac{110}{15} + \frac{12}{15} = c$

$\frac{122}{15} = c$

$8\frac{2}{15} = c$

27. $-2 = \frac{3}{10} + f$

$-2 - \frac{3}{10} = \frac{3}{10} + f - \frac{3}{10}$

$\frac{-20}{10} - \frac{3}{10} = f$

$\frac{-23}{10}$ or $-2\frac{3}{10} = f$

28. $\frac{7}{9}k = -\frac{5}{12}$

$\frac{9}{7}\left(\frac{7}{9}k\right) = \frac{9}{7}\left(\frac{-5}{12}\right)$

$k = -\frac{45}{84}$ or $-\frac{15}{28}$

29. $4.1p = 16.4$

$\frac{4.1p}{4.1} = \frac{16.4}{4.1}$

$p = 4$

30. $8 = \frac{2}{3}d$

$\frac{3}{2}(8) = \frac{3}{2}\left(\frac{2}{3}d\right)$

$12 = d$

31. $0.4y = 2$

$\frac{0.4y}{0.4} = \frac{2}{0.4}$

$y = 5$

32. $\frac{1}{5}t = 9$

$\frac{5}{1}\left(\frac{1}{5}t\right) = \frac{5}{1}(9)$

$t = 45$

33. $4 = -\frac{1}{8}q$

$-8(4) = -8\left(-\frac{1}{8}q\right)$

$-32 = q$

34. $\frac{1}{3}n = \frac{2}{9}$

$3\left(\frac{1}{3}n\right) = 3\left(\frac{2}{9}\right)$

$n = \frac{2}{3}$

35. $\frac{5}{8} = \frac{1}{2}r$

$2\left(\frac{5}{8}\right) = 2\left(\frac{1}{2}r\right)$

$\frac{15}{4} = r$

36. $\frac{2}{3}a = 6$

$\frac{3}{2}\left(\frac{2}{3}a\right) = \frac{3}{2}(6)$

$a = 9$

37. $b - 1\frac{1}{2} = 4\frac{1}{4}$

$b - 1\frac{1}{2} + 1\frac{1}{2} = 4\frac{1}{4} + 1\frac{1}{2}$

$b = (4 + 1) + \left(\frac{1}{4} + \frac{1}{2}\right)$

$b = 5 + \left(\frac{1}{4} + \frac{2}{4}\right)$

$b = 5\frac{3}{4}$

38. $7\frac{1}{2} = r - 5\frac{2}{3}$

$7\frac{1}{2} + 5\frac{2}{3} = r - 5\frac{2}{3} + 5\frac{2}{3}$

$(7 + 5) + \left(\frac{1}{2} + \frac{2}{3}\right) = r$

$12 + \left(\frac{3}{6} + \frac{4}{6}\right) = r$

$12\frac{7}{6} = r$

$13\frac{1}{6} = r$

39. $3\frac{3}{4} + n = 6\frac{5}{8}$

$3\frac{3}{4} + n - 3\frac{3}{4} = 6\frac{5}{8} - 3\frac{3}{4}$

$n = \frac{53}{8} - \frac{15}{4}$

$n = \frac{53}{8} - \frac{30}{8}$

$n = \frac{23}{8}$ or $2\frac{7}{8}$

40. $y + 1\frac{1}{3} = 3\frac{1}{18}$

$y + 1\frac{1}{3} - 1\frac{1}{3} = 3\frac{1}{18} - 1\frac{1}{3}$

$y = \frac{55}{18} - \frac{4}{3}$

$y = \frac{55}{18} - \frac{24}{18}$

$y = \frac{31}{18}$ or $1\frac{13}{18}$

41. $n - 1\frac{1}{4} = 12\frac{1}{4}$

$n - 1\frac{1}{4} + 1\frac{1}{4} = 12\frac{1}{4} + 1\frac{1}{4}$

$n = 13\frac{2}{4}$ or $13\frac{1}{2}$ inches

$x + \frac{1}{2} = 22$

$x + \frac{1}{2} - \frac{1}{2} = 22 - \frac{1}{2}$

$x = 21\frac{1}{2}$ inches

The old dimensions were $13\frac{1}{2}$ in. by $21\frac{1}{2}$ in.

42. $2.36 - 1.63 = 0.73$ million or 730,000 metric tons

43. If the price is marked $\frac{1}{3}$ off, then it is $\frac{2}{3}$ of the original price.

$\frac{2}{3}(24.99) = \frac{24.99}{3}$

$= \$16.66$

44. $2\frac{1}{2}x = 3\frac{3}{4}$

$\frac{5}{2}x = \frac{15}{4}$

$\frac{2}{5}\left(\frac{5}{2}x\right) = \frac{2}{5}\left(\frac{15}{4}\right)$

$x = \frac{\cancel{2}^{1}}{\cancel{5}_{1}} \cdot \frac{\cancel{15}^{3}}{\cancel{4}_{2}}$

$x = \frac{3}{2}$ or $1\frac{1}{2}$ cups

45. $\overset{1}{\cancel{3}}\ \cancel{\text{inches}} \cdot \dfrac{1 \text{ foot}}{\underset{4}{\cancel{12}}\ \cancel{\text{inches}}} = \dfrac{1}{4} \text{ foot}$

$$\frac{1}{4}(9) = \frac{9}{4}$$
$$= 2\frac{1}{4} \text{ feet}$$

46. $\dfrac{x+1}{x+4+1} = \dfrac{1}{2}$

$$\frac{x+1}{x+5} = \frac{1}{2}$$
$$1(x+5) = 2(x+1)$$
$$x+5 = 2x+2$$
$$x+5-x = 2x+2-x$$
$$5 = x+2$$
$$5-2 = x+2-2$$
$$3 = x$$

The original fraction is $\dfrac{x}{x+4}$ or $\dfrac{3}{3+4} = \dfrac{3}{7}$.

47. Equations with fractions can be written to represent the number of vibrations per second for different notes. To solve, multiply each side of the equation by the reciprocal of the fraction. Answers should include the following.

- For example, if n vibrations per second produce middle C, then $\frac{5}{4}n$ vibrations per second produce the note E above middle C.
- The equation $\frac{5}{3}n = 440$ represents the number of vibrations per second to produce middle C. To solve, multiply each side by the reciprocal of $\frac{5}{3}$, $\frac{3}{5}$.

48. A; $\dfrac{5}{6}z = \dfrac{3}{5}$

$$\frac{6}{5}\left(\frac{5}{6}z\right) = \left(\frac{3}{5}\right)\frac{6}{5}$$
$$z = \frac{18}{25}$$

49. D; $33\frac{3}{4} = \frac{1}{2}(9)h$

$$\frac{135}{4} = \frac{9}{2}h$$
$$\frac{2}{9}\left(\frac{135}{4}\right) = \left(\frac{9}{2}h\right)\frac{2}{9}$$
$$\frac{\overset{1}{\cancel{2}}}{\underset{1}{\cancel{9}}} \cdot \frac{\overset{15}{\cancel{135}}}{\underset{2}{\cancel{4}}} = h$$
$$\frac{15}{2} \text{ or } 7\frac{1}{2} = h$$

Page 248 Maintain Your Skills

50. mean $= \dfrac{2+8+5+18+3+5+6}{7}$

$$= \frac{47}{7} \approx 6.7$$

The mean is 6.7.

median: 2, 3, 5, <u>5</u>, 6, 8, 18

The median is 5.

5 is the mode, as it occurs twice.

51. mean $= \dfrac{11+12+12+14+16+11+15}{7}$

$$= \frac{91}{7} = 13$$

The mean is 13.

median: 11, 11, 12, <u>12</u>, 14, 15, 16

The median is 12.

11 and 12 are the modes, as they both occur twice.

52. mean $= \dfrac{0.9+0.5+0.7+0.4+0.3+0.2}{6}$

$$= \frac{3}{6} = 0.5$$

The mean is 0.5.

median: 0.2, 0.3, <u>0.4, 0.5</u>, 0.7, 0.9

$$\frac{0.4+0.5}{2} = 0.45$$

The median is 0.45.

There is no mode because each number in the set occurs once.

53. mean $= \dfrac{56+77+60+60+72+100}{6}$

$$= \frac{425}{6} \approx 70.8$$

The mean is 70.8.

median: 56, 60, <u>60, 72</u>, 77, 100

$$\frac{60+72}{2} = 66$$

The median is 66.

60 is the mode, as it occurs twice.

54. $\dfrac{3}{5} + \dfrac{1}{3} = \dfrac{3}{5} \cdot \dfrac{3}{3} + \dfrac{1}{3} \cdot \dfrac{5}{5}$

$$= \frac{9}{15} + \frac{5}{15}$$
$$= \frac{14}{15}$$

55. $\dfrac{5}{6} - \dfrac{1}{8} = \dfrac{5}{6} \cdot \dfrac{4}{4} - \dfrac{1}{8} \cdot \dfrac{3}{3}$

$$= \frac{20}{24} - \frac{3}{24}$$
$$= \frac{17}{24}$$

56. $-4\frac{1}{4} - \frac{1}{6} = \dfrac{-17}{4} - \dfrac{1}{6}$

$$= -\frac{17}{4} \cdot \frac{3}{3} - \frac{1}{6} \cdot \frac{2}{2}$$
$$= -\frac{51}{12} - \frac{2}{12}$$
$$= -\frac{53}{12} \text{ or } -4\frac{5}{12}$$

57. $\dfrac{5}{9} + \left(-\dfrac{1}{12}\right) = \dfrac{5}{9} \cdot \dfrac{4}{4} + \left(-\dfrac{1}{12}\right) \cdot \dfrac{3}{3}$

$$= \frac{20}{36} + \left(-\frac{3}{36}\right)$$
$$= \frac{17}{36}$$

58. $-3\frac{3}{4} + \left(-2\frac{1}{8}\right) = -\dfrac{15}{4} + \left(-\dfrac{17}{8}\right)$

$$= -\frac{15}{4} \cdot \frac{2}{2} + \left(-\frac{17}{8}\right)$$
$$= -\frac{30}{8} + \left(-\frac{17}{8}\right)$$
$$= -\frac{47}{8} \text{ or } -5\frac{7}{8}$$

59. $8\frac{9}{10} - 1\frac{1}{6} = \dfrac{89}{10} - \dfrac{7}{6}$

$$= \frac{89}{10} \cdot \frac{3}{3} - \frac{7}{6} \cdot \frac{5}{5}$$
$$= \frac{267}{30} - \frac{35}{30}$$
$$= \frac{232}{30}$$
$$= \frac{116}{15} \text{ or } 7\frac{11}{15}$$

60. $a - b = 9\frac{5}{6} - 1\frac{1}{6}$

$$= \frac{59}{6} - \frac{7}{6}$$
$$= \frac{52}{6}$$
$$= \frac{26}{3} \text{ or } 8\frac{2}{3}$$

61. 9 hours out of 24 hours;

$\frac{9}{24} = \frac{3 \cdot \cancel{3}^{1}}{2 \cdot 2 \cdot 2 \cdot \cancel{3}_{1}} = \frac{3}{8}$ of a day

62.
$$\begin{aligned} 3t - 6 &= 15 \\ 3t - 6 + 6 &= 15 + 6 \\ 3t &= 21 \\ \frac{3t}{3} &= \frac{21}{3} \\ t &= 7 \end{aligned}$$

63.
$$\begin{aligned} 8 &= \frac{k}{-2} + 5 \\ 8 - 5 &= \frac{k}{-2} + 5 - 5 \\ 3 &= \frac{k}{-2} \\ (-2)(3) &= (-2)\left(\frac{k}{-2}\right) \\ -6 &= k \end{aligned}$$

64.
$$\begin{aligned} 9n - 13n &= 4 \\ -4n &= 4 \\ \frac{4n}{-4} &= \frac{4}{-4} \\ n &= -1 \end{aligned}$$

65. $-18 \div 3 = -6$

66. $24 \div (-2) = -12$

67. $-20 \div (-4) = 5$

68. $-55 \div (-5) = 11$

69. $-12 \div 36 = \frac{-12}{36} = -\frac{1}{3}$

70. $9 \div (-81) = \frac{9}{-81} = -\frac{1}{9}$

Page 248 Practice Quiz 2

1. $8 = 2 \cdot 2 \cdot 2 \cdot 2^3$

$9 = 3 \cdot 3 = 3^2$

$\text{LCM} = 2^3 \cdot 3^2 = 72$

The LCM of 8 and 9 is 72.

2. $12 = 2 \cdot 2 \cdot 3 \cdot = 2^2 \cdot 3$

$30 = 2 \cdot 3 \cdot 5 = 2 \cdot 3 \cdot 5$

$\text{LCM} = 2^2 \cdot 3 \cdot 5 = 60$

The LCM of 12 and 30 is 60.

3. $2 = 2$

$10 = 2 \cdot 5$

$\text{LCM} = 2 \cdot 5 = 10$

The LCM of 2 and 10 is 10.

4. $6 = 2 \cdot 3 = 2 \cdot 3$

$8 = 2 \cdot 2 \cdot 2 = 2^3$

$\text{LCM} = 2^3 \cdot 3 = 24$

The LCM of 6 and 8 is 24.

5. $\frac{3}{4} + \frac{1}{16} = \frac{3}{4} \cdot \frac{4}{4} + \frac{1}{16} = \frac{12}{16} + \frac{1}{16} = \frac{13}{16}$

6. $-\frac{3}{7} + \frac{5}{14} = -\frac{3}{7} \cdot \frac{2}{2} + \frac{5}{14} = \frac{-6}{14} + \frac{5}{14} = -\frac{1}{14}$

7.
$$\begin{aligned} 1\tfrac{1}{5} - \left(-\tfrac{7}{15}\right) &= \tfrac{6}{5} - \left(-\tfrac{7}{15}\right) \\ &= \tfrac{6}{5} \cdot \tfrac{3}{3} - \left(-\tfrac{7}{15}\right) \\ &= \tfrac{18}{15} - \left(-\tfrac{7}{15}\right) \\ &= \tfrac{25}{15} \\ &= \tfrac{5}{3} \text{ or } 1\tfrac{2}{3} \end{aligned}$$

8.
$$\begin{aligned} 6\tfrac{3}{4} - 2\tfrac{1}{6} &= \tfrac{27}{4} - \tfrac{13}{6} \\ &= \tfrac{27}{4} \cdot \tfrac{3}{3} - \tfrac{13}{6} \cdot \tfrac{2}{2} \\ &= \tfrac{81}{12} - \tfrac{26}{12} \\ &= \tfrac{55}{12} \text{ or } 4\tfrac{7}{12} \end{aligned}$$

9. mean $= \frac{389}{12} \approx 32.4$

median: 22, 29, 29, 29, 30, $\underline{30, 31}$, 31, 35, 38, 40, 45

The median is $\frac{30 + 31}{2} = 30.5$.

29 is the mode, as it occurs 3 times.

10.
$$\begin{aligned} a - 1\tfrac{1}{3} &= 4\tfrac{1}{6} \\ a - 1\tfrac{1}{3} + 1\tfrac{1}{3} &= 4\tfrac{1}{6} + 1\tfrac{1}{3} \\ a &= 4\tfrac{1}{6} + 1\tfrac{1}{3} \\ a &= (4 + 1) + \left(\tfrac{1}{6} + \tfrac{1}{3}\right) \\ a &= 5 + \left(\tfrac{1}{6} + \tfrac{1}{3} \cdot \tfrac{2}{2}\right) \\ a &= 5 + \left(\tfrac{1}{6} + \tfrac{2}{6}\right) \\ a &= 5 + \tfrac{3}{6} \\ a &= 5\tfrac{1}{2} \end{aligned}$$

5-10 Arithmetic and Geometric Sequences

Page 249 How can sequences be used to make predictions?

a. 70 ft

b. 20, 45, 80, 125, 180, ?

(+ 25, + 35, + 45, + 55, + 65)

$180 + 65 = 245$

The braking distance of 70 mph is 245 ft.

c. 10 ft

d. Sample answer: the difference in braking distance increases by 10 ft for each 10 mph increase in speed: 25, 35, 45, 55, ...

Page 251 Check for Understanding

1. Arithmetic sequences have a common difference and the terms can be found by adding or subtracting. Geometric sequences have a common ratio and the terms can be found by multiplying or dividing.
2. The terms are decreasing; Sample answer: 8, 4, 2, 1; 0.5
3. 3, 7, 11, 15,... ($+4$, $+4$, $+4$)

   Arithmetic; the common difference is 4; the next 3 terms are 19, 23, and 27.
4. 1, 3, 9, 27,... ($\times 3$, $\times 3$, $\times 3$)

   Geometric; the common ratio is 3; the next 3 terms are 81, 243, and 729.
5. 6, 8, 12, 18,... ($+2$, $+4$, $+6$)

   neither arithmetic or geometric
6. 13, 8, 3, –2,... (-5, -5, -5)

   Arithmetic; the common difference is –5; the next 3 terms are –7, –12, and –17.
7. 3, $\frac{8}{3}$, $\frac{7}{3}$, 2, ...

   $\frac{9}{3}$, $\frac{8}{3}$, $\frac{7}{3}$, $\frac{6}{3}$ ($-\frac{1}{3}$, $-\frac{1}{3}$, $-\frac{1}{3}$)

   Arithmetic; the common difference is $-\frac{1}{3}$; the next 3 terms are $\frac{5}{3}$, $\frac{4}{3}$, and 1.
8. 48, 12, 3, $\frac{3}{4}$,... ($\times\frac{1}{4}$, $\times\frac{1}{4}$, $\times\frac{1}{4}$)

   Geometric; the common ratio is $\frac{1}{4}$; the next 3 terms are $\frac{3}{16}$, $\frac{3}{64}$, and $\frac{3}{256}$.
9. 20,000, 16,400, 13,448, 11,027.36 ($\times 0.82$, $\times 0.82$, $\times 0.82$)

   The car will be worth $11,027.36.

Pages 251–252 Practice and Apply

10. 2, 5, 8, 11,... ($+3$, $+3$, $+3$)

    Arithmetic; the common difference is 3; the next 3 terms are 14, 17, and 20.
11. −6, 5, 16, 27,... ($+11$, $+11$, $+11$)

    Arithmetic; the common difference is 11; the next 3 terms are 38, 49, and 60.
12. $\frac{1}{2}$, 1, 2, 4,... ($\times 2$, $\times 2$, $\times 2$)

    Geometric; the common ratio is 2; the next 3 terms are 8, 16, and 32.
13. 2, 6, 18, 54,... ($\times 3$, $\times 3$, $\times 3$)

    Geometric; the common ratio is 3; the next 3 terms are 162, 486, and 1458.
14. 18, 11, 4, −3,... (-7, -7, -7)

    Arithmetic; the common difference is −7; the next 3 terms are −10, −17, and −24.
15. 25, 22, 19, 16,... (-3, -3, -3)

    Arithmetic; the common difference is −3; the next 3 terms are 13, 10, and 7.
16. 4, 1, $\frac{1}{4}$, $\frac{1}{16}$,... ($\times\frac{1}{4}$, $\times\frac{1}{4}$, $\times\frac{1}{4}$)

    Geometric; the common ratio is $\frac{1}{4}$; the next 3 terms are $\frac{1}{64}$, $\frac{1}{256}$, and $\frac{1}{1024}$.
17. −5, 1, $-\frac{1}{5}$, $\frac{1}{25}$,... ($\times\left(-\frac{1}{5}\right)$, $\times\left(-\frac{1}{5}\right)$, $\times\left(-\frac{1}{5}\right)$)

    Geometric; the common ratio is $-\frac{1}{5}$; the next 3 terms are $-\frac{1}{125}$, $\frac{1}{625}$, and $-\frac{1}{3125}$.
18. $\frac{1}{2}$, 1, $\frac{3}{2}$, 2, ($+\frac{1}{2}$, $+\frac{1}{2}$, $+\frac{1}{2}$)

    Arithmetic; the common difference is $\frac{1}{2}$; the next 3 terms are $\frac{5}{2}$, 3, and $\frac{7}{2}$.
19. 0, $\frac{1}{6}$, $\frac{1}{3}$, $\frac{1}{2}$,...

    0, $\frac{1}{6}$, $\frac{2}{6}$, $\frac{3}{6}$,... ($+\frac{1}{6}$, $+\frac{1}{6}$, $+\frac{1}{6}$)

    Arithmetic; the common difference is $\frac{1}{6}$; the next 3 terms are $\frac{2}{3}$, $\frac{5}{6}$, and 1.
20. 0.75, 1.5, 2.25,... ($+0.75$, $+0.75$)

    Arithmetic; the common difference is 0.75; the next 3 terms are 3, 3.75, and 4.5.
21. 4.5, 4.0, 3.5 3.0,... (-0.5, -0.5, -0.5)

    Arithmetic; the common difference is −0.5; the next 3 terms are 2.5, 2.0, and 1.5.
22. 11, 14, 19, 26,... ($+3$, $+4$, $+5$)

    Neither
23. 17, 16, 14, 11,... (-1, -2, -3)

    Neither
24. 24, 12, 6, 3,... ($\times\frac{1}{2}$, $\times\frac{1}{2}$, $\times\frac{1}{2}$)

    Geometric; the common ratio is $\frac{1}{2}$; the next 3 terms are $\frac{3}{2}$, $\frac{3}{4}$, and $\frac{3}{8}$.

25. 18, −6, 2, $-\frac{2}{3}$...

$\times\left(-\frac{1}{3}\right)\times\left(-\frac{1}{3}\right)\times\left(-\frac{1}{3}\right)$

Geometric; the common ratio is $-\frac{1}{3}$; the next 3 terms are $\frac{2}{9}$, $-\frac{2}{27}$, and $\frac{2}{81}$.

26. 0.1, 0.3, 0.9, 2.7, ...

× 3 × 3 × 3

Geometric; the common ratio is 3; the next 3 terms are 8.1, 24.3, and 72.9.

27. $\frac{1}{2}$, $\frac{1}{4}$, $\frac{1}{8}$ $\frac{1}{16}$, ...

$\times\frac{1}{2}$ $\times\frac{1}{2}$ $\times\frac{1}{2}$

Geometric; the common ratio is $\frac{1}{2}$; the next 3 terms are $\frac{1}{32}$, $\frac{1}{64}$, and $\frac{1}{128}$.

28. 160, 120, 90 67.5, ...

× 0.75 × 0.75 × 0.75

67.5 feet

29a. Arithmetic; the common difference is $3.

29b. $6 + 3(15 - 1) = 6 + 42$
$= \$48$

30. Sample answer: 9, 10, 11, 12; 6, 9, 12, 15

31. Find the pattern, continue the sequence, and use the new values to make predictions. Answers should include the following.

- The difference between any two consecutive terms in an arithmetic sequence is the common difference. To find the next value in such a sequence, add the common difference to the last term. The ratio of any two consecutive terms in a geometric sequence is the common ratio. So, to find the next value in such a sequence, multiply the last term by the common ratio.
- Sequences occurring in nature include geysers spouting every few minutes, the arrangement of geese in migration patterns, and ocean tides.

32. C; 56, 48, 40, 32

−8 −8 −8

The next term is 24.

33. A; 4, 8, 12, 16, 20

+4 +4 +4 +4

The perimeter values form an arithmetic sequence.

34. $6 + (n - 1)(3) = a_1 + (n - 1)d$

35. $a_1 + (n - 1)d = 16 + (9 - 1)5$
$= 16 + 8(5)$
$= 16 + 40$
$= 56$

Page 252 Maintain Your Skills

Exercises 36–39 For checks see students' work.

36.
$$\frac{7}{6} = y + \frac{5}{12}$$
$$\frac{7}{6} - \frac{5}{12} = y + \frac{5}{12} - \frac{5}{12}$$
$$\frac{14}{12} - \frac{5}{12} = y$$
$$\frac{9}{12} = y$$
$$\frac{3}{4} = y$$

37.
$$k - 4 \cdot 1 = -9.38$$
$$k - 4.1 + 4.1 = -9.38 + 4.1$$
$$k = -5.28$$

38.
$$40.3 = 6.2x$$
$$\frac{40.3}{6.2} = \frac{6.2x}{6.2}$$
$$6.5 = x$$

39.
$$\frac{3}{4}b = 7\frac{1}{2}$$
$$\frac{4}{3}\left(\frac{3}{4}b\right) = \frac{4}{3}\left(7\frac{1}{2}\right)$$
$$b = \frac{4}{3} \cdot \frac{15}{2}$$
$$b = \frac{\cancel{4}^2}{\cancel{3}_1} \cdot \frac{\cancel{15}^5}{\cancel{2}_1}$$
$$b = 10$$

40. mean $= \frac{11 + 45 + 62 + 12 + 47 + 8 + 12 + 35}{8}$
$= \frac{232}{8} = 29$

The mean is 29.

median: 8, 11, 12, 12, 35, 45, 47, 62
$= \frac{12 + 35}{2}$
$= \frac{47}{2} = 23.5$

The median is 23.5.

12 is the mode, as it occurs twice.

41. mean $= \frac{2.3 + 3.6 + 4.1 + 3.6 + 2.9 + 3.0}{6}$
$= \frac{19.5}{6} \approx 3.3$

The mean is about 3.3.

median: 2.3, 2.9, 3.0, 3.6, 3.6, 4.1
$= \frac{3.0 + 3.6}{2}$
$= 3.3$

The median is 3.3.

3.6 is the mode, as it occurs twice.

42.
$$6a^2 = 2 \cdot 3 \cdot a^2$$
$$9a^3 = 3^2 \cdot a^3$$
$$\text{LCM} = 2 \cdot 3^2 \cdot a^3$$
$$= 18a^3$$

The LCD of $\frac{1}{6a^2}$ and $\frac{5}{9a^3}$ is $18a^3$.

43. $34\frac{3}{8} + 33\frac{7}{8} + 34\frac{5}{8}$
$= (34 + 33 + 34) + \left(\frac{3}{8} + \frac{7}{8} + \frac{5}{8}\right)$
$= 101 + \frac{15}{8}$
$= 101 + 1\frac{7}{8}$
$= 102\frac{7}{8}$ in.

44. $3^4 \cdot 3^2 = 3^{4+2}$
$= 3^6$

45. $\frac{b^5}{b^2} = b^{5-2}$
$= b^3$

46. $2x^3(5x^2) = 2.5x^{3+2}$
$= 10x^5$

47. 36: (2) · 2 · (3) · 3
42: (2) · (3) · 7
The GCF is 6.

48. 9: (3) · 3
24: 2 · 2 · 2 · (3)
The GCF is 3.

49. 60: 2 · 2 · (3) · (5)
45: (3) · 3 · (5)
30: 2 · (3) · (5)
The GCF is 15.

Page 253 Algebra Activity (Follow-Up of Lesson 5-10)

1. Sample answer: Most of them should be terms of the Fibonacci sequence.
2. Sample answer: Most of them should be terms of the Fibonacci sequence. Some groups may have different numbers based on the size of their object.
3. Sample answer: The number of rows is a term of the Fibonacci sequence.
4. 1, 1, 2, 3, 5, 8, 13, 21, 34, 55, 89, 144, 233, 377, 610
5. 1, 2, 1.5, 1.6666667, 1.6, 1.625, 1.6153846, 1.6190476, 1.6176471, 1.6181818, 1.6179775, 1.6180556, 1.6180258, 1.6180371
6. They get increasingly closer to 1.618.
7. The golden ratio is approximately 1:1.618. After the first few terms, the ratio of successive numbers in the Fibonacci sequence is the golden ratio.

Chapter 5 Study Guide and Review

Page 254 Vocabulary and Concept Check

1. rational
2. terminating
3. algebraic fraction
4. multiples
5. LCD
6. 1
7. reciprocal
8. mode
9. arithmetic

Pages 254–258 Lesson-by-Lesson Review

10.
```
   0.875
 8)7.000
  -6 4
     60
    -56
     40
    -40
      0
```
0.875 is a terminating decimal.

11.
```
    0.45
 20)9.00
   -8 0
     1 00
    -1 00
        0
```
0.45 is a terminating decimal.

12. $\frac{2}{3} \rightarrow 3\overline{)2.000...}$ (quotient 0.666...)
So, $\frac{2}{3} = 0.\overline{6}$.

13. $-\frac{7}{15} \rightarrow 15\overline{)7.0000...}$ (quotient 0.4666...)
So, $-\frac{7}{15} = 0.4\overline{6}$.

14. $8\frac{3}{25} = 8 + \frac{3}{25}$
```
    0.12
 25)3.00
   -2 5
     50
    -50
      0
```
$8\frac{3}{25} = 8 + 0.12 = 8.12$

15. $6\frac{4}{11} = 6 + \frac{4}{11}$
$\frac{4}{11} \rightarrow 11\overline{)4.0000...}$ (quotient 0.3636...)
$6\frac{4}{11} = 6 + 0.\overline{36}$
$= 6.\overline{36}$

16. $0.23 = \frac{23}{100}$

17. $0.6 = \frac{6}{10}$
$= \frac{3}{5}$

18. $-0.05 = -\frac{5}{100}$
$= -\frac{1}{20}$

19. $0.125 = \frac{125}{1000}$
$= \frac{1}{8}$

20. $2.36 = 2\frac{36}{100}$
$= 2\frac{9}{25}$

21. $4.44 = 4\frac{44}{100}$
$= 4\frac{11}{25}$

22. $-8.002 = -8\frac{2}{1000}$
$= -8\frac{1}{500}$

23. $N = 0.555...$
$10N = 10(0.555...)$
$10N = 5.555...$
$\underline{-(N = 0.555)}$
$\frac{9N}{9} = \frac{5}{9}$
$N = \frac{5}{9}$

Therefore, $0.555... = \frac{5}{9}$.

24. $N = 0.333...$
$10N = 10(0.333...)$
$10N = 3.333...$
$\underline{-(N = 0.333)}$
$\frac{9N}{9} = \frac{3}{9}$
$N = \frac{3}{9}$ or $\frac{1}{3}$

25. $N = 1.777...$
$10N = 10(1.777...)$
$10N = 17.777...$
$\underline{-(N = 1.777...)}$
$\frac{9N}{9} = \frac{16}{9}$
$N = \frac{16}{9}$ or $1\frac{7}{9}$

Therefore, $1.\overline{7} = 1\frac{7}{9}$.

26. $N = 0.7272...$
$100N = 100(0.7272...)$
$100N = 72.7272...$
$\underline{-(N = 0.7272...)}$
$\frac{99N}{99} = \frac{72}{99}$
$N = \frac{72}{99}$ or $\frac{8}{11}$

Therefore, $0.\overline{72} = \frac{8}{11}$.

27. $N = 3.3636...$
$100N = 100(3.3636...)$
$100N = 336.3636...$
$\underline{-(N = 3.3636...)}$
$\frac{99N}{99} = \frac{333}{99}$
$N = \frac{333}{99}$
$N = \frac{37}{11}$ or $3\frac{4}{11}$

Therefore, $3.\overline{36} = 3\frac{4}{11}$.

28. $\frac{2}{3} \cdot \frac{1}{8} = \frac{\overset{1}{\cancel{2}}}{3} \cdot \frac{1}{\underset{4}{\cancel{8}}}$
$= \frac{1}{12}$

29. $-\frac{7}{15} \cdot \frac{5}{9} = -\frac{7}{\underset{3}{\cancel{15}}} \cdot \frac{\overset{1}{\cancel{5}}}{9}$
$= -\frac{7}{27}$

30. $\frac{6}{11} \cdot \frac{2}{15} = \frac{\overset{2}{\cancel{6}}}{11} \cdot \frac{2}{\underset{5}{\cancel{15}}}$
$= \frac{4}{55}$

31. $8 \cdot \frac{4}{5} = \frac{8}{1} \cdot \frac{4}{5}$
$= \frac{32}{5}$ or $6\frac{2}{5}$

32. $-1\frac{5}{6} \cdot 9 = \frac{-11}{6} \cdot \frac{9}{1}$
$= \frac{-11}{\underset{2}{\cancel{6}}} \cdot \frac{\overset{3}{\cancel{9}}}{1}$
$= -\frac{33}{2}$ or $-16\frac{1}{2}$

33. $\frac{6}{7} \cdot \frac{14}{9} = \frac{\overset{2}{\cancel{6}}}{\underset{1}{\cancel{7}}} \cdot \frac{\overset{2}{\cancel{14}}}{\underset{3}{\cancel{9}}}$
$= \frac{4}{3}$ or $1\frac{1}{3}$

34. $\frac{8}{9} \cdot \frac{5}{12} = \frac{\overset{2}{\cancel{8}}}{9} \cdot \frac{5}{\underset{3}{\cancel{12}}}$
$= \frac{10}{27}$

35. $2\frac{1}{4} \cdot \left(-\frac{4}{3}\right) = \frac{9}{4} \cdot \frac{-4}{3}$
$= \frac{\overset{3}{\cancel{9}}}{\underset{1}{\cancel{4}}} \cdot \frac{\overset{1}{\cancel{-4}}}{\underset{1}{\cancel{3}}}$
$= -\frac{3}{1}$ or -3

36. $\frac{14}{15} \cdot 3\frac{2}{7} = \frac{14}{15} \cdot \frac{23}{7}$
$= \frac{\overset{2}{\cancel{14}}}{15} \cdot \frac{23}{\underset{1}{\cancel{7}}}$
$= \frac{46}{15}$ or $3\frac{1}{15}$

37. $2\frac{5}{6} \cdot 3\frac{1}{3} = \frac{17}{6} \cdot \frac{10}{3}$
$= \frac{17}{\underset{3}{\cancel{6}}} \cdot \frac{\overset{5}{\cancel{10}}}{3}$
$= \frac{85}{9}$ or $9\frac{4}{9}$

38. $\frac{ab}{4} \cdot \frac{2}{bc} = \frac{a\overset{1}{\cancel{b}}}{\underset{2}{\cancel{4}}} \cdot \frac{\overset{1}{\cancel{2}}}{\underset{1}{\cancel{b}}c}$
$= \frac{a}{2c}$

39. $\frac{x^2}{r} \cdot \frac{r}{x} = \frac{x \cdot x}{r} \cdot \frac{r}{x}$
$= \frac{x \cdot \overset{1}{\cancel{x}}}{\underset{1}{\cancel{r}}} \cdot \frac{\overset{1}{\cancel{r}}}{\underset{1}{\cancel{x}}}$
$= \frac{x}{1}$ or x

40. $\frac{4}{9} \div \frac{1}{3} = \frac{4}{\underset{3}{\cancel{9}}} \cdot \frac{\overset{1}{\cancel{3}}}{1}$
$= \frac{4}{3}$ or $1\frac{1}{3}$

41. $\frac{2}{5} \div \left(-\frac{1}{15}\right) = \frac{2}{\underset{1}{\cancel{5}}} \cdot \left(-\frac{\overset{3}{\cancel{15}}}{1}\right)$
$= -\frac{6}{1}$ or -6

42. $\frac{6}{13} \div \frac{8}{9} = \frac{\overset{3}{\cancel{6}}}{13} \cdot \frac{9}{\underset{4}{\cancel{8}}}$
$= \frac{27}{52}$

43. $5 \div \left(-1\frac{1}{3}\right) = \frac{5}{1} \div \left(-\frac{4}{3}\right)$
$= \frac{5}{1} \cdot \left(-\frac{3}{4}\right)$
$= -\frac{15}{4}$ or $-3\frac{3}{4}$

44. $\frac{11}{18} \div 4\frac{1}{2} = \frac{11}{18} \div \frac{9}{2}$
$= \frac{11}{\underset{9}{\cancel{18}}} \cdot \frac{\overset{1}{\cancel{2}}}{9}$
$= \frac{11}{81}$

45. $-3\frac{3}{5} \div \frac{6}{7} = -\frac{\overset{3}{\cancel{18}}}{5} \cdot \frac{7}{\underset{1}{\cancel{6}}}$
$= \frac{-21}{5}$ or $-4\frac{1}{5}$

46. $\frac{n}{8} \div \frac{n}{32} = \frac{\overset{1}{\cancel{n}}}{\underset{1}{\cancel{8}}} \cdot \frac{\overset{4}{\cancel{32}}}{\underset{1}{\cancel{n}}}$
$= \frac{4}{1}$ or 4

47. $\frac{2}{7x} \div \frac{3}{2} = \frac{2}{7x} \cdot \frac{2}{3}$
$= \frac{4}{21x}$

48. $\frac{5}{18} + \frac{11}{18} = \frac{5+11}{18}$
$= \frac{16}{18}$ or $\frac{8}{9}$

49. $\frac{7}{9} + \left(-\frac{2}{9}\right) = \frac{7+(-2)}{9}$
$= \frac{5}{9}$

50. $\frac{19}{20} - \frac{17}{20} = \frac{19-17}{20}$
$= \frac{2}{20}$ or $\frac{1}{10}$

51. $1\frac{16}{21} - \frac{9}{21} = \frac{37}{21} - \frac{9}{21}$
$= \frac{37-9}{21}$
$= \frac{28}{21}$
$= \frac{4}{3}$ or $1\frac{1}{3}$

52. $-\frac{12}{17} + \frac{10}{17} = \frac{-12+10}{17}$
$= -\frac{2}{17}$

53. $8\frac{3}{10} - 5\frac{7}{10} = \frac{83}{10} - \frac{57}{10}$
$= \frac{83-57}{10}$
$= \frac{26}{10}$
$= \frac{13}{5}$ or $2\frac{3}{5}$

54. $\frac{7t}{15} + \frac{t}{15} = \frac{7t+t}{15}$
$= \frac{8t}{15}$

55. $\frac{5}{3x} - \frac{1}{3x} = \frac{5-1}{3x}$
$= \frac{4}{3x}$

56. $4 = 2 \cdot 2 = 2^2$
$18 = 2 \cdot 3 \cdot 3 = 2 \cdot 3^2$
$\text{LCM} = 2^2 \cdot 3^2$
$= 36$
The LCM of 4 and 18 is 36.

57. $24 = 2 \cdot 2 \cdot 2 \cdot 3 = 2^3 \cdot 3$
$20 = 2 \cdot 2 \cdot 5 = 2^2 \cdot 5$
$\text{LCM} = 2^3 \cdot 3 \cdot 5$
$= 120$
The LCM of 24 and 20 is 120.

58. $4a = 2 \cdot 2 \cdot a = 2^2 \cdot a$
$6a = 2 \cdot 3 \cdot a = 2 \cdot 3 \cdot a$
$\text{LCM} = 2^2 \cdot 3 \cdot a$
$= 12a$
The LCM of $4a$ and $6a$ is $12a$.

59. $7c^2 = 7 \cdot c^2$
$21c = 3 \cdot 7 \cdot c$
$\text{LCM} = 3 \cdot 7 \cdot c^2$
$= 21c^2$
The LCM of $7c^2$ and $21c$ is $21c^2$.

60. The LCD is $2^2 \cdot 3$ or 24.
Rewrite fractions using the LCD.
$\frac{3}{8} \cdot \frac{3}{3} = \frac{9}{24}$ $\quad \frac{5}{12} \cdot \frac{2}{2} = \frac{10}{24}$
Since $9 < 10$, $\frac{9}{24} < \frac{10}{24}$. So, $\frac{3}{8} < \frac{5}{12}$.

61. The LCD is $3^2 \cdot 5$ or 45.
Rewrite fractions using the LCD.
$\frac{2}{9} \cdot \frac{5}{5} = \frac{10}{45}$ $\quad \frac{4}{15} \cdot \frac{3}{3} = \frac{12}{45}$
Since $10 < 12$, $\frac{10}{45} < \frac{12}{45}$. So, $\frac{2}{9} < \frac{4}{15}$.

62. The LCD is $2^2 \cdot 5$ or 20.
Rewrite fractions using the LCD.
$\frac{5}{20}$ $\quad \frac{1}{4} \cdot \frac{5}{5} = \frac{5}{20}$
Since $5 = 5$, $\frac{5}{20} = \frac{5}{20}$. So, $\frac{5}{20} = \frac{1}{4}$.

63. The LCD is $3 \cdot 7$ or 21.
Rewrite fractions using the LCD.
$\frac{3}{7} \cdot \frac{3}{3} = \frac{9}{21}$ $\quad \frac{8}{21}$
Since $9 > 8$, $\frac{9}{21} > \frac{8}{21}$. So, $\frac{3}{7} > \frac{8}{21}$.

64. $\frac{1}{3} + \frac{5}{6} = \frac{1}{3} \cdot \frac{2}{2} + \frac{5}{6}$
$= \frac{2}{6} + \frac{5}{6}$
$= \frac{7}{6}$ or $1\frac{1}{6}$

65. $\frac{11}{12} + \frac{3}{4} = \frac{11}{12} + \frac{3}{4} \cdot \frac{3}{3}$
$= \frac{11}{12} + \frac{9}{12}$
$= \frac{20}{12}$
$= \frac{5}{3}$ or $1\frac{2}{3}$

66. $\frac{7}{8} - \frac{5}{6} = \frac{7}{6} \cdot \frac{3}{3} - \frac{5}{6} \cdot \frac{4}{4}$
$= \frac{21}{24} - \frac{20}{24}$
$= \frac{1}{24}$

67. $3\frac{7}{12} - \frac{3}{4} = \frac{43}{12} - \frac{3}{4}$
$= \frac{43}{12} - \frac{3}{4} \cdot \frac{3}{3}$
$= \frac{43}{12} - \frac{9}{12}$
$= \frac{34}{12}$
$= \frac{17}{6}$ or $2\frac{5}{6}$

68. $-\frac{3}{7} + \left(-\frac{11}{14}\right) = \frac{-3}{7} \cdot \frac{2}{2} + \frac{-11}{14}$
$= \frac{-6}{14} + \frac{-11}{14}$
$= \frac{-17}{14}$ or $-1\frac{3}{14}$

69. $1\frac{2}{5} - \left(-\frac{1}{3}\right) = \frac{7}{5} - \left(\frac{-1}{3}\right)$
$= \frac{7}{5} \cdot \frac{3}{3} - \left(-\frac{1}{3}\right) \cdot \frac{5}{5}$
$= \frac{21}{15} - \left(\frac{-5}{15}\right)$
$= \frac{26}{15}$ or $1\frac{11}{15}$

70. $5\frac{1}{2} - 2\frac{2}{3} = \frac{11}{2} - \frac{8}{3}$
$= \frac{11}{2} \cdot \frac{3}{3} - \frac{8}{3} \cdot \frac{2}{2}$
$= \frac{33}{6} - \frac{16}{6}$
$= \frac{17}{6}$ or $2\frac{5}{6}$

71. $-2\frac{1}{6} + 5\frac{1}{3} = \frac{-13}{6} + \frac{16}{3}$
$= \frac{-13}{6} + \frac{16}{3} \cdot \frac{2}{2}$
$= \frac{-13}{6} + \frac{32}{6}$
$= \frac{19}{6}$ or $3\frac{1}{6}$

72. mean: $\frac{4 + 5 + 7 + 3 + 9 + 11 + 23 + 37}{8}$

≈ 12.4

median: 8

mode: none

73. mean: $\frac{3.6 + 7.2 + 9.0 + 5.2 + 7.2 + 6.5 + 3.6}{7}$

≈ 6.0

median: 6.5

mode: 3.6 and 7.2

74.
$$\frac{1}{2} = a + \frac{3}{8}$$
$$\frac{1}{2} - \frac{3}{8} = a + \frac{3}{8} - \frac{3}{8}$$
$$\frac{1}{2} \cdot \frac{4}{4} - \frac{3}{8} = a$$
$$\frac{4}{8} - \frac{3}{8} = a$$
$$\frac{1}{8} = a$$

Check: $\frac{1}{2} = a + \frac{3}{8}$
$$\frac{1}{2} \stackrel{?}{=} \frac{1}{8} + \frac{3}{8}$$
$$\frac{1}{2} \stackrel{?}{=} \frac{4}{8}$$
$$\frac{1}{2} = \frac{1}{2} \checkmark$$

75.
$$x - 1.5 = 1.75$$
$$x - 1.5 + 1.5 = 1.75 + 1.5$$
$$x = 3.25$$

Check: $x - 1.5 = 1.75$
$$3.25 - 1.5 \stackrel{?}{=} 1.75$$
$$1.75 = 1.75 \checkmark$$

76. $0.2t = 6$
$$\frac{0.2t}{0.2} = \frac{6}{0.2}$$
$$t = 30$$

Check: $0.2t = 6$
$$0.2(30) \stackrel{?}{=} 6$$
$$6 = 6 \checkmark$$

77.
$$2 = -\frac{4}{5}n$$
$$-\frac{5}{4}(2) = \left(-\frac{4}{5}n\right) - \frac{5}{4}$$
$$-\frac{5}{\cancel{4}_2} \cdot \frac{\cancel{2}^1}{1} = n$$
$$-\frac{5}{2} \text{ or } -2\frac{1}{2} = n$$

Check: $2 = -\frac{4}{5}n$
$$2 \stackrel{?}{=} -\frac{4}{5}\left(-\frac{5}{2}\right)$$
$$2 \stackrel{?}{=} -\frac{\cancel{4}^2}{\cancel{5}_1}\left(-\frac{\cancel{5}^1}{\cancel{2}_1}\right)$$
$$2 \stackrel{?}{=} \frac{2}{1}$$
$$2 = 2 \checkmark$$

78. 4, 9, 14, 19,... (+5, +5, +5)

Arithmetic; the common difference is 5; the next 3 terms are 19 + 5 or 24, 24 + 5 or 29, and 29 + 5 or 34.

79. 1, 3, 9, 27,... (×3, ×3, ×3)

Geometric; the common ratio is 3; the next 3 terms are 27 × 3 or 81, 81 × 3 or 243, and 243 × 3 or 729.

80. 32, 8, 2, $\frac{1}{2}$,... ($\times\frac{1}{4}$, $\times\frac{1}{4}$, $\times\frac{1}{4}$)

Geometric; the common ratio is $\frac{1}{4}$; the next 3 terms are $\frac{1}{2} \times \frac{1}{4}$ or $\frac{1}{8}$, $\frac{1}{8} \times \frac{1}{4}$ or $\frac{1}{32}$, and $\frac{1}{32} \times \frac{1}{4}$ or $\frac{1}{128}$.

81. −6, −5, −2, 3,... (+1, +3, +5)

Neither

Chapter 5 Practice Test

Page 259

1. The digits of a terminating decimal end, and the digits of a repeating decimal continue repeating.
2. Rename the fractions with a common denominator. Then add and simplify.
3. a sequence in which the quotient of any two consecutive terms is the same
4.
$$\begin{array}{r} 0.45 \\ 20\overline{)9.00} \\ -8\,0 \\ \hline 1\,00 \\ -1\,00 \\ \hline 0 \end{array}$$
0.45 is a terminating decimal.
5.
$$\begin{array}{r} 0.875 \\ 8\overline{)7.000} \\ -6\,4 \\ \hline 60 \\ -56 \\ \hline 40 \\ -40 \\ \hline 0 \end{array}$$
−0.875 is a terminating decimal.
6. $4\frac{2}{9} = 4 + \frac{2}{9}$
$$\frac{2}{9} \rightarrow \begin{array}{r} 0.222... \\ 9\overline{)2.000...} \end{array}$$
$$= 4 + 0.\overline{2}$$
$$= 4.\overline{2}$$
7. $0.24 = \frac{24}{100}$
$$= \frac{6}{25}$$
8. $5.06 = 5\frac{6}{100}$
$$= 5\frac{3}{50}$$

9. $N = 2.333...$
$10N = 10(2.333...)$
$10N = 23.333...$
$-(N = 2.333...)$
$\frac{9N}{9} = \frac{21}{9}$
$N = \frac{21}{9}$
$N = \frac{7}{3}$ or $2\frac{1}{3}$

Therefore, $-2.\overline{3} = -2\frac{1}{3}$.

10. $0.6 \bigcirc \frac{2}{3}$
$0.6 \bigcirc 0.\overline{6}$
$0.6 < 0.\overline{6}$
$0.6 < \frac{2}{3}$

11. $-1\frac{5}{8} \bigcirc -1.6$
$-1.625 \bigcirc -1.6$
$-1.625 < -1.6$
$-1\frac{5}{8} < -1.6$

12. $6 = 2 \cdot 3 = 2 \cdot 3$
$9 = 3 \cdot 3 = 3^2$
$\text{LCM} = 2 \cdot 3^2$
$= 18$

The LCD of $\frac{5}{6}$ and $\frac{2}{9}$ is 18.

13. $4a^2 = 2 \cdot 2 \cdot a^2 = 2^2 \cdot a^2$
$3ab = 3 \cdot a \cdot b = 3 \cdot a \cdot b$
$\text{LCM} = 2^2 \cdot 3 \cdot a^2 \cdot b$
$= 12a^2b$

The LCD of $\frac{9}{4a^2}$ and $\frac{2}{3ab}$ is $12a^2b$.

14. $\frac{5}{8} \cdot \frac{6}{11} = \frac{5}{\cancel{8}_4} \cdot \frac{\cancel{6}^3}{11}$
$= \frac{15}{44}$

15. $\frac{5}{8} + \frac{1}{8} = \frac{5+1}{8}$
$= \frac{6}{8}$
$= \frac{3}{4}$

16. $\frac{7}{9} \div \frac{4}{15} = \frac{7}{\cancel{9}_3} \cdot \frac{\cancel{15}^5}{4}$
$= \frac{35}{12}$ or $2\frac{11}{12}$

17. $3\frac{5}{6} + 1\frac{2}{9} = (3 + 1) + \left(\frac{5}{6} + \frac{2}{9}\right)$
$= 4 + \left(\frac{5}{6} \cdot \frac{3}{3} + \frac{2}{9} \cdot \frac{2}{2}\right)$
$= 4 + \left(\frac{15}{18} + \frac{4}{18}\right)$
$= 4\frac{19}{18}$
$= 5\frac{1}{18}$

18. $\frac{ab}{9} \div \frac{b}{3} = \frac{a\cancel{b}^1}{\cancel{9}_3} \cdot \frac{\cancel{3}^1}{\cancel{b}_1}$
$= \frac{a}{3}$

19. $\frac{11x}{3y} - \frac{8x}{3y} = \frac{11x-8x}{3y}$
$= \frac{3x}{3y}$
$= \frac{x}{y}$

20. mean: $\frac{20.5 + 18.6 + 16.3 + 4.8 + 19.1 + 17.3 + 20.5}{7}$
≈ 16.7

median: 18.6

mode: 20.5

21. 4.8 is an extreme value; mean with extreme value ≈ 16.7;

mean without extreme value:
$\frac{20.5 + 18.6 + 16.3 + 19.1 + 17.3 + 20.5}{6}$
≈ 18.7

It decreases the mean by 2.

22. $x + 4.3 = 9.8$
$x + 4.3 - 4.3 = 9.8 - 4.3$
$x = 5.5$

Check: $x + 4.3 = 9.8$
$5.5 + 4.3 \stackrel{?}{=} 9.8$
$9.8 = 9.8$ ✓

23. $12 = 0.75x$
$\frac{12}{0.75} = \frac{0.75x}{0.75}$
$16 = x$

Check: $12 = 0.75x$
$12 \stackrel{?}{=} 0.75(16)$
$12 = 12$ ✓

24. $3.1m = 12.4$
$\frac{3.1m}{3.1} = \frac{12.4}{3.1}$
$m = 4$

Check: $3.1m = 12.4$
$3.1(4) \stackrel{?}{=} 12.4$
$12.4 = 12.4$ ✓

25. $p - \frac{4}{5} = \frac{2}{3}$
$p - \frac{4}{5} + \frac{4}{5} = \frac{2}{3} + \frac{4}{5}$
$p = \frac{2}{3} \cdot \frac{5}{5} + \frac{4}{5} \cdot \frac{3}{3}$
$p = \frac{10}{15} + \frac{12}{15}$
$p = \frac{22}{15}$ or $1\frac{7}{15}$

Check: $p - \frac{4}{5} = \frac{2}{3}$
$\frac{22}{15} - \frac{4}{5} \stackrel{?}{=} \frac{2}{3}$
$\frac{22}{15} - \frac{4}{5} \cdot \frac{3}{3} \stackrel{?}{=} \frac{2}{3}$
$\frac{22}{15} - \frac{12}{15} \stackrel{?}{=} \frac{2}{3}$
$\frac{10}{15} \stackrel{?}{=} \frac{2}{3}$
$\frac{2}{3} = \frac{2}{3}$ ✓

26. $-\frac{3}{8} = \frac{5}{3}a$

$\frac{3}{5}\left(-\frac{3}{8}\right) = \frac{3}{5}\left(\frac{5}{3}a\right)$

$-\frac{9}{40} = a$

Check: $-\frac{3}{8} = \frac{5}{3}a$

$-\frac{3}{8} \stackrel{?}{=} \frac{\cancel{5}^1}{\cancel{3}_1}\left(-\frac{\cancel{9}^3}{\cancel{40}_8}\right)$

$-\frac{3}{8} = -\frac{3}{8}$ ✓

27. $y - 2\frac{1}{4} = 1\frac{5}{6}$

$y - 2\frac{1}{4} + 2\frac{1}{4} = 1\frac{5}{6} + 2\frac{1}{4}$

$y = (1 + 2) + \left(\frac{5}{6} + \frac{1}{4}\right)$

$y = 3 + \left(\frac{5}{6} \cdot \frac{2}{2} + \frac{1}{4} \cdot \frac{3}{3}\right)$

$y = 3 + \left(\frac{10}{12} + \frac{3}{12}\right)$

$y = 3 + \frac{13}{12}$

$y = 4\frac{1}{12}$

Check: $y - 2\frac{1}{4} = 1\frac{5}{6}$

$4\frac{1}{12} - 2\frac{1}{4} \stackrel{?}{=} 1\frac{5}{6}$

$\frac{49}{12} - \frac{9}{4} \stackrel{?}{=} 1\frac{5}{6}$

$\frac{49}{12} - \frac{9}{4} \cdot \frac{3}{3} \stackrel{?}{=} 1\frac{5}{6}$

$\frac{49}{12} - \frac{27}{12} \stackrel{?}{=} 1\frac{5}{6}$

$\frac{22}{12} \stackrel{?}{=} 1\frac{5}{6}$

$\frac{11}{6}$ or $1\frac{5}{6} = 1\frac{5}{6}$ ✓

28. 2, 8, 32, 128, ...

$\times 4 \quad \times 4 \quad \times 4$

Geometric; the common ratio is 4; the next 3 terms are 512, 2048, and 8192.

29. 5.5, 4.9, 4.3, 3.7, ...

$-0.6 \quad -0.6 \quad -0.6$

Arithmetic; the common difference is −0.6; the next 3 terms are 3.1, 2.5, and 1.9.

30. 1, 2, 4, 7, ...

$+1 \quad +2 \quad +3$

Neither

31. $d = rt$

$d = 65(6)$

$d = 390$ miles

32. $\frac{3}{4} \cdot \left(4\frac{2}{3}\right) = \frac{\cancel{3}^1}{\cancel{4}_2} \cdot \frac{\cancel{14}^7}{\cancel{3}_1}$

$= \frac{7}{2}$

$= 3\frac{1}{2}$ yd

33. A; $-6\frac{1}{4} + \left(-9\frac{3}{20}\right) = \frac{-25}{4} + \frac{-183}{20}$

$= \frac{-25}{4} \cdot \frac{5}{5} + \frac{-183}{20}$

$= \frac{-125}{20} + \frac{-183}{20}$

$= \frac{-308}{20}$

$= \frac{-77}{5}$ or $-15\frac{2}{5}$

Chapter 5 Standardized Test Practice

Pages 260–261

1. A; $20.00 - (3.00 + 0.30 + 0.02)$

$= 20.00 - (3.32)$

$= \$16.68$

2. A; Canada: 24

England: 28

Australia: 6

$24 + 28 + 6 = 58$

3. A; $-1 - 7(4) = -1 - 28$

$= -29°$

4. C; $\frac{6t^4}{18ts} = \frac{\cancel{6} \cdot \cancel{t} \cdot t \cdot t \cdot t}{3 \cdot \cancel{6} \cdot \cancel{t} \cdot s}$

$= \frac{t^3}{3s}$

5. B; $8 \times 10^{-2} = 8 \times 0.01$

$= 0.08$

6. C; $\frac{2}{3} + \frac{1}{4} + \frac{5}{6} = \frac{2}{3} \cdot \frac{4}{4} + \frac{1}{4} \cdot \frac{3}{3} + \frac{5}{6} \cdot \frac{2}{2}$

$= \frac{8}{12} + \frac{3}{12} + \frac{10}{12}$

$= \frac{11}{12} + \frac{10}{12}$

$= \frac{21}{12}$

$= \frac{7}{4}$ or $1\frac{3}{4}$

7. B; $\frac{1}{4}(2 - x) - x = \frac{1}{4}\left(2 - \frac{1}{2}\right) - \frac{1}{2}$

$= \frac{1}{4}\left(1\frac{1}{2}\right) - \frac{1}{2}$

$= \frac{1}{4} \cdot \frac{3}{2} - \frac{1}{2}$

$= \frac{3}{8} - \frac{1}{2}$

$= \frac{3}{8} - \frac{1}{2} \cdot \frac{4}{4}$

$= \frac{3}{8} - \frac{4}{8}$

$= -\frac{1}{8}$

8. B; $\frac{324 + 375 + 420 + 397}{4}$

$= 379$

9. D; 3, 6, 12, 24, ...

$\times 2 \quad \times 2 \quad \times 2$

10. A; 128, 32, 8, 2, ...

$\times\frac{1}{4} \quad \times\frac{1}{4} \quad \times\frac{1}{4}$

$2 \times \frac{1}{4} = \frac{1}{2}$; $\frac{1}{2} \times \frac{1}{4} = \frac{1}{8}$;

$\frac{1}{8} \times \frac{1}{4} = \frac{1}{32}$

11. 15 ~~feet~~ $\times \frac{12 \text{ inches}}{1 \text{ foot}} = 180$ in.

3 ~~feet~~ $\times \frac{12 \text{ inches}}{1 \text{ foot}} = 36$ in. + 4

$= 40$ inches

$40 \times 4 = 160$ inches

He can make 4 shelves.

12. The x-coordinate is 4.

The y-coordinate is –3.

The ordered pair is (4, –3).

13. $3x + 4 = 28$

$3x + 4 - 4 = 28 - 4$

$\frac{3x}{3} = \frac{24}{3}$

$x = 8$

14. $t = 8 \cdot 125$

15. $d = rt$

$275 = 55t$

$\frac{275}{55} = \frac{55t}{55}$

$5 = t$

Juan's trip will take 5 hours.

16. $4^2 \times 5^3 = 4 \cdot 4 \cdot 5 \cdot 5 \cdot 5$

$= 2 \cdot 2 \cdot 2 \cdot 2 \cdot 5 \cdot 5 \cdot 5$

17. $\overset{11}{\cancel{44}}$ ~~minutes~~ $\times \frac{1 \text{ day}}{\underset{360}{\cancel{1440}} \text{ ~~minutes~~}}$

$= \frac{11}{360}$ day

18. $6.526 \times 10^6 = 6{,}526{,}000$ lb.

19. $\frac{3}{8} = 0.375$

20. 3 : 3, 6, 9, (12), 15, 18, 21, (24), 27, 30, 33, (36), 39

4 : 4, 8, (12), 16, 20, (24), 28, 32, (36), 40

Floors 12, 24, and 36 are served by both elevators.

21. $2\frac{1}{2} - \frac{3}{4} = \frac{5}{2} - \frac{3}{4}$

$= \frac{5}{2} \cdot \frac{2}{2} - \frac{3}{4}$

$= \frac{10}{4} - \frac{3}{4}$

$= \frac{7}{4}$ or $1\frac{3}{4}$ cups

22. 307, 352, 367, 417, 419, 433

$\frac{367 + 417}{2} = 392$

The median number of tickets sold is 392.

23a. mean $= \frac{60 + 64 + 68 + 73 + 73 + 74 + 78}{7}$

$= \frac{490}{7}$ or 70

median = 73

mode = 73

23b. mean $= \frac{490 + 40}{8}$

$= \frac{530}{8}$ or 66.25

median $= \frac{68 + 73}{2}$ or 70.5

mode = 73

23c. The mean changed the most. To find the difference, subtract the mean for the 8 scores from the mean for the 7 scores. Do the same for the medians and the modes.

	7 scores	8 scores	Change
mean	70	66.25	70 − 66.25 = 3.75
median	73	70.5	73 − 70.5 = 2.5
mode	73	73	73 − 73 = 0

23d. The median best represents the team's eight scores, because it is less affected by the playoff game score, which is much lower than the other scores.

Chapter 6 Ratio, Proportion, and Percent

Page 263 Getting Started

1. $2 \times 12 = 24$
   2 ft = 24 in.
2. $4 \times 3 = 12$
   4 yd = 12 ft
3. $2 \times 5280 = 10{,}560$
   2 mi = 10,560 ft
4. $3 \times 60 = 180$
   3 h = 180 min
5. $8 \times 60 = 480$
   8 min = 480 s
6. $4 \times 16 = 64$
   4 lb = 64 oz
7. $2 \times 2000 = 4000$
   2T = 4000 lb
8. $5 \times 4 = 20$
   5 gal = 20 qt
9. $3 \times 2 = 6$
   3 pt = 6c
10. $3 \times 100 = 300$
    3m = 300 cm
11. $5.8 \times 100 = 580$
    5.8m = 580 cm
12. $2 \times 1000 = 2000$
    2 km = 2000 m
13. $5 \times 10 = 50$
    5 cm = 50 mm
14. $2.3 \times 1000 = 2300$
    2.3L = 2300 mL
15. $15 \times 1000 = 15{,}000$
    15 kg = 15,000 g

16. $$\begin{array}{r} 3.4 \\ \times\ 7 \\ \hline 23.8 \end{array}$$

17. $$\begin{array}{r} 6.1 \\ \times\ 8 \\ \hline 48.8 \end{array}$$

18. $$\begin{array}{r} 2.8 \\ \times\ 5.9 \\ \hline 252 \\ 140\ \\ \hline 16.52 \end{array}$$

19. $$\begin{array}{r} 1.6 \\ \times\ 8.4 \\ \hline 64 \\ 128\ \\ \hline 13.44 \end{array}$$

20. $$\begin{array}{r} 0.8 \\ \times\ 9.3 \\ \hline 24 \\ 72\ \\ \hline 7.44 \end{array}$$

21. $$\begin{array}{r} 0.6 \\ \times\ 0.3 \\ \hline 0.18 \end{array}$$

22. $$\begin{array}{r} 12.4 \\ \times\ 3.8 \\ \hline 992 \\ 372\ \\ \hline 47.12 \end{array}$$

23. $$\begin{array}{r} 15.2 \\ \times\ 0.2 \\ \hline 3.04 \end{array}$$

24. $\frac{4}{8} = \frac{\overset{1}{\cancel{2}} \times \overset{1}{\cancel{2}}}{2 \times \underset{1}{\cancel{2}} \times \underset{1}{\cancel{2}}} = \frac{1}{2}$

25. $\frac{5}{15} = \frac{1 \times \overset{1}{\cancel{5}}}{3 \times \underset{1}{\cancel{5}}} = \frac{1}{3}$

26. $\frac{6}{10} = \frac{3 \times \overset{1}{\cancel{2}}}{5 \times \underset{1}{\cancel{2}}} = \frac{3}{5}$

27. $\frac{12}{25}$ is simplified.

28. $\frac{22}{20} = \frac{11 \times \overset{1}{\cancel{2}}}{5 \times 2 \times \underset{1}{\cancel{2}}} = \frac{11}{10}$

29. $\frac{15}{16}$ is simplified.

30. $\frac{36}{42} = \frac{\overset{1}{\cancel{2}} \times 2 \times 3 \times \overset{1}{\cancel{3}}}{\underset{1}{\cancel{2}} \times \underset{1}{\cancel{3}} \times 7} = \frac{6}{7}$

31. $\frac{36}{48} = \frac{\overset{1}{\cancel{2}} \times \overset{1}{\cancel{2}} \times 3 \times \overset{1}{\cancel{3}}}{\underset{1}{\cancel{2}} \times \underset{1}{\cancel{2}} \times 2 \times 2 \times \underset{1}{\cancel{3}}} = \frac{3}{4}$

6-1 Ratios and Rates

Page 264 How are ratios used in paint mixtures?

a. Combination B; It has the same ratio as the original mixture but in a smaller amount.
b. For each part of blue paint, 2 parts yellow paint should be used.

Page 266 Check for Understanding

1. Sample answer:

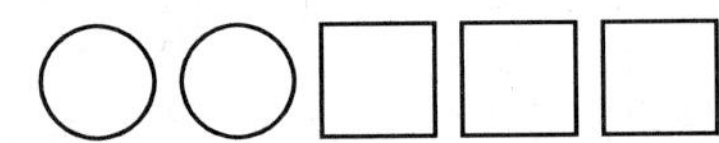

2. A ratio is a comparison of two numbers by division. A rate is a ratio of two measurements having different units of measure.
3. Sample answer: \$12 per person
4. $\frac{4}{10} \overset{\div 2}{=} \frac{2}{5}$ (÷2 above and below)

   The ratio of goals to attempts is $\frac{2}{5}$.
5. $\frac{15}{24} \overset{\div 3}{=} \frac{5}{8}$ (÷3 above and below)

   The ratio of dimes to coins is $\frac{5}{8}$.

6. $\frac{10 \text{ inches}}{3 \text{ feet}} = \frac{10 \text{ inches}}{36 \text{ inches}}$

$= \frac{5 \text{ inches}}{18 \text{ inches}}$

The ratio is $\frac{5}{18}$.

7. $\frac{5 \text{ feet}}{5 \text{ yards}} = \frac{5 \text{ feet}}{15 \text{ feet}}$

$= \frac{1 \text{ feet}}{3 \text{ feet}}$

The ratio is $\frac{1}{3}$.

8. $\frac{183 \text{ dollars}}{4 \text{ tickets}} = \frac{47.75 \text{ dollars}}{1 \text{ ticket}}$ (÷4)

= $45.75/ticket

9. $\frac{9 \text{ inches}}{12 \text{ hours}} = \frac{0.75 \text{ inches}}{1 \text{ hour}}$ (÷12)

= 0.75 inch/hour

10. $\frac{100 \text{ feet}}{14.5 \text{ seconds}} \approx \frac{6.9 \text{ feet}}{1 \text{ second}}$ (÷14.5)

≈ 6.9 feet/second

11. $\frac{254.1 \text{ mi}}{10.5 \text{ gal}} = \frac{24.2 \text{ mi}}{1 \text{ gal}}$ (÷10.5)

So, the unit rate is 24.2 miles per gallon.

12. $\frac{20 \text{ mi}}{1 \text{ h}} = \frac{20 \text{ mi}}{1 \text{ h}} \cdot \frac{5280 \text{ ft}}{1 \text{ mi}} \div \frac{60 \text{ min}}{1 \text{ h}}$

$= \frac{20 \text{ mi}}{1 \text{ h}} \cdot \frac{5280 \text{ ft}}{1 \text{ mi}} \cdot \frac{1 \text{ h}}{60 \text{ min}}$

$= \frac{20 \text{ mi}}{1 \text{ h}} \cdot \frac{\overset{1760}{5280} \text{ ft}}{1 \text{ mi}} \cdot \frac{1 \text{ h}}{\underset{\underset{1}{3}}{60} \text{ min}}$

$= \frac{1760 \text{ ft}}{\text{min}}$

So, 20 miles per hour is equivalent to 1760 feet per minute.

13. $\frac{16 \text{ cm}}{1 \text{ s}} = \frac{16 \text{ cm}}{1 \text{ s}} \cdot \frac{1 \text{ m}}{100 \text{ cm}} \div \frac{1 \text{ h}}{3600 \text{ s}}$

$= \frac{16 \text{ cm}}{1 \text{ s}} \cdot \frac{1 \text{ m}}{100 \text{ cm}} \cdot \frac{3600 \text{ s}}{1 \text{ h}}$

$= \frac{16 \text{ cm}}{1 \text{ s}} \cdot \frac{1 \text{ m}}{\underset{1}{100} \text{ cm}} \cdot \frac{\overset{36}{3600} \text{ s}}{1 \text{ h}}$

$= \frac{576 \text{ m}}{\text{h}}$

So, 16 centimeteres per second is equivalent to 576 meters per hour.

14. $\frac{6}{10} = \frac{3}{5}$ (÷2)

15. $\frac{6+2}{10+2} = \frac{8}{12}$

$\frac{8}{12} = \frac{2}{3}$ (÷4)

No; the ratio will be 2 to 3.

Pages 267–268 Practice and Apply

16. $\frac{6}{27} = \frac{2}{9}$ (÷3)

The ratio of ladybugs to insects is $\frac{2}{9}$.

17. $\frac{14}{35} = \frac{2}{5}$ (÷7)

The ratio of girls to boys is $\frac{2}{5}$.

18. $\frac{18}{45} = \frac{2}{5}$ (÷9)

The ratio of cups to cups is $\frac{2}{5}$.

19. $\frac{12}{28} = \frac{3}{7}$ (÷4)

The ratio of roses to flowers is $\frac{3}{7}$.

20. $\frac{7 \text{ cups}}{9 \text{ pints}} = \frac{7 \text{ cups}}{18 \text{ cups}}$

$= \frac{7 \text{ cups}}{18 \text{ cups}}$

The ratio is $\frac{7}{18}$.

21. $\frac{9 \text{ pounds}}{16 \text{ tons}} = \frac{9 \text{ pounds}}{32{,}000 \text{ pounds}}$

$= \frac{9 \text{ pounds}}{32{,}000 \text{ pounds}}$

The ratio is $\frac{9}{32{,}000}$.

22. $\frac{11 \text{ gallons}}{11 \text{ quarts}} = \frac{44 \text{ quarts}}{11 \text{ quarts}}$

$= \frac{4 \text{ quarts}}{1 \text{ quart}}$

The ratio is $\frac{4}{1}$.

23. $\frac{18 \text{ miles}}{18 \text{ yards}} = \frac{31{,}680 \text{ yards}}{18 \text{ yards}}$

$= \frac{1760 \text{ yards}}{1 \text{ yard}}$

The ratio is $\frac{1760}{1}$.

24. $\frac{15 \text{ dollars}}{123 \text{ dollars}} = \frac{5 \text{ dollars}}{41 \text{ dollars}}$

$= \frac{5 \text{ dollars}}{41 \text{ dollars}}$

The ratio is $\frac{5}{41}$.

25. $\frac{17}{118}$ is simplified.

The ratio of rubies to gems is $\frac{17}{118}$.

26. $\dfrac{155}{75} \overset{\div 5}{=} \dfrac{31}{15}$

The ratio of apples to oranges is $\frac{31}{15}$.

27. $\dfrac{321}{107} \overset{\div 107}{=} \dfrac{3}{1}$

The ratio of articles to magazines is $\frac{3}{1}$.

28. $\dfrac{3 \text{ dollars}}{6 \text{ cans}} \overset{\div 6}{=} \dfrac{0.5 \text{ dollar}}{1 \text{ can}}$

$= \$0.50/\text{can}$

29. $\dfrac{0.99 \text{ dollar}}{10 \text{ pencils}} \overset{\div 10}{\approx} \dfrac{0.1 \text{ dollars}}{1 \text{ pencil}}$

$\approx \$0.10/\text{pencil}$

30. $\dfrac{140 \text{ miles}}{6 \text{ gallons}} \overset{\div 6}{\approx} \dfrac{23.3 \text{ miles}}{1 \text{ gallon}}$

$\approx 23.3 \text{ mi/gal}$

31. $\dfrac{68 \text{ m}}{15 \text{ s}} \overset{\div 15}{\approx} \dfrac{4.5 \text{ m}}{1 \text{ s}}$

$\approx 4.5 \text{ m/sec}$

32. $\dfrac{19 \text{ yd}}{2.5 \text{ min}} \overset{\div 2.5}{=} \dfrac{7.6 \text{ yd}}{1 \text{ min}}$

$= 7.6 \text{ yd/min}$

33. $\dfrac{25 \text{ ft}}{3.2 \text{ h}} \overset{\div 3.2}{\approx} \dfrac{7.8 \text{ ft}}{1 \text{ h}}$

$\approx 7.8 \text{ ft/h}$

34. $\dfrac{236.7 \text{ mi}}{4.5 \text{ d}} \overset{\div 4.5}{=} \dfrac{52.6 \text{ mi}}{1 \text{ d}}$

$= 52.6 \text{ mi/d}$

35. $\dfrac{331.5 \text{ pages}}{8.5 \text{ weeks}} \overset{\div 8.5}{=} \dfrac{39 \text{ pages}}{1 \text{ week}}$

$= 39 \text{ pages/week}$

36. $\dfrac{40.5 \text{ dollars}}{18 \text{ issues}} \overset{\div 18}{=} \dfrac{2.25 \text{ dollars}}{1 \text{ issue}}$

$= \$2.25/\text{issue}$

$\dfrac{33.6 \text{ dollars}}{12 \text{ issues}} \overset{\div 12}{=} \dfrac{2.8 \text{ dollars}}{1 \text{ issue}}$

$= \$2.80/\text{issue}$

Since the 12-issue subscription costs $2.80 per issue, and the 18-issue subscription costs $2.25 per issue, the 12-issue subscription costs more per issue.

37. $\dfrac{2.2 \text{ dollars}}{6 \text{ cans}} \overset{\div 6}{\approx} \dfrac{0.37 \text{ dollar}}{1 \text{ can}}$

$\approx \$0.37/\text{can}$

$\dfrac{4.25 \text{ dollars}}{12 \text{ cans}} \approx \dfrac{0.35 \text{ dollar}}{1 \text{ can}}$

$\approx \$0.35/\text{can}$

The 6-pack of soda costs $0.37 per can. The 12-pack of soda costs $0.35 per can. So, the 12-pack is less expensive.

38. $\dfrac{45 \text{ mi}}{1 \text{ h}} = \dfrac{45 \text{ mi}}{1\text{h}} \cdot \dfrac{5280 \text{ ft}}{1 \text{ mi}} \div \dfrac{3600 \text{ s}}{1 \text{ h}}$

$= \dfrac{45 \text{ mi}}{1 \text{ h}} \cdot \dfrac{5280 \text{ ft}}{1 \text{ mi}} \cdot \dfrac{1 \text{ h}}{3600 \text{ s}}$

$= \dfrac{\overset{1}{\cancel{45}} \cancel{\text{mi}}}{1 \cancel{\text{h}}} \cdot \dfrac{\overset{66}{\cancel{5280}} \text{ ft}}{1 \cancel{\text{mi}}} \cdot \dfrac{1 \cancel{\text{h}}}{\underset{\underset{1}{\cancel{80}}}{\cancel{3600}} \text{ s}}$

$= \dfrac{66 \text{ ft}}{\text{s}}$

So, 45 miles per hour is equivalent to 66 feet per second.

39. $\dfrac{18 \text{ mi}}{1 \text{ h}} = \dfrac{18 \text{ mi}}{1 \text{ h}} \cdot \dfrac{5280 \text{ ft}}{1 \text{ mi}} \div \dfrac{3600 \text{ s}}{1 \text{ h}}$

$= \dfrac{18 \text{ mi}}{1 \text{ h}} \cdot \dfrac{5280 \text{ ft}}{1 \text{ mi}} \cdot \dfrac{1 \text{ h}}{3600 \text{ s}}$

$= \dfrac{\overset{1}{\cancel{18}} \cancel{\text{mi}}}{1 \cancel{\text{h}}} \cdot \dfrac{\overset{26.4}{\cancel{5280}} \text{ ft}}{1 \cancel{\text{mi}}} \cdot \dfrac{1 \cancel{\text{h}}}{\underset{\underset{1}{\cancel{200}}}{\cancel{3600}} \text{ s}}$

$= \dfrac{26.4 \text{ ft}}{\text{s}}$

So, 18 miles per hour is equivalent to 26.4 feet per second.

40. $\dfrac{26 \text{ cm}}{1 \text{ s}} = \dfrac{26 \text{ cm}}{1 \text{ s}} \cdot \dfrac{1 \text{ m}}{100 \text{ cm}} \div \dfrac{1 \text{ min}}{60 \text{ s}}$

$= \dfrac{26 \text{ cm}}{1 \text{ s}} \cdot \dfrac{1 \text{ m}}{100 \text{ cm}} \cdot \dfrac{60 \text{ s}}{1 \text{ min}}$

$= \dfrac{26 \cancel{\text{cm}}}{1 \cancel{\text{s}}} \cdot \dfrac{1 \text{ m}}{\underset{1}{\cancel{100}} \cancel{\text{cm}}} \cdot \dfrac{\overset{0.6}{\cancel{60}} \cancel{\text{s}}}{1 \text{ min}}$

$= \dfrac{15.6 \text{ m}}{\text{min}}$

So, 26 cm per second is equivalent to 15.6 meters per minute.

41. $\dfrac{32 \text{ cm}}{1 \text{ s}} = \dfrac{32 \text{ cm}}{1 \text{ s}} \cdot \dfrac{1 \text{ m}}{100 \text{ cm}} \div \dfrac{1 \text{ min}}{60 \text{ s}}$

$= \dfrac{32 \text{ cm}}{1 \text{ s}} \cdot \dfrac{1 \text{ m}}{100 \text{ cm}} \cdot \dfrac{60 \text{ s}}{1 \text{ min}}$

$= \dfrac{32 \cancel{\text{cm}}}{1 \cancel{\text{s}}} \cdot \dfrac{1 \text{ m}}{\underset{1}{\cancel{100}} \cancel{\text{cm}}} \cdot \dfrac{\overset{0.6}{\cancel{60}} \cancel{\text{s}}}{1 \text{ min}}$

$= \dfrac{19.2 \text{ m}}{\text{min}}$

So, 32 cm per second is equivalent to 19.2 meters per minute.

42. $$\begin{aligned}\frac{2.5\text{ qt}}{1\text{ min}} &= \frac{2.5\text{ qt}}{1\text{ min}} \cdot \frac{1\text{ gal}}{4\text{ qt}} \div \frac{1\text{ h}}{60\text{ min}}\\ &= \frac{2.5\text{ qt}}{1\text{ min}} \cdot \frac{1\text{ gal}}{4\text{ qt}} \cdot \frac{60\text{ min}}{1\text{ h}}\\ &= \frac{2.5\,\cancel{\text{qt}}}{1\,\cancel{\text{min}}} \cdot \frac{1\text{ gal}}{\underset{1}{\cancel{4}}\,\cancel{\text{qt}}} \cdot \frac{\overset{15}{\cancel{60}}\,\cancel{\text{min}}}{1\text{ h}}\\ &= \frac{37.5\text{ gal}}{\text{h}}\end{aligned}$$

So, 2.5 quarts per minute is equivalent to 37.5 gallons per hour.

43. $$\begin{aligned}\frac{4.8\text{ qt}}{1\text{ min}} &= \frac{4.8\text{ qt}}{1\text{ min}} \cdot \frac{1\text{ gal}}{4\text{ qt}} \div \frac{1\text{ h}}{60\text{ min}}\\ &= \frac{4.8\text{ qt}}{1\text{ min}} \cdot \frac{1\text{ gal}}{4\text{ qt}} \cdot \frac{60\text{ min}}{1\text{ h}}\\ &= \frac{4.8\,\cancel{\text{qt}}}{1\,\cancel{\text{min}}} \cdot \frac{1\text{ gal}}{\cancel{4}\,\cancel{\text{qt}}} \cdot \frac{\overset{15}{\cancel{60}}\,\cancel{\text{min}}}{1\text{ h}}\\ &= \frac{72\text{ gal}}{\text{h}}\end{aligned}$$

So, 4.8 quarts per minute is equivalent to 72 gallons per hour.

44. $$\begin{aligned}\frac{4\text{ c}}{1\text{ min}} &= \frac{4\text{ c}}{1\text{ min}} \cdot \frac{1\text{ qt}}{4\text{ c}} \div \frac{1\text{ h}}{60\text{ min}}\\ &= \frac{4\text{ c}}{1\text{ min}} \cdot \frac{1\text{ qt}}{4\text{ c}} \cdot \frac{60\text{ min}}{1\text{ h}}\\ &= \frac{\cancel{4}\,\cancel{\text{c}}}{1\,\cancel{\text{min}}} \cdot \frac{1\text{ qt}}{\cancel{4}\,\cancel{\text{c}}} \cdot \frac{60\,\cancel{\text{min}}}{1\text{ h}}\\ &= \frac{60\text{ qt}}{\text{h}}\end{aligned}$$

So, 4 cups per minute is equivalent to 60 quarts per hour.

45. $$\begin{aligned}\frac{7\text{ c}}{1\text{ min}} &= \frac{7\text{ c}}{1\text{ min}} \cdot \frac{1\text{ qt}}{4\text{ c}} \div \frac{1\text{ h}}{60\text{ min}}\\ &= \frac{7\text{ c}}{1\text{ min}} \cdot \frac{1\text{ qt}}{4\text{ c}} \cdot \frac{60\text{ min}}{1\text{ hr}}\\ &= \frac{7\,\cancel{\text{c}}}{1\,\cancel{\text{min}}} \cdot \frac{1\text{ qt}}{\cancel{4}\,\cancel{\text{c}}} \cdot \frac{\overset{15}{\cancel{60}}\,\cancel{\text{min}}}{1\text{ h}}\\ &= \frac{105\text{ qt}}{\text{h}}\end{aligned}$$

So, 7 cups per minute is equivalent to 105 quarts per hour.

46. Alaska:

÷570374

$$\frac{626{,}932\text{ people}}{570{,}374\text{ sq mi}} \approx \frac{1\text{ person}}{1\text{ sq mi}}$$

÷570374

≈ 1 person/sq mi

New York:

÷47224

$$\frac{18{,}976{,}457\text{ people}}{47{,}224\text{ sq mi}} \approx \frac{402\text{ people}}{1\text{ sq mi}}$$

÷47224

≈ 402 people/sq mi

Rhode Island:

÷1045

$$\frac{1{,}048{,}319\text{ people}}{1045\text{ sq mi}} \approx \frac{1003\text{ people}}{1\text{ sq mi}}$$

÷1045

≈ 1003 people/sq mi

Texas:

÷261914

$$\frac{20{,}851{,}820\text{ people}}{261{,}914\text{ sq mi}} \approx \frac{80\text{ people}}{1\text{ sq mi}}$$

÷261914

≈ 80 people/sq mi

Wyoming:

÷97105

$$\frac{493{,}782\text{ people}}{97{,}105\text{ sq mi}} \approx \frac{5\text{ people}}{1\text{ sq mi}}$$

÷97105

≈ 5 people/sq mi

47. 1015 + 1011 = 2026

÷3.5

$$\frac{2026\text{ miles}}{3.5\text{ hours}} \approx \frac{579\text{ miles}}{1\text{ hour}}$$

÷3.5

≈ 579 mi/h

The airplane traveled at 579 mi/h.

48. $\frac{5685\text{ dollars}}{285\text{ people}} \approx \frac{19.95\text{ dollars}}{1\text{ person}}$

≈ \$19.95/person/h

49. Original earnings:

$\frac{\text{Marty}}{\text{Spencer}} = \frac{3}{1}$

$\frac{3}{1} = \frac{6}{2} = \frac{9}{3} = \frac{12}{4}$ and so on

After Marty gives Spencer \$3:

$\frac{\text{Marty}}{\text{Spencer}} = \frac{1}{1}$

$\frac{9}{3} = \frac{9-3}{3+3} = \frac{6}{6} = \frac{1}{1}$

Therefore, Marty earned \$9.

50. Ratios are used in paint mixtures to achieve a desired color by using two or more different colors. Answers should include the following.

- If you want a darker shade of green, use a ratio in which there is a greater amount of blue paint compared to yellow paint. An example is 3 blue to 4 yellow.
- If you want a lighter shade of green, use a ratio in which there is a greater amount of yellow paint than blue paint. An example is 5 yellow to 2 blue.

51. C; $\frac{4}{3} = \frac{12}{9}$ (×3 on top and bottom)

52. A; $\frac{2.79\text{ dollars}}{4\text{ quarts}} \approx \frac{0.70\text{ dollars}}{1\text{ quart}}$ (÷4 on top and bottom)

≈ \$0.70/qt

53a. Answers will vary.

53b. The ratios should be close in value.

53c. Sample answer: Pyramid of Khufu in Giza, Egypt; The Taj Mahal in India; The Lincoln Memorial in Washington, D.C.

Page 268 Maintain Your Skills

54. $-3, 6, -12, 24, \ldots$ (each term $\times(-2)$)

Geometric; the common ratio is -2; the next 3 terms are -48, 96, and -192.

55. $12.1, 12.4, 12.7, 13, \ldots$ (each term $+0.3$)

Arithmetic; the common difference is 0.3; the next 3 terms are 13.3, 13.6, and 13.9.

56.
$$3.6 = x - 7.1$$
$$3.6 + 7.1 = x - 7.1 + 7.1$$
$$10.7 = x$$

57.
$$y + \frac{3}{4} = \frac{2}{3}$$
$$y + \frac{3}{4} - \frac{3}{4} = \frac{2}{3} - \frac{3}{4}$$
$$y = \frac{2}{3} \cdot \frac{4}{4} - \frac{3}{4} \cdot \frac{3}{3}$$
$$y = \frac{8}{12} - \frac{9}{12}$$
$$y = -\frac{1}{12}$$

58. $-4.8 = 6z$
$$\frac{-4.8}{6} = \frac{6z}{6}$$
$$-0.8 = z$$

59.
$$\frac{3}{8}w = 5$$
$$\frac{8}{3}\left(\frac{3}{8}w\right) = \left(\frac{5}{1}\right)\frac{8}{3}$$
$$w = \frac{40}{3} \text{ or } 13\frac{1}{3}$$

60.
$$1\frac{1}{7} \div \left(-\frac{4}{7}\right) = \frac{8}{7} \div \left(-\frac{4}{7}\right)$$
$$= \frac{\overset{2}{\cancel{8}}}{\underset{1}{\cancel{7}}} \cdot \frac{\overset{1}{\cancel{-7}}}{\underset{1}{\cancel{4}}}$$
$$= -2$$

61. $52{,}000{,}000 = 5.2 \times 10{,}000{,}000$
$$= 5.2 \times 10^7$$

62. $42{,}240 = 4.224 \times 10{,}000$
$$= 4.224 \times 10^4$$

63. $0.038 = 3.8 \div 100$
$$= 3.8 \times 10^{-2}$$

64. $8 \cdot (k + 3) \cdot (k + 3) = 8(k + 3)^2$

65. $10x = 300$
$$\frac{10x}{10} = \frac{300}{10}$$
$$x = 30$$

66. $25m = 225$
$$\frac{25m}{25} = \frac{225}{25}$$
$$m = 9$$

67. $8k = 320$
$$\frac{8k}{8} = \frac{320}{8}$$
$$k = 40$$

68. $192 = 4t$
$$\frac{192}{4} = \frac{4t}{4}$$
$$48 = t$$

69. $195 = 15w$
$$\frac{195}{15} = \frac{15w}{15}$$
$$13 = w$$

70. $231 = 33n$
$$\frac{231}{33} = \frac{33n}{33}$$
$$7 = n$$

Page 269 Reading Mathematics

1. Sample answer: difference comparison: The Houston Zoo has 6000 fewer animals than the Columbus Zoo; ratio comparison: The ratio of the size of the San Diego Zoo to the Houston Zoo is 100:55 or 20:11. The San Diego Zoo is about twice the size of the Houston Zoo.
2. difference comparison
3. ratio comparison
4. ratio comparison

6-2 Using Proportions

Page 270 How are proportions used in recipes?

a. Lemonade, grape juice, and orange juice are 12 oz each.

$$\frac{\text{juice}}{\text{water}} = \frac{12}{84} = \frac{12 \div 12}{84 \div 12} = \frac{1}{7} \text{ each juice}$$
$$\frac{\text{lemon-lime soda}}{\text{water}} = \frac{40}{84} = \frac{40 \div 4}{84 \div 4} = \frac{10}{21}$$

b.
$$\frac{\text{juice} \times 2}{\text{water} \times 2} = \frac{1 \times 2}{7 \times 2} = \frac{2}{14} = \frac{1}{7} \text{ each juice}$$
$$\frac{\text{lemon-lime soda} \times 2}{\text{water} \times 2} = \frac{10 \times 2}{21 \times 2} = \frac{20}{42} = \frac{10}{21}$$

c. Yes; each part was multiplied by the same number.

Page 272 Check for Understanding

1. *Proportion* is a statement of equality of two ratios.
2. Sample answer: $\frac{2}{3} = \frac{3}{4}$; $\frac{1}{2} = \frac{1}{4}$

3.
$$\frac{1}{4} \stackrel{?}{=} \frac{4}{16}$$
$$1 \cdot 16 \stackrel{?}{=} 4 \cdot 4$$
$$16 = 16$$

So, $\frac{1}{4} = \frac{4}{16}$.

Yes; the ratios form a proportion.

4.
$$\frac{2.1}{3.5} \stackrel{?}{=} \frac{3}{7}$$
$$2.1 \cdot 7 \stackrel{?}{=} 3.5 \cdot 3$$
$$14.7 \neq 10.5$$

So, $\frac{2.1}{3.5} \neq \frac{3}{7}$.

No; the ratios do not form a proportion.

5.
$$\frac{k}{35} = \frac{3}{7}$$
$$k \cdot 7 = 35 \cdot 3$$
$$7k = 105$$
$$\frac{7k}{7} = \frac{105}{7}$$
$$k = 15$$

6. $\frac{3}{t} = \frac{18}{24}$

$3 \cdot 24 = t \cdot 18$

$72 = 18t$

$\frac{72}{18} = \frac{18t}{18}$

$4 = t$

7. $\frac{10}{8.4} = \frac{5}{m}$

$10 \cdot m = 8.4 \cdot 5$

$10m = 42$

$\frac{10m}{10} = \frac{42}{10}$

$m = 4.2$

8. $\frac{\text{width of original}}{\text{length of original}} = \frac{\text{width of new photo}}{\text{length of new photo}}$

$\frac{3}{5} = \frac{x}{7}$

$3 \cdot 7 = 5 \cdot x$

$21 = 5x$

$\frac{21}{5} = \frac{5x}{5}$

$4\frac{1}{5} = x$

The width of the new photo is $4\frac{1}{5}$ in.

Pages 273–274 Practice and Apply

9. $\frac{2}{3} \stackrel{?}{=} \frac{8}{12}$

$2 \cdot 12 \stackrel{?}{=} 3 \cdot 8$

$24 = 24$

So, $\frac{2}{3} = \frac{8}{12}$.

Yes; the ratio forms a proportion.

10. $\frac{4}{2} \stackrel{?}{=} \frac{16}{5}$

$4 \cdot 5 \stackrel{?}{=} 2 \cdot 16$

$20 \neq 32$

So, $\frac{4}{2} \neq \frac{16}{5}$.

No; the ratios do not form a proportion.

11. $\frac{1.5}{5.0} \stackrel{?}{=} \frac{3}{9}$

$1.5 \cdot 9 \stackrel{?}{=} 5.0 \cdot 3$

$13.5 \neq 15$

So, $\frac{1.5}{5.0} \neq \frac{3}{9}$.

No; the ratios do not form a proportion.

12. $\frac{18}{2.4} \stackrel{?}{=} \frac{15}{2}$

$18 \cdot 2 \stackrel{?}{=} 2.4 \cdot 15$

$36 = 36$

So, $\frac{18}{2.4} = \frac{15}{2}$.

Yes; the ratios form a proportion.

13. $\frac{3.4}{1.6} \stackrel{?}{=} \frac{5.1}{2.4}$

$3.4 \cdot 2.4 \stackrel{?}{=} 1.6 \cdot 5.1$

$8.16 = 8.16$

So, $\frac{3.4}{1.6} = \frac{5.1}{2.4}$.

Yes; the ratios form a proportion.

14. $\frac{5.3}{15.9} \stackrel{?}{=} \frac{2.7}{8.1}$

$5.3 \cdot 8.1 \stackrel{?}{=} 15.9 \cdot 2.7$

$42.93 = 42.93$

So, $\frac{5.3}{15.9} = \frac{2.7}{8.1}$.

Yes; the ratios form a proportion.

15. $\frac{p}{6} = \frac{24}{36}$

$p \cdot 36 = 6 \cdot 24$

$36p = 144$

$\frac{36p}{36} = \frac{144}{36}$

$p = 4$

16. $\frac{w}{11} = \frac{14}{22}$

$w \cdot 22 = 11 \cdot 14$

$22w = 154$

$\frac{22w}{22} = \frac{154}{22}$

$w = 7$

17. $\frac{4}{10} = \frac{8}{a}$

$4 \cdot a = 10 \cdot 8$

$4a = 80$

$\frac{4a}{4} = \frac{80}{4}$

$a = 20$

18. $\frac{18}{12} = \frac{24}{q}$

$18 \cdot q = 12 \cdot 24$

$18q = 288$

$\frac{18q}{18} = \frac{288}{18}$

$q = 16$

19. $\frac{5}{h} = \frac{10}{30}$

$5 \cdot 30 = h \cdot 10$

$150 = 10h$

$\frac{150}{10} = \frac{10h}{10}$

$15 = h$

20. $\frac{51}{z} = \frac{17}{7}$

$51 \cdot 7 = z \cdot 17$

$357 = 17z$

$\frac{357}{17} = \frac{17z}{17}$

$21 = z$

21. $\frac{7}{45} = \frac{x}{9}$

$7 \cdot 9 = 45 \cdot x$

$63 = 45x$

$\frac{63}{45} = \frac{45x}{45}$

$1.4 = x$

22. $\frac{2}{15} = \frac{c}{72}$

$2 \cdot 72 = 15 \cdot c$

$144 = 15c$

$\frac{144}{15} = \frac{15c}{15}$

$9.6 = c$

23. $\frac{7}{5} = \frac{10.5}{b}$

$7 \cdot b = 5 \cdot 10.5$

$7b = 52.5$

$\frac{7b}{7} = \frac{52.5}{7}$

$b = 7.5$

24. $\frac{16}{7} = \frac{4.8}{h}$

$16 \cdot h = 7 \cdot 4.8$

$16h = 33.6$

$\frac{16h}{16} = \frac{33.6}{16}$

$h = 2.1$

25. $\frac{2}{9.4} = \frac{0.2}{v}$

$2 \cdot v = 9.4 \cdot 0.2$

$2v = 1.88$

$\frac{2v}{2} = \frac{1.88}{2}$

$v = 0.94$

26. $\frac{9}{7.2} = \frac{3.5}{k}$

$9 \cdot k = 7.2 \cdot 3.5$

$9k = 25.2$

$\frac{9k}{9} = \frac{25.2}{9}$

$k = 2.8$

27. $\frac{a}{0.28} = \frac{4}{1.4}$

$a \cdot 1.4 = 0.28 \cdot 4$

$1.4a = 1.12$

$\frac{1.4a}{1.4} = \frac{1.12}{1.4}$

$a = 0.8$

28. $\frac{3}{14} = \frac{15}{m - 3}$

$3(m - 3) = 14 \cdot 15$

$3m - 9 = 210$

$3m - 9 + 9 = 210 + 9$

$3m = 219$

$\frac{3m}{3} = \frac{219}{3}$

$m = 73$

29. $\frac{16}{x + 5} = \frac{4}{5}$

$16 \cdot 5 = (x + 5)4$

$80 = 4x + 20$

$80 - 20 = 4x + 20 - 20$

$60 = 4x$

$\frac{60}{4} = \frac{4x}{4}$

$15 = x$

30. $\frac{5.1}{1.7} = \frac{7.5}{d}$

$5.1 \cdot d = 1.7 \cdot 7.5$

$5.1d = 12.75$

$\frac{5.1d}{5.1} = \frac{12.75}{5.1}$

$d = 2.5$

31. $\frac{6.5}{1.3} = \frac{m}{5.2}$

$6.5 \cdot 5.2 = 1.3 \cdot m$

$33.8 = 1.3m$

$\frac{33.8}{1.3} = \frac{1.3m}{1.3}$

$26 = m$

32. $\frac{8 \text{ pencils}}{2 \text{ boxes}} = \frac{20 \text{ pencils}}{x \text{ boxes}}$

$\frac{8}{2} = \frac{20}{x}$

$8 \cdot x = 2 \cdot 20$

$8x = 40$

$\frac{8x}{8} = \frac{40}{8}$

$x = 5$

33. $\frac{12 \text{ glasses}}{3 \text{ crates}} = \frac{72 \text{ glasses}}{m \text{ crates}}$

$\frac{12}{3} = \frac{72}{m}$

$12 \cdot m = 3 \cdot 72$

$12m = 216$

$\frac{12m}{12} = \frac{216}{12}$

$m = 18$

34. $\frac{y \text{ dollars}}{5.4 \text{ gallons}} = \frac{14 \text{ dollars}}{3 \text{ gallons}}$

$\frac{y}{5.4} = \frac{14}{3}$

$y \cdot 3 = 5.4 \cdot 14$

$3y = 75.6$

$\frac{3y}{3} = \frac{75.6}{3}$

$y = 25.2$

35. $\frac{5 \text{ quarts}}{6.25 \text{ dollars}} = \frac{d \text{ quarts}}{8.75 \text{ dollars}}$

$\frac{5}{6.25} = \frac{d}{8.75}$

$5 \cdot 8.75 = 6.25 \cdot d$

$43.75 = 6.25d$

$\frac{43.75}{6.25} = \frac{6.25d}{6.25}$

$7 = d$

36. $\frac{1 \text{ m}}{3.28 \text{ ft}} = \frac{110 \text{ m}}{x \text{ ft}}$

37. $\frac{1}{3.28} = \frac{110}{x}$

$1 \cdot x = 3.28 \cdot 110$

$x = 360.8 \text{ ft}$

38. $\frac{\text{width of original}}{\text{length of original}} = \frac{\text{width of new}}{\text{length of new}}$

$\frac{8 \text{ inches}}{10 \text{ inches}} = \frac{4.5 \text{ inches}}{x \text{ inches}}$

$\frac{8}{10} = \frac{4.5}{x}$

$8 \cdot x = 10 \cdot 4.5$

$8x = 45$

$\frac{8x}{8} = \frac{45}{8}$

$x = 5\frac{5}{8} \text{ in.}$

39. $\frac{0.667 \text{ pound}}{1 \text{ dollar}} = \frac{14.99 \text{ pounds}}{x \text{ dollars}}$

$\frac{0.667}{1} = \frac{14.99}{x}$

$0.667 \cdot x = 1 \cdot 14.99$

$0.667x = 14.99$

$\frac{0.667x}{0.667} = \frac{14.99}{0.667}$

$x \approx \$22.47$

40. $\frac{3.481 \text{ pound}}{1 \text{ dollar}} = \frac{12.50 \text{ pounds}}{x \text{ dollars}}$

$\frac{3.481}{1} = \frac{12.50}{x}$

$3.481 \cdot x = 1 \cdot 12.50$

$3.481x = 12.50$

$\frac{3.481x}{3.481} = \frac{12.50}{3.481}$

$x \approx \$3.59$

41. $\frac{\text{raisins in original}}{\text{chocolate in original}} = \frac{\text{raisins in new}}{\text{chocolate in new}}$

$$\frac{1}{0.5} = \frac{6}{x}$$

$$1 \cdot x = 0.5 \cdot 6$$

$$x = 3 \text{ cups of chocolate}$$

$$\frac{\text{raisins in original}}{\text{peanuts in original}} = \frac{\text{raisins in new}}{\text{peanuts in new}}$$

$$\frac{1}{0.25} = \frac{6}{x}$$

$$1 \cdot x = 0.25 \cdot 6$$

$$x = 1.5 \text{ or } 1\tfrac{1}{2} \text{ cups of peanuts}$$

42. $5 \times 5 = 25$ sq ft $\quad 9.5 \times 9.5 = 90.25$ sq ft

$$\frac{1 \text{ pint}}{25 \text{ sq ft}} = \frac{x \text{ pints}}{90.25 \text{ sq ft}}$$

$$\frac{1}{25} = \frac{x}{90.25}$$

$$1 \cdot 90.25 = 25 \cdot x$$

$$90.25 = 25x$$

$$\frac{90.25}{25} = \frac{25x}{25}$$

$$3.61 \text{ or } 4 = x$$

4 pints must be purchased.

43. $\frac{a}{c} = \frac{b}{d}$; $\frac{b}{a} = \frac{d}{c}$; $\frac{c}{a} = \frac{d}{b}$

$ad = cb \quad bc = ad \quad cb = ad$

$ad = bc \quad ad = bc \quad ad = bc$

44. Proportions can be used in recipes to determine the amount of ingredients needed when increasing or decreasing a recipe. Answers should include the following.

- If you want to increase or decrease a recipe, you can use proportions to find the amounts needed so that the ratios remain the same.
- Adding 10 ounces to each ingredient in the punch recipe will not result in the same flavor because the ingredients are in proportion to each other. By adding the same amount to each ingredient, the ingredients are no longer proportional and the flavor is changed.

45. C; $\frac{\text{Jack's height}}{\text{Jack's shadow}} = \frac{\text{pole's height}}{\text{pole's shadow}}$

$$\frac{6}{3} = \frac{x}{12}$$

Page 274 Maintain Your Skills

46. $\frac{5 \text{ dollars}}{4 \text{ loaves}} = \frac{1.25 \text{ dollars}}{1 \text{ loaf}}$ (÷4 numerator and denominator)

$$= \$1.25/\text{loaf}$$

47. $\frac{183.4 \text{ miles}}{3.2 \text{ hours}} \approx \frac{57.3 \text{ miles}}{1 \text{ hour}}$ (÷3.2 numerator and denominator)

$$\approx 57.3 \text{ mph}$$

48. 2, 5, 8, 11, 14 (+3, +3, +3, +3)

$14 + 3 = 17$; $17 + 3 = 20$; $20 + 3 = 23$

49. $\frac{x}{5} \div \frac{x}{20} = \frac{x}{5} \cdot \frac{20}{x}$

$$= \frac{\cancel{x}}{\cancel{5}_1} \cdot \frac{\cancel{20}^4}{\cancel{x}}$$

$$= 4$$

50. $\frac{3y}{4} \div \frac{5y}{8} = \frac{3y}{4} \cdot \frac{8}{5y}$

$$= \frac{3\cancel{y}}{\cancel{4}_1} \cdot \frac{\cancel{8}^2}{5\cancel{y}}$$

$$= \frac{6}{5} \text{ or } 1\tfrac{1}{5}$$

51. $\frac{4z}{w} \div \frac{7yz}{w} = \frac{4z}{w} \cdot \frac{w}{7yz}$

$$= \frac{4\cancel{z}}{\cancel{w}} \cdot \frac{\cancel{w}}{7y\cancel{z}}$$

$$= \frac{4}{7y}$$

52. $5 \times 12 = 60$
5 ft = 60 in.

53. $8.5 \times 12 = 102$
8.5 ft = 102 in.

54. $36 \div 12 = 3$
36 in. = 3 ft

55. $78 \div 12 = 6.5$
78 in. = 6.5 ft

Page 275 Algebra Activity (Follow-Up of Lesson 6-2)

1–2. Answers will vary. The estimate should be close to the actual number.

3. Sample answer: More samples leads to an accurate estimate.

4. Sample answer: The ratio of tagged fish to actual fish should be approximately the same as the ratio of tagged fish to total recaptured fish in the samples.

6-3 Scale Drawings and Models

Page 276 How are scale drawings used in everyday life?

a. $2 \cdot 4 = 8$ ft

b. Sample answer: Suppose a map has a scale of 0.25 inch = 10 miles, and two towns are 1 inch apart. Since 1 inch is equivalent to four times 0.25 inch, and each of the 0.25 inch equals 10 miles, this tells you that the actual distance is 4×10 or 40 miles. Use a proportion.

$$\frac{0.25}{10} = \frac{1.0}{x}$$

Pages 278–279 Check for Understanding

1. 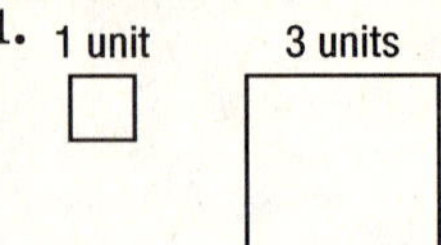

2. Luisa; since there are 24 inches in 2 feet, the ratio 1:24 is correct.

3. $\frac{1 \text{ inch}}{20 \text{ miles}} = \frac{2 \text{ inches}}{x \text{ miles}}$

$1 \cdot x = 20 \cdot 2$

$x = 40$

The distance is 40 miles.

4. $\frac{1 \text{ inch}}{20 \text{ miles}} = \frac{1.75 \text{ inches}}{x \text{ miles}}$

$1 \cdot x = 20 \cdot 1.75$

$x = 35$

The distance is 35 miles.

5. 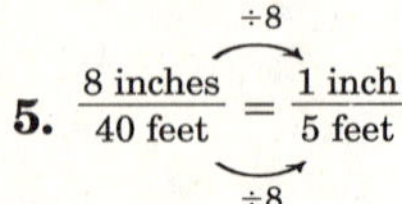

$\frac{8 \text{ inches}}{40 \text{ feet}} = \frac{1 \text{ inch}}{5 \text{ feet}}$ (÷8)

The scale is 1 in. = 5 ft.

6. $\frac{1 \text{ inch}}{5 \text{ feet}} = \frac{1 \text{ inch}}{60 \text{ inches}}$

The scale factor is $\frac{1}{60}$.

7. $\frac{0.5 \text{ in.}}{10 \text{ ft}} = \frac{x \text{ in.}}{15 \text{ ft}}$

$0.5 \cdot 15 = 10 \cdot x$

$7.5 = 10x$

$0.75 = x$

The drawing's length is 0.75 or $\frac{3}{4}$ inch.

$\frac{0.5 \text{ in.}}{10 \text{ ft}} = \frac{x \text{ in.}}{10 \text{ ft}}$

$0.5 \cdot 10 = 10 \cdot x$

$5 = 10x$

$0.5 = x$

The drawing's width is 0.5 or $\frac{1}{2}$ inch.

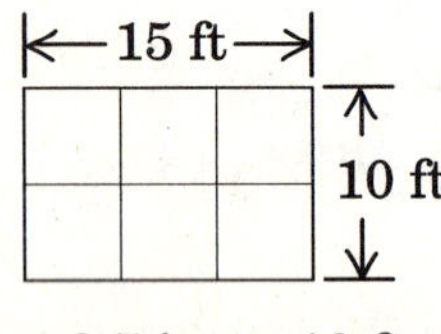

0.5 in. = 10 ft

Pages 279–280 Practice and Apply

8. $\frac{0.5 \text{ inch}}{3 \text{ feet}} = \frac{7 \text{ inches}}{x \text{ feet}}$

$0.5 \cdot x = 3 \cdot 7$

$0.5x = 21$

$\frac{0.5x}{0.5} = \frac{21}{0.5}$

$x = 42$ ft

9. $\frac{0.5 \text{ inch}}{3 \text{ feet}} = \frac{2 \text{ inches}}{x \text{ feet}}$

$0.5 \cdot x = 3 \cdot 2$

$0.5x = 6$

$\frac{0.5x}{0.5} = \frac{6}{0.5}$

$x = 12$ ft

10. $\frac{0.5 \text{ inch}}{3 \text{ feet}} = \frac{2.3 \text{ inches}}{x \text{ feet}}$

$0.5 \cdot x = 3 \cdot 2.3$

$0.5x = 6.9$

$\frac{0.5x}{0.5} = \frac{6.9}{0.5}$

$x = 13.8$ ft

11. $\frac{0.5 \text{ inch}}{3 \text{ feet}} = \frac{4.1 \text{ inches}}{x \text{ feet}}$

$0.5 \cdot x = 3 \cdot 4.1$

$0.5x = 12.3$

$\frac{0.5x}{0.5} = \frac{12.3}{0.5}$

$x = 24.6$ ft

12. $\frac{0.5 \text{ inch}}{3 \text{ feet}} = \frac{2.2 \text{ inches}}{x \text{ feet}}$

$0.5 \cdot x = 3 \cdot 2.2$

$0.5x = 6.6$

$\frac{0.5x}{0.5} = \frac{6.6}{0.5}$

$x = 13.2$ ft

13. $\frac{0.5 \text{ inch}}{3 \text{ feet}} = \frac{1.9 \text{ inches}}{x \text{ feet}}$

$0.5 \cdot x = 3 \cdot 1.9$

$0.5x = 5.7$

$\frac{0.5x}{0.5} = \frac{5.7}{0.5}$

$x = 11.4$ ft

14. $\frac{\frac{1}{2} \text{ inch}}{3 \text{ feet}} = \frac{3\frac{3}{4} \text{ inches}}{x \text{ feet}}$

$\frac{1}{2} \cdot x = 3 \cdot 3\frac{3}{4}$

$\frac{\cancel{2}^{1}}{1} \cdot \frac{1}{\cancel{2}_{1}}x = \frac{3}{1} \cdot \frac{15}{\cancel{4}_{2}} \cdot \frac{\cancel{2}^{1}}{1}$

$x = \frac{45}{2} = 22\frac{1}{2}$ ft

15. $\frac{\frac{1}{2} \text{ inch}}{3 \text{ feet}} = \frac{8\frac{1}{4} \text{ inches}}{x \text{ feet}}$

$\frac{1}{2}x = 3 \cdot 8\frac{1}{4}$

$\frac{2}{1} \cdot \frac{1}{2}x = \frac{3}{1} \cdot \frac{33}{\cancel{4}_{2}} \cdot \frac{\cancel{2}^{1}}{1}$

$x = \frac{99}{2} = 49\frac{1}{2}$ ft

16. $\frac{0.5 \text{ inch}}{3 \text{ feet}} = \frac{0.5 \text{ inch}}{36 \text{ inches}}$

$= \frac{1 \text{ inch}}{72 \text{ inches}}$

The scale factor is $\frac{1}{72}$.

17. $\frac{8 \text{ inches}}{1 \text{ foot}} = \frac{8 \text{ inches}}{12 \text{ inches}}$

$= \frac{2 \text{ inches}}{3 \text{ inches}}$

The scale factor is $\frac{2}{3}$.

18. $\frac{6 \text{ inches}}{210 \text{ feet}} = \frac{1 \text{ inch}}{x \text{ feet}}$

$6 \cdot x = 210 \cdot 1$

$6x = 210$

$\frac{6x}{6} = \frac{210}{6}$

$x = 35$

The scale is 1 in. = 35 ft.

19. $\frac{4.8 \text{ cm}}{1.2 \text{ cm}} = \frac{1 \text{ cm}}{x \text{ cm}}$

$$\begin{aligned} 4.8 \cdot x &= 1.2 \cdot 1 \\ 4.8x &= 1.2 \\ \frac{4.8x}{4.8} &= \frac{1.2}{4.8} \\ x &= 0.25 \end{aligned}$$

The scale is 1 cm = 0.25 cm.

20. $\frac{0.25 \text{ inch}}{2 \text{ feet}} = \frac{x \text{ inches}}{8 \text{ feet}}$

$$\begin{aligned} 0.25 \cdot 8 &= 2 \cdot x \\ 2 &= 2x \\ 1 &= x \end{aligned}$$

The drawing width is 1 inch.

$$\frac{0.25 \text{ inch}}{2 \text{ feet}} = \frac{x \text{ inches}}{16 \text{ feet}}$$

$$\begin{aligned} 0.25 \cdot 16 &= 2 \cdot x \\ 4 &= 2x \\ 2 &= x \end{aligned}$$

The drawing length is 2 inches.

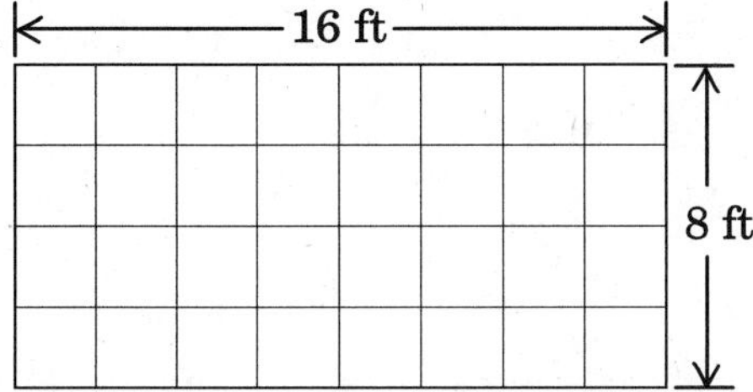

21. A scale factor less than 1 means that the drawing or model is drawn smaller than actual size. A scale factor of 1 means that the drawing or model is drawn actual size. A scale factor greater than 1 means the drawing or model is drawn larger than actual size.

22. In everyday life, scale drawings or models are used to represent objects that are too large or too small to be drawn or built at actual size. Answers should include the following.

- Sample answer: a model airplane, a globe, and a set of blueprints
- Sample answer: A scale drawing such as a map can help you find actual distance and a scale model such as a replica of a dinosaur can help to see the actual image of a dinosaur.

23. B; $\frac{6 \text{ inches}}{9 \text{ feet}} = \frac{6 \text{ inches}}{108 \text{ inches}}$

$$= \frac{1 \text{ inch}}{18 \text{ inches}}$$

$$= \frac{1}{18}$$

24. C; $\frac{1 \text{ inch}}{16 \text{ inch}} = \frac{5.8 \text{ feet}}{x \text{ feet}}$

$$\begin{aligned} 1 - x &= 16 \cdot 5.8 \\ x &= 92.8 \text{ feet} \end{aligned}$$

25a.
$$\begin{aligned} 2l + 2w &= 2(5) + 2(2) \\ &= 14 \\ 2l + 2w &= 2(10) + 2(4) \\ &= 28 \end{aligned}$$

$$\frac{14 \text{ inches}}{28 \text{ inches}} = \frac{1}{2}$$

25b.
$$\begin{aligned} lw &= 5(2) \\ &= 10 \\ lw &= 10(4) \\ &= 40 \end{aligned}$$

$$\frac{10 \text{ sq. cm}}{40 \text{ sq. cm}} = \frac{1}{4}$$

25c. The perimeter of a 3-inch by 5-inch rectangle is 16 inches. The area is 15 square inches. For a 6-inch by 10-inch rectangle, the perimeter should be double the 3-inch by 5-inch rectangle. That is, 16 × 2 or 32 inches. The area should be 4 times the area of the 3-inch by 5-inch rectangle. That is, 4 × 15 or 60 square inches. Since the perimeter of the 6-inch by 10-inch rectangle is 32 and its area is 60, the conjecture is true.

Page 280 Maintain Your Skills

26. $\frac{n}{20} = \frac{15}{50}$

$$\begin{aligned} n \cdot 50 &= 20 \cdot 15 \\ 50n &= 300 \\ \frac{50n}{50} &= \frac{300}{50} \\ n &= 6 \end{aligned}$$

27. $\frac{14}{32} = \frac{x}{8}$

$$\begin{aligned} 14 \cdot 8 &= 32 \cdot x \\ 112 &= 32x \\ \frac{112}{32} &= \frac{32x}{32} \\ 3.5 &= x \end{aligned}$$

28. $\frac{3}{2.2} = \frac{7.5}{y}$

$$\begin{aligned} 3 \cdot y &= 2.2 \cdot 7.5 \\ 3y &= 16.5 \\ \frac{3y}{3} &= \frac{16.5}{3} \\ y &= 5.5 \end{aligned}$$

29.
$$\begin{aligned} \frac{36 \text{ cm}}{1 \text{ s}} &= \frac{36 \text{ cm}}{1 \text{ s}} \cdot \frac{1 \text{ m}}{100 \text{ cm}} \div \frac{1 \text{ min}}{60 \text{ s}} \\ &= \frac{36 \text{ cm}}{1 \text{ s}} \cdot \frac{1 \text{ m}}{100 \text{ cm}} \cdot \frac{60 \text{ s}}{1 \text{ min}} \\ &= \frac{\overset{7.2}{\cancel{36}} \cancel{\text{cm}}}{1 \cancel{\text{s}}} \cdot \frac{1 \text{ m}}{\underset{\underset{1}{\cancel{5}}}{\cancel{100}} \cancel{\text{cm}}} \cdot \frac{\overset{3}{\cancel{60}} \cancel{\text{s}}}{1 \text{ min}} \\ &= \frac{21.6 \text{ m}}{\text{min}} \end{aligned}$$

So, 36 cm per second is equivalent to 21.6 meters per minute.

30.
$$\begin{aligned} \frac{66 \text{ gal}}{1 \text{ h}} &= \frac{66 \text{ gal}}{1 \text{ h}} \cdot \frac{4 \text{ qt}}{1 \text{ gal}} \div \frac{60 \text{ min}}{1 \text{ h}} \\ &= \frac{66 \text{ gal}}{1 \text{ h}} \cdot \frac{4 \text{ qt}}{1 \text{ gal}} \cdot \frac{1 \text{ h}}{60 \text{ min}} \\ &= \frac{\overset{4.4}{\cancel{66}} \cancel{\text{gal}}}{1 \cancel{\text{h}}} \cdot \frac{\overset{1}{\cancel{4}} \text{ qt}}{1 \cancel{\text{gal}}} \cdot \frac{1 \cancel{\text{h}}}{\underset{15}{\cancel{60}} \text{ min}} \\ &= \frac{4.4 \text{ qt}}{\text{min}} \end{aligned}$$

So, 66 gallons per hour is equivalent to 4.4 quarts per minute.

31. $1\frac{1}{4} + 4\frac{5}{6} = (1 + 4) + \left(\frac{1}{4} + \frac{5}{6}\right)$
$= 5 + \left(\frac{1}{4} \cdot \frac{3}{3} + \frac{5}{6} \cdot \frac{2}{2}\right)$
$= 5 + \left(\frac{3}{12} + \frac{10}{12}\right)$
$= 5 + \frac{13}{12}$
$= 6\frac{1}{12}$

32. $4^3 \cdot 4^5 = 4^{3+5}$
$= 4^8$

33. $3t^4 \cdot 6t = 3 \cdot 6 \cdot t^4 \cdot t$
$= 18t^{4+1}$
$= 18t^5$

34. $7^{14} \div 7^8 = 7^{14-8}$
$= 7^6$

35. $\frac{24\text{ m}^5}{18\text{ m}^2} = \frac{4 \times 6\text{ m}^{5-2}}{3 \times 6}$
$= \frac{4\text{m}^3}{3}$

36. $14x^2y = \boxed{1} \cdot 2 \cdot \boxed{7} \cdot 14 \cdot \boxed{x} \cdot x \cdot \boxed{y}$
$35xy^3 = \boxed{1} \cdot 5 \cdot \boxed{7} \cdot 35 \cdot \boxed{x} \cdot \boxed{y} \cdot y \cdot y$

37. $\frac{5}{100} = \frac{1 \cdot \overset{1}{\cancel{5}}}{2 \cdot 2 \cdot 5 \cdot \underset{1}{\cancel{5}}}$
$= \frac{1}{20}$

38. $\frac{25}{100} = \frac{1 \cdot \overset{1}{\cancel{5}} \cdot \overset{1}{\cancel{5}}}{2 \cdot 2 \cdot \underset{1}{\cancel{5}} \cdot \underset{1}{\cancel{5}}}$
$= \frac{1}{4}$

39. $\frac{40}{100} = \frac{\overset{1}{\cancel{2}} \cdot \overset{1}{\cancel{2}} \cdot 2 \cdot \overset{1}{\cancel{5}}}{\underset{1}{\cancel{2}} \cdot \underset{1}{\cancel{2}} \cdot 5 \cdot \underset{1}{\cancel{5}}}$
$= \frac{2}{5}$

40. $\frac{52}{100} = \frac{13 \cdot \overset{1}{\cancel{2}} \cdot \overset{1}{\cancel{2}}}{5 \cdot 5 \cdot \underset{1}{\cancel{2}} \cdot \underset{1}{\cancel{2}}}$
$= \frac{13}{25}$

41. $\frac{78}{100} = \frac{39 \cdot \overset{1}{\cancel{2}}}{50 \cdot \underset{1}{\cancel{2}}}$
$= \frac{39}{50}$

42. $\frac{75}{100} = \frac{3 \cdot \overset{1}{\cancel{5}} \cdot \overset{1}{\cancel{5}}}{2 \cdot 2 \cdot \underset{1}{\cancel{5}} \cdot \underset{1}{\cancel{5}}}$
$= \frac{3}{4}$

43. $\frac{82}{100} = \frac{41 \cdot \overset{1}{\cancel{2}}}{50 \cdot \underset{1}{\cancel{2}}}$
$= \frac{41}{50}$

44. $\frac{95}{100} = \frac{19 \cdot \overset{1}{\cancel{5}}}{20 \cdot \underset{1}{\cancel{5}}}$
$= \frac{19}{20}$

6-4 Fractions, Decimals, and Percents

Page 281 How are percents related to fractions and decimals?

a. $\frac{3}{4}; \frac{3}{5}; \frac{4}{5}$

b. $\frac{75}{100}; \frac{60}{100}; \frac{80}{100}$

c. circle

d. Part b; the fractions have a common denominator.

Page 283 Check for Understanding

1. Sample answer: write an equivalent fraction with a denominator of 100 or express the fraction as a decimal and then express the decimal as a percent. A fraction is greater than 100% if it is greater than 1. It is less than 1% if it is less than $\frac{1}{100}$.

2. Sample answer: Change $64\frac{1}{2}\%$ to 64.5% and then move the decimal point two places to the left and drop the percent sign.

3. $30\% = \frac{30}{100}$
$= \frac{3}{10}$
$30\% = 30\%$
$= 0.3$

4. $12\frac{1}{2}\% = \frac{12\frac{1}{2}}{100}$
$= 12\frac{1}{2} \div 100$
$= \frac{\overset{1}{\cancel{25}}}{2} \cdot \frac{1}{\underset{4}{\cancel{100}}}$
$= \frac{1}{8}$
$12\frac{1}{2}\% = 12.5\%$
$= 0.125$

5. $125\% = \frac{125}{100}$
$= \frac{5}{4}$ or $1\frac{1}{4}$
$125\% = 125\%$
$= 1.25$

6. $65\% = \frac{65}{100}$
$= \frac{13}{20}$
$65\% = 65\%$
$= 0.65$

7. $135\% = \frac{135}{100}$
$= \frac{27}{20}$ or $1\frac{7}{20}$
$135\% = 135\%$
$= 1.35$

8. $0.2\% = \frac{0.2}{100}$
$= \frac{0.2}{100} \cdot \frac{10}{10}$
$= \frac{2}{1000}$ or $\frac{1}{500}$
$0.2 = 0.2\%$
$= .002$

9. $0.45 = 0.45$
$= 45\%$

10. $1.3 = 1.3$
$= 130\%$

11. $0.008 = 0.008$
$= 0.8\%$

12. $\frac{1}{4} = \frac{25}{100}$
$= 25\%$

13. $\frac{12}{9} = 1.333...$
$\approx 133.3\%$

14. $\frac{3}{600} = 0.005$
$= 0.5\%$

15. $\frac{3}{5} = 0.60$
$= 60\%$
Since $60\% > 55\%$, more people get it from the daily newspaper.

Pages 284–285 Practice and Apply

16. $42\% = \frac{42}{100}$ $= \frac{21}{50}$; $42\% = 42\%$ $= 0.42$

17. $88\% = \frac{88}{100}$ $= \frac{22}{25}$; $88\% = 88\%$ $= 0.88$

18. $16\frac{2}{3}\% = \frac{16\frac{2}{3}}{100}$ $= 16\frac{2}{3} \div 100$ $= \frac{\cancel{50}^{1}}{3} \cdot \frac{1}{\cancel{100}_{2}}$ $= \frac{1}{6}$; $16\frac{2}{3}\% = 16.\overline{6}\%$ $= 0.16\overline{6}$

19. $87.5\% = \frac{87.5}{100}$ $= \frac{87.5}{100} \cdot \frac{10}{10}$ $= \frac{875}{1000}$ $= \frac{7}{8}$; $87.5\% = 87.5\%$ $= 0.875$

20. $150\% = \frac{150}{100}$ $= \frac{3}{2}$ or $1\frac{1}{2}$; $150\% = 150\%$ $= 1.5$

21. $350\% = \frac{350}{100}$ $= \frac{7}{2}$ or $3\frac{1}{2}$; $350\% = 350\%$ $= 3.5$

22. $18\% = \frac{18}{100}$ $= \frac{9}{50}$; $18\% = 18\%$ $= 0.18$

23. $61\% = \frac{61}{100}$; $61\% = 61\%$ $= 0.61$

24. $117\% = \frac{117}{100}$ $= 1\frac{17}{100}$; $117\% = 117\%$ $= 1.17$

25. $223\% = \frac{223}{100}$ $= 2\frac{23}{100}$; $223\% = 223\%$ $= 2.23$

26. $0.8\% = \frac{0.8}{100}$ $= \frac{0.8}{100} \cdot \frac{10}{10}$ $= \frac{8}{1000}$ $= \frac{1}{125}$; $0.8\% = 0.8\%$ $= 0.008$

27. $0.53\% = \frac{0.53}{100}$ $= \frac{0.53}{100} \cdot \frac{100}{100}$ $= \frac{53}{10{,}000}$; $0.53\% = 0.53\%$ $= 0.0053$

28. $0.51 = 0.51$ $= 51\%$

29. $0.09 = 0.09$ $= 9\%$

30. $3.21 = 3.21$ $= 321\%$

31. $2.7 = 2.7$ $= 270\%$

32. $0.0042 = 0.0042$ $= 0.42\%$

33. $0.0006 = 0.0006$ $= 0.06\%$

34. $\frac{7}{25} = \frac{28}{100}$ $= 28\%$

35. $\frac{9}{40} = 0.225$ $= 22.5\%$

36. $\frac{10}{3} = 3.\overline{3}$ $\approx 333.3\%$

37. $\frac{14}{8} = \frac{7}{4}$ $= \frac{175}{100}$ $= 175\%$

38. $\frac{15}{2500} = 0.006$ $= 0.6\%$

39. $\frac{20}{1200} = 0.01\overline{6}$ $= 1.7\%$

40. $46\% = \frac{46}{100}$ $= \frac{23}{50}$

41. $3.7\% = \frac{3.7}{100}$ $= \frac{3.7}{100} \cdot \frac{10}{10}$ $= \frac{37}{1000}$

42. $\frac{2}{5} = \frac{40}{100}$ $= 40\%$

Since $\frac{2}{5} = 40\%$, and 40% is greater than 22%, the group that said they perfer ketchup is larger.

43. $\frac{2}{5} = \frac{40}{100}$ $= 40\%$

$0.45 = 0.45$ $= 45\%$

$\frac{3}{8} = 0.375$ $= 37.5\%$

45%, or 0.45 is largest.

44. $\frac{3}{4} = \frac{75}{100}$ $= 75\%$

$0.70 = 0.70$ $= 70\%$

$\frac{4}{5} = \frac{80}{100}$ $= 80\%$

80%, or $\frac{4}{5}$ is largest.

45. $\frac{3}{16} = 0.1875$ $= 18.75\%$

$0.155 = 0.155$ $= 15.5\%$

$\frac{2}{15} = 0.1\overline{3}$ $\approx 13.3\%$

19% is the largest.

46. $\frac{10}{11} = 0.90\overline{90}$
$\approx 90.9\%$
$0.884 = 0.884$
$= 88.4\%$
$\frac{12}{14} \approx 0.857\%$
$= 85.7\%$
$\frac{10}{11}$ or 90.9% is the largest.

47. $\frac{2}{3} = 0.\overline{6}$
$\approx 66.7\%$
$0.69 = 0.69$
$= 69\%$
61%, 66.7% or $\frac{2}{3}$, 69% or 0.69

48. $\frac{2}{7} \approx 0.286$
$= 28.6\%$
$0.027 = 0.027$
$= 2.7\%$
2.7% or 0.027, 27%, 28.6% or $\frac{2}{7}$

49. $\frac{2}{5} = 0.4$

50. $\frac{2}{5} \times 25 = \frac{2}{\cancel{5}_1} \times \cancel{25}^5$
$= 10$ units in length
$A = lw$
$= 10(15)$
$= 150$ square units

51. There are only two possibilities that satisfy the conditions, $\frac{1}{4}$ and $\frac{2}{5}$.

52. Because they can be expressed in the form $\frac{a}{b}$ where a and b are integers and $b \neq 0$.

53. Percents are related to fractions and decimals because they can be expressed as them. Answers should include the following.

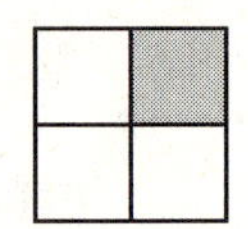

25%

30%

40%

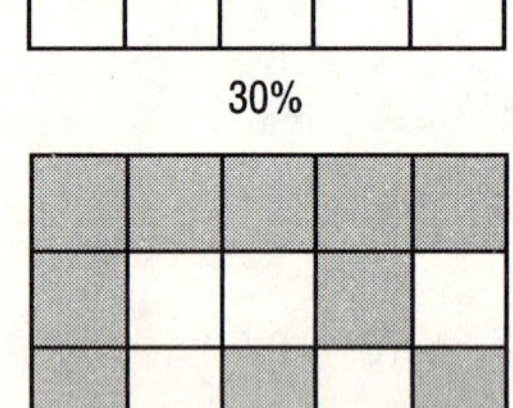

65%

- $25\% = \frac{1}{4} = 0.25$; $30\% = \frac{3}{10} = 0.3$; $40\% = \frac{2}{5} = 0.4$; $65\% = \frac{13}{20} = 0.65$

54. D; $\frac{4}{8} = 0.50$
$= 50\%$
$\frac{3}{9} = 0.3\overline{3}$
$\approx 33.3\%$
$\frac{1}{4} = 0.25$
$= 25\%$
$\frac{9}{15} = 0.60$
$= 60\%$

55. A; $85\% = \frac{85}{100}$
$= \frac{17}{20}$
17 to 20

Page 285 Maintain Your Skills

56. $\frac{3 \text{ inches}}{18 \text{ inches}} = \frac{1}{6}$

57. $\frac{2 \text{ inches}}{2 \text{ feet}} = \frac{2 \text{ inches}}{24 \text{ inches}}$
$= \frac{1}{12}$

58. $\frac{x}{54} = \frac{2}{3}$
$x \cdot 3 = 54 \cdot 2$
$3x = 108$
$\frac{3x}{3} = \frac{108}{3}$
$x = 36$

59. $\frac{4}{7} \cdot \frac{11}{12} = \frac{\cancel{4}^1}{7} \cdot \frac{11}{\cancel{12}_3}$
$= \frac{11}{21}$

60. $-\frac{3}{5} \cdot \frac{10}{18} = -\frac{\cancel{3}^1}{\cancel{5}_1} \cdot \frac{\cancel{\cancel{10}}^{\cancel{2}^1}}{\cancel{\cancel{18}}_{\cancel{6}_3}}$
$= -\frac{1}{3}$

61. $4 \cdot \frac{16}{52} = \frac{\cancel{4}^1}{1} \cdot \frac{16}{\cancel{52}_{13}}$
$= \frac{16}{13}$
$= 1\frac{3}{13}$

62. $5.6 \times 10^{-4} = 0.00056$

63. $21 = 3 \cdot 7$ composite

64. 47 is prime

65. $57 = 3 \cdot 19$ composite

66. $\frac{25}{4} = \frac{x}{100}$
$25 \cdot 100 = 4 \cdot x$
$2500 = 4x$
$\frac{2500}{4} = \frac{4x}{4}$
$625 = x$

67. $\frac{56}{7} = \frac{y}{100}$
$56 \cdot 100 = 7 \cdot y$
$5600 = 7y$
$\frac{5600}{7} = \frac{7y}{7}$
$800 = y$

68. $\frac{75}{8} = \frac{n}{100}$
$75 \cdot 100 = 8 \cdot n$
$7500 = 8n$
$\frac{7500}{8} = \frac{8n}{8}$
$937.5 = n$

69. $\frac{m}{10} = \frac{9.4}{100}$

$m \cdot 100 = 10 \cdot 9.4$

$100m = 94$

$\frac{100m}{100} = \frac{94}{100}$

$m = 0.94$

70. $\frac{h}{350} = \frac{46}{100}$

$h \cdot 100 = 350 \cdot 46$

$100h = 16{,}100$

$\frac{100h}{100} = \frac{16{,}100}{100}$

$h = 161$

71. $\frac{86.4}{k} = \frac{27}{100}$

$86.4 \cdot 100 = k \cdot 27$

$8640 = 27k$

$\frac{8640}{27} = \frac{27k}{27}$

$320 = k$

Pages 286–287 Algebra Activity (Preview of Lesson 6-5)

1. Exact answer: 60%

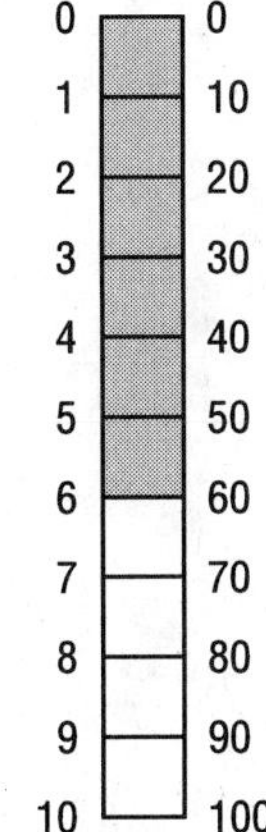

2. Exact answer: 90%

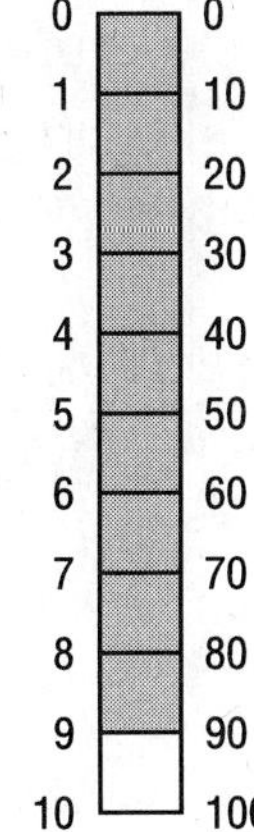

3. Exact answer: 40%

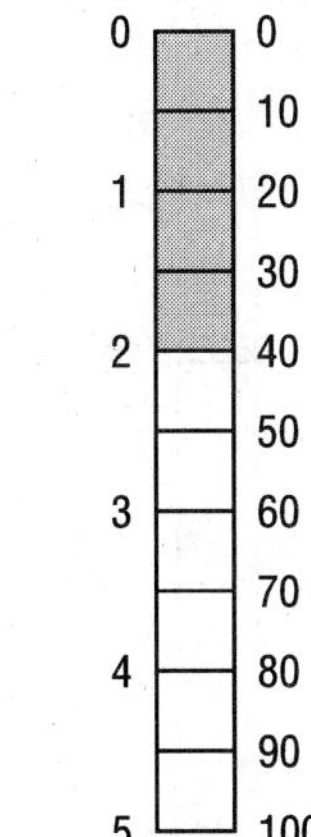

4. Exact answer: 75%

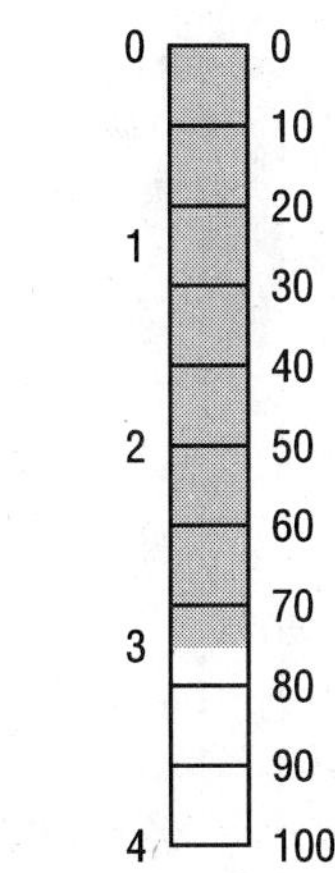

5. Exact answer: 45%

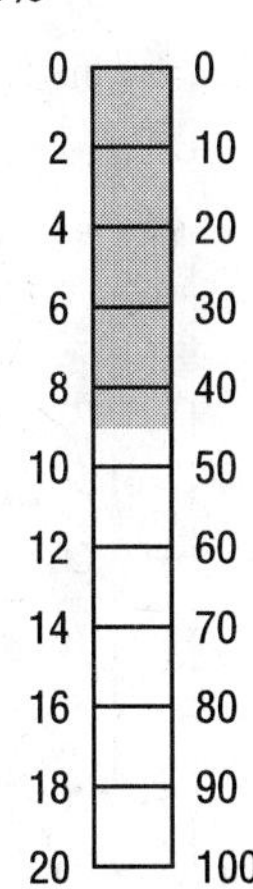

6. Exact answer: 16%

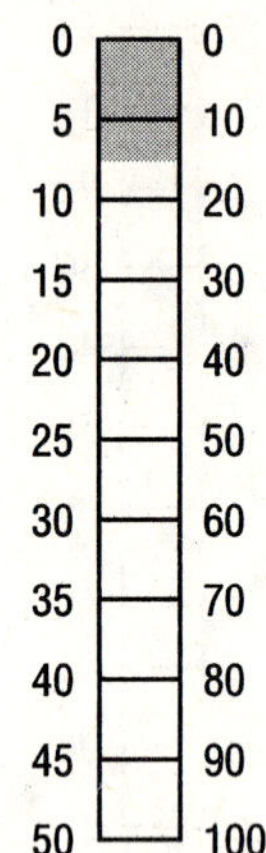

7. Exact answer: 25%

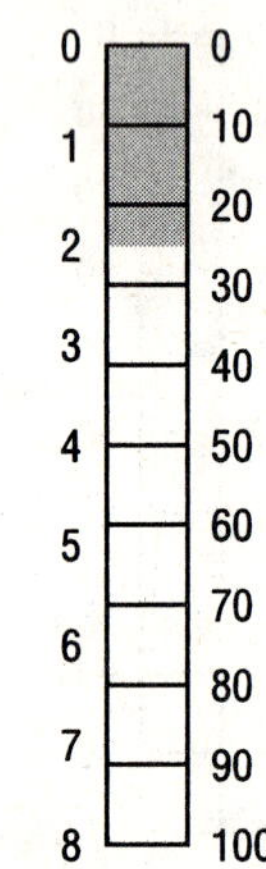

8. Exact answer: 37.5%

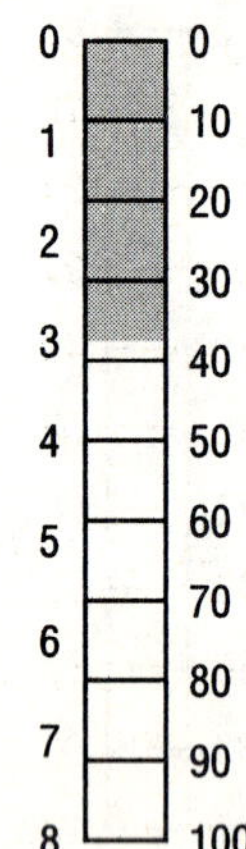

9. Exact answer: $66\frac{2}{3}\%$

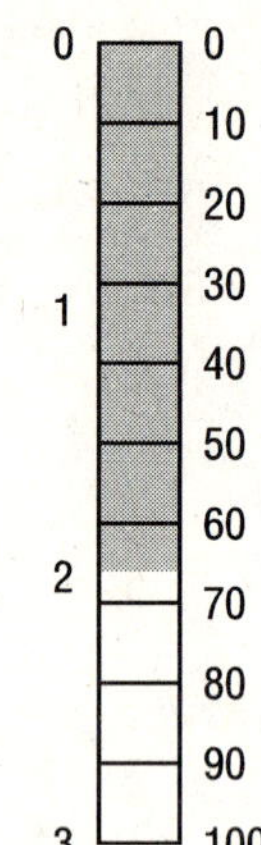

10. Exact answer: $55\frac{5}{9}\%$

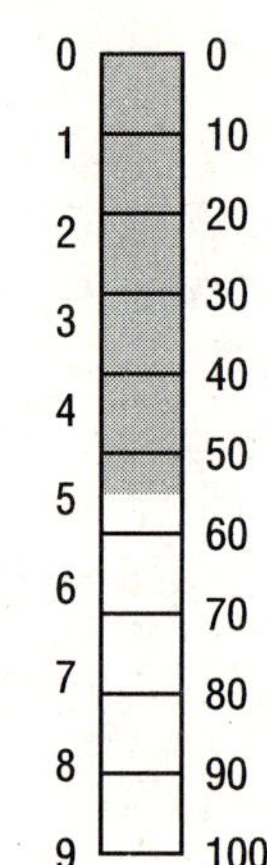

11. Exact answer: 5

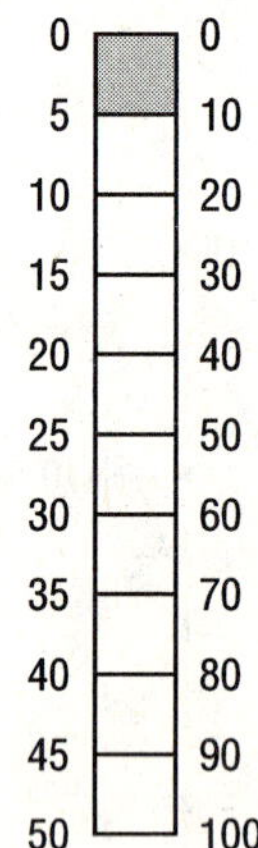

12. Exact answer: 12

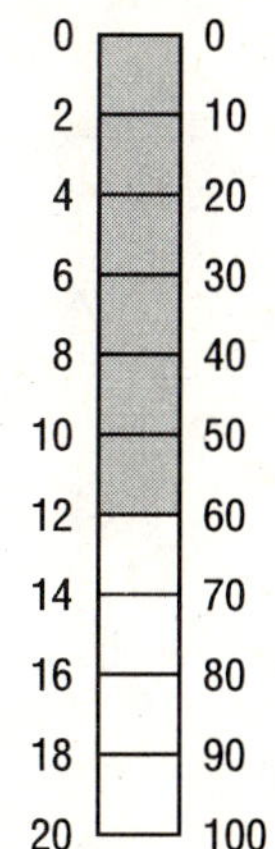

13. Exact answer: 36

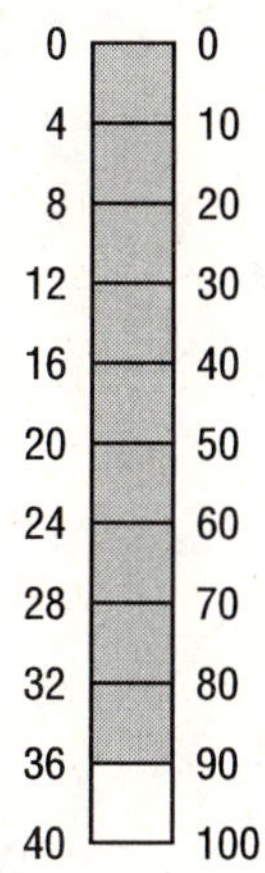

14. Exact answer: 3

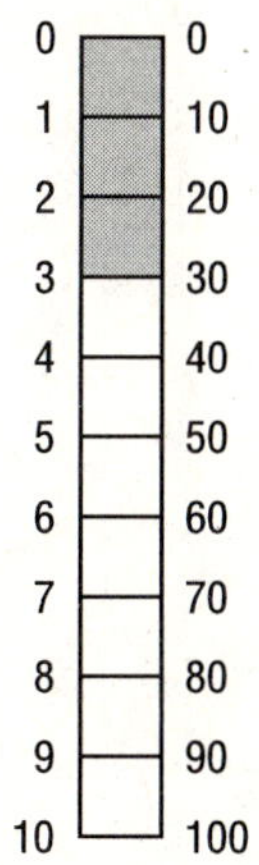

15. Exact answer: 5

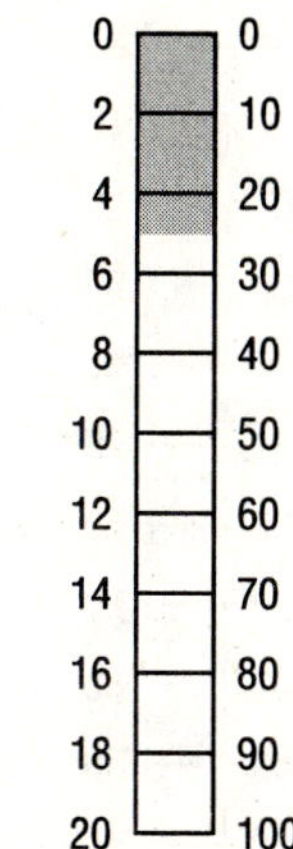

16. Exact answer: 30

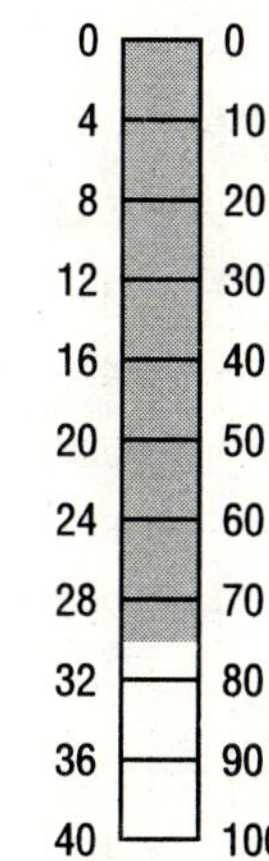

17. Exact answer: 10

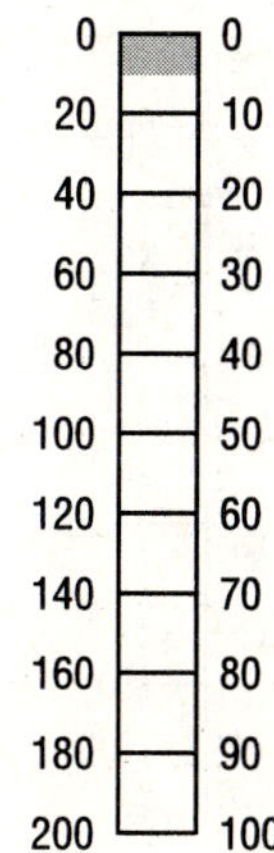

18. Exact answer: 425

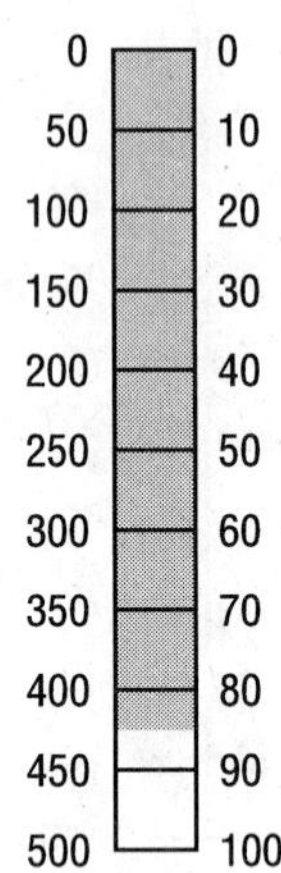

19. Exact answer: 4

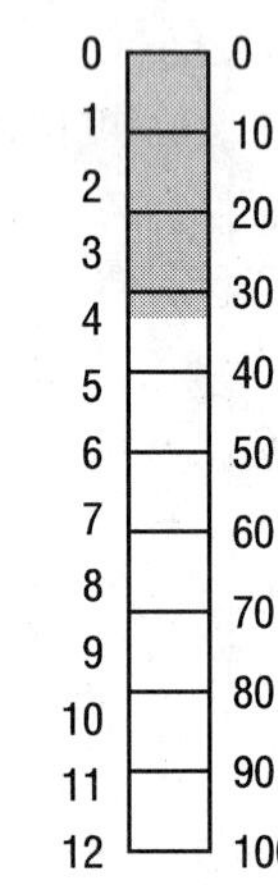

20. Exact answer: 6

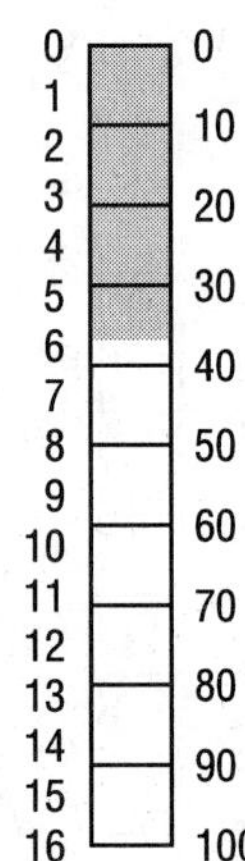

6-5 Using the Percent Proportion

Page 288 Why are percents important in real-world situations?

a. 3 to 4

b. $\frac{3}{4}$, 75%

Page 291 Check for Understanding

1. $\frac{\text{number correct}}{50} = \frac{\%}{100}$

2. Judie; in the sentence *what number is 35% of 21, what number* is the part, 21 is the base, and 35% is the percent. So, to solve the problem, you would use the proportion $\frac{n}{21} = \frac{35}{100}$.

3. $\frac{16}{40} = \frac{p}{100}$

$16 \cdot 100 = 40 \cdot p$

$1600 = 40p$

$\frac{1600}{40} = \frac{40p}{40}$

$40 = p$

So, 16 is 40% of 40.

4. $\frac{21}{b} = \frac{30}{100}$

$21 \cdot 100 = b \cdot 30$

$2100 = 30b$

$\frac{2100}{30} = \frac{30b}{30}$

$70 = b$

So, 21 is 30% of 70.

5. $\frac{a}{130} = \frac{80}{100}$

$a \cdot 100 = 130 \cdot 80$

$100a = 10{,}400$

$\frac{100a}{100} = \frac{10{,}400}{100}$

$a = 104$

So, 104 is 80% of 130.

6. $\frac{14}{5} = \frac{p}{100}$

$14 \cdot 100 = 5 \cdot p$

$1400 = 5p$

$\frac{1400}{5} = \frac{5p}{5}$

$280 = p$

So, 280% of 5 is 14.

7. $\frac{54}{90} = \frac{p}{100}$

$54 \cdot 100 = 90 \cdot p$

$5400 = 90p$

$\frac{5400}{90} = \frac{90p}{90}$

$60 = p$

So, 60% of the books are history books.

8. $\frac{a}{1074} = \frac{66}{100}$

$a \cdot 100 = 1074 \cdot 66$

$100a = 70{,}884$

$a = 708.84$ or 709

About 709 of the students do not clean their room because they do not like to clean.

Pages 291–292 Practice and Apply

9. $\frac{72}{160} = \frac{p}{100}$

$72 \cdot 100 = 160 \cdot p$

$7200 = 160p$

$\frac{7200}{160} = \frac{160p}{160}$

$45 = p$

So, 72 is 45% of 160.

10. $\frac{17}{85} = \frac{p}{100}$

$17 \cdot 100 = 85 \cdot p$

$1700 = 85p$

$\frac{1700}{85} = \frac{85p}{85}$

$20 = p$

So, 17 is 20% of 85.

11. $\frac{36}{b} = \frac{72}{100}$

$36 \cdot 100 = b \cdot 72$

$3600 = 72b$

$\frac{3600}{72} = \frac{72b}{72}$

$50 = b$

So, 36 is 72% of 50.

12. $\frac{27}{b} = \frac{90}{100}$

$27 \cdot 100 = b \cdot 90$

$2700 = 90b$

$\frac{2700}{90} = \frac{90b}{90}$

$30 = b$

So, 27 is 90% of 30.

13. $\frac{a}{175} = \frac{44}{100}$

$a \cdot 100 = 175 \cdot 44$

$100a = 7700$

$\frac{100a}{100} = \frac{7700}{100}$

$a = 77$

So, 77 is 44% of 175.

14. $\frac{a}{150} = \frac{84}{100}$

$a \cdot 100 = 150 \cdot 84$

$100a = 12{,}600$

$\frac{100a}{100} = \frac{12{,}600}{100}$

$a = 126$

So, 126 is 84% of 150.

15. $\frac{52.2}{145} = \frac{p}{100}$

$52.2 \cdot 100 = 145 \cdot p$

$5220 = 145p$

$\frac{5220}{145} = \frac{145p}{145}$

$36 = p$

So, 52.2 is 36% of 145.

16. $\frac{19.8}{36} = \frac{p}{100}$

$19.8 \cdot 100 = 36 \cdot p$

$1980 = 36p$

$\frac{1980}{36} = \frac{36p}{36}$

$55 = p$

So, 19.8 is 55% of 36.

17. $\frac{14}{b} = \frac{12.5}{100}$

$14 \cdot 100 = b \cdot 12.5$

$1400 = 12.5b$

$\frac{1400}{12.5} = \frac{12.5b}{12.5}$

$112 = b$

So, 14 is $12\frac{1}{2}$% of 112.

18. $\frac{36}{b} = \frac{8.75}{100}$

$36 \cdot 100 = b \cdot 8.75$

$3600 = 8.75b$

$\frac{3600}{8.75} = \frac{8.75b}{8.75}$

$411.4 \approx b$

So, 36 is about $8\frac{3}{4}$% of 411.4.

19. $\frac{7}{3500} = \frac{p}{100}$

$7 \cdot 100 = 3500 \cdot p$

$700 = 3500p$

$\frac{700}{3500} = \frac{3500p}{3500}$

$0.2 = p$

So, 7 is 0.2% of 3500.

20. $\frac{a}{750} = \frac{0.3}{100}$

$a \cdot 100 = 750 \cdot 0.3$

$100a = 225$

$\frac{100a}{100} = \frac{225}{100}$

$a = 2.25$

So, 2.25 is 0.3% of 750.

21. $\frac{12}{75} = \frac{p}{100}$

$12 \cdot 100 = 75 \cdot p$

$1200 = 75p$

$\frac{1200}{75} = \frac{75p}{75}$

$16 = p$

16% are parakeets.

22. $\frac{a}{50} = \frac{26}{100}$

$a \cdot 100 = 50 \cdot 26$

$100a = 1300$

$\frac{100a}{100} = \frac{1300}{100}$

$a = 13$

13 are angelfish.

23. $\frac{6{,}300{,}000}{9{,}013{,}400} = \frac{p}{100}$

$6{,}300{,}000 \cdot 100 = 9{,}013{,}400 \cdot p$

$630{,}000{,}000 = 9{,}013{,}400p$

$\frac{630{,}000{,}000}{9{,}013{,}400} = \frac{9{,}013{,}400p}{9{,}013{,}400}$

$70 \approx p$

The Icecap contains about 70% of the freshwater.

24. The total volume of the world's fresh and salt water is 326,038,400 mi^3.

$\frac{6{,}300{,}000}{326{,}038{,}400} = \frac{p}{100}$

$6{,}300{,}000 \cdot 100 = 326{,}038{,}400 \cdot p$

$630{,}000{,}000 = 326{,}038{,}400p$

$2 \approx p$

The Antarctic Icecap contains about 2% of the world's total water supply.

25. $\frac{a}{145} = \frac{18.5}{100}$

$a \cdot 100 = 145 \cdot 18.5$

$100a = 2682.5$

$a \approx 26.8$

A person who weighs 145 lb contains about 26.8 lb of carbon.

26. Since n is 25% of some number a, and 35% of some number b, we can write $\frac{n}{a} = \frac{25}{100}$ and $\frac{n}{b} = \frac{35}{100}$. By solving for n, we find that $n = \frac{25a}{100}$ and $n = \frac{35b}{100}$. Since $n = n$, we can write $\frac{25a}{100} = \frac{35b}{100}$. Solving for a, $a = \frac{35b}{25} = \frac{7b}{5} = \frac{7}{5}b = 1.4b$ results. So, $a > b$.

27. In real-world situations, percents are important because they show how something compares to the whole. Answers should include the following.
   - Sample answer: The outer layer of the new state quarters is an alloy of 3 parts copper to 1 part nickel.
   - There are 4 parts to the outer layer (copper, copper, copper, nickel). The outer layer is $\frac{3}{4}$ or 75% copper and $\frac{1}{4}$ or 25% nickel.

28. B; $\frac{15}{60} = \frac{p}{100}$
$$15 \cdot 100 = 60 \cdot p$$
$$1500 = 60p$$
$$25 = p$$
Altos make up 25% of the chorale.

Page 292 Maintain Your Skills

29. $42\% = \frac{42}{100} = \frac{21}{50}$

30. $56\% = \frac{56}{100} = \frac{14}{25}$

31. $120\% = \frac{120}{100} = \frac{6}{5}$ or $1\frac{1}{5}$

32. $\frac{0.5 \text{ inch}}{1.5 \text{ miles}} = \frac{1.75 \text{ inches}}{x \text{ miles}}$
$$0.5 \cdot x = 1.5 \cdot 1.75$$
$$0.5x = 2.625$$
$$\frac{0.5x}{0.5} = \frac{2.625}{0.5}$$
$$x = 5.25$$
The distance is 5.25 miles.

33. $\frac{2}{9} + \frac{5}{9} = \frac{2+5}{9} = \frac{7}{9}$

34. $\frac{11}{12} - \frac{3}{12} = \frac{11-3}{12} = \frac{8}{12} = \frac{2}{3}$

35. $2\frac{5}{8} + \frac{7}{8} = (2 + 0) + \left(\frac{5}{8} + \frac{7}{8}\right)$
$$= 2 + \frac{12}{8}$$
$$= 2 + \frac{3}{2}$$
$$= 2 + 1\frac{1}{2}$$
$$= 3\frac{1}{2}$$

36. $\frac{1}{2} \times 14 = \frac{1}{\cancel{2}_1} \times \frac{\cancel{14}^7}{1} = 7$

37. $\frac{1}{4} \times 32 = \frac{1}{\cancel{4}_1} \times \frac{\cancel{32}^8}{1} = 8$

38. $\frac{1}{5} \times 15 = \frac{1}{\cancel{5}_1} \times \frac{\cancel{15}^3}{1} = 3$

39. $\frac{2}{3} \times 9 = \frac{2}{\cancel{3}_1} \times \frac{\cancel{9}^3}{1} = 6$

40. $\frac{3}{4} \times 16 = \frac{3}{\cancel{4}_1} \times \frac{\cancel{16}^4}{1} = 12$

41. $\frac{5}{6} \times 30 = \frac{5}{\cancel{6}_1} \times \frac{\cancel{30}^5}{1} = 25$

Page 292 Practice Quiz 1

1. $\frac{3.29 \text{ dollars}}{24 \text{ cans}} \approx \frac{0.14 \text{ dollars}}{1 \text{ can}}$ (÷24)
$$\approx \$0.14/\text{can}$$

2. $\frac{3}{4} = \frac{x}{68}$
$$3 \cdot 68 = 4 \cdot x$$
$$204 = 4x$$
$$\frac{204}{4} = \frac{4x}{4}$$
$$51 = x$$

3. $\frac{4 \text{ feet}}{12{,}276 \text{ feet}} = \frac{1 \text{ foot}}{3069 \text{ feet}}$ (÷4)

So, 1 ft = 3,069 ft.

4. $352\% = 352\% = 3.52$

5. $\frac{a}{60} = \frac{32.5}{100}$
$$a \cdot 100 = 60 \cdot 32.5$$
$$100a = 1950$$
$$\frac{100a}{100} = \frac{1950}{100}$$
$$a = 19.5$$
So, 32.5% of 60 is 19.5.

6-6 Finding Percents Mentally

Page 293 How is estimation used when determining sale prices?

a. tennis racket: $\$68 - \left(\frac{1}{2} \cdot 68\right) = \$34$;

baseball cap: $\$15 - \left(\frac{1}{2} \cdot 15\right) = \$7.50$;

baseball glove: $\$35 - \left(\frac{1}{2} \cdot 35\right) = \$18.50$;

football: $\$44 - \left(\frac{1}{2} \cdot 44\right) = \$22$

b. $\frac{1}{2} = \frac{x}{100}$

$2x = 100$

$x = 50$ or 50%

c. Sample answer: Divide the regular price by 4 and then subtract the result from the original price.

Page 295 Check for Understanding

1. 18% is about 20% or $\frac{1}{5}$. 216 is about 220. $\frac{1}{5}$ of 220 is 44. So, 18% of 216 is about 44.

2. 2 of the 18 squares are shaded.

$\frac{2}{18}$ is about $\frac{2}{20}$ or $\frac{1}{10}$. $\frac{1}{10} = 10\%$. So, about 10% of the model is shaded.

3. See students' work.

Exercises 4–12 Sample answers are given.

4. 75% of 64 $= \frac{3}{4}$ of 64
$= 48$

5. 25% of 52 $= \frac{1}{4}$ of 52
$= 13$

6. $33\frac{1}{3}\%$ of 27 $= \frac{1}{3}$ of 27
$= 9$

7. 90% of 80 $= \frac{9}{10}$ of 80
$= 72$

8. 20% of 61—fraction method;

20% is equal to $\frac{1}{5}$.

61 is about 60.

$\frac{1}{5}$ of 60 is 12.

So, 20% of 61 is about 12.

9. 34% of 24—fraction method;

34% is about $33\frac{1}{3}\%$ or $\frac{1}{3}$.

$\frac{1}{3}$ of 24 is 8.

So, 34% of 24 is about 8.

10. $\frac{1}{2}\%$ of 396—1% method;

$\frac{1}{2}\% = \frac{1}{2} \times 1\%$

396 is almost 400.

1% of 400 is 4.

So, $\frac{1}{2}\%$ of 396 is about $\frac{1}{2} \times 4$ or 2.

11. 152% of 14—meaning of percent method;
152% means about 150 for every 100 or about 15 for every 10.
14 has 1 ten.
$15 \times 1 = 15$.
So, 152% of 14 is about 21.

12. 20% of 52.48

20% is equal to $\frac{1}{5}$.

52.48 is about 50.

$\frac{1}{5}$ of 50 is 10.

So, 20% of \$52.48 is about \$10.

Pages 296–297 Practice and Apply

13. 50% of 28 $= \frac{1}{2}$ of 28
$= 14$

14. 75% of 16 $= \frac{3}{4}$ of 16
$= 12$

15. 60% of 55 $= \frac{3}{5}$ of 55
$= 33$

16. 20% of 105 $= \frac{1}{5}$ of 105
$= 21$

17. $87\frac{1}{2}\%$ of 56 $= \frac{7}{8}$ of 56
$= 49$

18. $16\frac{2}{3}\%$ of 42 $= \frac{1}{6}$ of 42
$= 7$

19. $12\frac{1}{2}\%$ of 32 $= \frac{1}{8}$ of 32
$= 4$

20. $66\frac{2}{3}\%$ of \$24 $= \frac{2}{3}$ of \$24
$= \$16$

21. 200% of 45 $= 2 \times 45$
$= 90$

22. 150% of 54 = (100% + 50%) of 54
$= 54 + 27$
$= 81$

23. 125% of 300 = (100% + 25%) of 300
$= 300 + \frac{1}{4}$ of 300
$= 300 + 75$
$= 375$

24. 175% of 200 = (100% + 75%) of 200
$= 200 + \frac{3}{4}$ of 200
$= 200 + 150$
$= 350$

25. 25% of 7

25% is equal to $\frac{1}{4}$.

7 is about 8.

$\frac{1}{4}$ of 8 is 2.

So, about 2 billion \$5 bills were in circulation.

26. 20% of 7

20% is equal to $\frac{1}{5}$.

7 is about 10.

$\frac{1}{5}$ of 10 is 2.

So, about 2 billion $10 bills were in circulation.

27. 30% of 89—fraction method;

30% is equal to $\frac{3}{10}$.

89 is about 90.

$\frac{3}{10}$ of 90 is 27.

So, 30% of 89 is about 27.

28. 25% of 162—fraction method;

25% is equal to $\frac{1}{4}$.

162 is about 160.

$\frac{1}{4}$ of 160 is 40.

So, 25% of 162 is about 40.

29. 38% of 88—fraction method;

38% is about 40% or $\frac{2}{5}$.

88 is about 90.

$\frac{2}{5}$ of 90 is 36.

So, 38% of 88 is about 36.

30. 81% of 25—fraction method;

81% is about 80% or $\frac{4}{5}$.

$\frac{4}{5}$ of 25 is 20.

So, 81% of 25 is about 20.

31. $\frac{1}{4}$% of 806—1% method;

$\frac{1}{4}$% is $\frac{1}{4} \times 1\%$.

806 is about 800.
1% of 800 is 8.

So, $\frac{1}{4}$% of 806 is about $\frac{1}{4} \times 8$ or 2.

32. $\frac{1}{5}$% of 40—1% method;

$\frac{1}{5}$% is $\frac{1}{5} \times 1\%$.

1% of 40 is 0.4.

So, $\frac{1}{5}$% of 40 is about $\frac{1}{5} \times 0.4$ or 0.08.

33. 127% of 64—meaning of percent;
127% means about 130 for every 100 or 13 for every 10.
64 has 6 tens.
$13 \times 6 = 78$
So, 127% of 64 is about 78.

34. 140% of 95—meaning of percent;
140% means about 140 for every 100.
95 has 1 100.
$140 \times 1 = 140$.
So, 140% of 95 is about 140.

35. 295% of 145—meaning of percent;
295% means about 300 for every 100 or 30 for every 10.
145 has one 100 and about 5 tens.
$(300 \times 1) + (30 \times 5) = 300 + 150$ or 450.
So, 295% of 145 is about 450.

36.
$$50\% \text{ of } 1516 = \tfrac{1}{2} \text{ of } 1516 = 758$$

Pluto has a radius of 707, which is the closest to being half that of Mercury.

37.
$$\tfrac{1}{3} \text{ of Mars} = \tfrac{1}{3} \text{ of } 2107 \approx 702$$

The radius of Pluto is close to $\frac{1}{3}$ the radius of Mars.

$$\tfrac{1}{3} \text{ of Jupiter} = \tfrac{1}{3} \text{ of } 43450 = 14{,}483.\overline{3}$$

The radius of Neptune is close to $\frac{1}{3}$ the radius of Jupiter.

38. 330% of Saturn = 330% of 95.160
330% means about 330 for every 100.
95 has one 100.
$330 \times 1 = 330$
The mass of Jupiter is about 330% the mass of Saturn.

39. 35% of 88,633

35% is about $33\frac{1}{3}$% or $\frac{1}{3}$.

88,633 is about 90,000.

$\frac{1}{3}$ of 90,000 is 30,000.

About 30,000 miles of shoreline is located in Alaska.

40. 8.5% of 40,298

8.5% is about 10% or $\frac{1}{10}$.

40,298 is about 40,000.

$\frac{1}{10}$ of 40,000 is 4000.

About 4000 miles of coastline is located in California.

41. 7 out of 90
9 out of 90 is 10%.
So, about 10% of the calories are from fat.

42. 56% of 160

56% is about 50% or $\frac{1}{2}$.

$\frac{1}{2}$ of 160 is 80.

About 80 calories are from fat.

43.
$$\begin{aligned}
0.40D + 0.925R &= 0.68(D + R)\\
0.4D + 0.925R &= 0.68D + 0.68R\\
0.4D + 0.925R - 0.68D &= 0.68D + 0.68R - 0.68D\\
-0.28D + 0.925R - 0.925R &= 0.68R - 0.925R\\
-0.28D &= -0.245R\\
\frac{-0.28D}{-0.28R} &= \frac{-0.245R}{-0.28R}\\
\frac{D}{R} &= \frac{0.245}{0.28} \text{ or } 0.875\\
\frac{D}{R} &= \frac{7}{8}
\end{aligned}$$

44. By estimating the cost of an item and estimating the percent of the discount, the sale price can be determined. Answers should include the following.

- Sample answer: Estimating the cost of a book that was on sale for 25% off.
- Sample answer: Estimating the amount of paint needed to paint a room.

45. C; $\frac{3}{5} = 60\%$ $\qquad \frac{2}{3} = 66\frac{2}{3}\%$

64% is between 60% and $66\frac{2}{3}\%$.

46. A; 26% of 362

26% is about 25% or $\frac{1}{4}$.

362 is about 360.

$\frac{1}{4}$ of 360 is 90.

Page 297 Maintain Your Skills

47. $\frac{a}{75} = \frac{28}{100}$

$a \cdot 100 = 75 \cdot 28$

$100a = 2100$

$\frac{100a}{100} = \frac{2100}{100}$

$a = 21$

48. $\frac{37.8}{84} = \frac{p}{100}$

$37.8 \cdot 100 = 84 \cdot p$

$3780 = 84p$

$\frac{3780}{84} = \frac{84p}{84}$

$45 = p$

49. Maine:

$\frac{a}{35{,}387} = \frac{89.9}{100}$

$a \cdot 100 = 3{,}181{,}291.3$

$100a = 3{,}181{,}291.3$

$\frac{100a}{100} = \frac{3{,}181{,}291.3}{100}$

$a \approx 31{,}813$ sq mi

New Hampshire:

$\frac{a}{9351} = \frac{88.1}{100}$

$a \cdot 100 = 9351 \cdot 88.1$

$100a = 823{,}823$

$\frac{100a}{100} = \frac{823{,}823}{100}$

$a \approx 8238$ sq mi

West Virginia:

$\frac{a}{24231} = \frac{77.5}{100}$

$a \cdot 100 = 24{,}231 \cdot 77.5$

$100a = 1{,}877{,}902.5$

$\frac{100a}{100} = \frac{1{,}877{,}902.5}{100}$

$a \approx 18{,}779$ sq mi

Vermont:

$\frac{a}{9615} = \frac{75.7}{100}$

$a \cdot 100 = 9615 \cdot 75.7$

$100a = 727{,}855.5$

$\frac{100a}{100} = \frac{727{,}855.5}{100}$

$a \approx 7279$ sq mi

Alabama:

$\frac{a}{52{,}423} = \frac{66.9}{100}$

$a \cdot 100 = 52{,}423 \cdot 66.9$

$100a = 3{,}507{,}098.7$

$\frac{100a}{100} = \frac{3{,}507{,}098.7}{100}$

$a \approx 35{,}071$ sq mi

50. $0.27 = 27\%$

51. $1.6 = 160\%$

52. $0.008 = 0.8\%$

53. $77\% = 0.77$

54. $8\% = 0.08$

55. $421\% = 4.21$

56. $3.56\% = 0.0356$

57. $n + 4.7 = 13.6$

$n + 4.7 - 4.7 = 13.6 - 4.7$

$n = 8.9$

58. $x + \frac{5}{6} = 2\frac{3}{8}$

$x + \frac{5}{6} - \frac{5}{6} = 2\frac{3}{8} - \frac{5}{6}$

$x = \frac{19}{8} - \frac{5}{6}$

$x = \frac{19}{8} \cdot \frac{3}{3} - \frac{5}{6} \cdot \frac{4}{4}$

$x = \frac{57}{24} - \frac{20}{24}$

$x = \frac{37}{24}$

$x = 1\frac{13}{24}$

59. $\frac{3}{7}r = -9$

$\frac{7}{3}\left(\frac{3}{7}r\right) = (-9)\frac{7}{3}$

$r = \frac{-\overset{3}{\cancel{9}}}{1} \cdot \frac{7}{\underset{1}{\cancel{3}}}$

$r = -21$

60. $P = 2\ell + 2w$

$22 = 2(7) + 2w$

$22 = 14 + 2w$

$22 - 14 = 14 + 2w - 14$

$8 = 2w$

$\frac{8}{2} = \frac{2w}{2}$

$4 = w$

The width is 4 feet.

61. $10a = 5$

$\frac{10a}{10} = \frac{5}{10}$

$a = 0.5$

62. $20m = 4$

$\frac{20m}{20} = \frac{4}{20}$

$m = 0.2$

63. $60h = 15$

$\frac{60h}{60} = \frac{15}{60}$

$h = 0.25$

64. $28g = 1.4$

$\frac{28g}{28} = \frac{1.4}{28}$

$g = 0.05$

65. $80w = 5.6$

$\frac{80w}{80} = \frac{5.6}{80}$

$w = 0.07$

66. $125n = 15$

$\frac{125n}{125} = \frac{15}{125}$

$n = 0.12$

6-7 Using Percent Equations

Page 298 How is the percent proportion related to an equation?

a. Alabama: $\frac{x}{35} = \frac{4}{100}$; $x = \$1.40$

Connecticut: $\frac{x}{35} = \frac{6}{100}$; $x = \$2.10$

New Mexico: $\frac{x}{35} = \frac{5}{100}$; $x = \$1.65$

Texas: $\frac{x}{35} = \frac{6.25}{100}$; $x \approx \$2.19$

b. Alabama: 0.04; Connecticut: 0.06; New Mexico: 0.05; Texas: 0.0625

c. Alabama: $0.04 \times 35 = \$1.40$
Connecticut: $0.06 \times 35 = \$2.10$
New Mexico: $0.05 \times 35 = \$1.65$
Texas: $0.0625 \times 35 \approx \$2.19$

d. They are the same.

Page 300 Check for Understanding

1. Use the percent equation in any situation where the rate and base are known.

2. Discount is the amount by which the regular price of an item is reduced.

3. I = interest; p = principal; r = annual interest rate; t = time in years

4. $15 = n(60)$
$\frac{15}{60} = n$
$0.25 = n$
So, 15 is 25% of 60.

5. $30 = 0.6n$
$\frac{30}{0.6} = \frac{0.6n}{0.6}$
$50 = n$
So, 30 is 60% of 50.

6. $n = 0.2(110)$
$n = 22$
So, 20% of 110 is 22.

7. $12 = n(400)$
$\frac{12}{400} = n$
$0.03 = n$
So, 12 is 3% of 400.

8. $n = 0.2(268)$
$n = 53.60$
The discount is \$53.60.

9. $I = prt$
$I = 8000(0.06)(3.5)$
$I = 1680$
The interest is \$1680.

10. $n = 0.35(180)$
$n = 63$
$180 - 63 = 117$
The sale price is \$117.

11. $I = prt$
$252 = 2400(0.07)t$
$252 = 168t$
$\frac{252}{168} = \frac{168t}{168}$
$1.5 = t$
It will take 1.5 years.

Pages 301–302 Practice and Apply

12. $9 = n(25)$
$\frac{9}{25} = n$
$0.36 = n$
So, 9 is 36% of 25.

13. $38 = n(40)$
$\frac{38}{40} = n$
$0.95 = n$
So, 38 is 95% of 40.

14. $48 = 0.64n$
$\frac{48}{0.64} = \frac{0.64n}{0.64}$
$75 = n$
So, 48 is 64% of 75.

15. $27 = 0.54n$
$\frac{27}{0.54} = \frac{0.54n}{0.54}$
$50 = n$
So, 27 is 54% of 50.

16. $n = 0.12(72)$
$n = 8.64$
So, 12% of 72 is 8.64.

17. $n = 0.42(150)$
$n = 63$
So, 42% of 150 is 63.

18. $39.2 = n(112)$
$\frac{39.2}{112} = n$
$0.35 = n$
So, 39.2 is 35% of 112.

19. $49.5 = n(132)$
$\frac{49.5}{132} = n$
$0.375 = n$
So, 49.5 is 37.5% of 132.

20. $n = 0.375(89)$
$n = 33.375$
So, 37.5% of 89 is 33.375.

21. $n = 0.242(60)$
$n = 14.52$
So, 24.2% of 60 is 14.52.

22. $37.5 = n(30)$
$\frac{37.5}{30} = n$
$1.25 = n$
So, 37.5 is 125% of 30.

23. $43.6 = n(20)$
$\frac{43.6}{20} = n$
$2.18 = n$
So, 43.6 is 218% of 20.

24. $1.6 = n(400)$
$\frac{1.6}{400} = n$
$0.004 = n$
So, 1.6 is 0.4% of 400.

25. $1.35 = n(150)$
$\frac{1.35}{150} = n$
$0.009 = n$
So, 1.35 is 0.9% of 150.

26. $83.5 = 1.25n$
$\frac{83.5}{1.25} = \frac{1.25n}{1.25}$
$66.8 = n$
So, 83.5 is 125% of 66.8.

27. $17.6 = \frac{4}{3}n$
$\frac{3}{4}(17.6) = \frac{3}{4}\left(\frac{4}{3}n\right)$
$13.2 = n$
So, 17.6 is $133\frac{1}{3}$% of 13.2.

28. $n = 0.25(4.85)$
$n \approx 1.21$
$4.85 - 1.21 = 3.64$
The sale price is \$3.64.

29. $n = 0.15(29.99)$
$n \approx 4.50$
$29.99 - 4.50 = 25.49$
The sale price is \$25.49.

30. $n = 0.25(65)$
$n = 16.25$
The discount is \$16.25.

31. $n = 0.20(85)$
$n = 17$
The discount is \$17.

32. $n = 0.15(489)$
$n = 73.35$
The discount is \$73.35.

33. $n = 0.25(74)$
$n = 18.5$
The discount is \$18.50.

34. $I = prt$
$I = 5432(0.062)(3)$
$I \approx 1010.35$
The interest is \$1010.35.

35. $I = prt$
$I = 4500(0.055)(4.5)$
$I = 1113.75$
The interest is \$1113.75.

36. $I = prt$
$I = 3680(0.0675)(2.25)$
$I = 558.9$
The interest is \$558.90.

37. $I = prt$
$I = 2543(0.055)(1.75)$
$I \approx 244.76$
The interest is \$244.76

38. $I = prt$
$456 = 1600(r)(6)$
$456 = 9600r$
$\frac{456}{9600} = \frac{9600r}{9600}$
$0.0475 = r$
The rate is 4.75%.

39. $7 = 0.4375n$
$\frac{7}{0.4375} = \frac{0.4375n}{0.4375}$
$16 = n$
They played 16 games.

40. $n = 0.03(130{,}000)$
$n = 3900$
The commission fee is \$3900.

41. $n = 0.15(8)$
$n = 1.2$
The markup is \$1.20.

42. Yes; n% of $m = \frac{n}{100} \cdot m$ or $\frac{mn}{100}$ and m% of $n = \frac{m}{100} \cdot n$ or $\frac{mn}{100}$.

43. If you know two of the three values, you can use the percent proportion to solve for the missing value. Answers should include the following.

- To find the amount of tax on an item, you can use the percent proportion or the percent equation.
- For example, the following methods can be used to find 6% tax on 24.99.
  Method 1: Percent Proportion
  $\frac{x}{24.99} = \frac{6}{100}$
  Method 2: Percent Equation
  $n = 0.06(24.99)$
  Using either method, $x = 1.50$. The amount of tax is \$1.50.

44. C; $19.2 = n(320)$
$\frac{19.2}{320} = n$
$0.06 = n$
6%

45. B; $n = 0.3(150)$
$n = 45$
$150 - 45 = 105$
\$105

Page 302 Maintain Your Skills

46. 47% of 84—fraction method;
47% is about 50% or $\frac{1}{2}$.
84 is about 80.
$\frac{1}{2}$ of 80 is 40.
47% of 84 is about 40.

47. 126% of 198—meaning of percent;
126% is about 125% or 125 for every hundred.
198 is about 200.
200 has two 100s.
$125 \times 2 = 250$.
126% of 198 is about 250.

48. 9% of 514—1% method;
1% of 514 is about 5.
9% is $9 \times 1\%$.
$9 \times 5 = 45$.
So, 9% of 514 is about 45.

49. $\frac{a}{220} = \frac{55}{100}$
$a \cdot 100 = 220 \cdot 55$
$100a = 12{,}100$
$\frac{100a}{100} = \frac{12{,}100}{100}$
$a = 121$
121 is 55% of 220.

50. $\frac{50.88}{96} = \frac{p}{100}$
$50.88 \cdot 100 = 96 \cdot p$
$5088 = 96p$
$\frac{5088}{96} = \frac{96p}{96}$
$53 = p$
50.88 is 53% of 96.

51. mean $= \frac{52 + 60 + 41 + 36 + 33 + 24}{6}$
$= \frac{246}{6}$
$= 41$
The mean is 41.

52. 30: 1, 2, 3, 5, 6, 10, 15, 30

53. $P = 2\ell + 2w$
$P = 2(13) + 2(6)$
$P = 26 + 12$
$P = 38$ cm

54. $P = 2\ell + 2w$
$P = 2(25) + 2(11)$
$P = 50 + 22$
$P = 72$ in.

55. $(w - 3)8 = 8w - 24$

56. $0.58 = 0.58$
$= 58\%$

57. $0.89 = 0.89$
$= 89\%$

58. $0.125 = 0.125$
$= 12.5\%$

59. $1.56 = 1.56$
$= 156\%$

60. $2.04 = 2.04$
$= 204\%$

61. $0.224 = 0.224$
$= 22.4\%$

Page 303 Spreadsheet Investigation (Follow-Up of Lesson 6-7)

1. $I = prt$
$I = 1000(0.06)5$
$I = 300$
$1343.92 - 1000 = 343.92$
$343.92 - 300 = 43.92$
The simple interest, $300, is $43.92 less than the compound interest.

2. $1346.86

3. $115.00, $115.97, $116.08

4. The amount of interest increases.

6-8 Percent of Change

Page 304 How can percents help to describe a change in area?

a. $\frac{1}{3}$ or $33\frac{1}{3}\%$

b. $\frac{1}{5}$ or 20%

c. $\frac{1}{4}$ or 25%

d. $\frac{1}{6}$ or $16\frac{2}{3}\%$

e. The percent of change is different because the area of each original rectangle is different.

Page 306 Check for Understanding

1. If the amount increases, it is a percent of increase. If the amount decreases, it is a percent of decrease.

2. Sample answer: a temperature change from 65°F to 50°F.

3. Mark; he divided the difference of the new amount and the original amount by the original amount.

4. $67 - 50 = 17$
percent of change $= \frac{17}{50}$
$= 0.34$ or 34%
The percent of increase is 34%.

5. $18 - 45 = -27$
percent of change $= \frac{-27}{45}$
$= -0.6$ or -60%
The percent of decrease is 60%.

6. $55 - 80 = -25$
percent of change $= \frac{-25}{80}$
≈ -0.313 or -31.3%
The percent of decrease is 31.3%.

7. $251 - 228 = 23$
percent of change $= \frac{23}{228}$
≈ 0.101 or 10.1%
The percent of increase is 10.1%.

8. $367 - 356 = 11$
percent of change $= \frac{11}{356}$
≈ 0.031 or 3.1%
The percent of change was 3.1%.

9. A; $36{,}851 - 30{,}735 = 6116$
percent of change $= \frac{6116}{30{,}735}$
≈ 0.199 or 19.9%
The percent of change is 19.9%.

Pages 307–308 Practice and Apply

10. $36 - 25 = 11$

$$\text{percent of change} = \frac{11}{25}$$
$$= 0.44 \text{ or } 44\%$$

The percent of increase is 44%.

11. $27 - 10 = 17$

$$\text{percent of change} = \frac{17}{10}$$
$$= 1.7 \text{ or } 170\%$$

The percent of increase is 170%.

12. $51 - 68 = -17$

$$\text{percent of change} = \frac{-17}{68}$$
$$= -0.25 \text{ or } -25\%$$

The percent of decrease is 25%.

13. $44 - 50 = -6$

$$\text{percent of change} = \frac{-6}{50}$$
$$= -0.12 \text{ or } -12\%$$

The percent of decrease is 12%.

14. $120 - 135 = -15$

$$\text{percent of change} = \frac{-15}{135}$$
$$\approx -0.111 \text{ or } -11.1\%$$

The percent of decrease is 11.1%.

15. $243 - 257 = -14$

$$\text{percent of change} = \frac{-14}{257}$$
$$\approx -0.054 \text{ or } -5.4\%$$

The percent of decrease is 5.4%.

16. $421 - 365 = 56$

$$\text{percent of change} = \frac{56}{365}$$
$$\approx 0.153 \text{ or } 15.3\%$$

The percent of increase is 15.3%.

17. $762 - 289 = 473$

$$\text{percent of change} = \frac{473}{289}$$
$$\approx 1.64 \text{ or } 164\%$$

The percent of increase is 164%.

18. $3.8 - 6 = -2.2$

$$\text{percent of change} = \frac{-2.2}{6}$$
$$\approx -0.367 \text{ or } -36.7\%$$

The percent of change is -36.7%.

19. $4{,}447{,}100 - 4{,}040{,}587 = 406{,}513$

$$\text{percent of change} = \frac{-406{,}513}{4{,}040{,}587}$$
$$\approx -0.101 \text{ or } 10.1\%$$

The percent of change was -10.1%.

20. $\text{percent of change} = \frac{36}{24}$

$$= 1.5 \text{ or } 150\%$$

The percent of change is 150%.

21. $\text{percent of change} = \frac{-4}{28}$

$$\approx -0.143 \text{ or } -14.3\%$$

The percent of change is -14.3%.

22. $\text{percent of change} = \frac{\text{amt. of change}}{\text{original measurement}}$

$$-10\% = \frac{n}{2875}$$
$$2875(-0.1) = \left(\frac{n}{2875}\right)2875$$
$$-287.5 = n$$
$$\text{January} = 2875 - 287.5$$
$$= \$2587.50$$
$$-15\% = \frac{n}{2587.50}$$
$$2587.5(-0.15) = \left(\frac{n}{2587.5}\right)2587.5$$
$$-388.13 = n$$
$$\text{February} = 2587.50 - 388.13$$
$$= \$2199.38$$

The expenses should be $2199.38.

23. $3 - 2 = 1$

$$\text{percent of change} = \frac{1}{2}$$
$$= 0.5 \text{ or } 50\%$$
$$4.5 - 3 = 1.5$$
$$\text{percent of change} = \frac{1.5}{3}$$
$$= 0.5 \text{ or } 50\%$$

The enlargement should be 50% larger, or 150% of the original.

24. Suppose x represents the original amount. After the increase, it is $1.1x$. After the decrease, it is $0.9(1.1x)$ or $0.99x$. Since when x is positive, $0.99x < x$, the final amount is less than the original.

25. The amount by which a rectangle is increased or decreased can be represented by a percent. Answers should include the following.

- If the size of the new rectangle is greater than the size of the original rectangle, the percent of increase is greater than 100%.

Original Rectangle

Less than 100%

Greater than 100%

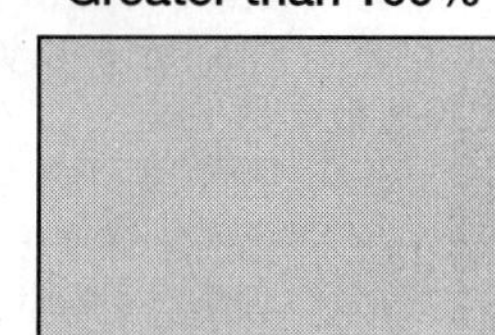

26. See students' work.

27. D; $49{,}080 - 53{,}322 = -4242$

$$\text{percent of change} = \frac{-4242}{53{,}322}$$
$$= -0.080 \text{ or } -8\%$$

28. A; Siberian Husky

$\frac{18,106 - 21,078}{21,078} = \frac{-2,972}{21,078}$

≈ -0.141 or -14.1%

Cocker Spaniel

$\frac{29,958 - 34,632}{34,632} = \frac{-4674}{34,632}$

≈ -0.135 or -13.5%

Golden Retreiver

$\frac{62,652 - 65,681}{65,681} = \frac{-3029}{65,681}$

≈ -0.0460 or -4.6%

Labrador Retriever

$\frac{157,897 - 157,936}{157,936} = \frac{-39}{157,936}$

≈ -0.0002 or -0.2%

The Siberian Husky had the largest percent of decrease.

Page 308 Maintain Your Skills

29. $n = 0.15(999)$
$n = 149.85$
The discount is \$149.85.

30. $I = prt$
$I = 1590(0.08)3$
$I = 381.60$
The interest for 3 years is \$381.60.

31. $n = 0.10(14.95)$
$n = 1.495$
$14.95 - 1.495 = 13.455$
The sale price is \$13.46.

32. 60% of 135—fraction method;
60% is $\frac{3}{5}$.
$\frac{3}{5} \times 135 = 81$.
So, 60% of 135 is 81.

33. 88% of 72—fraction method;
88% is about 90% or $\frac{9}{10}$.
72 is about 70.
$\frac{9}{10} \times 70 = 63$.
So, 88% of 72 is about 63.

34. 123% of 32—meaning of percent;
123% is 123 for every 100 or 12 for every 10.
32 has 3 tens.
$3 \times 12 = 36$.
So, 123% of 32 is about 36.

35. -8 is an integer and rational.

36. $1\frac{1}{4}$ or $\frac{5}{4}$ is rational.

37. -5.63 or $-\frac{563}{100}$ is rational.

38. $\frac{3}{4} = \frac{75}{100}$
$= 75\%$

39. $\frac{1}{5} = \frac{20}{100}$
$= 20\%$

40. $\frac{2}{3} = 0.\overline{6}$
$= 66.\overline{6}\%$
$= 66\frac{2}{3}\%$

41. $\frac{5}{6} = 0.8\overline{3}$
$= 83.\overline{3}\%$
$= 83\frac{1}{3}\%$

42. $\frac{3}{8} = 0.375$
$= 37.5\%$
$= 37\frac{1}{2}\%$

Page 308 Practice Quiz 2

1. 42% of 68—fraction method;
42% is about 40% or $\frac{2}{5}$.
68 is about 70.
$\frac{2}{5} \times 70 = 28$
So, 42% of 68 is about 28.

2. $66\frac{2}{3}\%$ of 34—fraction method;
$66\frac{2}{3}\%$ is $\frac{2}{3}$.
34 is about 33.
$\frac{2}{3} \times 33 = 22$.
So, $66\frac{2}{3}\%$ of 34 is about 22.

3. $n = 0.25(58)$
$= 14.5$
The discount is \$14.50.

4. $I = prt$
$I = 2500(0.04)(2.5)$
$I = \$250$

5. $2.45 - 0.95 = 1.5$
percent of change $= \frac{1.5}{0.95}$
≈ 1.58 or 158%
The percent change is 158%.

Page 309 Algebra Activity (Preview of Lesson 6-9)

1. Yes; strength: large enough population
2. No; weakness: the people in a library may read more than others
3. No; weakness: many of the fans may be Dolphin fans
4. Systematic; People may prefer the restaurant.
5. Find the percent of students in each grade that eat breakfast.
6a–d. See students' work.

6-9 Probability and Predictions

Page 310 How can probability help you make predictions?

a. $\frac{12}{100}$
b. 12%
c. $\frac{12}{100} = \frac{12 \div 4}{100 \div 4} = \frac{3}{25}$

d. There is a better chance of choosing an N because there are more Ns than Ds.

Page 312 Check for Understanding

1. The event will not happen.

2. Theoretical probability is what should occur. Experimental probability is what actually occurs.

3. Sample answer: Spinning the spinner shown and having it land on 4.

4. $P(5) = \frac{1}{10}$

The probability of choosing 5 is $\frac{1}{10}$ or 10%.

5. $P(\text{odd}) = \frac{5}{10}$ or $\frac{1}{2}$

The probability of choosing odd is $\frac{1}{2}$ or 50%.

6. $P(\text{less than } 3) = \frac{2}{10}$ or $\frac{1}{5}$

The probability of choosing less than 3 is $\frac{1}{5}$ or 20%.

7. $P(\text{greater than } 6) = \frac{4}{10}$ or $\frac{2}{5}$

The probability of choosing greater than 6 is $\frac{2}{5}$ or 40%.

8. There are 6 outcomes in which the sum is 2 or 6.

$P(\text{sum of 2 or 6}) = \frac{6}{36}$ or $\frac{1}{6}$

The probability of rolling a sum of 2 or 6 is $\frac{1}{6}$ or $16\frac{2}{3}$%.

9. There are 36 outcomes in which the sum is even or odd.

$P(\text{even or odd sum}) = \frac{36}{36}$ or 1.

The probability of rolling a sum that is even or odd is 1 or 100%.

10. $\frac{\text{number of times heads occur}}{\text{number of possible outcomes}} = \frac{14}{14 + 11}$ or $\frac{14}{25}$

The experimental probability of getting heads in this case is $\frac{14}{25}$ or 44%.

11. $\frac{a}{250} = \frac{30}{100}$

$100 \cdot a = 30 \cdot 250$

$100a = 7500$

$a = 75$

She can expect 75 to be red.

Pages 313–314 Practice and Apply

12. There is 1 possible outcome of 8.

$P(8) = \frac{1}{8}$ or 0.125

So, the probability of spinning an 8 is $\frac{1}{8}$ or 12.5%.

13. There are 2 possible outcomes of red.

$P(\text{red}) = \frac{2}{8}$

$= \frac{1}{4}$ or 0.25

So, the probability of spinning red is $\frac{1}{4}$ or 25%.

14. There are 4 possible outcomes of even numbers.

$P(\text{even}) = \frac{4}{8}$

$= \frac{1}{2}$ or 0.5

So the probability spinning even is $\frac{1}{2}$ or 50%.

15. There are 4 possible outcomes of prime numbers.

$P(\text{prime}) = \frac{4}{8}$

$= \frac{1}{2}$ or 0.5

So, the probability of spinning a prime number is $\frac{1}{2}$ or 50%.

16. There are 5 possible outcomes of spinning greater than 5.

$P(\text{greater than } 5) = \frac{5}{8}$ or 0.625

So, the probability of spinning greater than 5 is $\frac{5}{8}$ or 62.5%.

17. There are 0 possible outcomes of spinning less than 2.

$P(\text{less than } 2) = \frac{0}{8}$ or 0.

So, the probability of spinning less than 2 is 0.

18. There are 3 possible outcomes of spinning blue and 1 possible outcome of spinning 11.

$P(\text{blue or } 11) = \frac{4}{8}$

$= \frac{1}{2}$ or 0.5

So, the probability of spinning blue or 11 is $\frac{1}{2}$ or 50%.

19. There are 7 possible outcomes of not spinning yellow.

$P(\text{not yellow}) = \frac{7}{8}$ or 0.875

So, the probability of not spinning yellow is $\frac{7}{8}$ or 87.5%.

20. There are 6 possible outcomes of not spinning red.

$P(\text{not red}) = \frac{6}{8}$

$= \frac{3}{4}$ or 0.75

So, the probability of not spinning red is $\frac{3}{4}$ or 75%.

21. There are 4 possible outcomes of selecting blue.

$P(\text{blue}) = \frac{4}{18}$ or $\frac{2}{9}$

So, the probability of selecting blue is $\frac{2}{9}$ or 22.2%.

22. There are 5 possible outcomes of selecting yellow.

$P(\text{yellow}) = \frac{5}{18}$

So, the probability of selecting yellow is $\frac{5}{18}$ or 27.8%.

23. There are 11 possible outcomes of not selecting green.

$P(\text{not green}) = \frac{11}{18}$

So, the probability of not selecting green is $\frac{11}{18}$ or 61.1%.

24. There are 0 possible outcomes of selecting purple.

$P(\text{purple}) = \frac{0}{18}$ or 0

So, the probability of selecting purple is 0%.

25. There are 2 possible outcomes of selecting red and 4 possible outcomes of selecting blue.

$P(\text{red or blue}) = \frac{6}{18}$ or $\frac{1}{3}$

So, the probability of selecting red or blue is $\frac{1}{3}$ or 33.3%.

26. There are 4 possible outcomes of selecting blue and 5 possible outcomes of selecting yellow.

$P(\text{blue or yellow}) = \frac{9}{18}$ or $\frac{1}{2}$

So, the probability of selecting blue or yellow is $\frac{1}{2}$ or 50%.

27. There are 18 possible outcomes of not selecting orange.

$P(\text{not orange}) = \frac{18}{18}$ or 1

So, the probability of not selecting orange is 100%.

28. There are 12 possible outcomes of not selecting blue or red.

$P(\text{not blue or not red}) = \frac{12}{18}$ or $\frac{2}{3}$

So, the probability of not selecting blue or red is $\frac{2}{3}$ or $66\frac{2}{3}$%.

29. There are 2 outcomes of selecting January or April. There are 12 possible outcomes.

$P(\text{January or April}) = \frac{2}{12}$ or $\frac{1}{6}$

The probability of a calendar being turned to January of April is $\frac{1}{6}$.

30. There is no date of November 31, so the probability is 0.

31.

	5	6	7	8
1	(1, 5)	(1, 6)	(1, 7)	(1, 8)
2	(2, 5)	(2, 6)	(2, 7)	(2, 8)
3	(3, 5)	(3, 6)	(3, 7)	(3, 8)
4	(4, 5)	(4, 6)	(4, 7)	(4, 8)

There is 1 possible outcome of spinning 2 and 7.

$P(2,7) = \frac{1}{16}$

So, the probability of spinning 2 and 7 is $\frac{1}{16}$.

32. (See table in #31)
There are 4 possible outcomes of spinning 2 even numbers.

$P(\text{even, even}) = \frac{4}{16}$

$= \frac{1}{4}$

So, the probability of spinning 2 even numbers is $\frac{1}{4}$.

33. (See table in #31)
There are 4 possible outcomes of spinning a sum of 9.

$P(\text{sum of } 9) = \frac{4}{16}$

$= \frac{1}{4}$

So, the probability of spinning a sum of 9 is $\frac{1}{4}$.

34. (See table in #31)
There are 3 possible outcomes of spinning a 2 and a number greater than 5.

$P(2, \text{greater than } 5) = \frac{3}{16}$

So, the probability of spinning a 2 and a number greater than 5 is $\frac{3}{16}$.

35. $\frac{\text{number of times 19 and under}}{\text{number of possible outcomes}} = \frac{9}{180}$

$= \frac{1}{20}$ or 0.05

The experimental probability of a driver age 19 and under is 0.05 or 5%.

36. $\frac{\text{number of times age 40–49}}{\text{number of possible outcomes}} = \frac{37}{180}$

The experimental probability of a driver age 40–49 is 0.21 or 21%.

37. $\frac{a}{1200} = \frac{37}{100}$

$100 \cdot a = 37 \cdot 1200$

$100a = 44{,}400$

$a = 444$

You can expect 444 people to say they crave chocolate after dinner.

38. No; many of the common two-letter words like *at*, *in*, *of*, and *on* do not contain the letter E.

39. Once the probability or likeliness of something happening is known, then you can use the probability to make a prediction. For example, in football, if you know the number of field goals a player has made in the past, you can use the information to predict the number of field goals he/she will make in upcoming games. Answers should include the following.

- E: $\frac{12}{100}$ or 12%; A, I: $\frac{9}{100}$ or 9%; O: $\frac{8}{100}$ or 8%; N, R, T: $\frac{6}{100}$ or 6%; D, L, S, U: $\frac{4}{100}$ or 4%; G: $\frac{3}{100}$ or 3%; B, C, F, H, M, P, V, W, Y, blank: $\frac{2}{100}$ or 2%; J, K, Q, X, Z: $\frac{1}{100}$ or 1%
- See students' work.

40. D; There are 4 equal outcomes, 2 of which are 6. Three of the outcomes are even.

$P(\text{even}) = \frac{3}{4}$

Page 314 Maintain Your Skills

41. $79 - 32 = 47$

$\text{percent of change} = \frac{47}{32}$

≈ 1.469 or 146.9%

The percent of increase is 146.9%.

42. $7 = n(32)$

$\frac{7}{32} = n$

$0.219 \approx n$

So, 7 is 21.9% of 32.

43. $n = 0.285(84)$

$n \approx 23.9$

So, 23.9 is 28.5% of 84.

44. $7^2 \cdot 7^3 = 7^{2+3}$

$= 7^5$

45. $x^4 \cdot 2x = 2x^{4+1}$

$= 2x^5$

46. $\frac{8^{12}}{8^8} = 8^{12-8}$

$= 8^4$

47. $\frac{36n^4}{14n^2} = \frac{\overset{18}{\cancel{36}} \cdot n \cdot n \cdot \cancel{n} \cdot \cancel{n}}{\underset{7}{\cancel{14}} \cdot \cancel{n} \cdot \cancel{n}}$

$= \frac{18n^2}{7}$

Page 315 Graphing Calculator Investigation (Follow-Up of Lesson 6-9)

1–2. See students' answers.

3–4. To use the random number generator, let red = 1, let white = 2, and let blue = 3. See students' answers.

Chapter 6 Study Guide and Review

Page 316 Vocabulary and Concept Check

1. proportion
2. percent
3. scale factor
4. sample space
5. experimental probability

Pages 316–320 Lesson-by-Lesson Review

6. $\frac{9 \text{ students}}{33 \text{ students}} = \frac{3 \text{ students}}{11 \text{ students}}$

$= \frac{3}{11}$

7. $\frac{12 \text{ hits}}{16 \text{ times at bat}} = \frac{3 \text{ hits}}{4 \text{ times at bat}}$

$= \frac{3}{4}$

8. $\frac{30 \text{ hours}}{18 \text{ hours}} = \frac{5 \text{ hours}}{3 \text{ hours}}$

$= \frac{5}{3}$

9. $\frac{5 \text{ quarts}}{5 \text{ gallons}} = \frac{5 \text{ quarts}}{20 \text{ quarts}}$

$= \frac{1 \text{ quart}}{4 \text{ quarts}}$ or $\frac{1}{4}$

10. $\frac{10 \text{ inches}}{4 \text{ feet}} = \frac{10 \text{ inches}}{48 \text{ inches}}$

$= \frac{5 \text{ inches}}{24 \text{ inches}}$ or $\frac{5}{24}$

11. $\frac{2 \text{ tons}}{1800 \text{ pounds}} = \frac{4000 \text{ pounds}}{1800 \text{ pounds}}$

$= \frac{20 \text{ pounds}}{9 \text{ pounds}}$ or $\frac{20}{9}$

12. $\frac{n}{12} = \frac{4}{3}$

$n \cdot 3 = 12 \cdot 4$

$3n = 48$

$\frac{3n}{3} = \frac{48}{3}$

$n = 16$

13. $\frac{21}{x} = \frac{84}{120}$

$21 \cdot 120 = x \cdot 84$

$2520 = 84x$

$\frac{2520}{84} = \frac{84x}{84}$

$30 = x$

14. $\frac{9}{7} = \frac{22.5}{y}$

$9 \cdot y = 7 \cdot 22.5$

$9y = 157.5$

$\frac{9y}{9} = \frac{157.5}{9}$

$y = 17.5$

15. $\frac{5}{7.5} = \frac{0.6}{k}$

$5 \cdot k = 7.5 \cdot 0.6$

$5k = 4.5$

$\frac{5k}{5} = \frac{4.5}{5}$

$k = 0.9$

16. $\frac{1 \text{ in.}}{12 \text{ ft}} = \frac{0.9 \text{ in.}}{x \text{ ft}}$

$1 \cdot x = 12 \cdot 0.9$

$x = 10.8$

The actual length of the stateroom is 10.8 ft.

17. $\frac{1 \text{ in.}}{12 \text{ ft}} = \frac{3.8 \text{ in.}}{x \text{ ft}}$

$1 \cdot x = 12 \cdot 3.8$

$x = 45.6$

The actual length of the galley is 45.6 ft.

18. $\frac{1 \text{ in.}}{12 \text{ ft}} = \frac{6 \text{ in.}}{x \text{ ft}}$

$1 \cdot x = 12 \cdot 6$

$x = 72$

The actual length of the gym is 72 ft.

19. $35\% = \frac{35}{100}$ or $\frac{7}{20}$

$35\% = 35\%$ or 0.35

20. $42\% = \frac{42}{100}$ or $\frac{21}{50}$

$42\% = 42\%$ or 0.42

21. $8\% = \frac{8}{100}$ or $\frac{2}{25}$

$8\% = 8\%$ or 0.08

22. $19\% = \frac{19}{100}$

$19\% = 19\%$ or 0.19

23. $120\% = \frac{120}{100}$

$= \frac{6}{5}$ or $1\frac{1}{5}$

$120\% = 120\%$ or 1.2

24. $250\% = \frac{250}{100}$

$= \frac{5}{2}$ or $2\frac{1}{2}$

$250\% = 250\%$ or 2.5

25. $62.5\% = \frac{62.5}{100} \times \frac{10}{10}$
$= \frac{625}{1000}$ or $\frac{5}{8}$
$62.5\% = \underbrace{62.5}\%$ or 0.625

26. $8.8\% = \frac{8.8}{100} \times \frac{10}{10}$
$= \frac{88}{1000}$ or $\frac{11}{125}$
$8.8\% = \underbrace{\ 8}.8\%$ or 0.088%

27. $0.24 = 0.24$ or 24%

28. $0.03 = 0.03$ or 3%

29. $0.452 = 0.452$ or 45.2%

30. $1.9 = 1.9$ or 190%

31. $\frac{2}{5} = 0.4$ or 40%

32. $\frac{13}{22} \approx 0.591$ or 59.1%

33. $\frac{6}{80} = 0.075$ or 7.5%

34. $\frac{77}{225} \approx 0.342$ or 34.2%

35. $\frac{18}{45} = \frac{p}{100}$
$18 \cdot 100 = 45 \cdot p$
$1800 = 45p$
$40 = p$
So, 18 is 40% of 45.

36. $\frac{39}{60} = \frac{p}{100}$
$39 \cdot 100 = 60 \cdot p$
$3900 = 60p$
$65 = p$
So, 65% of 60 is 39.

37. $\frac{23}{b} = \frac{92}{100}$
$23 \cdot 100 = b \cdot 92$
$2300 = 92b$
$25 = b$
So, 23 is 92% of 25.

38. $\frac{a}{110} = \frac{74}{100}$
$a \cdot 100 = 110 \cdot 74$
$100a = 8140$
$a = 81.4$
So, 81.4 is 74% of 110.

39. $\frac{a}{62.5} = \frac{80}{100}$
$a \cdot 100 = 62.5 \cdot 80$
$100a = 5000$
$a = 50$
So, 50 is 80% of 62.5.

40. $\frac{36}{b} = \frac{15}{100}$
$36 \cdot 100 = b \cdot 15$
$3600 = 15b$
$240 = b$
So, 36 is 15% of 240.

41. 50% of $86 = \frac{1}{2}$ of 86
$= 43$
So, 50% of 86 is 43.

42. 20% of $55 = \frac{1}{5}$ of 55
$= 11$
So, 20% of 55 is 11.

43. 25% of $36 = \frac{1}{4}$ of 36
$= 9$
So, 25% of 36 is 9.

44. 40% of $75 = \frac{2}{5}$ of 75
$= 30$
So, 40% of 75 is 30.

45. $33\frac{1}{3}\%$ of $24 = \frac{1}{3}$ of 24
$= 8$
So, $33\frac{1}{3}\%$ of 24 is 8.

46. 90% of $60 = \frac{9}{10}$ of 60
$= 54$
So, 90% of 60 is 54.

47. 48% of 32—fraction method;
48% is about 50% or $\frac{1}{2}$.
$\frac{1}{2}$ of 32 is 16.
So, 48% of 32 is about 16.

48. 67% of 30—fraction method;
67% is about $66\frac{2}{3}\%$ or $\frac{2}{3}$.
$\frac{2}{3}$ of 30 is 20.
So, 67% of 30 is about 20.

49. 20% of 51—fraction method;
51 is about 50.
20% or $\frac{1}{5}$ of 50 is 10.
So, 20% of 51 is about 10.

50. 25% of 27—fraction method;
27 is almost 28.
25% or $\frac{1}{4}$ of 28 is 7.
So, 25% of 27 is about 7.

51. $\frac{1}{3}\%$ of 304—1% method;
1% of 304 is about 3.
$\frac{1}{3}\%$ of 304 is about $\frac{1}{3}$ of 3 or 1.
So, $\frac{1}{3}\%$ of 304 is about 1.

52. 147% of 200—meaning of percent;
147% means about 150 for every 100.
200 has 2 100's.
$150 \times 2 = 300$.
So, 147% of 200 is about 300.

53. $24 = n(50)$
$\frac{24}{50} = \frac{50n}{50}$
$0.48 = n$
So, 24 is 48% of 50.

54. $70 = 0.4n$
$\frac{70}{0.4} = \frac{0.4n}{0.4}$
$175 = n$
So, 70 is 40% of 175.

55. $n = 0.9(105)$
$n = 94.5$
So, 94.5 is 90% of 105.

56. $n = 0.125(68)$
$n = 8.5$
So, 8.5 is 12.5% of 68.

57. $56 = 0.28n$

$\frac{56}{0.28} = \frac{0.28n}{0.28}$

$200 = n$

So, 56 is 28% of 200.

58. $35.7 = n(17)$

$\frac{35.7}{17} = \frac{17n}{17}$

$210 = n$

So, 35.7 is 210% of 17.

59. percent of change $= \frac{12 - 40}{40}$

$= \frac{-28}{40}$

$= -0.7$ or -70%

The percent of decrease is 70%.

60. percent of change $= \frac{96 - 80}{80}$

$= \frac{16}{80}$

$= 0.2$ or 20%

The percent of increase is 20%.

61. percent of change $= \frac{54 - 29}{29}$

$= \frac{25}{29}$

≈ 0.826 or 86.2%

The percent of increase is about 86.2%.

62. percent of change $= \frac{77 - 80}{80}$

$= \frac{-3}{80}$

≈ 0.038 or 3.8%

The percent of decrease is about 3.8%.

63. Favorable outcomes: 5
Possible outcomes: 15

$P(\text{red}) = \frac{5}{15}$ or $\frac{1}{3}$

So, the probability of selecting red is $\frac{1}{3}$.

64. Favorable outcomes: 8
Possible outcomes: 15

$P(\text{green}) = \frac{8}{15}$

So, the probability of selecting green is $\frac{8}{15}$.

65. Favorable outcomes: 10
Possible outcomes: 15

$P(\text{blue or green}) = \frac{10}{15}$ or $\frac{2}{3}$

So, the probability of selecting blue or green is $\frac{2}{3}$.

66. Favorable outcomes: 13
Possible outcomes: 15

$P(\text{not blue}) = \frac{13}{15}$

So, the probability of not selecting blue is $\frac{13}{15}$.

67. Favorable outcomes: 0
Possible outcomes: 15

$P(\text{yellow}) = \frac{0}{15}$

So, the probability of selecting yellow is 0.

68. Favorable outcomes: 15
Possible outcomes: 15

$P(\text{green, red, or blue}) = \frac{15}{15}$ or 1.

So, the probability of selecting green, red, or blue is 1.

Chapter 6 Practice Test

Page 321

1. A ratio is a comparison of two numbers by division. A rate is a ratio of two measurements having diferent kinds of units. For example, 2 inches to 24 inches is a ratio, and 2 inches in 3 hours is a rate.

2. To write a fraction as a percent, first write the fraction as a decimal. Then write the decimal as a percent by moving the decimal 2 places to the right and adding the percent symbol.

3. $\frac{15 \text{ girls}}{40 \text{ students}} = \frac{3 \text{ girls}}{8 \text{ students}}$

$= \frac{3}{8}$

4. $\frac{6 \text{ feet}}{3 \text{ yards}} = \frac{6 \text{ feet}}{9 \text{ feet}}$

$= \frac{2 \text{ feet}}{3 \text{ feet}}$

$= \frac{2}{3}$

5. $\frac{145 \text{ mi}}{3 \text{ h}} \approx \frac{48.3 \text{ mi}}{1 \text{ h}}$ (÷3 numerator and denominator)

$= 48.3$ mi/h

6. $\frac{245 \text{ dollars}}{9 \text{ tickets}} \approx \frac{27.22 \text{ dollars}}{1 \text{ ticket}}$ (÷9 numerator and denominator)

$= \$27.22$/ticket

7. $\frac{15 \text{ mi}}{1 \text{ h}} = \frac{15 \text{ mi}}{1 \text{ h}} \cdot \frac{5280 \text{ ft}}{1 \text{ mi}} \div \frac{60 \text{ min}}{1 \text{ h}}$

$= \frac{\overset{1}{\cancel{15}} \cancel{\text{mi}}}{1 \cancel{\text{h}}} \cdot \frac{\overset{1320}{\cancel{5280}} \text{ ft}}{1 \cancel{\text{mi}}} \cdot \frac{1 \cancel{\text{h}}}{\underset{\underset{1}{\cancel{4}}}{\cancel{60}} \text{ min}}$

$= \frac{1320 \text{ ft}}{1 \text{ min}}$

$= 1320$ feet/min

8. $\frac{8.4}{y} = \frac{1.2}{1.1}$

$8.4 \cdot 1.1 = y \cdot 1.2$

$9.24 = 1.2y$

$\frac{9.24}{1.2} = \frac{1.2y}{1.2}$

$7.7 = y$

9. $36\% = \frac{36}{100}$ or $\frac{9}{25}$

$36\% = 36\%$ or 0.36

10. $52\% = \frac{52}{100}$ or $\frac{13}{25}$

$52\% = 52\%$ or 0.52

11. $225\% = \frac{225}{100}$

$= \frac{9}{4}$ or $2\frac{1}{4}$

$225\% = 225\%$ or 2.25

12. $315\% = \frac{315}{100}$

$= \frac{63}{20}$ or $3\frac{3}{20}$

$315\% = 315\%$ or 3.15

13. $0.6\% = \frac{0.6}{100} \times \frac{10}{10}$

$= \frac{6}{1000}$ or $\frac{3}{500}$

$0.6\% = 0.6\%$ or 0.006

14. $0.4\% = \frac{0.4}{100} \times \frac{10}{10}$

$= \frac{4}{1000}$ or $\frac{1}{250}$

$0.4\% = 0.4\%$ or 0.004

15. $0.47 = 0.47$ or 47%

16. $0.025 = 0.025$ or 2.5%

17. $5.38 = 5.38$ or 538%

18. $\frac{7}{20} = 0.35$ or 35%

19. $\frac{30}{22} \approx 1.36$ or 136%

20. $\frac{18}{4000} = 0.0045$ or 0.45%

21. $\frac{36}{80} = \frac{p}{100}$

$36 \cdot 100 = 80 \cdot p$

$3600 = 80p$

$45 = p$

So, 36 is 45% of 80.

22. $\frac{35.23}{b} = \frac{63}{100}$

$35.23 \cdot 100 = b \cdot 63$

$3523 = 63b$

$56 \approx b$

So, 35.23 is 63% of 56.

23. 25% of 82

25% is $\frac{1}{4}$.

82 is about 80.

$\frac{1}{4}$ of 80 is 20.

So, 25% of 82 is about 20.

24. 63% of 77

63% is about 60% or $\frac{6}{10}$.

77 is about 80.

$\frac{6}{10} \times 80 = 48$.

So, 63% of 77 is about 48.

25. $I = prt$

$I = 2700(0.04)(2.5)$

$I = 270$

The interest is \$270.

26. $n = 0.15(135)$

$n = 20.25$

The discount is \$20.25.

27. percent of change $= \frac{140 - 175}{175}$

$= \frac{-35}{175}$

$= -0.2$ or -20%

The percent of change is −20%.

28. Favorable outcomes: 5

Possible outcomes: 16

$P(\text{orange}) = \frac{5}{16}$

The probability of selecting an orange ball is $\frac{5}{16}$.

29. $\frac{1 \text{ in.}}{6 \text{ ft}} = \frac{8.5 \text{ in.}}{x \text{ ft}}$

$1 \cdot x = 6 \cdot 8.5$

$x = 51$

The swimming pool is 51 feet long.

30. A; $n = 0.54(350)$

$n = 189$

You can expect 189 to say they use it to track delivery.

Chapter 6 Standardized Test Practice

Pages 322–323

1. C; $x - y + z = -6 - 9 + (-3)$

$= -15 + (-3)$

$= -18$

2. D; $A = \ell w$

$= 16 \cdot 12$

$= 192$

3. C; $m - n$ has 2 terms.

It is not a monomial.

4. C; $\frac{8 \text{ apples}}{36 \text{ pieces}} = \frac{2 \text{ apples}}{9 \text{ pieces}}$

$= \frac{2}{9}$

5. B; $\frac{5}{4} = \frac{x}{27 - x}$

$5(27 - x) = 4x$

$135 - 5x = 4x$

$135 - 5x + 5x = 4x + 5x$

$135 = 9x$

$\frac{135}{9} = \frac{9x}{9}$

$15 = x$

There are 15 girls.

6. B; $\frac{1 \text{ in.}}{8 \text{ ft}} = \frac{13.5 \text{ in.}}{x \text{ ft}}$

$1 \cdot x = 8 \cdot 13.5$

$x = 108$ feet

7. A; Randy: $\frac{47}{50} = \frac{94}{100}$ or 94%

Eduardo: 91%

Kelli: $\frac{9}{10} = \frac{90}{100}$ or 90%

Randy had the highest score.

8. C; $\frac{200}{75 + 200 + 100 + 125} = \frac{200}{500}$

$= \frac{2}{5}$ or 40%

9. B; percent of increase $= \frac{3{,}530{,}000 - 823{,}000}{823{,}000}$

$= \frac{2{,}707{,}000}{823{,}000}$

≈ 3.29 or 329%

Basketball had an increase of about 325%.

10. $4(6.80) + 4(6.80) + 5(2 \cdot 6.80)$

$= 27.20 + 27.20 + 68$

$= 122.40$

Ana earned \$122.40.

11. Juan: $3.89 + 1.00 + 0.99$
$= 5.88$

Julia: $3.49 + 0.79 + 0.99$
$= 5.27$

$5.88 + 5.27 = 11.15$
The total cost of lunch was 11.15.

12. $4^4 = 256$

13.
$$\frac{3}{8}m = \frac{1}{4}$$
$$\frac{8}{3}\left(\frac{3}{8}m\right) = \left(\frac{1}{4}\right)\frac{8}{3}$$
$$m = \frac{1}{4} \cdot \frac{8}{3}$$
$$m = \frac{2}{3}$$

14.
$$\frac{1.49 \text{ dollars}}{8 \text{ buns}} \approx \frac{0.19 \text{ dollars}}{1\text{bun}} \quad (\div 9)$$
$= \$0.19$/per hamburger bun

15.
$$\frac{a}{70} = \frac{40}{100}$$
$a \cdot 100 = 70 \cdot 40$
$100a = 2800$
$a = 28$
So, 40% of 70 is 28.

16. $n = 0.3(69.99)$
$n \approx 21$
He saved \$21.

17. Favorable outcomes: 3
Possible outcomes: 8
$P(\text{even}) = \frac{3}{8}$
The probability it will stop on an even number is $\frac{3}{8}$.

18a. $n = 0.4(679)$
$= 271.6$
$679 - 271.6 = 407.40$
He will pay \$407.40 on Saturday.

18b. $n = 0.1(679)$
$= 67.9$
$407.4 - 67.9 = 339.5$
He will pay \$339.50 on Wednesday.

18c. $n = 0.4(679)$
$= 271.6$
He will save \$271.60 on Saturday.

19. From 1997 to 1998, the percent of change was 118%, from 1998 to 1999, the percent of change was 168%, and from 1999 to 2000, the percent of change was 213%.

Chapter 7 Equations and Inequalities

Page 327 Getting Started

1.
$$\begin{aligned} 2x + 5 &= 13 \\ 2x + 5 - 5 &= 13 - 5 \\ 2x &= 8 \\ \frac{2x}{2} &= \frac{8}{2} \\ x &= 4 \end{aligned}$$

Check:
$$\begin{aligned} 2x + 5 &= 13 \\ 2(4) + 5 &\stackrel{?}{=} 13 \\ 8 + 5 &\stackrel{?}{=} 13 \\ 13 &= 13 \checkmark \end{aligned}$$

The solution is 4.

2.
$$\begin{aligned} 4n - 3 &= 5 \\ 4n - 3 + 3 &= 5 + 3 \\ 4n &= 8 \\ \frac{4n}{4} &= \frac{8}{4} \\ n &= 2 \end{aligned}$$

Check:
$$\begin{aligned} 4n - 3 &= 5 \\ 4(2) - 3 &\stackrel{?}{=} 5 \\ 8 - 3 &\stackrel{?}{=} 5 \\ 5 &= 5 \checkmark \end{aligned}$$

The solution is 2.

3.
$$\begin{aligned} 16 &= 8 + \frac{d}{3} \\ 16 - 8 &= 8 - 8 + \frac{d}{3} \\ 8 &= \frac{d}{3} \\ 8 \cdot 3 &= \frac{d}{3} \cdot 3 \\ 24 &= d \end{aligned}$$

Check:
$$\begin{aligned} 16 &= 8 + \frac{d}{3} \\ 16 &\stackrel{?}{=} 8 + \frac{24}{3} \\ 16 &\stackrel{?}{=} 8 + 8 \\ 16 &= 16 \checkmark \end{aligned}$$

The solution is 24.

4.
$$\begin{aligned} \frac{c}{-4} + 3 &= -9 \\ \frac{c}{-4} + 3 - 3 &= -9 - 3 \\ \frac{c}{-4} &= -12 \\ \frac{c}{-4}(-4) &= -12(-4) \\ c &= 48 \end{aligned}$$

Check:
$$\begin{aligned} \frac{c}{-4} + 3 &= -9 \\ \frac{48}{-4} + 3 &\stackrel{?}{=} -9 \\ -12 + 3 &\stackrel{?}{=} -9 \\ -9 &= -9 \checkmark \end{aligned}$$

The solution is 48.

5. $-28 + (-16) = -44$

6. $17 + (-25) = -8$

7. $-13 + 24 = 11$

8. $36 + (-18) = 18$

9. $13 - 48 = 13 + (-48)$
$= -17$

10. $-16 - 7 = -16 + (-7)$
$= -23$

11. $4 - (-12) = 4 + 12$
$= 16$

12. $-23 - (-29) = -23 + 29$
$= 6$

13. $-19 - (-5) = -19 + 5$
$= -14$

14. $-6(8) = -48$

15. $-3 \cdot 5 = -15$

16. $-6(-25) = 150$

17. $2(-4)(-9) = [2(-4)](-9)$
$= -8(-9)$
$= 72$

18. $64 \div (-32) = -2$

19. $-15 \div 3 = -5$

20. $-12 \div (-3) = 4$

21. $-6 \div (-6) = 1$

22. $24 \div (-2) = -12$

Pages 328–329 Algebra Activity (Preview of Lesson 7-1)

1.

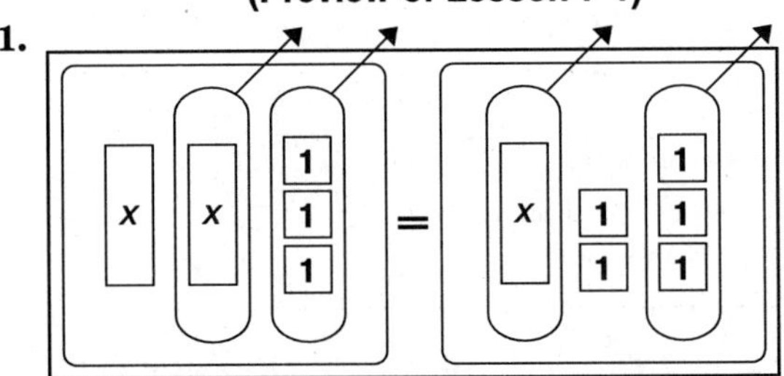

$2x + 3 = x + 5$

$x = 2$

2.

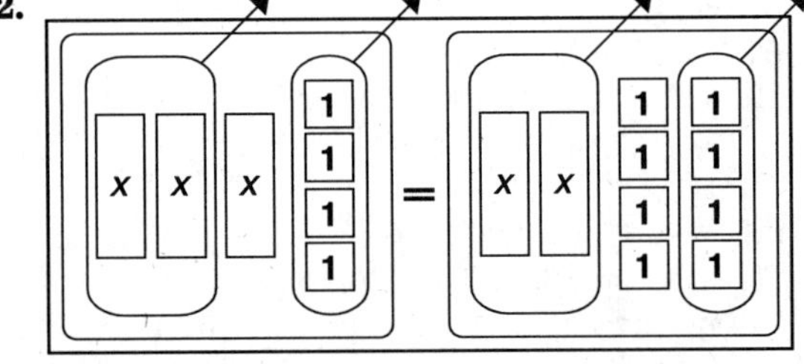

$3x + 4 = 2x + 8$

$x = 4$

3.

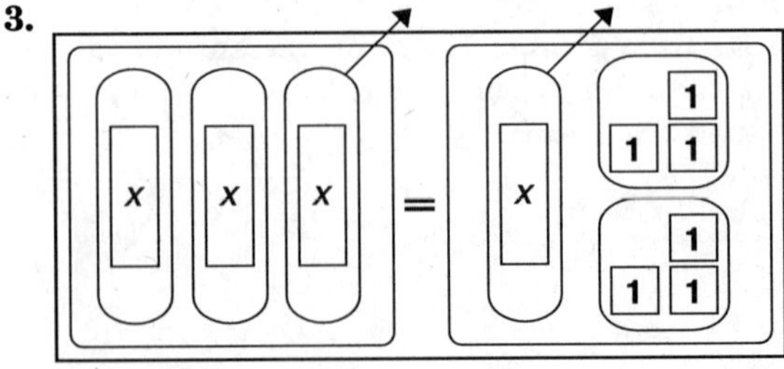

$3x = x + 6$

$x = 3$

4.

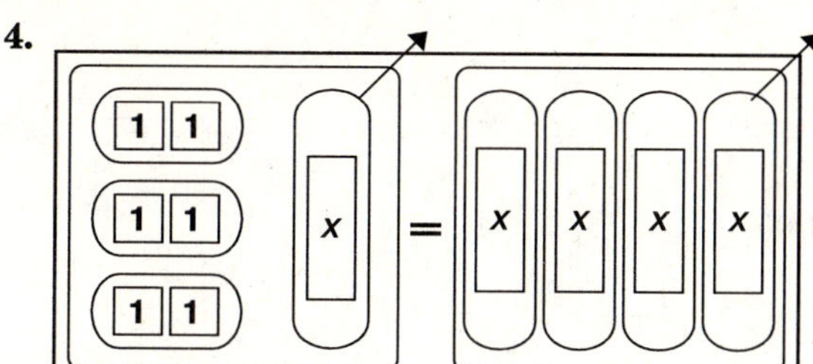

$6 + x = 4x$

$6 = 3x$

$2 = x$

5.

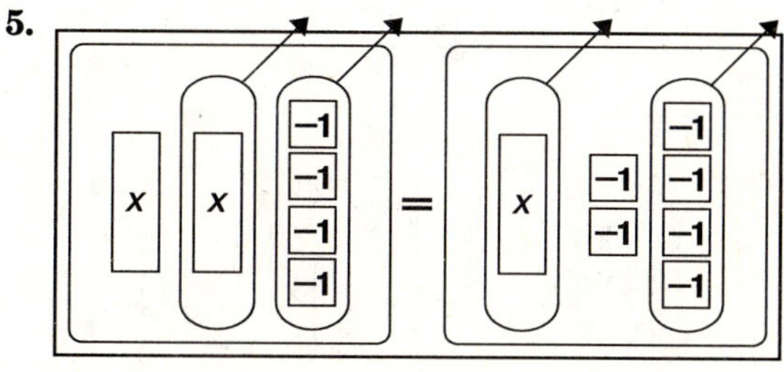

$2x - 4 = x - 6$

$x = -2$

6.

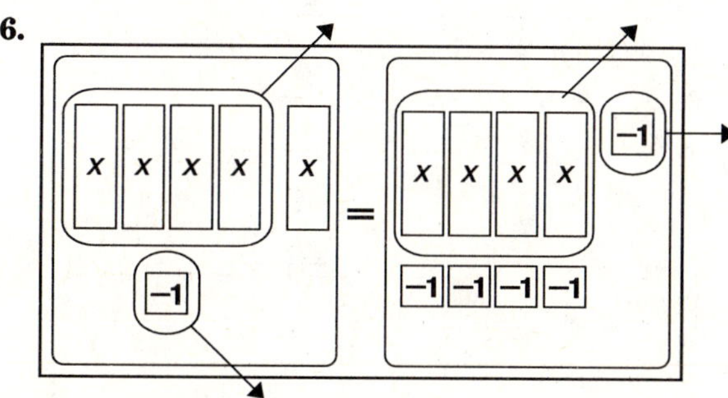

$5x - 1 = 4x - 5$

$x = -4$

7. subtraction

8. The value of x is the same on each side of the mat.

9.

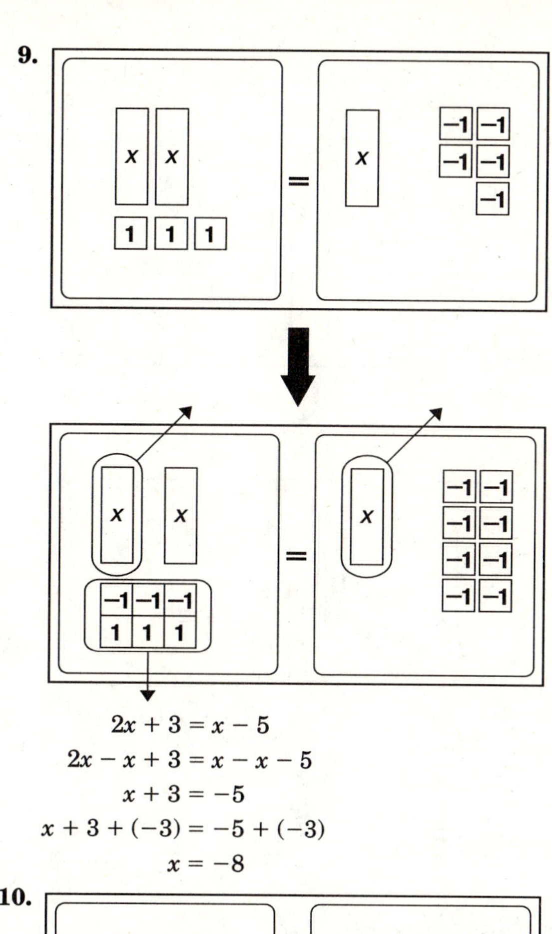

$2x + 3 = x - 5$

$2x - x + 3 = x - x - 5$

$x + 3 = -5$

$x + 3 + (-3) = -5 + (-3)$

$x = -8$

10.

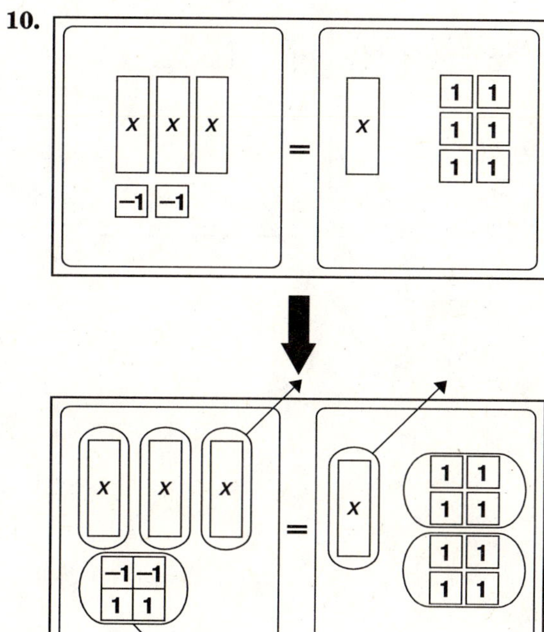

$3x - 2 = x + 6$

$3x - x - 2 = x - x + 6$

$2x - 2 = 6$

$2x - 2 + 2 = 6 + 2$

$2x = 8$

$x = 4$

11.

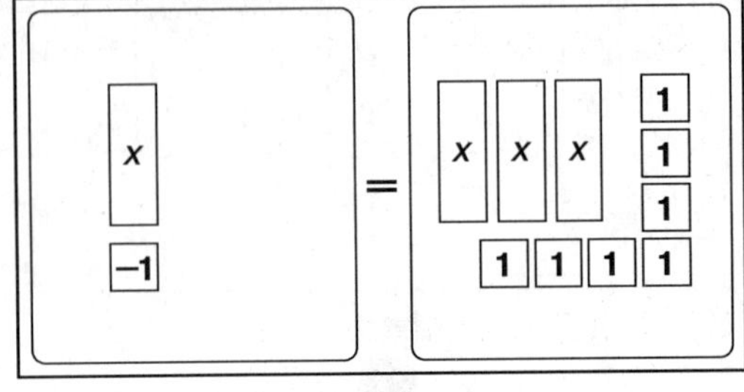

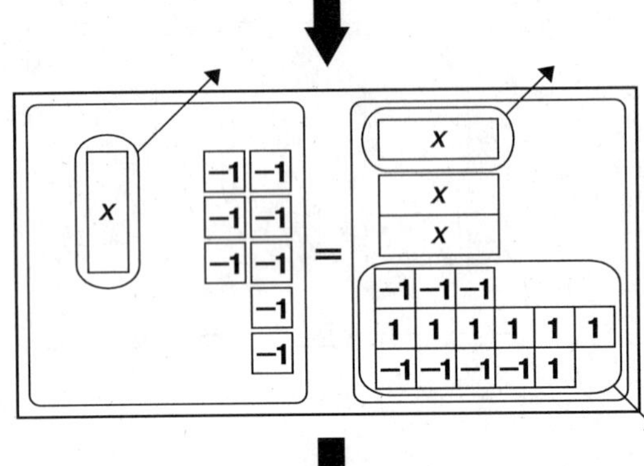

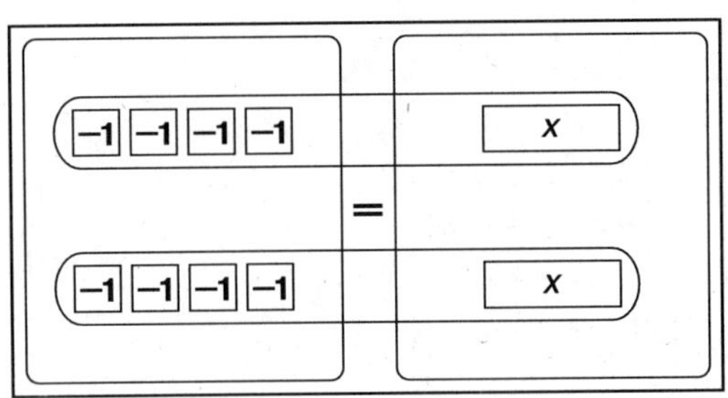

$$x - 1 = 3x + 7$$
$$x - x - 1 = 3x - x + 7$$
$$-1 = 2x + 7$$
$$-1 + (-7) = 2x + 7 + (-7)$$
$$-8 = 2x$$
$$-4 = x$$

12.

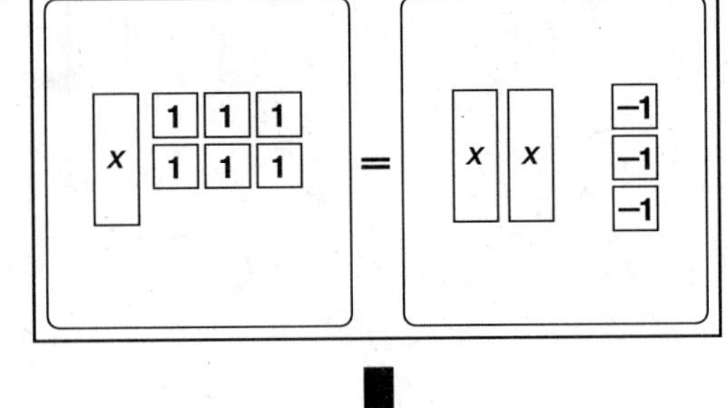

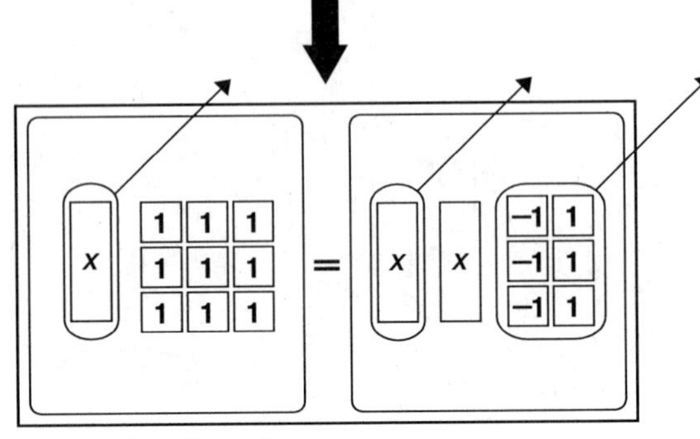

$$x + 6 = 2x - 3$$
$$x - x + 6 = 2x - x - 3$$
$$6 = x - 3$$
$$6 + 3 = x - 3 + 3$$
$$9 = x$$

13.

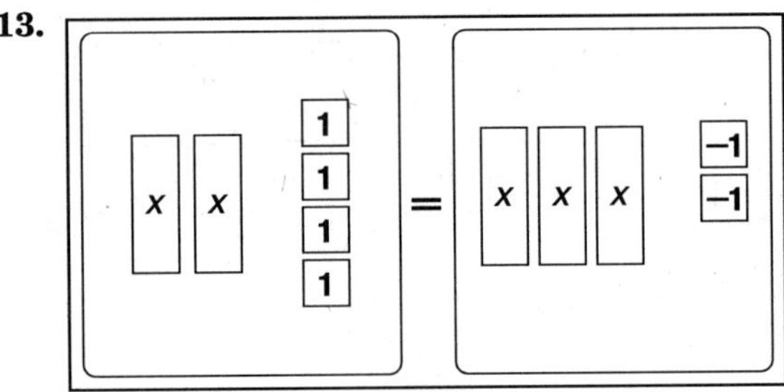

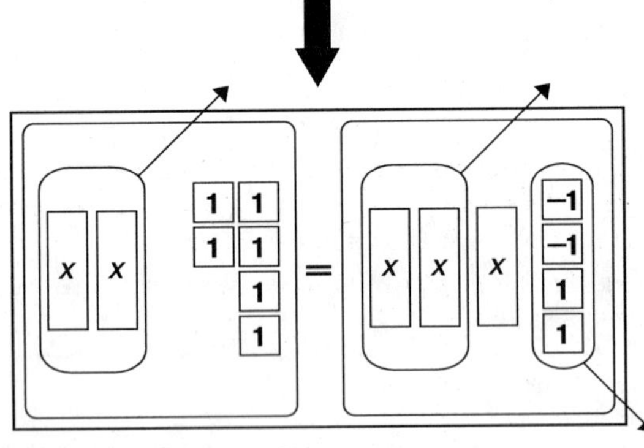

$$2x + 4 = 3x - 2$$
$$2x - 2x + 4 = 3x - 2x - 2$$
$$4 = x - 2$$
$$4 + 2 = x - 2 + 2$$
$$6 = x$$

14.

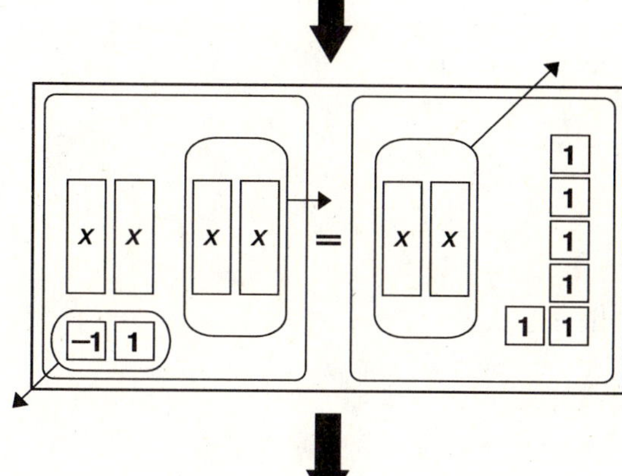

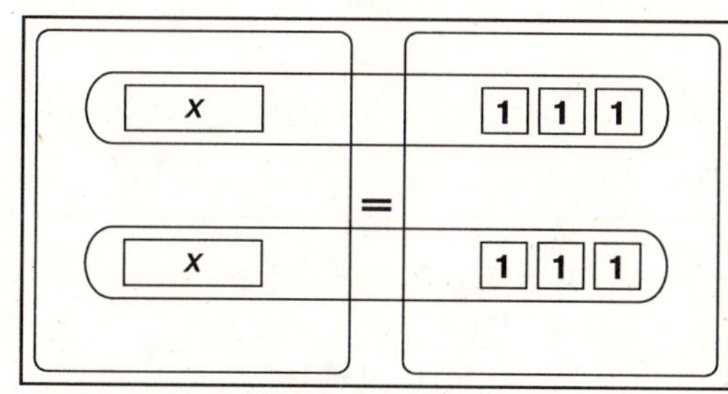

$$4x - 1 = 2x + 5$$
$$4x - 2x - 1 = 2x - 2x + 5$$
$$2x - 1 = 5$$
$$2x - 1 + 1 = 5 + 1$$
$$2x = 6$$
$$x = 3$$

15. You can remove either tile first. But it is a good idea to remove the 1-tiles first, because you are using the work backward strategy.

16. Add two 1-tiles to each side. Then add 2 x-tiles to each side. So, $x = 7$.

7-1 Solving Equations with Variables on Each Side

Page 330 How is solving equations with variables on each side like solving equations with variables on one side?

a. Sample answer: Remove one bag from each side. Then remove 3 blocks from each side. There must be 2 blocks in each bag.

b. Removing the same thing from each side keeps the scale in balance.

c. $2x + 3 = x + 5$

d. Subtract x from each side. Then subtract 3 from each side.

Page 332 Check for Understanding

1. Subtraction Property of Equality

2. Sample answer: $3x + 4 = 2x - 8$; subtract $2x$ from each side, subtract 4 from each side.

3.
$$4x - 8 = 5x$$
$$4x - 4x - 8 = 5x - 4x$$
$$-8 = x$$
Check: $4x - 8 = 5x$
$$4(-8) - 8 \stackrel{?}{=} 5(-8)$$
$$-32 - 8 \stackrel{?}{=} -40$$
$$-40 = -40 \checkmark$$
The solution is -8.

4.
$$12x = 2x + 40$$
$$12x - 2x = 2x - 2x + 40$$
$$10x = 40$$
$$\frac{10x}{10} = \frac{40}{10}$$
$$x = 4$$
Check: $12x = 2x + 40$
$$12(4) \stackrel{?}{=} 2(4) + 40$$
$$48 \stackrel{?}{=} 8 + 40$$
$$48 = 48 \checkmark$$
The solution is 4.

5.
$$4x - 1 = 3x + 2$$
$$4x - 3x - 1 = 3x - 3x + 2$$
$$x - 1 = 2$$
$$x - 1 + 1 = 2 + 1$$
$$x = 3$$
Check: $4x - 1 = 3x + 2$
$$4(3) - 1 \stackrel{?}{=} 3(3) + 2$$
$$12 - 1 \stackrel{?}{=} 9 + 2$$
$$11 = 11 \checkmark$$
The solution is 3.

6.
$$4k + 24 = 6k - 10$$
$$4k - 4k + 24 = 6k - 4k - 10$$
$$24 = 2k - 10$$
$$24 + 10 = 2k - 10 + 10$$
$$34 = 2k$$
$$\frac{34}{2} = \frac{2k}{2}$$
$$17 = k$$
Check: $4k + 24 = 6k - 10$
$$4(17) + 24 \stackrel{?}{=} 6(17) - 10$$
$$68 + 24 \stackrel{?}{=} 102 - 10$$
$$92 = 92 \checkmark$$
The solution is 17.

7. $n + 0.4 = -n + 1$

$$n + n + 0.4 = -n + n + 1$$
$$2n + 0.4 = 1$$
$$2n + 0.4 - 0.4 = 1 - 0.4$$
$$2n = 0.6$$
$$\frac{2n}{2} = \frac{0.6}{2}$$
$$n = 0.3$$

Check: $n + 0.4 = -n + 1$

$$0.3 + 0.4 \stackrel{?}{=} -(0.3) + 1$$
$$0.7 = 0.7 \checkmark$$

The solution is 0.3.

8. $3.1w + 5 = 0.8 + w$

$$3.1w - w + 5 = 0.8 + w - w$$
$$2.1w + 5 = 0.8$$
$$2.1w + 5 - 5 = 0.8 - 5$$
$$2.1w = -4.2$$
$$\frac{2.1w}{2.1} = \frac{-4.2}{2.1}$$
$$w = -2$$

Check: $3.1w + 5 = 0.8 + w$

$$3.1(-2) + 5 \stackrel{?}{=} 0.8 + (-2)$$
$$-6.2 + 5 \stackrel{?}{=} -1.2$$
$$-1.2 = -1.2 \checkmark$$

The solution is -2.

9. Let m represent the number of miles.

$$\underbrace{\text{\$25 a day plus \$0.45 a mile}}_{25 + 0.45m}$$

$$\underbrace{\text{\$40 a day plus \$0.25 a mile}}_{40 + 0.25m}$$

$$25 + 0.45m = 40 + 0.25m$$
$$25 + 0.45m - 0.25m = 40 + 0.25m - 0.25m$$
$$25 + 0.20m = 40$$
$$25 - 25 + 0.20m = 40 - 25$$
$$0.20m = 15$$
$$\frac{0.20m}{0.20} = \frac{15}{0.20}$$
$$m = 75$$

The daily costs are the same for 75 miles.

Pages 332–333 Practice and Apply

Exercises 10–27 For checks, see students' work.

10. $4x + 9 = 7x$

$$4x - 4x + 9 = 7x - 4x$$
$$9 = 3x$$
$$\frac{9}{3} = \frac{3x}{3}$$
$$3 = x$$

The solution is 3.

11. $6a = 26 + 4a$

$$6a - 4a = 26 + 4a - 4a$$
$$2a = 26$$
$$\frac{2a}{2} = \frac{26}{2}$$
$$a = 13$$

The solution is 13.

12. $3y + 16 = 5y$

$$3y - 3y + 16 = 5y - 3y$$
$$16 = 2y$$
$$\frac{16}{2} = \frac{2y}{2}$$
$$8 = y$$

The solution is 8.

13. $n - 14 = 3n$

$$n - n - 14 = 3n - n$$
$$-14 = 2n$$
$$\frac{-14}{2} = \frac{2n}{2}$$
$$-7 = n$$

The solution is -7.

14. $8 - 3c = 2c - 2$

$$8 - 3c + 3c = 2c + 3c - 2$$
$$8 = 5c - 2$$
$$8 + 2 = 5c - 2 + 2$$
$$10 = 5c$$
$$\frac{10}{5} = \frac{5c}{5}$$
$$2 = c$$

The solution is 2.

15. $3 - 4b = 10b + 10$

$$3 - 4b + 4b = 10b + 4b + 10$$
$$3 = 14b + 10$$
$$3 - 10 = 14b + 10 - 10$$
$$-7 = 14b$$
$$\frac{-7}{14} = \frac{14b}{14}$$
$$-0.5 = b$$

The solution is -0.5.

16. $7d - 13 = 3d + 7$

$$7d - 3d - 13 = 3d - 3d + 7$$
$$4d - 13 = 7$$
$$4d - 13 + 13 = 7 + 13$$
$$4d = 20$$
$$\frac{4d}{4} = \frac{20}{4}$$
$$d = 5$$

The solution is 5.

17. $2f - 6 = 7f + 24$

$$2f - 2f - 6 = 7f - 2f + 24$$
$$-6 = 5f + 24$$
$$-6 - 24 = 5f + 24 - 24$$
$$-30 = 5f$$
$$\frac{-30}{5} = \frac{5f}{5}$$
$$-6 = f$$

The solution is -6.

18.
$$
\begin{aligned}
-s + 4 &= 7s - 3\\
-s + s + 4 &= 7s + s - 3\\
4 &= 8s - 3\\
4 + 3 &= 8s - 3 + 3\\
7 &= 8s\\
\tfrac{7}{8} &= \tfrac{8s}{8}\\
0.875 &= s
\end{aligned}
$$
The solution is 0.875.

19.
$$
\begin{aligned}
4a - 2 &= 7a - 6\\
4a - 4a - 2 &= 7a - 4a - 6\\
-2 &= 3a - 6\\
-2 + 6 &= 3a - 6 + 6\\
4 &= 3a\\
\tfrac{4}{3} &= \tfrac{3a}{3}\\
\tfrac{4}{3} &= a
\end{aligned}
$$
The solution is $\frac{4}{3}$.

20.
$$
\begin{aligned}
12n - 24 &= -14n + 28\\
12n + 14n - 24 &= -14n + 14n + 28\\
26n - 24 &= 28\\
26n - 24 + 24 &= 28 + 24\\
26n &= 52\\
\tfrac{26n}{26} &= \tfrac{52}{26}\\
n &= 2
\end{aligned}
$$
The solution is 2.

21.
$$
\begin{aligned}
13y - 18 &= -5y + 36\\
13y + 5y - 18 &= -5y + 5y + 36\\
18y - 18 &= 36\\
18y - 18 + 18 &= 36 + 18\\
18y &= 54\\
\tfrac{18y}{18} &= \tfrac{54}{18}\\
y &= 3
\end{aligned}
$$
The solution is 3.

22.
$$
\begin{aligned}
12 + 1.5a &= 3a\\
12 + 1.5a - 1.5a &= 3a - 1.5a\\
12 &= 1.5a\\
\tfrac{12}{1.5} &= \tfrac{1.5a}{1.5}\\
8 &= a
\end{aligned}
$$
The solution is 8.

23.
$$
\begin{aligned}
12.6 - x &= 2x\\
12.6 - x + x &= 2x + x\\
12.6 &= 3x\\
\tfrac{12.6}{3} &= \tfrac{3x}{3}\\
4.2 &= x
\end{aligned}
$$
The solution is 4.2.

24.
$$
\begin{aligned}
2b + 6.2 &= 13.2 - 8b\\
2b + 8b + 6.2 &= 13.2 - 8b + 8b\\
10b + 6.2 &= 13.2\\
10b + 6.2 - 6.2 &= 13.2 - 6.2\\
10b &= 7\\
\tfrac{10b}{10} &= \tfrac{7}{10}\\
b &= 0.7
\end{aligned}
$$
The solution is 0.7.

25.
$$
\begin{aligned}
3c + 4.5 &= 7.2 - 6c\\
3c + 6c + 4.5 &= 7.2 - 6c + 6c\\
9c + 4.5 &= 7.2\\
9c + 4.5 - 4.5 &= 7.2 - 4.5\\
9c &= 2.7\\
\tfrac{9c}{9} &= \tfrac{2.7}{9}\\
c &= 0.3
\end{aligned}
$$
The solution is 0.3.

26.
$$
\begin{aligned}
12.4y + 14 &= 6y - 2\\
12.4y - 6y + 14 &= 6y - 6y - 2\\
6.4y + 14 &= -2\\
6.4y + 14 - 14 &= -2 - 14\\
6.4y &= -16\\
\tfrac{6.4y}{6.4} &= \tfrac{-16}{6.4}\\
y &= -2.5
\end{aligned}
$$
The solution is −2.5.

27.
$$
\begin{aligned}
4.3n - 1.6 &= 2.3n + 5.2\\
4.3n - 2.3n - 1.6 &= 2.3n - 2.3n + 5.2\\
2n - 1.6 &= 5.2\\
2n - 1.6 + 1.6 &= 5.2 + 1.6\\
2n &= 6.8\\
\tfrac{2n}{2} &= \tfrac{6.8}{2}\\
n &= 3.4
\end{aligned}
$$
The solution is 3.4.

28. Let x = the number.
$$
\begin{aligned}
2x &= 6x - 220\\
2x - 6x &= 6x - 6x - 220\\
-4x &= -220\\
\tfrac{-4x}{-4} &= \tfrac{-220}{-4}\\
x &= 55
\end{aligned}
$$
The number is 55.

29. Let y = the number.
$$
\begin{aligned}
3y - 14 &= y\\
3y - 3y - 14 &= y - 3y\\
-14 &= -2y\\
\tfrac{-14}{-2} &= \tfrac{-2y}{-2}\\
7 &= y
\end{aligned}
$$
The number is 7.

Exercises 30–33 For checks, see students' work.

30. $\frac{4}{5}y - 8 = \frac{2}{5}y + 16$

$\frac{4}{5}y - \frac{2}{5}y - 8 = \frac{2}{5}y - \frac{2}{5}y + 16$

$\frac{2}{5}y - 8 = 16$

$\frac{2}{5}y - 8 + 8 = 16 + 8$

$\frac{2}{5}y = 24$

$\frac{5}{2} \cdot \frac{2}{5}y = \frac{5}{2} \cdot 24$

$y = 60$

The solution is 60.

31. $\frac{3}{4}k + 16 = 2 - \frac{1}{8}k$

$\frac{3}{4}k + \frac{1}{8}k + 16 = 2 - \frac{1}{8}k + \frac{1}{8}k$

$\frac{6}{8}k + \frac{1}{8}k + 16 = 2$

$\frac{7}{8}k + 16 = 2$

$\frac{7}{8}k + 16 - 16 = 2 - 16$

$\frac{7}{8}k = -14$

$\frac{8}{7} \cdot \frac{7}{8}k = \frac{8}{7} \cdot (-14)$

$k = -16$

The solution is -16.

32. $\frac{x}{0.4} = 2x + 1.2$

Multiply both sides by 0.4.

$0.4\left(\frac{x}{0.4}\right) = 0.4(2x + 1.2)$

$x = 0.4(2x) + 0.4(1.2)$

$x = 0.8x + 0.48$

$x - 0.8x = 0.48$

$0.2x = 0.48$

$\frac{0.2x}{0.2} = \frac{0.48}{0.2}$

$x = 2.4$

The solution is 2.4.

33. $\frac{1}{3}b + 8 = \frac{1}{2}b - 4$

$\frac{1}{3}b - \frac{1}{2}b + 8 = \frac{1}{2}b - \frac{1}{2}b - 4$

$\frac{2}{6}b - \frac{3}{6}b + 8 = -4$

$-\frac{1}{6}b + 8 = -4$

$-\frac{1}{6}b + 8 - 8 = -4 - 8$

$-\frac{1}{6}b = -12$

$(-6) \cdot \left(-\frac{1}{6}b\right) = (-6)(-12)$

$b = 72$

The solution is 72.

34. Let x = the length of Louisiana's coastline.

$2x + 46 = x + 443$

$2x - x + 46 = x - x + 443$

$x + 46 = 443$

$x + 46 - 46 = 443 - 46$

$x = 397$

Louisiana's coastline is 397 miles long.

California's coastline:

$x + 443 = 397 + 443$

$= 840$

California's coastline is 840 miles long.

35. Let x = the number of minutes for local calls.

Cost of 1st carrier: $29.75 + 0.15x$

Cost of 2nd carrier: $19.95 + 0.29x$

$29.75 + 0.15x = 19.95 + 0.29x$

$29.75 + 0.15x - 0.15x = 19.95 + 0.29x - 0.15x$

$29.75 = 19.95 + 0.14x$

$29.75 - 19.95 = 19.95 - 19.95 + 0.14x$

$9.8 = 0.14x$

$\frac{9.8}{0.14} = \frac{0.14x}{0.14}$

$70 = x$

The monthly cost is the same for 70 minutes of local calls.

36. Let x = the number of seconds it will take to fill the bucket.

The first faucet can fill the bucket in 6 minutes, so it can fill $\frac{1}{6}$ of it in one minute. The second faucet can fill the bucket in 4 minutes, so it can fill $\frac{1}{4}$ of it in one minute. Together, they can fill it in x minutes, so they can fill $\frac{1}{x}$ of it in one minute.

$\frac{1}{6} + \frac{1}{4} = \frac{1}{x}$

Multiply both sides of the equation by $24x$.

$24x\left(\frac{1}{6} + \frac{1}{4}\right) = 24x\left(\frac{1}{x}\right)$

$4x + 6x = 24$

$10x = 24$

$\frac{10x}{10} = \frac{24}{10}$

$x = 2.4$

2.4 minutes is $2.4 \times 60 = 144$ seconds. Together, the two faucets can fill the bucket in 144 seconds.

37. $3(y + 7) = 4(y - 2)$

$3y + 21 = 4y - 8$

$3y - 3y + 21 = 4y - 3y - 8$

$21 = y - 8$

$21 + 8 = y - 8 + 8$

$29 = y$

38. In both cases, the goal is to isolate the variable. Anwers should include:

- Sample equations: variables on both sides: $-2x - 8 = 4x - 2$, variables on one side: $3x + 2 = 8$
- They are alike because the goal is to isolate the variable. However, in equations with variables on both sides, you have to add or subtract variables from each side as well as constant terms.

39. A; Let s = the amount of Olivia's sales.

Pay for Plan 1: $3 + 0.15s$

Pay for Plan 2: $4 + 0.10s$

$3 + 0.15s = 4 + 0.10s$

40. B; $3x - 1 = x + 3$

$$3x - x - 1 = x - x + 3$$
$$2x - 1 = 3$$
$$2x - 1 + 1 = 3 + 1$$
$$2x = 4$$
$$\frac{2x}{2} = \frac{4}{2}$$
$$x = 2$$

41. Substitute C for F.

$$C = \frac{9}{5}C + 32$$
$$C - \frac{9}{5}C = \frac{9}{5}C - \frac{9}{5}C + 32$$
$$\frac{5}{5}C - \frac{9}{5}C = 32$$
$$\frac{-4}{5}C = 32$$
$$\left(\frac{-5}{4}\right)\left(\frac{-4}{5}C\right) = \left(\frac{-5}{4}\right)(32)$$
$$C = -40$$

At 40°, the two scales are the same.

Page 333 Maintain Your Skills

42. $P(\text{T}) = \frac{\text{number of Ts}}{\text{number of possible letters}}$

$$= \frac{2}{10}$$
$$= 0.2 \text{ or } 20\%$$

43. percent of increase $= \frac{\text{amount of increase}}{\text{original price}}$

$$= \frac{90 - 80}{80}$$
$$= \frac{10}{80}$$
$$= 0.125 \text{ or } 12.5\%$$

44. You know that the part is 14 and the base is 20. Let n represent the percent.

$$14 = n(20)$$
$$\frac{14}{20} = n$$
$$\frac{7}{10} = n$$
$$n = 0.7 \text{ or } 70\%$$

45. You know that 36 is the percent and 18 is the base. Let n represent the part.

$$n = 0.36(18)$$
$$n = 6.48$$

46. You know that 1.5 is the part and 30 is the percent. Let n represent the base.

$$1.5 = 0.30n$$
$$\frac{1.5}{0.30} = \frac{0.30n}{0.30}$$
$$5 = n$$

47. You know that 140 is the percent and 50 is the base. Let n represent the part.

$$n = 1.40(50)$$
$$n = 70$$

48. $4(x - 8) = 4[x + (-8)]$

$$= 4x + 4(-8)$$
$$= 4x + (-32)$$
$$= 4x - 32$$

49. $3(2a + 9) = 3 \cdot 2a + 3 \cdot 9$

$$= 6a + 27$$

50. $5(12 - x) = 5[12 + (-x)]$

$$= 5 \cdot 12 + 5(-x)$$
$$= 60 - 5x$$

51. $2(1.2c + 14) = 2(1.2c) + 2 \cdot 14$

$$= 2.4c + 28$$

52. $8(-4k + 2.3) = 8(-4k) + 8(2.3)$

$$= -32k + 18.4$$

53. $\frac{1}{2}(n - 9) = \frac{1}{2}[n + (-9)]$

$$= \frac{1}{2}n + \frac{1}{2}(-9)$$
$$= \frac{1}{2}n - \frac{9}{2}$$

7-2 Solving Equations with Grouping Symbols

Page 334 Why is the Distributive Property important in solving equations?

a. t represents the time Josh travels.

b. She left 1 hour later than Josh.

c. $2t = 10(t - 1)$

Page 336 Check for Understanding

1. Multiply 4 times $(x - 1)$.
   Subtract $2x$ from each side.
   Add 4 to each side.
   Divide each side by 2.

2. Sample answer: $3x + 2 = 3x + 1$ has no solution; $2x + 4 = 2(x + 2)$ is an identity.

3.

$$3(a - 5) = 18$$
$$3a - 15 = 18$$
$$3a - 15 + 15 = 18 + 15$$
$$3a = 33$$
$$\frac{3a}{3} = \frac{33}{3}$$
$$a = 11$$

Check: $3(a - 5) = 18$

$$3(11 - 5) \stackrel{?}{=} 18$$
$$3(6) \stackrel{?}{=} 18$$
$$18 = 18 \checkmark$$

The solution is 11.

4.

$$32 = 4(x + 9)$$
$$32 = 4x + 36$$
$$32 - 36 = 4x + 36 - 36$$
$$-4 = 4x$$
$$\frac{-4}{4} = \frac{4x}{4}$$
$$-1 = x$$

Check: $32 = 4(x + 9)$

$$32 \stackrel{?}{=} 4(-1 + 9)$$
$$32 \stackrel{?}{=} 4(8)$$
$$32 = 32 \checkmark$$

The solution is -1.

5. $2(d + 6) = 3d - 1$
$2d + 12 = 3d - 1$
$2d - 2d + 12 = 3d - 2d - 1$
$12 = d - 1$
$12 + 1 = d - 1 + 1$
$13 = d$
Check: $2(d + 6) = 3d - 1$
$2(13 + 6) \stackrel{?}{=} 3(13) - 1$
$2(19) \stackrel{?}{=} 39 - 1$
$38 = 38$ ✓
The solution is 13.

6. $6(n - 3) = 4(n + 2.1)$
$6n - 18 = 4n + 8.4$
$6n - 4n - 18 = 4n - 4n + 8.4$
$2n - 18 = 8.4$
$2n - 18 + 18 = 8.4 + 18$
$2n = 26.4$
$\frac{2n}{2} = \frac{26.4}{2}$
$n = 13.2$
Check: $6(n - 3) = 4(n + 2.1)$
$6(13.2 - 3) \stackrel{?}{=} 4(13.2 + 2.1)$
$6(10.2) \stackrel{?}{=} 4(15.3)$
$61.2 = 61.2$ ✓
The solution is 13.2.

7. $12 - h = -h + 3$
$12 - h + h = -h + h + 3$
$12 = 3$
The sentence $12 = 3$ is never true, so the solution is $\varnothing$.

8. $3(2g + 4) = 6(g + 2)$
$6g + 12 = 6g + 12$
$6g + 12 - 12 = 6g + 12 - 12$
$6g = 6g$
$g = g$
The sentence $g = g$ is always true, so the solution set is all numbers.

9. Let ℓ = the length.
Let $\ell - 4$ = the width.
Perimeter = 2 times length + 2 times width
$20 = 2\ell + 2(\ell - 4)$
$20 = 2\ell + 2\ell - 8$
$20 = 4\ell - 8$
$20 + 8 = 4\ell - 8 + 8$
$28 = 4\ell$
$\frac{28}{4} = \frac{4\ell}{4}$
$7 = \ell$
The length is 7 ft.
Evaluate $\ell - 4$ to find the width.
$\ell - 4 = 7 - 4 = 3$.
The width is 3 ft.
$A = \ell w$
$A = 7 \cdot 3$
$A = 21$
The area is 21 ft^2.

Pages 337–338 Practice and Apply

Exercises 10–29 For checks, see students' work.

10. $3(g - 3) = 6$
$3g - 9 = 6$
$3g - 9 + 9 = 6 + 9$
$3g = 15$
$\frac{3g}{3} = \frac{15}{3}$
$g = 5$
The solution is 5.

11. $3(x + 1) = 21$
$3x + 3 = 21$
$3x + 3 - 3 = 21 - 3$
$3x = 18$
$\frac{3x}{3} = \frac{18}{3}$
$x = 6$
The solution is 6.

12. $5(2c + 7) = 80$
$10c + 35 = 80$
$10c + 35 - 35 = 80 - 35$
$10c = 45$
$\frac{10c}{10} = \frac{45}{10}$
$c = 4.5$
The solution is 4.5.

13. $6(3d + 5) = 75$
$18d + 30 = 75$
$18d + 30 - 30 = 75 - 30$
$18d = 45$
$\frac{18d}{18} = \frac{45}{18}$
$d = 2.5$
The solution is 2.5.

14. $3(a - 3) = 2(a + 4)$
$3a - 9 = 2a + 8$
$3a - 2a - 9 = 2a - 2a + 8$
$a - 9 = 8$
$a - 9 + 9 = 8 + 9$
$a = 17$
The solution is 17.

15. $3(s + 22) = 4(s + 12)$
$3s + 66 = 4s + 48$
$3s - 3s + 66 = 4s - 3s + 48$
$66 = s + 48$
$66 - 48 = s + 48 - 48$
$18 = s$
The solution is 18.

16. $4(x - 2) = 3(1.5 + x)$
$4x - 8 = 4.5 + 3x$
$4x - 3x - 8 = 4.5 + 3x - 3x$
$x - 8 = 4.5$
$x - 8 + 8 = 4.5 + 8$
$x = 12.5$
The solution is 12.5.

17. $3(a-1) = 4(a-1.5)$
$$3a - 3 = 4a - 6$$
$$3a - 3a - 3 = 4a - 3a - 6$$
$$-3 = a - 6$$
$$-3 + 6 = a - 6 + 6$$
$$3 = a$$
The solution is 3.

18. $2(3.5n + 6) = 2.5n - 2$
$$7n + 12 = 2.5n - 2$$
$$7n - 2.5n + 12 = 2.5n - 2.5n - 2$$
$$4.5n + 12 = -2$$
$$4.5n + 12 - 12 = -2 - 12$$
$$4.5n = -14$$
$$\frac{4.5n}{4.5} = \frac{-14}{4.5}$$
$$n = -3.\overline{1}$$
The solution is $-3.\overline{1}$.

19. $4.2x - 9 = 3(1.2x + 4)$
$$4.2x - 9 = 3.6x + 12$$
$$4.2x - 3.6x - 9 = 3.6x - 3.6x + 12$$
$$0.6x - 9 = 12$$
$$0.6x - 9 + 9 = 12 + 9$$
$$0.6x = 21$$
$$\frac{0.6x}{0.6} = \frac{21}{0.6}$$
$$x = 35$$
The solution is 35.

20. $4(f + 3) + 5 = 17 + 4f$
$$4f + 12 + 5 = 17 + 4f$$
$$4f + 17 = 17 + 4f$$
$$4f + 17 - 17 = 17 - 17 + 4f$$
$$4f = 4f$$
$$f = f$$
The sentence $f = f$ is always true, so the solution set is all numbers.

21. $3n + 4 = 5(n + 2) - 2n$
$$3n + 4 = 5n + 10 - 2n$$
$$3n + 4 = 3n + 10$$
$$3n - 3n + 4 = 3n - 3n + 10$$
$$4 = 10$$
The sentence $4 = 10$ is never true, so the solution set is $\emptyset$.

22. $8y - 3 = 5(y - 1) + 3y$
$$8y - 3 = 5y - 5 + 3y$$
$$8y - 3 = 8y - 5$$
$$8y - 8y - 3 = 8y - 8y - 5$$
$$-3 = -5$$
The sentence $-3 = -5$ is never true, so the solution set is $\emptyset$.

23. $2(x - 5) = 4x - 2(x + 5)$
$$2x - 10 = 4x - 2x - 10$$
$$2x - 10 = 2x - 10$$
$$2x - 10 + 10 = 2x - 10 + 10$$
$$2x = 2x$$
$$x = x$$
The sentence $x = x$ is always true, so the solution set is all numbers.

24. $\frac{1}{2}(2n - 5) = 4n - 1$
$$\frac{1}{2}(2n) - \frac{1}{2}(5) = 4n - 1$$
$$n - \frac{5}{2} = 4n - 1$$
$$n - n - \frac{5}{2} = 4n - n - 1$$
$$-\frac{5}{2} = 3n - 1$$
$$-\frac{5}{2} + 1 = 3n - 1 + 1$$
$$-\frac{5}{2} + \frac{2}{2} = 3n$$
$$\frac{-3}{2} = 3n$$
$$\frac{1}{3}\left(-\frac{3}{2}\right) = \frac{1}{3}(3n)$$
$$-\frac{1}{2} = n$$
$$-0.5 = n$$
The solution is -0.5.

25. $y - 2 = \frac{1}{3}(y + 6)$
$$y - 2 = \frac{1}{3}y + \frac{1}{3} \cdot 6$$
$$y - 2 = \frac{1}{3}y + 2$$
$$y - \frac{1}{3}y - 2 = \frac{1}{3}y - \frac{1}{3}y + 2$$
$$\frac{3}{3}y - \frac{1}{3}y - 2 = 2$$
$$\frac{2}{3}y - 2 = 2$$
$$\frac{2}{3}y - 2 + 2 = 2 + 2$$
$$\frac{2}{3}y = 4$$
$$\frac{3}{2}\left(\frac{2}{3}y\right) = \frac{3}{2}(4)$$
$$y = 6$$
The solution is 6.

26. $-3(4b - 10) = \frac{1}{2}(-24b + 60)$
$$-12b + 30 = \frac{1}{2}(-24b) + \frac{1}{2}(60)$$
$$-12b + 30 = -12b + 30$$
$$-12b + 30 - 30 = -12b + 30 - 30$$
$$-12b = -12b$$
$$b = b$$
The sentence $b = b$ is always true, so the solution is all numbers.

27. $\frac{3}{4}a + 4 = \frac{1}{4}(3a + 16)$

$\frac{3}{4}a + 4 = \frac{1}{4}(3a) + \frac{1}{4}(16)$

$\frac{3}{4}a + 4 = \frac{3}{4}a + 4$

$\frac{3}{4}a + 4 - 4 = \frac{3}{4}a + 4 - 4$

$\frac{3}{4}a = \frac{3}{4}a$

$a = a$

The sentence $a = a$ is always true, so the solution is all numbers.

28. $\frac{d}{0.4} = 2d + 1.24$

$0.4\left(\frac{d}{0.4}\right) = 0.4(2d + 1.24)$

$d = 0.4(2d) + 0.4(1.24)$

$d = 0.8d + 0.496$

$d - 0.8d = 0.8d - 0.8d + 0.496$

$0.2d = 0.496$

$\frac{0.2d}{0.2} = \frac{0.496}{0.2}$

$d = 2.48$

The solution is 2.48.

29. $\frac{a - 6}{12} = \frac{a - 2}{4}$

$12\left(\frac{a - 6}{12}\right) = 12\left(\frac{a - 2}{4}\right)$

$a - 6 = 3(a - 2)$

$a - 6 = 3a - 6$

$a - a - 6 = 3a - a - 6$

$-6 = 2a - 6$

$-6 + 6 = 2a - 6 + 6$

$0 = 2a$

$\frac{0}{2} = \frac{2a}{2}$

$0 = a$

The solution is 0.

30. Perimeter = 2 times length + 2 times width

$460 = 2(w + 30) + 2w$

$460 = 2w + 60 + 2w$

$460 = 4w + 60$

$460 - 60 = 4w + 60 - 60$

$400 = 4w$

$\frac{400}{4} = \frac{4w}{4}$

$100 = w$

Evaluate $w + 30$ to find the length.

$w + 30 = 100 + 30$ or 130

The dimensions of the rectangle are 100 ft by 130 ft.

31. Perimeter = 2 times length + 2 times width

$440 = 2(3w - 60) + 2w$

$440 = 6w - 120 + 2w$

$440 = 8w - 120$

$440 + 120 = 8w - 120 + 120$

$560 = 8w$

$\frac{560}{8} = \frac{8w}{8}$

$70 = w$

Evaluate $3w - 60$ to find the length.

$3w - 60 = 3(70) - 60$
$= 210 - 60$
$= 150$

The dimensions are 70 yd by 150 yd.

32. Perimeter = 2 times length + 2 times width

$11 = 2w + 2(2w - 2)$

$11 = 2w + 4w - 4$

$11 = 6w - 4$

$11 + 4 = 6w - 4 + 4$

$15 = 6w$

$\frac{15}{6} = \frac{6w}{6}$

$2.5 = w$

Evaluate $2w - 2$ to find the length.

$2w - 2 = 2(2.5) - 2$
$= 5 - 2$
$= 3$

The dimensions are 2.5 m by 3 m.

33. Let w = the width.
Let $3w + 4$ = the length.
Perimeter = 2 times length + 2 times width

$32 = 2(3w + 4) + 2w$

$32 = 6w + 8 + 2w$

$32 = 8w + 8$

$32 - 8 = 8w + 8 - 8$

$24 = 8w$

$\frac{24}{8} = \frac{8w}{8}$

$3 = w$

Evaluate $3w + 4$ to find the length.

$3w + 4 = 3(3) + 4$
$= 9 + 4$
$= 13$

The width is 3 ft and the length is 13 ft.

$A = \ell w$
$A = 13 \cdot 3$
$A = 39$

The area is 39 ft^2.

34. Let x = the first integer.
Let $x + 1$ = the second consecutive integer.
Let $x + 2$ = the third consecutive integer.

$3[(x) + (x + 1) + (x + 2)] = 72$

$3(3x + 3) = 72$

$9x + 9 = 72$

$9x + 9 - 9 = 72 - 9$

$9x = 63$

$\frac{9x}{9} = \frac{63}{9}$

$x = 7$

Evaluate $x + 1$ and $x + 2$ to find the next two consecutive integers.

$x + 1 = 7 + 1 = 8$
$x + 2 = 7 + 2 = 9$

The integers are 7, 8, and 9.

35. Perimeter of the triangle:

$x + x + 1 + x + 2$

Perimeter of the rectangle:

$2(x - 3) + 2(x + 1)$

Since, according to the problem, the perimeters are the same, set them equal to each other:

$$x + x + 1 + x + 2 = 2(x - 3) + 2(x + 1)$$
$$3x + 3 = 2x - 6 + 2x + 2$$
$$3x + 3 = 4x - 4$$
$$3x - 3x + 3 = 4x - 3x - 4$$
$$3 = x - 4$$
$$3 + 4 = x - 4 + 4$$
$$7 = x$$

The other sides of the triangle can be found by evaluating:

$x + 1 = 7 + 1 = 8$

$x + 2 = 7 + 2 = 9$

The sides of the triangle are 7, 8, and 9. The perimeter is 7 + 8 + 9 or 24.

The sides of the rectangle can be found by evaluating:

$x - 3 = 7 - 3 = 4$

$x + 1 = 7 + 1 = 8$

The sides of the rectangle are 4 and 8. Its perimeter is 2(4) + 2(8) or 24.

36. Let x = Kim's points.
Let $x + 5$ = Lynn's points.
Let $3(x + 5)$ = Camilla's points.

Kim's points + Lynn's points + Camilla's points = 2 times Jasmine's points

$$x + x + 5 + 3(x + 5) = 2(25)$$
$$x + x + 5 + 3x + 15 = 50$$
$$5x + 20 = 50$$
$$5x + 20 - 20 = 50 - 20$$
$$5x = 30$$
$$\frac{5x}{5} = \frac{30}{5}$$
$$x = 6$$

Kim made 6 points.
Evaluate $x + 5$ and $3(x + 5)$ to find Lynn's and Camilla's points:

$x + 5 = 6 + 5$ or 11

$$3(x + 5) = 3(6 + 5)$$
$$= 3(11)$$
$$= 33$$

Lynn made 11 points.
Camilla made 33 points.

37. Total square footage
= wall height × combined length
− 2 windows × 15 square feet
− 2 doors × 15 square feet
= 7(15 + 12 + 15 + 12) − 2(15) − 2(15)
= 7(54) − 30 − 30
= 378 − 60
= 318 square feet

For 2 coats of paint, we need to cover 2(318) or 636 square feet. Each gallon covers 350 square feet.

636 ÷ 350 = 1.82 gallons

Since we can't buy 1.82 gallons, we will need to buy 2 gallons.

38. Let x = cost of banana.

Let $\frac{1}{2}(x + 15)$ = cost of orange.

Let $x + \frac{1}{2}(x + 15) - 10$ = cost of apple.

cost of apple = cost of 2 oranges

$$x + \frac{1}{2}(x + 15) - 10 = 2\left[\frac{1}{2}(x + 15)\right]$$
$$x + \frac{1}{2}x + \frac{15}{2} - 10 = x + 15$$
$$x - x + \frac{1}{2}x + \frac{15}{2} - 10 = x - x + 15$$
$$\frac{1}{2}x + \frac{15}{2} - 10 = 15$$
$$\frac{1}{2}x + \frac{15}{2} - \frac{20}{2} = \frac{30}{2}$$
$$\frac{1}{2}x - \frac{5}{2} = \frac{30}{2}$$
$$\frac{1}{2}x - \frac{5}{2} + \frac{5}{2} = \frac{30}{2} + \frac{5}{2}$$
$$\frac{1}{2}x = \frac{35}{2}$$
$$2\left(\frac{1}{2}x\right) = 2\left(\frac{35}{2}\right)$$
$$x = 35$$

Bananas cost 15 cents.
Cost of oranges:

$$\frac{1}{2}(x + 15) = \frac{1}{2}(35 + 15)$$
$$= \frac{1}{2}(50)$$
$$= 25 \text{ cents}$$

Cost of apples:

$$x + \frac{1}{2}(x + 15) - 10 = 35 + \frac{1}{2}(35 + 15) - 10$$
$$= 35 + \frac{1}{2}(50) - 10$$
$$= 35 + 25 - 10$$
$$= 60 - 10$$
$$= 50 \text{ cents}$$

39. Many equations include grouping symbols. You must use the Distributive Property to correctly solve the equation. Answers should include the following.

- The Distributive Property states that $a(b + c) = ab + ac$.
- You use the Distributive Property to remove the grouping symbols when you are solving equations.

40. C; $2(3x - 1) = 10 + 2x$

$$6x - 2 = 10 + 2x$$
$$6x - 2x - 2 = 10 + 2x - 2x$$
$$4x - 2 = 10$$

41. D; Let x = the time Car X drives.
Let $x - 1$ = the time Car Y drives.

Use $d = rt$ to find the distance each car drives.

Distance of Car X: $55x$
Distance of Car Y: $60(x - 1)$

When Car Y catches up to Car X, their distances will be equal.

$55x = 60(x - 1)$

Page 338 Maintain Your Skills

42. $4x = 2x + 5$

$4x - 2x = 2x - 2x + 5$

$2x = 5$

$\frac{2x}{2} = \frac{5}{2}$

$x = \frac{5}{2}$ or 2.5

Check: $4x = 2x + 5$

$4(2.5) \stackrel{?}{=} 2(2.5) + 5$

$10 \stackrel{?}{=} 5 + 5$

$10 = 10$ ✓

The solution is 2.5.

43. $3x + 5 = 7 - 2x$

$3x + 2x + 5 = 7 - 2x + 2x$

$5x + 5 = 7$

$5x + 5 - 5 = 7 - 5$

$5x = 2$

$\frac{5x}{5} = \frac{2}{5}$

$x = \frac{2}{5}$ or 0.4

Check: $3x + 5 = 7 - 2x$

$3(0.4) + 5 \stackrel{?}{=} 7 - 2(0.4)$

$1.2 + 5 \stackrel{?}{=} 7 - 0.8$

$6.2 = 6.2$ ✓

The solution is 0.4.

44. $1.5x + 9 = 3x - 3$

$1.5x - 1.5x + 9 = 3x - 1.5x - 3$

$9 = 1.5x - 3$

$9 + 3 = 1.5x - 3 + 3$

$12 = 1.5x$

$\frac{12}{1.5} = \frac{1.5x}{1.5}$

$8 = x$

Check: $1.5x + 9 = 3x - 3$

$1.5(8) + 9 \stackrel{?}{=} 3(8) - 3$

$12 + 9 \stackrel{?}{=} 24 - 3$

$21 = 21$ ✓

The solution is 8.

45. $P(\text{girl's name}) = \frac{\text{number of girls' names}}{\text{number of possible names}}$

$= \frac{20}{50}$

$= 0.4$ or 40%

46. $10\overline{)4.0}$ with quotient 0.4

-40

0

So, $\frac{4}{10} = 0.4$.

47. $8\overline{)3.0}$ with quotient 0.375

-24

60

-56

40

-40

0

So, $\frac{3}{8} = 0.375$.

48. $3\overline{)1.0}$ with quotient 0.33...

So, $-\frac{1}{3} = -0.\overline{3}$.

49. $25\overline{)6.00}$ with quotient 0.24

-50

100

-100

0

So, $3\frac{6}{25} = 3 + 0.24$ or 3.024.

50. $11\overline{)5.000}$ with quotient .4545...

So, $4\frac{5}{11} = 4 + 0.\overline{45}$ or $4.\overline{45}$.

51. $x - 12 = 5 - 12$

$= -7$

52. $b + 11 = -15 + 11$

$= -4$

53. $4a = 4(-6)$

$= -24$

54. $2t + 8 = 2(-3) + 8$

$= -6 + 8$

$= 2$

55. $\frac{24}{c} = \frac{24}{-3}$

$= -8$

56. $\frac{3x}{4} + 2 = \frac{3(6)}{4} + 2$

$= \frac{18}{4} + 2$

$= 4.5 + 2$

$= 6.5$

Page 338 Practice Quiz 1

1. Let x = the number.

$2x = 5x - 150$

$2x - 5x = 5x - 5x - 150$

$-3x = -150$

$\frac{-3x}{-3} = \frac{-150}{-3}$

$x = 50$

The number is 50.

2. $6y + 42 = 4y$

$6y - 6y + 42 = 4y - 6y$

$42 = -2y$

$\frac{42}{-2} = \frac{-2y}{-2}$

$-21 = y$

Check: $6y + 42 = 4y$

$6(-21) + 42 \stackrel{?}{=} 4(-21)$

$-126 + 42 \stackrel{?}{=} -84$

$-84 = -84$ ✓

The solution is -21.

3. $7m - 12 = 2.5m + 2$

$7m - 2.5m - 12 = 2.5m - 2.5m + 2$

$4.5m - 12 = 2$

$4.5m - 12 + 12 = 2 + 12$

$4.5m = 14$

$\frac{4.5m}{4.5} = \frac{14}{4.5}$

$m = 3.\overline{1}$

Check: $7m - 12 = 2.5m + 2$
$7(3.\overline{1}) - 12 \stackrel{?}{=} 2.5(3.\overline{1}) + 2$
$21.\overline{7} - 12 \stackrel{?}{=} 7.\overline{7} + 2$
$9.\overline{7} = 9.\overline{7}$ ✓

The solution is $3.\overline{1}$.

4. $8(p - 4) = 2(2p + 1)$
$8p - 32 = 4p + 2$
$8p - 4p - 32 = 4p - 4p + 2$
$4p - 32 = 2$
$4p - 32 + 32 = 2 + 32$
$4p = 34$
$\frac{4p}{4} = \frac{34}{4}$
$p = 8.5$

Check: $8(p - 4) = 2(2p + 1)$
$8(8.5 - 4) \stackrel{?}{=} 2(2 \cdot 8.5 + 1)$
$8(4.5) \stackrel{?}{=} 2(17 + 1)$
$36 \stackrel{?}{=} 2(18)$
$36 = 36$ ✓

The solution is 8.5.

5. $b + 2(b + 5) = 3(b - 1) + 13$
$b + 2b + 10 = 3b - 3 + 13$
$3b + 10 = 3b + 10$
$3b + 10 - 10 = 3b + 10 - 10$
$3b = 3b$
$b = b$

The sentence $b = b$ is always true, so the solution is all numbers.

Page 339 Reading Mathematics

1. Sample answer: at most—no more than; at least—no less than
2. Sample answer: You must earn \$50 or more than \$50; $e \geq 50$.
3. Sample answer: The sum is 6 or less than 6; $a + b \leq 6$.
4. Sample answer: You drive 250 miles or more; $m \geq 250$.
5. Sample answer: You will hike 4 hours or less than 4 hours; $h \leq 4$.

7-3 Inequalities

Page 340 How can inequalities help you describe relationships?

a. Sample ages: 3, 4, 5; no
b. Sample heights: 42 in., 43 in., 44 in.; no
c. Sample speeds: 25 mph, 30 mph, 35 mps; yes

Pages 342–343 Check for Understanding

1. An inequality represents all numbers greater or less than a given number. A number line graph can represent all those numbers.

2. Sample answer: $x < 9$ means that the value of x is less than 9 (and does not include 9); $x > 9$ means that the value of x is greater than 9 (and does not include 9); $x \leq 9$ means that the value of x is 9 or less; $x \geq 9$ means that the value of x is 9 or greater.

3. Let n represent the number.
$n + 14 \geq 25$

4. Let n represent the number.
$5n < 65$

5. $n + 4 > 6$
$12 + 4 \stackrel{?}{>} 6$
$16 > 6$
This sentence is true.

6. $34 \leq 4r$
$34 \stackrel{?}{\leq} 4(8)$
$34 \not\leq 32$
This sentence is false.

7. $n > 3$

−2 −1 0 1 2 3 4 5 6

8. $p \leq 5$

−2 −1 0 1 2 3 4 5 6

9. $x < 7$

1 2 3 4 5 6 7 8 9

10. $x < 11$

11. $x \geq -20$

12. Let m represent the safe load.
$m \leq 3600$

Pages 343–344 Practice and Apply

13. Let f represent the number of fans.
$f > 18{,}000$

14. Let h represent the number of hours Kyle worked. Let $15h$ represent Kyle's earnings.
$15h \leq 60$

15. Let w represent the winner's time.
$86 \geq 2w$

16. Let s represent the beginning amount in the savings account.
$s - 75 < 500$

17. $18 - x > 4$
$18 - 12 \stackrel{?}{>} 4$
$6 > 4$
This sentence is true.

18. $14 + n < 23$
$14 + 8 \stackrel{?}{<} 23$
$22 < 23$
This sentence is true.

19. $5k > 35$
$5(7) \stackrel{?}{>} 35$
$35 \not> 35$
This sentence is false.

20. $16 \leq 3c$
$16 \overset{?}{\leq} 3(8)$
$16 \leq 24$
This sentence is true.

21. $\frac{x}{3} \geq 2$
$\frac{9}{3} \overset{?}{\geq} 2$
$3 \geq 2$
This sentence is true.

22. $\frac{14}{c} < 7$
$\frac{14}{2} \overset{?}{<} 7$
$7 \not< 7$
This sentence is false.

23. $a > 4$

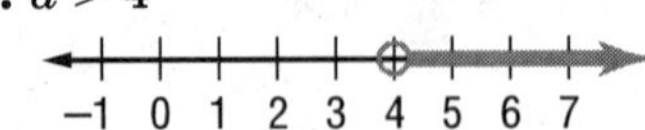

24. $x > 6$

3 4 5 6 7 8 9 10 11

25. $n < 11$

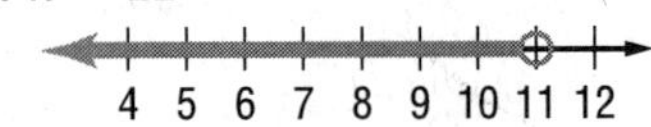

26. $x < 5$

−2 −1 0 1 2 3 4 5 6

27. $t \geq 9$

6 7 8 9 10 11 12 13 14

28. $b \geq 8$

6 7 8 9 10 11 12 13 14

29. $d \leq 5$

−2 −1 0 1 2 3 4 5 6

30. $w \leq 8$

2 3 4 5 6 7 8 9 10

31. $x > -4$

−6 −5 −4 −3 −2 −1 0 1 2

32. $n \geq -3$

−6 −5 −4 −3 −2 −1 0 1 2

33. $x \leq -5$

−8 −7 −6 −5 −4 −3 −2 −1 0

34. $x < -2$

−6 −5 −4 −3 −2 −1 0 1 2

35. $x > 13$

36. $x \geq -8$

37. $x \leq -3$

38. $x < 0$

39. $x \geq -32$

40. $x < 11$

41. Let m represent Inali's homework time.
$m \geq 5.4 + 1$
$m \geq 6.4$

42. Let j represent Anna's homework time.
$j \leq 6.8$

43. Let b represent the number of basketball programs.
$b + 14{,}600 > 30{,}000$
$b + 14{,}600 - 14{,}600 > 30{,}000 - 14{,}600$
$b > 15{,}400$

44. Sample answer: 0.7

45. Symmetric: If $a < b$, then $b < a$; not true.
Sample counterexample: $4 < 5$, but $5 \not< 4$.
Transitive: If $a < b$ and $b < c$, then $a < c$; true.

46. Inequalities describe numbers that are greater than or less than a given number. Answers should include the following.

- Sample answer: My brother's age is less than (<) my age. My age is less than or equal to (≤) the ages of the other students in my class. My house is more than 2 miles (>) from school. I drink at least 2 (≥) glasses of milk each day.
- Sample answer: My brother is younger than I am. The other students are my age or older. I live farther than 2 miles from school. I drink 2 or more glasses of milk each day.

47. B; Let n represent the number.
$n - 2 \leq 8$

48. A; $4 \leq x + 2$

49a. −3 −2 −1 0 1 2 3 4 5

49b. −3 −2 −1 0 1 2 3 4 5

Page 344 Maintain Your Skills

50.
$2(3 + x) = 14$
$6 + 2x = 14$
$6 - 6 + 2x = 14 - 6$
$2x = 8$
$\frac{2x}{2} = \frac{8}{2}$
$x = 4$
Check: $2(3 + x) = 14$
$2(3 + 4) \overset{?}{=} 14$
$2(7) \overset{?}{=} 14$
$14 = 14$ ✓
The solution is 4.

51.
$63 = 9(2y - 3)$
$63 = 18y - 27$
$63 + 27 = 18y - 27 + 27$
$90 = 18y$
$\frac{90}{18} = \frac{18y}{18}$
$5 = y$
Check: $63 = 9(2y - 3)$
$63 \overset{?}{=} 9(2 \cdot 5 - 3)$
$63 \overset{?}{=} 9(10 - 3)$
$63 \overset{?}{=} 9(7)$
$63 = 63$ ✓
The solution is 5.

52. $$3(n-1) = 1.5(n+2)$$
$$3n - 3 = 1.5n + 1.5(2)$$
$$3n - 3 = 1.5n + 3$$
$$3n - 1.5n - 3 = 1.5n - 1.5n + 3$$
$$1.5n - 3 = 3$$
$$1.5n - 3 + 3 = 3 + 3$$
$$1.5n = 6$$
$$\frac{1.5n}{1.5} = \frac{6}{1.5}$$
$$n = 4$$
Check: $3(n-1) = 1.5(n+2)$
$$3(4-1) \stackrel{?}{=} 1.5(4+2)$$
$$3(3) \stackrel{?}{=} 1.5(6)$$
$$9 = 9 \checkmark$$
The solution is 4.

53. Let n = the number.
$$4n - 6 = 3n + 2$$
$$4n - 3n - 6 = 3n - 3n + 2$$
$$n - 6 = 2$$
$$n - 6 + 6 = 2 + 6$$
$$n = 8$$

54. Arithmetic, with common difference 3; The next three terms are: $5 + 3 = 8$, $8 + 3 = 11$, and $11 + 3 = 14$.

55. Geometric, with common ratio -2; the next three terms are: $8(-2) = -16$, $-16(-2) = 32$, and $32(-2) = -64$.

56. Neither; the pattern is $+1, +2, +3, \ldots$. The next three terms are: $7 + 4 = 11$, $11 + 5 = 16$, and $16 + 6 = 22$.

57. $$x + 19 = 32$$
$$x + 19 - 19 = 32 - 19$$
$$x = 13$$

58. $$a + 7 = -3$$
$$a + 7 - 7 = -3 - 7$$
$$a = -10$$

59. $$26 + c = 19$$
$$26 - 26 + c = 19 - 26$$
$$c = -7$$

60. $$44 - c = 26$$
$$44 - 44 - c = 26 - 44$$
$$-c = -18$$
$$\frac{-1c}{-1} = \frac{-18}{-1}$$
$$c = 18$$

61. $$y - 9.7 = 10.1$$
$$y - 9.7 + 9.7 = 10.1 + 9.7$$
$$y = 19.8$$

62. $$r - 1.6 = -0.6$$
$$r - 1.6 + 1.6 = -0.6 + 1.6$$
$$r = 1$$

7-4 Solving Inequalities by Adding or Subtracting

Page 345 How is solving an inequality similar to solving an equation?

a. 3

b. Sample answer: Remove 2 blocks from each side. There are 3 blocks remaining on the right, so there must be 3 blocks in the bag.

c. 0, 1, or 2

d. $x < 3$

Page 347 Check for Understanding

1. Use addition to undo subtraction; use subtraction to undo addition.

2. Sample answer: Dylan; both solutions include 18, but the solution to the inequality also includes numbers greater than 18.

3. Sample answer: Joanna works part-time at a clothing store. Part-time workers must work fewer than 30 hours a week. If Joanna has already worked 8 hours this week, how many hours can she still work?

4. $$x + 3 < 8$$
$$x + 3 - 3 < 8 - 3$$
$$x < 5$$
Check: Replace x with any number less than 5.
$$x + 3 < 8$$
$$4 + 3 \stackrel{?}{<} 8$$
$$7 < 8 \checkmark$$
The solution is $x < 5$.

5. $$14 + y \geq 7$$
$$14 - 14 + y \geq 7 - 14$$
$$y \geq -7$$
Check: Replace y with any number greater than or equal to -7.
$$14 + y \geq 7$$
$$14 + (-5) \stackrel{?}{\geq} 7$$
$$9 \geq 7 \checkmark$$
The solution is $y \geq -7$.

6. $$-13 \geq 9 + b$$
$$-13 - 9 \geq 9 - 9 + b$$
$$-22 \geq b \text{ or } b \leq -22$$
Check: Replace b with any number less than or equal to -22.
$$-13 \geq 9 + b$$
$$-13 \stackrel{?}{\geq} 9 + (-30)$$
$$-13 \geq -21 \checkmark$$
The solution is $b \leq -22$.

7. $$a - 5 > 6$$
$$a - 5 + 5 > 6 + 5$$
$$a > 11$$
Check: Replace a with any number greater than 11.
$$a - 5 > 6$$
$$12 - 5 \stackrel{?}{>} 6$$
$$7 > 6 \checkmark$$
The solution is $a > 11$.

8. $c - (-2) \leq 3$
$c + 2 \leq 3$
$c + 2 - 2 \leq 3 - 2$
$c \leq 1$

Check: Replace c with any number less than or equal to 1.
$c - (-2) \leq 3$
$0 - (-2) \overset{?}{\leq} 3$
$2 \leq 3$ ✓

The solution is $c \leq 1$.

9. $-5 < t - 2$
$-5 + 2 < t - 2 + 2$
$-3 < t$ or $t > -3$

Check: Replace t with any number greater than -3.
$-5 < t - 2$
$-5 \overset{?}{<} 0 - 2$
$-5 < -2$ ✓

The solution is $t > -3$.

10. $h + 4 > 4$
$h + 4 - 4 > 4 - 4$
$h > 0$

−4 −3 −2 −1 0 1 2 3 4

11. $x - 6 \leq 4$
$x - 6 + 6 \leq 4 + 6$
$x \leq 10$

7 8 9 10 11 12 13 14 15

12. Let x represent the amount Chris needs to save.
$x + 62.50 \geq 100$
$x + 62.50 - 62.50 \geq 100 - 62.50$
$x \geq 37.50$
Chris needs to save at least \$37.50.

Pages 348–349 Practice and Apply

13. $p + 7 < 9$
$p + 7 - 7 < 9 - 7$
$p < 2$

Check: Replace p with any number less than 2.
$p + 7 < 9$
$0 + 7 \overset{?}{<} 9$
$7 < 9$ ✓

The solution is $p < 2$.

14. $t + 6 > -3$
$t + 6 - 6 > -3 - 6$
$t > -9$

Check: Replace t with any number greater than -9.
$t + 6 > -3$
$-8 + 6 \overset{?}{>} -3$
$-2 > -3$ ✓

The solution is $t > -9$.

15. $-14 \geq 8 + b$
$-14 - 8 \geq 8 - 8 + b$
$-22 \geq b$ or $b \leq -22$

Check: Replace b with any number less than or equal to -22.
$-14 \geq 8 + b$
$-14 \overset{?}{\geq} 8 + (-28)$
$-14 \geq -20$ ✓

The solution is $b \leq -22$.

16. $16 > -11 + k$
$16 + 11 > -11 + 11 + k$
$27 > k$ or $k < 27$

Check: Replace k with any number less than 27.
$16 > -11 + k$
$16 \overset{?}{>} -11 + 25$
$16 > 14$ ✓

The solution is $k < 27$.

17. $3 \geq -2 + y$
$3 + 2 \geq -2 + 2 + y$
$5 \geq y$ or $y \leq 5$

Check: Replace y with any number less than or equal to 5.
$3 \geq -2 + y$
$3 \overset{?}{\geq} -2 + (4)$
$3 \geq 2$ ✓

The solution is $y \leq 5$.

18. $25 < n + (-12)$
$25 + 12 < n + (-12) + 12$
$37 < n$ or $n > 37$

Check: Replace n with any number greater than 37.
$25 < n + (-12)$
$25 \overset{?}{<} 40 + (-12)$
$25 < 28$ ✓

The solution is $n > 37$.

19. $r - 5 \leq 2$
$r - 5 + 5 \leq 2 + 5$
$r \leq 7$

Check: Replace r with any number less than or equal to 7.
$r - 5 \leq 2$
$6 - 5 \overset{?}{\leq} 2$
$1 \leq 2$ ✓

The solution is $r \leq 7$.

20. $a - 6 < 13$
$a - 6 + 6 < 13 + 6$
$a < 19$

Check: Replace a with any number less than 19.
$a - 6 < 13$
$18 - 6 \overset{?}{<} 13$
$12 < 13$ ✓

The solution is $a < 19$.

21. $j - 8 \leq -12$
$j - 8 + 8 \leq -12 + 8$
$j \leq -4$

Check: Replace j with any number less than or equal to -4.
$j - 8 \leq -12$
$-5 - 8 \overset{?}{\leq} -12$
$-13 \leq -12$ ✓

The solution is $j \leq -4$.

22. $-8 > h - 1$
$-8 + 1 > h - 1 + 1$
$-7 > h$ or $h < -7$

Check: Replace h with any number less than -7.
$-8 > h - 1$
$-8 \overset{?}{>} -10 - 1$
$-8 > -11$ ✓

The solution is $h < -7$.

23. $22 > w - (-16)$
$22 > w + 16$
$22 - 16 > w + 16 - 16$
$6 > w$ or $w < 6$

Check: Replace w with any number less than 6.
$22 > w - (-16)$
$22 \overset{?}{>} 5 - (-16)$
$22 \overset{?}{>} 5 + 16$
$22 > 21$ ✓

The solution is $w < 6$.

24. $-30 \leq d + (-5)$
$-30 + 5 \leq d + (-5) + 5$
$-25 \leq d$ or $d \geq -25$

Check: Replace d with any number greater than or equal to -25.
$-30 \leq d + (-5)$
$-30 \overset{?}{\leq} -20 + (-5)$
$-30 \leq -25$ ✓

The solution is $d \geq -25$.

25. $1 + y \leq 2.4$
$1 - 1 + y \leq 2.4 - 1$
$y \leq 1.4$

Check: Replace y with any number less than or equal to 1.4.
$1 + y \leq 2.4$
$1 + 0 \overset{?}{\leq} 2.4$
$1 \leq 2.4$ ✓

The solution is $y \leq 1.4$.

26. $2.9 < c + 7$
$2.9 - 7 < c + 7 - 7$
$-4.1 < c$ or $c > -4.1$

Check: Replace c with any number greater than -4.1.
$2.9 < c + 7$
$2.9 \overset{?}{<} 0 + 7$
$2.9 < 7$ ✓

The solution is $c > -4.1$.

27. $f + (-4) \geq 1.4$
$f + (-4) + 4 \geq 1.4 + 4$
$f \geq 5.4$

Check: Replace f with any number greater than or equal to 5.4.
$f + (-4) \geq 1.4$
$6 + (-4) \overset{?}{\geq} 1.4$
$2 \geq 1.4$ ✓

The solution is $f \geq 5.4$.

28. $z + (-2) > -3.8$
$z + (-2) + 2 > -3.8 + 2$
$z > -1.8$

Check: Replace z with any number greater than -1.8.
$z + (-2) > -3.8$
$0 + (-2) \overset{?}{>} -3.8$
$-2 > -3.8$ ✓

The solution is $z > -1.8$.

29. $b - \frac{3}{4} < 2\frac{1}{2}$
$b - \frac{3}{4} < \frac{5}{2}$
$b - \frac{3}{4} + \frac{3}{4} < \frac{5}{2} + \frac{3}{4}$
$b < \frac{10}{4} + \frac{3}{4}$
$b < \frac{13}{4}$ or $3\frac{1}{4}$

Check: Replace b with any number less than $3\frac{1}{4}$.
$b - \frac{3}{4} < 2\frac{1}{2}$
$1\frac{3}{4} - \frac{3}{4} \overset{?}{<} 2\frac{1}{2}$
$1 < 2\frac{1}{2}$ ✓

The solution is $b < 3\frac{1}{4}$.

30. $g - 1\frac{2}{3} > 2\frac{1}{6}$
$g - \frac{5}{3} > \frac{13}{6}$
$g - \frac{5}{3} + \frac{5}{3} > \frac{13}{6} + \frac{5}{3}$
$g > \frac{13}{6} + \frac{10}{6}$
$g > \frac{23}{6}$ or $3\frac{5}{6}$

Check: Replace g with any number greater than $3\frac{5}{6}$.
$g - 1\frac{2}{3} > 2\frac{1}{6}$
$4\frac{2}{3} - 1\frac{2}{3} \overset{?}{>} 2\frac{1}{6}$
$3 > 2\frac{1}{6}$ ✓

The solution is $g > 3\frac{5}{6}$.

31. $n + 4 < 9$
$n + 4 - 4 < 9 - 4$
$n < 5$

1 2 3 4 5 6 7 8 9

32. $t + 7 > 12$
$t + 7 - 7 > 12 - 7$
$t > 5$

1 2 3 4 5 6 7 8 9

33. $p + (-5) > -3$
$p + (-5) + 5 > -3 + 5$
$p > 2$

1 2 3 4 5 6 7 8 9

34. $-3 + z > 2$
$-3 + 3 + z > 2 + 3$
$z > 5$

1 2 3 4 5 6 7 8 9

35. $-13 \geq x - 8$
$-13 + 8 \geq x - 8 + 8$
$-5 \geq x$ or $x \leq -5$

−7 −6 −5 −4 −3 −2 −1 0 1

36. $-32 \geq a + (-5)$
$-32 + 5 \geq a + (-5) + 5$
$-27 \geq a$ or $a \leq -27$

−30 −28 −26 −24 −22

37. $33 \leq m - (-6)$
$33 \leq m + 6$
$33 - 6 \leq m + 6 - 6$
$27 \leq m$ or $m \geq 27$

23 25 27 29 31

38. $k + 9 \geq -21$
$k + 9 - 9 \geq -21 - 9$
$k \geq -30$

−31 −29 −27 −25 −23

39. $1\frac{1}{4} + b < 3$
$\frac{5}{4} + b < 3$
$\frac{5}{4} - \frac{5}{4} + b < 3 - \frac{5}{4}$
$b < \frac{12}{4} - \frac{5}{4}$
$b < \frac{7}{4}$ or $1\frac{3}{6}$

0 $\frac{1}{2}$ 1 $1\frac{1}{2}$ 2

40. $3 \leq \frac{1}{2} + a$
$3 - \frac{1}{2} \leq \frac{1}{2} - \frac{1}{2} + a$
$2\frac{1}{2} \leq a$ or $a \geq 2\frac{1}{2}$

0 1 2 3 4

41. $4 \geq s - \frac{2}{3}$
$4 + \frac{2}{3} \geq s - \frac{2}{3} + \frac{2}{3}$
$4\frac{2}{3} \geq s$ or $s \leq 4\frac{2}{3}$

3 4 5

42. $-\frac{3}{4} < w - 1$
$-\frac{3}{4} + 1 < w - 1 + 1$
$-\frac{3}{4} + \frac{4}{4} < w$
$\frac{1}{4} < w$ or $w > \frac{1}{4}$

0 $\frac{1}{2}$ 1 $1\frac{1}{2}$ 2

43. Let x represent the allowable weight for passengers.
$x + 120 \leq 1100$
$x + 120 - 120 \leq 1100 - 120$
$x \leq 980$
The maximum allowable weight for passengers is 980 pounds.

44. Let y represent the amount the whale could grow.
$12 + y \leq 19$
$12 - 12 + y \leq 19 - 12$
$y \leq 7$
The whale could grow as much as 7 feet.

45. Let x represent the amount that the winds must increase.
$42 + x \geq 74$
$42 - 42 + x \geq 74 - 42$
$x \geq 32$
The winds must increase at least 32 mph to become a hurricane.

46. Let y represent the amount greater than hurricane speed.
$74 + y \geq 110$
$74 - 74 + y \geq 110 - 74$
$y \geq 36$
Wind speeds must be at least 36 mph greater than the slowest hurricane to be a major storm.

47. Always; subtracting x gives $-1 < 0$, which is always true.

48. Solving an inequality is similar to solving an equation because when you add or subtract the same thing on each side, the inequality remains true. Answers should include the following.

- If you subtract 3 blocks from each side, the balance will stay the same.
- Subtracting 3 blocks from each side of the scale is like subtracting 3 from each side of an inequality—the inequality remains true.

49. C; $t \leq 42$

50. C; $14 + t \leq 25$
$14 - 14 + t \leq 25 - 14$
$t \leq 11$
He can spend at most $11.

Page 349 Maintain Your Skills

51. $x - 5 \overset{?}{>} 4$
$9 - 5 \overset{?}{>} 4$
$4 \not> 4$
This sentence is false.

52. $9 + a \le 3$
$9 + (-7) \overset{?}{\le} 3$
$2 \le 3$
This sentence is true.

53. $\frac{x}{2} \ge 8$
$\frac{4}{2} \overset{?}{\ge} 8$
$2 \not\ge 8$
This sentence is false.

54. $6n < -4$
$6(-1) \overset{?}{<} -4$
$-6 < -4$
This sentence is true.

55. Let w = the width.
Let $2w + 3$ = the length.
Perimeter = 2 times length + 2 times width
$24 = 2(2w + 3) + 2w$
$24 = 4w + 6 + 2w$
$24 = 6w + 6$
$24 - 6 = 6w$
$18 = 6w$
$\frac{18}{6} = \frac{6w}{6}$
$3 = w$
Evaluate $2w + 3$ to find the length.
$2w + 3 = 2(3) + 3$
$= 6 + 3$
$= 9$
The dimensions are 3 cm by 9 cm.

56. $P = 2\ell + 2w$
$P = 2(19) + 2(8)$
$P = 38 + 16$
$P = 54$
The perimeter is 54 cm.
$A = \ell w$
$A = 19 \cdot 8$
$A = 152$
The area is 152 cm^2.

57. $4(2 + 8) = 4(2) + 4(8)$
$= 8 + 32$

58. $-2(n + 6) = -2n + (-2)(6)$
$= -2n - 12$

59. $5(x - 3.5) = 5x + 5(-3.5)$
$= 5x - 17.5$

60. $(9 - d)(-3c) = 9(-3c) + (-d)(-3c)$
$= -27c + 3cd$

61. $-15 - (-12) = -15 + 12$
$= -3$

62. $8 - (-5) = 8 + 5$
$= 13$

63. $-9 - 6 = -9 + (-6)$
$= -15$

64. $27 - 45 = 27 + (-45)$
$= -18$

65. $-7x = 14$
$\frac{-7x}{-7} = \frac{14}{-7}$
$x = -2$

66. $-3y = -27$
$\frac{-3y}{-3} = \frac{-27}{-3}$
$y = 9$

67. $5x = -20$
$\frac{5x}{5} = \frac{-20}{5}$
$x = -4$

68. $\frac{d}{-3} = -6$
$-3\left(\frac{d}{-3}\right) = -3(-6)$
$d = 18$

69. $\frac{c}{-4} = 12$
$-4\left(\frac{c}{-4}\right) = -4(12)$
$c = -48$

70. $\frac{a}{2} = -8$
$2\left(\frac{a}{2}\right) = 2(-8)$
$a = -16$

7-5 Solving Inequalities by Multiplying or Dividing

Page 350 How are inequalities used in studying space?

a. Yes; $150 > 25$

b. Earth; $300 > 67$, so $5(300) > 5(67)$.

Page 353 Check for Understanding

1. Multiply each side by -12 and reverse the inequality symbol.

2. Sample answer: $3x < 24$

3. Tamika is correct. She divided each side of the inequality by 9. Since 9 is a positive number, she did not reverse the inequality symbol.

4. $2x < 8$
$\frac{2x}{2} < \frac{8}{2}$
$x < 4$
The solution is $x < 4$. You can check this solution by substituting a number less than 4 into the inequality.

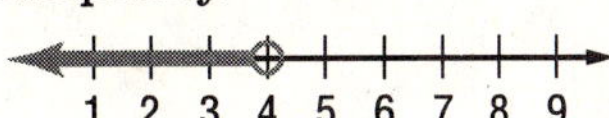

5. $3x \ge -6$
$\frac{3x}{3} \ge \frac{-6}{3}$
$x \ge -2$
The solution is $x \ge -2$. You can check this solution by substituting -2 or a number greater than -2 into the inequality.

−4 −3 −2 −1 0 1 2 3 4

6. 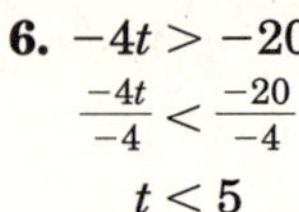

$-4t > -20$

$\frac{-4t}{-4} < \frac{-20}{-4}$

$t < 5$

The solution is $t < 5$. You can check this solution by substituting a number less than 5 into the inequality.

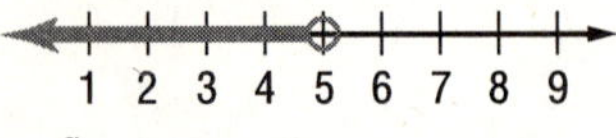

7. $\frac{a}{5} > 10$

$5\left(\frac{a}{5}\right) > 5(10)$

$a > 50$

The solution is $a > 50$. You can check this solution by substituting a number greater than 50 into the inequality.

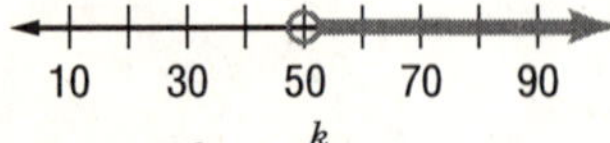

8. $-8 > \frac{k}{-0.4}$

$-8(-0.4) < \left(\frac{k}{-0.4}\right)(-0.4)$

$3.2 < k$ or $k > 3.2$

The solution is $k > 3.2$. You can check this solution by substituting a number greater than 3.2 into the inequality.

3 3.2 3.4 3.6 3.8

9. $\frac{m}{-7} \leq 1.2$

$-7\left(\frac{m}{-7}\right) \geq -7(1.2)$

$m \geq -8.4$

The solution is $m \geq -8.4$. You can check this solution by substituting -8.4 or a number greater than -8.4 into the inequality.

–9 –8.6 –8.2 –7.8 –7.4

10. $-\frac{s}{3} \leq -3.5$

$-3\left(-\frac{s}{3}\right) \geq -3(-3.5)$

$s \geq 10.5$

The solution is $s \geq 10.5$. You can check this solution by substituting 10.5 or a number greater than 10.5 into the inequality.

9 10 11 12 13

11. $36 \geq -\frac{1}{2}y$

$-2(36) \leq -2\left(-\frac{1}{2}y\right)$

$-72 \leq y$ or $y \geq -72$

The solution is $y \geq -72$. You can check this solution by substituting -72 or a number greater than -72 into the inequality.

–80 –76 –72 –68 –64

12. $-273 \geq -13z$

$\frac{-273}{-13} \leq \frac{-13z}{-13}$

$21 \leq z$ or $z \geq 21$

The solution is $z \geq 21$. You can check this solution by substituting 21 or a number greater than 21 into the inequality.

20 22 24 26 28

13. C; Let x represent the number of pizzas.

The average tip times the number of pizzas

1.50 × x

is at least the amount earned in tips.

$\geq$ 20

$1.50x \geq 20$

$\frac{1.50x}{1.50} \geq \frac{20}{1.50}$

$x \geq 13.\overline{33}$

She must deliver at least 14 pizzas.

Pages 353–354 Practice and Apply

14. $4x < 4$

$\frac{4x}{4} < \frac{4}{4}$

$x < 1$

The solution is $x < 1$. You can check this solution by substituting a number less than 1 into the inequality.

–4 –3 –2 –1 0 1 2 3 4

15. $7y > 63$

$\frac{7y}{7} > \frac{63}{7}$

$y > 9$

The solution is $y > 9$. You can check this solution by substituting a number greater than 9 into the inequality.

5 6 7 8 9 10 11 12 13

16. $13a \geq -26$

$\frac{13a}{13} \geq \frac{-26}{13}$

$a \geq -2$

The solution is $a \geq -2$. You can check this solution by substituting -2 or a number greater than -2 into the inequality.

–4 –3 –2 –1 0 1 2 3 4

17. $-15 \leq 5b$

$\frac{-15}{5} \leq \frac{5b}{5}$

$-3 \leq b$ or $b \geq -3$

The solution is $b \geq -3$. You can check this solution by substituting -3 or a number greater than -3 into the inequality.

–4 –3 –2 –1 0 1 2 3 4

18. $144 < 12d$

$\frac{144}{12} < \frac{12d}{12}$

$12 < d$ or $d > 12$

The solution is $d > 12$. You can check this solution by substituting a number greater than 12 into the inequality.

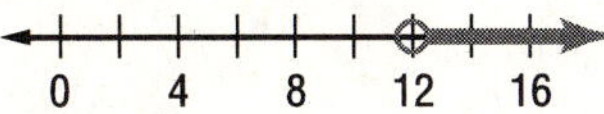

19. $15 \geq 3t$

$\frac{15}{3} \geq \frac{3t}{3}$

$5 \geq t$ or $t \leq 5$

The solution is $t \leq 5$. You can check this solution by substituting 5 or a number less than 5 into the inequality.

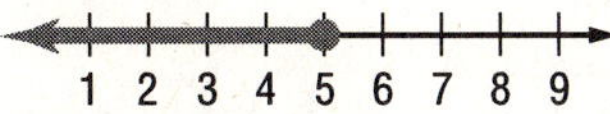

20. $\frac{p}{6} > 5$

$6\left(\frac{p}{6}\right) > 6(5)$

$p > 30$

The solution is $p > 30$. You can check this solution by substituting a number greater than 30 into the inequality.

10 30 50 70 90

21. $7 \geq \frac{h}{14}$

$14(7) \geq 14\left(\frac{h}{14}\right)$

$98 \geq h$ or $h \leq 98$

The solution is $h \leq 98$. You can check this solution by substituting 98 or a number less than 98 into the inequality.

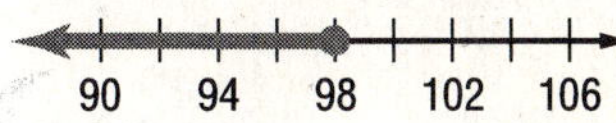

22. $-3m > -33$

$\frac{-3m}{-3} < \frac{-33}{-3}$

$m < 11$

The solution is $m < 11$. You can check this solution by substituting a number less than 11 into the inequality.

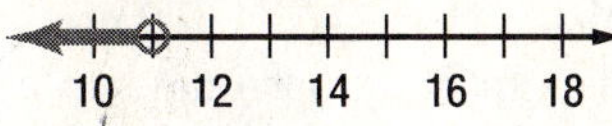

23. $-8z \leq -24$

$\frac{-8z}{-8} \geq \frac{-24}{-8}$

$z \geq 3$

The solution is $z \geq 3$. You can check this solution by substituting 3 or a number greater than 3 into the inequality.

1 2 3 4 5 6 7 8 9

24. $18 > -2g$

$\frac{18}{-2} < \frac{-2g}{-2}$

$-9 < g$ or $g > -9$

The solution is $g > -9$. You can check this solution by substituting a number greater than -9 into the inequality.

−10 −8 −6 −4 −2

25. $-8 \leq -4w$

$\frac{-8}{-4} \geq \frac{-4w}{-4}$

$2 \geq w$ or $w \leq 2$

The solution is $w \leq 2$. You can check this solution by substituting 2 or a number less than 2 into the inequality.

−4 −3 −2 −1 0 1 2 3 4

26. $6 > \frac{x}{-7}$

$-7(6) < -7\left(\frac{x}{-7}\right)$

$-42 < x$ or $x > -42$

The solution is $x > -42$. You can check this solution by substituting a number greater than -42 into the inequality.

−50 −46 −42 −38 −34

27. $\frac{r}{-2} < -2$

$-2\left(\frac{r}{-2}\right) > -2(-2)$

$r > 4$

The solution is $r > 4$. You can check this solution by substituting a number greater than 4 into the inequality.

1 2 3 4 5 6 7 8 9

28. $\frac{y}{-3} < -7$

$-3\left(\frac{y}{-3}\right) > -3(-7)$

$y > 21$

The solution is $y > 21$. You can check this solution by substituting a number greater than 21 into the inequality.

16 18 20 22 24

29. $\frac{k}{-2} < 9$

$-2\left(\frac{k}{-2}\right) > -2(9)$

$k > -18$

The solution is $k > -18$. You can check this solution by substituting a number greater than -18 into the inequality.

−20 −18 −16 −14 −12

30. $-6a > -78$

$\frac{-6a}{-6} < \frac{-78}{-6}$

$a < 13$

The solution is $a < 13$. You can check this solution by substituting a number less than 13 into the inequality.

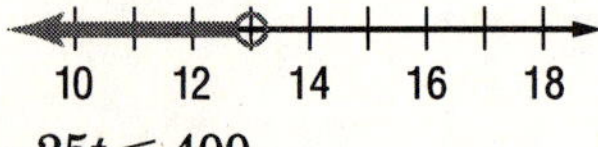

31. $-25t \leq 400$

$\frac{-25t}{-25} \geq \frac{400}{-25}$

$t \geq -16$

The solution is $t \geq -16$. You can check this solution by substituting -16 or a number greater than -16 into the inequality.

−20 −18 −16 −14 −12

32. $\frac{y}{4} \geq 2.4$

$4\left(\frac{y}{4}\right) \geq 4(2.4)$

$y \geq 9.6$

The solution is $y \geq 9.6$. You can check this solution by substituting 9.6 or a number greater than 9.6 into the inequality.

9.0 9.2 9.4 9.6 9.8

33. $\frac{n}{5} \leq 0.8$

$5\left(\frac{n}{5}\right) \leq 5(0.8)$

$n \leq 4$

The solution is $n \leq 4$. You can check this solution by substituting 4 or a number less than 4 into the inequality.

1 2 3 4 5 6 7 8 9

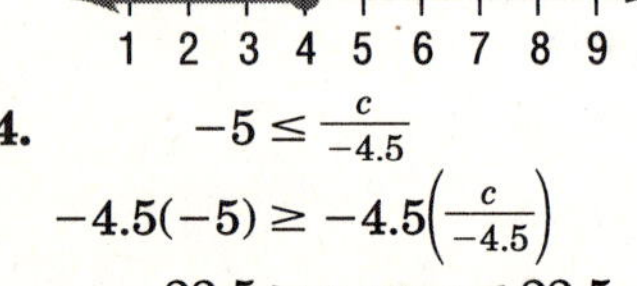

34. $-5 \leq \frac{c}{-4.5}$

$-4.5(-5) \geq -4.5\left(\frac{c}{-4.5}\right)$

$22.5 \geq c$ or $c \leq 22.5$

The solution is $c \leq 22.5$. You can check this solution by substituting 22.5 or a number less than 22.5 into the inequality.

21 22 23 24 25

35. $-19 > \frac{y}{-0.3}$

$-0.3(-19) < -0.3\left(\frac{y}{-0.3}\right)$

$5.7 < y$ or $y > 5.7$

The solution is $y > 5.7$. You can check this solution by substituting a number greater than 5.7 into the inequality.

5 5.2 5.4 5.6 5.8

36. $-\frac{1}{3}x \geq -9$

$-3\left(-\frac{1}{3}x\right) \leq -3(-9)$

$x \leq 27$

The solution is $x \leq 27$. You can check this solution by substituting 27 or a number less than 27 into the inequality.

25 27 29 31 33

37. $-36 < -\frac{1}{2}b$

$-2(-36) > -2\left(-\frac{1}{2}b\right)$

$72 > b$ or $b < 72$

The solution is $b < 72$. You can check this solution by substituting a number less than 72 into the inequality.

70 74 78 82 86

38a. Let s represent the amount left for shoes.

$24 + s < 100$

38b. $24 + s < 100$

$24 - 24 + s < 100 - 24$

$s < 76$

He can spend less than \$76 on shoes.

39a. Let m represent the number of minutes spent swimming.

Use $d = rt$.

rate · time > distance

$40m > 2000$

39b. $40m > 2000$

$\frac{40m}{40} > \frac{2000}{40}$

$m > 50$

She should swim at least 50 minutes.

40. Let n represent the integer.

$n \cdot (-7) < -84$

$\frac{n \cdot (-7)}{-7} > \frac{-84}{-7}$

$n > 12$

Since n cannot equal 12, the smallest integer is 13.

41. Inequalities can be used to compare the weights of objects on different planets. Answers should include the following.

- Comparing the weight of an astronaut in a space suit on Mars to the same astronaut on the moon: $113 > 50$.
- If you multiply or divide the astronaut's weight by the same number, the inequality comparing the weights on different planets would still be true.

42. D; length × width = Area

$x(4) < 36$

$4x < 36$

$\frac{4x}{4} < \frac{36}{4}$

$x < 9$

Since the length must be less than 9, 9 is not a possible length.

43. There are 4 quarts in a gallon, so there are $20 \cdot 4$ or 80 quarts in 20 gallons. Each pitcher holds 2 quarts, so she will need $80 \div 2$ or 40 pitchers.

Page 354 Maintain Your Skills

44.
$$-4 + x > 23$$
$$-4 + 4 + x > 23 + 4$$
$$x > 27$$
The solution is $x > 27$. You can check this solution by substituting a number greater than 27 into the inequality.

45.
$$c + 18 \leq -2$$
$$c + 18 - 18 \leq -2 - 18$$
$$c \leq -20$$
The solution is $c \leq -20$. You can check this solution by substituting -20 or a number less than -20 into the inequality.

46.
$$6 > n - 10$$
$$6 + 10 > n - 10 + 10$$
$$16 > n \text{ or } n < 16$$
The solution is $n < 16$. You can check this solution by substituting a number less than 16 into the inequality.

47. $2n \leq 14$

48. $\frac{1}{8} \cdot \frac{3}{4} = \frac{3}{32}$

49. $-\frac{3}{7} \cdot \frac{5}{9} = -\frac{15}{63}$
$$= -\frac{5}{21}$$

50. $2\frac{1}{2} \cdot \left(-\frac{5}{6}\right) = \frac{5}{2} \cdot \left(-\frac{5}{6}\right)$
$$= -\frac{25}{12} \text{ or } -2\frac{1}{12}$$

51. $\frac{ab}{2} \cdot \frac{4}{bc} = \frac{4ab}{2bc}$
$$= \frac{2a}{c}$$

52.
$$2x + 3 = 9$$
$$2x + 3 - 3 = 9 - 3$$
$$2x = 6$$
$$\frac{2x}{2} = \frac{6}{2}$$
$$x = 3$$

53.
$$5a - 6 = 14$$
$$5a - 6 + 6 = 14 + 6$$
$$5a = 20$$
$$\frac{5a}{5} = \frac{20}{5}$$
$$a = 4$$

54.
$$3n - 8 = -26$$
$$3n - 8 + 8 = -26 + 8$$
$$3n = -18$$
$$\frac{3n}{3} = \frac{-18}{3}$$
$$n = -6$$

55.
$$\frac{t}{3} + 5 = 2$$
$$\frac{t}{3} + 5 - 5 = 2 - 5$$
$$\frac{t}{3} = -3$$
$$3\left(\frac{t}{3}\right) = 3(-3)$$
$$t = -9$$

56.
$$\frac{c}{4} - 1 = 4$$
$$\frac{c}{4} - 1 + 1 = 4 + 1$$
$$\frac{c}{4} = 5$$
$$4\left(\frac{c}{4}\right) = 4(5)$$
$$c = 20$$

57.
$$\frac{d}{2} + 3 = 19$$
$$\frac{d}{2} + 3 - 3 = 19 - 3$$
$$\frac{d}{2} = 16$$
$$2\left(\frac{d}{2}\right) = 2(16)$$
$$d = 32$$

Page 354 Practice Quiz 2

1. $x < -3$

−4 −3 −2 −1 0 1 2 3 4

2. $y \geq 5$

1 2 3 4 5 6 7 8 9

3.
$$a - 26 \leq 14$$
$$a - 26 + 26 \leq 14 + 26$$
$$a \leq 40$$
The solution is $a \leq 40$. You can check this solution by substituting 40 or a number less than 40 into the inequality.

4.
$$46 + k > -8$$
$$46 - 46 + k > -8 - 46$$
$$k > -54$$
The solution is $k > -54$. You can check this solution by substituting a number greater than -54 into the inequality.

5.
$$115 \leq -9 + n$$
$$115 + 9 \leq -9 + 9 + n$$
$$124 \leq n \text{ or } n > 124$$
The solution is $n \geq 124$. You can check this solution by substituting 124 or a number greater than 124 into the inequality.

6. $2.5 > 5r$
$$\frac{2.5}{5} > \frac{5r}{5}$$
$$0.5 > r \text{ or } r < 0.5$$
The solution is $r < 0.5$. You can check this solution by substituting a number less than 0.5 into the inequality.

7.
$$\frac{r}{5} < -45$$
$$5\left(\frac{r}{5}\right) < 5(-45)$$
$$r < -225$$
The solution is $r < -225$. You can check this solution by substituting a number less than -225 into the inequality.

8. $-\frac{s}{8} < -80$

$-8\left(-\frac{s}{8}\right) > -8(-80)$

$s > 640$

The solution is $s > 640$. You can check this solution by substituting a number greater than 640 into the inequality.

9. $-12g \geq -84$

$\frac{-12g}{-12} \leq \frac{-84}{-12}$

$g \leq 7$

The solution is $g \leq 7$. You can check this solution by substituting 7 or a number less than 7 into the inequality.

10. $5w \geq -2$

$\frac{5w}{5} \geq \frac{-2}{5}$

$w \geq -0.4$

The solution is $w \geq -0.4$. You can check this solution by substituting -0.4 or a number greater than -0.4 into the inequality.

7-6 Solving Multi-Step Inequalities

Page 355 How are multi-step inequalities used in backpacking?

a. $3(p + c) < b$

b. $3(5 + c) < 120$

Page 357 Check for Understanding

1. Check the solution by replacing the variable with a number in the solution. If the inequality is true, the solution checks.

2. $2x + 3 < 7$

$2x + 3 - 3 < 7 - 3$

$2x < 4$

$\frac{2x}{2} < \frac{4}{2}$

$x < 2$

3. Jerome is correct. By the Distributive Property, $2(2y + 3) = 4y + 6$, not $4y + 3$.

4. $3x + 4 \leq 31$

$3x + 4 - 4 \leq 31 - 4$

$3x \leq 27$

$\frac{3x}{3} \leq \frac{27}{3}$

$x \leq 9$

The solution is $x \leq 9$. You can check this solution by substituting 9 or a number less than 9 into the inequality.

5 6 7 8 9 10 11 12 13

5. $2n + 5 > 11 - n$

$2n + n + 5 > 11 - n + n$

$3n + 5 > 11$

$3n + 5 - 5 > 11 - 5$

$3n > 6$

$\frac{3n}{3} > \frac{6}{3}$

$n > 2$

The solution is $n > 2$. You can check this solution by substituting a number greater than 2 into the inequality.

1 2 3 4 5 6 7 8 9

6. $y + 1 \geq 4y + 4$

$y - y + 1 \geq 4y - y + 4$

$1 \geq 3y + 4$

$1 - 4 \geq 3y + 4 - 4$

$-3 \geq 3y$

$\frac{-3}{3} \geq \frac{3y}{3}$

$-1 \geq y$ or $y \leq -1$

The solution is $y \leq -1$. You can check this solution by substituting -1 or a number less than -1 into the inequality.

−4 −3 −2 −1 0 1 2 3 4

7. $16 - 2c < 14$

$16 - 16 - 2c < 14 - 16$

$-2c < -2$

$\frac{-2c}{-2} > \frac{-2}{-2}$

$c > 1$

The solution is $c > 1$. You can check this solution by substituting a number greater than 1 into the inequality.

−4 −3 −2 −1 0 1 2 3 4

8. $-6.1n \geq 3.9n + 5$

$-6.1n - 3.9n \geq 3.9n - 3.9n + 5$

$-10n \geq 5$

$\frac{-10n}{-10} \leq \frac{5}{-10}$

$n \leq -\frac{1}{2}$ or $n \leq -0.5$

The solution is $n \leq -0.5$. You can check this solution by substituting -0.5 or a number less than -0.5 into the inequality.

−1 0 1 2 3

9. $-4 \leq \frac{x}{4} - 6$

$-4 + 6 \leq \frac{x}{4} - 6 + 6$

$2 \leq \frac{x}{4}$

$4(2) \leq 4\left(\frac{x}{4}\right)$

$8 \leq x$ or $x \geq 8$

The solution is $x \geq 8$. You can check this solution by substituting 8 or a number greater than 8 into the inequality.

1 2 3 4 5 6 7 8 9

10. $-3(b-1) > 18$
$-3b + 3 > 18$
$-3b + 3 - 3 > 18 - 3$
$-3b > 15$
$\frac{-3b}{-3} < \frac{15}{-3}$
$b < -5$

The solution is $b < -5$. You can check this solution by substituting a number less than -5 into the inequality.

−7 −6 −5 −4 −3 −2 −1 0 1

11. $\frac{1}{2}(2d + 3) < -8$
$\frac{1}{2}(2d) + \frac{1}{2}(3) < -8$
$d + \frac{3}{2} < -8$
$d + \frac{3}{2} - \frac{3}{2} < -8 - \frac{3}{2}$
$d < -\frac{16}{2} - \frac{3}{2}$
$d < -\frac{19}{2}$ or $d < -9\frac{1}{2}$

The solution is $d < -9\frac{1}{2}$. You can check this solution by substituting a number less than $-9\frac{1}{2}$ into the inequality.

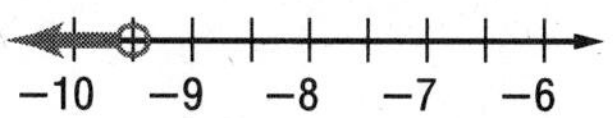

12. Let x represent the number of minutes Dante can talk.
$10 + 0.05x < 30$
$10 - 10 + 0.05x < 30 - 10$
$0.05x < 20$
$\frac{0.05x}{0.05} < \frac{20}{0.05}$
$x < 400$

Since he must talk less than 400 minutes, he can talk at most 399 minutes.

Pages 357–359 Practice and Apply

13. $2x + 8 > 24$
$2x + 8 - 8 > 24 - 8$
$2x > 16$
$\frac{2x}{2} > \frac{16}{2}$
$x > 8$

The solution is $x > 8$. You can check this solution by substituting a number greater than 8 into the inequality.

1 2 3 4 5 6 7 8 9

14. $3y - 1 \le 5$
$3y - 1 + 1 \le 5 + 1$
$3y \le 6$
$\frac{3y}{3} \le \frac{6}{3}$
$y \le 2$

The solution is $y \le 2$. You can check this solution by substituting 2 or a number less than 2 into the inequality.

−4 −3 −2 −1 0 1 2 3 4

15. $3 + 4c > -13$
$3 - 3 + 4c > -13 - 3$
$4c > -16$
$\frac{4c}{4} > \frac{-16}{4}$
$c > -4$

The solution is $c > -4$. You can check this solution by substituting a number greater than -4 into the inequality.

−4 −3 −2 −1 0 1 2 3 4

16. $9 + 2p \le 15$
$9 - 9 + 2p \le 15 - 9$
$2p \le 6$
$\frac{2p}{2} \le \frac{6}{2}$
$p \le 3$

The solution is $p \le 3$. You can check this solution by substituting 3 or a number less than 3 into the inequality.

−4 −3 −2 −1 0 1 2 3 4

17. $3x - 2 > 10 - x$
$3x + x - 2 > 10 - x + x$
$4x - 2 > 10$
$4x - 2 + 2 > 10 + 2$
$4x > 12$
$\frac{4x}{4} > \frac{12}{4}$
$x > 3$

The solution is $x > 3$. You can check this solution by substituting a number greater than 3 into the inequality.

−4 −3 −2 −1 0 1 2 3 4

18. $c - 1 < 3c + 5$
$c - c - 1 < 3c - c + 5$
$-1 < 2c + 5$
$-1 - 5 < 2c + 5 - 5$
$-6 < 2c$
$\frac{-6}{2} < \frac{2c}{2}$
$-3 < c$ or $c > -3$

The solution is $c > -3$. You can check this solution by substituting a number greater than -3 into the inequality.

−4 −3 −2 −1 0 1 2 3 4

19. $4 - 3k \le 19$
$4 - 4 - 3k \le 19 - 4$
$-3k \le 15$
$\frac{-3k}{-3} \ge \frac{15}{-3}$
$k \ge -5$

The solution is $k \ge -5$. You can check this solution by substituting -5 or a number greater than -5 into the inequality.

−7 −6 −5 −4 −3 −2 −1 0 1

20.
$$16 - 4n > 20$$
$$16 - 16 - 4n > 20 - 16$$
$$-4n > 4$$
$$\frac{-4n}{-4} < \frac{4}{-4}$$
$$n < -1$$
The solution is $n < -1$. You can check this solution by substituting a number less than -1 into the inequality.

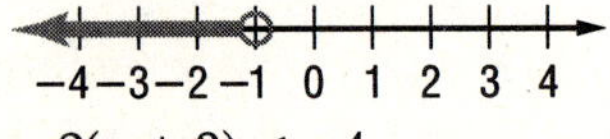

21.
$$2(n + 3) < -4$$
$$2n + 6 < -4$$
$$2n + 6 - 6 < -4 - 6$$
$$2n < -10$$
$$\frac{2n}{2} < \frac{-10}{2}$$
$$n < -5$$
The solution is $n < -5$. You can check this solution by substituting a number less than -5 into the inequality.

−7 −6 −5 −4 −3 −2 −1 0 1

22.
$$2(d + 1) > 16$$
$$2d + 2 > 16$$
$$2d + 2 - 2 > 16 - 2$$
$$2d > 14$$
$$\frac{2d}{2} > \frac{14}{2}$$
$$d > 7$$
The solution is $d > 7$. You can check this solution by substituting a number greater than 7 into the inequality.

1 2 3 4 5 6 7 8 9

23.
$$8 + 3b \le 2(9 - b)$$
$$8 + 3b \le 18 - 2b$$
$$8 + 3b + 2b \le 18 - 2b + 2b$$
$$8 + 5b \le 18$$
$$8 - 8 + 5b \le 18 - 8$$
$$5b \le 10$$
$$\frac{5b}{5} \le \frac{10}{5}$$
$$b \le 2$$
The solution is $b \le 2$. You can check this solution by substituting 2 or a number less than 2 into the inequality.

−4 −3 −2 −1 0 1 2 3 4

24.
$$\frac{m}{2} + 9 \ge 5$$
$$\frac{m}{2} + 9 - 9 \ge 5 - 9$$
$$\frac{m}{2} \ge -4$$
$$2\left(\frac{m}{2}\right) \ge 2(-4)$$
$$m \ge -8$$
The solution is $m \ge -8$. You can check this solution by substituting -8 or a number greater than -8 into the inequality.

−10 −8 −6 −4 −2

25.
$$2 + 0.3y \ge 11$$
$$2 - 2 + 0.3y \ge 11 - 2$$
$$0.3y \ge 9$$
$$\frac{0.3y}{0.3} \ge \frac{9}{0.3}$$
$$y \ge 30$$
The solution is $y \ge 30$. You can check this solution by substituting 30 or a number greater than 30 into the inequality.

10 30 50 70 90

26.
$$0.5a - 1.4 \le 2.1$$
$$0.5a - 1.4 + 1.4 \le 2.1 + 1.4$$
$$0.5a \le 3.5$$
$$\frac{0.5a}{0.5} \le \frac{3.5}{0.5}$$
$$a \le 7$$
The solution is $a \le 7$. You can check this solution by substituting 7 or a number less than 7 into the inequality.

1 2 3 4 5 6 7 8 9

27.
$$\frac{1}{2}(6 - c) > 5$$
$$\frac{1}{2}(6) - \frac{1}{2}(c) > 5$$
$$3 - \frac{1}{2}c > 5$$
$$3 - 3 - \frac{1}{2}c > 5 - 3$$
$$-\frac{1}{2}c > 2$$
$$-2\left(-\frac{1}{2}c\right) < -2(2)$$
$$c < -4$$
The solution is $c < -4$. You can check this solution by substituting a number less than -4 into the inequality.

−10 −8 −6 −4 −2

28.
$$\frac{2}{3}(9 - x) < 3$$
$$\frac{2}{3}(9) - \frac{2}{3}(x) < 3$$
$$\frac{18}{3} - \frac{2}{3}x < 3$$
$$6 - \frac{2}{3}x < 3$$
$$6 - 6 - \frac{2}{3}x < 3 - 6$$
$$-\frac{2}{3}x < -3$$
$$-\frac{3}{2}\left(-\frac{2}{3}x\right) > -\frac{3}{2}(-3)$$
$$x > \frac{9}{2} \text{ or } x > 4.5$$
The solution is $x > 4.5$. You can check this solution by substituting a number greater than 4.5 into the inequality.

1 2 3 4 5 6 7 8 9

29. Let n represent the number or numbers.

$$4n - 6 > 2n + 8$$
$$4n - 2n - 6 > 2n - 2n + 8$$
$$2n - 6 > 8$$
$$2n - 6 + 6 > 8 + 6$$
$$2n > 14$$
$$\frac{2n}{2} > \frac{14}{2}$$
$$n > 7$$

Any number greater than 7 is a solution.

30. Let n represent the number or numbers.

$$\frac{1}{2}(n + 6) < 25$$
$$\frac{1}{2}n + \frac{1}{2}(6) < 25$$
$$\frac{1}{2}n + 3 < 25$$
$$\frac{1}{2}n + 3 - 3 < 25 - 3$$
$$\frac{1}{2}n < 22$$
$$2\left(\frac{1}{2}n\right) < 2(22)$$
$$n < 44$$

Any number less than 44 is a solution.

31.

$$1.3n + 6.7 \geq 3.1n - 1.4$$
$$1.3n - 3.1n + 6.7 \geq 3.1n - 3.1n - 1.4$$
$$-1.8n + 6.7 \geq -1.4$$
$$-1.8n + 6.7 - 6.7 \geq -1.4 - 6.7$$
$$-1.8n \geq -8.1$$
$$\frac{-1.8n}{-1.8} \leq \frac{-8.1}{-1.8}$$
$$n \leq 4.5$$

The solution is $n \leq 4.5$. You can check this solution by substituting 4.5 or a number less than 4.5 into the inequality.

1 2 3 4 5 6 7 8 9

32.

$$-5a + 3 > 3a + 23$$
$$-5a - 3a + 3 > 3a - 3a + 23$$
$$-8a + 3 > 23$$
$$-8a + 3 - 3 > 23 - 3$$
$$-8a > 20$$
$$\frac{-8a}{-8} < \frac{20}{-8}$$
$$a < -2.5$$

The solution is $a < -2.5$. You can check this solution by substituting a number less than -2.5 into the inequality.

−4 −3 −2 −1 0 1 2 3 4

33.

$$-5(t + 4) \geq 3(t - 4)$$
$$-5t - 20 \geq 3t - 12$$
$$-5t - 3t - 20 \geq 3t - 3t - 12$$
$$-8t - 20 \geq -12$$
$$-8t - 20 + 20 \geq -12 + 20$$
$$-8t \geq 8$$
$$\frac{-8t}{-8} \leq \frac{8}{-8}$$
$$t \leq -1$$

The solution is $t \leq -1$. You can check this solution by substituting -1 or a number less than -1 into the inequality.

−4 −3 −2 −1 0 1 2 3 4

34.

$$8x - (x - 5) > x + 17$$
$$8x - x + 5 > x + 17$$
$$7x + 5 > x + 17$$
$$7x - x + 5 > x - x + 17$$
$$6x + 5 > 17$$
$$6x + 5 - 5 > 17 - 5$$
$$6x > 12$$
$$\frac{6x}{6} > \frac{12}{6}$$
$$x > 2$$

The solution is $x > 2$. You can check this solution by substituting a number greater than 2 into the inequality.

1 2 3 4 5 6 7 8 9

35.

$$\frac{c + 8}{4} < \frac{5 - c}{9}$$
$$36\left(\frac{c + 8}{4}\right) < 36\left(\frac{5 - c}{9}\right)$$
$$9(c + 8) < 4(5 - c)$$
$$9c + 72 < 20 - 4c$$
$$9c + 4c + 72 < 20 - 4c + 4c$$
$$13c + 72 < 20$$
$$13c + 72 - 72 < 20 - 72$$
$$13c < -52$$
$$\frac{13c}{13} < -\frac{52}{13}$$
$$c < -4$$

The solution is $c < -4$. You can check this solution by substituting a number less than -4 into the inequality.

−10 −8 −6 −4 −2

36.

$$\frac{2(n + 1)}{7} \geq \frac{n + 4}{5}$$
$$35\left[\frac{2(n + 1)}{7}\right] \geq 35\left(\frac{n + 4}{5}\right)$$
$$5 \cdot 2(n + 1) \geq 7(n + 4)$$
$$10n + 10 \geq 7n + 28$$
$$10n - 7n + 10 \geq 7n - 7n + 28$$
$$3n + 10 \geq 28$$
$$3n + 10 - 10 \geq 28 - 10$$
$$3n \geq 18$$
$$\frac{3n}{3} \geq \frac{18}{3}$$
$$n \geq 6$$

The solution is $n \geq 6$. You can check this solution by substituting 6 or a number greater than 6 into the inequality.

1 2 3 4 5 6 7 8 9

37. Let c represent the number of candy bars.

$$0.55c + 0.35 \leq 2$$
$$0.55c + 0.35 - 0.35 \leq 2 - 0.35$$
$$0.55c \leq 1.65$$
$$\frac{0.55c}{0.55} \leq \frac{1.65}{0.55}$$
$$c \leq 3$$

You can buy at most 3 candy bars.

38. Let p represent the number of points Nate can get on the fifth test.

$$\frac{\text{total of points}}{\text{number of tests}} \geq \text{average score}$$
$$\frac{85 + 91 + 89 + 93 + p}{5} \geq 90$$
$$\frac{358 + p}{5} \geq 90$$
$$5\left(\frac{358 + p}{5}\right) \geq 5(90)$$
$$358 + p \geq 450$$
$$358 - 358 + p \geq 450 - 358$$
$$p \geq 92$$

Nate must score at least 92 points.

39. Let s represent the number of subscriptions.

$$2s + 10 \geq 40$$
$$2s + 10 - 10 \geq 40 - 10$$
$$2s \geq 30$$
$$\frac{2s}{2} \geq \frac{30}{2}$$
$$s \geq 15$$

You must sell at least 15 subscriptions.

40. Let s represent the monthly sales.

$$1500 + 0.04s \geq 5000$$
$$1500 - 1500 + 0.04s \geq 5000 - 1500$$
$$0.04s \geq 3500$$
$$\frac{0.04s}{0.04} \geq \frac{3500}{0.04}$$
$$s \geq 87{,}500$$

The agent must have at least \$87,500 in monthly sales.

41.
$$30 + 0.05x > 20 + 0.10x$$
$$30 + 0.05x - 0.05x > 20 + 0.10x - 0.05x$$
$$30 > 20 + 0.05x$$
$$30 - 20 > 20 - 20 + 0.05x$$
$$10 > 0.05x$$
$$\frac{10}{0.05} > \frac{0.05x}{0.05}$$
$$200 > x \text{ or } x < 200$$

Sample explanation: The inequality finds at what mileage Able's charge is greater than Baker's charge.

42. Let t represent the number of hours spent on the trail. Use rate · time = distance.

$$3t \geq 18$$
$$\frac{3t}{3} \geq \frac{18}{3}$$
$$t \geq 6 \text{ hours walking}$$

Since you spend 1 hour for lunch, you should expect to spend at least 7 hours on the trail.

43. Let x represent the number of minutes she uses the phone.

Cost for FoneCom > Cost for Best Phone

$$15 + 0.10x > 20 + 0.05x$$
$$15 + 0.10x - 0.05x > 20 + 0.05x - 0.05x$$
$$15 + 0.05x > 20$$
$$15 - 15 + 0.05x > 20 - 15$$
$$0.05x > 5$$
$$\frac{0.05x}{0.05} > \frac{5}{0.05}$$
$$x > 100$$

Miko uses her phone more than 100 minutes each month.

44. Let x represent the number of programs the Club must sell. Use Revenue − Cost = Profit.

Revenue = amount from selling programs
= $1x$ or x

Cost = total cost to make 400 programs
= 60 + 0.20(400)
= 60 + 80
= 140

$$x - 140 \geq 200$$
$$x - 140 + 140 \geq 200 + 140$$
$$x \geq 340$$

The Club must sell at least 340 programs.

45.
$$10 - 2|k| > 4$$
$$10 - 10 - 2|k| > 4 - 10$$
$$-2|k| > -6$$
$$\frac{-2|k|}{-2} < \frac{-6}{-2}$$
$$|k| < 3$$

$k < 3$ and $k > -3$. Since k is an integer, $k = \{-2, -1, 0, 1, 2\}$.

46. Multi-step inequalities are used to find the amount of weight a person can carry in proportion to that person's body weight. Answers should include the following.

- Multi-step inequalities are inequalities that involve more than one operation.
- $c < 35$

47. D; $5 + 2n < 10$

48. D; Let s represent the score on her sixth test.

$$\frac{\text{total of scores}}{\text{number of scores}} = \text{average score}$$
$$\frac{85 + 84 + 90 + 95 + 88 + s}{6} > 88$$
$$\frac{442 + s}{6} > 88$$
$$6\left(\frac{442 + s}{6}\right) > 6(88)$$
$$442 + s > 528$$
$$442 - 442 + s > 528 - 442$$
$$s > 86$$

49.
$$-10 < 3x + 5 < 8$$
$$-10 - 5 < 3x + 5 - 5 < 8 - 5$$
$$-15 < 3x < 3$$
$$\frac{-15}{3} < \frac{3x}{3} < \frac{3}{3}$$
$$-5 < x < 1$$

Page 359 Maintain Your Skills

50.
$$20 < -9 + k$$
$$20 + 9 < -9 + 9 + k$$
$$29 < k \text{ or } k > 29$$

The solution is $k > 29$. You can check this solution by substituting a number greater than 29 into the inequality.

51.
$$22 \leq -15 + y$$
$$22 + 15 \leq -15 + 15 + y$$
$$37 \leq y \text{ or } y \geq 37$$

The solution is $y \geq 37$. You can check this solution by substituting 37 or a number greater than 37 into the inequality.

52. $6x < -27$

$\frac{6x}{6} < \frac{-27}{6}$

$x < -4.5$

The solution is $x < -4.5$. You can check this solution by substituting a number less than -4.5 into the inequality.

53. $-5n \geq -25$

$\frac{-5n}{-5} \leq \frac{-25}{-5}$

$n \leq 5$

The solution is $n \leq 5$. You can check this solution by substituting 5 or a number less than 5 into the inequality.

54. $\frac{n}{-4} \leq -11$

$-4\left(\frac{n}{-4}\right) \geq -4(-11)$

$n \geq 44$

The solution is $n \geq 44$. You can check this solution by substituting 44 or a number greater than 44 into the inequality.

55. $\frac{a}{-3} > 6.2$

$-3\left(\frac{a}{-3}\right) < -3(6.2)$

$a < -18.6$

The solution is $a < -18.6$. You can check this solution by substituting a number less than -18.6 into the inequality.

56. Let p represent the percent.

$\frac{12}{20} = \frac{p}{100}$

$12 \cdot 100 = p \cdot 20$

$1200 = 20p$

$\frac{1200}{20} = \frac{20p}{20}$

$60 = p$

60% of the class are boys.

57. $\frac{1}{200} = 0.005$

$= 0.5\%$

58. $\frac{5 \text{ dollars}}{2 \text{ loaves}} = \frac{2.5 \text{ dollars}}{1 \text{ loaf}}$ (÷2, ÷2)

The unit rate is \$2.50 a loaf.

59. $\frac{200 \text{ miles}}{12 \text{ gallons}} = \frac{16.\overline{6} \text{ miles}}{1 \text{ gallon}}$ (÷12, ÷12)

The unit rate is $16.\overline{6}$ mpg.

60. $\frac{24 \text{ meters}}{4 \text{ seconds}} = \frac{6 \text{ meters}}{1 \text{ second}}$ (÷4, ÷4)

The unit rate is 6 meters per second.

61. $\frac{11.25 \text{ dollars}}{9 \text{ issues}} = \frac{1.25 \text{ dollars}}{1 \text{ issue}}$ (÷9, ÷9)

The unit rate is \$1.25 an issue.

62.

$$\begin{aligned} P &= 2(\ell + w) \\ 49.6 &= 2(18.4 + w) \\ 49.6 &= 36.8 + 2w \\ 49.6 - 36.8 &= 36.8 - 36.8 + 2w \\ 12.8 &= 2w \\ \frac{12.8}{2} &= \frac{2w}{2} \\ 6.4 &= w \end{aligned}$$

The width is 6.4 feet.

63.

$$\begin{aligned} A &= \ell w \\ 30.6 &= \ell \cdot (5.1) \\ \frac{30.6}{5.1} &= \frac{\ell \cdot (5.1)}{5.1} \\ 6 &= \ell \end{aligned}$$

The length is 6 meters.

Chapter 7 Study Guide and Review

Page 360 Vocabulary and Concept Check

1. true
2. true
3. false; identity
4. false; Multiplication
5. false; inequality
6. true
7. true
8. true
9. false; is geater than or equal to
10. true

Page 360–362 Lesson-by-Lesson Review

Exercises 11–15 For checks, see students' work.

11.

$$\begin{aligned} 2a + 9 &= 5a \\ 2a - 2a + 9 &= 5a - 2a \\ 9 &= 3a \\ 3 &= a \end{aligned}$$

The solution is 3.

12.

$$\begin{aligned} x - 4 &= 3x \\ x - x - 4 &= 3x - x \\ -4 &= 2x \\ -2 &= x \end{aligned}$$

The solution is -2.

13.

$$\begin{aligned} 3y - 8 &= y \\ 3y - 3y - 8 &= y - 3y \\ -8 &= -2y \\ 4 &= y \end{aligned}$$

The solution is 4.

14.

$$\begin{aligned} 19t &= 26 + 6t \\ 19t - 6t &= 26 + 6t - 6t \\ 13t &= 26 \\ \frac{13t}{13} &= \frac{26}{13} \\ t &= 2 \end{aligned}$$

The solution is 2.

15. $2 + 7n = 8 + n$
$2 + 7n - n = 8 + n - n$
$2 + 6n = 8$
$2 - 2 + 6n = 8 - 2$
$6n = 6$
$n = 1$
The solution is 1.

16. $5 + 6t = 10t - 7$
$5 + 6t - 10t = 10t - 10t - 7$
$5 - 4t = -7$
$5 - 5 - 4t = -7 - 5$
$-4t = -12$
$t = 3$
The solution is 3.

17. $-r + 4.2 = 8.8r + 14$
$-r - 8.8r + 4.2 = 8.8r - 8.8r + 14$
$-9.8r + 4.2 = 14$
$-9.8r + 4.2 - 4.2 = 14 - 4.2$
$-9.8r = 9.8$
$r = -1$
The solution is −1.

18. $12 + 1.5x = 9x$
$12 + 1.5x - 1.5x = 9x - 1.5x$
$12 = 7.5x$
$\frac{12}{7.5} = \frac{7.5x}{7.5}$
$1.6 = x$
The solution is 1.6.

19. $5b - 1 = 2.5b - 4$
$5b - 2.5b - 1 = 2.5b - 2.5b - 4$
$2.5b - 1 = -4$
$2.5b - 1 + 1 = -4 + 1$
$2.5b = -3$
$\frac{2.5b}{2.5} = \frac{-3}{2.5}$
$b = -1.2$
The solution is −1.2.

20. $4(k + 1) = 16$
$4k + 4 = 16$
$4k + 4 - 4 = 16 - 4$
$4k = 12$
$k = 3$

21. $2(n - 5) = 8$
$2n - 10 = 8$
$2n - 10 + 10 = 8 + 10$
$2n = 18$
$n = 9$

22. $11 + 2q = 2(q + 4)$
$11 + 2q = 2q + 8$
$11 + 2q - 2q = 2q - 2q + 8$
$11 = 8$
The sentence 11 = 8 is never true, so the solution is ∅.

23. $\frac{1}{2}(t + 8) = \frac{3}{4}t$
$\frac{1}{2}t + 4 = \frac{3}{4}t$
$\frac{1}{2}t - \frac{1}{2}t + 4 = \frac{3}{4}t - \frac{1}{2}t$
$4 = \frac{3}{4}t - \frac{2}{4}t$
$4 = \frac{1}{4}t$
$4(4) = 4\left(\frac{1}{4}t\right)$
$16 = t$

24. $4(x + 2.5) = 3(7 + x)$
$4x + 4(2.5) = 21 + 3x$
$4x + 10 = 21 + 3x$
$4x - 3x + 10 = 21 + 3x - 3x$
$x + 10 = 21$
$x + 10 - 10 = 21 - 10$
$x = 11$

25. $3(x + 1) - 5 = 3x - 2$
$3x + 3 - 5 = 3x - 2$
$3x - 2 = 3x - 2$
$3x - 2 + 2 = 3x - 2 + 2$
$3x = 3x$
$x = x$
The sentence $x = x$ is always true, so the solution is all numbers.

26. $x + 4 > 9$
$12 + 4 \overset{?}{>} 9$
$16 > 9$ ✓
This sentence is true.

27. $15 \le 5n$
$15 \overset{?}{\le} 5(3)$
$15 \le 15$ ✓
This sentence is true.

28. $3n + 1 \ge 14$
$3(4) + 1 \overset{?}{\ge} 14$
$12 + 1 \overset{?}{\ge} 14$
$13 \ngeq 14$
This sentence is false.

29. $b - 9 \ge 8$
$b - 9 + 9 \ge 8 + 9$
$b \ge 17$

14 16 18 20 22 24

30. $x + 4.8 \le 2$
$x + 4.8 - 4.8 \le 2 - 4.8$
$x \le -2.8$

−3 −2.6 −2.2 −1.8 −1.4 −1

31. $t + \frac{1}{2} < 4$
$t + \frac{1}{2} - \frac{1}{2} < 4 - \frac{1}{2}$
$t < \frac{8}{2} - \frac{1}{2}$
$t < \frac{7}{2}$ or $t < 3\frac{1}{2}$

0 1 2 3 4 5

32. $\frac{n}{4} < 6$
$4\left(\frac{n}{4}\right) < 4(6)$
$n < 24$

20 22 24 26 28

33. $\frac{k}{1.7} \le 3$
$1.7\left(\frac{k}{1.7}\right) \le 1.7(3)$
$k \le 5.1$

−1 0 1 2 3 4 5 6 7 8 9

34. $0.5x > 3.2$

$$\frac{0.5x}{0.5} > \frac{3.2}{0.5}$$

$$x > 6.4$$

2 4 6 8 10 12

35. $-56 \geq 8y$

$$\frac{-56}{8} \geq \frac{8y}{8}$$

$$-7 \geq y$$

−12 −10 −8 −6 −4

36. $9 > \frac{x}{-4}$

$$-4(9) < -4\left(\frac{x}{-4}\right)$$

$$-36 < x \text{ or } x > -36$$

−40 −38 −36 −34 −32

37. $-\frac{5}{6}a \leq 2$

$$-\frac{6}{5}\left(-\frac{5}{6}a\right) \geq -\frac{6}{5}(2)$$

$$a \geq -\frac{12}{5} \text{ or } -2\frac{2}{5}$$

−4 −2 0 2 4 6

38. $2x - 3 > 19$

$$2x - 3 + 3 > 19 + 3$$

$$2x > 22$$

$$\frac{2x}{2} > \frac{22}{2}$$

$$x > 11$$

39. $5n + 4 \leq 24$

$$5n + 4 - 4 \leq 24 - 4$$

$$5n \leq 20$$

$$\frac{5n}{5} \leq \frac{20}{5}$$

$$n \leq 4$$

40. $6 \geq \frac{r}{7} + 1$

$$6 - 1 \geq \frac{r}{7} + 1 - 1$$

$$5 \geq \frac{r}{7}$$

$$7(5) \geq 7\left(\frac{r}{7}\right)$$

$$35 \geq r \text{ or } r \leq 35$$

41. $\frac{t}{-2} + 15 < 21$

$$\frac{t}{-2} + 15 - 15 < 21 - 15$$

$$\frac{t}{-2} < 6$$

$$-2\left(\frac{t}{-2}\right) > -2(6)$$

$$t > -12$$

42. $3(a + 8.4) > 30$

$$3a + 25.2 > 30$$

$$3a + 25.2 - 25.2 > 30 - 25.2$$

$$3a > 4.8$$

$$\frac{3a}{3} > \frac{4.8}{3}$$

$$a > 1.6$$

43. $\frac{1}{4} + 2b < 13 + 5b$

$$\frac{1}{4} + 2b - 5b < 13 + 5b - 5b$$

$$\frac{1}{4} - 3b < 13$$

$$\frac{1}{4} - \frac{1}{4} - 3b < 13 - \frac{1}{4}$$

$$-3b < \frac{52}{4} - \frac{1}{4}$$

$$-3b < \frac{51}{4}$$

$$-\frac{1}{3}(-3b) > -\frac{1}{3}\left(\frac{51}{4}\right)$$

$$b > -\frac{17}{4} \text{ or } b > -4\frac{1}{4}$$

Chapter 7 Practice Test

Page 363

1. An open circle is used when the inequality symbol is $>$ or $<$; a closed circle is used when the inequality symbol is $\leq$ or $\geq$.
2. When an inequality is multiplied or divided by a negative number the direction of the inequality symbol is reversed.
3. $7x - 3 = 10x$

$$7x - 7x - 3 = 10x - 7x$$

$$-3 = 3x$$

$$-1 = x$$

Check: $7x - 3 = 10x$

$$7(-1) - 3 \stackrel{?}{=} 10(-1)$$

$$-7 - 3 \stackrel{?}{=} -10$$

$$-10 = -10 \checkmark$$

The solution is −1.

4. $p - 9 = 4p$

$$p - p - 9 = 4p - p$$

$$-9 = 3p$$

$$-3 = p$$

Check: $p - 9 = 4p$

$$-3 - 9 \stackrel{?}{=} 4(-3)$$

$$-12 = -12 \checkmark$$

The solution is −3.

5. $2.3n - 8 = 1.2n + 3$

$$2.3n - 1.2n - 8 = 1.2n - 1.2n + 3$$

$$1.1n - 8 = 3$$

$$1.1n - 8 + 8 = 3 + 8$$

$$1.1n = 11$$

$$\frac{1.1n}{1.1} = \frac{11}{1.1}$$

$$n = 10$$

Check: $2.3n - 8 = 1.2n + 3$

$$2.3(10) - 8 \stackrel{?}{=} 1.2(10) + 3$$

$$23 - 8 \stackrel{?}{=} 12 + 3$$

$$15 = 15 \checkmark$$

The solution is 10.

6. $\frac{3}{8}y - 5 = \frac{5}{8}y - 3$

$\frac{3}{8}y - \frac{5}{8}y - 5 = \frac{5}{8}y - \frac{5}{8}y - 3$

$-\frac{2}{8}y - 5 = -3$

$-\frac{1}{4}y - 5 + 5 = -3 + 5$

$-\frac{1}{4}y = 2$

$-4\left(-\frac{1}{4}y\right) = -4(2)$

$y = -8$

Check: $\frac{3}{8}y - 5 = \frac{5}{8}y - 3$

$\frac{3}{8}(-8) - 5 \stackrel{?}{=} \frac{5}{8}(-8) - 3$

$-3 - 5 \stackrel{?}{=} -5 - 3$

$-8 = -8$ ✓

The solution is -8.

7. $6 + 2(x - 4) = 2(x - 1)$

$6 + 2x - 8 = 2x - 2$

$2x - 2 = 2x - 2$

$2x - 2 + 2 = 2x - 2 + 2$

$2x = 2x$

$x = x$

The sentence $x = x$ is always true, so the solution is all numbers.

8. $2(6 - 5d) = 8$

$12 - 10d = 8$

$12 - 12 - 10d = 8 - 12$

$-10d = -4$

$\frac{-10d}{-10} = \frac{-4}{-10}$

$d = 0.4$

Check: $2(6 - 5d) = 8$

$2(6 - 5(0.4)) \stackrel{?}{=} 8$

$2(6 - 2) \stackrel{?}{=} 8$

$2(4) \stackrel{?}{=} 8$

$8 = 8$ ✓

The solution is 0.4.

9. $8(2x - 9) = 4(5 + 4x)$

$16x - 72 = 20 + 16x$

$16x - 16x - 72 = 20 + 16x - 16x$

$-72 = 20$

The sentence $-72 = 20$ is never true, so the solution is $\varnothing$.

10. $4(a + 3) = 20$

$4a + 12 = 20$

$4a + 12 - 12 = 20 - 12$

$4a = 8$

$a = 2$

Check: $4(a + 3) = 20$

$4(2 + 3) \stackrel{?}{=} 20$

$4(5) \stackrel{?}{=} 20$

$20 = 20$ ✓

The solution is 2.

11. $\frac{1}{3}(9b + 1) = b - 1$

$\frac{1}{3}(9b) + \frac{1}{3}(1) = b - 1$

$3b + \frac{1}{3} = b - 1$

$3b - b + \frac{1}{3} = b - b - 1$

$2b + \frac{1}{3} = -1$

$2b + \frac{1}{3} - \frac{1}{3} = -1 - \frac{1}{3}$

$2b = -\frac{3}{3} - \frac{1}{3}$

$2b = -\frac{4}{3}$

$\frac{1}{2}(2b) = \frac{1}{2}\left(-\frac{4}{3}\right)$

$b = -\frac{2}{3}$

Check: $\frac{1}{3}(9b + 1) = b - 1$

$\frac{1}{3}\left(9\left(-\frac{2}{3}\right) + 1\right) \stackrel{?}{=} -\frac{2}{3} - 1$

$\frac{1}{3}(-6 + 1) \stackrel{?}{=} -\frac{2}{3} - \frac{3}{3}$

$\frac{1}{3}(-5) \stackrel{?}{=} -\frac{5}{3}$

$-\frac{5}{3} = -\frac{5}{3}$ ✓

The solution is $-\frac{2}{3}$.

12. Let x represent the number.

$3x + 8 = x - 4$

$3x - x + 8 = x - x - 4$

$2x + 8 = -4$

$2x + 8 - 8 = -4 - 8$

$2x = -12$

$x = -6$

The number is -6.

13. Let x represent the number.

$x \cdot 5 = x + 12$

$5x = x + 12$

$5x - x = x - x + 12$

$4x = 12$

$x = 3$

The number is 3.

14. Perimeter = twice the length + twice the width

$22 = 2(2w + 3.5) + 2w$

$22 = 4w + 7 + 2w$

$22 = 6w + 7$

$22 - 7 = 6w + 7 - 7$

$15 = 6w$

$\frac{15}{6} = \frac{6w}{6}$

$2.5 = w$

Evaluate $2w + 3.5$ to find the length.

$2w + 3.5 = 2(2.5) + 3.5$

$= 5 + 3.5$

$= 8.5$

The dimensions are 2.5 ft by 8.5 ft.

15. Let s represent the number of shirts.

$4s \geq 120$

16. $x > 4$

17. $x \leq -1$

18. $-4 \geq p - 2$

$-4 + 2 \geq p - 2 + 2$

$-2 \geq p$ or $p \leq -2$

The solution is $p \leq -2$. You can check this solution by substituting -2 or a number less than -2 into the inequality.

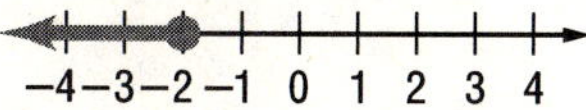

19. $3x \geq 15$

$\frac{3x}{3} \geq \frac{15}{3}$

$x \geq 5$

The solution is $x \geq 5$. You can check this solution by substituting 5 or a number greater than 5 into the inequality.

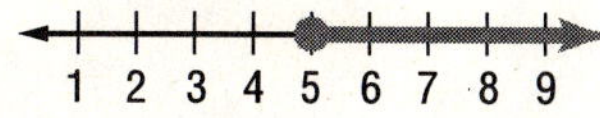

20. $-42 < -0.6x$

$\frac{-42}{-0.6} > \frac{-0.6x}{-0.6}$

$70 > x$ or $x < 70$

The solution is $x < 70$. You can check this solution by substituting a number less than 70 into the inequality.

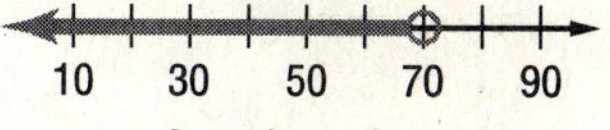

21. $c - 3 \leq 4c + 9$

$c - c - 3 \leq 4c - c + 9$

$-3 \leq 3c + 9$

$-3 - 9 \leq 3c + 9 - 9$

$-12 \leq 3c$

$-4 \leq c$ or $c \geq -4$

The solution is $c \geq -4$. You can check this solution by substituting -4 or a number greater than -4 into the inequality.

−10 −8 −6 −4 −2

22. $7(3 - 2b) > 5b + 2$

$21 - 14b > 5b + 2$

$21 - 14b - 5b > 5b - 5b + 2$

$21 - 19b > 2$

$21 - 21 - 19b > 2 - 21$

$-19b > -19$

$\frac{-19b}{-19} < \frac{-19}{-19}$

$b < 1$

The solution is $b < 1$. You can check this solution by substituting a number less than 1 into the inequality.

−4 −3 −2 −1 0 1 2 3 4

23. $\frac{1}{2}(a + 4) > \frac{1}{4}(a - 8)$

$\frac{1}{2}a + \frac{1}{2}(4) > \frac{1}{4}a - \frac{1}{4}(8)$

$\frac{1}{2}a + 2 > \frac{1}{4}a - 2$

$\frac{1}{2}a - \frac{1}{4}a + 2 > \frac{1}{4}a - \frac{1}{4}a - 2$

$\frac{1}{4}a + 2 > -2$

$\frac{1}{4}a + 2 - 2 > -2 - 2$

$\frac{1}{4}a > -4$

$4\left(\frac{1}{4}a\right) > 4(-4)$

$a > -16$

The solution is $a > -16$. You can check this solution by substituting a number greater than -16 into the inequality.

−20 −18 −16 −14 −12

24. Let x represent the number of cookies.

Revenue − Cost = Profit

Revenue = amount from cookies sold $= 0.95x$

Cost = fixed cost + cost for cookies $= 300 + 0.45x$

$0.95x - (300 + 0.45x) \geq 100$

$0.95x - 300 - 0.45x \geq 100$

$0.50x - 300 \geq 100$

$0.50x - 300 + 300 \geq 100 + 300$

$0.50x \geq 400$

$\frac{0.50x}{0.50} \geq \frac{400}{0.50}$

$x \geq 800$

At least 800 cookies must be sold.

25. D; Let h represent the number of hours Danny works each week.

pay rate · hours worked = total pay

$6.50 \cdot h \geq 100$

$6.50h \geq 100$

Chapter 7 Standardized Test Practice

Pages 364–365

1. B; $c = 0.30w + 6$

$c = 0.30(8) + 6$

$c = 2.4 + 6$

$c = 8.4$

The lowest possible cost is $8.40.

2. C; $\frac{2}{3}$ × number of students on first bus

$+ \frac{3}{4}$ × number of students on second bus

= total who ate at Hamburger Haven

$\frac{2}{3}(36) + \frac{3}{4}(36) = 24 + 27$

$= 51$

51 students ate at Hamburger Haven.

3. B; $\frac{5}{9}(360) = 200$

$360 - 200 = 160$

Shanté had $160 left.

4. C; $\frac{170 \text{ beats}}{1 \text{ min}} \cdot \frac{60 \text{ min}}{1 \text{ h}} = 10{,}200$ beats per hour.

5. C; The rectangle has 8 shaded and 15 total squares, so $\frac{8}{15}$ or about 53% of it shaded. Circle C has just over 50% shaded.

6. D; 40% = 0.40, so 40% > 0.04.

7. C; Let x represent the number of visitors likely to attend a movie.
$x = 0.15(8700)$
$x = 1305$
1305 visitors are likely to attend a movie.

8. B; $1536 + (0.05)(1536) = 1612.8$
There are about 1600 students this year.

9. B;
$$\begin{aligned} 5(x+2) &= 40 \\ 5x + 10 &= 40 \\ 5x + 10 - 10 &= 40 - 10 \\ 5x &= 30 \\ \frac{5x}{5} &= \frac{30}{5} \\ x &= 6 \end{aligned}$$

10. D; $\frac{x}{3} < 5$
$$3\left(\frac{x}{3}\right) < 3(5)$$
$$x < 15$$

11.
$$\begin{aligned} 16 + 18 \div 2 \times 3 &= 16 + (18 \div 2) \times 3 \\ &= 16 + 9 \times 3 \\ &= 16 + 27 \\ &= 43 \end{aligned}$$

12. $\frac{-25 \text{ points}}{5 \text{ days}} = -5$ points per day

13. $\frac{1}{5 \times 5 \times 5 \times 5} = \frac{1}{5^4} = 5^{-4}$

14.
$$\begin{aligned} \frac{5}{12} - \frac{3}{8} &= \frac{5 \cdot 2}{12 \cdot 2} - \frac{3 \cdot 3}{8 \cdot 3} \\ &= \frac{10}{24} - \frac{9}{24} \\ &= \frac{1}{24} \end{aligned}$$

15. List the temperatures from lowest to highest:
41°, 45°, 53°, 57°, 62°
↑
median
The median high temperature is 53°.

16. Ratio = $\frac{\text{total amount of white}}{\text{total amount of paint}}$
$= \frac{5}{8}$
The ratio is 5:8.

17. $P(\text{red}) = \frac{\text{total red pencils}}{\text{total number of pencils}}$
$$\begin{aligned} \frac{3}{7} &= \frac{r}{42} \\ 3 \cdot 42 &= 7r \\ 126 &= 7r \\ \frac{126}{7} &= r \\ 18 &= r \end{aligned}$$
There are 18 red pencils.

18. part = percent · base
$$\begin{aligned} 350 &= p \cdot 625 \\ \frac{350}{625} &= \frac{p \cdot 625}{625} \\ 0.56 &= p \end{aligned}$$
56% of the total cost was paid for with the grant.

19. percent discount = $\frac{\text{original price} - \text{sale price}}{\text{original price}}$
Let x represent the original price of the jacket.
$$\begin{aligned} 0.40 &= \frac{x - (x - 64)}{x} \\ x(0.40) &= x\left(\frac{x - (x - 64)}{x}\right) \\ 0.40x &= x - (x - 64) \\ 0.40x &= x - x + 64 \\ 0.40x &= 64 \\ \frac{0.40x}{0.40} &= \frac{64}{0.40} \\ x &= 160 \end{aligned}$$
The original price of the jacket was $160.

20.
$$\begin{aligned} 8x - 12 &= 5x + 6 \\ 8x - 5x - 12 &= 5x - 5x + 6 \\ 3x - 12 &= 6 \\ 3x - 12 + 12 &= 6 + 12 \\ 3x &= 18 \\ x &= 6 \end{aligned}$$

21. Perimeter = twice the length + twice the width
$$\begin{aligned} 88 &= 2(5w - 4) + 2w \\ 88 &= 10w - 8 + 2w \\ 88 &= 12w - 8 \\ 88 + 8 &= 12w - 8 + 8 \\ 96 &= 12w \\ \frac{96}{12} &= \frac{12w}{12} \\ 8 &= w \end{aligned}$$
The width is 8 meters.

22. hourly pay × number of hours = total pay
Let h represent the number of hours he will have to work.
$$\begin{aligned} 8h &\geq 1200 \\ \frac{8h}{8} &\geq \frac{1200}{8} \\ h &\geq 150 \end{aligned}$$
He will need to work at least 150 hours.

23a. $0.01x + 0.22x + 0.34x + 2.80 = 14$

23b. the number of each kind of stamp

23c.
$$\begin{aligned} 0.01x + 0.22x + 0.34x + 2.80 &= 14.20 \\ (0.01 + 0.22 + 0.34)x + 2.80 &= 14.20 \\ 0.57x + 2.80 &= 14.20 \\ 0.57x + 2.80 - 2.80 &= 14.20 - 2.80 \\ 0.57x &= 11.40 \\ \frac{0.57x}{0.57} &= \frac{11.40}{0.57} \\ x &= 20 \end{aligned}$$

23d. The customer bought 20 of each kind of stamp.

24a.
$$\begin{aligned} \frac{p}{100} &= \frac{3}{50} \\ 50p &= 3 \cdot 100 \\ 50p &= 300 \\ \frac{50p}{50} &= \frac{300}{50} \\ p &= 6 \end{aligned}$$
6% do not renew their subscriptions.

24b. If 3 out of 50 do *not* renew, then 47 out of 50 *do* renew.
$\frac{47}{50} = 0.94$ or 94%

24c. $0.94(24{,}000) = 22{,}560$

Chapter 8 Functions and Graphing

Page 367 Getting Started

1.

x	y
0	4
−3	3

domain = {0, −3}; range = {4, 3}

2.

x	y
−5	11
2	1

domain = {−5, 2}; range = {11, 1}

3.

x	y
6	8
7	10
8	12

domain = {6, 7, 8}; range = {8, 10, 12}

4.

x	y
1	−9
5	12
−3	−10

domain = {1, 5, −3}; range = {−9, 12, −10}

5.

x	y
−8	5
7	−1
6	1
1	−2

domain = {−8, 7, 6, 1}; range = {5, −1, 1, −2}

6. For (−3, 0), start at the origin. Move left 3 units. Then move 0 units up or down. The point is *C*.
7. For (3, −2), start at the origin. Move right 3 units. Then move down 2 units. The point is *E*.
8. For (−4, −2), start at the origin. Move left 4 units. Then move down 2 units. The point is *D*.
9. For (0, 4), start at the origin. Move 0 units left or right. Then move up 4 units. The point is *A*.
10. For (4, 6), start at the origin. Move right 4 units. Then move up 6 units. The point is *B*.
11. For (4, 0), start at the origin. Move right 4 units. Then move 0 units up or down. The point is *F*.
12. $8y \geq 25$
$8(4) \overset{?}{\geq} 25$
$32 \geq 25$ ✓
This sentence is true.
13. $18 < t + 12$
$18 \overset{?}{<} 10 + 12$
$18 < 22$ ✓
This sentence is true.
14. $n - 15 > 7$
$20 - 15 \overset{?}{>} 7$
$5 \not> 7$
This sentence is false.
15. $5 \geq 2x + 3$
$5 \overset{?}{\geq} 2(1) + 3$
$5 \overset{?}{\geq} 2 + 3$
$5 \geq 5$ ✓
Although the inequality $5 > 5$ is false, the equation $5 = 5$ is true. Therefore, the sentence is true.
16. $12 \leq \frac{2}{3}n$
$12 \overset{?}{\leq} \frac{2}{3}(9)$
$12 \not\leq 6$
This sentence is false.
17. $\frac{1}{2}x - 5 < 0$
$\frac{1}{2}(8) - 5 \overset{?}{<} 0$
$4 - 5 \overset{?}{<} 0$
$-1 < 0$ ✓
This sentence is true.

Page 368 Alegbra Activity (Preview of Lesson 8-1)

1. −5, −3, −1, 1, 3, 5, 7, 9, 11
2. $y = 2x + 3$
3. Write an equation and work backward to solve for the input value.
4. False; the output values depend on the input values.
5. $y = 2x$; $y = x + 4$
6. Sample answer:

Input	Output
−5	−9
−4	−7
−3	−5
−2	−3
−1	−1
0	1
1	3
2	5
3	7
4	9

rule: × 2 + 1

8-1 Functions

Page 369 How can the relationship between actual temperatures and windchill temperatures be a function?

a.

b. Sample answer: As the actual temperature increases, the windchill temerature also increases.

c. $-46°F$; when the actual temperature decreases, the windchill temperature also decreases.

Page 371 Check for Understanding

1. Sample answer: a set of ordered pairs: {(1, 2), (4, 3), (−2, −1), (−3, 3)}

a table:

x	y
1	2
4	3
−2	−1
−3	3

a graph:

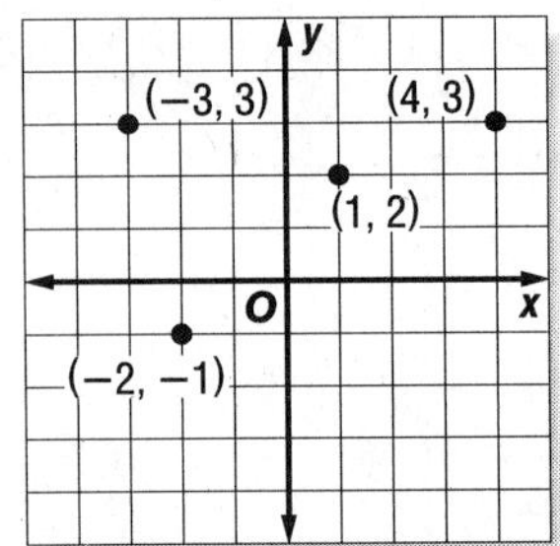

2. Sample answer: Check that each domain value is paired with only one range value or use the vertical line test.

3. Sample answer: This graph does not represent a function because when x equals 1, there are two y values, 0 and 2.

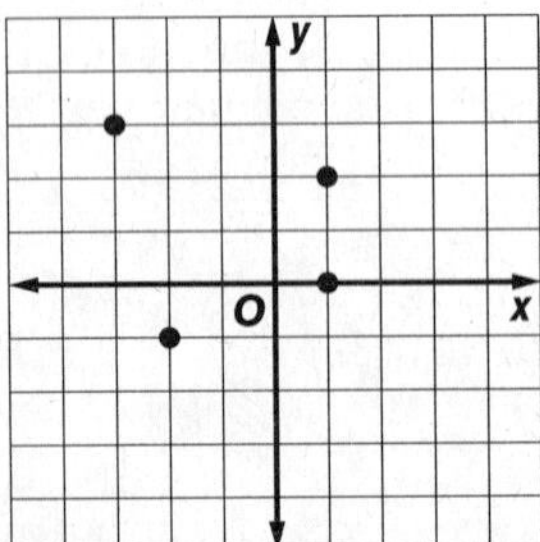

4. No; 13 is paired with 5 and 10.

5. Yes; each x value is paired with only one y value.

6. Yes; each domain value is paired with only one range value.

7. No; 5 is paired with 4 and 14.

8. No; a vertical line passes through more than one point.

9. Yes; any vertical line passes through no more than one point of the graph.

10. Yes; each wind speed is paired with only one windchill temperature.

11. As wind speed increases, the windchill temperature decreases.

Pages 371–373 Practice and Apply

12. Yes; each x value is paired with only one y value.

13. Yes; each x value is paired with only one y value.

14. No; 6 in the domain is paired with 21 and 22 in the range.

15. No; 5 in the domain is paired with −4 and −1 in the range.

16. Yes; each domain value is paired with only one range value.

17. No; −2 in the domain is paired with 5 and 1 in the range.

18. No; the x value 0 is paired with y values 4 and 10; the x value 11 is paired with y values 6 and 8.

19. Yes; each x value is paired with only one y value.

20. Yes; any vertical line passes through no more than one point of the graph.

21. No; a vertical line passes through more than one point.

22. Yes; any vertical line passes through no more than one point of the graph.

23. Yes; any vertical line passes through no more than one point of the graph.

24. Yes; each year is only paired with one value for the number of farms.

25. Generally, as the years progress, the number of farms decreases. An exception is in the year 2000.

26. Yes; each year is paired with only one value for the average size of farms.

27. Generally, as the years progress, the size of farms increases. An exception is in the year 2000.

28. No; 24 in the domain is paired with 163 and 168 in the range.

29. Generally, as foot length increases, height increases.

30. Always; every function can be written as a set of ordered pairs.

31. Sometimes; a relation that has a member of the domain paired with more than one member in the range is not a function.

32a. Yes

32b. Sometimes; the inverse of {(1, 5), (2, 10)} is also a function, but the inverse of {(1, 5), (2, 5)} is not a function.

33. For a given wind speed, there is only one windchill temperature for each actual temperature. So, the relationship between actual temperatures and windchill temperatures is a function. Answers should include the following.

- For a given wind speed, as the actual temperature increases, the windchill temperature increases.
- Since the relationship between actual temperatures and windchill temperatures is a function, there cannot be two different windchill temperatures for the same actual temperature when the wind speed remains the same.

34. D; (2, 18) causes the value of 2 to be paired with both 11 and 18.

35. A; The data represent a function, since each x value is paired with only one y value.

Page 373 Maintain Your Skills

36. $4y > 24$

$\frac{4y}{4} > \frac{24}{4}$

$y > 6$

The solution is $y > 6$. You can check this solution by substituting a number greater than 6 into the inequality.

37. $\frac{a}{3} < -7$

$3\left(\frac{a}{3}\right) < 3(-7)$

$a < -21$

The solution is $a < -21$. You can check this solution by substituting a number less than -21 into the inequality.

38. $18 \geq -2k$

$\frac{18}{-2} \leq \frac{-2k}{-2}$

$-9 \leq k$ or $k \geq -9$

The solution is $k \geq -9$. You can check this solution by substituting -9 or a number greater than -9 into the inequality.

39. $2x + 5 < 17$

$2x + 5 - 5 < 17 - 5$

$2x < 12$

$\frac{2x}{2} < \frac{12}{2}$

$x < 6$

The solution is $x < 6$. You can check this solution by substituting a number less than 6 into the inequality.

40. $2t - 3 \geq 1.4t + 6$

$2t - 1.4t - 3 \geq 1.4t - 1.4t + 6$

$0.6t - 3 \geq 6$

$0.6t - 3 + 3 \geq 6 + 3$

$0.6t \geq 9$

$\frac{0.6t}{0.6} \geq \frac{9}{0.6}$

$t \geq 15$

The solution is $t \geq 15$. You can check this solution by substituting 15 or a number greater than 15 into the inequality.

41. $12r - 4 > 7 + 12r$

$12r - 12r - 4 > 7 + 12r - 12r$

$-4 > 7$

The sentence $-4 > 7$ is never true, so the solution is $\varnothing$.

42. Let n represent the percent.

$10 = n \cdot 50$

$\frac{10}{50} = \frac{n \cdot 50}{50}$

$0.2 = n$

So, 10 is 20% of 50.

43. Let n represent the part.

$n = 0.15(120)$

$n = 18$

So, 18 is 15% of 120.

44. Let n represent the part.

$n = 0.95(256)$

$n = 243.2$

So, 243.2 is 95% of 256.

45. Let n represent the base.

$46.5 = 0.62n$

$\frac{46.5}{0.62} = \frac{0.62n}{0.62}$

$75 = n$

So, 46.5 is 62% of 75.

46. arithmetic, with common difference -20; 40, 20, 0

47. $3x + 1 = 3(4) + 1$

$= 12 + 1$

$= 13$

48. $2y = 2(-1)$

$= -2$

49. $y + 6 = -1 + 6$

$= 5$

50. $-5x = -5(4)$

$= -20$

51. $2x - 8 = 2(4) - 8$

$= 8 - 8$

$= 0$

52. $3y - 4 = 3(-1) - 4$

$= -3 - 4$

$= -7$

Page 374 Graphing Calculator Investigation (Preview of Lesson 8-2)

1a. Enter Y = −2X + 4 in the Y= list.

KEYSTROKES: [Y=] −2 [X,T,θ,n] [+] 4

Use TBLSET to select *Ask* for the independent variable and *Auto* for the dependent variable. Then enter any value for the domain.

KEYSTROKES: [2nd] [TBLSET] [▼] [▼] [▶] [ENTER] [▼] [ENTER]

Access the table.

KEYSTROKES: [2nd] [TABLE]

Enter the domain values.

KEYSTROKES: −2 [ENTER] −1 [ENTER] 0 [ENTER] 1 [ENTER] 2 [ENTER]

The range is {8, 6, 4, 2, 0}.

X	Y1	
-2	8	
-1	6	
0	4	
1	2	
2	0	

X=

1b. As X increases by 1 unit, Y decreases by 2 units.

1c. Greater than 8; the pattern of the range values indicates that if the domain value decreases by 1, then the range value increases by 2.

2a. Y = 60X

2b.

Time x	Distance y
0	0
1	60
3.5	210
10	600

2c. 1 h and 3.5 h

2d. Sample answer: Input X values between 1 and 3.5 until you get a Y value closer to 150.

3a. \$1.90, \$2.30, \$2.70, \$3.10, \$6.30

3b. $y = 1.50 + 0.25x$
$y = 1.50 + 0.25(12)$
$y = 1.50 + 3.00$
$y = 4.50$
The cost is \$4.50 for pencils and 12 plain folders. It cost \$6.30 for pencils and 12 fancy folders (from part (a)), so Serena saves \$6.30 − \$4.50 = \$1.80.

8-2 Linear Equations in Two Variables

Page 375 How can linear equations represent a function?

a.

Number of Cans (x)	$1.50x$	Cost (y)
1	1.50(1)	1.50
2	1.50(2)	3.00
3	1.50(3)	4.50
4	1.50(4)	6.00

b.

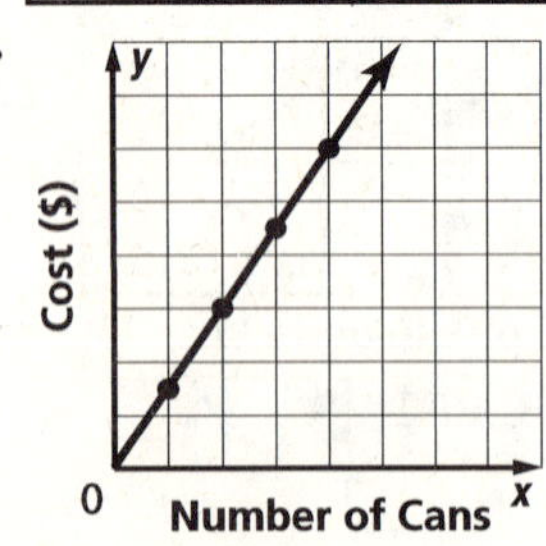

c. $y = 1.5x$ or $1.50x$

Pages 377–378 Check for Understanding

1. Sample answer: Infinitely many values can be substituted for x, or the domain.

2. Sample answer: $y = x + 6$

3.

x	$x + 5$	y
−3	−3 + 5	2
−1	−1 + 5	4
0	0 + 5	5
1	1 + 5	6

(−3, 2), (−1, 4), (0, 5), (1, 6)

4. Choose four values for x and substitute them into $y = x + 8$.
Sample answer:

x	$y = x + 8$	y	(x, y)
−1	$y = -1 + 8 = 7$	7	(−1, 7)
0	$y = 0 + 8 = 8$	8	(0, 8)
1	$y = 1 + 8 = 9$	9	(1, 9)
2	$y = 2 + 8 = 10$	10	(2, 10)

Four solutions are (−1, 7), (0, 8), (1, 9) and (2, 10).

5. Choose four values for x and substitute them into $y = 4x$.
Sample answer:

x	$y = 4x$	y	(x, y)
−1	$y = 4(-1) = -4$	−4	(−1, −4)
0	$y = 4(0) = 0$	0	(0, 0)
1	$y = 4(1) = 4$	4	(1, 4)
2	$y = 4(2) = 8$	8	(2, 8)

Four solutions are (−1, −4), (0, 0), (1, 4), and (2, 8).

6. Choose four values for x and substitute them into $y = 2x - 7$.
Sample answer:

x	$y = 2x - 7$	y	(x, y)
−1	$y = 2(-1) - 7 = -9$	−9	(−1, −9)
0	$y = 2(0) - 7 = -7$	−7	(0, −7)
1	$y = 2(1) - 7 = -5$	−5	(1, −5)
2	$y = 2(2) - 7 = -3$	−3	(2, −3)

Four solutions are (−1, −9), (0, −7), (1, −5), and (2, −3).

7. First, rewrite the equation by solving for y:
$-5x + y = 6$
$-5x + 5x + y = 6 + 5x$
$y = 6 + 5x$
Choose four values for x and substitute them into $y = 6 + 5x$.
Sample answer:

x	$y = 6 + 5x$	y	(x, y)
−1	$y = 6 + 5(-1) = 1$	1	(−1, 1)
0	$y = 6 + 5(0) = 6$	6	(0, 6)
1	$y = 6 + 5(1) = 11$	11	(1, 11)
2	$y = 6 + 5(2) = 16$	16	(2, 16)

Four solutions are (−1, 1), (0, 6), (1, 11), and (2, 16).

8. Sample ordered pairs: $(-2, 1)$, $(-1, 2)$, $(0, 3)$, $(1, 4)$

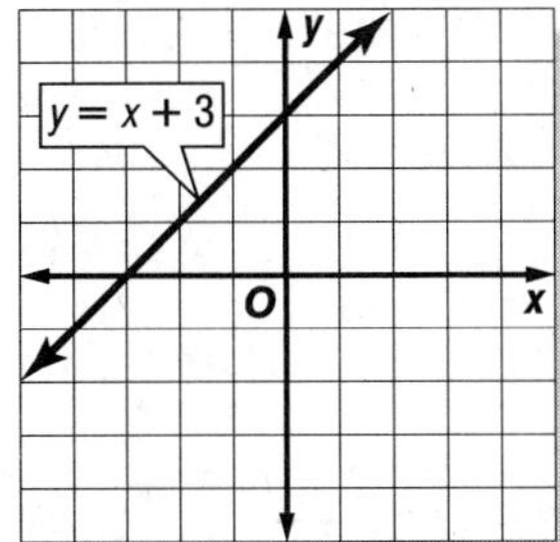

9. Sample ordered pairs: $(-1, -3)$, $(0, -1)$, $(1, 1)$, $(2, 3)$

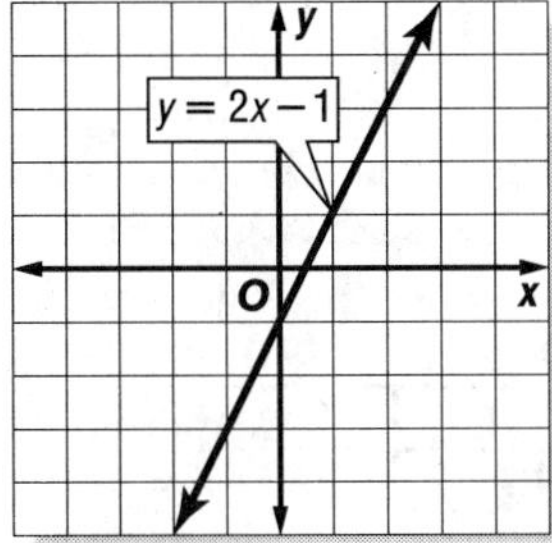

10. First, rewrite the equation by solving for y:

$$x + y = 5$$
$$x - x + y = 5 - x$$
$$y = 5 - x$$

Sample ordered pairs: $(-1, 6)$, $(0, 5)$, $(1, 4)$, $(2, 3)$

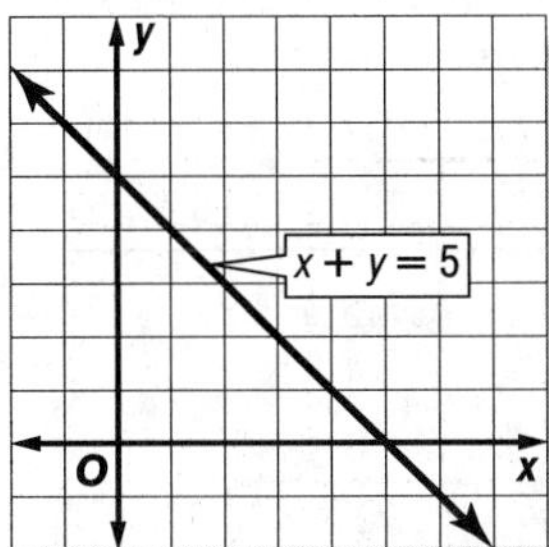

11. Sample answer: (1, 186,000) means that light travels 186,000 miles in 1 second. (2, 372,000) means that light travels 372,000 miles in 2 seconds.

Pages 378–379 Practice and Apply

12.

x	$x - 9$	y
-1	$-1 - 9$	-10
0	$0 - 9$	-9
4	$4 - 9$	-5
7	$7 - 9$	-2

$(-1, -10)$, $(0, -9)$, $(4, -5)$, $(7, -2)$

13.

x	$2x + 6$	y
-4	$2(-4) + 6$	-2
0	$2(0) + 6$	6
2	$2(2) + 6$	10
4	$2(4) + 6$	14

$(-4, -2)$, $(0, 6)$, $(2, 10)$, $(4, 14)$

14. Choose four values for x and substitute them into $y = x + 2$.
Sample answer:

x	$y = x + 2$	y	(x, y)
-1	$y = -1 + 2 = 1$	1	$(-1, 1)$
0	$y = 0 + 2 = 2$	2	$(0, 2)$
1	$y = 1 + 2 = 3$	3	$(1, 3)$
2	$y = 2 + 2 = 4$	4	$(2, 4)$

Four solutions are $(-1, 1)$, $(0, 2)$, $(1, 3)$, and $(2, 4)$.

15. Choose four values for x and substitute them into $y = x - 7$.
Sample answer:

x	$y = x - 7$	y	(x, y)
-1	$y = -1 - 7 = -8$	-8	$(-1, -8)$
0	$y = 0 - 7 = -7$	-7	$(0, -7)$
1	$y = 1 - 7 = -6$	-6	$(1, -6)$
2	$y = 2 - 7 = -5$	-5	$(2, -5)$

Four solutions are $(-1, -8)$, $(0, -7)$, $(1, -6)$, and $(2, -5)$.

16. Choose four values for x and substitute them into $y = 3x$.
Sample answer:

x	$y = 3x$	y	(x, y)
-1	$y = 3(-1) = -3$	-3	$(-1, -3)$
0	$y = 3(0) = 0$	0	$(0, 0)$
1	$y = 3(1) = 3$	3	$(1, 3)$
2	$y = 3(2) = 6$	6	$(2, 6)$

Four solutions are $(-1, -3)$, $(0, 0)$, $(1, 3)$, and $(2, 6)$.

17. Choose four values for x and substitute them into $y = -5x$.
Sample answer:

x	$y = -5x$	y	(x, y)
-1	$y = -5(-1) = 5$	5	$(-1, 5)$
0	$y = -5(0) = 0$	0	$(0, 0)$
1	$y = -5(1) = -5$	-5	$(1, -5)$
2	$y = -5(2) = -10$	-10	$(2, -10)$

Four solutions are $(-1, 5)$, $(0, 0)$, $(1, -5)$, and $(2, -10)$.

18. Choose four values for x and substitute them into $y = 2x - 3$.
Sample answer:

x	$y = 2x - 3$	y	(x, y)
-1	$y = 2(-1) - 3 = -5$	-5	$(-1, -5)$
0	$y = 2(0) - 3 = -3$	-3	$(0, -3)$
1	$y = 2(1) - 3 = -1$	-1	$(1, -1)$
2	$y = 2(2) - 3 = 1$	1	$(2, 1)$

Four solutions are $(-1, -5)$, $(0, -3)$, $(1, -1)$, and $(2, 1)$.

19. Choose four values for x and substitute them into $y = 3x + 1$.
Sample answer:

x	$y = 3x + 1$	y	(x, y)
-1	$y = 3(-1) + 1 = -2$	-2	$(-1, -2)$
0	$y = 3(0) + 1 = 1$	1	$(0, 1)$
1	$y = 3(1) + 1 = 4$	4	$(1, 4)$
2	$y = 3(2) + 1 = 7$	7	$(2, 7)$

Four solutions are $(-1, -2)$, $(0, 1)$, $(1, 4)$, and $(2, 7)$.

20. First, solve the equation for y:
$$x + y = 9$$
$$x - x + y = 9 - x$$
$$y = 9 - x$$
Choose four values for x and substitute them into $y = 9 - x$.
Sample answer:

x	$y = 9 - x$	y	(x, y)
-1	$y = 9 - (-1) = 10$	10	$(-1, 10)$
0	$y = 9 - 0 = 9$	9	$(0, 9)$
1	$y = 9 - 1 = 8$	8	$(1, 8)$
2	$y = 9 - 2 = 7$	7	$(2, 7)$

Four solutions are $(-1, 10)$, $(0, 9)$, $(1, 8)$, and $(2, 7)$.

21. First, solve the equation for y:
$$x + y = -6$$
$$x - x + y = -6 - x$$
$$y = -6 - x$$
Choose four values for x and substitute them into $y = -6 - x$.
Sample answer:

x	$y = -6 - x$	y	(x, y)
-1	$y = -6 - (-1) = -5$	-5	$(-1, -5)$
0	$y = -6 - 0 = -6$	-6	$(0, -6)$
1	$y = -6 - 1 = -7$	-7	$(1, -7)$
2	$y = -6 - 2 = -8$	-8	$(2, -8)$

Four solutions are $(-1, -5)$, $(0, -6)$, $(1, -7)$, and $(2, -8)$.

22. First, solve the equation for y:
$$4x + y = 2$$
$$4x - 4x + y = 2 - 4x$$
$$y = 2 - 4x$$
Choose four values for x and substitute them into $y = 2 - 4x$
Sample answer:

x	$y = 2 - 4x$	y	(x, y)
-1	$y = 2 - 4(-1) = 6$	6	$(-1, 6)$
0	$y = 2 - 4(0) = 2$	2	$(0, 2)$
1	$y = 2 - 4(1) = -2$	-2	$(1, -2)$
2	$y = 2 - 4(2) = -6$	-6	$(2, -6)$

Four solutions are $(-1, 6)$, $(0, 2)$, $(1, -2)$, and $(2, -6)$.

23. First, solve the equation for y:
$$3x - y = 10$$
$$3x - 3x - y = 10 - 3x$$
$$-y = 10 - 3x$$
$$y = -10 + 3x$$
Choose four values for x and substitute them into $y = -10 + 3x$.
Sample answer:

x	$y = -10 + 3x$	y	(x, y)
-1	$y = -10 + 3(-1) = -13$	-13	$(-1, -13)$
0	$y = -10 + 3(0) = -10$	-10	$(0, -10)$
1	$y = -10 + 3(1) = -7$	-7	$(1, -7)$
2	$y = -10 + 3(2) = -4$	-4	$(2, -4)$

Four solutions are $(-1, -13)$, $(0, -10)$, $(1, -7)$, and $(2, -4)$.

24. All y values are 8, so choose any four x values.
Sample answer: $(-1, 8)$, $(0, 8)$, $(1, 8)$, $(2, 8)$

25. All x values are -1, so choose any four y values.
Sample answer: $(-1, 0)$, $(-1, 1)$, $(-1, 2)$, $(-1, 3)$

26. $(8, 4.96)$ means that 8 kilometers is approximately equal to 4.96 miles.

27. $y = 0.62x$
$y = 0.62(10)$
$y = 6.2$ miles

28. $y = 0.7(220 - x)$
20-year-old: $y = 0.7(220 - 20)$
$y = 0.7(200)$
$y = 140$ beats per minute
50-year-old: $y = 0.7(220 - 50)$
$y = 0.7(170)$
$y = 119$ beats per mintue

29. Quadrant I; a person cannot have a negative age or heart rate.

30. Sample answer:

x	$y = x + 2$	y	(x, y)
-1	$y = -1 + 2$	1	$(-1, 1)$
0	$y = 0 + 2$	2	$(0, 2)$
1	$y = 1 + 2$	3	$(1, 3)$
2	$y = 2 + 2$	4	$(2, 4)$

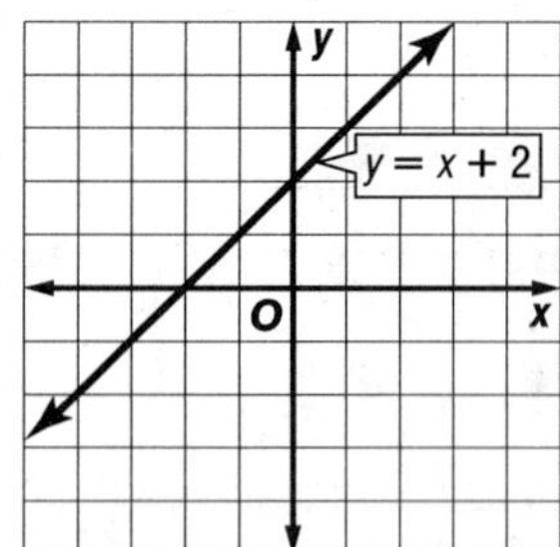

31. Sample answer:

x	$y = x + 5$	y	(x, y)
-2	$y = -2 + 5$	3	$(-2, 3)$
-1	$y = -1 + 5$	4	$(-1, 4)$
0	$y = 0 + 5$	5	$(0, 5)$
1	$y = 1 + 5$	6	$(1, 6)$

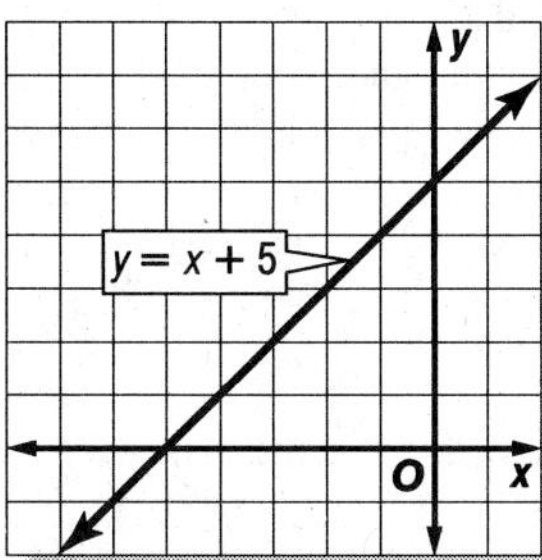

32. Sample answer:

x	$y = x - 4$	y	(x, y)
−1	$y = -1 - 4$	−5	(−1, −5)
0	$y = 0 - 4$	−4	(0, −4)
1	$y = 1 - 4$	−3	(1, −3)
2	$y = 2 - 4$	−2	(2, −2)

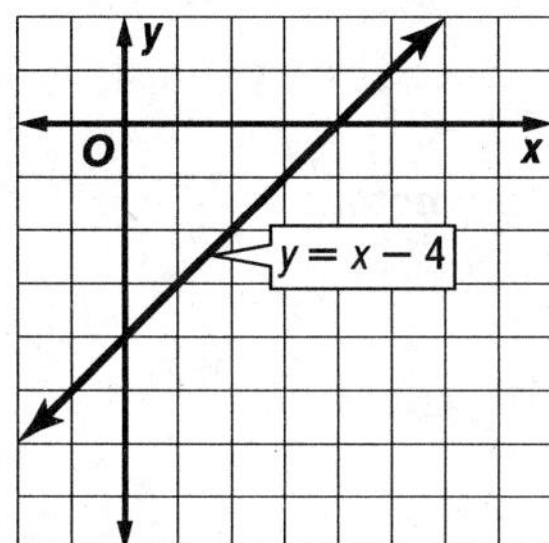

33. Sample answer:

x	$y = -x - 6$	y	(x, y)
−4	$y = -(-4) - 6$	−2	(−4, −2)
−3	$y = -(-3) - 6$	−3	(−3, −3)
−2	$y = -(-2) - 6$	−4	(−2, −4)
−1	$y = -(-1) - 6$	−5	(−1, −5)

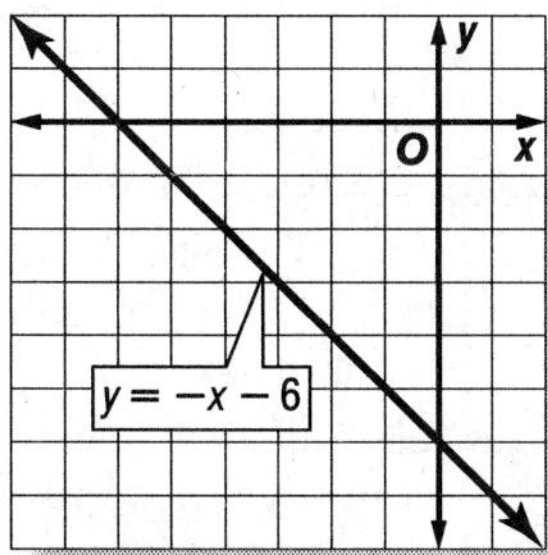

34. Sample answer:

x	$y = -2x + 2$	y	(x, y)
−1	$y = -2(-1) + 2$	4	(−1, 4)
0	$y = -2(0) + 2$	2	(0, 2)
1	$y = -2(1) + 2$	0	(1, 0)
2	$y = -2(2) + 2$	−2	(2, −2)

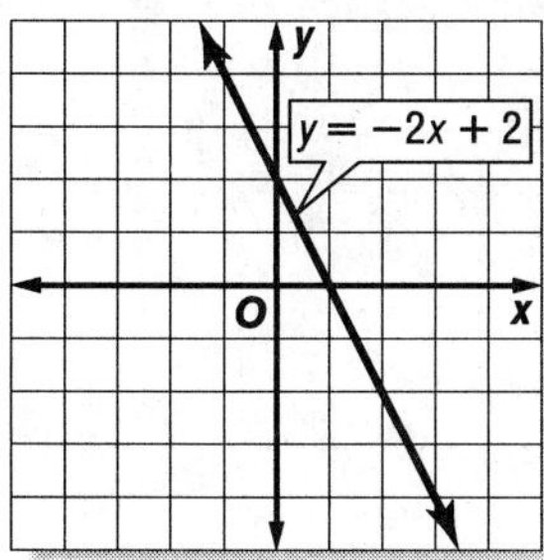

35. Sample answer:

x	$y = 3x - 4$	y	(x, y)
−1	$y = 3(-1) - 4$	−7	(−1, −7)
0	$y = 3(0) - 4$	−4	(0, −4)
1	$y = 3(1) - 4$	−1	(1, −1)
2	$y = 3(2) - 4$	2	(2, 2)

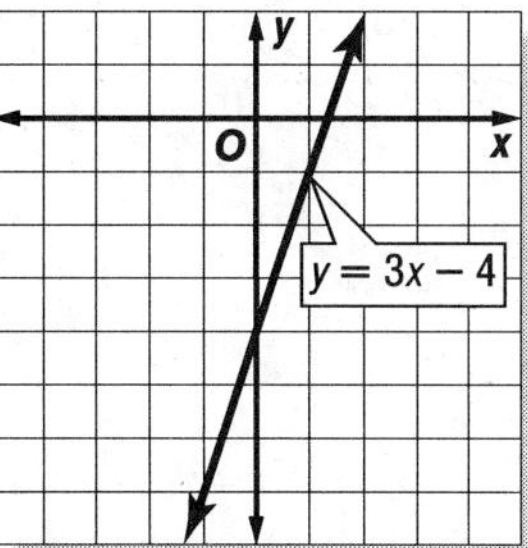

36. First, solve the equation for y.

$$x + y = 1$$
$$x - x + y = 1 - x$$
$$y = 1 - x$$

Sample answer:

x	$y = 1 - x$	y	(x, y)
−1	$y = 1 - (-1)$	2	(−1, 2)
0	$y = 1 - 0$	1	(0, 1)
1	$y = 1 - (1)$	0	(1, 0)
2	$y = 1 - (2)$	−1	(2, −1)

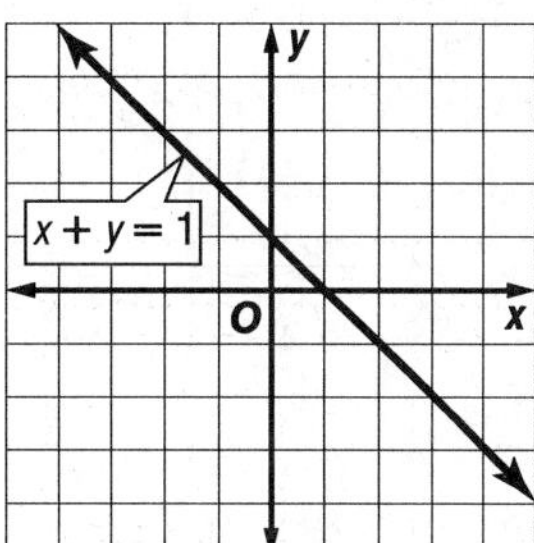

37. First, solve the equation for y.

$$x - y = 6$$
$$x - x - y = 6 - x$$
$$-y = 6 - x$$
$$y = -6 + x$$

Sample answer:

x	$y = -6 + x$	y	(x, y)
0	$y = -6 + 0$	−6	(0, −6)
1	$y = -6 + 1$	−5	(1, −5)
2	$y = -6 + 2$	−4	(2, −4)
3	$y = -6 + 3$	−3	(3, −3)

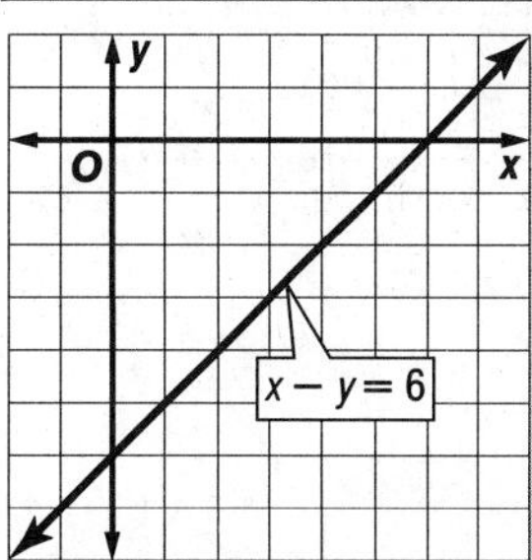

38. First, solve the equation for y.

$$2x + y = 5$$
$$2x - 2x + y = 5 - 2x$$
$$y = 5 - 2x$$

Sample answer:

x	$y = 5 - 2x$	y	(x, y)
−1	$y = 5 - 2(-1)$	7	(−1, 7)
0	$y = 5 - 2(0)$	5	(0, 5)
1	$y = 5 - 2(1)$	3	(1, 3)
2	$y = 5 - 2(2)$	1	(2, 1)

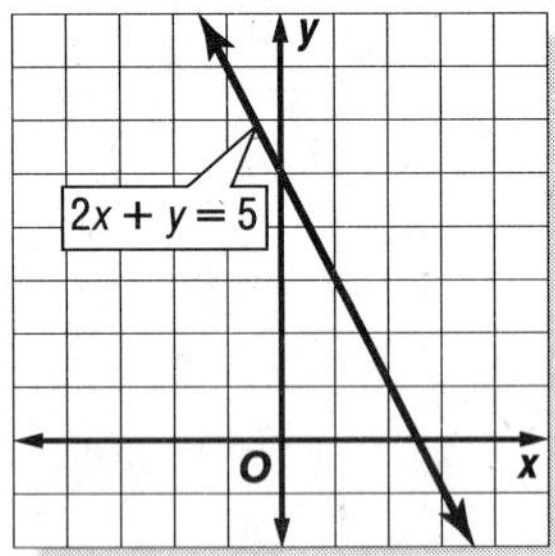

39. First, solve the equation for y.

$$3x - y = 7$$
$$3x - 3x - y = 7 - 3x$$
$$-y = 7 - 3x$$
$$y = -7 + 3x$$

Sample answer:

x	$y = -7 + 3x$	y	(x, y)
0	$y = -7 + 3(0)$	−7	(0, −7)
1	$y = -7 + 3(1)$	−4	(1, −4)
2	$y = -7 + 3(2)$	−1	(2, −1)
3	$y = -7 + 3(3)$	2	(3, 2)

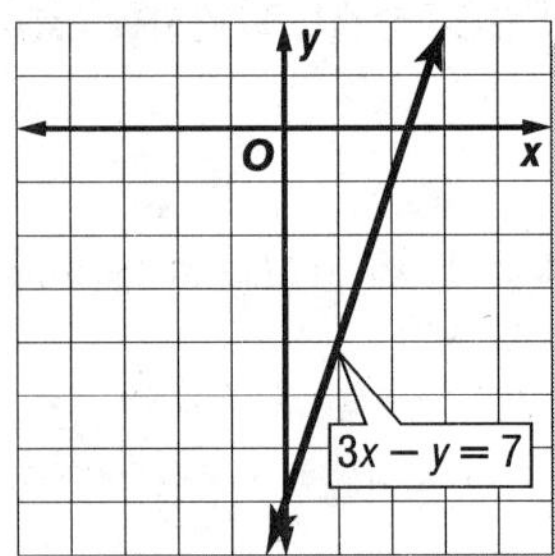

40. All ordered pairs have an x-coordinate of 2. This is a vertical line. Sample ordered pairs: (2, −3), (2, 0), (2, 1), (2, 4)

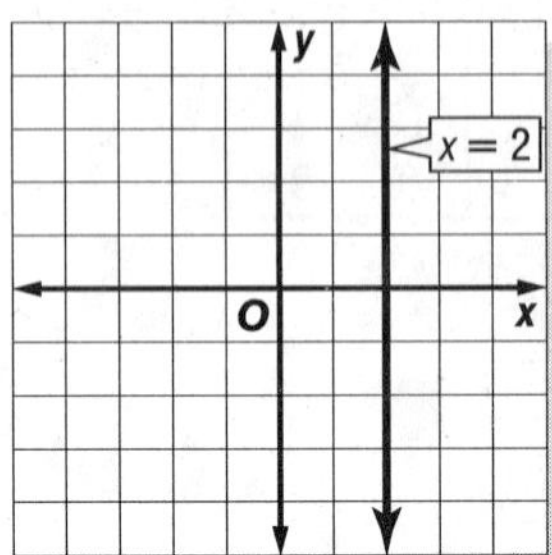

41. All ordered pairs have a y-coordinate of −3. This is a horizontal line. Sample ordered pairs: (−4, −3), (−1, −3), (2, −3), (3, −3)

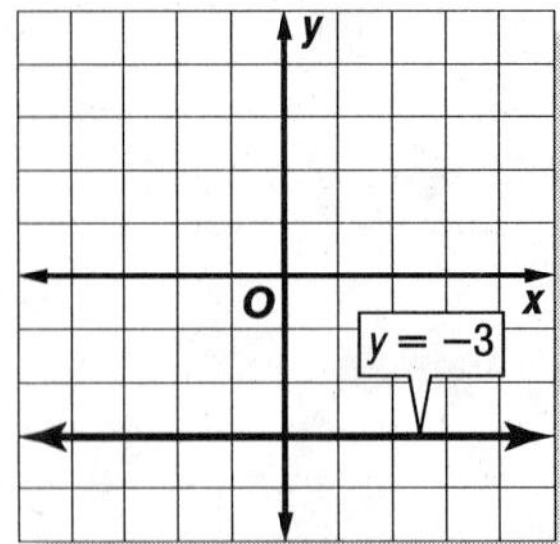

42. Sample answer:

s	$P = 4s$	P	(s, P)
1	$P = 4(1)$	4	(1, 4)
2	$P = 4(2)$	8	(2, 8)
3	$P = 4(3)$	12	(3, 12)

43. Sample answer:

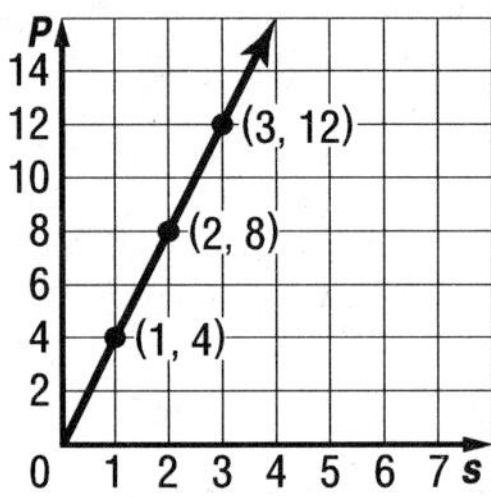

44. Length cannot have a negative value.

45. Yes; the points lie on a straight line.

46. No; the points do not lie on a straight line.

47. Yes; the points lie on a straight line.

48. Yes; the graph is a line.

49. No; the exponent of x is not 1.

50. Yes; the graph is a horizontal line.

51. Sample answer: In the first table, as the x values increase by 1, the y values increase by 2. In the second table, as the x values increase by 1, the y values do not change by a constant amount.

52. Linear equations use variables to show the relationship between the domain values and the range values of a function. Answers should include the following.

- Functions can be represented using a table, a graph, a verbal description, or an equation.
- The equation $y = 0.49x$ represents the cost of x pounds of bananas at \$0.49 per pound.

53. B; Substitute the first coordinate pair in the table into each equation choice. Choice A, $11 \neq -2 + 5$, can be eliminated. Next, substitute the second coordinate pair into the remaining equations. Choice C, $5 \neq -5(0) + 1$, and Choice D, $5 \neq 0 + 13$, can be eliminated, leaving Choice B.

54. C; First, solve the equation for y.

$$2x - y = 4$$
$$2x - 2x - y = 4 - 2x$$
$$-y = 4 - 2x$$
$$y = -4 + 2x$$

When $x = 1, y = -4 + 2(1) = -4 + 2$ or -2.
When $x = 3, y = -4 + 2(3) = -4 + 6$ or 2.
The points are $(1, -2)$ and $(3, 2)$.

Page 379 Maintain Your Skills

55. Yes; each x value is paired with only one y value.

56. Yes; each x value is paired with only one y value.

57. No; 11 in the domain is paired with 8 and 21 in the range.

58. No; -0.1 in the domain is paired with 5 and -5 in the range.

59.
$$3x + 4 < 16$$
$$3x + 4 - 4 < 16 - 4$$
$$3x < 12$$
$$\frac{3x}{3} < \frac{12}{3}$$
$$x < 4$$

The solution is $x < 4$. You can check this solution by substituting a number less than 4 into the inequality.

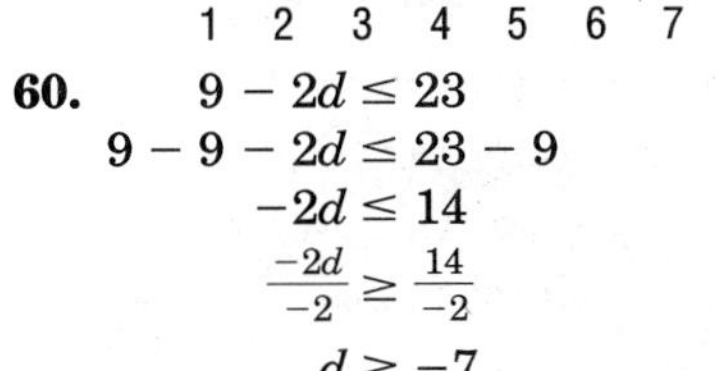

60.
$$9 - 2d \leq 23$$
$$9 - 9 - 2d \leq 23 - 9$$
$$-2d \leq 14$$
$$\frac{-2d}{-2} \geq \frac{14}{-2}$$
$$d \geq -7$$

The solution is $d \geq -7$. You can check this solution by substituting -7 or a number greater than -7 into the inequality.

−10 −9 −8 −7 −6 −5 −4

61.
$$a \div b = \frac{4}{7} \div \frac{2}{3}$$
$$= \frac{\overset{2}{\cancel{4}}}{7} \cdot \frac{3}{\underset{1}{\cancel{2}}}$$
$$= \frac{6}{7}$$

62.
$$y = 5x - 3$$
$$y = 5(0) - 3$$
$$y = 0 - 3$$
$$y = -3$$

63.
$$-x + y = 3$$
$$-(0) + y = 3$$
$$0 + y = 3$$
$$y = 3$$

64.
$$x + 2y = 12$$
$$0 + 2y = 12$$
$$2y = 12$$
$$y = 6$$

65.
$$4x - 5y = -20$$
$$4(0) - 5y = -20$$
$$0 - 5y = -20$$
$$-5y = -20$$
$$y = 4$$

Page 380 Reading Mathematics

1. Sample answer: One meaning of the word *function* is purpose. In mathematics, the purpose of a function is to assign each value of the domain exactly one range value.

2. Sample answer: x represents the domain; y represents the range.

3.

x	$f(x) = 3x + 5$	$f(x)$
0	$f(0) = 3(0) + 5$	5
1	$f(1) = 3(1) + 5$	8
2	$f(2) = 3(2) + 5$	11
3	$f(3) = 3(3) + 5$	14

4a.
$$f(2) = 4(2) - 1$$
$$= 8 - 1$$
$$= 7$$

4b.
$$f(-3) = 4(-3) - 1$$
$$= -12 - 1$$
$$= -13$$

4c.
$$f\left(\frac{1}{2}\right) = 4\left(\frac{1}{2}\right) - 1$$
$$= 2 - 1$$
$$= 1$$

5.
$$f(x) = -2x + 5$$
$$-7 = -2x + 5$$
$$-7 - 5 = -2x + 5 - 5$$
$$-12 = -2x$$
$$\frac{-12}{-2} = \frac{-2x}{-2}$$
$$6 = x$$

8-3 Graphing Linear Equations Using Intercepts

Page 381 How can intercepts be used to represent real-life information?

a. (0, 32); a temperature of 0°C equals 32°F.

b. (−18, 0); a temperature of approximately −18°C equals 0°F.

Page 384 Check for Understanding

1. To find the x-intercept, let $y = 0$ and solve for x. To find the y-intercept, let $x = 0$ and solve for y.

2. Sample answer:

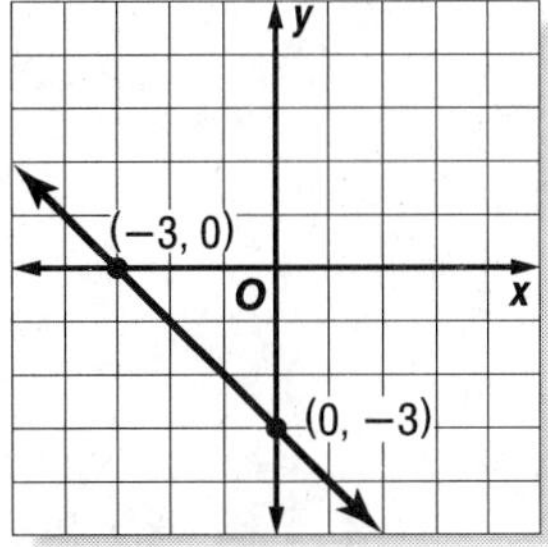

3. The graph crosses the x-axis at -1. The x-intercept is -1. The graph crosses the y-axis at -3. The y-intercept is -3.

4. The graph crosses the x-axis at 5. The x-intercept is 5. The graph does not cross the y-axis. There is no y-intercept.

5. To find the x-intercept, let $y = 0$.
$y = x + 4$
$0 = x + 4$
$-4 = x$
The x-intercept is -4.
To find the y-intercept, let $x = 0$.
$y = x + 4$
$y = 0 + 4$
$y = 4$
The y-intercept is 4.

6. Note that $y = 7$ is the same as $0x + y = 7$. There is no x-intercept, since $0x + 0 = 7$ has no solution. The y-intercept is 7.

7. To find the x-intercept, let $y = 0$.
$2x + 3y = 6$
$2x + 3(0) = 6$
$2x = 6$
$x = 3$
The x-intercept is 3.
To find the y-intercept, let $x = 0$.
$2x + 3y = 6$
$2(0) + 3y = 6$
$3y = 6$
$y = 2$
The y-intercept is 2.

8. Find the x-intercept.
$y = x + 1$
$0 = x + 1$
$-1 = x$
Find the y-intercept.
$y = x + 1$
$y = 0 + 1$
$y = 1$
Graph the points $(-1, 0)$ and $(0, 1)$ and draw a line through them.

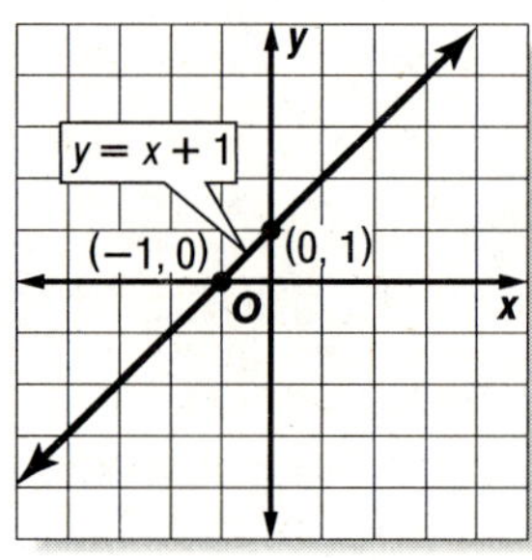

9. Find the x-intercept.
$x - 2y = 6$
$x - 2(0) = 6$
$x = 6$
Find the y-intercept.
$x - 2y = 6$
$0 - 2y = 6$
$-2y = 6$
$y = -3$
Graph the points $(6, 0)$ and $(0, -3)$ and draw a line through them.

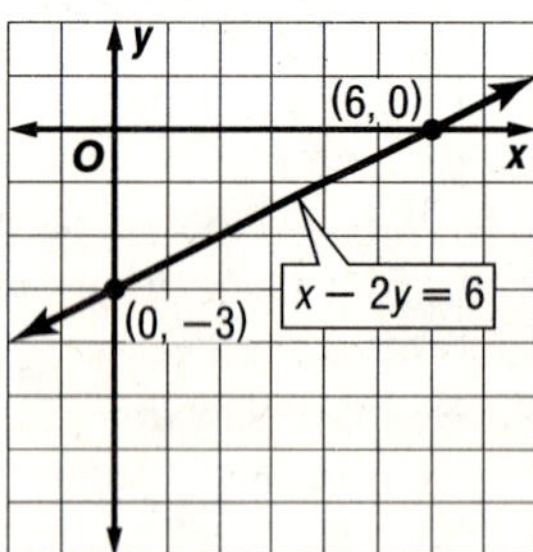

10. Note that $x = -1$ is the same as $x + 0y = -1$. The x-intercept is $(-1, 0)$, and there is no y-intercept.

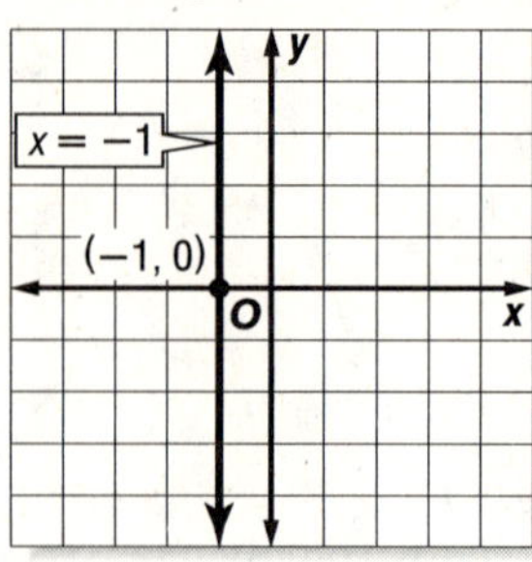

11. Find the x-intercept.
$y = 6x + 3$
$0 = 6x + 3$
$0 - 3 = 6x + 3 - 3$
$-3 = 6x$
$-\frac{3}{6} = \frac{6x}{6}$
$-\frac{1}{2} = x$
Find the y-intercept.
$y = 6x + 3$
$y = 6(0) + 3$
$y = 3$
Plot the points $\left(-\frac{1}{2}, 0\right)$ and $(0, 3)$ and draw a line through them.
The y-intercept 3 represents the base fee of \$3.

Pages 384–385 Practice and Apply

12. The graph crosses the x-axis at -3. The x-intercept is -3. The graph crosses the y-axis at 2. The y-intercept is 2.

13. The graph crosses the x-axis at 1. The x-intercept is 1. The graph crosses the y-axis at 1. The y-intercept is 1.

14. The graph crosses the x-axis at -4. The x-intercept is -4. The graph crosses the y-axis at -6. The y-intercept is -6.

15. The graph does not cross the x-axis. There is no x-intercept. The graph crosses the y-axis at -5. The y-intercept is -5.

16. To find the x-intercept, let $y = 0$.
$y = x - 1$
$0 = x - 1$
$1 = x$
The x-intercept is 1.
To find the y-intercept, let $x = 0$.
$y = x - 1$
$y = 0 - 1$
$y = -1$
The y-intercept is -1.

17. To find the x-intercept, let $y = 0$.
$y = x + 5$
$0 = x + 5$
$-5 = x$
The x-intercept is -5.

To find the y-intercept, let $x = 0$.
$y = x + 5$
$y = 0 + 5$
$y = 5$
The y-intercept is 5.

18. Note that $x = 9$ is the same as $x + 0y = 9$. The x-intercept is 9. There is no y-intercept, since $0 + 0y = 9$ has no solution.

19. Note that $y + 4 = 0$ is the same as $y = -4$ or $0x + y = -4$. There is no x-intercept, since $0x + 0 = -4$ has no solution. The y-intercept is -4.

20. To find the x-intercept, let $y = 0$.
$y = 2x + 10$
$0 = 2x + 10$
$0 - 10 = 2x + 10 - 10$
$-10 = 2x$
$\frac{-10}{2} = \frac{2x}{2}$
$-5 = x$
The x-intercept is -5.
To find the y-intercept, let $x = 0$.
$y = 2x + 10$
$y = 2(0) + 10$
$y = 10$
The y-intercept is 10.

21. To find the x-intercept, let $y = 0$.
$x - 2y = 8$
$x - 2(0) = 8$
$x = 8$
The x-intercept is 8.
To find the y-intercept, let $x = 0$.
$x - 2y = 8$
$0 - 2y = 8$
$-2y = 8$
$y = -4$
The y-intercept is -4.

22. To find the x-intercept, let $y = 0$.
$y = -3x - 12$
$0 = -3x - 12$
$0 + 12 = -3x - 12 + 12$
$12 = -3x$
$-4 = x$
The x-intercept is -4.
To find the y-intercept, let $x = 0$.
$y = -3x - 12$
$y = -3(0) - 12$
$y = -12$
The y-intercept is -12.

23. To find the x-intercept, let $y = 0$.
$4x + 5y = 20$
$4x + 5(0) = 20$
$4x = 20$
$x = 5$
The x-intercept is 5.
To find the y-intercept, let $x = 0$.
$4x + 5y = 20$
$4(0) + 5y = 20$
$5y = 20$
$y = 4$
The y-intercept is 4.

24. To find the x-intercept, let $y = 0$.
$6x + 7y = 12$
$6x + 7(0) = 12$
$6x = 12$
$x = 2$
The x-intercept is 2.
To find the y-intercept, let $x = 0$.
$6x + 7y = 12$
$6(0) + 7y = 12$
$7y = 12$
$y = \frac{12}{7}$
The y-intercept is $\frac{12}{7}$.

25. Find the x-intercept.
$y = x + 2$
$0 = x + 2$
$-2 = x$
Find the y-intercept.
$y = x + 2$
$y = 0 + 2$
$y = 2$
Graph the points $(-2, 0)$ and $(0, 2)$ and draw a line through them.

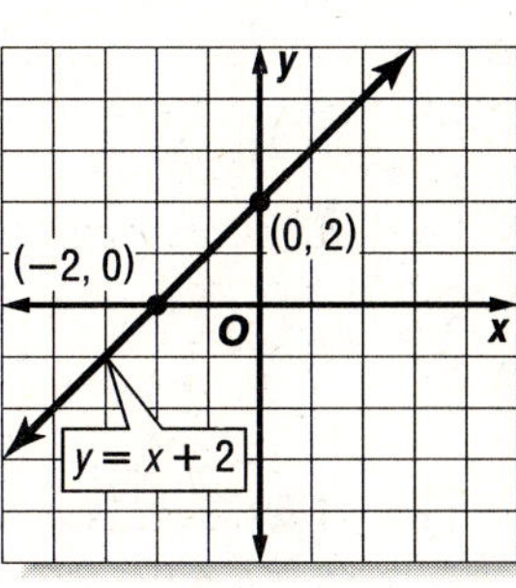

26. Find the x-intercept.
$y = x - 3$
$0 = x - 3$
$3 = x$
Find the y-intercept.
$y = x - 3$
$y = 0 - 3$
$y = -3$
Graph the points $(3, 0)$ and $(0, -3)$ and draw a line through them.

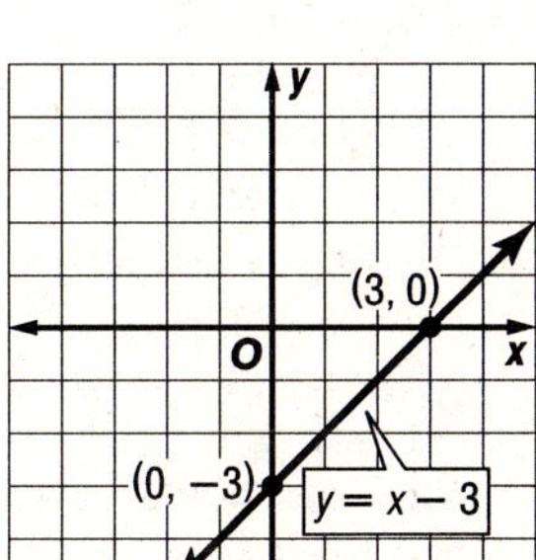

27. Find the x-intercept.
$x + y = 4$
$x + 0 = 4$
$x = 4$
Find the y-intercept.
$x + y = 4$
$0 + y = 4$
$y = 4$
Graph the points $(4, 0)$ and $(0, 4)$ and draw a line through them.

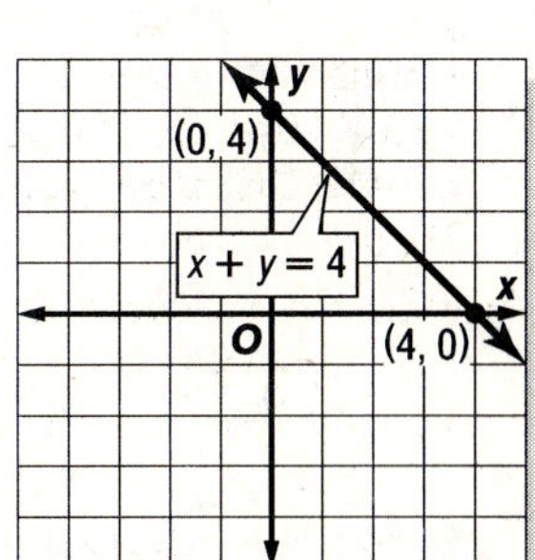

28. Find the x-intercept.
$y = 5x + 5$
$0 = 5x + 5$
$-5 = 5x$
$-1 = x$
Find the y-intercept.
$y = 5x + 5$
$y = 5(0) + 5$
$y = 5$
Graph the points $(-1, 0)$ and $(0, 5)$ and draw a line through them.

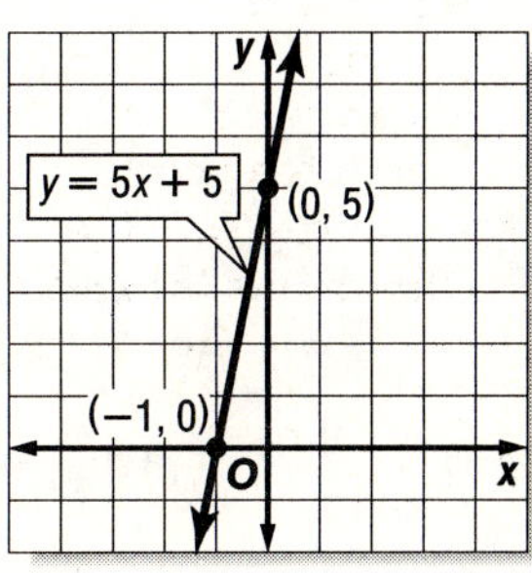

29. Find the x-intercept.
$y = -2x + 4$
$0 = -2x + 4$
$-4 = -2x$
$2 = x$
Find the y-intercept.
$y = -2x + 4$
$y = -2(0) + 4$
$y = 4$
Graph the points (2, 0) and (0, 4) and draw a line through them.

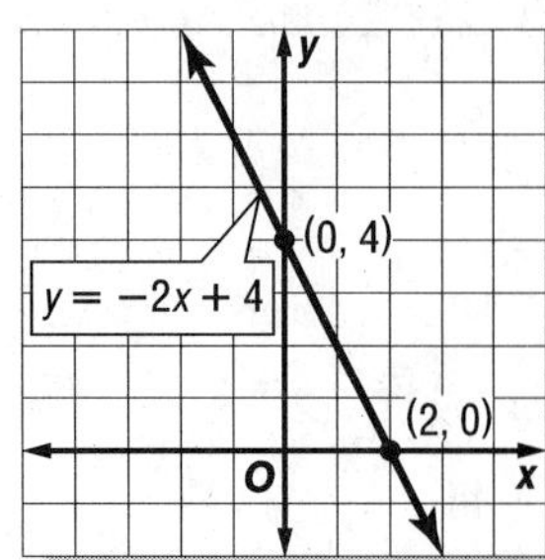

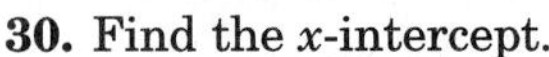

30. Find the x-intercept.
$x + 2y = -6$
$x + 2(0) = -6$
$x = -6$
Find the y-intercept.
$x + 2y = -6$
$0 + 2y = -6$
$2y = -6$
$y = -3$
Graph the points $(-6, 0)$ and $(0, -3)$ and draw a line through them.

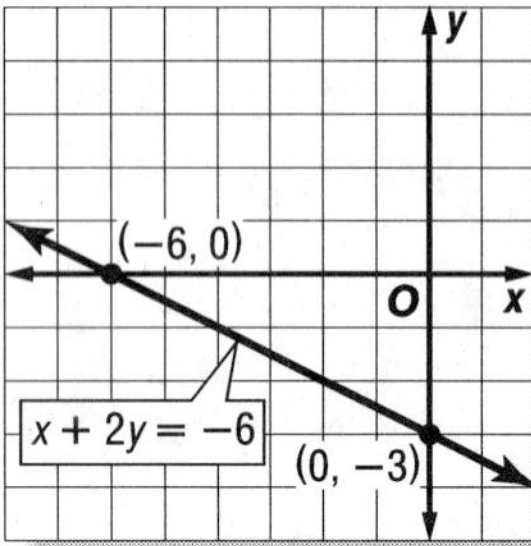
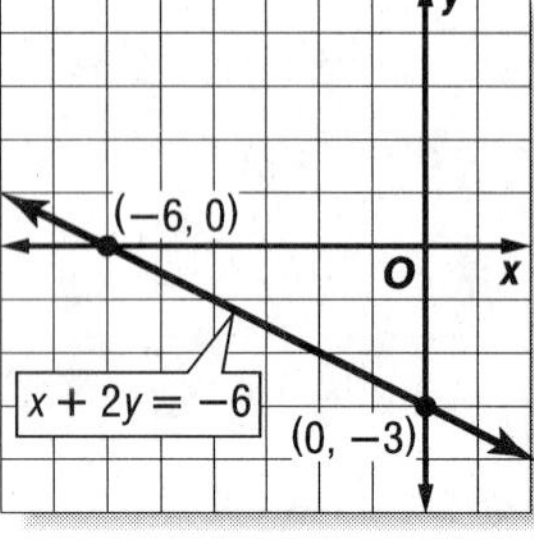

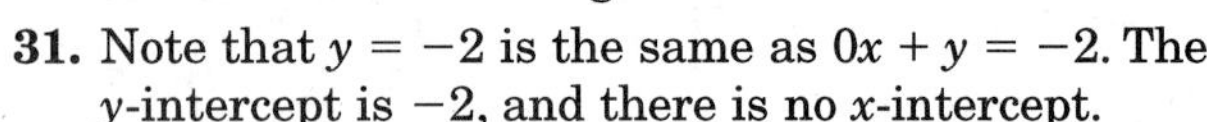

31. Note that $y = -2$ is the same as $0x + y = -2$. The y-intercept is -2, and there is no x-intercept.

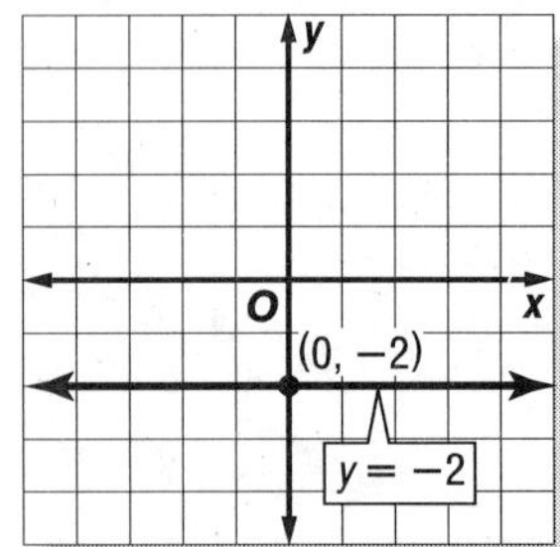

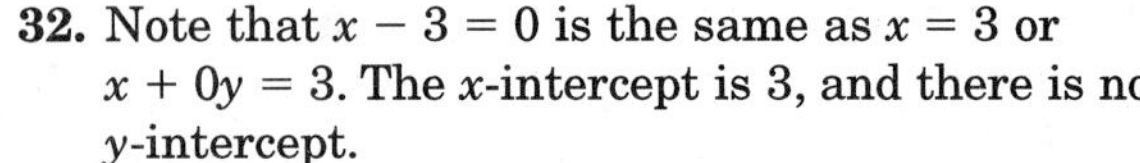

32. Note that $x - 3 = 0$ is the same as $x = 3$ or $x + 0y = 3$. The x-intercept is 3, and there is no y-intercept.

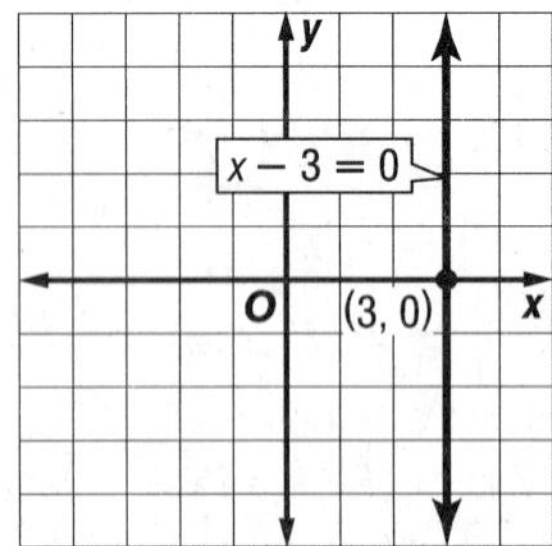

33. Find the x-intercept.
$3x + 6y = 18$
$3x + 6(0) = 18$
$3x = 18$
$x = 6$
Find the y-intercept.
$3x + 6y = 18$
$3(0) + 6y = 18$
$6y = 18$
$y = 3$
Graph the points (6, 0) and (0, 3) and draw a line through them.

34. Find the x-intercept.
$y = 8x + 24$
$0 = 8x + 24$
$-24 = 8x$
$-3 = x$
Find the y-intercept.
$y = 8x + 24$
$y = 8(0) + 24$
$y = 24$
Graph the points $(-3, 0)$ and (0, 24) and draw a line through them.
Note that the number of people can only be positive integers. Therefore, the graph must start at the y-intercept, and only points for integer coordinates should be plotted.
The y-intercept 24 represents the setup fee.

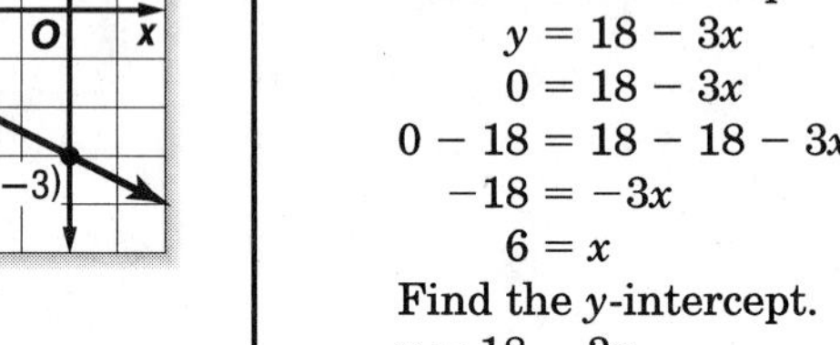

35. Find the x-intercept.
$y = 18 - 3x$
$0 = 18 - 3x$
$0 - 18 = 18 - 18 - 3x$
$-18 = -3x$
$6 = x$
Find the y-intercept.
$y = 18 - 3x$
$y = 18 - 3(0)$
$y = 18$
Graph the points (6, 0) and (0, 18) and draw a line through them.
The x-intercept 6 represents the number of books that she can buy with no money left over. The y-intercept 18 represents the money she has before she buys any books.

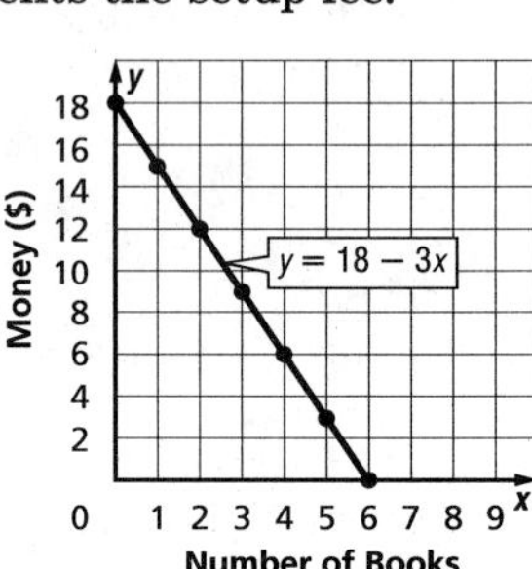

36. Let $\ell = 0$.
$50 = 2\ell + 2w$
$50 = 2(0) + 2w$
$50 = 2w$
$25 = w$
Let $w = 0$
$50 = 2\ell + 2w$
$50 = 2\ell + 2(0)$
$50 = 2\ell$
$25 = \ell$
The intercepts are 25 and 25. They represent one dimension when the other is 0, which is impossible for a rectangle.

37. The x- and y-intercept are both 0. Therefore, the line passes through the origin. Since two points are needed to graph a line, $y = 2x$ cannot be graphed using only the intercepts.

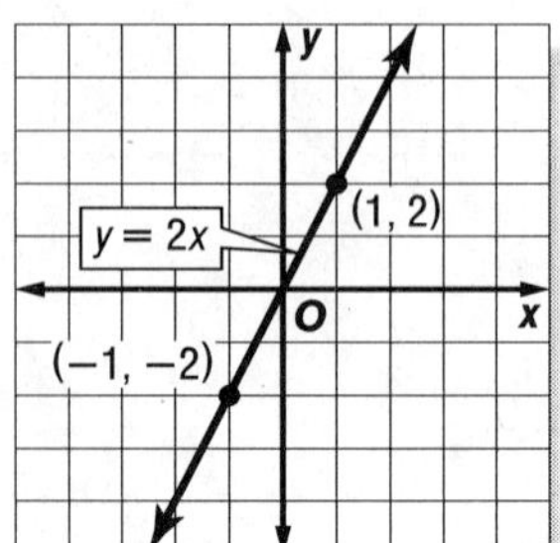

38. The intercepts can be used to show values at time 0, temperature 0, and so on. Answers should include the following.

- Sample graph:

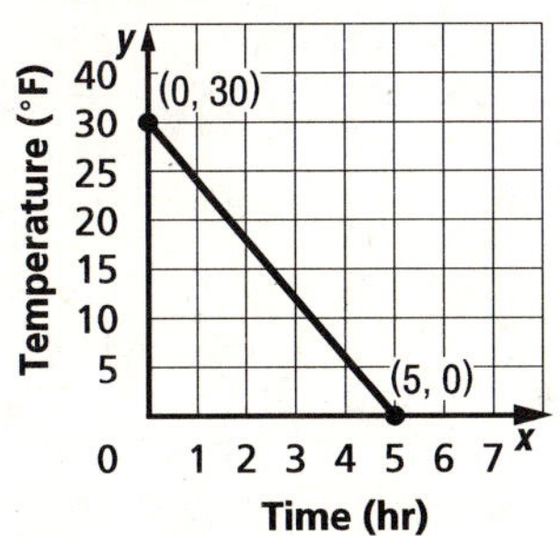

- The x-intercept 5 means that after 5 hours the temperature is 0°F. The y-intercept 30 means that at time 0, the temperature was 30°F.

39. B; Let $y = 0$ in $y = 8x - 32$.

$$0 = 8x - 32$$
$$0 + 32 = 8x - 32 + 32$$
$$32 = 8x$$
$$\frac{32}{8} = \frac{8x}{8}$$
$$4 = x$$

The x-intercept is 4.

40. C; To find the y-intercept of $x + 3y = 6$, let $x = 0$.

$$0 + 3y = 6$$
$$3y = 6$$
$$y = 2$$

$x + 3y = 6$ does not have a y-intercept of 3.

Page 385 Maintain Your Skills

41. Choose four values for x and substitute them into $y = 2x + 7$.

Sample answer:

x	$y = 2x + 7$	y	(x, y)
-1	$y = 2(-1) + 7 = 5$	5	$(-1, 5)$
0	$y = 2(0) + 7 = 7$	7	$(0, 7)$
1	$y = 2(1) + 7 = 9$	9	$(1, 9)$
2	$y = 2(2) + 7 = 11$	11	$(2, 11)$

Four solutions are $(-1, 5)$, $(0, 7)$, $(1, 9)$, and $(2, 11)$.

42. Choose four values for x and substitute them into $y = -3x + 1$.

Sample answer:

x	$y = -3x + 1$	y	(x, y)
-1	$y = -3(-1) + 1 = 4$	4	$(-1, 4)$
0	$y = -3(0) + 1 = 1$	1	$(0, 1)$
1	$y = -3(1) + 1 = -2$	-2	$(1, -2)$
2	$y = -3(2) + 1 = -5$	-5	$(2, -5)$

Four solutions are $(-1, 4)$, $(0, 1)$, $(1, -2)$, and $(2, -5)$.

43. First solve the equation for y.

$$4x - y = -5$$
$$4x - 4x - y = -5 - 4x$$
$$-y = -5 - 4x$$
$$y = 5 + 4x$$

Choose four values for x and substitute them into $y = 5 + 4x$.

Sample answer:

x	$y = 5 + 4x$	y	(x, y)
-1	$y = 5 + 4(-1) = 1$	1	$(-1, 1)$
0	$y = 5 + 4(0) = 5$	5	$(0, 5)$
1	$y = 5 + 4(1) = 9$	9	$(1, 9)$
2	$y = 5 + 4(2) = 13$	13	$(2, 13)$

Four solutions are $(-1, 1)$, $(0, 5)$, $(1, 9)$, and $(2, 13)$.

44. Yes, because each x-value is paired with only one y-value.

45. No, because the domain values of -4.3 is paired with 16, 15, and 14 in the range.

46.
$$y + 3 < 5$$
$$y + 3 - 3 < 5 - 3$$
$$y < 2$$

47.
$$-2 + n > 10$$
$$-2 + 2 + n > 10 + 2$$
$$n > 12$$

48.
$$7 \leq x + 8$$
$$7 - 8 \leq x + 8 - 8$$
$$-1 \leq x \text{ or } x \geq -1$$

49. $0.028 = 2.8\%$

50. $-11 - 13 = -11 + (-13)$
$= -24$

51. $15 - 31 = 15 + (-31)$
$= -16$

52. $-26 - (-26) = -26 + 26$
$= 0$

53. $9 - (-16) = 9 + 16$
$= 25$

Page 386 Algebra Activity (Preview of Lesson 8-4)

1. The slope increased.
2. The slope decreased.
3. $\frac{25}{18}$ because it is greater than $\frac{18}{25}$; the greater the slope, the farther the toy car traveled.
4. The car will travel farther if the slope is greater than the other slopes and less far if the slope is less than the other slopes.

8-4 Slope

Page 387 How is slope used to describe roller coasters?

a. $\frac{56}{42} = \frac{4}{3}$

b. $\frac{56 + 14}{42} = \frac{70}{42}$
$= \frac{5}{3}$

The hill is steeper than the original hill.

Pages 389–390 Check for Understanding

1. Sample answer: horizontals do not rise, so slope $= \frac{\text{rise}}{\text{run}} = \frac{0}{\text{run}}$ or 0.

2. Sample answer:

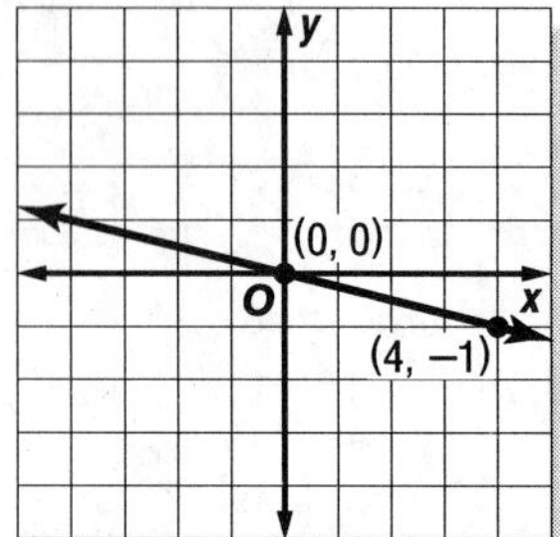

3. Mike; Chloe should have subtracted −2 from 11 in the denominator.

4. $\text{slope} = \frac{\text{rise}}{\text{run}}$
$= \frac{-24}{30}$
$= -\frac{4}{5}$

5. $m = \frac{y_2 - y_1}{x_2 - x_1}$
$m = \frac{0 - (-2)}{-3 - 0}$
$m = \frac{2}{-3}$ or $-\frac{2}{3}$

6. $m = \frac{y_2 - y_1}{x_2 - x_1}$
$m = \frac{2 - 2}{1 - (-1)}$
$m = \frac{0}{2}$
$m = 0$

7. $m = \frac{y_2 - y_1}{x_2 - x_1}$
$m = \frac{6 - 4}{4 - 3}$
$m = \frac{2}{1}$ or 2

8. $m = \frac{y_2 - y_1}{x_2 - x_1}$
$m = \frac{10 - 0}{-8 - (-8)}$
$m = \frac{10}{-8 + 8}$
$m = \frac{10}{0}$ or undefined

9. $m = \frac{y_2 - y_1}{x_2 - x_1}$
$m = \frac{-1 - (-1)}{9 - 7}$
$m = \frac{-1 + 1}{2}$
$m = \frac{0}{2}$ or 0

10. $m = \frac{y_2 - y_1}{x_2 - x_1}$
$m = \frac{-3 - (-4)}{-8 - (-6)}$
$m = \frac{-3 + 4}{-8 + 6}$
$m = \frac{1}{-2}$ or $-\frac{1}{2}$

11. B; The slopes for each can be found by $\frac{\text{rise}}{\text{run}}$ or $\frac{\text{height}}{\text{length}}$.
Ramp 1: $\frac{6}{8} = \frac{3}{4}$
Ramp 2: $\frac{10}{4} = 2.5$
Ramp 3: $\frac{5}{3} = 1.\overline{6}$
Ramp 4: $\frac{8}{4} = 2$
The steepest ramp is the one with the greatest slope, ramp 2.

Pages 390–391 Practice and Apply

12. $\text{slope} = \frac{\text{rise}}{\text{run}}$ or $\frac{\text{riser}}{\text{tread}}$
$= \frac{8}{12}$
$= \frac{2}{3}$

13. $\text{slope} = \frac{\text{rise}}{\text{run}}$ or $\frac{\text{height of ladder}}{\text{distance from house}}$
$= \frac{16}{4}$
$= 4$

14. $m = \frac{y_2 - y_1}{x_2 - x_1}$
$m = \frac{3 - 0}{1 - (-2)}$
$m = \frac{3}{1 + 2}$
$m = \frac{3}{3}$
$m = 1$

15. $m = \frac{y_2 - y_1}{x_2 - x_1}$
$m = \frac{-4 - (-1)}{3 - (-1)}$
$m = \frac{-4 + 1}{3 + 1}$
$m = \frac{-3}{4}$ or $-\frac{3}{4}$

16. $m = \frac{y_2 - y_1}{x_2 - x_1}$
$m = \frac{-5 - (-1)}{-6 - (-6)}$
$m = \frac{-5 + 1}{-6 + 6}$
$m = \frac{-4}{0}$ or undefined

17. $m = \frac{y_2 - y_1}{x_2 - x_1}$
$m = \frac{-5 - (-5)}{3 - 0}$
$m = \frac{-5 + 5}{3}$
$m = \frac{0}{3}$ or 0

18. $m = \frac{y_2 - y_1}{x_2 - x_1}$
$m = \frac{4 - (-3)}{5 - 1}$
$m = \frac{4 + 3}{4}$
$m = \frac{7}{4}$

19. $m = \frac{y_2 - y_1}{x_2 - x_1}$
$m = \frac{-2 - (-3)}{5 - 4}$
$m = \frac{-2 + 3}{1}$
$m = 1$

20. $m = \frac{y_2 - y_1}{x_2 - x_1}$
$m = \frac{4 - (-1)}{-3 - 5}$
$m = \frac{4 + 1}{-8}$
$m = \frac{5}{-8}$ or $-\frac{5}{8}$

21. $m = \frac{y_2 - y_1}{x_2 - x_1}$

$m = \frac{9 - 6}{-5 - (-3)}$

$m = \frac{3}{-5 + 3}$

$m = \frac{3}{-2}$

22. $m = \frac{y_2 - y_1}{x_2 - x_1}$

$m = \frac{6 - 6}{-1 - 2}$

$m = \frac{0}{-3}$ or 0

23. $m = \frac{y_2 - y_1}{x_2 - x_1}$

$m = \frac{8 - (-4)}{-9 - (-9)}$

$m = \frac{8 + 4}{-9 + 9}$

$m = \frac{12}{0}$ or undefined

24. $m = \frac{y_2 - y_1}{x_2 - x_1}$

$m = \frac{2.1 - 1.6}{0.5 - 0}$

$m = \frac{0.5}{0.5}$

$m = 1$

25. $m = \frac{y_2 - y_1}{x_2 - x_1}$

$m = \frac{6 - 5\frac{1}{4}}{2\frac{1}{2} - 3\frac{1}{2}}$

$m = \frac{\frac{3}{4}}{-1}$

$m = -\frac{3}{4}$

26. '99–'00; it has the steepest positive slope.

27. It decreased; it has a negative slope.

28. $m = \frac{y_2 - y_1}{x_2 - x_1}$

$m = \frac{b - 0}{a - 0}$

$m = \frac{b}{a}$

The slope is the y-coordinate of C divided by the x-coordinate of C, when the line passes through the origin (0, 0).

29. Slope can be used to describe the steepness of roller coaster hills. Answers should include the following.
   - Slope is the steepness of a line or incline. It is the ratio of the rise to the run.
   - An increase in rise with no change in run makes a roller coaster hill steeper. An increase in run with no change in rise makes a roller coaster less steep.

30. D; Graph A has a negative slope, graph B has an undefined slope, and graph C has a slope of 0.

31. A; $m = \frac{y_2 - y_1}{x_2 - x_1}$

$m = \frac{-5 - (-2)}{3 - 9}$

$m = \frac{-5 + 2}{-6}$

$m = \frac{-3}{-6}$ or $\frac{1}{2}$

Page 391 Maintain Your Skills

32. To find the x-intercept, let $y = 0$.

$y = x + 8$
$0 = x + 8$
$-8 = x$

The x-intercept is -8.

To find the y-intercept, let $x = 0$.

$y = x + 8$
$y = 0 + 8$
$y = 8$

The y-intercept is 8.

33. To find the x-intercept, let $y = 0$.

$y = -3x + 6$
$0 = -3x + 6$
$0 - 6 = -3x + 6 - 6$
$-6 = -3x$
$2 = x$

The x-intercept is 2.

To find the y-intercept, let $x = 0$

$y = -3x + 6$
$y = -3(0) + 6$
$y = 6$

The y-intercept is 6.

34. To find the x-intercept, let $y = 0$.

$4x - y = 12$
$4x - 0 = 12$
$4x = 12$
$x = 3$

The x-intercept is 3.

To find the y-intercept, let $x = 0$.

$4x - y = 12$
$4(0) - y = 12$
$-y = 12$
$y = -12$

The y-intercept is -12.

35. Choose four values for x and substitute them into $y = 2x + 5$.

Sample answer:

x	$y = 2x + 5$	y	(x, y)
-1	$y = 2(-1) + 5 = 3$	3	$(-1, 3)$
0	$y = 2(0) + 5 = 5$	5	(0, 5)
1	$y = 2(1) + 5 = 7$	7	(1, 7)
2	$y = 2(2) + 5 = 9$	9	(2, 9)

Four solutions are $(-1, 3)$, (0, 5), (1, 7), and (2, 9).

36. Choose four values for x and substitute them into $y = -3x$.

Sample answer:

x	$y = -3x$	y	(x, y)
-1	$y = -3(-1) = 3$	3	$(-1, 3)$
0	$y = -3(0) = 0$	0	(0, 0)
1	$y = -3(1) = -3$	-3	$(1, -3)$
2	$y = -3(2) = -6$	-6	$(2, -6)$

Four solutions are $(-1, 3)$, (0, 0), $(1, -3)$, and $(2, -6)$.

37. First, solve the equation for y.

$$x + y = 7$$
$$x - x + y = 7 - x$$
$$y = 7 - x$$

Choose four values for x and substitute them into $y = 7 - x$.

Sample answer:

x	$y = 7 - x$	y	(x, y)
−1	$y = 7 - (-1) = 8$	8	(−1, 8)
0	$y = 7 - (0) = 7$	7	(0, 7)
1	$y = 7 - (1) = 6$	6	(1, 6)
2	$y = 7 - (2) = 5$	5	(2, 5)

Four solutions are (−1, 8), (0, 7), (1, 6), and (2, 5).

38. $y = 5x$

39. $y = -2x$

40. $y = 0.25x$

41. $y = \frac{1}{3}x$

Page 392 Algebra Activity (Preview of Lesson 8-5)

1. See students' work.
2. Line 2 has a steeper negative slope.
3. number of measures removed
4. The height of water decreases.
5. Using a $\frac{1}{8}$-cup; the cup is larger than the tablespoon, so it emptied the glass at a faster rate.
6. The steeper the slope, the faster the rate.
7. Sample answer: teaspoon: slope is less steep than the other graphs because the rate of emptying the glass is slower; $\frac{1}{4}$-cup: steeper slope than the other graphs because the rate of emptying the glass is faster.

8a. Sample answer:

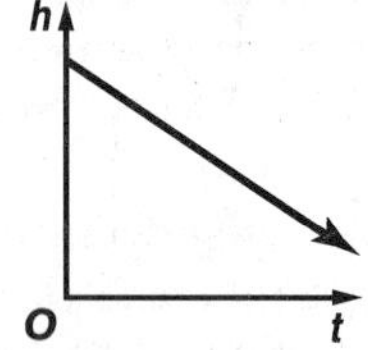

8b. Sample answer:

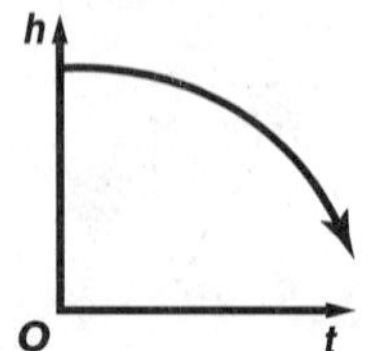

8c. Sample answer:

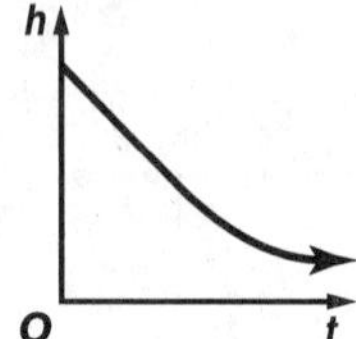

8-5 Rate of Change

Page 393 How are slope and speed related?

a. Let n = increase in time each hour

$$110 + n = 165$$
$$110 - 110 + n = 165 - 110$$
$$n = 55$$
$$165 + n = 220$$
$$165 - 165 + n = 220 - 165$$
$$n = 55$$

For every 1-hour increase in time, the change in distance is 55 miles.

b. $m = \frac{y_2 - y_1}{x_2 - x_1}$

$m = \frac{165 - 110}{3 - 2}$

$m = \frac{55}{1}$ or 55

The slope is 55.

c. Slope is equal to the speed of the car.

Pages 395–396 Check for Understanding

1. The slope is 60, the rate of change is 60 units for every 1 unit, and the constant of variation is 60.
2. Sample answer:

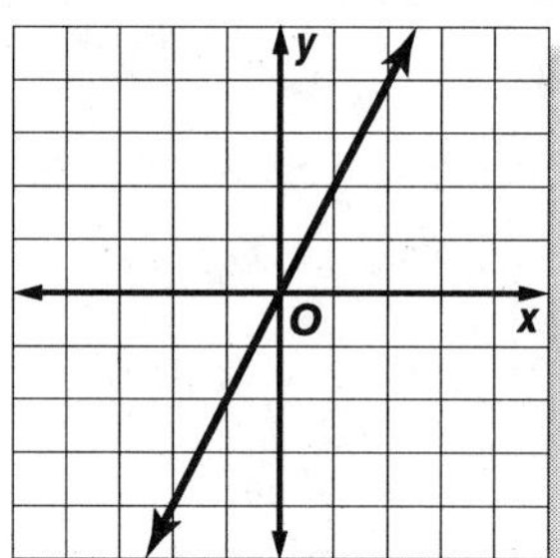

rate of change = 2

3. Justin; any linear function, including direct variations, has a rate of change.
4. rate of change $= \frac{\text{change in } y}{\text{change in } x}$

$= \frac{35 - 25}{0 - 2}$

$= \frac{10}{-2}$

$= -5$

There is a decrease of 5 gal/min.

5. rate of change $= \frac{\text{change in } y}{\text{change in } x}$

$= \frac{12 - 0}{1 - 0}$

$= \frac{12}{1}$

$= 12$

There is an increase of $12 per hour.

6. Find k:

$y = kx$

$5 = k \cdot (-15)$

$\frac{5}{-15} = \frac{k \cdot (-15)}{-15}$

$-\frac{1}{3} = k$

Write the equation:

$y = kx$

$y = -\frac{1}{3}x$

7. Find k: $y = kx$, $24 = k(4)$, $6 = k$

Write the equation: $y = kx$, $y = 6x$

8a. Find k: $y = kx$, $8 = k \cdot 25$, $\frac{8}{25} = \frac{k \cdot 25}{25}$, $0.32 = k$

Write the equation: $y = kx$, $y = 0.32x$

8b. Replace x with 60.
$y = 0.32x$
$y = 0.32(60)$
$y = 19.2$
The length is about 19.2 cm.

Pages 396–397 Practice and Apply

9. rate of change $= \frac{\text{change in } y}{\text{change in } x} = \frac{12 - 0}{1 - 0} = \frac{12}{1} = 12$
There is an increase of 12 in./ft.

10. rate of change $= \frac{\text{change in } y}{\text{change in } x} = \frac{16 - 20}{400 - 0} = \frac{-4}{400} = \frac{-1}{100}$
There is a decrease of $\frac{1}{100}$°C/m or 0.01°C/m.

11. rate of change $= \frac{\text{change in } y}{\text{change in } x} = \frac{58 - 56}{0 - 1} = \frac{2}{-1} = -2$
There is a decrease of 2°F/min.

12. rate of change $= \frac{\text{change in } y}{\text{change in } x} = \frac{0 - 25}{0.0 - 0.5} = \frac{-25}{-0.5} = 50$
There is an increase of 50 mi/h.

13. Sample answer: The population of wild condors decreased from 1966 to 1990. Then the population increased from 1990 to 1996. The population of condors in captivity increased slowly from 1966 to 1982, then increased more rapidly from 1982 to 1996.

	Rate of Change (number per year)	
Interval	Condors in the Wild	Condors in Captivity
1966–1982	−2.25	0.1875
1982–1990	−3	4.625
1990–1992	0	8
1992–1994	1.5	14.5
1994–1996	12.5	3.5

14. Find k: $y = kx$, $8 = k(4)$, $2 = k$

Write the equation: $y = kx$, $y = 2x$

15. Find k: $y = kx$, $-30 = k(6)$, $-5 = k$

Write the equation: $y = kx$, $y = -5x$

16. Find k: $y = kx$, $9 = k(24)$, $\frac{9}{24} = k$, $\frac{3}{8} = k$

Write the equation: $y = kx$, $y = \frac{3}{8}x$

17. Find k: $y = kx$, $7.5 = k(10)$, $\frac{7.5}{10} = k$, $0.75 = k$

Write the equation: $y = kx$, $y = 0.75x$

18. Let y represent the cost of cheese.
Let x represent the number of pounds.
Find k: $y = kx$, $8.40 = k(2)$, $\frac{8.40}{2} = k$, $4.20 = k$
Write the equation: $y = kx$, $y = 4.20x$
Substitute 3.5 for x in the equation.
$y = 4.20(3.5)$
$y = 14.7$
The cost is \$14.70.

19. Let y represent the number of centimeters.
Let x represent the number of inches.
Find k: $y = kx$, $2.54 = k(1)$, $2.54 = k$
Write the equation: $y = kx$, $y = 2.54x$

20. A horizontal line represents a zero rate of change because there is no change in y. A vertical line represents no passage of time, so a rate of change does not make sense.

21. A line representing the relationship between time and distance has a slope that is equal to the speed. Answers should include the following.
- Sample drawing:

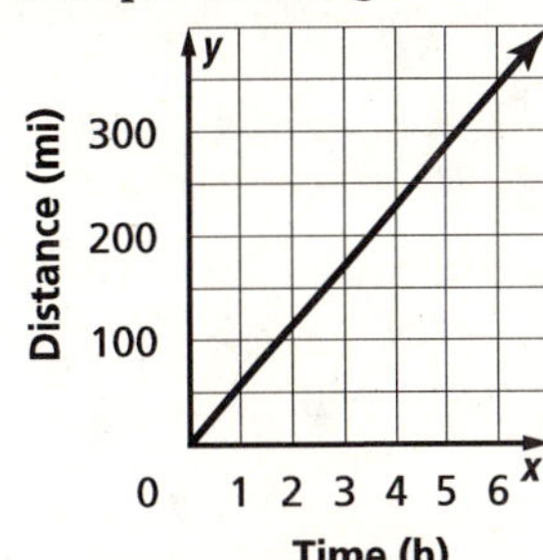

- As speed increases, the slope of the graph becomes steeper.

22. A; positive slope

23. B; $y = 1$ does not match the form $y = kx$.

Page 397 Maintain Your Skills

24. $m = \frac{y_2 - y_1}{x_2 - x_1}$

$m = \frac{5 - 4}{3 - (-4)}$

$m = \frac{1}{3 + 4}$

$m = \frac{1}{7}$

25. $m = \frac{y_2 - y_1}{x_2 - x_1}$

$m = \frac{0 - 6}{-1 - 2}$

$m = \frac{-6}{-3}$

$m = 2$

26. Find the x-intercept.

$y = x + 5$

$0 = x + 5$

$-5 = x$

Find the y-intercept.

$y = x + 5$

$y = 0 + 5$

$y = 5$

Graph the points $(-5, 0)$ and $(0, 5)$ and draw a line through them.

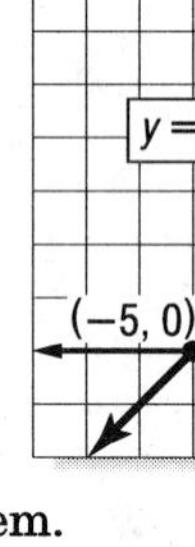

27. Find the x-intercept.

$y = -x + 1$

$0 = -x + 1$

$-1 = -x$

$1 = x$

Find the y-intercept.

$y = -x + 1$

$y = -(0) + 1$

$y = 1$

Graph the points $(1, 0)$ and $(0, 1)$ and draw a line through them.

28. Find the x-intercept.

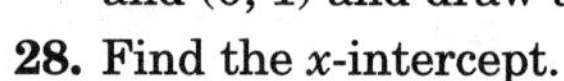

$2x + y = 4$

$2x + 0 = 4$

$2x = 4$

$x = 2$

Find the y-intercept.

$2x + y = 4$

$2(0) + y = 4$

$y = 4$

Graph the points $(2, 0)$ and $(0, 4)$ and draw a line through them.

29. 20% is $\frac{1}{5}$.

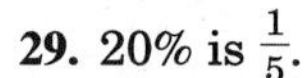

72 is about 70.

$\frac{1}{5}$ of 70 is 14.

20% of 72 is about 14.

30. $x + y = 6$

$x - x + y = 6 - x$

$y = 6 - x$ or $y = -x + 6$

31. $3x + y = 1$

$3x - 3x + y = 1 - 3x$

$y = 1 - 3x$ or $y = -3x + 1$

32. $-x + 5y = 10$

$-x + x + 5y = 10 + x$

$5y = 10 + x$

$\frac{5y}{5} = \frac{10 + x}{5}$

$y = 2 + \frac{1}{5}x$ or $y = \frac{1}{5}x + 2$

Page 397 Practice Quiz 1

1. No; 1 is paired with 2 and −3.
2. Yes; each x value is paired with only one y value.
3. Sample answer:

x	$y = x - 4$	y	(x, y)
0	$y = 0 - 4$	−4	(0, −4)
1	$y = 1 - 4$	−3	(1, −3)
2	$y = 2 - 4$	−2	(2, −2)
3	$y = 3 - 4$	−1	(3, −1)

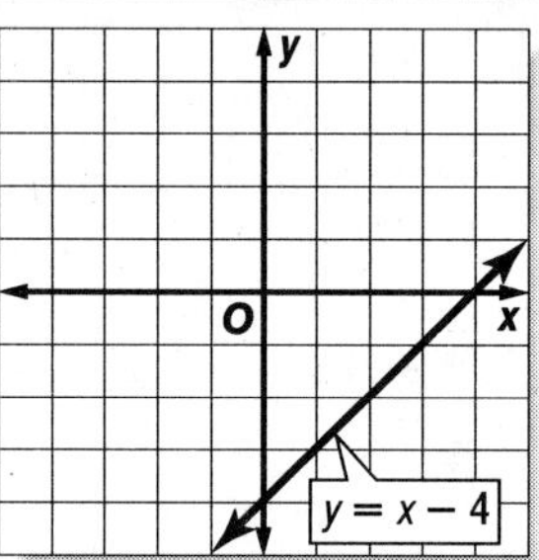

4. Sample answer:

x	$y = 2x + 3$	y	(x, y)
−3	$y = 2(-3) + 3$	−3	(−3, −3)
−2	$y = 2(-2) + 3$	−1	(−2, −1)
−1	$y = 2(-1) + 3$	1	(−1, 1)
0	$y = 2(0) + 3$	3	(0, 3)

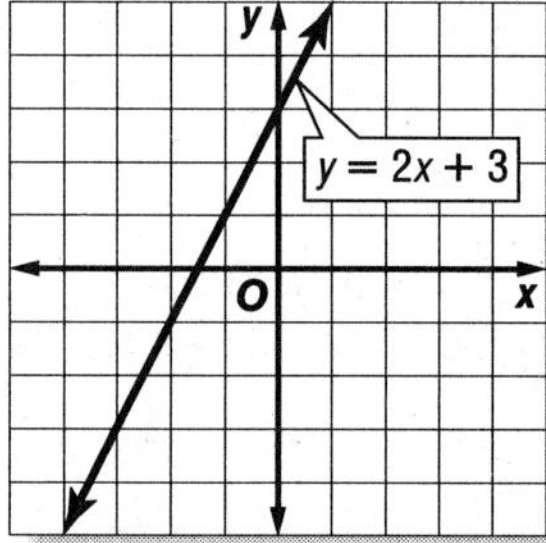

5. To find the x-intercept, let $y = 0$.

$y = x + 9$

$0 = x + 9$

$-9 = x$

The x-intercept is −9.

To find the y-intercept, let $x = 0$.

$y = x + 9$

$y = 0 + 9$

$y = 9$

The y-intercept is 9.

6. To find the x-intercept, let $y = 0$.

$x + 2y = 12$

$x + 2(0) = 12$

$x = 12$

The x-intercept is 12.

To find the y-intercept, let $x = 0$.

$x + 2y = 12$
$0 + 2y = 12$
$2y = 12$
$y = 6$

The y-intercept is 6.

7. To find the x-intercept, let $y = 0$.

$4x - 5y = 20$
$4x - 5(0) = 20$
$4x = 20$
$x = 5$

The x-intercept is 5.

To find the y-intercept, let $x = 0$.

$4x - 5y = 20$
$4(0) - 5y = 20$
$-5y = 20$
$y = -4$

The y-intercept is -4.

8. $m = \frac{y_2 - y_1}{x_2 - x_1}$

$m = \frac{0 - 4}{0 - 1}$

$m = \frac{-4}{-1}$

$m = 4$

9. $m = \frac{y_2 - y_1}{x_2 - x_1}$

$m = \frac{-6 - 4}{3 - (-2)}$

$m = \frac{-10}{5}$

$m = -2$

10. $m = \frac{y_2 - y_1}{x_2 - x_1}$

$m = \frac{2 - 2}{5 - 0}$

$m = \frac{0}{5}$ or 0

8-6 Slope-Intercept Form

Page 398 How can knowing the slope and y-intercept help you graph an equation?

a.

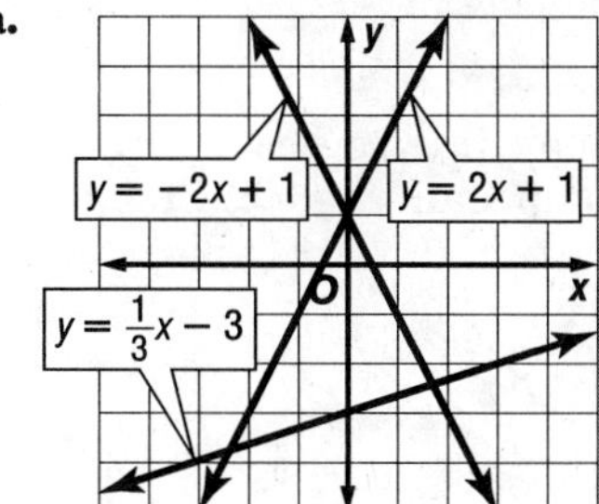

b.

Equation	Slope	y-intercept
$y = 2x + 1$	2	1
$y = \frac{1}{3}x - 3$	$\frac{1}{3}$	-3
$y = -2x + 1$	-2	1

c. The slope is the coefficient of x; the y-intercept is the constant.

Page 400 Check for Understanding

1. a, since the numerator of the fraction represents the change in y.

2. Sample answer:

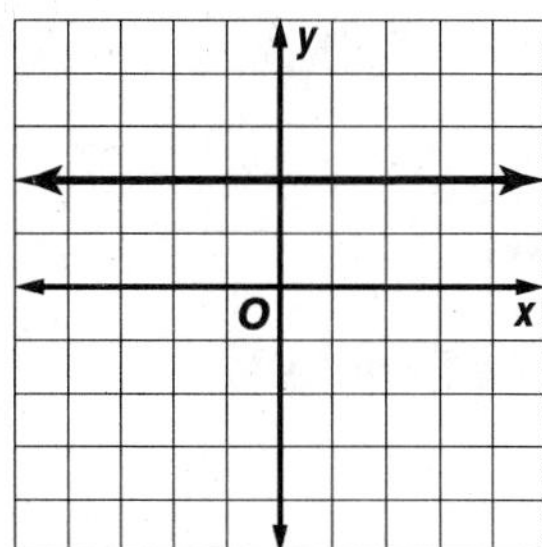

slope = 0

3. Alex; the equation in slope-intercept form is $y = -\frac{1}{2}x + 4$.

4. $y = mx + b$

$y = 1x + 8$

slope = 1; y-intercept = 8

5. Write the equation in the form $y = mx + b$:

$x + y = 0$
$x - x + y = 0 - x$
$y = -x$
$y = -x + 0$

slope = -1; y-intercept = 0

6. Write the equation in the form $y = mx + b$:

$x + 3y = 6$
$x - x + 3y = 6 - x$
$3y = 6 - x$
$\frac{3y}{y} = \frac{6 - x}{3}$
$y = 2 - \frac{1}{3}x$
$y = -\frac{1}{3}x + 2$

slope = $-\frac{1}{3}$; y-intercept = 2

7. The slope is $\frac{1}{4}$ and the y-intercept is 1. Graph the point (0, 1). Then go up 1 and right 4. Connect these points.

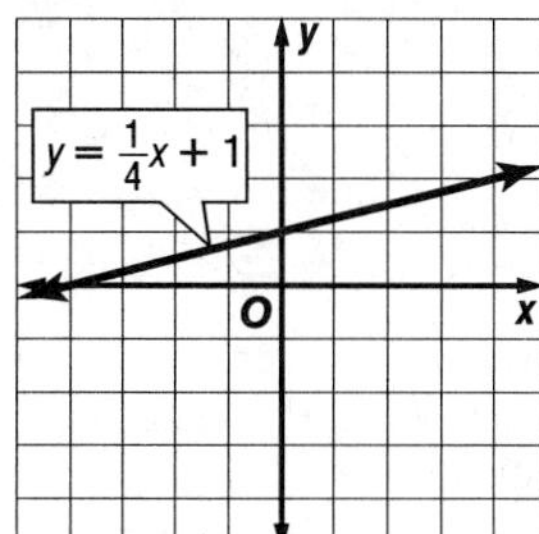

8. Write the equation in the form $y = mx + b$:

$3x + y = 2$
$3x - 3x + y = 2 - 3x$
$y = 2 - 3x$
$y = -3x + 2$

The slope is -3 and the y-intercept is 2. Graph the point (0, 2). Then go down 3 and right 1. Connect these points.

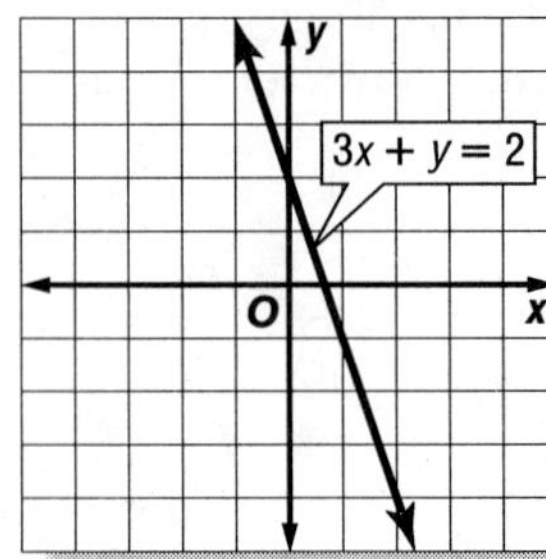

9. Write the equation in the form $y = mx + b$:

$$x - 2y = 4$$
$$x - x - 2y = 4 - x$$
$$-2y = 4 - x$$
$$\frac{-2y}{-2} = \frac{4 - x}{-2}$$
$$y = -2 + \frac{1}{2}x$$
$$y = \frac{1}{2}x - 2$$

The slope is $\frac{1}{2}$ and the y-intercept is -2. Graph the point $(0, -2)$. Then go up 1 and right 2. Connect these points.

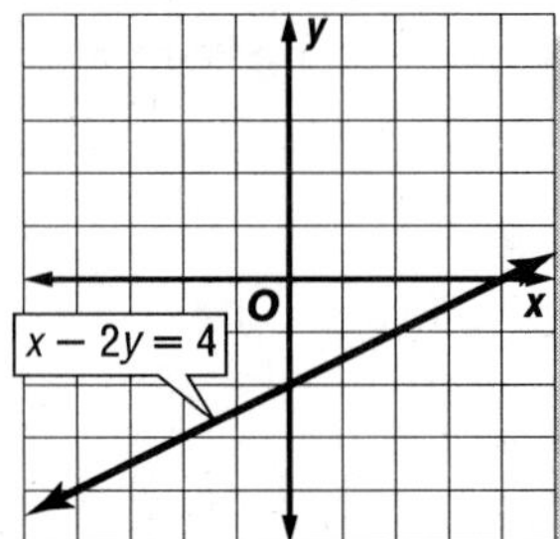

10. The slope is 1.5 and the y-intercept is 25. Graph the point $(0, 25)$. Then go up 1.5 and right 1. Connect these points.

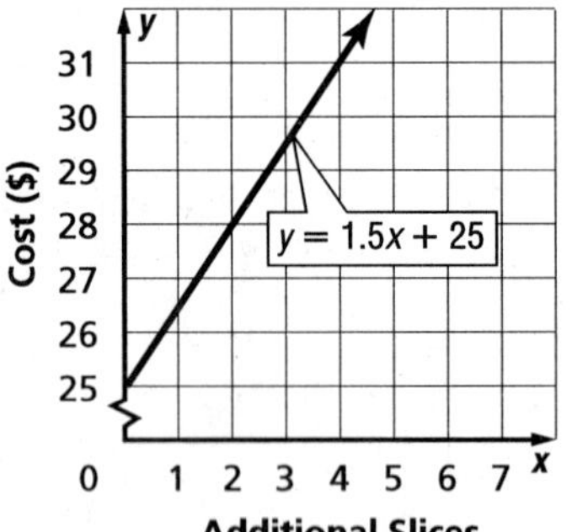

11. The y-intercept 25 represents the charge for a basic cake. Slope 1.5 represents the cost per additional slice.

Pages 400–401 Practice and Apply

12. slope $= 1$; y-intercept $= 2$
13. slope $= 2$; y-intercept $= -4$
14. Write the equation in the form $y = mx + b$:

$$x + y = -3$$
$$x - x + y = -3 - x$$
$$y = -3 - x$$
$$y = -x - 3$$

slope $= -1$; y-intercept $= -3$

15. Write the equation in the form $y = mx + b$:

$$2x + y = -3$$
$$2x - 2x + y = -3 - 2x$$
$$y = -3 - 2x$$
$$y = -2x - 3$$

slope $= -2$; y-intercept $= -3$

16. Write the equation in the form $y = mx + b$:

$$5x + 4y = 20$$
$$5x - 5x + 4y = 20 - 5x$$
$$4y = 20 - 5x$$
$$\frac{4y}{4} = \frac{20 - 5x}{4}$$
$$y = 5 - \frac{5}{4}x$$
$$y = -\frac{5}{4}x + 5$$

slope $= -\frac{5}{4}$; y-intercept $= 5$

17. Write the equation in the form $y = mx + b$:

$y = 4$

$y = 0x + 4$

slope $= 0$; y-intercept $= 4$

18. Graph the point $(0, 1)$. Then go up 3 and right 1. Connect the points.

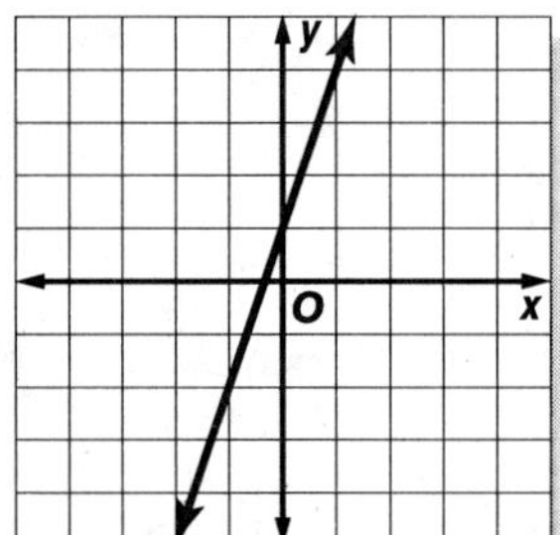

19. Graph the point $(0, -1)$. Then go down 3 and right 2. Connect the points.

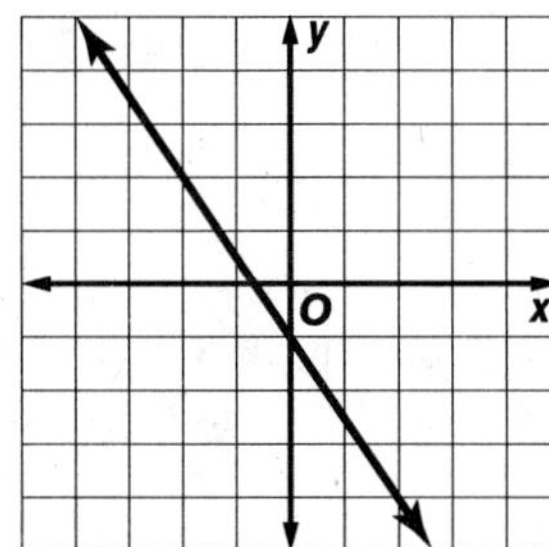

20. The slope is 1 and the y-intercept is 5. Graph the point $(0, 5)$. Then go up 1 and right 1. Connect the points.

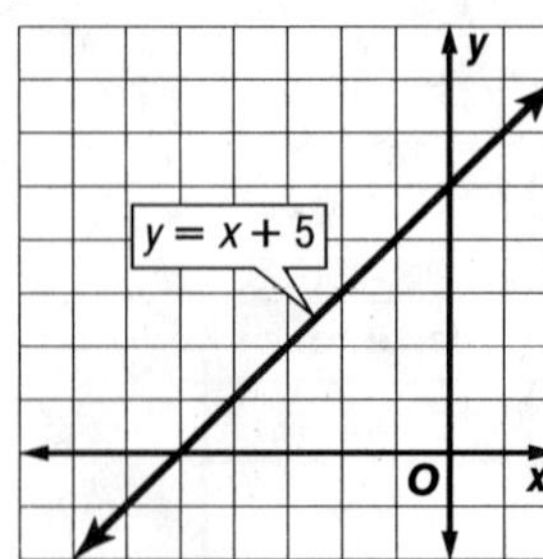

21. The slope is -1 and the y-intercept is 6. Graph the point (0, 6). Then go down 1 and right 1. Connect the points.

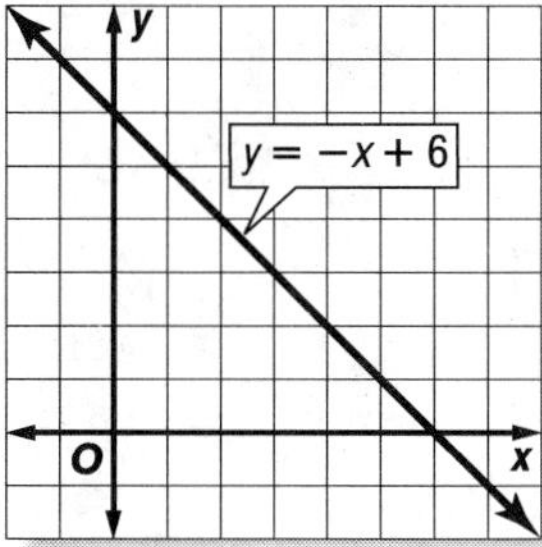

22. The slope is 2 and the y-intercept is -3. Graph the point $(0, -3)$. Then go up 2 and right 1. Connect these points.

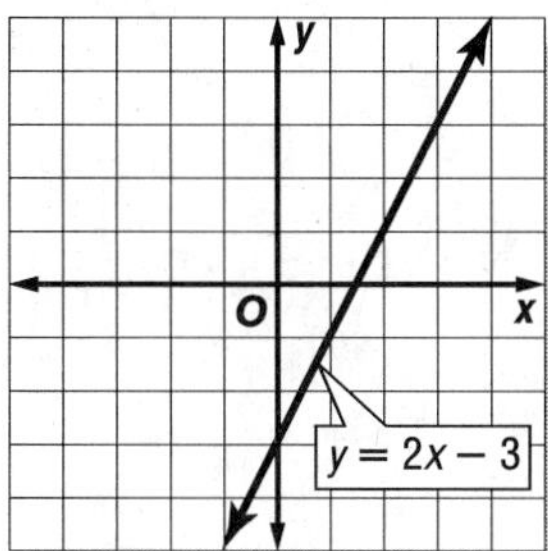

23. The slope is $\frac{3}{4}$ and the y-intercept is 2. Graph the point (0, 2). Then go up 3 and right 4. Connect these points.

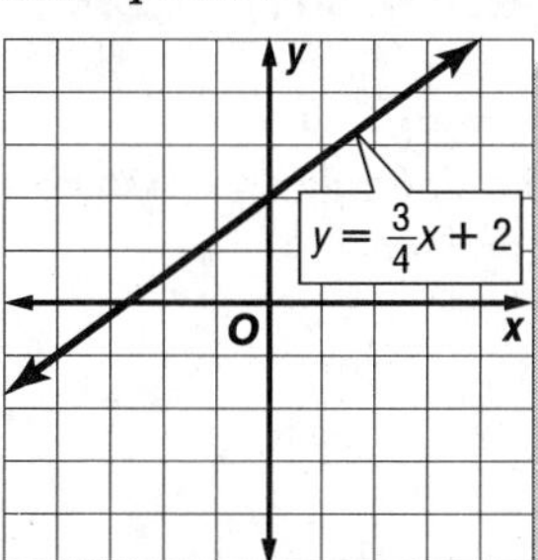

24. Write the equation in the form $y = mx + b$:

$$x + y = -3$$
$$x - x + y = -3 - x$$
$$y = -3 - x$$
$$y = -x - 3$$

The slope is -1 and the y-intercept is -3. Graph the point $(0, -3)$. Then go down 1 and right 1. Connect these points.

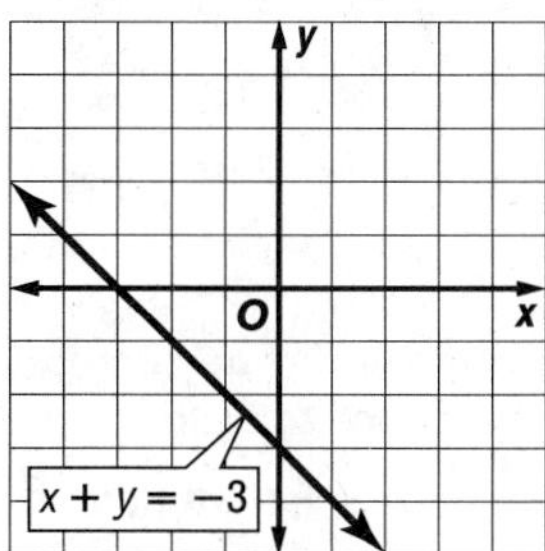

25. Write the equation in the form $y = mx + b$:

$$x + y = 0$$
$$x - x + y = 0 - x$$
$$y = -x$$
$$y = -x + 0$$

The slope is -1 and the y-intercept is 0. Graph the point (0, 0). Then go down 1 and right 1. Connect these points.

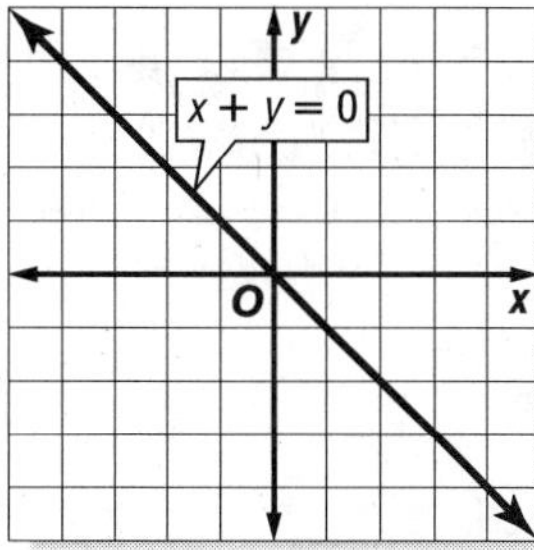

26. Write the the equation in the form $y = mx + b$:

$$-2x + y = -1$$
$$-2x + 2x + y = -1 + 2x$$
$$y = 2x - 1$$

The slope is 2 and the y-intercept is -1. Graph the point $(0, -1)$. Then go up 2 and right 1. Connect these points.

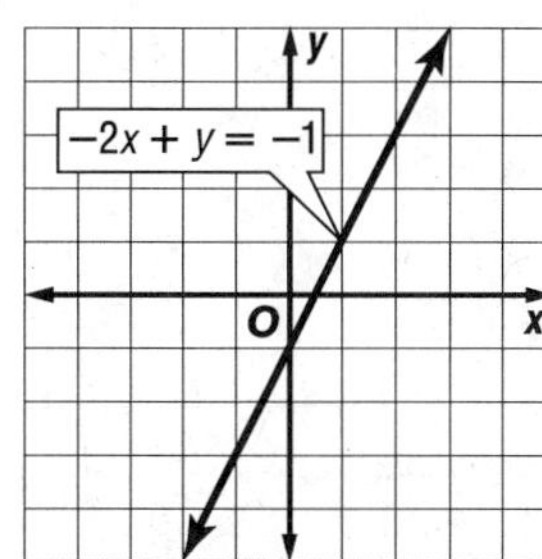

27. Write the equation in the form $y = mx + b$:

$$5x + y = -3$$
$$5x - 5x + y = -3 - 5x$$
$$y = -5x - 3$$

The slope is -5 and the y-intercept is -3. Graph the point $(0, -3)$. Then go down 5 and right 1. Connect these points.

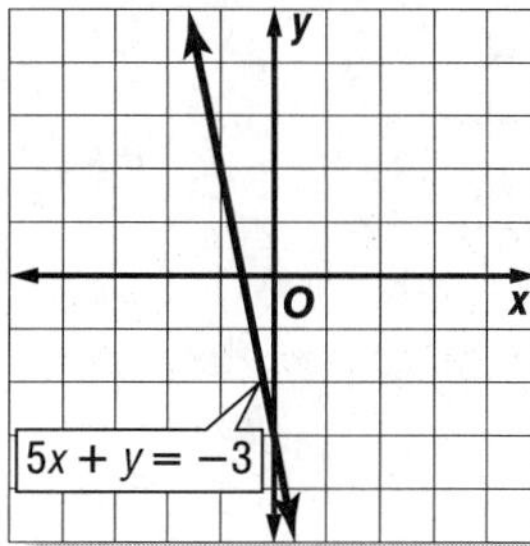

28. Write the equation in the form $y = mx + b$:

$$x - 3y = -6$$
$$x - x - 3y = -6 - x$$
$$-3y = -6 - x$$
$$\frac{-3y}{-3} = \frac{-6 - x}{-3}$$
$$y = 2 + \frac{1}{3}x$$
$$y = \frac{1}{3}x + 2$$

The slope is $\frac{1}{3}$ and the y-intercept is 2. Graph the points (0, 2). Then go up 1 and right 3. Connect these points.

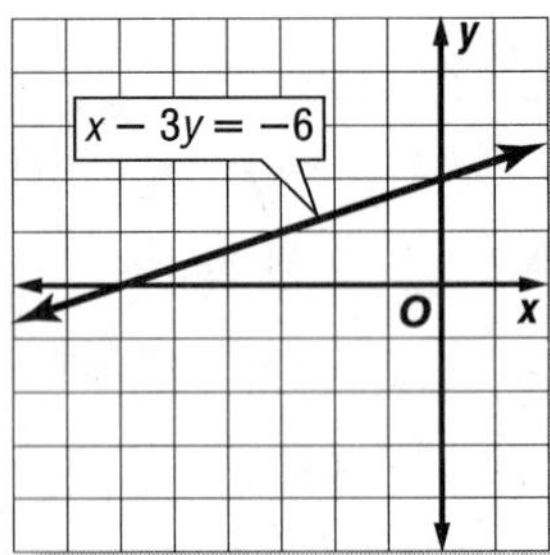

29. Write the equation in the form $y = mx + b$:

$$2x + 3y = 12$$
$$2x - 2x + 3y = 12 - 2x$$
$$3y = 12 - 2x$$
$$\frac{3y}{3} = \frac{12 - 2x}{3}$$
$$y = 4 - \frac{2}{3}x$$
$$y = -\frac{2}{3}x + 4$$

The slope is $-\frac{2}{3}$ and the y-intercept is 4. Graph the point (0, 4). Then go down 2 and right 3. Connect these points.

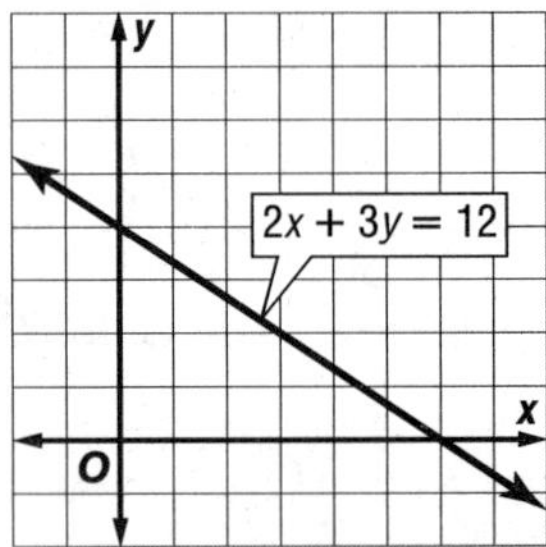

30. Write the equation in the form $y = mx + b$:

$$3x + 4y = 12$$
$$3x - 3x + 4y = 12 - 3x$$
$$4y = 12 - 3x$$
$$\frac{4y}{4} = \frac{12 - 3x}{4}$$
$$y = 3 - \frac{3}{4}x$$
$$y = -\frac{3}{4}x + 3$$

The slope is $-\frac{3}{4}$ and the y-intercept is 3. Graph the point (0, 3). Then go down 3 and right 4. Connect these points.

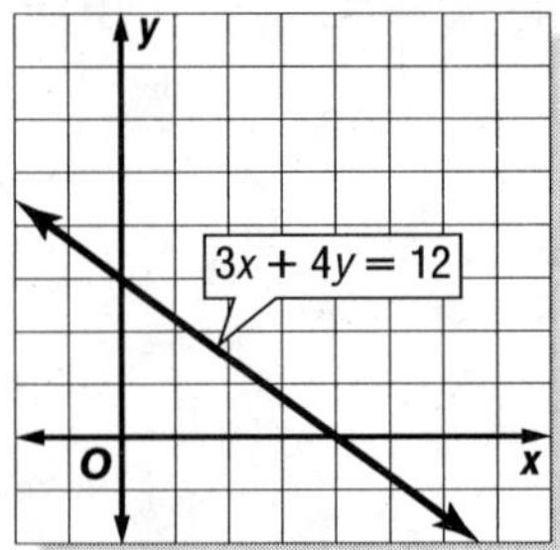

31. Write the equation in the form $y = mx + b$:

$y = -3$

$y = 0x - 3$

The slope is 0 and the y-intercept is -3. This is a horizontal line.

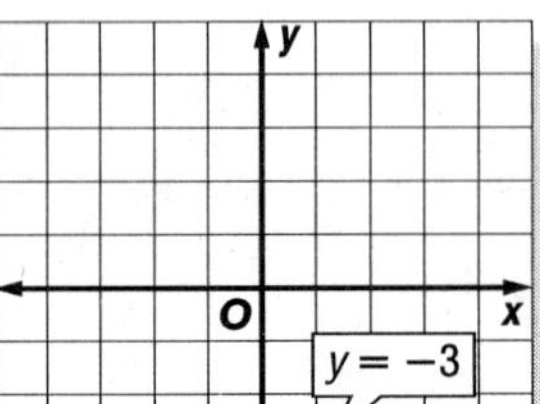

32. Write the the equation in the form $y = mx + b$:

$y = 300 - 50x$

$y = -50x + 300$

The slope is -50 and the y-intercept is 300. Graph (0, 300). Then go down 50 and right 1. Connect these points.

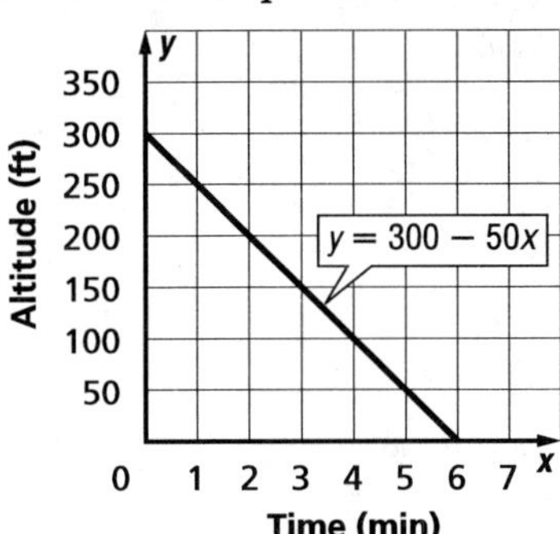

33. The slope of -50 is the descent, in feet per minute. The y-intercept of 300 is the initial altitude of 300 feet.

34. To find the x-intercept, let $y = 0$.

$$y = 300 - 50x$$
$$0 = 300 - 50x$$
$$0 + 50x = 300 - 50x + 50x$$
$$50x = 300$$
$$\frac{50x}{50} = \frac{300}{50}$$
$$x = 6$$

The x-intercept is 6. It represents the time in minutes that it takes the glider to land.

35. $-\frac{b}{m}$; replace y with 0 in $y = mx + b$ and solve for x.

36. Graph the y-intercept point. Then use the slope to locate a second point on the line. Draw a line through the two points. Answers should include the following.

- When an equation is written in slope-intercept form $y = mx + b$, the slope is m and the y-intercept is b.
- To write an equation of a line, subsitute the known values for m and b in $y = mx + b$.

37. B; Solve $2x + 3y = 6$ for y:

$$2x + 3y = 6$$
$$2x - 2x + 3y = 6 - 2x$$
$$3y = 6 - 2x$$
$$\frac{3y}{3} = \frac{6 - 2x}{3}$$
$$y = 2 - \frac{2}{3}x$$
$$y = -\frac{2}{3}x + 2$$

38. D; Write the equation in the form $y = mx + b$:

$$-x + 2y = 6$$
$$-x + x + 2y = 6 + x$$
$$2y = 6 + x$$
$$\frac{2y}{2} = \frac{6 + x}{2}$$
$$y = 3 + \frac{1}{2}x$$
$$y = \frac{1}{2}x + 3$$

slope $= \frac{1}{2}$; y-intercept $= 3$

Page 401 Maintain Your Skills

39. Find k:

$$y = kx$$
$$-36 = k(9)$$
$$-4 = k$$

Write the equation:

$$y = kx$$
$$y = -4x$$

40. Find k:

$$y = kx$$
$$5 = k(25)$$
$$\frac{5}{25} = k$$
$$\frac{1}{5} = k$$

Write the equation:

$$y = kx$$
$$y = \frac{1}{5}x$$

41. $m = \frac{y_2 - y_1}{x_2 - x_1}$

$$m = \frac{7 - 1}{6 - 3}$$
$$m = \frac{6}{3}$$
$$m = 2$$

42. $m = \frac{y_2 - y_1}{x_2 - x_1}$

$$m = \frac{5 - 5}{8 - (-2)}$$
$$m = \frac{0}{10} \text{ or } 0$$

42. $m = \frac{y_2 - y_1}{x_2 - x_1}$

$$m = \frac{-4 - 4}{0 - 2}$$
$$m = \frac{-8}{-2}$$
$$m = 4$$

44.

$$4(r - 3) = 8$$
$$4r - 12 = 8$$
$$4r - 12 + 12 = 8 + 12$$
$$4r = 20$$
$$r = 5$$

45. Let x represent the number.

$$6x = 2x + 28$$
$$6x - 2x = 2x - 2x + 28$$
$$4x = 28$$
$$x = 7$$

46. $2(18) - 1 = 36 - 1$
$= 35$

47. $(-2 - 4) \div 10 = -6 \div 10$
$= \frac{-6}{10}$
$= -\frac{3}{5}$

48. $-1(6) + 8 = -6 + 8$
$= 2$

49. $5 - 8(-3) = 5 + (-8)(-3)$
$= 5 + 24$
$= 29$

50. $(9 + 6) \div 3 = 15 \div 3$
$= 5$

51. $3 - (-2)(4) = 3 - (-8)$
$= 3 + 8$
$= 11$

Page 403 Graphing Calculator Investigation (Follow-Up of Lesson 8-6)

1. They all have the same slope, different y-intercepts.
2. It shifts the graph vertically c units.
3. $y = 3x + 2$
4. $y = 3x$
5. $y = x + 6, y = x, y = x - 4, y = x - 7$
6. As the coefficient of x increases, the line becomes steeper.
7. $y = 1.4x$, because $1.4 > 0.4$
8. The graphs have opposite slopes; the graph of $y = -4x$ falls from left to right and the graph of $y = 4x$ rises from left to right.
9. If the coefficient is negative, the graph has a negative slope. If the coefficient is positive, the graph has a positive slope.
10. As the absolute value of the coefficient of x increases, the line becomes steeper.
11. 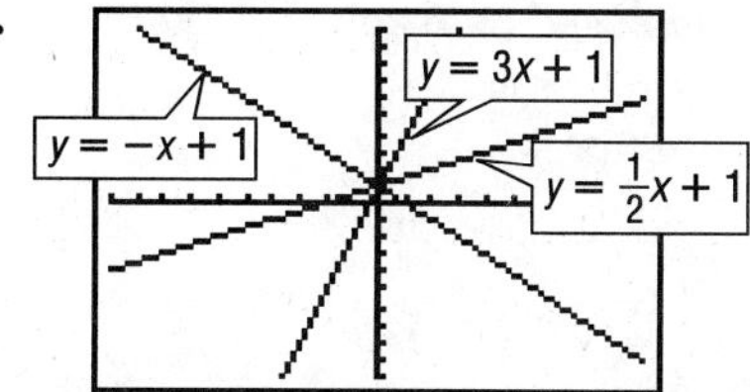

12a. same slope, different y-intercepts

12b. Same y-intercepts, opposite slopes; $y = 2x - 3$ slopes up from left to right and $y = -2x - 3$ slopes down from left to right.

12c. The graph of $y = 0.5x + 3$ is less steep and has a different y-intercept.

13. Sample answer: $y = -4x$

8-7 Writing Linear Equations

Page 404 How can you model data with a linear equation?

a.

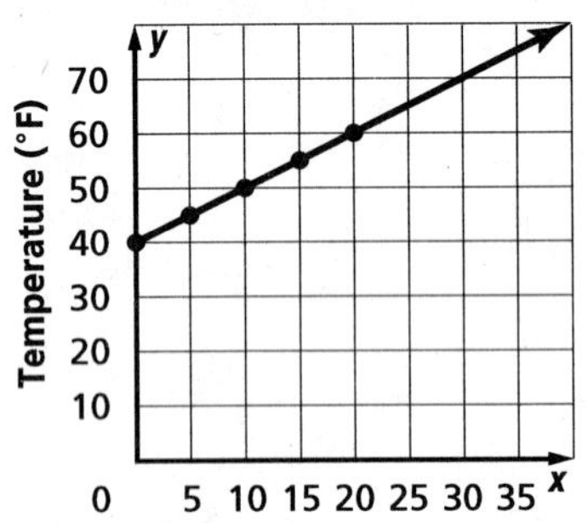

b. $m = \frac{y_2 - y_1}{x_2 - x_1}$

$m = \frac{60 - 55}{20 - 15}$

$m = \frac{5}{5}$

$m = 1$

The slope, 1, represents the rate of change, 1 degree for every 1 chirp; the y-intercept represents the temperature, 40°F, at which the crickets are not chirping.

c. Substitute the slope, 1, and y-intercept, 40, into the slope-intercept form of the equation.

$y = mx + b$

$y = 1x + 40$

$y = x + 40$

The temperature equals the number of chirps in 15 seconds plus 40.

Page 407 Check for Understanding

1. Sample answer: Find the y-intercept b and another point on the line. Use the points to determine the slope m. Then substitute these values in $y = mx + b$ and write the equation.

2. Sample answer: State the slope and y-intercept. Include a graph whose y-intercept point and another point on the line are both labeled. Show how the slope was used to graph the line.

3. $y = mx + b$

$y = \frac{1}{2}x + 1$

4. $y = mx + b$

$y = 0x + (-7)$

$y = -7$

5. The y-intercept is 3. From (0, 3), you can go down 2 units and right 1 unit to another point on the line. So, the slope is $\frac{-2}{1}$, or -2.

$y = mx + b$

$y = -2x + 3$

6. The y-intercept is -4. From $(0, -4)$, you can go up 2 units and right 3 units to another point on the line. So, the slope is $\frac{2}{3}$.

$y = mx + b$

$y = \frac{2}{3}x + (-4)$

$y = \frac{2}{3}x - 4$

7. Find the slope m.

$m = \frac{y_2 - y_1}{x_2 - x_1}$

$m = \frac{3 - 2}{4 - 2}$, or $\frac{1}{2}$

Find the y-intercept b. Use the slope and the coordinates of either point.

$y = mx + b$

$3 = \frac{1}{2}(4) + b$

$1 = b$

Substitute the slope and y-intercept.

$y = mx + b$

$y = \frac{1}{2}x + 1$

8. Find the slope m.

$m = \frac{y_2 - y_1}{x_2 - x_1}$

$m = \frac{4 - (-4)}{-1 - 3}$ or -2

Find the y-intercept b. Use the slope and the coordinates of either point.

$y = mx + b$

$-4 = -2(3) + b$

$2 = b$

Substitute the slope and y-intercept.

$y = mx + b$

$y = -2x + 2$

9. Find the slope m. Use the coordinates of any two points.

$m = \frac{y_2 - y_1}{x_2 - x_1}$

$m = \frac{5 - 2}{8 - 4}$ or $\frac{3}{4}$

Find the y-intercept b. Use the slope and the coordinates of any point.

$y = mx + b$

$-1 = \frac{3}{4}(0) + b$

$-1 = b$

Substitute the slope and y-intercept.

$y = mx + b$

$y = \frac{3}{4}x + (-1)$

$y = \frac{3}{4}x - 1$

10a. Find the slope m.

$m = \frac{\text{change in } y}{\text{change in } x}$

$m = \frac{10}{1}$

$m = 10$

Find the y-intercept b.

$(x, y) = (\text{hours, cost})$

$= (0, 50)$

When the hours are 0, the cost is \$50. So, the y-intercept is 50.

Write the equation.

$y = mx + b$

$y = 10x + 50$

10b. $y = 10x + 50$
$y = 10(8) + 50$
$y = 130$
The cost of renting the pavilion for 8 hours is $130.

Pages 407–408 Practice and Apply

11. $y = mx + b$
$y = 2x + 6$

12. $y = mx + b$
$y = -4x + 1$

13. $y = mx + b$
$y = 0x + 5$
$y = 5$

14. $y = mx + b$
$y = 1x + (-2)$
$y = x - 2$

15. $y = mx + b$
$y = -\frac{1}{3}x + 8$

16. $y = mx + b$
$y = \frac{2}{5}x + 0$
$y = \frac{2}{5}x$

17. The y-intercept is 3. From (0, 3), you can go down 2 units and left 1 unit to another point on the line. So, the slope is $\frac{-2}{-1}$, or 2.
$y = mx + b$
$y = 2x + 3$

18. The y-intercept is 6. From (0, 6), you can go down 1 unit and right 1 unit to another point on the line. So, the slope is $\frac{-1}{1}$, or -1.
$y = mx + b$
$y = -1x + 6$
$y = -x + 6$

19. The y-intercept is -2.5. The line is horizontal, so the slope is 0.
$y = mx + b$
$y = 0x + (-2.5)$
$y = -2.5$

20. The y-intercept is -5. From $(0, -5)$, you can go up 4 units and right 1 unit to another point on the line. So, the slope is $\frac{4}{1}$, or 4.
$y = mx + b$
$y = 4x + (-5)$
$y = 4x - 5$

21. The y-intercept is 0. From (0, 0), you can go down 1 unit and right 2 units to another point on the line. So, the slope is $-\frac{1}{2}$.
$y = mx + b$
$y = -\frac{1}{2}x + 0$
$y = -\frac{1}{2}x$

22. The y-intercept is 2. From (0, 2), you can go down 1 unit and left 4 units to another point on the line. So, the slope is $-\frac{1}{4}$.
$y = mx + b$
$y = -\frac{1}{4}x + 2$

23. Find the slope m.
$m = \frac{y_2 - y_1}{x_2 - x_1}$
$m = \frac{2 - (-1)}{1 - (-2)}$ or 1
Find the y-intercept b. Use the slope and the coordinates of either point.
$y = mx + b$
$2 = 1(1) + b$
$1 = b$
Substitute the slope and y-intercept.
$y = mx + b$
$y = 1x + 1$
$y = x + 1$

24. Find the slope m.
$m = \frac{y_2 - y_1}{x_2 - x_1}$
$m = \frac{-1 - 3}{4 - (-4)}$ or $-\frac{1}{2}$
Find the y-intercept b. Use the slope and the coordinates of either point.
$y = mx + b$
$-1 = -\frac{1}{2}(4) + b$
$1 = b$
Substitute the slope and y-intercept.
$y = mx + b$
$y = -\frac{1}{2}x + 1$

25. Find the slope m.
$m = \frac{y_2 - y_1}{x_2 - x_1}$
$m = \frac{1 - 0}{-1 - 0}$ or -1
Find the y-intercept b. Use the slope and the coordinates of either point.
$y = mx + b$
$1 = (-1)(-1) + b$
$0 = b$
Substitute the slope and y-intercept.
$y = mx + b$
$y = -1x + 0$
$y = -x$

26. Find the slope m.
$m = \frac{y_2 - y_1}{x_2 - x_1}$
$m = \frac{-16 - 2}{-8 - 4}$ or $\frac{3}{2}$
Find the y-intercept b. Use the slope and the coordinates of either point.
$y = mx + b$
$2 = \frac{3}{2}(4) + b$
$-4 = b$
Substitute the slope and y-intercept.
$y = mx + b$
$y = \frac{3}{2}x + (-4)$
$y = \frac{3}{2}x - 4$

27. Find the slope m.

$m = \frac{y_2 - y_1}{x_2 - x_1}$

$m = \frac{7 - 7}{-9 - 8}$ or 0

Find the y-intercept b. Use the slope and the coordinates of either point.

$y = mx + b$
$7 = 0(8) + b$
$7 = b$

Substitute the slope and y-intercept.

$y = mx + b$
$y = 0x + 7$
$y = 7$

28. Find the slope m.

$m = \frac{y_2 - y_1}{x_2 - x_1}$

$m = \frac{2 - (-6)}{3 - 5}$ or -4

Find the y-intercept b. Use the slope and the coordinates of either point.

$y = mx + b$
$-6 = -4(5) + b$
$14 = b$

Substitute the slope and y-intercept.

$y = mx + b$
$y = -4x + 14$

29. Find the slope m. Use the coordinates of any two points.

$m = \frac{y_2 - y_1}{x_2 - x_1}$

$m = \frac{5 - 1}{2 - 1}$ or 4

Find the y-intercept b. Use the slope and the coordinates of any point.

$y = mx + b$
$5 = 4(2) + b$
$-3 = b$

Substitute the slope and y-intercept.

$y = mx + b$
$y = 4x + (-3)$
$y = 4x - 3$

30. Find the slope m. Use the coordinates of any two points.

$m = \frac{y_2 - y_1}{x_2 - x_1}$

$m = \frac{5 - 7}{-1 - (-3)}$ or -1

Find the y-intercept b. Use the slope and the coordinates of any point.

$y = mx + b$
$7 = -1(-3) + b$
$4 = b$

Substitute the slope and y-intercept.

$y = mx + b$
$y = -1x + 4$
$y = -x + 4$

31. Find the slope m. Use the coordinates of any two points.

$m = \frac{y_2 - y_1}{x_2 - x_1}$

$m = \frac{1088 - 0}{1 - 0}$ or 1088

Find the y-intercept b. Use the slope and the coordinates of any point.

$y = mx + b$
$0 = 1088(0) + b$
$0 = b$

Substitute the slope and y-intercept.

$y = mx + b$
$y = 1088x + 0$
$y = 1088x$

The speed of sound is 1088 feet per second at 0°C.

32. Sound travels at 1088 feet each second, from your answer to Exercise 31.

$\frac{1088 \text{ ft}}{1 \text{ s}} \cdot \frac{60 \text{ s}}{1 \text{ min}} \cdot \frac{1 \text{ mi}}{5280 \text{ ft}} = \frac{12.4 \text{ mi}}{1 \text{ min}}$

In 1 minute, sound travels 12.4 miles.

33a. Since Kim bought the CD player at a 30% discount, she paid 70% of the full price or $0.70c$.

Kim paid 6% tax, so she paid 106% of the discounted price, or $0.70c \times 1.06$.

Kim sells for 50% of what she paid, or $0.50(0.70c \times 1.06)$. So, $d = 0.50(0.70c \times 1.06)$ or $d = 0.371c$.

33b. $d = 0.371(50)$
$d = 18.55$
Kim sold it for $18.55.

34. Write the data as ordered pairs and graph the points. Then draw a line through the points. Find the slope and y-intercept of the line and substitute the values in $y = mx + b$. Answers should include the following.

- The y-intercept is the y value in the table when the corresponding x value is 0. Use two pairs of x- and y-coordinates in the table to find the slope.

35. C; Find the slope m.

$m = \frac{y_2 - y_1}{x_2 - x_1}$

$m = \frac{2 - (-2)}{0 - 2}$ or -2

Find the y-intercept b. Use the slope and the coordinates of either point.

$y = mx + b$
$2 = -2(0) + b$
$2 = b$

Substitute the slope and y-intercept.

$y = mx + b$
$y = -2x + 2$

36. C; Find the slope m. Use the coordinates of any two points.

$m = \frac{y_2 - y_1}{x_2 - x_1}$

$m = \frac{8 - 6}{-8 - (-4)}$ or $-\frac{1}{2}$

Find the y-intercept b. Use the slope and the coordinates of any point.

$y = mx + b$

$6 = -\frac{1}{2}(-4) + b$

$4 = b$

Substitute the slope and y-intercept.

$y = mx + b$

$y = -\frac{1}{2}x + 4$

Page 408 Maintain Your Skills

37. slope = 6; y-intercept = 7

38. slope = −1; y-intercept = 4

39. Solve the equation for y to write it as $y = mx + b$:

$-3x + y = -2$

$-3x + 3x + y = -2 + 3x$

$y = 3x - 2$

slope = 3; y-intercept = −2

40. Find k: Write the equation:

$y = kx$ $\quad y = kx$

$14 = k(35)$ $\quad y = \frac{2}{5}x$

$\frac{14}{35} = k$

$\frac{2}{5} = k$

41. As the x values increase, the y values increase, so the set of points would show a positive relationship.

8-8 Best-Fit Lines

Page 409 How can a line be used to predict life expectancy for future generations?

a. 83

b. Sample answer: Medical improvements could change the accuracy of the model. Also, the line continues to go up, but people's ages will not increase forever.

Pages 410–411 Check for Understanding

1. Sample answer: Use a ruler to extend the line so that it passes through the x value for which you want to predict. Locate the x value on the line and determine the corresponding y value. Or, write an equation for the best-fit line and substitute the desired value of x to find the corresponding value of y.

2. Sample answer:

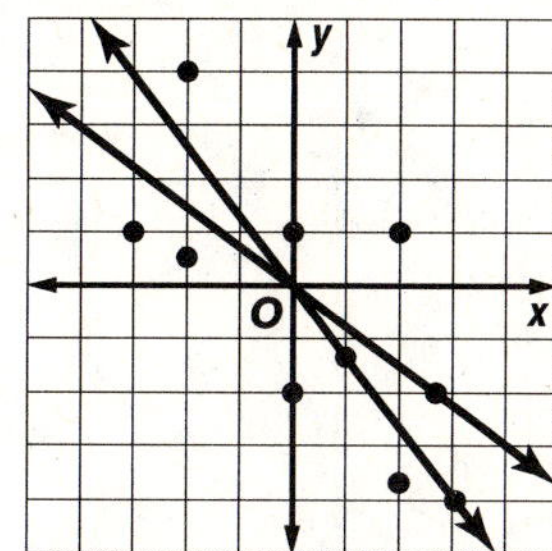

3. Sample answer:

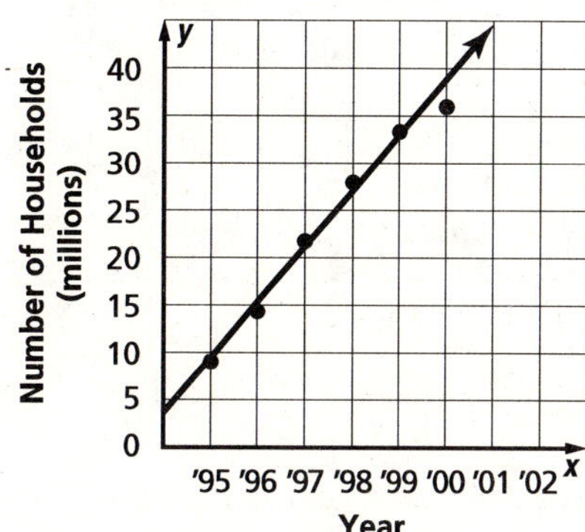

4. Extend the line so that you can find the y value for an x value of 2005.
Sample answer: about 65 million

5. First, select two points on the line and find the slope.

$m = \frac{y_2 - y_1}{x_2 - x_1}$

$m = \frac{400 - 350}{4 - 1.5}$ or 20

Next, find the y-intercept.

$y = mx + b$

$400 = 20(4) + b$

$320 = b$

Write the equation.

$y = mx + b$

$y = 20x + 320$

6. The year 2008 is 15 years since 1993. Substitute 15 for x.

$y = 20x + 320$

$y = 20(15) + 320$

$y = 620$

Travelers will spend about \$620 billion in 2008.

Pages 411–413 Practice and Apply

7. Sample answer:

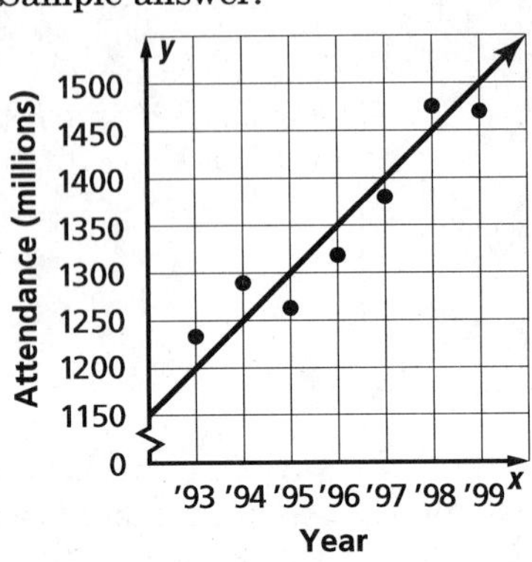

8. Extend the line so that you can find the y value for an x value of 2005.
Sample answer: 1700 million

9. Sample answer:

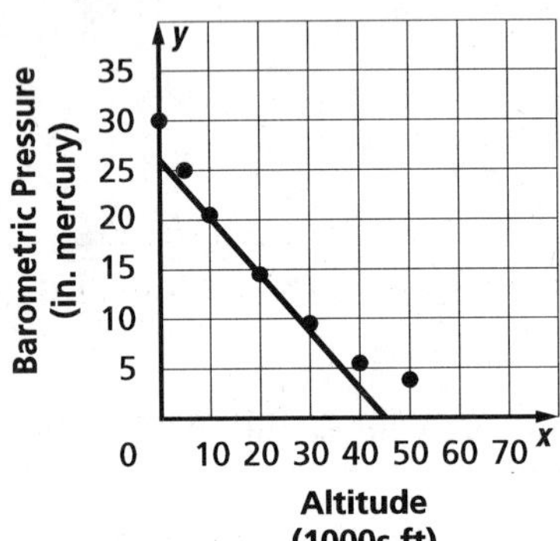

10. Sample answer:
First, select two points on the line and find the slope. Use (0, 27) and (10,000, 21).

$m = \frac{y_2 - y_1}{x_2 - x_1}$

$m = \frac{27 - 21}{0 - 10{,}000}$ or -0.0006

Next, find the y-intercept.

$y = mx + b$

$27 = -0.0006(0) + b$

$27 = b$

Write the equation.

$y = mx + b$

$y = -0.0006x + 27$

Substitute 60,000 into the equation for x.

$y = -0.0006(60{,}000) + 27$

$y = -9$

This is not reasonable because barometric pressure cannot be negative.

11. No, the equation gives a negative value for barometric pressure, which is not possible. Also, the data in the scatter plot do not appear to be linear.

12. Sample answer:

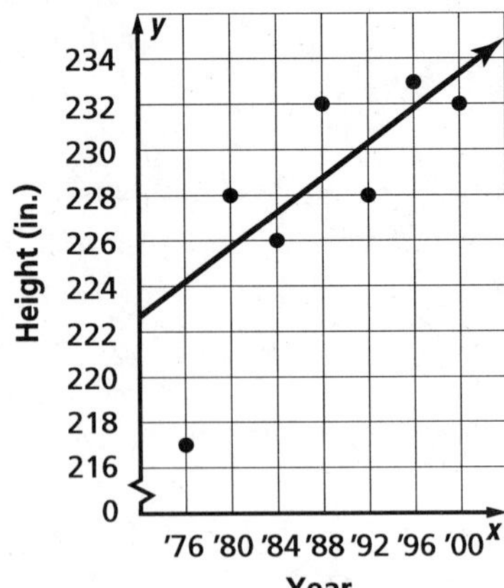

13. Extend the line so that you can find the y value for an x value of 2008.
Sample answer: 238 in.

14. Sample answer:

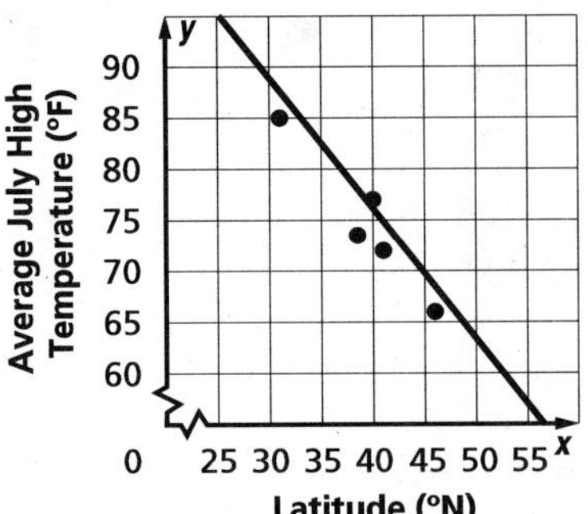

15. As latitude increases, temperature decreases.

16. Sample answer: $y = -1.33x + 127.8$

17. Substitute 50 for x in your equation from Exercise 16. Sample answer:

$y = -1.33(50) + 127.8$

$y = 61.3°F$

18. Sample answer: The slopes are the same because the rate of change in the percent of schools is the same per year whether you use the year or the number of years since 1996. The y-intercepts are different, depending on which scale you use.

19. The data describing the life expectancy for past generations can be displayed using a scatter plot. Then a line is drawn as close to as many of the points as possible. Then the line can be extended and used to predict the life expectancy for future generations. Answers should include the following.
   - A best-fit line is a line drawn as close to as many of the data points as possible.
   - Although the points may not be exactly linear, a best-fit line can be used to approximate the data set.

20. B; Locate 7 on the x-axis. Move up to the graph and locate the y value corresponding to 7. The y value is 6.

21. A; A best-fit line is close to most of the data points.

Page 413 Maintain Your Skills

22. $y = mx + b$
$y = 3x + 5$

23. $y = mx + b$
$y = -2x + 2$

24. The y-intercept is -1. From the point $(0, -1)$, you can go up 3 units and right 2 units to another point on the line. So, the slope is $\frac{3}{2}$.
$y = mx + b$
$y = \frac{3}{2}x + (-1)$
$y = \frac{3}{2}x - 1$

25. The y-intercept is -4. This is a horzontal line, so the slope is 0.
$y = mx + b$
$y = 0x + (-4)$
$y = -4$

26. The slope is 1. The y-intercept is -2. Graph the point $(0, -2)$. Then go up 1 unit and right 1 unit. Connect these points.

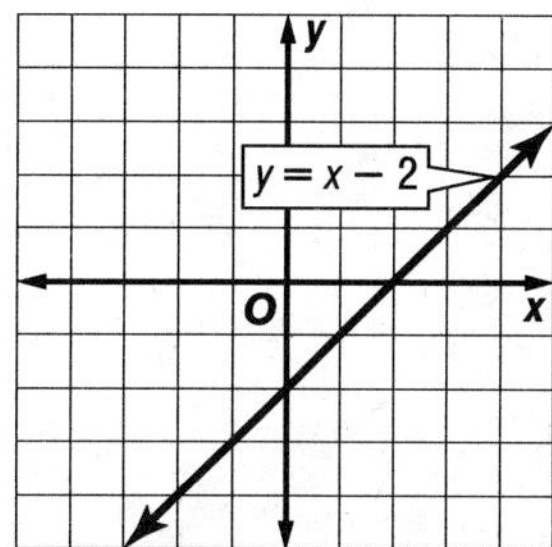

27. The slope is -1. The y-intercept is 3. Graph the point $(0, 3)$. Then go down 1 unit and right 1 unit. Connect these points.

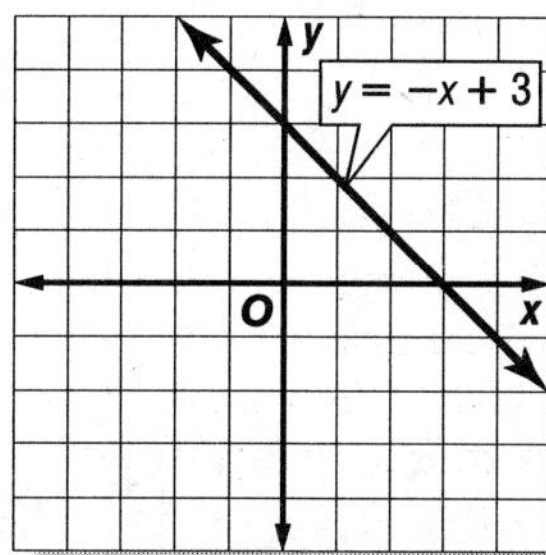

28. The slope is $\frac{1}{2}$. The y-intercept is 0. Graph the point $(0, 0)$. Then go up a unit and right 2 units. Connect these points.

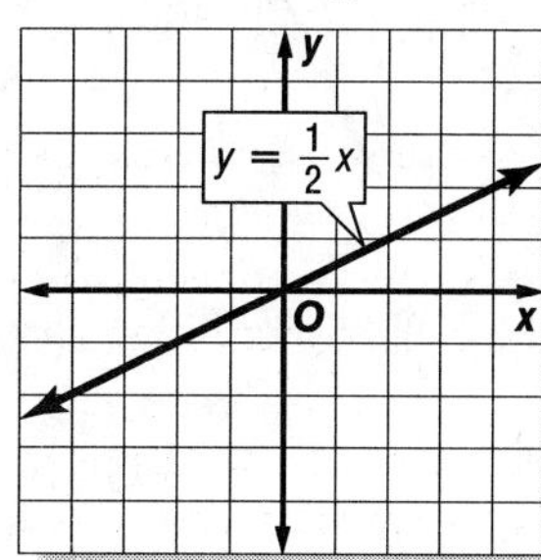

29.
$$3n - 11 \le 10$$
$$3n - 11 + 11 \le 10 + 11$$
$$3n \le 21$$
$$n \le 7$$
The solution is $n \le 7$. You can check this solution by substituting 7 or a number less than 7 into the inequality.

30.
$$8(x + 1) < 16$$
$$8x + 8 < 16$$
$$8x + 8 - 8 < 16 - 8$$
$$8x < 8$$
$$x < 1$$
The solution is $x < 1$. You can check this solution by substituting a number less than 1 into the inequality.

31.
$$5d + 2 > d - 4$$
$$5d - d + 2 > d - d - 4$$
$$4d + 2 > -4$$
$$4d + 2 - 2 > -4 - 2$$
$$4d > -6$$
$$d > \frac{-6}{4}$$
$$d > -\frac{3}{2}$$
The solution is $d > -\frac{3}{2}$. You can check this solution by substituting a number greater than $-\frac{3}{2}$ into the inequality.

32. Let d represent the number of days left.
$d \le 12$

33.
$$\frac{a}{3} = \frac{16}{24}$$
$$a \cdot 24 = 16 \cdot 3$$
$$24a = 48$$
$$a = \frac{48}{24}$$
$$a = 2$$

34.
$$\frac{5}{10} = \frac{15}{x}$$
$$5x = 15 \cdot 10$$
$$5x = 150$$
$$x = \frac{150}{5}$$
$$x = 30$$

35.
$$\frac{2}{16} = \frac{n}{36}$$
$$2 \cdot 36 = n \cdot 16$$
$$72 = 16n$$
$$\frac{72}{16} = n$$
$$4.5 = n$$

36.
$$xy = \frac{\overset{4}{\cancel{8}}}{\underset{3}{\cancel{9}}} \cdot \frac{\overset{4}{\cancel{12}}}{\underset{15}{\cancel{30}}}$$
$$= \frac{16}{45}$$

37. $y = x + 1$
$y = 2 + 1$
$y = 3$

38. $y = x + 5$
$y = -1 + 5$
$y = 4$

39. $y = x - 4$
$y = 3 - 4$
$y = -1$

40. $x + y = 2$
$0 + y = 2$
$y = 2$

41. $x + y = -1$
$1 + y = -1$
$y = -2$

42. $x + y = 0$
$4 + y = 0$
$y = -4$

8-9 Solving Systems of Equations

Page 414 How can a system of equations be used to compare data?

a. Income = hourly rate · number of hours worked + bonus.
Let y = income and let x = number of hours worked.
Job A: $y = 10x + 50$
Job B: $y = 15x$

b. $y = 10x + 50$: slope = 10 and y-intercept = 50
$y = 15x$: slope = 15 and y-intercept is 0, or the origin.

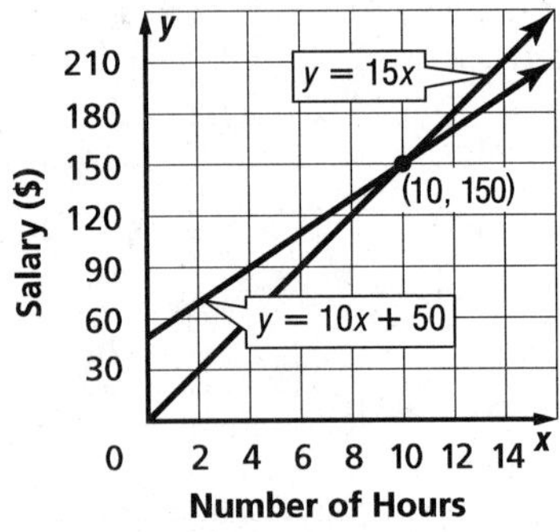

c. (10, 150); For 10 hours worked, the salary is the same for both jobs, $150.

Pages 416–417 Check for Understanding

1. Sample answer: A group of two or more equations form a system of equations. The solution is the ordered pair that satisfies all the equations in the system. If the equations are graphed, the solution is the coordinates of the point where the graphs intersect. The system has infinitely many solutions if the equations are the same and the graphs coincide.

2. Sample answers:
one solution

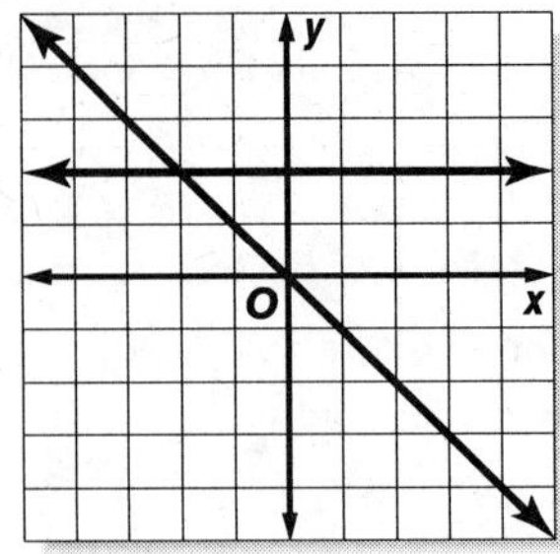

no solution

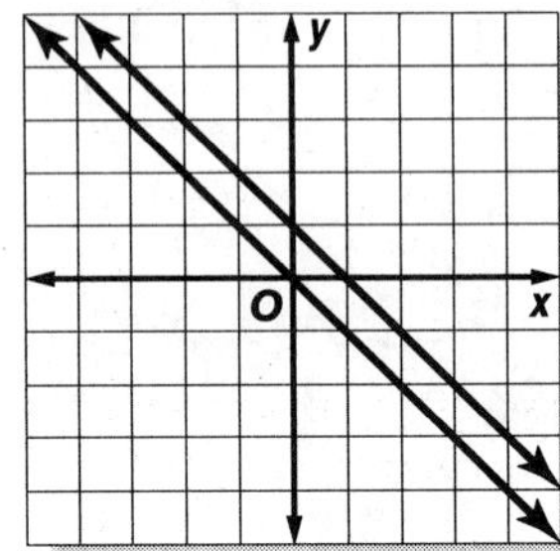

infinitely many solutions

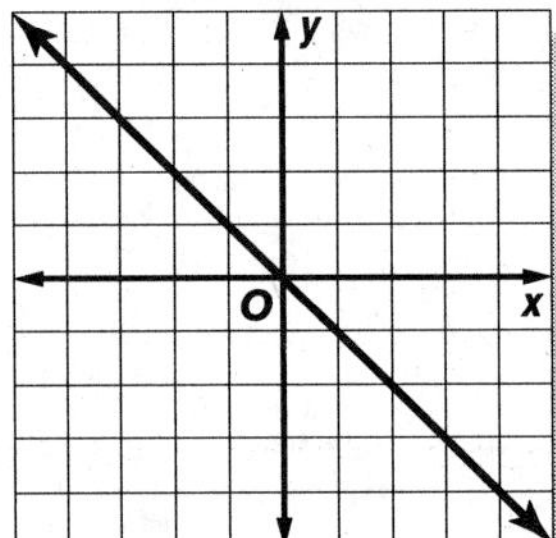

3. The graphs appear to intersect at $(-5, -4)$.
Check:
$y = -\frac{2}{5}x - 6$
$-4 \stackrel{?}{=} -\frac{2}{5}(-5) - 6$
$-4 \stackrel{?}{=} 2 - 6$
$-4 = -4$ ✓

4. $y = 2x + 1$
$y = -x + 1$
The graphs appear to intersect at (0, 1).

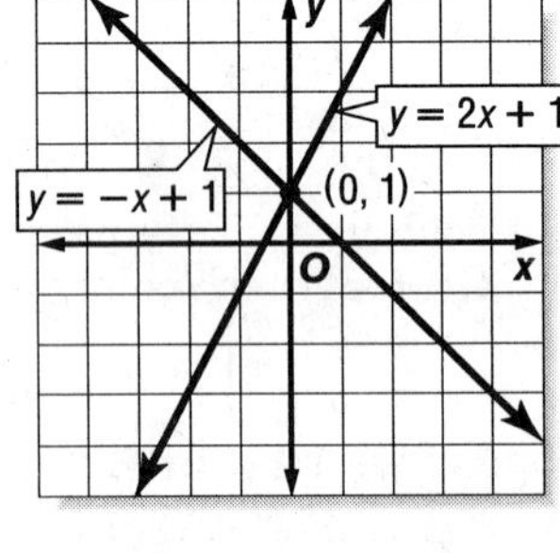

Check:
$y = 2x + 1$
$1 \stackrel{?}{=} 2(0) + 1$
$1 = 1$ ✓
$y = -x + 1$
$1 \stackrel{?}{=} -0 + 1$
$1 = 1$ ✓

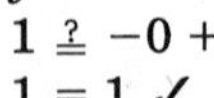

The solution of the system of equations is (0, 1).

5. $x + y = 4$
$x + y = 2$
The graphs appear to be parallel lines. Since there is no coordinate pair that is a solution to both equations, there is no solution of this system of equations.

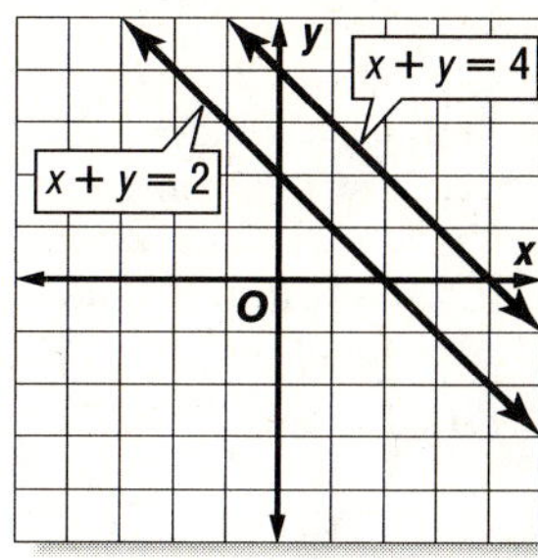

6. $y = 3x - 4$
$x = 0$
Replace x with 0 in the first equation.
$y = 3(0) - 4$
$y = -4$
The solution of this system of equations is $(0, -4)$. You can check the solution by graphing. The graphs appear to intersect at $(0, -4)$, so the solution is correct.

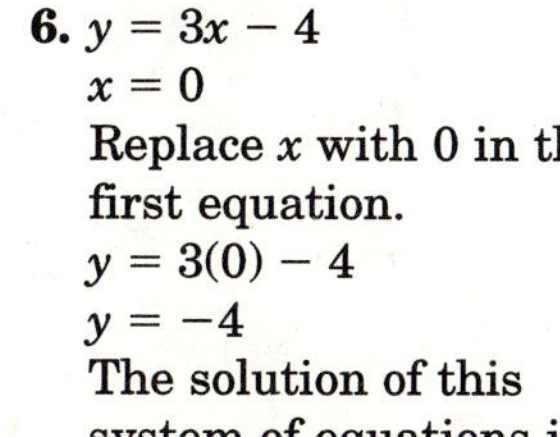

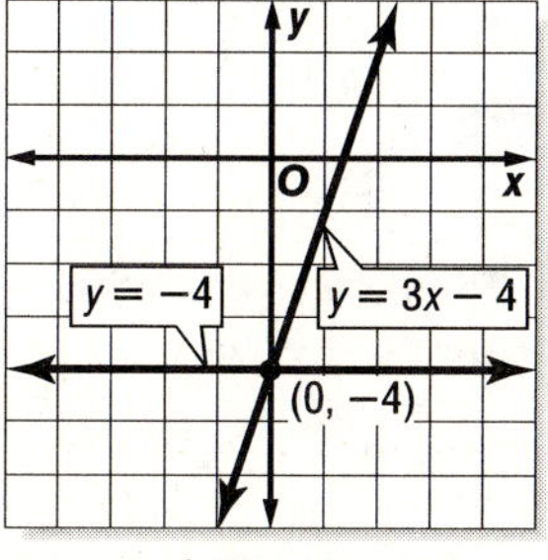

7. $x + y = 8$
$y = 6$
Replace y with 6 in the first equation.
$x + 6 = 8$
$x = 2$
The solution of this system of equations is $(2, 6)$. You can check the solution by graphing. The graphs appear to intersect at $(2, 6)$, so the solution is correct.

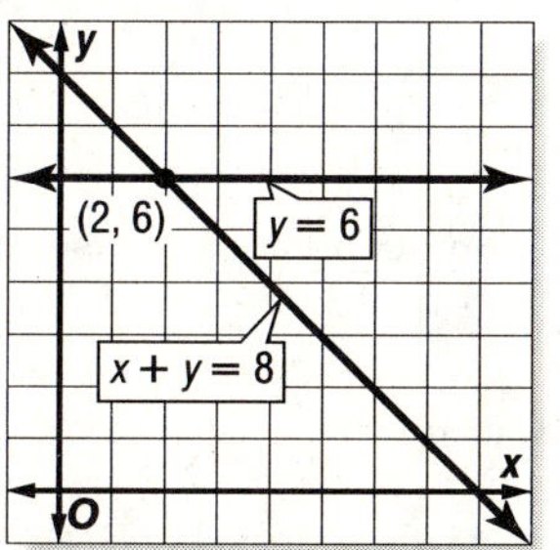

8. Let x represent the length of the garden and y represent the width.
$2 \cdot \text{length} + 2 \cdot \text{width} = \text{perimeter}$

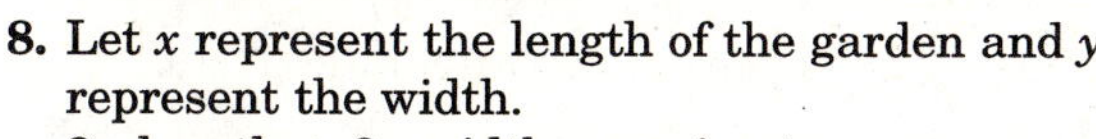

$2x + 2y = 40$
$y = 7$
Replace y with 7 in the first equation.
$2x + 2y = 40$
$2x + 2(7) = 40$
$2x + 14 = 40$
$2x = 40$
$2x = 26$
$x = 13$
The solution of this system of equations is $(13, 7)$. The length of the garden is 13 feet.

Pages 417–418 Practice and Apply

9. The graphs appear to intersect at $(3, -2)$.

10. The graphs appear to be parallel lines. Since there is no coordinate pair that is a solution to both equations, there is no solution of this system of equations.

11. Both equations have the same graph. Any ordered pair on the graph will satisfy both equations. Therefore, there are infinitely many solutions of this system of equations.

12. $y = -x$
$y = x + 2$
The graphs appear to intersect at $(-1, 1)$.
Check:
$y = -x \qquad y = x + 2$
$1 \stackrel{?}{=} -(-1) \qquad 1 \stackrel{?}{=} -1 + 2$
$1 = 1$ ✓ $\qquad 1 = 1$ ✓
The solution of the system of equations is $(-1, 1)$.

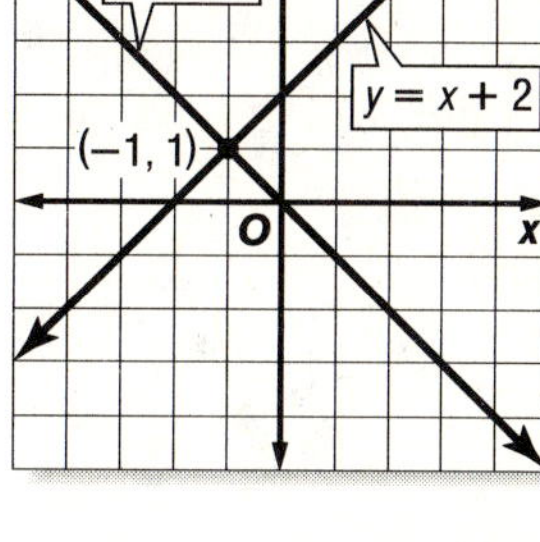

13. $x + y = 6$
$y = x$
The graphs appear to intersect at $(3, 3)$.
Check:
$x + y = 6 \qquad y = x$
$3 + 3 \stackrel{?}{=} 6 \qquad 3 = 3$ ✓
$6 = 6$ ✓
The solution of the system of equations is $(3, 3)$.

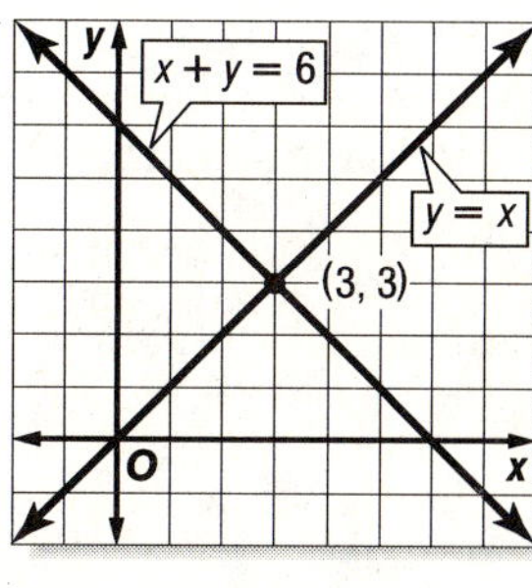

14. $2x + y = 1$
$y = -2x + 5$
The graphs appear to be parallel lines. Since there is no coordinate pair that is a solution to both equations, there is no solution of this system of equations.

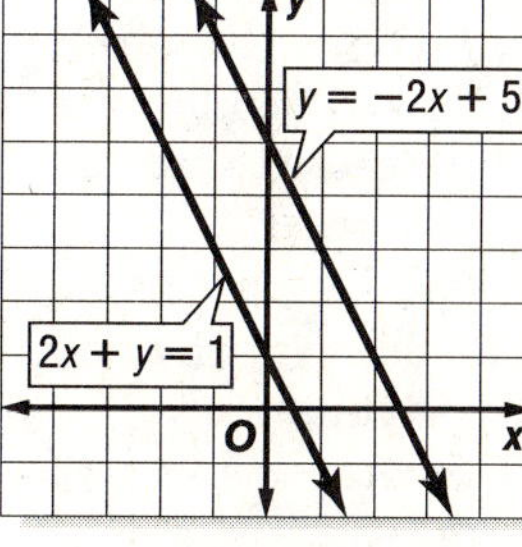

15. $y = \frac{2}{3}x$
$2x - 3y = 0$
Both equations have the same graph. Any ordered pair on the graph will satisfy both equations. Therefore, there are infinitely many solutions of this system of equations.

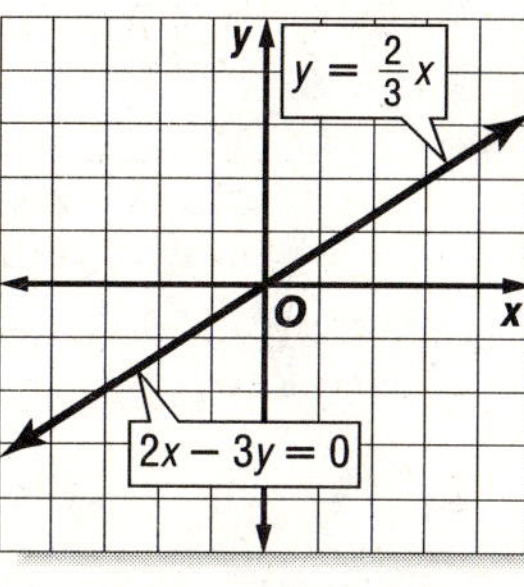

16. $x + y = -4$
$x - y = -4$
The graphs appear to intersect at $(-4, 0)$.
Check:
$x + y = -4$
$-4 + 0 \stackrel{?}{=} -4$
$-4 = -4$ ✓
$x - y = -4$
$-4 - 0 \stackrel{?}{=} -4$
$-4 = -4$ ✓
The solution of this system of equations is $(-4, 0)$.

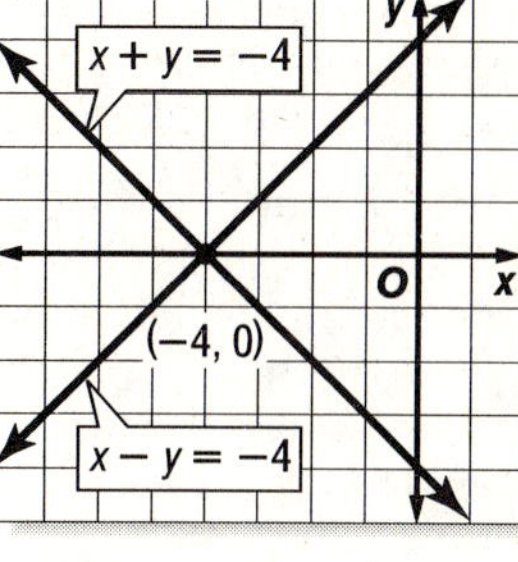

17. $y = -\frac{1}{2}x + 3$
$y = -2x$
The graphs appear to intersect at $(-2, 4)$.
Check:
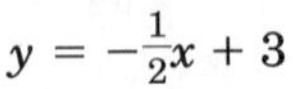

$y = -\frac{1}{2}x + 3$
$4 \stackrel{?}{=} -\frac{1}{2}(-2) + 3$
$4 \stackrel{?}{=} 1 + 3$
$4 = 4$ ✓
$y = -2x$
$4 \stackrel{?}{=} -2(-2)$
$4 = 4$ ✓
The solution of this system of equations is $(-2, 4)$.

18. $y = x + 1$
$x = 3$
Replace x with 3 in the first equation.
$y = x + 1$
$y = 3 + 1$
$y = 4$
The solution of this system of equations is $(3, 4)$. You can check this solution by graphing. The graphs appear to intersect at $(3, 4)$, so the solution is correct.

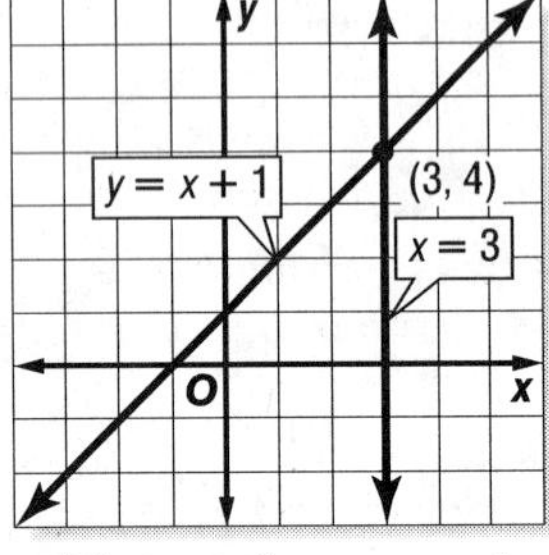

19. $y = x + 2$
$y = 0$
Replace y with 0 in the first equation.
$y = x + 2$
$0 = x + 2$
$-2 = x$
The solution of this system of equations is $(-2, 0)$. You can check this solution by graphing. The graphs appear to intersect at $(-2, 0)$, so the solution is correct.

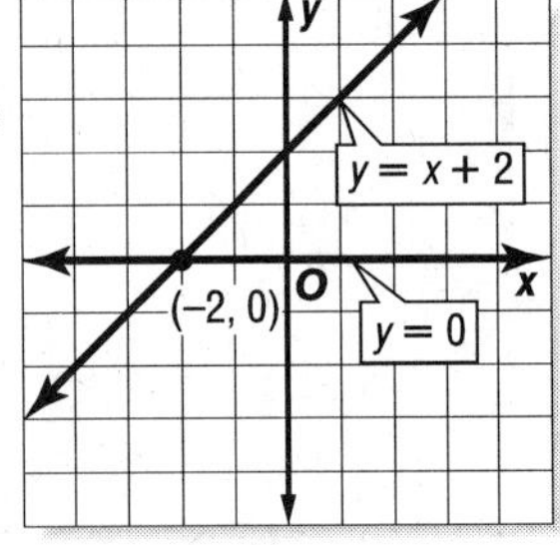

20. $y = 2x + 7$
$y = -5$
Replace y with -5 in the first equation.
$y = 2x + 7$
$-5 = 2x + 7$
$-12 = 2x$
$-6 = x$
The solution of this system of equations is $(-6, -5)$. You can check the solution by graphing. The graphs appear to intersect at $(-6, -5)$, so the solution is correct.

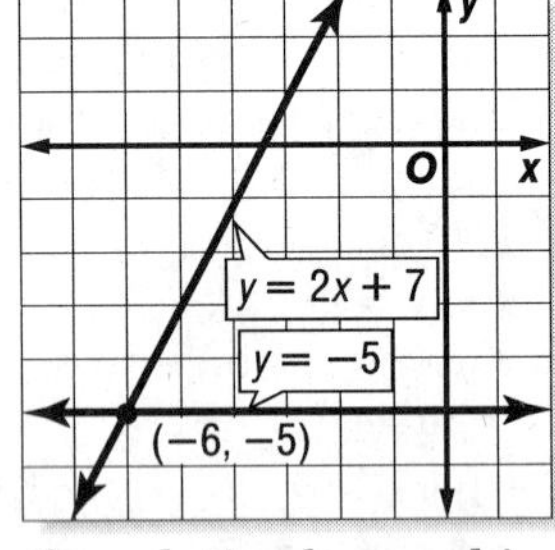

21. $x + y = 6$
$x = -4$
Replace x with -4 in the first equation.
$x + y = 6$
$-4 + y = 6$
$y = 10$
The solution of this system of equations is $(-4, 10)$. You can check the solution by graphing. The graphs appear to intersect at $(-4, 10)$, so the solution is correct.

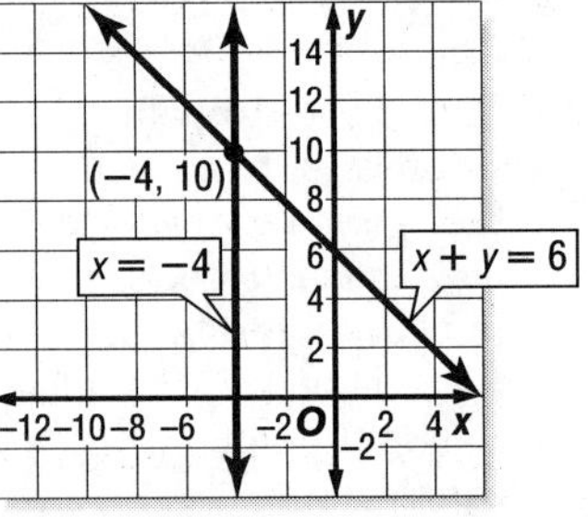

22. $2x + 3y = 5$
$y = x$
Replace y with x in the first equation.
$2x + 3y = 5$
$2x + 3x = 5$
$5x = 5$
$x = 1$
Replace x with 1 in the second equation.
$y = x$
$y = 1$
The solution of this system of equations is $(1, 1)$. You can check the solution by graphing. The graphs appear to intersect at $(1, 1)$, so the solution is correct.

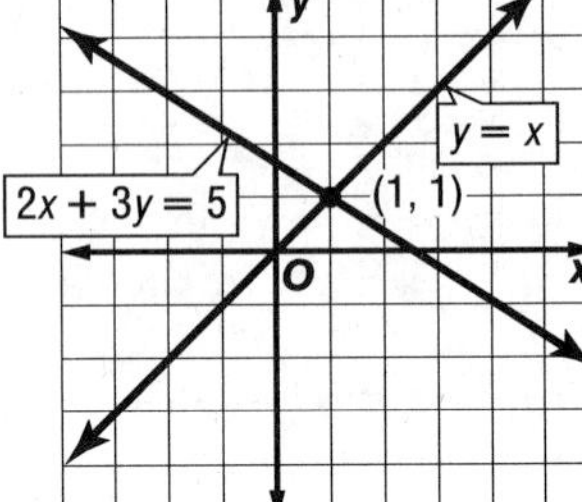

23. $x + y = 9$
$y = 2x$
Replace y with $2x$ in the first equation.
$x + y = 9$
$x + 2x = 9$
$3x = 9$
$x = 3$
Replace x with 3 in the second equation.
$y = 2x$
$y = 2(3)$
$y = 6$
The solution of this system of equations is $(3, 6)$. You can check the solution by graphing. The graphs appear to intersect at $(3, 6)$, so the solution is correct.

24. Site A: $y = 5.00 + 1.00x$ or $y = 5 + x$
Site B: $y = 2.00 + 1.50x$ or $y = 2 + 1.5x$

25. The graph of the system shows the solution is $(6, 11)$. An order of 6 pounds will cost the same, $11.00, for both sites.

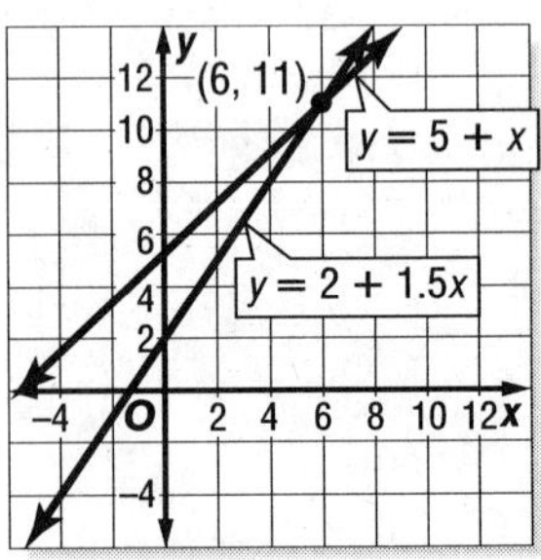

26. Replace x with 8 in each equation.

Site A: $y = 5 + x$
$y = 5 + 8$
$y = 13$

Site B: $y = 2 + 1.5x$
$y = 2 + 1.5(8)$
$y = 2 + 12$
$y = 14$

Site A would be less expensive, since $13.00 < $14.00.

27a. Sample answer:

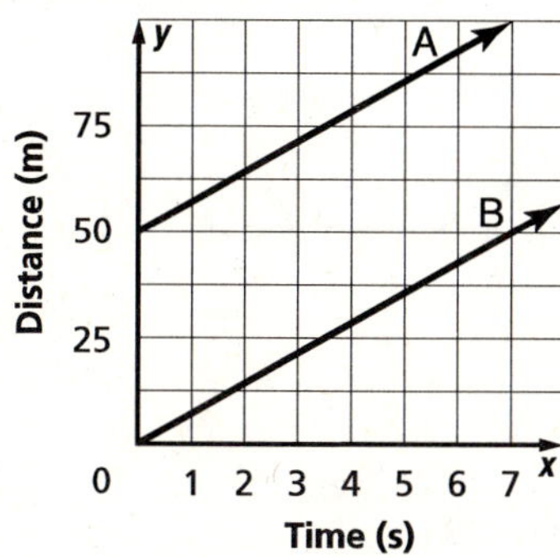

No; since slope represents rate, the slopes of the graphs are the same. Therefore, the runners will never meet.

27b. Sample answer:

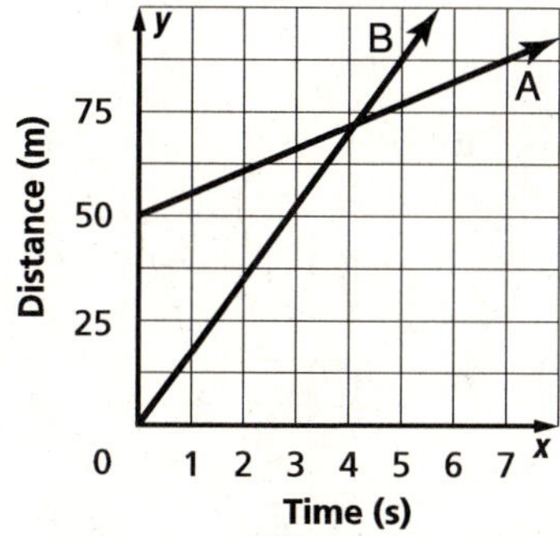

The line representing runner B has a steeper slope than the line representing runner A. The intersection point represents the time and distance at which runner B catches up to runner A.

28. Each set of data can be represented by an equation. All the equations in the system can be graphed and analyzed to compare the data. Answers should include the following.

- Equations can be used to represent incomes of different employees, where y represents income, m is the hourly rate, x is the number of hours, and b is the bonus.
- A solution (x, y) represents a situation in which the number of hours worked and the income is the same for all the employees.

29. C; $x + y = 6$
$y = 3x$

30. C; The system
$x + y = 1$
$y = x + 7$
is graphed below. The solution appears to be $(-3, 4)$.

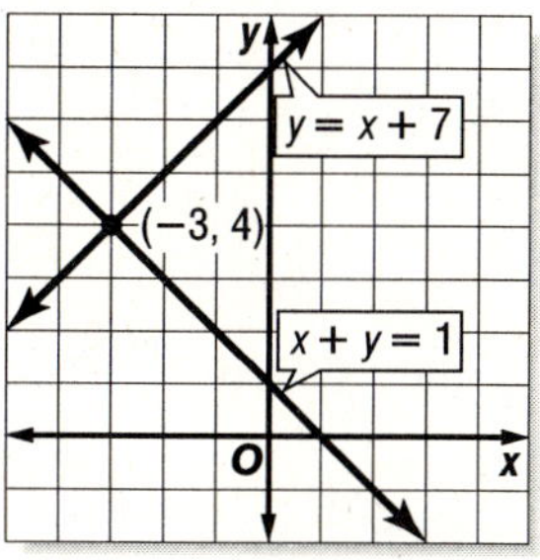

Page 418 Maintain Your Skills

31. positive slope

32. Find the slope m.

$m = \frac{y_2 - y_1}{x_2 - x_1}$

$m = \frac{7 - 1}{3 - 0}$ or 2

Find the y-intercept b. Use the slope and the coordinates of either point.

$y = mx + b$
$7 = 2(3) + b$
$1 = b$

Substitute the slope and y-intercept.

$y = mx + b$
$y = 2x + 1$

33. Find the slope m.

$m = \frac{y_2 - y_1}{x_2 - x_1}$

$m = \frac{-3 - 6}{1 - (-2)}$ or -3

Find the y-intercept b. Use the slope and the coordinates of either point.

$y = mx + b$
$6 = -3(-2) + b$
$0 = b$

Substitute the slope and y-intercept.

$y = mx + b$
$y = -3x + 0$
$y = -3x$

34. Find the slope m.

$m = \frac{y_2 - y_1}{x_2 - x_1}$

$m = \frac{-4 - 0}{-8 - 8}$ or $\frac{1}{4}$

Find the y-intercept b. Use the slope and the coordinates of either point.

$y = mx + b$
$0 = \frac{1}{4}(8) + b$
$-2 = b$

Substitute the slope and y-intercept.

$y = mx + b$
$y = \frac{1}{4}x + (-2)$
$y = \frac{1}{4}x - 2$

35. $1 < x + 3$
$1 \overset{?}{<} 0 + 3$
$1 < 3$ ✓
Yes, 0 is a solution.

36. $9 > t - 5$
$9 \overset{?}{>} 1 - 5$
$9 > -4$ ✓
Yes, 1 is a solution.

37. $2y \geq 2$
$2(-1) \overset{?}{\geq} 2$
$-2 \ngeq 2$
No, -1 is not a solution.

38. $14 < 6 - n$
$14 \overset{?}{<} 6 - 0$
$14 \nless 6$
No, 0 is not a solution.

39. $35 > 12 + k$
$35 \overset{?}{>} 12 + 12$
$36 > 24$ ✓
Yes, 12 is a solution.

40. $5n + 1 \leq 0$
$5(-2) + 1 \overset{?}{\leq} 0$
$-10 + 1 \overset{?}{\leq} 0$
$-9 \leq 0$ ✓
Yes, -2 is a solution.

Page 418 Practice Quiz 2

1. slope $= -1$; y-intercept $= 8$
2. slope $= 3$; y-intercept $= -5$
3. Write it in the form $y = mx + b$:
$x + 2y = 6$
$x - x + 2y = 6 - x$
$2y = 6 - x$
$y = \frac{6 - x}{2}$
$y = 3 - \frac{1}{2}x$
$y = -\frac{1}{2}x + 3$
slope $= -\frac{1}{2}$; y-intercept $= 3$
4. $y = mx + b$
$y = 6x + (-7)$
$y = 6x - 7$
5. $y = mx + b$
$y = 0x + 1$
$y = 1$
6. $y = mx + b$
$y = 1x + 0$
$y = x$
7. Sample answer:

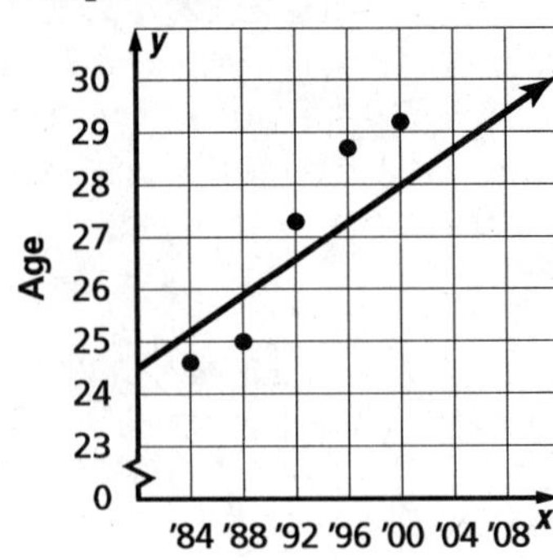

8. $y = x - 1$
$y = 2$
Replace y with 2 in the first equation.
$y = x - 1$
$2 = x - 1$
$3 = x$
The solution of this system is (3, 2).
9. $y = x + 5$
$x = 0$
Replace x with 0 in the first equation.
$y = x + 5$
$y = 0 + 5$
$y = 5$
The solution of this system is (0, 5).
10. $y = 2x + 4$
$y = -4$
Replace y with -4 in the first equation.
$y = 2x + 4$
$-4 = 2x + 4$
$-8 = 2x$
$-4 = x$
The solution of this system is $(-4, -4)$.

8-10 Graphing Inequalites

Page 419 How can shaded regions on a graph model inequalities?

a. $y > 2x + 1$
$2 \overset{?}{>} 2(-4) + 1$
$2 \overset{?}{>} -8 + 1$
$2 > -7$

$y > 2x + 1$
$1 \overset{?}{>} 2(3) + 1$
$1 \overset{?}{>} 6 + 1$
$1 \ngtr 7$

b. $y < 2x + 1$
$2 \overset{?}{<} 2(-4) + 1$
$2 \overset{?}{<} -8 + 1$
$2 \nless -7$

$y < 2x + 1$
$1 \overset{?}{<} 2(3) + 1$
$1 \overset{?}{<} 6 + 1$
$1 < 7$

c. shaded area

Page 421 Check for Understanding

1. $y < -x + 3$
2. Sample answer: Choose a test point not on the boundary line and substitute its coordinates in the inequality. If the result is true, shade that side of the boundary. If the result is false, shade the other side of the boundary.

3a. Sample answer: (0, −1), (1, 0), (2, 1)

3b. Sample answer: (0, 0), (1, 1), (2, 2)

4. Sample answer: $x + y < 6$

5. $y > x - 1$

Graph $y = x - 1$. Draw a dashed line since the boundary is not part of the graph.

Test (0, 0): $y > x - 1$

$0 \overset{?}{>} 0 - 1$

$0 > -1$ ✓

Thus, the graph is all points in the region above the boundary.

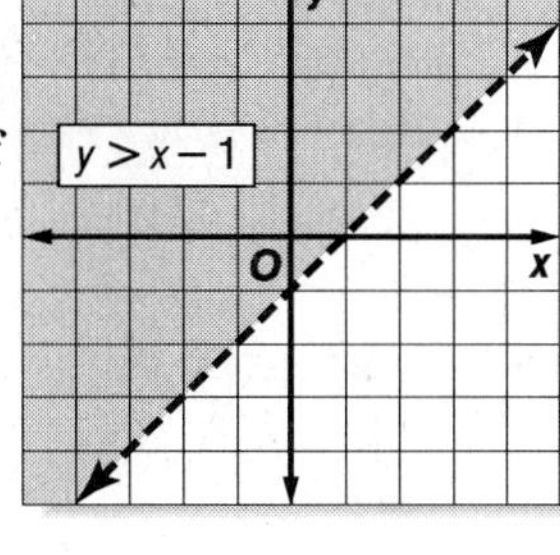

6. $y \le 2$

Graph $y = 2$. Draw a solid line since the boundary is part of the graph.

Test (0, 0): $y \le 2$

$0 \le 2$ ✓

Thus, the graph is all points in the region below the boundary.

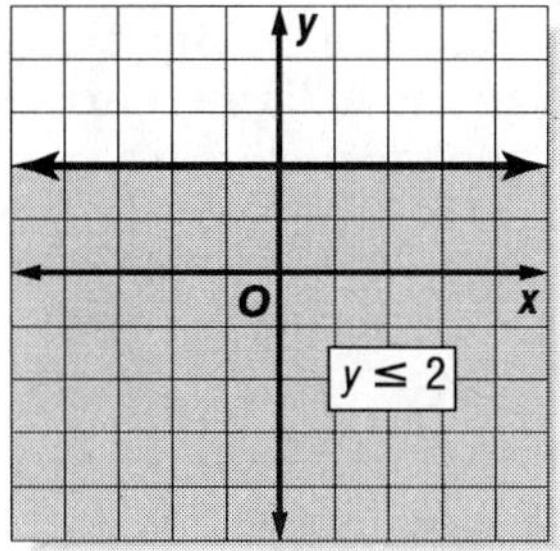

7. $y \ge -3x + 2$

Graph $y = -3x + 2$. Draw a solid line since the boundary is part of the graph.

Test (0, 0): $y \ge -3x + 2$

$0 \overset{?}{\ge} -3(0) + 2$

$0 \not\ge 2$

(0, 0) is not a solution, so shade the other half plane.

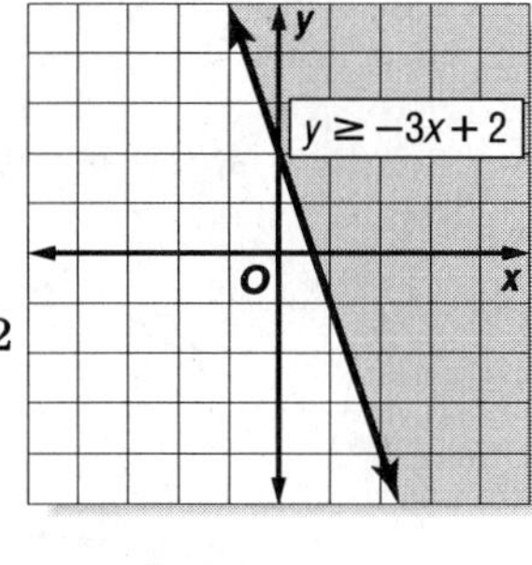

8. Let x represent the number of adult tickets. Let y represent the number of children's tickets. $25x + 15y \le 630$

9. $25x + 15y \le 630$

Graph $25x + 15y = 630$. Draw a solid line since the boundary is part of the graph.

Test (0, 0): $25x + 15y \le 630$

$25(0) + 15(0) \overset{?}{\le} 630$

$0 \le 630$ ✓

Thus, the graph is all points in the region below the boundary.

Sample answer: 5 adults, 30 children; 10 adults, 22 children; 15 adults, 15 children

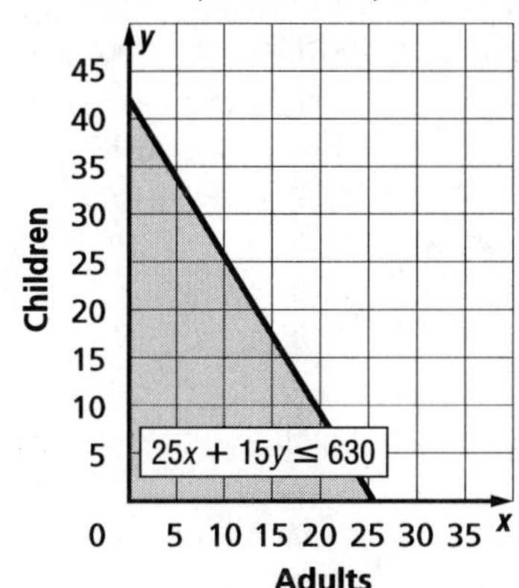

Pages 421–422 Practice and Apply

10. $y \ge x$

Graph $y = x$. Draw a solid line since the boundary is part of the graph.

Test (2, −1): $y \ge x$

$-1 \not\ge 2$

(2, −1) is not a solution, so shade the other half plane.

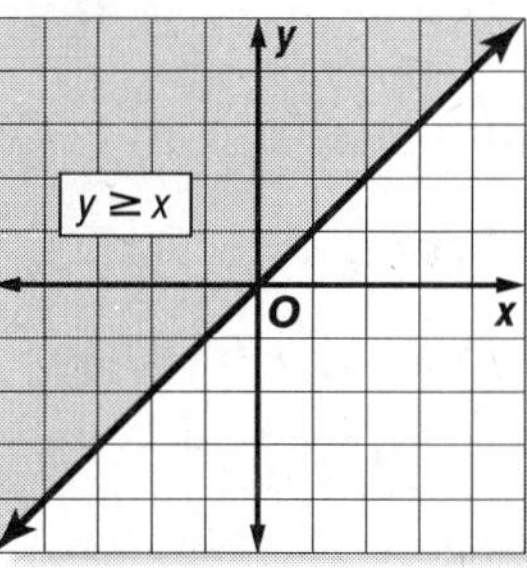

11. $y < x$

Graph $y = x$. Draw a dashed line since the boundary is not part of the graph.

Test (2, −1): $y < x$

$-1 < 2$ ✓

Thus, the graph is all points in the region below the boundary.

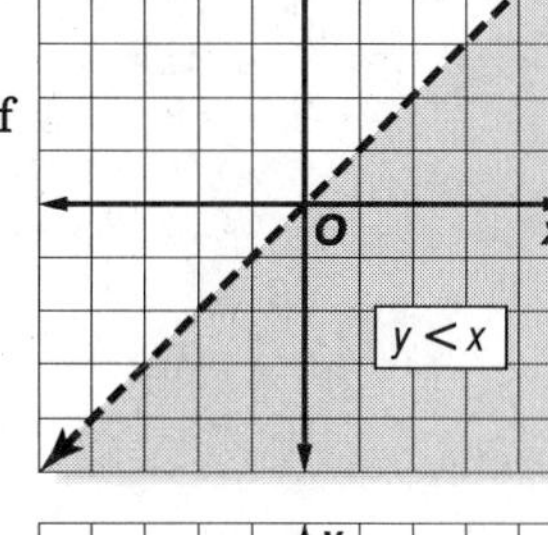

12. $y < 1$

Graph $y = 1$. Draw a dashed line since the boundary is not part of the graph.

Test (0, 0): $y < 1$

$0 < 1$ ✓

Thus, the graph is all points in the region below the boundary.

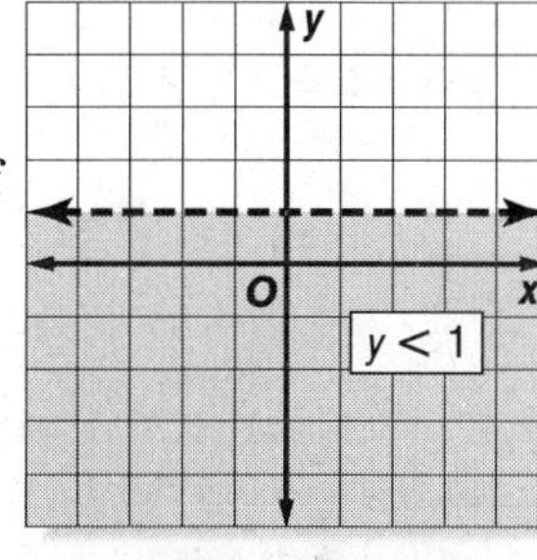

13. $y \ge -3$

Graph $y = -3$. Draw a solid line since the boundary is part of the graph.

Test (0, 0): $y \ge -3$

$0 \ge -3$ ✓

Thus, the graph is all points in the region above the boundary.

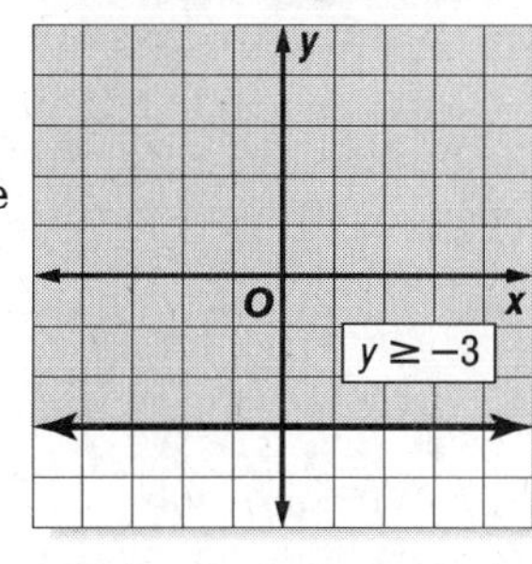

14. $y > x - 1$

Graph $y = x - 1$. Draw a dashed line since the boundary is not part of the graph.

Test (0, 0): $y > x - 1$

$0 \overset{?}{>} 0 - 1$

$0 > -1$ ✓

Thus, the graph is all points in the region above the boundary.

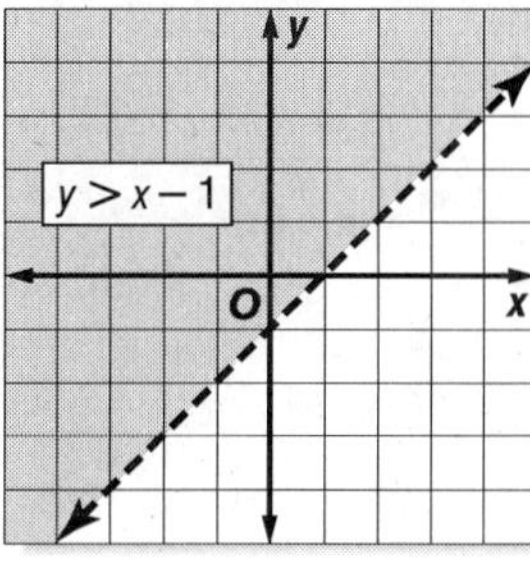

15. $y \leq x + 4$

Graph $y = x + 4$. Draw a solid line since the boundary is part of the graph.

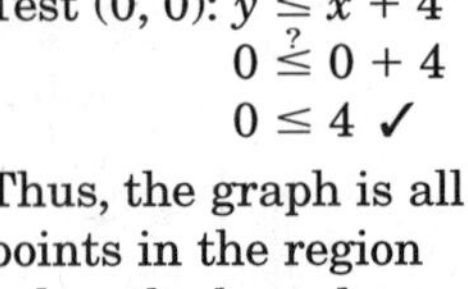

Test (0, 0): $y \leq x + 4$

$0 \overset{?}{\leq} 0 + 4$

$0 \leq 4$ ✓

Thus, the graph is all points in the region below the boundary.

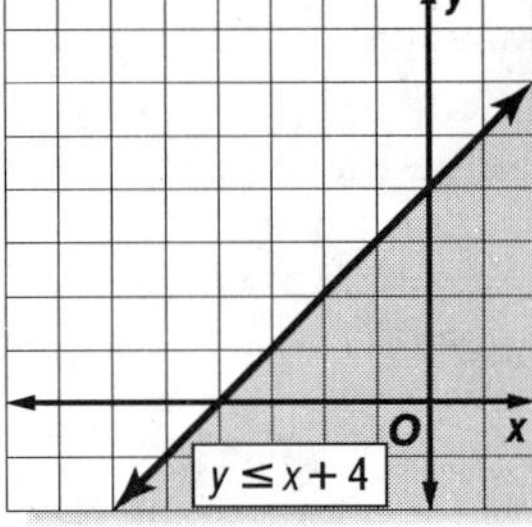

16. $y \leq -x$

Graph $y = x$. Draw a solid line since the boundary is part of the graph.

Test (2, 3): $y \leq -x$

$3 \not\leq -2$

(2, 3) is not a solution, so shade the other half plane.

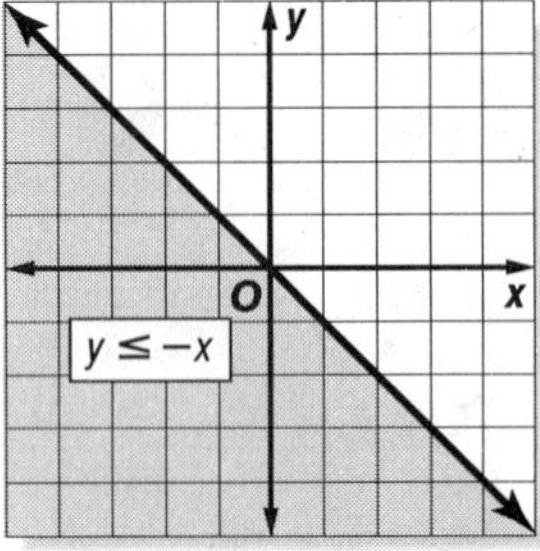

17. $y \geq 0$

Graph $y = 0$. Draw a solid line since the boundary is part of the graph.

Test (0, −4): $y \geq 0$

$-4 \not\geq 0$

(0, −4) is not a solution, so shade the other half plane.

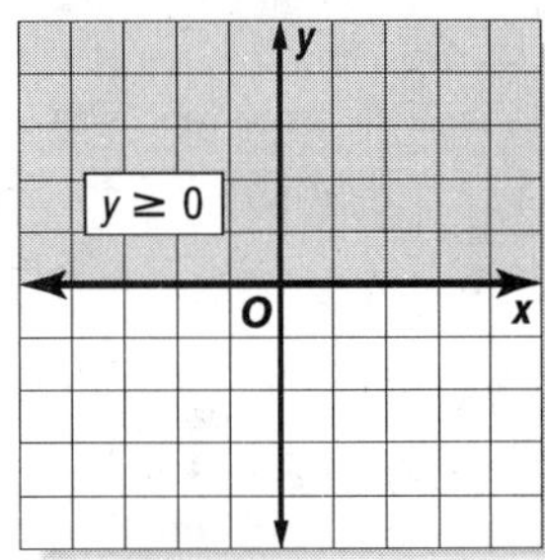

18. $y > 2x + 3$

Graph $y = 2x + 3$. Draw a dashed line since the boundary is not part of the graph.

Test (0, 0): $y > 2x + 3$

$0 \overset{?}{>} 2(0) + 3$

$0 \not> 3$

(0, 0) is not a solution, so shade the other half plane.

y > 2x + 3

19. $y < 2x - 4$

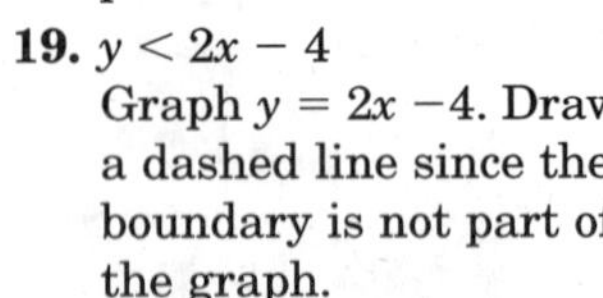

Graph $y = 2x - 4$. Draw a dashed line since the boundary is not part of the graph.

Test (0, 0): $y < 2x - 4$

$0 \overset{?}{<} 2(0) - 4$

$0 \not< -4$

(0, 0) is not a solution, so shade the other half plane.

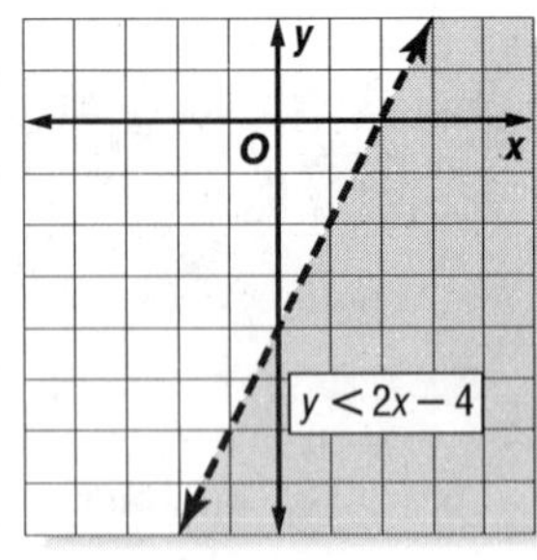

20. $y < \frac{1}{2}x + 2$

Graph $y = \frac{1}{2}x + 2$.

Draw a dashed line since the boundary is not part of the graph.

Test (0, 0): $y < \frac{1}{2}x + 2$

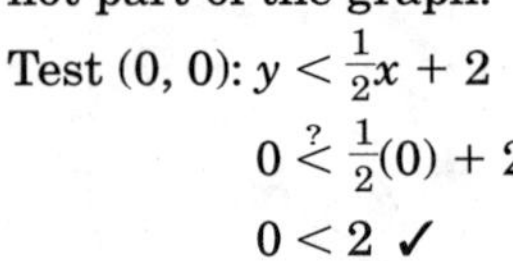

$0 \overset{?}{<} \frac{1}{2}(0) + 2$

$0 < 2$ ✓

Thus, the graph is all points in the region below the boundary.

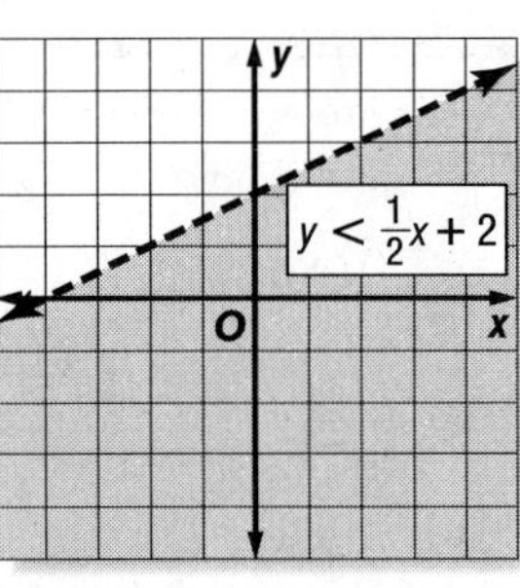

21. $y \geq -\frac{1}{3}x - 1$

Graph $y = -\frac{1}{3}x - 1$.

Draw a solid line since the boundary is part of the graph.

Test (0, 0): $y \geq -\frac{1}{3}x - 1$

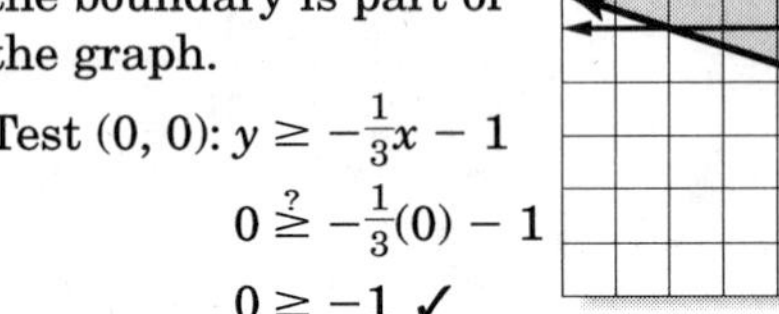

$0 \overset{?}{\geq} -\frac{1}{3}(0) - 1$

$0 \geq -1$ ✓

Thus, the graph is all points in the region above the boundary.

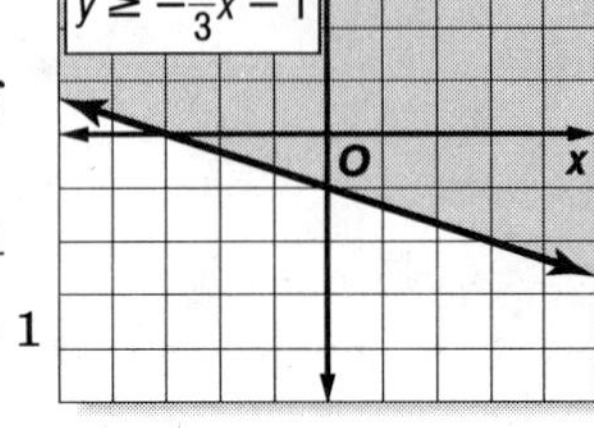

22. $y \geq x + 3000$

23. $y \geq x + 3000$

Graph $y = x + 3000$. Draw a solid line since the boundary is part of the graph.

Test (0, 0): $y \geq x + 3000$

$0 \overset{?}{\geq} 0 + 3000$

$0 \not\geq 3000$

(0, 0) is not a solution, so shade the other half plane.

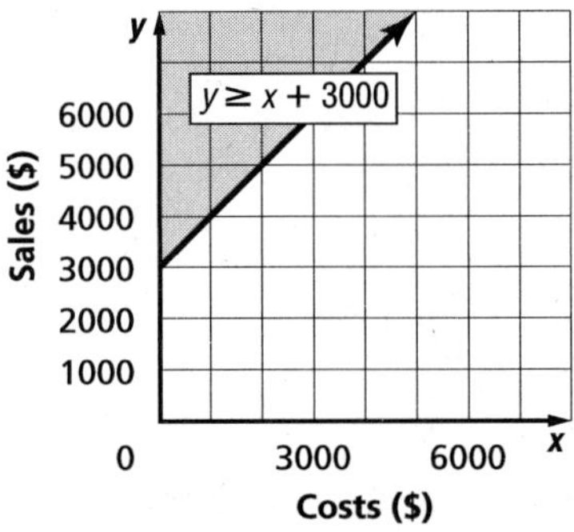

24. Above the line; (1000, 5000) is a solution of $y \geq x + 3000$.

25. No; the number of sales and the costs cannot be negative.

26. Sample answer: (1000, 5000), (4000, 7000)

27. Let x represent the number of small baskets. Let y represent the number of large baskets. Convert 24 hours to minutes:

$24 \text{ h} \cdot \frac{60 \text{ min}}{1 \text{ h}} = 1440 \text{ min}$

$10x + 25y \leq 1440$

28. $10x + 25y \leq 1440$
Graph $10x + 25y = 1440$. Draw a solid line since the boundary is part of the graph.
Test (0, 0): $10x + 25y \leq 1440$
$10(0) + 25(0) \overset{?}{\leq} 1440$
$0 \leq 1440$ ✓
Thus, the graph is all points in the region below the boundary.

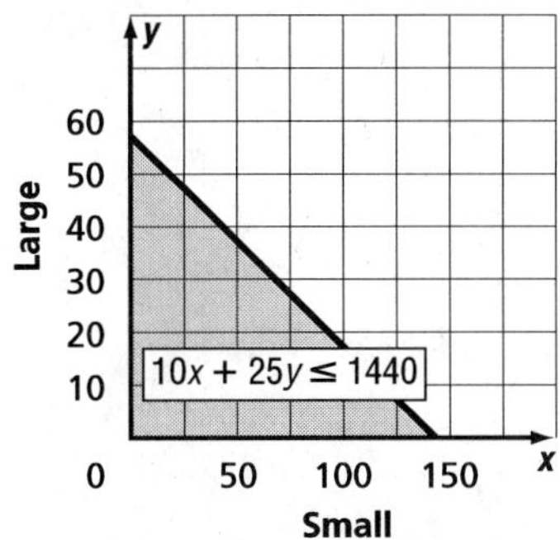

29. Sample answer: 25 small, 45 large; 50 small, 30 large; 100 small, 10 large

30. Let x represent the number of Super Rafts.
Let y represent the number of Econo Rafts.
$100x + 40y \geq 1500$

31. $100x + 40y \geq 1500$
Graph $100x + 40y = 1500$. Draw a solid line since the boundary is part of the graph.
Test (0, 0): $100x + 40y \geq 1500$
$100(0) + 40(0) \overset{?}{\geq} 1500$
$0 \ngeq 1500$
(0, 0) is not a solution, so shade the other half plane.

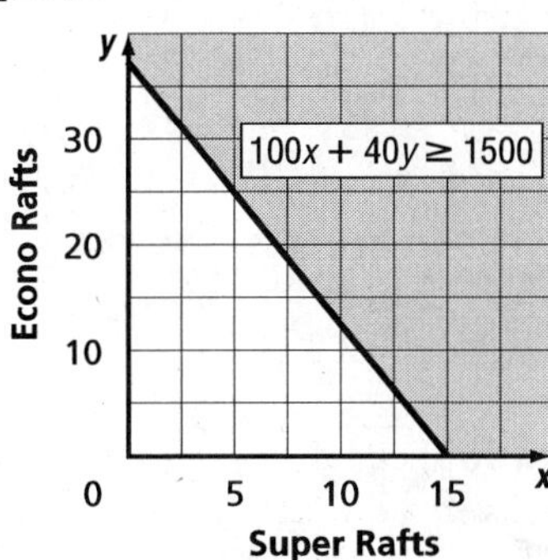

32. Sample answer: 5 Super, 30 Econo; 10 Super, 15 Econo; 15 Super, 5 Econo

33a. $y > x - 1, y \leq -\frac{3}{2}x + 2$

33b. Sample answers: (0, 2), (−1, 2), (−3, −2)

34. To model an inequality, graph the related equation and choose a point not on the line. If the inequality is true when the coordinates of the point are substituted, then that region is shaded. If the inequality is false, then the region on the other side of the line is shaded. Answers should include the following.

- Any point in the shaded region is a solution of the inequality.

35. C; Substitute −9 for x and 0 for y in the inequality.
$y - 8 > x$
$0 - 8 \overset{?}{>} -9$
$-8 > -9$ ✓

36. A; $y > x$, since the inequality does not include an equal sign.

Page 422 Maintain Your Skills

37. $y = x + 3$
$x = 0$
The graphs appear to intersect at (0, 3).

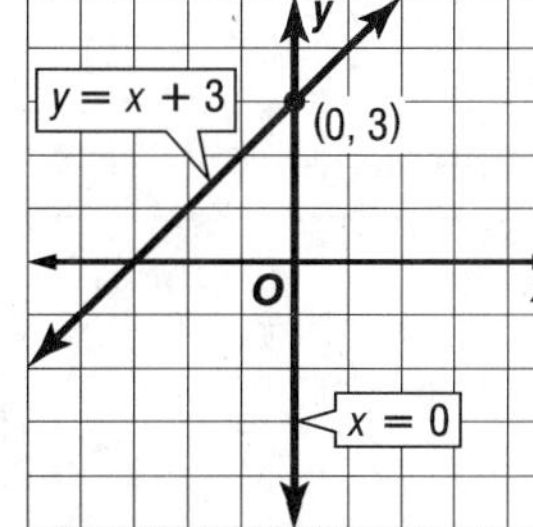

Check:
$y = x + 3$ $\quad x = 0$
$3 \overset{?}{=} 0 + 3$ $\quad 0 = 0$ ✓
$3 = 3$ ✓
The solution is (0, 3).

38. $y = -x + 2$
$y = x + 4$
The graphs appear to intersect at (−1, 3).

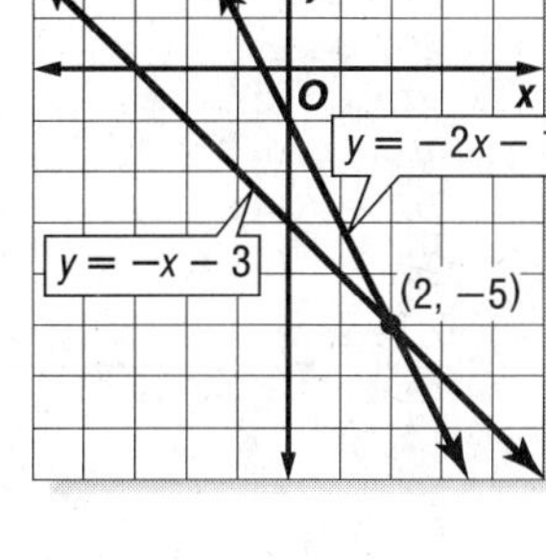

Check:
$y = -x + 2$
$3 \overset{?}{=} -(-1) + 2$
$3 = 3$ ✓
$y = x + 4$
$3 \overset{?}{=} -1 + 4$
$3 = 3$ ✓
The solution is (−1, 3).

39. $y = -2x - 1$
$y = -x - 3$
The graphs appear to intersect at (2, −5).

Check:
$y = -2x - 1$
$-5 \overset{?}{=} -2(2) - 1$
$-5 = -5$ ✓
$y = -x - 3$
$-5 \overset{?}{=} -2 - 3$
$-5 = -5$ ✓
The solution is (2, −5).

40. Sample answer: Make a scatter plot of the data and draw a best-fit line. Then use the line to make a prediction.

41.
$$\begin{array}{r} 0.4 \\ 5\overline{)2.0} \\ \underline{-2\,0} \\ 0 \end{array}$$
$\frac{2}{5} = 0.4$

42.
$$\begin{array}{r} 0.7 \\ 10\overline{)7.0} \\ \underline{-7\,0} \\ 0 \end{array}$$
$3\frac{7}{10} = 3 + 0.7 = 3.7$

43. $\begin{array}{r} 0.\overline{5} \\ 9\overline{)5.0} \\ \underline{-4\,5} \end{array}$

$-\frac{5}{9} = 0.\overline{5}$

Page 423 Graphing Calculator Investigation (Follow-Up of Lesson 8-10)

1. The graphs have the same boundary line, but opposite sides of the line are shaded.

2a.

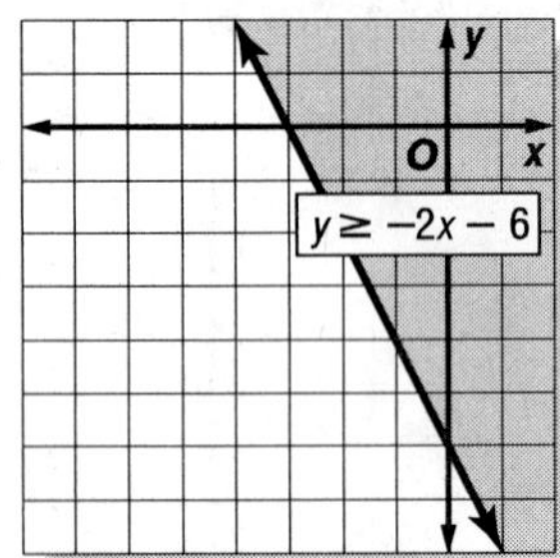

2b. lower bound: $-2x - 6$; upper bound: Ymax or 10

2c. Sample answer: (0, 0), (1, 5), (−1, 4), (−2, 2)

3.

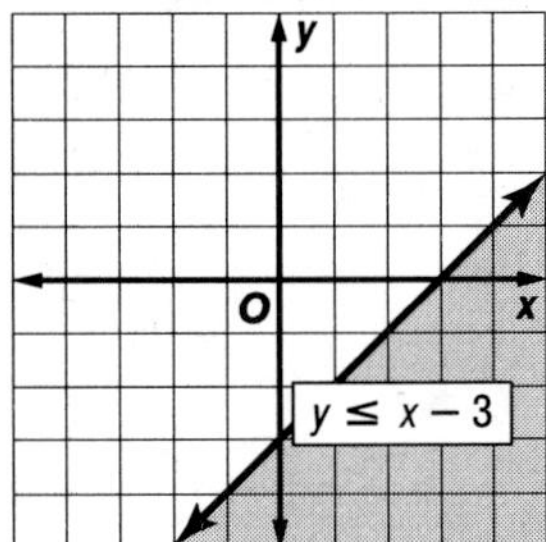

4.

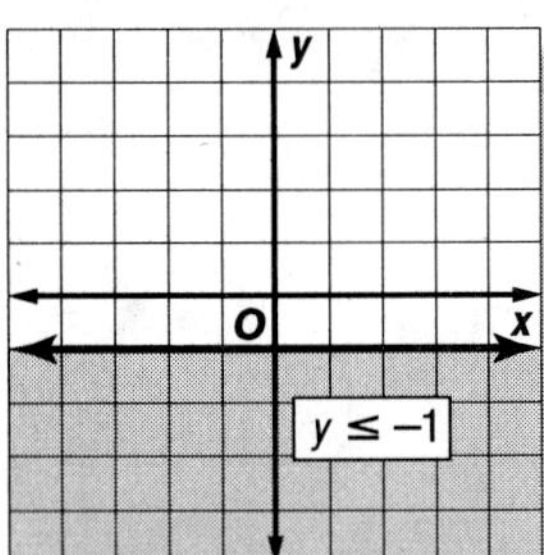

5.

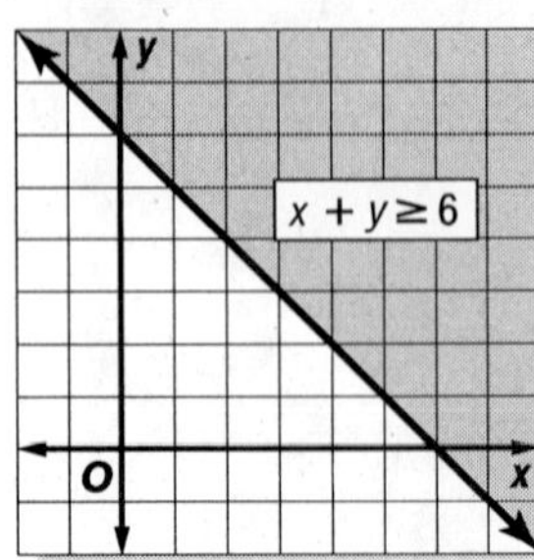

6.

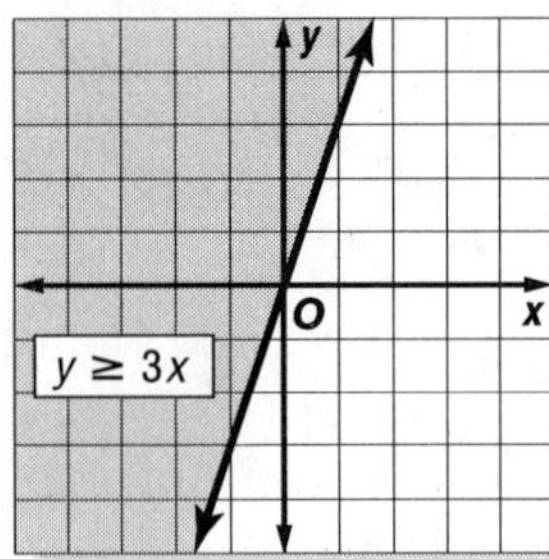

7.

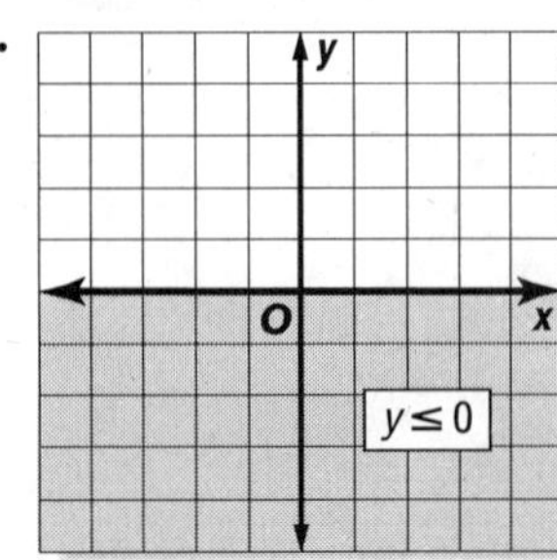

8.

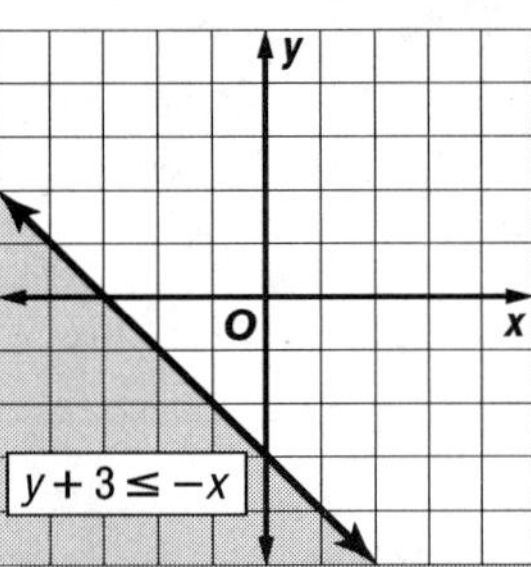

9.

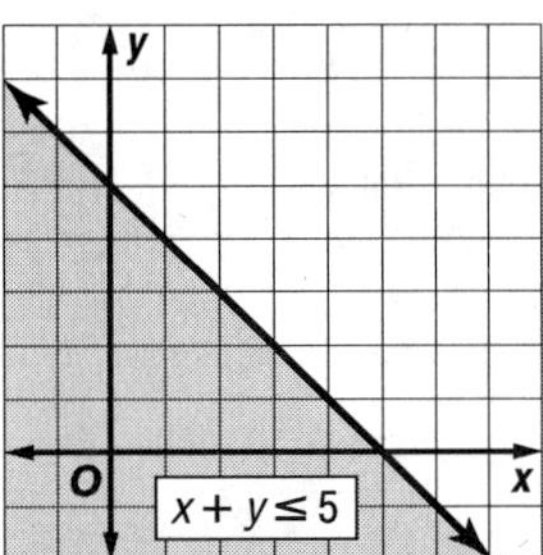

10.

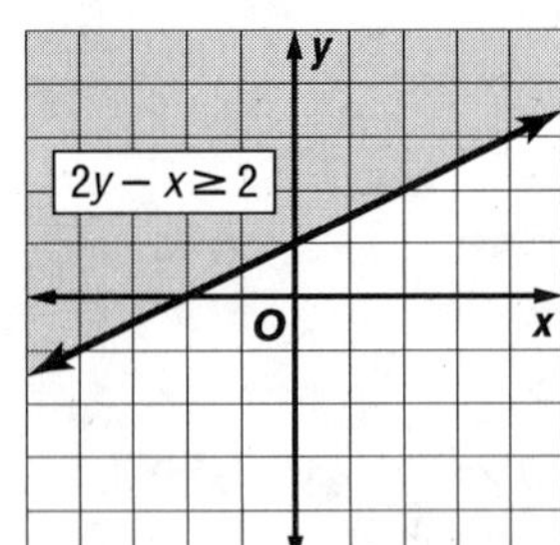

Chapter 8 Study Guide and Review

Page 424 Vocabulary and Concept Check

1. d; function
2. b; boundary
3. h; slope
4. j; system of equations
5. a; best-fit line
6. g; rate of change

7. f; linear equation
8. i; substitution
9. c; direct variation
10. e; half plane

Pages 424–428 Lesson-by-Lesson Review

11. Yes; each x value is paired with only one y value.
12. No; 11.8 in the domain is paired with -9 and 3.8 in the range.
13. Yes; each x value is paired with only one y value.
14. Yes; each x value is paired with only one y value.
15. Sample ordered pairs: $(-4, 0)$, $(-3, 1)$, $(-2, 2)$, and $(-1, 3)$

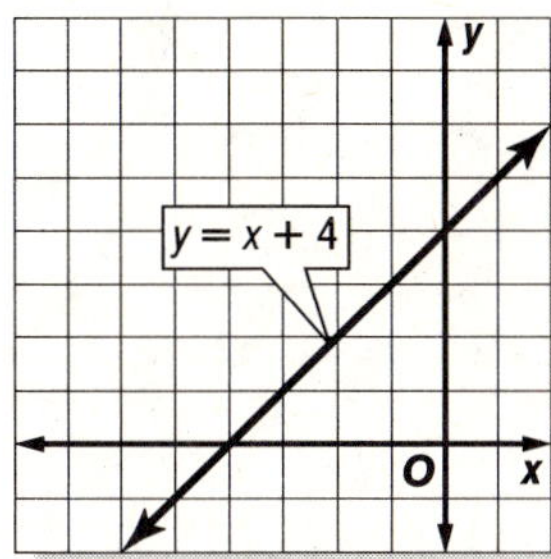

16. Sample ordered pairs: $(-1, -3)$, $(0, -2)$, $(1, -1)$, and $(2, 0)$

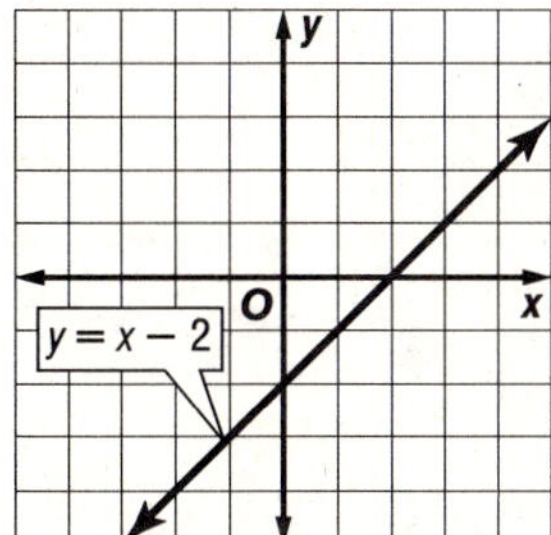

17. Sample ordered pairs: $(-2, 2)$, $(-1, 1)$, $(0, 0)$, and $(1, -1)$

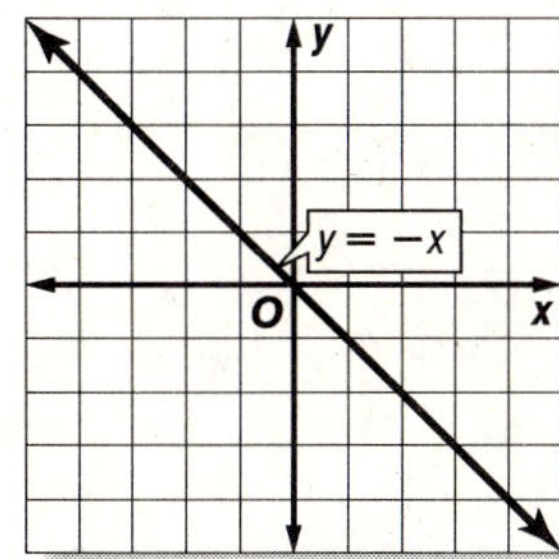

18. Sample ordered pairs: $(-1, -2)$, $(0, 0)$, $(1, 2)$, and $(2, 4)$

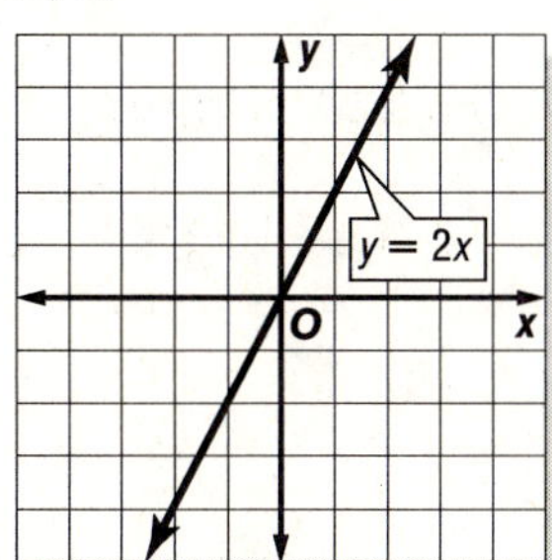

19. Sample ordered pairs: $(-2, -4)$, $(-1, -1)$, $(0, 2)$, and $(1, 5)$

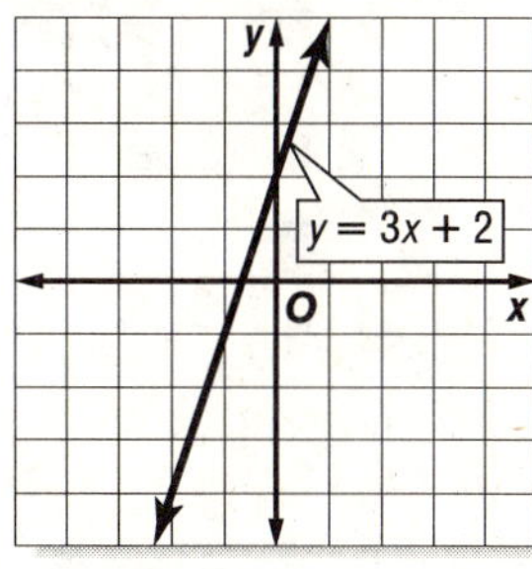

20. Sample ordered pairs: $(-2, 0)$, $(-1, -2)$, $(0, -4)$, and $(1, -6)$

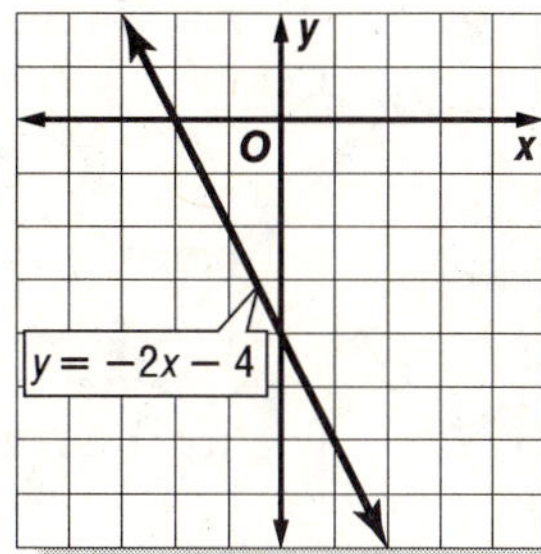

21. Sample ordered pairs: $(0, 4)$, $(1, 3)$, $(2, 2)$, and $(3, 1)$

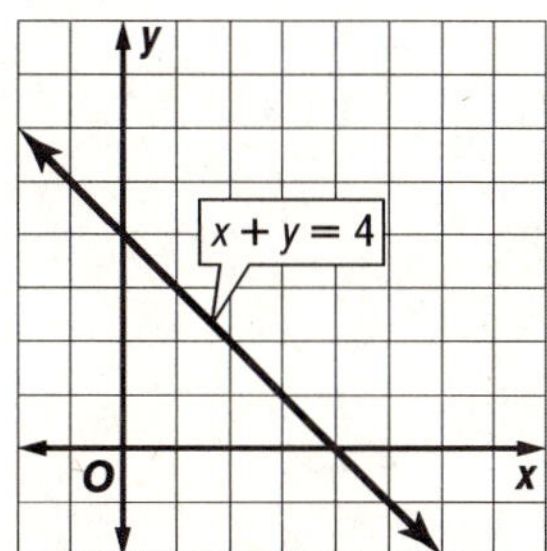

22. Sample ordered pairs: $(-2, 1)$, $(-1, 2)$, $(0, 3)$, and $(1, 4)$

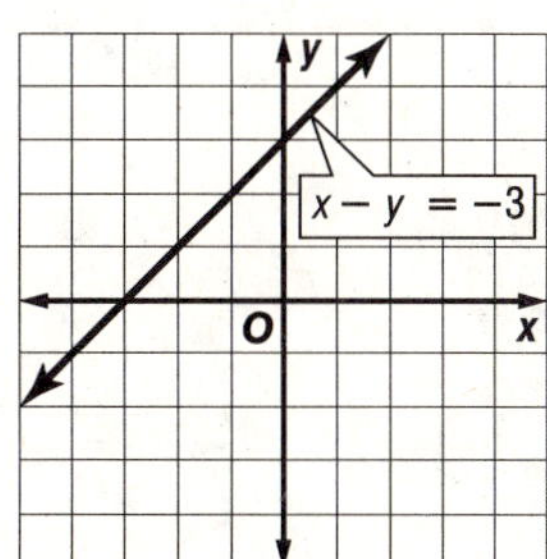

23. Find the x-intercept.

$y = x + 2$
$0 = x + 2$
$-2 = x$

Find the y-intercept.

$y = x + 2$
$y = 0 + 2$
$y = 2$

Graph the points $(-2, 0)$ and $(0, 2)$ and draw a line through them.

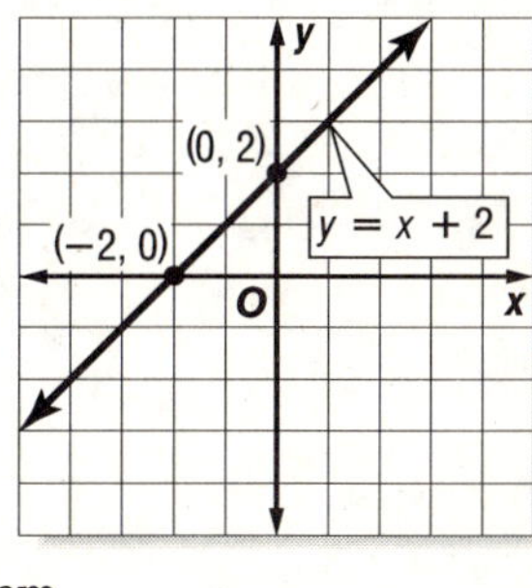

24. Find the x-intercept.
$y = x + 6$
$0 = x + 6$
$-6 = x$
Find the y-intercept.
$y = x + 6$
$y = 0 + 6$
$y = 6$
Graph the points $(-6, 0)$ and $(0, 6)$ and draw a line through them.

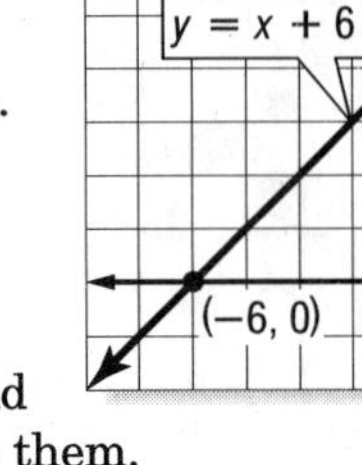

25. Find the x-intercept.
$y = -x - 3$
$0 = -x - 3$
$3 = -x$
$-3 = x$
Find the y-intercept.
$y = -x - 3$
$y = -0 - 3$
$y = -3$
Graph the points $(-3, 0)$ and $(0, -3)$ and draw a line through them.

26. Find the x-intercept.
$y = x - 1$
$0 = x - 1$
$1 = x$
Find the y-intercept.
$y = x - 1$
$y = 0 - 1$
$y = -1$
Graph the points $(1, 0)$ and $(0, -1)$ and draw a line through them.

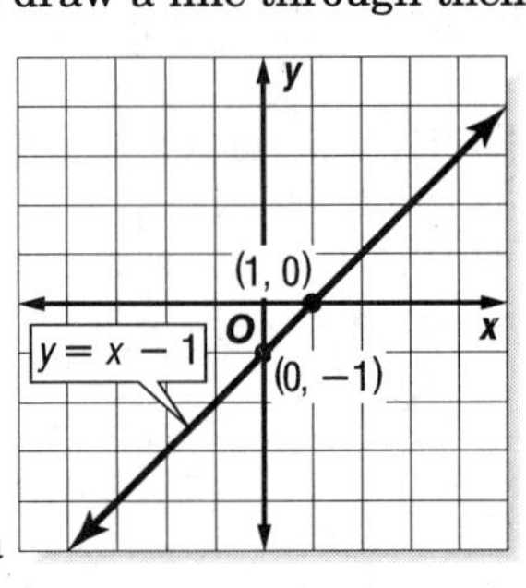

27. Note that $x = -4$ is the same as $x + 0y = -4$. The x-intercept is $(-4, 0)$, and there is no y-intercept.

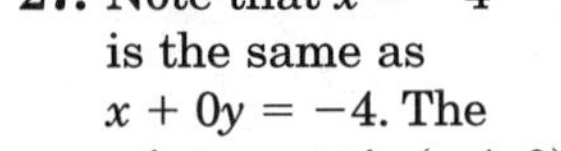

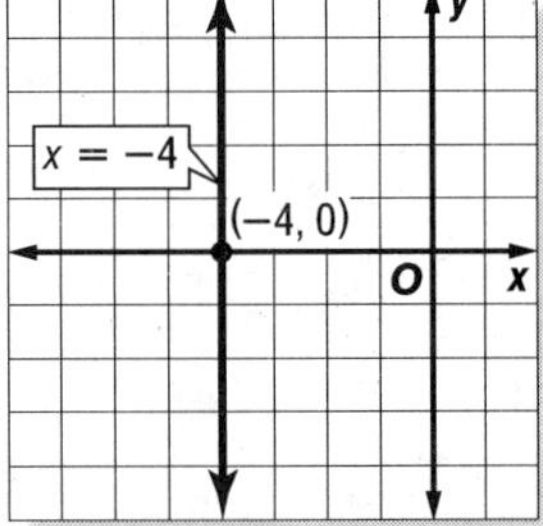

28. Note that $y = 5$ is the same as $0x + y = 5$. The y-intercept is $(0, 5)$, and there is no x-intercept.

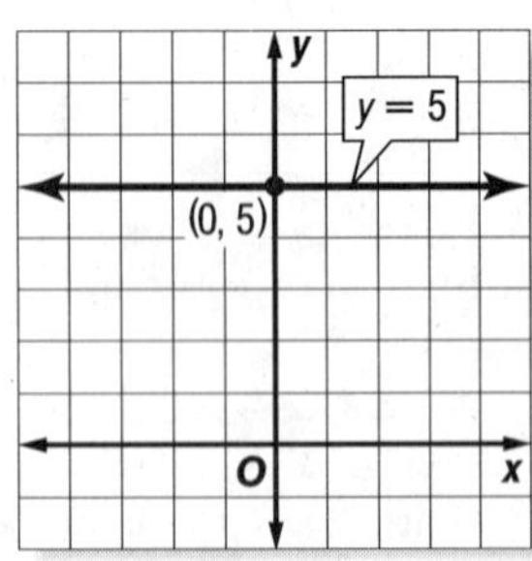

29. Find the x-intercept.
$x + y = -2$
$x + 0 = -2$
$x = -2$
Find the y-intercept.
$x + y = -2$
$0 + y = -2$
$y = -2$
Graph the points $(-2, 0)$ and $(0, -2)$ and draw a line through them.

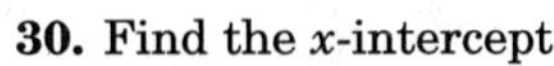

30. Find the x-intercept.
$3x + y = 6$
$3x + 0 = 6$
$3x = 6$
$x = 2$
Find the y-intercept.
$3x + y = 6$
$3(0) + y = 6$
$y = 6$
Graph the points $(2, 0)$ and $(0, 6)$ and draw a line through them.

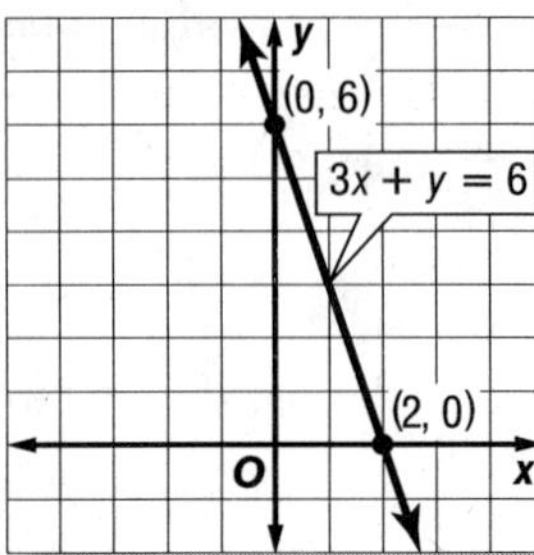

31. $m = \frac{y_2 - y_1}{x_2 - x_1}$
$m = \frac{5 - 4}{4 - 3}$ or 1

32. $m = \frac{y_2 - y_1}{x_2 - x_1}$
$m = \frac{7 - 8}{6 - 2}$ or $-\frac{1}{4}$

33. $m = \frac{y_2 - y_1}{x_2 - x_1}$
$m = \frac{-4 - 3}{-1 - 7}$ or $\frac{7}{8}$

34. $m = \frac{y_2 - y_1}{x_2 - x_1}$
$m = \frac{6 - 10}{4 - 2}$ or -2

35. $m = \frac{y_2 - y_1}{x_2 - x_1}$
$m = \frac{9 - 5}{-1 - (-1)}$
$m = \frac{4}{0}$ or undefined

36. $m = \frac{y_2 - y_1}{x_2 - x_1}$
$m = \frac{8 - 8}{-3 - 0}$
$m = \frac{0}{-3}$ or 0

37. rate of change $= \frac{\text{change in } y}{\text{change in } x}$
$= \frac{8 - 0}{1 - 0}$
$= 8$
There is an increase of 8 m/s.

38. rate of change $= \frac{\text{change in } y}{\text{change in } x}$
$= \frac{43 - 45}{2 - 1}$
$= -\frac{2}{1}$
$= -2$
There is a decrease of 2°F/h.

39. The slope is 1 and the y-intercept is 4. Graph the point (0, 4). Then go up 1 and right 1. Connect these points.

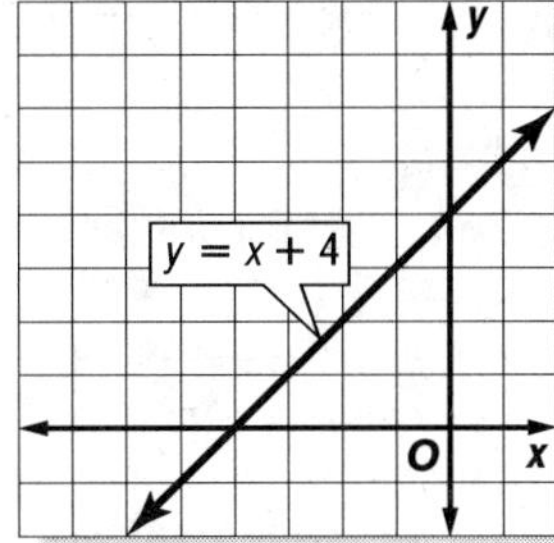

40. The slope is −2 and the y-intercept is 1. Graph the point (0, 1). Then go down 2 and right 1. Connect these points.

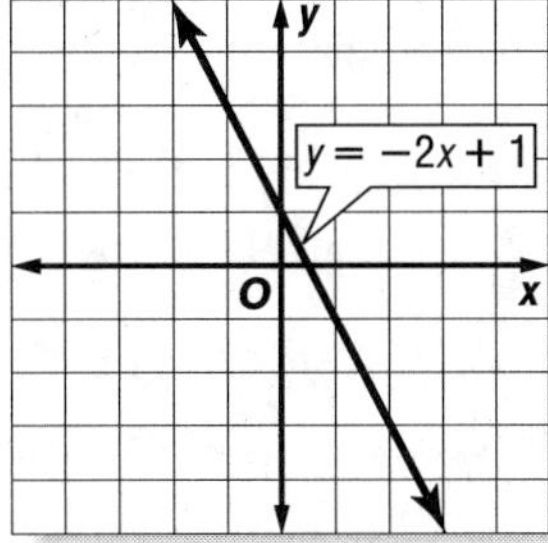

41. The slope is $\frac{1}{3}$ and the y-intercept is −2. Graph the point (0, −2). Then go up 1 and right 3. Connect these points.

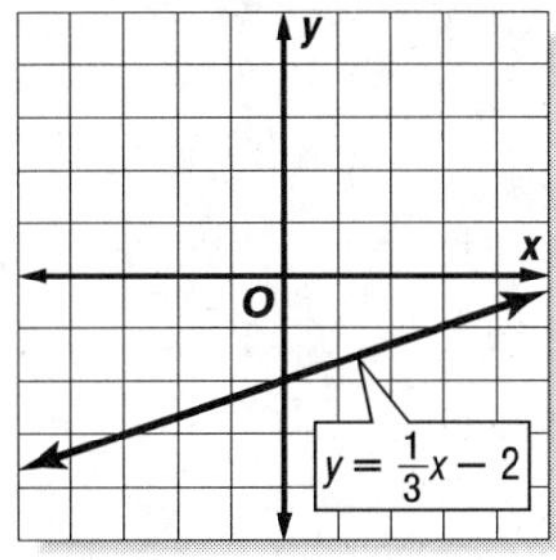

42. Write the equation in the form $y = mx + b$:

$$x + y = -5$$
$$x - x + y = -5 - x$$
$$y = -x - 5$$

The slope is −1 and the y-intercept is −5. Graph (0, −5). Then go down 1 and right 1. Connect these points.

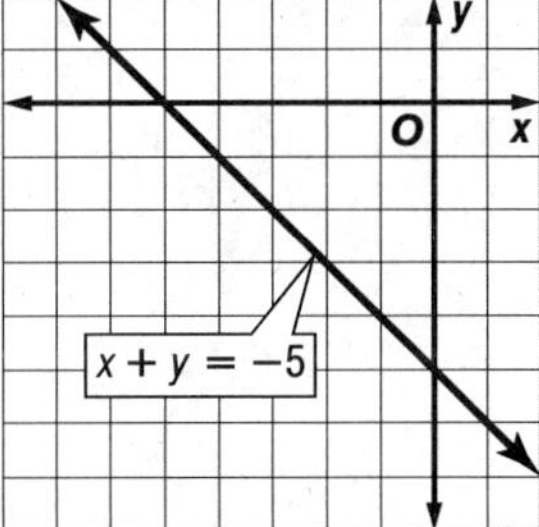

43. $y = mx + b$
$y = -1x + 3$
$y = -x + 3$

44. $y = mx + b$
$y = 6x + (-3)$
$y = 6x - 3$

45. Sample answer:

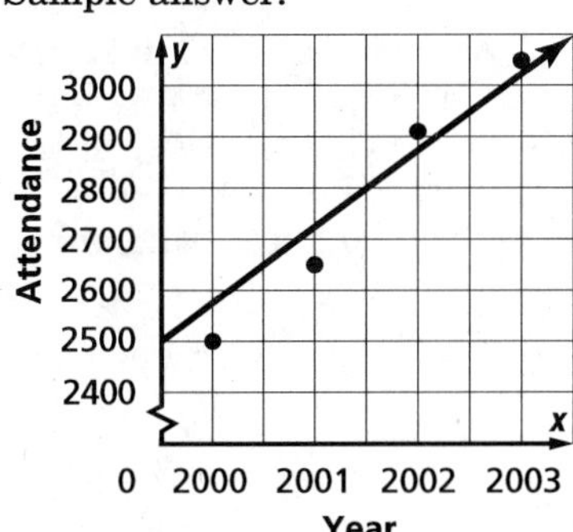

46. Extend the line so that you can find the y value for an x value of 2008.
Sample answer:
The y value for 2008 is about 3600.

47. The graphs appear to intersect at (3, 3).
Check:
$y = x$ $\quad$ $y = 3$
$3 = 3$ ✓ $\quad$ $3 = 3$ ✓
The solution of the system of equations is (3, 3).

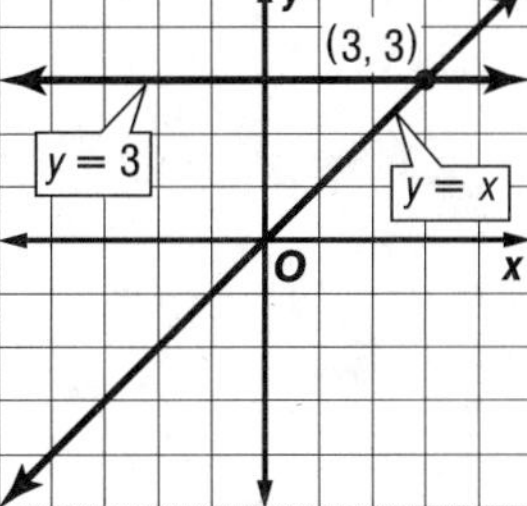

48. The graphs appear to intersect at (−2, 0).
Check:
$y = 2x + 4$
$0 \stackrel{?}{=} 2(-2) + 4$
$0 \stackrel{?}{=} -4 + 4$
$0 = 0$ ✓
$x + y = -2$
$-2 + 0 \stackrel{?}{=} -2$
$-2 = -2$ ✓

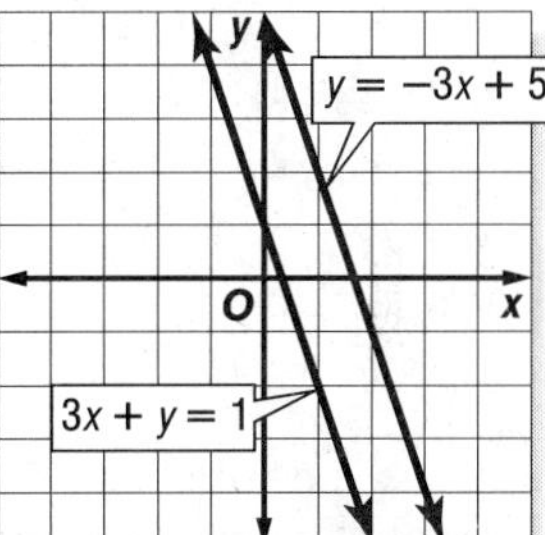

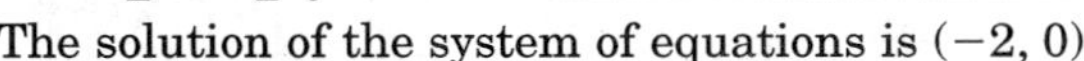
The solution of the system of equations is (−2, 0).

49. The graphs appear to be parallel lines. Since there is no coordinate pair that is a solution to both equations, there is no solution of this system of equations.

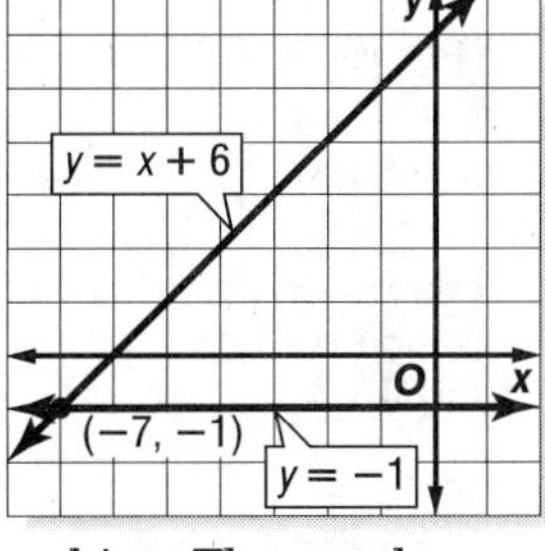

50. $y = x + 6$
$y = -1$
Replace y with −1 in the first equation.
$y = x + 6$
$-1 = x + 6$
$-7 = x$
The solution of this system of equations is (−7, −1). You can check this solution by graphing. The graphs appear to intersect at (−7, −1), so the solution is correct.

51. $y = x$
$x = 4$
Replace x with 4 in the first equation.
$y = x$
$y = 4$
The solution of this system of equations is (4, 4). You can check this solution by graphing. The graphs appear to intersect at (4, 4), so the solution is correct.

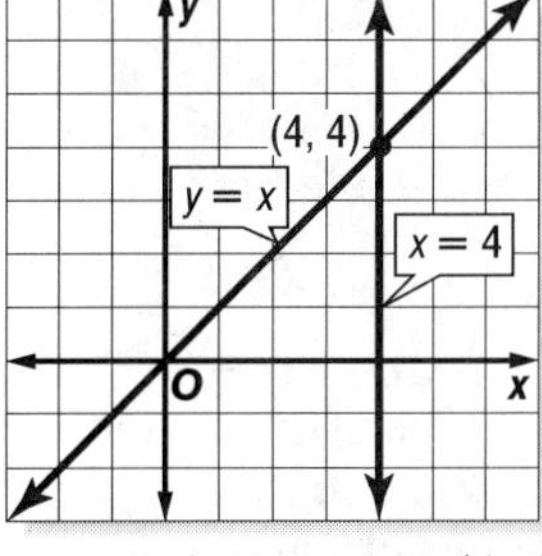

52. $y = 2x - 3$
$y = 0$
Replace y with 0 in the first equation.
$y = 2x - 3$
$0 = 2x - 3$
$3 = 2x$
$\frac{3}{2} = x$

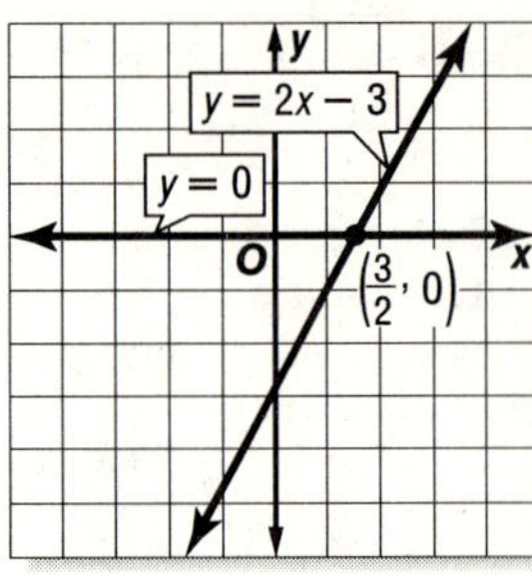

The solution of this system of equations is $\left(\frac{3}{2}, 0\right)$. You can check this solution by graphing. The graphs appear to intersect at $\left(\frac{3}{2}, 0\right)$, so the solution is correct.

53. $y \leq 3x - 1$
Graph $y = 3x - 1$.
Draw a solid line since the boundary is part of the graph.
Test (0, 0): $y \leq 3x - 1$
$0 \overset{?}{\leq} 3(0) - 1$
$0 \not\leq -1$
(0, 0) is not a solution, so shade the other half plane.

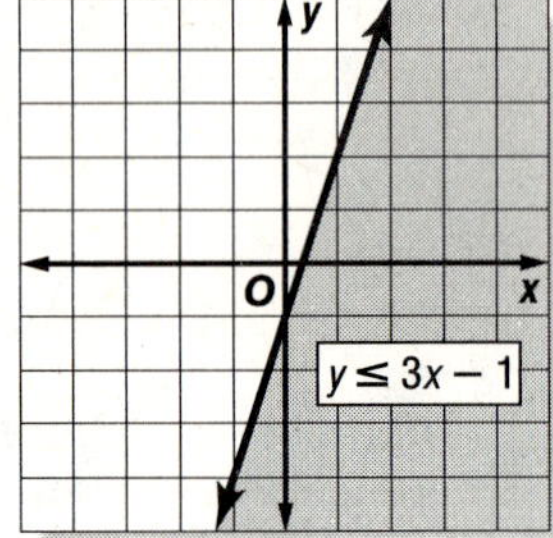

54. $y > \frac{1}{2}x - 3$
Graph $y = \frac{1}{2}x - 3$.
Draw a dashed line since the boundary is not part of the graph.
Test (0, 0): $y > \frac{1}{2}x - 3$
$0 \overset{?}{>} \frac{1}{2}(0) - 3$
$0 > -3$ ✓

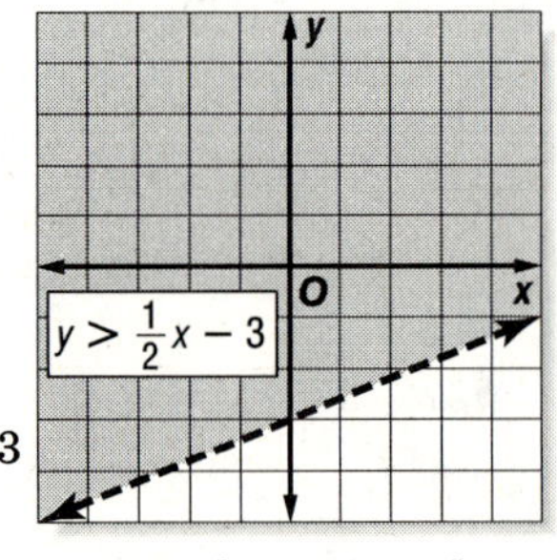

Thus, the graph is all points in the region above the boundary.

55. $y \geq -x + 4$
Graph $y = -x + 4$. Draw a solid line since the boundary is part of the graph.
Test (0, 0): $y \geq -x + 4$
$0 \overset{?}{\geq} -(0) + 4$
$0 \not\geq 4$
(0, 0) is not a solution, so shade the other half plane.

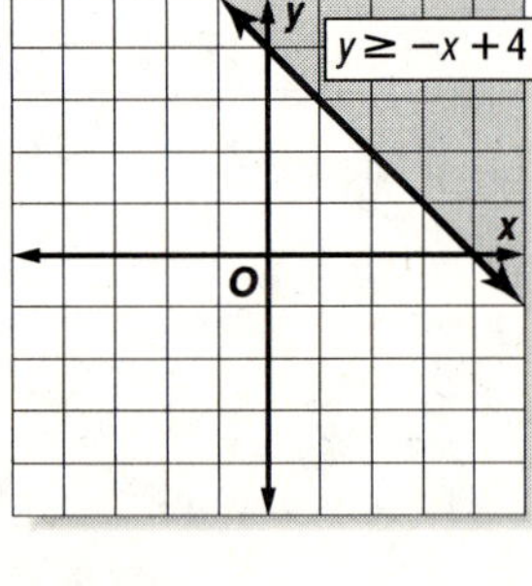

56. $y < -2x$
Graph $y = -2x$. Draw a dashed line since the boundary is not part of the graph.
Test (−3, −1):

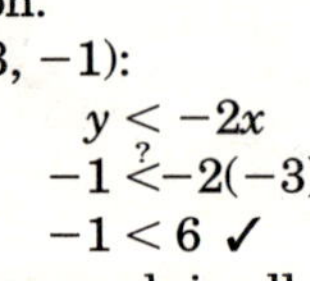

$y < -2x$
$-1 \overset{?}{<} -2(-3)$
$-1 < 6$ ✓

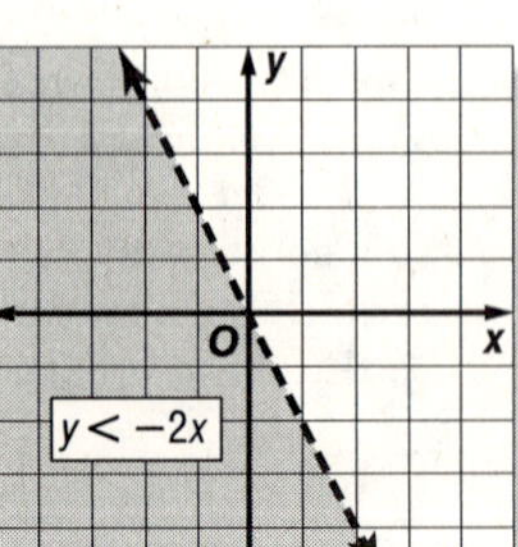

Thus, the graph is all points in the region below the boundary.

Chapter 8 Practice Test

Page 429

1. To find the x-intercept, let $y = 0$ in the equation and solve for x. To find the y-intercept, let $x = 0$ in the equation and solve for y.
2. Choose a point above or below the boundary line and substitute its coordinates in the inequality. If the result is true, shade that side of the boundary. If the result is false, shade the other side of the boundary.
3. Sample answer: $y = x$ and $y = -x + 2$; the solution (1, 1) is a solution of both equations.
4. No; −3 in the domain is paired with 4 and 6 in the range.
5. Yes; each x value is paired with only one y value.
6. Sample ordered pairs: (−2, −3), (−1, −1), (0, 1), and (1, 3)

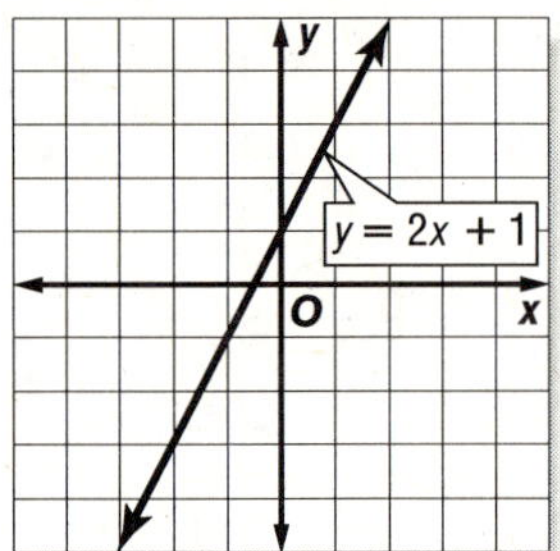

7. Sample ordered pairs: (−1, 7), (0, 4), (1, 1), and (2, −2)

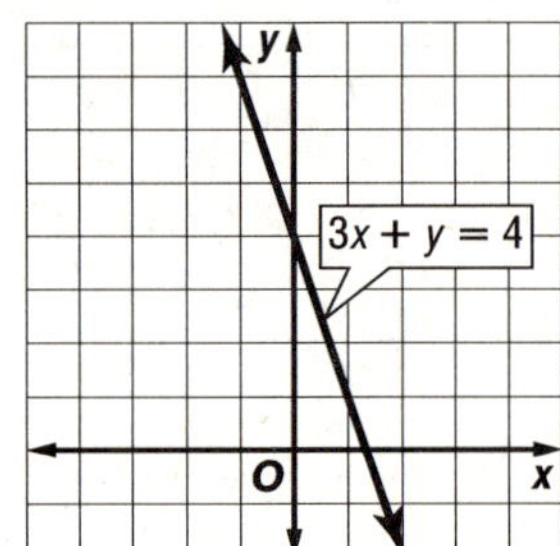

8. Find the x-intercept.
$y = x + 3$
$0 = x + 3$
$-3 = x$
The x-intercept is −3.
Find the y-intercept.
$y = x + 3$
$y = 0 + 3$
$y = 3$
The y-intercept is 3.
Graph the points (−3, 0) and (0, 3) and draw a line through them.

9. Find the x-intercept.
$2x - y = 4$
$2x - 0 = 4$
$2x = 4$
$x = 2$
The x-intercept is 2.
Find the y-intercept.
$2x - y = 4$
$2(0) - y = 4$
$-y = 4$
$y = -4$
The y-intercept is -4. Graph the points (2, 0) and (0, −4) and draw a line through them.

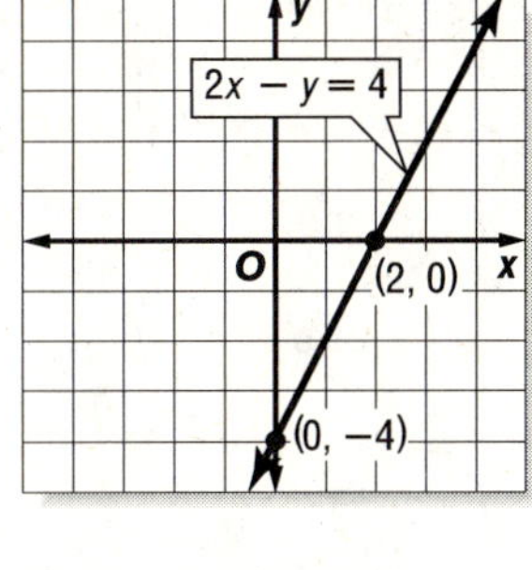

10. $m = \frac{y_2 - y_1}{x_2 - x_1}$
$m = \frac{11 - 5}{4 - 2}$ or 3

11. $m = \frac{y_2 - y_1}{x_2 - x_1}$
$m = \frac{-3 - 5}{6 - (-4)}$ or $-\frac{4}{5}$

12. rate of change $= \frac{\text{change in money earned}}{\text{change in hours worked}}$
$= \frac{11.00 - 5.50}{2 - 1}$
$= 5.50$
There is an increase of \$5.50/h.

13. The slope is $\frac{2}{3}$ and the y-intercept is -4. Graph the point (0, −4). Then go up 2 and right 3. Connect these points.

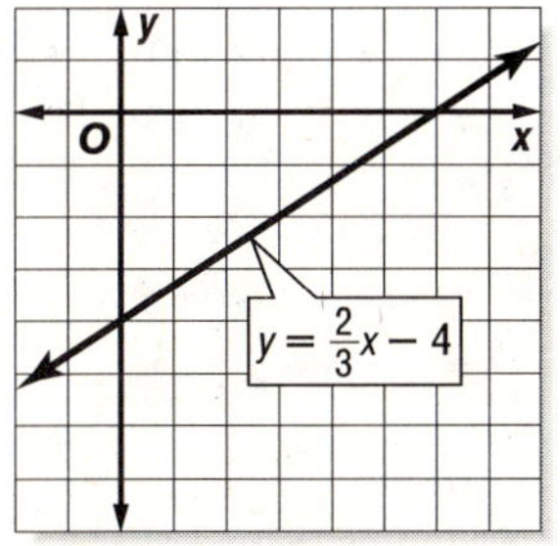

14. Write the equation in the form $y = mx + b$:
$2x + 4y = 12$
$2x - 2x + 4y = 12 - 2x$
$4y = 12 - 2x$
$y = \frac{12 - 2x}{4}$
$y = 3 - \frac{1}{2}x$
$y = -\frac{1}{2}x + 3$

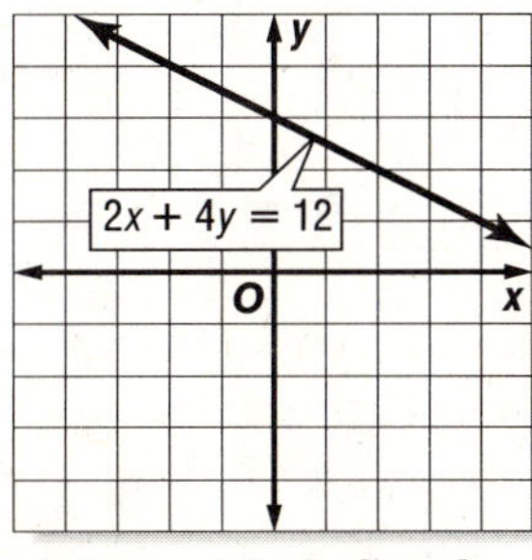

The slope is $-\frac{1}{2}$ and the y-intercept is 3. Graph the point (0, 3). Then go down 1 and right 2. Connect these points.

15. $y = mx + b$
$y = \frac{3}{8}x + (-2)$
$y = \frac{3}{8}x - 2$

16. The graphs appear to intersect at (1, −2).
Check:
$2x - y = 4$
$2(1) - (-2) \stackrel{?}{=} 4$
$2 + 2 \stackrel{?}{=} 4$
$4 = 4$ ✓

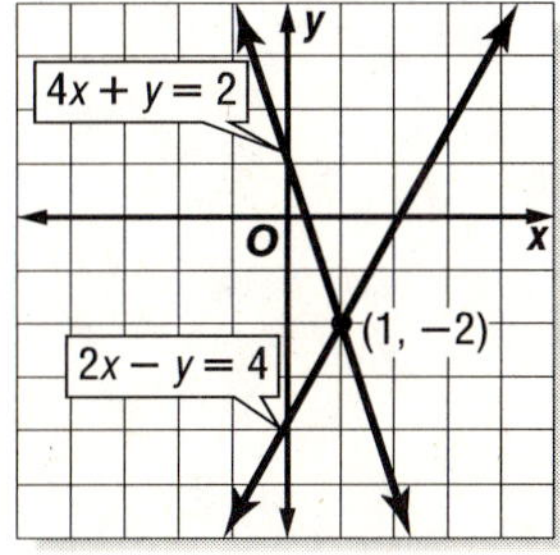

$4x + y = 2$
$4(1) + (-2) \stackrel{?}{=} 2$
$4 - 2 \stackrel{?}{=} 2$
$2 = 2$ ✓
The solution of the system of equations is (1, −2).

17. $y > 2x - 1$
Graph $y = 2x - 1$.
Draw a dashed line since the boundary is not part of the graph.
Test (0, 0): $y > 2x - 1$
$0 \stackrel{?}{>} 2(0) - 1$
$0 > -1$ ✓
Thus, the graph is all points in the region above the boundary.

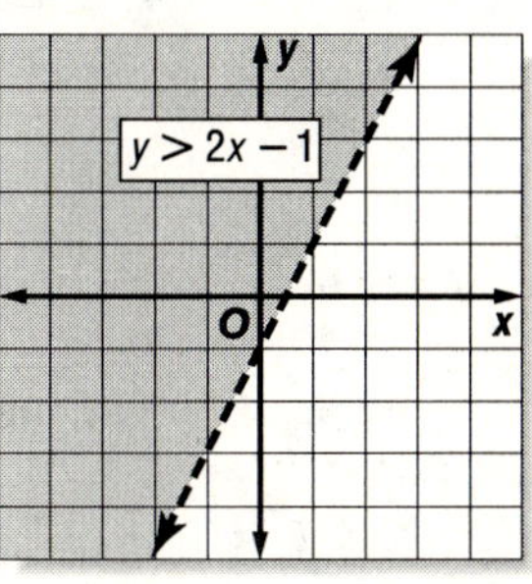

18.

y
20
15
10
5
0
10 20 30 40 x
Number of Tomatoes
Height (in.)

19. Extend the line so that you can find the y value for an x value of 43.
Sample answer: 20

20. D; $y \geq -2x + 5$
$-1 \stackrel{?}{\geq} -2(7) + 5$
$-1 \stackrel{?}{\geq} -14 + 5$
$-1 \geq -9$ ✓
(−1, 7) is a solution.

Chapter 8 Standardized Test Practice

Pages 430–431

1. C; $5.5 \text{ ft} = 5.5 \text{ ft} \cdot \frac{12 \text{ in.}}{1 \text{ ft}}$
$= 66$ in.
$66 \div 10 = 6.6$ in. each piece

2. D; $-\frac{3}{9} = -\frac{1}{3}$ and $-\frac{9}{3} = -3$, so $-\frac{3}{9} > -\frac{9}{3}$.

3. B; $P(\text{yellow}) = \frac{\text{number of favorable outcomes}}{\text{number of possible outcomes}}$
$= \frac{\text{2 yellow sides}}{\text{6 sides of a cube}}$
$= \frac{1}{3}$

4. D; Use the ratio $\frac{\text{quarts of paint}}{\text{square feet of wall}}$ to write the proportion.
$\frac{3}{175} = \frac{x}{700}$

5. C; 30% of 12 ounces $= 0.30 \times 12$
$= 3.6$
$3.6 + 12 = 15.6$ ounces

6. B; The percentage of garbage that is not paper is $16 + 25 + 7 + 15$ or 63%.
$0.63 \times 6000 = 3780$ pounds

7. D; Let n represent an integer.
Let $n + 1$ represent the next greater integer.

$$n + n + 1 > 51$$
$$2n + 1 > 51$$
$$2n > 50$$
$$n > 25$$

Since the integer must be greater than 25, 26 is the only possible solution from the choices.

8. B; y appears to increase by 2 for each 1 unit increase in x.

x	y
1	3
2	5
3	7
4	9
5	11
6	13

9. B; The y-intercept is 8 and the slope is 2.
$y = mx + b$
$y = 2x + 8$

10. C; Check $(5, -1)$ in each equation.

$$2x + 3y = 7 \qquad 3x - 3y = 18$$
$$2(5) + 3(-1) \stackrel{?}{=} 7 \qquad 3(5) - 3(-1) \stackrel{?}{=} 18$$
$$10 - 3 \stackrel{?}{=} 7 \qquad 15 + 3 \stackrel{?}{=} 18$$
$$7 = 7 \checkmark \qquad 18 = 18 \checkmark$$

11. Find 4 on the x-axis, and move up to the point and then left to the y-axis. It appears to have a y value of 13, or $13.

12.
$$P = 2\ell + 2w$$
$$24 = 2(8) + 2w$$
$$24 = 16 + 2w$$
$$24 - 16 = 16 - 16 + 2w$$
$$8 = 2w$$
$$4 = w$$
$$A = \ell w$$
$$A = 8 \cdot 4$$
$$A = 32$$

13. $y = \frac{1}{2}x + 5$

14. $0.09357 = 9.357 \times 10^{-2}$

15.
$$y = 4x - 3$$
$$-\frac{3}{2} = 4x - 3$$
$$-\frac{3}{2} + 3 = 4x - 3 + 3$$
$$-\frac{3}{2} + \frac{6}{2} = 4x$$
$$\frac{3}{2} = 4x$$
$$\frac{1}{4}\left(\frac{3}{2}\right) = \frac{1}{4}(4x)$$
$$\frac{3}{8} = x$$

16. $\frac{3}{5} = 0.60$
$= 60\%$

17.
$$68.2 + 68.9 + 67.5 + 71.7 + x > 345.4$$
$$276.3 + x > 345.4$$
$$276.3 - 276.3 + x > 345.4 - 276.3$$
$$x > 69.1$$

18. Use $d = rt$.
30 min = 0.5 h and 15 min = 0.25 h
Her distance in the first 30 minutes is:
$d = rt$
$d = 30(0.5)$
$d = 15$ miles
Her distance in the next hour and fifteen minutes is:
$d = rt$
$d = 56(1.25)$
$d = 70$ miles
She traveled a total of $15 + 70$, or 85 miles.

19.
$$m = \frac{y_2 - y_1}{x_2 - x_1}$$
$$m = \frac{5 - 4}{2 - 0}$$
$$m = \frac{1}{2}$$

20. The graphs appear to intersect at $(4, 5)$.

21a. X: $c = 0 + 0.24m$
Y: $c = 15.95 + 0.08m$
Z: $c = 25.95 + 0.04m$

21b. X: $c = 0.24(100) = \$24.00$
Y: $c = 15.95 + 0.08(100) = \$23.95$
Z: $c = 25.95 + 0.04(100) = \$29.95$

21c. Plan Y

21d. X: $c = 0.24(300) = \$72.00$
Y: $c = 15.95 + 0.08(300) = \$39.95$
Z: $c = 25.95 + 0.04(300) = \$37.95$

21e. Plan Z

Chapter 9 Real Numbers and Right Triangles

Page 435 Getting Started

1. $<$
2. $>$
3. $<$
4. $>$
5. $>$
6. $<$
7. $<$
8. $>$
9. $3x = 24$
   $\frac{3x}{3} = \frac{24}{3}$
   $x = 8$
10. $7y = 49$
    $\frac{7y}{7} = \frac{49}{7}$
    $y = 7$
11. $120 = 2n$
    $\frac{120}{2} = \frac{2n}{2}$
    $60 = n$
12. $54 = 6a$
    $\frac{54}{6} = \frac{6a}{6}$
    $9 = a$
13. $90 = 10m$
    $\frac{90}{10} = \frac{10m}{10}$
    $9 = m$
14. $144 = 12m$
    $\frac{144}{12} = \frac{12m}{12}$
    $12 = m$
15. $15d = 165$
    $\frac{15d}{15} = \frac{165}{15}$
    $d = 11$
16. $182 = 14w$
    $\frac{182}{14} = \frac{14w}{14}$
    $13 = w$
17. $(3 - 1)^2 + (4 - 2)^2 = (2)^2 + (2)^2$
    $= 4 + 4$
    $= 8$
18. $(5 - 2)^2 + (6 - 3)^2 = (3)^2 + (3)^2$
    $= 9 + 9$
    $= 18$
19. $(4 - 7)^2 + (3 - 8)^2 = (-3)^2 + (-5)^2$
    $= 9 + 25$
    $= 34$
20. $(8 - 2)^2 + (3 - 9)^2 = (6)^2 + (-6)^2$
    $= 36 + 36$
    $= 72$
21. $(2 - 6)^2 + [(-8) - 1]^2 = (-4)^2 + (-9)^2$
    $= 16 + 81$
    $= 97$
22. $(-7 - 2)^2 + [3 - (-4)]^2 = (-9)^2 + (7)^2$
    $= 81 + 49$
    $= 130$

9-1 Squares and Square Roots

Page 436 How are square roots related to factors?

x	x^2
5	25
7	49
13	169
15	225
2.8	8
3.5	12
8.1	65
10.5	110

a. The first four are whole numbers and the last four are not.

b. To find an exact answer, determine what number times itself is equal to each value of x.

c. Choose a decimal value that, when multiplied by itself, is approximately equal to the value of x.

Page 438 Check for Understanding

1. A positive number squared results in a positive number, and a negative number squared results in a positive number.
2. Sample answer: 225, 256, 324
3. Sample answer: $-\sqrt{1.69}$

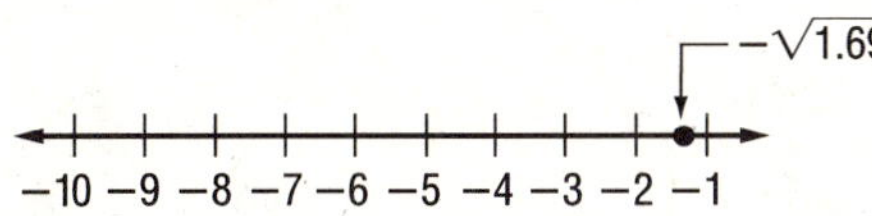

4. Since $7^2 = 49$, $\sqrt{49} = 7$.
5. Since $8^2 = 64$, $-\sqrt{64} = -8$.
6. not possible
7. 3.9
8. −5.7
9. $\sqrt{66}$
   $64 < 66 < 81$
   $\sqrt{64} < \sqrt{66} < \sqrt{81}$
   $8 < \sqrt{66} < 9$
   $\sqrt{66} \approx 8$
10. $-\sqrt{103}$
    $-121 < -103 < -100$
    $-\sqrt{121} < -\sqrt{103} < -\sqrt{100}$
    $-11 < -\sqrt{103} < -10$
    $-\sqrt{103} \approx -10$
11. $D = 1.22 \times \sqrt{A}$
    $= 1.22 \times \sqrt{1353}$
    $\approx 1.22 \times 36.78$
    ≈ 44.8716
    Ryan can see about 44.9 miles.

Pages 439–440 Practice and Apply

12. Since $4^2 = 16$, $\sqrt{16} = 4$.

13. Since $6^2 = 36$, $\sqrt{36} = 6$.

14. Since $1^2 = 1$, $-\sqrt{1} = -1$.

15. Since $5^2 = 25$, $-\sqrt{25} = -5$.

16. not possible

17. not possible

18. Since $10^2 = 100$, $\sqrt{100} = 10$.

19. Since $14^2 = 196$, $\sqrt{196} = 14$.

20. Since $16^2 = 256$, $\pm\sqrt{256} = 16, -16$.

21. Since $18^2 = 324$, $\pm\sqrt{324} = 18, -18$.

22. Since $(0.9)^2 = 0.81$, $\sqrt{0.81} = 0.9$.

23. Since $(1.5)^2 = 2.25$, $\sqrt{2.25} = 1.5$.

24. 3.9

25. 7.5

26. −6.6

27. −9.3

28. 13.4

29. 15.8

30. −0.9

31. −1.7

32. −31.6

33. $x = \sqrt{5000}$
$x = 70.7$

34. $49 < 54 < 64$
$\sqrt{49} < \sqrt{54} < \sqrt{64}$
$7 < \sqrt{54} < 8$
The number $\sqrt{54}$ lies between 7 and 8 on a number line.

35. $64 < 79 < 81$
$\sqrt{64} < \sqrt{79} < \sqrt{81}$
$8 < \sqrt{79} < 9$
$\sqrt{79} \approx 9$

36. $81 < 95 < 100$
$\sqrt{81} < \sqrt{95} < \sqrt{100}$
$9 < \sqrt{95} < 10$
$\sqrt{95} \approx 10$

37. $-64 < -54 < -49$
$-\sqrt{64} < -\sqrt{54} < -\sqrt{49}$
$-8 < -\sqrt{54} < -7$
$-\sqrt{54} \approx -7$

38. $-144 < -125 < -121$
$-\sqrt{144} < -\sqrt{125} < -\sqrt{121}$
$-12 < -\sqrt{125} < -11$
$-\sqrt{125} \approx -11$

39. $196 < 200 < 225$
$\sqrt{196} < \sqrt{200} < \sqrt{225}$
$14 < \sqrt{200} < 15$
$\sqrt{200} \approx 14$

40. $361 < 396 < 400$
$\sqrt{361} < \sqrt{396} < \sqrt{400}$
$19 < \sqrt{396} < 20$
$\sqrt{396} \approx 20$

41. $-289 < -280 < -256$
$-\sqrt{289} < -\sqrt{280} < -\sqrt{256}$
$-17 < -\sqrt{280} < -16$
$-\sqrt{280} \approx -17$

42. $-529 < -490 < -484$
$-\sqrt{529} < -\sqrt{490} < -\sqrt{484}$
$-23 < -\sqrt{490} < -22$
$-\sqrt{490} \approx -22$

43. $-9 < -5.25 < -4$
$-\sqrt{9} < -\sqrt{5.25} < -\sqrt{4}$
$-3 < -\sqrt{5.25} < -2$
$-\sqrt{5.25} \approx -2$

44. $-25 < -17.3 < -16$
$-\sqrt{25} < -\sqrt{17.3} < -\sqrt{16}$
$-5 < -\sqrt{17.3} < -4$
$-\sqrt{17.3} \approx -4$

45. $36 < 38.75 < 49$
$\sqrt{36} < \sqrt{38.75} < \sqrt{49}$
$6 < \sqrt{38.75} < 7$
$\sqrt{38.75} \approx 6$

46. $121 < 140.57 < 144$
$\sqrt{121} < \sqrt{140.57} < \sqrt{144}$
$11 < \sqrt{140.57} < 12$
$\sqrt{140.57} \approx 12$

47. $D = 1.22 \times \sqrt{A}$
$= 1.22 \times \sqrt{148}$
$\approx 1.22 \times 12.17$
≈ 14.8474
A person can see about 14.8 miles.

48. $D = 1.22 \times \sqrt{A}$
$= 1.22 \times \sqrt{255}$
$\approx 1.22 \times 15.97$
≈ 19.4834
A person can see about 19.5 miles.

49. 9; Since $64 < 65 < 81$, $\sqrt{64} < \sqrt{65} < \sqrt{81}$. Thus, it follows that $8 < \sqrt{65} < 9$.
So, 9 is greater than $\sqrt{65}$.

50. $\sqrt{120}$; Since $100 < 120 < 121$, $\sqrt{100} < \sqrt{120} < \sqrt{121}$. Thus, it follows that $10 < \sqrt{120} < 11$.
So, $\sqrt{120}$ is less than 11.

51a. 10.4 in.

51b. 14.2 cm

51c. 8.4 m

52. $2 + \sqrt{81} \times 3 = 2 + 9 \times 3$
$= 2 + 27$
$= 29$

53. $\sqrt{256} \div 8 \times 5 = 16 \div 8 \times 5$
$= 2 \times 5$
$= 10$

54. $(9 - \sqrt{36}) + (16 - \sqrt{100}) = (9 - 6) + (16 - 10)$
$= 3 + 6$
$= 9$

55. $(\sqrt{1} + 7) \div (\sqrt{121} - 7) = (1 + 7) \div (11 - 7)$
$= 8 \div 4$
$= 2$

56. about 182.4 m

57. Sample answer: A number that has a rational square root will have an ending digit of 0, 1, 4, 5, 6, or 9. The last digit is the ending digit in one of the squares from 1–100. There are just six ending digits.

58. Square roots are related to factors in that if a number is a perfect square, the square root of the number is factor of the number. Answers should include the following.

- Sample answer: 121
- Sample answer: 120

59. D; $-5 < -\sqrt{17} < -4$.

60. A; $-\sqrt{361} = -19$

61. Sample answer: addition and subtraction

62a. 64

62b. 100

62c. 169

63. a

Page 440 Maintain Your Skills

64.

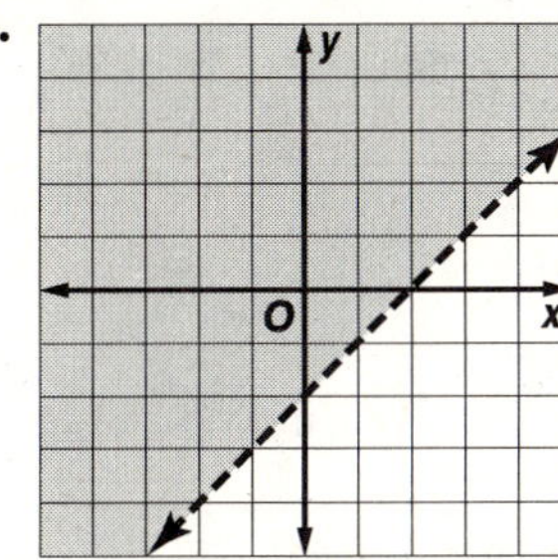

65.

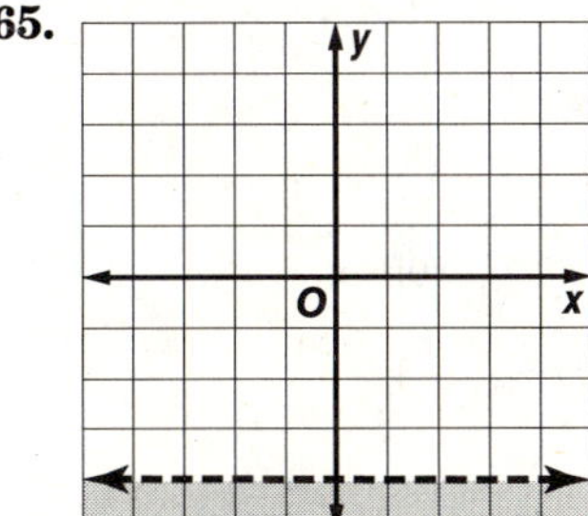

66.

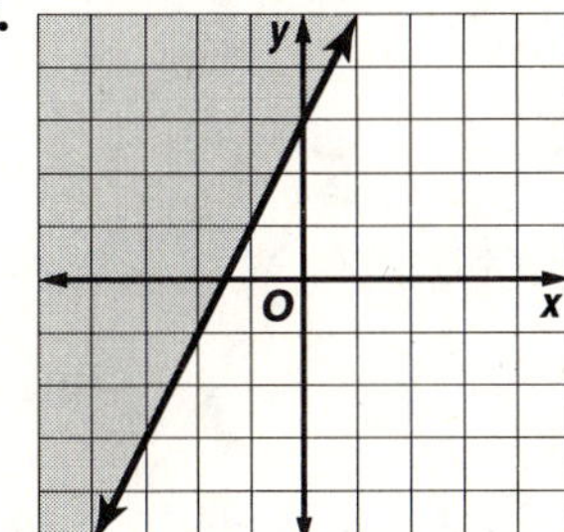

67. $y = x + 2$
$x = 5$
Replace x with 5:
$y = 5 + 2$
$y = 7$
The solution of this system is (5, 7).

68. $y = x + 3$
$y = 0$
Replace y with 0:
$0 = x + 3$
$0 - 3 = x + 3 - 3$
$-3 = x$
The solution of this system is $(-3, 0)$.

69. $y = 2x + 5$
$y = -3$
Replace y with -3:
$-3 = 2x + 5$
$-3 - 5 = 2x + 5 - 5$
$\frac{-8}{2} = \frac{2x}{2}$
$-4 = x$
The solution of this system is $(-4, -3)$.

70. No, because 4 is paired with both -1 and 2.

71. $\frac{1}{6} + \frac{3}{6} = \frac{4}{6}$
$= \frac{2}{3}$

72. $\frac{n}{9} = \frac{15}{27}$
$\frac{27n}{27} = \frac{135}{27}$
$n = 5$

73. $\frac{4}{b} = \frac{16}{36}$
$\frac{144}{16} = \frac{16b}{16}$
$9 = b$

74. $\frac{7}{8.4} = \frac{0.5}{x}$
$\frac{7x}{7} = \frac{4.2}{7}$
$x = 0.6$

75. It can be written as a fraction.

76. It can be written as $\frac{3}{2}$.

77. It can be written as $\frac{3}{4}$.

78. It can be written as $\frac{8}{9}$.

79. It can be written as $\frac{6}{1}$.

80. It can be written as $-\frac{7}{1}$.

9-2 The Real Number System

Page 441 How can squares have lengths that are not rational numbers?

a. $\frac{1}{2}$ unit^2

b. square

c. 2 square units

d. $\sqrt{2}$

Page 443 Check for Understanding

1. Whereas rational numbers can be expressed in the form $\frac{a}{b}$, where a and b are integers and b does not equal 0, irrational numbers cannot.

2. Sample answer: $-\sqrt{25}$

3. N, W, Z, Q

4. Q

5. Q

6. I

7. $6\frac{4}{5} = 6.8$
$\sqrt{48} = 6.92820323$
Since $6.8 < 6.92820323$, $6\frac{4}{5} < \sqrt{48}$.

8. $-\sqrt{74} = -8.602325267$
$-8.\overline{4} = -8.44444...$
Since $-8.602325267 < -8.\overline{4}$, $-\sqrt{74} < -8.\overline{4}$.

9. $3.\overline{7} = 3.7777...$
$3\frac{3}{5} = 3.6$
$\sqrt{13} = 3.605551275$
$\frac{10}{3} = 3.3333...$
From least to greatest, the order is $\frac{10}{3}$, $3\frac{3}{5}$, $\sqrt{13}$, $3.\overline{7}$.

10. $y^2 = 25$
$\sqrt{y^2} = \sqrt{25}$
$y = \sqrt{25}$ or $y = -\sqrt{25}$
$y = 5$ or $y = -5$

11. $m^2 = 74$
$\sqrt{m^2} = \sqrt{74}$
$m = \sqrt{74}$ or $m = -\sqrt{74}$
$m \approx 8.6$ or $m \approx -8.6$

12. $r = \sqrt{\frac{A}{\pi}}$
$r = \sqrt{\frac{40}{\pi}}$
$r \approx \sqrt{12.7324}$
$r \approx 3.6$ ft

Pages 444–445 Practice and Apply

13. N, W, Z, Q
14. N, W, Z, Q
15. Q
16. Q
17. Q
18. Q
19. I
20. I
21. I
22. I
23. Z, Q
24. Z, Q
25. Q
26. Q
27. Z, Q
28. Z, Q
29. always
30. never
31. always
32. sometimes

33. $5\frac{1}{4} = 5.25$
$\sqrt{26} = 5.099019514$
Since $5.25 > 5.099019514$, $5\frac{1}{4} > \sqrt{26}$.

34. $\sqrt{80} = 8.94427191$
9.2
Since $8.94427191 < 9.2$, $\sqrt{80} < 9.2$.

35. -3.3
$-\sqrt{10} = -3.16227766$
Since $-3.3 < -3.16227766$, $-3.3 < -\sqrt{10}$.

36. $-\sqrt{18} = -4.242640687$
$-4\frac{3}{8} = -4.375$
Since $-4.242640687 > -4.375$, $-\sqrt{18} > -4\frac{3}{8}$.

37. $1\frac{1}{2} = 1.5$
$\sqrt{2.25} = 1.5$
$1\frac{1}{2} = \sqrt{2.25}$

38. $-\sqrt{6.25} = -2.5$
$-\frac{5}{2} = -2.5$
$-\sqrt{6.25} = -\frac{5}{2}$

39. $5\frac{1}{4} = 5.25$
$2.\overline{1} = 2.1111...$
$\sqrt{4} = 2$
$\frac{6}{5} = 1.2$
From least to greatest, the order is $\frac{6}{5}$, $\sqrt{4}$, $2.\overline{1}$, $5\frac{1}{4}$.

40. $4.\overline{23} = 4.2323...$
$4\frac{2}{3} = 4.6666...$
$\sqrt{18} = 4.242640687$
$\sqrt{16} = 4$
From least to greatest, the order is $\sqrt{16}$, $4.\overline{23}$, $\sqrt{18}$, $4\frac{2}{3}$.

41. -10
-1.05
$-\sqrt{105} = -10.24695077$
$-10\frac{1}{2} = -10.5$
From least to greatest, the order is -10.5, $-\sqrt{105}$, -10, -1.05.

42. $-\sqrt{14} = -3.741657387$
$-4\frac{1}{10} = -4.1$
-3.8
$-\frac{17}{4} = -4.25$
From least to greatest, the order is $-\frac{17}{4}$, $-4\frac{1}{10}$, -3.8, $-\sqrt{14}$.

43. Sample answer: $\sqrt{4}$ and $\sqrt{49}$

44. Sample answer: $\frac{1}{2}$ and -4.5

45. $a^2 = 49$
$\sqrt{a^2} = \sqrt{49}$
$a = \sqrt{49}$ or $a = -\sqrt{49}$
$a = 7$ or $a = -7$

46. $d^2 = 81$
$\sqrt{d^2} = \sqrt{81}$
$d = \sqrt{81}$ or $d = -\sqrt{81}$
$d = 9$ or $d = -9$

47. $y^2 = 22$
$\sqrt{y^2} = \sqrt{22}$
$y = \sqrt{22}$ or $y = -\sqrt{22}$
$y \approx 4.7$ or $y \approx -4.7$

48. $p^2 = 63$
$\sqrt{p^2} = \sqrt{63}$
$p = \sqrt{63}$ or $p = -\sqrt{63}$

49. $144 = w^2$
$\sqrt{144} = \sqrt{w^2}$
$\sqrt{144} = w$ or $-\sqrt{144} = w$
$w = 12$ or $w = -12$

50. $289 = m^2$
$\sqrt{289} = \sqrt{m^2}$
$\sqrt{289} = m$ or $-\sqrt{289} = m$
$m = 17$ or $m = -17$

51. $127 = b^2$
$\sqrt{127} = \sqrt{b^2}$
$\sqrt{127} = b$ or $-\sqrt{127} = b$
$b \approx 11.3$ or $b \approx -11.3$

52. $300 = h^2$
$\sqrt{300} = \sqrt{h^2}$
$\sqrt{300} = h$ or $-\sqrt{300} = h$
$h \approx 17.3$ or $h \approx -17.3$

53. $x^2 = 1.69$
$\sqrt{x^2} = \sqrt{1.69}$
$x = \sqrt{1.69}$ or $x = -\sqrt{1.69}$
$x = 1.3$ or $x = -1.3$

54. $0.0016 = q^2$
$\sqrt{0.0016} = \sqrt{q^2}$
$\sqrt{0.0016} = q$ or $-\sqrt{0.0016} = q$
$q = 0.04$ or $q = -0.04$

55. $n^2 = 3.56$
$\sqrt{n^2} = \sqrt{3.56}$
$n = \sqrt{3.56}$ or $n = -\sqrt{3.56}$
$n \approx 1.9$ or $n \approx -1.9$

56. $0.0058 = k^2$
$\sqrt{0.0058} = \sqrt{k^2}$
$\sqrt{0.0058} = k$ or $-\sqrt{0.0058} = k$
$k \approx 0.1$ or $k \approx -0.1$

57. $(-a)^2 = 144$
$\sqrt{(-a)^2} = \sqrt{144}$
$|-a| = \sqrt{144}$ or $|-a| = -\sqrt{144}$
$a = 12$ or $a = -12$

58. $x^2 - 4^2 = \sqrt{15^2}$
$x^2 - 16 = 15$
$x^2 - 16 + 16 = 15 + 16$
$x^2 = 31$
$\sqrt{x^2} = \sqrt{31}$
$x = \sqrt{31}$ or $x = -\sqrt{31}$
$x \approx 5.6$ or $x \approx -5.6$

59. $\sqrt{256} = m^2$
$16 = m^2$
$\sqrt{16} = \sqrt{m^2}$
$\sqrt{16} = m$ or $-\sqrt{16} = m$
$m = 4$ or $m = -4$

60. $t^2 = \frac{d^3}{216}$
$t^2 = \frac{7^3}{216}$
$t^2 = \frac{343}{216}$
$\sqrt{t^2} = \sqrt{\frac{343}{216}}$
$t = \sqrt{1.58796}$
$t \approx 1.3$ h
The storm will last about 1.3 h.

61. $x^2 = 324$
$\sqrt{x^2} = \sqrt{324}$
$x = 18$
Each side of the room is 18 feet long.
18 ft ÷ 6 in. = 18 ft ÷ 0.5 ft
= 36
There will be 36 tiles in each row.

62. Irrational; see students' explanations.

63. If a square has an area that is not a perfect square, the lengths of the sides will be irrational. Answers should include the following.

Area = 56 in^2	Area = 64 in^2

64. A

65. B; Zero is a whole number but *not* a natural number.

66. $P = a + b + c$ $\quad s = \frac{1}{2}P$
$P = 5 + 8 + 12$ $\quad s = \frac{1}{2}(25)$
$P = 25$ $\quad s = 12.5$
$A = \sqrt{s(s-a)(s-b)(s-c)}$
$A = \sqrt{12.5(12.5-5)(12.5-8)(12.5-12)}$
$A = \sqrt{210.938}$
$A \approx 14.5\ m^2$

67. $P = a + b + c$ $\quad s = \frac{1}{2}P$
$P = 65 + 82 + 95$ $\quad s = \frac{1}{2}(242)$
$P = 242$ $\quad s = 121$
$A = \sqrt{s(s-a)(s-b)(s-c)}$
$A = \sqrt{121(121-65)(121-82)(121-95)}$
$A = \sqrt{6870864}$
$A \approx 2621.2$ ft^2

68. No, because triangles can have dimensions that are irrational numbers. Thus, the area will be irrational. For example, a triangle with a base of 2 inches and a height of $\sqrt{3}$ inches has an area of $\frac{1}{2} \cdot 2 \cdot \sqrt{3}$ or $\sqrt{3}$ square inches.

Page 445 Maintain Your Skills

69. $49 < 54 < 64$
$\sqrt{49} < \sqrt{54} < \sqrt{64}$
$7 < \sqrt{54} < 8$
$\sqrt{54} \approx 7$

70. $-144 < -126 < -121$
$-\sqrt{144} < -\sqrt{126} < -\sqrt{121}$
$-12 < -\sqrt{126} < -11$
$-\sqrt{126} \approx -11$

71. $4 < 8.67 < 9$
$\sqrt{4} < \sqrt{8.67} < \sqrt{9}$
$2 < \sqrt{8.67} < 3$
$\sqrt{8.67} \approx 3$

72. Sample answer: (0, −2), (1, −3), (2, −4)

73. Sample answer: (0, 5), (1, 6), (−6, 2)

74. Sample answer: (0, −3), (1, −2), (2, −3)

75. $3x + 2 > 17$
$3x + 2 - 2 > 17 - 2$
$\frac{3x}{3} > \frac{15}{3}$
$x > 5$

76. $-2y + 9 \leq 3$
$-2y + 9 - 9 \leq 3 - 9$
$\frac{-2y}{-2} \leq \frac{-6}{-2}$
$y \geq 3$

77. $\frac{8}{15} = 0.5\overline{3}$ or \$0.53/cupcake

78. $\frac{120}{4.3} = 27.9$ mi/gal

79.

3:00

80.

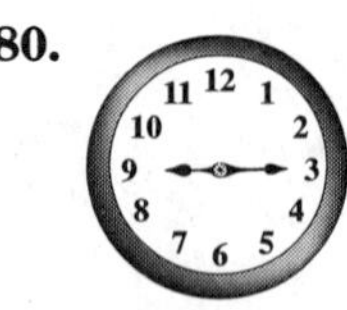

9:15

81.

10:00

82.

2:30

83.

7:45

84.

8:05

Page 446 Reading Mathematics

1. ray: A ray of sunshine appeared through a cloud.
degree: Today's temperature is 10 degrees colder than yesterday's temperature.
acute: Neena is experiencing acute stomach pains.
obtuse: His answer to the question was vague and obtuse.

2a. See students' work.
2b. See students' work.
2c. See students' work.
3. equilateral, scalene, isosceles

9-3 Angles

Page 447 How are angles used in circle graphs?

a. in booths
b. 26
c. in booths

Page 449 Check for Understanding

1. N; $\overrightarrow{NM}$, $\overrightarrow{NP}$; $\angle 1$, $\angle MNP$, $\angle PNM$, $\angle N$

2. Sample answer:

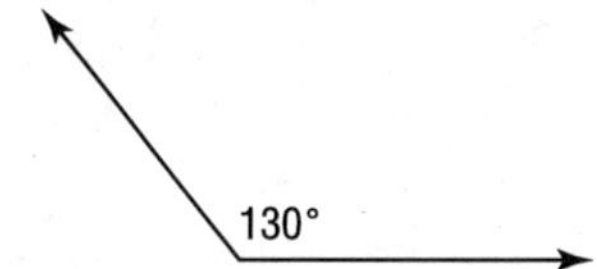

3. $m\angle CED < 20°$, so $\angle CED$ is acute.
4. $m\angle BED = 90°$, so $\angle BED$ is right.
5. $m\angle CEB < 70°$, so $\angle CEB$ is acute.
6. $m\angle AED > 125°$, so $\angle AED$ is obtuse.
7.

55°

$55° < 90°$, so the angle is acute.

8.

140°

$144° > 90°$, so the angle is obtuse.

9. straight angle

Page 450 Practice and Apply

10. 30°; $m\angle XZY < 90°$, so $\angle XZY$ is acute.
11. 15°; $m\angle SZT < 90°$, so $\angle SZT$ is acute.
12. 180°; $m\angle SZY > 90°$, so $\angle SZY$ is obtuse.
13. 85°; $m\angle UZX < 90°$, so $\angle UZX$ is acute.
14. 105°; $m\angle TZW > 90°$, so $\angle TZW$ is obtuse.
15. 135°; $m\angle XZT > 90°$, so $\angle XZT$ is obtuse.
16. 25°; $m\angle UZV < 90°$, so $\angle UZV$ is acute.
17. 60°; $m\angle WZU < 90°$, so $\angle WZU$ is acute.
18. 45°
19. 90°
20. 60°
21.

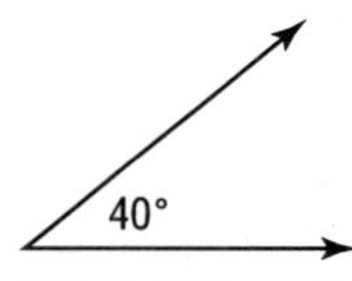

$40° < 90°$, so the angle is acute.

22.

70°

$70° < 90°$, so the angle is acute.

23.

65°

$65° < 90°$, so the angle is acute.

24. 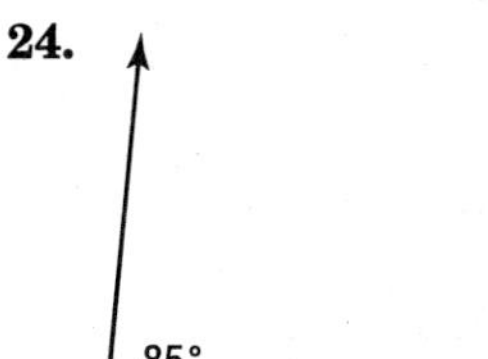

$85° < 90°$, so the angle is acute.

25. 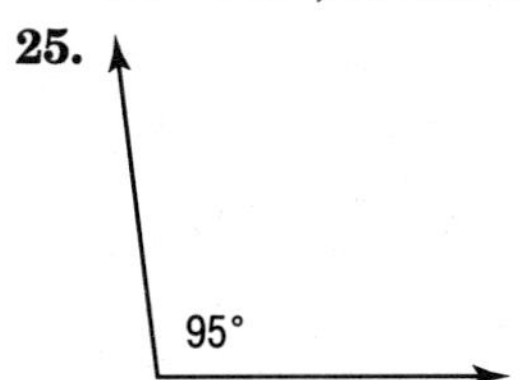

$95° > 90°$, so the angle is obtuse.

26. 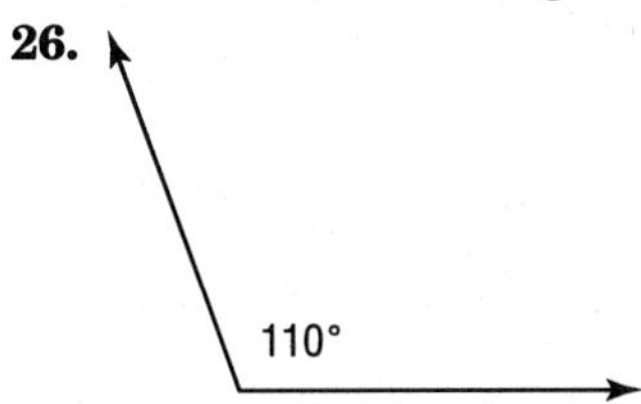

$110° > 90°$, so the angle is obtuse.

27. 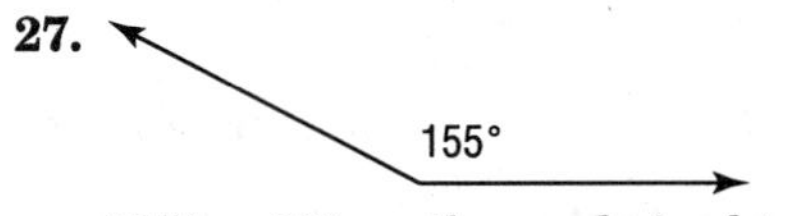

$155° > 90°$, so the angle is obtuse.

28. 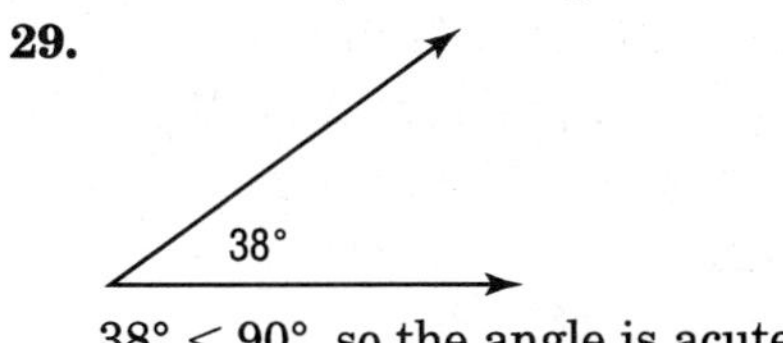

$140° > 90°$, so the angle is obtuse.

29. 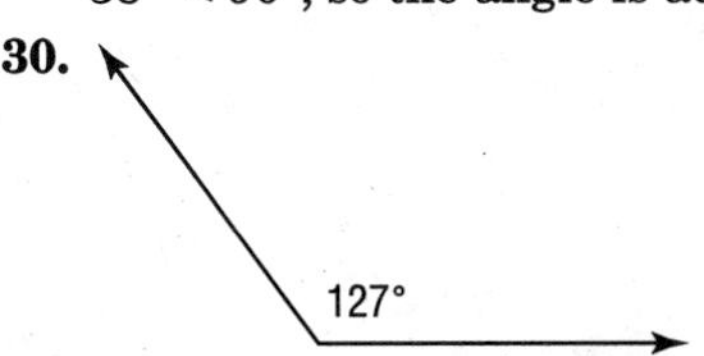

$38° < 90°$, so the angle is acute.

30.

127°

$127° > 90°$, so the angle is obtuse.

31a. $m\angle JKL > 90°$, so $\angle JKL$ is obtuse.

31b. $m\angle RST = 90°$, so $\angle RST$ is right.

31c. $m\angle DOT < 90°$, so $\angle DOT$ is acute.

32. Moderate: obtuse; all other angles: acute

33. Intense: about 68°; Moderate: about 104°; Light: 72°; No standard routine: about 47°; Don't exercise regularly: about 68°

34. 29% of 500 = 145 **35.** 24

36. The sides of each section on a circle graph represent an angle. Answers should include the following.

- In booths: 180°; On Internet: about 86°; By mail: about 83°; Other: about 11°
- $0 < p < 25$; $25 < p < 50$

37. C; $110° > 90°$, so the angle is obtuse.

38. A

Page 451 Maintain Your Skills

39. Z, Q

40. Q

41. I

42. Q

43.
$$\begin{aligned} 16 &< 18 < 25 \\ \sqrt{16} &< \sqrt{18} < \sqrt{25} \\ 4 &< \sqrt{18} < 5 \\ \sqrt{18} &\approx 4 \end{aligned}$$

44.
$$\begin{aligned} 64 &< 79 < 81 \\ \sqrt{64} &< \sqrt{79} < \sqrt{81} \\ 8 &< \sqrt{79} < 9 \\ \sqrt{79} &\approx 9 \end{aligned}$$

45. $x - 7 \geq 22$

46.
$$\begin{aligned} 18 + 57 + x &= 180 \\ 75 + x &= 180 \\ 75 - 75 + x &= 180 - 75 \\ x &= 105 \end{aligned}$$

47.
$$\begin{aligned} x + 27 + 54 &= 180 \\ x + 81 &= 180 \\ x + 81 - 81 &= 180 - 81 \\ x &= 99 \end{aligned}$$

48.
$$\begin{aligned} 85 + x + 24 &= 180 \\ x + 109 &= 180 \\ x + 109 - 109 &= 180 - 109 \\ x &= 71 \end{aligned}$$

49.
$$\begin{aligned} x + x + x &= 180 \\ 3x &= 180 \\ \frac{3x}{3} &= \frac{180}{3} \\ x &= 60 \end{aligned}$$

50.
$$\begin{aligned} 2x + 3x + 4x &= 180 \\ 9x &= 180 \\ \frac{9x}{9} &= \frac{180}{9} \\ x &= 20 \end{aligned}$$

51.
$$\begin{aligned} 2x + 3x + 5x &= 180 \\ 10x &= 180 \\ \frac{10x}{10} &= \frac{180}{10} \\ x &= 18 \end{aligned}$$

Page 451 Practice Quiz 1

1. Since $6^2 = 36$, $\sqrt{36} = 6$
2. Since $13^2 = 169$, $-\sqrt{169} = -13$
3. $$\begin{aligned} m^2 &= 68 \\ \sqrt{m^2} &= \sqrt{68} \\ m = \sqrt{68} &\text{ or } m = -\sqrt{68} \\ m \approx 8.2 &\text{ or } m \approx -8.2 \end{aligned}$$
4. $83° < 90°$, so the angle is acute.
5. $115° > 90°$, so the angle is obtuse.

Page 452 Spreadsheet Investigation (Follow-Up of Lesson 9-3)

1. $P(\text{Blue}) = \frac{1}{2}$, so you would expect $\frac{1}{2}$ of 20 or 10 Blue. $P(\text{Yellow}) = \frac{1}{4}$, so you would expect $\frac{1}{4}$ of 20 or 5 Yellow. Similarly, you would expect 5 Red.
2. See students' work.
3. The central angle for the Blue section should be a straight angle because $\frac{1}{2}$ of 360° is 180°. The central angle for the Red or Yellow section should be a right angle because $\frac{1}{4}$ of 360° is 90°.
4. See students' work.
5. They look the same.

9-4 Triangles

Page 453 How do the angles of a triangle relate to each other?

a. See students' work.
b. See students' work.
c. See students' work.
d. See students' work.
e. The sum of the measures is 180°.

Page 455 Check for Understanding

1. Whereas an isosceles triangle has at least two sides congruent, an equilateral triangle has three sides congruent.
2. Sample answer: yield sign.
3. Sample answer:

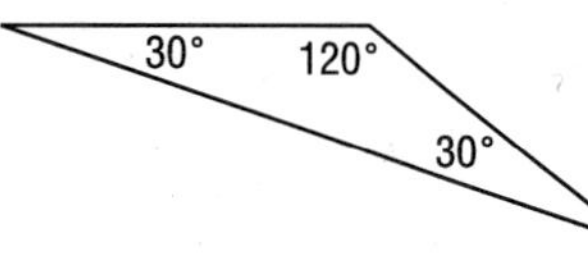

4. $$\begin{aligned} x + 72 + 83 &= 180 \\ x + 155 &= 180 \\ x + 155 - 155 &= 180 - 155 \\ x &= 25\text{; acute} \end{aligned}$$
5. $$\begin{aligned} x + 61 + 29 &= 180 \\ x + 90 &= 180 \\ x + 90 - 90 &= 180 - 90 \\ x &= 90\text{; right} \end{aligned}$$
6. $$\begin{aligned} x + 48 + 27 &= 180 \\ x + 75 &= 180 \\ x + 75 - 75 &= 180 - 75 \\ x &= 105\text{; obtuse} \end{aligned}$$
7. Right isosceles because the triangle has a right angle and two congruent sides
8. obtuse scalene because the triangle has an obtuse angle and no congruent sides
9. acute isosceles because the triangle has all acute angles and two congruent sides
10. $$\begin{aligned} 2x + 4x + 9x &= 180 \\ 15x &= 180 \\ \frac{15x}{15} &= \frac{180}{15} \\ x &= 12 \end{aligned}$$
Since $x = 12$, $2x = 2(12)$ or 24, $4x = 4(12)$ or 48, and $9x = 9(12)$ or 108, the measures of the angles are 24°, 48°, and 108°.

Pages 456–457 Practice and Apply

11. $$\begin{aligned} x + 90 + 63 &= 180 \\ x + 153 &= 180 \\ x + 153 - 153 &= 180 - 153 \\ x &= 27 \text{ or } 27° \end{aligned}$$
right because it has a right angle
12. $$\begin{aligned} x + 57 + 68 &= 180 \\ x + 125 &= 180 \\ x + 125 - 125 &= 180 - 125 \\ x &= 55 \text{ or } 55° \end{aligned}$$
acute because all the angles are acute
13. $$\begin{aligned} x + 32 + 36 &= 180 \\ x + 68 &= 180 \\ x + 68 - 68 &= 180 - 68 \\ x &= 112 \text{ or } 112° \end{aligned}$$
obtuse because the triangle has one obtuse angle
14. $$\begin{aligned} x + 45 + 33 &= 180 \\ x + 78 &= 180 \\ x + 78 - 78 &= 180 - 78 \\ x &= 102 \text{ or } 102° \end{aligned}$$
obtuse because the triangle has one obtuse angle
15. $$\begin{aligned} x + 28 + 62 &= 180 \\ x + 90 &= 180 \\ x + 90 - 90 &= 180 - 90 \\ x &= 90 \text{ or } 90° \end{aligned}$$
right because the triangle has a right angle
16. $$\begin{aligned} x + 43 + 71 &= 180 \\ x + 114 &= 180 \\ x + 114 - 114 &= 180 - 114 \\ x &= 66 \text{ or } 66° \end{aligned}$$
acute because the triangle has all acute angles
17. $$\begin{aligned} x + 3x + 5x &= 180 \\ 9x &= 180 \\ \frac{9x}{9} &= \frac{180}{9} \\ x &= 20 \end{aligned}$$
Since $x = 20$, $3x = 3(20)$ or 60, and $5x = 5(20)$ or 100. The measures of the angles are 20°, 60°, and 100°.
18. $$\begin{aligned} 7x + 7x + 22x &= 180 \\ 36x &= 180 \\ \frac{36x}{36} &= \frac{180}{36} \\ x &= 5 \end{aligned}$$
Since $x = 5$, $7x = 7(5)$ or 35, and $22x = 22(5)$ or 110, the measures of the angles are 35°, 35°, and 110°.
19. acute equilateral because the triangle has all sides congruent
20. obtuse isosceles because the triangle has one obtuse angle and two sides congruent

21. right scalene because the triangle has a right angle and no congruent sides
22. acute isosceles because the triangle has all acute angles and two sides congruent
23. acute scalene because the triangle has all acute angles and no congruent sides
24. obtuse isosceles because the triangle has an obtuse angle and two congruent sides
25. always
26. sometimes
27. obtuse because the triangle appears to have an obtuse angle
28. right because the triangle appears to have a right angle
29. acute because the triangle appears to have all acute angles
30.

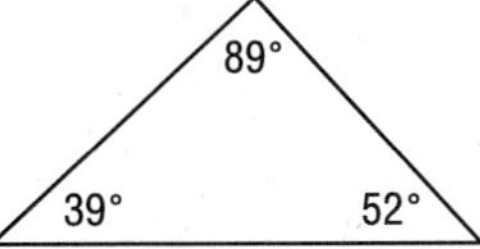

31. not possible
32.

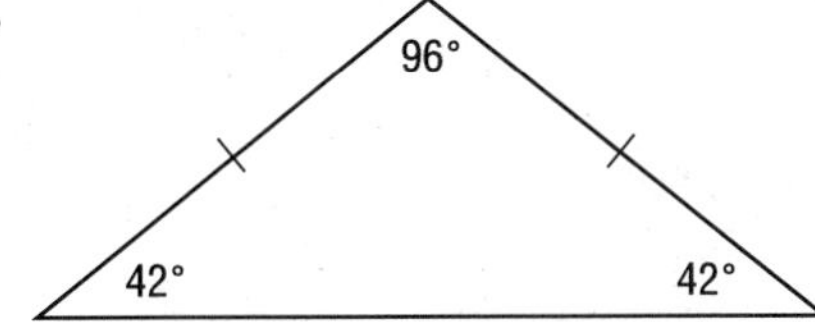

33. not possible
34. $x + 5x + 3x = 180$

$$9x = 180$$
$$\frac{9x}{9} = \frac{180}{9}$$
$$x = 20$$

Since $x = 20$, $5x = 5(20)$ or 100, and $3x = 3(20)$ or 60, the measures of the angles are 20°, 60°, and 100°.

35. $x + 85 + (x + 5) = 180$

$$2x + 90 = 180$$
$$2x + 90 - 90 = 180 - 90$$
$$2x = 90$$
$$\frac{2x}{2} = \frac{90}{2}$$
$$x = 45$$

Since $x = 45$, $x + 5 = 45 + 5$ or 50, the measures of the angles are 45°, 85°, and 50°.

36. $2x + (2x + 15) + 7x = 180$

$$11x + 15 = 180$$
$$11x + 15 - 15 = 180 - 15$$
$$11x = 165$$
$$x = 15$$

Since $x = 15$, $2x = 2(15)$ or 30, $2x + 15 = 2(15) + 15$ or 45, and $7x = 7(15)$ or 105, the measures of the angles are 30°, 45°, 105°.

37. 10, 15, 21
38. The angles of a triangle have a sum of 180°. Answers should include the following.
    - The angles of a triangle can be folded to form a straight angle. Since the measure of a straight angle is 180°, the sum of the angles of a triangle is 180°.
    - Sample answer:

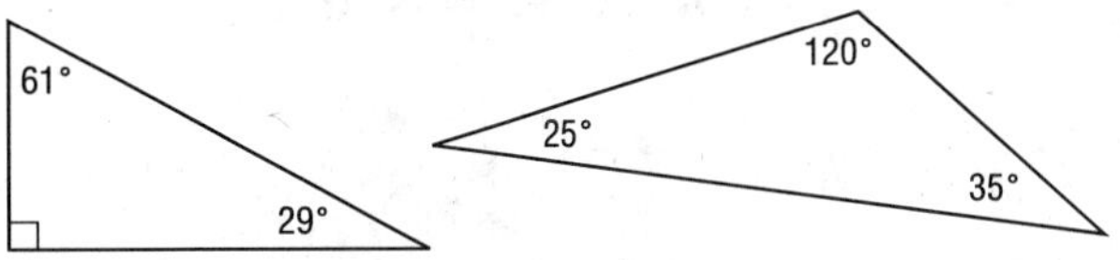

39. B; $m\angle R + 110 + 32 = 180$

$$m\angle R + 142 = 180$$
$$m\angle R + 142 - 142 = 180 - 142$$
$$m\angle R = 38$$

40. C

Page 457 Maintain Your Skills

41. $m\angle ADC = 180°$, so $\angle ADC$ is straight.
42. $m\angle ADE < 90°$, so $\angle ADE$ is acute.
43. $m\angle CDE > 90°$, so $\angle CDE$ is obtuse.
44. $m\angle EDF < 90°$, so $\angle EDF$ is acute.
45. $m^2 = 81$

$$\sqrt{m^2} = \sqrt{81}$$
$$m = \sqrt{81} \text{ or } -\sqrt{81}$$
$$m = 9 \text{ or } -9$$

46. $196 = y^2$

$$\sqrt{196} = \sqrt{y^2}$$
$$\sqrt{196} \text{ or } -\sqrt{196} = y$$
$$14 \text{ or } -14 = y$$

47. $84 = p^2$

$$\sqrt{84} = \sqrt{p^2}$$
$$\sqrt{84} \text{ or } -\sqrt{84} = p$$
$$9.2 \text{ or } -9.2 \approx p$$

48. $\frac{26}{x} = \frac{25}{100}$

$$25x = 2600$$
$$\frac{25x}{25} = \frac{2600}{25}$$
$$x = 104$$

26 is 25% of 104.

49. $12^2 = 12 \cdot 12 = 144$
50. $15^2 = 15 \cdot 15 = 225$
51. $18^2 = 18 \cdot 18 = 324$
52. $24^2 = 24 \cdot 24 = 576$
53. $27^2 = 27 \cdot 27 = 729$
54. $31^2 = 31 \cdot 31 = 961$

Pages 458–459 Algebra Activity (Preview of Lesson 9-5)

1. 8 units^2
2. 6 units^2
3. 10 units^2
4. 17 units^2
5. The sum of the areas of squares A and B is equal to the area of square C.

6a. Yes; $3^2 + 4^2 = 5^2$

6b. No; $5^2 + 7^2 \neq 9^2$

6c. No; $6^2 + 9^2 \neq 12^2$

6d. Yes; $7^2 + 24^2 = 25^2$

7. Sample answer: 10, 24, 26 and 12, 16, 20

9-5 Pythagorean Theorem

Page 460 How do the sides of a right triangle relate to each other?

a. $3 \cdot 3 = 9$ units^2, $4 \cdot 4 = 16$ units^2, $5 \cdot 5 =$ 25 units^2

b. The area of the large square is equal to the sum of the areas of the two smaller squares.

c. The area of the large square is equal to the sum of the areas of the two smaller squares.

Page 462 Check for Understanding

1. Sample answer: 8, 15, 17

2. Marcus; Allyson incorrectly substitutes 15 for *b* in the equation. Since the side measuring 15 units is opposite the right angle, it is a hypotenuse. So, 15 should be substituted for *c* in the equation.

3.
$$\begin{aligned} c^2 &= a^2 + b^2 \\ c^2 &= 15^2 + 20^2 \\ c^2 &= 225 + 400 \\ c^2 &= 625 \\ \sqrt{c^2} &= \sqrt{625} \\ c &= 25 \end{aligned}$$

4.
$$\begin{aligned} c^2 &= a^2 + b^2 \\ c^2 &= 6^2 + 12^2 \\ c^2 &= 36 + 144 \\ c^2 &= 180 \\ \sqrt{c^2} &= \sqrt{180} \\ c &\approx 13.4 \end{aligned}$$

5.
$$\begin{aligned} c^2 &= a^2 + b^2 \\ 17^2 &= 8^2 + b^2 \\ 289 &= 64 + b^2 \\ 289 - 64 &= 64 - 64 + b^2 \\ 225 &= b^2 \\ \sqrt{225} &= \sqrt{b^2} \\ 15 &= b \end{aligned}$$

6.
$$\begin{aligned} c^2 &= a^2 + b^2 \\ 25^2 &= a^2 + 24^2 \\ 625 &= a^2 + 576 \\ 625 - 576 &= a^2 + 576 - 576 \\ 49 &= a^2 \\ \sqrt{49} &= \sqrt{a^2} \\ 7 &= a \end{aligned}$$

7. No;
$$\begin{aligned} c^2 &= a^2 + b^2 \\ 8^2 &\stackrel{?}{=} 5^2 + 7^2 \\ 64 &\stackrel{?}{=} 25 + 49 \\ 64 &\neq 74 \end{aligned}$$

8. Yes;
$$\begin{aligned} c^2 &= a^2 + b^2 \\ 26^2 &\stackrel{?}{=} 10^2 + 24^2 \\ 676 &\stackrel{?}{=} 100 + 576 \\ 676 &= 676 \end{aligned}$$

9. B;
$$\begin{aligned} c^2 &= a^2 + b^2 \\ 55^2 &= 40^2 + b^2 \\ 3025 &= 1600 + b^2 \\ 3025 - 1600 &= 1600 - 1600 + b^2 \\ 1425 &= b^2 \\ \sqrt{1425} &= \sqrt{b^2} \\ 37.7 &\approx b \end{aligned}$$

Pages 463–464 Practice and Apply

10.
$$\begin{aligned} c^2 &= a^2 + b^2 \\ c^2 &= 6^2 + 8^2 \\ c^2 &= 36 + 64 \\ c^2 &= 100 \\ \sqrt{c^2} &= \sqrt{100} \\ c &= 10 \end{aligned}$$

11.
$$\begin{aligned} c^2 &= a^2 + b^2 \\ c^2 &= 10^2 + 24^2 \\ c^2 &= 100 + 576 \\ c^2 &= 676 \\ \sqrt{c^2} &= \sqrt{676} \\ c^2 &= 26 \end{aligned}$$

12.
$$\begin{aligned} c^2 &= a^2 + b^2 \\ c^2 &= 5^2 + 9^2 \\ c^2 &= 25 + 81 \\ c^2 &= 106 \\ \sqrt{c^2} &= \sqrt{106} \\ c &\approx 10.3 \end{aligned}$$

13.
$$\begin{aligned} c^2 &= a^2 + b^2 \\ c^2 &= 6^2 + 11^2 \\ c^2 &= 36 + 121 \\ c^2 &= 157 \\ \sqrt{c^2} &= \sqrt{157} \\ c &\approx 12.5 \end{aligned}$$

14.
$$\begin{aligned} c^2 &= a^2 + b^2 \\ c^2 &= 2.7^2 + 7.2^2 \\ c^2 &= 7.29 + 51.84 \\ c^2 &= 59.13 \\ \sqrt{c^2} &= \sqrt{59.13} \\ c &\approx 7.7 \end{aligned}$$

15.
$$\begin{aligned} c^2 &= a^2 + b^2 \\ c^2 &= 12.8^2 + 13.9^2 \\ c^2 &= 163.84 + 193.21 \\ c^2 &= 357.05 \\ \sqrt{c^2} &= \sqrt{357.05} \\ c &\approx 18.9 \end{aligned}$$

16.
$$\begin{aligned} c^2 &= a^2 + b^2 \\ c^2 &= 40^2 + 40^2 \\ c^2 &= 1600 + 1600 \\ c^2 &= 3200 \\ \sqrt{c^2} &= \sqrt{3200} \\ c &\approx 56.6 \text{ or } 56.6 \text{ ft} \end{aligned}$$

17.
$$\begin{aligned} c^2 &= a^2 + b^2 \\ 35^2 &= a^2 + 26^2 \\ 1225 &= a^2 + 676 \\ 1225 - 676 &= a^2 + 676 - 676 \\ 549 &= a^2 \\ \sqrt{549} &= \sqrt{a^2} \\ 23 &\approx a \text{ or } 23 \text{ in.} \end{aligned}$$

18.
$$\begin{aligned} c^2 &= a^2 + b^2 \\ 41^2 &= 9^2 + b^2 \\ 1681 &= 81 + b^2 \\ 1681 - 81 &= 81 - 81 + b^2 \\ 1600 &= b^2 \\ \sqrt{1600} &= \sqrt{b^2} \\ 40 &= b \end{aligned}$$

19.
$$\begin{aligned} c^2 &= a^2 + b^2 \\ 37^2 &= a^2 + 35^2 \\ 1369 &= a^2 + 1225 \\ 1369 - 1225 &= a^2 + 1225 - 1225 \\ 144 &= a^2 \\ \sqrt{144} &= \sqrt{a^2} \\ 12 &= a \end{aligned}$$

20.
$$\begin{aligned} c^2 &= a^2 + b^2 \\ 19^2 &= a^2 + 12^2 \\ 361 &= a^2 + 144 \\ 361 - 144 &= a^2 + 144 - 144 \\ 217 &= a^2 \\ \sqrt{217} &= \sqrt{a^2} \\ 14.7 &\approx a \end{aligned}$$

21.
$$\begin{aligned} c^2 &= a^2 + b^2 \\ 14^2 &= 7^2 + b^2 \\ 196 &= 49 + b^2 \\ 196 - 49 &= 49 - 49 + b^2 \\ 147 &= b^2 \\ \sqrt{147} &= \sqrt{b^2} \\ 12.1 &\approx b \end{aligned}$$

22.
$$\begin{aligned} c^2 &= a^2 + b^2 \\ 61^2 &= 27^2 + b^2 \\ 3721 &= 729 + b^2 \\ 3721 - 729 &= 729 - 729 + b^2 \\ 2992 &= b^2 \\ \sqrt{2992} &= \sqrt{b^2} \\ 54.7 &\approx b \end{aligned}$$

23.
$$\begin{aligned} c^2 &= a^2 + b^2 \\ 82^2 &= a^2 + 73^2 \\ 6724 &= a^2 + 5329 \\ 6724 - 5329 &= a^2 + 5329 - 5329 \\ 1395 &= a^2 \\ \sqrt{1395} &= \sqrt{a^2} \\ 37.3 &\approx a \end{aligned}$$

24.
$$\begin{aligned} c^2 &= a^2 + b^2 \\ 22^2 &= a^2 + (\sqrt{123})^2 \\ 484 &= a^2 + 123 \\ 484 - 123 &= a^2 + 123 - 123 \\ 361 &= a^2 \\ \sqrt{361} &= \sqrt{a^2} \\ 19 &= a \end{aligned}$$

25.
$$\begin{aligned} c^2 &= a^2 + b^2 \\ 31^2 &= (\sqrt{177})^2 + b^2 \\ 961 &= 177 + b^2 \\ 961 - 177 &= 177 - 177 + b^2 \\ 784 &= b^2 \\ \sqrt{784} &= \sqrt{b^2} \\ 28 &= b \end{aligned}$$

26.
$$\begin{aligned} c^2 &= a^2 + b^2 \\ 28^2 &= 17^2 + x^2 \\ 784 &= 289 + x^2 \\ 784 - 289 &= 289 - 289 + x^2 \\ 495 &= x^2 \\ \sqrt{495} &= \sqrt{x^2} \\ 22.2 &\approx x \end{aligned}$$

27.
$$\begin{aligned} c^2 &= a^2 + b^2 \\ 45^2 &= x^2 + 30^2 \\ 2025 &= x^2 + 900 \\ 2025 - 900 &= x^2 + 900 - 900 \\ 1125 &= x^2 \\ \sqrt{1125} &= \sqrt{x^2} \\ 33.5 &\approx x \end{aligned}$$

28. No; $c^2 = a^2 + b^2$
$$\begin{aligned} 9^2 &\stackrel{?}{=} 5^2 + 8^2 \\ 81 &\stackrel{?}{=} 25 + 64 \\ 81 &\neq 89 \end{aligned}$$

29. Yes; $c^2 = a^2 + b^2$
$$\begin{aligned} 34^2 &\stackrel{?}{=} 16^2 + 30^2 \\ 1156 &\stackrel{?}{=} 256 + 900 \\ 1156 &= 1156 \end{aligned}$$

30. Yes; $c^2 = a^2 + b^2$
$$\begin{aligned} 30^2 &\stackrel{?}{=} 18^2 + 24^2 \\ 900 &\stackrel{?}{=} 324 + 576 \\ 900 &= 900 \end{aligned}$$

31. No; $c^2 = a^2 + b^2$
$$\begin{aligned} 32^2 &\stackrel{?}{=} 24^2 + 28^2 \\ 1024 &\stackrel{?}{=} 576 + 784 \\ 1024 &\neq 1360 \end{aligned}$$

32. Yes; $c^2 = a^2 + b^2$
$$\begin{aligned} \sqrt{57}^2 &\stackrel{?}{=} \sqrt{21}^2 + 6^2 \\ 57 &\stackrel{?}{=} 21 + 36 \\ 57 &= 57 \end{aligned}$$

33. No; $c^2 = a^2 + b^2$
$$\begin{aligned} \sqrt{177}^2 &\stackrel{?}{=} 11^2 + \sqrt{55}^2 \\ 177 &\stackrel{?}{=} 121 + 55 \\ 177 &\neq 176 \end{aligned}$$

34.
$$\begin{aligned} c^2 &= a^2 + b^2 \\ c^2 &= 6^2 + 6^2 \\ c^2 &= 36 + 36 \\ c^2 &= 72 \\ \sqrt{c^2} &= \sqrt{72} \\ c &\approx 8.5 \end{aligned}$$
The length of $\overline{AB}$ is about 8.5 in.

35.
$$\begin{aligned} c^2 &= a^2 + b^2 & P &= 2\ell + 2w \\ \sqrt{128}^2 &= a^2 + b^2 \ (a = b) & P &= 2 \cdot 8 + 2 \cdot 8 \\ \sqrt{128}^2 &\stackrel{?}{=} 8^2 + 8^2 & P &= 16 + 16 \\ 128 &\stackrel{?}{=} 64 + 64 & P &= 32 \\ 128 &= 128 \end{aligned}$$
If $\overline{AB}$ measures $\sqrt{128}$, the length of each side is 4 inches and the perimeter is 32 inches.

36.
$$\begin{aligned} c^2 &= a^2 + b^2 \\ 17^2 &= 8^2 + b^2 \\ 289 &= 64 + b^2 \\ 289 - 64 &= 64 - 64 + b^2 \\ 225 &= b^2 \\ \sqrt{225} &= \sqrt{b^2} \\ 15 &= b \end{aligned}$$
The length of $\overline{BC}$ is 15.

37. There is enough information to find the lengths of the legs. Since the right triangle is an isosceles right triangle, we know that the lengths of the legs are equal. So, in the Pythagorean Theorem, we can say that $a = b$. In addition we know that $c = 8$.

$c^2 = a^2 + b^2$	Pythagorean Theroem
$8^2 = a^2 + a^2$	$a = b$ and $c = 8$
$8^2 = 2a^2$	Add a^2 and a^2.
$64 = 2a^2$	Evaluate 8^2.
$\frac{64}{2} = \frac{2a^2}{2}$	Divide each side by 2.
$32 = a^2$	Simplify.
$\sqrt{32} = a$	Take the square root of each side.

So, if a triangle is an isosceles right triangle and the hypotenuse is 8 inches, the length of each leg is $\sqrt{32}$ inches or about 5.7 inches.

38. The sum of the squares of the lengths of the legs is equal to the square of the length of the hypotenuse. Answers should include the following.

- 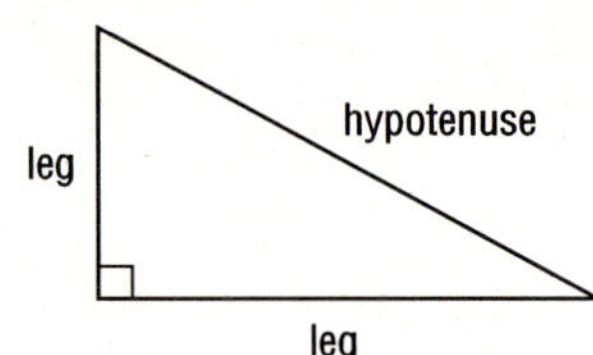

- An example of a set of numbers that represents the measures of the lengths of the legs and hypotenuse of a right triangle is 15, 20, 25.

39. D; $c^2 = a^2 + b^2$

$$10^2 \stackrel{?}{=} 8^2 + 6^2$$
$$100 \stackrel{?}{=} 64 + 36$$
$$100 = 100$$

40. B;

$$c^2 = a^2 + b^2$$
$$12^2 = 9^2 + x^2$$
$$144 = 81 + x^2$$
$$144 - 81 = 81 - 81 + x^2$$
$$63 = x^2$$
$$\sqrt{63} = \sqrt{x^2}$$
$$7.9 \approx x$$

41.

$$c^2 = a^2 + b^2$$
$$c^2 = 15^2 + 8^2$$
$$c^2 = 225 + 64$$
$$c^2 = 289$$
$$\sqrt{c^2} = \sqrt{289}$$
$$c = 17$$

The length of $\overline{BD}$ is 17 units.

42.

$$c^2 = a^2 + b^2$$
$$c^2 = 17^2 + 10^2$$
$$c^2 = 289 + 100$$
$$c^2 = 389$$
$$\sqrt{c^2} = \sqrt{389}$$
$$c \approx 19.7 \text{ units}$$

The length of $\overline{FD}$ is about 19.7 units.

43. See students' work.

Page 464 Maintain Your Skills

44.

$$x + 46 + 33 = 180$$
$$x + 79 = 180$$
$$x + 79 - 79 = 180 - 79$$
$$x = 101$$

The measure of the angle is 101°. Since the triangle has an obtuse angle, the triangle is obtuse.

45.

$$x + 63 + 27 = 180$$
$$x + 90 = 180$$
$$x + 90 - 90 = 180 - 90$$
$$x = 90$$

The measure of the angle is 90°. Since the triangle has a right angle, the triangle is a right triangle.

46.

$$x + 48 + 54 = 180$$
$$x + 102 = 180$$
$$x + 102 - 102 = 180 - 102$$
$$x = 78$$

The measure of the angle is 78°. Since the triangle has all acute angles, the triangle is acute.

47. See students' work.

48.

$$x + 4 < 12$$
$$x + 4 - 4 < 12 - 4$$
$$x < 8$$

49.

$$-15 \le n - 6$$
$$-15 + 6 \le n - 6 + 6$$
$$-9 \le n$$

50. $(2 + 6)^2 + (-5 + 6)^2 = (8)^2 + (1)^2$

$$= 64 + 1$$
$$= 65$$

51. $(-4 + 3)^2 + (0 - 2)^2 = (-1)^2 + (-2)^2$

$$= 1 + 4$$
$$= 5$$

52. $[3 + (-1)]^2 + (8 - 4)^2 = (2)^2 + (4)^2$

$$= 4 + 16$$
$$= 20$$

Page 465 Algebra Activity (Follow-Up of Lesson 9-5)

1. $5 = 2^2 + 1^2$

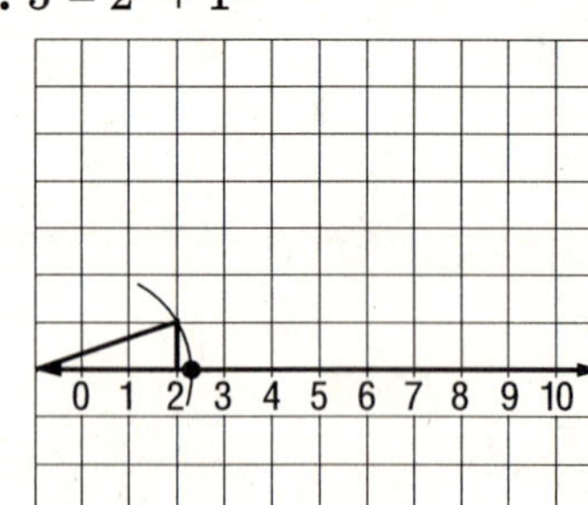

2. $20 = 4^2 + 2^2$

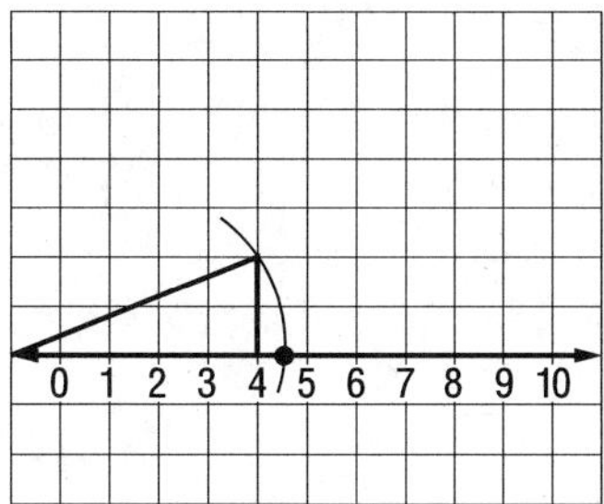

3. $45 = 3^2 + 6^2$

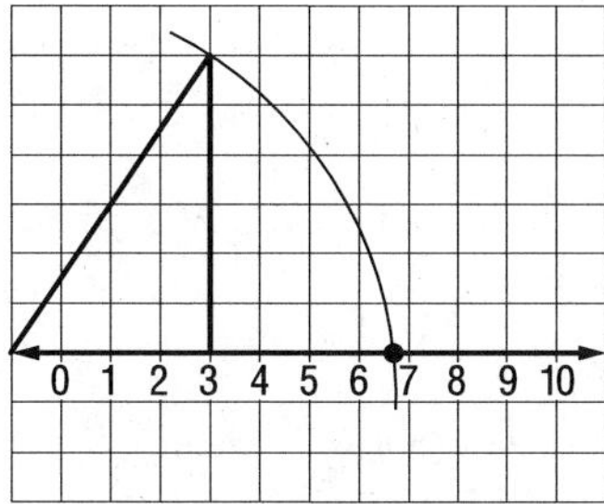

4. $97 = 9^2 + 4^2$

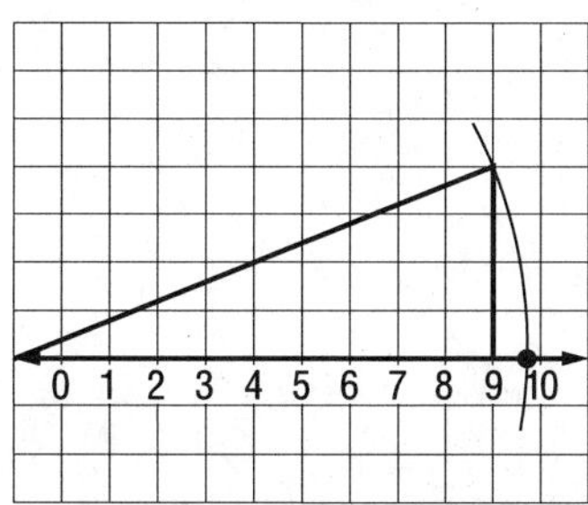

5. Sample answer: First, find two numbers whose square roots have a sum of 34. Since $3^2 + 5^2 = 34$, draw a right triangle with legs 3 units and 5 units long. Then, open the compass to the length of the hypotenuse. With the tip of the compass at 0, draw an arc that intersects the number line. Or, use the procedure in the activity to find the lengths of $\sqrt{3}$ and $\sqrt{5}$ and then use those lengths as the legs of a right triangle.
6. Sample answer: Since $(\sqrt{2})^2 + 1^2 = 3$, use $\sqrt{2}$ as one leg of a right triangle and 1 unit as the other leg. Then follow the procedure in the activity to locate the graph of $\sqrt{3}$.

9-6 The Distance and Midpoint Formulas

Page 466 How is the Distance Formula related to the Pythagorean Theorem?

a. (3, 3)
b. 7 units
c. 3 units
d. right
e. Pythagorean Theorem
f. $c^2 = a^2 + b^2$
$c^2 = 7^2 + 3^2$
$c^2 = 49 + 9$
$c^2 = 58$
$c = \sqrt{58}$ or ≈ 7.62 units

Page 468 Check for Understanding

1. the point halfway between two endpoints of a line segment
2. Sample answer: To find the distance between any two points with coordinates (x_1, y_1) and (x_2, y_2), first find the difference between x_2 and x_1, and the difference between y_2 and y_1. Then square each difference and take the square root of the sum of the squares.
3. Sample answer:

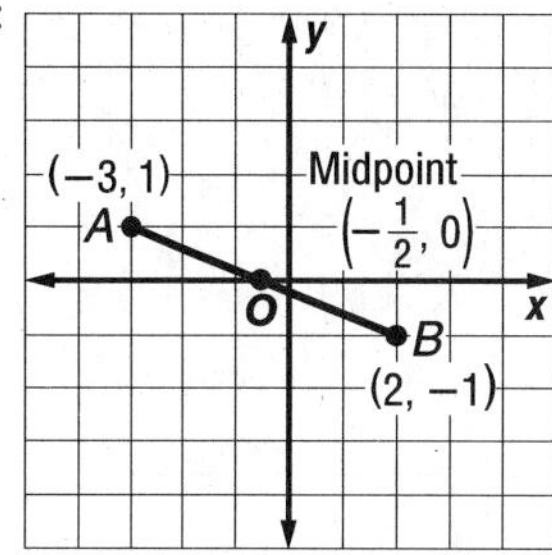

$$\text{midpoint} = \left(\frac{x_1 + x_2}{2}, \frac{y_1 + y_2}{2}\right) = \left(\frac{-3 + 2}{2}, \frac{1 + (-1)}{2}\right) = \left(-\frac{1}{2}, 0\right)$$

The coordinates of the midpoint of $\overline{AB}$ are $\left(-\frac{1}{2}, 0\right)$.

4. $d = \sqrt{(x_2 - x_1)^2 + (y_2 - y_1)^2}$
$AB = \sqrt{[8 - (-1)]^2 + (-6 - 3)^2}$
$AB = \sqrt{9^2 + (-9)^2}$
$AB = \sqrt{81 + 81}$
$AB = \sqrt{162}$
$AB \approx 12.7$
The distance between points A and B is about 12.7 units.
5. $d = \sqrt{(x_2 - x_1)^2 + (y_2 - y_1)^2}$
$MN = \sqrt{(-6 - 4)^2 + [-7 - (-2)]^2}$
$MN = \sqrt{(-10)^2 + (-5)^2}$
$MN = \sqrt{100 + 25}$
$MN = \sqrt{125}$
$MN \approx 11.2$
The distance between points M and N is about 11.2 units.
6. $\text{midpoint} = \left(\frac{x_1 + x_2}{2}, \frac{y_1 + y_2}{2}\right)$
$\text{midpoint } \overline{BC} = \left(\frac{4 + (-2)}{2}, \frac{1 + 5}{2}\right) = (1, 3)$
The coordinates of the midpoint of $\overline{BC}$ are (1, 3).
7. $\text{midpoint} = \left(\frac{x_1 + x_2}{2}, \frac{y_1 + y_2}{2}\right)$
$\text{midpoint } \overline{RS} = \left(\frac{3 + 1}{2}, \frac{-6 + (-4)}{2}\right) = (2, -5)$
The coordinates of the midpoint of $\overline{RS}$ are (2, −5).

8. Side $\overline{EF}$: $E(1, 4)$, $F(-3, 0)$

$$d = \sqrt{(x_2 - x_1)^2 + (y_2 - y_1)^2}$$
$$EF = \sqrt{(-3 - 1)^2 + (0 - 4)^2}$$
$$EF = \sqrt{(-4)^2 + (-4)^2}$$
$$EF = \sqrt{16 + 16}$$
$$EF = \sqrt{32}$$

Side $\overline{FG}$: $F(-3, 0)$, $G(4, -1)$

$$d = \sqrt{(x_2 - x_1)^2 + (y_2 - y_1)^2}$$
$$FG = \sqrt{[4 - (-3)]^2 + (-1 - 0)^2}$$
$$FG = \sqrt{(7)^2 + (-1)^2}$$
$$FG = \sqrt{49 + 1}$$
$$FG = \sqrt{50}$$

Side $\overline{EG}$: $E(1, 4)$, $G(4, -1)$

$$d = \sqrt{(x_2 - x_1)^2 + (y_2 - y_1)^2}$$
$$EG = \sqrt{(4 - 1)^2 + (-1 - 4)^2}$$
$$EG = \sqrt{3^2 + (-5)^2}$$
$$EG = \sqrt{9 + 25}$$
$$EG = \sqrt{34}$$
$$P = \sqrt{32} + \sqrt{50} + \sqrt{34} \approx 5.7 + 7.1 + 5.8$$
$$\approx 18.6$$

The perimeter is about 18.6 units.

Pages 469–470 Practice and Apply

9. $d = \sqrt{(x_2 - x_1)^2 + (y_2 - y_1)^2}$
$$JK = \sqrt{(-1 - 5)^2 + [3 - (-4)]^2}$$
$$JK = \sqrt{(-6)^2 + (7)^2}$$
$$JK = \sqrt{36 + 49}$$
$$JK = \sqrt{85}$$
$$JK \approx 9.2$$

The distance between points J and K is about 9.2 units.

10. $d = \sqrt{(x_2 - x_1)^2 + (y_2 - y_1)^2}$
$$CD = \sqrt{[6 - (-7)]^2 + (-4 - 2)^2}$$
$$CD = \sqrt{13^2 + (-6)^2}$$
$$CD = \sqrt{169 + 36}$$
$$CD = \sqrt{205}$$
$$CD \approx 14.3$$

The distance between points C and D is about 14.3 units.

11. $d = \sqrt{(x_2 - x_1)^2 + (y_2 - y_1)^2}$
$$EF = \sqrt{[9 - (-1)]^2 + [-4 - (-2)]^2}$$
$$EF = \sqrt{10^2 + (-2)^2}$$
$$EF = \sqrt{100 + 4}$$
$$EF = \sqrt{104}$$
$$EF \approx 10.2$$

The distance between points E and F is about 10.2 units.

12. $d = \sqrt{(x_2 - x_1)^2 + (y_2 - y_1)^2}$
$$VW = \sqrt{(-3 - 8)^2 + [-5 - (-5)]^2}$$
$$VW = \sqrt{(-11)^2 + (0)^2}$$
$$VW = \sqrt{121 + 0}$$
$$VW = \sqrt{121}$$
$$VW = 11$$

The distance between points V and W is 11 units.

13. $d = \sqrt{(x_2 - x_1)^2 + (y_2 - y_1)^2}$
$$ST = \sqrt{[6 - (-9)]^2 + (-7 - 0)^2}$$
$$ST = \sqrt{15^2 + (-7)^2}$$
$$ST = \sqrt{225 + 49}$$
$$ST = \sqrt{274}$$
$$ST \approx 16.6$$

The distance between points S and T is about 16.6 units.

14. $d = \sqrt{(x_2 - x_1)^2 + (y_2 - y_1)^2}$
$$MN = \sqrt{(-7 - 0)^2 + (-8 - 0)^2}$$
$$MN = \sqrt{(-7)^2 + (-8)^2}$$
$$MN = \sqrt{49 + 64}$$
$$MN = \sqrt{113}$$
$$MN \approx 10.6$$

The distance between points M and N is about 10.6 units.

15. $d = \sqrt{(x_2 - x_1)^2 + (y_2 - y_1)^2}$
$$QR = \sqrt{\left(2 - 5\tfrac{1}{4}\right)^2 + \left(6\tfrac{1}{2} - 3\right)^2}$$
$$QR = \sqrt{\left(\tfrac{8}{4} - \tfrac{21}{4}\right)^2 + \left(\tfrac{13}{2} - \tfrac{6}{2}\right)^2}$$
$$QR = \sqrt{\left(-\tfrac{13}{4}\right)^2 + \left(\tfrac{7}{2}\right)^2}$$
$$QR = \sqrt{10\tfrac{9}{16} + 12\tfrac{1}{4}}$$
$$QR = \sqrt{10\tfrac{9}{16} + 12\tfrac{4}{16}}$$
$$QR = \sqrt{22\tfrac{13}{16}}$$
$$QR \approx 4.8$$

The distance between points Q and R is about 4.8 units.

16. $d = \sqrt{(x_2 - x_1)^2 + (y_2 - y_1)^2}$
$$AB = \sqrt{\left[-8\tfrac{3}{4} - \left(-2\tfrac{1}{2}\right)\right]^2 + \left(-6\tfrac{1}{4} - 0\right)^2}$$
$$AB = \sqrt{\left(-\tfrac{35}{4} + \tfrac{10}{4}\right)^2 + \left(-6\tfrac{1}{4}\right)^2}$$
$$AB = \sqrt{\left(-\tfrac{25}{4}\right)^2 + \left(-6\tfrac{1}{4}\right)^2}$$
$$AB = \sqrt{39\tfrac{1}{6} + 39\tfrac{1}{16}}$$
$$AB = \sqrt{78\tfrac{1}{8}}$$
$$AB \approx 8.8$$

The distance between points A and B is about 8.8 units.

17. Side $\overline{XY}$: $X(-2, 3)$, $Y(3, 0)$
$$d = \sqrt{(x_2 - x_1)^2 + (y_2 - y_1)^2}$$
$$XY = \sqrt{[3 - (-2)]^2 + (0 - 3)^2}$$
$$XY = \sqrt{5^2 + (-3)^2}$$
$$XY = \sqrt{25 + 9}$$
$$XY = \sqrt{34}$$

Side $\overline{XZ}$: $X(-2, 3)$, $Z(-2, -4)$
$$d = \sqrt{(x_2 - x_1)^2 + (y_2 - y_1)^2}$$
$$XZ = \sqrt{[-2 - (-2)]^2 + (-4 - 3)^2}$$
$$XZ = \sqrt{0^2 + (-7)^2}$$
$$XZ = \sqrt{49}$$

Side $\overline{YZ}$: $Y(3, 0), Z(-2, -4)$

$$d = \sqrt{(x_2 - x_1)^2 + (y_2 - y_1)^2}$$
$$YZ = \sqrt{(-2 - 3)^2 + (-4 - 0)^2}$$
$$YZ = \sqrt{(-5)^2 + (-4)^2}$$
$$YZ = \sqrt{25 + 16}$$
$$YZ = \sqrt{41}$$
$$P = \sqrt{34} + \sqrt{49} + \sqrt{41} \approx 5.8 + 7 + 6.4 \approx 19.2$$

The perimeter is about 19.2 units.

18. Side $\overline{AB}$: $A(4, 4), B(1, -4)$

$$d = \sqrt{(x_2 - x_1)^2 + (y_2 - y_1)^2}$$
$$AB = \sqrt{(1 - 4)^2 + (-4 - 4)^2}$$
$$AB = \sqrt{(-3)^2 + (-8)^2}$$
$$AB = \sqrt{9 + 64}$$
$$AB = \sqrt{73}$$

Side $\overline{AC}$: $A(4, 4), C(-2, -2)$

$$d = \sqrt{(x_2 - x_1)^2 + (y_2 - y_1)^2}$$
$$AC = \sqrt{(-2 - 4)^2 + (-2 - 4)^2}$$
$$AC = \sqrt{(-6)^2 + (-6)^2}$$
$$AC = \sqrt{36 + 36}$$
$$AC = \sqrt{72}$$

Side $\overline{BC}$: $B(1, -4), C(-2, -2)$

$$d = \sqrt{(x_2 - x_1)^2 + (y_2 - y_1)^2}$$
$$BC = \sqrt{(-2 - 1)^2 + [-2 - (-4)]^2}$$
$$BC = \sqrt{(-3)^2 + (2)^2}$$
$$BC = \sqrt{9 + 4}$$
$$BC = \sqrt{13}$$
$$P = \sqrt{73} + \sqrt{72} + \sqrt{13} \approx 8.5 + 8.5 + 3.6 \approx 20.6$$

The perimeter is about 20.6 units.

19. $$\text{midpoint} = \left(\frac{x_1 + x_2}{2}, \frac{y_1 + y_2}{2}\right)$$
$$\text{midpoint } \overline{XY} = \left(\frac{-2 + 4}{2}, \frac{3 + 3}{2}\right) = (-1, 3)$$

The coordinates of the midpoint of $\overline{XY}$ are $(-1, 3)$.

20. $$\text{midpoint} = \left(\frac{x_1 + x_2}{2}, \frac{y_1 + y_2}{2}\right)$$
$$\text{midpoint } \overline{RS} = \left(\frac{1 + 1}{2}, \frac{4 + (-3)}{2}\right) = \left(1, \frac{1}{2}\right)$$

The coordinates of the midpoint of $\overline{RS}$ are $\left(1, \frac{1}{2}\right)$.

21. $$\text{midpoint} = \left(\frac{x_1 + x_2}{2}, \frac{y_1 + y_2}{2}\right)$$
$$\text{midpoint } \overline{AB} = \left(\frac{6 + 2}{2}, \frac{1 + (-5)}{2}\right) = (4, -2)$$

The coordinates of the midpoint of $\overline{AB}$ are $(4, -2)$.

22. $$\text{midpoint} = \left(\frac{x_1 + x_2}{2}, \frac{y_1 + y_2}{2}\right)$$
$$\text{midpoint } \overline{JK} = \left(\frac{-3 + 7}{2}, \frac{5 + 9}{2}\right) = (2, 7)$$

The coordinates of the midpoint of $\overline{JK}$ are $(2, 7)$.

23. $$\text{midpoint} = \left(\frac{x_1 + x_2}{2}, \frac{y_1 + y_2}{2}\right)$$
$$\text{midpoint } \overline{MN} = \left(\frac{-1 + 5}{2}, \frac{-3 + 7}{2}\right) = (2, 2)$$

The coordinates of the midpoint of $\overline{MN}$ are $(2, 2)$.

24. $$\text{midpoint} = \left(\frac{x_1 + x_2}{2}, \frac{y_1 + y_2}{2}\right)$$
$$\text{midpoint } \overline{CD} = \left(\frac{-4 + 6}{2}, \frac{9 + (-5)}{2}\right) = (1, 2)$$

The coordinates of the midpoint of $\overline{CD}$ are $(1, 2)$.

25. $$\text{midpoint} = \left(\frac{x_1 + x_2}{2}, \frac{y_1 + y_2}{2}\right)$$
$$\text{midpoint } \overline{TU} = \left(\frac{10 + (-4)}{2}, \frac{-3 + (-5)}{2}\right) = (3, -4)$$

The coordinates of the midpoint of $\overline{TU}$ are $(3, -4)$.

26. $$\text{midpoint} = \left(\frac{x_1 + x_2}{2}, \frac{y_1 + y_2}{2}\right)$$
$$\text{midpoint } \overline{PQ} = \left(\frac{6 + (-4)}{2}, \frac{11 + (-3)}{2}\right) = (1, 4)$$

The coordinates of the midpoint of $\overline{PQ}$ are $(1, 4)$.

27. $$\text{midpoint} = \left(\frac{x_1 + x_2}{2}, \frac{y_1 + y_2}{2}\right)$$
$$\text{midpoint } \overline{FG} = \left(\frac{15 + 8}{2}, \frac{-4 + (-6)}{2}\right) = \left(11\tfrac{1}{2}, -5\right)$$

The coordinates of the midpoint of $\overline{FG}$ are $\left(11\frac{1}{2}, -5\right)$.

28. $$\text{midpoint} = \left(\frac{x_1 + x_2}{2}, \frac{y_1 + y_2}{2}\right)$$
$$\text{midpoint } \overline{EF} = \left(\frac{-12 + (-3)}{2}, \frac{-5 + (-4)}{2}\right) = \left(-7\tfrac{1}{2}, -4\tfrac{1}{2}\right)$$

The coordinates of the midpoint of $\overline{EF}$ are $\left(-7\frac{1}{2}, -4\frac{1}{2}\right)$.

29. Yes; $\overline{PM}$ and $\overline{MN}$ have equal measures.

30. Yes; none of the measures of the sides are equal.

31. $$\text{midpoint} = \left(\frac{x_1 + x_2}{2}, \frac{y_1 + y_2}{2}\right)$$
$$8 = \left(\frac{x_1 + x_2}{2}\right) \text{ and } -9 = \left(\frac{y_1 + y_2}{2}\right)$$
$$8 = \left(\frac{x_1 + 18}{2}\right) \text{ and } -9 = \left(\frac{y_1 + (-21)}{2}\right)$$
$$8 = \left(\frac{-2 + 18}{2}\right) \text{ and } -9 = \left(\frac{3 + (-21)}{2}\right)$$
$$x_1 = -2 \qquad x_2 = 3$$

The coordinates of A are $(-2, 3)$.

32. Both the Distance Formula and the Pythagorean Theorem can be used to find distance between any two points on the coordinate system. Answers should include the following.

- First, use the two points. Draw vertical and horizontal lines so that a right triangle is formed. Then determine the lengths of the legs. Replace the lengths of the legs in the Pythagorean Theorem to determine the length of the the hypotenuse.

Pythagorean Theorem
$c^2 = a^2 + b^2$
$c^2 = 5^2 + 4^2$
$c^2 = 25 + 16$
$c^2 = 41$
$c \approx 6.4$

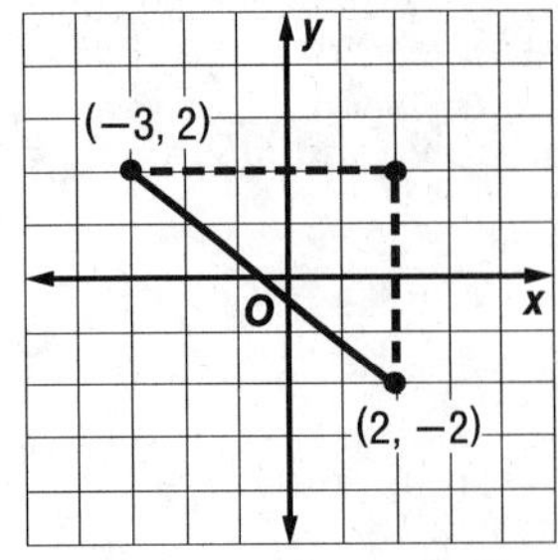

- The values of $(x_2 - x_1)$ and $(y_2 - y_1)$ equal the lengths of the legs of a right triangle.

33. C; midpoint $= \left(\frac{x_1 + x_2}{2}, \frac{y_1 + y_2}{2}\right)$

midpoint $\overline{HG} = \left(\frac{-2 + 8}{2}, \frac{0 + 6}{2}\right)$
$= (3, 3)$

34. B; $d = \sqrt{(x_2 - x_1)^2 + (y_2 - y_1)^2}$
$MN = \sqrt{[2 - (-5)]^2 + [3 - (-3)]^2}$

Page 470 Maintain Your Skills

35. $c^2 = a^2 + b^2$
$c^2 = 6^2 + 7^2$
$c^2 = 36 + 49$
$c^2 = 85$
$\sqrt{c^2} = \sqrt{85}$
$c \approx 9.2$
The length of the hypotenuse is about 9.2 ft.

36. $c^2 = a^2 + b^2$
$c^2 = 9^2 + 24^2$
$c^2 = 81 + 576$
$c^2 = 657$
$\sqrt{c^2} = \sqrt{657}$
$c \approx 25.6$
The length of the hypotenuse is about 25.6 yd.

37. $c^2 = a^2 + b^2$
$c^2 = 12^2 + 33^2$
$c^2 = 144 + 1089$
$c^2 = 1233$
$\sqrt{c^2} = \sqrt{1233}$
$c \approx 35.1$
The length of the hypotenuse is about 35.1 km.

38. $x + 4x + 5x = 180$
$10x = 180$
$\frac{10x}{10} = \frac{180}{10}$
$x = 18$
Since $x = 18$, $4x = 4(18)$ or 72, and $5x = 5(18)$ or 90. The measures of the angles are 18°, 72°, and 90°.

39. $y = mx + b$
$y = 2x + 1$
slope $= 2$, y-intercept $= 1$

40. $x + y = -4$
$x - x + y = -4 - x$
$y = -x - 4$
slope $= -1$, y-intercept $= -4$

41. $y = mx + b$
$4x + y = -6$
$4x - 4x + y = -4x - 6$
$y = -4x - 6$
slope $= -4$, y-intercept $= -6$

42. $\frac{x}{85} = \frac{56}{100}$
$100x = 4760$
$\frac{100x}{100} = \frac{4760}{100}$
$x = 47.6$
56% of 85 is 47.6.

43. $\frac{4}{16} = \frac{7}{x}$
$4x = 112$
$\frac{4x}{4} = \frac{112}{4}$
$x = 28$

44. $\frac{a}{15} = \frac{12}{60}$
$60a = 180$
$a = 3$

45. $\frac{84}{m} = \frac{52}{13}$
$1092 = 52m$
$\frac{1092}{52} = \frac{52m}{52}$
$21 = m$

46. $\frac{2.8}{h} = \frac{4.2}{12}$
$33.6 = 4.2h$
$\frac{33.6}{4.2} = \frac{4.2h}{4.2}$
$8 = h$

47. $\frac{3.4}{85} = \frac{2.36}{n}$
$3.4n = 200.6$
$\frac{3.4n}{3.4} = \frac{200.6}{3.4}$
$n = 59$

48. $\frac{k}{5.5} = \frac{111}{15}$
$15k = 610.5$
$\frac{15k}{15} = \frac{610.5}{15}$
$k = 40.7$

Page 470 Practice Quiz 2

1. obtuse isosceles because the triangle has an obtuse angle and two congruent sides
2. right scalene because the triangle has a right angle and no congruent sides
3. Yes; $34^2 = 30^2 + 16^2$
4. $d = \sqrt{(x_2 - x_1)^2 + (y_2 - y_1)^2}$
$AB = \sqrt{[7 - (-4)]^2 + [-9 - (-1)]^2}$
$AB = \sqrt{11^2 + (-8)^2}$
$AB = \sqrt{121 + 64}$
$AB = \sqrt{185}$
$AB \approx 13.6$
The distance between points A and B is about 13.6 units.

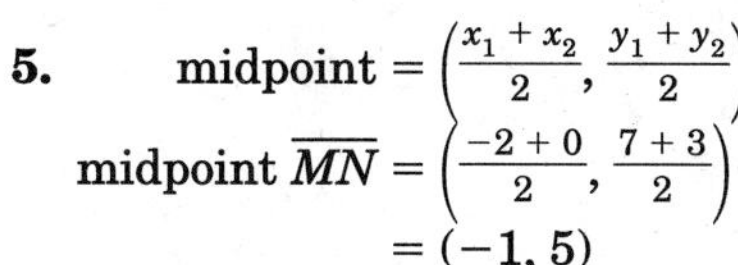

5. $\text{midpoint} = \left(\frac{x_1 + x_2}{2}, \frac{y_1 + y_2}{2}\right)$

$\text{midpoint } \overline{MN} = \left(\frac{-2 + 0}{2}, \frac{7 + 3}{2}\right)$

$= (-1, 5)$

The coordinates of the midpoint of $\overline{MN}$ are $(-1, 5)$.

9-7 Similar Triangles and Indirect Measurement

Page 471 How can similar triangles be used to create patterns?

a. They are the same.

b. They are shorter.

Page 473 Check for Understanding

1. Sample answer:

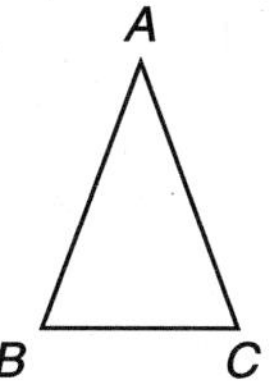

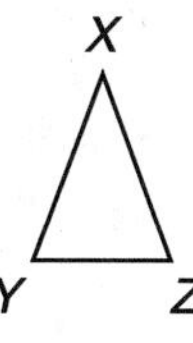

corresponding sides: $\frac{AB}{XY} = \frac{BC}{YZ} = \frac{AC}{XZ}$

2. Sample answer: A method used to find measurements that are difficult to measure directly.

3. $\frac{ST}{DE} = \frac{RS}{CD}$

$\frac{x}{15} = \frac{6}{9}$

$9 \cdot x = 15 \cdot 6$

$9x = 90$

$x = 10$

The value of x is 10.

4. $\frac{BC}{PN} = \frac{AB}{NM}$

$\frac{x}{6} = \frac{6}{2}$

$2 \cdot x = 6 \cdot 6$

$2x = 36$

$x = 18$

The value of x is 18.

5. $\frac{DE}{AB} = \frac{DC}{CB}$

$\frac{x}{18} = \frac{2}{6}$

$6 \cdot x = 18 \cdot 2$

$6x = 36$

$x = 6$

The distance from Austintown to North Jackson is 6 km.

6. $\frac{10}{x} = \frac{8}{40}$

$8 \cdot x = 10 \cdot 40$

$8x = 400$

$x = 50$

The tree is 50 feet tall.

Pages 474–475 Practice and Apply

7. $\frac{GJ}{XZ} = \frac{GH}{XY}$

$\frac{x}{4} = \frac{10}{5}$

$5 \cdot x = 4 \cdot 10$

$5x = 40$

$x = 8$

The value of x is 8.

8. $\frac{JM}{AC} = \frac{JK}{AB}$

$\frac{x}{6} = \frac{12}{8}$

$8 \cdot x = 6 \cdot 12$

$8x = 72$

$x = 9$

The value of x is 9.

9. $\frac{HG}{DE} = \frac{EG}{CD}$

$\frac{x}{18} = \frac{20}{15}$

$15 \cdot x = 18 \cdot 20$

$15x = 360$

$x = 24$

The value of x is 24.

10. $\frac{JM}{JK} = \frac{RP}{QP}$

$\frac{x}{9} = \frac{9}{6}$

$6 \cdot x = 9 \cdot 9$

$6x = 81$

$x = 13.5$

The value of x is 13.5.

11. $\frac{EB}{EC} = \frac{AB}{CD}$

$\frac{x}{3} = \frac{7}{4.2}$

$4.2 \cdot x = 3 \cdot 7$

$4.2x = 21$

$x = 5$

The value of x is 5.

12. $\frac{SK}{UV} = \frac{RK}{RV}$

$\frac{x}{9} = \frac{2}{7.5}$

$7.5 \cdot x = 9 \cdot 2$

$7.5x = 18$

$x = 2.4$

The value of x is 2.4.

13. always

14. sometimes

15. $\frac{x}{51} = \frac{64}{48}$

$48 \cdot x = 51 \cdot 64$

$48x = 3264$

$x = 68$

The pavilion is 68 yards from the log cabin.

16. $\frac{28}{84} = \frac{35}{x}$

$x \cdot 28 = 84 \cdot 35$

$28x = 2940$

$x = 105$

The gorillas are 105 meters away from the cheetahs.

17. $\frac{3.2}{x} = \frac{8}{15}$

$8 \cdot x = 3.2(15)$
$8x = 48$
$x = 6$

The baby giraffe is 6 feet tall.

18. $\frac{31.5}{126} = \frac{19}{x}$

$31.5 \cdot x = 126 \cdot 19$
$31.5x = 2394$
$x = 76$

The Ferris wheel is 76 feet tall.

19. Sample answer: 4 units by 6 units by 8 units and 1 unit, 1.5 units, 2 units

20. By increasing or decreasing the size of similar triangles, patterns can be formed. Answers should include the following.
- All of the triangles used in Sierpinski's triangle are similar.
- Sample answer:

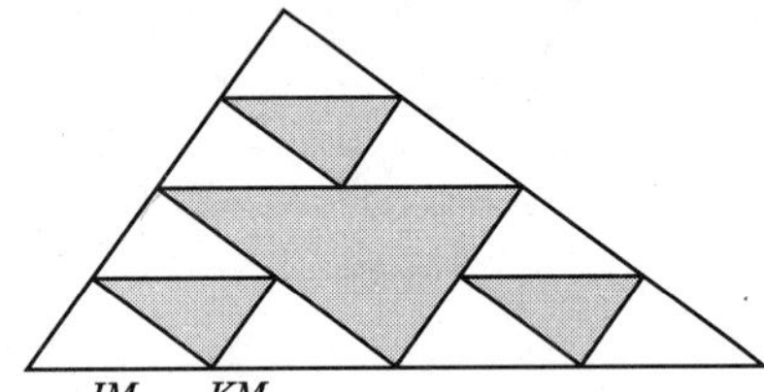

21. A; $\frac{JM}{RT} = \frac{KM}{ST}$

$\frac{x}{5.5} = \frac{2}{3}$
$3 \cdot x = 2 \cdot 5.5$
$3x = 11$
$x \approx 3.7$

22. B; $\frac{MN}{RS} = \frac{MP}{x}$, let $x = RT$

$\frac{4}{2} = \frac{6}{x}$
$4 \cdot x = 2 \cdot 6$
$4x = 12$
$x = 3$

The length of $\overline{RT}$ is 3 units. The coordinates of point T are $(1, -1)$.

Page 475 Maintain Your Skills

23. $d = \sqrt{(x_2 - x_1)^2 + (y_2 - y_1)^2}$

$ST = \sqrt{(0 - 2)^2 + (6 - 3)^2}$
$ST = \sqrt{(-2)^2 + (3)^2}$
$ST = \sqrt{4 + 9}$
$ST = \sqrt{13}$
$ST \approx 3.6$

The distance between points S and T is about 3.6 units.

24. $d = \sqrt{(x_2 - x_1)^2 + (y_2 - y_1)^2}$

$EF = \sqrt{[3 - (-1)]^2 + (-2 - 1)^2}$
$EF = \sqrt{4^2 + (-3)^2}$
$EF = \sqrt{16 + 9}$
$EF = \sqrt{25}$
$EF = 5$

The distance between points E and F is 5 units.

25. $d = \sqrt{(x_2 - x_1)^2 + (y_2 - y_1)^2}$

$WV = \sqrt{(-3 - 4)^2 + [-5 - (-6)]^2}$
$WV = \sqrt{(-7)^2 + (1)^2}$
$WV = \sqrt{49 + 1}$
$WV = \sqrt{50}$
$WV \approx 7.1$

The distance between points W and V is about 7.1 units.

26.
$c^2 = a^2 + b^2$
$34^2 = 8^2 + b^2$
$1156 = 64 + b^2$
$1156 - 64 = 64 - 64 + b^2$
$1092 = b^2$
$\sqrt{1092} = \sqrt{b^2}$
$33.0 \approx b$

27.
$c^2 = a^2 + b^2$
$82^2 = a^2 + 27^2$
$6724 = a^2 + 729$
$6724 - 729 = a^2 + 729 - 729$
$5995 = a^2$
$\sqrt{5995} = \sqrt{a^2}$
$77.4 \approx a$

28. Find the slope.

$m = \frac{(y_2 - y_1)}{(x_2 - x_1)}$
$m = \frac{[11 - (-1)]}{(-2 - 1)}$
$m = \frac{12}{-3}$
$m = -4$

Find the y-intercept.

$y = mx + b$
$-1 = -4(1) + b$
$-1 = -4 + b$
$3 = b$

Substitute the slope and y-intercept.

$y = mx + b$
$y = -4x + 3$

The equation of the line is $y = -4x + 3$.

29. $25\% = \frac{25}{100} = \frac{1}{4}$

30. $87\frac{1}{2}\% = \frac{87.5}{100} = \frac{7}{8}$

31. $150\% = \frac{150}{100} = 1\frac{1}{2}$

32. $\frac{12}{16} = 16\overline{)12.00} = 0.75$

$$\begin{array}{r} 0.75 \\ 16\overline{)12.00} \\ \underline{11\,2} \\ 80 \\ \underline{80} \\ 0 \end{array}$$

33. $\frac{10}{14} = 14\overline{)10.00000...} \approx 0.7143$

$$\begin{array}{r} 0.71428 \\ 14\overline{)10.00000...} \\ \underline{9\,8} \\ 20 \\ \underline{14} \\ 60 \\ \underline{56} \\ 40 \\ \underline{28} \\ 120 \\ \underline{112} \\ 80 \end{array}$$

34. $\frac{20}{25} = 25\overline{)20.0} = 0.8$

$$\begin{array}{r} 0.8 \\ 25\overline{)20.0} \\ \underline{20\ 0} \\ 0 \end{array}$$

35. $\frac{9}{40} = 40\overline{)9.000} = 0.225$

$$\begin{array}{r} 0.225 \\ 40\overline{)9.000} \\ \underline{8\ 0} \\ 1\ 00 \\ \underline{80} \\ 200 \\ \underline{200} \\ 0 \end{array}$$

Page 476 Algebra Activity (Preview of Lesson 9-8)

Sample answer:

	30° angle	60° angle
Length (mm) of opposite leg	20	35
Length (mm) of adjacent leg	35	20
Length (mm) of hypotenuse	40	40
Ratio 1	$\frac{20}{40}$, or 0.5	$\frac{35}{40}$, or 0.875
Ratio 2	$\frac{35}{40}$, or 0.875	$\frac{20}{40}$, or 0.5
Ratio 3	$\frac{20}{35}$, or 0.57	$\frac{35}{20}$, or 1.75

1. See students' work.
2. The ratios of the sides of any 30°-60°-90° triangle are about 0.50, 0.875, and 1.75.
3. See students' work. The ratios of the sides of any 45°-45°-90° triangle are about 0.71, 0.71, and 1.00.

9-8 Sine, Cosine, and Tangent Ratios

Page 477 How are ratios in right triangles used in the real world?

a. right
b. $\overline{AC}$
c. acute
d. $\overline{AB}$
e. $\overline{BC}$

Page 479 Check for Understanding

1. The sine ratio compares the measure of the leg opposite the angle to the measure of the hypotenuse. The cosine ratio compares the measure of the leg adjacent to the angle to the measure of the hypotenuse. The tangent ratio compares the measure of the leg opposite the angle to the measure of the leg adjacent to the angle.
2. Susan; the sine ratio is the ratio of the opposite side to the hypotenuse. The length of the side opposite the 15° angle is x m. The length of the hypotenuse is 580 m. So, $\sin 15° = \frac{x}{580}$.
3. $\sin N = \frac{\text{measure of leg opposite } \angle N}{\text{measure of hypotenuse}}$
   $= \frac{15}{17}$
   $\sin N \approx 0.8824$
4. $\cos N = \frac{\text{measure of the leg adjacent to } \angle N}{\text{measure of hypotenuse}}$
   $= \frac{8}{17}$
   $\cos N \approx 0.4706$
5. $\tan N = \frac{\text{measure of leg opposite } \angle N}{\text{measure of leg adjacent to } \angle N}$
   $= \frac{15}{8}$
   $\tan N = 1.875$
6. $\sin 52° = 0.7880$
7. $\cos 92° = 0.9455$
8. $\tan 76° = 4.0108$
9. $\sin 53° = \frac{x}{32}$
   $32(\sin 53°) = 32 \cdot \frac{x}{32}$
   $25.6 \approx x$
10. $\sin 21° = \frac{x}{73}$
    $73(\sin 21°) = 73 \cdot \frac{x}{73}$
    $26.1 \approx x$
11. $\tan 33° = \frac{x}{12}$
    $12(\tan 33°) = 12 \cdot \frac{x}{12}$
    $7.8 \approx x$
12. $\sin 50° = \frac{x}{50}$
    $50(\sin 50°) = 50 \cdot \frac{x}{50}$
    $38.3 \approx x$
    The kite is about 38.3 yards above the ground.

Pages 480–481 Practice and Apply

13. $\sin N = \frac{12}{13}$
    $\sin N \approx 0.9231$
14. $\sin K = \frac{5}{13}$
    $\sin K \approx 0.3846$
15. $\cos C = \frac{21}{29}$
    $\cos C \approx 0.7241$
16. $\cos A = \frac{20}{29}$
    $\cos A \approx 0.6897$
17. $\tan N = \frac{12}{5}$
    $\tan N \approx 2.4$
18. $\tan C = \frac{20}{21}$
    $\tan C \approx 0.9524$

19. $\sin R = \frac{8}{\sqrt{80}}$

$\sin R \approx 0.8944$

$\cos R = \frac{4}{\sqrt{80}}$

$\cos R \approx 0.44721$

$\tan R = \frac{8}{4}$

$\tan R = 2.0$

20. $\sin X = \frac{7}{\sqrt{74}}$

$\sin X \approx 0.8137$

$\cos X = \frac{5}{\sqrt{74}}$

$\cos X \approx 0.5812$

$\tan X = \frac{7}{5}$

$\tan X = 1.4$

21. $\sin 6° \approx 0.1045$

22. $\sin 51° \approx 0.7771$

23. $\cos 31° \approx 0.8572$

24. $\cos 87° \approx 0.0523$

25. $\tan 12° \approx 0.2126$

26. $\tan 66° \approx 2.2460$

27. $\sin 18° = \frac{x}{22}$

$22(\sin 18°) = 22 \cdot \frac{x}{22}$

$6.8 \approx x$

28. $\sin 25° = \frac{x}{8}$

$8(\sin 25°) = 8 \cdot \frac{x}{8}$

$3.4 \approx x$

29. $\cos 41° = \frac{x}{33}$

$33(\cos 41°) = 33 \cdot \frac{x}{33}$

$24.9 \approx x$

30. $\cos 73° = \frac{x}{16}$

$16(\cos 73°) = 16 \cdot \frac{x}{16}$

$4.7 \approx x$

31. $\tan 47° = \frac{x}{91}$

$91(\tan 47°) = 91 \cdot \frac{x}{91}$

$97.6 \approx x$

32. $\tan 62° = \frac{x}{13}$

$13(\tan 62°) = 13 \cdot \frac{x}{13}$

$24.4 \approx x$

33. $\tan 40° = \frac{x}{25}$

$25(\tan 40°) = 25 \cdot \frac{x}{25}$

$21.0 \approx x$

The flagpole is about 21 meters tall.

34. $\sin 6° = \frac{x}{20}$

$20(\sin 6°) = 20 \cdot \frac{x}{20}$

$2.1 \approx x$

The height of the ramp is about 2.1 feet.

35. $\tan 26.5° = \frac{x}{85}$

$85(\tan 26.5°) = 85 \cdot \frac{x}{85}$

$42.4 \approx x$

$42.4 + 1.6 = 44$

The building is about 44 meters tall.

36. $\cos 17° = \frac{x}{210}$

$210(\cos 17°) = 210 \cdot \frac{x}{210}$

$200.8 \approx x$

The base of the ramp is about 200.8 yards long.

37. $\sin 22° = \frac{9}{x}$

$x\sin 22° = x \cdot \frac{9}{x}$

$x(0.3746) = 9$

$\frac{x(0.3746)}{0.3746} = \frac{9}{0.3746}$

$x \approx 24.0$

38. $\tan 30° = \frac{8}{x}$

$x\tan 30° = x \cdot \frac{8}{x}$

$x\tan 30° = 8$

$x(0.5774) = 8$

$\frac{x(0.5774)}{0.5774} = \frac{8}{0.5774}$

$x \approx 13.9$

39. $\cos 42° = \frac{30}{x}$

$x\cos 42° = x \cdot \frac{30}{x}$

$x\cos 42° = 30$

$x(0.7431) = 30$

$\frac{x(0.7431)}{0.7431} = \frac{30}{0.7431}$

$x \approx 40.4$

40.

x	30°	45°	60°
$\sin x$	0.5	0.7071	0.866
$\cos x$	0.866	0.7071	0.5
$\tan x$	0.5774	1	1.7321

41. $\sin 45° = \cos 45°$; $\sin 60° = \cos 30°$; $\sin 30° = \cos 60°$

42. 45°

43. To find heights of buildings. Answers should include the following.

- If two of the three measures of the sides of a right triangle are known, you can use the Pythagorean Theorem to find the measure of the third side. If the measure of an acute angle and the length of one side of a right triangle are known, trigonometric ratios can be used to find the missing measures.
-

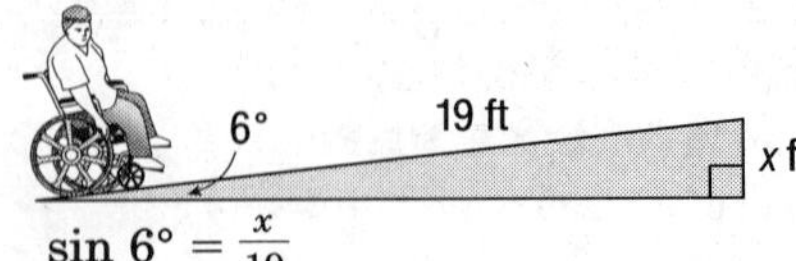

$\sin 6° = \frac{x}{19}$

$19(\sin 6°) = x$

$2.0 \approx x$

The height of the ramp is about 2 feet.

44. D; $\cos 58° = \frac{x}{5}$

$5(\cos 58°) = 5 \cdot \frac{x}{5}$

$2.6496 \approx x$

45. C; $\tan \angle P = \frac{8}{21}$

≈ 0.4

Page 481 Maintain Your Skills

46. $\frac{40}{22} = \frac{x}{3.3}$

$3.3 \cdot 40 = 22 \cdot x$

$132 = 22x$

$\frac{132}{22} = \frac{22x}{22}$

$6 = x$

The height of the fence is 6 feet.

47. $\text{midpoint} = \left(\frac{x_1 + x_2}{2}, \frac{y_1 + y_2}{2}\right)$

$\text{midpoint } \overline{AC} = \left(\frac{3+0}{2}, \frac{-4+5}{2}\right)$

$= \left(1\frac{1}{2}, \frac{1}{2}\right)$

The coordinates of the midpoint of $\overline{AC}$ are $\left(1\frac{1}{2}, \frac{1}{2}\right)$.

48. $\text{midpoint} = \left(\frac{x_1 + x_2}{2}, \frac{y_1 + y_2}{2}\right)$

$\text{midpoint } \overline{MN} = \left(\frac{-2+6}{2}, \frac{1+(-9)}{2}\right)$

$= (2, -4)$

The coordinates of the midpoint of $\overline{MN}$ are $(2, -4)$.

49. $\text{midpoint} = \left(\frac{x_1 + x_2}{2}, \frac{y_1 + y_2}{2}\right)$

$\text{midpoint } \overline{RS} = \left(\frac{-3+(-3)}{2}, \frac{0+(-8)}{2}\right)$

$= (-3, -4)$

The coordinates of the midpoint of $\overline{RS}$ are $(-3, -4)$.

50. $5m < 5$

$\frac{5m}{5} < \frac{5}{5}$

$m < 1$

51. $\frac{a}{-2} > 3$

$-2 \cdot \frac{a}{-2} < 3 \cdot (-2)$

$a < -6$

52. $-4x \geq -16$

$\frac{-4x}{-4} \leq \frac{-16}{-4}$

$x \leq 4$

Page 482 Graphing Calculator Investigation (Follow-Up of Lesson 9-8)

1. $\cos \angle B = \frac{\text{adjacent}}{\text{hypotenuse}}$

$\cos \angle B = \frac{7}{13.5}$

To the nearest degree, the measure of $\angle B$ is 59°.

$\sin \angle C = \frac{\text{opposite}}{\text{hypotenuse}}$

$\sin \angle C = \frac{7}{13.5}$

To the nearest degree, the measure of $\angle A$ is 31°.

2. $\sin \angle F = \frac{\text{opposite}}{\text{hypotenuse}}$

$\sin \angle F = \frac{17}{39}$

To the nearest degree, the measure of $\angle F$ is 26°.

$\cos \angle D = \frac{\text{adjacent}}{\text{hypotenuse}}$

$\cos \angle D = \frac{17}{39}$

To the nearest degree, the measure of $\angle D$ is 64°.

3. Find the measure of the angle at the flower garden.

$\tan \angle\text{flower garden} = \frac{19}{46}$

$\tan \angle\text{flower garden} \approx 22°$

Then, find the measure of the missing angle.

$22 + 90 + x = 180$

$102 + x = 180$

$102 - 102 + x = 180 - 102$

$x = 68$

The measures of the angles formed by these three park features are 22°, 68°, and 90°.

Chapter 9 Study Guide and Review

Page 483 Vocabulary and Concept Check

1. real numbers
2. acute angle
3. obtuse
4. equilateral triangle
5. hypotenuse
6. similar
7. trigonometric ratio
8. ray

Pages 483–486 Lesson-by-Lesson Review

9. Since $6^2 = 36$, $\sqrt{36} = 6$.

10. Since $10^2 = 100$, $\sqrt{100} = 10$.

11. Since $9^2 = 81$, $-\sqrt{81} = -9$.

12. Since $\pm 11^2 = 121$, $\pm\sqrt{121} = \pm 11$.

13. not possible

14. Since $15^2 = 225$, $-\sqrt{225} = -15$.

15. $n^2 = 81$

$\sqrt{n^2} = \sqrt{81}$

$n = \sqrt{81}$ or $n = -\sqrt{81}$

$n = 9$ or $n = -9$

16. $t^2 = 38$

$\sqrt{t^2} = \sqrt{38}$

$t = \sqrt{38}$ or $t = -\sqrt{38}$

$t \approx 6.2$ or $t \approx -6.2$

17. $y^2 = 1.44$

$\sqrt{y^2} = \sqrt{1.44}$

$y = \sqrt{1.44}$ or $y = -\sqrt{1.44}$

$y = 1.2$ or $y = -1.2$

18. $7.5 = r^2$

$\sqrt{7.5} = \sqrt{r^2}$

$\sqrt{7.5} = r$ or $-\sqrt{7.5} = r$

$2.7 \approx r$ or $-2.7 \approx r$

19. $m\angle FGH = 90°$
Since $m\angle FGH = 90°$, $\angle FGH$ is right.

20. $m\angle QRS = 115°$
Since $m\angle QRS > 90°$, $\angle QRS$ is obtuse.

21. $m\angle WXY = 35°$
Since $m\angle WXY < 90°$, $\angle WXY$ is acute.

22. acute equilateral because the triangle has all acute angles and all sides congruent

23. obtuse isosceles because the triangle has an obtuse angle and two sides congruent

24. right scalene because the triangle has a right angle and no congruent sides

25.
$$c^2 = a^2 + b^2$$
$$15^2 = 6^2 + b^2$$
$$225 = 36 + b^2$$
$$225 - 36 = 36 - 36 + b^2$$
$$189 = b^2$$
$$\sqrt{189} = \sqrt{b^2}$$
$$13.7 \approx b$$

26.
$$c^2 = a^2 + b^2$$
$$7^2 = a^2 + 2^2$$
$$49 = a^2 + 4$$
$$49 - 4 = a^2 + 4 - 4$$
$$45 = a^2$$
$$\sqrt{45} = \sqrt{a^2}$$
$$6.7 \approx a$$

27.
$$c^2 = a^2 + b^2$$
$$24^2 = 18^2 + b^2$$
$$576 = 324 + b^2$$
$$576 - 324 = 324 - 324 + b^2$$
$$252 = b^2$$
$$\sqrt{252} = \sqrt{b^2}$$
$$15.9 \approx b$$

28.
$$d = \sqrt{(x_2 - x_1)^2 + (y_2 - y_1)^2}$$
$$d = \sqrt{(2 - 0)^2 + (7 - 9)^2}$$
$$d = \sqrt{2^2 + (-2)^2}$$
$$d = \sqrt{4 + 4}$$
$$d = \sqrt{8}$$
$$d \approx 2.8$$
The distance between points J and K is about 2.8 units.

29.
$$d = \sqrt{(x_2 - x_1)^2 + (y_2 - y_1)^2}$$
$$d = \sqrt{[3 - (-5)]^2 + (6 - 1)^2}$$
$$d = \sqrt{8^2 + 5^2}$$
$$d = \sqrt{64 + 25}$$
$$d = \sqrt{89}$$
$$d \approx 9.4$$
The distance between points A and B is about 9.4 units.

30.
$$d = \sqrt{(x_2 - x_1)^2 + (y_2 - y_1)^2}$$
$$d = \sqrt{(3 - 8)^2 + [3 - (-4)]^2}$$
$$d = \sqrt{(-5)^2 + 7^2}$$
$$d = \sqrt{25 + 49}$$
$$d = \sqrt{74}$$
$$d \approx 8.6$$
The distance between points W and Y is about 8.6 units.

31.
$$\text{midpoint} = \left(\frac{x_1 + x_2}{2}, \frac{y_1 + y_2}{2}\right)$$
$$\text{midpoint } \overline{MN} = \left(\frac{8 + (-2)}{2}, \frac{0 + 10}{2}\right)$$
$$= (3, 5)$$
The coordinates of the midpoint of $\overline{MN}$ are (3, 5).

32.
$$\text{midpoint} = \left(\frac{x_1 + x_2}{2}, \frac{y_1 + y_2}{2}\right)$$
$$\text{midpoint } \overline{CD} = \left(\frac{5 + (-7)}{2}, \frac{9 + 3}{2}\right)$$
$$= (-1, 6)$$
The coordinates of the midpoint of $\overline{CD}$ are (−1, 6).

33.
$$\text{midpoint} = \left(\frac{x_1 + x_2}{2}, \frac{y_1 + y_2}{2}\right)$$
$$\text{midpoint } \overline{QR} = \left(\frac{6 + (-6)}{2}, \frac{-8 + 4}{2}\right)$$
$$= (0, -2)$$
The coordinates of the midpoint of $\overline{QR}$ are (0, −2).

34.
$$\frac{LM}{TU} = \frac{MN}{UV}$$
$$\frac{24}{8} = \frac{21}{x}$$
$$x \cdot 24 = 8 \cdot 21$$
$$24x = 168$$
$$\frac{24x}{24} = \frac{168}{24}$$
$$x = 7$$

35.
$$\frac{AB}{HJ} = \frac{BC}{JK}$$
$$\frac{x}{15} = \frac{6}{9}$$
$$9 \cdot x = 15 \cdot 6$$
$$9x = 90$$
$$\frac{9x}{9} = \frac{90}{9}$$
$$x = 10$$

36. $\sin R = \frac{15}{25} = 0.6$

37. $\tan S = \frac{20}{15} \approx 1.3333$

38. $\tan R = \frac{15}{20} = 0.75$

Chapter 9 Practice Test

Page 487

1. Sample answer: 1; 5; $\sqrt{27}$; $\frac{5}{8}$; −6

2. To classify a triangle by its angles and sides, first determine whether the triangle is acute, right, or obtuse. Then determine whether the triangle is scalene, isosceles, or equilateral.

3. Since $9^2 = 81$, $\sqrt{81} = 9$.

4. Since $11^2 = 121$, $-\sqrt{121} = -11$.

5. Since $7^2 = 49$, $\pm\sqrt{49} = -7, 7$.

6.
$$-49 < -42 < -36$$
−42 is closer to −36 than to −49.
$$-\sqrt{42} < -\sqrt{36}$$
$$-\sqrt{42} < -6$$
$$-\sqrt{42} \approx -6$$

7.
$$x^2 = 100$$
$$\sqrt{x^2} = \sqrt{100}$$
$$x = \sqrt{100} \text{ or } x = -\sqrt{100}$$
$$x = 10 \quad \text{or } x = -10$$

8. $w^2 = 38$
$\sqrt{w^2} = \sqrt{38}$
$w = \sqrt{38}$ or $w = -\sqrt{38}$
$w \approx 6.2$ or $w \approx -6.2$

9. $m\angle CAB = 65°$
Since $m\angle CAB < 90°$, $\angle CAB$ is acute.

10. $m\angle M + 87° + 32° = 180°$
$m\angle M + 119 = 180$
$m\angle M + 119 - 119 = 180 - 119$
$m\angle M = 61$
The measure of $\angle M$ is 61°.

11. acute scalene because $\triangle MNP$ has all acute angles and no sides congruent

12. $c^2 = a^2 + b^2$
$c^2 = 6^2 + 8^2$
$c^2 = 36 + 64$
$c^2 = 100$
$\sqrt{c^2} = \sqrt{100}$
$c = \sqrt{100}$
$c = 10$

13. $c^2 = a^2 + b^2$
$32^2 = 15^2 + b^2$
$1024 = 225 + b^2$
$1024 - 225 = 225 - 225 + b^2$
$799 = b^2$
$\sqrt{799} = \sqrt{b^2}$
$28.3 \approx b$

14. Starting point (0, 0)
Ending point (4, −7)
$d = \sqrt{(x_2 - x_1)^2 + (y_2 - y_1)^2}$
$d = \sqrt{(4 - 0)^2 + (-7 - 0)^2}$
$d = \sqrt{4^2 + (-7)^2}$
$d = \sqrt{16 + 49}$
$d = \sqrt{65}$
$d \approx 8.1$
Brandon is about 8.1 miles from his starting point.

15. $d = \sqrt{(x_2 - x_1)^2 + (y_2 - y_1)^2}$
$d = \sqrt{(-5 - 3)^2 + (2 - 8)^2}$
$d = \sqrt{(-8)^2 + (-6)^2}$
$d = \sqrt{64 + 36}$
$d = \sqrt{100}$
$d = 10$
The distance between points A and B is 10 units.
midpoint $\overline{AB} = \left(\frac{x_1 + x_2}{2}, \frac{y_1 + y_2}{2}\right)$
midpoint $\overline{AB} = \left(\frac{3 + (-5)}{2}, \frac{8 + 2}{2}\right)$
$= (-1, 5)$
The coordinates of the midpoint of $\overline{AB}$ are $(-1, 5)$.

16. $\frac{x}{45} = \frac{50}{36}$
$36 \cdot x = 45 \cdot 50$
$36x = 2250$
$\frac{36x}{36} = \frac{2250}{36}$
$x = 62.5$
The distance from the playground to the swimming pool is 62.5 ft.

17. $\sin P = \frac{12}{13} \approx 0.9231$

18. $\tan M = \frac{5}{12} \approx 0.4167$

19. $\cos P = \frac{5}{13} \approx 0.3846$

20. C; $-49 > \quad -59 \quad > -64$
$-\sqrt{49} > -\sqrt{59} > -\sqrt{64}$
$-7 > -\sqrt{59} > -8$

Chapter 9 Standardized Test Practice

Pages 488–489

1. D; $m + n = n + m$ is true because of the commutative property of addition.

2. A; $|-8| + |5| = 13$

3. B; $A = \ell \cdot w$
$54 = \ell \cdot 9$
$\frac{54}{9} = \frac{9\ell}{9}$
$6 = \ell$
The length of the rectangle is 6 feet.

4. A; $(-2)^5 = (-2)(-2)(-2)(-2)(-2) = -32$

5. D; $\frac{1}{2} + x = \frac{5}{6}$
$\frac{1}{2} - \frac{1}{2} + x = \frac{5}{6} - \frac{1}{2}$
$x = \frac{5}{6} - \frac{1}{2}$
$x = \frac{5}{6} - \frac{3}{6}$
$x = \frac{2}{6} = \frac{1}{3}$

6. B; \$15.95 + \$6.95 = \$22.90
22.90 + (22.90 × 0.06) = total cost of books
22.90 + 1.37 = \$24.27
\$25.00 − \$24.27 = \$0.73
Elisa received \$0.73 in change.

7. A; $2s - t = s + 3t$ if $s = \frac{1}{2}$
$2\left(\frac{1}{2}\right) - t = \frac{1}{2} + 3t$
$1 - t = \frac{1}{2} + 3t$
$1 - t + t = \frac{1}{2} + 3t + t$
$1 = \frac{1}{2} + 4t$
$1 - \frac{1}{2} = \frac{1}{2} - \frac{1}{2} + 4t$
$\frac{1}{2} = 4t$
$\frac{\frac{1}{2}}{4} = \frac{4t}{4}$
$\frac{1}{8} = t$

8. C; The pattern is add consecutive multiples of 2 starting with 2. If $x = 5$ then $y = 17$ because $5 + 12 = 17$.

9. D; slope $= \frac{\text{change in } y}{\text{change in } x} = \frac{2 - 4}{2 - 1} = \frac{-2}{1} = -2$

10. B; $\sqrt{7} \approx 2.6$

11. D; $\frac{AB}{DE} = \frac{BC}{FE}$, $\frac{x}{12} = \frac{5}{8}$
$8 \cdot x = 12 \cdot 5$
$8x = 60$
$\frac{8x}{8} = \frac{60}{8}$
$x = 7.5$

12. $3 \cdot 8 + n = 27$
$$24 + n = 27$$
$$24 - 24 + n = 27 - 24$$
$$n = 3$$

13. $1.4 \times 10^{-5}, 1.04 \times 10^{-2}, 4.0 \times 10^2$

14. $\frac{6 \text{ mi}}{4 \text{ h}} = \frac{6 \text{ mi}}{240 \text{ min}} \cdot \frac{5280 \text{ ft}}{1 \text{ mi}} = \frac{31680 \text{ ft}}{240 \text{ min}} = 132 \text{ ft/min}$

15. $\frac{11}{14} \approx 0.7857$
$$0.7857 \times 100 = 78.6\%$$

16. $\frac{4}{8}$ or $\frac{1}{2}$

17. $y \leq 4x - 3$ if $y = 9$
$$9 \leq 4x - 3$$
$$9 + 3 \leq 4x - 3 + 3$$
$$12 \leq 4x$$
$$\frac{12}{4} \leq \frac{4x}{4}$$
$$3 \leq x$$

18. $y < \frac{1}{2}x$

19. slope $= \frac{\text{change in } y}{\text{change in } x} = \frac{1 - 3}{[1 - (-2)]} = -\frac{2}{3}$

20. Since $14^2 = 196$, $\sqrt{196} = 14$.

21. obtuse since the angle formed by the hands of the clock $> 90°$

22. $x + 90 + 22 = 180$
$$x + 112 = 180$$
$$x + 112 - 112 = 180 - 112$$
$$x = 68$$
$m\angle X$ is 68°.

23. $d = \sqrt{(x_2 - x_1)^2 + (y_2 - y_1)^2}$
$d = \sqrt{[6 - (-5)]^2 + (-3 - 4)^2}$
$d = \sqrt{11^2 + (-7)^2}$
$d = \sqrt{121 + 49}$
$d = \sqrt{170}$
$d \approx 13.0$ units

24a. Sample answer:

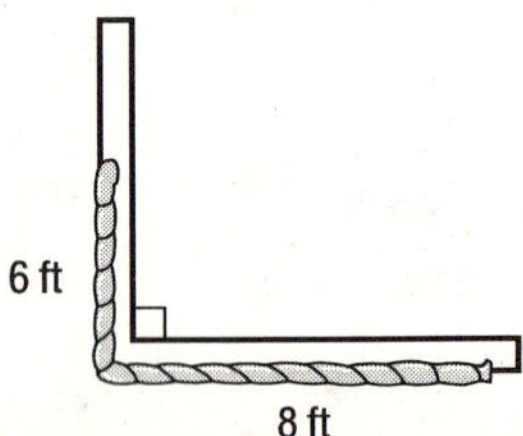

24b. $d = \sqrt{(x_2 - x_1)^2 + (y_2 - y_1)^2}$ when (0, 0) and (8, −6)
$d = \sqrt{(8 - 0)^2 + (-6 - 0)^2}$
$d = \sqrt{8^2 + (-6)^2}$
$d = \sqrt{64 + 36}$
$d = \sqrt{100}$
$d = 10$
The distance between the ends is 10 feet.

24c. Sample answer:

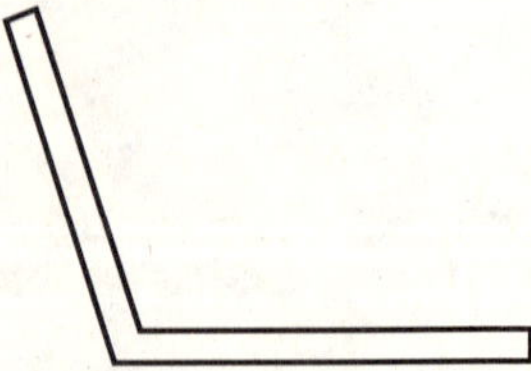

24d. No; If the measure of the angle between the walls is greater than 90°, then the length of the side opposite it will be greater than the side opposite the 90° angle.

Chapter 10 Two-Dimensional Figures

Page 491 Getting Started

1. $$\begin{aligned} x + 46 &= 90 \\ x + 46 - 46 &= 90 - 46 \\ x &= 44 \end{aligned}$$

2. $$\begin{aligned} x + 35 &= 180 \\ x + 35 - 35 &= 180 - 35 \\ x &= 145 \end{aligned}$$

3. $$\begin{aligned} 2x - 12 &= 90 \\ 2x - 12 + 12 &= 90 + 12 \\ 2x &= 102 \\ \frac{2x}{2} &= \frac{102}{2} \\ x &= 51 \end{aligned}$$

4. $$\begin{aligned} 3x - 24 &= 180 \\ 3x - 24 + 24 &= 180 + 24 \\ 3x &= 204 \\ \frac{3x}{3} &= \frac{204}{3} \\ x &= 68 \end{aligned}$$

5. $$\begin{aligned} 2x + 34 &= 90 \\ 2x + 34 - 34 &= 90 - 34 \\ 2x &= 56 \\ \frac{2x}{2} &= \frac{56}{2} \\ x &= 28 \end{aligned}$$

6. $$\begin{aligned} 4x + 44 &= 180 \\ 4x + 44 - 44 &= 180 - 44 \\ 4x &= 136 \\ \frac{4x}{4} &= \frac{136}{4} \\ x &= 34 \end{aligned}$$

7. $$\begin{aligned} 5x + 165 &= 360 \\ 5x + 165 - 165 &= 360 - 165 \\ 5x &= 195 \\ \frac{5x}{5} &= \frac{195}{5} \\ x &= 39 \end{aligned}$$

8. $$\begin{aligned} 4x + 184 &= 360 \\ 4x + 184 - 184 &= 360 - 184 \\ 4x &= 176 \\ \frac{4x}{4} &= \frac{176}{4} \\ x &= 44 \end{aligned}$$

9. $(5.5)(8) = 44$

10. $(7.5)(3.4) = 25.5$

11. $(6.3)(11.4) = 71.82$
≈ 71.8

12. $\frac{1}{2}(8)(2.5) = 10$

13. $\frac{1}{2}(4.3)(5.8) = 12.47$
≈ 12.5

14. $(3.14)(7) = 21.98$
≈ 22.0

15. $(2)(3.14)(1.7) = 10.676$
≈ 10.7

16. $2(3.1)(3.14) = 19.468$
≈ 19.5

17. $$\begin{aligned} 5\tfrac{1}{2} + 4\tfrac{2}{3} &= \frac{11}{2} + \frac{14}{3} \\ &= \frac{33}{6} + \frac{28}{6} \\ &= \frac{61}{6} \\ &= 10\tfrac{1}{6} \end{aligned}$$

18. $$\begin{aligned} 2\tfrac{1}{3} + 3\tfrac{3}{4} &= \frac{7}{3} + \frac{15}{4} \\ &= \frac{28}{12} + \frac{45}{12} \\ &= \frac{73}{12} \\ &= 6\tfrac{1}{12} \end{aligned}$$

19. $$\begin{aligned} 1\tfrac{3}{8} + 2\tfrac{1}{2} &= \frac{11}{8} + \frac{5}{2} \\ &= \frac{11}{8} + \frac{20}{8} \\ &= \frac{31}{8} \\ &= 3\tfrac{7}{8} \end{aligned}$$

20. $$\begin{aligned} 6\tfrac{1}{4} + 1\tfrac{5}{6} &= \frac{25}{4} + \frac{11}{6} \\ &= \frac{75}{12} + \frac{22}{12} \\ &= \frac{97}{12} \\ &= 8\tfrac{1}{12} \end{aligned}$$

21. $$\begin{aligned} 3\tfrac{5}{8} + 1\tfrac{3}{4} &= \frac{29}{8} + \frac{7}{4} \\ &= \frac{29}{8} + \frac{14}{8} \\ &= \frac{43}{8} \\ &= 5\tfrac{3}{8} \end{aligned}$$

22. $$\begin{aligned} 2\tfrac{3}{5} + 4\tfrac{7}{10} &= \frac{13}{5} + \frac{47}{10} \\ &= \frac{26}{10} + \frac{47}{10} \\ &= \frac{73}{10} \\ &= 7\tfrac{3}{10} \end{aligned}$$

23. $$\begin{aligned} 2\tfrac{2}{3} + 3\tfrac{5}{9} &= \frac{8}{3} + \frac{32}{9} \\ &= \frac{24}{9} + \frac{32}{9} \\ &= \frac{56}{9} \\ &= 6\tfrac{2}{9} \end{aligned}$$

24. $$\begin{aligned} 5\tfrac{2}{3} + 3\tfrac{4}{5} &= \frac{17}{3} + \frac{19}{5} \\ &= \frac{85}{15} + \frac{57}{15} \\ &= \frac{142}{15} \\ &= 9\tfrac{7}{15} \end{aligned}$$

10-1 Line and Angle Relationships

Page 492 How are parallel lines and angles related?

a. See students' work.

b. See students' work.

c. Sample answer: $m\angle 1 = 110°$, $m\angle 2 = 70°$, $m\angle 3 = 110°$, $m\angle 4 = 70°$, $m\angle 5 = 110°$, $m\angle 6 = 70°$, $m\angle 7 = 110°$, $m\angle 8 = 70°$

d. There are just two different measures.

e. Sample answer: The angles across from each other; the angles in the same position on each of the horizontal lines; the angles on opposite sides inside the horizontal lines; the angles on opposite sides above and below the parallel lines

f. Their sum is 180°.

Page 495 Check for Understanding

1. Complementary angles have a sum of 90° and supplementary angles have a sum of 180°.

2. Sample answer:

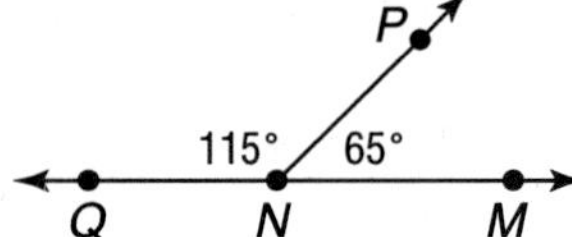

3. Since $\angle 1$ and $\angle 2$ are vertical angles, they are congruent. So, $m\angle 2 = 56°$.

4. Since $\angle 1$ and $\angle 3$ are corresponding angles, they are congruent. So, $m\angle 3 = 56°$.

5. Since $\angle 3$ and $\angle 4$ are supplementary, the sum of their measures is 180°.
$180 - 56 = 124$. So, $m\angle 4 = 124°$.

6. Since $x°$ and 140° are vertical angles, they are equal. So, $x = 140$.

7. Since $x°$ and 152° are supplementary angles, the sum of their measures is 180°.
$180 - 152 = 28$. So, $x = 28$.

8. Find the value of x.

$$m\angle N + m\angle M = 180°$$
$$3x + 2x = 180°$$
$$5x = 180°$$
$$x = 36°$$

Replace x with 36 to find the measure of each angle.

$m\angle N = 3x = 3(36) = 108$

$m\angle M = 2x = 2(36) = 72$

So, $m\angle N = 108°$ and $m\angle M = 72°$.

9. D; Since $\angle A$ and $\angle B$ are complementary,

$$m\angle A + m\angle B = 90°$$
$$m\angle A + m\angle B = 90°$$
$$m\angle A + 26° = 90°$$
$$m\angle A + 26° - 26° = 90° - 26°$$
$$m\angle A = 64°$$

Pages 496–497 Practice and Apply

10. Since $\angle 1$ and $\angle 4$ are alternate exterior angles, they are congruent. So, $m\angle 1 = 53°$.

11. Since $\angle 4$ and $\angle 5$ are supplementary, the sum of their measures is 180°.
$180 - 53 = 127$. So, $m\angle 5 = 127°$.

12. Since $\angle 7$ and $\angle 5$ are alternate exterior angles, they are congruent. So, $m\angle 7 = 127°$.

13. Since $\angle 1$ and $\angle 8$ are vertical angles, they are congruent. So, $m\angle 8 = 53°$.

14. Since $\angle 2$ and $\angle 4$ are vertical angles, they are congruent. So, $m\angle 2 = 53°$.

15. Since $\angle 3$ and $\angle 7$ are vertical angles, they are congruent. So, $m\angle 3 = 127°$.

16. Since $x°$ and 45° are vertical angles, they are congruent. So, $x = 45$.

17. Since $x°$ and 148° are vertical angles, they are congruent. So, $x = 148$.

18. Since $x°$ and 31° are complementary, the sum of their measures is 90°.

$$x + 31 = 90$$
$$x + 31 - 31 = 90 - 31$$
$$x = 59$$

19. Since $x°$ and 5° are supplementary angles, the sum of their measures is 180°.

$$x + 5 = 180$$
$$x + 5 - 5 = 180 - 5$$
$$x = 175$$

20. Since $8x°$ and $4x°$ are supplementary, the sum of their measures is 180°.

$$8x + 4x = 180$$
$$12x = 180$$
$$\frac{12x}{12} = \frac{180}{12}$$
$$x = 15$$

21. Since $5x°$ and $5x°$ are complementary, the sum of their measures is 90°.

$$5x + 5x = 90$$
$$10x = 90$$
$$\frac{10x}{10} = \frac{90}{10}$$
$$x = 9$$

22. Since $\angle A$ and $\angle B$ are complementary,

$$m\angle A + m\angle B = 90°$$
$$m\angle A + m\angle B = 90°$$
$$m\angle A + 17° = 90$$
$$m\angle A + 17° - 17° = 90° - 17°$$
$$m\angle A = 73°$$

23. Since angles P and Q are supplementary,

$$m\angle P + m\angle Q = 180°$$
$$m\angle P + m\angle Q = 180°$$
$$m\angle P + 139° = 180°$$
$$m\angle P + 139° - 139° = 180° - 139°$$
$$m\angle P = 41°$$

24. Find the value of x.

$$m\angle J + m\angle K = 90°$$
$$(x - 9) + (x + 5) = 90°$$
$$2x - 4 = 90°$$
$$2x = 94°$$
$$x = 47°$$

Replace x with 47 to find the measure of each angle.

$m\angle J = x - 9 = 47 - 9 = 38$ $\quad$ $m\angle K = x + 5 = 47 + 5 = 52$

So, $m\angle J = 38°$ and $m\angle K = 52°$.

25. Find the value of x.

$$m\angle E + m\angle F = 180°$$
$$(2x + 15) + (5x - 38) = 180°$$
$$7x - 23 = 180°$$
$$7x = 203°$$
$$\frac{7x}{7} = \frac{203°}{7}$$
$$x = 29°$$

Replace x with 29 to find the measure of each angle.

$m\angle E = 2x + 15 = 2(29) + 15 = 73$ $\quad$ $m\angle F = 5x - 38 = 5(29) - 38 = 107$

So, $m\angle E = 73°$ and $m\angle F = 107°$.

26. Since $\angle 3$ and $\angle 4$ are alternate interior angles, they are congruent. So, $m\angle 3 = 105°$.
Since $\angle 2$ and $\angle 3$ are vertical angles, they are congruent. So, $m\angle 2 = 105°$.
Since $\angle 5$ and $\angle 6$ are alternate exterior angles, they are congruent. So, $m\angle 5 = 75°$.
Since $\angle 7$ and $\angle 6$ are vertical angles, they are congruent. So, $m\angle 7 = 75°$.
Since $\angle 8$ and $\angle 2$ are supplementary, $m\angle 8 = 180° - m\angle 2 = 180 - 105 = 75$. Therefore, $m\angle 8 = 75°$.

27. $\overleftrightarrow{XY}$ and $\overleftrightarrow{BC}$ are parallel.
Since $\angle YXB$ and $\angle YXA$ are supplementary,

$$m\angle YXB + m\angle YXA = 180°$$
$$68° + m\angle YXA = 180°$$
$$m\angle YXA = 112°$$

Since $\angle YXA$ and $\angle XBC$ are corresponding angles, they are congruent. So, $m\angle XBC = 112°$.

28. $m\angle YXB + m\angle XBC = 68° + 112° = 180°$

So, they are supplementary angles.

29. Since $\angle 2$ and $\angle 4$ are vertical angles, they are congruent.

$$m\angle 2 = m\angle 4$$
$$2x + 3 = 4x - 7$$
$$2x - 2x + 3 = 4x - 2x - 7$$
$$3 = 2x - 7$$
$$10 = 2x$$
$$5 = x$$

30. Since $\angle 8$ and $\angle 5$ are supplementary, their sum is 180°.

$$m\angle 8 + m\angle 5 = 180°$$
$$(4x - 32) + (5x + 50) = 180$$
$$9x + 18 = 180$$
$$9x = 162$$
$$x = 18$$

31. Since $\angle 7$ and $\angle 3$ are corresponding angles, they are congruent.

$$m\angle 7 = m\angle 3$$
$$10x + 15 = 7x + 42$$
$$10x - 7x + 15 = 7x - 7x + 42$$
$$3x + 15 = 42$$
$$3x = 27$$
$$x = 9$$

32. Let x represent the measure of the angle. Then, $180 - x$ = the supplement of the angle and $90 - x$ = the complement of the angle.

$$180 - x = 4(90 - x) - 15$$
$$180 - x = 360 - 4x - 15$$
$$180 - x = 345 - 4x$$
$$180 - x + 4x = 345 - 4x + 4x$$
$$180 + 3x = 345$$
$$3x = 165$$
$$x = 55°$$

33. They are supplementary.

34. Angles and parallel lines are related in that when two parallel lines are cut by a transversal, angles are formed. Answers should include the following.

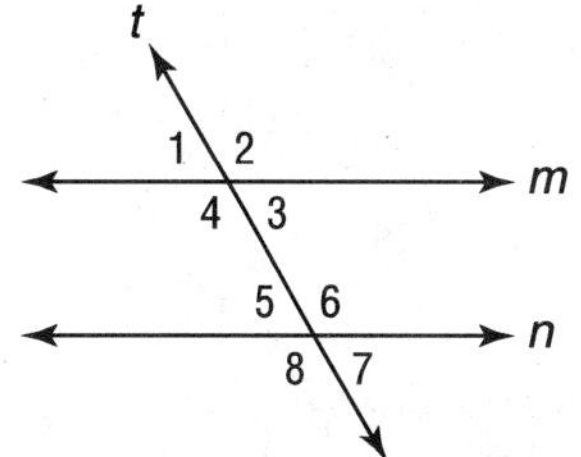

- congruent angles:
  $\angle 1, \angle 3, \angle 5, \angle 7$ and $\angle 2, \angle 4, \angle 6, \angle 8$
  supplementary angles:
  $\angle 1$ and $\angle 2$, $\angle 1$ and $\angle 4$, $\angle 3$ and $\angle 4$, $\angle 2$ and $\angle 3$, $\angle 5$ and $\angle 6$, $\angle 6$ and $\angle 7$, $\angle 8$ and $\angle 7$, $\angle 5$ and $\angle 8$, $\angle 1$ and $\angle 6$, $\angle 1$ and $\angle 8$, $\angle 2$ and $\angle 5$, $\angle 2$ and $\angle 7$, $\angle 3$ and $\angle 6$, $\angle 3$ and $\angle 8$, $\angle 4$ and $\angle 5$, $\angle 4$ and $\angle 7$

35. A; The alternate interior angle to the 45° angle, formed by extending the top rail, is supplemental to x.
So, $x + 45° = 180°$
$x = 135°$.

36. C; Since the second vertical post is parallel to the first vertical post, the 45° angle is a corresponding angle to the angle formed by the second post. So, the angle is 45°.

37. Sample answer: Both graphs intersect to form right angles.

38. They are negative reciprocals of each other.

39. Sample answer: The slopes of the graphs of perpendicular lines are negative reciprocals of each other.

Page 497 Maintain Your Skills

40. COS 21 ENTER 0.9335804265
$\cos 21° = 0.9336$

41. SIN 63 ENTER 0.8910065242
$\sin 63° = 0.8910$

42. TAN 38 ENTER 0.7812856265
$\tan 38° = 0.7813$

43. Since the triangles are similar, their sides are proportional.
$\frac{15}{10} = \frac{x}{5}$
$15 \cdot 5 = 10x$
$75 = 10x$
$7.5 = x$

44. $6a + (-18)a = [6 + (-18)]a$
$= -12a$

45. $-5m + (-4)m = [-5 + (-4)]m$
$= -9m$

46.

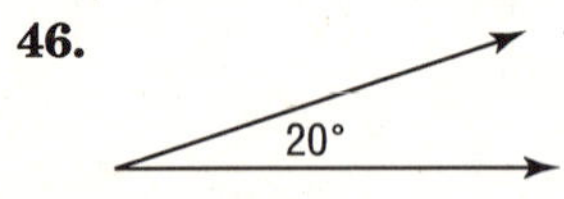

47.

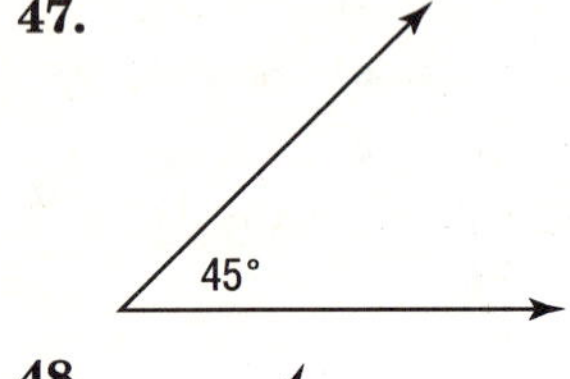

48.

49.

50.

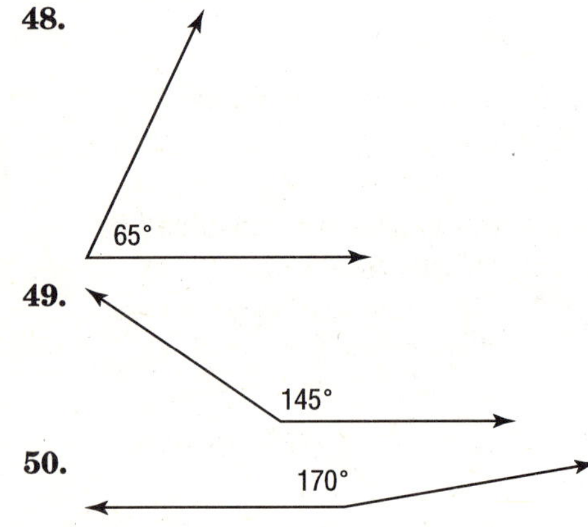

Pages 498–499 Algebra Activity (Follow-Up of Lesson 10-1)

1. See students' work.
2. See students' work.
3. to divide into two equal parts
4. See students' work.
5. They are the same.
6. $\overrightarrow{NQ}$ divides $\angle MNP$ into two equal angles.

10-2 Congruent Triangles

Page 500 Where are congruent triangles present in nature?

a. See students' work.

b. $AB = DF, AC = DE, BC = FE$

c. $m\angle A = m\angle D, m\angle B = m\angle F, m\angle C = m\angle E$

d. Since the angles have the same measures and the sides have the same length, the triangles have the same size and shape.

Page 502 Check for Understanding

1. They have the same size and shape.
2. Sample answer:

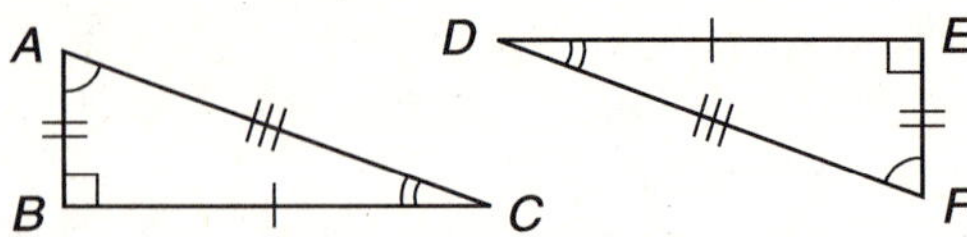

3. $\angle J \cong \angle C, \angle K \cong \angle B, \angle M \cong \angle G, \overline{KM} \cong \overline{BG}, \overline{MJ} \cong \overline{GC}, \overline{KJ} \cong \overline{BC}; \triangle KMJ \cong \triangle BGC$
4. $\angle C \cong \angle D, \angle B \cong \angle B, \angle E \cong \angle A, \overline{CB} \cong \overline{DB}, \overline{BE} \cong \overline{BA}, \overline{CE} \cong \overline{DA}; \triangle CBE \cong \triangle DBA$
5. $\angle J$ corresponds to $\angle M$, so $\angle J \cong \angle M$.
6. $\angle A$ corresponds to $\angle K$, so $\angle A \cong \angle K$.
7. J corresponds to M, and D corresponds to N, so $\overline{JD} \cong \overline{MN}$.
8. A corresponds to K, and N corresponds to J, so $\overline{AN} \cong \overline{KD}$.
9. $\overline{TC}$ corresponds to $\overline{RC}$, so $\overline{TC} \cong \overline{RC}$. Since $TC = 15$ feet, $RC = 15$ feet.

Pages 503–504 Practice and Apply

10. $\angle D \cong \angle P, \angle F \cong \angle Q, \angle E \cong \angle R, \overline{DF} \cong \overline{PQ}, \overline{FE} \cong \overline{QR}, \overline{DE} \cong \overline{PR}; \triangle DFE \cong \triangle PQR$
11. $\angle K \cong \angle N, \angle J \cong \angle P, \angle M \cong \angle M, \overline{KJ} \cong \overline{NP}, \overline{JM} \cong \overline{PM}, \overline{KM} \cong \overline{NM}; \triangle KJM \cong \triangle NPM$
12. $\angle D \cong \angle E, \angle B \cong \angle B, \angle A \cong \angle C, \overline{DB} \cong \overline{EB}, \overline{BA} \cong \overline{BC}, \overline{DA} \cong \overline{EC}, \triangle DBA \cong \triangle EBC$
13. $\angle Z \cong \angle S, \angle W \cong \angle T, \angle Y \cong \angle R, \overline{ZW} \cong \overline{ST}, \overline{WY} \cong \overline{TR}, \overline{ZY} \cong \overline{SR}; \triangle ZWY \cong \triangle STR$
14. always
15. sometimes; for example, two noncongruent triangles with sides of lengths 3, 5, and 7, and 4, 5, and 6 both have perimeters of 15.
16. Since $\triangle TRU \cong \triangle SRU$, $TU \cong SU$. Since $TU + SU = 32$ feet, each equals 16 feet.
17. Since $\triangle TRU \cong \triangle SRU$, $\angle T \cong \angle S$. So, $\angle T = 30°$.
18. $\angle G$ corresponds to $\angle D$, so $\angle G \cong \angle D$.
19. $\angle C$ corresponds to $\angle F$, so $\angle C \cong \angle F$.
20. $\angle M$ corresponds to $\angle R$, so $\angle M \cong \angle R$.
21. $\angle Q$ corresponds to $\angle A$, so $\angle Q \cong \angle A$.
22. H corresponds to B, and G corresponds to D, so $\overline{HG} \cong \overline{BD}$.
23. B corresponds to H, and C corresponds to F, so $\overline{BC} \cong \overline{HF}$.
24. D corresponds to G, and C corresponds to F, so $\overline{DC} \cong \overline{GF}$.
25. G corresponds to D, and F corresponds to C, so $\overline{GF} \cong \overline{DC}$.

26. G corresponds to D, F corresponds to C, and H corresponds to B, so $\angle GFH \cong \angle DCB$.
27. $\triangle ABC \cong \triangle DEC$, and $\overline{BC} \cong \overline{EC}$, so $x = 20$.
28. $\triangle JKM \cong \triangle RST$, and $\overline{JM} \cong \overline{RT}$, so $x = 18$.
29. $\overline{AC} \cong \overline{XZ}$, so $3x + 12 = 30$.

$$3x + 12 = 30$$
$$3x = 18$$
$$x = 6$$

30. Sample answer: $\triangle WCQ$
31. Sample answer: $\triangle ZQN$
32. Sample answer: $\triangle CYP$
33. Sample answer: $\triangle WQO$
34. $\triangle ABH \cong \triangle IJG$; $\triangle ACG \cong \triangle IEF$
35. Congruent triangles can be found on objects in nature like leaves and animals. Answers should include the following.
    - Congruent triangles are triangles with the same angle measures and the same side lengths.
    - Sample answer: A bird's wings when extended is an example of congruent angles. Another example of congruent angles is the wings of a butterfly.
36. D; C corresponds to H, A corresponds to E, and D corresponds to F, so $\angle CAD \cong \angle HEF$.
37. D; $\overline{UY} \cong \overline{XY}$, so $UY = 26$.

Pages 504 Maintain Your Skills

38. Since angles P and Q are supplementary,

$$m\angle P + m\angle Q = 180°$$
$$m\angle P + 129° = 180°$$
$$m\angle P = 51°$$

39. [COS] 83 [ENTER] 0.1218693434, so cos 83° is about 0.1219.
40. [SIN] 39.7 [ENTER] 0.6387678175, so sin 39.7° is about 0.6388.
41. [TAN] 49.2 [ENTER] 1.15851115, so tan 49.2° is about 1.1585.
42. $2\frac{1}{2} = \frac{5}{2}$, so the multiplicative inverse is $\frac{2}{5}$, since $\frac{5}{2} \cdot \frac{2}{5} = 1$.
43.

$$-5a - 6 = 24$$
$$-5a - 6 + 6 = 24 + 6$$
$$-5a = 30$$
$$\frac{-5a}{-5} = \frac{30}{-5}$$
$$a = -6$$

44–49.

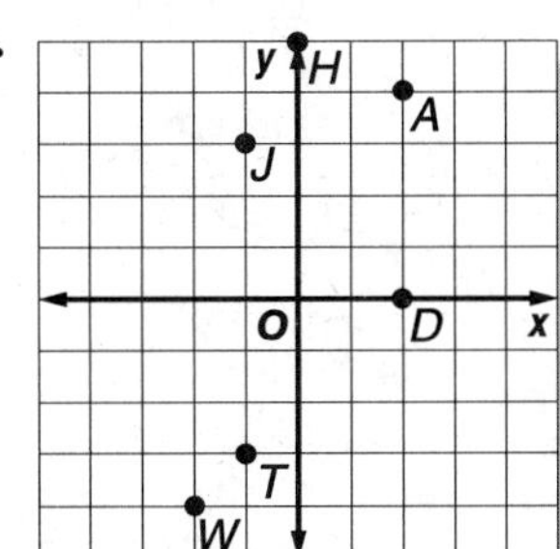

Page 505 Algebra Activity (Preview of Lesson 10-3)

1.

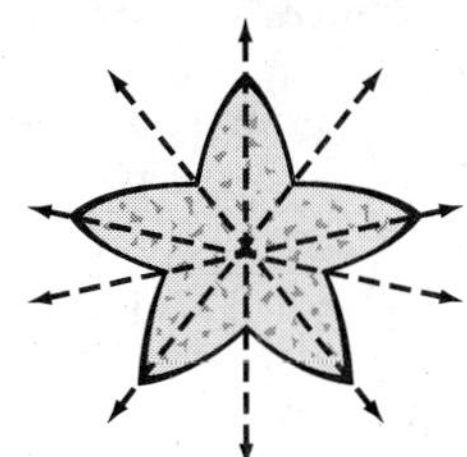

2. none
3.
4. yes
5. yes
6. no
7. Sample answer: capital letter H, yield sign, square

10-3 Transformations on the Coordinate Plane

Page 506 How are transformations involved in recreational activities?

a. Sample answer: The skateboard lands on an opposite position from where it started.
b. Sample answer: a back-and-forth motion
c. Sample answer: A scooter moves from one position to another position.

Page 509 Check for Understanding

1. The figure is moved 5 units to the right and 2 units down.

2. Sample answer:

vertex	5 right	translation	**90° counterclockwise rotation** (switch coordinates; multiply the first by -1)
$A(-5, -1)$	$+ (5, 0) \rightarrow$	$A'(0, -1)$	$\rightarrow (-1, 0) \rightarrow (1, 0)$
$B(-5, -4)$	$+ (5, 0) \rightarrow$	$B'(0, -4)$	$\rightarrow (-4, 0) \rightarrow (4, 0)$
$C(-3, -4)$	$+ (5, 0) \rightarrow$	$C'(2, -4)$	$\rightarrow (-4, 2) \rightarrow (4, 2)$

The coordinates of the vertices of $A'B'C'$ after rotation are $A'(1, 0)$, $B'(4, 0)$, and $C'(4, 2)$.

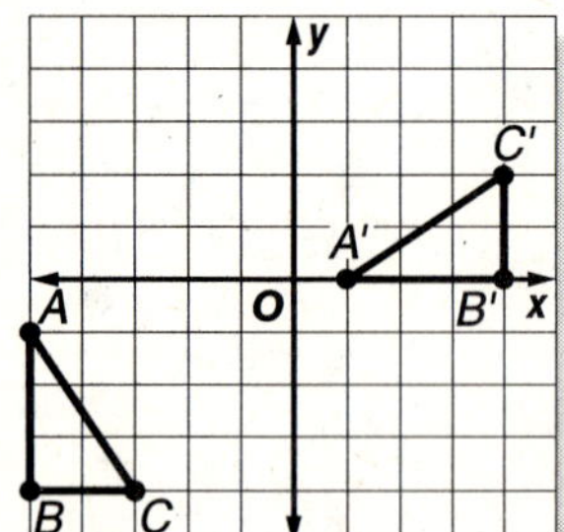

3.

vertex	4 right, 2 down	translation
$R(-5, 3)$	$+ (4, -2) \rightarrow$	$R'(-1, 1)$
$S(-1, 3)$	$+ (4, -2) \rightarrow$	$S'(3, 1)$
$T(-1, 0)$	$+ (4, -2) \rightarrow$	$T'(3, -2)$
$U(-5, 0)$	$+ (4, -2) \rightarrow$	$U'(-1, -2)$

The coordinates of the vertices of $R'S'T'U'$ are $R'(-1, 1)$, $S'(3, 1)$, $T'(3, -2)$, and $U'(-1, -2)$.

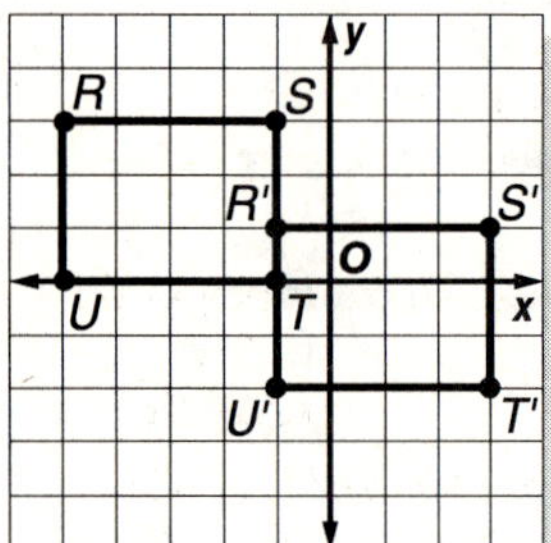

4. To find the coordinates, use the same y-coordinate and multiply the x-coordinate by -1.

vertex	reflection
$W(-4, 3) \rightarrow (-1 \cdot -4, 3)$	$\rightarrow W'(4, 3)$
$X(-1, 5) \rightarrow (-1 \cdot -1, 5)$	$\rightarrow X'(1, 5)$
$Z(-1, -2) \rightarrow (-1 \cdot -1, -2)$	$\rightarrow Z'(1, -2)$
$Y(-3, -1) \rightarrow (-1 \cdot -3, -1)$	$\rightarrow Y'(3, -1)$

The coordinates of the vertices of the reflected figure are $W'(4, 3)$, $X'(1, 5)$, $Z'(1, -2)$, and $Y'(3, -1)$.

5. To rotate the figure, switch the coordinates of each vertex and multiply the first by -1.
$A(-1, 2) \rightarrow A'(-2, -1)$
$B(3, 3) \rightarrow B'(-3, 3)$
$C(4, -2) \rightarrow C'(2, 4)$

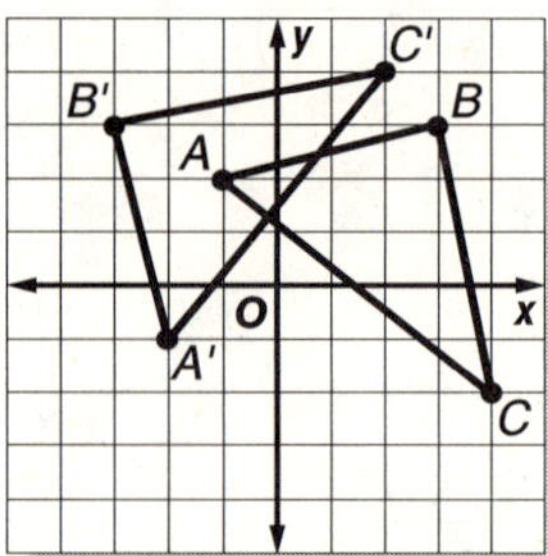

6. translation or reflection

Pages 510–511 Practice and Apply

7.

vertex	2 right, 3 up	translation
$A(-4, 2)$	$+ (2, 3) \rightarrow$	$A'(-2, 5)$
$B(-1, 0)$	$+ (2, 3) \rightarrow$	$B'(1, 3)$
$C(-4, -3)$	$+ (2, 3) \rightarrow$	$C'(-2, 0)$

The coordinates of the vertices are $A'(-2, 5)$, $B'(1, 3)$, and $C'(-2, 0)$.

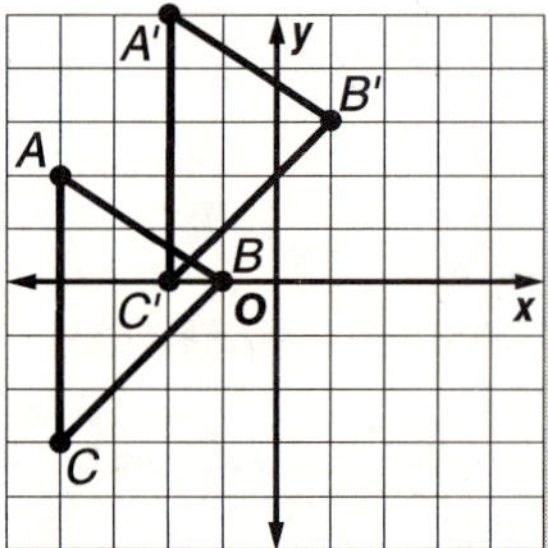

8.

vertex	left 4, up 3	translation
$H(0, 0)$	$+ (-4, 3) \rightarrow$	$H'(-4, 3)$
$J(4, -1)$	$+ (-4, 3) \rightarrow$	$J'(0, 2)$
$I(4, -3)$	$+ (-4, 3) \rightarrow$	$I'(0, 0)$
$K(0, -4)$	$+ (-4, 3) \rightarrow$	$K'(-4, -1)$

The coordinates of the vertices are $H'(-4, 3)$, $J'(0, 2)$, $I'(0, 0)$, and $K'(-4, -1)$.

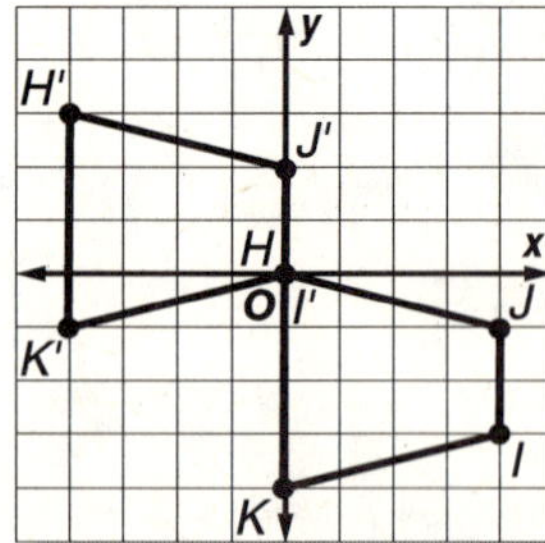

9.

vertex	right 5, up $2\frac{1}{2}$	translation
$D(-4, 1)$	$+ \left(5, 2\frac{1}{2}\right) \rightarrow$	$D'\left(1, 3\frac{1}{2}\right)$
$E(-2, 2)$	$+ \left(5, 2\frac{1}{2}\right) \rightarrow$	$E'\left(3, 4\frac{1}{2}\right)$
$F(-2, -3)$	$+ \left(5, 2\frac{1}{2}\right) \rightarrow$	$F'\left(3, -\frac{1}{2}\right)$
$G(-4, -4)$	$+ \left(5, 2\frac{1}{2}\right) \rightarrow$	$G'\left(1, -1\frac{1}{2}\right)$

The coordinates of the vertices are $D'\left(1, 3\frac{1}{2}\right)$, $E'\left(3, 4\frac{1}{2}\right)$, $F'\left(3, -\frac{1}{2}\right)$, and $G\left(1, -1\frac{1}{2}\right)$.

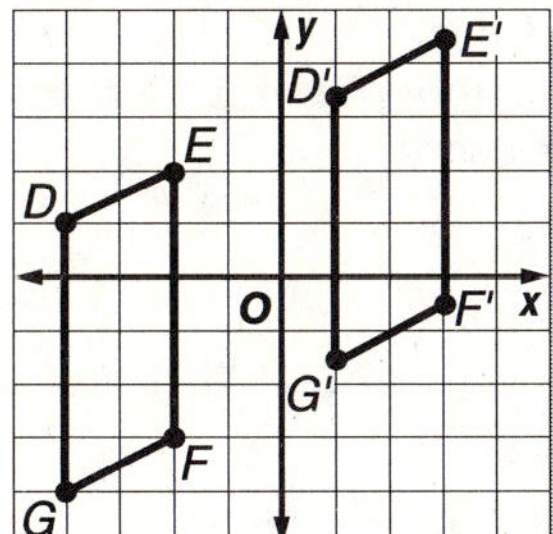

10.

vertex	4 down	translation
$D(1, 2)$	$+ (0, -4) \rightarrow$	$D'(1, -2)$
$E(1, 4)$	$+ (0, -4) \rightarrow$	$E'(1, 0)$
$F(-1, 2)$	$+ (0, -4) \rightarrow$	$F'(-1, -2)$
$G(-4, 4)$	$+ (0, -4) \rightarrow$	$G'(-4, 0)$

The coordinates of the vertices are $D'(1, -2)$, $E'(1, 0)$, $F'(-1, -2)$, and $G'(-4, 0)$.

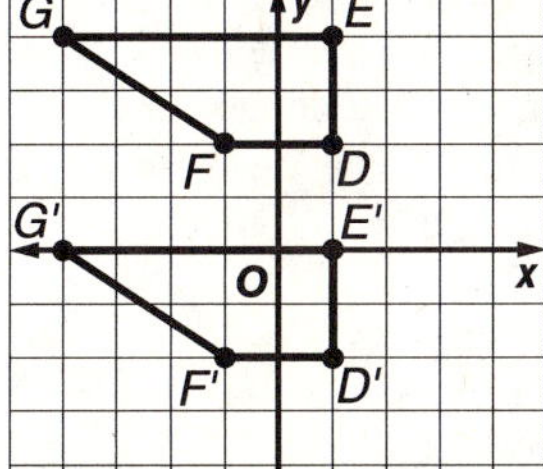

11. Use the same x-coordinate and multiply the y-coordinate by -1.

vertex		reflection
$R(-4, 3)$	$\rightarrow (-4, -1 \cdot 3)$	$\rightarrow R'(-4, -3)$
$S(0, 3)$	$\rightarrow (0, -1 \cdot 3)$	$\rightarrow S'(0, -3)$
$T(-4, -1)$	$\rightarrow (-4, -1 \cdot -1)$	$\rightarrow T'(-4, 1)$

The coordinates of the vertices are $R'(-4, -3)$, $S'(0, -3)$, and $T'(-4, 1)$.

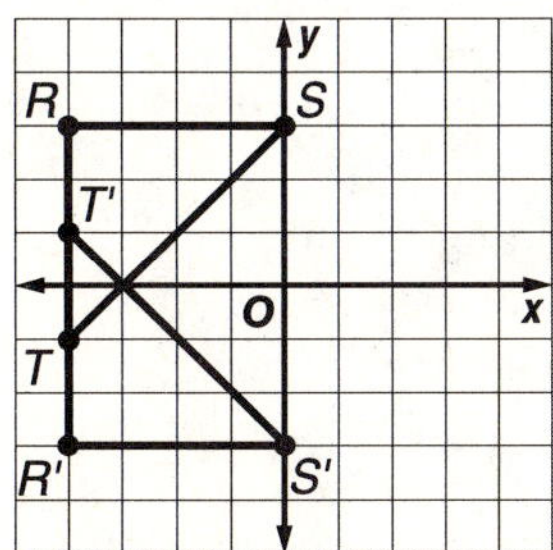

12. Use the same y-coordinate and multiply the x-coordinate by -1.

vertex		reflection
$A(1, 3)$	$\rightarrow (-1 \cdot 1, 3)$	$\rightarrow A'(-1, 3)$
$B(4, 4)$	$\rightarrow (-1 \cdot 4, 4)$	$\rightarrow B'(-4, 4)$
$C(4, 1)$	$\rightarrow (-1 \cdot 4, 1)$	$\rightarrow C'(-4, 1)$
$D(2, -1)$	$\rightarrow (-1 \cdot 2, -1)$	$\rightarrow D'(-2, -1)$

The coordinates of the vertices are $A'(-1, 3)$, $B'(-4, 4)$, $C'(-4, 1)$, and $D'(-2, -1)$.

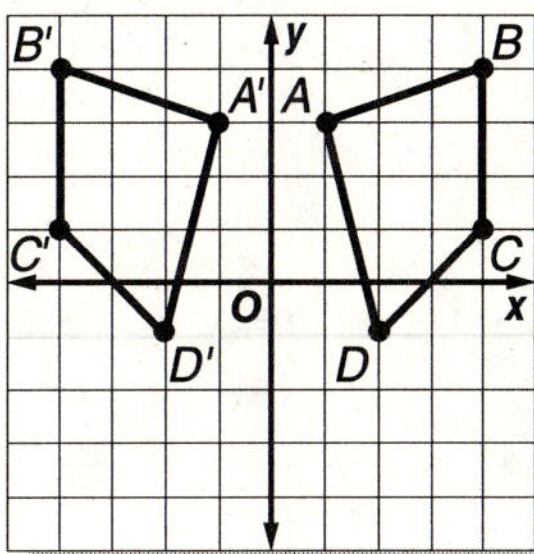

13. Use the same x-coordinate and multiply the y-coordinate by -1.

vertex		reflection
$M(-1, 2)$	$\rightarrow (-1, -1 \cdot 2)$	$\rightarrow M'(-1, -2)$
$N(4, 4)$	$\rightarrow (4, -1 \cdot 4)$	$\rightarrow N'(4, -4)$
$O(3, 2)$	$\rightarrow (3, -1 \cdot 2)$	$\rightarrow O'(3, -2)$
$P(3, 0)$	$\rightarrow (3, -1 \cdot 0)$	$\rightarrow P'(3, 0)$
$Q(0, 0)$	$\rightarrow (0, -1 \cdot 0)$	$\rightarrow Q'(0, 0)$

The coordinates of the vertices are $M'(-1, -2)$, $N'(4, -4)$, $O'(3, -2)$, $P'(3, 0)$, and $Q'(0, 0)$.

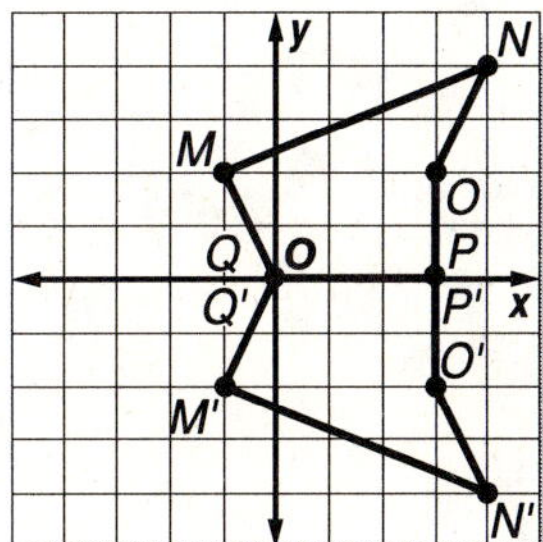

14. Use the same y-coordinate and multiply the x-coordinate by -1.

vertex		reflection
$W(-3, -3)$	$\rightarrow (-1 \cdot -3, -3)$	$\rightarrow W'(3, -3)$
$X(0, -4)$	$\rightarrow (-1 \cdot 0, -4)$	$\rightarrow X'(0, -4)$
$Y(4, -2)$	$\rightarrow (-1 \cdot 4, -2)$	$\rightarrow Y'(-4, -2)$
$Z(2, -1)$	$\rightarrow (-1 \cdot 2, -1)$	$\rightarrow Z'(-2, -1)$

The coordinates of the vertices are $W'(3, -3)$, $X'(0, -4)$, $Y'(-4, -2)$, and $Z'(-2, -1)$.

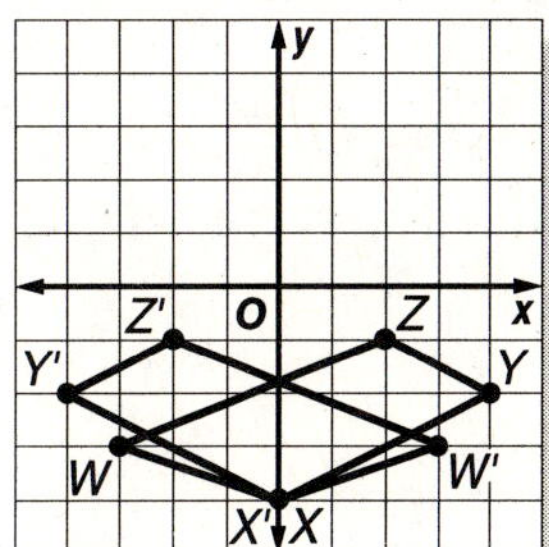

15. Switch the coordinates of each point and then multiply the first by -1.
$A(1, -1) \rightarrow A'(1, 1)$
$B(3, 0) \rightarrow B'(0, 3)$
$C(5, -2) \rightarrow C'(2, 5)$
$D(3, -4) \rightarrow D'(4, 3)$
$E(1, -4) \rightarrow E'(4, 1)$
The coordinates of the vertices are $A'(1, 1)$, $B'(0, 3)$, $C'(2, 5)$, $D'(4, 3)$, and $E'(4, 1)$.

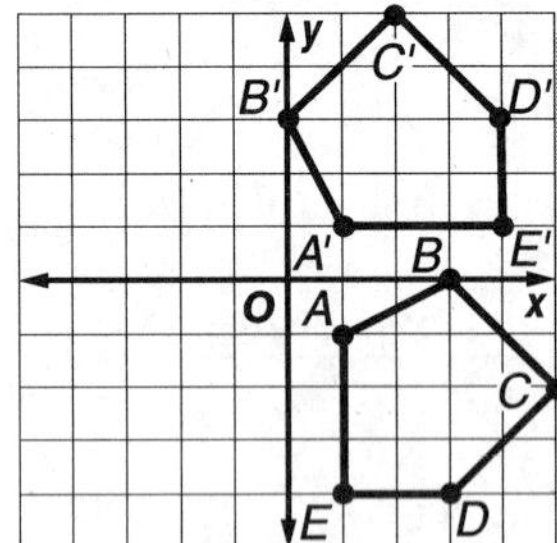

16. Multiply both the coordinates of each point by -1.
$A(1, -1) \rightarrow A'(-1, 1)$
$B(3, 0) \rightarrow B'(-3, 0)$
$C(5, -2) \rightarrow C'(-5, 2)$
$D(3, -4) \rightarrow D'(-3, 4)$
$E(1, -4) \rightarrow E'(-1, 4)$

17. Switch the coordinates of each point and then multiply the second by -1.
$A(1, -1) \rightarrow A'(-1, -1)$
$B(3, 0) \rightarrow B'(0, -3)$
$C(5, -2) \rightarrow C'(-2, -5)$
$D(3, -4) \rightarrow D'(-4, -3)$
$E(1, -4) \rightarrow E'(-4, -1)$
The coordinates of the vertices are $A'(-1, -1)$, $B'(0, -3)$, $C'(-2, -5)$, $D'(-4, -3)$, and $E'(-4, -1)$.

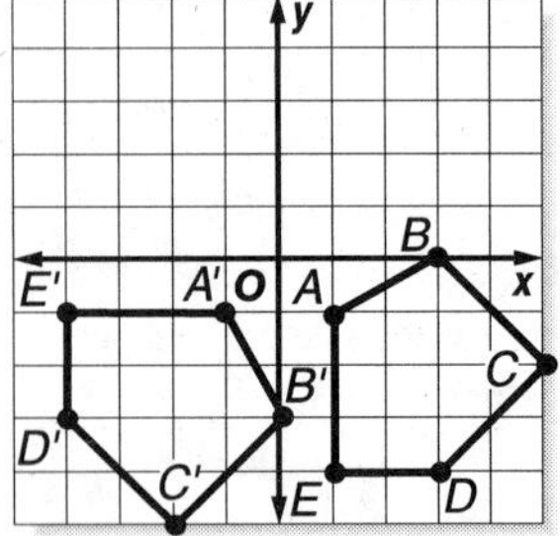

18. rotation
19. translation
20. reflection
21. Sample answer:

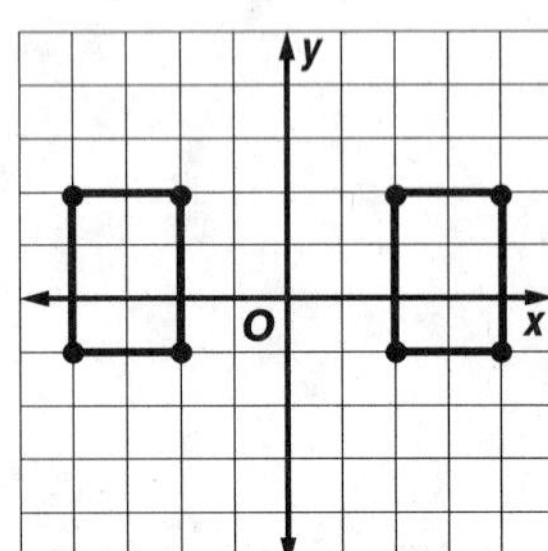

The image of the figure's reflection is the same as the image of its translation to the right 6 units.

22. translation

23. reflection

24. To get the vertices of the rotated image, the coordinates were switched and the first coordinate was multiplied by -1. To get the coordinates before the rotation, multiply the first coordinate by -1, then switch the coordinates.
$A'(3, 2) \rightarrow A'(2, -3)$
$B'(0, 4) \rightarrow B'(4, 0)$
$C'(5, 5) \rightarrow C'(5, -5)$
The coordinates of the vertices of $A'B'C'$ are $A'(2, -3)$, $B'(4, 0)$, and $C'(5, -5)$.

25. Since many of the movements used in recreational activities involve rotating, sliding, and flipping, they are examples of transformations. Answers should include the following.
- a translation is a slide, a reflection is a mirror-image, and a rotation is a turn.
- sample answer: the swing represents a rotation, the scooter represents a translation, and the skateboard represents reflection.

26. B; The five coordinates of the translated image can be found by adding -4 to each x-coordinate, and -3 to each y-coordinate.
$(1, 2) + (-4, -3) \rightarrow (-3, -1)$
$(2, 4) + (-4, -3) \rightarrow (-2, 1)$
$(5, 1) + (-4, -3) \rightarrow (1, -2)$
$(3, 1) + (-4, -3) \rightarrow (-1, -2)$
$(2, 0) + (-4, -3) \rightarrow (-2, -3)$
Choice B is *not* one of the vertices.

27. C; $(x, y) \rightarrow (x - 4, y - 3)$ means to subtract 4 from each x-coordinate and to subtract 3 from each y-coordinate.

Page 511 Maintain Your Skills

28. $\angle D$ corresponds to $\angle A$, so $\angle D \cong \angle A$.

29. A corresponds to D and C corresponds to F, so $\overline{AC} \cong \overline{DF}$.

30. D corresponds to A and E corresponds to B, so $\overline{DE} \cong \overline{AB}$.

31. x and 110° are vertical angles, so they are congruent. So, $x = 110°$.

32. x and 26° are complementary angles, so
$x + 26 = 90.$
$x + 26 = 90$
$x = 64°$

33. $x - 3.4 \geq 6.2$
$x - 3.4 + 3.4 \geq 6.2 + 3.4$
$x \geq 9.6$

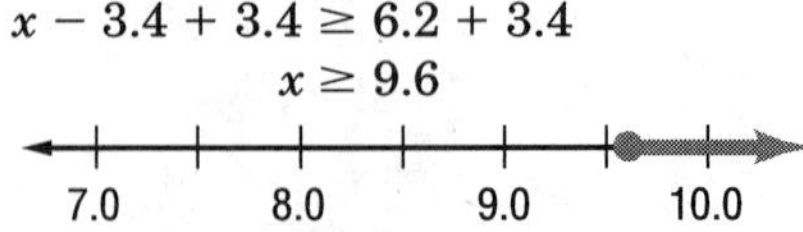

34. $|-4| - |3| = 4 - 3$
$= 1$

35. $2x + 134 = 360$
$2x + 134 - 134 = 360 - 134$
$2x = 226$
$\frac{2x}{2} = \frac{226}{2}$
$x = 113$

36. $3x + 54 = 360$
$3x + 54 - 54 = 360 - 54$
$3x = 306$
$\frac{3x}{3} = \frac{306}{3}$
$x = 102$

37. $5x + 125 = 360$
$5x + 125 - 125 = 360 - 125$
$5x = 235$
$\frac{5x}{5} = \frac{235}{5}$
$x = 47$

38. $4x + 92 = 360$
$4x + 92 - 92 = 360 - 92$
$4x = 268$
$\frac{4x}{4} = \frac{268}{4}$
$x = 67$

39. $2x + 148 = 360$
$2x + 148 - 148 = 360 - 148$
$2x = 212$
$\frac{2x}{2} = \frac{212}{2}$
$x = 106$

40. $6x + 102 = 360$
$6x + 102 - 102 = 360 - 102$
$6x = 258$
$\frac{6x}{6} = \frac{258}{6}$
$x = 43$

Page 512 Algebra Activity (Follow-up of Lesson 10-3)

1. The corresponding angles have the same measure.
2. The lengths of the sides of the enlarged trapezoid are twice the length of the corresponding sides of the smaller trapezoid.
3. 2:1; The ratios are the same.
4. See students' work. The results are not the same. By multiplying the coordinates of trapezoid *ABCD* by a scale factor of one-half, the resulting figure is a trapezoid that has sides that are half the size of the original trapezoid.
5. If the scale factor is greater than 1, the image is an enlargement. If the scale factor is less than 1, the image is a reduction.
6. Whereas translations, reflections, and rotations produce congruent figures, dilations produce enlargements or reductions of the figure.
7. Multiply the coordinates of each vertex by 3.
$M(1, 2) \rightarrow M'(3, 6)$
$N(2, 0) \rightarrow N'(6, 0)$
$M'(3, 6); N'(6, 0);$

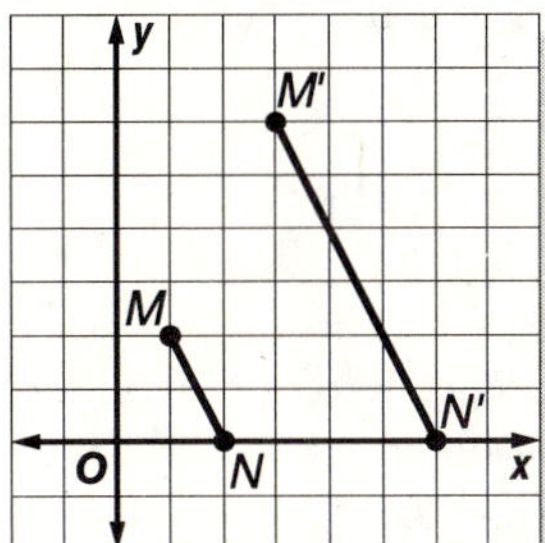

8. Multiply the coordinates of each vertex by $\frac{1}{4}$.
$A(-4, 4) \rightarrow A'(-1, 1)$
$B(4, 2) \rightarrow B'\left(1, \frac{1}{2}\right)$
$C(4, 8) \rightarrow C'(1, 2)$
$A'(-1, 1); B'\left(1, \frac{1}{2}\right); C'(1, 2);$

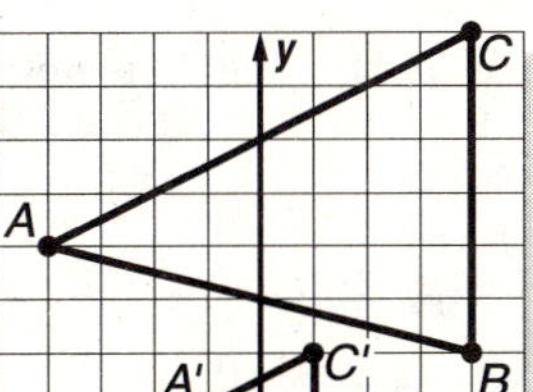

9. Multiply the coordinates of each vertex by $1\frac{1}{2}$.
$D(-4, -2) \rightarrow D'(-6, -3)$
$E(-5, -5) \rightarrow E'(-7.5, -7.5)$
$F(0, -4) \rightarrow F'(0, -6)$
$G(-1, -1) \rightarrow G'(-1.5, -1.5)$
$D'(-6, -3); E'(-7.5, -7.5); F'(0, -6);$
$G'(-1.5, -1.5);$

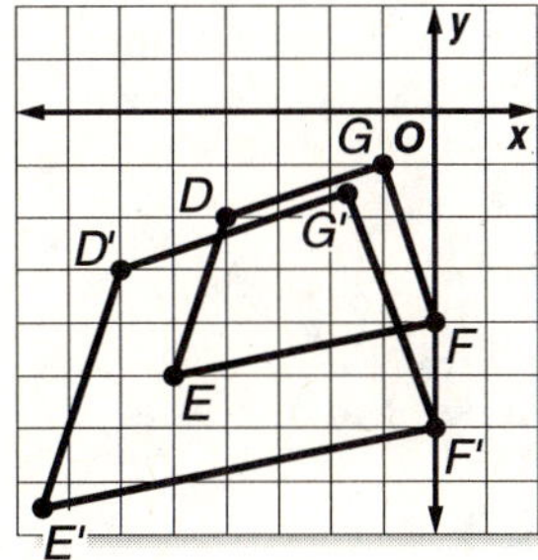

10. reflection
11. dilation
12. translation

10-4 Quadrilaterals

Page 513 How are quadrilaterals used in design?

a. squares, rectangles, trapezoids

b. Sample answer: The center of the circle is one square brick. The first row contains small trapezoids. The next four rows contain larger trapezoids. The next three rows contain trapezoids and squares that alternate.

Page 515 Check for Understanding

1. Sample answer: A textbook is an example if a quadrilateral; the tiles in a shuffleboard scoring region are examples of parallelograms; a "dead end" road sign is an example of a rhombus; and a floppy disk is an example of a square.
2. A rectangle has four sides, four right angles, opposite sides parallel and opposite sides congruent.

3. The sum of the measures of the angles is 360°.
$x + 64 + 118 + 68 = 360$
$x + 250 = 360$
$x = 110$
The value of x is 110, so the missing angle is 110°.
4. The sum of the measures of the angles is 360°.
$x + 2x + 100 + 125 = 360$
$3x + 225 = 360$
$3x = 135$
$x = 45$
The value of x is 45, so the missing angles are 45° and 2(45) or 90°.
5. The quadrilateral has four right angles, but the four sides are not congruent. It is a rectangle.
6. The shape has four sides, but no congruent sides or angles, and no parallel sides. It is a quadrilateral.
7. Each of the quadrilaterals has at least one set of opposite sides parallel. The shapes are trapezoids and parallelograms.

Pages 516–517 Practice and Apply

8. The sum of the measures of the angles is 360°.
$x + 115 + 65 + 109 = 360$
$x + 289 = 360$
$x = 71$
The value of x is 71, so the missing angle is 71°.
9. The sum of the measures of the angles is 360°.
$x + 52 + 110 + 96 = 360$
$x + 258 = 360$
$x = 102$
The value of x is 102, so the missing angle is 102°.
10. The sum of the measures of the angles is 360°.
$x + 128 + 3x + 120 = 360$
$4x + 248 = 360$
$4x = 112$
$x = 28$
The value of x is 28, so the missing angles are 28° and 3(28) or 84°.
11. The sum of the measures of the angles is 360°.
$x + x + 120 + 2x = 360$
$4x + 210 = 360$
$4x = 240$
$x = 60$
The value of x is 60, so the missing angles are 60°, 60°, and 2(60) or 120°.
12. The sum of the measures of the angles is 360°.
$x + 2x + (x + 5) + 135 = 360$
$4x + 140 = 360$
$4x = 220$
$x = 55$
The value of x is 55, so the missing angles are 55°, 2(55) or 110°, and 55 + 5 or 60°.
13. The sum of the measures of the angles is 360°.
$x + 90 + (2x - 20) + (x + 10) = 360$
$4x + 80 = 360$
$4x = 280$
$x = 70$
The value of x is 70, so the missing angles are 70°, 2(70) − 20 or 120°, and 70 + 10 or 80°.
14. Sample answer: A baking sheet; it is a rectangle because it is a parallelogram with 4 right angles.
15. Sample answer: A chessboard; it is a square because it is a parallelogram with 4 congruent sides and 4 right angles.
16. The quadrilateral has four right angles, but the four sides are not congruent. It is a rectangle.
17. The quadrilateral has four right angles and four congruent sides. It is a square.
18. The quadrilateral has four right angles, and opposite sides are congruent. All four sides are not congruent, so it is a rectangle.
19. The quadrilateral has opposite sides parallel and opposite sides congruent. It is a parallelogram.
20. The shape has four sides, but no congruent sides or angles, and no parallel sides. It is a quadrilateral.
21. The quadrilateral has four congruent sides, but no right angles. It is a rhombus.
22. Sample answer: The artist used triangles, quadrilaterals, and a few shapes having five and six sides. Some of the quadrilaterals are trapezoids.
23. always
24. sometimes
25. sometimes
26. always
27. See students' drawings; quadrilateral.
28. See students' drawings; parallelogram.

29a. Yes; a rhombus is equilateral but may not be equiangular.

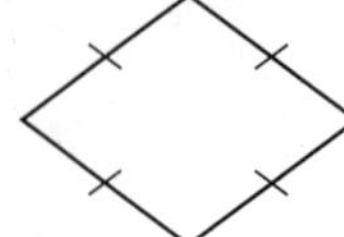

29b. Yes; a rectangle is equiangular but may not be equilateral.

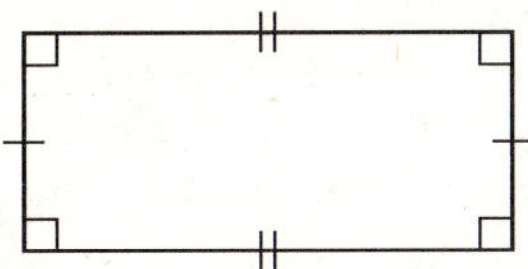

30. Many designs contain patterns formed by using shapes such as quadrilaterals. Answers should include the following.

- Sample answer: Many quilt designs contain quadrilaterals. One such example is shown.

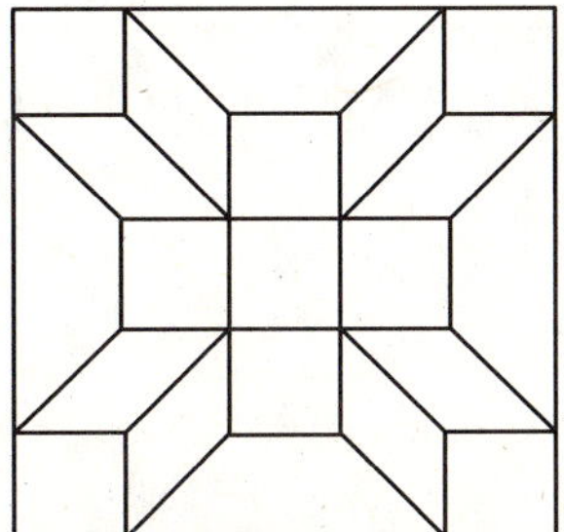

- The quadrilaterals used are trapezoids, squares, and parallelograms.

31. B; It is a parallelogram with all four sides congruent, so it is a rhombus.

32. The sum of the measures of the angles is 360°.

$90 + 82 + 41 + x = 360$

$213 + x = 360$

$x = 147$

Page 517 Maintain Your Skills

33.

vertex	4 down	translation
$D(1, 2)$	$+ (0, -4) \rightarrow$	$D'(1, -2)$
$E(1, 4)$	$+ (0, -4) \rightarrow$	$E'(1, 0)$
$F(-4, 4)$	$+ (0, -4) \rightarrow$	$F'(-4, 0)$
$G(-2, 2)$	$+ (0, -4) \rightarrow$	$G'(-2, -2)$

The coordinates of the vertices are $D'(1, -2)$, $E'(1, 0)$, $F'(-4, 0)$, and $G'(-2, -2)$.

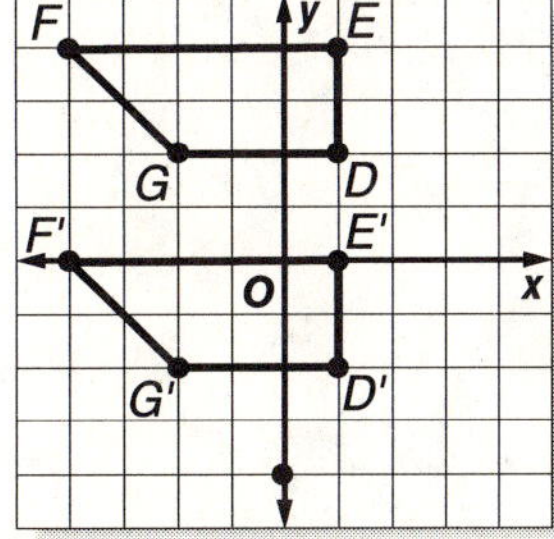

34. $\angle K$ corresponds to $\angle D$, so $\angle K \cong \angle D$.

35. Q corresponds to M, and N corresponds to A, so $\overline{QN} \cong \overline{MA}$.

36. $45(0.20) = \$9$

37. $(3)(4.8) = 14.4$

38. $(5.4)(6) = 32.4$

39. $(9.2)(3.1) = 28.52$

40. $(10.5)(5.7) = 59.85$

Page 517 Practice Quiz 1

1. Since $\angle A$ and $\angle B$ are complementary,

$m\angle A + m\angle B = 90°$

$m\angle A + 55 = 90$

$m\angle A = 35°$

2. $\angle D$ corresponds to $\angle A$, so $\angle D \cong \angle A$.

3. To reflect a point over the x-axis, use the same x-coordinate and multiply y-coordinate by -1.

vertex	reflection
$Q(3, 3) \rightarrow (3, -1 \cdot 3)$	$\rightarrow Q'(3, -3)$
$R(5, 6) \rightarrow (5, -1 \cdot 6)$	$\rightarrow R'(5, -6)$
$S(7, 3) \rightarrow (7, -1 \cdot 3)$	$\rightarrow S'(7, -3)$

The coordinates of the vertices of the reflected figure are $Q'(3, -3)$, $R'(5, -6)$, and $S'(7, -3)$.

4. The sum of the measures of the angles is 360°.

$95 + 25 + 175 + x = 360$

$295 + x = 360$

$x = 65$

The value of x is 65, so the missing angle is 65°.

5. The sum of the measures of the angles is 360°.

$90 + 90 + 2x + x = 360$

$180 + 3x = 360$

$3x = 180$

$x = 60$

The value of x is 60, so the missing angles are 60° and 2(60) or 120°.

Pages 518–519 Algebra Activity (Preview of Lesson 10-5)

1. 11 sq units
2. about 4 sq units
3. about 5.5 sq units
4. Sample answer: Count whole-squares and half-squares.
5. See students' work.
6. 3 sq units
7. 6 sq units
8. 4 sq units
9. See students' work.
10. See students' work.
11. 4.5 sq units
12. 6 sq units
13. 4 sq units
14. See students' work.

10-5 Area: Parallelograms, Triangles, and Trapezoids

Page 520 How is the area of a parallelogram related to the area of a rectangle?

a. parallelogram
b. They are the same, 18 square units.
c. length and width or height and base

Page 523 Check for Understanding

1. Sample answer:

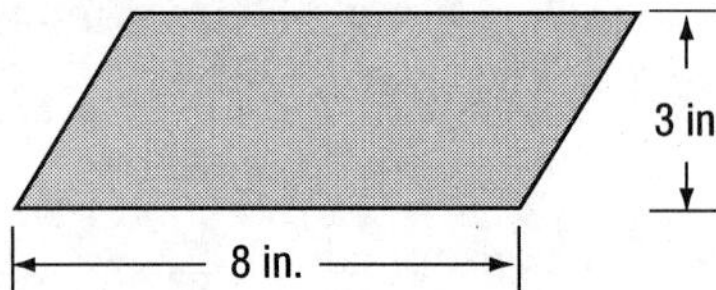

2. The altitude of a triangle is a line segment perpendicular to the base with endpoints on the base and the vertex opposite the base.

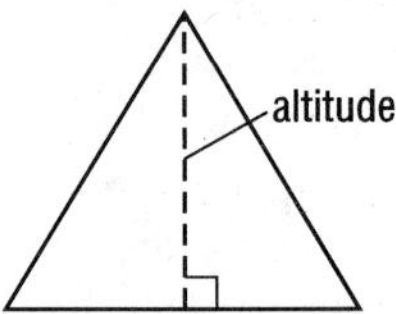

3. The base is 4 ft. The height is 2 ft.
$A = bh$
$A = 4 \cdot 2$
$A = 8 \text{ ft}^2$
The area is 8 ft^2.

4. The base is 3 cm. The height is 5.4 cm.
$A = \frac{1}{2}bh$
$A = \frac{1}{2}(3)(5.4)$
$A = 8.1 \text{ cm}^2$
The area is 8.1 cm^2.

5. The height is 6 m. The bases are 15 m and 5 m.
$A = \frac{1}{2}h(a + b)$
$A = \frac{1}{2} \cdot 6(15 + 5)$
$A = \frac{1}{2} \cdot 6 \cdot 20$
$A = 60 \text{ m}^2$
The area is 60 m^2.

6. The height is $18 - 12 = 6$ inches. The bases are 9 in. and 7 in.
$A = \frac{1}{2}h(a + b)$
$A = \frac{1}{2} \cdot 6(9 + 7)$
$A = \frac{1}{2} \cdot 6 \cdot 16$
$A = 48$ in.
The area is 48 in^2.

Pages 524–525 Practice and Apply

7. The base is 5.5 m. The height is 2 m.
$A = bh$
$A = (5.5)(2)$
$A = 11 \text{ m}^2$
The area is 11 m^2.

8. The base is 7.4 in. The height is 3.5 in.
$A = bh$
$A = (7.4)(3.5)$
$A = 25.9 \text{ in}^2$
The area is 25.9 in^2.

9. The base is 15 cm. The height is 12 cm.
$A = \frac{1}{2}bh$
$A = \frac{1}{2} \cdot 15 \cdot 12$
$A = 90 \text{ cm}^2$
The area is 90 cm^2.

10. The base is 6 in. The height is 7.2 in.
$A = \frac{1}{2}bh$
$A = \frac{1}{2}(6)(7.2)$
$A = 21.6 \text{ in}^2$
The area is 21.6 in^2.

11. The bases are 9.9 cm and 6.7 cm. The height is 5.3 cm.
$A = \frac{1}{2}h(a + b)$
$A = \frac{1}{2} \cdot (5.3)(9.9 + 6.7)$
$A = \frac{1}{2} \cdot (5.3)(16.6)$
$A = 43.99 \text{ cm}^2$
The area is 43.99 cm^2.

12. The bases are 20 ft and 14 ft. The height is $9\frac{1}{5}$ ft or $\frac{46}{5}$ ft.
$A = \frac{1}{2}h(a + b)$
$A = \frac{1}{2} \cdot \left(\frac{46}{5}\right)(20 + 14)$
$A = \frac{1}{2} \cdot \left(\frac{46}{5}\right)(34)$
$A = 156.4 \text{ ft}^2$
The area is 156.4 ft^2.

13. $A = \frac{1}{2}bh$
$A = \frac{1}{2} \cdot 8 \cdot 7$
$A = 28 \text{ in}^2$

14. $A = \frac{1}{2}h(a + b)$
$A = \frac{1}{2} \cdot 2(3 + 6)$
$A = \frac{1}{2} \cdot 2(9)$
$A = 9 \text{ cm}^2$

15. $A = bh$
$A = (3.8)(6)$
$A = 22.8 \text{ yd}^2$

16. $A = \frac{1}{2}bh$
$A = \frac{1}{2}(9)(3.2)$
$A = 14.4 \text{ ft}^2$

17. $A = \frac{1}{2}h(a + b)$
$A = \frac{1}{2} \cdot (3.5)(10 + 11)$
$A = 36.75 \text{ m}^2$

18. $A = bh$
$A = (5.6)(4.5)$
$A = 25.2 \text{ km}^2$

19. Oregon is in the approximate shape of a parallelogram with a base of 332 mi and a height of 287 mi.

$A = bh$

$A = 332 \cdot 287$

$A = 95{,}284 \text{ mi}^2$

The area is about 95,284 mi^2.

20. Arkansas is in the approximate shape of a trapezoid with bases 270 mi and 165 mi, and a height of 235 mi.

$A = \frac{1}{2}h(a + b)$

$A = \frac{1}{2} \cdot 235(270 + 165)$

$A = \frac{1}{2} \cdot 235(435)$

$A = 51{,}112.5 \text{ mi}^2$

The area is about 51,113 mi^2.

21. The figure is made up of a rectangle and a trapezoid.

Area of the rectangle	**Area of the trapezoid**
$A = bh$	$A = \frac{1}{2}h(a + b)$
$A = 15 \cdot 8$	$A = \frac{1}{2} \cdot 2(15 + 12)$
$A = 120\ km^2$	$A = \frac{1}{2} \cdot 2(27)$
	$A = 27 \text{ km}^2$

The total area is 120 + 27 or 147 km^2.

22. The figure is made up of a triangle and a trapezoid.

Area of the triangle	**Area of the trapezoid**
$A = \frac{1}{2}bh$	$A = \frac{1}{2}h(a + b)$
$A = \frac{1}{2} \cdot 6 \cdot 8$	$A = \frac{1}{2} \cdot 4(9 + 3)$
$A = 24 \text{ m}^2$	$A = \frac{1}{2} \cdot 4(12)$
	$A = 24 \text{ m}^2$

The total area is 24 + 24 or 48 m^2.

23. The figure is made up of a rectangle and a triangle. The height of the triangle is 11 − 8 or 3 ft.

Area of the rectangle	**Area of the triangle**
$A = bh$	$A = \frac{1}{2}bh$
$A = 8 \cdot 6$	$A = \frac{1}{2} \cdot 6 \cdot 3$
$A = 48 \text{ ft}^2$	$A = 9 \text{ ft}^2$

The total area is 48 + 9 or 57 ft^2.

24. The figure is a trapezoid with bases 125 ft and 100 ft. The height is 84 ft.

$A = \frac{1}{2}h(a + b)$

$A = \frac{1}{2} \cdot 84(125 + 100)$

$A = \frac{1}{2} \cdot 84(225)$

$A = 9270 \text{ ft}^2$

Area of figure	−	Area of patio	=	Area of lawn
9450	−	$15 \cdot 12$	=	Area of lawn
9450	−	180	=	9270 ft^2

The area of the lawn is 9270 ft^2.

25. Each bag covers 2000 square feet, so $9270 \div 2000 = 4.635$ bags. Since she can only buy whole bags, she needs 5 bags.

26. $A = bh$

$36.8 = b \cdot (9.2)$

$\frac{36.8}{9.2} = \frac{b \cdot (9.2)}{9.2}$

$4 = b$

The base is 4m.

27. $A = \frac{1}{2}bh$

$20 = \frac{1}{2}(2.5)h$

$20 = 1.25h$

$\frac{20}{1.25} = \frac{1.25h}{1.25}$

$16 = h$

The height is 16 in.

28. $A = \frac{1}{2}h(a + b)$

$54 = \frac{1}{2}h(16 + 8)$

$54 = \frac{1}{2}h(24)$

$54 = 12h$

$\frac{54}{12} = \frac{12h}{12}$

$4.5 = h$

The height is 4.5 ft.

29. The land is in the approximate shape of a trapezoid with bases 90.54 ft and 57.62 ft. The height is 144.34 ft.

$A = \frac{1}{2}h(a + b)$

$A = \frac{1}{2}(144.34)(90.54 + 57.62)$

$A = \frac{1}{2}(144.34)(148.16)$

$A = 10{,}692.7072 \text{ ft}^2$

Let p represent the percent of an acre.

$p = \frac{10{,}692.7072}{43{,}560} \cdot 100\%$

$p = 24.5\%$

30. $A = \frac{1}{2}h(a + b)$; in a triangle, one "base" = 0, so $A = \frac{1}{2}h(0 + b)$, then $A = \frac{1}{2}hb$ or $A = \frac{1}{2}bh$. In a parallelogram, the bases are congruent, so $a = b$. Thus, $A = \frac{1}{2}h(b + b)$ or $A = \frac{1}{2}h(2b)$, then $A = bh$.

31. The area of a parallelogram is found by multiplying the base and the height of the parallelogram. The area of a rectangle is found by multiplying the length and the width of the rectangle. Since in a parallelogram, the base is the length of the parallelogram, and the height is the width of the parallelogram, both areas are found by multiplying the length and the width. Answers should include the following.

- Parallelograms and rectangles are similar in that they are quadrilaterals with opposite sides parallel and opposite sides congruent. They are different in that rectangles always have 4 right angles.
- 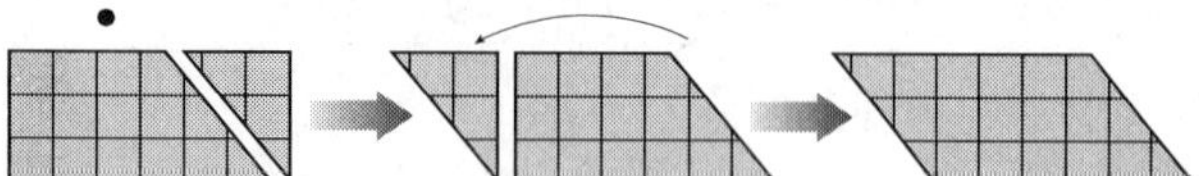

32. C; The area of the rectangle is:
$A = bh$
$A = 3(1.6)$
$A = 4.8 \text{ m}^2$
The rectangle does *not* have an area of 48 m^2.

33. B: The area of a square is $A = s \cdot s$, where s is the length of a side.
Area of Square x
$A = s \cdot s$
$9 = s \cdot s$, so s is 3 ft.
Square x has sides of length 3 ft, so square y has sides of length 6 feet since they are twice as long.
Area of Square y
$A = s \cdot s$
$A = 6 \cdot 6$
$A = 36 \text{ ft}^2$

Pages 525 Maintain Your Skills

34. The sum of the measures of the angles is 360°.
$x + 60 + 120 + 60 = 360$
$x + 240 = 360$
$x = 120$
The value of x is 120, so the missing angle is 120°.

35. The sum of the measures of the angles is 360°.
$x + 4x + 110 + 130 = 360$
$5x + 240 = 360$
$5x = 120$
$x = 24$
The value of x is 24, so the missing angles are 24° and 5(24) or 120°.

36. This translation can be written as the ordered pair (−3, −4). To find the coordinates of the translated image, add −3 to each x-coordinate and add −4 to each y-coordinate.

vertex	3 left, 4 down		translation
$M(-1, 1)$	$+ (-3, -4)$	→	$M'(-4, -3)$
$N(5, 4)$	$+ (-3, -4)$	→	$N'(2, 0)$
$P(4, 1)$	$+ (-3, -4)$	→	$P'(1, -3)$

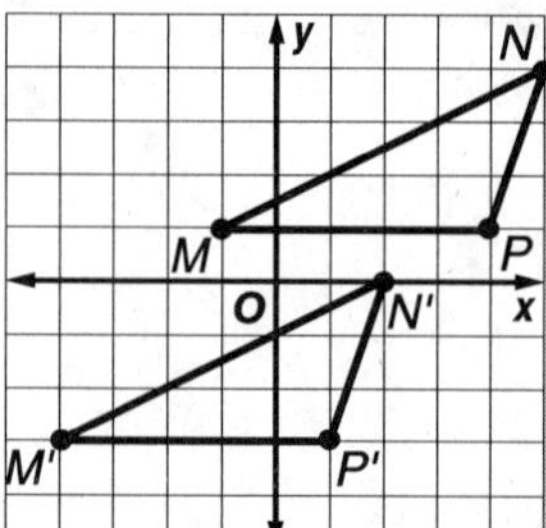

The coordinates of the vertices of $\triangle M'N'P'$ are $M'(-4, -3)$, $N'(2, 0)$, and $P'(1, -3)$.

37. $(0.25)(120) = 30$

38. $(0.75)(160) = 120$

39. $(0.40)(65) = 26$

40. $(5 - 2)180 = 3(180)$
$= 540$

41. $(7 - 2)180 = 5(180)$
$= 900$

42. $(10 - 2)\ 180 = 8(180)$
$= 1440$

43. $(9 - 2)180 = 7(180)$
$= 1260$

Page 526 Reading Mathematics

1. Sample answer: quad; A leap day is a quadrennial event.
pent; Many military personnel work at the Pentagon.
hex; An insect is an example of a hexapod.
hept; Heptarchy was the name given to the seven kingdoms of England during the 7th and 8th centuries.
oct; The giant octopus found in the Pacific Ocean may have a diameter greater than 30 feet.
dec; Dan O'Brien is the current world record holder in the decathlon with 8891 points.

2. Sample answer: quad; quadrant—an arc of 90° that is one quarter of a circle.
pent; pentahedron—a three-dimensional solid with five faces.
hex; hexagram—a two-dimensional figure that has equilateral triangles on each side of a regular hexagon.
hept; heptagon—a two-dimensional, closed figure with seven sides.
oct; octal—a number system with a base of 8.
dec; decare—a metric unit of area equal to 0.2471 acre.

3. Sample answer: quad; quadriceps—a muscle located on the front of the thigh that is divided into four parts.
pent; pentamerous—divided into or consisting of five parts.
hex; hexahydrate—a chemical compound with six molecules of water.
hept; heptameter—a line of verse consisting of seven metrical feet.
oct; octogenarian—a person whose age is in the eighties.
dec; decasyllabic—consisting of ten syllables or composed of verses of ten syllables.

10-6 Polygons

Page 527 How are polygons used in tessellations?

a. The first figure uses squares. The second figure uses triangles. The third figure uses hexagons.

b. 360

c. Yes; for each tessellation shown, the sum of the measures of the angles that surround a vertex is 360°.

d. The sum is 360°.

Page 529 Check for Understanding

1.

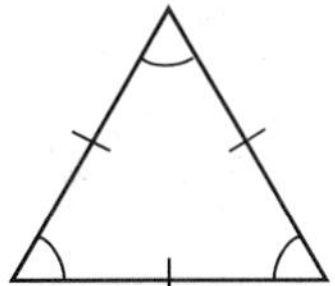

2. It has four congruent angles and four congruent sides.

3. The number of triangles is 2 less than the number of sides.

4. The polygon has 5 sides. It is a pentagon. All sides are not congruent, so it is not regular.

5. The polygon has 8 sides. It is an octagon. All the sides and angles are congruent, so it is regular.

6. A nonagon has 9 sides. Therefore, $n = 9$.
$(n - 2)180 = (9 - 2)180$
$= 7(180)$ or 1260
The sum of the measures of the angles is 1260°.

7. Find the sum of the measures of the angles. A heptagon has 7 sides. Therefore, $n = 7$.
$(n - 2)180 = (7 - 2)180$
$= 5(180)$ or 900
The sum of the measures of the interior angles is 900°.
Divide the sum by 7 to find the measure of one angle.
$900 \div 7 = 128.6°$
So, the measure of each interior angle is 128.6°.

8. square, hexagon, 12-gon

Pages 530–531 Practice and Apply

9. The polygon has 6 sides. It is a hexagon. All sides and angles are congruent, so it is regular.

10. The polygon has 7 sides. It is a heptagon. All sides are not congruent, so it is not regular.

11. The polygon has 9 sides. It is a nonagon. All sides are not congruent, so it is not regular.

12. The polygon has 5 sides. It is a pentagon. All sides and angles are congruent, so it is regular.

13. The polygon has 10 sides. It is a decagon. All sides are congruent, but all angles are not congruent. It is not regular.

14. The polygon has 10 sides. It is a decagon. All sides are not congruent, so it is not regular.

15. A pentagon has 5 sides. Therefore, $n = 5$.
$(n - 2)180 = (5 - 2)180$
$= 3(180)$
$= 540$
The sum of the measures of the angles is 540°.

16. An octagon has 8 sides. Therefore, $n = 8$.
$(n - 2)180 = (8 - 2)180$
$= 6(180)$
$= 1080$
The sum of the measures of the angles is 1080°.

17. A decagon has 10 sides. Therefore, $n = 10$.
$(n - 2)180 = (10 - 2)180$
$= 8(180)$
$= 1440$
The sum of the measures of the angles is 1440°.

18. A hexagon has 6 sides. Therefore, $n = 6$.
$(n - 2)180 = (6 - 2)180$
$= 4(180)$
$= 720$
The sum of the measures of the angles is 720°.

19. An 18-gon has 18 sides. Therefore, $n = 18$.
$(n - 2)180 = (18 - 2)180$
$= 16(180)$
$= 2880$
The sum of the measures of the angles is 2880°.

20. A 23-gon has 23 sides. Therefore, $n = 23$.
$(n - 2)180 = (23 - 2)180$
$= 21(180)$
$= 3780$
The sum of the measures of the angles is 3780°.

21. Sample answer: hexagon, triangle, decagon, quadrilateral, and pentagon

22. No, it is not a polygon since it has curved sides.

23. Find the sum of the measures of the angles. A nonagon has 9 sides. Therefore, $n = 9$.
$(n - 2)180 = (9 - 2)180$
$= 7(180)$ or 1260°
The sum of the measures of the angles is 1260°.
Divide the sum by 9 to find the measure of one angle.
$1260 \div 9 = 140$
So, the measure of an interior angle is 140°.

24. Find the sum of the measures of the angles. A pentagon has 5 sides. Therefore, $n = 5$.
$(n - 2)180 = (5 - 2)180$
$= 3(180)$ or $540°$
The sum of the measures of the angles is 540°. Divide the sum by 5 to find the measure of one angle.
$540 \div 5 = 108$
So, the measure of an interior angle is 108°.

25. Find the sum of the measures of the angles. An octagon has 8 sides. Therefore, $n = 8$.
$(n - 2)180 = (8 - 2)180$
$= 6(180)$ or $1080°$
The sum of the measures of the angles is 1080°. Divide the sum by 8 to find the measure of one angle.
$1080 \div 8 = 135$
So, the measure of an interior angle is 135°.

26. Find the sum of the measures of the angles. A decagon has 10 sides.
Therefore, $n = 10$.
$(n - 2)180 = (10 - 2)180$
$= 8(180)$ or 1440
The sum of the measures of the angles is 1440°. Divide the sum by 10 to find the measure of one angle.
$1440 \div 10 = 144$
So, the measure of one interior angle is 144°.

27. Find the sum of the measures of the angles. A 12-gon has 12 sides. Therefore, $n = 12$.
$(n - 2)180 = (12 - 2)180$
$= 10(180)$ or 1800
The sum of the measures of the angles is 1800°. Divide the sum by 12 to find the measure of one angle.
$1800 \div 12 = 150$
So, the measure of an interior angle is 150°.

28. Find the sum of the measures of the angles. A 25-gon has 25 sides. Therefore, $n = 25$.
$(n - 2)180 = (25 - 2)180$
$= 23(180)$ or 4140
The sum of the measures of the angles is 4140°. Divide the sum by 25 to find the measure of one angle.
$4140 \div 25 = 165.6$
So, the measure of an interior angle is 165.6°.

29. octagons, squares

30. hexagons, triangles, squares

31. The sides of the squares and the hexagons have the same lengths as the sides of the 12-gon. There are 36 sides on the perimeter of the design, and each side is 5 cm long, so the perimeter is 36(5) or 180 cm long.

32. A regular pentagon has 5 equal sides. The perimeter is 5(4.2) or 21 ft.

33. A regular nonagon has 9 equal sides. The perimeter is $9\left(6\frac{1}{2}\right)$ or $9(6.5)$ or 58.5 in.

34. Sample answer:

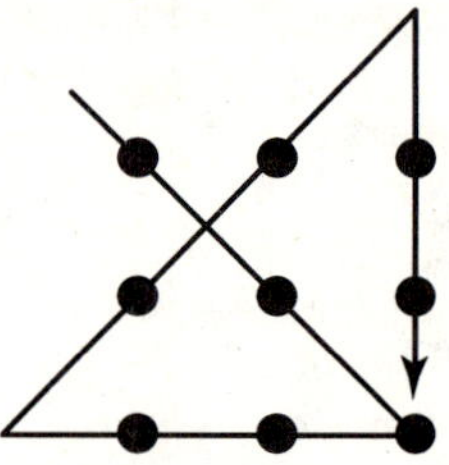

35. In tessellations, polygons are fit together to create a pattern such that there are no gaps or spaces. Answers should include the following.
Sample answer:

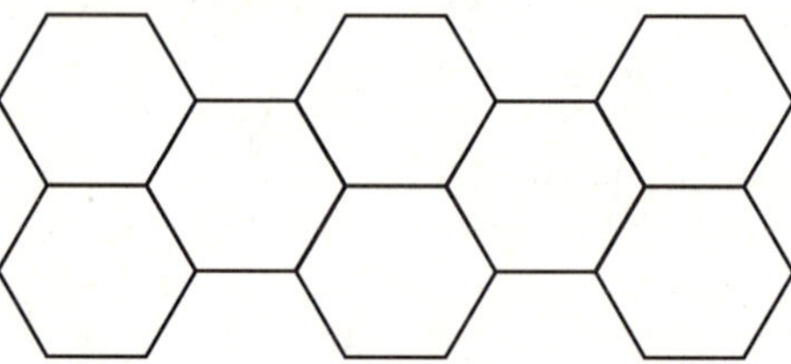

Sample answer:

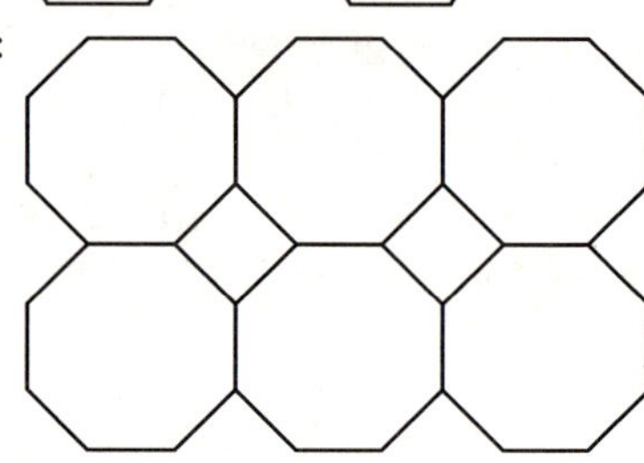

36. A triangle has 3 sides. Therefore, $n = 3$.
$$\frac{180(n-2)}{n} = \frac{180(3-2)}{3}$$
$$= \frac{180(1)}{3} \text{ or } 60$$
The measure of an interior angle in a regular triangle is 60°.

37. B; The square has all sides congruent and all angles congruent.

38. A regular octagon has 8 equal interior angles, so it has 8 equal exterior angles. The sum of the exterior angles is 360°, so each exterior angle is $360 \div 8$ or 45°.

39. A regular triangle has 3 equal interior angles, so it has 3 equal exterior angles. The sum of the exterior angles is 360°, so each exterior angle is $360 \div 3$ or 120°.

40. A regular nonagon has 9 equal interior angles, so it has 9 equal exterior angles. The sum of the exterior angles is 360°, so each exterior angle is $360 \div 9$ or 40°.

41. A regular hexagon has 6 equal interior angles, so it has 6 equal exterior angles. The sum of the exterior angles is 360°, so each exterior angle is $360 \div 6$ or 60°.

42. A regular decagon has 10 equal interior angles, so it has 10 equal exterior angles. The sum of the exterior angles is 360°, so each exterior angle is $360 \div 10$ or 36°.

43. A regular 12-gon has 12 equal interior angles, so it has 12 equal exterior angles. The sum of the exterior angles is 360°, so each exterior angle is $360 \div 12$ or 30°.

Page 531 Maintain Your Skills

44. $A = \frac{1}{2}bh$

$A = \frac{1}{2} \cdot 9 \cdot 6$

$A = 27 \text{ in}^2$

45. $A = \frac{1}{2}h(a + b)$

$A = \frac{1}{2} \cdot 3(4 + 8)$

$A = \frac{1}{2} \cdot 3 \cdot 12$

$A = 18 \text{ cm}^2$

46. The quadrilateral has four congruent sides, but no right angles. It is a rhombus.

47. The figure has four sides, but no congruent sides or angles. It is a quadrilateral.

48. The quadrilateral has opposite sides parallel, opposite sides congruent, and four right angles. It is a rectangle.

49. $4.6x + 2.5x + 9.3x$

$= (4.6 + 2.5 + 9.3)x$

$= 16.4x$

50. $\pi \cdot 4.3 \approx 13.50... \approx 13.5$

51. $2 \cdot \pi \cdot 5.4 \approx 33.92... \approx 33.9$

52. $\pi \cdot 4^2 \approx 50.26... \approx 50.3$

53. $\pi(2.4)^2 \approx 18.09... \approx 18.1$

Page 532 Algebra Activity (Follow-Up of Lesson 10-6)

1.

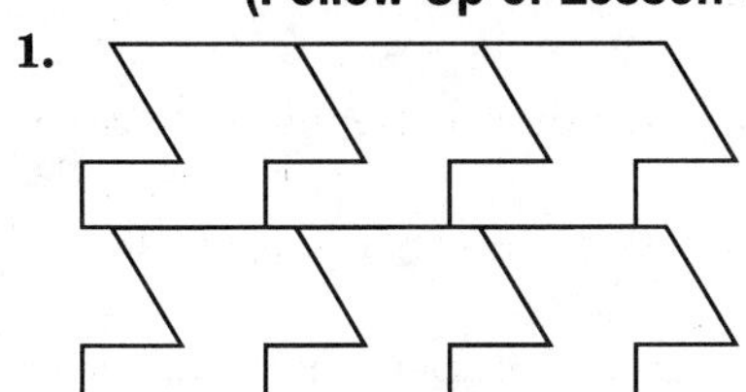

2.

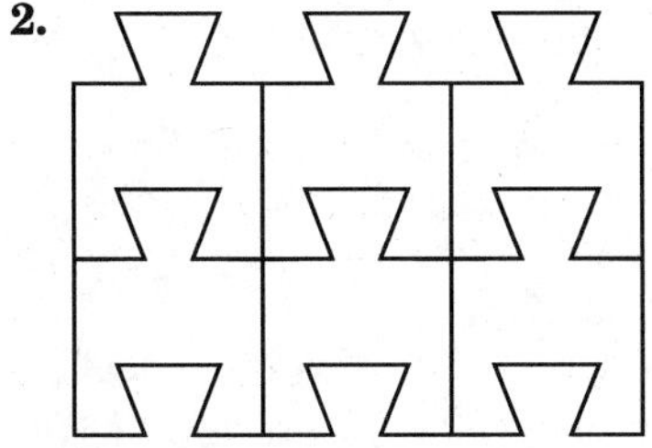

3.

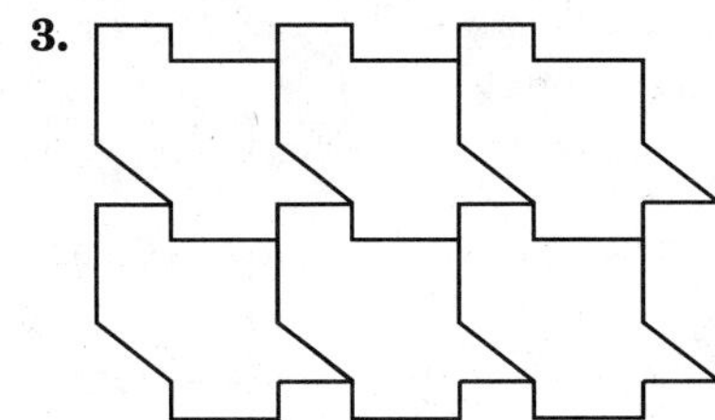

4.

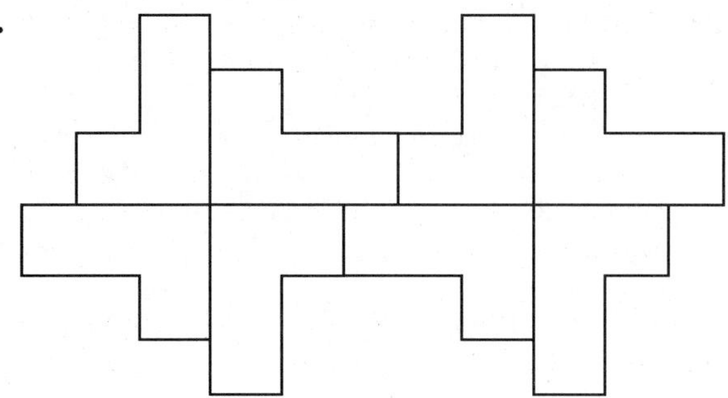

5.

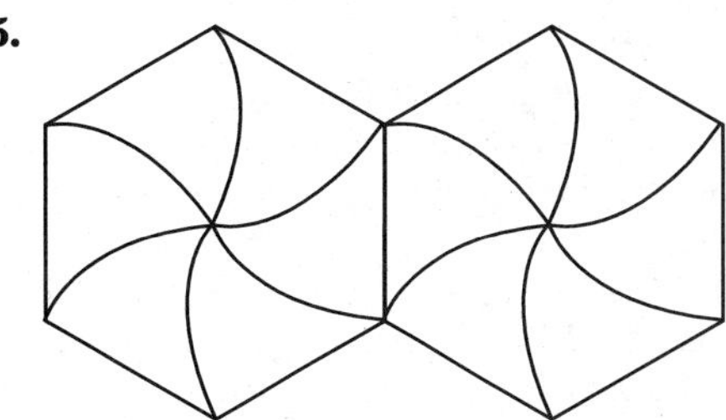

6.

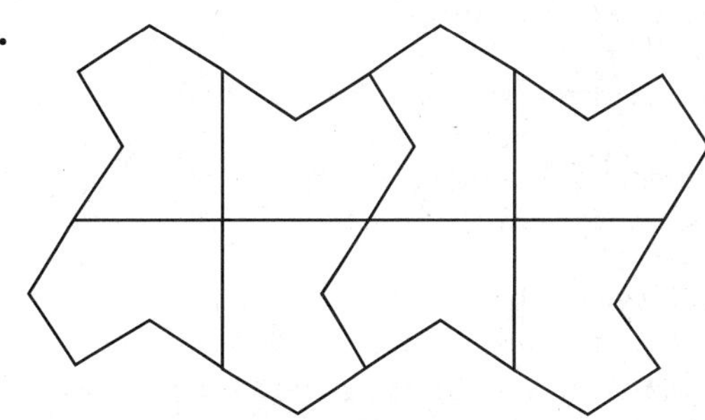

7. See students' work.

10-7 Circumference and Area: Circles

Page 533 How are circumference and diameter related?

a. See students' work.

b. See students' work.

c. The results are about 3.

Page 535–536 Check for Understanding

1. Multiply 2 times π times the radius.

2. Sample answer:

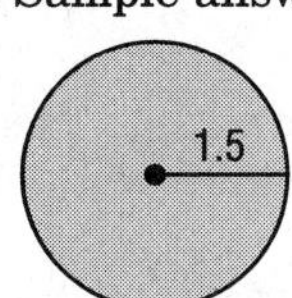

$A = \pi r^2$

$A = \pi(1.5)^2$

$A \approx 7.1$

3. Mark; since the diameter of the circle is 7 units, its radius is 3.5 units. Thus, the area of the circle is $\pi \cdot (3.5)^2$ or 38.5 square units.

4. $C = \pi d$
$C = \pi \cdot 4$
$C \approx 12.6$
The circumference is about 12.6 in.
$A = \pi r^2$
$A = \pi \cdot 2^2$
$A = \pi \cdot 4$
$A \approx 12.6$
The area is about 12.6 in^2.

5. $C = 2\pi r$
$C = 2 \cdot \pi \cdot 8$
$C \approx 50.3$
The circumference is about 50.3 m.
$A = \pi r^2$
$A = \pi \cdot 8^2$
$A = \pi \cdot 64$
$A \approx 201.1$
The area is about 201.1 m^2.

6. $C = 2\pi r$
$C = 2 \cdot \pi \cdot 5$
$C \approx 31.4$
The circumference is about 31.4 mi^2.
$A = \pi r^2$
$A = \pi \cdot 5^2$
$A = \pi \cdot 25$
$A \approx 78.5$
The area is about 78.5 mi^2.

7. $C = 2\pi r$
$C = 2 \cdot \pi \cdot 1.3$
$C \approx 8.2$
The circumference is about 8.2 km.
$A = \pi r^2$
$A = \pi \cdot (1.3)^2$
$A = \pi \cdot 1.69$
$A \approx 5.3$
The area is about 5.3 km^2.

8. $C = \pi d$
$C = \pi \cdot 6.1$
$C \approx 19.2$
The circumference is about 19.2 cm.
$A = \pi r^2$
$A = \pi \cdot (3.05)^2$
$A = \pi \cdot 9.3025$
$A \approx 29.2$
The area is about 29.2 cm^2.

9. $A = \pi r^2$
$A = \pi \cdot (1.7)^2$
$A = \pi \cdot 2.89$
$A \approx 9.1$
The area is about 9.1 mi^2.

Pages 536–538 Practice and Apply

10. $C = \pi d$
$C = \pi \cdot 6$
$C \approx 18.8$
The circumference is about 18.8 cm.
$A = \pi r^2$
$A = \pi \cdot 3^2$
$A = \pi \cdot 9$
$A \approx 28.3$
The area is about 28.3 cm^2.

11. $C = \pi d$
$C = \pi \cdot 13$
$C \approx 40.8$
The circumference is about 40.8 in.
$A = \pi r^2$
$A = \pi \cdot (6.5)^2$
$A = \pi \cdot 42.25$
$A \approx 132.7$
The area is about 132.7 in^2.

12. $C = 2\pi r$
$C = 2\pi \cdot 10$
$C \approx 62.8$
The circumference is about 62.8 m.
$A = \pi r^2$
$A = \pi \cdot 10^2$
$A = \pi \cdot 100$
$A \approx 314.2$
The area is about 314.2 m^2.

13. $C = 2\pi r$
$C = 2 \cdot \pi \cdot 21$
$C = 131.9$
The circumference is about 131.9 km.
$A = \pi r^2$
$A = \pi \cdot 21^2$
$A = \pi \cdot 441$
$A = 1385.4$
The area is about 1385.4 km^2.

14. $C = \pi d$
$C = \pi \cdot 9.5$
$C \approx 29.8$
The circumference is about 29.8 ft.
$A = \pi r^2$
$A = \pi \cdot (4.75)^2$
$A = \pi \cdot 22.5625$
$A = 70.9$
The area is about 70.9 ft^2.

15. $C = 2\pi r$
$C = 2 \cdot \pi \cdot 12.7$
$C \approx 79.8$
The circumference is about 79.8 m.
$A = \pi r^2$
$A = \pi \cdot (12.7)^2$
$A = \pi \cdot 161.29$
$A \approx 506.7$
The area is about 506.7 m^2.

16. $C = 2\pi r$
$C = 2 \cdot \pi \cdot 4.5$
$C \approx 28.3$
The circumference is about 28.3 m.
$A = \pi r^2$
$A = \pi \cdot (4.5)^2$
$A = \pi \cdot 20.25$
$A = 63.6$
The area is about 63.6 m^2.

17. $C = \pi d$
$C = \pi \cdot 7.3$
$C \approx 22.9$
The circumference is about 22.9 cm.
$A = \pi r^2$
$A = \pi \cdot (3.65)^2$
$A = 41.9$
The area is about 41.9 cm^2.

18. $C = \pi d$
$C = \pi \cdot (7.8)$
$C \approx 24.5$
The circumference is about 24.5 ft.
$A = \pi r^2$
$A = \pi \cdot (3.9)^2$
$A = \pi \cdot 15.21$
$A \approx 47.8$
The area is about 47.8 ft^2.

19. The radius is $15\frac{3}{8}$ or 15.375 m.
$C = 2\pi r$
$C = 2 \cdot \pi \cdot (15.375)$
$C \approx 96.6$
The circumference is about 96.6 in.
$A = \pi r^2$
$A = \pi \cdot (15.375)^2$
$A = \pi \cdot (236.390625)$
$A \approx 742.6$
The area is about 742.6 in^2.

20. $C = \pi d$
$25.8 = \pi d$
$\frac{25.8}{\pi} = \frac{\pi d}{\pi}$
$8.2 \approx d$
The diameter is about 8.2 in.

21. $C = 2\pi r$
$9.2 = 2\pi r$
$\frac{9.2}{2\pi} = \frac{2\pi r}{2\pi}$
$1.5 \approx r$
The radius is about 1.5 m.

22. $A = \pi r^2$
$254.5 = \pi r^2$
$\frac{254.5}{\pi} = \frac{\pi r^2}{\pi}$
$81.0 \approx r^2$
$9 \approx r$
The radius is about 9 in.

23. $A = \pi r^2$
$132.7 = \pi r^2$
$\frac{132.7}{\pi} = \frac{\pi r^2}{\pi}$
$42.2 \approx r^2$
$6.5 \approx r$
The radius is about 6.5 m, so the diameter is about 2(6.5) or 13 m.

24. In one rotation, the bicycle travels the length of its circumference.
$C = \pi d$
$C = \pi \cdot 27$
$C \approx 84.82$
The bicycle travels 10 · (84.82) or 848.2 inches.

25. The distance to the center of the earth is the radius of the earth.
$C = 2\pi r$
$25{,}000 = 2\pi r$
$\frac{25{,}000}{2\pi} = \frac{2\pi r}{2\pi}$
$3979 \approx r$
The distance to the center of the earth is about 3979 miles.

26. c; $A = \pi r^2$
$A = \pi \cdot 4^2$
$A = \pi \cdot 16$
$A \approx 50.3$ units^2

27. d; $C = \pi d$
$C = \pi \cdot 7$
$C \approx 21.99$ or 22.0 units

28. b; $A = \pi r^2$
$A = \pi \cdot (1.5)^2$
$A = \pi \cdot 2.25$
$A \approx 7.1$ units^2

29. a; $C = 2\pi r$
$C = 2 \cdot \pi \cdot 6$
$C \approx 37.7$ units

30. $A = \pi r^2$
$A = \pi \cdot (150)^2$
$A = \pi \cdot 22{,}500$
$A \approx 70{,}685.8$
The area is about 70,685.8 ft^2.

31. **Astrodome**
$A = \pi r^2$
$A = \pi \cdot (108.2)^2$
$A \approx 36{,}779.4$

Pantheon
$A = \pi r^2$
$A = \pi \cdot (21.35)^2$
$A = 1432.0$

Divide the area of the Astrodome by the area of the Pantheon:
$36{,}779.4 \div 1432.0 \approx 26$
The Astrodome covers about 26 times more area than the Pantheon.

32. Find the area of the circle graph:
$A = \pi r^2$
$A = \pi \cdot (4.5)^2$
$A \approx 63.6$ in^2
The graph will cover about 63.6 in^2.

33. Are safe: 40% of 63.6 in^2
$0.40 \cdot (63.6) = 25.4$ in^2
Are unsafe: 38% of 63.6 in^2
$0.38 \cdot (63.6) = 24.2$ in^2
Don't know: 22% of 63.6 in^2
$0.22(63.6) = 14.0$ in^2

34.

d	$C = \pi d$
1	$C = \pi(1) \approx 3.14$
2	$C = \pi(2) \approx 6.28$
3	$C = \pi(3) \approx 9.42$
4	$C = \pi(4) \approx 12.57$

$$\text{slope} = \frac{y_2 - y_1}{x_2 - x_1} = \frac{2\pi - \pi}{2 - 1} = \frac{\pi}{1} \text{ or } \pi$$

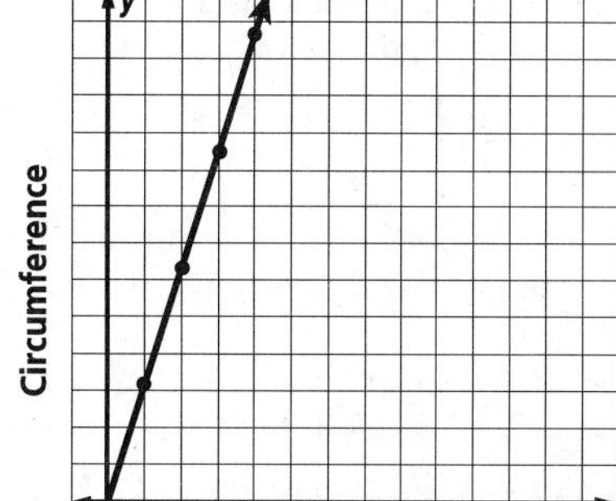

35.

r	$C = 2\pi r$	$A = \pi r^2$
1	$C = 2\pi(1) \approx 2\pi$	$A = \pi(1)^2 \approx \pi$
2	$C = 2\pi(2) \approx 4\pi$	$A = \pi(2)^2 \approx 4\pi$
3	$C = 2\pi(3) \approx 6\pi$	$A = \pi(3)^2 \approx 9\pi$
4	$C = 2\pi(4) \approx 8\pi$	$A = \pi(4)^2 \approx 16\pi$

Since 16π is twice 8π, the numerical value of the area is twice the numerical value of the circumference. The radius is 4.

36. The circumference of a circle is about 3 times its diameter. Answers should include the following.
- Since the circumference is about 3 times the diameter, the ratio describing the relationship would be about 3 to 1.
- As the diameter increases, the circumference increases. As the diameter decreases, the circumference decreases.

37. C; The diameter is 16 units, so the radius is 8 units.
$A = \pi(8)^2$
$A = \pi \cdot 64$
$A \approx 201.1 \text{ units}^2$

38.
$C = 2\pi r$
$18.8 = 2\pi r$
$\frac{18.8}{2\pi} = \frac{2\pi r}{2\pi}$
$2.992 = r$
$A = \pi r^2$
$A = \pi \cdot (2.992)^2$
$A \approx 28.1$
The area is 28.1 m^2.

39. The arc has the same measure as the measure of the central angle, 125°.

40. $\angle 1$ is supplemental to the 125° angle, so $\angle 1 = 180° - 125°$ or 55°.

41. Since minor arc *PQ* has measure 125°, major arc *QP* has measure 360° − 125° or 235°.

42. $\angle 2$ is supplemental to $\angle 1$, and $\angle 1$ equals 55° from Exercise 40.
So, $\angle 2 = 180° - 55°$ or 125°.

43. This arc has the same measure as the measure of the central angle, $\angle 2$. $\angle 2$ equals 125° from exercise 42, so the measure of minor arc *TR* is 125°.

44. Minor arc *RQ* has a central angle which is a vertical angle to $\angle 1$, so its measure is 55°. The minor arc *RQ* has the same measure as its central angle of 55°, so its measure is 55°.

45. $\angle PSQ$ intercepts minor arc *PQ*, so its measure is $\frac{1}{2}(125°)$ or 62.5°.

46. Minor arc *SR* can be found by subtracting the measure of minor arc *TR* from the measure of minor arc *ST*.
Minor arc *ST* has a measure of 2(15°) or 30°, because the inscribed angle is half the measure of the arc.
Minor arc *TR* has a measure of 125°, from Exercise 43.
The measure of minor arc *SR* is 125° − 30° or 95°.

47. Sample answer: $\overline{SP}, \overline{SQ}, \overline{RP}$.

Page 538 Maintain Your Skills

48. A hexagon has 6 sides. Therefore, $n = 6$.
$(n - 2)180 = (6 - 2)180$
$= 4(180)$
$= 720°$
The sum of the measures of the interior angles is 720°. Divide the sum by 6 to find the measure of an interior angle.
$720 \div 6 = 120°$
The measure of an interior angle is 120°.

49. A decagon has 10 sides. Therefore, $n = 10$.
$(n - 2)180 = (10 - 2)180$
$= 8(180)$
$= 1440°$
The sum of the measures of the angles is 1440°. Divide the sum by 10 to find the measure of an interior angle.
$1440° \div 10 = 144°$
The measure of an interior angle is 144°.

50. An octagon has 8 sides. Therefore, $n = 8$.
$(n - 2)180 = (8 - 2)180$
$= 6(180)$
$= 1080°$
The sum of the measures of the angles is 1080°. Divide the sum by 8 to find the measure of an interior angle.
$1080° \div 8 = 135°$
The measure of an interior angle is 135°.

51. $A = \frac{1}{2}h(a + b)$
$A = \frac{1}{2} \cdot 2(20 + 18)$
$A = \frac{1}{2} \cdot 2(38)$
$A = 38 \text{ m}^2$

52. $A = bh$
$A = 6 \cdot 8$
$A = 48 \text{ km}^2$

53.
$2x - 7 > 5x + 14$
$2x - 5x - 7 > 5x - 5x + 14$
$-3x - 7 > 14$
$-3x - 7 + 7 > 14 + 7$
$-3x > 21$
$\frac{-3x}{-3} < \frac{21}{-3}$
$x < -7$

54. $200 + 43.9 = 243.9$

55. $23.6 + 126.9 = 150.5$

56. $345.14 + 23.8 = 368.94$

Page 538 Practice Quiz 2

1. The figure is a triangle with base 17.1 cm and height 13.2 cm.
$A = \frac{1}{2}bh$
$A = \frac{1}{2}(17.1)(13.2)$
$A = 112.86$
The area is 112.86 cm^2.

2. The figure is a parallelogram with base 11 in. and height 7 in.
$A = bh$
$A = 11 \cdot 7$
$A = 77$
The area is 77 in^2.

3. The figure is a trapezoid with height 4 m and bases 10 m and 6 m.
$A = \frac{1}{2}h(a + b)$
$A = \frac{1}{2} \cdot 4(10 + 6)$
$A = \frac{1}{2} \cdot 4(16)$
$A = 32$
The area is 32 m^2.

4. A 15-gon has 15 sides, so $n = 15$.
$(n - 2)180 = (15 - 2)180$
$= 13(180)$
$= 2340$
The sum of the measures of the interior angles is 2340°.

5. $C = 2\pi r$
$C = 2 \cdot \pi \cdot 4.7$
$C \approx 29.5$
The circumference is about 29.5 in.
$A = \pi r^2$
$A = \pi \cdot (4.7)^2$
$A = \pi \cdot 22.09$
$A \approx 69.4$
The area is about 69.4 in^2.

10-8 Area: Irregular Figures

Page 539 How can polygons help to find the area of an irregular figure?

a. two trapezoids, a triangle, and a rectangle

b. Add the areas to find the total area.

c. Area of large trapezoid:
$A = \frac{1}{2}h(a + b)$
$A = \frac{1}{2} \cdot 210(213.3 + 546.7)$
$A = \frac{1}{2} \cdot 210(760)$
$A = 79{,}800$ mi^2
Area of triangle:
$A = \frac{1}{2}bh$
$A = \frac{1}{2} \cdot (160 + 133.3) \cdot 280$
$A = \frac{1}{2} \cdot 293.3 \cdot 280$
$A = 41{,}062$ mi^2
Area of small trapezoid:
$A = \frac{1}{2}h(a + b)$
$A = \frac{1}{2} \cdot 160(40 + 160)$
$A = \frac{1}{2} \cdot 160(200)$
$A = 16{,}000$ mi^2
Area of rectangle:
$A = bh$
$A = 133.3 \cdot 160$
$A = 21{,}328$ mi^2

d. Total area $= 79{,}800 + 41{,}062 + 16{,}000 + 21{,}328$
$= 158{,}190$ mi^2

Page 541 Check for Understanding

1. Sample answer:

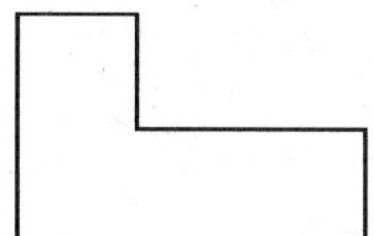
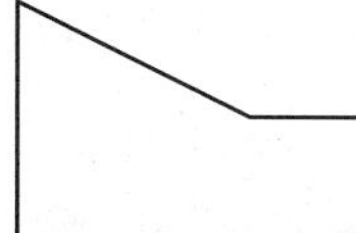

2. Sample answer: First, separate the figure into a rectangle and two triangles. Then find the area of each of the figures. Then find the sum of their areas.

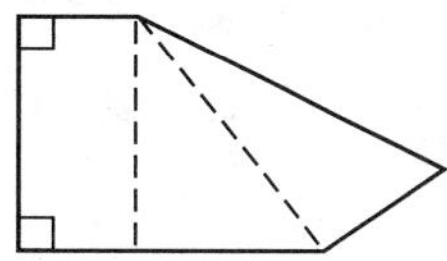

3. First, separate the figure into a rectangle and a triangle.
Area of rectangle:
$A = bh$
$A = 9 \cdot 4$
$A = 36$
Area of triangle:
$A = \frac{1}{2}bh$
$A = \frac{1}{2} \cdot 9 \cdot 3$
$A = 13.5$
The area of the figure is $36 + 13.5$ or 49.5 yd^2.

4. First, separate the figure into a square and a semicircle.
Area of square:
$A = bh$
$A = 5 \cdot 5$
$A = 25$
Area of semicircle:
$A = \frac{1}{2}\pi r^2$
$A = \frac{1}{2} \cdot \pi \cdot (2.5)^2$
$A \approx 9.8$
The area of the figure is about $25 + 9.8$ or 34.8 in^2.

5. Find the area of the deck.
Area of small trapezoid:
$A = \frac{1}{2}h(a + b)$
$A = \frac{1}{2} \cdot 10(7 + 18)$
$A = \frac{1}{2} \cdot 10(25)$
$A = 125$

Area of large trapezoid:
$A = \frac{1}{2}h(a + b)$
$A = \frac{1}{2} \cdot 12(9 + 20)$
$A = \frac{1}{2} \cdot 12(29)$
$A = 174$
The area of the deck is 125 + 174 or 299 ft^2.
Find the number of gallons of stain needed.
$299 \div 200 = 1.495$.
So, they will need to buy 2 gallons of stain.

6. The total cost is 2 · (19.95) or $39.90.

Pages 542–543 Practice and Apply

7. First, separate the figure into two rectangles, by drawing a line horizontally to make a 12 by 4 and a 4 by 6 rectangle.
Area of small rectangle:
$A = bh$
$A = 4 \cdot 6$
$A = 24$
Area of large rectangle:
$A = bh$
$A = 12 \cdot 4$
$A = 48$
The area of the figure is 24 + 48 or 72 ft^2.

8. Area of square:
$A = bh$
$A = 3 \cdot 3$
$A = 9$
Area of triangle:
$A = \frac{1}{2}bh$
$A = \frac{1}{2} \cdot (4.2 + 3) \cdot 5$
$A = \frac{1}{2}(7.2) \cdot 5$
$A = 18$
The area of the figure is 9 + 18 or 27 m^2.

9. Area of triangle:
$A = \frac{1}{2}bh$
$A = \frac{1}{2} \cdot (3.8) \cdot (9.5)$
$A \approx 18.1$
Area of parallelogram:
$A = bh$
$A = (9.5) \cdot 4$
$A = 38$
The area of the figure is 18.1 + 38 or about 56.1 cm^2.

10. Area of rectangle:
$A = bh$
$A = 10 \cdot (4.3)$
$A = 43$
Area of semicircle:
$A = \frac{1}{2}\pi r^2$
$A = \frac{1}{2} \cdot \pi \cdot (2.5)^2$
$A \approx 9.8$
The area of the figure is 43 + 9.8 or about 52.8 km^2.

11. Area of square:
$A = bh$
$A = (3.2) \cdot (3.2)$
$A \approx 10.3$
Area of two semicircles:
$A = 2 \cdot \left(\frac{1}{2}\pi r^2\right)$
$A = 2 \cdot \left(\frac{1}{2} \cdot \pi \cdot 1.6^2\right)$
$A \approx 8.0$
The area of the figure is 10.3 + 8.0 or about 18.3 in^2.

12. First, separate the figure into a rectangle, a square, and a triangle by drawing horizontal lines to make a 17 by 3 rectangle, a 9 by 9 square, and a triangle with base 9 and height 4.
Area of rectangle:
$A = bh$
$A = 17 \cdot 3$
$A = 51$
Area of square:
$A = bh$
$A = 9 \cdot 9$
$A = 81$
Area of triangle:
$A = \frac{1}{2}bh$
$A = \frac{1}{2} \cdot 9 \cdot 4$
$A = 18$
The area of the figure is 51 + 81 + 18 or 150 m^2.

13. First, separate the figure into two identical trapezoids and a rectangle by drawing two horizontal lines to make an 8 by 3 rectangle between the two trapezoids.
Area of trapezoid:
$A = \frac{1}{2}h(a + b)$
$A = \frac{1}{2} \cdot (1.5) \cdot (8 + 6)$
$A = 10.5$
Area of rectangle:
$A = bh$
$A = 8 \cdot 3$
$A = 24$
The area of the figure is 10.5 + 24 + 10.5 or 45 ft^2.

14. Area of square:
$A = bh$
$A = 8 \cdot 8$
$A = 64$
Area of semicircle:
$A = \frac{1}{2}\pi r^2$
$A = \frac{1}{2} \cdot \pi \cdot 2.8^2$
$A \approx 12.3$
The area of the figure is 64 + 12.3 or 76.3 m^2.

15. Area of rectangle:
$A = bh$
$A = (3.5) \cdot (2.8)$
$A = 9.8$

Area of semicircle:
$A = \frac{1}{2}\pi r^2$
$A = \frac{1}{2} \cdot \pi \cdot 7^2$
$A \approx 77.0$
The area of the figure is 9.8 + 77 or 86.8 yd^2.

16. Area of square:
$A = bh$
$A = 15 \cdot 15$
$A = 225$
Area of rectangle:
$A = bh$
$A = 8 \cdot 4$
$A = 32$
Area of semicircle:
$A = \frac{1}{2}\pi r^2$
$A = \frac{1}{2} \cdot \pi \cdot 4^2$
$A \approx 25.1$
The area of the non-shaded figure is 32 + 25.1 or about 57.1 square units. The area of the shaded figure is 225 − 57.1 or 167.9 square units.

17. Area of trapezoid:
$A = \frac{1}{2}h(a + b)$
$A = \frac{1}{2} \cdot 12 \cdot (10 + 14)$
$A = 144$
Area of rectangle:
$A = bh$
$A = 5 \cdot 10$
$A = 50$
Area of square:
$A = bh$
$A = 3 \cdot 3$
$A = 9$
The area of the non-shaded figure is 50 + 9 or 59 square units. The area of the shaded figure is 144 − 59 or 85 square units.

18. Area of circle:
$A = \pi r^2$
$A = \pi \cdot 11^2$
$A \approx 380.1$
Area of square:
$A = bh$
$A = 7.7$
$A = 49$
The area of the shaded figure is 380.1 − 49 or about 331.1 square units.

19. First, separate the figure into a rectangle and two semicircles.
Area of rectangle:
$A = bh$
$A = 100 \cdot 50$
$A = 5000$
Area of circle:
$A = \pi r^2$
$A = \pi \cdot 25^2$
$A \approx 1963.5$
The area of the grass region is 5000 + 1963.5 or about 6963.5 yd^2.

20. Area of outer square:
$A = bh$
$A = 12 \cdot 12$
$A = 144$
Area of inner square:
$A = 6 \cdot 6$
$A = 36$
The area of the sidewalk is 144 − 36 or 108 ft^2.

21. Sample answer: Separate the area into a rectangle and a trapezoid.

22. Sample answer: 62,500 mi^2

23. 68,679 mi^2

24. The outer and inner figures are made up of rectangles and semicircles.
Area of outer rectangle:
$A = bh$
$A = 30 \cdot 24$
$A = 720$
Area of outer semicircle:
$A = \frac{1}{2}\pi r^2$
$A = \frac{1}{2} \cdot \pi \cdot 12^2$
$A \approx 226.2$
The area of the outer region is 720 + 226.2 or about 946.2 ft^2.
Area of inner rectangle:
$A = bh$
$A = 26 \cdot 16$
$A = 416$
Area of inner semicircle:
$A = \frac{1}{2}\pi r^2$
$A = \frac{1}{2} \cdot \pi \cdot 8^2$
$A \approx 100.5$
The area of the inner region is 416 + 100.5 or about 516.5 ft^2.
The area of the patio is 946.2 − 516.5 or about 429.7 ft^2.

25. You can use polygons to find the area of an irregular figure by finding the area of each individual polygon and then finding the total area of the irregular figure. Answers should include the following.
- An example of an irregular figure:

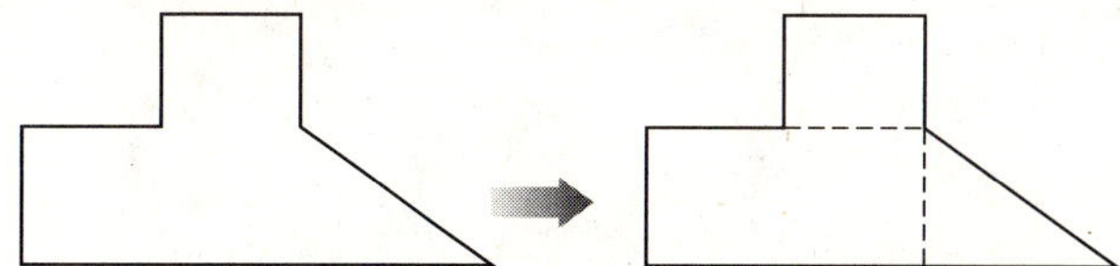

- To find the area of the irregular figure shown above, the figure can be separated into two rectangles and a triangle.

26. C; Area of shaded square:
$A = bh$
$A = 4 \cdot 4$
$A = 16$
Area of shaded triangle:
$A = \frac{1}{2}bh$
$A = \frac{1}{2} \cdot 7 \cdot 3$
$A = 10.5$
The area of the figure is 16 + 10.5 or 26.5 square units. Since each square unit represents 5 square feet, the area is 26.5 · 5 or 132.5 ft^2.

27. C; Area of outer square:
$A = bh$
$A = 10 \cdot 10$
$A = 100$ square units
Since each square unit represents 5 square feet, the area is 5 · 100 or 500 ft^2.
The area of the nonshaded region is 500 − 132.5 or 367.5 ft^2.

Page 543 Maintain Your Skills

28. $C = \pi d$
$C = \pi \cdot 8.5$
$C \approx 26.7$
The circumference is about 26.7 ft.
$A = \pi r^2$
$A = \pi \cdot (4.25)^2$
$A \approx 56.7$
The area is about 56.7 ft^2.

29. $C = 2\pi r$
$C = 2 \cdot \pi \cdot 7$
$C \approx 44.0$
The circumference is about 44.0 cm.
$A = \pi r^2$
$A = \pi \cdot 7^2$
$A \approx 153.9$
The area is about 153.9 cm^2.

30. $C = \pi d$
$C = \pi \cdot 19$
$C \approx 59.7$
The circumference is about 59.7 in.
$A = \pi r^2$
$A = \pi \cdot (9.5)^2$
$A \approx 283.5$
The area is about 283.5 in^2.

31. A pentagon has 5 sides, so $n = 5$.
$(n - 2)180 = (5 - 2)180$
$= 3 \cdot (180)$
$= 540°$

32. A quadrilateral has 4 sides, so $n = 4$.
$(n - 2) \cdot 180 = (4 - 2) \cdot 180$
$= 2 \cdot (180)$
$= 360°$

33. An octagon has 8 sides, so $n = 8$.
$(n - 2) \cdot 180 = (8 - 2) \cdot 180$
$= 6 \cdot (180)$
$= 1080°$

34.

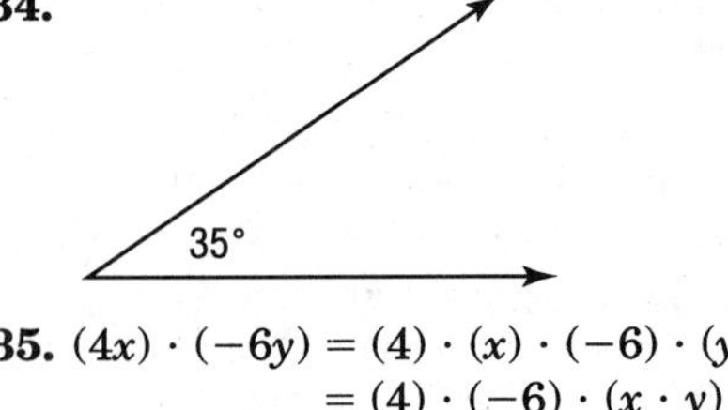

35. $(4x) \cdot (-6y) = (4) \cdot (x) \cdot (-6) \cdot (y)$
$= (4) \cdot (-6) \cdot (x \cdot y)$
$= -24xy$

Chapter 10 Study Guide and Review

Page 544 Vocabulary and Concept Check

1. supplementary
2. rhombus
3. corresponding angles
4. rotation
5. equilateral

Pages 544–548 Lesson-by-Lesson Review

6. Since ∠1 and ∠5 are corresponding angles, they are congruent. So, $m\angle 5 = 109°$.

7. Since ∠1 and ∠3 are vertical angles, they are congruent. So, $m\angle 3 = 109°$.

8. Since ∠1 and ∠2 are supplementary, the sum of their measures is 180°.
180 − 109 = 71. So, $m\angle 2 = 71°$.

9. Since ∠2 and ∠6 are corresponding angles, they are congruent. So, $m\angle 6 = 71°$.

10. $\angle F$ corresponds to $\angle Q$, so $\angle F \cong \angle Q$.

11. $\angle S$ corresponds to $\angle H$, so $\angle S \cong \angle H$.

12. $\angle R$ corresponds to $\angle G$, so $\angle R \cong \angle G$.

13. GH corresponds to RS, so $\overline{GH} \cong \overline{RS}$.

14. HF corresponds to SQ, so $\overline{HF} \cong \overline{SQ}$.

15. RQ corresponds to GF, so $\overline{RQ} \cong \overline{GF}$.

16. This translation can be written (4, −2).

vertex	4 right, 2 down		translation
$C(0, 2)$	+ (4, −2)	→	$C'(4, 0)$
$D(2, 0)$	+ (4, −2)	→	$D'(6, -2)$
$F(-1, -3)$	+ (4, −2)	→	$F'(3, -5)$
$G(-3, -1)$	+ (4, −2)	→	$G'(1, -3)$

The coordinates of the vertices of the rectangle are $C'(4, 0)$, $D'(6, -2)$, $F'(3, -5)$, and $G'(1, -3)$.

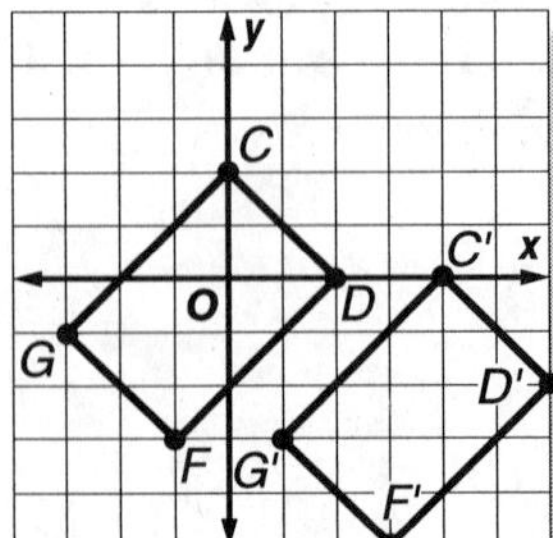

17. Use the same y-coordinate and multiply the x-coordinate by -1.

vertex		reflection
$H(-1, 4)$	$\rightarrow (-1 \cdot -1, 4)$	$\rightarrow H'(1, 4)$
$I(-4, -2)$	$\rightarrow (-1 \cdot -4, -2)$	$\rightarrow I'(4, -2)$
$J(-2, -1)$	$\rightarrow (-1 \cdot -2, -1)$	$\rightarrow J'(2, -1)$

The coordinates of the vertices of the triangle are $H'(1, 4)$, $I'(4, -2)$, and $J'(2, -1)$.

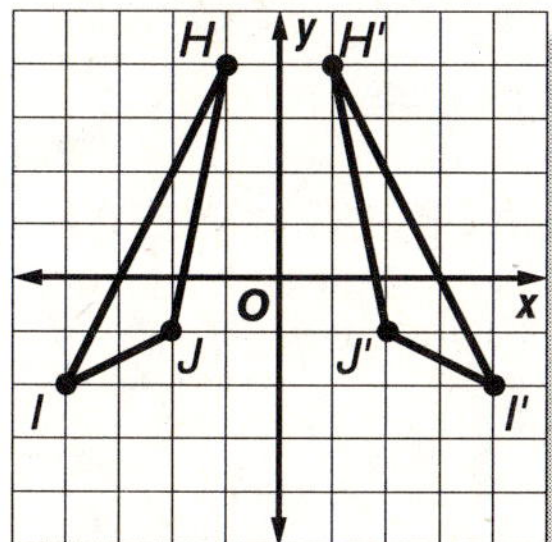

18. Switch the coordinates of each vertex and multiply the second by -1.

$N(1, 3) \rightarrow N'(3, -1)$
$P(3, 0) \rightarrow P'(0, -3)$
$Q(1, -1) \rightarrow Q'(-1, -1)$

The coordinates of the vertices of $N'(3, -1)$, $P'(0, -3)$, and $Q'(-1, -1)$.

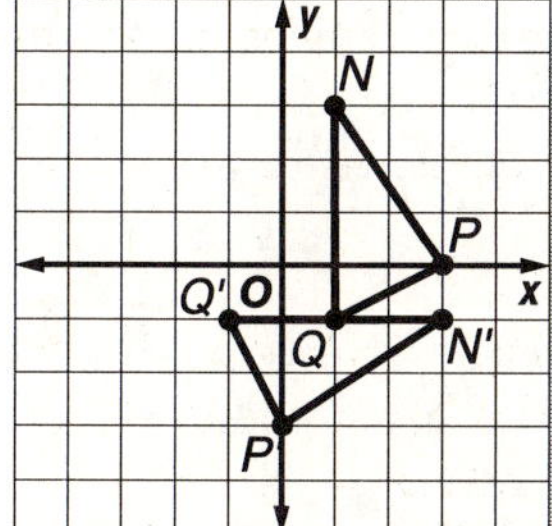

19. $x + 66 + 115 + 67 = 360$
$x + 248 = 360$
$x = 112$
The value of x is 112. So, the missing angle measure is 112°.

20. $78 + 90 + 2x + x = 360$
$168 + 3x = 360$
$3x = 192$
$x = 64$
The value of x is 64. So, the missing angle measures are 64° and 2(64) or 128°.

21. $110 + 110 + x + x = 360$
$220 + 2x = 360$
$2x = 140$
$x = 70$
The value of x is 70. So, the missing angles are each 70°.

22. $A = bh$
$A = 9 \cdot 13$
$A = 117$
The area of the parallelogram is 117 in^2.

23. $A = \frac{1}{2}bh$
$A = \frac{1}{2} \cdot 4 \cdot \frac{11}{2}$
$A = 11$
The area of the triangle is 11 yd^2.

24. $A = \frac{1}{2}h(a + b)$
$A = \frac{1}{2} \cdot (6.2) \cdot (8.7 + 5.0)$
$A = \frac{1}{2} \cdot (6.2) \cdot (13.7)$
$A = 42.47$
The area of the trapezoid is 42.47 m^2.

25. The polygon has 6 sides. It is a hexagon.
$(n - 2) \cdot 80 = (6 - 2) \cdot 180$
$= 4 \cdot (180)$ or 720
The sum of the measures of the angles is 720°.

26. The polygon has 7 sides. It is a heptagon.
$(n - 2) \cdot 180 = (7 - 2) \cdot 180$
$= 5 \cdot (180)$ or 900
The sum of the measures of the angles is 900°.

27. The polygon has 10 sides. It is a decagon.
$(n - 2) \cdot 180 = (10 - 2) \cdot 180$
$= 8 \cdot (180)$ or 1440
The sum of the measures of the angles is 1440°.

28. $C = 2\pi r$
$C = 2 \cdot \pi \cdot s$
$C \approx 31.4$
The circumference is about 31.4 cm.
$A = \pi r^2$
$A = \pi \cdot 5^2$
$A = \pi \cdot 25$
$A \approx 78.5$
The area is about 78.5 cm^2.

29. $C = 2\pi r$
$C = 2 \cdot \pi \cdot 2.1$
$C \approx 13.2$
The circumference is about 13.2 m.
$A = \pi r^2$
$A = \pi \cdot 2.1^2$
$A = \pi \cdot 4.41$
$A \approx 13.9$
The area is about 13.9 m^2.

30. $C = \pi d$
$C = \pi \cdot 18$
$C \approx 56.5$
The circumference is about 56.5 ft.
$A = \pi r^2$
$A = \pi \cdot 9^2$
$A = \pi \cdot 81$
$A \approx 254.5$
The area is about 254.5 ft^2.

31. Area of square:
$A = s^2$
$A = 7^2$ or 49
Area of trapezoid:
$A = \frac{1}{2}h(a + b)$
$A = \frac{1}{2} \cdot 6 \cdot (7 + 16)$
$A = \frac{1}{2} \cdot 6 \cdot 23$
$A = 69$
The area of the figure is 49 + 69 or 118 in^2.

32. Area of triangle:

$A = \frac{1}{2}bh$

$A = \frac{1}{2} \cdot (11 - 5.2) \cdot 9.4$

$A = \frac{1}{2} \cdot 5.8 \cdot 9.4$

$A = 27.26$

Area of rectangle:

$A = bh$

$A = (5.2) \cdot (9.4)$ or 48.88

The area of the figure is 27.26 + 48.88 or about 76.1 m^2.

33. Area of triangle:

$A = \frac{1}{2}bh$

$A = \frac{1}{2} \cdot 30 \cdot 34$

$A = 510$

Area of semicircle:

$A = \frac{1}{2}\pi r^2$

$A = \frac{1}{2} \cdot \pi \cdot 15^2$

$A = \frac{1}{2} \cdot \pi \cdot 225$

$A \approx 353.4$

The area of the figure is 510 + 353.4 or about 863.4 cm^2.

Chapter 10 Practice Test

Page 549

1a.

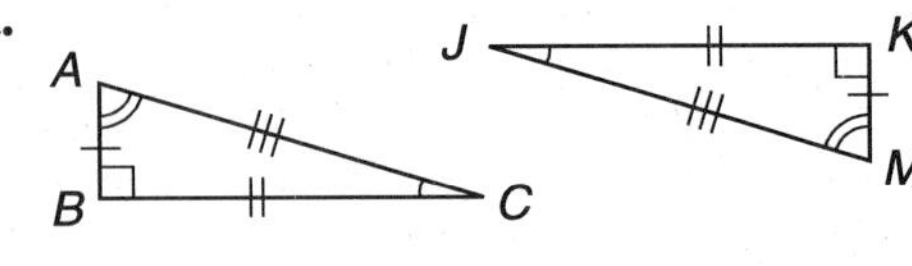

1b.

1c.

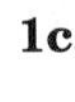

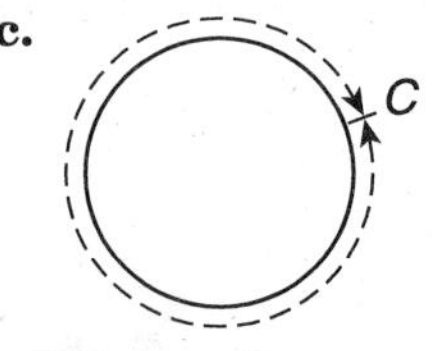

2. Whereas the sum of complementary angles is 90°, the sum of supplementary angles is 180°.
3. Since ∠5 and ∠6 are vertical angles, they are congruent. So, $m\angle 6 = 58°$.
4. Since ∠5 and ∠1 are corresponding angles, they are congruent. So, $m\angle 1 = 58°$. Since ∠1 and ∠7 are supplementary angles, the sum of their measures is 180°. 180 − 58 = 122. So, $m\angle 7 = 122°$.
5. Since ∠5 and ∠4 are alternate interior angles, they are congruent. So, $m\angle 4 = 58°$.
6. Since ∠7 and ∠3 are vertical angles, they are congruent. So, $m\angle 3 = 122°$.
7. Since $\angle P$ corresponds to $\angle M$, $\angle P \cong \angle M$.
8. Since $\overline{RS}$ corresponds to $\overline{NO}$, $\overline{RS} \cong \overline{NO}$.
9. Since $\angle MNO$ corresponds to $\angle PRS$, $\angle MNO \cong \angle PRS$.
10. Use the same y-coordinate and multiply the x-coordinate by −1.

vertex		reflection
$D(1, 3)$	$\to (-1 \cdot 1, 3)$	$\to D'(-1, 3)$
$E(4, 4)$	$\to (-1 \cdot 4, 4)$	$\to E'(-4, 4)$
$F(4, -2)$	$\to (-1 \cdot 4, -2)$	$\to F'(-4, -2)$
$G(1, -1)$	$\to (-1 \cdot 1, -1)$	$\to G'(-1, -1)$

The coordinates of the vertices are $D'(-1, 3)$, $E'(-4, 4)$, $F'(-4, -2)$, and $G'(-1, -1)$.

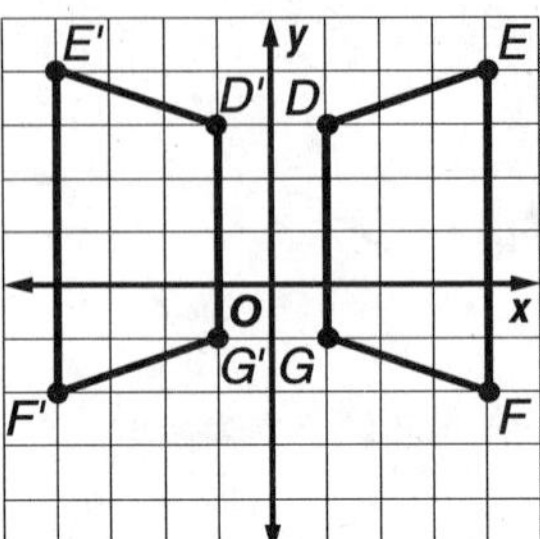

11. Switch the coordinates of each point and multiply the second by −1.

$E(2, -1) \to E'(-1, -2)$

$F(4, -1) \to F'(-1, -4)$

$H(4, -5) \to H'(-5, -4)$

$G(2, -3) \to G'(-3, -2)$

The coordinates of the vertices are $E'(-1, -2)$, $F'(-1, -4)$, $H'(-5, -4)$, and $G'(-3, -2)$.

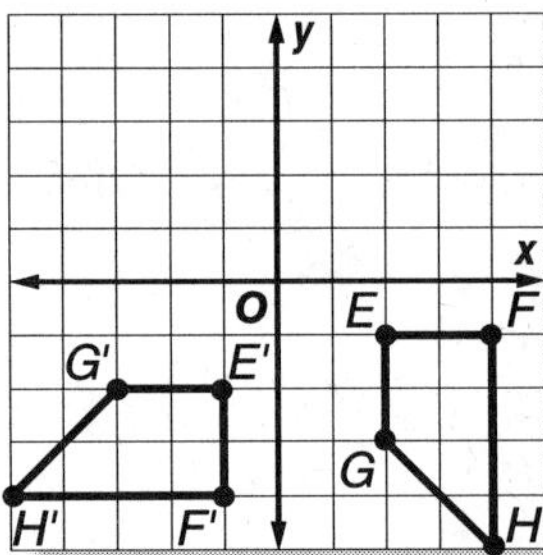

12. $x + 70 + 121 + 63 = 360$

$x + 254 = 360$

$x = 106$

The value of x is 106. So, the missing angle measure is 106°.

13. $(x - 11) + (2x - 1) + x + 40 = 360$

$x + 2x + x + (-11) + (-1) + 40 = 360$

$4x + 28 = 360$

$4x = 332$

$x = 83$

The value of x is 83. So, the missing angles are 83 − 11 or 72°, 2 · (83) − 1 or 165°, and 83°.

14. $A = \frac{1}{2}bh$

$A = \frac{1}{2} \cdot 21 \cdot 16$

$A = 168$

The area is 168 ft^2.

15. $A = bh$

$A = 7 \cdot (2.5)$ or 17.5

The area is 17.5 ft^2.

16. The polygon has 6 sides. It is a hexagon.
$(n - 2) \cdot 180 = (6 - 2) \cdot 180$
$= 4 \cdot (180)$ or 720
The sum of the measures of the interior angles is 720°.

17. The polygon has 8 sides. It is an octagon.
$(n - 2) \cdot 180 = (8 - 2) \cdot 180$
$= 6 \cdot (180)$ or 1080
The sum of the measures of the interior angles is 1080°.

18. $A = \pi r^2$
$A = \pi \cdot 3^2$
$A = \pi \cdot 9$
$A \approx 28.3$
The area is about 28.3 mi^2.
$C = 2\pi r$
$C = 2 \cdot \pi \cdot 3$
$C \approx 18.8$
The circumference is about 18.8 mi.

19. $A = \pi r^2$
$A = \pi \cdot 5^2$
$A = \pi \cdot 25$
$A \approx 78.5$
The area is about 78.5 in^2.
$C = \pi d$
$C = \pi \cdot 10$
$C \approx 31.4$
The circumference is about 31.4 in.

20. C; $C = \pi d$
$54.8 = \pi d$
$\frac{54.8}{\pi} = \frac{\pi d}{\pi}$
$17.4 \approx d$
The circumference is about 17.4 m.

Chapter 10 Standardized Test Practice

Pages 550–551

1. B; $\frac{5^5}{5^3} = 5^{5-3}$
$= 5^2$

2. D; $\frac{16}{4} = \frac{x}{3}$

3. C; Since 2% are expected to be defective, 100% − 2% or 98% should not be defective.
$750 \cdot (0.98) = 735$
There should be 735 that are not defective.

4. B; There are 6 sides on a cube. Since $\frac{1}{3} = \frac{2}{6}$, there should be 2 sides painted blue.

5. B; $C = A - 2B$

6. C; $c^2 = a^2 + b^2$
$c^2 = 36^2 + 48^2$
$c^2 = 1296 + 2304$
$c^2 = 3600$
$c = 60$
The hypotenuse is 60 units long.

7. A; Since the rectangles are parallel, $\angle 1$ and $\angle 2$ are congruent alternate interior angles. So, $m\angle 1 - m\angle 2 = 0$.

8. A; The sum of the measures of the angles in a quadrilateral is 360°.
$100 + 100 + 90 + m\angle D = 360$
$290 + m\angle D = 360$
$m\angle D = 70$

9. A; Since the side of the square is 3 in, the radius of the circle is $\frac{1}{2} \cdot (3)$ or 1.5 in.
$A = \pi r^2$
$A = \pi \cdot 1.5^2$
$A = \pi \cdot 2.25$
$A \approx 7$
The area of the circle is about 7 in^2.

10. $\frac{14x + 6}{5x - 3} = \frac{14 \cdot 3 + 6}{5 \cdot 3 - 3}$
$= \frac{42 + 6}{15 - 3}$
$= \frac{48}{12}$ or 4

11. The median is $\frac{22 + 24}{2}$ or 23. The mode is 24. Since $24 > 23$, the mode is greater.

12. $0.83 \cdot 512 = 424.96$
About 425 students use e-mail.

13. $n + 2$

14. $4x + 7 < 39$
$4x < 32$
$x < 8$

15. Write the equation in the form $y = mx + b$.
$3x + y = 5$
$y = -3x + 5$
The slope is −3.

16. This translation can be written (4, −2).

vertex	2 left, 1 down	translation
$W(0, 4)$	$+ (-2, -1) \rightarrow$	$W'(-2, 3)$
$X(4, 3)$	$+ (-2, -1) \rightarrow$	$X'(2, 2)$
$Y(4, 1)$	$+ (-2, -1) \rightarrow$	$Y'(2, 0)$
$Z(1, 2)$	$+ (-2, -1) \rightarrow$	$Z'(-1, 1)$

The coordinates of the vertices are $W'(-2, 3)$, $X'(2, 2)$, $Y'(2, 0)$, and $Z'(-1, 1)$.

17. The obtuse angle in the second triangle corresponds to the 105° angle in the first triangle. The angle sum in a triangle is 180°.
$x + 35 + 105 = 180$
$x + 140 = 180$
$x = 40$
The value of x is 40.

18. The quadrilateral has 4 congruent sides, but no right angles. It is a rhombus.

19. First, separate the figure into a 2 ft by 6 ft rectangle and a 12 ft by 3 ft rectangle.
Area of smaller rectangle:
$A = bh$
$A = 2 \cdot 6$ or 12
Area of larger rectangle:
$A = bh$
$A = 12 \cdot 3$ or 36
The area of the figure, without the cut-out is 12 + 36 or 48 ft^2.
Area of rectangular cut-out:
$A = bh$
$A = 3 \cdot 2$ or 6
The area of the counter top is 48 − 6 or 42 ft^2.

20a. $J(-3, 4), K(-2, 1), M(-1, 2)$

20b–d.

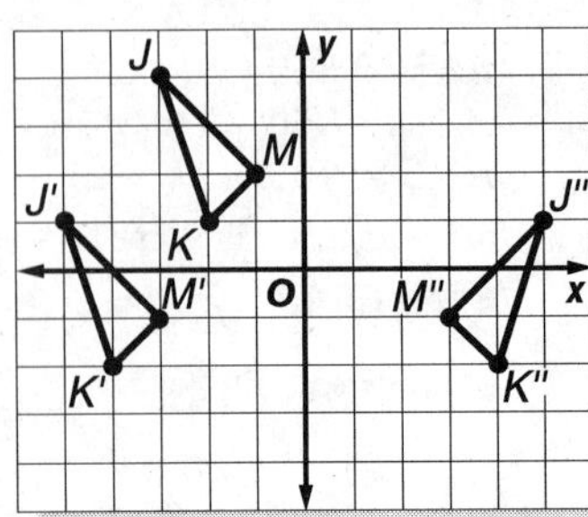

20c. This translation can be written as $(-2, -3)$.

vertex	2 left, 3 down	translation
$J(-3, 4)$	$+ (-2, -3) \rightarrow$	$J'(-5, 1)$
$K(-2, 1)$	$+ (-2, -3) \rightarrow$	$K'(-4, -2)$
$M(-1, 2)$	$+ (-2, -3) \rightarrow$	$M'(-3, -1)$

20d. Use the same y-coordinate and multiply the x-coordinate by -1.

vertex		translation
$J'(-5, 1)$	$\rightarrow (-1 \cdot -5, 1)$	$\rightarrow J''(5, 1)$
$K'(-4, -2)$	$\rightarrow (-1 \cdot -4, -2)$	$\rightarrow K''(4, -2)$
$M'(-3, -1)$	$\rightarrow (-1 \cdot -3, -1)$	$\rightarrow M''(3, -1)$

20e–f.

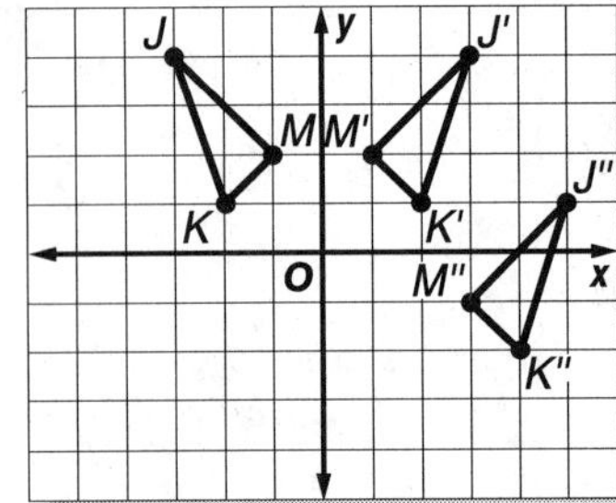

20e. Use the same y-coordinate and multiply the x-coordinate by -1.

vertex		translation
$J(-3, 4)$	$\rightarrow (-1 \cdot -3, 4)$	$\rightarrow J'(3, 4)$
$K(-2, 1)$	$\rightarrow (-1 \cdot -2, 1)$	$\rightarrow K'(2, 1)$
$M(-1, 2)$	$\rightarrow (-1 \cdot -1, 2)$	$\rightarrow M'(1, 2)$

20f. This translation can be written as $(2, -3)$.

vertex	2 right, 3 down	translation
$J'(3, 4)$	$+ (2, -3) \rightarrow$	$J''(5, 1)$
$K'(2, 1)$	$+ (2, -3) \rightarrow$	$K''(4, -2)$
$M'(1, 2)$	$+ (2, -3) \rightarrow$	$M''(3, -1)$

20g. The two images are the same because the order in which these operations are performed does not matter. Thus, the result is the same.

Chapter 11 Three-Dimensional Figures

Page 553 Getting Started

1. Yes; it is a triangle.
2. Not a polygon
3. Not a polygon
4. Yes; it is a quadrilateral.
5. $8.5 \cdot 2 = 17$
6. $3.2(3.2)10 = 102.4$
7. $\frac{1}{2} \cdot 14 = \frac{1}{\cancel{2}_1} \cdot \frac{\cancel{14}^7}{1}$
   $= 7$
8. $\frac{1}{2}(6.4)(5) = \frac{1}{\cancel{2}_1} \cdot \frac{\cancel{6.4}^{3.2}}{1} \cdot \frac{5}{1}$
   $= 16$
9. $\frac{1}{3}(50)(9.3) = \frac{1}{\cancel{3}_1} \cdot \frac{50}{1} \cdot \frac{\cancel{9.3}^{3.1}}{1}$
   $= 155$
10. $\frac{1}{3}\left(\frac{1}{2} \cdot 3 \cdot 8\right) = \frac{1}{\cancel{3}_1} \cdot \frac{1}{\cancel{2}_1} \cdot \frac{\cancel{3}^1}{1} \cdot \frac{\cancel{8}^4}{1}$
    $= 4$
11. $\frac{3}{8} \stackrel{?}{=} \frac{9}{24}$
    $3 \cdot 24 \stackrel{?}{=} 8 \cdot 9$
    $72 = 72$
    Yes, $\frac{3}{8}$ and $\frac{9}{24}$ form a proportion.
12. $\frac{7}{2} \stackrel{?}{=} \frac{14}{6}$
    $7 \cdot 6 \stackrel{?}{=} 2 \cdot 14$
    $42 \neq 28$
    No, $\frac{7}{2}$ and $\frac{14}{6}$ do not form a proportion.
13. $\frac{18}{32} \stackrel{?}{=} \frac{9}{16}$
    $18 \cdot 16 \stackrel{?}{=} 32 \cdot 9$
    $288 = 288$
    Yes, $\frac{18}{32}$ and $\frac{9}{16}$ form a proportion.
14. $\frac{12}{15} \stackrel{?}{=} \frac{4}{5}$
    $12 \cdot 5 \stackrel{?}{=} 15 \cdot 4$
    $60 = 60$
    Yes, $\frac{12}{15}$ and $\frac{4}{5}$ form a proportion.
15. $\frac{1.2}{5} \stackrel{?}{=} \frac{6}{25}$
    $1.2 \cdot 25 \stackrel{?}{=} 5 \cdot 6$
    $30 = 30$
    Yes, $\frac{1.2}{5}$ and $\frac{6}{25}$ form a proportion.
16. $\frac{1.6}{2} \stackrel{?}{=} \frac{3.6}{6}$
    $1.6 \cdot 6 \stackrel{?}{=} 2 \cdot 3.6$
    $9.6 \neq 7.2$
    No, $\frac{1.6}{2}$ and $\frac{3.6}{6}$ do not form a proportion.

Pages 554–555 Geometry Activity (Preview of Lesson 11-1)

1. Sample answer:

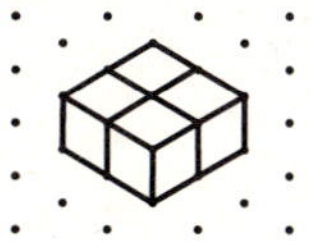

2. Sample answer:

3. Sample answer:

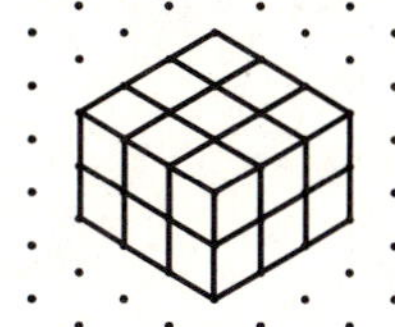

4. Sample answer:

5. Sample answer:

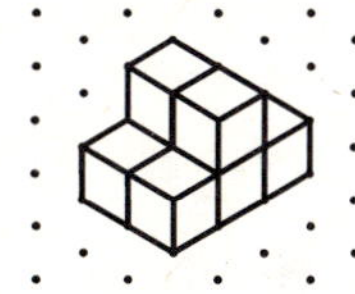

6. Sample answer:

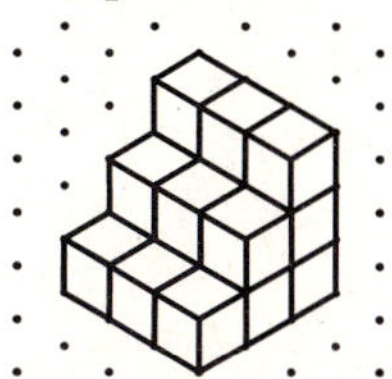

7. Sample answer:

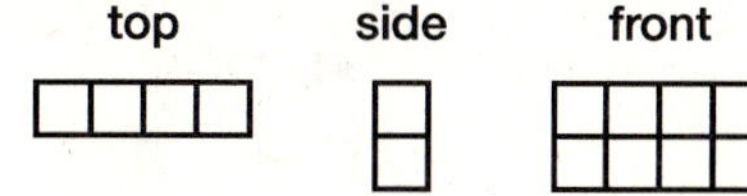

8. Sample answer:

top side front

9. Sample answer:

top side front

10. Sample answer:

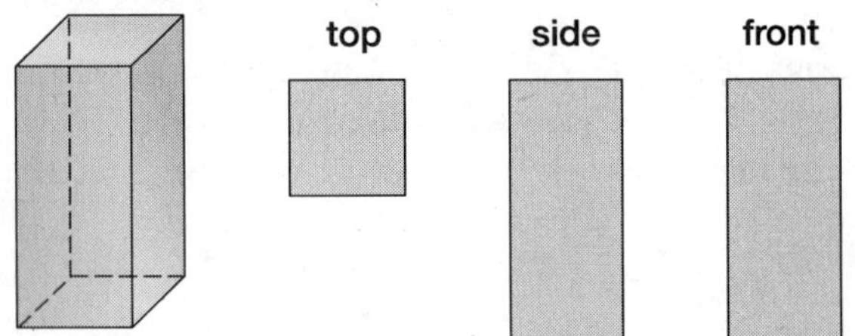

11. Sample answer:

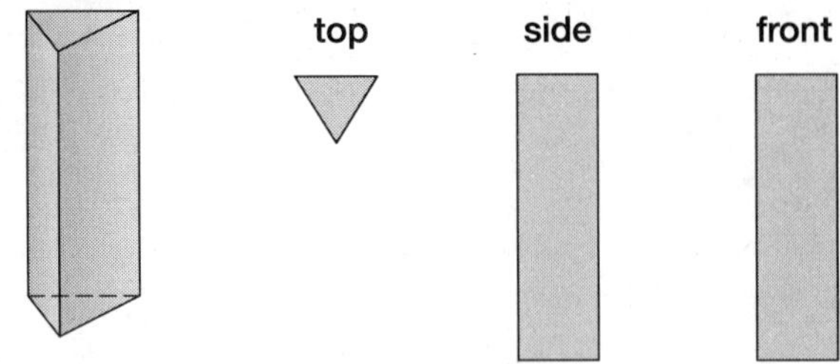

12. Sample answer:

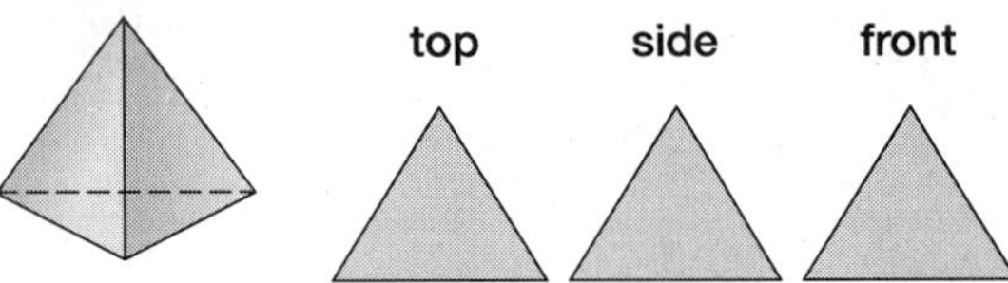

13. Sample answer:

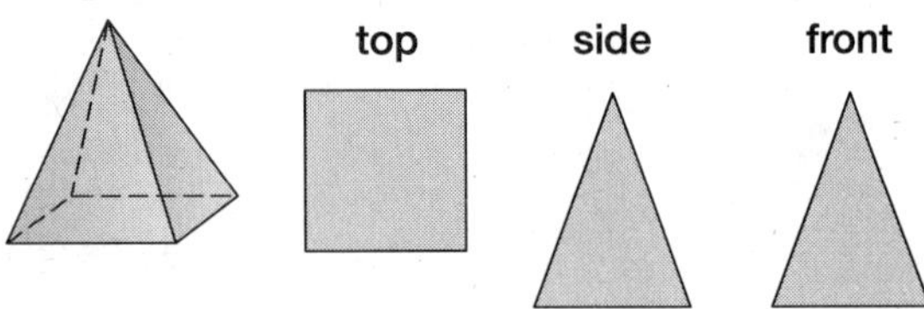

14. Sample answer:

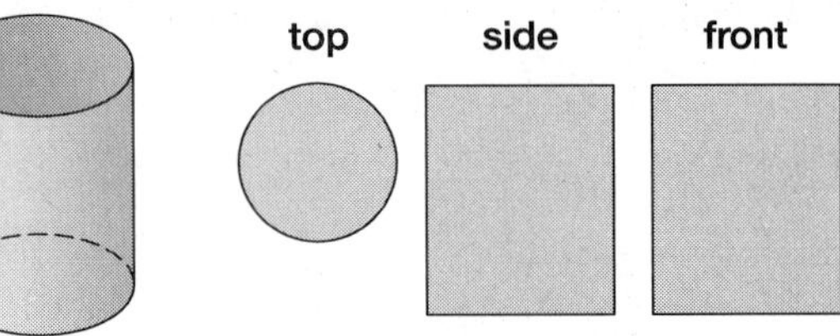

15. Sample answer:

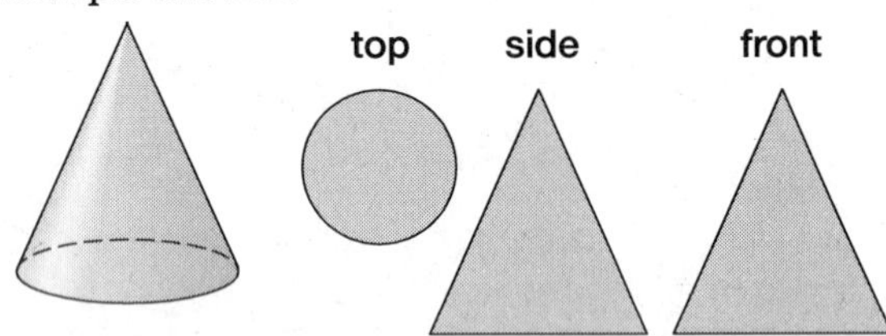

11-1 Three-Dimensional Figures

Page 556 How are 2-dimensional figures related to 3-dimensional figures?

a. Square

b. Pyramid, triangle; Trade center, rectangle

c. Sample answer: When a three-dimensional figure has faces that are different polygons, the polygon you see depends on your point of view.

Page 559 Check for Understanding

1. Five planes form a square pyramid because the solid has five faces. An edge is formed when two planes intersect; a vertex is formed when three or four planes intersect.

2. Sample answer: On a television, the top front edge and a bottom side edge form skew lines. The top and bottom front edges do not form skew lines because they are parallel. See students' drawings.

3. The figure is a triangular pyramid since it has one triangular base.
The faces are *RST*, *QRS*, *QST*, and *QRT*; any one of the faces can be considered a base.
The edges are $\overline{QR}$, $\overline{QS}$, $\overline{QT}$, $\overline{RT}$, $\overline{RS}$, and $\overline{ST}$.
The vertices are *Q*, *R*, *S*, and *T*.

4. The figure is a rectangular prism.
The bases are *ABCD*, *EHGF*, or *ABHE*, *DCGF* or *ADFE*, and *BCGH*.
The faces are *ABCD*, *EHGF*, *ABHE*, *DCGF*, *ADFE*, and *BCGH*.
The edges are $\overline{AB}$, $\overline{BC}$, $\overline{CD}$, $\overline{DA}$, $\overline{EF}$, $\overline{FG}$, $\overline{GH}$, $\overline{HE}$, $\overline{AE}$, $\overline{BH}$, $\overline{CG}$, and $\overline{DF}$.
The vertices are *A*, *B*, *C*, *D*, *E*, *F*, *G*, and *H*.

5. intersecting

6. $\overline{MS}$ and $\overline{MT}$ are skew to $\overline{PO}$.

7.

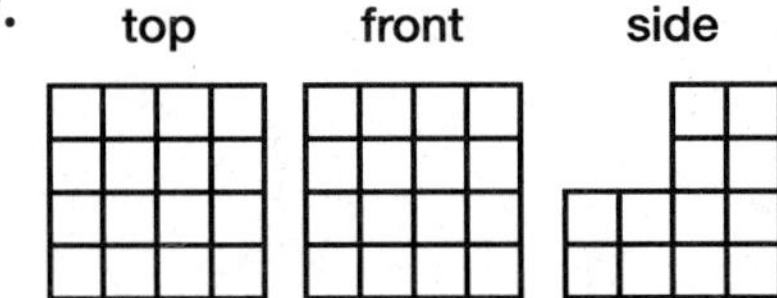

8. The steps are 4 units high.
4×4 inches $= 16$ inches
$\overset{4}{\cancel{16}}\ \cancel{\text{in.}} \cdot \frac{1 \text{ ft}}{\underset{3}{\cancel{12}}\ \cancel{\text{in.}}} = \frac{4}{3}$ or $1\frac{1}{3}$ ft

Pages 559–561 Practice and Apply

9. The figure has two parallel, congruent bases that are rectangles, *LMNP* and *QRST*, or *LPTQ* and *MNSR*, or *PNST* and *LMRQ*, so it is a rectangular prism.
faces: *LMNP*, *QRST*, *LPTQ*, *MNSR*, *PNST*, LMRQ
edges: $\overline{LM}$, $\overline{MN}$, $\overline{NP}$, $\overline{PL}$, $\overline{LQ}$, $\overline{MR}$, $\overline{NS}$, $\overline{PT}$, $\overline{QR}$, $\overline{RS}$, $\overline{ST}$, $\overline{TQ}$
vertices: *L*, *M*, *N*, *P*, *Q*, *R*, *S*, *T*

10. The figure has two parallel, congruent bases that are triangles, *FGH* and *IJK*, so it is a triangular prism.
faces: *FGH*, *IJK*, *HKIF*, *GHKJ*, *FGJI*
edges: $\overline{FG}$, $\overline{GH}$, $\overline{HF}$, $\overline{IJ}$, $\overline{JK}$, $\overline{KI}$, $\overline{FI}$, $\overline{HK}$, $\overline{GJ}$
vertices: *F*, *G*, *H*, *I*, *J*, *K*

11. The figure has one triangular base, *WXY*, *WYZ*, *WZX*, or *XYZ*, so it is a triangular pyramid.
edges: $\overline{WX}$, $\overline{WY}$, $\overline{WZ}$, $\overline{XZ}$, $\overline{XY}$, $\overline{YZ}$
vertices: *W*, *X*, *Y*, *Z*

12. The figure has one rectangular base, $BCDE$, so it is a rectangular pyramid.
faces: $ABC, ACD, ADE, ABE, BCDE$
edges: $\overline{AB}, \overline{AC}, \overline{AD}, \overline{AE}, \overline{BC}, \overline{CD}, \overline{DE}, \overline{EB}$
vertices: A, B, C, D, E
13. Sample answer: $\overline{WR}$
14. Sample answer: $\overline{WP}$, $\overline{ZS}$, $\overline{WX}$ and $\overline{ZY}$ are skew to $\overline{QR}$.
15. skew
16. Sample answer: $\overline{PS}$
17. top
18. top side front

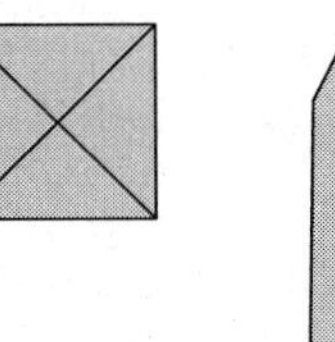
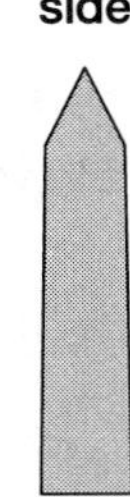
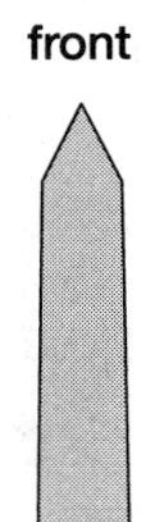

19. They are rectangular and pentagonal prisms.
20. Sample answer: using angles and shading to portray depth
21. See students' work.
22. Sometimes; two planes may be parallel and not intersect at all.
23. Never; only three or more planes can intersect in a single point.
24. Never; every pair of vertices on a pyramid has a common edge.
25. Sometimes; three planes may intersect in a line. Or, the planes may be parallel and not intersect at all.
26. Sample answer:

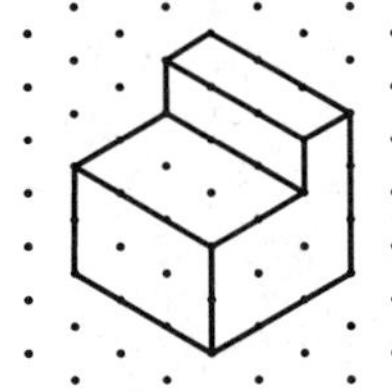

The figure has bilateral symmetry because it can be divided into two matching halves. It does not have rotational symmetry because when rotated less than 360°, it never looks exactly as it does in its original position.

27. Two-dimensional figures form three-dimensional figures. Answers should include the following.
    - Two-dimensional figures have length and width and therefore lie in a single plane. Three-dimensional figures have length, width, and depth.
    - Two-dimensional figures form the faces of three-dimensional figures.
28. A; The intersection is a point.
29. D; Figure D has 2 faces that are squares; the others do not.

Page 561 Maintain Your Skills

30. Area of rectangle:
$A = \ell \cdot w$
$A = 9 \cdot 4$
$A = 36$
Area of left triangle:
$A = \frac{1}{2}b \cdot h$
$A = \frac{1}{2}(4) \cdot 2$
$A = 4$
Area of right triangle:
$A = \frac{1}{2}b \cdot h$
$A = \frac{1}{2}(2) \cdot 2$
$A = 2$
Area of the polygon:
A = (area of rectangle) − (area of left triangle) − (area of right triangle)
$A = 36 - 4 - 2$
$A = 30$
The area is 30 cm^2.

31. $C = 2\pi r$ $\quad A = \pi r^2$
$C = 2\pi(6)$ $\quad A = \pi(6)^2$
$C \approx 37.7$ cm $\quad A \approx 113.1$ cm^2

32. $\sin 35° \approx 0.5736$
33. $\sin 30° = 0.5000$
34. $\cos 280° \approx 0.1736$
35. $c + 4 < 12$
$c + 4 - 4 < 12 - 4$
$c < 8$
To check the solution, try any number less than 8.
Check: $c + 4 < 12$
$7 + 4 \stackrel{?}{<} 12$
$11 < 12$ ✓

36. $7 \geq t - 2$
$7 + 2 \geq t - 2 + 2$
$9 \geq t$
To check the solution, try any number less than or equal to 9.
Check: $7 \geq t - 2$
$7 \stackrel{?}{\geq} 8 - 2$
$7 \geq 6$ ✓

37. $-26 < n + (-15)$
$-26 - (-15) < n + (-15) - (-15)$
$-11 < n$
To check the solution, try any number greater than −11.
Check: $-26 < n + (-15)$
$-26 \stackrel{?}{<} -10 + (-15)$
$-26 < -25$ ✓

38. $k + (-4) \geq 3.8$
$k + (-4) - (-4) \geq 3.8 - (-4)$
$k \geq 7.8$
To check the solution, try any number greater than or equal to 7.8.
Check: $k + (-4) \geq 3.8$
$8 + (-4) \stackrel{?}{\geq} 3.8$
$4 \geq 3.8$ ✓

39. $y - \frac{1}{4} < 1\frac{1}{2}$

$y - \frac{1}{4} + \frac{1}{4} < 1\frac{1}{2} + \frac{1}{4}$

$y < \frac{3}{2} + \frac{1}{4}$

$y < \frac{3}{2} \cdot \frac{2}{2} + \frac{1}{4}$

$y < \frac{6}{4} + \frac{1}{4}$

$y < \frac{7}{4}$ or $y < 1\frac{3}{4}$

To check the solution, try any number less than $\frac{7}{4}$.

Check: $y - \frac{1}{4} < 1\frac{1}{2}$

$\frac{6}{4} - \frac{1}{4} \overset{?}{<} 1\frac{1}{2}$

$\frac{5}{4} \overset{?}{<} 1\frac{1}{2}$

$1\frac{1}{4} < 1\frac{1}{2}$ ✓

40. $3\frac{1}{5} > a - \frac{3}{10}$

$3\frac{1}{5} + \frac{3}{10} > a - \frac{3}{10} + \frac{3}{10}$

$\frac{16}{5} + \frac{3}{10} > a$

$\frac{16}{5} \cdot \frac{2}{2} + \frac{3}{10} > a$

$\frac{32}{10} + \frac{3}{10} > a$

$\frac{35}{10} > a$

$\frac{7}{2} > a$ or $a < 3\frac{1}{2}$

To check the solution, try any number less than $\frac{7}{2}$.

Check: $3\frac{1}{5} > a - \frac{3}{10}$

$3\frac{1}{5} \overset{?}{>} \frac{6}{2} - \frac{3}{10}$

$3\frac{1}{5} \overset{?}{>} \frac{6}{2} \cdot \frac{5}{5} - \frac{3}{10}$

$3\frac{1}{5} \overset{?}{>} \frac{30}{10} - \frac{3}{10}$

$3\frac{1}{5} > \frac{27}{10}$ or $2\frac{7}{10}$ ✓

41. $A = \frac{1}{2}b \cdot h$

$A = \frac{1}{2}(4)(7)$

$A = 14 \text{ in}^2$

42. $A = \frac{1}{2}b \cdot h$

$A = \frac{1}{2}(10)(9)$

$A = 45 \text{ ft}^2$

43. $A = \frac{1}{2}b \cdot h$

$A = \frac{1}{2}(6.5)(2)$

$A = 6.5 \text{ cm}^2$

44. $A = \frac{1}{2}b \cdot h$

$A = \frac{1}{2}(0.4)(1.3)$

$A = 0.26 \text{ m}^2$

Page 562 Geometry Activity (Preview of Lesson 11-2)

1. cylinder; triangular prism
2. same height; 5 in.
3. same; 8 in.
4. circular: $\approx 5 \text{ in}^2$; triangular: $\approx 3 \text{ in}^2$; square: 4 in^2
5. cylinder
6. Yes; the greater the base area, the greater the volume.

11-2 Volume: Prisms and Cylinders

Page 563 How is volume related to area?

a. Sample answers:

Prism	Length (units)	Width (units)	Height (units)	Area of Base (units^2)
1	6	1	4	6
2	6	2	2	12
3	4	3	2	12
4	8	1	3	8

b. Volume equals the product of the length, width, and height.

c. $V = Bh$

Page 565–566 Check for Understanding

1. Sample answer: Finding how much sand is needed to fill a child's rectangular sandbox; measure the length, width, and height, and then multiply to find the volume.
2. Marissa; a cube has three dimensions, and when each dimension is multiplied by 2, the volume is multiplied by $2 \times 2 \times 2$ or 8.
3. $V = Bh$
   $V = (\ell \cdot w)h$
   $V = (4 \cdot 5.1)9$
   $V = 183.6$
   The volume is 183.6 cm^3.
4. $V = Bh$
   $V = \left(\frac{1}{2} \cdot 7 \cdot 8\right)h$
   $V = \left(\frac{1}{2} \cdot 7 \cdot 8\right)15$
   $V = 420$
   The volume is 420 in^3.
5. $V = \pi r^2 h$
   $V = \pi \cdot 8^2 \cdot 8$
   $V \approx 1608.5$
   The volume is about 1608.5 ft^3.
6. $V = Bh$
   $V = (\ell \cdot w)h$
   $V = (6 \cdot 6)9$
   $V = 324$
   The volume is 324 in^3.

7. $V = \pi r^2 h$
$V = \pi \cdot 3^2 \cdot 10$
$V \approx 282.7$
The volume is about 282.7 yd^3.

8. $V = Bh$
$V = \ell \cdot w \cdot h$
$60.3 = 3 \cdot 1.5 \cdot h$
$60.3 = 4.5h$
$13.4 = h$
The height of the prism is 13.4 m.

9. $V \leq \pi r^2 h$
$1{,}000{,}000 \leq \pi \cdot 40^2 \cdot h$
$1{,}000{,}000 \leq 5026.5h$
$198.9 \leq h$
The height should be at least 198.9 feet.

10. D; $V_1 = Bh$ | $V_2 = Bh$
$V_1 = (\ell \cdot w)h$ | $V_2 = (\ell \cdot w)h$
$V_1 = (2 \cdot 2)1.5$ | $V_2 = (5 \cdot 2)1$
$V_1 = 6$ | $V_2 = 10$
$V_1 + V_2 = 6 + 10$
$= 16 \text{ in}^3$
The volume is 16 in^3.

Pages 566–567 Practice and Apply

11. $V = Bh$
$V = (\ell \cdot w)h$
$V = (4 \cdot 16)8$
$V = 512$
The volume is 512 cm^3.

12. $V = Bh$
$V = (\ell \cdot w)h$
$V = (7 \cdot 2.6)4.5$
$V = 81.9$
The volume is 81.9 m^3.

13. $V = Bh$
$V = \left(\frac{1}{2} \cdot 11 \cdot 8\right)h$
$V = \left(\frac{1}{2} \cdot 11 \cdot 8\right)17$
$V = 748$
The volume is 748 in^3.

14. $V = Bh$
$V = \left(\frac{1}{2} \cdot 10 \cdot 8\right)h$
$V = \left(\frac{1}{2} \cdot 10 \cdot 8\right)15$
$V = 600$
The volume is 600 in^3.

15. $V = \pi r^2 h$
$V = \pi \cdot 2^2 \cdot 7$
$V \approx 88.0$
The volume is about 88 ft^3.

16. $r = \frac{1}{2}(2.7)$
$r = 1.35$
$V = \pi r^2 h$
$V = \pi \cdot 1.35^2 \cdot 30$
$V \approx 171.8$
The volume is about 171.8 m^3.

17. $V = Bh$
$V = (\ell \cdot w)h$
$V = (3 \cdot 5)15$
$V = 225$
The volume is 225 mm^3.

18. $V = Bh$
$V = \left(\frac{1}{2} \cdot 8 \cdot 15\right)h$
$V = \left(\frac{1}{2} \cdot 8 \cdot 15\right)6.5$
$V = 390$
The volume is 390 in^3.

19. $r = \frac{1}{2}(2.6)$
$r = 1.3$
$V = \pi r^2 h$
$V = \pi \cdot 1.3^2 \cdot 3.5$
$V \approx 18.6$
The volume is about 18.6 m^3.

20. $V = Bh$
$V = 25 \cdot 1.5$
$V = 37.5$
The volume is 37.5 m^3.

21. $V = \ell \cdot w \cdot h$
$83.3 = 4.2 \cdot 3.2 \cdot h$
$83.3 = 13.44h$
$6.2 \approx h$
The height of the prism is about 6.2 m.

22. $V = \pi r^2 h$
$28.3 = \pi \cdot 2^2 \cdot h$
$28.3 = 12.57h$
$2.3 \approx h$
The height of the cylinder is about 2.3 ft.

23. $1 \text{ ft}^3 = 12 \cdot 12 \cdot 12 = 1728 \text{ in}^3$

24. $1 \text{ cm}^3 = 10 \cdot 10 \cdot 10 = 1000 \text{ mm}^3$

25. $1 \text{ m}^3 = 100 \cdot 100 \cdot 100 = 1{,}000{,}000 \text{ cm}^3$

26. $V = B \cdot h$
$V = \ell \cdot w \cdot h$
$V = 2 \cdot 3 \cdot 2$
$V = 12 \text{ cm}^3$
$\text{mass} = 12 \cancel{\text{cm}^3} \cdot \frac{19.29 \text{ g}}{\cancel{\text{cm}^3}} = 231.48 \text{ g}$

27. $1.2 \text{ ft}^3 = 1.2 \cancel{\text{ft}^3} \cdot \frac{1728 \text{ in}^3}{1 \cancel{\text{ft}^3}}$
$= 2073.6 \text{ in}^3$
$V = Bh$
$V = \ell \cdot w \cdot h$
$2073.6 = 10 \cdot 18 \cdot h$
$2073.6 = 180h$
$11.5 \approx h$
The microwave is about 11.5 inches deep.

28. $r = \frac{1}{2}(33.3)$
$r = 16.65$
$V = \pi r^2 h$
$V = \pi(16.65)^2(61.1)$
$V \approx 53213.2$
$53213.2 \text{ mm}^3 = \frac{53213.2 \cancel{\text{mm}^3}}{1} \cdot \frac{1 \text{ cm}^3}{1000 \cancel{\text{mm}^3}}$
$\approx 53.2 \text{ cm}^3$
A size D battery has a volume of about 53.2 cm^3.

$r = \frac{1}{2}(25.5)$
$r = 12.75$
$V = \pi r^2 h$
$V = \pi(12.75)^2(50)$
$V \approx 25535.3$

$$25535.3 \text{ mm}^3 = \frac{25535.3 \cancel{\text{mm}^3}}{1} \cdot \frac{1 \text{ cm}^3}{1000 \cancel{\text{mm}^3}} \approx 25.5 \text{ cm}^3$$

A size C battery has a volume of about 25.5 cm^3.

$r = \frac{1}{2}(14.5)$
$r = 7.25$
$V = \pi r^2 h$
$V = \pi(7.25)^2(50.5)$
$V \approx 8339$

$$8339 \text{ mm}^3 = \frac{8339 \cancel{\text{mm}^3}}{1} \cdot \frac{1 \text{ cm}^3}{1000 \cancel{\text{mm}^3}} \approx 8.3 \text{ cm}^3$$

A size AA battery has a volume of about 8.3 cm^3.

$r = \frac{1}{2}(10.5)$
$r = 5.25$
$V = \pi r^2 h$
$V = \pi(5.25)^2(44.5)$
$V \approx 3853.3$

$$3853.3 \text{ mm}^3 = \frac{3853.3 \cancel{\text{mm}^3}}{1} \cdot \frac{1 \text{ cm}^3}{1000 \cancel{\text{mm}^3}} \approx 3.9 \text{ cm}^3$$

A size AAA battery has a volume of about 3.9 cm^3.

29. $C = 2\pi r$

$8.5 = 2\pi r$	$11 = 2\pi r$
$1.4 \approx r$	$1.8 \approx r$
$V = \pi r^2 h$	$V = \pi r^2 h$
$V = \pi(1.4)^2(11)$	$V = \pi(1.8)^2(8.5)$
$V \approx 67.7$ in^3	$V \approx 86.5$ in^3

The volume will be greater if the height is $8\frac{1}{2}$ inches. By using the formula for circumference, you can find the radius and volume of each cylinder. If the height is $8\frac{1}{2}$ inches, the volume is 86.5 in^3; if the height is 11 inches, the volume is 67.7 in^3.

30. Volume of a solid equals the area of the solid's base times its height. Answers should include the following.

- Area is a measure of figures on a flat plane having two dimensions, length and width. Volume is a measure of objects in space having three dimensions, length, width, and height.
- The dimensions that are used to find area are also used to find volume, so the formula for volume includes the formula for area.

31. C; $V = \ell \cdot w \cdot h$
$V = 18.79 \cdot 18.79 \cdot 18.79$
$V \approx 20 \cdot 20 \cdot 20$
$V \approx 8000$ mm^3

32. A; $r = \frac{1}{2}(2.5)$
$r = 1.25$
The figure is half of a cylinder.
$V = Bh$
$V = \left(\frac{1}{2}\pi r^2\right)h$
$V = \frac{1}{2}\pi(1.25)^2 \cdot 10$
$V = \frac{1}{2}\pi \cdot 1.5625 \cdot 10$
$V \approx 24.5$ ft^3

Page 567 Maintain Your Skills

33. Sample answer: $\overline{QT}$ and $\overline{YZ}$

34. First, approximate the curved parts of the figure with straight lines.
The figure becomes a rectangle with a triangle missing from its top left corner and a second triangle missing from its top right corner.

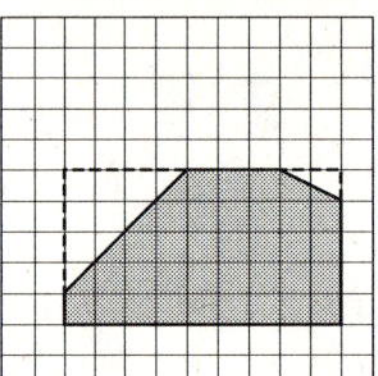

Area of the rectangle:
$A = \ell \cdot w$
$A = 9 \cdot 5$
$A = 45$
Area of left triangle:
$A = \frac{1}{2}bh$
$A = \frac{1}{2} \cdot 4 \cdot 4$
$A = 8$
Area of right triangle:
$A = \frac{1}{2}bh$
$A = \frac{1}{2} \cdot 2 \cdot 1$
$A = 1$
Area $= 45 - 8 - 1 = 36$
The area of the figure is approximately 36 square units.

35. $x + 5 > -3$
$x + 5 - 5 > -3 - 5$
$x > -8$
To check the solution, try any number greater than -8.
Check: $x + 5 > -3$
$-7 + 5 \overset{?}{>} -3$
$-2 > -3$ ✓

36. $k + (-9) \geq 1.8$
$k + (-9) - (-9) \geq 1.8 - (-9)$
$k \geq 10.8$
To check the solution, try any number greater than or equal to 10.8.
Check: $k + (-9) \geq 1.8$
$11 + (-9) \overset{?}{\geq} 1.8$
$2 \geq 1.8$ ✓

37. $\frac{1}{3}\cdot 5\cdot 15 = \frac{1}{\cancel{3}_1}\cdot\frac{5}{1}\cdot\frac{\cancel{15}^{5}}{1}$
$= 25$

38. $\frac{1}{3}\cdot 4\cdot 9 = \frac{1}{\cancel{3}_1}\cdot\frac{4}{1}\cdot\frac{\cancel{9}^{3}}{1}$
$= 12$

39. $\frac{1}{3}\cdot 2\cdot 2\cdot 3 = \frac{1}{\cancel{3}_1}\cdot\frac{2}{1}\cdot\frac{2}{1}\cdot\frac{\cancel{3}^{1}}{1}$
$= 4$

40. $\frac{1}{3}\cdot 3\cdot 4\cdot 8 = \frac{1}{\cancel{3}_1}\cdot\frac{\cancel{3}^{1}}{1}\cdot\frac{4}{1}\cdot\frac{8}{1}$
$= 32$

41. $\frac{1}{3}\cdot 2^2\cdot 21 = \frac{1}{\cancel{3}_1}\cdot\frac{4}{1}\cdot\frac{\cancel{21}^{7}}{1}$
$= 28$

42. $\frac{1}{3}\cdot 3^2\cdot 10 = \frac{1}{\cancel{3}_1}\cdot\frac{\cancel{9}^{3}}{1}\cdot\frac{10}{1}$
$= 30$

11-3 Volume: Pyramids and Cones

Page 568 How is the volume of a pyramid related to the volume of a prism?

a. The areas of the bases and the heights of both solids are the same.

b. 3 times

c. $\frac{1}{3}$

Page 570 Check for Understanding

1. The base is a circle.
2. Prism $V = Bh$; cylinder $V = \pi r^2 h = Bh$; pyramid $V = \frac{1}{3}Bh$; cone $V = \frac{1}{3}\pi r^2 h = \frac{1}{3}Bh$; B = area of base, h = height, r = radius; the formula for the volume of a pyramid and a cone is one-third times the formula for the volume of a prism and a cylinder, respectively.
3. Sample answer:

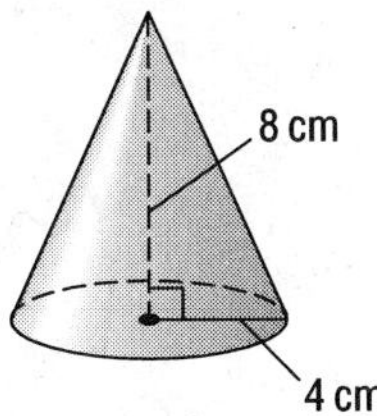

$V = \frac{1}{3}\pi r^2 h$
$= \frac{1}{3}\pi(4)^2 8$
$\approx 134.0\ \text{cm}^3$
Since $V \approx 134.0\ \text{cm}^3$,
$100\ \text{cm}^3 < V < 1000\ \text{cm}^3$.

4. $V = \frac{1}{3}Bh$
$V = \frac{1}{3}(5\cdot 4)h$
$V = \frac{1}{3}(5\cdot 4)15$
$V = 100$
The volume is 100 m^3.

5. $V = \frac{1}{3}\pi r^2 h$
$V = \frac{1}{3}\pi\cdot 3^2\cdot 4$
$V \approx 37.7$
The volume is about 37.7 cm^3.

6. $V = \frac{1}{3}Bh$
$V = \frac{1}{3}(48)(10)$
$V = 160$
The volume is 160 in^3.

7. $V = \frac{1}{3}Bh$
$V = \frac{1}{3}(9\cdot 7)h$
$V = \frac{1}{3}(9\cdot 7)18$
$V = 378$
The volume is 378 ft^3.

8. $V = \frac{1}{3}\pi r^2 h$
$V = \frac{1}{3}\pi\cdot 4^2\cdot 6.5$
$V \approx 108.9$
The volume is about 108.9 mm^3.

9. $V = \frac{1}{3}Bh$
$V = \frac{1}{3}(756\cdot 756)h$
$V = \frac{1}{3}(756\cdot 756)481$
$V = 91{,}636{,}272$
The volume was 91,636,272 ft^3.

Pages 571–572 Practice and Apply

10. $V = \frac{1}{3}Bh$
$V = \frac{1}{3}(4\cdot 4)h$
$V = \frac{1}{3}(4\cdot 4)6$
$V = 32$
The volume is 32 ft^3.

11. $V = \frac{1}{3}Bh$
$V = \frac{1}{3}(10\cdot 10.3)h$
$V = \frac{1}{3}(10\cdot 10.3)12$
$V = 412$
The volume is 412 cm^3.

12. $V = \frac{1}{3}Bh$
$V = \frac{1}{3}(40.6)9.2$
$V \approx 124.5$
The volume is about 124.5 mm^3.

13. $V = \frac{1}{3}Bh$

$V = \frac{1}{3}\left(\frac{1}{2} \cdot 5 \cdot 12\right)h$

$V = \frac{1}{3}\left(\frac{1}{2} \cdot 5 \cdot 12\right)15$

$V = 150$

The volume is 150 in^3.

14. $V = \frac{1}{3}\pi r^2 h$

$V = \frac{1}{3}\pi \cdot 10^2 \cdot 6$

$V \approx 628.3$

The volume is about 628.3 in^3.

15. $r = \frac{1}{2}(5)$

$r = 2.5$

$V = \frac{1}{3}\pi r^2 h$

$V = \frac{1}{3}\pi \cdot 2.5^2 \cdot 12$

$V \approx 78.5$

The volume is about 78.5 m^3.

16. $V = \frac{1}{3}Bh$

$V = \frac{1}{3}(5 \cdot 5)h$

$V = \frac{1}{3}(5 \cdot 5)6$

$V = 50$

The volume is 50 in^3.

17. $V = \frac{1}{3}Bh$

$V = \frac{1}{3} \cdot 125 \cdot 6.5$

$V \approx 270.8$

The volume is about 270.8 cm^3.

18. $V = \frac{1}{3}\pi r^2 h$

$V = \frac{1}{3}\pi \cdot 3^2 \cdot 14$

$V \approx 131.9$

The volume is about 131.9 yd^3.

19. $r = \frac{1}{2} \cdot 12$

$r = 6$

$V = \frac{1}{3}\pi r^2 h$

$V = \frac{1}{3}\pi \cdot 6^2 \cdot 15$

$V \approx 565.5$

The volume is about 565.5 m^3.

20. The bottom solid is a cube:

$V = Bh$

$V = 20 \cdot 20 \cdot 20$

$V = 8000$

The top solid is a square pyramid:

$V = \frac{1}{3}Bh$

$V = \frac{1}{3}(20 \cdot 20)15$

$V = 2000$

Add the volumes together:

$8000 + 2000 = 10{,}000$

The volume is 10,000 ft^3.

21. The bottom solid is a cylinder:

$V = \pi r^2 h$

$V = \pi \cdot 2^2 \cdot 3$

$V \approx 37.7$

The top solid is a cone:

$V = \frac{1}{3}\pi r^2 h$

$V = \frac{1}{3}\pi \cdot 2^2 \cdot 1.5$

$V \approx 6.3$

Add the volumes together:

$37.7 + 6.3 = 44.0$

The volume is about 44.0 m^3.

22. The bottom solid is a cone:

$V = \frac{1}{3}\pi r^2 h$

$V = \frac{1}{3}\pi \cdot 5^2 \cdot 20$

$V \approx 523.6$

The top solid is a cone:

$V = \frac{1}{3}\pi r^2 h$

$V = \frac{1}{3}\pi \cdot 5^2 \cdot 8$

$V \approx 209.4$

Add the volumes together:

$523.6 + 209.4 = 733.0$

The volume is about 733.0 cm^3.

23a. $r = \frac{1}{2} \cdot 1.5$

$r = 0.75$

$V = \frac{1}{3}\pi r^2 h$

$V = \frac{1}{3}\pi \cdot 0.75^2 \cdot 4$

$V \approx 2.4$

The volume is about 2.4 ft^3.

23b. $\frac{131 \text{ lb}}{\cancel{ft^3}} \cdot 2.4 \cancel{ft^3} = 314.4 \text{ lb}$

It weighs about 314.4 lb.

24. Standard:

$r = \frac{1}{2} \cdot 48$

$r = 24$

$V = \frac{1}{3}\pi r^2 h$

$V = \frac{1}{3}\pi \cdot 24^2 \cdot 40$

$V \approx 24{,}127.4 \text{ mm}^3$

buchner:

$r = \frac{1}{2} \cdot 34$

$r = 17$

$V = \frac{1}{3}\pi r^2 h + \pi r^2 h$

$V = \frac{1}{3}\pi \cdot 17^2 \cdot 18 + \pi \cdot 17^2 \cdot 20$

$V \approx 5447.5 + 18{,}158.4$

$V \approx 23{,}605.9 \text{ mm}^3$

The standard funnel has the greatest volume.

25. The volume of a pyramid is one-third the volume of a prism with the same base and height. Answers should include the following.

- The height of the pyramid and prism are equal. The bases of the pyramid and prism are squares with equal side lengths. Therefore, their base areas are equal.
- The formula for the volume of a pyramid is one-third times the formula for the volume of a prism.

26a. $V = \frac{1}{3}\pi r^2 h$

$\frac{1}{3}\pi r^2 2h = 2\left(\frac{1}{3}\pi r^2 h\right) = 2V$

The volume doubles.

26b. $V = \frac{1}{3}\pi r^2 h$

$\frac{1}{3}\pi(2r)^2 h = \frac{1}{3}\pi 4r^2 h = 4V$

The volume quadruples.

27. B; $V = \frac{1}{3}Bh$

$V = \frac{1}{3}(4.9 \cdot 3)7$

$V \approx \frac{1}{3} \cdot 5 \cdot 3 \cdot 7$

$V \approx \frac{1}{\cancel{3}_1} \cdot \frac{5}{1} \cdot \frac{\cancel{3}^1}{1} \cdot \frac{7}{1}$

$V \approx 35 \text{ cm}^3$

28. D; If a cone and a cylinder have the same base area, the cone will have $\frac{1}{3}$ the volume of the cylinder.

29. $V = \frac{4}{3}\pi r^3$

$V = \frac{4}{3}\pi \cdot 6^3$

$V \approx 904.8$

The volume is about 904.8 cm^3.

30. $V = \frac{4}{3}\pi r^3$

$V = \frac{4}{3}\pi \cdot 1.4^3$

$V \approx 11.5$

The volume is about 11.5 in^3.

31. $r = \frac{1}{2} \cdot 5.9$

$r = 2.95$

$V = \frac{4}{3}\pi r^3$

$V = \frac{4}{3}\pi \cdot 2.95^3$

$V \approx 107.5$

The volume is about 107.5 mm^3.

Page 572 Maintain Your Skills

32. $V = Bh$

$V = \ell \cdot w \cdot h$

$V = 4 \cdot 8 \cdot 2$

$V = 64$

The volume is 64 cm^3.

33. $r = \frac{1}{2} \cdot 1.6$

$r = 0.8$

$V = \pi r^2 h$

$V = \pi \cdot 0.8^2 \cdot 5$

$V \approx 10.1$

The volume is about 10.1 in^3.

34. The figure has one triangular base, *DCG*, *DGF*, *DFC*, or *CGF*, so it is a triangular pyramid.

edges: $\overline{DC}, \overline{DG}, \overline{DF}, \overline{CG}, \overline{GF}, \overline{FC}$

vertices: *C*, *D*, *F*, *G*

35. $d = \sqrt{(x_2 - x_1)^2 + (y_2 - y_1)^2}$

$d = \sqrt{(-2 - 3)^2 + (1 - 7)^2}$

$d = \sqrt{(-5)^2 + (-6)^2}$

$d = \sqrt{25 + 36}$

$d = \sqrt{61}$

$d \approx 7.8$

36. $4.9 \cdot 5.1 \cdot 3 \approx 5 \cdot 5 \cdot 3$

$= 75$

37. $2 \cdot 1.7 \cdot 9 \approx 2 \cdot 2 \cdot 9$

$= 36$

38. $2 \cdot \pi \cdot 6.8 \approx 2 \cdot 3 \cdot 7$

$= 42$

Page 572 Practice Quiz 1

1. triangular prism
2. rectangular pyramid
3. $V = \pi r^2 h$

   $V = \pi \cdot 2^2 \cdot 1$

   $V \approx 12.6$

   The volume is about 12.6 cm^3.
4. $V = Bh$

   $V = 42 \cdot 18$

   $V = 756$

   The volume is 756 ft^3.
5. $r = \frac{1}{2} \cdot 420$

   $r = 210$

   $V = \frac{1}{3}\pi r^2 h$

   $V = \frac{1}{3}\pi \cdot 210^2 \cdot 250$

   $V \approx 11{,}545{,}353$

   The volume removed was about 11,545,353 ft^3.

11-4 Surface Area: Prisms and Cylinders

Page 573 How is the surface area of a solid different from its volume?

a. Box A: $A = 8(8)$

$= 64$

There are six faces with the same area.

$6(64) = 384 \text{ in}^2$

Box B: $A = 15(10)$

$= 150$

$A = 10(12)$

$= 120$

$A = 15(12)$

$= 180$

There are 2 faces with each area.

$2(150) + 2(120) + 2(180) = 900 \text{ in}^2$

Box C: $A = 20(14)$
$= 280$
$A = 14(10)$
$= 140$
$A = 20(10)$
$= 200$
There are 2 faces with each area.
$2 \times 280 + 2 \times 140 + 2 \times 200 = 1240 \text{ in}^3$

b. Box A: $V = \ell \cdot w \cdot h$
$V = 8 \cdot 8 \cdot 8$
$V = 512 \text{ in}^3$
Box B: $V = \ell \cdot w \cdot h$
$V = 15 \cdot 10 \cdot 12$
$V = 1800 \text{ in}^3$
Box C: $V = \ell \cdot w \cdot h$
$V = 20 \cdot 14 \cdot 10$
$V = 2800 \text{ in}^3$
No, for each box the volume is different from the surface area.

Pages 575–576 Check for Understanding

1. Sample answer: The surface of a solid is two-dimensional. Its area is the sum of the face areas, which are given in square units.

2. Sample answer: Find the surface area of a cedar chest and a cylinder-shaped potato chip container.

3. $S = 2\ell w + 2\ell h + 2wh$
$S = 2(5)(5) + 2(5)(5) + 2(5)(5)$
$S = 150$
The surface area is 150 ft^2.

4. bottom: $5 \cdot 10 = 50$
left side: $3 \cdot 10 = 30$
right side: $4 \cdot 10 = 40$
two bases: $2\left(\frac{1}{2} \cdot 3 \cdot 4\right) = 12$
$50 + 30 + 40 + 12 = 132$
The surface area is 132 cm^2.

5. $r = \frac{1}{2} \cdot 14$
$r = 7$
$S = 2\pi r^2 + 2\pi rh$
$S = 2\pi(7)^2 + 2\pi(7)(6)$
$S \approx 571.8$
The surface area is about 571.8 in^2.

6. $S = 2\ell w + 2\ell h + 2wh$
$S = 2(3)(2) + 2(3)(1) + 2(2)(1)$
$S = 22$
The surface area is 22 cm^2.

7. $S = 2\pi r^2 + 2\pi rh$
$S = 2\pi(4)^2 + 2\pi(4)(1.6)$
$S \approx 140.7$
The surface area is about 140.7 mm^2.

8. Box A: $S = 2\ell w + 2\ell h + 2wh$
$S = 2(9)(4) + 2(9)(5) + 2(4)(5)$
$S = 166 \text{ in}^2$
Box B: $S = 2\ell w + 2\ell h + 2wh$
$S = 2(5)(5) + 2(5)(6) + 2(5)(6)$
$S = 145 \text{ in}^2$
Because $166 \text{ in}^2 > 145 \text{ in}^2$, Box A will require more canvas.

Pages 576–577 Practice and Apply

9. $S = 2\ell w + 2\ell h + 2wh$
$S = 2(12)(7) + 2(12)(3) + 2(7)(3)$
$S = 282$
The surface area is 282 in^2.

10. $S = 2\ell w + 2\ell h + 2wh$
$S = 2(14)(3.5) + 2(14)(3.5) + 2(3.5)(3.5)$
$S = 220.5$
The surface area is 220.5 m^2.

11. bottom: $8 \cdot 9 = 72$
left side: $9 \cdot 10 = 90$
right side: $6 \cdot 9 = 54$
two bases: $2\left(\frac{1}{2} \cdot 8 \cdot 6\right) = 48$
$72 + 90 + 54 + 48 = 264$
The surface area is 264 m^2.

12. left side: $6 \cdot 10 = 60$
right side: $6 \cdot 10 = 60$
back side: $6 \cdot 10 = 60$
two bases: $2\left(\frac{1}{2} \cdot 6 \cdot 5.2\right) = 31.2$
$60 + 60 + 60 + 31.2 = 211.2$
The surface area is 211.2 cm^2.

13. $S = 2\pi r^2 + 2\pi rh$
$S = 2\pi(10)^2 + 2\pi(10)(20)$
$S \approx 1885.0$
The surface area is about 1885.0 ft^2.

14. $r = \frac{1}{2} \cdot 9$
$r = 4.5$
$S = 2\pi r^2 + 2\pi rh$
$S = 2\pi(4.5)^2 + 2\pi(4.5)(1)$
$S \approx 155.5$
The surface area is about 155.5 in^2.

15. $S = 2\ell w + 2\ell h + 2wh$
$S = 2(7)(7) + 2(7)(7) + 2(7)(7)$
$S = 294$
The surface area is 294 ft^2.

16. $S = 2\ell w + 2\ell h + 2wh$
$S = 2(6.2)(4) + 2(6.2)(8.5) + 2(4)(8.5)$
$S = 223$
The surface area is 223 cm^2.

17. $S = 2\pi r^2 + 2\pi rh$
$S = 2\pi(5)^2 + 2\pi(5)(15)$
$S \approx 628.3$
The surface area is about 628.3 in^2.

18. $r = \frac{1}{2} \cdot 4$
$r = 2$
$S = 2\pi r^2 + 2\pi rh$
$S = 2\pi(2)^2 + 2\pi(2)(20)$
$S \approx 276.5$
The surface area is about 276.5 m^2.

19. The top is half of a cylinder.
$r = \frac{1}{2} \cdot 4$
$r = 2$
$S = \frac{1}{2}(2\pi r^2 + 2\pi rh)$
$S = \frac{1}{2}(2\pi(2)^2 + 2\pi(2)(12))$
$S \approx 88.0$

The bottom is a rectangular prism without one side.
$S = 2\ell w + 2\ell h + 2wh - \ell w$
$S = 2(12)(4) + 2(12)(4) + 2(4)(4) - (12)(4)$
$S = 176$
Add the top area and bottom area.
$88.0 + 176 = 264.0$
The surface area was about 264.0 in^2.

20a. $V = \ell \cdot w \cdot h$
$4600 = 24 \cdot 12 \cdot h$
$4600 = 288h$
$16 = h$
The height of the aquarium is about 16 in.

20b. left side: $12 \cdot 16 = 192$
right side: $12 \cdot 16 = 192$
front side: $24 \cdot 16 = 384$
back side: $24 \cdot 16 = 384$
base: $24 \cdot 12 = 288$
$192 + 192 + 384 + 384 + 288 = 1440$
$1440 \text{ in}^2 = 1440 \cancel{\text{in}^2} \cdot \frac{1 \text{ ft}^2}{144 \cancel{\text{in}^2}}$
$= 10 \text{ ft}^2$
10 ft^2 of glass was needed.

20c. side: $9 \cdot 16 = 144$
base: 392.4
$S = 8(144) + 392.4 = 1544.4$
The sides of both tanks have the same surface area, 1152 in^2. However, the area of the octagonal base is 392.4 in^2, and the area of the rectangular base is 288 in^2. So, the total surface area of the octagonal tank is greater.

21. Since there is only one base, the formula would be:
$r = \frac{1}{2} \cdot 24$
$r = 12$
$S = \pi r^2 + 2\pi rh$
$S = \pi(12)^2 + 2\pi(12)(4)$
$S \approx 754$
The area of the liner is 754 ft^2.

22. The figure is a cylinder with a rectangular prism inserted at its center.
Cylinder:
$r = \frac{1}{2} \cdot 16$
$r = 8$
$S = \pi r^2 + 2\pi rh$
$S = \pi(8)^2 + 2\pi(8)(4)$
$S \approx 402$
Rectangular Prism:
bottom $16 \cdot 12 = 192$
side $4 \cdot 12 = 48$
side $4 \cdot 12 = 48$
$402 + 192 + 48 + 48 = 690$
The area of the liner is about 690 ft^2.

23. $S = 2\ell w + 2\ell h + 2wh$
$= 2(2\ell)(2w) + 2(2\ell)(2h) + 2(2w)(2h)$
$= 8\ell w + 8\ell h + 8wh$
$= 4(2\ell w + 2\ell h + 2wh)$
$= 4\,s$
The surface area is 4 times greater.

24. Surface area is the boundary around a solid. Volume is the space inside a solid. Answers should include the following.
- The formula for surface area involves the product of two dimensions and is given in square units. The formula for volume involves the product of three dimensions and is given in cubic units.
- Surface area describes the size of the total surfaces of a solid. Volume describes the capacity or amount of space contained inside a solid.

25. D; $r = \frac{1}{2} \cdot 15$
$r = 7.5$
$S = 2\pi r^2 + 2\pi rh$
$S = 2\pi(7.5)^2 + 2\pi(7.5)(2)$
$S \approx 447.7 \text{ cm}^2$

26. B; $S = 2\ell w + 2\ell h + 2wh$
$S = 2(10)(4) + 2(10)(6) + 2(4)(6)$
$S = 248 \text{ in}^2$
A 2-inch square has an area of 4 in^2.
$248 \text{ in}^2 \div 4 \text{ in}^2 = 62 \text{ in}^2$

27. square or rectangle

28. circle

29. rectangle

Page 577 Maintain Your Skills

30. $V = \frac{1}{3}Bh$
$V = \frac{1}{3}(6 \cdot 5) \cdot 7$
$V = 70$
The volume is 70 ft^3.

31. $r = \frac{1}{2} \cdot 6$
$r = 3$
$V = \pi r^2 h$
$V = \pi(3)^2(20)$
$V \approx 565.5$
The volume is about 565.5 in^3.

32. $10.3(8) = 82.4$

33. $3.9(3.9) = 15.21$

34. $12.3(9.2)(6) = 678.96$

35. $\frac{1}{2} \cdot 2.6 = \frac{1}{\cancel{2}_1} \cdot \frac{\cancel{2.6}^{1.3}}{1}$
$= 1.3$

36. $\frac{1}{2} \cdot 82 \cdot 90 = \frac{1}{\cancel{2}_1} \cdot \frac{\cancel{82}^{41}}{1} \cdot \frac{90}{1}$
$= 3690$

37. $\frac{1}{2}\left(6\frac{1}{2}\right) = \frac{1}{2} \cdot \frac{13}{2}$
$= \frac{13}{4}$
$= 3\frac{1}{4}$

11-5 Surface Area: Pyramids and Cones

Page 578 How is surface area important in architecture?

a. $A = \frac{1}{2}b \cdot h$

$A = \frac{1}{2} \cdot 230 \cdot 120$

$A = \frac{1}{\cancel{2}_1} \cdot \frac{\cancel{230}^{115}}{1} \cdot \frac{120}{1}$

$A = 13{,}800 \text{ ft}^2$

b. Find the area of all the sides and add.

Page 580 Check for Understanding

1. Slant height is the altitude of a triangular face of a pyramid; it is the length from the vertex of a cone to the edge of its base. Height of a pyramid or cone is the altitude of the whole solid.

2. Use $A = \frac{1}{2}bh$ to find the area of each face. Then multiply by the number of faces.

3. Sample answer: An architect might use the formulas to calculate the amount of materials needed for parts of a structure.

4. Area of each lateral face:

$A = \frac{1}{2}bh$

$A = \frac{1}{2}(4)(6.3)$

$A = 12.6$

There are 4 faces, so the lateral area is 4(12.6) or 50.4 ft^2.

Area of base:

$A = s^2$

$A = 4^2$

$A = 16$

$S = 50.4 + 16$

The surface area of the pyramid is 66.4 ft^2.

5. Area of each lateral face:

$A = \frac{1}{2}bh$

$A = \frac{1}{2}(4)(6)$

$A = 12$

There are 3 lateral faces, so the lateral area is 3(12) or 36 m^2.

Area of base:

$A = 6.9$

$S = 36 + 6.9$

The surface area of the pyramid is 42.9 m^2.

6. $S = \pi r\ell + \pi r^2$

$S = \pi(5)(13) + \pi(5)^2$

$S \approx 282.7$

The surface area of the cone is about 282.7 cm^2.

7. $A = \frac{1}{2}bh$

$A = \frac{1}{2}(8)(14)$

$A = 56$

There are 6 lateral faces, so the lateral area is 6(56) or 336 ft^2.

It will take 336 ft^2 of roofing.

Pages 580–582 Practice and Apply

8. Area of lateral face:

$A = \frac{1}{2}bh$

$A = \frac{1}{2}(8)(9)$

$A = 36$

There are 4 lateral faces, so the lateral area is 4(36) or 144 m^2.

Area of base:

$A = s^2$

$A = 8^2$ or 64

$S = 144 + 64$

$S = 208$

The surface area of the pyramid is 208 m^2.

9. Area of lateral face:

$A = \frac{1}{2}bh$

$A = \frac{1}{2}(5.5)(6)$

$A = 16.5$

There are 4 lateral faces, so the lateral area is 4(16.5) or 66 in^2.

Area of base:

$A = s^2$

$A = (5.5)^2$ or 30.25

$S = 66 + 30.25$

$S \approx 96.3$

The surface area of the pyramid is about 96.3 in^2.

10. Area of lateral face:

$A = \frac{1}{2}bh$

$A = \frac{1}{2}(8)(6.9)$

$A = 27.6$

There are 3 lateral faces, so the lateral area is 3(27.6) or 82.8 ft^2.

Area of base:

$A = \frac{1}{2}bh$

$A = \frac{1}{2}(8)(6.9)$

$A = 27.6$

$S = 82.8 + 27.6$

The surface area of the pyramid is 110.4 ft^2.

11. Area of lateral face:

$A = \frac{1}{2}bh$

$A = \frac{1}{2}(6)(5.2)$

$A = 15.6$

There are 3 lateral faces, so the lateral area is 3(15.6) or 46.8 in^2.

Area of base:

$A = \frac{1}{2}bh$

$A = \frac{1}{2}(6)(5.2)$

$A = 15.6$

$S = 46.8 + 15.6$

The surface area of the pyramid is 62.4 in^2.

12. $S = \pi r\ell + \pi r^2$

$S = \pi(5)(10) + \pi(5)^2$

$S \approx 235.6$

The surface area of the cone is about 235.6 cm^2.

13. $r = \frac{1}{2}(10)$
$r = 5$
$S = \pi r\ell + \pi r^2$
$S = \pi(5)(16.6) + \pi(5)^2$
$S \approx 339.3$
The surface area of the cone is about 339.3 in^2.

14. $S = \pi r\ell + \pi r\ell$
$S = \pi(5)(15) + \pi(5)(15)$
$S \approx 471.2$
The surface area of the solid is about 471.2 cm^2.

15. Area of lateral face:
$A = \frac{1}{2}bh$
$A = \frac{1}{2}(8.3)(10.6)$
$A = 43.99$
There are 6 faces, so the lateral area is 6(43.99) or 263.9 m^2.
Area of base:
$A = 48$
$S = 263.9 + 48$
$S = 311.9$
The surface area of the pyramid is 311.9 m^2.

16. Top Pyramid:
$A = \frac{1}{2}bh$
$A = \frac{1}{2}(5)(12.3)$
$A = 30.75$
There are 4 faces, so the lateral area is 4(30.75) or 123 in^2.
Bottom Pyramid:
$A = \frac{1}{2}bh$
$A = \frac{1}{2}(5)(15.2)$
$A = 38$
There are 4 faces, so the lateral area is 4(38) or 152 in^2.
$S = 123 + 152$
$S = 275$
The surface area of the solid is 275 in^2.

17. $S = \pi r\ell + \pi r^2$
$S = \pi(7.5)(14) + \pi(7.5)^2$
$S \approx 506.6$
The surface area of the cone is 506.6 mm^2.

18. Area of lateral face:
$A = \frac{1}{2}bh$
$A = \frac{1}{2}(9)(8)$
$A = 36$
There are 4 faces, so the lateral area is 4(36) or 144 yd^2.
Area of base:
$A = s^2$
$A = 9^2$ or 81
$S = 144 + 81$
$S = 225$
The surface area of the pyramid is 225 yd^2.

19. Style 8M:
$S = \pi r\ell$
$S = \pi(4)(8)$
$S \approx 101 \text{ in}^2$
Style 65M:
$S = \pi r\ell$
$S = \pi(3.5)(6.5)$
$S \approx 72 \text{ in}^2$
Style 8M has 29 in^2 more plastic.

20. $S = \pi r\ell$
$S = \pi(8)(23)$
$S \approx 578.0$
The lateral surface area is 578 ft^2.
$578 \div 100 = 5.78$
Six squares would be needed.

21. $S = \pi r\ell$
$S = \pi(9)(12)$
$S \approx 339.3$
The lateral surface area is 339.3 ft^2.
$339.3 \div 100 = 3.393$
Four squares would be needed.

22. Bar:
$S = 2\ell w + 2\ell h + 2wh$
$S = 2(13)(2) + 2(13)(1) + 2(2)(1)$
$S = 82$
Sinkers:
$r = \frac{1}{2} \cdot 1$
$r = \frac{1}{2}$
$S = \pi r\ell + \pi r^2$
$S = \pi\left(\frac{1}{2}\right)(1.1) + \pi\left(\frac{1}{2}\right)^2$
$S \approx 2.51$
$100 \times 2.51 = 251$
The sinkers have a surface area of about 251 in^2. The original bar has a surface area of 82 in^2. The surface area has more than tripled.

23. Many building materials are priced and purchased by square footage. Architects use surface area when designing buildings. Answers should include the following.
- Surface area is used in covering building exteriors and in designing interiors.
- It is important to know surface areas so the amounts and costs of building materials can be estimated.

24. D; $S = \pi r\ell + \pi r^2$
$S = \pi(7)(11.4) + \pi(7)^2$
$S \approx 404.6 \text{ cm}^2$

25. C; $S = \frac{1}{2}bh$
$S = \frac{1}{2}(5)(7)$
$S = 17.5 \text{ in}^2$
There are 4 faces, so the lateral area is 4(17.5) or 70 in^2.

26. $S = 4\pi r^2$
$S = 4\pi(8)^2$
$S \approx 804.2$
The surface area of the sphere is about 804.2 cm^2.

27. $S = 4\pi r^2$
$S = 4\pi(21)^2$
$S \approx 5541.8$
The surface area of the sphere is about 5541.8 mm^2.

28. $S = 4\pi r^2$
$S = 4\pi(1.3)^2$
$S \approx 21.2$
The surface area of the sphere is about 21.2 in^2.

Page 582 Maintain Your Skills

29. $S = 2\ell w + 2\ell h + 2wh$
$S = 2(2)(1) + 2(2)(0.5) + 2(1)(0.5)$
$S = 7$
The surface area of the prism is 7 ft^2.
30. $S = 2\pi r^2 + 2\pi rh$
$S = 2\pi(4)^2 + 2\pi(4)(13.8)$
$S \approx 447.4$
The surface area of the cylinder is about 447.4 cm^2.
31. $V = \frac{1}{3}\pi r^2 h$
$V = \frac{1}{3}\pi(2)^2(6)$
$V \approx 25.1$
The volume of the cone is about 25.1 in^3.
32. The solution is $(-1, 4)$.
33. There are infinitely many solutions.
34. $\frac{1}{6} = \frac{x}{24}$
$1 \cdot 24 = 6 \cdot x$
$24 = 6x$
$4 = x$
35. $\frac{9}{15} = \frac{n}{5}$
$9 \cdot 5 = 15 \cdot n$
$45 = 15n$
$3 = n$
36. $\frac{t}{7} = \frac{40}{56}$
$t \cdot 56 = 7 \cdot 40$
$56t = 280$
$t = 5$
37. $\frac{1.6}{y} = \frac{9.6}{18}$
$1.6 \cdot 18 = y \cdot 9.6$
$28.8 = 9.6y$
$3 = y$
38. $\frac{4}{3.2} = \frac{w}{20}$
$4 \cdot 20 = 3.2 \cdot w$
$80 = 3.2w$
$25 = w$
39. $\frac{18}{21} = \frac{2.7}{n}$
$18 \cdot n = 21 \cdot 2.7$
$18n = 56.7$
$n = 3.15$

Page 583 Geometry Activity (Preview of Lesson 11-6)

1. 8
2. 1 unit^2
3. 4 units^2
4. 1 unit^3
5. 8 units^3
6. 27
7. 9 units^2
8. 27 units^3
9.

Scale Factor	Side Length	Area of a Face	Volume
1	1	1	1
2	2	4	8
3	3	9	27

10. 4 times greater; 9 times greater
11. x^2; $6x^2$
12. 2^3 or 8 times greater; 3^3 or 27 times greater
13. x^3
14. Surface area $= 6x^2$
$= 6(4)^2$
$= 96$ units^2
Volume $= x^3$
$= 4^3$
$= 64$ units^3
15. See students' work.

11-6 Similar Solids

Page 584 How can linear dimensions be used to identify similar solids?

a. $87 \times 8.5 = 739.5$ inches long
$= 739.5 \text{ in.} \cdot \frac{1 \text{ ft}}{12 \text{ in.}}$
≈ 61.63 ft long
$87 \times 1 = 87$ inches wide
$= 87 \text{ in.} \cdot \frac{1 \text{ ft}}{12 \text{ in.}}$
$= 7.25$ ft wide
b. $87 \times 7 = 609$ inches long
$= 609 \text{ in.} \cdot \frac{1 \text{ ft}}{12 \text{ in.}}$
$= 50.75$ ft long
c. It is 87 times greater.

Pages 586–587 Check for Understanding

1. Sample answer:

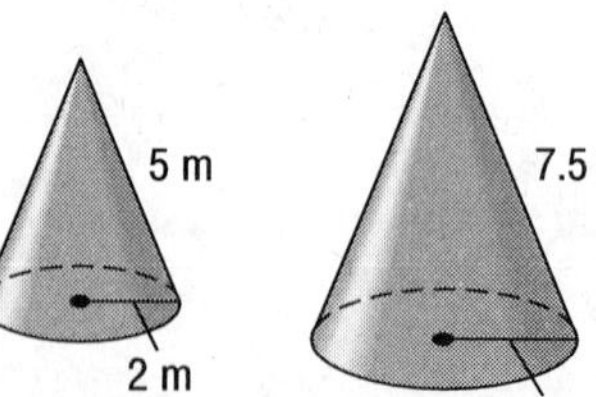

The cones are similar because the ratios comparing their radii and slant heights are equal: $\frac{2}{3} = \frac{5}{7.5}$.
2. Multiply the surface area of the smaller cylinder by the square of the scale factor.

3. $\frac{1}{4} \stackrel{?}{=} \frac{3}{8}$

$1(8) \stackrel{?}{=} 4(3)$

$8 \neq 12$

The corresponding measures are not proportional, so the solids are not similar.

4. $\frac{2}{3} \stackrel{?}{=} \frac{4}{6}$

$2(6) \stackrel{?}{=} 3(4)$

$12 = 12$

The radii and heights are proportional, so the cones are similar.

5. $\frac{45}{6} = \frac{250}{x}$

$45x = 6(250)$

$x = 33\frac{1}{3}$

The height of the cylinder is $33\frac{1}{3}$ ft.

6. $\frac{30}{45} = \frac{x}{75}$ $\qquad$ $\frac{30}{45} = \frac{24}{y}$

$30(75) = 45x$ $\qquad$ $30y = 45(24)$

$50 = x$ $\qquad$ $y = 36$

The width of the small box is 50 cm, and the length of the large box is 36 cm.

7. $\frac{1}{1.5} = \frac{350}{x}$

$1 \cdot x = 1.5 \cdot 350$

$x = 525$

The building is 525 m tall.

8. $\frac{1 \text{ cm}}{1.5 \text{ m}} = \frac{1 \text{ cm}}{1.5 \cancel{\text{m}}} \cdot \frac{1 \cancel{\text{m}}}{100 \text{ cm}}$

$= \frac{1 \text{ cm}}{150 \text{ cm}}$

$= \frac{1}{150}$

9. $V = \ell \cdot w \cdot h$

$V = 60 \cdot 42 \cdot 350$

$V = 882{,}000 \text{ cm}^3$

$\frac{\text{volume of model}}{\text{volume of building}} = \frac{(1 \text{ cm})^3}{(1.5 \text{ m})^3}$

$= \frac{1 \text{ cm}^3}{3.375 \text{ m}^3}$

So, the volume of the building in cubic meters is 3.375 times the volume of the model in cubic centimeters.

$3.375 \cdot 882{,}000 = 2{,}976{,}750 \text{ m}^3$

Pages 587–588 Practice and Apply

10. $\frac{8}{10} \stackrel{?}{=} \frac{60}{72}$

$8(72) \stackrel{?}{=} 10(60)$

$576 \neq 600$

The corresponding measures are not proportional, so the solids are not similar.

11. $\frac{5}{9} \stackrel{?}{=} \frac{2.5}{4.5}$

$5(4.5) \stackrel{?}{=} 9(2.5)$

$22.5 = 22.5$ ✓

$\frac{5}{9} \stackrel{?}{=} \frac{10}{18}$

$5 \cdot 18 \stackrel{?}{=} 9 \cdot 10$

$90 = 90$ ✓

The sides are proportional, so the solids are similar.

12. $\frac{5}{6} \stackrel{?}{=} \frac{10}{6}$

$5(6) \stackrel{?}{=} 6(10)$

$30 \neq 60$

The corresponding measures are not proportional, so the solids are not similar.

13. $\frac{8}{4} \stackrel{?}{=} \frac{10}{5}$

$8(5) \stackrel{?}{=} 4(10)$

$40 = 40$ ✓

$\frac{8}{4} \stackrel{?}{=} \frac{6}{3}$

$8(3) \stackrel{?}{=} 4(6)$

$24 = 24$ ✓

The measurements are proportional, so the cones are similar.

14. $\frac{4}{12} = \frac{x}{15.3}$

$4(15.3) = 12x$

$5.1 = x$

The height of the cylinder is 5.1 ft.

15. $\frac{15}{5} = \frac{y}{6}$ $\qquad$ $\frac{15}{5} = \frac{21}{x}$

$15(6) = 5y$ $\qquad$ $15x = 5(21)$

$18 = y$ $\qquad$ $x = 7$

The height of the large box is 18 m, and the length of the small box is 7 m.

16. Always; same shape, lengths are proportional
17. Sometimes; bases must be the same polygon, and the corresponding side lengths must be propotional.
18. Never; different shapes
19. Always; same shape, diameters or radii are proportional
20. Area of lateral side:

$A = \frac{1}{2}b \cdot h$

$A = \frac{1}{2}(110)(88.5)$

$A = 4867.5$

There are 4 faces, so the lateral area is 4(4867.5) or 19,470 m^2.

Area of the base:

$A = \ell \cdot w$

$A = 110^2$

$A = 12{,}100$

$S = 19{,}470 + 12{,}100$

$S = 31{,}570$

The surface area of the original pyramid is 31,570 m^2.

Ratio of surface areas $= \frac{(4 \text{ m})^2}{(2 \text{ cm})^2}$

$= \frac{16\text{m}^2}{4 \text{ cm}^2}$

$\frac{16 \text{ m}^2}{4 \text{ cm}^2} = \frac{31{,}570 \text{ m}^2}{x \text{ cm}^2}$

$16x = 4(31{,}570)$

$x = 7892.5$

You will need 7892.5 cm^2 of surface material.

21. $\frac{4 \text{ m}}{2 \text{ cm}} = \frac{4 \cancel{\text{m}}}{2 \cancel{\text{cm}}} \cdot \frac{100 \cancel{\text{cm}}}{1 \cancel{\text{m}}}$

$= \frac{400}{2}$

$= \frac{200}{1}$

$\frac{\text{volume of pyramid}}{\text{volume of model}} = \frac{200^3}{1^3}$

$= \frac{8{,}000{,}000}{1}$

The volume is 8,000,000 times greater.

22. Sample answer: The prisms are similar only if all the dimensions decreased proportionately.

23. If the ratios of corresponding linear dimensions are equal, the solids are similar. Answers should include the following.

- For example, if the ratio comparing the heights of two cylinders equals the ratio comparing their radii, then the cylinders are similar.
- A cone and a prism are not similar.

24. C; $\frac{4}{6} \stackrel{?}{=} \frac{3}{4.5} \stackrel{?}{=} \frac{2}{3}$

$4(4.5) \stackrel{?}{=} 6(3) \quad 3(3) \stackrel{?}{=} 4.5(2)$

$18 = 18$ ✓ $\quad 9 = 9$ ✓

Prisms A and B are similar.

$\frac{4}{28} \stackrel{?}{=} \frac{3}{21} \stackrel{?}{=} \frac{2}{14}$

$4(21) \stackrel{?}{=} 28(3) \quad 3(14) \stackrel{?}{=} 21(2)$

$84 = 84$ ✓ $\quad 42 = 42$ ✓

Prism D is similar to Prism A and, therefore, also similar to Prism B.

$\frac{6}{5} \stackrel{?}{=} \frac{4.5}{4} \stackrel{?}{=} \frac{3}{2}$

$6(4) \stackrel{?}{=} 5(4.5) \quad 2(4.5) \stackrel{?}{=} 4(3)$

$24 \neq 22.5 \quad 9 \neq 12$

Prism C is not similar to Prism B and, therefore, not similar to Prism A or D.

25. C; $S = \pi r\ell + \pi r^2$

$= \pi(2r)(2\ell) + \pi(2r)^2$

$= \pi 4r\ell + \pi 4r^2$

$= 4(\pi r\ell + \pi r^2)$

$= 4\,S$

The surface area is quadrupled.

Page 588 Maintain Your Skills

26. $r = \frac{1}{2}(10)$

$r = 5$

$S = \pi r\ell + \pi r^2$

$S = \pi(5)(13) + \pi(5)^2$

$S \approx 282.7$

The surface area of the cone is about 282.7 in^2.

27. front side $5 \cdot 10 = 50$

back side $5 \cdot 10 = 50$

bottom side $6 \cdot 10 = 60$

two bases $2\left(\frac{1}{2} \cdot 6 \cdot 4\right) = 24$

$50 + 50 + 60 + 24 = 184$

The surface area of the prism is 184 ft^2.

28. $S = 2\pi r^2 + 2\pi rh$

$S = 2\pi(14)^2 + 2\pi(14)(22)$

$S \approx 3166.7$

The surface area of the cylinder is about 3166.7 m^2.

29. $m\angle J + m\angle K = 90°$

$25° + m\angle K = 90°$

$25° + m\angle K - 25° = 90° - 25°$

$m\angle K = 65°$

30. $r - 3.5 = 8$

$r - 3.5 + 3.5 = 8 + 3.5$

$r = 11.5$

Check: $r - 3.5 = 8$

$11.5 - 3.5 \stackrel{?}{=} 8$

$8 = 8$ ✓

31. $\frac{2}{3} + y = \frac{1}{9}$

$\frac{2}{3} + y - \frac{2}{3} = \frac{1}{9} - \frac{2}{3}$

$y = \frac{1}{9} - \frac{2}{3} \cdot \frac{3}{3}$

$y = \frac{1}{9} - \frac{6}{9}$

$y = \frac{-5}{9}$

Check: $\frac{2}{3} + y = \frac{1}{9}$

$\frac{2}{3} + \frac{-5}{9} \stackrel{?}{=} \frac{1}{9}$

$\frac{2}{3} \cdot \frac{3}{3} + \frac{-5}{9} \stackrel{?}{=} \frac{1}{9}$

$\frac{6}{9} + \frac{-5}{9} \stackrel{?}{=} \frac{1}{9}$

$\frac{1}{9} = \frac{1}{9}$ ✓

32. $\frac{1}{4}a = 6$

$\frac{4}{1}\left(\frac{1}{4}a\right) = \frac{4}{1}(6)$

$a = 24$

Check: $\frac{1}{4}a = 6$

$\frac{1}{4} \cdot 24 \stackrel{?}{=} 6$

$\frac{1}{\cancel{4}_1} \cdot \frac{\cancel{24}^6}{1} \stackrel{?}{=} 6$

$6 = 6$ ✓

33. $13.28 + 6.05 = 19.33 \approx 19.3$

34. $8.99 - 1.2 = 7.79 \approx 7.8$

35. $2.4 \cdot 2.5 = 6.0$

36. $55 \div 3.8 = 14.47 \ldots \approx 14.5$

37. $6 + 1.9 + 1.45 = 9.35 \approx 9.4$

38. $6.7(0.3)(1.8) = 3.618 \approx 3.6$

Page 588 Practice Quiz 2

1. $S = 2\ell w + 2\ell h + 2wh$

$S = 2(7)(4) + 2(7)(5) + 2(4)(5)$

$S = 166$

The surface area of the prism is 166 mm^2.

2. $S = 2\pi r^2 + 2\pi rh$

$S = 2\pi(4)^2 + 2\pi(4)(12)$

$S \approx 402.1$

The surface area of the cylinder is about 402.1 in^2.

3. $S = \pi r\ell + \pi r^2$

$S = \pi(1)(2) + \pi(1)^2$

$S \approx 9.4$

The surface area of the cone is about 9.4 m^2.

4. Area of lateral face:

$A = \frac{1}{2}bh$

$A = \frac{1}{2}(10)(8.25)$

$A = 41.25$

There are 4 lateral faces, so the lateral area is 4(41.25) or 165 in^2.

Area of base:
$A = s^2$
$A = 10^2$
$A = 100$
$S = 165 + 100$
$S = 265$
The surface area of the pyramid is 265 in^2.

5. $\frac{24}{16} \stackrel{?}{=} \frac{21}{14}$
$24(14) \stackrel{?}{=} 16(21)$
$336 = 336$ ✓
Yes, they are similar. Corresponding dimensions are proportional.

Page 589 Reading Mathematics

1. Sample answer: Accuracy is how close a measurement is to the true value. Precision is how exact a measurement is.
2. Sample answer: A certain analytical balance is precise to the nearest 0.01 milligram.
3. See students' work.
4. meters
5. tenth of a pound
6. gram
7. eighth of an inch
8. Sample answer: Accuracy, because it is more important to have approximately the correct amount of cement than it is to have a very precise amount.

11-7 Precision and Significant Digits

Page 590 Why are all measurements really approximations?

a. See students' work.
b. Sample answer: The third ruler; it gives the most precise measurements.

Page 592 Check for Understanding

1. Josh; the first two 0s are not significant because they are placeholders for the decimal point. The 0 between 2 and 5 is significant because it is between two significant digits and shows the actual value in the thousandths place.
2a. cm ruler
2b. surveyor's tools
2c. 12-ft tape measure
2d. yardstick
3. Sample answer: 0.012 or 2500
4. The precision unit is one ounce.
5. 3 significant digits
6. 1 significant digit
7. 3 significant digits

8. $\begin{array}{r} 14.38 \\ +\ 5.7 \\ \hline 20.08 \end{array}$

The least precise measurement, 5.7 cm, has 1 decimal place. So, round 20.08 to 1 decimal place, 20.1.

9. $\begin{array}{r} 15.273 \\ -\ 8.2 \\ \hline 7.073 \end{array}$

The least precise measurement, 8.2 L, has 1 decimal place. So, round 7.073 to 1 decimal place, 7.1 L.

10. $\begin{array}{r} 3.147 \\ \times\ 1.8 \\ \hline 5.6646 \end{array}$

The answer cannot have more significant digits than the measurement 1.88 mm. So, round 5.6646 mm^2 to 2 significant digits, 5.7 mm^2.

11. $\begin{array}{r} 60.42 \\ \times\ 9.012 \\ \hline 544.50504 \end{array}$

The answer cannot have more significant digits than the measurements given. So, round 544.50504 in^2 to 4 significant digits, 544.5 in^2.

12. $\begin{array}{r} 7.85 \\ \times\ 13.0 \\ \hline 102.050 \end{array}$

The answer cannot have more significant digits than the measurements given. So, round 102.050 ft^2 to 3 significant digits, 102 ft^2.

Pages 593–594 Practice and Apply

13. The precision unit is $\frac{1}{32}$ inch.
14. The precision unit is 0.5 centimeter.
15. 3 significant digits
16. 1 significant digit
17. 2 significant digits
18. 3 significant digits
19. 1 significant digit
20. 2 significant digits
21. 4 significant digits
22. 2 siginficant digits

23. $\begin{array}{r} 27 \\ +\ 18.2 \\ \hline 45.2 \end{array}$

The least precise measurement, 27 in., has 0 decimal places. So, round 45.2 to no decimal places, 45 in.

24. $\begin{array}{r} 6.75 \\ -\ 3.2 \\ \hline 3.55 \end{array}$

The least precise measurement, 3.2 mm, has 1 decimal place. So, round 3.55 to 1 decimal place, 3.6 mm.

25. $\begin{array}{r} 0.4 \\ \times\ 5.1 \\ \hline 2.04 \end{array}$

The answer cannot have more significant digits than the measurement 0.4 ft. So, round 2.04 ft^2 to 1 siginificant digit, 2 ft^2.

26. $\begin{array}{r} 7.30 \\ \times\ 1.61 \\ \hline 11.753 \end{array}$

The answer cannot have more significant digits than the measurements given. So, round 11.753 yd^2 to 3 significant digits, 11.8 yd^2.

27. $\begin{array}{r} 29.307 \\ 4.23 \\ +\ 50.93 \\ \hline 84.467 \end{array}$

The least precise measurements, 4.23 and 50.93, have 2 decimal places. So, round 84.467 to 2 decimal places, 84.47 m.

28. $\begin{array}{r} 127.2 \\ +\ 42.3 \\ \hline 169.5 \end{array}$ $\quad$ $\begin{array}{r} 169.5 \\ -\ 5.7 \\ \hline 163.8 \end{array}$

The measurements all have 1 decimal place. So, round 163.8 to 1 decimal place, 163.8 g.

29. $\begin{array}{r} 50.2 \\ -\ 0.75 \\ \hline 49.45 \end{array}$

The least precise measurement, 50.2, has 1 decimal place. So, round 49.45 to 1 decimal place, 49.5 cm.

30. $\begin{array}{r} 18.160 \\ -\ 15 \\ \hline 3.160 \end{array}$

The least precise measurement, 15, has no decimal places. So, round 3.160 to no decimal places, 3 L.

31. $\begin{array}{r} 5.327 \\ \times\ 4.8 \\ \hline 25.5696 \end{array}$

The answer cannot have more significant digits than the measurement 4.8 m. So, round 25.5696 m^2 to 2 significant digits, 26 m^2.

32. $\begin{array}{r} 4.397 \\ \times\ 2.01 \\ \hline 8.83797 \end{array}$

The answer cannot have more significant digits than the measurement 2.01 cm. So, round 8.83797 cm^2 to 3 significant digits, 8.84 cm^2.

33. b; The precision unit of ruler a is $\frac{1}{8}$ inch. The precision unit of ruler b is $\frac{1}{16}$ inch.

34. 0.004 mm, 0.40 mm, 0.4 mm, 40 mm

35. No, the numbers are estimated to the nearest 0.1 million.

36. 2; 3

37. 12,400,000; 3

38. $\begin{array}{r} 26.375 \\ \times\ 15.5 \\ \hline 408.8125 \end{array}$ $\quad$ $\begin{array}{r} 25 \\ \times\ 15.5 \\ \hline 387.5 \end{array}$

$409 - 388 = 21$
The error will cause her to underestimate the area of the shelf by about 21 square inches.

39. Not necessarily, because the actual size of the 2.0 mm wrench can range from 1.95 mm to 2.05 mm, and the actual size of the 2 mm bolt can range from 1.5 mm to 2.5 mm. So, the bolts could be larger than the corresponding wrenches.

40. Measurements are accurate to the nearest precision unit. A smaller precision unit gives a more accurate measure. Answers should include the following.

- The smallest unit of measure that is used determines the precision of a measurement.
- For example, when a pharmacist is determining the amount of medicine to put in a prescription, an exact solution is needed. When a landscaper is determining how much mulch is needed for a client's yard, an approximate solution is sufficient.

41. A; 12 mm is most precise. The precision unit is 1 mm.

42. D; The answer cannot have more significant digits than the measurement 0.1 mm. So, round to 1 significant digit, 0.3 mm^2.

43. $\begin{aligned} \text{greatest possible error} &= \tfrac{1}{2} \cdot \text{precision unit} \\ &= \tfrac{1}{2} \cdot 0.1 \text{ mi} \\ &= 0.05 \text{ mi} \end{aligned}$

The possible actual distance traveled is 0.05 miles less than or 0.05 miles more than 132.8 or between 132.75 and 132.85 miles.

Page 594 Maintain Your Skills

44. $\frac{14}{12} \stackrel{?}{=} \frac{8}{6}$

$14(6) \stackrel{?}{=} 12(8)$
$84 \neq 96$
The corresponding measures are not proportional, so the solids are not similar.

45. $S = \pi r^2 + \pi r\ell$
$S = \pi(1.3)^2 + \pi(1.3)(5.2)$
$S \approx 26.5$
The surface area of the cone is about 26.5 cm^2.

46. $S = \pi r^2 + \pi r\ell$
$S = \pi(7)^2 + \pi(7)(9)$
$S \approx 351.9$
The surface area of the cone is about 351.9 mm^2.

47. Area of lateral face:
$A = \frac{1}{2}b \cdot h$
$A = \frac{1}{2}(1.8)(3.2)$
$A = 2.88$
There are 4 lateral faces, so the lateral area is 4(2.88) or 11.52 m^2.

Area of base:
$A = s^2$
$A = 1.8^2$
$A = 3.24$
$S = 11.52 + 3.24$
$S = 14.76$
The surface area of the pyramid is about 14.8 m^2.

Chapter 11 Study Guide and Review

Page 595 Vocabulary and Concept Check

1. False; lateral area
2. True
3. False; slant height
4. True
5. False; cylinder
6. True
7. True
8. True

Pages 595–598 Lesson-by-Lesson Review

9. The figure has two parallel, congruent bases that are rectangles, $QRST$ and $UVWX$, $QTXU$ and $RSWV$, or $QRVU$ and $TSWX$, so it is a rectangular prism.
faces: $QRST$, $UVWX$, $QTXU$, $RSWV$, $QRVU$, $TSWX$
edges: $\overline{QR}, \overline{RS}, \overline{ST}, \overline{TQ}, \overline{UV}, \overline{VW}, \overline{WX}, \overline{XU}, \overline{QU}, \overline{TX}, \overline{SW}, \overline{RV}$
vertices: Q, R, S, T, U, V, W, X

10. The figure has two parallel, congruent bases that are triangles, GHJ and CDF, so it is a triangular prism.
faces: GHJ, CDF, $CGJF$, $CGHD$, $FDHJ$
edges: $\overline{CD}, \overline{DF}, \overline{FC}, \overline{GH}, \overline{HJ}, \overline{JG}, \overline{CG}, \overline{DH}, \overline{FJ}$
vertices: C, D, F, G, H, J

11. The figure has one rectangular base, $YNPZ$, so it is a rectangular pyramid.
faces: AYZ, AZP, ANP, ANY, $YNPZ$
edges: $\overline{AY}, \overline{AZ}, \overline{AP}, \overline{AN}, \overline{YZ}, \overline{ZP}, \overline{PN}, \overline{NY}$
vertices: A, Y, Z, P, N

12. $V = \pi r^2 h$
$V = \pi \cdot 3.4^2 \cdot 6$
$V \approx 217.9$
The volume is about 217.9 m^3.

13. $V = \ell \cdot w \cdot h$
$V = 0.8 \cdot 0.5 \cdot 1.9$
$V \approx 0.8$
The volume is about 0.8 mm^3.

14. $V = Bh$
$V = \left(\frac{1}{2} \cdot 7 \cdot 5\right)11$
$V = 192.5$
The volume is 192.5 cm^3.

15. $V = \frac{1}{3}Bh$
$V = \frac{1}{3}(2)^2 \cdot 3$
$V = 4$
The volume is 4 ft^3.

16. $V = \frac{1}{3}Bh$
$V = \frac{1}{3}(7.5)5.1$
$V \approx 12.8$
The volume is about 12.8 m^3.

17. $V = \frac{1}{3}\pi r^2 h$
$V = \frac{1}{3} \cdot \pi \cdot 9^2 \cdot 20.3$
$V \approx 1721.9$
The volume is about 1721.9 cm^3.

18. $S = 2\ell w + 2\ell h + 2wh$
$S = 2(10)(3) + 2(10)(5) + 2(3)(5)$
$S = 190$
The surface area is 190 in^2.

19. top $15 \cdot 13.1 = 196.5$
bottom $11 \cdot 15 = 165$
right side $7.2 \cdot 15 = 108$
two bases $2\left(\frac{1}{2} \cdot 11 \cdot 7.2\right) = 79.2$
$S = 196.5 + 165 + 108 + 79.2$
$S = 548.7$
The surface area is 548.7 mm^2.

20. $S = 2\pi r^2 + 2\pi rh$
$S = 2\pi \cdot (4.5)^2 + 2 \cdot \pi \cdot 4.5 \cdot 12$
$S \approx 466.5$
The surface area is about 466.5 cm^2.

21. Area of lateral faces:
$A = \frac{1}{2}(3)(6)$
$A = 9$
Area of base:
$A = 3^2$
$A = 9$
$S = 4(9) + 9$
$S = 45$
The surface area is 45 in^2.

22. $S = \pi r\ell + \pi r^2$
$S = \pi(14)(18.1) + \pi(14)^2$
$S \approx 1411.8$
The surface area is about 1411.8 cm^2.

23. $S = \pi r\ell + \pi r^2$
$S = \pi(1.75)(5) + \pi(1.75)^2$
$S \approx 37.1$
The surface area is about 37.1 in^2.

24. $\frac{14.5}{13.3} \stackrel{?}{=} \frac{6}{5}$
$14.5(5) \stackrel{?}{=} 13.3(6)$
$72.5 \neq 79.8$
The measurements are not proportional, so the cylinders are not similar.

25. $\frac{11.25}{18} \stackrel{?}{=} \frac{20}{32}$
$11.25(32) \stackrel{?}{=} 18(20)$
$360 = 360$ ✓
The measurements are proportional, so the solids are similar.

26. $\frac{x}{28} = \frac{9}{15}$
$15x = 9(28)$
$x = 16.8$
The slant height is 16.8 in.

27. $\begin{array}{r} 10.3 \\ +\ 8.7 \\ \hline 19.0 \end{array}$

The measurements all have 1 decimal place. So, round 19.0 to 1 decimal place, 19.0 cm.

28. $\begin{array}{r} 25.71 \\ -\ 11.2 \\ \hline 14.51 \end{array}$

The least precise measurement, 11.2, has 1 decimal place. So, round 14.51 to 1 decimal place, 14.5 kg.

29. $\begin{array}{r} 0.04 \\ +\ 0.9 \\ \hline 0.94 \end{array}$

The least precise measurement, 0.9, has 1 decimal place. So, round 0.94 to 1 decimal place, 0.9 m.

30. $\begin{array}{r} 5.186 \\ \times\ 1.5 \\ \hline 7.779 \end{array}$

The answer cannot have more significant digits than the measurement 1.5. So, round 7.779 to 2 significant digits, 7.8 in^2.

31. $\begin{array}{r} 32.0 \\ \times\ 30.4 \\ \hline 972.8 \end{array}$

The answer cannot have more significant digits than the measurements given. So, round 972.8 to 3 significant digits, 973 ft^2.

32. $\begin{array}{r} 80.51 \\ -\ 6.01 \\ \hline 74.50 \end{array}$

The measurements all have 2 decimal places. So, round 74.50 to two decimal places, 74.50 g.

Chapter 11 Practice Test

Page 599

1. A prism has two parallel congruent bases. A pyramid has one base.
2. Sample answer: Similar solids have the same shape, and the corresponding linear measures are proportional.
3. Sample answer: 0.123 or 1230
4. The figure has two parallel, congruent bases that are rectangular, *ABCD* and *EFGH*, *ABFE* and *DCGH*, or *ADHE* and *BCGF*, so it is a rectangular prism.
   faces: *ABCD, EFGH, ABFE, CDGH, ADHE, BCGF*
   edges: $\overline{AB}, \overline{BC}, \overline{CD}, \overline{DA}, \overline{EF}, \overline{FH}, \overline{HG}, \overline{HE}, \overline{AE}, \overline{BF}, \overline{CG}, \overline{DH}$
   vertices: *A, B, C, D, E, F, G, H*
5. The figure has one rectangular base, *MNOP*, so it is a rectangular pyramid.
   faces: *OMN, LNO, LOP, LMP, MNOP*
   edges: $\overline{LM}, \overline{LN}, \overline{LO}, \overline{LP}, \overline{MN}, \overline{NO}, \overline{OP}, \overline{PM}$
   vertices: *L, M, N, O, P*
6. $V = \pi r^2 h$
   $V = \pi(1.7)^2(8)$
   $V \approx 72.6$
   The volume is about 72.6 mm^3.
7. $V = \frac{1}{3}Bh$
   $V = \frac{1}{3}(14)(8)(5)$
   $V \approx 186.7$
   The volume is about 186.7 in^3.
8. $V = \ell \cdot w \cdot h$
   $V = 9.2(9.2)(9.2)$
   $V \approx 778.7$
   The volume is about 778.7 cm^3.
9. $V = \frac{1}{3}\pi r^2 h$
   $V = \frac{1}{3}\pi(13)^2(31)$
   $V \approx 5486.3$
   The volume is about 5486.3 ft^3.
10. $S = 2\pi r^2 + 2\pi rh$
    $S = 2\pi(10)^2 + 2\pi(10)(40)$
    $S \approx 3141.6$
    The surface area is about 3141.6 cm^2.
11. Area of lateral face:
    $A = \frac{1}{2} \cdot 5 \cdot 8.2$
    $A = 20.5$
    There are four faces, so the lateral area is 4(20.5) or 82 ft^2.
    Area of base:
    $A = 5^2$
    $A = 25$
    $S = 82 + 25$
    $S = 107$
    The surface area is 107 ft^2.
12. $S = \pi r^2 + \pi r\ell$
    $S = \pi(8)^2 + \pi(8)(20)$
    $S \approx 703.7$
    The surface area is about 703.7 mm^2.
13. $\frac{3}{5} \stackrel{?}{=} \frac{8}{12}$
    $3(12) \stackrel{?}{=} 5(8)$
    $36 \neq 40$
    The corresponding measures are not proportional, so the solids are not similar.
14. $\frac{x}{4} = \frac{135}{15}$
    $15x = 4(135)$
    $x = 36$
    The radius is 36 in.
15. $\frac{x}{21} = \frac{8}{12}$ $\qquad$ $\frac{6}{y} = \frac{8}{12}$
    $12x = 8(21)$ $\qquad$ $8y = 6(12)$
    $x = 14$ $\qquad$ $y = 9$
    The length of the small box is 14 meters, and the width of the large box is 9 m.
16. 2 significant digits
17. 2 significant digits
18. $\begin{array}{r} 37.65 \\ -\ 12.9 \\ \hline 24.75 \end{array}$

    The least precise measurement, 12.9 cm, has 1 decimal place. So, round 24.75 to 1 decimal place, 24.8 cm.

19. $\begin{array}{r} 6.8 \\ \times\ 3.875 \\ \hline 26.3500 \end{array}$

The answer cannot have more significant digits than the measurement 6.8 ft. So, round 26.3500 to 2 significant digits, 26 ft^2.

20. C; $\frac{50 \text{ ft}}{3 \text{ in.}} = \frac{x \text{ ft}}{15 \text{ in.}}$

$50(15) = 3x$

$250 = x$

The length is 250 ft.

Chapter 11 Standardized Test Practice

Pages 600–601

1. D; $-3x + 7 = -29$

$-3x + 7 - 7 = -29 - 7$

$\frac{-3x}{-3} = \frac{-36}{-3}$

$x = 12$

2. A; $22 \times 18 = 396$ in^2

$24 \times 16 = 384$ in^2

$26 \times 14 = 364$ in^2

$28 \times 12 = 336$ in^2

A is the greatest.

3. C; mean $= \frac{\$105.52 + \$98.26 + \$101.29 + \$91.73}{4}$

$= \$99.20$

4. There were 10 girls and 5 boys or 15 children who chose Brand Z. Because 5 is not half of 15, choice D is not true.

5. C; $\frac{2}{12} = \frac{x}{30}$

$2(30) = 12x$

$5 = x$

5 pounds of meat will be needed.

6. B; 50% of $\frac{3}{4} = \frac{1}{2} \times \frac{3}{4}$

$= \frac{3}{8}$

$\frac{3}{8}$ of the population voted.

7. B; $\text{Money to spend} - \text{1 bag popcorn} - \text{1.00 per video game}$

$= 10.00 - 3.25 - 1.00n$

8. B; $\frac{8}{36} = \frac{5}{n}$

$8n = 36(5)$

$n = 22.5$

The tree is 22.5 feet tall.

9. The range, or y-values, is {1, 2, 8, 11}.

10. $-9s(4t) = -36st$

11. $A = \ell w$

$308 = 22(w)$

$14 = w$

The width is 14 ft.

12. $x^{-2} = 3^{-2}$

$= \frac{1}{3^2}$

$= \frac{1}{9}$

13. $3.45 \times 10^3 = 3{,}450$

$5.87 \times 10^2 = 587$

3.45×10^3 is greater.

14. $\frac{5}{6} - \frac{1}{5} - \frac{1}{30} = \frac{5}{6} \cdot \frac{5}{5} - \frac{1}{5} \cdot \frac{6}{6} - \frac{1}{30}$

$= \frac{25}{30} - \frac{6}{30} - \frac{1}{30}$

$= \frac{18}{30} = \frac{3}{5}$

15. $300 - 240 = 60$

$\frac{60}{300} = \frac{n}{100}$

$300n = 60(100)$

$n = 20$

20% of the budgeted amount was left.

16. $y + 5 = 2x$

$y + 5 = 2(0)$

$y + 5 = 0$

$y + 5 - 5 = 0 - 5$

$y = -5$

The y-intercept is -5.

17. $m\angle P + 35° + 35° = 180°$

$m\angle P + 70° = 180°$

$m\angle P + 70° - 70° = 180° - 70°$

$m\angle P = 110°$

18. $C = 2\pi r$

$C = 2\pi(9)$

$C \approx 56.5$

The circumference is about 56.5 cm.

19. $V = \ell \cdot w \cdot h$

$V = 7(20)(4)$

$V = 560$

Each step is 560 in^3. Six steps will be 6(560) or 3360 in^3.

20. $\frac{x}{24.1} = \frac{45}{18}$

$18x = 24.1(45)$

$x = 60.3$

The height is about 60.3 cm.

21a. 32 feet long ÷ 3 feet per box = $10.\overline{6}$ boxes, so the boxes can be stored 10 boxes long.

8 feet wide ÷ 2 feet per box = 4 boxes, so the boxes can be stored 4 boxes wide.

10 feet high ÷ 2 feet per box = 5 boxes, so the boxes can be stored 5 boxes high.

$10 \times 4 \times 5 = 200$ boxes

The company can store 200 boxes.

21b. $V = \ell \cdot w \cdot h$

$V = 3(2)(2)$

$V = 12$ ft^3 per box

$12(200) = 2400$ cubic feet

The volume of the boxes is 2400 ft^3.

21c. $V = \ell \cdot w \cdot h$

$V = 32(8)(10)$

$V = 2560$ cubic feet

The volume of the storage area is 2560 ft^3.

21d. $2560 - 2400 = 160$

160 ft^3 of space is not filled.

10 ft
5 boxes
8 ft
4 boxes
32 ft
10 boxes
Leftover space:
2 ft x 8 ft x 10 ft

Chapter 12 More Statistics and Probability

Page 605 Getting Started

1. mean $= \frac{10 + 15 + 23}{3}$
$= \frac{48}{3}$
$= 16$
median $= 15$
no mode

2. mean $= \frac{21 + 24 + 24 + 24 + 42 + 48}{6}$
$= \frac{183}{6}$
$= 30.5$
median $= 24$
mode $= 24$

3. mean $= \frac{3.2 + 5.1 + 6.5 + 6.5}{4}$
$= \frac{21.3}{4}$
≈ 5.3
median $= \frac{5.1 + 6.5}{2}$
$= \frac{11.6}{2}$
$= 5.8$
mode $= 6.5$

4. mean $= \frac{2.2 + 4.3 + 5.4 + 3.2 + 4.8 + 5.4 + 6.2 + 8.1}{8}$
$= \frac{39.6}{8}$
≈ 5.0
To find the median, first order the numbers from smallest to largest:
2.2, 3.2, 4.3, 4.8, 5.4, 5.4, 6.2, 8.1
median $= \frac{4.8 + 5.4}{2}$
$= 5.1$
mode $= 5.4$

5. $P(\text{green}) = \frac{2}{6}$ or $\frac{1}{3}$

6. $P(4) = \frac{1}{6}$

7. 2, 4, and 6 are even numbers.
$P(\text{even}) = \frac{3}{6}$ or $\frac{1}{2}$

8. $P(\text{not blue}) = \frac{4}{6}$ or $\frac{2}{3}$

9. 2, 3, and 5 are prime numbers.
$\text{P(prime)} = \frac{3}{6}$ or $\frac{1}{2}$

10. 4 and 6 are composite numbers.
$P(\text{composite}) = \frac{2}{6}$ or $\frac{1}{2}$

11. $\frac{1}{6} + \frac{1}{3} = \frac{1}{6} + \frac{1 \cdot 2}{3 \cdot 2}$
$= \frac{1}{6} + \frac{2}{6}$
$= \frac{3}{6}$ or $\frac{1}{2}$

12. $\frac{2}{3} - \frac{4}{9} = \frac{2 \cdot 3}{3 \cdot 3} - \frac{4}{9}$
$= \frac{6}{9} - \frac{4}{9}$
$= \frac{2}{9}$

13. $\frac{3}{4} \times \frac{1}{6} = \frac{3 \times 1}{4 \times 6}$
$= \frac{3}{24}$ or $\frac{1}{8}$

14. $\frac{5}{8} \times \frac{3}{4} = \frac{5 \times 3}{8 \times 4}$
$= \frac{15}{32}$

15. $\frac{3}{8} \times \frac{4}{5} \times \frac{5}{9} = \frac{3 \times 4 \times 5}{8 \times 5 \times 9}$
$= \frac{60}{360}$ or $\frac{1}{6}$

16. $\frac{1}{2} \times \frac{5}{6} \times \frac{3}{4} = \frac{1 \times 5 \times 3}{2 \times 6 \times 4}$
$= \frac{15}{48}$ or $\frac{5}{16}$

17. $\frac{5}{12} + \frac{1}{12} - \frac{1}{3} = \frac{5}{12} + \frac{1}{12} - \frac{1 \cdot 4}{3 \cdot 4}$
$= \frac{5}{12} + \frac{1}{12} - \frac{4}{12}$
$= \frac{2}{12}$ or $\frac{1}{6}$

18. $\frac{3}{8} + \frac{1}{2} - \frac{1}{4} = \frac{3}{8} + \frac{1 \cdot 4}{2 \cdot 4} - \frac{1 \cdot 2}{4 \cdot 2}$
$= \frac{3}{8} + \frac{4}{8} - \frac{2}{8}$
$= \frac{5}{8}$

12-1 Stem-and-Leaf Plots

Page 606 How can stem-and-leaf plots help you understand an election?

a. Sample answer: Even though the intervals are the same, the data are not distributed evenly as the number of pieces of data in each interval are not the same.

b. Sample answer: You can see how the data are distributed.

Page 608 Check for Understanding

1. Sample answer: The age of the youngest President at the time of his inauguration was 42 and the age of the oldest President to be inaugurated was 69. However, most of the Presidents were 50 to 59 years old at the time of their inauguration.

2. The stems will be the tens digits for this data. The stems are: 3, 4, 5, and 6.

3. The stems will be the tens digits for this data. The stems are: 0, 1, 2, 3, and 4.
Arrange the leaves, the ones digits, with their corresponding stems, in order from least to greatest.

Stem	Leaf
0	6 7
1	2 5 5
2	0
3	5
4	0 1

2 | 0 = 20

4. The stems will be the tens digits, or the first two digits in the three-digit numbers.
The stems are 2 to 12.
Arrange the leaves, the ones digits, with their corresponding stems, in order from least to greatest.

Stem	Leaf
2	2 3 9
3	
4	2 2 2 4
5	
6	1
7	
8	2
9	
10	3
11	
12	8

12 | 8 = 128

5. The lowest score is 50. The highest score is 99.
6. The median, or the number in the middle, is $\frac{77 + 78}{2}$ or 77.5
7. Sample answer: The lowest score was 50. The highest score was 99. Most of the scores were in the 70–79 interval.
8. The chicken sandwiches have at most 20 fat grams. The burgers have at most 36 fat grams.
9. Chicken; whereas chicken sandwiches have 8–20 grams of fat, burgers have 10–36 grams of fat.

Pages 609–611 Practice and Apply

10.

Stem	Leaf
0	1 2 2 2 6
1	0 3
2	1
3	1

2 | 1 = 21

11.

Stem	Leaf
1	3
2	6 8
3	4 9
4	
5	2 3
6	2 2 7
7	7 9
8	4

7 | 7 = 77

12.

Stem	Leaf
5	2 4 4 8 8 9 9 9
6	0 0 0 0 0

5 | 8 = 58

13.

Stem	Leaf
0	6 6 8
1	0 1 4 6 8
2	0 2 5
3	2 7
4	
5	0

3 | 7 = 37

14.

Stem	Leaf
46	8
47	2 5 5 6 8 8
48	1 7
49	3 6
50	1 8
51	1 7
52	4 6
53	0 4
54	0
55	0
56	
57	6

53 | 4 = 53.4

15.

Stem	Leaf
40	0 0 5
41	
42	5
43	
44	
45	0 6
46	
47	
48	0 4 7
49	
50	
51	0
52	0
⋮	
62	5
63	
⋮	
76	4
77	
78	9

76 | 4 = 764

16. Always; back-to-back stem-and-leaf plots are used only when representing two sets of data.
17. Never; a basic stem-and-leaf plot can have only one key.
18.

Stem	Leaf
2	7 7 7 8 8
3	0 1 2 5

3 | 1 = 31

19. The greatest percent of people in a city who exercise daily is 35%.
20. In 5 of the 9 cities, fewer than 30% of the people exercise daily.
21. Sample answer: In most populated U.S. cities, about 27 to 35% of the people exercise daily.
22. The greatest number of games won by a Big Ten Conference team was 24.

23. The least number of games won by a Big East Conference team was 5.

24. There are 14 teams in the Big East Conference.

25. The average number of games won by the teams in the Big East Conference is less than the average number of games won by the teams in the Big Ten Conference.

26. See students' work.

27. See students' work.

28. Stem-and-leaf plots can help you understand an election by allowing you to see how the number of electors in the U.S. is distributed. Answers should include the following.

Stem	Leaf
0	3 3 3 3 3 3 3 3 4 4 4 4 4 4 5 5 5 5
	6 6 7 7 7 8 8 8 8 8 8 9 9
1	0 0 1 1 1 1 2 2 3 3 4 5 8
2	1 2 3 5
3	2 3
4	
5	4

  3|2 = 32
- Most of the states in the U.S. have 0 to 10 electors. However, there are some states that have 10–19, 20–29, 30–39 and 50–59 electors.
- A presidential candidate might use the display to determine the importance of each state when campaigning.

29a. Sample answer: It would be easier to find the median in a stem-and-leaf plot because the data are arranged in order from least to greatest.

29b. Sample answer: It would be easier to find the mean in a table because once you find the sum, it may be easier to count the number of items in a table.

29c. Sample answer: It would be easier to find the mode in a stem-and-leaf plot because the value or values that occur most often are grouped together.

30. B; The stems are the tens digits. {1, 2, 3, 4, 5}

31. C; Average protein in legumes, nuts, and seeds

$= \frac{5 + 6 + 9 + 14 + 15 + 18 + 39}{7}$

$= \frac{106}{7}$ or about 15 grams

Average protein in dairy products

$= \frac{2 + 2 + 5 + 8 + 8 + 7 + 7 + 9 + 10 + 26}{10}$

$= \frac{84}{10}$ or 8.4 grams

The average amount of protein in legumes, nuts, and seeds is more than the average amount in dairy products.

Page 611 Maintain Your Skills

32. Perimeter = sum of side lengths

$= 12.38 + 7.5 + 6.185$

$= 26.065$

≈ 26.1 in.

33. No; the ratios of the sides are not equal: $\frac{6 \text{ in.}}{10 \text{ in.}} \neq \frac{4 \text{ in.}}{6 \text{ in.}}$ and $\frac{6 \text{ in.}}{5 \text{ in.}} \neq \frac{4 \text{ in.}}{3 \text{ in.}}$.

34. Yes; the ratios of radius to height are equal: $\frac{8}{9.8} = \frac{12}{14.7}$

35. $C = 2\pi r$ $A = \pi r^2$

$C = 2 \cdot \pi \cdot 10$ $A = \pi \cdot 10^2$

$C \approx 62.8$ ft $A = \pi \cdot 100$

$A \approx 314.2 \text{ ft}^2$

36. $0.36 = \frac{36}{100}$ or 36%

37. $2.47 = \frac{247}{100}$ or 247%

38. $0.019 = \frac{1.9}{100}$ or 1.9%

39. $0.0065 = \frac{0.65}{100}$ or 0.65%

40. $\frac{6}{25} = \frac{6 \cdot 4}{25 \cdot 4}$

$= \frac{24}{100}$ or 24%

41. $$\begin{array}{r} 0.571 \\ 7\overline{)4.0} \\ -3\,5 \\ \hline 50 \\ -49 \\ \hline 10 \\ -7 \\ \hline 3 \end{array}$$

$\frac{4}{7} = 0.571$

$= \frac{57.1}{100}$ or 57.1%

42. $$\begin{array}{r} 1.875 \\ 8\overline{)15.0} \\ -8 \\ \hline 7\,0 \\ -6\,4 \\ \hline 60 \\ -56 \\ \hline 40 \\ -40 \\ \hline 0 \end{array}$$

$\frac{15}{8} = 1.875$

$= \frac{187.5}{100}$ or 187.5%

43. $$\begin{array}{r} 0.016 \\ 1500\overline{)24.00} \\ -15\,00 \\ \hline 9\,000 \\ -9\,000 \\ \hline 0 \end{array}$$

$\frac{24}{1500} = 0.016$

$= \frac{1.6}{100}$ or 1.6%

44. Order the data from least to greatest.
21, 23, 28, 35, 45
The median, or middle value, is 28.

45. Order the data from least to greatest.
1,2, 4, 6, 6, 9, 13, 15, 18
The median, or middle value, is 6.

46. Order the data from least to greatest.
39, 45, 50, 54, 64, 78
The median, or middle value, is $\frac{50 + 54}{2}$ or 52.

47. Order the data from least to greatest.
0.2, 0.3, 0.4, 0.8, 1.3, 1.8, 2.1, 2.6
The median, or middle value, is $\frac{0.8 + 1.3}{2}$ or about 1.1

12-2 Measures of Variation

Page 612 Why are measures of variation important in interpreting data?

a. 173 mph

b. 142 mph

c. $173 - 142 = 31$ mph

d. The fastest winning average speed and the slowest winning average speed are within 31 miles per hour of each other.

Page 614 Check for Understanding

1. The range describes how the entire set of data is distributed, while the interquartile range describes how the middle half of the data is distributed.

2. The lower quartile is the median of the lower half of the set of data. The upper quartile is the median of the upper half of the set of data.

3. Sample answer: {8, 9, 13, 25, 26, 26, 26, 27, 28, 30, 35, 40}

4. The greatest value is 48, and the least value is 12. So, the range is $48 - 12$ or 36.
To find the interquartile range, first list the data from least to greatest. Then find the median and the upper and lower quartiles.

lower half (12 26 32) | upper half (35 41 48)

12 26 32 35 41 48

LQ = 26 ↑, median ↑ (between 32 and 35), UQ = 41 ↑

The interquartile range is $41 - 26$ or 15.

5. The greatest value is 99, and the least value is 72. So, the range is $99 - 72$ or 27.
To find the interquartile range, first list the data from least to greatest. Then find the median and the upper and lower quartiles.

lower half (72 73 76 76) | upper half (80 80 81 99)

72 73 76 76 79 80 80 81 99

median ↑ 79

$LQ = \frac{73 + 76}{2}$ or 74.5 $\quad UQ = \frac{80 + 81}{2}$ or 80.5

The interquartile range is $80.5 - 74.5$ or 6.

6. List the data from least to greatest. The median value divides the data in half.
10, 11, 16, 17, 24, 25, 154, 1416, 5832
↑ median (24)
Earth has the median day length.

7. The median day length is 24 hours.

8. Sample answer: Since the length of a day ranges from $5832 - 10$ or 5822 hours, the lengths of days for the planets vary greatly.

Pages 615–616 Practice and Apply

9. The greatest value is 92, and the least value is 15. So, the range is $92 - 15$ or 77.
To find the interquartile range, first list the data from least to greatest. Then find the median and the upper and lower quartiles.

lower half (15 34 43) | upper half (65 73 92)

15 34 43 64 65 73 92

LQ = 34 ↑, median = 64 ↑, UQ = 73 ↑

The interquartile range is $73 - 34$ or 39.

10. The greatest value is 25, and the least value is 1. So, the range is $25 - 1$ or 24.
To find the interquartile range, first list the data from least to greatest. Then find the median and the upper and lower quartiles.

lower half (1 5 6 8) | upper half (9 9 13 25)

1 5 6 8 9 9 13 25

median ↑ (between 8 and 9)

$LQ = \frac{5 + 6}{2}$ or 5.5 $\quad UQ = \frac{9 + 13}{2} = 11$

The interquartile range is $11 - 5.5$ or 5.5.

11. The greatest value is 82, and the least value is 32. So, the range is $82 - 32$ or 50.
To find the interquartile range, first list the data from least to greatest. Then find the median and upper and lower quartiles.

lower half (32 55 65 68 69) | upper half (70 74 75 77 82)

32 55 65 68 69 70 74 75 77 82

LQ = 65 ↑, median ↑ (between 69 and 70), UQ = 75 ↑

The interquartile range is $75 - 65$ or 10.

12. The greatest value is 91, and the least value is 10. So, the range is $91 - 10$ or 81.
To find the interquartile range, first list the data from least to greatest. Then find the median and the upper and lower quartiles.

lower half (10 21 25 39 43) | upper half (55 58 76 89 91)

10 21 25 39 43 44 55 58 76 89 91

LQ = 25 ↑, median = 44 ↑, UQ = 76 ↑

The interquartile range is $76 - 25$ or 51.

13. The greatest value is 26, and the least value is 1. So, the range is $26 - 1$ or 25.
To find the interquartile range, first list the data from least to greatest. Then find the median and the upper and lower quartiles.

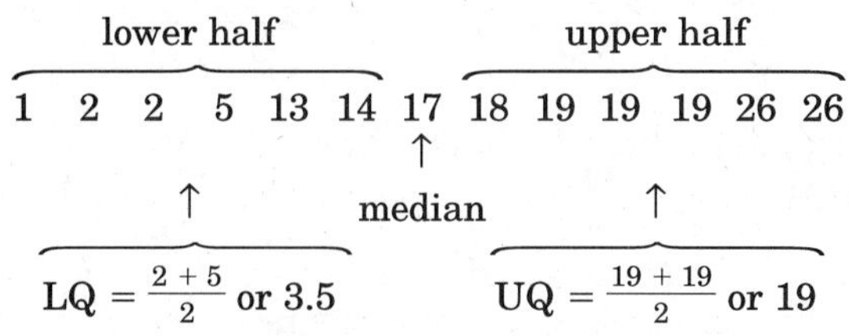

The interquartile range is $19 - 3.5$ or 15.5.

14. The greatest value is 69, and the least value is 40. So, the range is 69 − 40 or 29.
To find the interquartile range, first list the data from least to greatest. Then find the median and the upper and lower quartiles.

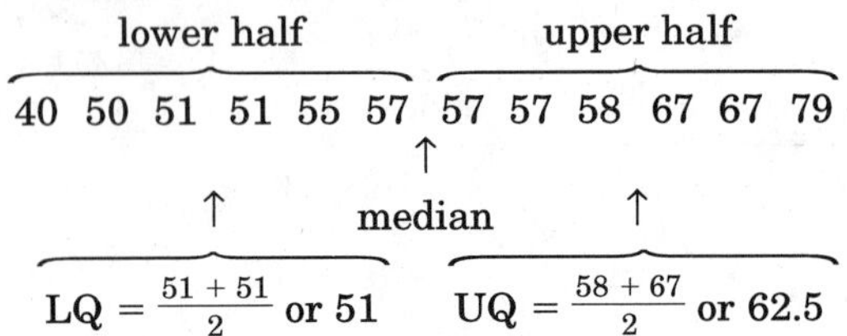

The interquartile range is 62.5 − 51 or 11.5.

15. To find the interquartile range, first list the data from least to greatest. Then find the median and the upper and lower quartiles.

lower half — upper half

119 204 210 213 215 226 362

LQ (204) — median (213) — UQ (226)

The interquartile range is 226 − 204 or 22.

16. First, list the data from least to greatest.

lower half — upper half

21.2 27.5 28.5 29.3 30.1 30.2 31.6 35.3

median

$LQ = \frac{27.5 + 28.5}{2}$ or 28 $\quad UQ = \frac{30.2 + 31.6}{2}$ or 30.9

The middle half of the data ranges from the lower quartile to the upper quartile. So, the range of the middle half is 30.9 − 28 or 2.9.

17. To find the interquartile range, first list each set of data from least to greatest. Then find the median and the upper and lower quartiles of each set.

February

lower half — upper half

24 30 34 36 39 41 44 44 45 46 51 54 62

median

$LQ = \frac{34 + 36}{2}$ or 35 $\quad UQ = \frac{46 + 51}{2}$ or 48.5

The interquartile range for the February data is 48.5 − 35 or 13.5.

July

lower half — upper half

68 70 73 75 76 77 79 80 82 82 82 82 83

median

$LQ = \frac{73 + 75}{2}$ or 74 $\quad UQ = \frac{82 + 82}{2}$ or 82

The interquartile range for the July data is 82 − 74 or 8.

18. Sample answer: July; the temperatures are more tightly clustered around the median.

19. **National League:**
The greatest value is 73 and the least value is 31. So, the range is 73 − 31 or 42.
To find the interquartile range, first list the data from least to greatest. Then find the median and the upper and lower quartiles.

lower half

31, 35, 36, 36, 36, 37, 37, 37, 38, 38, 38, 39, 39, 40,
↑ LQ

40, 40, 40, 40, 41, 43, 44, 44, 44, 45, 45, 46, 46, 47,
↑ median

upper half

47, 47, 48, 48, 48, 49, 49, 49, 50, 52, 52, 65, 70, 73
↑ UQ

The median is $\frac{44 + 44}{2}$ or 44.
The upper quartile is 48.
The lower quartile is 38.
The interquartile range is 48 − 38 or 10.

American League:
The greatest value is 61, and the least value is 22. So, the range is 61 − 22 or 39.
To find the interquartile range, first list the data from least to greatest. Then find the median and the upper and lower quartiles.

lower half

22, 32, 32, 32, 32, 33, 36, 36, 37, 39, 39, 39, 40, 40,
↑

LQ

40, 40, 41, 42, 43, 43, 44, 44, 44, 44, 45, 45, 46, 46,
↑ median

upper half

47, 48, 48, 49, 49, 49, 49, 50, 51, 52, 52, 56, 56, 61
↑ UQ

The median is $\frac{44 + 44}{2}$ or 44.
The upper quartile is 49.
The lower quartile is 39.
The interquartile range is 49 − 39 or 10.

20. Sample answer: Since the measures of variation are about the same, we can conclude that the number of home runs hit by the leaders in each league is consistent. That is, neither league exceeds the other.

21a. Sample answer:
{43, 49, 50, 50, 58, 60, 60, 66, 70, 70, 71, 78}
LQ — median — UQ

21b. Sample answer:
{15, 18, 20, 20, 44, 60, 60, 64, 70, 70, 75, 79}
LQ — median — UQ

21c. Sample answer: The first set of data has a smaller interquartile range, thus, the data in the first set are more tightly clustered around the median and the data in the second are more spread out over the range.

22. They allow us to see how the data are distributed. Answers should include the following.

- 156.5; 31; 11
- The average speed was 156.5 miles per hour. The speeds varied by 31 miles per hour. The speeds of the cars in the middle of the data set are close in value.

23. C; Order the data from least to greatest.

42, 43, 49, 54, 55, 67, 73, 78, 82, 91, 93, 94

↑

median $= \frac{67 + 73}{2}$ or 70

24. B; Order the data from least to greatest.

14,083, 14,150, 14,165, 14,196, 14,197, 14,238,

↑ LQ (14,196)

14,246, 14,264, 14,265, 14,269, 14,286, 14,309,

↑ median (14,264) ↑ UQ (14,309)

14,361, 14,420, 14,433

The interquartile range is 14,309 − 14,196 or 113. Since the interquartile range is small, the heights of the mountains in Colorado are clustered around the median height.

Page 616 Maintain Your Skills

25.

Stem	Leaf
0	9
1	2 4 5 8
2	1 7
3	7

3 | 7 = $37

26. 27.08 mm + 6.5 mm = 33.58 mm ≈ 33.6 mm

27. $V = \frac{1}{3}\pi r^2 h$

$V = \frac{1}{3} \cdot \pi \cdot 7^2 \cdot 9$

$V = \frac{1}{3} \cdot \pi \cdot 49 \cdot 9$

$V \approx 461.8 \text{ cm}^3$

28. $V = \frac{1}{3}\pi r^2 h$

$V = \frac{1}{3} \cdot \pi \cdot (4.2)^2 \cdot (6.5)$

$V = \frac{1}{3} \cdot \pi \cdot (17.64) \cdot (6.5)$

$V \approx 120.1 \text{ yd}^3$

29. $C = 2\pi r$

$9.82 = 2\pi r$

$\frac{9.82}{2\pi} = \frac{2\pi r}{2\pi}$

$1.6 \text{ ft} \approx r$

30. {4.3, 4.8, 4.9, 5.0, 5.3, 5.6}

31. {0.2, 0.3, 0.6, 0.8, 1.2, 1.4, 1.5}

32. {40.6, 45.2, 45.4, 46.0, 50.7}

33. {9.8, 9.9, 10.5, 10.9, 11.2, 11.4}

12-3 Box-and-Whisker Plots

Page 617 How can box-and-whisker plots help you interpret data?

a. Order the data from least to greatest for each city.

Tampa, FL

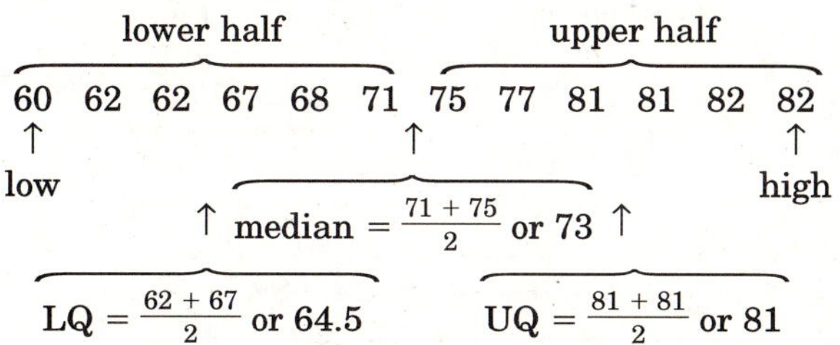

low, 60; high, 82; median, 73; upper quartile, 81; lower quartile, 64.5

Caribou, ME

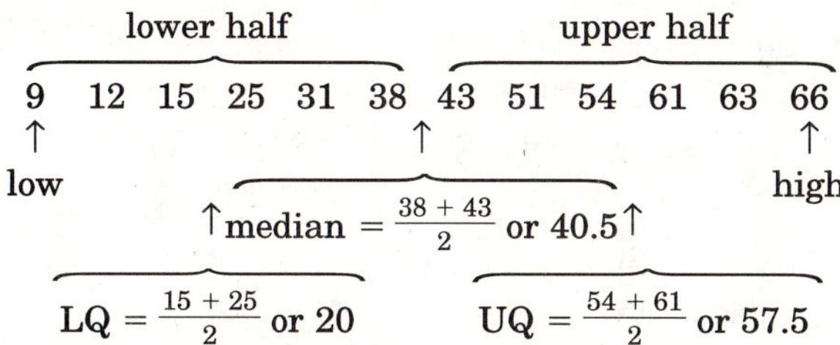

low, 9; high, 66; median, 40.5; upper quartile, 57.5; lower quartile, 20

b. 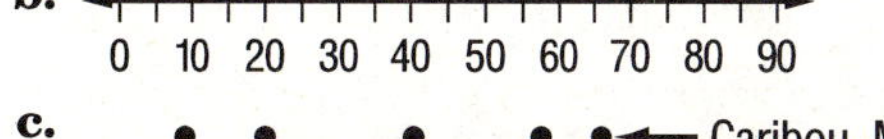

c. 

d. Sample answer: The average monthly temperatures vary greatly in Caribou, ME, while the average monthly temperatures in Tampa, FL, are more consistent.

Pages 619–620 Check for Understanding

1. lower quartile and least value; upper quartile and greatest value

2. A box-and-whisker plot separates the data into fourths.

3. Sample answer: {28, 30, 52, 68, 90, 92}; The lower quartile is 30 and the upper quartile is 90, so the box will be long. The lower extreme is 28, which is very close to the lower quartile of 30. The upper extreme is 92, which is very close to the upper quartile of 90. So, the whiskers will be short.

4. The median is 25. The extremes are 19 and 40. The lower quartile is 22 and the upper quartile is 34.

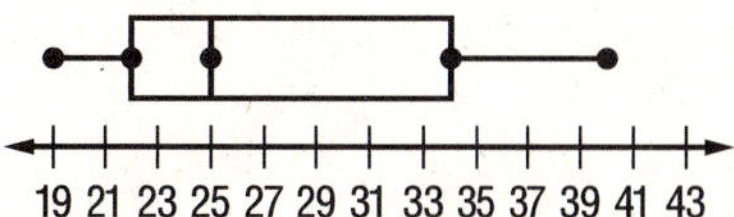

5. The median is 29.5. The extremes are 15 and 50. The lower quartile is 22 and the upper quartile is 32.

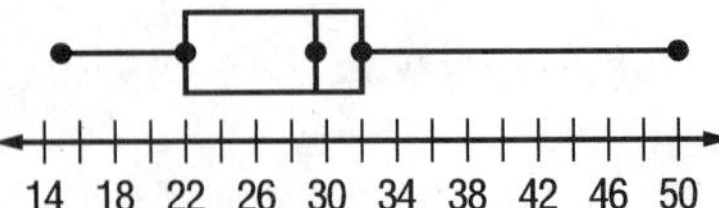

6. The median is 134. The extremes are 129 and 161. The lower quartile is 131 and the upper quartile is 14.9.

Summer Olympic Games 1924–2000
Winning Times for Men's Marathon

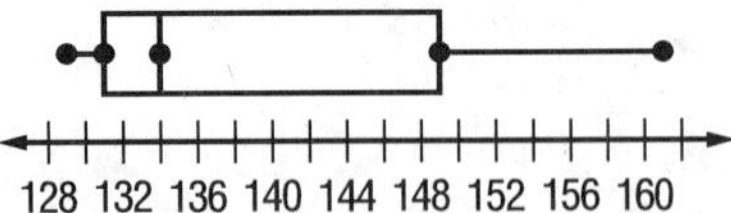

7. Sample answer: The length of the box-and-whisker plot shows that the winning times of the men's marathons are not concentrated around a certain time.
8. Since SUVs average the least miles per gallon, they tend to be less fuel-efficient.
9. The most fuel-efficient SUV and the least fuel-efficent sedan both average 22 miles per gallon.

Pages 620–621 Practice and Apply

10. The median is 88. The extremes are 35 and 100. The lower quartile is 61 and the upper quartile is 97.

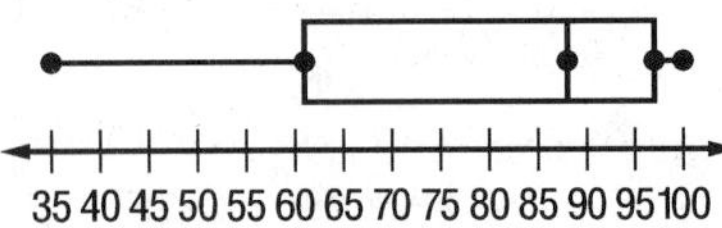

11. The median is 90. The extremes are 60 and 130. The lower quartile is 68 and the upper quartile is 104.

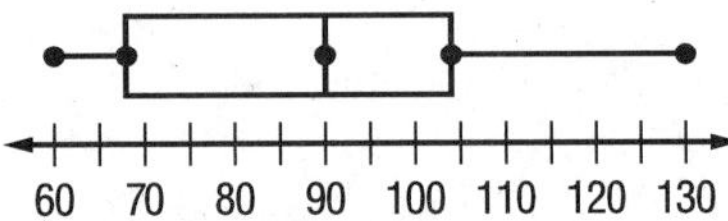

12. The median is 67. The extremes are 40 and 81. The lower quartile is 55 and the upper quartile is 77.

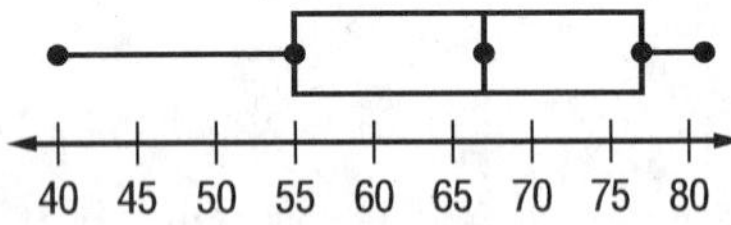

13. The median is 113.5. The extremes are 55 and 174. The lower quartile is 95 and the upper quartile is 162.

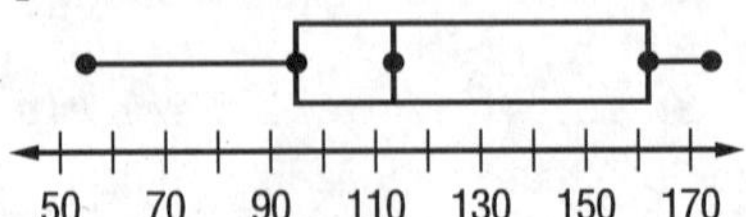

14. The median is 6.7. The extremes are 5.8 and 7.9. The lower quartile is 6.1 and the upper quartile is 7.35.

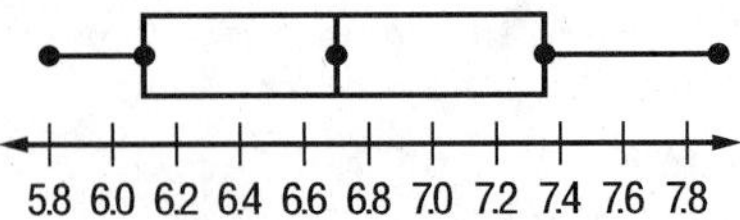

15. The median is 30.6. The extremes are 26.8 and 37.1. The lower quartile is 29.65 and the upper quartile is 32.75.

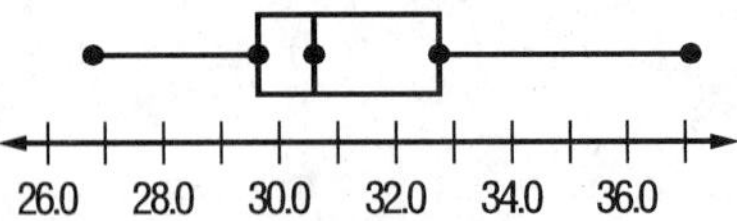

16. The highest quiz score was 96.
17. 50% of the students scored between 80 and 96.
18. Sample answer: Based on the plot, the students' overall scores are between 63 and 96.
19. Sample answer: The least number of games won for NFC is 3 and the least number of games won for the AFC is 1. The most games won for the NFC is 12 and the most games won for the AFC is 13. In addition, for both conferences, the median number of games won is about 9.
20. Sample answer: {60, 60, 60, 60, 60, 70, 75, 80, 85, 85, 85, 85}
    The minimum value and the lower quartile are both 60. The maximum value and the upper quartile are both 85. So, there are no whiskers.
21. A box-and-whisker plot would clearly display any upper and lower extreme temperatures and the median temperature. Answers should include the following.
    -
    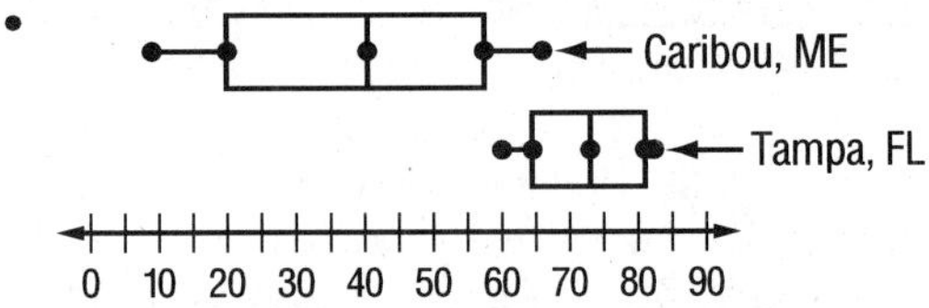

    - Sample answer: Tampa has a median temperature of 73 and Caribou has a median temerature of 40.5. Whereas the highest temperature for Tampa is 82, the highest average temperature for Caribou is 66.
    - Sample answer: You can easily see how the temperatures vary.
22. A; The least of the values for highest wind speed is 40 mph.
23. C; 75% of the data range from the minimum to the upper quartile, 40 to 70 mph.
24. Order the data from least to greatest.
    56, 65, 67, 70, 72, 75, 89
    The interquartile range is 75 − 65 or 10. The outliers are the values more than 1.5(10) or 15 from the quartiles.
    65 − 15 = 50 75 + 15 = 90
    The limits of the outliers are 50 and 90. So, there are no outliers.

25. Order the data from least to greatest.
4, 27, [29], 30, 30, 31, 35, 36, [37], 38, 46
The interquartile range is 37 − 29 or 8. The outliers are the values more than 1.5(8) or 12 from the quartiles.
29 − 12 = 17 37 + 12 = 49
The limits for the outliers are 17 and 49. So, there is one outlier, 4.

Page 621 Maintain Your Skills

26. The greatest value is 4.8, and the least value is 1.4. So, the range is 4.8 − 1.4 or 3.4.
To find the interquartile range, first list the data from least to greatest. Then find the median and the upper and lower quartiles.

lower half (1.4 2.1 2.4), upper half (2.7 3.9 4.8)

1.4 2.1 2.4 2.7 3.9 4.8
LQ = 2.1, median, UQ = 3.9

The interquartile range is 3.9 − 2.1 or 1.8.

27.

Stem	Leaf
1	4
2	1 4 7
3	9
4	8

2|4 = 2.4

28. 3 + 8 + 9 + 5 or 25 people
29. 9 + 5 or 14 people

Page 622 Graphing Calculator Investigation (Follow-Up of Lesson 12-3)

1. Class A: The least value is 25 and the greatest value is 40. The lower quartile is 27 and the upper quartile is 39. The median value is 33.
Class B: The least value is 16 and the greatest value is 40. The lower quartile is 19 and the upper quartile is 26. The median value is 23.
2. Class A: The interquartile range is 39 − 27 or 12.
Class B: The interquartile range is 26 − 19 or 7.
3. Yes; a point appears outside the box-and-whisker plot.
4a. 0%; there are no students under age 25.
4b. 25%; there are 25% of the students between the minimum and the lower quartile, or ages 16 to 19.
5. Sample answer: Class B because there would be more members around my age.

12-4 Histograms

Page 623 How are histograms similar to frequency tables?

a. a state
b. the totals in the row of tally marks
c. They are equal.

Page 625 Check for Understanding

1. Sample answer: Because the intervals are continuous.
2. Sample answer: By displaying the data in a histogram, it is much easier to see how the data in the intervals compare to each other.
3.

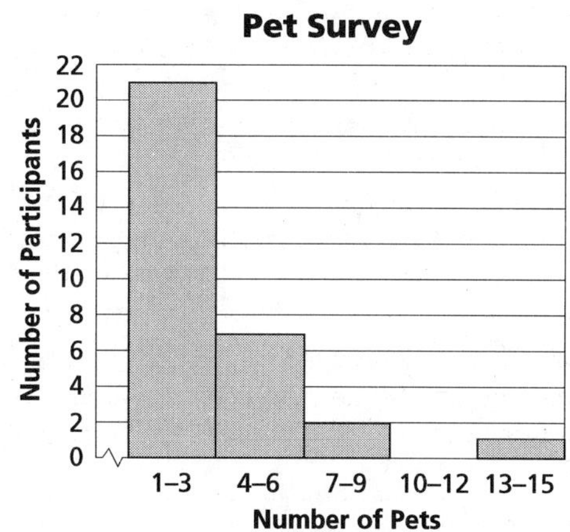

4.

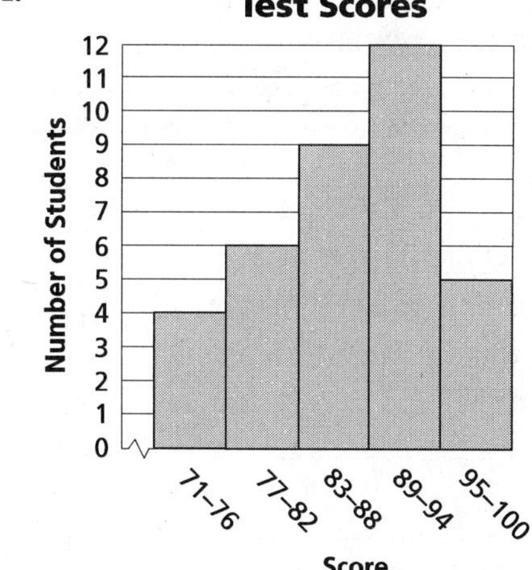

5. The data show the number of states that have a certain number of roller coasters.
6. The interval from 0–9 has the most roller coasters.
7. The numbers between 10 and 34 are omitted.
8. No; the graph shows that 34 states have anywhere from 0 to 9 roller coasters but it does not indicate how many states have none.
9. The number of states with two or more national parks is 8 + 1 + 2 or 11. The number of states with two or more national monuments is 11 + 1 + 2 + 1 + 1 or 16. There are more states with two or more national monuments.

10. The number of states with either one or no national parks is 38. The number of states with either one or no national monuments is 34. So, 38 − 34 or 4 more states have either one or no national parks then either one or no national monuments.

Pages 626–628 Practice and Apply

11.

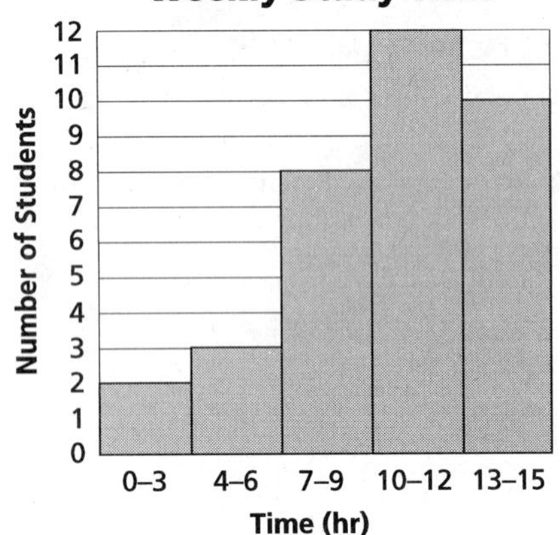

12.

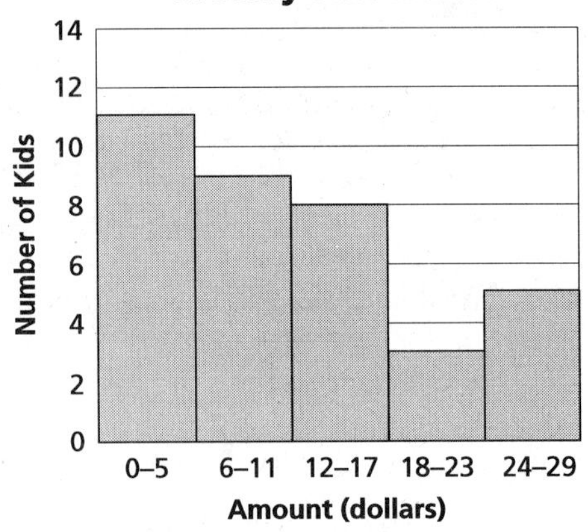

13.

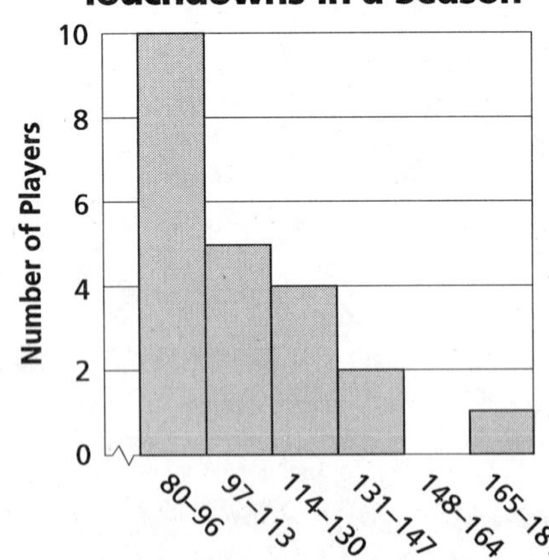

14.

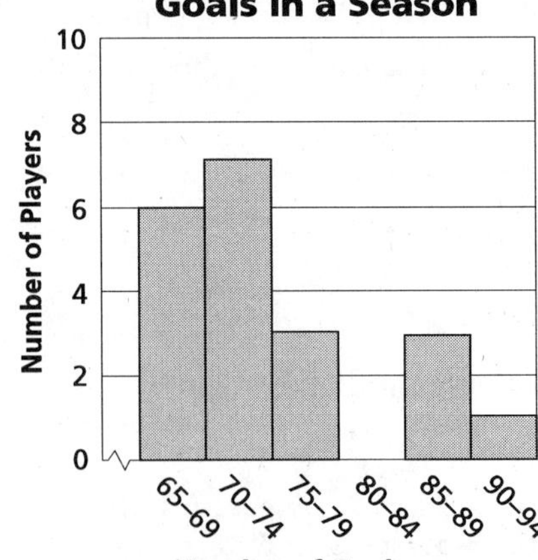

15. Since 3 restaurants sell sandwiches costing \$1.50–\$1.99, 6 restaurants sell sandwiches costing \$2.00–\$2.49, and 4 restaurants sell sandwiches costing \$2.50–\$2.99, 3 + 6 + 4 or 13 restaurants sell sandwiches costing under \$3.

16. There were 3 + 6 + 4 + 2 + 1 or 16 restaurants surveyed.

17. There were 6 restaurants selling sandwiches costing between \$2.00 and \$2.49, and there were 16 restaurants surveyed, so the percent is $\frac{6}{16} \cdot 100\%$ or 37.5%.

18. Dallas has a total of 13 + 7 + 5 + 3 + 1 + 1 or 30 tall buildings. Los Angeles has a total of 10 + 9 + 4 + 6 + 1 + 1 or 31 tall buildings. So, Los Angeles has more tall buildings.

19. Dallas has 5 + 3 + 1 + 1 or 10 buildings over 600 feet. Los Angeles has 4 + 6 + 1 + 1 or 12 buildings over 600 feet. So, Dallas has fewer buildings over 600 feet.

20. Sample answer: Whereas Dallas, Texas, has more buildings with a height of 400-499 feet, Los Angeles, California, has more buildings classified as tall buildings.

21. True; the median is shown inside the box.

22. True; the range is found by subtracting the smallest value from the largest value.

23. False; histograms show only the total number within a range of values, not individual values.

24. False; a box-and-whisker plot shows only the minimum, maximum, lower quartile, upper quartile, and median.

25. True; you can find the total number of values, then find the lower half and the upper half of the data. The median will be between these two halves.

26. Histograms and frequency tables are similar in that they both categorize data using intervals. Answers should include the following.

•

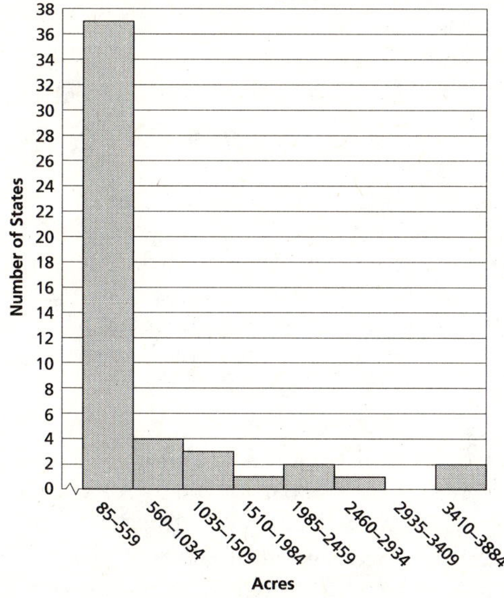

Size of U.S. Farms		
Acres	**Tally**	**Frequency**
85-559	~~||||~~ ~~||||~~ ~~||||~~ ~~||||~~ ~~||||~~ ~~||||~~ ~~||||~~ ||	37
560-1034	||||	4
1035-1509	|||	3
1510-1984	|	1
1985-2459	||	2
2460-2934	|	1
2935-3409		0
3410-3884	||	2

- In the histogram, the size of the farm is shown on the horizontal axis and the number of states that contain a farm of that size is shown on the vertical axis. The bars are used to show the number of states that contain a farm of the given size. In the frequency table, the size of the farm is listed in the first column and the number of states that contain a farm of the given size is listed in the third column. The middle column is used for tallying the data.

27. A; 18 or 19.

28. C; The total number of students is 6 + 8 + 6 + 4 or 24 students.

29. The absolute frequencies are 2, 9, 17, 12, 7, and 2. The relative frequencies are $\frac{2}{49}$, $\frac{9}{49}$, $\frac{12}{49}$, $\frac{7}{49}$ or $\frac{1}{7}$, and $\frac{2}{49}$. The cumulative frequencies are 2, 2 + 9 or 11, 2 + 9 + 17 or 28, 2 + 9 + 17 + 12 or 40, 2 + 9 + 17 + 12 + 7 or 47, and 2 + 9 + 17 + 12 + 7 + 2 or 49.

30. The absolute frequencies are 11, 20, 21, 12, 1, and 1. The relative frequencies are $\frac{11}{66}$ or $\frac{1}{6}$, $\frac{20}{66}$ or $\frac{2}{33}$, $\frac{21}{66}$ or $\frac{7}{22}$, $\frac{12}{66}$ or $\frac{2}{11}$, $\frac{1}{66}$, and $\frac{1}{66}$. The cumulative frequencies are 11, 11 + 20 or 31, 11 + 20 + 21 or 52, 11 + 20 + 21 + 12 or 64, 11 + 20 + 21 + 12 + 1 or 65, and 11 + 20 + 21 + 12 + 1 + 1 or 66.

31.

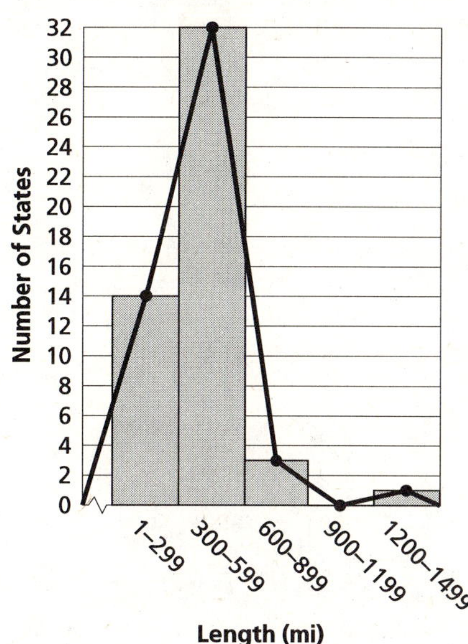

Source: *The World Almanac*

Page 628 Maintain Your Skills

32. 75% of the data in a box-and-whisker plot lie between the minimum and the upper quartile, 78. So, 75% of the presidents died by the time they were 78 years old.

33. The greatest value is 90, and the least value is 46. So, the range is 90 − 46 or 44.
The upper quartile is 78, and the lower quartile is 63. So, the interquartile range is 78 − 63 or 15.

34. Brand D, since it has the tallest bar.

35. Sample answer: The cost of Brand B appears to be twice the cost of Brand C.

Page 628 Practice Quiz 1

1.

Stem	Leaf
1	6 8 9 9
2	0 2 6 8 8 8
3	0 2
4	0 2 3
5	5
6	
7	
8	6 9

5|5 = 55

2. The greatest value is 89, and the least value is 16. So, the range is 89 − 16 or 73.
The upper quartile is 42 and the lower quartile is 20. The interquartile range is 42 − 20 or 22.

3.

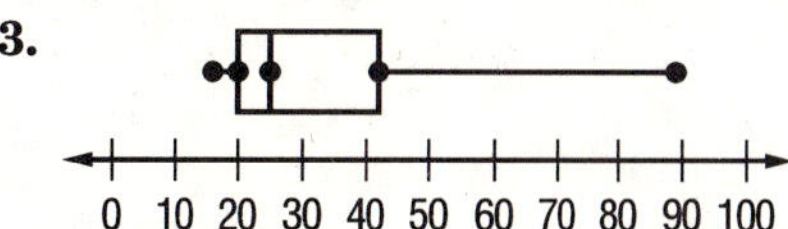

4. The values of \$16, \$18, \$19, and \$19 were less than \$20. So, $\frac{4}{18} \cdot 100\%$ or about 22% of the costs were less than \$20.

5.

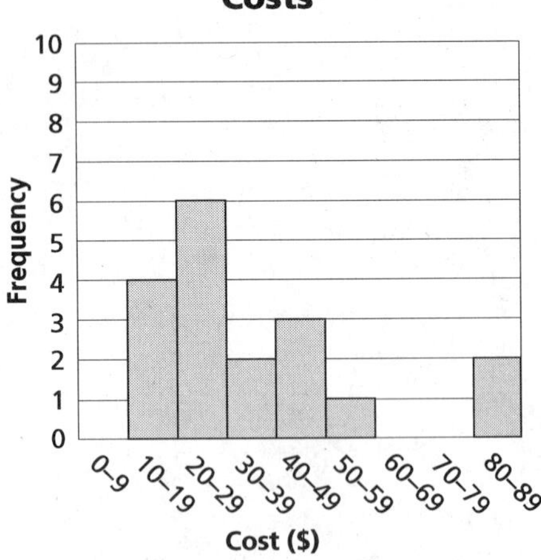

Page 629 Graphing Calculator Investigation (Follow-Up of Lesson 12-4)

1.

Interval	Frequency
40 up to 45	2
45 up to 50	6
50 up to 55	13
55 up to 60	12
60 up to 65	7
65 up to 70	3
70 up to 75	0

2. It contains all the possible ages.
3. It determines interval size by the scale factor of x.
4. There were 6 + 13 + 12 + 7 or 38 presidents at least 45, but less than 65.
5. The percent is $\frac{38}{43} \cdot 100\%$ or about 88%.
6. No; the histogram shows the number of presidents for a range of ages, not the number of presidents for individual ages.
7. See students' work.

12-5 Misleading Statistics

Page 630 How can graphs be misleading?

a. Yes; the data are the same, but the scales are different.
b. Graph B
c. Graph A
d. Sample answer: The graphs differ in that Graph A shows a gradual increase and decrease over the year; Graph B shows a drastic increase and decrease. They are similar in that each graph contains labeled axes and a title, and they contain the same dots.

Page 631 Check for Understanding

1. Sample answer: inconsistent vertical scale and break in vertical scale
2. See students' work.
3. Graph A has a break in the vertical scale.
4. Graph A because the increase appears more drastic.
5. From the vertical scale, you can see that the number of area codes in 1999 is about 1.5 times the number of area codes in 1996. The graph is misleading because the drawing of the phone for 1999 is about 3 times the size of the phone for 1996. Also, there is a break in the vertical scale.

Pages 632–633 Practice and Apply

6. Graph A, since the bar for movie A appears dramatically taller than the others.
7. Graph B, since the bar for movie C is about the same height as the others.
8. The vertical scale of graph A goes from 0 to 8.0, whereas the vertical scale of graph B goes from 4.0 to 8.0. In addition, the actual distance between vertical scales is less than the distance used in graph B.
9. Graph B; the vertical scale used makes the decrease in the unemployment rates appear more drastic.
10. Yes; the intervals are inconsistent and the horizontal scale has a break in it.

11a. Sample answer:

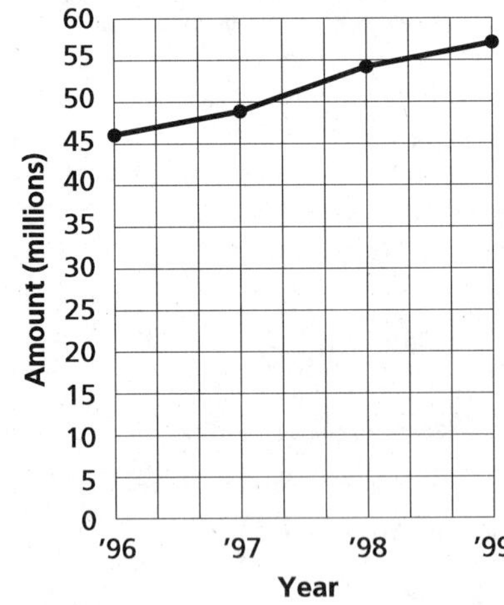

11b. Sample answer:

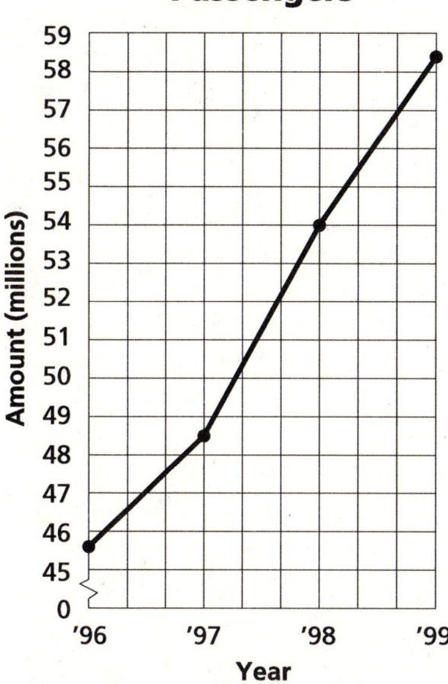

12. The use of inconsistent scales, broken scales, and expanded or shortened axes will make the graph misleading. Answers should include the following.

•

Cookie Sales

Amount Made (millions)
700
600
500
400
300
A B C D E
Type of Cookie

- This graph is misleading because the vertical axis does not start at zero. To redo the graph so that it is not misleading, you can redraw the vertical axis so that it does include zero and the intervals of the vertical axis are equal.

13. B; The amount spent in 1994 was almost \$1700. The amount spent in 1998 was about \$2025. Since 1700(1.2) = 2040, the amount spent in 1998 was about 1.2 times the amount spent in 1994.

Page 633 Maintain Your Skills

14.

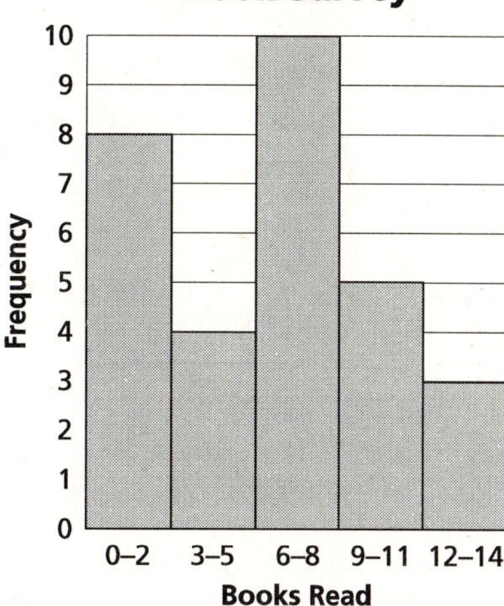

15. The median is 45. The extremes are 38 and 56. The lower quartile is 41.5 and the upper quartile is 52.5.

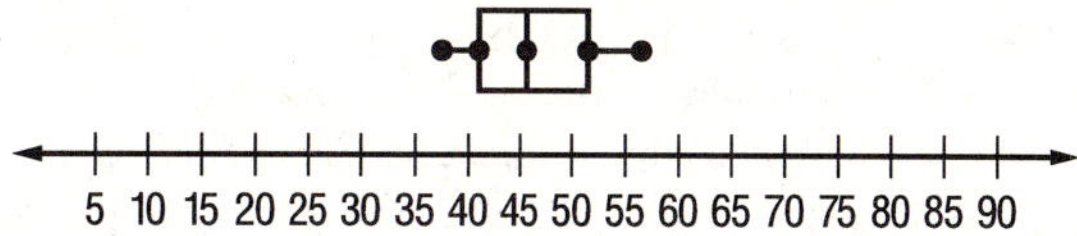

16. $A = \frac{1}{2}bh$

$A = \frac{1}{2} \cdot 6 \cdot 4.2$

$A = 12.6 \text{ ft}^2$

17. $A = \frac{1}{2}h(a + b)$

$A = \frac{1}{2} \cdot 5.8(4 + 3)$

$A = \frac{1}{2} \cdot 5.8 \cdot 7$

$A = 20.3 \text{ m}^2$

18. $P(\text{yellow}) = \frac{3}{12}$ or $\frac{1}{4}$; 25%

19. $P(\text{blue}) = \frac{2}{12}$ or $\frac{1}{6}$; $16\frac{2}{3}\%$

20. $P(\text{not purple}) = \frac{5}{12}$; $41\frac{2}{3}\%$

21. $P(\text{not blue}) = \frac{10}{12}$ or $\frac{5}{6}$; $83\frac{1}{3}\%$

22. $P(\text{blue or purple}) = \frac{9}{12}$ or $\frac{3}{4}$; 75%

23. $P(\text{yellow or not blue}) = \frac{10}{12}$ or $\frac{5}{6}$; $83\frac{1}{3}\%$

Page 634 Reading Mathematics

1. Sample answer: This situation may result in bias. Many people may not be at home during these hours, and thus the survey will not include a portion of the population.
2. Sample answer: Since many cities offer public transportation, people living in urban households may be more apt to use public transportaion. Thus, this may be biased.
3. Sample answer: The survey may be biased. It is a voluntary response and those who call in may have strong opinions.
4. Sample answer: This survey is biased due to the wording of the question.

12-6 Counting Outcomes

Page 635 How can you count the number of skateboard designs that are available from a catalog?

a. See students' work.

b. See students' work.

c. Alien, Eagle; Alien, Cloud; Alien, Red Hot; Birdman, Eagle; Birdman, Cloud; Birdman, Red Hot;
Candy, Eagle; Candy, Cloud; Candy, Red Hot; Radical, Eagle, Radical, Cloud; Radical, Red Hot; Trickster, Eagle; Trickster, Cloud; Trickster, Red Hot

d. 15

Page 637 Check for Understanding

1. Sample answer: Both methods find the number of outcomes. Using the Fundamental Counting Principle is faster and uses less space; using a tree diagram shows what each outcome is.
2. Sample answer: Choosing an outfit from 3 pairs of shorts and 4 T-shirts.
3. First find the number of outcomes possible. The possible outcomes are taco-chicken, taco-beef, taco-bean, burrito-chicken, burrito-beef, and burrito-bean, for a total of 6 outcomes. Two of the outcomes have chicken filling, so the probability of chicken filling is $\frac{2}{6}$ or $\frac{1}{3}$.
4. 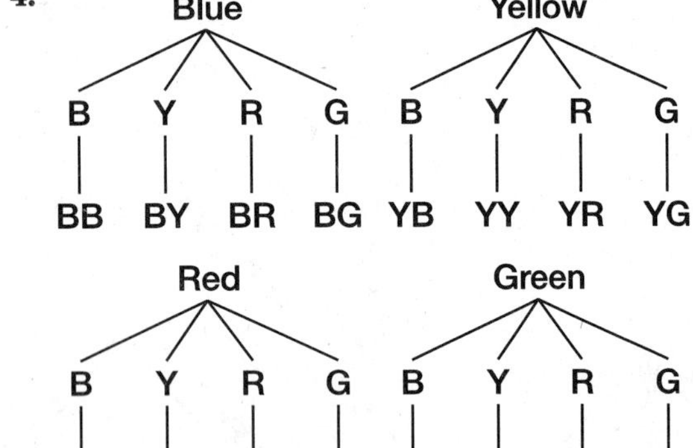

There are $4 \cdot 4$ or 16 possible outcomes.

5. $P(\text{two blues}) = \frac{\text{number of favorable outcomes}}{\text{number of possible outcomes}} = \frac{1}{16}$
6. Use the Fundamental Counting Principle.

$$\underbrace{2}_{\substack{\text{number of}\\ \text{possible}\\ \text{outcomes}\\ \text{of coin toss}}} \times \underbrace{6}_{\substack{\text{number of}\\ \text{possible}\\ \text{outcomes of}\\ \text{number cube}}} = \underbrace{12}_{\substack{\text{total}\\ \text{number of}\\ \text{outcomes}}}$$

There are 12 possible outcomes.

7. List the possible outcomes.

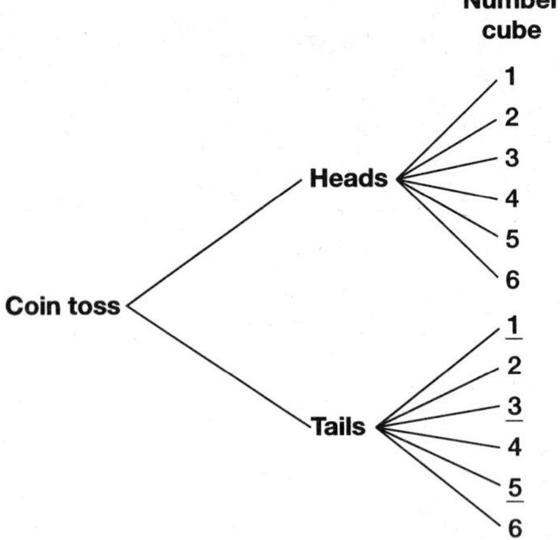

There are 3 favorable outcomes of tails and an odd number.

$$P(\text{tails, odd number}) = \frac{\text{number of favorable outcomes}}{\text{number of possible outcomes}}$$
$$= \frac{3}{12} \text{ or } \frac{1}{4}$$

8. Each coin has 2 possible outcomes. So, the number of outcomes possible is $2 \cdot 2 \cdot 2 \cdot 2$ or 16.
9. There are 3 choices of bread and 3 choices of beverage. The number of outcomes is $3 \cdot 3$ or 9.

Pages 638–639 Practice and Apply

10. 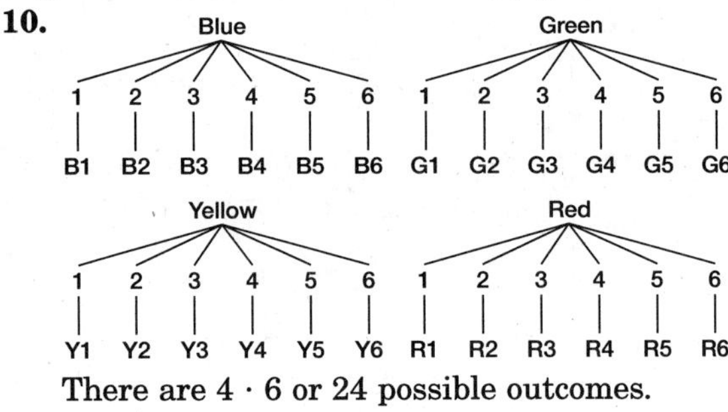

There are $4 \cdot 6$ or 24 possible outcomes.

11. 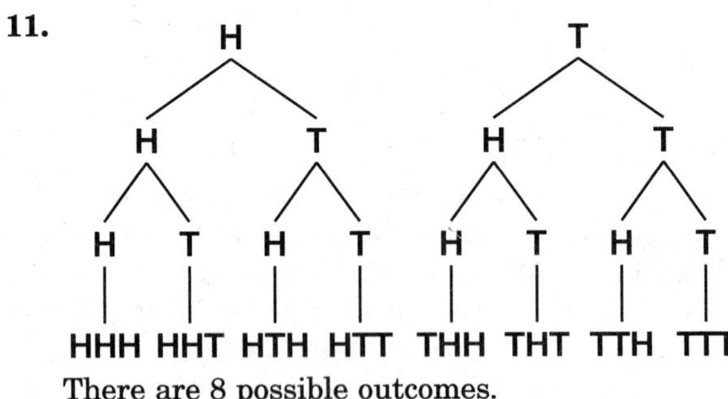

There are 8 possible outcomes.

12.

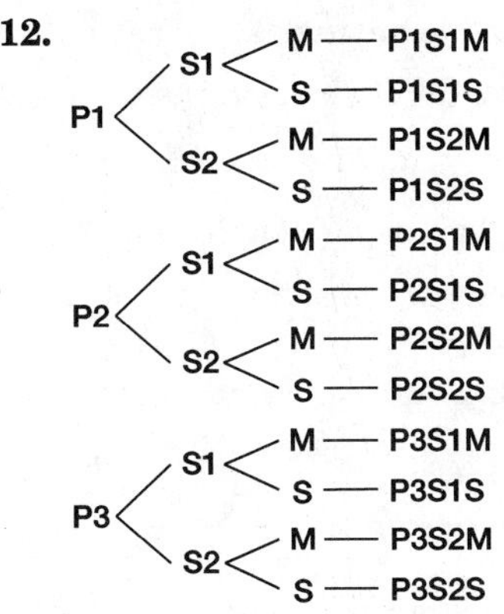

There are 12 possible outcomes.

13.

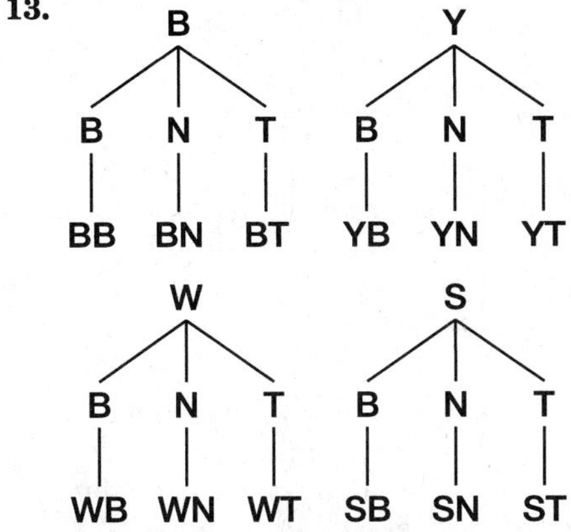

There are 12 possible outcomes.

14. Use the Fundamental Counting Principle.

number of sizes		number of colors		number of possible outcomes
4	×	4	=	16

There are 16 possible outcomes.

15. Use the Fundamental Counting Principle.

number of possible outcomes of first roll		number of possible outcomes of second roll		number of possible outcomes
6	×	6	=	36

There are 36 possible outcomes.

16. Use the Fundamental Counting Principle.

number of possible outcomes of first coin toss		number of possible outcomes of second coin toss		number of possible outcomes of number cube		number of possible outcomes
2	×	2	×	6	=	24

There are 24 possible outcomes.

17. Use the Fundamental Counting Principle.

number of door choices		number of engine choices		number of color choices		number of possible outcomes
2	×	2	×	6	=	24

There are 24 possible outcomes.

18. Use the Fundamental Counting Principle.
Each question has 2 possible choices.
$2 \times 2 \times 2 \times 2 \times 2 = 32$
There are 32 possible outcomes.

19. Use the Fundamental Counting Principle.
Each question has 4 possible choices.
$4 \times 4 \times 4 \times 4 \times 4 = 1024$
There are 1024 possible outcomes.

20. First find the number of outcomes.

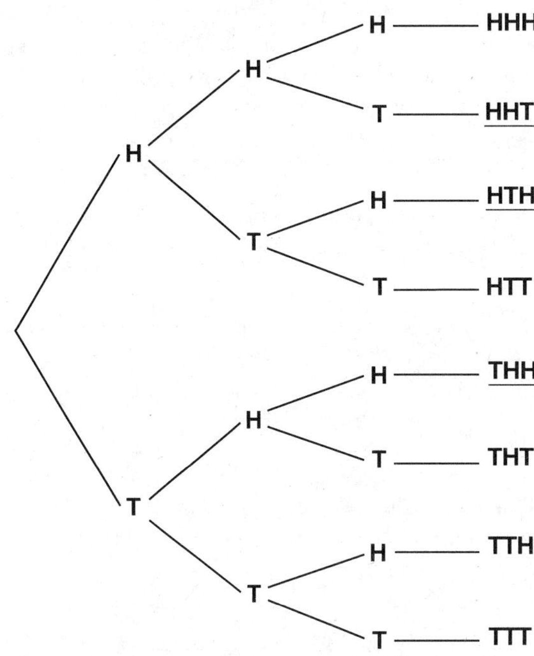

There are 8 possible outcomes. There are three outcomes that have two heads and one tail.

$$P(\text{two heads, one tail}) = \frac{\text{number of favorable outcomes}}{\text{number of possible outcomes}}$$
$$= \frac{3}{8}$$

The probability is $\frac{3}{8}$.

21. First find the number of outcomes.

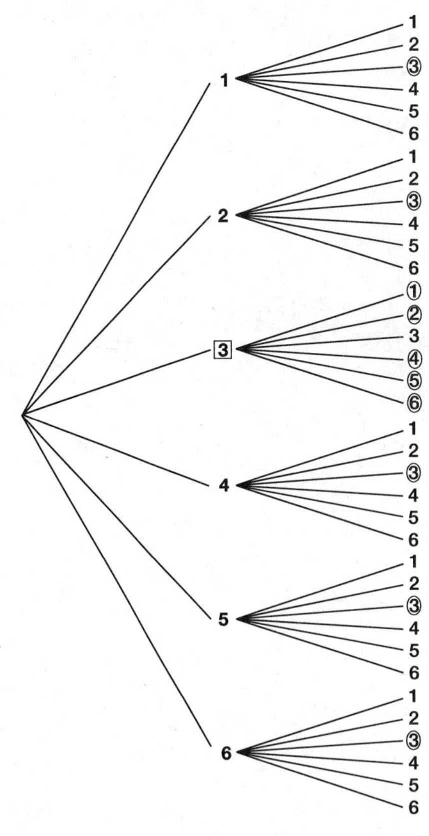

There are 36 possible outcomes. There are 10 outcomes that have a 3 on exactly one of the number cubes.

$P(\text{one } 3) = \frac{10}{36}$ or $\frac{5}{18}$

The probability is $\frac{5}{18}$.

22. First, find the number of possible outcomes. Use the Fundamental Counting Principle.

choices for the 1st die		choices for the 2nd die		choices for the 3rd die		number of possible outcomes
8	×	8	×	8	=	512

There are 512 possible outcomes. Since there is only one way to roll three 7s, the probability is $\frac{1}{512}$.

23. First, find the number of possible outcomes. Use the Fundamental Counting Principle.

choices for the 1st digit		choices for the 2nd digit		choices for the 3rd digit		choices for the 4th digit		choices for the 5th digit		total number of outcomes
10	×	10	×	10	×	10	×	10	=	100,000

There are 100,000 possible outcomes. There is 1 winning number. So, the probability of winning with one ticket is $\frac{1}{100{,}000}$.

24. Use the Fundamental Counting Principle.

number of choices of decks		number of choices of trucks		number of choices of wheels		number of possible outcomes
10	×	8	×	12	=	960

There are 960 different deluxe skateboards possible.

25. First find the number of possible outcomes.

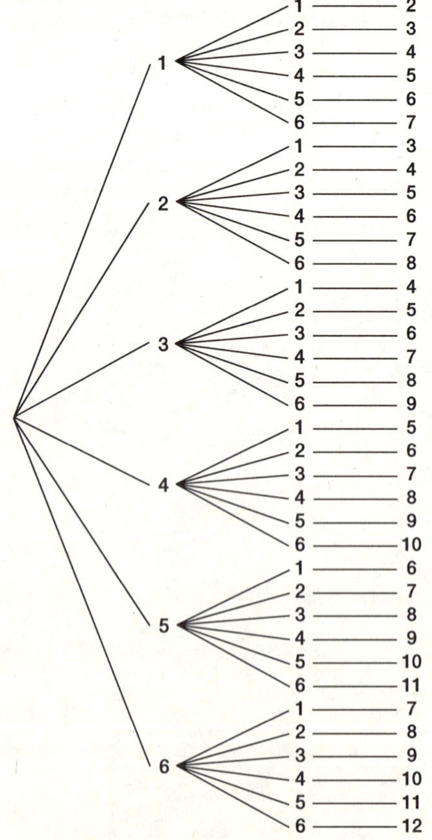

There are 66 or 36 possible outcomes. There are 18 possible "even" sums. So, $P(\text{even sum}) = \frac{18}{36}$ or $\frac{1}{2}$. There are 18 possible "odd" sums. So, $P(\text{odd sum}) = \frac{18}{36}$ or $\frac{1}{2}$. The player should choose either even or odd, since the probability of rolling either even or odd is one-half.

26a. Use the Fundamental Counting Principle.

choices for processor		choices for RAM		choices for external drive		choices for color		number of possible outcomes
2	×	3	×	2	×	5	=	60

There are 60 possible customized computers that include the deluxe printer.

26b. Use the Fundamental Counting Principle.

choices for external drive		choices for printer		choices for color		number of possible outcomes
2	×	4	×	5	=	40

There are 40 possible customized computers that include 256 MB of RAM and a high-speed processor.

27. Sample answer: any two letters followed by any four digits

28. Answers may include making a list pairing each deck choice with each wheel choice, making a tree diagram, multiplying the number of choices for deck times the number of choices for wheels. Answers should include the following.

- The strategy used by the student to count the possible designs is explained.
- The number of designs is the same as the product of the number of choices.
- The number of skateboards would double.

29. D; Use the Fundamental Counting Principle.

choices for 1st letter		choices for 2nd letter		choices for 3rd letter		choices for 4th letter		number of possible outcomes
25	×	25	×	25	×	25	=	390,625

There are 390,625 possible passwords.

30. D; Use the Fundamental Counting Principle.

choices for sweater		choices for pants		choices for shoes		number of possible outcomes
6	×	4	×	3	=	72

There are 72 possible outfits.

Page 639 Maintain Your Skills

31. Sample answer: vertical scale that does not start at zero

32. 10 years

33. 11–20 has the greatest number since its bar is tallest.

34. Since 1 animal has a lifespan 21–30 years, 1 animal has a lifespan 31–40 years, and 1 animal has a lifespan 91–100 years, 1 + 1 + 1 or 3 animals have a lifespan more than 20 years.

35. The slope is 3 and the y-intercept is 1. Graph the point (0, 1). Then go up 3 and right 1. Draw a line through the points.

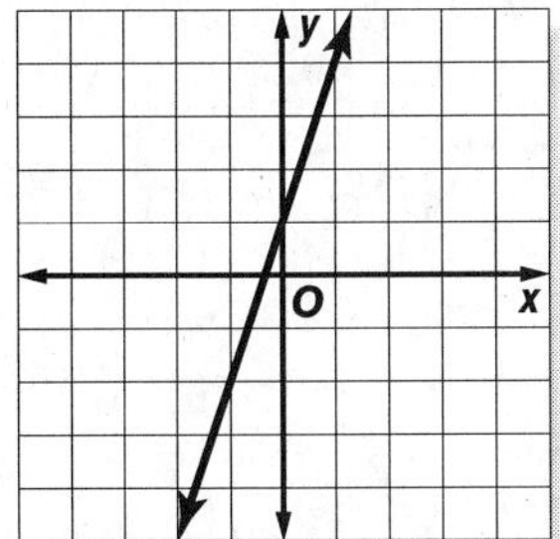

36. The slope is $\frac{1}{2}$ and the y-intercept is -2. Graph the point $(0, -2)$. Then go up 1 and right 2. Draw a line through the points.

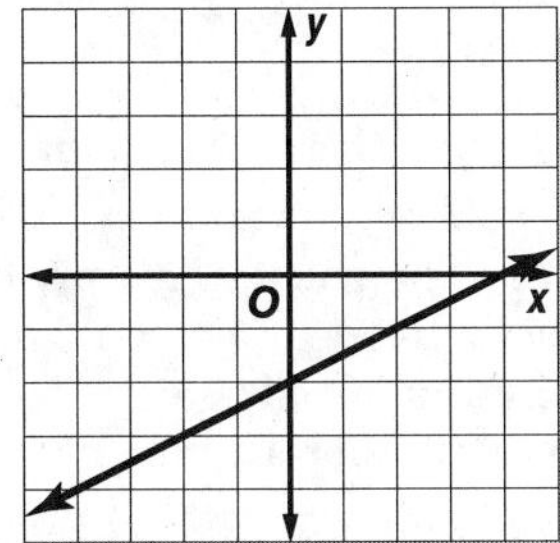

37. Write the equation in the form $y = mx + b$.
$y = 0x + 5$
The slope is 0, and the y-intercept is 5. This is a horizontal line.

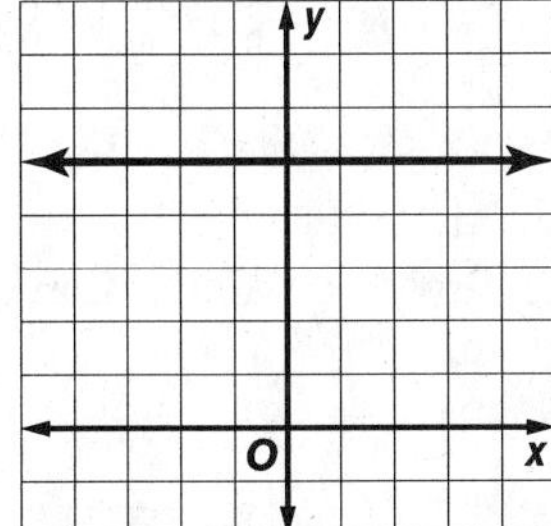

38. Write the equation in the form $y = mx + b$.

$$2x - y = 4$$
$$2x - 2x - y = 4 - 2x$$
$$-y = 4 - 2x$$
$$\frac{-y}{-1} = \frac{4 - 2x}{-1}$$
$$y = 2x - 4$$

The slope is 2 and the y-intercept is -4. Graph the point $(0, -4)$. Then go up 2 and right 1. Draw a line through the points.

39. $\frac{6 \cdot 5}{2 \cdot 1} = \frac{15}{1}$ or 15

40. $\frac{5 \cdot 4 \cdot 3}{3 \cdot 2 \cdot 1} = \frac{10}{1}$ or 10

41. $\frac{8 \cdot 7}{2 \cdot 1} = \frac{28}{1}$ or 28

42. $\frac{6 \cdot 5 \cdot 4 \cdot 3}{4 \cdot 3 \cdot 2 \cdot 1} = \frac{15}{1}$ or 15

Page 640 Algebra Activity (Follow-Up of Lesson 12-6)

Step 1 Tails; H, T; T, H; T, T

Step 2

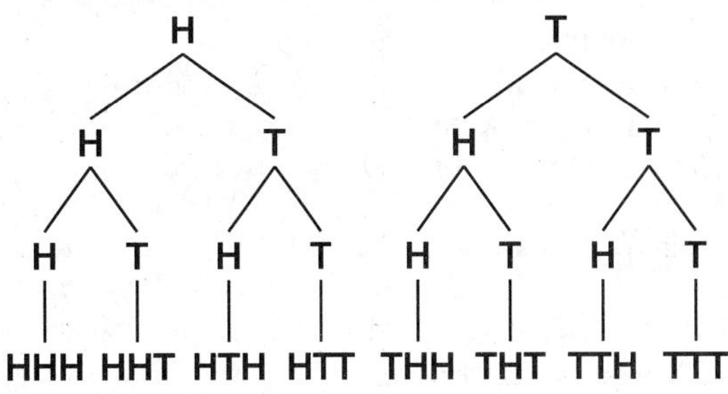

Step 3

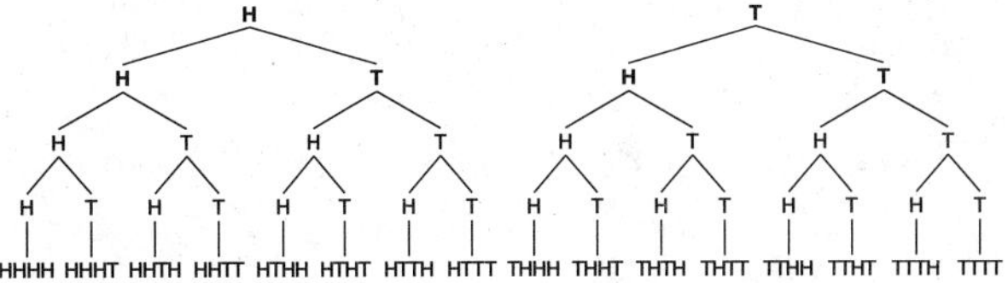

1. 4; 2
2. $P(\text{two heads}) = \frac{1}{4}$; $P(\text{one head, one tail}) = \frac{2}{4}$ or $\frac{1}{2}$; $P(\text{two tails}) = \frac{1}{4}$
3. 8; 3; 3
4. $P(\text{three heads}) = \frac{1}{8}$; $P(\text{two heads, one tail}) = \frac{3}{8}$; $P(\text{one head, two tails}) = \frac{3}{8}$; $P(\text{three tails}) = \frac{1}{8}$
5. 16; 4; 6; 4; 4
6. $P(\text{four heads}) = \frac{1}{16}$; $P(\text{three heads, one tail}) = \frac{4}{16}$ or $\frac{1}{4}$; $P(\text{two heads, two tails}) = \frac{6}{16}$ or $\frac{3}{8}$; $P(\text{four tails}) = \frac{1}{16}$
7. The sum of the numbers in each row is the number of outcomes for tossing a number of coins one less than the row number. The numbers in each row are the numerators for the probabilities for outcomes for the coin tossing connected with that row.
8. The numbers in the row are 1, 5, 10, 10, 5, 1. The number of outcomes is 32. The probabilities are: $P(\text{5 heads}) = \frac{1}{32}$, $P(\text{4 heads, 1 tail}) = \frac{5}{32}$, $P(\text{3 heads, 2 tails}) = \frac{10}{32}$, $P(\text{2 heads, 3 tails}) = \frac{10}{32}$, $P(\text{1 head, 4 tails}) = \frac{5}{32}$, $P(\text{5 tails}) = \frac{1}{32}$.
9. See students' work.

12-7 Permutations and Combinations

Page 641 Why is order sometimes important when determining outcomes?

a. Leonora-Michael, Leonora-Ned, Leonora-Olivia, Leonora-Patrick, Michael-Leonora, Michael-Ned, Michael-Olivia, Michael-Patrick, Ned-Michael, Ned-Leonora, Ned-Olivia, Ned-Patrick, Olivia-Michael, Olivia-Ned, Olivia-Leonora, Olivia-Patrick, Patrick-Michael, Patrick-Ned, Patrick-Olivia, Patrick-Leonora

b. The results are the same.

c. Leonora-Michael, Leonora-Ned, Leonora-Olivia, Leonora-Patrick, Michael-Ned, Michael-Olivia, Michael-Patrick, Ned-Olivia, Ned-Patrick, Olivia-Patrick

d. The answer in part c equals the answer in part a divided by 2.

Page 643 Check for Understanding

1. Sample answer: How many 3-digit numbers can be made from the digits 1, 2, 3, and 4 if no digit is repeated?
2. Both expressions have 5, 4, and 3 as factors; 5! also has factors of 2 and 1.
3. Sarah; five CDs from a collection of 30 is a combination because order is not important.
4. $6! = 6 \cdot 5 \cdot 4 \cdot 3 \cdot 2 \cdot 1$
   $= 720$
5. permutation; $P(5, 5) = 5 \cdot 4 \cdot 3 \cdot 2 \cdot 1$ or 120 ways
6. permutation; $P(8, 4) = 8 \cdot 7 \cdot 6 \cdot 5$ or 1680 programs
7. combination; $C(9, 3) = \frac{P(9, 3)}{3!} = \frac{9 \cdot 8 \cdot 7}{3 \cdot 2 \cdot 1}$ or 84 ways
8. combination; $C(12, 6) = \frac{P(12, 6)}{6!} = \frac{12 \cdot 11 \cdot 10 \cdot 9 \cdot 8 \cdot 7}{6 \cdot 5 \cdot 4 \cdot 3 \cdot 2 \cdot 1}$ or 924 ways
9. Since order is not important, this is a combination.
   $C(12, 3) = \frac{P(12, 3)}{3!} = \frac{12 \cdot 11 \cdot 10}{3 \cdot 2 \cdot 1}$ or 220 pizzas

Pages 644–645 Practice and Apply

10. permutation; $P(6, 6) = 6 \cdot 5 \cdot 4 \cdot 3 \cdot 2 \cdot 1$ or 720 ways
11. permutation; $P(4, 3) = 4 \cdot 3 \cdot 2$ or 24 flags
12. combination; $C(10, 4) = \frac{P(10, 4)}{4!} = \frac{10 \cdot 9 \cdot 8 \cdot 7}{4 \cdot 3 \cdot 2 \cdot 1}$ or 210 ways
13. combination; $C(15, 2) = \frac{P(15, 2)}{2!} = \frac{15 \cdot 14}{2 \cdot 1}$ or 105 ways
14. permutation; $P(3, 3) = 3 \cdot 2 \cdot 1$ or 6 numbers
15. permutation; $P(12, 3) = 12 \cdot 11 \cdot 10$ or 1320 ways
16. combination; $C(52, 5) = \frac{P(52, 5)}{5!} = \frac{52 \cdot 51 \cdot 50 \cdot 49 \cdot 48}{5 \cdot 4 \cdot 3 \cdot 2 \cdot 1}$ or 2,598,960 hands
17. combination; $C(14, 3) = \frac{P(14, 3)}{3!} = \frac{14 \cdot 13 \cdot 12}{3 \cdot 2 \cdot 1}$ or 364 ways
18. $7! = 7 \cdot 6 \cdot 5 \cdot 4 \cdot 3 \cdot 2 \cdot 1$ or 5040
19. $8! = 8 \cdot 7 \cdot 6 \cdot 5 \cdot 4 \cdot 3 \cdot 2 \cdot 1$ or 40,320
20. $10! = 10 \cdot 9 \cdot 8 \cdot 7 \cdot 6 \cdot 5 \cdot 4 \cdot 3 \cdot 2 \cdot 1$ or 3,628,800
21. $11! = 11 \cdot 10 \cdot 9 \cdot 8 \cdot 7 \cdot 6 \cdot 5 \cdot 4 \cdot 3 \cdot 2 \cdot 1$ or 39,916,800
22. Since the line segment between 1 and 7 is the same as the line segment between 7 and 1, this is a combination. Find the combination of 12 points taken 2 at a time.

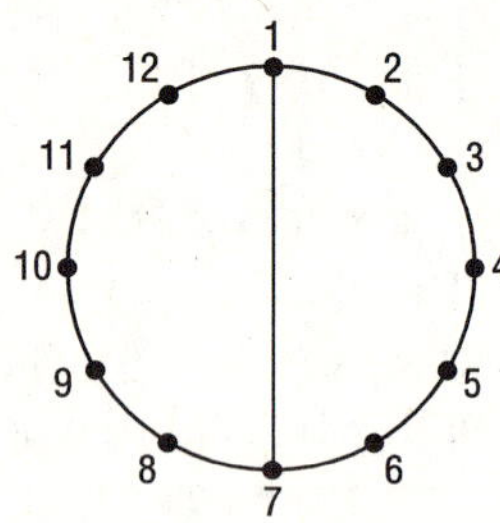

   $C(12, 2) = \frac{P(12, 2)}{2!}$
   $= \frac{12 \cdot 11}{2 \cdot 1}$ or 66

   There can be 66 line segments drawn between any two of the points.
23. Since a handshake between person 1 and person 2 is the same as a handshake between person 2 and person 1, this is a combination. Find the combination of 9 people taken 2 at a time.
   $C(9, 2) = \frac{P(9, 2)}{2!} = \frac{9 \cdot 8}{2 \cdot 1}$ or 36
   There will be 36 handshakes.
24. Since order is important, this is a permutation. Find the permutation of 14 coasters taken 8 at a time.
   $P(14, 8) = 14 \cdot 13 \cdot 12 \cdot 11 \cdot 10 \cdot 9 \cdot 8 \cdot 7$
   $= 121{,}080{,}960$
   There are 121,080,960 ways to ride the eight coasters if order is important.
25. If order is not important, this is a combination of 14 coasters taken 8 at a time.
   $C(14, 8) = \frac{P(14, 8)}{8!}$
   $= \frac{14 \cdot 13 \cdot 12 \cdot 11 \cdot 10 \cdot 9 \cdot 8 \cdot 7}{8 \cdot 7 \cdot 6 \cdot 5 \cdot 4 \cdot 3 \cdot 2 \cdot 1}$ or 3003
   There are 3003 ways to ride the eight coasters if order is not important.
26. There are 26 letters of the alphabet and there are 10 digits for the numbers 0 through 9.

26 choices for the first letter
25 choices for the second letter
24 choices for the third letter
9 digit choices since 0 cannot be the first number
10 choices for the second digit
10 choices for the third digit
10 choices for the fourth digit

$26 \cdot 25 \cdot 24 \cdot 9 \cdot 10 \cdot 10 \cdot 10 = 140{,}400{,}000$
There are 140,400,000 license plates possible.

27. $C(5, 3) = \frac{P(5, 3)}{3!} = \frac{5 \cdot 4 \cdot 3}{3 \cdot 2 \cdot 1}$ or 10 combinations

28. There are 10 possible combinations of 3 roses selected from 5 color choices. If the color choices are only pink, white, and yellow, this is the combination of 3 roses taken from 3 choices.

$C(3, 3) = \frac{P(3, 3)}{3!} = \frac{3 \cdot 2 \cdot 1}{3 \cdot 2 \cdot 1}$ or 1 possible choice

There is 1 way to select the 3 roses, and 10 possible combinations, so the probability is $\frac{1}{10}$.

29. If the color choices are pink, white, yellow, and orange, there are 4 color choices.

$C(4, 3) = \frac{P(4, 3)}{3!} = \frac{4 \cdot 3 \cdot 2}{3 \cdot 2 \cdot 1}$ or 4 possible choices

There are 4 ways to select the 3 roses and 10 possible combinations (including red), so the probability is $\frac{4}{10}$ or $\frac{2}{5}$.

30. Always; since $C(x, y) = \frac{P(x, y)}{y!}$ and neither x nor y equals 1 and $x \neq y$, the value of $P(x, y)$ is divided by a number larger than 1, so $C(x, y)$ must be less than $P(x, y)$.

31. When order is not important, duplicate arrangements are not included in the number of arrangements. Answers should include the following.

- When order was not important there were half as many pairs.
- Order is important when you arrange things in a line; order is not important when you choose a group of things.

32. D; There are 3 choices for president, 2 choices for secretary, and 3 choices for treasurer. So, there are $3 \cdot 2 \cdot 3$ or 18 different slates of officers.

33. C; There are 4 choices for the last four digits. Since none of the digits are repeated, there are 4 choices for the first digit, 3 choices for the second digit, 2 choices for the third digit, and 1 choice for the last digit. The greatest possible number of calls is $4 \cdot 3 \cdot 2 \cdot 1$ or 24 calls.

Page 645 Maintain Your Skills

34. 36 outcomes;

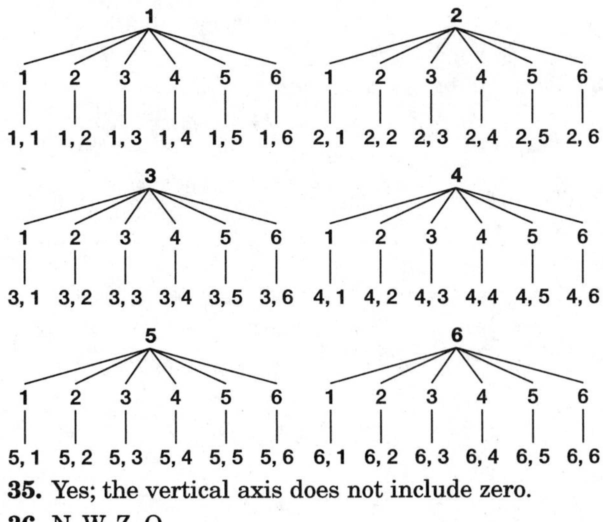

35. Yes; the vertical axis does not include zero.

36. N, W, Z, Q

37. Q

38. Z, Q

39. I

40. 16:14 or $\frac{16}{14} = \frac{8 \cdot 2}{7 \cdot 2} = \frac{8}{7}$ or 8:7

41. 4:32 or $\frac{4}{32} = \frac{1 \cdot 4}{8 \cdot 4} = \frac{1}{8}$ or 1:8

42. 4:48 or $\frac{4}{48} = \frac{1 \cdot 4}{12 \cdot 4} = \frac{1}{12}$ or 1:12

43. 44:8 or $\frac{44}{8} = \frac{11 \cdot 4}{2 \cdot 4} = \frac{11}{2}$ or 11:2

44. 45:55 or $\frac{45}{55} = \frac{9 \cdot 5}{11 \cdot 5} = \frac{9}{11}$ or 9:11

Page 645 Practice Quiz 2

1. Sample answer: Bars are different widths.

2.

Shorts	blue			tan		
Shirt	red	white	yellow	red	white	yellow
Outfit	B, R	B, W	B, Y	T, R	T, W	T, Y

3. Use the Fundamental Counting Principle. Each question has 2 possible choices.
$2 \times 2 \times 2 \times 2 \times 2 \times 2 \times 2 \times 2$ or 256
There are 256 possible outcomes.

4. There are 5 different letters. So, there are 5 choices for the first letter, 4 choices for the second, 3 choices for the third, 2 choices for the fourth, and 1 choice for the last letter.
$5 \cdot 4 \cdot 3 \cdot 2 \cdot 1$ or 120
There are 120 ways to arrange the letters.

5. There are 6 vertices on a hexagon. Since a line segment drawn between 1 and 3 is the same as a line segment drawn between 3 and 1, this is a combination. Find the combination of 6 vertices taken 2 at a time.

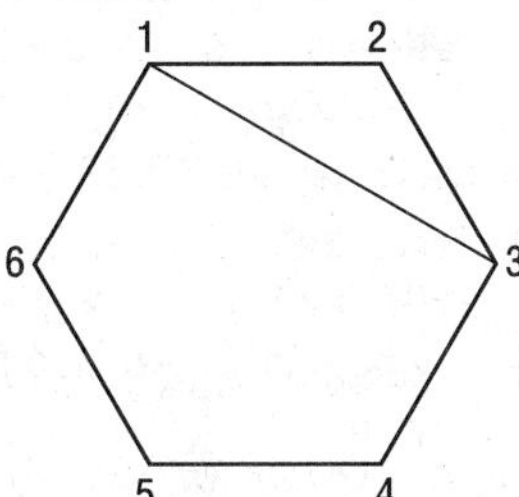

$C(6, 2) = \frac{P(6, 2)}{2!} = \frac{6 \cdot 5}{2 \cdot 1}$ or 15

There can be 15 line segments drawn between the vertices.

12-8 Odds

Page 646 How are odds related to probability?

a. See students' work.

b. See students' work; theoretical probability is $\frac{5}{12}$.

c. See students' work; based on theoretical probability, the ratio is 5:7

Page 648 Check for Understanding

1. Write a ratio comparing the ways the event can occur to the ways the event cannot occur.

2. Sample answer: Prize contests list the odds of winning the grand prize.

3. Hoshi; the probability of rolling a 2 is 1 out of 6. The odds of rolling a 2 are 1 in 5.
4. There are 6 possible outcomes. There is 1 number less than 2. There are 6 − 1 or 5 numbers that are not less than 2.
   Odds of rolling a number less than 2 = 1:5
5. There are 6 possible outcomes. There are 2 numbers that are multiples of 3. They are 3 and 6. There are 6 − 2 or 4 numbers that are not multiples of 3.
   Odds of rolling a multiple of 3 = 2:4 or 1:2
6. There are 6 possible outcomes. There are 3 numbers greater than 3. They are 4, 5, and 6. There are 6 − 3 or 3 numbers that are not greater than 3.
   Odds of rolling a number greater than 3 = 3:3 or 1:1
7. There are 6 possible outcomes. There are 5 numbers that are not 5. They are 1, 2, 3, 4, and 6. There are 6 − 5 or 1 number that is a 5.
   Odds of rolling a number that is not a 5 = 5:1
8. D; There are 8 cards available. There are 2 instant winner cards, so there are 8 − 2 or 6 cards that are not instant winners.
   Odds against the next card being an instant winner = 6:2 or 3:1

Pages 648–649 Practice and Apply

9. There are 6 spaces on the spinner. There are 2 yellow spaces.
   There are 6 − 2 or 4 spaces that are not yellow.
   Odds of spinning yellow = 2:4 or 1:2
10. There are no orange spaces on the spinner. The odds of spinning orange are 0.
11. There are 6 spaces on the spinner. There are 4 spaces that are *not* green, and 2 spaces that are green.
    Odds of spinning a color that is not green = 4:2 or 2:1
12. There are 6 spaces on the spinner. There are 3 spaces that are *not* red or green.
    There are 3 spaces that are red or green.
    Odds of spinning a color that is not red or green = 3:3 or 1:1
13. There are 6 · 6 or 36 sums possible for rolling a pair of number cubes. There are 18 sums that are odd. There are 36 − 18 or 18 sums that are not odd.
    Odds of rolling an odd sum = 18:18 or 1:1
14. There are 6 · 6 or 36 sums possible for rolling a pair of number cubes. There are 18 sums that are even.
    There are 36 − 18 or 18 sums that are not even.
    Odds of rolling an even sum = 18:18 or 1:1
15. There are 6 · 6 or 36 sums possible for rolling a pair of number cubes. There are 9 sums that are multiples of 4.
    There are 36 − 9 or 27 sums that are not multiples of 4.
    Odds of rolling a sum that is a multiple of 4 = 9:27 or 1:3
16. There are no sums less than 2. So, the odds are 0.
17. There are 6 · 6 or 36 sums possible for rolling a pair of number cubes. There are 15 sums that are prime numbers.
    There are 36 − 15 or 21 sums that are not prime numbers.
    Odds of rolling a sum that is a prime number = 15:21 or 5:7
18. There are 6 · 6 or 36 sums possible for rolling a pair of number cubes. There are 21 sums that are composite numbers.
    There are 36 − 21 or 15 sums that are not composite numbers.
    Odds of rolling a sum that is a composite number = 21:15 or 7:5
19. There are 6 · 6 or 36 sums possible for rolling a pair of number cubes. There are 5 sums of 8, so there are 36 − 5 or 31 sums that are not 8.
    Odds of rolling a sum that is *not* 8 = 31:5
20. There are 6 · 6 or 36 sums possible for rolling a pair of number cubes. There are 7 sums of 9 or 10, so there are 36 − 7 or 29 sums that are not 9 or 10.
    Odds of rolling a sum that is not 9 or 10 = 29:7
21. There are 6 · 6 or 36 sums possible for rolling a pair of number cubes. There are 2 sums of 6 with a 4 on one number cube. They are (4, 2) and (2, 4). There are 36 − 2 or 34 sums that are not 6 with a 4 on one number cube.
    Odds of rolling a 6 with a 4 on one number cube = 2:34 or 1:17
22. There are 6 · 6 or 36 sums possible for rolling a pair of number cubes. There are 30 sums that are even or greater than 6.
    There are 36 − 30 or 6 sums that are not even or greater than 6.
    Odds of rolling an even sum or a sum greater than 6 = 30:6 or 5:1
23. There are 6 · 6 or 36 sums possible for rolling a pair of number cubes. There are 16 sums of 6, 7, or 8, so there are 36 − 16 or 20 sums that are not 6, 7, or 8.
    Odds of rolling a sum that is not 6, 7, or 8 = 20:16 or 5:4
24. There are 2 red queens.
    There are 52 − 2 or 50 cards that are not red queens.
    Odds of selecting a red queen = 2:50 or 1:25

25. There are 13 cards that are diamonds, and 3 aces that are not diamonds. So, there are 13 + 3 or 16 cards that are diamonds or aces. There are 52 − 16 or 36 cards that are *not* diamonds or aces.
Odds of *not* selecting a diamond or an ace

$$= \underbrace{36}_{\text{number of cards that are not a diamond or an ace}} : \underbrace{16}_{\text{number of cards that are diamonds or aces}} \text{ or } 9:4$$

26. There is 1 eight-child family with all girls in 256 possible eight-child families. There are 256 − 1 or 255 eight-child families that are not all girls.
Odds in favor of an eight-girl family = 1:255
Odds against an eight-girl family = 255:1

27. There is 1 13-child family with all boys and 8191 13-child families without all boys. So, there are 8191 + 1 or 8192 13-child families in all.
The probability of a 13-child family having all boys is $\frac{1}{8192}$.

28. There are 54 out of 100 men who believe in aliens. There are 100 − 54 or 46 men who do not believe in aliens.
Odds that a man believes in aliens = 54:46 or 27:23

29. There are 47 out of 100 women who do not believe in aliens.
There are 100 − 47 or 53 women who do believe in aliens or who are not sure.
Odds that a woman does *not* believe in aliens = 47:53

30. 54% of men believe in aliens. So, 0.54 · 500 or 270 men in a group of 500 can be expected to believe in aliens.

31. No; 512 is the total number of possible outcomes, so it could be a number in the odds of winning. The odds are 8:512 − 8 or 8:504.

32. Odds and probability both express the likelihood of an event happening. Answers should include the following.

- The sum of the numbers in the odds ratio is the total number of outcomes in the probability ratio; the first number in the odds ratio is the same as the numerator in the probability ratio.
- You can find the odds of an event by using the numerator of its probability as the first number in the odds ratio and subtracting that number from the denominator to find the second number in the odds ratio.

33. A; There are 40 − 25 or 15 cards greater than 25. There are 40 − 15 or 25 cards that are not greater than 25.
Odds that the card drawn is greater than 25 = 15:25 or 3 to 5

34. D; Judie did not make a basket 25 − 15 or 10 times.
Odds *against* Judie making a basket the next time she shoots the ball = 10:15 or 2 to 3

Page 649 Maintain Your Skills

35.

father sits in aisle seat		three choices for 2nd seat		two choices for 3rd seat		one choice for 4th seat	
↓		↓		↓		↓	
1	·	3	·	2	·	1	or 6 ways

36. Use the Fundamental Counting Principle.

$$\underbrace{6}_{\text{choices for first cube}} \cdot \underbrace{6}_{\text{choices for second cube}} \cdot \underbrace{6}_{\text{choices for third cube}} = \underbrace{216}_{\text{number of possible outcomes}}$$

There are 216 possible outcomes.

37.
$$\begin{aligned} 2a - 3 &\geq 9 \\ 2a - 3 + 3 &\geq 9 + 3 \\ 2a &\geq 12 \\ \frac{2a}{2} &\geq \frac{12}{2} \\ a &\geq 6 \end{aligned}$$

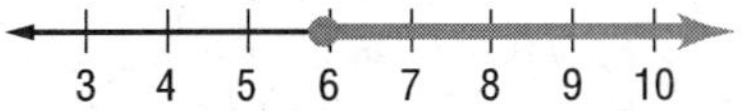

38.
$$\begin{aligned} 4c + 4 &> 32 \\ 4c + 4 - 4 &> 32 - 4 \\ 4c &> 28 \\ \frac{4c}{4} &> \frac{28}{4} \\ c &> 7 \end{aligned}$$

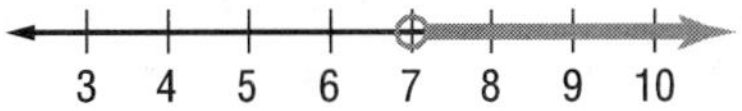

39.
$$\begin{aligned} -2y + 3 &< 9 \\ -2y + 3 - 3 &< 9 - 3 \\ -2y &< 6 \\ \frac{-2y}{-2} &> \frac{6}{-2} \\ y &> -3 \end{aligned}$$

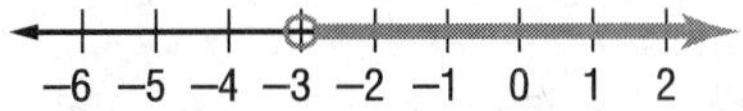

40. $\frac{1}{6} \cdot \frac{1}{3} = \frac{1 \cdot 1}{6 \cdot 3} = \frac{1}{18}$

41. $\frac{2}{3} \cdot \frac{3}{6} = \frac{\overset{1}{\cancel{2}} \cdot \overset{1}{\cancel{3}}}{\underset{1}{\cancel{3}} \cdot \underset{3}{\cancel{6}}} = \frac{1}{3}$

42. $\frac{1}{3} \cdot \frac{1}{3} \cdot \frac{1}{3} = \frac{1 \cdot 1 \cdot 1}{3 \cdot 3 \cdot 3} = \frac{1}{27}$

43. $\frac{3}{8} \cdot \frac{2}{7} \cdot \frac{1}{6} = \frac{\overset{1}{\cancel{3}} \cdot \overset{1}{\cancel{2}} \cdot 1}{8 \cdot 7 \cdot \underset{1}{\cancel{6}}} = \frac{1}{56}$

12-9 Probability of Compound Events

Page 650 How are compound events related to simple events?

a. See students' work; theoretical probability is $\frac{1}{4}$.

b. Yes; by not replacing the first counter drawn, you affect the possibilities for the second draw.

Page 653 Check for Understanding

1. Independent and dependent events are similar because both are a connection of two or more simple events and the probability of the compound event is found by multiplying the probabilities of each simple event. They are different because the second event in a dependent event is influenced by the outcome of the first event. Therefore, the probability of the second event used in calculating the probability of the compound event is dependent on the outcome of the first event.
2. Sample answer: choosing an odd or composite number when rolling a die
3. The result of the 1st event must be taken into account.
4. The events are independent since a roll of the number cube does not affect the outcome of the spinner.
   There are 3 ways to roll an odd number, 1, 3, or 5, and there are 6 ways to roll a number cube.
   So, the probability of rolling an odd number is $\frac{3}{6}$ or $\frac{1}{2}$.
   There is only one way to spin a B on the spinner and there are 4 ways to spin. So, the probability of spinning a B is $\frac{1}{4}$.
   P(an odd number and a B)
   $= P(\text{an odd number}) \cdot P(\text{a B}) = \frac{1}{2} \cdot \frac{1}{4}$ or $\frac{1}{8}$
5. The events are independent since a roll of the number cube does not affect the outcome of the spinner.
   There are 2 ways to roll a composite number, 4 or 6, and there are 6 ways to roll a number cube. So, the probability of rolling a composite number is $\frac{2}{6}$ or $\frac{1}{3}$.
   There is only one way to spin a vowel, A, and there are 4 ways to spin. So, the probability of spinning a vowel is $\frac{1}{4}$.
   P(a composite number and a vowel)
   $= P(\text{a composite number}) \cdot P(\text{a vowel})$
   $= \frac{1}{3} \cdot \frac{1}{4}$ or $\frac{1}{12}$
6. $P(5 \text{ and } 2) = \frac{1}{8} \cdot \frac{1}{7}$
7. $P(\text{two odd numbers}) = \frac{4}{8} \cdot \frac{3}{7}$
8. The events are mutually exclusive because a card cannot be both a diamond and a club. There are 13 diamonds and 13 clubs in a standard deck of cards.
   $P(\text{diamond or club}) = \frac{13}{52} + \frac{13}{52} = \frac{26}{52}$ or $\frac{1}{2}$
9. The events are mutually exclusive, since a book cannot be both a history book and an animal book at the same time.
   $P(\text{history book or animal book}) = \frac{5}{12} + \frac{4}{12}$
   $= \frac{9}{12}$ or $\frac{3}{4}$
10. The events are mutually exclusive since he cannot roll both a sum of 5 and a sum of 8 at the same time.
    $P(5 \text{ or } 8) = \frac{4}{36} + \frac{5}{36} = \frac{9}{36}$ or $\frac{1}{4}$

Pages 653–655 Practice and Apply

11. The events are independent since each roll of the number cube does not affect the outcome of the spin.
    There is 1 way to roll a 3 and there are 6 ways to roll the number cube.
    There is 1 way to spin an E and there are 5 possible spins.
    $P(3 \text{ and E}) = P(3) \cdot P(\text{E}) = \frac{1}{6} \cdot \frac{1}{5}$ or $\frac{1}{30}$
12. The events are independent since each roll of the number cube does not affect the outcome of the spin.
    There are 3 ways to roll an even number and there are 6 ways to roll a number cube.
    There is 1 way to spin an A and there are 5 possible spins.
    P(an even number and A)
    $= P(\text{an even number}) \cdot P(\text{A}) = \frac{3}{6} \cdot \frac{1}{5}$ or $\frac{1}{10}$
13. The events are independent since each roll of a number cube does not affect the outcome of the spin.
    There are 3 ways to roll a prime number, 2, 3, or 5, and there are 6 ways to roll a number cube.
    There are 2 ways to spin a vowel, A or E, and there are 5 possible spins.
    P(a prime number and a vowel)
    $= P(\text{a prime number}) \cdot P(\text{a vowel}) = \frac{3}{6} \cdot \frac{2}{5}$ or $\frac{1}{5}$
14. The events are independent since each roll of a number cube does not affect the outcome of the spin.
    There are 3 ways to roll an odd number, 1, 3, or 5, and there are 6 ways to roll a number cube.
    There are 3 ways to roll a consonant, B, C, or D, and there are 5 possible spins.
    P(an odd number and a consonant)
    $= P(\text{an odd number}) \cdot P(\text{a consonant})$
    $= \frac{3}{6} \cdot \frac{3}{5}$ or $\frac{3}{10}$
15. $P(\text{two yellow marbles}) = \frac{2}{14} \cdot \frac{1}{13}$
    $= \frac{2}{182}$ or $\frac{1}{91}$
16. $P(\text{two blue marbles}) = \frac{5}{14} \cdot \frac{4}{13}$
    $= \frac{20}{182}$ or $\frac{10}{91}$
17. $P(\text{blue and green}) = \frac{5}{14} \cdot \frac{4}{13}$
    $= \frac{20}{182}$ or $\frac{10}{91}$
18. $P(\text{yellow and red}) = \frac{2}{14} \cdot \frac{3}{13}$
    $= \frac{6}{182}$ or $\frac{3}{91}$
19. $P(\text{blue and yellow and red}) = \frac{5}{14} \cdot \frac{2}{13} \cdot \frac{3}{12}$
    $= \frac{30}{2184}$ or $\frac{5}{364}$

20. $P(\text{three green}) = \frac{4}{14} \cdot \frac{3}{13} \cdot \frac{2}{12}$
$= \frac{24}{2184}$ or $\frac{1}{91}$

21. The events are mutually exclusive because the roll cannot be both 3 and an even number at the same time.
$P(3 \text{ or even}) = P(3) + P(\text{even})$
$= \frac{1}{8} + \frac{4}{8}$
$= \frac{5}{8}$

22. The events are mutually exclusive because the roll cannot be both 6 and a prime number at the same time.
$P(6 \text{ or prime}) = P(6) + P(\text{prime})$
$= \frac{1}{8} + \frac{4}{8}$
$= \frac{5}{8}$

23. The events are mutually exclusive because the card drawn cannot be both 3 and a multiple of 2 at the same time.
$P(3 \text{ or multiple of } 2) = P(3) + P(\text{multiple of } 2)$
$= \frac{1}{7} + \frac{2}{7}$
$= \frac{3}{7}$

24. The events are mutually exclusive because the card drawn cannot be both 4 and greater than 5 at the same time.
$P(4 \text{ or greater than } 5) = P(4) + P(\text{greater than } 5)$
$= \frac{1}{7} + \frac{3}{7}$
$= \frac{4}{7}$

25. The events are mutually exclusive because the card drawn cannot be both odd and even at the same time.
$P(\text{odd or even}) = P(\text{odd}) + P(\text{even})$
$= \frac{5}{7} + \frac{2}{7}$
$= \frac{7}{7}$ or 1

26. The events are mutually exclusive because the card drawn cannot be both 2 and 6 at the same time.
$P(2 \text{ or } 6) = P(2) + P(6)$
$= 0 + \frac{1}{7}$ or $\frac{1}{7}$

27. The events are independent, since drawing the first marble does not affect the outcome of drawing the second marble.
$P(\text{red and blue}) = P(\text{red}) \cdot P(\text{blue})$
$= \frac{3}{9} \cdot \frac{6}{9}$
$= \frac{18}{81}$ or $\frac{2}{9}$

28. The events are independent since each birth does not affect the outcome of the next birth.
$P(\text{four girls}) = P(\text{girl}) \cdot P(\text{girl}) \cdot P(\text{girl}) \cdot P(\text{girl})$
$= \frac{1}{2} \cdot \frac{1}{2} \cdot \frac{1}{2} \cdot \frac{1}{2}$
$= \frac{1}{16}$

29. $P(\text{games and studying}) = P(\text{games}) \cdot P(\text{studying})$
$= (0.32) \cdot (0.51)$
≈ 0.16

30. $P(\text{browse and e-mail}) = P(\text{browse}) \cdot P(\text{e-mail})$
$= (0.54) \cdot (0.51)$
≈ 0.28 or 28%

31a. Sample answer: 3 red, 2 white, and 4 blue

31b. Sample answer: The numerator must be 24. Any combination of 3, 2, and 4 will have a probability of $\frac{1}{21}$.

32. The probability of compound events is based on the probability of each simple event. Answers should include the following.
- The probability for two draws is the product of the probabilities of each single draw.
- Independent events do not affect one another; dependent events do.

33. B; $P(\text{blue and yellow}) = \frac{2}{10} \cdot \frac{1}{9}$
$= \frac{1}{45}$

34. D; There are 4 jacks, 4 queens, and 4 kings in a standard deck of cards.
$P(\text{jack, queen, or king})$
$= P(\text{jack}) + P(\text{queen}) + P(\text{king})$
$= \frac{4}{52} + \frac{4}{52} + \frac{4}{52}$
$= \frac{12}{52}$ or $\frac{3}{13}$

35. $P(\text{green or even})$
$= P(\text{green}) + P(\text{even}) - P(\text{green and even})$
$= \frac{2}{6} + \frac{3}{6} - \frac{1}{6}$
$= \frac{5}{6} - \frac{1}{6}$
$= \frac{4}{6}$ or $\frac{2}{3}$

Page 655 Maintain Your Skills

36. There are 26 red cards in a standard deck of 52 cards.
There are 52 − 26 or 26 cards that are not red.
Odds of selecting a red card = 26:26 or 1:1

37. There are 4 jacks, 4 queens, and 4 kings in a standard deck of cards. So, there are 4 + 4 + 4 or 12 jacks, queens, and kings.
There are 52 − 12 or 40 cards that are not jacks, queens, or kings.
Odds of selecting a jack, queen, or king = 12:40 or 3:10

38. $C(8, 3) = \frac{P(8, 3)}{3!}$
$= \frac{8 \cdot 7 \cdot 6}{3 \cdot 2 \cdot 1}$ or 56

39. There are 26 letter choices and 10 digit choices.

choices for 1st letter		choices for 2nd letter		choices for 3rd letter		choices for 1st digit		choices for 2nd digit		choices for 3rd digit
↓		↓		↓		↓		↓		↓
26	·	26	·	26	·	10	·	10	·	10

= 17,576,000 license plates

40. Since ∠1 and the 65° angle form a right angle, their sum is 90°.
$m\angle 1 + 65 = 90$
$m\angle 1 = 25°$

41. Since $\angle 2$ and the 65° angle are corresponding angles, they are congruent. So, $m\angle 2 = 65°$

42. Since $\angle 3$ and $\angle 2$ are supplementary, their sum is 180°.

$m\angle 3 + m\angle 2 = 180°$
$m\angle 3 + 65° = 180°$
$m\angle 3 = 115°$

43. Since $\angle 4$ and $\angle 2$ are vertical angles, they are congruent. So, $m\angle 4 = 65°$.

44. Since $\angle 5$ and $\angle 3$ are vertical angles, they are congruent. So, $m\angle 5 = 115°$.

45. Since $\angle 6$ and the 90° angle are vertical angles, they are congruent. So, $m\angle 6 = 90°$.

46. $a^2 + b^2 = c^2$
$6^2 + 8^2 = c^2$
$36 + 64 = c^2$
$100 = c^2$
$10 = c$

47. $a^2 + b^2 = c^2$
$7^2 + 40^2 = c^2$
$49 + 1600 = c^2$
$1649 = c^2$
$40.6 \approx c$

48. $a^2 + b^2 = c^2$
$8^2 + b^2 = 15^2$
$64 + b^2 = 225$
$b^2 = 161$
$b \approx 12.7$

49. $a^2 + b^2 = c^2$
$a^2 + 63^2 = 65^2$
$a^2 + 3969 = 4225$
$a^2 = 256$
$a = 16$

50. 40% of the 625 students = $(0.40) \cdot 625$ or 250 students. So, about 250 students are expected to have Type A blood.

Pages 656–657 Algebra Activity (Follow-Up of Lesson 12-9)

1. Sample answer: No; none of the simulations results in a passing grade.

2a. See students' work.

2b. Sample answer: at least 6

3. See students' work. The probability should be about 56%.

4. See students' work.

5. 0.64 or 64%

6. 40 red, 10 blue; red represents a basket; blue a miss

7. See students' work. The theoretical probability and the experimental probability should be about the same.

8. No; the data suggest the number varies from machine to machine.

Chapter 12 Study Guide and Review

Page 658 Vocabulary and Concept Check

1. b; permutation
2. e; odds
3. c; mutually exclusive events
4. a; combination
5. d; lower quartile

Pages 658–662 Lesson-by-Lesson Review

6.

Stem	Leaf
5	5 6 7 8 9 9
6	0 1 1 2 2 5

6 | 1 = 61 in.

7.

Stem	Leaf
4	5 5
5	0 0 5 5
6	0 0 0 0 0
7	5 5 5
8	5 5

7 | 5 = 75¢

8.

Stem	Leaf
11	0 2 4 8 9
12	0 3 4 4 8 9
13	
14	0 0 2 5 6

12 | 4 = 124

9. The greatest value is 50 and the least value is 27. So, the range is $50 - 27$ or 23.
To find the interquartile range, first list the data from least to greatest. Then find the median and the upper and lower quartiles.

lower half (27 38 39), upper half (42 45 50)

27 38 39 41 42 45 50

LQ = 38, median = 41, UQ = 45

The interquartile range is $45 - 38$ or 7.

10. The greatest value is 11 and the least value is 1. So, the range is $11 - 1$ or 10.
To find the interquartile range, first list the data from least to greatest. Then find the median and the upper and lower quartiles.

lower half (1 3 4 4 5), upper half (6 7 8 8 11)

1 3 4 4 5 5 6 7 8 8 11

LQ = 4, median = 5, UQ = 8

The interquartile range is $8 - 4$ or 4.

11. The greatest value is 80 and the least value is 58. So, the range is 80 − 58 or 22.
To find the interquartile range, first list the data from least to greatest. Then find the median and the upper and lower quartiles.

lower half: 58 62 64 65 | upper half: 70 72 74 80

median between 65 and 70; LQ between 62 and 64; UQ between 72 and 74

$LQ = \frac{62 + 64}{2}$ or 63 $\quad UQ = \frac{72 + 74}{2} = 73$

The interquartile range is 73 − 63 or 10.

12. The greatest value is 49 and the least value is 24. So, the range is 49 − 24 or 25.
The median is 36. The lower quartile is 30 and the upper quartile is 48. So, the interquartile range is 48 − 30 or 18.

13. The highest temperature was 80°F.

14. The median is 62. So, half the marathons were held on days having a high temperature of 62°F or less.

15. The lower quartile is 54.5 and the upper quartile is 72.5. So, 50% of the marathons were held on days that had a high temperature between 54.5°F and 72.5°F.

16. **Books Read in a Month**

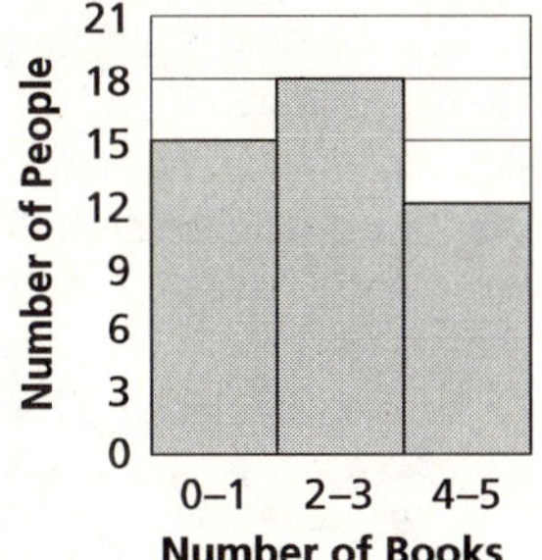

17. Graph A

18. Graph B

19. Use the Fundamental Counting Principle.

outcomes for 1st coin		outcomes for 2nd coin		outcomes for 3rd coin		outcomes for 4th coin		total possible outcomes
2	×	2	×	2	×	2	=	16

There are 16 possible outcomes.

20. Use the Fundamental Counting Principle.

choices of sizes		choices of styles		choices of colors		total possible choices
2	×	3	×	3	=	18

There are 18 possible outcomes for tennis shoes.

21. The order is not important, so this is a combination.
$C(14, 3) = \frac{P(14, 3)}{3!} = \frac{14 \cdot 13 \cdot 12}{3 \cdot 2 \cdot 1}$ or 364 ways

22. The order is important, so this is a permutation. Use the Fundamental Counting Principle.

choices for 1st digit		choices for 2nd digit		choices for 3rd digit		choices for 4th digit		choices for 5th digit
↓		↓		↓		↓		↓
10	·	10	·	10	·	10	·	10

= 100,000 codes

23. The order is not important, so this is a combination.
$C(7, 2) = \frac{P(7, 2)}{2!} = \frac{7 \cdot 6}{2 \cdot 1}$ or 21 ways

24. There are 8 spins possible. There is 1 way to spin yellow and there are 8 − 1 or 7 ways to spin a color that is not yellow.
Odds of spinning yellow = 1:7

25. There are 8 spins possible. There are 3 ways to spin blue and there are 8 − 3 or 5 ways to spin a color that is not blue.
Odds of spinning blue = 3:5

26. There are 8 spins possible. There are 6 ways to spin a color that is not green and there are 2 ways to spin green.
Odds of spinning a color that is not green = 6:2 or 3:1

27. There are 8 spins possible. There are 2 ways to spin white or red, and there are 8 − 2 or 6 ways to spin a color that is not white or red.
Odds of spinning white or red = 2:6 or 1:3

28. The events are mutually exclusive.
$P(\text{odd number or } 2) = P(\text{odd number}) + P(2)$
$= \frac{3}{6} + \frac{1}{6}$
$= \frac{4}{6}$ or $\frac{2}{3}$

29. $P(3 \text{ and } 6) = \frac{1}{6} \cdot \frac{1}{5}$ or $\frac{1}{30}$

30. The events are independent.
$P(4 \text{ and } 2) = P(4) \cdot P(2)$
$= \frac{1}{6} \cdot \frac{1}{6}$
$= \frac{1}{36}$

Chapter 12 Practice Test

Page 663

1. permutation: the number of ways 7 different plants can be planted in a row; combination: the number of ways 6 different candy bars can be chosen from 10 different candy bars

2. They cannot happen at the same time.

3.

Stem	Leaf
5	8 9
6	0 2 4 7 8 9
7	0 3 4 5 5 6 8
8	1

5|9 = 59 in.

4. The median value is $\frac{69 + 70}{2}$ or 69.5.

5. Most of the heights occur from 70 inches to 79 inches.

6. The greatest value is 99 and the least value is 71. So, the range is 99 − 71 or $28.

7. The median is 84. The extremes are 71 and 99. The lower quartile is 77 and the upper quartile is 96.

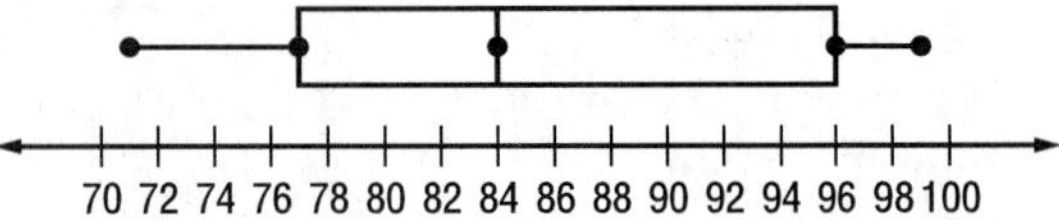

8. The upper quartile is 96 and the lower quartile is 77. So, the interquartile range is 96 − 77 or $19.

9.

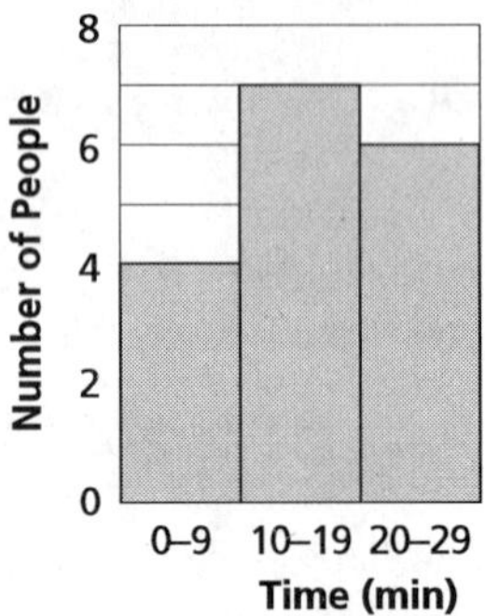

10. Use the Fundamental Counting Principle.

number of choices of meat		number of choices of vegetables		total possible outcomes	
4	×	3	=	12	outcomes

11. Order is important, so this is a permutation.
$P(7, 7) = 7 \cdot 6 \cdot 5 \cdot 4 \cdot 3 \cdot 2 \cdot 1$ or 5040 ways

12. Order is not important, so this is a combination.
$C(12, 3) = \frac{P(12, 3)}{3!} = \frac{12 \cdot 11 \cdot 10}{3 \cdot 2 \cdot 1}$ or 220 ways

13. $4! = 4 \cdot 3 \cdot 2 \cdot 1$ or 24

14. There are 10 possible rolls. There are 6 numbers greater than 4, and there are 10 − 6 or 4 numbers that are not greater than 4.
Odds of rolling a number greater than 4 = 6:4 or 3:2

15. $P(\text{four 4's}) = P(\text{4 on 1st roll}) \cdot P(\text{4 on 2nd roll}) \cdot P(\text{4 on 3rd roll}) \cdot P(\text{4 on 4th roll})$
$= \frac{1}{6} \cdot \frac{1}{6} \cdot \frac{1}{6} \cdot \frac{1}{6}$ or $\frac{1}{1296}$

16. $P(\text{5 or even}) = P(5) + P(\text{even})$
$= \frac{1}{7} + \frac{3}{7} = \frac{4}{7}$

17. $P(\text{even or 1}) = P(\text{even}) + P(1)$
$= \frac{3}{7} + \frac{1}{7} = \frac{4}{7}$

18. $P(\text{odd or even}) = P(\text{odd}) + P(\text{even})$
$= \frac{4}{7} + \frac{3}{7} = \frac{7}{7}$ or 1

19. $P(\text{2 or greater than 5}) = P(2) + P(\text{greater than 5})$
$= \frac{1}{7} + \frac{2}{7} = \frac{3}{7}$

20. C; 70° is the lower quartile and 95° is the upper extreme. So, 75% of the daily high temperatures range from 70° to 95°.

Chapter 12 Standardized Test Practice

Pages 664–665

1. B; $18.40 ÷ 20 = $0.92 for each disk

2. B; $8 \cdot 3 + 4 \cdot (-3) = 24 + (-12)$
$= 12$ points

3. D; 2(45)(0.89)

4. D;
$$\frac{5}{8} + m = \frac{3}{4}$$
$$\frac{5}{8} - \frac{5}{8} + m = \frac{3}{4} - \frac{5}{8}$$
$$m = \frac{6}{8} - \frac{5}{8}$$
$$m = \frac{1}{8}$$

5. C; $6.80 + (0.05) \cdot (6.80) = 6.80 + 0.34$
$= \$7.14$

6. C; It took Juliet 14 minutes more to walk 3 miles than to walk 2 miles. To walk 4 miles, the best estimate is 42 + 14 or 56 minutes.

7. B; Find the equation of the line through (4, 0) and (8, 8).
slope $= \frac{y_2 - y_1}{x_2 - x_1} = \frac{8 - 0}{8 - 4}$ or 2
Use the the slope-intercept equation, $y = mx + b$, and the point (4, 0) to find b.
$y = mx + b$
$0 = 2 \cdot 4 + b$
$0 = 8 + b$
$-8 = b$
The equation of the line is $y = 2x - 8$. The point (6, 4) lies on the line, since when $x = 6$, $y = 2 \cdot 6 - 8$ or 4.

8. C; Since the volume is 27 in^3, each side is 3 inches. The surface area is 6 · (area of one side) or 6(3 · 3) or 54 in^2.

9. B; Use the Fundamental Counting Principle.

choices for 1st digit		choices for 2nd digit		choices for 3rd digit		choices for 4th digit		total number of choices
4	·	3	·	2	·	1	=	24

10. C; $P(\text{red, then blue}) = \frac{4}{9} \cdot \frac{3}{8}$
$= \frac{12}{72}$ or $\frac{1}{6}$

11. $(0.3)^4 = (0.3) \cdot (0.3) \cdot (0.3) \cdot (0.3)$
$= 0.0081$

12. $16 \div \frac{1}{4} = 16 \cdot \frac{4}{1}$ or 64 patties

13. The translation can be written (−3, 0).

endpoint		**3 left, 0 up 3**		**new endpoint**
(1, 4)	+	(−3, 0)	→	(1 + −3, 4 + 0) or (−2, 4)
(3, 1)	+	(−3, 0)	→	(3 + −3, 1 + 0) or (0, 1)

14. $A = \frac{1}{2}h(a + b)$
$A = \frac{1}{2} \cdot 6 \cdot (8 + 14)$
$A = \frac{1}{2} \cdot 6 \cdot 22$ or 66 in^2

15. $V = \frac{1}{3}\pi r^2 h$
$V = \frac{1}{3} \cdot \pi \cdot 9^2 \cdot 12$
$V = \frac{1}{3} \cdot \pi \cdot 81 \cdot 12$ or about 1017.9 ft^3

16. First, find the circumference of the can.

$C = \pi d$

$C = \pi \cdot 8$ or about 25 cm

The width of the label is the same as the height of the can, 12 cm. The length of the label is the same as the circumference of the can, 25 cm.

17. The median price of the scooters is about $105.

18. The lower quartile is about 90, so about 25% of the scooters cost less than $90.

19. $C(5, 2) = \frac{P(5, 2)}{2!} = \frac{5 \cdot 4}{2 \cdot 1}$ or 10

20a.

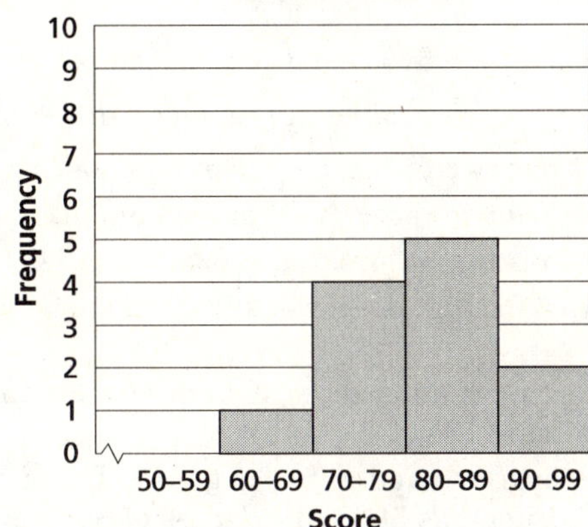

20b. The interval 80–89 contains the greatest number of test scores.

21a. PS, PO, PG, PM, PB, SO, SG, SM, SB, OG, OM, OB, GM, GB, MB

21b. Since order is not important, this is a combination.

$C(6, 2) = \frac{P(6, 2)}{2!} = \frac{6 \cdot 5}{2 \cdot 1}$ or 15 pizzas

21c. There are 15 possible pizzas. There are 9 pizzas that have pepperoni or sausage.

$P(\text{pepperoni or sausage}) = \frac{9}{15} = \frac{3}{5}$ or 60%

Chapter 13 Polynomials and Nonlinear Functions

Page 667 Getting Started

1. 1; one term is the product of a number and variables.

2. 2; two terms are variables or numbers.

3. 2; two terms are the products of numbers and variables.

4. 3; three terms are numbers, variables, or their products.

5. 0; this term is a quotient of a number and a variable, not a product of a number and a variable.

6. 4; four terms are products of numbers and variables, or numbers.

7. $5(a + 4) = 5 \cdot a + 5 \cdot 4$
$= 5a + 20$

8. $2(3y - 8) = 3[3y + (-8)]$
$= 2(3y) + (2)(-8)$
$= 6y - 16$

9. $-4(1 + 8n) = -4 \cdot 1 + (-4)(8n)$
$= -4 - 32n$

10. $6(x + 2y) = 6 \cdot x + 6 \cdot 2y$
$= 6x + 12y$

11. $(9b - 9c)3 = [9b + (-9c)]3$
$= 9b \cdot 3 + (-9c)(3)$
$= 27b - 27c$

12. $5(q - 2r + 3s) = 5[q + (-2r) + 3s]$
$= 5 \cdot q + (5)(-2r) + 5 \cdot 3s$
$= 5q - 10r + 15s$

13. Yes; each of the variables are raised to exponents of 1.

14. No; the variable x is raised to an exponent higher than 1.

15. Yes; each of the variables are raised to exponents of 1.

Page 668 Reading Mathematics

1. Each group has 2, 3, or many items.

2. *Bi* means two, *tri* means three, *poly* means many.

3. Sample answer: *binomial*, an expression with two terms; *trinomial*, an expression with three terms; *polynomial*, an expression with many terms

4. Sample answer: $x + y$; $a^2 + a = 1$; $x^3 = x^2 + x + 1$

5. Sample answer: *bicolored* means two-colored; a *trident* is a spear with three prongs; *polysyllable* means having many syllables.

13-1 Polynomials

Page 669 How are polynomials used to approximate real-world data?

a. 9

b. plus and minus signs

Pages 670–671 Check for Understanding

1. The degree of a monomial is the sum of the exponents of its variables. The degree of a polynomial is the same as the degree of the term with the greatest degree.

2. Sample answer: $2x + 1, x - y, x^2 + x$; they are sums or differences of two monomials.

3. Tanisha; the degree of a binomial is the degree of the term with the greater degree.

4. Yes; it has one term, so it is a monomial.

5. Yes; it has one term, so it is a monomial.

6. No; there is a variable in the denominator of $\frac{1}{x}$.

7. Yes; it has two terms, so it is a binomial.

8. Yes; it has two terms, so it is a binomial.

9. Yes; it has three terms, so it is a trinomial.

10. The variable b has degree 2. So, the degree of $4b^2$ is 2.

11. The degree of a nonzero constant is 0. So, the degree of 121 is 0.

12. x^3 has degree 3 and y^2 has degree 2. The sum of the exponents is 5. So, the degree of $8x^3y^2$ is 5.

13.

Term	Degree
$3x$	1
5	0

The greatest degree is 1. So, the degree of $3x + 5$ is 1.

14.

Term	Degree
r^3	3
$7r$	1

The greatest degree is 3. So, the degree of $r^3 + 7r$ is 3.

15.

Term	Degree
d^2	2
c^4	4

The greatest degree is 4. So, the degree of $d^2 + c^4$ is 4.

16. $A = \ell \cdot w$
$\ell = x$
$w = x - y$
$A = x(x - y)$ or $A = x^2 - xy$

17. 2; x^2 has degree 2 and xy has degree 2, so $x^2 - xy$ has degree 2.

Pages 671–672 Practice and Apply

18. Yes; it has one term, so it is a monomial.

19. Yes; it has two terms, so it is a binomial.

20. Yes; it has two terms, so it is a binomial.

21. Yes; it has one term, so it is a monomial.

22. No; there is a variable under a radical sign.

23. No; there is a variable in the denominator of $-\frac{2}{k}$.

24. Yes; it has two terms, so it is a binomial.

25. Yes; it has three terms, so it is a trinomial.

26. Yes; it has three terms, so it is a trinomial.

27. No; there is a variable under the radical sign of $\sqrt{y}$.

28. No; there is a variable in the denominator of $\frac{ab}{c}$.

29. Yes; it has three terms, so it is a trinomial.

30. The degree of a nonzero constant is 0, so the degree of 3 is 0.

31. The degree of a nonzero constant is 0, so the degree of 56 is 0.

32. a has degree 1 and b has degree 1. The degree of ab is 1 + 1 or 2.

33. c^3 has degree 3. So, the degree of $12c^3$ is 3.

34. x has degree 1, y has degree 1, and z^2 has degree 2. So xyz^2 has degree 1 + 1 + 2 or 4.

35. s^4 has degree 4 and t has degree 1. So, $9s^4t$ has degree 4 + 1 or 5.

36.

Term	Degree
2	0
$-8n$	1

The greatest degree is 1. So, the degree of $2 - 8n$ is 1.

37.

Term	Degree
g^5	5
$5h$	1

The greatest degree is 5. So, the degree of $g^5 + 5h$ is 5.

38.

Term	Degree
x^2	2
$3x$	1
2	0

The greatest degree is 2. So, the greatest degree of $x^2 + 3x + 2$ is 2.

39.

Term	Degree
$4y^3$	3
$6y^2$	2
$-5y$	1
-1	0

The greatest degree is 3. So, the degree of $4y^3 + 6y^2 - 5y - 1$ is 3.

40.

Term	Degree
d^2	2
c^4d^2	6

The greatest degree is 6. So, the degree of $d^2 + c^4d^2$ is 6.

41.

Term	Degree
x^3	3
$-x^2y^3$	5
8	0

The greatest degree is 5. So, the degree of $x^3 - x^2y^3 + 8$ has degree 5.

42. Sometimes; $x^3 + xy + 5$ has degree 3; $x + xy + 5$ has degree 2.

43. Always; any number is a monomial.

44. $-0.006t^4$ has degree 4; $0.140t^3$ has degree 3; $-0.53t^2$ has degree 2; and $1.79t$ has degree 1. So, the polynomial has degree 4.

45. The perimeter is the distance around the garden.

$$\begin{aligned}\text{Perimeter} &= x + y + z + y + x + xy \\ &= (x + x) + (y + y) + z + xy \\ &= 2x + 2y + z + xy\end{aligned}$$

46.

Term	Degree
$2x$	1
$2y$	1
z	1
xy	2

The greatest degree is 2. So, the degree of $2x + 2y + z + xy$ is 2.

47. See students' work.

48.

Term	Degree
a^{x+3}	$x + 3$
$x^{x-2}b^3$	$(x - 2) + 3$ or $x + 1$
b^{x+2}	$x + 2$

The greatest degree is $x + 3$. So, the degree is $x + 3$.

49. Polynomials approximate real-world data by using variables to represent quantities that are related. Answers should include the following.

- Heat index is found by using a polynomial in which one variable represents the percent humidity and another variable represents the temperature.
- Heat index cannot be approximated using a linear equation because the values do not change at a constant rate.

50. D; $7x + 2x$ is a monomial because the terms can be combined, $7x + 2x = 9x$.

51. C;

Term	Degree
$4x^3$	3
xy	2
$-y^2$	2

The greatest degree is 3. So, the degree of $4x^3 + xy - y^2$ is 3.

Page 672 Maintain Your Skills

52. Inclusive; an odd number and a number greater than 3 can occur at the same time.
There are 5 numbers on a number cube that are odd or greater than 3: 1, 3, 4, 5, and 6.
$P(\text{odd or greater than 3}) = \frac{5}{6}$

53. Mutually exclusive; the number 5 and an even number cannot occur at the same time.

$$\begin{aligned}P &= P(5) + P(\text{even}) \\ &= \tfrac{1}{6} + \tfrac{3}{6} \\ &= \tfrac{4}{6} \text{ or } \tfrac{2}{3}\end{aligned}$$

54. There are 6 possible outcomes. There are 4 numbers greater than 2 and there are 2 numbers that are not greater than 2.
Odds of rolling a number greater than 2 = 4:2 or 2:1.

55. $V = \frac{1}{3}Bh$

$V = \frac{1}{3}(6 \cdot 5) \cdot 7$

$V = 70$

The volume is 70 in^3.

56. $V = \frac{1}{3}\pi r^2 h$

$= \frac{1}{3} \cdot \pi \cdot 4^2 \cdot 11.3$

≈ 189.3

The volume is about 189.3 m^3.

57. $(x + 4) + 2x = (x + 2x) + 4$

58. $3x^2 - 1 + x^2 = (3x^2 + x^2) - 1$

59. $(6n + 2) + (3n + 5) = (6n + 3n) + (2 + 5)$

60. $(a + 2b) + (3a + b) = (a + 3a) + (2b + b)$

61. $(s + t) + (5s - 3t) = (s + 5s) + (t - 3t)$

62. $(x^2 + 4x) + (7x^2 - 3x) = (x^2 + 7x^2) + (4x - 3x)$

Page 673 Algebra Activity (Follow-Up of Lesson 13-1)

1.

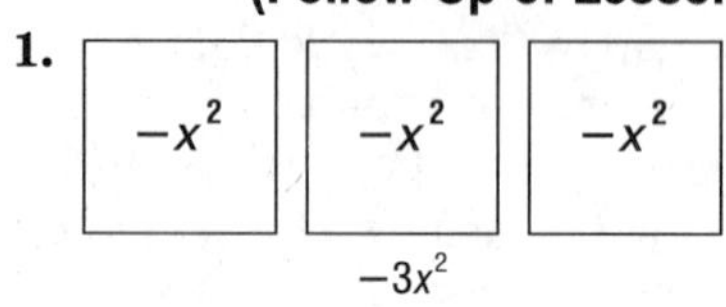

2.

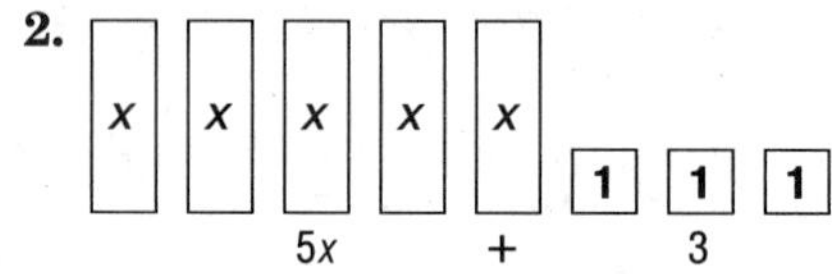

3.

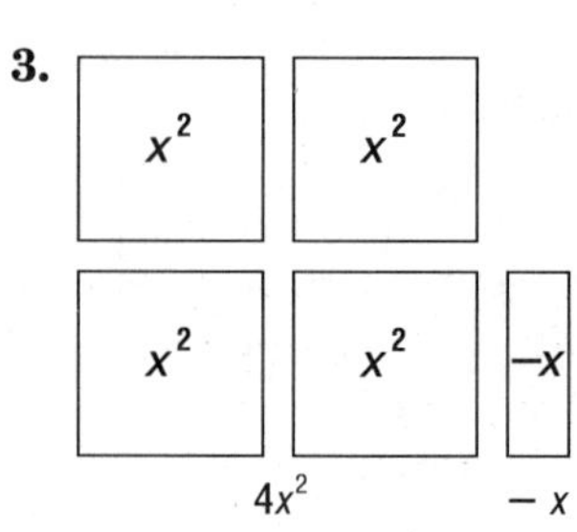

4.

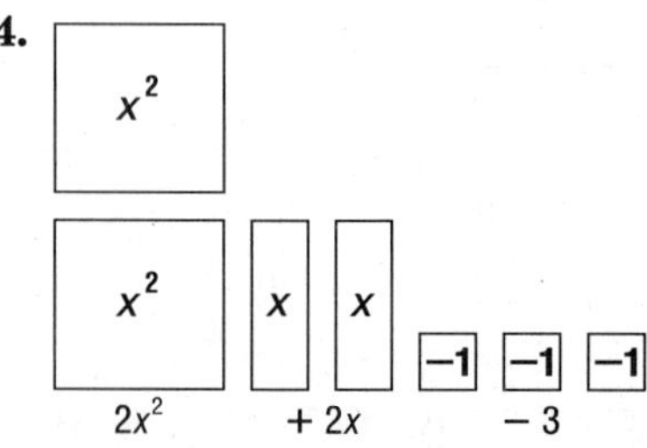

5. A monomial has tiles of one size, a binomial has tiles of two sizes, and a trinomial has tiles of three sizes.

6. $-x^2 + 3x - 5$

7. The degree of a polynomial is determined by the size of the largest algebra tile.

13-2 Adding Polynomials

Page 674 How can you use algebra tiles to add polynomials?

a. $x^2 - 2x + 2$

b. Model the polynomials.

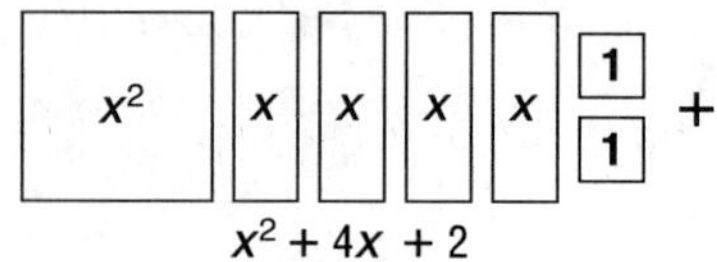

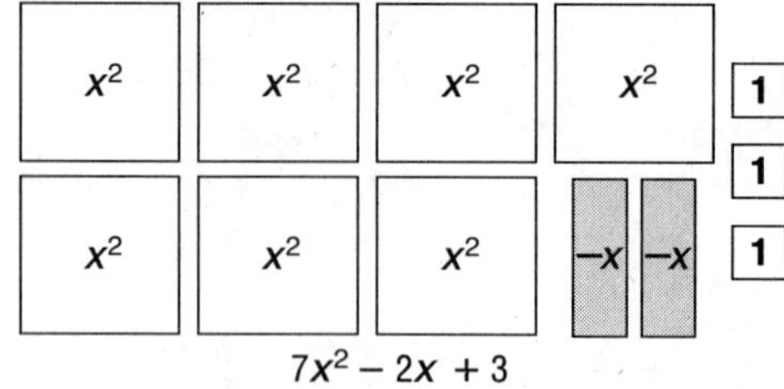

Combine the tiles that have the same shape. Remove zero pairs.

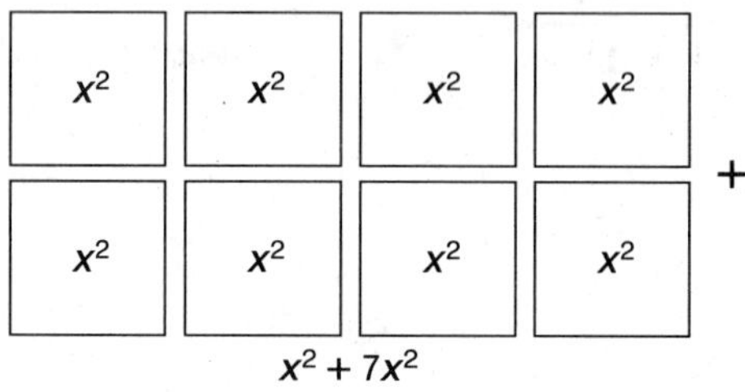

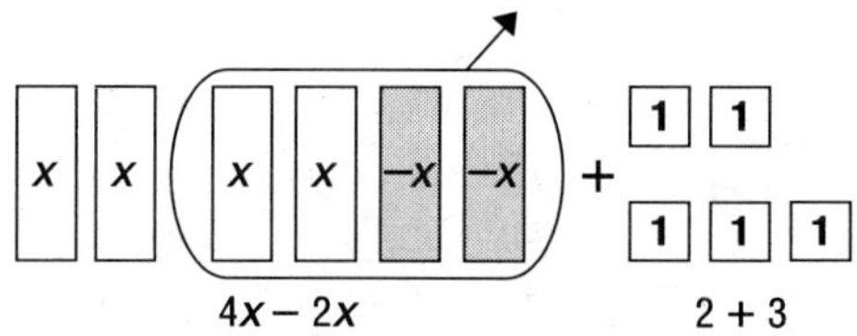

$(x^2 + 4x + 2) + (7x^2 - 2x + 3) = 8x^2 + 2x + 5$

c. The concept of the zero pairs is the same, but there are tiles that represent different terms in the polynomials.

Page 676 Check for Understanding

1. x^2 and $2x^2$; $5x$ and $-4x$; 2 and 7

2. Sample answer: $3x + 1$ and $4x + x^2$

3. Hai; the terms have the same variables in a different order.

4.
$$\begin{array}{rr} & 4x + 5 \\ (+) & \underline{-x - 3} \\ & 3x + 2 \end{array}$$

5.
$$\begin{array}{rr} & 3a^2 - 9a + 6 \\ (+) & \underline{4a^2 \qquad - 2} \\ & 7a^2 - 9a + 4 \end{array}$$

6. $(x + 3) + (2x + 5) = (x + 2x) + (3 + 5)$
$= 3x + 8$

7. $(13x - 7y) + 3y = 13x + (-7y + 3y)$
$= 13x - 4y$

8. $(2x^2 + 5x) + (9 - 7x) = 2x^2 + (5x - 7x) + 9$
$= 2x^2 - 2x + 9$

9. $(3x^2 - 2x + 1) + (x^2 + 5x - 3)$
$= (3x^2 + x^2) + (-2x + 5x) + [1 + (-3)]$
$= 4x^2 + 3x - 2$

10. $P = 2\ell + 2w$
$P = 2(2x + 7) + 2(x + 5)$
$= 4x + 14 + 2x + 10$
$= (4x + 2x) + (14 + 10)$
$= 6x + 24$

Pages 676–677 Practice and Apply

11. $\begin{array}{r} -5x + 4 \\ (+)\ \underline{8x - 1} \\ 3x + 3 \end{array}$

12. $\begin{array}{r} 7b - 5 \\ (+)\ \underline{-9b + 8} \\ -2b + 3 \end{array}$

13. $\begin{array}{r} 10x^2 + 5xy + 7y^2 \\ (+)\ \underline{x^2 \qquad - 3y^2} \\ 11x^2 + 5xy + 4y^2 \end{array}$

14. $\begin{array}{r} 4a^3 + a^2 + 8a - 8 \\ (+)\ \underline{2a^2 \qquad + 6} \\ 4a^3 + 3a^2 + 8a - 2 \end{array}$

15. $(3x + 9) + (x + 5) = (3x + x) + (9 + 5)$
$= 4x + 14$

16. $(4x + 3) + (x - 1) = (4x + x) + [3 + (-1)]$
$= 5x + 2$

17. $(6y - 5r) + (2y + 7r) = (6y + 2y) + (-5r + 7r)$
$= 8y + 2r$

18. $(8m - 2n) + (3m + n) = (8m + 3m) + (-2n + n)$
$= 11m - n$

19. $(x^2 + y) + (4x^2 + xy) = (x^2 + 4x^2) + xy + y$
$= 5x^2 + xy + y$

20. $(3a^2 + b^2) + (3a + b^2) = 3a^2 + (b^2 + b^2) + 3a$
$= 3a^2 + 2b^2 + 3a$

21. $(5x^2 + 6x + 4) + (2x^2 + 3x + 1)$
$= (5x^2 + 2x^2) + (6x + 3x) + (4 + 1)$
$= 7x^2 + 9x + 5$

22. $(-2x^2 + x - 5) + (x^2 - 3x + 2)$
$= (-2x^2 + x^2) + (x - 3x) + (-5 + 2)$
$= -x^2 - 2x - 3$

23. $(3a + 5b) + (2a - 9b) = (3a + 2a) + (5b - 9b)$
$= 5a - 4b$
$= 5(-3) - 4(4)$
$= -15 - 16$
$= -31$

24. $(a^2 + 7b^2) + (5 - 3b^2) + (2a^2 - 7)$
$= (a^2 + 2a^2) + (7b^2 - 3b^2) + (5 - 7)$
$= 3a^2 + 4b^2 - 2$
$= 3(-3)^2 + 4(4)^2 - 2$
$= 3(9) + 4(16) - 2$
$= 27 + 64 - 2$
$= 89$

25. $(3a + 5b - 4c) + (2a - 3b + 7c) + (-a + 4b - 2c)$
$= (3a + 2a - a) + (5b - 3b + 4b) +$
$(-4c + 7c - 2c)$
$= 4a + 6b + c$
$= 4(-3) + 6(4) + 2$
$= -12 + 24 + 2$
$= 14$

26. $x^0 + (2x - 30)^0 + (x - 14)^0$
$= (x + 2x + x)^0 + [-30 + (-14)]^0$
$= (4x - 44)^0$
The sum of the measures of the angles is $(4x - 44)^0$.

27. $4x - 44 = 180$
$4x - 44 + 44 = 180 + 44$
$4x = 224$
$\frac{4x}{4} = \frac{224}{4}$
$x = 56$
The value of x is 56.

28. If $x = 56$, then $2x - 30 = 2(56) - 30$ or 82, and $x - 14 = 56 - 14$ or 42. The measures of the angles are 56°, 42°, and 82°.

29. Let x = the hourly pay. Then $23x - 12$ represents Jason's pay for the week.

30. Let x = the hourly pay. Then $19x - 10$ represents Will's pay for the week.

31. $(23x - 12) + (19x - 10)$
$= (23x + 19x) + [-12 + (-10)]$
$= 42x - 22$

32.

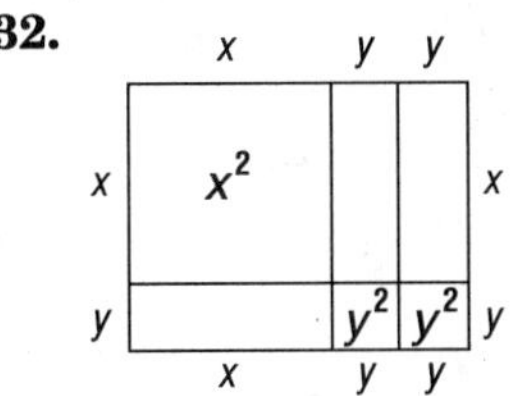

The length is $x + 2y$ and the width is $x + y$. So, the perimeter is $2(x + 2y) + 2(x + y)$ or $4x + 6y$.

33. Use algebra tiles to model each polynomial and combine the tiles that have the same size and shape. Answers should include the following.
- Algebra tiles that represent like terms have the same size and shape.
- When adding polynomials, a red tile and a white tile that have the same size and shape are zero pairs and may be removed. The result is the sum of the polynomials.

34. C; a^2 and b^2 do not contain the same variables.

35. B; $(11x + 2y) + (x - 5y) = (11x + x) + (2y - 5y)$
$= 12x - 3y$

Page 677 Maintain Your Skills

36. a^3 has degree 3 and b has degree 1, so the degree of a^3b is 3 + 1 or 4.

37.

Term	Degree
$3x$	1
$-5y$	1
z^2	2

The greatest degree is 2. So, the degree of $3x - 5y + z^2$ is 2.

38.

Term	Degree
c^2	2
$-7c^3y^4$	7

The greatest degree is 7. S, the degree of $c^2 - 7c^3y^4$ is 7.

39. The events are mutually exclusive because a 2 and a jack cannot be drawn at the same time. There are 4 2s and 4 jacks.
$P(2 \text{ or jack}) = P(2) + P(\text{jack})$
$= \frac{4}{52} + \frac{4}{52}$
$= \frac{8}{52}$ or $\frac{2}{13}$

40. The events are not mutually exclusive, since a card can be both 10 and red. There are 28 cards that are either 10 or red, since there are 4 10s and 26 red cards, but 2 cards that are both 10 and red.
$P(10 \text{ or red}) = \frac{28}{52}$ or $\frac{7}{13}$

41. The events are mutually exclusive since a card cannot be both an ace and a black 7 at the same time. There are 4 aces and 2 black 7s.
$P(\text{ace or black } 7) = P(\text{ace}) + P(\text{black } 7)$
$= \frac{4}{52} + \frac{2}{52}$
$= \frac{6}{52}$ or $\frac{3}{26}$

42. Write a proportion comparing lengths and widths.
$\frac{16}{4} \stackrel{?}{=} \frac{24}{6}$
$4 = 4$ ✓
The lengths and widths are proportional.
Write a proportion comparing heights and widths.
$\frac{9}{4} \stackrel{?}{=} \frac{13.5}{6}$
$2.25 = 2.5$
The corresponding dimensions are proportional.

43. $15c - 26 = 15c + (-26)$

44. $x^2 - 7 = x^2 + (-7)$

45. $1 - 2x = 1 + (-2x)$

46. $6b - 3a^2 = 6b + (-3a^2)$

47. $(n + rt) - r^2 = (n + rt) + (-r^2)$

48. $(s + t) - 2s = (s + t) + (-2s)$

13-3 Subtracting Polynomials

Page 678 How is subtracting polynomials similar to subtracting measurements?

a. 89 − 85 = 4 degrees; 35.4 − 27.3 = 8.1 minutes.

b. Subtract the degrees and subtract the minutes.

c.

Station 1	162°	16′	32″
Station 5	− 68°	8′	2″
	94°	8′	30″

The difference in longitude is 94°8′30″.

Page 680 Check for Understanding

1. Subtracting one polynomial from another is the same as adding the additive inverse.

2. Sample answer: $2x^2 + 4x + 1$ and $x^2 + 2x + 5$

3.
$$\begin{array}{rr} & r^2 + 5r \\ (-) & r^2 + \ \ r \\ \hline & 4r \end{array}$$

4. The additive inverse of $x^2 - 1$ is $-x^2 + 1$.
$$\begin{array}{rr} & 3x^2 + 5x + 4 \\ (-) & x^2 \qquad\ - 1 \\ \hline \end{array} \rightarrow \begin{array}{rr} & 3x^2 + 5x + 4 \\ (+) & -x^2 \qquad\ + 1 \\ \hline & 2x^2 + 5x + 5 \end{array}$$

5. $(9x + 5) - (4x + 3) = (9x + 5) + (-4x - 3)$
$= (9x - 4x) + (5 - 3)$
$= 5x + 2$

6. $(2x + 4) - (-x + 5) = (2x + 4) + (x - 5)$
$= (2x + x) + (4 - 5)$
$= 3x - 1$

7. $(3x^2 + x) - (8 - 2x) = (3x^2 + x) + (-8 + 2x)$
$= 3x^2 + (x + 2x) - 8$
$= 3x^2 + 3x - 8$

8. $(6a^2 - 3a + 9) - (7a^2 + 5a - 1)$
$= (6a^2 - 3a + 9) + (-7a^2 - 5a + 1)$
$= (6a^2 - 7a^2) + (-3a - 5a) + (9 + 1)$
$= -a^2 - 8a + 10$

9. perimeter − the lengths of three sides = the length of the missing side
$(16x + 1) - [(2x - 3) + (5x + 2) + (2x - 3)]$
$= (16x + 1) - [(2x + 5x + 2x) + (-3 + 2 - 3)]$
$= 16x + 1 - (9x - 4)$
$= 16x + 1 + (-9x + 4)$
$= (16x - 9x) + (1 + 4)$
$= 7x + 5$
One length of the missing base is $7x + 5$ units.

Pages 680–681 Practice and Apply

10.
$$\begin{array}{rr} & 8k + 9 \\ (-) & k + 2 \\ \hline & 7k + 7 \end{array}$$

11. The additive inverse of $n^2 - 5n$ is $-n^2 + 5n$.
$$\begin{array}{rr} & -n^2 + 1n \\ (-) & n^2 - 5n \\ \hline \end{array} \rightarrow \begin{array}{rr} & -n^2 + 1n \\ (+) & -n^2 + 5n \\ \hline & -2n^2 + 6n \end{array}$$

12. The additive inverse of $-3a^2 + 5a - 7$ is $3a^2 - 5a + 7$.
$$\begin{array}{rr} & 5a^2 + 9a - 12 \\ (-) & -3a^2 + 5a - \ \ 7 \\ \hline \end{array} \rightarrow \begin{array}{rr} & 5a^2 + 9a - 12 \\ (+) & 3a^2 - 5a + \ \ 7 \\ \hline & 8a^2 + 4a - \ \ 5 \end{array}$$

13. The additive inverse of $5y^2 + 2y - 7$ is $-5y^2 - 2y + 7$.
$$\begin{array}{rr} & 6y^2 - 5y + 3 \\ (-) & 5y^2 + 2y - 7 \\ \hline \end{array} \rightarrow \begin{array}{rr} & 6y^2 - 5y + \ \ 3 \\ (+) & -5y^2 - 2y + \ \ 7 \\ \hline & y^2 - 7y + 10 \end{array}$$

14. The additive inverse of $-3xy + 2y^2$ is $3xy - 2y^2$.
$$\begin{array}{rr} & 5x^2 - 4xy \\ (-) & -3xy + 2y^2 \\ \hline \end{array} \rightarrow \begin{array}{rr} & 5x^2 - 4xy \\ (+) & 3xy - 2y^2 \\ \hline & 5x^2 - \ \ xy - 2y^2 \end{array}$$

15. The additive inverse of $-6w^2 + 2w - 3$ is $6w^2 - 2w + 3$.
$$\begin{array}{rr} & 9w^2 \qquad\ + 7 \\ (-) & -6w^2 + 2w - 3 \\ \hline \end{array} \rightarrow \begin{array}{rr} & 9w^2 \qquad\ + \ \ 7 \\ (+) & 6w^2 - 2w + \ \ 3 \\ \hline & 15w^2 - 2w + 10 \end{array}$$

16. $(3x + 4) - (x + 2) = (3x + 4) + (-x - 2)$
$= (3x - x) + (4 - 2)$
$= 2x + 2$

17. $(7x + 5) - (3x + 2) = (7x + 5) + (-3x - 2)$
$= (7x - 3x) + (5 - 2)$
$= 4x + 3$

18. $(2y + 5) - (y + 8) = (2y + 5) + (-y - 8)$
$= (2y - y) + (5 - 8)$
$= y - 3$

19. $(3t - 2) - (5t - 4) = (3t - 2) + (-5t + 4)$
$= (3t - 5t) + (-2 + 4)$
$= -2t + 2$

20. $(2x + 3y) - (x - y) = (2x + 3y) + (-x + y)$
$= (2x - x) + (3y + y)$
$= x + 4y$

21. $(a^2 + 6b^2) - (-2a^2 + 4b^2)$
$= (a^2 + 6b^2) + (2a^2 - 4b^2)$
$= (a^2 + 2a^2) + (6b^2 - 4b^2)$
$= 3a^2 + 2b^2$

22. $(x^2 + 6x) - (3x^2 + 7) = (x^2 + 6x) + (-3x^2 - 7)$
$= (x^2 - 3x^2) + 6x - 7$
$= -2x^2 + 6x - 7$

23. $(9n^2 - 8) - (n + 4) = (9n^2 - 8) + (-n - 4)$
$= 9n^2 - n + (-8 - 4)$
$= 9n^2 - n - 12$

24. $(6x^2 + 3x + 9) - (2x^2 + 8x + 1)$
$= (6x^2 + 3x + 9) + (-2x^2 - 8x - 1)$
$= (6x^2 - 2x^2) + (3x - 8x) + (9 - 1)$
$= 4x^2 - 5x + 8$

25. $(3x^2 - 5xy + 7y^2) - (x^2 - 3xy + 4y^2)$
$= (3x^2 - 5xy + 7y^2) + (-x^2 + 3xy - 4y^2)$
$= (3x^2 - x^2) + (-5xy + 3xy) + (7y^2 - 4y^2)$
$= 2x^2 - 2xy + 3y^2$

26. area of the picture − area of the frame = the amount Alyssa must trim
$(2x^2 + 11x + 12) - (2x^2 + 5x + 2)$
$= (2x^2 + 11x + 12) + (-2x^2 - 5x - 2)$
$= (2x^2 - 2x^2) + (11x - 5x) + (12 - 2)$
$= 0 + 6x + 10$ or $6x + 10$
Alyssa must trim $6x + 10$ square units from the picture.

27. Let x = the highest temperature.
Let y = the lowest temperature.
Difference between highest and lowest temperature = 68° more than the sum of the temperatures.
$x - y = 68 + (x + y)$
$x - x - y + y = 68 + (x + y) - x + y$
$0 = 68 + x + y - x + y$
$0 - 68 = (x - x) + (y + y)$
$-68 = 2y$
$-34 = y$
The record low temperature in North Carolina is −34°F.

28. Since $(A + B) + (A - B) = (A + A) + (B - B) = 2A$, substitute for $A + B$ and $A - B$ in the equation:
$2A = (A + B) + (A - B)$
$2A = (3x^2 + 2x - 2) + (-x^2 + 4x - 8)$
$2A = (3x^2 - x^2) + (2x + 4x) + (-2 - 8)$
$2A = 2x^2 + 6x - 10$
$\frac{2A}{2} = \frac{2x^2 + 6x - 10}{2}$
$A = x^2 + 3x - 5$

Since $A = x^2 + 3x - 5$, and $A + B = 3x^2 + 2x - 2$, then substitute for A in $A + B$.
$(x^2 + 3x - 5) + B = 3x^2 + 2x - 2$
$B = 3x^2 + 2x - 2 - (x^2 + 3x - 5)$
$B = 3x^2 + 2x - 2 + (-x^2 - 3x + 5)$
$B = (3x^2 - x^2) + (2x - 3x) + (-2 + 5)$
$B = 2x^2 - x + 3$
So, $A = x^2 + 3x - 5$ and $B = 2x^2 - x + 3$.

29. In subtracting polynomials and in subtracting measurements, like parts are subtracted. Answers should include the following.
- To subtract measurements with two or more units, subtract the like units. To subtract polynomials with two or more terms, subtract the like terms.
- For example, to subtract 1 foot 5 inches from 3 feet 8 inches, subtract the feet 3 − 1 and subtract the inches 8 − 5. The difference is 2 feet 3 inches.

30. A; $(5x - 7) - (3x - 4) = (5x - 7) + (-3x + 4)$
$= (5x - 3x) + (-7 + 4)$
$= 2x - 3$

31. B; $(-1)(-4h^2 - hk - k^2) = 4h^2 + hk + k^2$

Page 681 Maintain Your Skills

32. $(2x - 3) + (x - 1) = (2x + x) + (-3 - 1)$
$= 3x - 4$

33. $(11x + 2y) + (x - 5y) = (11x + x) + (2y - 5y)$
$= 12x - 3y$

34. $(5x^2 - 7x + 9) + (3x^2 + 4x - 6)$
$= (5x^2 + 3x^2) + (-7x + 4x) + (9 - 6)$
$= 8x^2 - 3x + 3$

35. $(4t - t^2) + (8t + 2) = -t^2 + (4t + 8t) + 2$
$= -t^2 + 12t + 2$

36. No; there is a variable in the denominator.

37. Yes; it has two terms, so it is a binomial.

38. Yes; it has three terms, so it is a trinomial.

39.

Stem	Leaf
5	4 9
6	4 6 8
7	0 1 1 2
8	5 9
9	1

$5|4 = 54$

40. $x(3x) = x \cdot 3 \cdot x$
$= 3 \cdot x \cdot x$
$= 3x^2$

41. $(2y)(4y) = 2 \cdot y \cdot 4 \cdot y$
$= (2 \cdot 4)(y \cdot y)$
$= 8y^2$

42. $(t^2)(6t) = t \cdot t \cdot 6 \cdot t$
$= 6t^3$

43. $(4m)(m^2) = 4 \cdot m \cdot m \cdot m$
$= 4m^3$

44. $(w^2)(-3w) = w \cdot w \cdot (-3) \cdot w$
$= -3w^3$

45. $(2r^2)(5r^3) = 2 \cdot r \cdot r \cdot 5 \cdot r \cdot r \cdot r$
$= 2 \cdot 5 \cdot r \cdot r \cdot r \cdot r \cdot r$
$= 10r^5$

Page 681 Practice Quiz 1

1. c has degree 1 and d^3 has degree 3, so the degree of cd^3 is $1 + 3$ or 4.
2. a has degree 1 and $-4a^2$ has degree 2. The greatest degree is 2. So, the degree of $a - 4a^2$ is 2.
3. x^2y has degree 3, $7x^2$ has degree 2, and -21 has degree 0. The greatest degree is 3. So, the degree of $x^2y + 7x^2 - 21$ is 3.
4. $(2x - 8) + (x - 7) = (2x + x) + (-8 - 7)$
$= 3x - 15$
5. $(4x + 5) - (2x + 3) = (4x + 5) + (-2x - 3)$
$= (4x - 2x) + (5 - 3)$
$= 2x + 2$
6. $(5d^2 - 3) - (2d^2 - 7) = (5d^2 - 3) + (-2d^2 + 7)$
$= (5d^2 - 2d^2) + (-3 + 7)$
$= 3d^2 + 4$
7. $(3r + 6s) + (5r - 9s) = (3r + 5r) + (6s - 9s)$
$= 8r - 3s$
8. $(x^2 + 4x + 2) + (7x^2 - 2x + 3)$
$= (x^2 + 7x^2) + (4x - 2x) + (2 + 3)$
$= 8x^2 + 2x + 5$
9. $(9x - 4y) - (12x - 9y) = (9x - 4y) + (-12x + 9y)$
$= (9x - 12x) + (-4y + 9y)$
$= -3x + 5y$
10. perimeter − lengths of two sides = length of the third side
$8x + 3y - [(4x - y) + (x + 2y)]$
$= 8x + 3y + [-4x + y - x - 2y]$
$= (8x - 4x - x) + (3y + y - 2y)$
$= 3x + 2y$
The third side is $3x + 2y$ cm long.

Page 682 Algebra Activity (Preview of Lesson 13-4)

1. False;

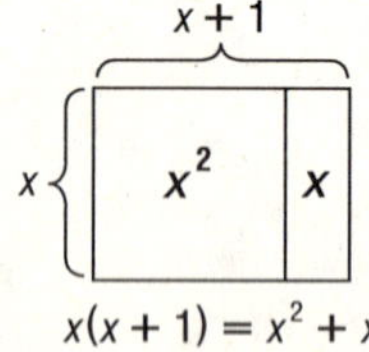

$x(x + 1) = x^2 + x$

2. True;

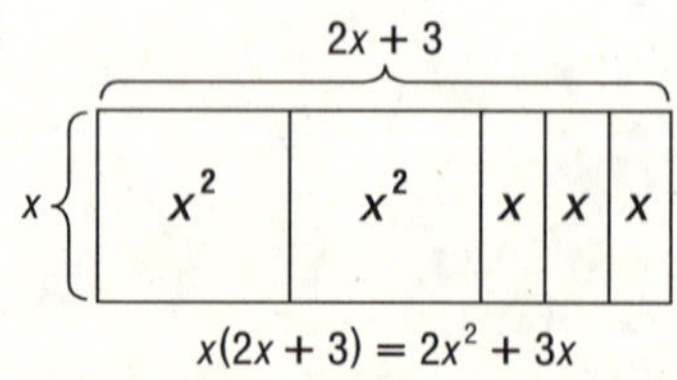

$x(2x + 3) = 2x^2 + 3x$

3. True;

$(x + 2)2x = 2x^2 + 4x$

4. False;

$2x(3x + 1) = 6x^2 + 2x$

5.

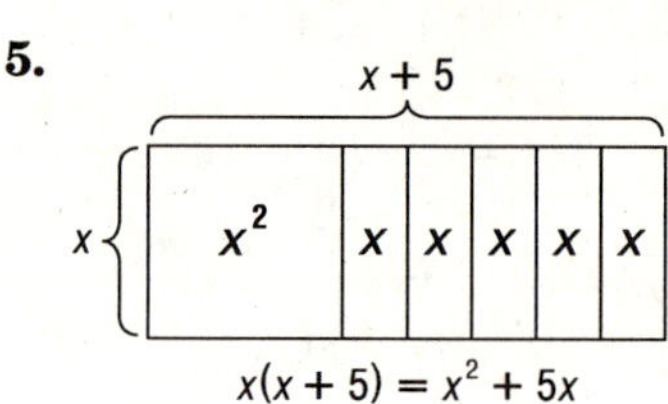

$x(x + 5) = x^2 + 5x$

6.

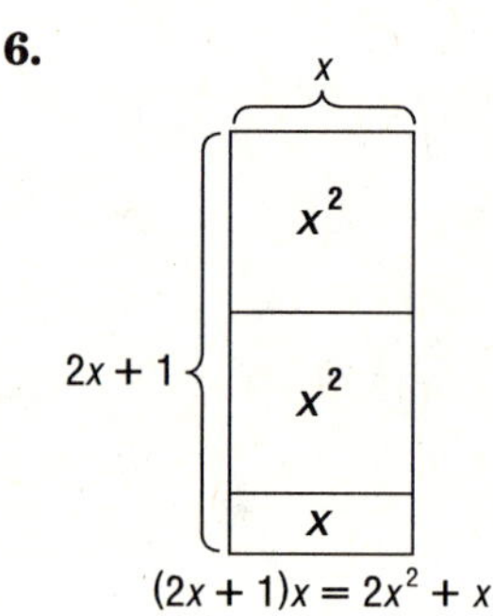

$(2x + 1)x = 2x^2 + x$

7.

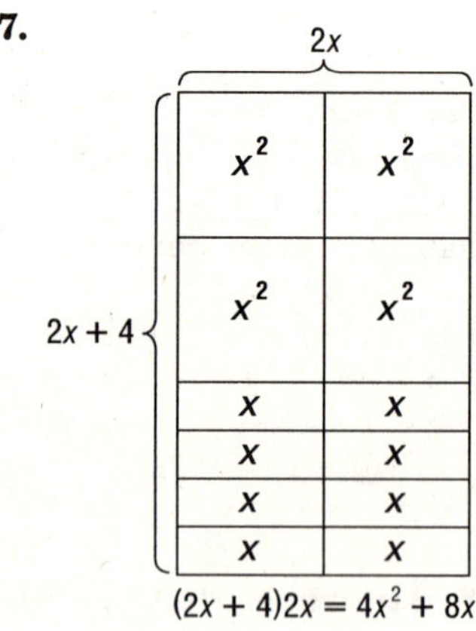

$(2x + 4)2x = 4x^2 + 8x$

8.

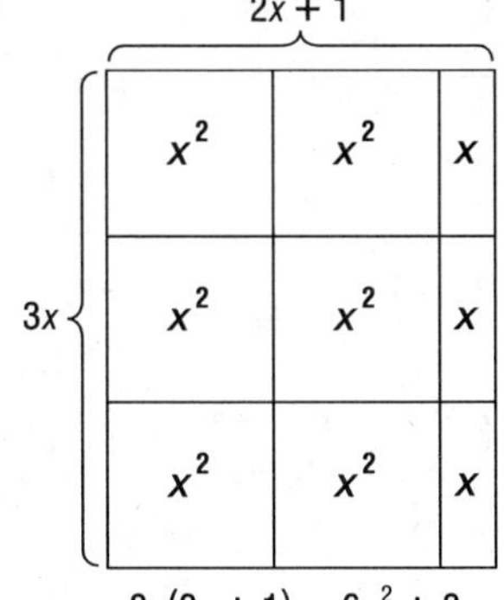

$3x(2x + 1) = 6x^2 + 3x$

9a. Original garden has length x and width x.
New garden has length $2x$ and width $x + 3$.
$A = \ell \cdot w$
$A = 2x(x + 3)$ or $2x^2 + 6x$

9b. Let $x = 10$
$A = 2x(x + 3)$ so $A = 2 \cdot 10(10 + 3)$
$= 20(13)$
$= 260$
The area of the new plot is 260 ft^2.

10. $(x + 3)(2x + 3) = 2x^2 + 3x + 6x + 9$
$= 2x^2 + 9x + 9$

13-4 Multiplying a Polynomial by a Monomial

Page 683 How is the Distributive Property used to multiply a polynomial by a monomial?

a. $A = \ell \cdot w$
$= w(2w - 52)$

b. $w(2w - 52) = w(2w) + w(-52)$
$= 2w^2 - 52w$

c. Since the volume of a prism is $V = Bh$, use the Distributive Property to multiply $2w^2 - 52w$ by w.
$(2w^2 - 52w)w = (2w^2)w - (52w)w$
$= 2w^3 - 52w^2$

Pages 684–685 Check for Understanding

1. False; the order in which numbers or terms are multiplied does not change the product, by the Commutative Property of Multiplication; $x(2x + 3) = 2x^2 + 3x$ and $(2x + 3)x = 2x^2 + 3x$.

2. Multiply x^3 by $4x$ and multiply -7 by $4x$. Then add.

3. Sample answer: $2x(x + 1) = 2x^2 + 2x$

4. $(5y - 4)3 = 5y(3) - 4(3)$
$= 15y - 12$

5. $a(a + 4) = a(a) + a(4)$
$= a^2 + 4a$

6. $t(7t + 8) = t(7t) + t(8)$
$= 7t^2 + 8t$

7. $(3x - 7)4x = 3x(4x) - 7(4x)$
$= 12x^2 - 28x$

8. $a(2a + b) = a(2a) + a(b)$
$= 2a^2 + ab$

9. $-5(3x^2 - 7x + 9) = -5(3x^2) + (-5)(-7x) + (-5)(9)$
$= -15x^2 + 35x - 45$

10. $P = 2(\ell + w)$
$228 = 2[(6 + 2w) + w]$
$228 = 2(6 + 3w)$
$228 = 2(6) + 2(3w)$
$228 = 12 + 6w$
$216 = 6w$
$36 = w$
The width is 36 feet, and the length is $6 + 2w = 6 + 2(36)$ or 78 feet.

Pages 685–686 Practice and Apply

11. $7(2n + 5) = 7(2n) + 7(5)$
$= 14n + 35$

12. $(1 + 4b)6 = 1(6) + 4b(6)$
$= 6 + 24b$

13. $t(t - 9) = t(t) - t(9)$
$= t^2 - 9t$

14. $(x + 5)x = x(x) + 5(x)$
$= x^2 + 5x$

15. $-a(7a + 6) = -a(7a) - a(6)$
$= -7a^2 - 6a$

16. $y(3 + 2y) = y(3) + y(2y)$
$= 3y + 2y^2$

17. $4n(10 + 2n) = 4n(10) + 4n(2n)$
$= 40n + 8n^2$

18. $-3x(6x - 4) = -3x(6x) + (-3x)(-4)$
$= -18x^2 + 12x$

19. $3y(y^2 - 2) = 3y(y^2) - 3y(2)$
$= 3y^3 - 6y$

20. $ab(a^2 + 7) = ab(a^2) + ab(7)$
$= a^3b + 7ab$

21. $5x(x + y) = 5x(x) + 5x(y)$
$= 5x^2 + 5xy$

22. $4m(m^2 - m) = 4m(m^2) - 4m(m)$
$= 4m^3 - 4m^2$

23. $7(-2x^2 + 5x - 11) = 7(-2x^2) + 7(5x) - 7(11)$
$= -14x^2 + 35x - 77$

24. $-3y(6 - 9y + 4y^2)$
$= -3y(6) + (-3y)(-9y) - 3y(4y^2)$
$= -18y + 27y^2 - 12y^3$

25. $4c(c^3 + 7c - 10) = 4c(c^3) + 4c(7c) - 4c(10)$
$= 4c^4 + 28c^2 - 40c$

26. $6x^2(-2x^3 + 8x + 1) = 6x^2(-2x^3) + 6x^2(8x) + 6x^2(1)$
$= -12x^5 + 48x^3 + 6x^2$

27. $30 = 6(-2w + 3)$
$30 = 6(-2w) + 6(3)$
$30 = -12w + 18$
$12 = -12w$
$-1 = w$

28. $-3(2a - 12) = 3a - 45$
$-3(2a) + (-3)(-12) = 3a - 45$
$-6a + 36 = 3a - 45$
$-6a + 6a + 36 = 3a + 6a - 45$
$36 = 9a - 45$
$36 + 45 = 9a - 45 + 45$
$81 = 9a$
$9 = a$

29. Highschool: $P = 2(\ell + w)$
$268 = 2[(2w - 16) + w]$
$268 = 2(3w - 16)$
$268 = 2(3w) - 2(16)$
$268 = 6w - 32$
$300 = 6w$
$50 = w$

The width is 50 feet and the length is $2w - 16 = 2(50) - 16$ or 84 feet.

College: $P = 2(\ell + w)$
$288 = 2\{[(2w - 16) + 10] + w\}$
$288 = 2(3w - 6)$
$288 = 2(3w) - 2(6)$
$288 = 6w - 12$
$300 = 6w$
$50 = w$

The width is 50 feet and the length is $(2w - 16) + 10 = (2 \cdot 50 - 16) + 10$ or 94 feet.

30a. Area of left and right sides
$= 2 \cdot (2x + 2y)(6x - 2y)$
$= 2 \cdot (12x^2 + 8xy - 4y^2)$
$= 2(12x^2) + 2(8xy) - 2(4y^2)$
$= 24x^2 + 16xy - 8y^2$
Area of front and back sides
$= 2 \cdot (2x)(2x + 2y)$
$= 4x(2x + 2y)$
$= 4x(2x) + 4x(2y)$
$= 8x^2 + 8xy$
The top and the bottom are the same, made up of 4 flaps.
Areas of left and right flaps (top and bottom)
$= 4 \cdot (x)(6x - 2y)$
$= 4x(6x - 2y)$
$= 4x(6x) - 4x(2y)$
$= 24x^2 - 8xy$
Areas of front and back flaps (top and bottom)
$= 4 \cdot (x)(2x)$
$= 8x^2$
Area of cardboard used to make the box
$= (24x^2 + 16xy - 8y^2) + (8x^2 + 8xy) + (24x^2 - 8xy) + 8x^2$
$= (24x^2 + 8x^2 + 24x^2 + 8x^2) + (16xy + 8xy - 8xy) + (-8y^2)$
$= 64x^2 + 16xy - 8y^2$

30b. $64x^2 + 16xy - 8y^2$
$= 64(1.2)^2 + 16(1.2)(0.1) - 8(0.1)^2$
$= 64 \cdot 1.44 + 1.92 - 8 \cdot 0.01$
$= 92.16 + 1.92 - 0.08$
$= 94$
The total amount of cardboard to make the box is 94 m^2.

31. $(a + b)(c + d) = ac + ad + bc + bd$;

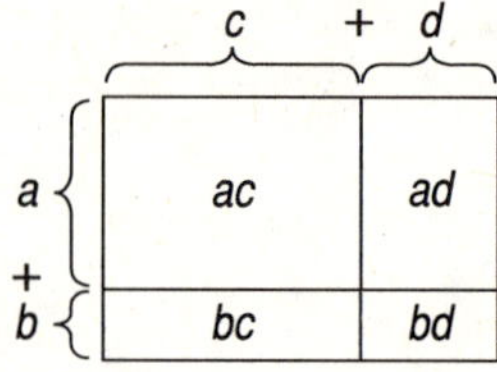

32. To use the Distributive Property, multiply each term of the polynomial by the monomial. Answers should include the following.

- The Distributive Property combines multiplication and addition when used to find the product of a polynomial and a monomial.
- To find the product of $3x$ and $x + 2$, first find $3x \cdot x$. The find $3x \cdot 2$. Finally, add the products, $3x^2 + 6x$.

33. D; $2x(x - 8) = 2x(x) - 2x(8)$
$= 2x^2 - 16x$

34. A; Choose an answer choice to check.
$A = \ell \cdot w$
Let $\ell = 18$.
$252 = 18 \cdot x$
$\frac{252}{18} = \frac{18}{18} \cdot x$
$14 = x$
Substitute 14 for x in the original equation.
$252 = (2x - 10)(x)$
$252 \stackrel{?}{=} (2 \cdot 14 - 10)(14)$
$252 \stackrel{?}{=} (18)(14)$
$252 = 252$ ✓

Page 686 Maintain Your Skills

35. $(2x - 1) + 5x = (2x + 5x) - 1$
$= 7x - 1$

36. $(9a + 3a^2) + (a + 4) = (9a + a) + 3a^2 + 4$
$= 10a + 3a^2 + 4$

37. $(y^2 + 6y + 2) + (3y^2 - 8y + 12)$
$= (y^2 + 3y^2) + (6y - 8y) + (2 + 12)$
$= 4y^2 - 2y + 14$

38. $(4x - 7) - (2x + 2) = (4x - 7) + (-2x - 2)$
$= (4x - 2x) + (-7 - 2)$
$= 2x - 9$

39. $(9x + 8y) - (x - 3y) = (9x + 8y) + (-x + 3y)$
$= (9x - x) + (8y + 3y)$
$= 8x + 11y$

40. $(13n^2 + 6n + 5) - (6n^2 + 5)$
$= (13n^2 + 6n + 5) + (-6n^2 - 5)$
$= (13n^2 - 6n^2) + 6n + (5 - 5)$
$= 7n^2 + 6n$

41. Sample answer: The scales are labeled inconsistently or the bars on a bar graph are different widths.

42. translation

43. reflection

44.

x	$4x$	(x, y)
0	0	(0, 0)
1	4	(1, 4)
2	8	(2, 8)
3	12	(3, 12)

45.

x	$2x^2 - 3$	(x, y)
0	-3	$(0, -3)$
1	-1	$(1, -1)$
2	5	(2, 5)
3	15	(3, 15)

46.

x	$x^3 + 1$	(x, y)
0	1	(0, 1)
1	2	(1, 2)
2	9	(2, 9)
3	28	(3, 28)

13-5 Linear and Nonlinear Functions

Page 687 How can you determine whether a function is linear?

a. $A = \ell w$
$A = (40 - 2x) \cdot x$
$A = x(40 - 2x)$ or $40x - 2x^2$

b. $A = x(40 - 2x)$

If $x = 6$, then $A = 6 \cdot (40 - 2 \cdot 6)$.
$= 6 \cdot (40 - 12)$
$= 6 \cdot 28$
$= 168$

If $x = 8$, then $A = 8 \cdot (40 - 2 \cdot 8)$.
$= 8 \cdot (40 - 16)$
$= 8 \cdot 24$
$= 192$

If $x = 10$, then $A = 10 \cdot (40 - 2 \cdot 10)$.
$= 10 \cdot (40 - 20)$
$= 10 \cdot 20$
$= 200$

If $x = 12$, then $A = 12 \cdot (40 - 2 \cdot 12)$.
$= 12 \cdot (40 - 24)$
$= 12 \cdot 16$
$= 192$

If $x = 14$, then $A = 14 \cdot (40 - 2 \cdot 14)$.
$= 14 \cdot (40 - 28)$
$= 14 \cdot 12$
$= 168$

The areas of the decks are 168 ft^2, 192 ft^2, 200 ft^2, 192 ft^2, and 168 ft^2.

c.

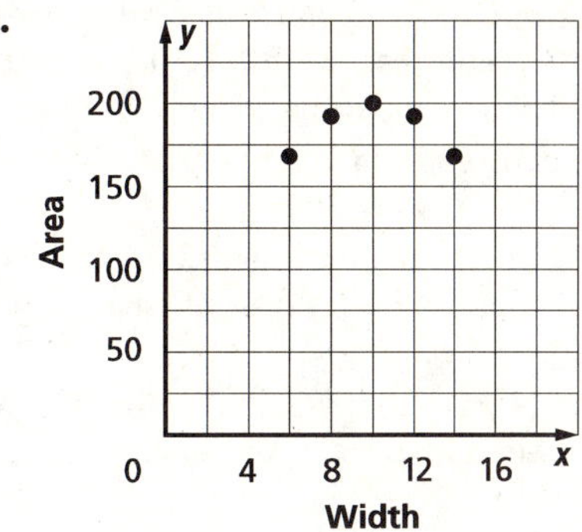

No; the connected points fall along a curve.

Page 689 Check for Understanding

1. Sample answer: Determine whether an equation can be written in the form $y = mx + b$ or look for a constant rate of change in a table of values.
2. Exponential; for the long term, growth of profits would occur at a faster rate.
3. Sample answer: population growth
4. Linear; graph is a straight line.
5. Nonlinear; graph is a curve.
6. Linear; equation can be written as $y = \frac{1}{5}x + 0$.
7. Nonlinear; equation cannot be written as $y = mx + b$.
8. Nonlinear; rate of change is not constant, since as x increases by 2, y changes by different amounts each time.
9. Linear; rate of change is constant, since as x increases by 1, y increases by 3.
10. B; equation cannot be written as $y = mx + b$.

Pages 690–691 Practice and Apply

11. Nonlinear; graph is a curve.
12. Linear; graph is a straight line.
13. Linear; graph is a straight line.
14. Nonlinear; graph is a curve.
15. Nonlinear; graph is a curve.
16. Nonlinear; graph is a curve.
17. Linear; equation can be written as $y = 0.9x + 0$.
18. Nonlinear; equation cannot be written as $y = mx + b$.
19. Linear; equation can be written as $y = \frac{3}{4}x + 0$.
20. Linear; equation can be written as $y = -\frac{2}{3}x + 4$.
21. Nonlinear; equation cannot be written as $y = mx + b$.
22. Nonlinear; equation cannot be written as $y = mx + b$.
23. Linear; rate of change is constant, since as x increases by 2, y decreases by 6.
24. Nonlinear; rate of change is not constant, since as x increases by 1, y increases by different amounts each time.

25. Nonlinear; rate of change is not constant, since as x increases by 2, y changes by different amounts each time.

26. Linear; rate of change is constant, since as x increases by 1, y decreases by 2.

27. Nonlinear; the points (year, applications) would lie on a curved line, not on a straight line. Or, the rate of change is not constant.

28. No, a vertical line based on a rule such as $x = 3$ is not a function because there is more than one y value for the x value 3.

29. No, the difference between the years varies, so the change is not constant.

30. A function is linear if it has a constant rate of change. This can be determined from graphs, equations, or tables of values. Answers should include the following.
    - Functions can be represented using graphs, equations, or tables.
    - A graph that is a straight line represents a linear function. An equation that can be written in the form $y = mx + b$ is a linear function. If a table of values shows a constant rate of change, the function is linear.

31. A; equation can be written in the form $y = mx + b$ as $y = \frac{1}{2}x + 0$.

32. C; equation cannot be written as $y = mx + b$.

Page 691 Maintain Your Skills

33. $t(4 + 9t) = t(4) + t(9t)$
$= 4t + 9t^2$

34. $5n(-1 + 3n) = 5n(-1) + 5n(3n)$
$= -5n + 15n^2$

35. $(a - 2b)ab = a(ab) - 2b(ab)$
$= a^2b - 2ab^2$

36. $(2x + 7) - (x - 1) = (2x + 7) + (-x + 1)$
$= (2x - x) + (7 + 1)$
$= x + 8$

37. $(4x + y) - (5x + y) = (4x + y) + (-5x - y)$
$= (4x - 5x) + (y - y)$
$= -x$

38. $(6a - a^2) - (8a + 3) = (6a - a^2) + (-8a - 3)$
$= (6a - 8a) - a^2 - 3$
$= -2a - a^2 - 3$

39. $65° < 90°$, so the angle is acute.

40. Sample answer:

x	$-x$	(x, y)
−2	$-(-2) = 2$	(−2, 2)
0	$-(0) = 0$	(0, 0)
2	$-(2) = -2$	(2, −2)

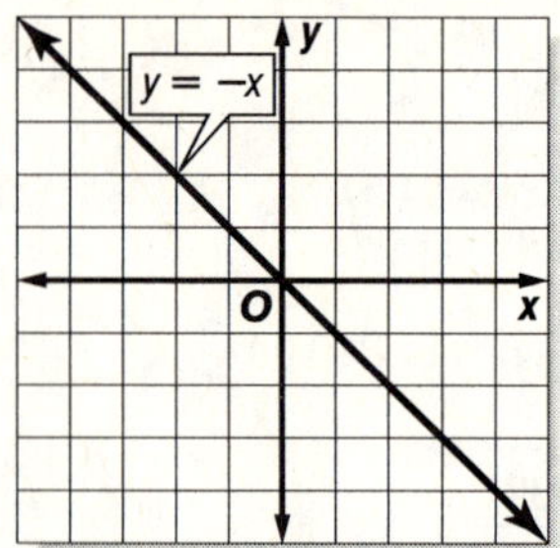

41. Sample answer:

x	$x - 4$	(x, y)
−2	$-2 - 4 = -6$	(−2, −6)
0	$0 - 4 = -4$	(0, −4)
2	$2 - 4 = -2$	(2, −2)

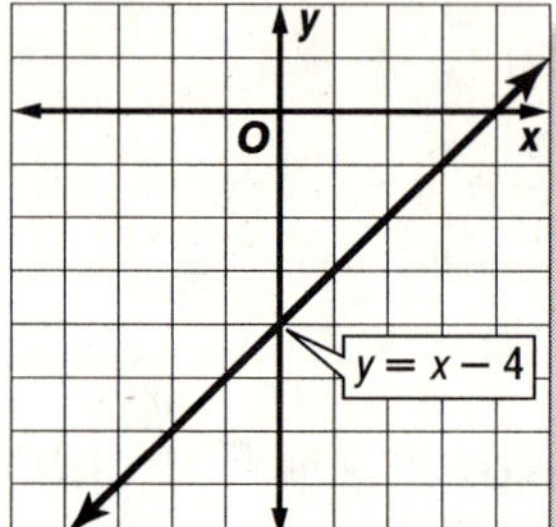

42. Sample answer:

x	$2x + 2$	(x, y)
−2	$2(-2) + 2 = -2$	(−2, −2)
−1	$2(-1) + 2 = 0$	(−1, 0)
0	$2(0) + 2 = 2$	(0, 2)

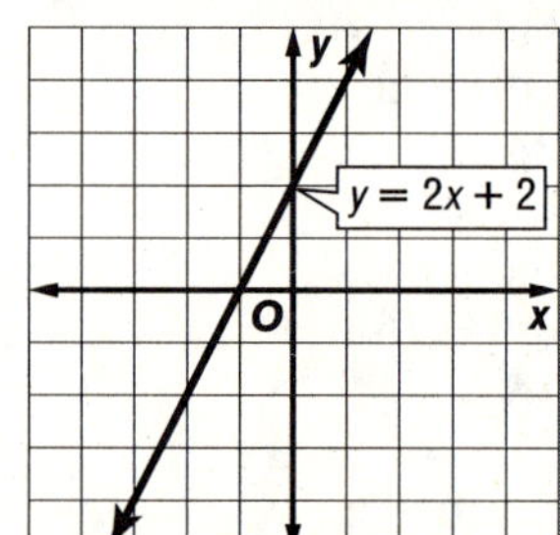

43. Sample answer:

x	$-\frac{1}{2}x + 3$	(x, y)
−2	$-\frac{1}{2}(-2) + 3 = 4$	(−2, 4)
0	$-\frac{1}{2}(0) + 3 = 3$	(0, 3)
2	$-\frac{1}{2}(2) + 3 = 2$	(2, 2)

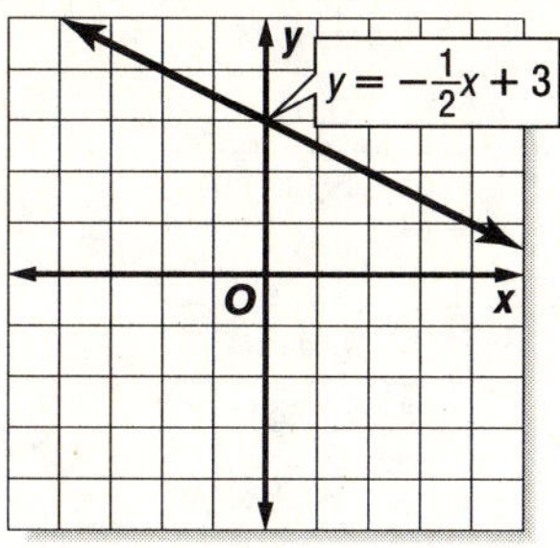

Page 691 Practice Quiz 2

1. $c(2c^2 - 8) = c(2c^2) - c(8)$
   $= 2c^3 - 8c$
2. $(4x + 2)3x = 4x(3x) + 2(3x)$
   $= 12x^2 + 6x$
3. $a^2(5 + a + 2a^2) = a^2(5) + a^2(a) + a^2(2a^2)$
   $= 5a^2 + a^3 + 2a^4$
4. Linear; equation can be written as $y = 9x + 0$.
5. Nonlinear; equation cannot be written as $y = mx + b$.

13-6 Graphing Quadratic and Cubic Functions

Page 692 How are functions, formulas, tables, and graphs related?

a. V = a^3

b.

a	a^3	(a, V)
0.0	0.0	(0.0, 0.0)
0.5	0.125	(0.5, 0.125)
1.0	1.0	(1.0, 1.0)
1.5	3.375	(1.5, 3.375)
2.0	8.0	(2.0, 8.0)

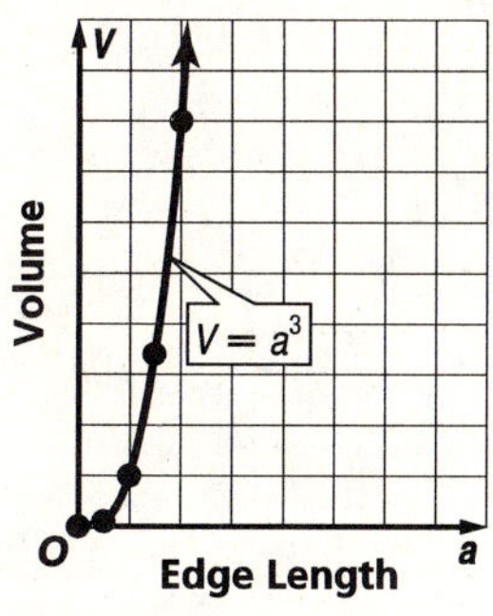

Page 694 Check for Understanding

1. Sample answer: The graph of $y = nx^2$ has line symmetry and the graph of $y = nx^3$ does not.
2. Sample answer: A function is quadratic if its equation contains one independent variable whose greatest exponent is 2.
3. Sample answer: $y = x^2 + 3$; make a table of values and plot the points.

4.

x	$y = x^2$	(x, y)
−2	$(-2)^2 = 4$	(−2, 4)
−1	$(-1)^2 = 1$	(−1, 1)
0	$(0)^2 = 0$	(0, 0)
1	$(1)^2 = 1$	(1, 1)
2	$(2)^2 = 4$	(2, 4)

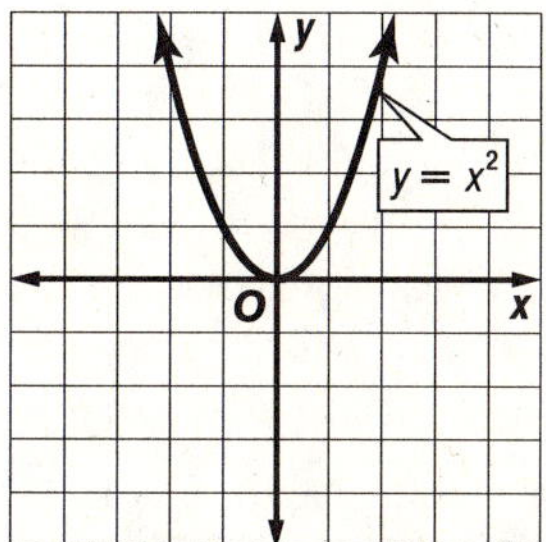

5.

x	$y = -2x^2$	(x, y)
−2	$-2(-2)^2 = -8$	(−2, −8)
−1	$-2(-1)^2 = -2$	(−1, −2)
0	$-2(0)^2 = 0$	(0, 0)
1	$-2(1)^2 = -2$	(1, −2)
2	$-2(2)^2 = -8$	(2, −8)

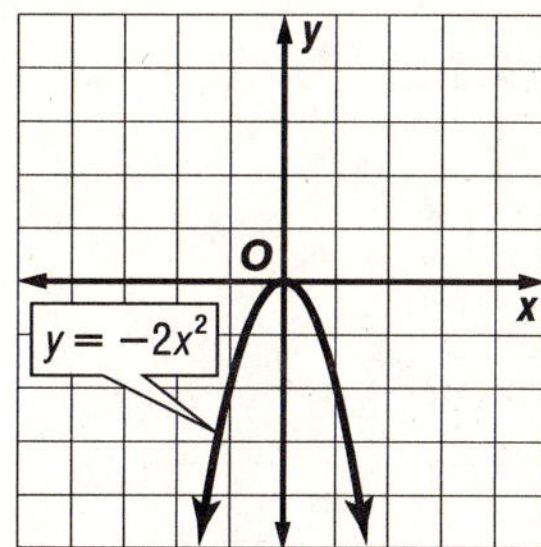

6.

x	$y = x^2 + 1$	(x, y)
−2	$(-2)^2 + 1 = 5$	(−2, 5)
−1	$(-1)^2 + 1 = 2$	(−1, 2)
0	$(0)^2 + 1 = 1$	(0, 1)
1	$(1)^2 + 1 = 2$	(1, 2)
2	$(2)^2 + 1 = 5$	(2, 5)

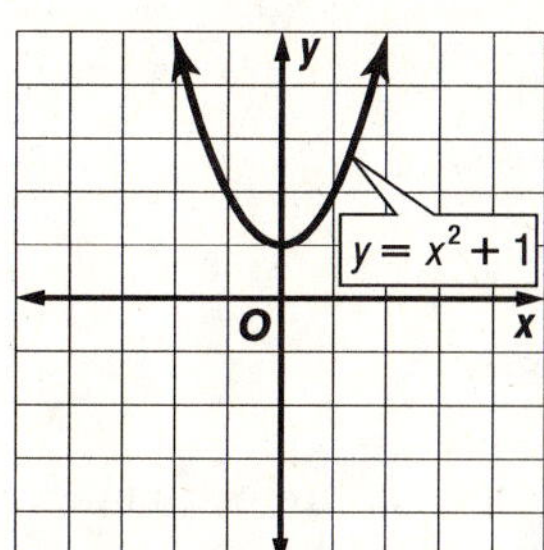

7.

x	$y = -x^3$	(x, y)
-2	$-(-2)^3 = 8$	$(-2, 8)$
-1	$-(-1)^3 = 1$	$(-1, 1)$
0	$-(0)^3 = 0$	$(0, 0)$
1	$-(1)^3 = -1$	$(1, -1)$
2	$-(2)^3 = -8$	$(2, -8)$

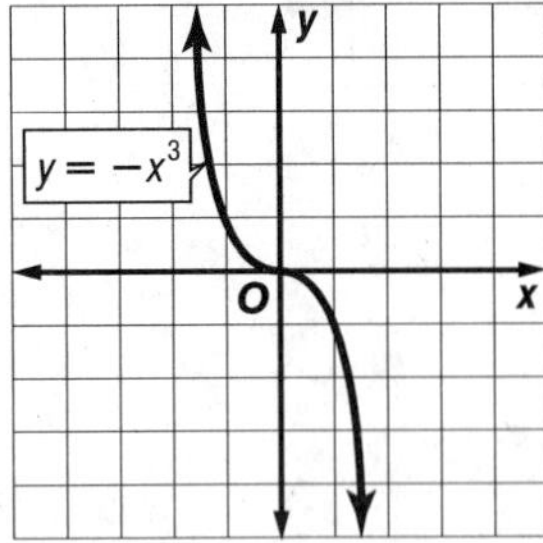

8.

x	$y = 0.5x^3$	(x, y)
-2	$0.5(-2)^3 = -4$	$(-2, -4)$
-1	$0.5(-1)^3 = -0.5$	$(-1, -0.5)$
0	$0.5(0)^3 = 0$	$(0, 0)$
1	$0.5(1)^3 = 0.5$	$(1, 0.5)$
2	$0.5(2)^3 = 4$	$(2, 4)$

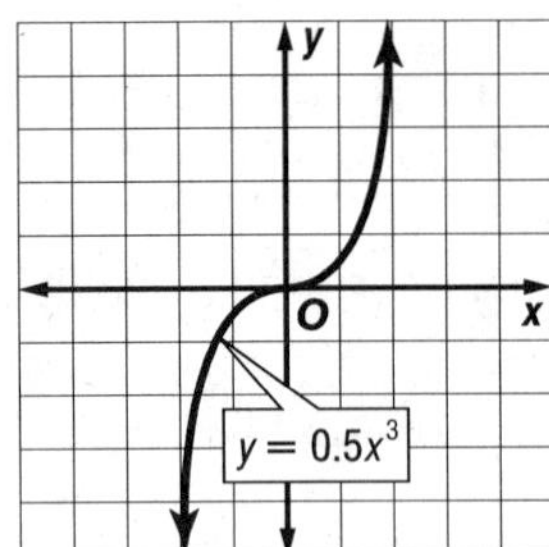

9.

x	$y = x^3 - 2$	(x, y)
-2	$(-2)^3 - 2 = -10$	$(-2, -10)$
-1	$(-1)^3 - 2 = -3$	$(-1, -3)$
0	$(0)^3 - 2 = -2$	$(0, -2)$
1	$(1)^3 - 2 = -1$	$(1, -1)$
2	$(2)^3 - 2 = 6$	$(2, 6)$

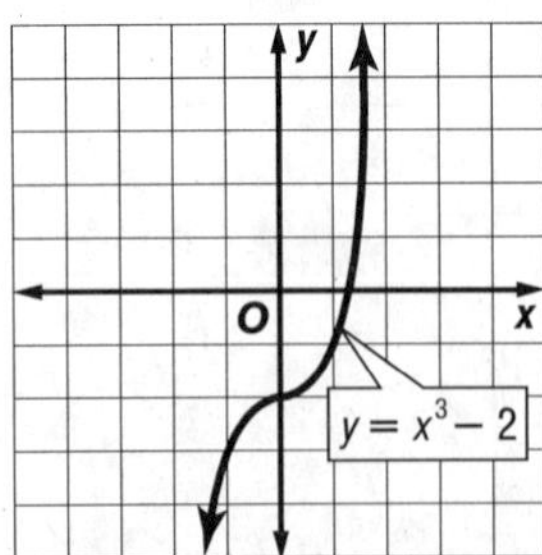

10a. Each side of the cube is a square, with side length a units. The area of each side is a^2. A cube has 6 sides, so the surface area of the cube is $S = 6a^2$.

10b.

a	$S = 6a^2$	(a, S)
0	$6(0)^2 = 0$	$(0, 0)$
0.5	$6(0.5)^2 = 1.5$	$(0.5, 1.5)$
1	$6(1)^2 = 6$	$(1, 6)$
1.5	$6(1.5)^2 = 13.5$	$(1.5, 13.5)$
2	$6(2)^2 = 24$	$(2, 24)$

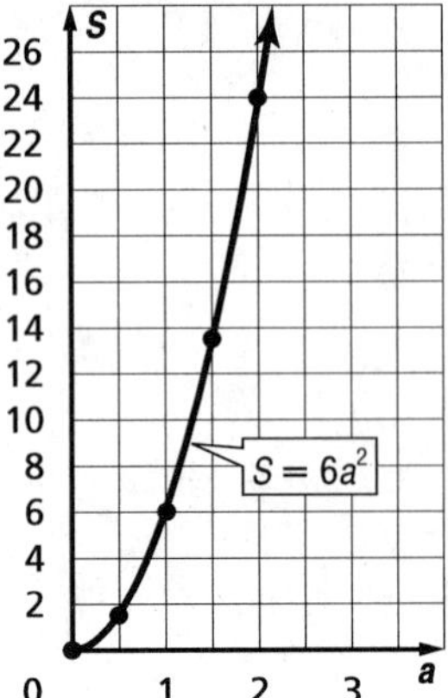

Pages 694–696 Practice and Apply

11.

x	$y = 3x^2$	(x, y)
-2	$3(-2)^2 = 12$	$(-2, 12)$
-1	$3(-1)^2 = 3$	$(-1, 3)$
0	$3(0)^2 = 0$	$(0, 0)$
1	$3(1)^2 = 3$	$(1, 3)$
2	$3(2)^2 = 12$	$(2, 12)$

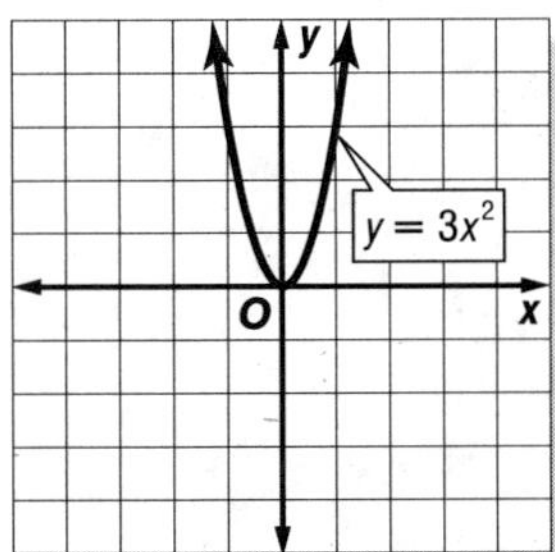

12.

x	$y = 0.5x^2$	(x, y)
-2	$0.5(-2)^2 = 2$	$(-2, 2)$
-1	$0.5(-1)^2 = 0.5$	$(-1, 0.5)$
0	$0.5(0)^2 = 0$	$(0, 0)$
1	$0.5(1)^2 = 0.5$	$(1, 0.5)$
2	$0.5(2)^2 = 2$	$(2, 2)$

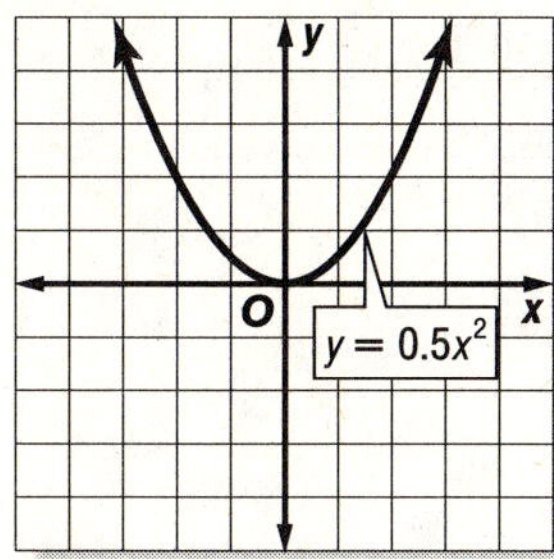

13.

x	$y = -x^2$	(x, y)
-2	$-(-2)^2 = -4$	$(-2, -4)$
-1	$-(-1)^2 = -1$	$(-1, -1)$
0	$-(0)^2 = 0$	$(0, 0)$
1	$-(1)^2 = -1$	$(1, -1)$
2	$-(2)^2 = -4$	$(2, -4)$

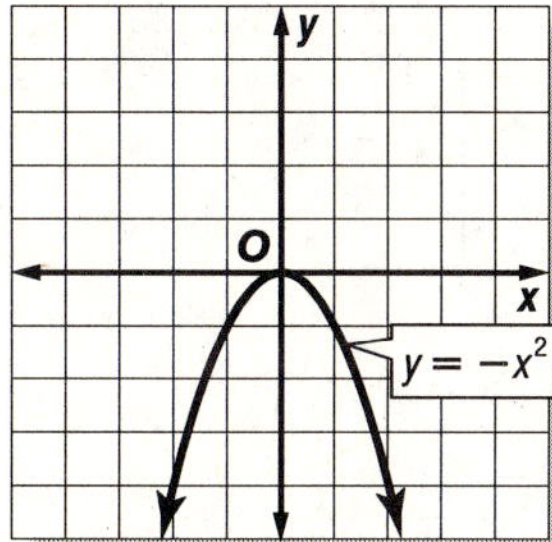

14.

x	$y = 3x^3$	(x, y)
-2	$3(-2)^3 = -24$	$(-2, -24)$
-1	$3(-1)^3 = -3$	$(-1, -3)$
0	$3(0)^3 = 0$	$(0, 0)$
1	$3(1)^3 = 3$	$(1, 3)$
2	$3(2)^3 = 24$	$(2, 24)$

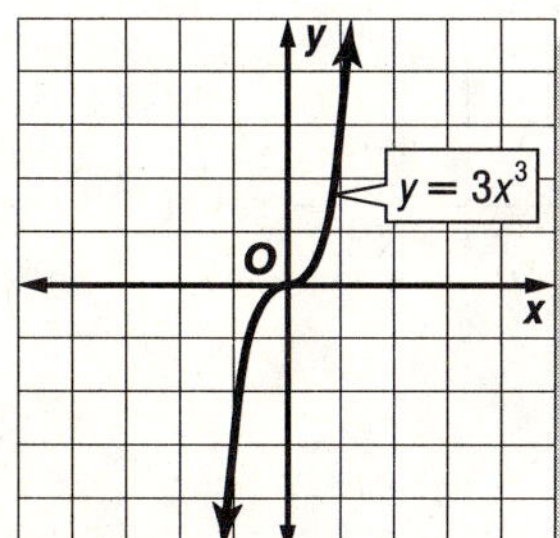

15.

x	$y = -2x^3$	(x, y)
-2	$-2(-2)^3 = 16$	$(-2, 16)$
-1	$-2(-1)^3 = 2$	$(-1, 2)$
0	$-2(0)^3 = 0$	$(0, 0)$
1	$-2(1)^3 = -2$	$(1, -2)$
2	$-2(2)^3 = -16$	$(2, -16)$

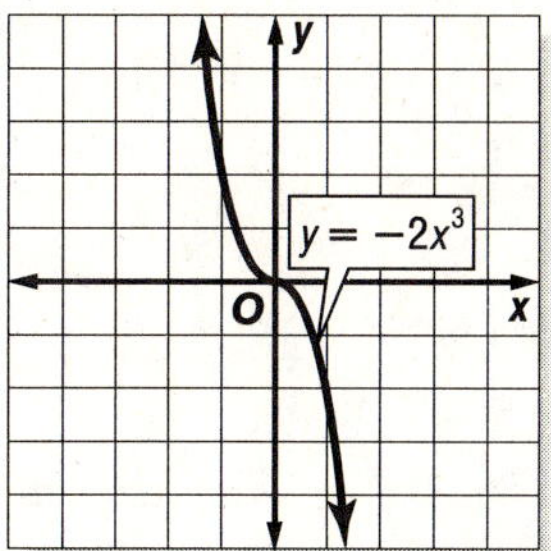

16.

x	$y = -0.5x^2$	(x, y)
-2	$-0.5(-2)^2 = -2$	$(-2, -2)$
-1	$-0.5(-1)^2 = -0.5$	$(-1, -0.5)$
0	$-0.5(0)^2 = 0$	$(0, 0)$
1	$-0.5(1)^2 = -0.5$	$(1, -0.5)$
2	$-0.5(2)^2 = -2$	$(2, -2)$

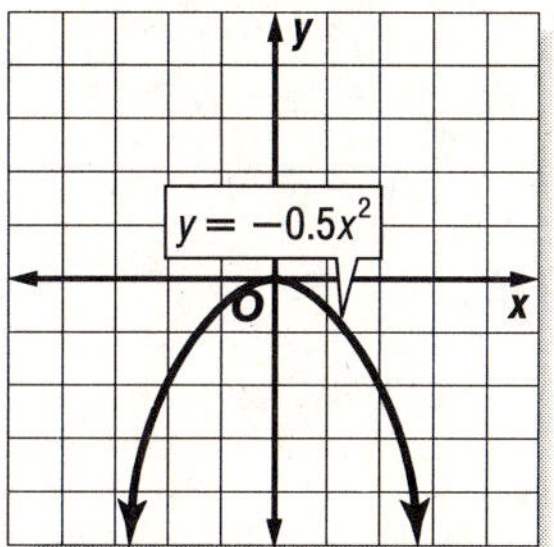

17.

x	$y = 2x^3$	(x, y)
-2	$2(-2)^3 = -16$	$(-2, -16)$
-1	$2(-1)^3 = -2$	$(-1, -2)$
0	$2(0)^3 = 0$	$(0, 0)$
1	$2(1)^3 = 2$	$(1, 2)$
2	$2(2)^3 = 16$	$(2, 16)$

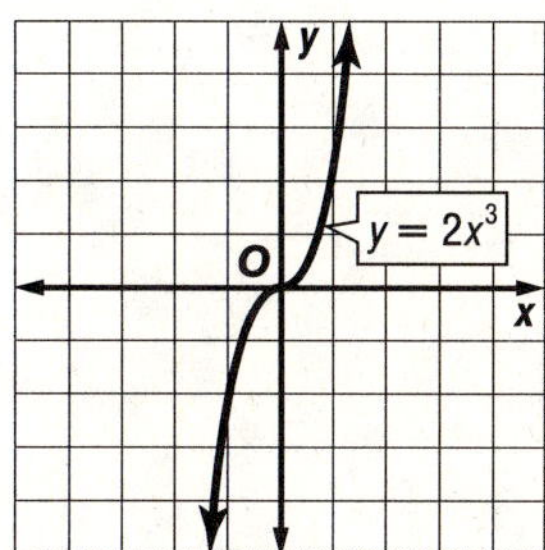

18.

x	$y = 0.1x^3$	(x, y)
-2	$0.1(-2)^3 = -0.8$	$(-2, -0.8)$
-1	$0.1(-1)^3 = -0.1$	$(-1, -0.1)$
0	$0.1(0)^3 = 0$	$(0, 0)$
1	$0.1(1)^3 = 0.1$	$(1, 0.1)$
2	$0.1(2)^3 = 0.8$	$(2, 0.8)$

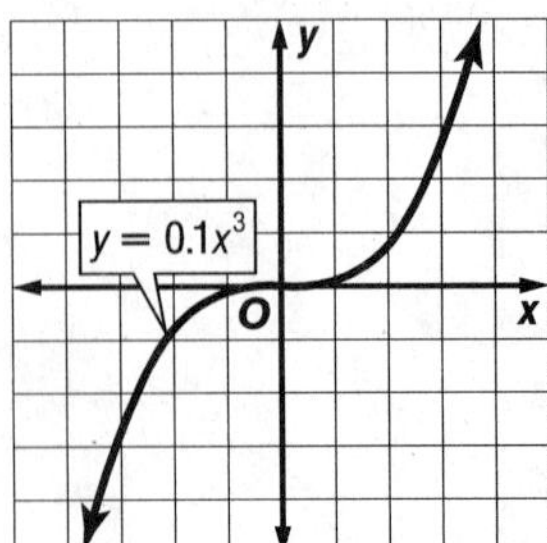

19.

x	$y = x^3 + 1$	(x, y)
-2	$(-2)^3 + 1 = -7$	$(-2, -7)$
-1	$(-1)^3 + 1 = 0$	$(-1, 0)$
0	$(0)^3 + 1 = 1$	$(0, 1)$
1	$(1)^3 + 1 = 2$	$(1, 2)$
2	$(2)^3 + 1 = 9$	$(2, 9)$

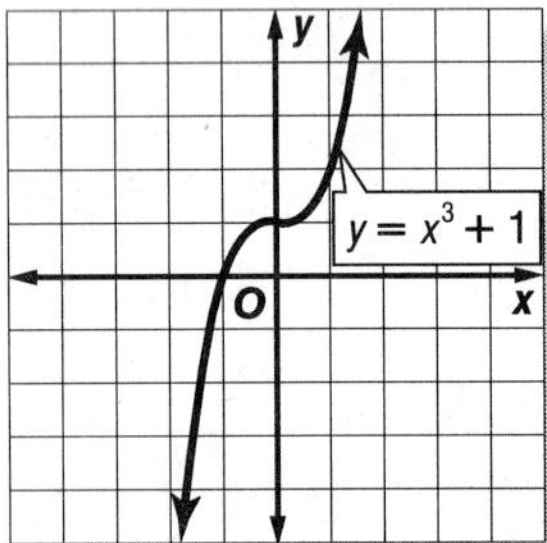

20.

x	$y = x^2 - 3$	(x, y)
-2	$(-2)^2 - 3 = 1$	$(-2, 1)$
-1	$(-1)^2 - 3 = -2$	$(-1, -2)$
0	$(0)^2 - 3 = -3$	$(0, -3)$
1	$(1)^2 - 3 = -2$	$(1, -2)$
2	$(2)^2 - 3 = 1$	$(2, 1)$

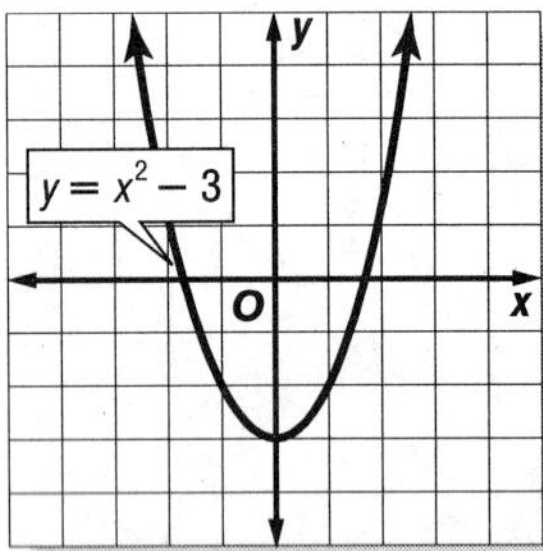

21.

x	$y = \frac{1}{2}x^2 + 1$	(x, y)
-2	$\frac{1}{2}(-2)^2 + 1 = 3$	$(-2, 3)$
-1	$\frac{1}{2}(-1)^2 + 1 = 1.5$	$(-1, 1.5)$
0	$\frac{1}{2}(0)^2 + 1 = 1$	$(0, 1)$
1	$\frac{1}{2}(1)^2 + 1 = 1.5$	$(1, 1.5)$
2	$\frac{1}{2}(2)^2 + 1 = 3$	$(2, 3)$

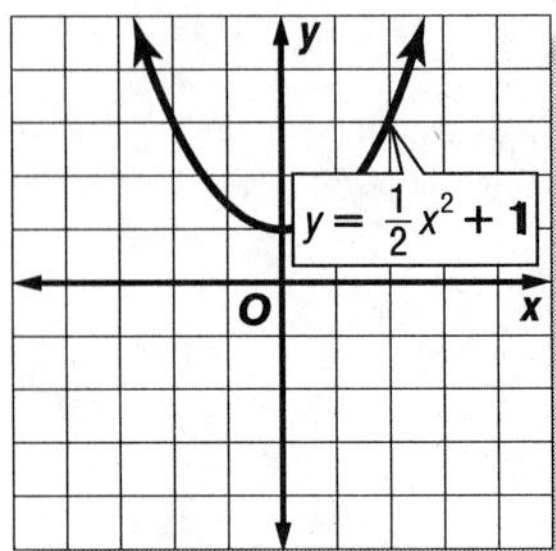

22.

x	$y = \frac{1}{3}x^3 + 2$	(x, y)
-3	$\frac{1}{3}(-3)^3 + 2 = -7$	$(-3, -7)$
-1	$\frac{1}{3}(-1)^3 + 2 \approx 1.7$	$(-1, 1.7)$
0	$\frac{1}{3}(0)^3 + 2 = 2$	$(0, 2)$
1	$\frac{1}{3}(1)^3 + 2 \approx 2.3$	$(1, 2.3)$
3	$\frac{1}{3}(3)^3 + 2 = 11$	$(3, 11)$

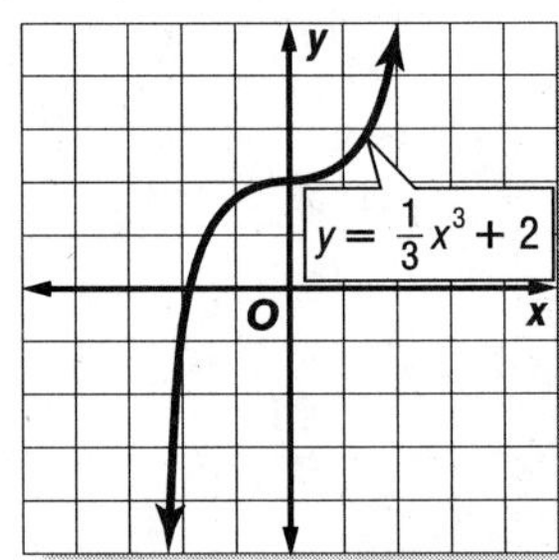

23.

x	$y = x^2 - 4$	(x, y)
-2	$(-2)^2 - 4 = 0$	$(-2, 0)$
-1	$(-1)^2 - 4 = -3$	$(-1, -3)$
0	$(0)^2 - 4 = -4$	$(0, -4)$
1	$(1)^2 - 4 = -3$	$(1, -3)$
2	$(2)^2 - 4 = 0$	$(2, 0)$

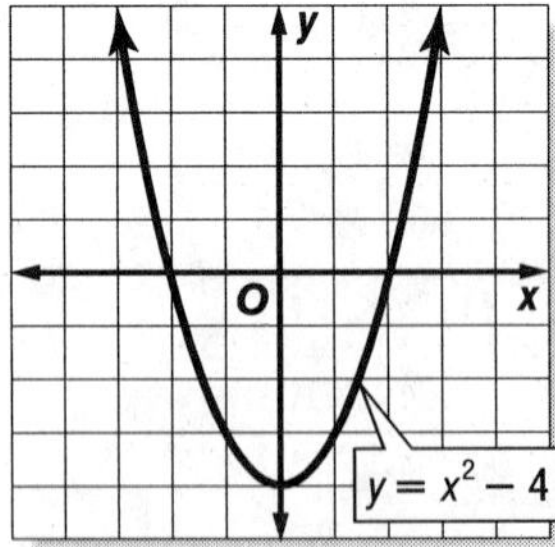

x	$y = -4x^3$	(x, y)
-2	$-4(-2)^3 = 32$	$(-2, 32)$
-1	$-4(-1)^3 = 4$	$(-1, 4)$
0	$-4(0)^3 = 0$	$(0, 0)$
1	$-4(1)^3 = -4$	$(1, -4)$
2	$-4(2)^3 = -32$	$(2, -32)$

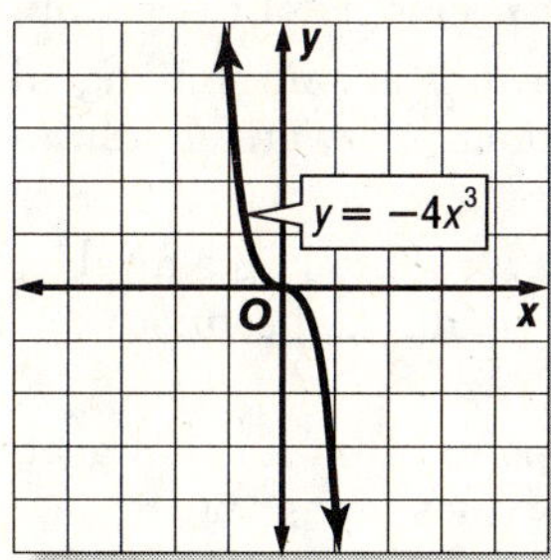

Both equations are functions because every value of x is paired with a unique value of y.

24.

x	$y = x^2$	(x, y)
0	$(0)^2 = 0$	(0, 0)
1	$(1)^2 = 1$	(1, 1)
2	$(2)^2 = 4$	(2, 4)

x	$y = x^3$	(x, y)
0	$(0)^3 = 0$	(0, 0)
1	$(1)^3 = 1$	(1, 1)
2	$(2)^3 = 8$	(2, 8)

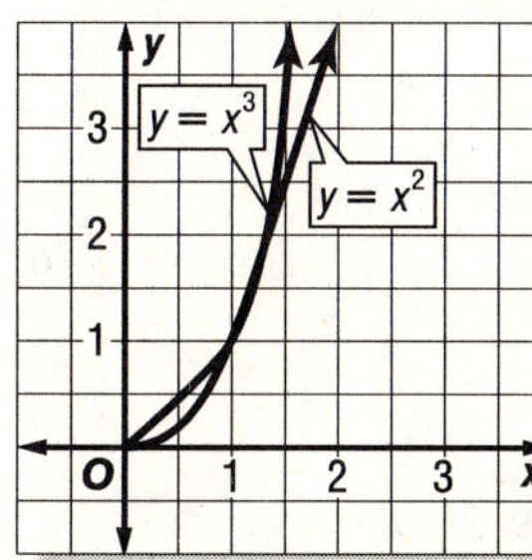

For $x > 1$, $y = x^3$ shows faster growth; the graph has a steeper upward slope.

25.

x	$y = -x^2 + 7$	(x, y)
−2	$-(-2)^2 + 7 = 3$	(−2, 3)
−1	$-(-1)^2 + 7 = 6$	(−1, 6)
0	$-(0)^2 + 7 = 7$	(0, 7)
1	$-(1)^2 + 7 = 6$	(1, 6)
2	$-(2)^2 + 7 = 3$	(2, 3)

The maximum point is (0, 7).

26.

x	$y = x^2 - 6$	(x, y)
−2	$(-2)^2 - 6 = -2$	(−2, −2)
−1	$(-1)^2 - 6 = -5$	(−1, −5)
0	$(0)^2 - 6 = -6$	(0, −6)
1	$(1)^2 - 6 = -5$	(1, −5)
2	$(2)^2 - 6 = -2$	(2, −2)

The minimum point is (0, −6).

27. Similar shape; $y = 3x^2$ is more narrow.

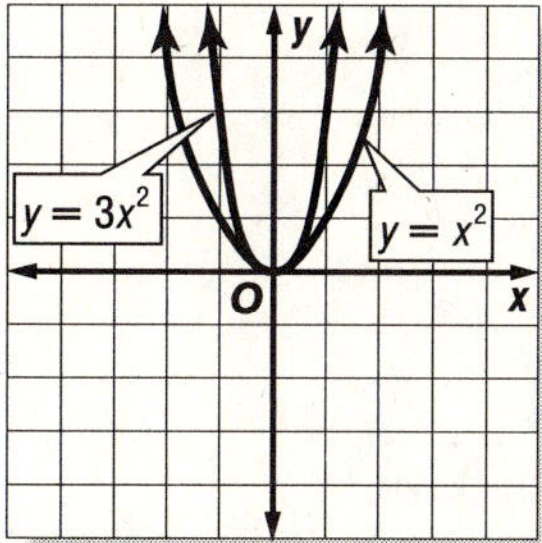

28. Similar shape; $y = 2x^3$ is more narrow.

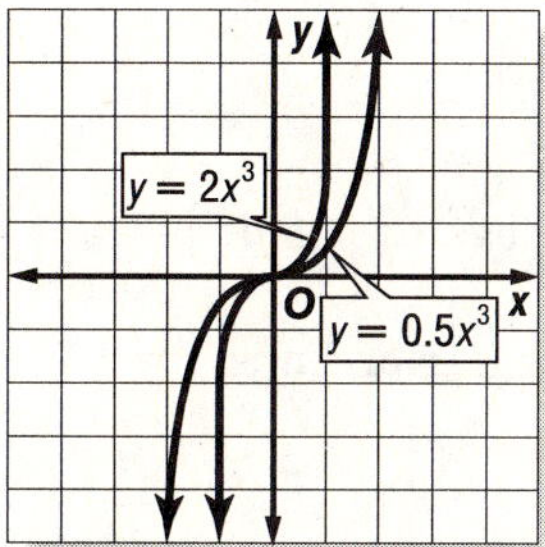

29. Same shape; $y = -2x^2$ is $y = 2x^2$ reflected over the x-axis.

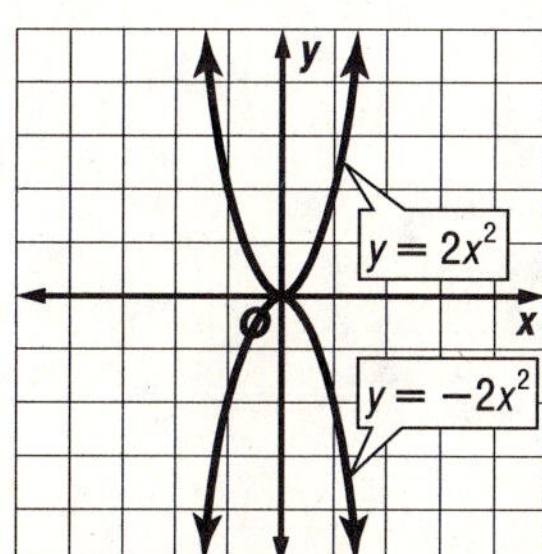

30. Same shape; $y = x^3 - 3$ is translated 3 units down from $y = x^3$.

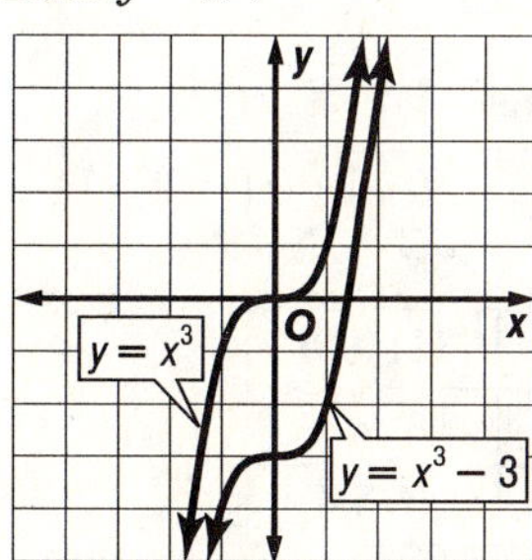

31. $A = \ell \cdot w$
$A = (50 - x)x$
$A = 50x - x^2$

32.

x	$y = 50x - x^2$	(x, A)
0	$50(0) - 0^2 = 0$	(0, 0)
10	$50(10) - 10^2 = 400$	(10, 400)
20	$50(20) - 20^2 = 600$	(20, 600)
30	$50(30) - 30^2 = 600$	(30, 600)
40	$50(40) - 40^2 = 400$	(40, 400)
50	$50(50) - 50^2 = 0$	(50, 0)

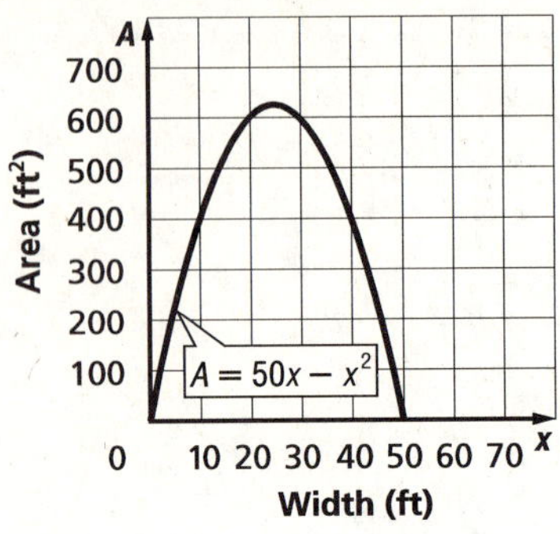

33. The greatest area occurs when x is 25. Find the length when x is 25:

length $= 50 - x$
$= 50 - 25$ or 25

The dimensions of the dog pen should be 25 ft by 25 ft.

34. The volume of a rectangular prism is $V = Bh$ or $V = \ell \cdot w \cdot h$.

$V = \ell \cdot w \cdot h$
$V = s \cdot s \cdot 2$
$V = 2s^2$

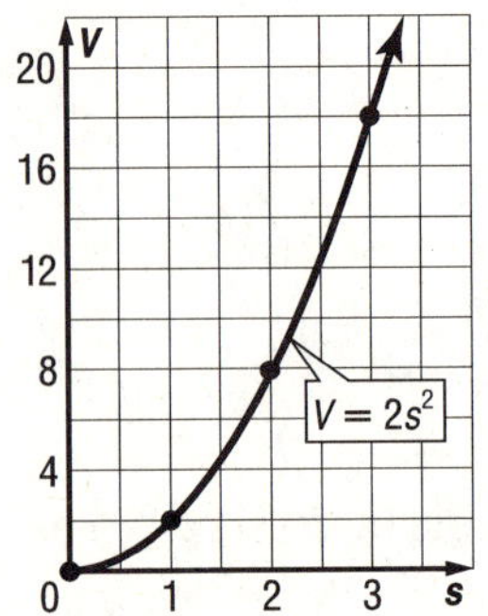

35. The volume of a cylinder is $V = \pi r^2 h$.

$V = \pi r^2 h$
$V = \pi \cdot r^2 \cdot (0.2)$
$V = 0.2\pi r^2$ $\quad V \approx 0.6r^2$

r	$V = 0.6r^2$	(r, V)
0	$0.6(0)^2 = 0$	(0, 0)
1	$0.6(1)^2 = 0.6$	(1, 0.6)
2	$0.6(2)^2 = 2.4$	(2, 2.4)
3	$0.6(3)^2 = 5.4$	(3, 5.4)
4	$0.6(4)^2 = 9.6$	(4, 9.6)

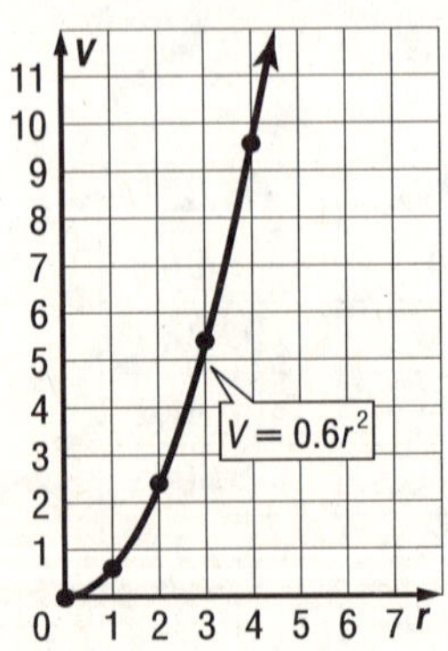

36. The solutions are the x-intercepts of the graph.

37. Formulas, tables, and graphs are interchangeable ways to represent functions. Answers should include the following.

- To make a graph, use a rule to make a table of values. Then plot the points and connect them to make a graph.
- To write a rule, find points that lie on a graph and make a table of values using the coordinates. Look for a pattern and write a rule that describes the pattern.

38. C; Equations A and B represent quadratic functions. The graph is of a cubic function. The ordered pairs for $y = 4x^3$ coincide with the points on the graph.

x	$y = 4x^3$	(x, y)
−1	$4(-1)^3 = -4$	(−1, −4)
0	$4(0)^3 = 0$	(0, 0)
1	$4(1)^3 = 4$	(1, 4)

39. C; The highest point on the graph occurs at a cost of $3.

40. Sample answer: 15 units^2

41. Sample answer: 18 units^2

Page 696 Maintain Your Skills

42. Nonlinear; the points on the graph do not lie on a straight line.

43. $(2x - 4)5 = 2x(5) - 4(5)$
$= 10x - 20$

44. $n(n + 6) = n(n) + n(6)$
$= n^2 + 6n$

45. $3y(8 - 7y) = 3y(8) - 3y(7y)$
$= 24y - 21y^2$

46. Find the slope m.

$m = \frac{y_2 - y_1}{x_2 - x_1}$

$m = \frac{9 - 6}{0 - 3}$

$m = \frac{3}{-3}$ or -1

Find the y-intercept b. Use the slope and the coordinates of either point.

$y = mx + b$
$9 = -1(0) + b$
$9 = b$

Substitute the slope and y-intercept.

$y = mx + b$
$y = -1x + 9$
$y = -x + 9$

47. Find the slope m.

$m = \frac{y_2 - y_1}{x_2 - x_1}$

$m = \frac{-7 - 5}{-1 - 2}$

$m = \frac{-12}{-3}$ or 4

Find the y-intercept b. Use the slope and the coordinate of either point.

$y = mx + b$

$5 = 4(2) + b$

$-3 = b$

Substitute the slope and y-intercept.

$y = mx + b$

$y = 4x + (-3)$

$y = 4x - 3$

48. Find the slope m.

$m = \frac{y_2 - y_1}{x_2 - x_1}$

$m = \frac{6 - (-3)}{8 - (-4)}$

$m = \frac{9}{12}$ or $\frac{3}{4}$

Find the y-intercept b. Use the slope and the coordinate of either point.

$y = mx + b$

$6 = \frac{3}{4}(8) + b$

$0 = b$

Substitute the slope and y-intercept.

$y = mx + b$

$y = \frac{3}{4}x + 0$

$y = \frac{3}{4}x$

Page 697 Graphing Calculator Investigation (Follow-Up of Lesson 13-6)

1.

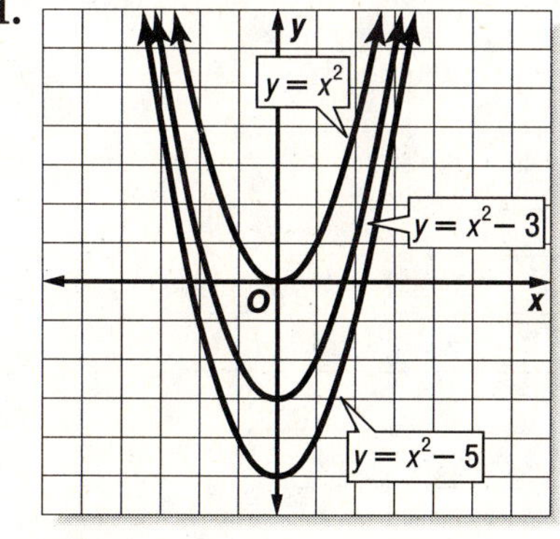

All the parabolas have the same shape. The graph of $y = x^2 - 5$ is shifted 5 units down from the graph of $y = x^2$. The graph of $y = x^2 - 3$ is shifted 3 units down from the graph of $y = x^2$.

2. It shifts the graph vertically c units.

3. $y = x^2 - 8, y = x^2 + 1, y = x^2 + 6$

4.

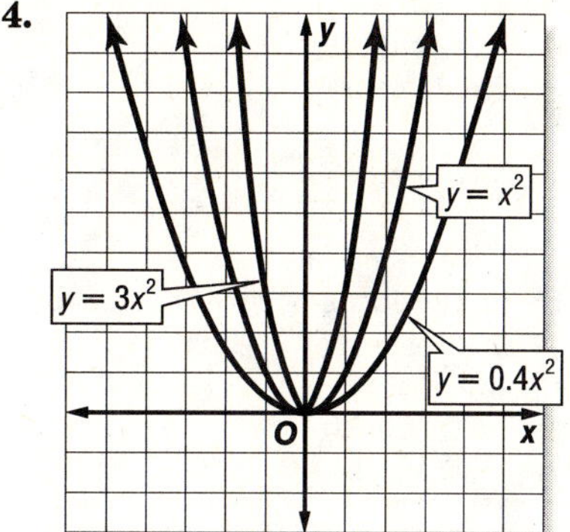

As the coefficient of x^2 increases, the parabola is narrower.

Chapter 13 Study Guide and Review

Page 698 Vocabulary and Concept Check

1. trinomial
2. power
3. cubic
4. quadratic
5. like terms
6. cubic
7. curve
8. Distributive

Pages 698–700 Lesson-by-Lesson Review

9. Yes; it has two terms, so it is a binomial.
10. Yes; it has one term, so it is a monomial.
11. Yes; it has one term, so it is a monomial.
12. No; there is a variable in the denominator of $\frac{6}{a}$.
13. Yes; it has three terms, so it is a trinomial.
14. Yes; it has two terms, so it is a binomial.
15. No; there is a variable under a radical sign.
16. Yes; it has three terms, so it is a trinomial.
17. The variable x has degree 1, so the degree of $2x$ is 1.
18. x has degree 1 and y has degree 1, so the degree of $5xy$ is 1 + 1 or 2.
19. a^2 has degree 2 and b has degree 1, so the degree of $3a^2b$ is 2 + 1 or 3.

20.

Term	Degree
n^2	2
-4	0

The greatest degree is 2. So, the degree of $n^2 - 4$ is 2.

21.

Term	Degree
x^6	6
y^6	6

The greatest degree is 6. So, the degree of $x^6 + y^6$ is 6.

22.

Term	Degree
$2xy$	2
$6yz^2$	3

The greatest degree is 3. So, the degree of $2xy + 6yz^2$ is 3.

23.

Term	Degree
$2x^5$	5
$9x$	1
1	0

The greatest degree is 5. So, the degree of $2x^5 + 9x + 1$ is 5.

24.

Term	Degree
x^2	2
xy^2	3
$-y^4$	4

The greatest degree is 4. So, the degree of $x^2 + xy^2 - y^4$ is 4.

25. $$\begin{array}{rr} & 3b + 8 \\ (+) & \underline{5b - 5} \\ & 8b + 3 \end{array}$$

26. $$\begin{array}{rr} & 2x^2 + 3x - 4 \\ (+) & \underline{6x^2 - x + 5} \\ & 8x^2 + 2x + 1 \end{array}$$

27. $$\begin{array}{rr} & 4y^2 + 2y + 3 \\ (+) & \underline{y^2 - 7} \\ & 5y^2 + 2y - 4 \end{array}$$

28. $(9m - 3n) + (10m + 4n)$
$= (9m + 10m) + (-3n + 4n)$
$= 19m + n$

29. $(-3y^2 + 2) + (4y^2 - y - 3)$
$= (-3y^2 + 4y^2) - y + (2 - 3)$
$= y^2 - y - 1$

30. The additive inverse of $3a^2 - 10$ is $-3a^2 + 10$.

$$\begin{array}{rr} & a^2 + 15 \\ (-) & \underline{3a^2 - 10} \end{array} \rightarrow \begin{array}{rr} & a^2 + 15 \\ (+) & \underline{-3a^2 + 10} \\ & -2a^2 + 25 \end{array}$$

31. The additive inverse of $x^2 + 2x - 4$ is $-x^2 - 2x + 4$.

$$\begin{array}{rr} & 4x^2 - 2x + 3 \\ (-) & \underline{x^2 + 2x - 4} \end{array} \rightarrow \begin{array}{rr} & 4x^2 - 2x + 3 \\ (+) & \underline{-x^2 - 2x + 4} \\ & 3x^2 - 4x + 7 \end{array}$$

32. The additive inverse of $2y^2 + 6$ is $-2y^2 - 6$.

$$\begin{array}{rr} & 18y^2 + 3y - 1 \\ (-) & \underline{2y^2 + 6} \end{array} \rightarrow \begin{array}{rr} & 18y^2 + 3y - 1 \\ (+) & \underline{-2y^2 - 6} \\ & 16y^2 + 3y - 7 \end{array}$$

33. $(x + 8) - (2x + 7) = (x + 8) + (-2x - 7)$
$= (x - 2x) + (8 - 7)$
$= -x + 1$

34. $(3n^2 + 7) - (n^2 - n + 4)$
$= (3n^2 + 7) + (-n^2 + n - 4)$
$= (3n^2 - n^2) + n + (7 - 4)$
$= 2n^2 + n + 3$

35. $5(4t - 2) = 5(4t) - 5(2)$
$= 20t - 10$

36. $(2x + 3y)7 = 2x(7) + 3y(7)$
$= 14x + 21y$

37. $k(6k + 3) = k(6k) + k(3)$
$= 6k^2 + 3k$

38. $4d(2d - 5) = 4d(2d) - 4d(5)$
$= 8d^2 - 20d$

39. $-2a(9 - a^2) = -2a(9) + (-2a)(-a^2)$
$= -18a + 2a^3$

40. $6(2x^2 + xy + 3y^2) = 6(2x^2) + 6(xy) + 6(3y^2)$
$= 12x^2 + 6xy + 18y^2$

41. Linear; graph is a straight line.

42. Linear; equation can be written as $y = \frac{1}{2}x + 0$.

43. Nonlinear; rate of change is not constant, since as x increases by 2, y changes by different amounts each time.

44.

x	$y = x^2 + 2$	(x, y)
-2	$(-2)^2 + 2 = 6$	$(-2, 6)$
-1	$(-1)^2 + 2 = 3$	$(-1, 3)$
0	$(0)^2 + 2 = 2$	$(0, 2)$
1	$(1)^2 + 2 = 3$	$(1, 3)$
2	$(2)^2 + 2 = 6$	$(2, 6)$

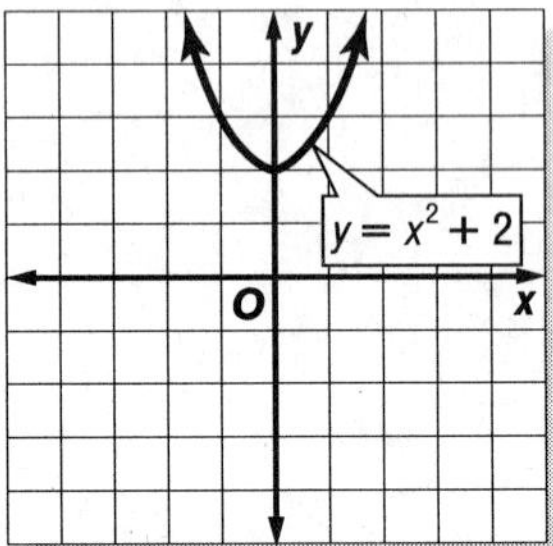

45.

x	$y = -3x^2$	(x, y)
-2	$-3(-2)^2 = -12$	$(-2, -12)$
-1	$-3(-1)^2 = -3$	$(-1, -3)$
0	$-3(0)^2 = 0$	$(0, 0)$
1	$-3(1)^2 = -3$	$(1, -3)$
2	$-3(2)^2 = -12$	$(2, -12)$

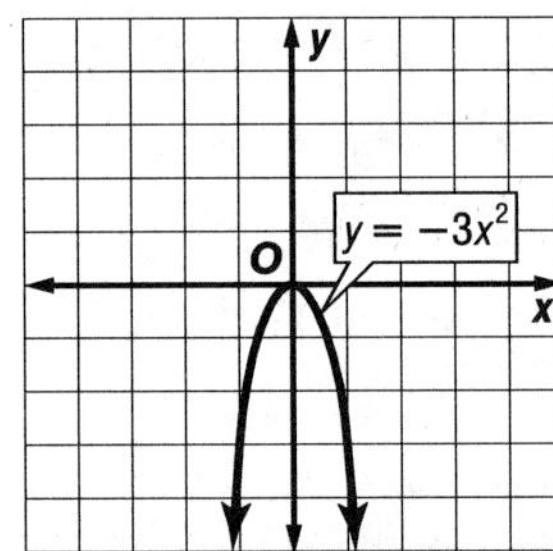

46.

x	$y = x^3 - 2$	(x, y)
-2	$(-2)^3 - 2 = -10$	$(-2, -10)$
-1	$(-1)^3 - 2 = -3$	$(-1, -3)$
0	$(0)^3 - 2 = -2$	$(0, -2)$
1	$(1)^3 - 2 = -1$	$(1, -1)$
2	$(2)^3 - 2 = 6$	$(2, 6)$

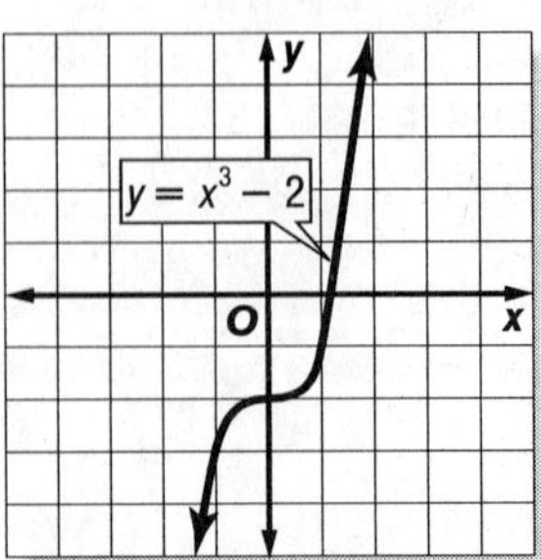

47.

x	$y = x^3 + 1$	(x, y)
−2	$(-2)^3 + 1 = -7$	(−2, −7)
−1	$(-1)^3 + 1 = 0$	(−1, 0)
0	$(0)^3 + 1 = 1$	(0, 1)
1	$(1)^3 + 1 = 2$	(1, 2)
2	$(2)^3 + 1 = 9$	(2, 9)

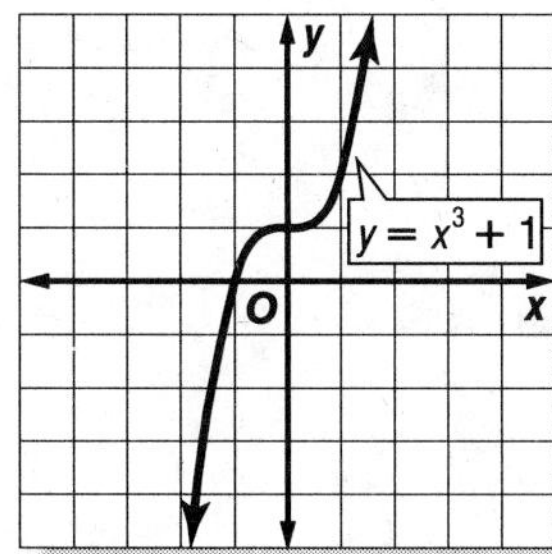

48.

x	$y = -x^3$	(x, y)
−2	$-(-2)^3 = 8$	(−2, 8)
−1	$-(-1)^3 = 1$	(−1, 1)
0	$-(0)^3 = 0$	(0, 0)
1	$-(1)^3 = -1$	(1, −1)
2	$-(2)^3 = -8$	(2, −8)

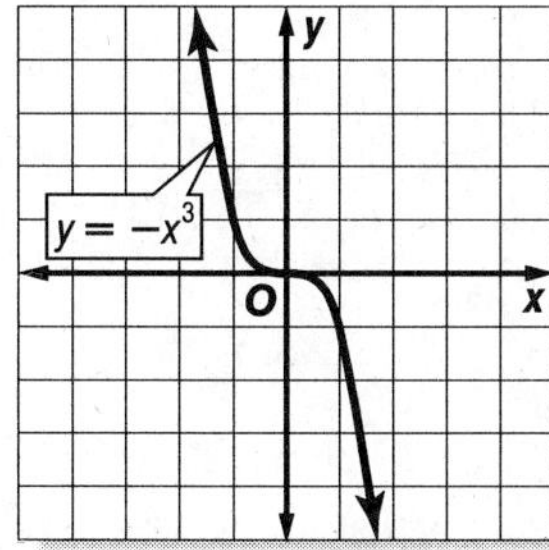

49.

x	$y = 2x^2 + 4$	(x, y)
−2	$2(-2)^2 + 4 = 12$	(−2, 12)
−1	$2(-1)^2 + 4 = 6$	(−1, 6)
0	$2(0)^2 + 4 = 4$	(0, 4)
1	$2(1)^2 + 4 = 6$	(1, 6)
2	$2(2)^2 + 4 = 12$	(2, 12)

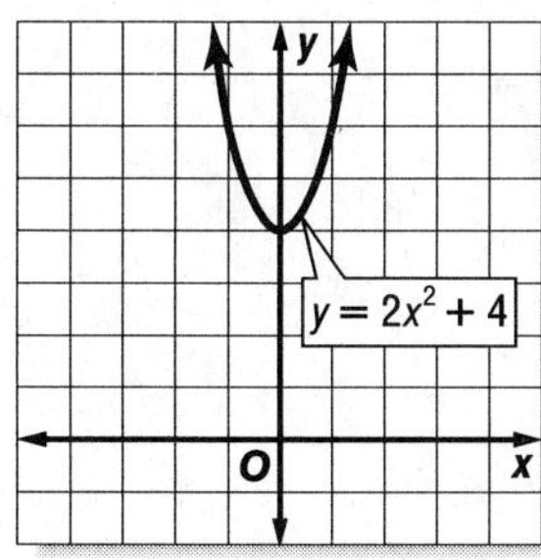

Chapter 13 Practice Test

Page 701

1. an algebraic expression that contains one or more monomials
2. Add the exponents of the variables.
3. See students' drawings; the linear function should be a straight line and the nonlinear function should be a curve.
4. Yes; there are three terms, so it is a trinomial.
5. No; there is a variable in the denominator of $\frac{5}{m}$.
6. Yes; there is one term, so it is a monomial.
7. a has degree 1 and b^3 has degree 3, so the degree of $5ab^3$ is 1 + 3 or 4.
8.

Term	Degree
w^5	5
$-3w^3y^4$	7
1	0

The greatest degree is 7. So, the degree of $w^5 - 3w^3y^4 + 1$ is 7.

9. $(5y + 8) + (-2y + 3) = (5y - 2y) + (8 + 3)$
$= 3y + 11$
10. $(5a - 2b) + (-4a + 5b) = (5a - 4a) + (-2b + 5b)$
$= a + 3b$
11. $(-3m^3 + 5m - 9) + (7m^3 - 2m^2 + 4)$
$= (-3m^3 + 7m^3) - 2m^2 + 5m + (-9 + 4)$
$= 4m^3 - 2m^2 + 5m - 5$
12. $(6p + 5) - (3p - 8) = (6p + 5) + (-3p + 8)$
$= (6p - 3p) + (5 + 8)$
$= 3p + 13$
13. $(5w - 3x) - (6w + 4x) = (5w - 3x) + (-6w - 4x)$
$= (5w - 6w) + (-3x - 4x)$
$= -w - 7x$
14. $(-2s^2 + 4s - 7) - (6s^2 - 7s - 9)$
$= (-2s^2 + 4s - 7) + (-6s^2 + 7s + 9)$
$= (-2s^2 - 6s^2) + (4s + 7s) + (-7 + 9)$
$= -8s^2 + 11s + 2$
15. $x(3x - 5) = x(3x) - x(5)$
$= 3x^2 - 5x$
16. $-5a(a^2 - b^2) = -5a(a^2) + 5a(b^2)$
$= -5a^3 + 5ab^2$
17. $6p(-2p^2 + 3p - 4) = 6p(-2p^2) + 6p(3p) - 6p(4)$
$= -12p^3 + 18p^2 - 24p$
18. Nonlinear; graph is a curve.
19. Linear; equation can be written as $y = \frac{5}{6}x - \frac{1}{3}$.
20. Nonlinear; rate of change is not constant, since as x increases by 2, y changes by different amounts each time.
21.

x	$y = 2x^2$	(x, y)
−2	$2(-2)^2 = 8$	(−2, 8)
−1	$2(-1)^2 = 2$	(−1, 2)
0	$2(0)^2 = 0$	(0, 0)
1	$2(1)^2 = 2$	(1, 2)
2	$2(2)^2 = 8$	(2, 8)

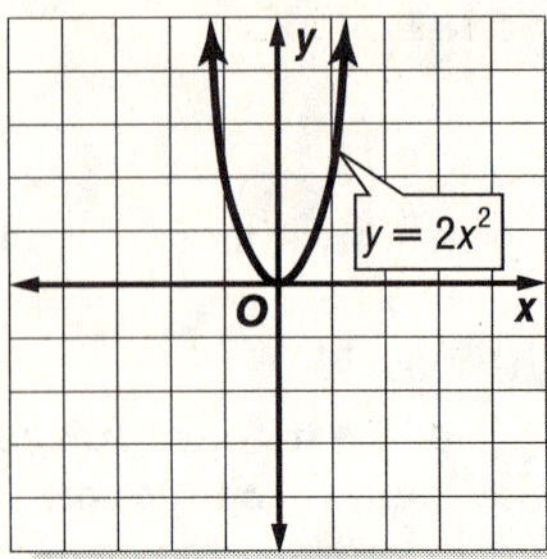

22.

x	$y = \frac{1}{2}x^3$	(x, y)
-2	$\frac{1}{2}(-2)^3 = -4$	$(-2, -4)$
-1	$\frac{1}{2}(-1)^3 = -0.5$	$(-1, -0.5)$
0	$\frac{1}{2}(0)^3 = 0$	$(0, 0)$
1	$\frac{1}{2}(1)^3 = 0.5$	$(1, 0.5)$
2	$\frac{1}{2}(2)^3 = 4$	$(2, 4)$

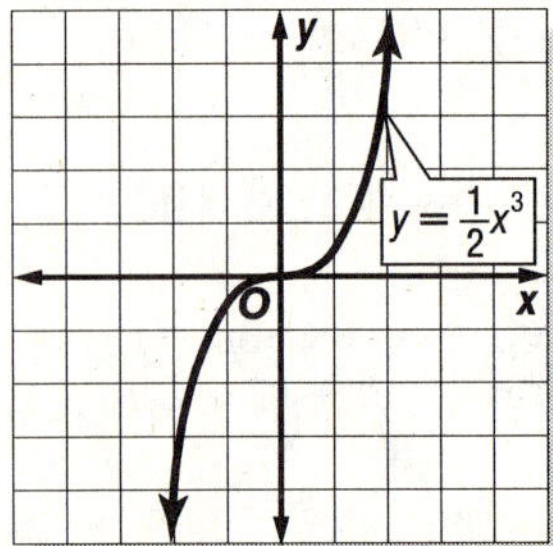

23.

x	$y = -x^2 + 3$	(x, y)
-2	$-(-2)^2 + 3 = -1$	$(-2, -1)$
-1	$-(-1)^2 + 3 = 2$	$(-1, 2)$
0	$-(0)^2 + 3 = 3$	$(0, 3)$
1	$-(1)^2 + 3 = 2$	$(1, 2)$
2	$-(2)^2 + 3 = -1$	$(2, -1)$

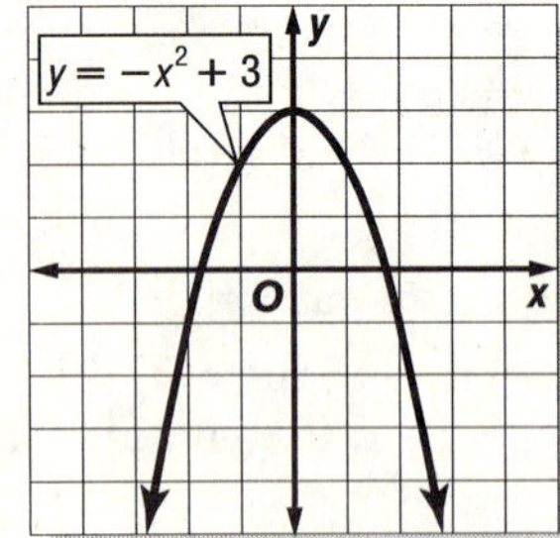

24a. $P = 2(\ell + w)$
$= 2[(12 - 2x) + (3x - 9)]$
$= 2[(-2x + 3x) + (12 - 9)]$
$= 2(x + 3)$
$= 2x + 6$
The expression for the perimeter of the rectangle is $2x + 6$ in.

24b. Let $P = 14$ in.
$P = 2x + 6$
$14 = 2x + 6$
$8 = 2x$
$4 = x$
The value of x is 4.

25. D; Let w = the width of the garden and let $4w - 5$ = the length of the garden.
$P = 2(\ell + w)$
$40 = 2[(4w - 5) + w]$
$40 = 2(5w - 5)$
$40 = 10w - 10$
$50 = 10w$
$5 = w$
The length of the garden is $4w - 5 = 4(5) - 5$ or 15 feet.

Chapter 13 Standardized Test Practice

Pages 702–703

1. B; $n + n + n = 3n$

2. D; Let m = the number Miguel sold. Then $m - 4$ = the number Connor sold, and $3(m - 4)$ = the number Kylie sold.

3. B;

$$\text{Average} = \frac{\text{miles on day 1} + \text{miles on day 2} + \text{miles on day 3}}{3}$$
$$400 = \frac{460 + 360 + x}{3}$$
$$400 \cdot 3 = \frac{460 + 360 + x}{3} \cdot 3$$
$$1200 = 460 + 360 + x$$
$$1200 = 820 + x$$
$$380 = x$$

Melissa's family drove 380 miles on the third day.

4. B; Let s = the side length of a square. Then its perimeter is $4s$. The ratio of the side to the perimeter is $\frac{s}{4s}$ or $\frac{1}{4}$.

5. B; $\frac{3}{5} = \frac{9}{T}$ and $\frac{3}{5} = \frac{S}{35}$
$3T = 45$ $\quad$ $105 = 5S$
$T = 15$ $\quad$ $21 = S$

6. C; $\angle 2$ and $\angle 5$ are supplementary angles, so the sum of their measures is 180°.

7. D; Use the same x-coordinate and multiply the y-coordinate by -1.
$(4, -6) \rightarrow (4, -1 \cdot -6) \rightarrow (4, 6)$

8. A; $(4x^2 + 6x - 3) - (2x^2 - 3x + 7)$
$= (4x^2 + 6x - 3) + (-2x^2 + 3x - 7)$
$= (4x^2 - 2x^2) + (6x + 3x) + (-3 - 7)$
$= 2x^2 + 9x - 10$

9. B;

x	$y = -2x$	(x, y)
-2	$-2(-2) = 4$	$(-2, 4)$
-1	$-2(-1) = 2$	$(-1, 2)$
0	$-2(0) = 0$	$(0, 0)$
1	$-2(1) = -2$	$(1, -2)$
2	$-2(2) = -4$	$(2, -4)$

$y = -2x$ includes all the ordered pairs in the table.

10. Order the data from least to greatest.
7 7 8 8 8 8 8 9 9 9 9 9 10 10 11 11 12 12 14 14 16
The number that occurs most often is the mode. The mode is 9.

11. $\frac{360}{\text{hr}} \times 4 \text{ hrs} = 1440$ containers
$\frac{10}{\text{min}} \times 240 \text{ min} = 2400$ containers
(240 min = 4 hours)
Both machines can make 1440 + 2400 or 3840 containers in 4 hours.

12. 165 is being compared to 275. So, 165 is the part and 275 is the base. Let p represent the percent.
$\frac{165}{275} = \frac{p}{100}$
$165 \cdot 100 = 275 \cdot p$
$\frac{16500}{275} = \frac{275p}{275}$
$60 = p$
So, 60% of 275 is 165.

13. Cost of 8 computers = 8(850)
Cost of p printers = $325p$
Cost of computers + cost of printers ≤ amount available to spend
$8(850) + 325p \leq 8500$

14. The graph crosses the y-axis at -3. The y-intercept is -3.

15. The vertical line through the kite divides it into two identical triangles with a base of 35 in. and a height of 9 in.
Area of kite $= 2 \cdot \left(\frac{1}{2}bh\right)$
$= 2 \cdot \left(\frac{1}{2} \cdot 35 \cdot 9\right)$
$= 315 \text{ in}^2$

16. Volume of the tank $= \ell \cdot w \cdot h$
Two-thirds the volume of the tank
$= \frac{2}{3}(\ell \cdot w \cdot h)$
$= \frac{2}{3}\left(20 \cdot 8\frac{1}{2} \cdot 20\right)$
$= \frac{2}{3}(3400)$
$\approx 2267 \text{ in}^3$

17. $s + t = (3x^2 - 2x - 1) + (-2x^2 + x + 2)$
$= (3x^2 - 2x^2) + (-2x + x) + (-1 + 2)$
$= x^2 - x + 1$

18. Let w = the width of the field. Then
$2w + 40$ = the length of the field
$P = 2(\ell + w)$
$1040 = 2[(2w + 40) + w]$
$1040 = 2(3w + 40)$
$1040 = 2(3w) + 2(40)$
$1040 = 6w + 80$
$960 = 6w$
$160 = w$
The length of the field is $2w + 40 = 2(160) + 40$ or 360 feet.

19a. The diameter is 12 inches, so the radius is 6 inches or 0.5 feet.
When $h = 6$, $V = 3.14(0.5)^2(6)$ or 4.7 ft^3.
When $h = 5$, $V = 3.14(0.5)^2(5)$ or 3.9 ft^3.
When $h = 4$, $V = 3.14(0.5)^2(4)$ or 3.1 ft^3.
When $h = 3$, $V = 3.14(0.5)^2(3)$ or 2.4 ft^3.
When $h = 2$, $V = 3.14(0.5)^2(2)$ or 1.6 ft^3.
The total volume is 4.7 + 3.9 + 3.1 + 2.4 + 1.6 or 15.7 ft^3.

19b. Let w = the weight of the sculpture.
$\text{Density} = \frac{\text{weight}}{\text{volume}}$
$12 = \frac{w}{15.7}$
$12 \cdot 15.7 = \frac{w}{15.7} \cdot 15.7$
$188.4 = w$
The weight of the sculpture is 188.4 lb.

20a–b.

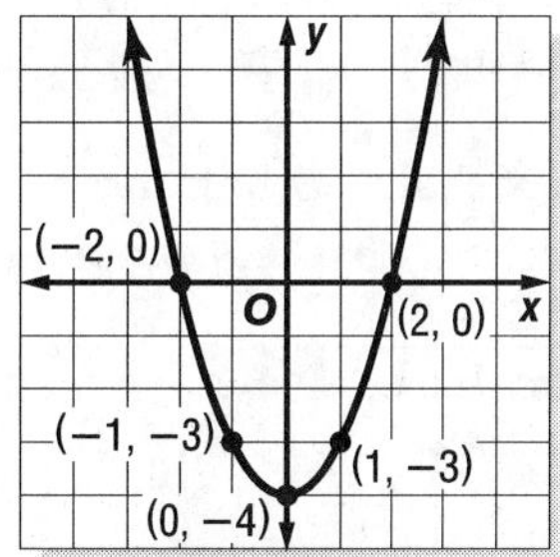

20c. The rule is square the first term and subtract 4 to get the second term. So, the quadratic function is $y = x^2 - 4$.

Page 706 Problem-Solving Strategy: Solve a Simpler Problem

1. The numbers between 1 and 1000 can be grouped into pairs whose sums are all 1001. Examples of such pairs are 1 and 1000, 2 and 999, and 500 and 501. There are 500 such pairs.

   $$1 + 2 + 3 + \cdots + 999 + 1000 = 50 \times 1001$$
   $$= 500{,}500$$

2. If the board contained only 1 of the small squares, the board would have only 1 square. If the board contained 4 of the smaller squares arranged in a 2 by 2 fashion, the board would contain the 2 by 2 square as well as the 4 small squares. Thus it would contain 1 + 4 or 5 squares. Similarly, if the board contained 9 of the smaller squares arranged in a 3 by 3 fashion, the square would contain one 3 by 3 square, four 2 by 2 squares, and 9 small squares, or $1 + 4 + 9 = 14$ squares.

   In the same manner, the board containing 64 small squares arranged in an 8 by 8 fashion will contain $1 + 4 + 9 + 16 + 25 + 36 + 49 + 64$ or 204 squares.

3. One link is needed to join 2 pieces of chain. Two links are needed to join 3 pieces of chain since 2 of the three pieces are joined by one link and then the third piece is joined by the second. In general, the number of links needed is one less than the number of pieces of chain.

   The join 30 pieces of chain, 29 links are needed.

4. Three people can pick six baskets of apples in one hour.

   First find how many baskets each person can pick in one hour. Divide 6 baskets by 3 people.

   $6 \div 3 = 2$

   So, each person can pick 2 baskets in one hour. To find how many baskets each person can pick in one-half hour, multiply 2 by $\frac{1}{2}$.

   $2 \times \frac{1}{2} = 1$

   Now find the number of baskets 12 people can pick in one-half hour by multiplying by 12.

   $1 \times 12 = 12$

   So, 12 people can pick 12 baskets of apples in one-half hour.

5. Three designers can make 12 shirts in 2 hours.

   First find how many shirts each designer can make in 2 hours. Divide 12 shirts by 3 designers.

   $12 \div 3 = 4$

   So, each designer can make 4 shirts in 2 hours. To find how many shirts each designer can make in 1 hour, divide 4 by 2.

   $4 \div 2 = 2$

   Now find the number of shirts each designer can make in 8 hours by multiplying 2 by 8.

   $2 \times 8 = 16$

   To find the number of designers needed, divide 112 by 16.

   $112 \div 16 = 7$

   Seven designers are needed to complete the orders in 8 hours.

Page 707 Problem-Solving Strategy: Work Backward

1. Katie now has \$2.25. Add \$1.75 to undo the amount spent on ice cream.

   \$2.25 + \$1.75 = \$4.00

   Multiply by 2 to undo the half she spent on a ticket.

   2 × \$4.00 = \$8.00

   Katie's allowance is \$8.00.

2. Michele has \$14 left. Multiply by 2 to undo the half she spent on a book.

   2 × \$14 = \$28

   Add \$24 to undo the amount spent on the ticket.

   \$28 + \$24 = \$52

   Multiply by 2 to undo the half she spent on clothing.

   2 × \$52 = \$104

   Add \$15 to undo the amount she put in savings.

   \$104 + \$15 = \$109

   Michele's paycheck was \$109.

3. The final answer is 41. Subtract 5 to undo adding 5.

   $41 - 5 = 36$

   Divide by 3 to undo multiplying by 3.

   $36 \div 3 = 12$

   The number is 12.

4. One-third of Mr. Jackson's stock is worth \$2700. Multiply by 3 to find the value of Mr. Jackson's stock.

   3 × \$2700 = \$8100

   Since Mrs. Jackson owns an equal number of shares, multiply by 2 to find the combined value of their stock.

   2 × \$8100 = \$16,200

   Before the sale, the value of their stock was \$16,200.

5. There are 1600 bacteria after 3 days or 72 hours. Divide by 2 to find the number of bacteria after 60 hours.

   $1600 \div 2 = 800$

   Divide by 2 again to find the number of bacteria after 48 hours.

   $800 \div 2 = 400$

   Divide by 2 again to find the number of bacteria after 36 hours.

   $400 \div 2 = 200$

Divide by 2 again to find the number of bacteria after 24 hours.

$200 \div 2 = 100$

Divide by 2 again to find the number of bacteria after 12 hours.

$100 \div 2 = 50$

Divide by 2 again to find the number of bacteria at the beginning.

$50 \div 2 = 25$

There were 25 bacteria at the beginning of the first day.

6. Maria has 3 pieces of gum. Multiply by 3 to find how many pieces Lisa had.

$3 \times 3 = 9$

Multiply by 2 to find how many pieces Bob had.

$9 \times 2 = 18$

Multiply by 4 to find the number of pieces Masao had.

$18 \times 4 = 72$

Masao had 72 pieces of bubble gum in the beginning.

7. Carla needs to catch a bus at 7:30. She needs a total of 30 + 30 + 15 or 75 minutes to get dressed, eat breakfast, and walk to the bus stop. Subtract 75 minutes or 1 hour 15 minutes from 7:30.

7:30 − 1:15 = 6:15

Carla needs to get up at 6:15 A.M.

8. Maria rented 10 DVDs. Subtract 4 to find the number Cole rented.

$10 - 4 = 6$

Subtract 4 to find the number Paloma rented.

$6 - 4 = 2$

Multiply 6, the number Cole rented, by 3 to find the number Justin rented.

$6 \times 3 = 18$

Justin rented 18 DVDs, Cole rented 6, Maria rented 10, and Paloma rented 2.

Page 708 Problem-Solving Strategy: Make a Table or List

1.

Quarters	Dimes	Nickels
2	0	0
1	2	1
1	1	3
1	0	5
0	5	0
0	4	2
0	3	4
0	2	6
0	1	8
0	0	10

There are 10 ways to make change for the half-dollar.

2.

Red	Blue
1	1
1	2
1	3
1	4
1	5
1	6
2	1
2	2
2	3
2	4
2	5

Red	Blue
3	1
3	2
3	3
3	4
4	1
4	2
4	3
5	1
5	2
6	1

The sum of each combination is less than 8. There are 21 ways to roll a sum less than 8.

3. List the coins you could draw.

1. penny, nickel
2. penny, dime
3. penny, quarter
4. nickel, dime
5. nickel, quarter
6. dime, quarter

Since each pair represents a different amount of money, 6 different amounts are possible.

4. List factorizations of 48.

1. 1×48
2. 2×24
3. 3×16
4. 4×12
5. 6×8

Six combinations of side lengths are possible.

5.

Given	Received
8	5
6	4
5	3
12	9

We see from the table that Malcolm gave up a total of 31 of his cards and gained 21 in return. Thus, Malcolm has 55 − 31 + 21 or 45 cards.

Page 709 Problem-Solving Strategy: Guess and Check

1. Try 27 and 29.
$27 \times 29 = 783$
The integers are 27 and 29.

2. Try 12 for Paula and 4 for Courtney.
$12 = 3 \times 4$ and $12 + 4 = 2(4 + 4)$
Paula is 12, and Courtney is 4.

3. Try 4.
$4 \times 5 \times 6 = 120$
The number is 4.

4. Try 5 postcards and 5 letters.
5(\$0.21) + 5(\$0.34) = \$2.75
He can send 5 postcards and 5 letters.

5. Try 14 in fingers, 8 in wrist, and 5 in palm.
$14 = 8 + 6$ and $5 = 8 - 3$
There are 14 bones in the fingers, 8 in the wrist, and 5 in the palm.

6. Sample answer: Try 125 candy bars and 105 pretzels.
   125($0.25) + 105($0.30) = $62.75
   They sold 125 candy bars and 105 pretzels.
7. Try 10.
   10($0.25) + 10($0.10) + 10($0.05) = $4.00
   Luis has 10 of each coin.
8. Sample answer: Try 3 $5-bills, 2 $10-bills, and 7 $20-bills.
   3($5) + 2($10) + 7($20) = $175
   Kelsey has 3 $5-bills, 2 $10-bills, and 7 $20-bills.
9. Try two 5-card and two 3-card packages.
   2(5) + 2(3) = 16
   You should buy two 5-card and two 3-card packages.

Page 710 Comparing and Ordering Decimals

1. 4.05 ● 4.45
   The digits in the tenths place are not the same.
   0 tenths $<$ 4 tenths, so $4.05 < 4.45$.
2. 2.26 ● 2.28
   The digits in the hundredths place are not the same.
   6 hundredths $<$ 8 hundredths, so $2.26 < 2.28$.
3. 3.005 ● 3.05
   The digits in the hundredths place are not the same.
   0 hundredths $<$ 5 hundredths, so $3.005 < 3.05$.
4. 8.7 ● 82.1
   The digits in the tens place are not the same.
   0 tens $<$ 8 tens, so $8.7 < 82.1$.
5. 6.2 ● 6.008
   The digits in the tenths place are not the same.
   2 tenths $>$ 0 tenths, so $6.2 > 6.008$.
6. 15.601 ● 16.9
   The digits in the ones place are not the same.
   5 ones $<$ 6 ones, so $15.601 < 16.9$.
7. 1.9 ● 1.96
   The digits in the hundredths place are not the same.
   0 hundredths $<$ 6 hundredths, so $1.9 < 1.96$.
8. 8.9 ● 7.99
   The digits in the ones place are not the same.
   8 ones $>$ 7 ones, so $8.9 > 7.99$.
9. 0.66 ● 0.582
   The digits in the tenths place are not the same.
   6 tenths $>$ 5 tenths, so $0.66 > 0.582$.
10. 7.14 ● 7.2
    The digits in the tenths place are not the same.
    1 tenth $<$ 2 tenths, so $7.14 < 7.2$.
11. 0.048 ● 0.11
    The digits in the tenths place are not the same.
    0 tenths $<$ 1 tenth, so $0.048 < 0.11$.
12. 10.1 ● 1.01
    The digits in the tens place are not the same.
    1 ten $>$ 0 tens, so $10.1 > 1.01$.
13. 32.1 ● 3.215
    The digits in the tens place are not the same.
    3 tens $>$ 0 tens, so $32.1 > 3.215$.
14. 1.098 ● 2
    The digits in the ones place are not the same.
    1 one $<$ 2 ones, so $1.098 < 2$.
15. 9.1 ● 9.005
    The digits in the tenths place are not the same.
    1 tenth $>$ 0 tenths, so $9.1 > 9.005$.
16. 16.8 ● 16.791
    The digits in the tenths place are not the same.
    8 tenths $>$ 7 tenths, so $16.8 > 16.791$.
17. 0.943 ● 0.4991
    The digits in the tenths place are not the same.
    9 tenths $>$ 4 tenths, so $0.943 > 0.4991$.
18. 0.117 ● 0.95
    The digits in the tenths place are not the same.
    1 tenth $<$ 9 tenths, so $0.117 < 0.95$.
19. {0.2, 0.01, 0.6}
    0.01 is less than both 0.2 and 0.6.
    0.2 is less than 0.6.
    Thus, the order from least to greatest is 0.01, 0.2, 0.6.
20. {1.2, 2.4, 0.04, 2.2}
    $0.04 < 1.2 < 2.2 < 2.4$
    Thus, the order from least to greatest is 0.04, 1.2, 2.2, 2.4.
21. {3.5, 0.6, 2.06, 0.28}
    $0.28 < 0.6 < 2.06 < 3.5$
    Thus, the order from least to greatest is 0.28, 0.6, 2.06, 3.5.
22. {0.8, 0.07, 1.001, 0.392}
    $0.07 < 0.392 < 0.8 < 1.001$
    Thus, the order from least to greatest is 0.07, 0.392, 0.8, 1.001.
23. {7.06, 7.026, 7.061, 7.009, 7.1}
    $7.009 < 7.026 < 7.06 < 7.061 < 7.1$
    Thus, the order from least to greatest is 7.009, 7.026, 7.06, 7.061, 7.1.
24. {0.82, 0.98, 0.103, 0.625, 0.809}
    $0.103 < 0.625 < 0.809 < 0.82 < 0.98$
    Thus, the order from least to greatest is 0.103, 0.625, 0.809, 0.82, 0.98.

Page 711 Rounding Decimals

1. 3.2
   3 is in the ones place. 2 is to the right of 3.
   $2 < 5$
   So, 3.2 rounded to the nearest whole number is 3.
2. 64.8
   4 is in the ones place. 8 is to the right of 4.
   $8 > 5$
   So, 64.8 rounded to the nearest whole number is 65.
3. 50.57
   0 is in the ones place. 5 is to the right of 0.
   $5 = 5$
   So, 50.57 rounded to the nearest whole number is 51.
4. 16.08
   6 is in the ones place. 0 is to the right of 6.
   $0 < 5$
   So, 16.08 rounded to the nearest whole number is 16.

5. 41.29
1 is in the ones place. 2 is to the right of 1.
$2 < 5$
So, 41.29 rounded to the nearest whole number is 41.

6. 38.726
8 is in the ones place. 7 is to the right of 8.
$7 > 5$
So, 38.726 rounded to the nearest whole number is 39.

7. 74.455
4 is in the ones place. 4 is to the right of 4.
$4 < 5$
So, 74.455 rounded to the nearest whole number is 74.

8. 86.299
6 is in the ones place. 2 is to the right of 6.
$2 < 5$
So, 86.299 rounded to the nearest whole number is 86.

9. 79.603
9 is in the ones place. 6 is to the right of 9.
$6 > 5$
So, 79.603 rounded to the nearest whole number is 80.

10. 16.57
5 is in the tenths place. 7 is to the right of 5.
$7 > 5$
So, 16.57 rounded to the nearest tenth is 16.6.

11. 1.05
0 is in the tenths place. 5 is to the right of 0.
$5 = 5$
So, 1.05 rounded to the nearest tenth is 1.1.

12. 43.827
8 is in the tenths place. 2 is to the right of 8.
$2 < 5$
So, 43.827 rounded to the nearest tenth is 43.8.

13. 53.865
8 is in the tenths place. 6 is to the right of 8.
$6 > 5$
So, 53.865 rounded to the nearest tenth is 53.9.

14. 80.349
3 is in the tenths place. 4 is to the right of 3.
$4 < 5$
So, 80.349 rounded to the nearest tenth is 80.3.

15. 24.731
7 is in the tenths place. 3 is to the right of 7.
$3 < 5$
So, 24.731 rounded to the nearest tenth is 24.7.

16. 49.5463
5 is in the tenths place. 4 is to the right of 5.
$4 < 5$
So, 49.5463 rounded to the nearest tenth is 49.5.

17. 131.9884
9 is in the tenths place. 8 is to the right of 9.
$8 > 5$
So, 131.9884 rounded to the nearest tenth is 132.0.

18. 68.3553
3 is in the tenths place. 5 is to the right of 3.
$5 = 5$
So, 68.3553 rounded to the nearest tenth is 68.4.

19. 62.624
2 is in the hundredths place. 4 is to the right of 2.
$4 < 5$
So, 62.624 rounded to the nearest hundredth is 62.62.

20. 44.138
3 is in the hundredths place. 8 is to the right of 3.
$8 > 5$
So, 44.138 rounded to the nearest hundredth is 44.14.

21. 85.5639
6 is in the hundredths place. 3 is to the right of 6.
$3 < 5$
So, 85.5639 rounded to the nearest hundredth is 85.56.

22. 105.3582
5 is in the hundredths place. 8 is to the right of 5.
$8 > 5$
So, 105.3582 rounded to the nearest hundredth is 105.36.

23. 99.9862
8 is in the hundredths place. 6 is to the right of 8.
$6 > 5$
So, 99.9862 rounded to the nearest hundredth is 99.99.

24. 24.8715
7 is in the hundredths place. 1 is to the right of 7.
$1 < 5$
So, 24.8715 rounded to the nearest hundredth is 24.87.

25. 458.7625
6 is in the hundredths place. 2 is to the right of 6.
$2 < 5$
So, 458.7625 rounded to the nearest hundredth is 458.76.

26. 206.6244
2 is in the hundredths place. 4 is to the right of 2.
$4 < 5$
So, 206.6244 rounded to the nearest hundredth is 206.62.

27. 153.2965
9 is in the hundredths place. 6 is to the right of 9.
$6 > 5$
So, 153.2965 rounded to the nearest hundredth is 153.30.

28. $40.29
0 is in the ones place. 2 is to the right of 0.
$2 < 5$
So, \$40.29 rounded to the nearest dollar is \$40.

29. $72.50
2 is in the ones place. 5 is to the right of 2.
$5 = 5$
So, \$72.50 rounded to the nearest dollar is \$73.

30. $36.82
6 is in the ones place. 8 is to the right of 6.
$8 > 5$
So, \$36.82 rounded to the nearest dollar is \$37.

Page 712 Estimating Sums and Differences of Decimals

1. 12.5 → 13
+ 44.8 + 45
58

2. 8.6 → 9
+ 11.9 + 12
21

3. 34.32 → 34
+ 19.51 + 20
54

4. 15.9 → 16
+ 20.32 + 20
36

5. 32.00 → 32
− 29.75 − 30
2

6. 125.8 → 126
− 22.4 − 22
104

7. 159.7 → 160
− 124.8 − 125
35

8. 8.890 → 9
+ 15.98 + 16
25

9. 0.7 → 1
+ 1.663 + 2
3

10. 52.4 → 52
− 21.01 − 21
31

11. 26.55 → 27
− 10 − 10
17

12. 2.79 3
5.9 → 6
+ 0.02 + 0
9

13. 42.1 42
16.25 → 16
+ 8.96 + 9
67

14. 209.5 → 210
− 110 −110
100

15. 18 → 18
− 12.49 − 12
6

16. $6.89 → $7.00
+ 1.20 + 1.00
$8.00

17. $5.72 → $ 6.00
+ 4.35 + 4.00
$10.00

18. $1.68 → $2.00
− 0.99 − 1.00
$1.00

19. $5.00 → $5.00
− 2.56 − 3.00
$2.00

20. $20.00 → $20.00
− 15.34 − 15.00
$ 5.00

21. $12.86 → $13.00
+ 3.33 + 3.00
$16.00

22. $4.99 → $5.00
+ 3.29 + 3.00
$8.00

23. $50.00 → $50.00
− 39.89 − 40.00
$10.00

24. $92.30 → $92.00
− 40.00 − 40.00
$52.00

25. $16.39 → $16.00
− 11.80 − 12.00
$ 4.00

26. $84.99 → $85.00
+ 5.52 + 6.00
$91.00

27. $132.62 → $133.00
− 45.81 − 46.00
$ 87.00

28. $20.19 $20.00
3.60 → 4.00
+ 5.08 + 5.00
$29.00

29. $4.80 $ 5.00
7.65 → 8.00
+ 2.59 + 3.00
$16.00

30. $325.44 → $325.00
+ 125.10 + 125.00
$450.00

31. 37.19 → 37
− 16.49 − 16
21

Seattle has about 21 inches more precipitation than Spokane each year.

32. $45.45 $ 45
45.19 → 45
+ 44.95 + 45
$135

The club collected about $135.

Page 713 Adding and Subtracting Decimals

1. 42.3 → $\overset{1}{4}2.30$
+ 0.81 + 0.81
43.11

2. 5.86
− 1.51
4.35

3. 13 → $1\overset{2}{\not{3}}.\overset{9}{\not{0}}\overset{9}{\not{0}}\overset{1}{}0$
− 0.324 − 0.324
12.676

4. 2.3
+ 1.1
3.4

5. 11.5
+ 4.2
15.7

6. 9.5
− 8.3
1.2

7. 24.8 − 3.6 = 21.2

8. 3.57 − 2.17 = 1.40

9. 7.43 − 5.34 = 2.09

10. 6.40 + 7.36 = 13.76

11. 15.20 + 0.16 = 15.36

12. 7.97 − 4.29 = 3.68

13. 8.70 + 0.64 = 9.34

14. 56.88 − 12.35 = 44.53

15. 4.192 + 1.255 = 5.447

16. 14.6 + 20.81 → 14.60 + 20.81 = 35.41

17. 5.2 − 3.01 → 5.20 − 3.01 = 2.19

18. 1.9 − 1.65 → 1.90 − 1.65 = 0.25

19. 6.38 − 1.1 → 6.38 − 1.10 = 5.28

20. 4.86 − 0.3 → 4.86 − 0.30 = 4.56

21. 9.43 + 1.8 → 9.43 + 1.80 = 11.23

22. 70.3 + 7.03 → 70.30 + 7.03 = 77.33

23. 0.5 + 1.674 → 0.500 + 1.674 = 2.174

24. 25 − 8.3 → 25.0 − 8.3 = 16.7

25. 18 − 12.31 → 18.00 − 12.31 = 5.69

26. 2.85 + 23.6 → 2.85 + 23.60 = 26.45

27. 0.8 + 9.612 → 0.800 + 9.612 = 10.412

28. 6.8 + 5.09 + 0.03 → 6.80 + 5.09 + 0.03 = 11.92

29. 0.5 + 2.41 + 6.7 → 0.50 + 2.41 + 6.70 = 9.61

30. 0.563 + 5.8 + 6.89 → 0.563 + 5.800 + 6.890 = 13.253

31. 41.30 + 0.28 + 6.15 = 47.73

32. 4.52 + 0.167 + 12.9 → 4.520 + 0.167 + 12.900 = 17.587

33. 23.4 + 9.865 + 18.26 → 23.400 + 9.865 + 18.260 = 51.525

34. 27.38 + 6.8 → 27.38 + 6.80 = 34.18

35. \$26.59 + 1.80 + 13 → \$26.59 + 1.80 + 13.00 = \$41.39

36. 42.05 − 11.621 → 42.050 − 11.621 = 30.429

37. \$115 − 102.90 → \$115.00 − 102.90 = \$ 12.10

38. \$ 6.50 + 37.99 + 13.79 = \$58.28

Page 714 Estimating Products and Quotients of Decimals

1. Sample answer:
9.2 × 4.89 → 9 × 5 = 45

2. Sample answer:
$6.75 \times 5.25 \rightarrow 7 \times 5 = 35$

3. Sample answer:
$12.19 \div 3.8 \rightarrow 12 \div 4 = 3$

4. Sample answer:
$39.79 \div 4.61 \rightarrow 40 \div 5 = 8$

5. Sample answer:
$11.2 \times 6.25 \rightarrow 11 \times 6 = 66$

6. Sample answer:
$15.2 \div 2.7 \rightarrow 15 \div 3 = 5$

7. Sample answer:
$47.2 \div 5.1 \rightarrow 45 \div 5 = 9$

8. Sample answer:
$16.53 \div 8.36 \rightarrow 16 \div 8 = 2$

9. Sample answer:
$4.32(107.6) \rightarrow 4 \times 100 = 400$

10. Sample answer:
$26 \times 10.9 \rightarrow 26 \times 10 = 260$

11. Sample answer:
$73.2 \div 6.99 \rightarrow 70 \div 7 = 10$

12. Sample answer:
$19.1(21.60) \rightarrow 20 \times 20 = 400$

13. Sample answer:
$4.6 \times 8.3 \rightarrow 5 \times 8 = 40$

14. Sample answer:
$5.12 \times 5.9 \rightarrow 5 \times 6 = 30$

15. Sample answer:
$7.5 \div 4.2 \rightarrow 8 \div 4 = 2$

16. Sample answer:
$9.27 \div 3.31 \rightarrow 9 \div 3 = 3$

17. Sample answer:
$19.8(2.6) \rightarrow 20 \times 3 = 60$

18. Sample answer:
$41.75 \div 6 \rightarrow 42 \div 6 = 7$

19. Sample answer:
$36.24 \div 8.7 \rightarrow 36 \div 9 = 4$

20. Sample answer:
$5.85 \times 7.55 \rightarrow 6 \times 8 = 48$

21. Sample answer:
$8.1 \div 2.2 \rightarrow 8 \div 2 = 4$

22. Sample answer:
$7.9(9.12) \rightarrow 8 \times 9 = 72$

23. Sample answer:
$6.1 \div 2.1 \rightarrow 6 \div 2 = 3$

24. Sample answer:
$9 \times 96.42 \rightarrow 9 \times 100 = 900$

25. Sample answer:
$13 \times 9.1 \rightarrow 13 \times 10 = 130$

26. Sample answer:
$10.1 \div 4.7 \rightarrow 10 \div 5 = 2$

27. Sample answer:
$28.6(5) \rightarrow 30 \times 5 = 150$

28. Sample answer:
$21 \div 7.6 \rightarrow 21 \div 7 = 3$

29. Sample answer:
$81 \div 10.5 \rightarrow 80 \div 10 = 8$

30. Sample answer:
$52.7 \div 5.3 \rightarrow 50 \div 5 = 10$

31. Sample answer:
$47.74 \times 2 \rightarrow 50 \times 2 = 100$

32. Sample answer:
$204.5 \times 3 \rightarrow 200 \times 3 = 600$

33. Sample answer:
$41.79 \div 7.23 \rightarrow 42 \div 7 = 6$

34. Sample answer:
$106.25 \times 1.8 \rightarrow 100 \times 2 = 200$
It can travel about 200 miles.

Page 715 Multiplying and Dividing Decimals

1.
$$\begin{array}{r} 1.2 \\ \underline{\times\ 3} \\ 3.6 \end{array}$$

2.
$$\begin{array}{r} 3.4 \\ \underline{\times\ 8} \\ 27.2 \end{array}$$

3.
$$\begin{array}{r} 7.2 \\ \underline{\times\ 0.2} \\ 1.44 \end{array}$$

4.
$$\begin{array}{r} 6.1 \\ \underline{\times\ 1.4} \\ 2\ 44 \\ \underline{6\ 1} \\ 8.54 \end{array}$$

5.
$$\begin{array}{r} 0.7 \\ 0.9\overline{)0.6\ 3} \\ \underline{6\ 3} \\ 0 \end{array}$$

6.
$$\begin{array}{r} 2\ 1 \\ 0.4\overline{)8.4} \\ \underline{8} \\ 0\ 4 \\ \underline{4} \\ 0 \end{array}$$

7.
$$\begin{array}{r} 0.06 \\ \underline{\times\ \ \ 3} \\ 0.18 \end{array}$$

8.
$$\begin{array}{r} 52.5 \\ 0.8\overline{)42.0\ 0} \\ \underline{40} \\ 2\ 0 \\ \underline{1\ 6} \\ 4\ 0 \\ \underline{4\ 0} \\ 0 \end{array}$$

9.
$$\begin{array}{r} 8.2 \\ \underline{\times\ 3.9} \\ 7\ 38 \\ \underline{24\ 6} \\ 31.98 \end{array}$$

10.
$$\begin{array}{r} 3.1 \\ \underline{\times\ 0.2} \\ 0.62 \end{array}$$

11.
$$\begin{array}{r} 9\ 0 \\ 0.3\overline{)27.0} \\ \underline{27} \\ 0\ 0 \end{array}$$

12. 0.4)64.0 = 16 0
4
24
24
0 0

13. 2)0.4 = 0.2
4
0

14. 0.16)14.40 = 90
14 4
00
0
0

15. 15.6
× 3 8
124 8
468
592.8

16. 0.03)0.51 = 17
3
21
21
0

17. 5.7
× 3.8
4 56
17 1
21.66

18. 7.07
× 4
28.28

19. 1.25
× 12
2 50
12 5
15.00

20. 100)62.900 = 0.629
60 0
2 90
2 00
900
900
0

21. 6.5
× 0.13
1 95
6 5
0.845

22. 14.9
× 0.56
894
7 45
8.344

23. 1.2)0.3 84 = 0.32
3 6
24
24
0

24. 1.05)4.20 = 4
4 20
0

25. 2.8)25.9 00 = 9.25
25 2
7 0
5 6
1 40
1 40
0

26. 3.01
× 0.47
2107
1 204
1.4147

27. 1.01
× 6.2
202
6 06
6.262

28. 0.375)9.000 = 24
7 50
1 500
1 500
0

29. 0.25)50.00 = 2 00
50
0 00

30. 3.2)500.0 00 = 15 6.25
32
180
160
20 0
19 2
8 0
6 4
1 60
1 60
0

31. 7.09
× 0.001
0.00709

32. 6.32
× 0.81
632
5 056
5.1192

33.
```
           1 460
0.002 )2.920
           2
           0 9
             8
             12
             12
             00
```

34.
```
     13.6
   × 9.15
      680
     1 36
   122 4
   124.440
```

35.
```
          6 55
0.11 )72.05
        66
        6 0
        5 5
          55
          55
           0
```

36.
```
  128.46
 ×    50
 6423.00
```
You would receive 6423 pesetas for \$50.

Page 716 Estimating Sums and Differences of Fractions and Mixed Numbers

1. $\frac{9}{10}$ rounds to 1.
2. $\frac{1}{8}$ rounds to 0.
3. $\frac{13}{25}$ rounds to $\frac{1}{2}$.
4. $\frac{3}{14}$ rounds to 0.
5. $\frac{9}{15}$ rounds to $\frac{1}{2}$.
6. $\frac{78}{81}$ rounds to 1.
7. Sample answer:
   $\frac{8}{9} + \frac{1}{4} \rightarrow 1 + 0 = 1$
8. Sample answer:
   $\frac{14}{15} + \frac{5}{6} \rightarrow 1 + 1 = 2$
9. Sample answer:
   $\frac{47}{90} + \frac{3}{24} \rightarrow \frac{1}{2} + 0 = \frac{1}{2}$
10. Sample answer:
    $\frac{11}{12} + \frac{4}{9} \rightarrow 1 + \frac{1}{2} = 1\frac{1}{2}$
11. Sample answer:
    $\frac{15}{16} + 9\frac{3}{4} \rightarrow 1 + 10 = 11$
12. Sample answer:
    $1\frac{5}{12} + \frac{7}{18} \rightarrow 1\frac{1}{2} + \frac{1}{2} = 2$
13. Sample answer:
    $5\frac{10}{11} + \frac{3}{5} \rightarrow 6 + \frac{1}{2} = 6\frac{1}{2}$
14. Sample answer:
    $21\frac{8}{9} + 6\frac{4}{25} \rightarrow 22 + 6 = 28$
15. Sample answer:
    $32\frac{3}{56} + 18\frac{2}{75} \rightarrow 32 + 18 = 50$
16. Sample answer:
    $\frac{4}{5} - \frac{1}{10} \rightarrow 1 - 0 = 1$
17. Sample answer:
    $\frac{7}{9} - \frac{13}{18} \rightarrow 1 - 1 = 0$
18. Sample answer:
    $\frac{9}{10} - \frac{3}{8} \rightarrow 1 - \frac{1}{2} = \frac{1}{2}$
19. Sample answer:
    $5\frac{1}{5} - 2\frac{3}{4} \rightarrow 5 - 3 = 2$
20. Sample answer:
    $8\frac{3}{5} - 2\frac{1}{8} \rightarrow 8\frac{1}{2} - 2 = 6\frac{1}{2}$
21. Sample answer:
    $16\frac{34}{35} - 3\frac{1}{6} \rightarrow 17 - 3 = 14$
22. Sample answer:
    $35\frac{7}{8} - 4\frac{1}{2} \rightarrow 36 - 4\frac{1}{2} = 31\frac{1}{2}$
23. Sample answer:
    $15\frac{4}{9} + 13\frac{9}{11} \rightarrow 15\frac{1}{2} + 14 = 29\frac{1}{2}$
24. Sample answer:
    $140\frac{4}{5} - 120\frac{2}{15} \rightarrow 141 - 120 = 21$
25. Sample answer:
    $4\frac{1}{2} - \frac{5}{6} \rightarrow 4\frac{1}{2} - 1 = 3\frac{1}{2}$
    $4\frac{1}{2}$ minutes is about $3\frac{1}{2}$ minutes longer than $\frac{5}{6}$ minute.
26. Sample answer:
    $3\frac{3}{10} + 2\frac{4}{5} + 3\frac{1}{3} \rightarrow 3 + 3 + 3 = 9$
27. Sample answer:
    $19\frac{3}{4} - 10\frac{7}{8} \rightarrow 20 - 11 = 9$
    $19\frac{3}{4}$ inches is about 9 inches longer than $10\frac{7}{8}$ inches.
28. Sample answer:
    $7\frac{1}{3} + 6\frac{4}{5} + 6\frac{3}{4} + 7\frac{1}{10} + 6\frac{15}{16} \rightarrow$
    $7 + 7 + 7 + 7 + 7 = 35$
29. Sample answer:
    $63\frac{5}{8} - 62\frac{1}{4} \rightarrow 63\frac{1}{2} - 62 = 1\frac{1}{2}$
    A board that is $63\frac{5}{8}$ inches long is about $1\frac{1}{2}$ inches longer than a board that is $62\frac{1}{4}$ inches long.

Page 717 Estimating Products and Quotients of Fractions and Mixed Numbers

1. Sample answer:
   $\frac{1}{4} \cdot 11 \rightarrow \frac{1}{4} \cdot 12 = 3$
2. Sample answer:
   $\frac{1}{3}(20) \rightarrow \frac{1}{3}(21) = 7$
3. Sample answer:
   $\frac{1}{3} \times 14 \rightarrow \frac{1}{3} \times 15 = 5$

4. Sample answer:
$\frac{1}{4}(15) \rightarrow \frac{1}{4}(16) = 4$
5. Sample answer:
$\frac{7}{15} \times 120 \rightarrow \frac{1}{2} \times 120 = 60$
6. Sample answer:
$\frac{11}{20}(60) \rightarrow \frac{1}{2}(60) = 30$
7. Sample answer:
$\frac{31}{40} \cdot 100 \rightarrow \frac{3}{4} \cdot 100 = 75$
8. Sample answer:
$\frac{6}{13} \times 150 \rightarrow \frac{1}{2} \times 150 = 75$
9. Sample answer:
$\frac{1}{5}(44) \rightarrow \frac{1}{5}(45) = 9$
10. Sample answer:
$1\frac{5}{6} \cdot 30 \rightarrow 2 \cdot 30 = 60$
11. Sample answer:
$2\frac{1}{4} \cdot 22 \rightarrow 2 \cdot 22 = 44$
12. Sample answer:
$4\frac{4}{5} \times 24 \rightarrow 5 \times 25 = 125$
13. Sample answer:
$5\frac{7}{8} \div 2 \rightarrow 6 \div 2 = 3$
14. Sample answer:
$8\frac{1}{4} \div 4 \rightarrow 8 \div 4 = 2$
15. Sample answer:
$14\frac{6}{7} \div 3 \rightarrow 15 \div 3 = 5$
16. Sample answer:
$50 \div 4\frac{7}{8} \rightarrow 50 \div 5 = 10$
17. Sample answer:
$61 \div 2\frac{4}{5} \rightarrow 60 \div 3 = 20$
18. Sample answer:
$148 \div 3\frac{1}{4} \rightarrow 150 \div 3 = 50$
19. Sample answer:
$79 \div 1\frac{9}{10} \rightarrow 80 \div 2 = 40$
20. Sample answer:
$75 \div 2\frac{11}{16} \rightarrow 75 \div 3 = 25$
21. Sample answer:
$88 \div 2\frac{1}{8} \rightarrow 88 \div 2 = 44$
22. Sample answer:
$3\frac{1}{2} \times 2\frac{1}{4} \rightarrow 4 \times 2 = 8$
About 8 cups of flour are needed.
23. Sample answer:
$17\frac{1}{2} \div 2\frac{3}{4} \rightarrow 18 \div 3 = 6$
Mario can place about 6 photographs on the poster board.
24. Sample answer:
$18\frac{1}{2} \times 3 \rightarrow 18 \times 3 = 54$
The circumference of the hoop is about 54 inches.

Page 719 Converting Measurements within the Metric System

1. liter
2. kilometer
3. millimeter
4. millimeter
5. centimeter
6. gram
7. meter
8. centimeter
9. millimeter
10. liter
11. millimeter
12. milligram
13. kilogram
14. kilogram
15. 5 km = __?__ m
5 × 1000 = 5000
5 km = 5000 m
16. 3.5 cm = __?__ mm
3.5 × 10 = 35
3.5 cm = 35 mm
17. 6 L = __?__ mL
6 × 1000 = 6000
6 L = 6000 mL
18. 370 mL = __?__ L
370 ÷ 1000 = 0.37
370 mL = 0.37 L
19. 20 mm = __?__ cm
20 ÷ 10 = 2
20 mm = 2 cm
20. 4000 g = __?__ kg
4000 ÷ 1000 = 4
4000 g = 4 kg
21. 18 cm = __?__ mm
18 × 10 = 180
18 cm = 180 mm
22. 0.75 L = __?__ mL
0.75 × 1000 = 750
0.75 L = 750 mL
23. 935 cm = __?__ m
935 ÷ 100 = 9.35
935 cm = 9.35 m
24. 210 mm = __?__ cm
210 ÷ 10 = 21
210 mm = 21 cm
25. 65 g = __?__ kg
65 ÷ 1000 = 0.065
65 g = 0.065 kg
26. 2 m __?__ cm
2 × 100 = 200
2 m = 200 cm
27. 52.9 kg = __?__ g
52.9 × 1000 = 52,900
52.9 kg = 52,900 g
28. 800 m = __?__ km
800 ÷ 1000 = 0.8
800 m = 0.8 km

29. 9.05 kg = ___?___ g
9.05 × 1000 = 9050
9.05 kg = 9050 g

30. 0.62 km = ___?___ m
0.62 × 1000 = 620
0.62 km = 620 m

31. 1250 mL = ___?___ L
1250 ÷ 1000 = 1.25
1250 mL = 1.25 L

32. 20,000 mg = ___?___ g
20,000 ÷ 1000 = 20
20,000 mg = 20 g

33. 3100 m = ___?___ km
3100 ÷ 1000 = 3.1
3100 m = 3.1 km

34. 2.6 m = ___?___ cm
2.6 × 100 = 260
2.6 m = 260 cm

35. 36 mg = ___?___ g
36 ÷ 1000 = 0.036
36 mg = 0.036 g

36. 7 mm = ___?___ cm
7 ÷ 10 = 0.7
7 mm = 0.7 cm

37. 0.085 L = ___?___ mL
0.085 × 1000 = 85
0.085 L = 85 mL

38. 125.9 g = ___?___ kg
125.9 ÷ 1000 = 0.1259
125.9 g = 0.1259 kg

39. 1.56 kg = ___?___ g
1.56 × 1000 = 1560
1.56 kg = 1560 g

40. 0.09 L = ___?___ mL
0.09 × 1000 = 90
0.09 L = 90 mL

41. 10 km = ___?___ m
10 × 1000 = 10,000
10 km = 10,000 m

42. 0.58 m = ___?___ cm
0.58 × 100 = 58
0.58 m = 58 cm

43. 355 mL = ___?___ L
355 ÷ 1000 = 0.355
355 mL = 0.355 L

Page 721 Converting Measurements within the Customary System

1. 5 ft = ___?___ in.
5 × 12 = 60
5 ft = 60 in.

2. 2 gal = ___?___ qt
2 × 4 = 8
2 gal = 8 qt

3. 96 oz = ___?___ lb
96 ÷ 16 = 6
96 oz = 6 lb

4. 2 T = ___?___ lb
2 × 2000 = 4000
2 T = 4000 lb

5. 9 ft = ___?___ yd
9 ÷ 3 = 3
9 ft = 3 yd

6. 6 c = ___?___ pt
6 ÷ 2 = 3
6 c = 3 pt

7. 2 mi = ___?___ ft
2 × 5280 = 10,560
2 mi = 10,560 ft

8. 72 in. = ___?___ ft
72 ÷ 12 = 6
72 in. = 6 ft

9. 3 lb = ___?___ oz
3 × 16 = 48
3 lb = 48 oz

10. 7 yd = ___?___ ft
7 × 3 = 21
7 yd = 21 ft

11. 32 fl oz = ___?___ c
32 ÷ 8 = 4
32 fl oz = 4 c

12. 15,840 ft = ___?___ mi
15,840 ÷ 5280 = 3
15,840 ft = 3 mi

13. 2 qt = ___?___ pt
2 × 2 = 4
2 qt = 4 pt

14. 5 pt = ___?___ c
5 × 2 = 10
5 pt = 10 c

15. 16 qt = ___?___ gal
16 ÷ 4 = 4
16 qt = 4 gal

16. 3000 lb = ___?___ T
3000 ÷ 2000 = 1.5
3000 lb = 1.5 T

17. 6 pt = ___?___ qt
6 ÷ 2 = 3
6 pt = 3 qt

18. 8 pt = ___?___ c
8 × 2 = 16
8 pt = 16 c

19. 14 pt = ___?___ qt
14 ÷ 2 = 7
14 pt = 7 qt

20. 8 yd = ___?___ ft
8 × 3 = 24
8 yd = 24 ft

21. 5 gal = ___?___ qt
5 × 4 = 20
5 gal = 20 qt

22. 36 qt = ___?___ gal
36 ÷ 4 = 9
36 qt = 9 gal

23. 5 c = ___?___ fl oz
5 × 8 = 40
5 c = 40 fl oz

24. 120 in. = __?__ ft
120 ÷ 12 = 10
120 in. = 10 ft

25. 30 in. = __?__ ft
30 ÷ 12 = 2.5
30 in. = 2.5 ft

26. 6.5 lb = __?__ oz
6.5 × 16 = 104
6.5 lb = 104 oz

27. 12 oz = __?__ lb
12 ÷ 16 = 0.75
12 oz = 0.75 lb

28. 1 yd = __?__ in.
1 yd = 3 ft
3 × 12 = 36
1 yd = 36 in.

29. 1 lb = 16 oz
1 T = 2000 lb
16 × 2000 = 32,000
1 T = 32,000 oz

30. 1 gal = 4 qt
4 × 2 = 8
4 qt = 8 pt
8 × 2 = 16
8 pt = 16 c
1 gal = 16 c

Page 723 Displaying Data in Graphs

1. Sample answer: A bar graph would show the different frequencies of the data.
2. Sample answer: A circle graph would show how each category compares to the whole class.
3. Sample answer: A double bar graph would make a side-by-side comparison of the two sets of data.
4. Sample answer: A line graph would show the wage's increase during those years.
5. Sample answer: A double line graph would show the changes in boys' and girls' participation over time.
6. Sample answer: The data would be best displayed using a line graph because it is easier to see the increase in events over time.
7. Sample answer: It shows how each age group makes up the whole group of Internet users.
8. Sample answer: A bar graph would also compare the different age groups, but a circle graph is more useful in comparing the parts to the whole.

Extra Practice

Page 724 Lesson 1-1

1a. The table shows the weight of a letter and the respective cost to mail it to another country. We need to find how much it will cost to mail a 5.5-ounce letter.

1b. Use the information in the table to solve the problem. Look for a pattern in the costs. Extend the pattern to find the cost for a 5.5-ounce letter.

1c. It costs $0.95 for the first ounce, and $0.39 for each additional half-ounce. To mail a 5.5-ounce letter, it would cost $0.95 for the first ounce and 9 × $0.39 or $3.51 for the 4.5 additional ounces. Therefore, the answer is $0.95 + $3.51 = $4.46.

1d. See students' work.

2a. Since the problem says "about," estimation is the best method of computation.

2b. See students' work.
Sample answer: about $30 million

3. 3 8 13 18 23 ?
(+5 +5 +5 +5 +5)

Assuming the pattern continues, the next term is 23 + 5 or 28.

4. 32 29 26 23 20 ?
(−3 −3 −3 −3 −3)

Assuming the pattern continues, the next term is 20 − 3 = 17.

5. 6 7 9 12 16 ?
(+1 +2 +3 +4 +5)

Assuming the pattern continues, the next term is 16 + 5 or 21.

Page 724 Lesson 1-2

1. $8 + 7 + 12 \div 4 = 8 + 7 + 3$
$= 18$

2. $20 \div 4 - 5 + 12 = 5 - 5 + 12$
$= 12$

3. $(25 \cdot 3) + (10 \cdot 3) = 75 + 30$
$= 105$

4. $36 \div 6 + 7 - 6 = 6 + 7 - 6$
$= 7$

5. $30 \cdot (6 - 4) = 30 \cdot 2$
$= 60$

6. $(40 \cdot 2) - (6 \cdot 11) = 80 - 66$
$= 14$

7. $\frac{86 - 11}{11 + 4} = (86 - 11) \div (11 + 4)$
$= 75 \div 15$
$= 5$

8. $\frac{12 + 84}{11 + 13} = (12 + 84) \div (11 + 13)$
$= 96 \div 24$
$= 4$

9. $\frac{5 \cdot 5 + 5}{5 \cdot 5 - 15} = (5 \cdot 5 + 5) \div (5 \cdot 5 - 15)$
$= (25 + 5) \div (25 - 15)$
$= 30 \div 10$
$= 3$

10. $(19 - 8)4 = (11)4$
$= 44$

11. $75 - 5(2 \cdot 6) = 75 - 5(12)$
$= 75 - 60$
$= 15$

12. $81 \div 27 \times 6 - 2 = 3 \times 6 - 2$
$= 18 - 2$
$= 16$

13. thirty-two divided by the product of four and two
32 ÷ (4 × 2)
Expression: $32 \div (4 \times 2) = 32 \div 8$
$= 4$

14. *Phrase*: three increased by nine
Key Word: increased
Expression: 3 + 9

15. *Phrase*: fifteen divided by three
Key Words: divided by
Expression: 15 ÷ 3

16. *Phrase*: six less than ten
Key Words: less than
Expression: 10 − 6

Page 724 Lesson 1-3

1. $ba - ac = (4)(2) - (2)(3)$
$= 8 - 6$
$= 2$

2. $4b + a \cdot a = (4)(4) + 2 \cdot 2$
$= 16 + 4$
$= 20$

3. $11 \cdot c - ab = 11(3) - (2)(4)$
$= 33 - 8$
$= 25$

4. $4b - (a + c) = (4)(4) - (2 + 3)$
$= 16 - 5$
$= 11$

5. $7(a + b) - c = 7(2 + 4) - 3$
$= 7(6) - 3$
$= 42 - 3$
$= 39$

6. $8a + 8b = 8(2) + 8(4)$
$= 16 + 32$
$= 48$

7. $\frac{8(a + b)}{4c} = 8(a + b) \div 4c$
$= 8(2 + 4) \div 4(3)$
$= 8(6) \div 12$
$= 48 \div 12$
$= 4$

8. $36 - 12c = 36 - 12(3)$
$= 36 - 36$
$= 0$

9. $\frac{9(b + a)}{c - 1} = 9(b + a) \div (c - 1)$
$= 9(4 + 2) \div (3 - 1)$
$= 9(6) \div 2$
$= 54 \div 2$
$= 27$

10. $abc - bc = (2)(4)(3) - (4)(3)$
$= 24 - 12$
$= 12$

11. $28 - bc + a = 28 - (4)(3) + 2$
$= 28 - 12 + 2$
$= 18$

12. $a(b - c) = 2(4 - 3)$
$= 2(1)$
$= 2$

13. $\underbrace{\text{nine}}_{9}\ \underbrace{\text{more than}}_{+}\ \underbrace{a}_{a}$
The expression is $a + 9$.

14. $\underbrace{k}_{k}\ \underbrace{\text{less}}_{-}\ \underbrace{\text{eleven}}_{11}$
The expression is $k - 11$.

15. $\underbrace{\text{three}}_{3}\ \underbrace{\text{times}}_{\times}\ \underbrace{p}_{p}$
The expression is $3p$.

16. Let n represent the number.
$\underbrace{\text{some number}}_{n}\ \underbrace{\text{product}}_{\times}\ \underbrace{\text{five}}_{5}$
The expression is $5n$.

17. Let s represent Shelly's score.
$\underbrace{\text{twice}}_{2}\ \underbrace{\text{Shelly's score}}_{s}\ \underbrace{\text{decreased by}}_{-}\ \underbrace{18}_{18}$
The expression is $2s - 18$.

18. $\underbrace{16}_{16}\ \underbrace{\text{quotient}}_{\div}\ \underbrace{n}_{n}$
The expression is $16 \div n$.

Page 725 Lesson 1-4

1. A number or variable was multiplied by 1. This is the Multiplicative Identity Property.
2. The grouping of the numbers or variables changed. This is the Associative Property of Addition.
3. The grouping of the numbers or variables changed. This is the Associative Property of Multiplication.
4. The order of the numbers or variables changed. This is the Commutative Property of Multiplication.
5. The number was multiplied by zero. This is the Multiplicative Property of Zero.
6. The order of the numbers or variables changed. This is the Commutative Property of Addition.
7. $6 + 8 + 14 = (6 + 14) + 8$
$= 20 + 8$
$= 28$

8. $5 \cdot 18 \cdot 2 = (5 \cdot 2) \cdot 18$
$= 10 \cdot 18$
$= 180$

9. $0(13 \cdot 6) = (0 \cdot 13)6$
$= 0 \cdot 6$
$= 0$

10. $8 + 4 + 12 + 16 = (8 + 12) + (4 + 16)$
$= 20 + 20$
$= 40$

11. $8 \cdot 20 \cdot 10 = (8 \cdot 10) \cdot 20$
$= 80 \cdot 20$
$= 1600$

12. $4 \cdot 14 \cdot 5 = (4 \cdot 5) \cdot 14$
$= 20 \cdot 14$
$= 280$

13. $(12 + x) + 9 = x + (12 + 9)$
$= x + 21$

14. $2 \cdot (6 \cdot x) = (2 \cdot 6)x$
$= 12x$

15. $(5 \cdot m) \cdot 3 = (5 \cdot 3) \cdot m$
$= 15m$

Page 725 Lesson 1-5

1.

Value for f	$16 - f = 11$	True or False?
3	$16 - 3 \stackrel{?}{=} 11$	false
5	$16 - 5 \stackrel{?}{=} 11$	true ✓
7	$16 - 7 \stackrel{?}{=} 11$	false

Therefore, the solution of $16 - f = 11$ is 5.

2.

Value for m	$9 = \frac{72}{m}$	True or False?
8	$9 \stackrel{?}{=} \frac{72}{8}$	true ✓
9	$9 \stackrel{?}{=} \frac{72}{9}$	false
11	$9 \stackrel{?}{=} \frac{72}{11}$	false

Therefore, the solution of $9 = \frac{72}{m}$ is 8.

3.

Value for b	$4b + 1 = 17$	True or False?
3	$4(3) + 1 \stackrel{?}{=} 17$	false
4	$4(4) + 1 \stackrel{?}{=} 17$	true ✓
5	$4(5) + 1 \stackrel{?}{=} 17$	false

Therefore, the solution of $4b + 1 = 17$ is 4.

4.

Value for r	$17 + r = 25$	True or False?
6	$17 + 6 \stackrel{?}{=} 25$	false
7	$17 + 7 \stackrel{?}{=} 25$	false
8	$17 + 8 \stackrel{?}{=} 25$	true ✓

Therefore, the solution of $17 + r = 25$ is 8.

5.

Value for n	$9 = 7n - 12$	True or False?
3	$9 \stackrel{?}{=} 7(3) - 12$	true ✓
5	$9 \stackrel{?}{=} 7(5) - 12$	false
7	$9 \stackrel{?}{=} 7(7) - 12$	false

Therefore, the solution of $9 = 7n - 12$ is 3.

6.

Value for q	$67 = 98 - q$	True or False?
21	$67 \stackrel{?}{=} 98 - 21$	false
26	$67 \stackrel{?}{=} 98 - 26$	false
31	$67 \stackrel{?}{=} 98 - 31$	true ✓

Therefore, the solution of $67 = 98 - q$ is 31.

7. $13 - u = 7$
$13 - 6 = 7$
$u = 6$

8. $23 = w + 6$
$23 = 17 + 6$
$w = 17$

9. $88 + y = 96$
$88 + 8 = 96$
$y = 8$

10. $9z = 45$
$9(5) = 45$
$z = 5$

11. $88 = 11d$
$88 = 11(8)$
$d = 8$

12. $5t = 0$
$5(0) = 0$
$t = 0$

13. $13g = 39$
$13(3) = 39$
$g = 3$

14. $\frac{x}{2} = 8$
$\frac{16}{2} = 8$
$x = 16$

15. $\frac{84}{h} = 12$
$\frac{84}{7} = 12$
$h = 7$

16. Let x = the number.
The sum of a number and 8 is 14.
$x + 8 = 14$
$x + 8 = 14$
$6 + 8 = 14$
$x = 6$

17. Let n = the number.
A number less than 12 is 50.
$n - 12 = 50$
$n - 12 = 50$
$62 - 12 = 50$
$n = 62$

18. Let n = the number.
The product of a number and 10 is 70.
$10n = 70$
$10n = 70$
$10(7) = 70$
$n = 7$

19. Let n = the number.
A number divided by three is nine.
$n \div 3 = 9$
$n \div 3 = 9$
$27 \div 3 = 9$
$n = 27$

Page 725 Lesson 1-6

1. Start at the origin and move 9 units to the right and 7 units up. The ordered pair (9, 7) is for point P.
2. Start at the origin and move 5 units to the right and 5 units up. The ordered pair (5, 5) is for point N.
3. Start at the origin and move 3 units to the right and 1 unit up. The ordered pair (3, 1) is for point Q.
4. Start at the origin and move 2 units to the right and 7 units up. The ordered pair (2, 7) is for point B.
5. Start at the origin and move 8 units to the right and 4 units up. The ordered pair (8, 4) is for point S.
6. Start at the origin and move 4 units to the right. Since the y-coordinate is 0, do not move up or down. The ordered pair (4, 0) is for point T.
7. The x-coordinate of R is 4, and the y-coordinate is 8. The ordered pair for point R is (4, 8).
8. The x-coordinate of P is 9, and the y-coordinate is 7. The ordered pair for point P is (9, 7).
9. The x-coordinate of W is 5, and the y-coordinate is 1. The ordered pair for point W is (5, 1).
10. The x-coordinate of C is 0, and the y-coordinate is 6. The ordered pair for point C is (0, 6).
11. The x-coordinate of D is 0, and the y-coordinate is 8. The ordered pair for point D is (0, 8).
12. The x-coordinate of F is 3, and the y-coordinate is 4. The ordered pair for point F is (3, 4).

13.

x	y
3	6
4	9
5	1

domain = {3, 4, 5};
range = {6, 9, 1}

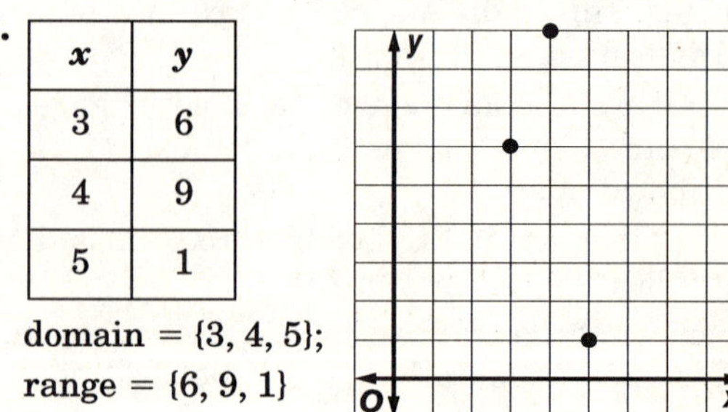

14.

x	y
2	1
4	4
6	7
4	3

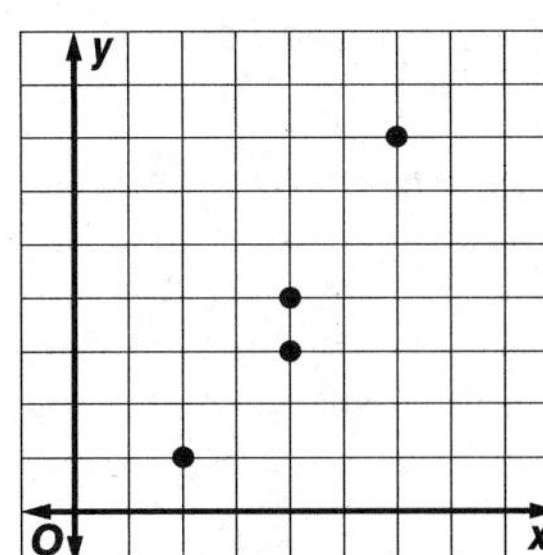

domain = {2, 4, 6};
range = {1, 4, 7, 3}
(Note: 4 is a repeated value in the domain.)

Page 726 Lesson 1-7

1. As the speed of the airplane increases, the miles traveled increase; positive relationship.

2. A person's shoe size is not affected by their weight; no relationship.

3. As the outside temperature decreases, the heating bill increases; negative relationship.

4.

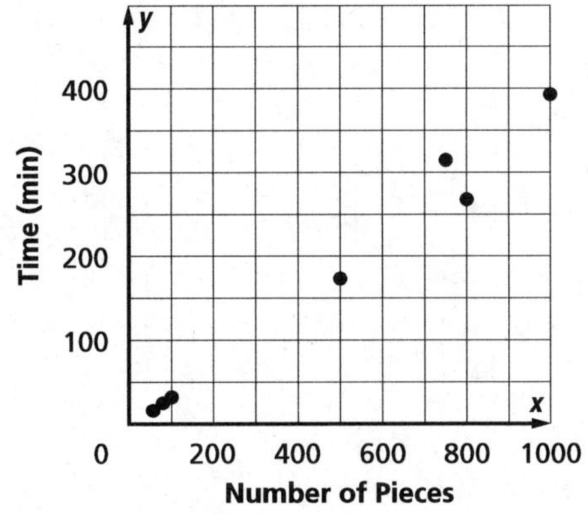

5. As the time increases, the number of pieces increases; positive relationship.

6. By looking at the pattern in the graph, we can predict that a puzzle with 650 pieces will take about 260 minutes.

Page 726 Lesson 2-1

1. $-4 > -8$

2. $-6 < 3$

3. $0 > -5$

4. $-12 < -9$

5. $12 > -25$

6. $3 > -7$

7. $0 > -2$

8. $-15 < 12$

9. $5 > -7$

10. $|6| > -2$, since $|6| = 6$

11. $-2 < |-3|$, since $|-3| = 3$.

12. $|-7| > |-4|$, since $|-7| = 7$ and $|-4| = 4$.

13. $\{-5, -1, 2\}$

14. $\{-9, -2, 0, 5, 8\}$

15. $\{-86, -34, 0, 21, 100\}$

16. $\{-43, -40, -1, 8, 16, 27\}$

17. $\{-23, -15, 0, 24, 75\}$

18. $\{-6, -5, 6, 18\}$

19. $|-3| + |9| = 3 + 9$
$= 12$

20. $|-18| - |5| = 18 - 5$
$= 13$

21. $|12 + 7| = |19|$
$= 19$

22. $-|6| = -6$

23. $|-8| + |4| = 8 + 4$
$= 12$

24. $-|-20| = -(20)$
$= -20$

25. $|15 - 12| = |3|$
$= 3$

26. $|8 + 9| = |17|$
$= 17$

27. $-|4| \cdot |-5| = -4 \cdot 5$
$= -20$

28. $|-6| \cdot |8| = 6 \cdot 8$
$= 48$

29. $-|12| \cdot |9| = -12 \cdot 9$
$= -108$

30. $-\|-16| + |-22\| = -|16 + 22|$
$= -|38|$
$= -(38)$
$= -38$

Page 726 Lesson 2-2

1. $5 + (-6) = -1$

2. $-17 + 24 = 7$

3. $15 + (-29) = -14$

4. $-6 + 13 = 7$

5. $50 + (-14) = 36$

6. $-21 + (-4) = -25$

7. $30 + (-7) = 23$

8. $(-3) + (-10) = -13$

9. $-15 + 26 = 11$

10. $-17 + 4 + (-2) = -17 + (-2) + 4$
$= [-17 + (-2)] + 4$
$= -19 + 4$
$= -15$

11. $50 + (-16) + (-11) = 50 + [(-16) + (-11)]$
$= 50 + (-27)$
$= 23$

12. $-17 + 8 + (-14) = -17 + (-14) + 8$
$= [-17 + (-14)] + 8$
$= -31 + 8$
$= -23$

13. $-11 + 15 + (-6) = -11 + (-6) + 15$
$= [-11 + (-6)] + 15$
$= -17 + 15$
$= -2$

14. $23 + (-64) = -41$

15. $-1 + 14 + (-13) = -1 + (-13) + 14$
$= [-1 + (-13)] + 14$
$= -14 + 14$
$= 0$

16. $33 + (-18) + 7 = 33 + 7 + (-18)$
$= (33 + 7) + (-18)$
$= 40 + (-18)$
$= 22$

17. $-75 + (-13) = -88$

18. $26 + 14 + (-71) = (26 + 14) + (-71)$
$= 40 + (-71)$
$= -31$

19. $8 + (-9) + (-1) = 8 + [(-9) + (-1)]$
$= 8 + (-10)$
$= -2$

20. $-16 + (-12) + 13 = [-16 + (-12)] + 13$
$= -28 + 13$
$= -15$

21. $35 + (-60) = -25$

22. $12 + (-20) + 16 = 12 + 16 + (-20)$
$= (12 + 16) + (-20)$
$= 28 + (-20)$
$= 8$

23. $100 + (-54) + (-17) = 100\ [(-54) + (-17)]$
$= 100 + (-71)$
$= 29$

24. $11 + (-22) + (-33) = 11 + [(-22) + (-33)]$
$= 11 + (-55)$
$= -44$

Page 727 Lesson 2-3

1. $8 - 17 = 8 + (-17)$
$= -9$

2. $-15 - 3 = -15 + (-3)$
$= -18$

3. $10 - 21 = 10 + (-21)$
$= -11$

4. $20 - (-5) = 20 + 5$
$= 25$

5. $5 - (-9) = 5 + 9$
$= 14$

6. $-12 - (-7) = -12 + 7$
$= -5$

7. $-19 - (-6) = -19 + 6$
$= -13$

8. $-16 - (-23) = -16 + 23$
$= 7$

9. $-56 - 32 = -56 + (-32)$
$= -88$

10. $-49 - (-52) = -49 + 52$
$= 3$

11. $-6 - 9 - (-7) = -6 + (-9) + 7$
$= -15 + 7$
$= -8$

12. $-6 - (-10) = -6 + 10$
$= 4$

13. $17 - 33 = 17 + (-33)$
$= -16$

14. $-21 - 19 = -21 + (-19)$
$= -40$

15. $12 - (-24) = 12 + 24$
$= 36$

16. $-35 - (-18) = -35 + 18$
$= -17$

17. $-54 - 27 = -54 + (-27)$
$= -81$

18. $32 - (-18) = 32 + 18$
$= 50$

19. $-26 - (-41) = -26 + 41$
$= 15$

20. $99 - (-1) = 99 + 1$
$= 100$

21. $-12 - (-25) = -12 + 25$
$= 13$

22. $18 - (-43) = 18 + 43$
$= 61$

23. $-66 - 13 = -66 + (-13)$
$= -79$

24. $54 - 100 = 54 + (-100)$
$= -46$

25. $y - z = -8 - (-3)$
$= -8 + 3$
$= -5$

26. $3 - z = 3 - (-3)$
$= 3 + 3$
$= 6$

27. $y - 5 = -8 - 5$
$= -8 + (-5)$
$= -13$

28. $x - y = 6 - (-8)$
$= 6 + 8$
$= 14$

29. $14 - y - x = 14 - (-8) - 6$
$= 14 + 8 - 6$
$= 22 - 6$
$= 16$

30. $6 + x - z = 6 + 6 - (-3)$
$= 12 + 3$
$= 15$

31. $y + z + w = -8 + (-3) + 4$
$= -11 + 4$
$= -7$

32. $w - z + 11 = 4 - (-3) + 11$
$= 4 + 3 + 11$
$= 7 + 11$
$= 18$

Page 727 Lesson 2-4

1. $-4(2) = -8$

2. $-8(-5) = 40$

3. $13(-4) = -52$

4. $-5 \cdot 6 \cdot 10 = [-5 \cdot 6] \cdot 10$
$= -30 \cdot 10$
$= -300$

5. $-6(-2)(-14) = [-6(-2)](-14)$
$= 12(-14)$
$= -168$

6. $18(-3)(6) = [18(-3)](6)$
$= -54(6)$
$= -324$

7. $4(-10)(-3) = [4(-10)](-3)$
$= -40(-3)$
$= 120$

8. $-9(3)(2) = [-9(3)](2)$
$= -27(2)$
$= -54$

9. $12(-8) = -96$

10. $-3 \cdot 5x = (-3 \cdot 5)x$
$= -15x$

11. $7(-8m) = [7 \cdot (-8)] \cdot m$
$= -56m$

12. $-10(-3k) = [-10(-3)] \cdot k$
$= 30k$

13. $-4y(-8z) = (-4)(y)(-8)(z)$
$= [(-4)(-8)](y \cdot z)$
$= 32yz$

14. $(-2r)(-3s) = (-2)(r)(-3)(s)$
$= [(-2)(-3)](r \cdot s)$
$= 6rs$

15. $6(-2m)(3n) = (6)(-2)(m)(3)(n)$
$= [6(-2)(3)](m \cdot n)$
$= [-12(3)](m \cdot n)$
$= -36mn$

16. $-6t = -6(15)$
$= -90$

17. $7p = 7(-9)$
$= -63$

18. $-4k = -4(-16)$
$= 64$

19. $aw = (0)(-72)$
$= 0$

20. $dk = (-12)(11)$
$= -132$

21. $st = (-8)(-10)$
$= 80$

22. $3hp = 3(9)(-3)$
$= [3(9)](-3)$
$= 27(-3)$
$= -81$

23. $-5bc = -5(-6)(2)$
$= [-5(-6)](2)$
$= 30(2)$
$= 60$

24. $-4wx = -4(-1)(-8)$
$= [-4(-1)](-8)$
$= 4(-8)$
$= -32$

Page 727 Lesson 2-5

1. $-36 \div 9 = -4$
2. $112 \div (-8) = -14$
3. $-72 \div 2 = -36$
4. $-26 \div (-13) = 2$
5. $-144 \div 6 = -24$
6. $-180 \div (-10) = 18$
7. $304 \div (-8) = -38$
8. $-216 \div (-9) = 24$
9. $80 \div (-5) = -16$
8. $-105 \div 15 = -7$
11. $120 \div (-30) = -4$
12. $-200 \div (-8) = 25$
13. $42 \div (-6) = -7$
14. $144 \div (-12) = -12$
15. $-360 \div 9 = -40$
16. $-84 \div (-6) = 14$
17. $125 \div (-5) = -25$
18. $180 \div (-15) = -12$
19. $-400 \div 20 = -20$
20. $72 \div (-9) = -8$
21. $-156 \div (-2) = 78$

22. $25 \div x = 25 \div (-5)$
$= -5$

23. $-42 \div w = -42 \div 7$
$= -6$

24. $3 \div y = 3 \div (-3)$
$= -1$

25. $2x \div z = 2(-5) \div 2$
$= [2(-5)] \div 2$
$= -10 \div 2$
$= -5$

26. $-3x \div y = -3(-5) \div (-3)$
$= [-3(-5)] \div (-3)$
$= 15 \div (-3)$
$= -5$

27. $x \div (-1) = (-5) \div (-1)$
$= 5$

28. $xyz \div 10 = (-5)(-3)(2) \div 10$
$= [(-5)(-3)](2) \div 10$
$= 15(2) \div 10$
$= [15(2)] \div 10$
$= 30 \div 10$
$= 3$

29. $yz \div 2 = (-3)(2) \div 2$
$= [(-3)(2)] \div 2$
$= -6 \div 2$
$= -3$

30. $\frac{3y}{-3} = \frac{3(-3)}{-3}$
$= \frac{[3(-3)]}{-3}$
$= \frac{-9}{-3}$
$= -9 \div (-3)$
$= 3$

31. $\frac{6-y}{y} = \frac{6-(-3)}{-3}$
$= \frac{6+3}{-3}$
$= \frac{9}{-3}$
$= 9 \div (-3)$
$= -3$

32. $\frac{w}{-7} = \frac{7}{-7}$
$= 7 \div (-7)$
$= -1$

33. $\frac{w-x}{y} = \frac{7-(-5)}{-3}$
$= \frac{7+5}{-3}$
$= \frac{12}{-3}$
$= 12 \div (-3)$
$= -4$

Page 728 Lesson 2-6

1. Start at the origin, and move 6 units to the left and 8 units up. The ordered pair for point D is $(-6, 8)$.
2. Start at the origin, and move 1 unit to the right and 2 units down. The ordered pair for point J is $(1,-2)$.

3. Start at the origin, and move 9 units to the right and 2 units up. The ordered pair for point C is (9, 2).

4. Start at the origin, and move 1 unit to the right and 4 units up. The ordered pair for point L is (1, 4).

5. Start at the origin, and move 3 units to the left and 4 units down. The ordered pair for point B is $(-3,-4)$.

6. Start at the origin, and move 2 units to the right and 5 units up. The ordered pair for point N is (2, 5).

7. Start at the origin, and move 3 units to the right. Do not move up or down. The ordered pair for point K is (3, 0).

8. Start at the origin, and move 5 units to the right and 1 unit down. The ordered pair for point M is $(5,-1)$.

9–14.

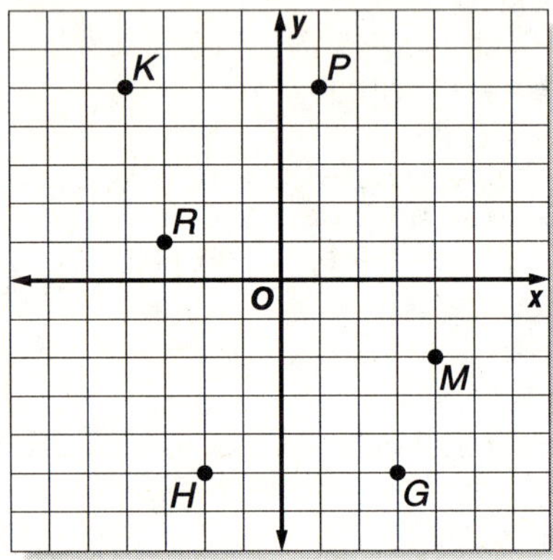

9. Start at the origin. Move 2 units to the left. Then move 5 units down and draw a dot. Point H is in Quadrant III.

10. Start at the origin. Move 1 unit to the right. Then move 5 units up and draw a dot. Point P is in Quadrant I.

11. Start at the origin. Move 3 units to the left. Then move 1 unit up and draw a dot. Point R is in Quadrant II.

12. Start at the origin. Move 4 units to the right. Then move 2 units down and draw a dot. Point M is in Quadrant IV.

13. Start at the origin. Move 4 units to the left. Then move 5 units up and draw a dot. Point K is in Quadrant II.

14. Start at the origin. Move 3 units to the right. Then move 5 units down and draw a dot. Point F is in Quadrant IV.

Page 728 Lesson 3-1

1. $2(4 + 5) = 2 \cdot 4 + 2 \cdot 5$
$= 8 + 10$
$= 18$

2. $4(5 + 3) = 4 \cdot 5 + 4 \cdot 3$
$= 20 + 12$
$= 32$

3. $3(7 - 6) = 3[7 + (-6)]$
$= 3(7) + 3(-6)$
$= 21 + (-18)$
$= 3$

4. $(2 + 5)9 = 2 \cdot 9 + 5 \cdot 9$
$= 18 + 45$
$= 63$

5. $(10 - 4)3 = [10 + (-4)]3$
$= 10(3) + (-4)(3)$
$= 30 + (-12)$
$= 18$

6. $-6(1 + 3) = -6 \cdot 1 + (-6 \cdot 3)$
$= -6 + (-18)$
$= -24$

7. $3(m + 4) = 3m + 3 \cdot 4$
$= 3m + 12$

8. $(y + 7)5 = y \cdot 5 + 7 \cdot 5$
$= 5y + 35$

9. $-6(x + 3) = -6x + (-6)(3)$
$= -6x - 18$

10. $(p - 4)5 = [p + (-4)]5$
$= 5p + 5(-4)$
$= 5p - 20$

11. $-3(s - 9) = -3[s + (-9)]$
$= -3s + (-3)(-9)$
$= -3s + 27$

12. $5(x + y) = 5x + 5y$

13. $b(c + 3d) = bc + b \cdot 3d$
$= bc + 3bd$

14. $(a - b)(-5) = [a + (-b)](-5)$
$= (-5)a + (-5)(-b)$
$= -5a + 5b$

15. $-6(v - 3w) = (-6)[v + (-3w)]$
$= -6v + (-6)(-3w)$
$= -6v + 18w$

16. $5(x + 12) = 5x + 5 \cdot 12$
$= 5x + 60$

17. $(m - 6)(4) = [m + (-6)](4)$
$= 4m + 4(-6)$
$= 4m - 24$

18. $-2(a - b) = -2[a + (-b)]$
$= -2a + (-2)(-b)$
$= -2a + 2b$

19. $(8 - m)(-3) = [8 + (-m)](-3)$
$= (8)(-3) + (-m)(-3)$
$= -24 + 3m$

20. $8(p + 3q) = 8p + 8 \cdot 3q$
$= 8p + 24q$

21. $(2x + 3y)(4) = 2x \cdot (4) + 3y \cdot (4)$
$= 8x + 12y$

22. $-5(9 - z) = -5[9 + (-z)]$
$= -5(9) + (-5)(-z)$
$= -45 + 5z$

23. $21(k - 3) = 21[k + (-3)]$
$= 21k + 21(-3)$
$= 21k - 63$

24. $(7 - 2h)(-3) = [7 + (-2h)](-3)$
$= 7(-3) + (-3)(-2h)$
$= -21 + 6h$

Page 728 Lesson 3-2

1. terms: 3, $4x$, x; like terms: $4x$, x
2. terms: $5n$, 2, $-3n$; like terms: $5n$, $-3n$
3. terms: 6, 1, $7y$; like terms: 6, 1
4. terms: $2c$, c, $8d$; like terms: $2c$, c
5. terms: $3a$, $-a$, b; like terms: none
6. terms: 2, $6k$, 7, $-5k$; like terms: 2, 7, $6k$, $-5k$

7. $8k + 2k + 7 = (8 + 2)k + 7$
$= 10k + 7$

8. $3 + 2b + b = 3 + (2 + 1)b$
$= 3 + 3b$

9. $t + 2t = 1t + 2t$
$= (1 + 2)t$
$= 3t$

10. $9(3 + 2x) = 9 \cdot 3 + 9 \cdot 2x$
$= 27 + 18x$

11. $4(y + 2) - 2 = 4y + 4 \cdot 2 - 2$
$= 4y + 8 - 2$
$= 4y + 6$

12. $(6 + 3e)4 = 6 \cdot 4 + 3e \cdot 4$
$= 24 + 12e$

13. $4 + 9c + 3(c + 2) = 4 + 9c + 3c + 3 \cdot 2$
$= 4 + (9 + 3)c + 6$
$= 4 + 6 + 12c$
$= 10 + 12c$

14. $5(7 + 2s) + 3(s + 4) = 5 \cdot 7 + 5 \cdot 2s + 3s + 3 \cdot 4$
$= 35 + 10s + 3s + 12$
$= (35 + 12) + 10s + 3s$
$= 47 + (10 + 3)s$
$= 47 + 13s$ or $13s + 47$

15. $9(f + 2) + 14f = 9f + 9 \cdot 2 + 14f$
$= 9f + 14f + 18$
$= (9 + 14)f + 18$
$= 23f + 18$

16. $5a - 9a = (5 - 9)a$
$= -4a$

17. $-6 + 4x + 9 - 2x = -6 + 9 + 4x - 2x$
$= 3 + (4 - 2)x$
$= 3 + 2x$

18. $6a + 11 + (-15) + 9a = 6a + 9a + 11 + (-15)$
$= (6 + 9)a + [11 + (-15)]$
$= 15a + (-4)$
$= 15a - 4$

19. $2(8w - 7) = 2 \cdot 8w - 2 \cdot 7$
$= 16w - 14$

20. $3(2d + 5) + 4d = 3 \cdot 2d + 3 \cdot 5 + 4d$
$= 6d + 15 + 4d$
$= 6d + 4d + 15$
$= (6 + 4)d + 15$
$= 10d + 15$

21. $2 + 4p - 6(p - 2) = 2 + 4p - 6p - 6(-2)$
$= 2 + (4 - 6)p + 12$
$= 2 + 12 - 2p$
$= 14 - 2p$ or $-2p + 14$

22. $-3(b + 4) = -3b - 3 \cdot 4$
$= -3b - 12$

23. $-6 + 3s + 11 - 5s = (-6 + 11) + 3s - 5s$
$= 5 + (3 - 5)s$
$= 5 - 2s$ or $-2s + 5$

24. $3(x - 5) + 7(x + 2) = 3x + 3(-5) + 7x + 7 \cdot 2$
$= 3x - 15 + 7x + 14$
$= 3x + 7x - 15 + 14$
$= 10x - 1$

25. $3q - r + q + 6r = 3q + 1q - 1r + 6r$
$= (3 + 1)q + [(-1) + 6]r$
$= 4q + 5r$

26. $8(r + 1) + 7 = 8r + 8 \cdot 1 + 7$
$= 8r + 8 + 7$
$= 8r + 15$

27. $3p - 2(p + 6q) = 3p - 2p - 2(6q)$
$= (3 - 2)p - 12q$
$= 1p - 12q$ or $p - 12q$

28. $a + 2b + 4a = 1a + 4a + 2b$
$= (1 + 4)a + 2b$
$= 5a + 2b$

29. $9x - 12 + 12 = 9x + [(-12) + 12]$
$= 9x + 0$
$= 9x$

30. $1 + g + 5g - 2 = [1 + (-2)] + 1g + 5g$
$= -1 + (1 + 5)g$
$= -1 + 6g$ or $6g - 1$

Page 729 Lesson 3-3

Exercises 1–21 For checks, see students' work.

1. $y + 49 = 26$
$y + 49 - 49 = 26 - 49$
$y = -23$
The solution is -23.

2. $d + 31 = -24$
$d + 31 - 31 = -24 - 31$
$d = -55$
The solution is -55.

3. $q - 8 = 16$
$q - 8 + 8 = 16 + 8$
$q = 24$
The solution is 24.

4. $x - 16 = 32$
$x - 16 + 16 = 32 + 16$
$x = 48$
The solution is 48.

5. $40 = a + 12$
$40 - 12 = a + 12 - 12$
$28 = a$
The solution is 28.

6. $b + 12 = -1$
$b + 12 - 12 = -1 - 12$
$b = -13$
The solution is -13.

7. $21 = u + 6$
$21 - 6 = u + 6 - 6$
$15 = u$
The solution is 15.

8. $-52 = p + 5$
$-52 - 5 = p + 5 - 5$
$-57 = p$
The solution is -57.

9. $-14 = 5 - g$
$-14 - 5 = 5 - 5 - g$
$-19 = -g$
$\frac{-19}{-1} = \frac{-1g}{-1}$
$19 = g$
The solution is 19.

10. $121 = k + (-12)$
$121 + 12 = k + (-12) + 12$
$133 = k$
The solution is 133.

11. $-234 = m - 94$
$-234 + 94 = m - 94 + 94$
$-140 = m$
The solution is -140.

12. $110 = x + 25$
$110 - 25 = x + 25 - 25$
$85 = x$
The solution is 85.

13. $f - 7 = 84$
$f - 7 + 7 = 84 + 7$
$f = 91$
The solution is 91.

14. $y - 864 = 652$
$y - 864 + 864 = 652 + 864$
$y = 1516$
The solution is 1516.

15. $475 + z = -18$
$475 - 475 + z = -18 - 475$
$z = -493$
The solution is -493.

16. $x + 12 = -9$
$x + 12 - 12 = -9 - 12$
$x = -21$
The solution is -21.

17. $15 - h = 11$
$15 - 15 - h = 11 - 15$
$-h = -4$
$\frac{-1h}{-1} = \frac{-4}{-1}$
$h = 4$
The solution is 4.

18. $16 = p + 21$
$16 - 21 = p + 21 - 21$
$-5 = p$
The solution is -5.

19. $-13 + t = -2$
$-13 + 13 + t = -2 + 13$
$t = 11$
The solution is 11.

20. $86 = x + 43$
$86 - 43 = x + 43 - 43$
$43 = x$
The solution is 43.

21. $y - 11 = -14$
$y - 11 + 11 = -14 + 11$
$y = -3$
The solution is -3.

22. $-6 + x = 8$
$-6 + 6 + x = 8 + 6$
$x = 14$

23. $x - 3 = -5$
$x - 3 + 3 = -5 + 3$
$x = -2$

24. $n + 7 = -9$
$n + 7 - 7 = -9 - 7$
$n = -16$

25. $w - 8 = 5$
$w - 8 + 8 = 5 + 8$
$w = 13$

Page 729 Lesson 3-4

1. $-y = -32$
$\frac{-1y}{-1} = \frac{-32}{-1}$
$y = 32$
The solution is 32.

2. $7r = -56$
$\frac{7r}{7} = \frac{-56}{7}$
$r = -8$
The solution is -8.

3. $\frac{t}{-3} = 12$
$\frac{t}{-3}(-3) = 12(-3)$
$t = -36$
The solution is -36.

4. $4 = \frac{s}{-14}$
$4(-14) = \frac{s}{-14}(-14)$
$-56 = s$
The solution is -56.

5. $\frac{b}{47} = -2$
$\frac{b}{47}(47) = -2(47)$
$b = -94$
The solution is -94.

6. $64 = -4n$
$\frac{64}{-4} = \frac{-4n}{-4}$
$-16 = n$
The solution is -16.

7. $-144 = 12q$
$\frac{-144}{12} = \frac{12q}{12}$
$-12 = q$
The solution is -12.

8. $\frac{r}{11} = -12$
$\frac{r}{11}(11) = -12(11)$
$r = -132$
The solution is -132.

9. $-5g = -385$
$\frac{-5g}{-5} = \frac{-385}{-5}$
$g = 77$
The solution is 77.

10. $-16x = -176$
$$\frac{-16x}{-16} = \frac{-176}{-16}$$
$$x = 11$$
The solution is 11.

11. $$-21 = \frac{y}{-4}$$
$$-21(-4) = \frac{y}{-4}(-4)$$
$$84 = y$$
The solution is 84.

12. $-372 = 31k$
$$\frac{-372}{31} = \frac{31k}{31}$$
$$-12 = k$$
The solution is -12.

13. $$84 = \frac{k}{5}$$
$$84(5) = \frac{k}{5}(5)$$
$$420 = k$$
The solution is 420.

14. $-b = 19$
$$\frac{-1b}{-1} = \frac{19}{-1}$$
$$b = -19$$
The solution is -19.

15. $$\frac{v}{112} = -9$$
$$\frac{v}{112}(112) = -9(112)$$
$$v = -1008$$
The solution is -1008.

16. $-3x = -27$
$$\frac{-3x}{-3} = \frac{-27}{-3}$$
$$x = 9$$
The solution is 9.

17. $$\frac{p}{-12} = 4$$
$$\frac{p}{-12}(-12) = 4(-12)$$
$$p = -48$$
The solution is -48.

18. $5q = -100$
$$\frac{5q}{5} = \frac{-100}{5}$$
$$q = -20$$
The solution is -20.

19. $$\frac{d}{11} = -8$$
$$\frac{d}{11}(11) = -8(11)$$
$$d = -88$$
The solution is -88.

20. $-9n = -45$
$$\frac{-9n}{-9} = \frac{-45}{-9}$$
$$n = 5$$
The solution is 5.

21. $125 = -25z$
$$\frac{125}{-25} = \frac{-25z}{-25}$$
$$-5 = z$$
The solution is -5.

22. $8m = -40$
$$\frac{8m}{8} = \frac{-40}{8}$$
$$m = -5$$
The solution is -5.

23. $$\frac{w}{-3} = 27$$
$$\frac{w}{-3}(-3) = 27(-3)$$
$$w = -81$$
The solution is -81.

24. $6e = -24$
$$\frac{6e}{6} = \frac{-24}{6}$$
$$e = -4$$
The solution is -4.

Page 729 Lesson 3-5

1. $$3t - 13 = 2$$
$$3t - 13 + 13 = 2 + 13$$
$$3t = 15$$
$$\frac{3t}{3} = \frac{15}{3}$$
$$t = 5$$
The solution is 5.

2. $$-8j - 7 = 57$$
$$-8j - 7 + 7 = 57 + 7$$
$$-8j = 64$$
$$\frac{-8j}{-8} = \frac{64}{-8}$$
$$j = -8$$
The solution is -8.

3. $$9d - 5 = 4$$
$$9d - 5 + 5 = 4 + 5$$
$$9d = 9$$
$$\frac{9d}{9} = \frac{9}{9}$$
$$d = 1$$
The solution is 1.

4. $$6 - 3w = -27$$
$$6 - 6 - 3w = -27 - 6$$
$$-3w = -33$$
$$\frac{-3w}{-3} = \frac{-33}{-3}$$
$$w = 11$$
The solution is 11.

5. $$\frac{k}{6} + 8 = 12$$
$$\frac{k}{6} + 8 - 8 = 12 - 8$$
$$\frac{k}{6} = 4$$
$$\frac{k}{6}(6) = 4(6)$$
$$k = 24$$
The solution is 24.

6. $$-4 = \frac{q}{8} - 19$$
$$-4 + 19 = \frac{q}{8} - 19 + 19$$
$$15 = \frac{q}{8}$$
$$15(8) = \frac{q}{8}(8)$$
$$120 = q$$
The solution is 120.

7. $15 - \frac{n}{7} = 13$

$15 - 15 - \frac{n}{7} = 13 - 15$

$-\frac{n}{7} = -2$

$-\frac{n}{7}(-7) = -2(-7)$

$n = 14$

The solution is 14.

8. $44 = -4 + 8p$

$44 + 4 = -4 + 4 + 8p$

$48 = 8p$

$\frac{48}{8} = \frac{8p}{8}$

$6 = p$

The solution is 6.

9. $21 - h = -32$

$21 - 21 - 1h = -32 - 21$

$-1h = -53$

$\frac{-1h}{-1} = \frac{-53}{-1}$

$h = 53$

The solution is 53.

10. $-19 = 11b - (-3)$

$-19 = 11b + 3$

$-19 - 3 = 11b + 3 - 3$

$-22 = 11b$

$\frac{-22}{11} = \frac{11b}{11}$

$-2 = b$

The solution is -2.

11. $6 = 20 + \frac{x}{3}$

$6 - 20 = 20 - 20 + \frac{x}{3}$

$-14 = \frac{x}{3}$

$-14(3) = \frac{x}{3}(3)$

$-42 = x$

The solution is -42.

12. $9 + 3a = -3$

$9 - 9 + 3a = -3 - 9$

$3a = -12$

$\frac{3a}{3} = \frac{-12}{3}$

$a = -4$

The solution is -4.

13. $2x - 8 = 10$

$2x - 8 + 8 = 10 + 8$

$2x = 18$

$\frac{2x}{2} = \frac{18}{2}$

$x = 9$

The solution is 9.

14. $\frac{m}{4} - 6 = 10$

$\frac{m}{4} - 6 + 6 = 10 + 6$

$\frac{m}{4} = 16$

$\frac{m}{4}(4) = 16(4)$

$m = 64$

The solution is 64.

15. $-12 + 3p = 3$

$-12 + 12 + 3p = 3 + 12$

$3p = 15$

$\frac{3p}{3} = \frac{15}{3}$

$p = 5$

The solution is 5.

16. $-18 = 6a - 6$

$-18 + 6 = 6a - 6 + 6$

$-12 = 6a$

$\frac{-12}{6} = \frac{6a}{6}$

$-2 = a$

The solution is -2.

17. $\frac{t}{-3} + 11 = 23$

$\frac{t}{-3} + 11 - 11 = 23 - 11$

$\frac{t}{-3} = 12$

$\frac{t}{-3}(-3) = 12(-3)$

$t = -36$

The solution is -36.

18. $3 + 2v = 11$

$3 - 3 + 2v = 11 - 3$

$2v = 8$

$\frac{2v}{2} = \frac{8}{2}$

$v = 4$

The solution is 4.

19. $16 = \frac{k}{3} - 11$

$16 + 11 = \frac{k}{3} - 11 + 11$

$27 = \frac{k}{3}$

$27(3) = \frac{k}{3}(3)$

$81 = k$

The solution is 81.

20. $-6g - 12 = -60$

$-6g - 12 + 12 = -60 + 12$

$-6g = -48$

$\frac{-6g}{-6} = \frac{-48}{-6}$

$g = 8$

The solution is 8.

21. $15 - 4c = -21$

$15 - 15 - 4c = -21 - 15$

$-4c = -36$

$\frac{-4c}{-4} = \frac{-36}{-4}$

$c = 9$

The solution is 9.

Page 730 Lesson 3-6

1. $3x - 5 = 13$

$3x - 5 + 5 = 13 + 5$

$3x = 18$

$\frac{3x}{3} = \frac{18}{3}$

$x = 6$

The number is 6.

2. $2p + 9 = 17$
$2p + 9 - 9 = 17 - 9$
$2p = 8$
$\frac{2p}{2} = \frac{8}{2}$
$p = 4$
The number is 4.

3. $4z + 10 = 46$
$4z + 10 - 10 = 46 - 10$
$4z = 36$
$\frac{4z}{4} = \frac{36}{4}$
$z = 9$
The number is 9.

4. $\frac{a}{-8} - 5 = -2$
$\frac{a}{-8} - 5 + 5 = -2 + 5$
$\frac{a}{-8} = 3$
$\frac{a}{-8}(-8) = 3(-8)$
$a = -24$
The number is -24.

5. $2n + 3 = 11$
$2n + 3 - 3 = 11 - 3$
$2n = 8$
$\frac{2n}{2} = \frac{8}{2}$
$n = 4$
The number is 4.

6. $\frac{w}{6} + 2 = -5$
$\frac{w}{6} + 2 - 2 = -5 - 2$
$\frac{w}{6} = -7$
$\frac{w}{6}(6) = -7(6)$
$w = -42$
The number is -42.

7. $-3x - 9 = 27$
$-3x - 9 + 9 = 27 + 9$
$-3x = 36$
$\frac{-3x}{-3} = \frac{36}{-3}$
$x = -12$
The number is -12.

Page 730 Lesson 3-7

1. $d = rt$
$366 = r \cdot 3$
$\frac{366}{3} = \frac{r \cdot 3}{3}$
$122 = r$

2. $S = (n - 2) \cdot 180$
$S = (8 - 2) \cdot 180$
$S = 6 \cdot 180$
$S = 1080$

3. $A = bh$
$36 = b \cdot 12$
$\frac{36}{12} = \frac{b \cdot 12}{12}$
$3 = b$

4. $P = 4s$
$108 = 4s$
$\frac{108}{4} = \frac{4s}{4}$
$27 = s$

5. $V = \ell wh$
$V = 27 \cdot 5 \cdot 2$
$V = 270$

6. $h = 69 + 2F$
$h = 69 + 2 \cdot 42$
$h = 69 + 84$
$h = 153$

7. $P = 2(\ell + w)$
$P = 2(23 + 9)$
$P = 2(32)$
$P = 64$
The perimeter is 64 cm.
$A = \ell w$
$A = 23 \cdot 9$
$A = 207$
The area is 207 cm^2.

8. $P = 2(\ell + w)$
$P = 2(16 + 14)$
$P = 2(30)$
$P = 60$
The perimeter is 60 ft.
$A = \ell w$
$A = 16 \cdot 14$
$A = 224$
The area is 224 ft^2.

9. $P = 2(\ell + w)$
$P = 2(31 + 3)$
$P = 2(34)$
$P = 68$
The perimeter is 68 m.
$A = \ell w$
$A = 31 \cdot 3$
$A = 93$
The area is 93 m^2.

10. $P = 2(\ell + w)$
$P = 2(7 + 7)$
$P = 2(14)$
$P = 28$
The perimeter is 28 m.
$A = \ell w$
$A = 7 \cdot 7$
$A = 49$
The area is 49 m^2.

11. $A = \ell w$
$126 = 9w$
$\frac{126}{9} = \frac{9w}{9}$
$14 = w$
The width is 14 ft.

12. $A = \ell w$

$108 = \ell \cdot 18$

$\frac{108}{18} = \frac{\ell \cdot 18}{18}$

$6 = \ell$

The length is 6 in.

13. $P = 2(\ell + w)$

$P = 2\ell + 2w$

$68 = 2 \cdot 13 + 2w$

$68 = 26 + 2w$

$68 - 26 = 26 - 26 + 2w$

$42 = 2w$

$\frac{42}{2} = \frac{2w}{2}$

$21 = w$

The width is 21 yd.

14. $A = \ell w$

$168 = \ell \cdot 12$

$\frac{168}{12} = \frac{\ell \cdot 12}{12}$

$14 = \ell$

The length is 14 cm.

15. $P = 2(\ell + w)$

$P = 2\ell + 2w$

$114 = 2\ell + 2 \cdot 3$

$114 = 2\ell + 6$

$114 - 6 = 2\ell + 6 - 6$

$108 = 2\ell$

$\frac{108}{2} = \frac{2\ell}{2}$

$54 = \ell$

The length is 54 m.

16. $P = 2(\ell + w)$

$P = 2\ell + 2w$

$50 = 2\ell + 2(10)$

$50 = 2\ell + 20$

$50 - 20 = 2\ell + 20 - 20$

$30 = 2\ell$

$\frac{30}{2} = \frac{2\ell}{2}$

$15 = \ell$

The length is 15 m.

17. $A = \ell w$

$96 = 12w$

$\frac{96}{12} = \frac{12w}{12}$

$8 = w$

The width is 8 in.

Page 730 Lesson 4-1

1.

Number	Divisible?	Reason
2	yes	The ones digit is 8, and 8 is divisible by 2.
3	no	The sum of the digits is 9 + 8 or 17, and 17 is not divisible by 3.
5	no	The ones digit is not 0 or 5.
6	no	98 is not divisible by 2 and 3.
10	no	The ones digit is not 0.

So, 98 is divisible by 2.

2.

Number	Divisible?	Reason
2	no	The ones digit is 3, and 3 is not divisible by 2.
3	yes	The sum of the digits is 2 + 4 + 3 or 9, and 9 is divisible by 3.
5	no	The ones digit is not 0 or 5.
6	no	243 is not divisible by 2 and 3.
10	no	The ones digit is not 0.

So, 243 is divisible by 3.

3.

Number	Divisible?	Reason
2	yes	The ones digit is 0, and 0 is divisible by 2.
3	no	The sum of the digits is 8 + 0 + 0 or 8, and 8 is not divisible by 3.
5	yes	The ones digit is 0 or 5.
6	no	800 is not divisible by 2 and 3.
10	yes	The ones digit is 0.

So, 800 is divisible by 2, 5, and 10.

4.

Number	Divisible?	Reason
2	yes	The ones digit is 2, and 2 is divisible by 2.
3	yes	The sum of the digits is 2 + 5 + 2 or 9, and 9 is divisible by 3.
5	no	The ones digit is not 0 or 5.
6	yes	252 is divisible by 2 and 3.
10	no	The ones digit is not 0.

So, 252 is divisible by 2, 3, and 6.

5.

Number	Divisible?	Reason
2	no	The ones digit is 5, and 5 is not divisible by 2.
3	yes	The sum of the digits is 1 + 0 + 5 or 6, and 6 is divisible by 3.
5	yes	The ones digit is 0 or 5.
6	no	105 is not divisible by 2 and 3.
10	no	The ones digit is not 0.

So, 105 is divisible by 3 and 5.

6.

Number	Divisible?	Reason
2	yes	The ones digit is 0, and 0 is divisible by 2.
3	yes	The sum of the digits is 2 + 1 + 0 or 3, and 3 is divisible by 3.
5	yes	The ones digit is 0 or 5.
6	yes	210 is divisible by 2 and 3.
10	yes	The ones digit is 0.

So, 210 is divisible by 2, 3, 5, 6, and 10.

7.

Number	Divisible?	Reason
2	no	The ones digit is 5, and 5 is not divisible by 2.
3	yes	The sum of the digits is 2 + 2 + 5 or 9, and 9 is divisible by 3.
5	yes	The ones digit is 0 or 5.
6	no	225 is not divisible by 2 and 3.
10	no	The ones digit is not 0.

So, 225 is divisible by 3 and 5.

8.

Number	Divisible?	Reason
2	yes	The ones digit is 0, and 0 is divisible by 2.
3	yes	The sum of the digits is 1 + 8 + 0 or 9, and 9 is divisible by 3.
5	yes	The ones digit is 0 or 5.
6	yes	180 is divisible by 2 and 3.
10	yes	The ones digit is 0.

So, 180 is divisible by 2, 3, 5, 6, and 10.

9. $1 \cdot 77 = 77$
$7 \cdot 11 = 77$
The factors of 77 are 1, 7, 11, and 77.

10. $1 \cdot 42 = 42$
$2 \cdot 21 = 42$
$3 \cdot 14 = 42$
$6 \cdot 7 = 42$
The factors of 42 are 1, 2, 3, 6, 7, 14, 21, and 42.

11. $1 \cdot 81 = 81$
$3 \cdot 27 = 81$
$9 \cdot 9 = 81$
The factors of 81 are 1, 3, 9, 27, and 81.

12. $1 \cdot 132 = 132$
$2 \cdot 66 = 132$
$3 \cdot 44 = 132$
$4 \cdot 33 = 132$
$6 \cdot 22 = 132$
$11 \cdot 12 = 132$
The factors of 132 are 1, 2, 3, 4, 6, 11, 12, 22, 33, 44, 66, and 132.

13. Yes; it is a product of a number and a variable.
14. No; it has two terms involving subtraction.
15. Yes; it is a variable.
16. No; two terms are added.
17. Yes; a number
18. No; $2(x + 9) = 2x + 18$, two terms are added.

Page 731 Lesson 4-2

1. $8 \cdot 8 \cdot 8 \cdot 8 = 8^4$
2. $9 = 9^1$
3. $(-6)(-6)(-6)(-6)(-6) = (-6)^5$
4. $(y \cdot y \cdot y) \cdot (y \cdot y \cdot y \cdot y) = y \cdot y \cdot y \cdot y \cdot y \cdot y \cdot y$
$= y^7$
5. $a \cdot b \cdot b = (a) \cdot (b \cdot b)$
$= ab^2$
6. $4 \cdot 4 \cdot 4 \cdot 4 \cdot x \cdot x \cdot x \cdot y = (4 \cdot 4 \cdot 4 \cdot 4) \cdot (x \cdot x \cdot x) \cdot (y)$
$= 4^4x^3y$
7. $3q \cdot 3q \cdot 3q \cdot 3q \cdot 3q \cdot 3q = (3q)^6$
8. $n \cdot n \cdot n \cdot n \cdot n \cdot n \cdot n \cdot n \cdot n \cdot n \cdot n \cdot n \cdot n \cdot n \cdot n \cdot n \cdot n = n^{17}$
9. $(x + y)(x + y) = (x + y)^2$
10. $56 = 50 + 6$
$= (5 \times 10^1) + (6 \times 10^0)$
11. $231 = 200 + 30 + 1$
$= (2 \times 10^2) + (3 \times 10^1) + (1 \times 10^0)$
12. $4075 = 4000 + 0 + 70 + 5$
$= (4 \times 10^3) + (0 \times 10^2) + (7 \times 10^1) + (5 \times 10^0)$
13. $3m^2 = 3(3)^2$
$= 3(9)$
$= 27$
14. $n^0 + m = (2)^0 + 3$
$= 1 + 3$
$= 4$
15. $7^4 = 7 \cdot 7 \cdot 7 \cdot 7$
$= 2401$
16. $-5^3 = -1 \cdot 5^3$
$= -1 \cdot 5 \cdot 5 \cdot 5$
$= -125$
17. $p^3 = (-4)^3$
$= (-4) \cdot (-4) \cdot (-4)$
$= -64$
18. $2(m - p)^2 = 2[3 - (-4)]^2$
$= 2(3 + 4)^2$
$= 2(7)^2$
$= 2 \cdot 49$
$= 98$
19. $-2n^3 + m = -2(2)^3 + 3$
$= -2[(2) \cdot (2) \cdot (2)] + 3$
$= -2(8) + 3$
$= -16 + 3$
$= -13$
20. $m - p^2 = 3 - (-4)^2$
$= 3 - [(-4) \cdot (-4)]$
$= 3 - 16$
$= -13$
21. $(m + n + p)^3 = [3 + 2 + (-4)]^3$
$= (5 - 4)^3$
$= (1)^3$
$= 1$
22. $5p - m^2 = 5(-4) - (3)^2$
$= -20 - 9$
$= -29$
23. $(n + p)^4 = [2 + (-4)]^4$
$= (2 - 4)^4$
$= (-2)^4$
$= (-2) \cdot (-2) \cdot (-2) \cdot (-2)$
$= 16$
24. $(m - n)^8 = (3 - 2)^8$
$= (1)^8$
$= 1$

Page 731 Lesson 4-3

1. composite; $57 = 1 \times 57$
$57 = 3 \times 19$

2. composite; $369 = 1 \times 369$
$369 = 3 \times 123$
$369 = 9 \times 41$

3. composite; $116 = 1 \times 116$
$116 = 2 \times 58$
$116 = 4 \times 29$

4. composite; $125 = 1 \times 125$
$125 = 5 \times 25$

5. prime; $83 = 1 \times 83$ is the only factorization.

6. composite; $99 = 1 \times 99$
$99 = 3 \times 33$
$99 = 9 \times 11$

7. composite; $91 = 1 \times 91$
$91 = 7 \times 13$

8. prime; $79 = 1 \times 79$ is the only factorization.

9. 21
/\
$7 \cdot 3$
The prime factorization of 21 is $3 \cdot 7$.

10. 44
/\
$4 \cdot 11$
/\
$2 \cdot 2$
The prime factorization of 44 is $2 \cdot 2 \cdot 11$ or $2^2 \cdot 11$.

11. 51
/\
$3 \cdot 17$
The prime factorization of 51 is $3 \cdot 17$.

12. 65
/\
$5 \cdot 13$
The prime factorization of 65 is $5 \cdot 13$.

13. 30
/\
$6 \cdot 5$
/\
$2 \cdot 3$
The prime factorization of 30 is $2 \cdot 3 \cdot 5$.

14. 28
/\
$4 \cdot 7$
/\
$2 \cdot 2$
The prime factorization of 28 is $2 \cdot 2 \cdot 7$ or $2^2 \cdot 7$.

15. 117
/\
$9 \cdot 13$
/\
$3 \cdot 3$
The prime factorization of 117 is $3 \cdot 3 \cdot 13$ or $3^2 \cdot 13$.

16. 88
/\
$8 \cdot 11$
/\
$4 \cdot 2$
/\
$2 \cdot 2$
The prime factorization of 88 is $2 \cdot 2 \cdot 2 \cdot 11$ or $2^3 \cdot 11$.

17. 54
/\
$6 \cdot 9$
/\ /\
$2 \cdot 3$ $3 \cdot 3$
The prime factorization of 54 is $2 \cdot 3 \cdot 3 \cdot 3$ or $2 \cdot 3^3$.

18. 32
/\
$8 \cdot 4$
/\ /\
$4 \cdot 2$ $2 \cdot 2$
/\
$2 \cdot 2$
The prime factorization of 32 is $2 \cdot 2 \cdot 2 \cdot 2 \cdot 2$ or 2^5.

19. 300
/\
$3 \cdot 100$
/\
$2 \cdot 50$
/\
$2 \cdot 25$
/\
$5 \cdot 5$
The prime factorization of 300 is $2 \cdot 2 \cdot 3 \cdot 5 \cdot 5$ or $2^2 \cdot 3 \cdot 5^2$.

20. 210
/\
$2 \cdot 105$
/\
$5 \cdot 21$
/\
$3 \cdot 7$
The prime factorization of 210 is $2 \cdot 3 \cdot 5 \cdot 7$.

21. $40 = 2 \cdot 2 \cdot 2 \cdot 5$
22. $630a = 2 \cdot 3 \cdot 3 \cdot 5 \cdot 7 \cdot a$
23. $187 = 11 \cdot 17$
24. $310 = 2 \cdot 5 \cdot 31$
25. $510 = 2 \cdot 3 \cdot 5 \cdot 17$
26. $1589 = 7 \cdot 227$
27. $-18ab^2 = -1 \cdot 2 \cdot 3 \cdot 3 \cdot a \cdot b \cdot b$
28. $-117x^3 = -1 \cdot 3 \cdot 3 \cdot 13 \cdot x \cdot x \cdot x$
29. $105j^2k^5 = 3 \cdot 5 \cdot 7 \cdot j \cdot j \cdot k \cdot k \cdot k \cdot k \cdot k$

Page 731 Lesson 4-4

1. $27 = \boxed{3} \cdot \boxed{3} \cdot 3$
$45 = \boxed{3} \cdot \boxed{3} \cdot 5$
The GCF of 27 and 45 is $3 \cdot 3$ or 9.

2. $30 = 2 \cdot 3 \cdot 5$
$12 = 2 \cdot 2 \cdot 3$
The GCF of 30 and 12 is $2 \cdot 3$ or 6.

3. $16 = 2 \cdot 2 \cdot 2 \cdot 2$
$40 = 2 \cdot 2 \cdot 2 \cdot 5$
$28 = 2 \cdot 2 \cdot 7$
The GCF of 16, 40, and 28 is $2 \cdot 2$ or 4.

4. $18 = 2 \cdot 3 \cdot 3$
$17 = 17$
$15 = 3 \cdot 5$
The GCF of 18, 17, and 15 is 1.

5. $112 = 2 \cdot 2 \cdot 2 \cdot 2 \cdot 7$
$216 = 2 \cdot 2 \cdot 2 \cdot 3 \cdot 3 \cdot 3$
The GCF of 112 and 216 is $2 \cdot 2 \cdot 2$ or 8.

6. $120 = 2 \cdot 2 \cdot 2 \cdot 3 \cdot 5$
$245 = 5 \cdot 7 \cdot 7$
The GCF of 120 and 245 is 5.

7. $84k = 2 \cdot 2 \cdot 3 \cdot 7 \cdot k$
$108k^2 = 2 \cdot 2 \cdot 3 \cdot 3 \cdot 3 \cdot k \cdot k$
The GCF of $84k$ and $108k^2$ is $2 \cdot 2 \cdot 3 \cdot k$ or $12k$.

8. $135ab = 3 \cdot 3 \cdot 3 \cdot 5 \cdot a \cdot b$
$171b = 3 \cdot 3 \cdot 19 \cdot b$
The GCF of $135ab$ and $171b$ is $3 \cdot 3 \cdot b$ or $9b$.

9. $185fg = 5 \cdot 37 \cdot f \cdot g$
$74f^2g = 2 \cdot 37 \cdot f \cdot f \cdot g$
The GCF of $185fg$ and $74f^2g$ is $37fg$.

10. $44m = 2 \cdot 2 \cdot 11 \cdot m$
$60n = 2 \cdot 2 \cdot 3 \cdot 5 \cdot n$
The GCF of $44m$ and $60n$ is $2 \cdot 2$ or 4.

11. $90gh = 2 \cdot 3 \cdot 3 \cdot 5 \cdot g \cdot h$
$225k = 3 \cdot 3 \cdot 5 \cdot 5 \cdot k$
The GCF of $90gh$ and $225k$ is $3 \cdot 3 \cdot 5$ or 45.

12. $8 = 2 \cdot 2 \cdot 2$
$28h = 2 \cdot 2 \cdot 7 \cdot h$
The GCF of 8 and $28h$ is $2 \cdot 2$ or 4.

13. $16w = 2 \cdot 2 \cdot 2 \cdot 2 \cdot w$
$28w^3 = 2 \cdot 2 \cdot 7 \cdot w \cdot w \cdot w$
The GCF of $16w$ and $28w^3$ is $2 \cdot 2 \cdot w$ or $4w$.

14. $24a = 2 \cdot 2 \cdot 2 \cdot 3 \cdot a$
$30ab = 2 \cdot 3 \cdot 5 \cdot a \cdot b$
$66a^2 = 2 \cdot 3 \cdot 11 \cdot a \cdot a$
The GCF of $24a$, $30ab$, and $66a^2$ is $2 \cdot 3 \cdot a$ or 6a.

15. $13z = 13 \cdot z$
$39yz = 3 \cdot 13 \cdot y \cdot z$
$52y = 2 \cdot 2 \cdot 13 \cdot y$
The GCF of $13z$, $39yz$, and $52y$ is 13.

16. $3m = 3 \cdot m$
$12 = 2 \cdot 2 \cdot 3$
The GCF is 3.
$3m + 12 = 3(m) + 3(4)$
$= 3(m + 4)$
So, $3m + 12 = 3(m + 4)$.

17. $5x = 5 \cdot x$
$15 = 3 \cdot 5$
The GCF is 5.
$5x + 15 = 5(x) + 5(3)$
$= 5(x + 3)$
So, $5x + 15 = 5(x + 3)$.

18. $4 = 2 \cdot 2$
$8b = 2 \cdot 2 \cdot 2 \cdot b$
The GCF is 4.
$4 + 8b = 4(1) + 4(2b)$
$= 4(1 + 2b)$
So, $4 + 8b = 4(1 + 2b)$.

19. $7x = 7 \cdot x$
$21 = 3 \cdot 7$
The GCF is 7.
$7x + 21 = 7(x) + 7(3)$
$= 7(x + 3)$
So, $7x + 21 = 7(x + 3)$.

20. $2a = 2 \cdot a$
$100 = 2 \cdot 2 \cdot 5 \cdot 5$
The GCF is 2.
$2a + 100 = 2(a) + 2(50)$
$= 2(a + 50)$
So, $2a + 100 = 2(a + 50)$.

21. $42 = 2 \cdot 3 \cdot 7$
$-14b = -1 \cdot 2 \cdot 7 \cdot b$
The GCF is $2 \cdot 7$ or 14.
$42 - 14b = 14(3) - 14(b)$
$= 14(3 - b)$
So, $42 - 14b = 14(3 - b)$.

22. $5f = 5 \cdot f$
$25 = 5 \cdot 5$
The GCF is 5.
$5f - 25 = 5(f) - 5(5)$
$= 5(f - 5)$
So, $5f - 25 = 5(f - 5)$.

23. $11p = 11 \cdot p$
$-66 = -1 \cdot 2 \cdot 3 \cdot 11$
The GCF is 11.
$11p - 66 = 11(p) - 11(6)$
$= 11(p - 6)$
So, $11p - 66 = 11(p - 6)$.

24. $7y = 7 \cdot y$
$-21 = -1 \cdot 3 \cdot 7$
The GCF is 7.
$7y - 21 = 7(y) - 7(3)$
$= 7(y - 3)$
So, $7y - 21 = 7(y - 3)$.

25. $48 = 2 \cdot 2 \cdot 2 \cdot 2 \cdot 3$
$12s = 2 \cdot 2 \cdot 3 \cdot s$
The GCF is $2 \cdot 2 \cdot 3$ or 12.
$48 + 12s = 12(4) + 12(s)$
$= 12(4 + s)$
So, $48 + 12s = 12(4 + s)$.

26. $18 = 2 \cdot 3 \cdot 3$
$-2w = -1 \cdot 2 \cdot w$
The GCF is 2.
$18 - 2w = 2(9) - 2(w)$
$= 2(9 - w)$
So, $18 - 2w = 2(9 - w)$.

27. $24k = 2 \cdot 2 \cdot 2 \cdot 3$
$96 = 2 \cdot 2 \cdot 2 \cdot 2 \cdot 2 \cdot 3$
The GCF is $2 \cdot 2 \cdot 2 \cdot 3$ or 24.
$24k + 96 = 24(k) + 24(4)$
$= 24(k + 4)$
So, $24k + 96 = 24(k + 4)$.

28. $2y = \boxed{2} \cdot y$
$14 = \boxed{2} \cdot 7$
The GCF is 2.
$2y + 14 = 2(y) + 2(7)$
$= 2(y + 7)$
So, $2y + 14 = 2(y + 7)$.

29. $42 = 2 \cdot 3 \cdot \boxed{7}$
$-7b = -1 \cdot \boxed{7} \cdot b$
The GCF is 7.
$42 - 7b = 7(6) - 7(b)$
$= 7(6 - b)$
So, $42 - 7b = 7(6 - b)$.

30. $13w = \boxed{13} \cdot w$
$39 = 3 \cdot \boxed{13}$
The GCF is 13.
$13w + 39 = 13(w) + 13(3)$
$= 13(w + 3)$
So, $13w + 39 = 13(w + 3)$.

Page 732 Lesson 4-5

1. $\frac{3}{54} = \frac{\overset{1}{\cancel{3}}}{2 \cdot \underset{1}{\cancel{3}} \cdot 3 \cdot 3}$
$= \frac{1}{18}$

2. 3 and 16 have no common factors.
$\frac{3}{16}$ is simplified.

3. $\frac{6}{58} = \frac{\overset{1}{\cancel{2}} \cdot 3}{\underset{1}{\cancel{2}} \cdot 29}$
$= \frac{3}{29}$

4. $\frac{15}{55} = \frac{3 \cdot \overset{1}{\cancel{5}}}{11 \cdot \underset{1}{\cancel{5}}}$
$= \frac{3}{11}$

5. $\frac{10}{90} = \frac{\overset{1}{\cancel{2}} \cdot \overset{1}{\cancel{5}}}{\underset{1}{\cancel{2}} \cdot 3 \cdot 3 \cdot \underset{1}{\cancel{5}}}$
$= \frac{1}{9}$

6. 20 and 49 have no common factors.
$\frac{20}{49}$ is simplified.

7. $\frac{8}{20} = \frac{\overset{1}{\cancel{2}} \cdot \overset{1}{\cancel{2}} \cdot 2}{\underset{1}{\cancel{2}} \cdot \underset{1}{\cancel{2}} \cdot 5}$
$= \frac{2}{5}$

8. $\frac{99}{9} = \frac{\overset{1}{\cancel{3}} \cdot \overset{1}{\cancel{3}} \cdot 11}{\underset{1}{\cancel{3}} \cdot \underset{1}{\cancel{3}}}$
$= 11$

9. $\frac{18}{54} = \frac{\overset{1}{\cancel{2}} \cdot \overset{1}{\cancel{3}} \cdot \overset{1}{\cancel{3}}}{\underset{1}{\cancel{2}} \cdot \underset{1}{\cancel{3}} \cdot \underset{1}{\cancel{3}} \cdot 3}$
$= \frac{1}{3}$

10. 21 and 64 have no common factors.
$\frac{21}{64}$ is simplified.

11. $\frac{40}{76} = \frac{\overset{1}{\cancel{2}} \cdot \overset{1}{\cancel{2}} \cdot 2 \cdot 5}{\underset{1}{\cancel{2}} \cdot \underset{1}{\cancel{2}} \cdot 19}$
$= \frac{10}{19}$

12. $\frac{49}{56} = \frac{\overset{1}{\cancel{7}} \cdot 7}{\underset{1}{\cancel{7}} \cdot 8}$
$= \frac{7}{8}$

13. $\frac{22}{66} = \frac{\overset{1}{\cancel{2}} \cdot \overset{1}{\cancel{11}}}{\underset{1}{\cancel{2}} \cdot 3 \cdot \underset{1}{\cancel{11}}}$
$= \frac{1}{3}$

14. $\frac{42}{49} = \frac{2 \cdot 3 \cdot \overset{1}{\cancel{7}}}{\underset{1}{\cancel{7}} \cdot 7}$
$= \frac{6}{7}$

15. $\frac{110}{200} = \frac{\overset{1}{\cancel{2}} \cdot \overset{1}{\cancel{5}} \cdot 11}{\underset{1}{\cancel{2}} \cdot 2 \cdot 2 \cdot \underset{1}{\cancel{5}} \cdot 5}$
$= \frac{11}{20}$

16. $\frac{b}{b^4} = \frac{\overset{1}{\cancel{b}}}{\underset{1}{\cancel{b}} \cdot b \cdot b \cdot b}$
$= \frac{1}{b^3}$

17. $\frac{16p}{24p} = \frac{\overset{1}{\cancel{2}} \cdot \overset{1}{\cancel{2}} \cdot \overset{1}{\cancel{2}} \cdot 2 \cdot \overset{1}{\cancel{p}}}{\underset{1}{\cancel{2}} \cdot \underset{1}{\cancel{2}} \cdot \underset{1}{\cancel{2}} \cdot 3 \cdot \underset{1}{\cancel{p}}}$
$= \frac{2}{3}$

18. $\frac{21x^2y}{81y} = \frac{\overset{1}{\cancel{3}} \cdot 7 \cdot x \cdot x \cdot \overset{1}{\cancel{y}}}{\underset{1}{\cancel{3}} \cdot 3 \cdot 3 \cdot 3 \cdot \underset{1}{\cancel{y}}}$
$= \frac{7x^2}{27}$

19. $\frac{32d^2}{6d} = \frac{\overset{1}{\cancel{2}} \cdot 2 \cdot 2 \cdot 2 \cdot 2 \cdot \overset{1}{\cancel{d}} \cdot d}{\underset{1}{\cancel{2}} \cdot 3 \cdot \underset{1}{\cancel{d}}}$
$= \frac{16d}{3}$

20. $\frac{72ab}{8b} = \frac{\overset{1}{\cancel{2}} \cdot \overset{1}{\cancel{2}} \cdot \overset{1}{\cancel{2}} \cdot 3 \cdot 3 \cdot a \cdot \overset{1}{\cancel{b}}}{\underset{1}{\cancel{2}} \cdot \underset{1}{\cancel{2}} \cdot \underset{1}{\cancel{2}} \cdot \underset{1}{\cancel{b}}}$
$= 9a$

21. $\frac{120z^3x}{18zx} = \frac{\overset{1}{\cancel{2}} \cdot 2 \cdot 2 \cdot \overset{1}{\cancel{3}} \cdot 5 \cdot \overset{1}{\cancel{z}} \cdot z \cdot z \cdot \overset{1}{\cancel{x}}}{\underset{1}{\cancel{2}} \cdot \underset{1}{\cancel{3}} \cdot 3 \cdot \underset{1}{\cancel{z}} \cdot \underset{1}{\cancel{x}}}$
$= \frac{20z^3}{3}$

22. There are 3 feet in 1 yard and 12 inches in 1 foot. Therefore, there are 36 inches in 1 yard.

$\frac{14}{36} = \frac{\overset{1}{\cancel{2}} \cdot 7}{\underset{1}{\cancel{2}} \cdot 2 \cdot 3 \cdot 3} = \frac{7}{18}$

So, 14 inches is $\frac{7}{18}$ of a yard.

23. There are 24 hours in one day.

$$\frac{9}{24} = \frac{\overset{1}{\cancel{3}} \cdot 3}{2 \cdot 2 \cdot 2 \cdot \underset{1}{\cancel{3}}} = \frac{3}{8}$$

So, 9 hours is $\frac{3}{8}$ of a day.

Page 732 Lesson 4-6

1. $r^4 \cdot r^2 = r^{4+2}$
   $= r^6$

2. $\frac{2^9}{2^3} = 2^{9-3}$
   $= 2^6$

3. $\frac{b^{18}}{b^5} = b^{18-5}$
   $= b^{13}$

4. $12^3 \cdot 12^8 = 12^{3+8}$
   $= 12^{11}$

5. $x \cdot x^9 = x^{1+9}$
   $= x^{10}$

6. $(2s^6)(4s^2) = (2 \cdot 4)(s^6 \cdot s^2)$
   $= (8)(s^{6+2})$
   $= 8s^8$

7. $w^3 \cdot w^4 \cdot w^2 = (w^3 \cdot w^4) \cdot w^2$
   $= w^{3+4} \cdot w^2$
   $= w^7 \cdot w^2$
   $= w^{7+2}$
   $= w^9$

8. $(-2)^2(-2)^5(-2) = [(-2)^2(-2)^5](-2)$
   $= (-2)^{2+5}(-2)^1$
   $= (-2)^7(-2)^1$
   $= (-2)^{7+1}$
   $= (-2)^8$

9. $\frac{4^7}{4^6} = 4^{7-6}$
   $= 4^1$
   $= 4$

10. $3(f^{17})(f^2) = 3(f^{17+2})$
    $= 3f^{19}$

11. $(5k)^2 \cdot k^7 = (5k)(5k) \cdot k^7$
    $= (5 \cdot 5)(k \cdot k) \cdot k^7$
    $= 25k^2 \cdot k^7$
    $= 25k^{2+7}$
    $= 25k^9$

12. $\frac{6m^8}{3m^2} = \frac{2m^8}{m^2}$
    $= 2m^{8-2}$
    $= 2m^6$

13. $(3x^4)(-6x) = [3 \cdot (-6)](x^4 \cdot x)$
    $= -18x^{4+1}$
    $= -18x^5$

14. $(4k^4)(-3k)^3 = (4k^4)(-3k)(-3k)(-3k)$
    $= [4(-3)(-3)(-3)](k^4)(k \cdot k \cdot k)$
    $= -108 \cdot k^4 \cdot k^3$
    $= -108k^{4+3}$
    $= -108k^7$

15. $\left(\frac{42}{-6}\right)\left(\frac{g^{10}}{g^3}\right) = (-7)(g^{10-3})$
    $= -7g^7$

Page 732 Lesson 4-7

1. $y^{-9} = \frac{1}{y^9}$

2. $m^{-4} = \frac{1}{m^4}$

3. $5^{-3} = \frac{1}{5^3}$

4. $2^{-7} = \frac{1}{2^7}$

5. $6^{-3} = \frac{1}{6^3}$

6. $a^{-11} = \frac{1}{a^{11}}$

7. $\frac{1}{p^4} = p^{-4}$

8. $\frac{1}{b^9} = b^{-9}$

9. $\frac{1}{5^3} = 5^{-3}$

10. $\frac{1}{7^4} = 7^{-4}$

11. $\frac{1}{15^2} = 15^{-2}$

12. $\frac{1}{25} = \frac{1}{5^2} = 5^{-2}$

13. $\frac{1}{c^7} = c^{-7}$

14. $\frac{1}{64} = \frac{1}{4^3}$ or $\frac{1}{8^2}$ or $\frac{1}{2^6}$
    $= 4^{-3}$ or 8^{-2} or 2^{-6}

15. $0.01 = \frac{1}{100}$ or $0.01 = \frac{1}{100}$
    $= \frac{1}{10^2}$ $\quad = 100^{-1}$
    $= 10^{-2}$

16. $0.00001 = \frac{1}{100{,}000}$
    $= \frac{1}{10^5}$
    $= 10^{-5}$

17. $0.0001 = \frac{1}{10{,}000}$ or $0.0001 = \frac{1}{10{,}000}$
    $= \frac{1}{10^4}$ $\quad = \frac{1}{100 \cdot 100}$
    $= 10^{-4}$ $\quad = 100^{-2}$

18. $0.001 = \frac{1}{1000}$
    $= \frac{1}{10^3}$
    $= 10^{-3}$

19. $0.1 = \frac{1}{10}$
    $= \frac{1}{10^1}$
    $= 10^{-1}$

20. $0.000001 = \frac{1}{1{,}000{,}000}$ or $0.000001 = \frac{1}{1{,}000{,}000}$
    $= \frac{1}{10^6}$ $\quad = \frac{1}{100 \cdot 100 \cdot 100}$
    $= 10^{-6}$ $\quad = 100^{-3}$

21. $x^{-2} = (3)^{-2}$
    $= \frac{1}{3^2}$
    $= \frac{1}{9}$

22. $9^y = 9^{-2}$
    $= \frac{1}{9^2}$
    $= \frac{1}{81}$

23. $y^{-3} = (-2)^{-3}$
$= \frac{1}{(-2)^3}$
$= \frac{1}{(-2)(-2)(-2)}$
$= -\frac{1}{8}$

24. $x^{-3} = (3)^{-3}$
$= \frac{1}{(3)^3}$
$= \frac{1}{27}$

25. $y^{-4} = (-2)^{-4}$
$= \frac{1}{(-2)^4}$
$= \frac{1}{(-2)(-2)(-2)(-2)}$
$= \frac{1}{16}$

26. $(xy)^{-2} = [(3) \cdot (-2)]^{-2}$
$= (-6)^{-2}$
$= \frac{1}{(-6)^2}$
$= \frac{1}{(-6)(-6)}$
$= \frac{1}{36}$

Page 733 Lesson 4-8

1. $9040 = 9.04 \times 1{,}000$
$= 9.04 \times 10^3$

2. $0.015 = 1.5 \times 0.01$
$= 1.5 \times 10^{-2}$

3. $6{,}180{,}000 = 6.18 \times 1{,}000{,}000$
$= 6.18 \times 10^6$

4. $27{,}210{,}000 = 2.721 \times 10{,}000{,}000$
$= 2.721 \times 10^7$

5. $0.00004637 = 4.637 \times 0.00001$
$= 4.637 \times 10^{-5}$

6. $0.00546 = 5.46 \times 0.001$
$= 5.46 \times 10^{-3}$

7. $500{,}300{,}100 = 5.003001 \times 100{,}000{,}000$
$= 5.003001 \times 10^8$

8. $-0.0000032 = -3.2 \times 0.000001$
$= -3.2 \times 10^{-6}$

9. $-9.5 \times 10^{-3} = -9.5 \times 0.001$
$= -0.0095$

10. $8.245 \times 10^{-4} = 8.245 \times 0.0001$
$= 0.0008245$

11. $8.2 \times 10^4 = 8.2 \times 10{,}000$
$= 82{,}000$

12. $-9.102040 \times 10^2 = -9.102040 \times 100$
$= -910.204$

13. $4.02 \times 10^3 = 4.02 \times 1000$
$= 4020$

14. $1.6 \times 10^{-2} = 1.6 \times 0.01$
$= 0.016$

15. $2.41023 \times 10^6 = 2.41023 \times 1{,}000{,}000$
$= 2{,}410{,}230$

16. $4.21 \times 10^{-5} = 4.21 \times 0.00001$
$= 0.0000421$

Page 733 Lesson 5-1

1. $10\overline{)6.0}$ = 0.6
$-6\,0$
0

So, $\frac{6}{10} = 0.6$.

2. $25\overline{)4.00}$ = 0.16
$-2\,5$
$1\,50$
$-1\,50$
0

So, $\frac{4}{25} = 0.16$.

3. $8\overline{)1.000}$ = 0.125
-8
20
-16
40
-40
0

So, $-\frac{1}{8} = -0.125$.

4. $1\frac{3}{4} = 1 + \frac{3}{4}$
$= 1 + 0.75$
$= 1.75$

5. $\frac{5}{6} \rightarrow 6\overline{)5.000\ldots}$ = 0.833 . . .
So, $\frac{5}{6} = 0.8\overline{3}$.

6. $20\overline{)9.00}$ = 0.45
$-8\,0$
$1\,00$
$-1\,00$
0

So, $\frac{9}{20} = 0.45$.

7. $-4\frac{7}{12} = -\left(4 + \frac{7}{12}\right)$
$\frac{7}{12} \rightarrow 12\overline{)7.0000\ldots}$ = 0.5833 . . .
$-4\frac{7}{12} = -(4 + 0.58\overline{3})$
$= -4.58\overline{3}$

8. $\frac{8}{11} \rightarrow 11\overline{)8.0000}\ldots$ (quotient $0.7272\ldots$)

So, $\frac{8}{11} = 0.\overline{72}$.

9. $3\frac{4}{18} = 3 + \frac{4}{18}$

$\frac{4}{18} \rightarrow 18\overline{)4.000}\ldots$ (quotient $0.222\ldots$)

$3\frac{4}{18} = 3 + 0.\overline{2}$

$= 3.\overline{2}$

10.
$$\begin{array}{r} 0.1875 \\ 16\overline{)3.0000} \\ -\underline{1\,6} \\ 1\,40 \\ -\underline{1\,28} \\ 120 \\ -\underline{112} \\ 80 \\ -\underline{80} \\ 0 \end{array}$$

So, $-\frac{3}{16} = -0.1875$.

11. $8\frac{36}{44} = 8 + \frac{36}{44}$

$\frac{36}{44} \rightarrow 44\overline{)36.0000}\ldots$ (quotient $0.8181\ldots$)

$8\frac{36}{44} = 8 + 0.\overline{81}$

$= 8.\overline{81}$

12.
$$\begin{array}{r} 0.4 \\ 15\overline{)6.0} \\ -\underline{6\,0} \\ 0 \end{array}$$

So, $\frac{6}{15} = 0.4$.

13. $\frac{7}{8} \bigcirc \frac{5}{6}$

$0.875 \bigcirc 0.8\overline{3}$

$0.875 > 0.8\overline{3}$

14. $0.04 \bigcirc \frac{5}{9}$

$0.04 \bigcirc 0.\overline{5}$

$0.04 < 0.\overline{5}$

15. $\frac{1}{3} \bigcirc \frac{2}{7}$

$0.\overline{3} \bigcirc 0.\overline{285714}$

$0.3 > 0.\overline{285714}$

16. $\frac{3}{5} \bigcirc \frac{12}{20}$

$0.6 \bigcirc 0.6$

$0.6 = 0.6$

17. $\frac{1}{2} \bigcirc 0.75$

$0.5 \bigcirc 0.75$

$0.5 < 0.75$

18. $0.3 \bigcirc \frac{1}{3}$

$0.3 \bigcirc 0.\overline{3}$

$0.3 < 0.\overline{3}$

19. $\frac{2}{3} \bigcirc 0.64$

$0.\overline{6} \bigcirc 0.64$

$0.\overline{6} > 0.64$

20. $\frac{2}{20} \bigcirc 0.10$

$0.1 \bigcirc 0.10$

$0.1 = 0.10$

21. $0.\overline{5} \bigcirc \frac{5}{9}$

$0.\overline{5} \bigcirc 0.\overline{5}$

$0.\overline{5} = 0.\overline{5}$

22. $2.\overline{1} \bigcirc 2\frac{1}{10}$

$2.\overline{1} \bigcirc 2.1$

$2.\overline{1} > 2.1$

23. $3\frac{7}{8} \bigcirc 3.78$

$3.875 \bigcirc 3.78$

$3.875 > 3.78$

24. $-\frac{6}{7} \bigcirc -\frac{5}{6}$

$-0.\overline{857142} \bigcirc -0.8\overline{3}$

$-0.\overline{857142} < -0.8\overline{3}$

Page 733 Lesson 5-2

1. $3\frac{4}{5} = \frac{19}{5}$

2. $-1\frac{2}{9} = -\frac{11}{9}$

3. $15 = \frac{15}{1}$

4. $2\frac{3}{8} = \frac{19}{8}$

5. $-13 = -\frac{13}{1}$

6. $2\frac{6}{7} = \frac{20}{7}$

7. $36 = \frac{36}{1}$

8. $-1\frac{3}{5} = -\frac{8}{5}$

9. $0.6 = \frac{6}{10}$

$= \frac{3}{5}$

10. $0.05 = \frac{5}{100}$

$= \frac{1}{20}$

11. $0.38 = \frac{38}{100}$

$= \frac{19}{50}$

12. $4.12 = 4\frac{12}{100}$

$= 4\frac{3}{25}$

13. $0.375 = \frac{375}{1000}$

$= \frac{3}{8}$

14. $-3.24 = -3\frac{24}{100}$

$= -3\frac{6}{25}$

15. $N = 0.222$

$10N = 2.222$

$\underline{-(N = 0.222)}$

$9N = 2$

$\frac{9N}{9} = \frac{2}{9}$

$N = \frac{2}{9}$

Therefore, $0.222\ldots = \frac{2}{9}$.

16. $N = 0.444\ldots$
$10N = 4.444$
$-(N = 0.444)$
$9N = 4$
$\frac{9N}{9} = \frac{4}{9}$
$N = \frac{4}{9}$
Therefore, $-0.\overline{4} = -\frac{4}{9}$.

17. Because $-4\frac{2}{5} = -\frac{22}{5}$, it is a rational number.

18. 6 is a natural number, a whole number, an integer and a rational number.

19. Because $3\frac{1}{3} = \frac{10}{3}$, it is a rational number.

20. -10 is an integer and a rational number.

21. Because $5.9 = 5\frac{9}{10} = \frac{59}{10}$, it is a rational number only.

22. Because $-\frac{3}{1} = -3$, it is an integer and a rational number.

23. Because $\frac{16}{8} = 2$, it is a natural number, a whole number, an integer, and a rational number.

24. 7.02002000 . . . is a nonterminating nonrepeating decimal, so it is not a rational number.

Page 734 Lesson 5-3

1. $\frac{2}{5} \cdot \frac{3}{16} = \frac{\cancel{2}^{1}}{5} \cdot \frac{3}{\cancel{16}_{8}}$
$= \frac{3}{40}$

2. $3\frac{1}{4} \cdot \frac{2}{11} = \frac{13}{4} \cdot \frac{2}{11}$
$= \frac{13}{\cancel{4}_{2}} \cdot \frac{\cancel{2}^{1}}{11}$
$= \frac{13}{22}$

3. $\frac{3}{5}\left(-\frac{5}{12}\right) = \frac{\cancel{3}^{1}}{\cancel{5}_{1}} \cdot -\frac{\cancel{5}^{1}}{\cancel{12}_{4}}$
$= -\frac{1}{4}$

4. $\frac{5}{8} \cdot \frac{2}{3} = \frac{5}{\cancel{8}_{4}} \cdot \frac{\cancel{2}^{1}}{3}$
$= \frac{5}{12}$

5. $-\frac{9}{10} \cdot \frac{5}{24} = \frac{-\cancel{9}^{3}}{\cancel{10}_{2}} \cdot \frac{\cancel{5}^{1}}{\cancel{24}_{8}}$
$= -\frac{3}{16}$

6. $\frac{1}{7} \cdot \frac{21}{22} = \frac{1}{\cancel{7}_{1}} \cdot \frac{\cancel{21}^{3}}{22}$
$= \frac{3}{22}$

7. $\frac{4}{5} \cdot \frac{1}{8} = \frac{\cancel{4}^{1}}{5} \cdot \frac{1}{\cancel{8}_{2}}$
$= \frac{1}{10}$

8. $2\frac{2}{6} \cdot 6\frac{2}{7} = \frac{14}{6} \cdot \frac{44}{7}$
$= \frac{\cancel{14}^{2}}{\cancel{6}_{3}} \cdot \frac{\cancel{44}^{22}}{\cancel{7}_{1}}$
$= \frac{44}{3}$ or $14\frac{2}{3}$

9. $2\left(-\frac{7}{12}\right) = \frac{2}{1} \cdot \left(\frac{-7}{12}\right)$
$= \frac{\cancel{2}^{1}}{1} \cdot \left(\frac{-7}{\cancel{12}_{6}}\right)$
$= -\frac{7}{6}$ or $-1\frac{1}{6}$

10. $1\frac{3}{7}\left(-9\frac{4}{5}\right) = \frac{10}{7} \cdot \left(\frac{-49}{5}\right)$
$= \frac{\cancel{10}^{2}}{\cancel{7}_{1}} \cdot \frac{-\cancel{49}^{7}}{\cancel{5}_{1}}$
$= -\frac{14}{1}$ or -14

11. $-\frac{6}{7}\left(-\frac{6}{7}\right) = \frac{-6}{7} \cdot \left(\frac{-6}{7}\right)$
$= \frac{36}{49}$

12. $\frac{6c}{10} \cdot \frac{2}{c} = \frac{6\cancel{c}}{\cancel{10}_{5}} \cdot \frac{\cancel{2}^{1}}{\cancel{c}_{1}}$
$= \frac{6}{5}$ or $1\frac{1}{5}$

13. $\frac{p^3}{4} \cdot \frac{12}{p} = \frac{pp\cancel{p}}{\cancel{4}_{1}} \cdot \frac{\cancel{12}^{3}}{\cancel{p}_{1}}$
$= \frac{3p^2}{1}$ or $3p^2$

14. $\frac{ab}{9} \cdot \frac{3}{b^2} = \frac{ab}{9} \cdot \frac{3}{b \cdot b}$
$= \frac{a\cancel{b}}{\cancel{9}_{3}} \cdot \frac{\cancel{3}^{1}}{\cancel{b} \cdot b}$
$= \frac{a}{3b}$

15. $\frac{4x}{3y} \cdot \frac{12y^4}{x^2} = \frac{4x}{3y} \cdot \frac{12yyyy}{xx}$
$= \frac{4\cancel{x}}{\cancel{3}\cancel{y}_{1}} \cdot \frac{\cancel{12}^{4}\cancel{y}yyy}{\cancel{x}x}$
$= \frac{16y^3}{x}$

16. $\frac{5}{12}$ yard $\cdot \frac{36 \text{ inches}}{1 \text{ yard}} = \frac{5 \cancel{\text{yd}}}{\cancel{12}_{1}} \cdot \frac{\cancel{36}^{3} \text{ in.}}{1 \cancel{\text{yd}}}$
$= 15$ in.

17. $\frac{1}{5}$ hour $\cdot \frac{60 \text{ minutes}}{1 \text{ hour}} = \frac{1 \cancel{\text{hr}}}{\cancel{5}_{1}} \cdot \frac{\cancel{60}^{12} \text{ min}}{1 \cancel{\text{hr}}}$
$= 12$ min.

18. $\frac{3}{4}$ pound $\cdot \frac{16 \text{ ounces}}{1 \text{ pound}} = \frac{3 \cancel{\text{lb}}}{\cancel{4}_{1}} \cdot \frac{\cancel{16}^{4} \text{ oz}}{1 \cancel{\text{lb}}}$
$= 12$ oz

19. $\frac{7}{8}$ day $\cdot \frac{24 \text{ hours}}{1 \text{ day}} = \frac{7 \cancel{\text{day}}}{\cancel{8}_{1}} \cdot \frac{\cancel{24}^{3} \text{ hr}}{1 \cancel{\text{day}}}$
$= 21$ hr

Page 734 Lesson 5-4

1. $\frac{4}{7}\left(\frac{7}{4}\right) = 1$
The multiplicative inverse of $\frac{4}{7}$ is $\frac{7}{4}$.

2. $-\frac{5}{9}\left(-\frac{9}{5}\right) = 1$
The multiplicative inverse of $-\frac{5}{9}$ is $-\frac{9}{5}$.

3. $\frac{1}{4}\left(\frac{4}{1}\right) = 1$
The multiplicative inverse of $\frac{1}{4}$ is 4.

4. $5\frac{3}{8} = \frac{43}{8}$
$\frac{43}{8}\left(\frac{8}{43}\right) = 1$
The multiplicative inverse of $5\frac{3}{8}$ is $\frac{8}{43}$.

5. $6\left(\frac{1}{6}\right) = 1$

The multiplicative inverse of 6 is $\frac{1}{6}$.

6. $-18\left(-\frac{1}{18}\right) = 1$

The multiplicative inverse of -18 is $-\frac{1}{18}$.

7. $\frac{7}{10}\left(\frac{10}{7}\right) = 1$

The multiplicative inverse of $\frac{7}{10}$ is $\frac{10}{7}$.

8. $2.35 = 2\frac{35}{100}$

$= 2\frac{7}{20}$

$= \frac{47}{20}$

$\frac{47}{20}\left(\frac{20}{47}\right) = 1$

The multiplicative inverse of 2.35 is $\frac{20}{47}$.

9. $\frac{4}{5} \div \frac{2}{5} = \frac{\overset{2}{\cancel{4}}}{\underset{1}{\cancel{5}}} \cdot \frac{\overset{1}{\cancel{5}}}{\underset{1}{\cancel{2}}}$

$= \frac{2}{1}$ or 2

10. $-\frac{1}{3} \div \frac{6}{7} = \frac{-1}{3} \cdot \frac{7}{6}$

$= -\frac{7}{18}$

11. $\frac{4}{9} \div \frac{1}{5} = \frac{4}{9} \cdot \frac{5}{1}$

$= \frac{20}{9}$ or $2\frac{2}{9}$

12. $\frac{2}{3} \div \frac{1}{9} = \frac{2}{\underset{1}{\cancel{3}}} \cdot \frac{\overset{3}{\cancel{9}}}{1}$

$= \frac{6}{1}$ or 6

13. $\frac{4}{5} \div \left(-\frac{8}{15}\right) = \frac{\overset{1}{\cancel{4}}}{\underset{1}{\cancel{5}}} \cdot \frac{-\overset{3}{\cancel{15}}}{\underset{2}{\cancel{8}}}$

$= -\frac{3}{2}$ or $-1\frac{1}{2}$

14. $\frac{1}{12} \div \frac{3}{4} = \frac{1}{\underset{3}{\cancel{12}}} \cdot \frac{\overset{1}{\cancel{4}}}{3}$

$= \frac{1}{9}$

15. $\frac{3}{4} \div \frac{15}{16} = \frac{\overset{1}{\cancel{3}}}{\underset{1}{\cancel{4}}} \cdot \frac{\overset{4}{\cancel{16}}}{\underset{5}{\cancel{15}}}$

$= \frac{4}{5}$

16. $16 \div 1\frac{7}{8} = \frac{16}{1} \div \frac{15}{8}$

$= \frac{16}{1} \cdot \frac{8}{15}$

$= \frac{128}{15}$ or $8\frac{8}{15}$

17. $2\frac{1}{6} \div \left(-1\frac{1}{5}\right) = \frac{13}{6} \div \left(-\frac{6}{5}\right)$

$= \frac{13}{6} \cdot \frac{-5}{6}$

$= -\frac{65}{36}$ or $-1\frac{29}{36}$

18. $-11 \div 3\frac{1}{7} = \frac{-11}{1} \div \frac{22}{7}$

$= \frac{-\overset{1}{\cancel{11}}}{1} \cdot \frac{7}{\underset{2}{\cancel{22}}}$

$= -\frac{7}{2}$ or $-3\frac{1}{2}$

19. $\frac{8}{45} \div \frac{10}{27} = \frac{\overset{4}{\cancel{8}}}{\underset{5}{\cancel{45}}} \cdot \frac{\overset{3}{\cancel{27}}}{\underset{5}{\cancel{10}}}$

$= \frac{12}{25}$

20. $-22 \div \left(-5\frac{1}{2}\right) = \frac{-22}{1} \div \left(\frac{-11}{2}\right)$

$= \frac{-\overset{2}{\cancel{22}}}{1} \cdot \left(\frac{-2}{\underset{1}{\cancel{11}}}\right)$

$= \frac{4}{1}$ or 4

21. $\frac{w}{5} \div \frac{w}{35} = \frac{\overset{1}{\cancel{w}}}{\underset{1}{\cancel{5}}} \cdot \frac{\overset{7}{\cancel{35}}}{\underset{1}{\cancel{w}}}$

$= \frac{7}{1}$ or 7

22. $\frac{ab}{12} \div \frac{b}{16} = \frac{a\cancel{b}}{\underset{3}{\cancel{12}}} \cdot \frac{\overset{4}{\cancel{16}}}{\cancel{b}}$

$= \frac{4a}{3}$

23. $\frac{21y}{8x^2} \div \frac{7y}{16x} = \frac{\overset{3}{\cancel{21y}}}{\underset{1}{\cancel{8x}}x} \cdot \frac{\overset{2}{\cancel{16x}}}{\underset{1}{\cancel{7y}}}$

$= \frac{6}{x}$

Page 734 Lesson 5-5

1. $\frac{2}{7} + \frac{3}{7} = \frac{2 + 3}{7}$

$= \frac{5}{7}$

2. $\frac{8}{15} - \frac{4}{15} = \frac{8 - 4}{15}$

$= \frac{4}{15}$

3. $\frac{3}{7} + \frac{4}{7} = \frac{3 + 4}{7}$

$= \frac{7}{7}$ or 1

4. $-\frac{8}{9} + \frac{1}{9} = \frac{-8 + 1}{9}$

$= \frac{-7}{9}$ or $-\frac{7}{9}$

5. $\frac{5}{6} - \frac{1}{6} = \frac{5 - 1}{6}$

$= \frac{4}{6}$ or $\frac{2}{3}$

6. $\frac{7}{12} - \frac{5}{12} = \frac{7 - 5}{12}$

$= \frac{2}{12}$ or $\frac{1}{6}$

7. $\frac{5}{12} + \frac{11}{12} = \frac{5 + 11}{12}$

$= \frac{16}{12}$

$= \frac{4}{3}$ or $1\frac{1}{3}$

8. $-\frac{3}{14} - \frac{5}{14} = \frac{(-3) - 5}{14}$

$= -\frac{8}{14}$ or $-\frac{4}{7}$

9. $3\frac{1}{4} + \left(-\frac{3}{4}\right) = \frac{13}{4} + \left(-\frac{3}{4}\right)$

$= \frac{13 + (-3)}{4}$

$= \frac{10}{4}$

$= \frac{5}{2}$ or $2\frac{1}{2}$

10. $\frac{3}{8} - \left(-1\frac{1}{8}\right) = \frac{3}{8} - \left(\frac{-9}{8}\right)$

$= \frac{3 - (-9)}{8}$

$= \frac{12}{8}$

$= \frac{3}{2}$ or $1\frac{1}{2}$

11. $4\frac{9}{10} - 1\frac{1}{10} = \frac{49}{10} - \frac{11}{10}$
$= \frac{49 - 11}{10}$
$= \frac{38}{10}$
$= \frac{19}{5}$ or $3\frac{4}{5}$

12. $-5\frac{3}{5} + \left(-2\frac{1}{5}\right) = \frac{-28}{5} + \left(\frac{-11}{5}\right)$
$= \frac{-28 + (-11)}{5}$
$= \frac{-39}{5}$ or $-7\frac{4}{5}$

13. $\frac{n}{5} + \frac{3n}{5} = \frac{n + 3n}{5}$
$= \frac{4n}{5}$

14. $\frac{15}{k} - \frac{8}{k} = \frac{15 - 8}{k}$
$= \frac{7}{k}, k \neq 0$

15. $12\frac{7}{8}s - 7\frac{3}{8}s = \frac{103s}{8} - \frac{59s}{8}$
$= \frac{103s - 59s}{8}$
$= \frac{44s}{8}$
$= \frac{11}{2}s$ or $5\frac{1}{2}s$

16. $-6\frac{4}{9}t - 3\frac{2}{9}t = \frac{-58t}{9} - \frac{29t}{9}$
$= \frac{-58t - 29t}{9}$
$= \frac{-87t}{9}$
$= \frac{-29}{3}t$ or $-9\frac{2}{3}t$

17. $6\frac{1}{4}g + \left(-6\frac{3}{4}g\right) = \frac{25g}{4} + \frac{-27g}{4}$
$= \frac{25g + (-27g)}{4}$
$= \frac{-2g}{4}$
$= -\frac{1}{2}g$

18. $7\frac{2}{5}n - \left(-4\frac{2}{5}n\right) = \frac{37n}{5} - \left(\frac{-22n}{5}\right)$
$= \frac{37n - (-22n)}{5}$
$= \frac{59n}{5}$ or $11\frac{4}{5}n$

Page 735 Lesson 5-6

1. $30 = 2 \cdot 3 \cdot 5 = 2 \cdot 3 \cdot 5$
$18 = 2 \cdot 3 \cdot 3 = 2 \cdot 3^2$
$\text{LCM} = 2 \cdot 3^2 \cdot 5$
$= 90$
The LCM of 30 and 18 is 90.

2. $4 = 2 \cdot 2 = 2^2$
$16 = 2 \cdot 2 \cdot 2 \cdot 2 = 2^4$
$\text{LCM} = 2^4$
$= 16$
The LCM of 4 and 16 is 16.

3. $3m = 3 \cdot m = 3 \cdot m$
$12 = 2 \cdot 2 \cdot 3 = 2^2 \cdot 3$
$\text{LCM} = 2^2 \cdot 3 \cdot m$
$= 12\,m$
The LCM of $3m$ and 12 is $12m$.

4. $6a = 2 \cdot 3 \cdot a = 2 \cdot 3 \cdot a$
$17a^5 = 17 \cdot a^5 = 17a^5$
$\text{LCM} = 2 \cdot 3 \cdot 17 \cdot a^5$
$= 102a^5$
The LCM of $6a$ and $17a^5$ is $102a^5$.

5. $2 = 2$
$5 = 5$
$7 = 7$
$\text{LCM} = 2 \cdot 5 \cdot 7$
$= 70$
The LCM of 2, 5, and 7 is 70.

6. $9x^2y = 3 \cdot 3 \cdot x^2 \cdot y = 3^2 \cdot x^2 \cdot y$
$12xy^3 = 2 \cdot 2 \cdot 3 \cdot x \cdot y^3 = 2^2 \cdot 3 \cdot x \cdot y^3$
$\text{LCM} = 2^2 \cdot 3^2 \cdot x^2 \cdot y^3$
$= 36x^2y^3$
The LCM of $9x^2y$ and $12xy^3$ is $36x^2y^3$.

7. $5 = 5$
$25 = 5 \cdot 5 = 5^2$
$\text{LCM} = 5^2$
$= 25$
The LCD of $\frac{2}{5}$ and $\frac{6}{25}$ is 25.

8. $12 = 2 \cdot 2 \cdot 3 = 2^2 \cdot 3$
$5 = 5$
$\text{LCM} = 2^2 \cdot 3 \cdot 5$
$= 60$
The LCD of $\frac{3}{12}$ and $\frac{4}{5}$ is 60.

9. $6 = 2 \cdot 3 = 2 \cdot 3$
$9 = 3 \cdot 3 = 3^2$
$\text{LCM} = 2 \cdot 3^2$
$= 18$
The LCD of $\frac{4}{6}$ and $\frac{7}{9}$ is 18.

10. $9 = 3 \cdot 3 = 3^2$
$12 = 2 \cdot 2 \cdot 3 = 2^2 \cdot 3$
$\text{LCM} = 2^2 \cdot 3^2$
$= 36$
The LCD of $\frac{5}{9}$ and $\frac{7}{12}$ is 36.

11. $4 = 2 \cdot 2 = 2^2$
$6 = 2 \cdot 3 = 2 \cdot 3$
$\text{LCM} = 2^2 \cdot 3$
$= 12$
The LCD of $\frac{1}{4}$ and $\frac{5}{6}$ is 12.

12. $20 = 2 \cdot 2 \cdot 5 = 2^2 \cdot 5$
$8 = 2 \cdot 2 \cdot 2 = 2^3$
$\text{LCM} = 2^3 \cdot 5$
$= 40$
The LCD of $\frac{11}{20}$ and $\frac{3}{8}$ is 40.

13. $10p = 2 \cdot 5 \cdot p$
$5p^3 = 5 \cdot p^3$
$\text{LCM} = 2 \cdot 5 \cdot p^3$
$= 10p^3$
The LCD of $\frac{3}{10p}$ and $\frac{7}{5p^3}$ is $10p^3$.

14. $a^2 = a^2$
$3a^4 = 3 \cdot a^4$
$\text{LCM} = 3 \cdot a^4$
$= 3a^4$
The LCD of $\frac{1}{a^2}$ and $\frac{2}{3a^4}$ is $3a^4$.

15. $\frac{2}{3} = \frac{2 \cdot 2^2}{2^3 \cdot 3} = \frac{8}{12}$

$\frac{3}{4} = \frac{3 \cdot 3}{2^2 \cdot 3} = \frac{9}{12}$

Since $\frac{8}{12} < \frac{9}{12}$, then $\frac{2}{3} < \frac{3}{4}$.

16. $-\frac{5}{8} = \frac{-5 \cdot 5}{2^3 \cdot 5} = \frac{-25}{40}$

$-\frac{3}{5} = \frac{-3 \cdot 2^3}{2^3 \cdot 5} = \frac{-24}{40}$

Since $-\frac{25}{40} < -\frac{24}{40}$, then $-\frac{5}{8} < -\frac{3}{5}$.

17. $\frac{4}{6} = \frac{4 \cdot 2}{2^2 \cdot 3} = \frac{8}{12}$

$\frac{7}{12} = \frac{7}{2^2 \cdot 3} = \frac{7}{12}$

Since $\frac{8}{12} > \frac{7}{12}$, then $\frac{4}{6} > \frac{7}{12}$.

18. $\frac{11}{18} = \frac{11 \cdot 3}{2 \cdot 3^3} = \frac{33}{54}$

$\frac{33}{54} = \frac{33}{2 \cdot 3^3} = \frac{33}{54}$

Since $\frac{33}{54} = \frac{33}{54}$, then $\frac{11}{18} = \frac{33}{54}$.

19. $\frac{4}{19} = \frac{4 \cdot 2}{2 \cdot 19} = \frac{8}{38}$

$\frac{8}{38} = \frac{8}{2 \cdot 19} = \frac{8}{38}$

Since $\frac{8}{38} = \frac{8}{38}$, then $\frac{4}{19} = \frac{8}{38}$.

20. $\frac{9}{15} = \frac{9 \cdot 2}{2 \cdot 3 \cdot 5} = \frac{18}{30}$

$\frac{1}{2} = \frac{1 \cdot 3 \cdot 5}{2 \cdot 3 \cdot 5} = \frac{15}{30}$

Since $\frac{18}{30} > \frac{15}{30}$, then $\frac{9}{15} > \frac{1}{2}$.

Page 735 Lesson 5-7

1. $\frac{1}{5} + \frac{2}{7} = \frac{1}{5} \cdot \frac{7}{7} + \frac{2}{7} \cdot \frac{5}{5}$
$= \frac{7}{35} + \frac{10}{35}$
$= \frac{17}{35}$

2. $\frac{4}{5} + \frac{7}{9} = \frac{4}{5} \cdot \frac{9}{9} + \frac{7}{9} \cdot \frac{5}{5}$
$= \frac{36}{45} + \frac{35}{45}$
$= \frac{71}{45}$ or $1\frac{26}{45}$

3. $\frac{1}{9} - \frac{7}{12} = \frac{1}{9} \cdot \frac{4}{4} - \frac{7}{12} \cdot \frac{3}{3}$
$= \frac{4}{36} - \frac{21}{36}$
$= -\frac{17}{36}$

4. $\frac{8}{11} - \frac{4}{5} = \frac{8}{11} \cdot \frac{5}{5} - \frac{4}{5} \cdot \frac{11}{11}$
$= \frac{40}{55} - \frac{44}{55}$
$= -\frac{4}{55}$

5. $\frac{7}{12} - \left(-\frac{4}{11}\right) = \frac{7}{12} \cdot \frac{11}{11} - \left(-\frac{4}{11}\right) \cdot \frac{12}{12}$
$= \frac{77}{132} - \left(-\frac{48}{132}\right)$
$= \frac{125}{132}$

6. $-\frac{9}{14} + \frac{15}{16} = -\frac{9}{14} \cdot \frac{8}{8} + \frac{15}{16} \cdot \frac{7}{7}$
$= -\frac{72}{112} + \frac{105}{112}$
$= \frac{33}{112}$

7. $-\frac{3}{8} + \left(-1\frac{5}{12}\right) = -\frac{3}{8} + \left(-\frac{17}{12}\right)$
$= -\frac{3}{8} \cdot \frac{3}{3} + \left(-\frac{17}{12}\right) \cdot \frac{2}{2}$
$= -\frac{9}{24} + \left(-\frac{34}{24}\right)$
$= -\frac{43}{24}$ or $-1\frac{19}{24}$

8. $-\frac{2}{15} - 3\frac{1}{5} = -\frac{2}{15} - \frac{16}{5}$
$= -\frac{2}{15} - \frac{16}{5} \cdot \frac{3}{3}$
$= -\frac{2}{15} - \frac{48}{15}$
$= -\frac{50}{15}$
$= -\frac{10}{3}$ or $-3\frac{1}{3}$

9. $-5\frac{1}{3} + \left(-\frac{1}{6}\right) = -\frac{16}{3} + \left(-\frac{1}{6}\right)$
$= -\frac{16}{3} \cdot \frac{2}{2} + \left(-\frac{1}{6}\right)$
$= -\frac{32}{6} + \left(-\frac{1}{6}\right)$
$= -\frac{33}{6}$
$= -\frac{11}{2}$ or $-5\frac{1}{2}$

10. $3\frac{2}{5} + 2\frac{4}{7} = \frac{17}{5} + \frac{18}{7}$
$= \frac{17}{5} \cdot \frac{7}{7} + \frac{18}{7} \cdot \frac{5}{5}$
$= \frac{119}{35} + \frac{90}{35}$
$= \frac{209}{35}$ or $5\frac{34}{35}$

11. $-4\frac{1}{8} + 2\frac{5}{9} = -\frac{33}{8} + \frac{23}{9}$
$= -\frac{33}{8} \cdot \frac{9}{9} + \frac{23}{9} \cdot \frac{8}{8}$
$= -\frac{297}{72} + \frac{184}{72}$
$= -\frac{113}{72}$ or $-1\frac{41}{72}$

12. $-3\frac{3}{7} - 5\frac{1}{14} = -\frac{24}{7} - \frac{71}{14}$
$= -\frac{24}{7} \cdot \frac{2}{2} - \frac{71}{14}$
$= -\frac{48}{14} - \frac{71}{14}$
$= -\frac{119}{14}$
$= -\frac{17}{2}$ or $-8\frac{1}{2}$

13. $11\frac{3}{5} - \left(-6\frac{5}{8}\right) = \frac{58}{5} - \left(-\frac{53}{8}\right)$
$= \frac{58}{5} \cdot \frac{8}{8} - \left(-\frac{53}{8}\right) \cdot \frac{5}{5}$
$= \frac{464}{40} - \left(-\frac{265}{40}\right)$
$= \frac{729}{40}$ or $18\frac{9}{40}$

14. $\frac{5}{14} + \frac{2}{21} = \frac{5}{14} \cdot \frac{3}{3} + \frac{2}{21} \cdot \frac{2}{2}$
$= \frac{15}{42} + \frac{4}{42}$
$= \frac{19}{42}$

15. $2\frac{1}{7} - 3\frac{1}{3} = \frac{15}{7} - \frac{10}{3}$
$= \frac{15}{7} \cdot \frac{3}{3} - \frac{10}{3} \cdot \frac{7}{7}$
$= \frac{45}{21} - \frac{70}{21}$
$= -\frac{25}{21}$ or $-1\frac{4}{21}$

Page 735 Lesson 5-8

1. mean $= \frac{82 + 79 + 93 + 91 + 95}{5}$
$= \frac{440}{5} = 88$
The mean is 88.
To find the median, order the numbers from least to greatest.
79, 82, <u>91</u>, 93, 95
The median is 91.
There is no mode because each number occurs only once.

2. mean $= \frac{88 + 85 + 76 + 94 + 85 + 97}{6}$
$= \frac{525}{6} = 87.5$
The mean is 87.5.
To find the median, order the numbers from least to greatest.
76, 85, 85, 88, 94, 97
$\frac{85 + 88}{2} = 86.5$
The median is 86.5.
Since 85 occurs twice, it is the mode.

3. mean $= \frac{23 + 32 + 19 + 27 + 41 + 21 + 26 + 32 + 23}{9}$
$= \frac{244}{9} \approx 27.1$
The mean is 27.1.
To find the median, order the numbers from least to greatest.
19, 21, 23, 23, <u>26</u>, 27, 32, 32, 41
The median is 26.
Since 23 and 32 occur twice, they are the modes.

4. mean $= \frac{7.4 + 8.3 + 6.1 + 5.4 + 6.8 + 7.1 + 8.0 + 9.2}{8}$
$= \frac{58.3}{8} \approx 7.3$
The mean is 7.3.
To find the median, order the numbers from least to greatest.
5.4, 6.1, 6.8, 7.1, 7.4, 8.0, 8.3, 9.2
$\frac{7.1 + 7.4}{2} = 7.3$
The median is 7.3.
There is no mode because each number occurs only once.

5. mean
$= \frac{0.57 + 12.81 + 12.6 + 0.96 + 6.1 + 14.3 + 4.1 + 12.81 + 0.96}{9}$
$= \frac{65.21}{9} \approx 7.2$
The mean is 7.2.
To find the median, order the numbers from least to greatest.
0.57, 0.96, 0.96, 4.1, <u>6.1</u>, 12.6, 12.81, 12.81, 14.3
The median is 6.1.
Since they both occur twice, the modes are 0.96 and 12.81.

6. mean $= \frac{8 + 11(4) + 12(3) + 13 + 14(2)}{11}$
$= \frac{129}{11} \approx 11.7$
The mean is 11.7.
There are 11 numbers, so the 6th number, or 12, is the median.
You can see from the data set that 11 occurs most often in the data set. The mode is 11.

7. mean $= \frac{0.2(2) + (0.3)4 + 0.4(3) + 0.5(3) + 0.6 + 0.8}{14}$
$= \frac{5.7}{14} \approx 0.4$
The mean is 0.4.
There are 14 numbers, so the mean of the 7th and 8th terms is the median.
$\frac{0.4 + 0.4}{2} = 0.4$
The median is 0.4.
The number 0.3 occurs the most often in the data set, so the mode is 0.3.

8. mean:
$= \frac{538.8 + 138.9 + 941 + 756.6 + 7372.4 + 11{,}669.3 + 1147.9 + 1023.6 + 42.1 + 2997.2 + 30.7 + 4063.8 + 27.7}{13}$
≈ 2365.4
median: 27.7, 30.7, 42.1, 138.9, 538.8, 756.6, <u>941</u>, 1023.6, 1147.9, 2997.2, 4063.8, 7372.4, 11,669.3
= 941
mode: none

Page 736 Lesson 5-9

Exercises 1–15 For checks, see students' work.

1. $a - 4.86 = 7.2$
$a - 4.86 + 4.86 = 7.2 + 4.86$
$a = 12.06$

2. $n + 6.98 = 10.3$
$n + 6.98 - 6.98 = 10.3 - 6.98$
$n = 3.32$

3. $87.64 = f - (-8.5)$
$87.64 + (-8.5) = f - (-8.5) + (-8.5)$
$79.14 = f$

4. $x - \frac{2}{5} = -\frac{8}{15}$
$x - \frac{2}{5} + \frac{2}{5} = \left(-\frac{8}{15}\right) + \frac{2}{5}$
$x = -\frac{8}{15} + \frac{2}{5} \cdot \frac{3}{3}$
$x = -\frac{8}{15} + \frac{6}{15}$
$x = -\frac{2}{15}$

5. $3\frac{3}{4} + m = 6\frac{5}{8}$
$3\frac{3}{4} + m - 3\frac{3}{4} = 6\frac{5}{8} - 3\frac{3}{4}$
$m = \frac{53}{8} - \frac{15}{4}$
$m = \frac{53}{8} - \frac{15}{4} \cdot \frac{2}{2}$
$m = \frac{53}{8} - \frac{30}{8}$
$m = \frac{23}{8}$ or $2\frac{7}{8}$

6. $4\frac{1}{6} = r + 6\frac{1}{4}$
$4\frac{1}{6} - 6\frac{1}{4} = r + 6\frac{1}{4} - 6\frac{1}{4}$
$\frac{25}{6} - \frac{25}{4} = r$
$\frac{25}{6} \cdot \frac{2}{2} - \frac{25}{4} \cdot \frac{3}{3} = r$
$\frac{50}{12} - \frac{75}{12} = r$
$-\frac{25}{12} = r$
$-2\frac{1}{12} = r$

7. $7\frac{1}{3} = c - \frac{4}{5}$

$7\frac{1}{3} + \frac{4}{5} = c - \frac{4}{5} + \frac{4}{5}$

$\frac{22}{3} + \frac{4}{5} = c$

$\frac{22}{3} \cdot \frac{5}{5} + \frac{4}{5} \cdot \frac{3}{3} = c$

$\frac{110}{15} + \frac{12}{15} = c$

$\frac{122}{15} = c$

$8\frac{2}{15} = c$

8. $-4.62 = h + (-9.4)$

$-4.62 - (-9.4) = h + (-9.4) - (-9.4)$

$4.78 = h$

9. $w - 1\frac{1}{5} = \frac{2}{9}$

$w - 1\frac{1}{5} + 1\frac{1}{5} = \frac{2}{9} + 1\frac{1}{5}$

$w = \frac{2}{9} + \frac{6}{5}$

$w = \frac{2}{9} \cdot \frac{5}{5} + \frac{6}{5} \cdot \frac{9}{9}$

$w = \frac{10}{45} + \frac{54}{45}$

$w = \frac{64}{45}$

$w = 1\frac{19}{45}$

10. $\frac{2}{3}w = \frac{1}{6}$

$\frac{3}{2}\left(\frac{2}{3}w\right) = \left(\frac{1}{6}\right)\frac{3}{2}$

$w = \frac{1}{\cancel{6}_2} \cdot \frac{\cancel{3}^1}{2}$

$w = \frac{1}{4}$

11. $6 = -\frac{3}{4}x$

$-\frac{4}{3}(6) = -\frac{4}{3}\left(-\frac{3}{4}x\right)$

$-\frac{4}{\cancel{3}_1} \cdot \frac{\cancel{6}^2}{1} = x$

$-8 = x$

12. $-0.5m = -10$

$\frac{-0.5m}{-0.5} = \frac{-10}{-0.5}$

$m = 20$

13. $-\frac{1}{9}t = 7$

$-\frac{9}{1}\left(-\frac{1}{9}t\right) = -\frac{9}{1}\left(\frac{7}{1}\right)$

$t = -63$

14. $5\frac{2}{3} = y - \frac{1}{8}$

$\frac{17}{3} + \frac{1}{8} = y - \frac{1}{8} + \frac{1}{8}$

$\frac{17}{3} \cdot \frac{8}{8} + \frac{1}{8} \cdot \frac{3}{3} = y$

$\frac{136}{24} + \frac{3}{24} = y$

$\frac{139}{24} = y$

$5\frac{19}{24} = y$

15. $-14.8 = -7.1 + t$

$-14.8 + 7.1 = -7.1 + t + 7.1$

$-7.7 = t$

Page 736 Lesson 5-10

1. 3.5, 4.3, 5.1, . . .

+ 0.8 + 0.8

Arithmetic; the common difference is 0.8; the next 3 terms are 5.1 + 0.8 or 5.9, 5.9 + 0.8 or 6.7, and 6.7 + 0.8 or 7.5.

2. 125, 75, 45, . . .

× 0.6 × 0.6

Geometric; the common ratio is 0.6; the next 3 terms are 45 × 0.6 or 27, 27 × 0.6 or 16.2, and 16.2 × 0.6 or 9.72.

3. 5, 10, 20, . . .

× 2 × 2

Geometric; the common ratio is 2; the next 3 terms are 20 × 2 or 40, 40 × 2 or 80, and 80 × 2 or 160.

4. 2401, 49, 7, . . .

÷ 49 ÷ 7

Neither

5. $\frac{1}{2}, \frac{5}{6}, 1\frac{1}{6}, \ldots$

$+\frac{1}{3}$ $+\frac{1}{3}$

Arithmetic; the common difference is $\frac{1}{3}$; the next 3 terms are $1\frac{1}{6} + \frac{1}{3}$ or $1\frac{1}{2}$, $1\frac{1}{2} + \frac{1}{3}$ or $1\frac{5}{6}$, and $1\frac{5}{6} + \frac{1}{3}$ or $2\frac{1}{6}$.

6. $-\frac{4}{5}, 2, -5, 12\frac{1}{2}, \ldots$

$\times\frac{-5}{2}$ $\times\frac{-5}{2}$ $\times\frac{-5}{2}$

Geometric; the common ratio is $-\frac{5}{2}$; the next 3 terms are $12\frac{1}{2} \times \left(-\frac{5}{2}\right)$ or $-31\frac{1}{4}$, $-31\frac{1}{4} \times \left(-\frac{5}{2}\right)$ or $78\frac{1}{8}$, and $78\frac{1}{8} \times \left(-\frac{5}{2}\right)$ or $-195\frac{5}{16}$.

7. $\frac{1}{4}, \frac{1}{2}, 1, 2, \ldots$

× 2 × 2 × 2

Geometric; the common ratio is 2; the next 3 terms are 2 × 2 or 4, 4 × 2 or 8, and 8 × 2 or 16.

8. 23, 18, 13, . . .

−5 −5

Arithmetic; the common difference is −5; the next 3 terms are 13 − 5 or 8, 8 − 5 or 3, and 3 − 5 or −2.

9. 45, 43, 39, 33

−2 −4 −6

Neither

10. 2, 4, 8, 16, . . .

× 2 × 2 × 2

Geometric; the common ratio is 2; the next 3 terms are 16 × 2 or 32, 32 × 2 or 64, and 64 × 2 or 128.

11. 100, 75, 50, . . .

$-25 \quad -25$

Arithmetic; the common difference is −25; the next 3 terms are 50 − 25 or 25, 25 − 25 or 0, and 0 − 25 or −25.

12. $\frac{1}{5}$, 1, 5, 25, . . .

$\times 5 \quad \times 5 \quad \times 5$

Geometric; the common ratio is 5; the next 3 terms are 25 × 5 or 125, 125 × 5 or 625, and 625 × 5 or 3125.

Page 736 Lesson 6-1

1. $\frac{15}{40} = \frac{3}{8}$ (÷ 5)

The ratio of vans to vehicles is $\frac{3}{8}$.

2. $\frac{6}{14} = \frac{3}{7}$ (÷ 2)

The ratio of pens to pencils is $\frac{3}{7}$.

3. $\frac{12}{18} = \frac{2}{3}$ (÷ 6)

The ratio of dolls to toys is $\frac{2}{3}$.

4. $\frac{8}{36} = \frac{2}{9}$ (÷ 4)

The ratio of red crayons to crayons is $\frac{2}{9}$.

5. $\frac{18}{45} = \frac{2}{5}$ (÷ 9)

The ratio of boys to students is $\frac{2}{5}$.

6. $\frac{30}{6} = \frac{5}{1}$ (÷ 6)

The ratio of birds to birds is $\frac{5}{1}$.

7. $\frac{98}{14} = \frac{7}{1}$ (÷ 14)

The ratio of ants to ladybugs is $\frac{7}{1}$.

8. $\frac{140}{12} = \frac{35}{3}$ (÷ 4)

The ratio of dogs to cats is $\frac{35}{3}$.

9. $\frac{321}{96} = \frac{107}{32}$ (÷ 3)

The ratio of pennies to dimes is $\frac{107}{32}$.

10. $\frac{3 \text{ cups}}{3 \text{ quarts}} = \frac{3 \text{ cups}}{12 \text{ cups}}$

$= \frac{1 \text{ cup}}{4 \text{ cups}}$

The ratio of cups to quarts is $\frac{1}{4}$.

11. $\frac{343.8 \text{ miles}}{9 \text{ gallons}} = \frac{38.2 \text{ miles}}{1 \text{ gallon}}$ (÷ 9)

$= 38.2$ mi/gal

12. $\frac{7.95 \text{ dollars}}{5 \text{ pounds}} = \frac{1.59 \text{ dollars}}{1 \text{ pound}}$ (÷ 5)

= \$1.59/lb

13. $\frac{52 \text{ dollars}}{8 \text{ tickets}} = \frac{6.5 \text{ dollars}}{1 \text{ ticket}}$ (÷ 8)

= \$6.50/ticket

14. $\frac{43.92 \text{ dollars}}{4 \text{ CDs}} = \frac{10.98 \text{ dollars}}{1 \text{ CD}}$ (÷ 4)

≈ \$11/CD

15. $\frac{450 \text{ miles}}{8 \text{ hours}} \approx \frac{56.3 \text{ miles}}{1 \text{ hour}}$ (÷ 8)

= 56.3 mi/h

16. $\frac{3.96 \text{ dollars}}{12 \text{ cans}} \approx \frac{0.3 \text{ dollars}}{1 \text{ can}}$ (÷ 12)

= \$0.3/can

17. $\frac{3.84 \text{ dollars}}{64 \text{ ounces}} = \frac{0.06 \text{ dollars}}{1 \text{ ounce}}$ (÷ 64)

= \$0.06/oz

18. $\frac{200 \text{ yards}}{32.3 \text{ seconds}} \approx \frac{6.2 \text{ yards}}{1 \text{ second}}$ (÷ 32.3)

= 6.2 yd/s

19. $\frac{3.98 \text{ dollars}}{4 \text{ notebooks}} = \frac{\$0.995}{\text{notebook}}$ (÷ 4 numerator and denominator)

$\frac{4.99 \text{ dollars}}{5 \text{ notebooks}} = \frac{\$0.998}{\text{notebook}}$

Since the 4-pack costs $0.995 per notebook and the 5-pack costs $0.998 per notebook, the 5-pack costs more per notebook.

20. $\frac{70 \text{ miles}}{1 \text{ hour}} = \frac{70 \text{ mi}}{1 \text{ h}} \cdot \frac{5280 \text{ ft}}{1 \text{ mi}} \div \frac{3600 \text{ s}}{1 \text{ h}}$

$= \frac{70 \text{ mi}}{1 \text{ h}} \cdot \frac{\overset{1.47}{5280} \text{ ft}}{1 \text{ mi}} \cdot \frac{1 \text{ h}}{\underset{1}{3600} \text{ s}}$

≈ 103 ft/s

Page 737 Lesson 6-2

1. $\frac{7}{k} = \frac{49}{63}$

$7 \cdot 63 = k \cdot 49$

$441 = 49k$

$\frac{441}{49} = \frac{49k}{49}$

$9 = k$

2. $\frac{s}{4.8} = \frac{30.6}{28.8}$

$s \cdot 28.8 = 4.8 \cdot 30.6$

$28.8\,s = 146.88$

$\frac{28.8\,s}{28.8} = \frac{146.88}{28.8}$

$s = 5.1$

3. $\frac{6}{11} = \frac{19.2}{g}$

$6 \cdot g = 11 \cdot 19.2$

$6g = 211.2$

$\frac{6g}{6} = \frac{211.2}{6}$

$g = 35.2$

4. $\frac{8}{13} = \frac{b}{65}$

$8 \cdot 65 = 13 \cdot b$

$520 = 13b$

$\frac{520}{13} = \frac{13b}{13}$

$40 = b$

5. $\frac{x}{12} = \frac{26}{24}$

$x \cdot 24 = 12 \cdot 26$

$24x = 312$

$\frac{24x}{24} = \frac{312}{24}$

$x = 13$

6. $\frac{21}{p} = \frac{3}{9}$

$21 \cdot 9 = p \cdot 3$

$189 = 3p$

$\frac{189}{3} = \frac{3p}{3}$

$63 = p$

7. $\frac{6.5}{8} = \frac{w}{20}$

$6.5 \cdot 20 = 8 \cdot w$

$130 = 8w$

$\frac{130}{8} = \frac{8w}{8}$

$16.25 = w$

8. $\frac{10}{4.21} = \frac{7}{y}$

$10 \cdot y = 4.21 \cdot 7$

$10y = 29.47$

$\frac{10y}{10} = \frac{29.47}{10}$

$y = 2.947$

9. $\frac{6 \text{ plums}}{1 \text{ dollar}} = \frac{10 \text{ plums}}{d \text{ dollars}}$

$\frac{6}{\$1.00} = \frac{10}{d}$

$6 \cdot d = 1 \cdot 10$

$6d = 10$

$\frac{6d}{6} = \frac{10}{6}$

$d \approx \$1.67$

10. $\frac{8 \text{ gallons}}{9.36 \text{ dollars}} = \frac{f \text{ gallons}}{17.55 \text{ dollars}}$

$\frac{8}{\$9.36} = \frac{f}{\$17.55}$

$8 \cdot 17.55 = 9.36 \cdot f$

$140.4 = 9.36f$

$\frac{140.4}{9.36} = \frac{9.36f}{9.36}$

$15 = f$

11. $\frac{3 \text{ packages}}{53.67 \text{ dollars}} = \frac{7 \text{ packages}}{m \text{ dollars}}$

$\frac{3}{\$53.67} = \frac{7}{m}$

$3 \cdot m = 53.67 \cdot 7$

$3m = 375.69$

$\frac{3m}{3} = \frac{375.69}{3}$

$m = \$125.23$

12. $\frac{10 \text{ cards}}{7.5 \text{ dollars}} = \frac{p \text{ cards}}{18 \text{ dollars}}$

$\frac{10}{\$7.50} = \frac{p}{\$18.00}$

$10 \cdot 18 = 7.5 \cdot p$

$180 = 7.5p$

$\frac{180}{7.5} = \frac{7.5p}{7.5}$

$24 = p$

13. $\frac{12 \text{ cookies}}{3 \text{ dollars}} = \frac{16 \text{ cookies}}{5 \text{ dollars}}$

$\frac{12}{\$3.00} = \frac{16}{s}$

$12 \cdot s = 3 \cdot 16$

$12s = 48$

$\frac{12s}{12} = \frac{48}{12}$

$s = \$4.00$

14. $\frac{6 \text{ toy cars}}{4.5 \text{ dollars}} = \frac{c \text{ toy cars}}{6.75 \text{ dollars}}$

$\frac{6}{\$4.50} = \frac{c}{\$6.75}$

$6 \cdot 6.75 = 4.5 \cdot c$

$40.5 = 4.5c$

$\frac{40.5}{4.5} = \frac{4.5c}{4.5}$

$9 = c$

Page 737 Lesson 6-3

1. $\frac{0.5 \text{ inch}}{4 \text{ feet}} = \frac{5 \text{ inches}}{x \text{ feet}}$
$0.5 \cdot x = 4 \cdot 5$
$0.5x = 20$
$\frac{0.5x}{0.5} = \frac{20}{0.5}$
$x = 40$ feet

2. $\frac{0.5 \text{ inch}}{4 \text{ feet}} = \frac{1.75 \text{ inches}}{x \text{ feet}}$
$0.5 \cdot x = 4 \cdot 1.75$
$0.5x = 7$
$\frac{0.5x}{0.5} = \frac{7}{0.5}$
$x = 14$ feet

3. $\frac{0.5 \text{ inch}}{4 \text{ feet}} = \frac{7.5 \text{ inches}}{x \text{ feet}}$
$0.5 \cdot x = 4 \cdot 7.5$
$0.5x = 30$
$\frac{0.5x}{0.5} = \frac{30}{0.5}$
$x = 60$ feet

4. $\frac{0.5 \text{ inch}}{4 \text{ feet}} = \frac{9.25 \text{ inches}}{x \text{ feet}}$
$0.5 \cdot x = 4 \cdot 9.25$
$0.5x = 37$
$\frac{0.5x}{0.5} = \frac{37}{0.5}$
$x = 74$ feet

5. $\frac{0.5 \text{ inch}}{4 \text{ feet}} = \frac{12.2 \text{ inches}}{x \text{ feet}}$
$0.5 \cdot x = 4 \cdot 12.2$
$0.5x = 48.8$
$\frac{0.5x}{0.5} = \frac{48.8}{0.5}$
$x = 97.6$ feet

6. $\frac{0.5 \text{ inch}}{4 \text{ feet}} = \frac{1.3 \text{ inches}}{x \text{ feet}}$
$0.5 \cdot x = 4 \cdot 1.3$
$0.5x = 5.2$
$\frac{0.5x}{0.5} = \frac{5.2}{0.5}$
$x = 10.4$ feet

Page 737 Lesson 6-4

1. $0.42 = 0.42$
$= 42\%$

2. $0.06 = 0.06$
$= 6\%$

3. $1.35 = 1.35$
$= 135\%$

4. $0.001 = 0.001$
$= 0.1\%$

5. $0.99 = 0.99$
$= 99\%$

6. $3.6 = 3.6$
$= 360\%$

7. $0.8 = 0.8$
$= 80\%$

8. $0.0052 = 0.0052$
$= 0.52\%$

9. $0.00009 = 0.00009$
$= 0.009\%$

10. $\frac{17}{50} = \frac{34}{100}$
$= 34\%$

11. $\frac{9}{25} = \frac{36}{100}$
$= 36\%$

12. $\frac{12}{8} = 1.5$
$= 150\%$

13. $\frac{7}{40} = 0.175$
$= 17.5\%$

14. $\frac{11}{33} = 0.333\ldots$
$\approx 33.3\%$

15. $\frac{36}{27} = 1.333\ldots$
$\approx 133.3\%$

16. $32\% = \frac{32}{100}$
$= \frac{8}{25}$
$32\% = 32\%$
$= 0.32$

17. $15\% = \frac{15}{100}$
$= \frac{3}{20}$
$15\% = 15\%$
$= 0.15$

18. $88\frac{1}{2}\% = \frac{88\frac{1}{2}}{100}$
$= 88\frac{1}{2} \div 100$
$= \frac{177}{2} \cdot \frac{1}{100}$
$= \frac{177}{200}$
$88\frac{1}{2}\% = 88.5\%$
$= 0.885$

19. $250\% = \frac{250}{100}$
$= \frac{5}{2}$ or $2\frac{1}{2}$
$250\% = 250\%$
$= 2.5$

20. $21\% = \frac{21}{100}$
$21\% = 21\%$
$= 0.21$

21. $64\% = \frac{64}{100}$
$= \frac{16}{25}$
$64\% = 64\%$
$= 0.64$

22. $25\% = \frac{25}{100}$
$= \frac{1}{4}$
$25\% = 25\%$
$= 0.25$

23. $131\% = \frac{131}{100}$
$131\% = 131\%$
$= 1.31$

24. $72.5\% = \frac{72.5}{100}$
$= \frac{72.5}{100} \cdot \frac{10}{10}$
$= \frac{725}{1000}$
$= \frac{29}{40}$
$72.5\% = 72.5\%$
$= 0.725$

25. $66\frac{2}{3}\% = \frac{66\frac{2}{3}}{100}$

$= 66\frac{2}{3} \div 100$

$= \frac{\overset{2}{\cancel{200}}}{3} \cdot \frac{1}{\underset{1}{\cancel{100}}}$

$= \frac{2}{3}$

$66\frac{2}{3}\% = 66.\overline{6}\%$

$= 0.\overline{6}$

26. $0.06\% = \frac{0.06}{100}$

$= \frac{0.06}{100} \cdot \frac{100}{100}$

$= \frac{6}{10{,}000}$

$= \frac{3}{5000}$

$0.06\% = 0.06\%$

$= 0.0006$

27. $315\% = \frac{315}{100}$

$= \frac{63}{20}$ or $3\frac{3}{20}$

$315\% = 315\%$

$= 3.15$

Page 738 Lesson 6-5

1. $\frac{a}{134} = \frac{81}{100}$

$a \cdot 100 = 134 \cdot 81$

$100a = 10854$

$\frac{100a}{100} = \frac{10854}{100}$

$a \approx 108.5$

So, 108.5 is 81% of 134.

2. $\frac{52.08}{b} = \frac{21}{100}$

$52.08 \cdot 100 = b \cdot 21$

$5208 = 21b$

$\frac{5208}{21} = \frac{21b}{21}$

$248 = b$

So, 52.08 is 21% of 248.

3. $\frac{11.18}{86} = \frac{p}{100}$

$11.18 \cdot 100 = 86 \cdot p$

$1118 = 86p$

$\frac{1118}{86} = \frac{86p}{86}$

$13\% = p$

So, 11.18 is 13% of 86.

4. $\frac{a}{312} = \frac{120}{100}$

$a \cdot 100 = 312 \cdot 120$

$100a = 37440$

$\frac{100a}{100} = \frac{37440}{100}$

$a = 374.4$

So, 374.4 is 120% of 312.

5. $\frac{140}{400} = \frac{p}{100}$

$140 \cdot 100 = 400 \cdot p$

$14000 = 400p$

$\frac{14000}{400} = \frac{400p}{400}$

$35\% = p$

So, 140 is 35% of 400.

6. $\frac{430.2}{b} = \frac{60}{100}$

$430.2 \cdot 100 = b \cdot 60$

$43020 = 60b$

$\frac{43020}{60} = \frac{60b}{60}$

$717 = b$

So, 430.2 is 60% of 717.

7. $\frac{32}{80} = \frac{p}{100}$

$32 \cdot 100 = 80 \cdot p$

$3200 = 80p$

$\frac{3200}{80} = \frac{80p}{80}$

$40\% = p$

So, 32 is 40% of 80.

8. $\frac{a}{125} = \frac{15}{100}$

$a \cdot 100 = 125 \cdot 15$

$100a = 1875$

$\frac{100a}{100} = \frac{1875}{100}$

$a \approx 18.8$

So, 18.8 is about 15% of 125.

9. $\frac{22}{110} = \frac{p}{100}$

$22 \cdot 100 = 110 \cdot p$

$2200 = 110p$

$\frac{2200}{110} = \frac{110p}{110}$

$20\% = p$

So, 22 is 20% of 110.

10. $\frac{9.4}{b} = \frac{40}{100}$

$9.4 \cdot 100 = b \cdot 40$

$940 = 40b$

$\frac{940}{40} = \frac{40b}{40}$

$23.5 = b$

So, 9.4 is 40% of 23.5.

11. $\frac{a}{95} = \frac{41.5}{100}$

$a \cdot 100 = 95 \cdot 41.5$

$100a = 3942.5$

$\frac{100a}{100} = \frac{3942.5}{100}$

$a \approx 39.4$

So, 39.4 is about 41.5% of 39.4.

12. $\frac{17.92}{112} = \frac{p}{100}$

$17.92 \cdot 100 = 112 \cdot p$

$1792 = 112p$

$\frac{1792}{112} = \frac{112p}{112}$

$16\% = p$

So, 17.92 is 16% of 112.

13. $\frac{28}{50} = \frac{p}{100}$

$28 \cdot 100 = 50 \cdot p$

$2800 = 50p$

$\frac{2800}{50} = \frac{50p}{50}$

$56\% = p$

56% of the cans are chicken noodle soup.

14. $\frac{a}{20} = \frac{60}{100}$

$a \cdot 100 = 20 \cdot 60$

$100a = 1200$

$\frac{100a}{100} = \frac{1200}{100}$

$a = 12$ boys

12 of the students are boys.

Page 738 Lesson 6-6

1. 40% of $60 = \frac{2}{5}$ of 60
$= 24$

2. 25% of $72 = \frac{1}{4}$ of 72
$= 18$

3. 50% of $96 = \frac{1}{2}$ of 96
$= 48$

4. $33\frac{1}{3}\%$ of $24 = \frac{1}{3}$ of 24
$= 8$

5. 150% of $42 = \frac{3}{2}$ of 42
$= 63$

6. $37\frac{1}{2}\%$ of $80 = \frac{3}{8}$ of 80
$= 30$

7. 200% of $125 = 2 \times 125$
$= 250$

8. $66\frac{2}{3}\%$ of $45 = \frac{2}{3}$ of 45
$= 30$

9. 60% of 49
Sample answer: fraction method
60% is $\frac{3}{5}$.
49 is about 50.
$\frac{3}{5}$ of 50 is 30.
So, 60% of 49 is about 30.

10. 19% of 41
Sample answer: fraction method
19% is about 20% or $\frac{1}{5}$.
41 is about 40.
$\frac{1}{5}$ of 40 is 8.
So, 19% of 41 is about 8.

11. 82% of 60
Sample answer: fraction method
82% is about 80% or $\frac{8}{10}$.
$\frac{8}{10}$ of 60 is 48.
So, 82% of 60 is about 48.

12. 125% of 81
Sample answer: meaning of percent
125% means 125 for every 100 or about 13 for every 10.
81 has about 8 tens.
$8 \times 13 = 104$
So, 125% of 81 is about 104.

13. $\frac{1}{2}\%$ of 502
Sample answer: 1% method
$\frac{1}{2}\% = \frac{1}{2} \times 1\%$
502 is about 500.
1% of 500 is 5.
So, $\frac{1}{2}\%$ of 502 is about $\frac{1}{2} \times 5$ or 2.5.

14. 31% of 19
Sample answer: fraction method
31% is about 30% or $\frac{3}{10}$.
19 is about 20.
$\frac{3}{10}$ of 20 is 6.
So, 31% of 19 is about 6.

Page 738 Lesson 6-7

1. $9.28 = n(58)$
$\frac{9.28}{58} = n$
$0.16 = n$
So, 9.28 is 16% of 58.

2. $n = 0.43(110)$
$n = 47.3$
So, 47.3 is 43% of 110.

3. $90 = 0.80(n)$
$\frac{90}{0.8} = n$
$112.5 = n$
So, 80% of 112.5 is 90.

4. $n = 0.61(524)$
$n = 319.64$
So, 319.64 is 61% of 524.

5. $126 = n(90)$
$\frac{126}{90} = n$
$1.4 = n$
So, 126 is 140% of 90.

6. $109.2 = 0.52(n)$
$\frac{109.2}{0.52} = n$
$210 = n$
So, 52% of 210 is 109.2.

7. $29.76 = 0.62(n)$
$\frac{29.76}{0.62} = n$
$48 = n$
So, 62% of 48 is 29.76.

8. $54 = n(90)$
$\frac{54}{90} = n$
$0.6 = n$
So, 54 is 60% of 90.

9. $n = 0.78(125)$
$n = 97.5$
So, 78% of 125 is 97.5.

10. $n = 0.002(12)$
$n = 0.024$
So, 0.2% of 12 is 0.024.

11. $49.5 = 0.66(n)$
$\frac{49.5}{0.66} = n$
$75 = n$
So, 66% of 75 is 49.5.

12. $36.45 = n(81)$
$\frac{36.45}{81} = n$
$0.45 = n$
So, 36.45 is 45% of 81.

13. $n = 0.2(35)$
$n = 7$
The discount is $7.

14. $n = 0.25(108)$
$n = 27$
The discount is $27.

15. $I = prt$
$I = 1585(0.06)(5)$
$I = 475.5$
The interest is $475.50.

16. $I = prt$
$I = 2934(0.0575)(3.5)$
$I \approx 590.47$
The interest is about $590.47.

17. $n = 0.15(29.99)$
$n = 4.50$
$29.99 - 4.50 = \$25.49$
The sale price is $25.49.

Page 739 Lesson 6-8

1. percent of decrease; $42 - 56 = -14$
percent of change $= \frac{-14}{56}$
$= -0.25$ or -25%
The percent of change is 25%.

2. percent of increase; $29.64 - 26 = 3.64$
percent of change $= \frac{3.64}{26}$
$= 0.14$ or 14%
The percent of change is 14%.

3. percent of increase; $37.18 - 22 = 15.18$
percent of change $= \frac{15.18}{22}$
$= 0.69$ or 69%
The percent of change is 69%.

4. percent of decrease; $85.25 - 137.50 = -52.25$
percent of change $= \frac{-52.25}{137.50}$
$= -0.38$ or -38%
The percent of change is 38%.

5. percent of increase; $955.50 - 455 = 500.50$
percent of change $= \frac{500.50}{455}$
$= 1.1$ or 110%
The percent of change is 110%.

6. percent of increase; $15 - 3 = 12$
percent of change $= \frac{12}{3}$
$= 4$ or 400%
The percent of change is 400%.

7. percent of increase; $765.51 - 750.75 = 14.76$
percent of change $= \frac{14.76}{750.75}$
≈ 0.02 or 2%
The percent of change is about 2%.

8. percent of decrease; $476.50 - 953 = -476.50$
percent of change $= \frac{-476.50}{953}$
$= -0.5$ or -50%
The percent of change is 50%.

9. percent of increase; $379.69 - 101.25 = 278.44$
percent of change $= \frac{278.44}{101.25}$
≈ 2.75 or 275%
The percent of change is about 275%.

10. percent of increase; $842.27 - 836 = 6.27$
percent of change $= \frac{6.27}{836}$
$= 0.0075$ or 0.75%
The percent of change is $\frac{3}{4}\%$.

11. percent of increase; $24 - 18 = 6$
percent of change $= \frac{6}{18}$
$= 0.\overline{3}$ or $33\frac{1}{3}\%$
The percent of change is $33\frac{1}{3}\%$.

12. percent of decrease; $100 - 250 = -150$
percent of change $= \frac{-150}{250}$
$= -0.6$ or -60%
The percent of change is 60%.

13. percent of decrease; $92 - 107.50 = -15.5$
percent of change $= \frac{-15.5}{107.5}$
≈ -0.144 or -14.4%
The percent of change is about 14.4%.

14. percent of increase; $394.20 - 365 = 29.2$
percent of change $= \frac{29.2}{365}$
$= 0.08$ or 8%
The percent of change is 8%.

15. percent of change $= \frac{-25}{340}$
≈ -0.074 or -7.4%.
The percent of decrease is about 7.4%.

Page 739 Lesson 6-9

1. There are 3 possible outcomes of selecting green.
$P(\text{green}) = \frac{3}{15}$
$= \frac{1}{5}$
So, the probability of selecting green is $\frac{1}{5}$ or 20%.

2. There are 4 possible outcomes of selecting blue.
$P(\text{blue}) = \frac{4}{15}$
So, the probability of selecting blue is $\frac{4}{15}$ or 27%.

3. There are 6 possible outcomes of selecting red.
$P(\text{red}) = \frac{6}{15}$
$= \frac{2}{5}$
So, the probability of selecting red is $\frac{2}{5}$ or 40%.

4. There are 2 possible outcomes of selecting yellow.
$P(\text{yellow}) = \frac{2}{15}$
So, the probability of selecting yellow is $\frac{2}{15}$ or 13%.

5. There are 12 possible outcomes of not selecting green.

$P(\text{not green}) = \frac{12}{15}$

$= \frac{4}{5}$

So, the probability of not selecting green is $\frac{4}{5}$ or 80%.

6. There are no possible outcomes of selecting white.

$P(\text{white}) = \frac{0}{15}$

$= 0$

So, the probability of selecting white is 0 or 0%.

7. There are 4 possible outcomes of selecting blue and 6 possible outcomes of selecting red.

$P(\text{blue or red}) = \frac{10}{15}$

$= \frac{2}{3}$

So, the probability of selecting blue or red is $\frac{2}{3}$ or 67%.

8. There are 13 possible outcomes of not selecting yellow.

$P(\text{not yellow}) = \frac{13}{15}$

So, the probability of not selecting yellow is $\frac{13}{15}$ or 87%.

9. There are 6 possible outcomes of selecting colors other than red and green.

$P(\text{neither red nor green}) = \frac{6}{15}$

$= \frac{2}{5}$

So, the probability of selecting neither red nor green is $\frac{2}{5}$ or 40%.

10. There are 6 possible outcomes of selecting red and 2 possible outcomes of selecting yellow.

$P(\text{red or yellow}) = \frac{8}{15}$

So, the probability of selecting red or yellow is $\frac{8}{15}$ or 53%.

11. There are 15 possible outcomes of selecting colors other than orange.

$P(\text{orange}) = \frac{15}{15}$

$= 1$

So, the probability of not selecting orange is 1 or 100%.

12. There are 9 possible outcomes of selecting colors other than blue or yellow.

$P(\text{neither blue nor yellow}) = \frac{9}{15}$

$= \frac{3}{5}$

So, the probability of selecting neither blue nor yellow is $\frac{3}{5}$ or 60%.

13. There are 9 possible outcomes of selecting colors other than red.

$P(\text{not red}) = \frac{9}{15}$

$= \frac{3}{5}$

So, the probability of not selecting red is $\frac{3}{5}$ or 60%.

14. There are 10 possible outcomes of selecting colors other than green or yellow.

$P(\text{not green or yellow}) = \frac{10}{15}$

$= \frac{2}{3}$

So, the probability of not selecting green or yellow is $\frac{2}{3}$ or 67%.

15. There are 36 possible outcomes and 10 outcomes of rolling a sum greater than 8.

$P(\text{sum greater than 8}) = \frac{10}{36}$

$= \frac{5}{18}$

So, the probability of rolling a sum greater than 8 is $\frac{5}{18}$ or about 28%.

16. $\frac{a}{45} = \frac{20}{100}$

$a \cdot 100 = 45.20$

$100a = 900$

$a = 9$

You can expect 9 cookies to be sugar cookies.

Page 739 Lesson 7-1

Exercises 1–20 For checks, see students' work.

1. $-7h - 5 = 4 - 4h$

$-7h + 4h - 5 = 4 - 4h + 4h$

$-3h - 5 = 4$

$-3h - 5 + 5 = 4 + 5$

$-3h = 9$

$h = -3$

The solution is -3.

2. $5t - 8 = 3t + 12$

$5t - 3t - 8 = 3t - 3t + 12$

$2t - 8 = 12$

$2t - 8 + 8 = 12 + 8$

$2t = 20$

$t = 10$

The solution is 10.

3. $m + 2m + 1 = 7$

$3m + 1 = 7$

$3m + 1 - 1 = 7 - 1$

$3m = 6$

$m = 2$

The solution is 2.

4. $2y + 5 = 6y + 25$

$2y - 2y + 5 = 6y - 2y + 25$

$5 = 4y + 25$

$5 - 25 = 4y + 25 - 25$

$-20 = 4y$

$-5 = y$

The solution is -5.

5. $3z - 1 = 23 - 3z$

$3z + 3z - 1 = 23 - 3z + 3z$

$6z - 1 = 23$

$6z - 1 + 1 = 23 + 1$

$6z = 24$

$z = 4$

The solution is 4.

6. $$\begin{aligned} 5a - 5 &= 7a - 19 \\ 5a - 5a - 5 &= 7a - 5a - 19 \\ -5 &= 2a - 19 \\ -5 + 19 &= 2a - 19 + 19 \\ 14 &= 2a \\ 7 &= a \end{aligned}$$
The solution is 7.

7. $$\begin{aligned} 5x + 12 &= 3x - 6 \\ 5x - 3x + 12 &= 3x - 3x - 6 \\ 2x + 12 &= -6 \\ 2x + 12 - 12 &= -6 - 12 \\ 2x &= -18 \\ x &= -9 \end{aligned}$$
The solution is -9.

8. $$\begin{aligned} 3x - 5 &= 7x + 7 \\ 3x - 7x - 5 &= 7x - 7x + 7 \\ -4x - 5 &= 7 \\ -4x - 5 + 5 &= 7 + 5 \\ -4x &= 12 \\ x &= -3 \end{aligned}$$
The solution is -3.

9. $$\begin{aligned} 5c + 9 &= 8c \\ 5c - 5c + 9 &= 8c - 5c \\ 9 &= 3c \\ 3 &= c \end{aligned}$$
The solution is 3.

10. $$\begin{aligned} 3p &= 4 - 9p \\ 3p + 9p &= 4 - 9p + 9p \\ 12p &= 4 \\ \frac{12p}{12} &= \frac{4}{12} \\ p &= \frac{1}{3} \end{aligned}$$
The solution is $\frac{1}{3}$.

11. $$\begin{aligned} 6z + 5 &= 4z - 7 \\ 6z - 4z + 5 &= 4z - 4z - 7 \\ 2z + 5 &= -7 \\ 2z + 5 - 5 &= -7 - 5 \\ 2z &= -12 \\ z &= -6 \end{aligned}$$
The solution is -6.

12. $$\begin{aligned} 2a + 4.2 &= 3a - 1.6 \\ 2a - 3a + 4.2 &= 3a - 3a - 1.6 \\ -1a + 4.2 &= -1.6 \\ -1a + 4.2 - 4.2 &= -1.6 - 4.2 \\ -1a &= -5.8 \\ a &= 5.8 \end{aligned}$$
The solution is 5.8.

13. $$\begin{aligned} 3.21 - 7y &= 10y - 1.89 \\ 3.21 - 7y + 7y &= 10y + 7y - 1.89 \\ 3.21 &= 17y - 1.89 \\ 3.21 + 1.89 &= 17y - 1.89 + 1.89 \\ 5.1 &= 17y \\ \frac{5.1}{17} &= \frac{17y}{17} \\ 0.3 &= y \end{aligned}$$
The solution is 0.3.

14. $$\begin{aligned} 1.9s + 6 &= 3.1 - s \\ 1.9s + 1s + 6 &= 3.1 - 1s + 1s \\ 2.9s + 6 &= 3.1 \\ 2.9s + 6 - 6 &= 3.1 - 6 \\ 2.9s &= -2.9 \\ \frac{2.9s}{2.9} &= \frac{-2.9}{2.9} \\ s &= -1 \end{aligned}$$
The solution is -1.

15. $$\begin{aligned} 12b - 5 &= 3b \\ 12b - 12b - 5 &= 3b - 12b \\ -5 &= -9b \\ \frac{-5}{-9} &= \frac{-9b}{-9} \\ \frac{5}{9} &= b \end{aligned}$$
The solution is $\frac{5}{9}$.

16. $$\begin{aligned} 9 + 11a &= -5a + 21 \\ 9 + 11a + 5a &= -5a + 5a + 21 \\ 9 + 16a &= 21 \\ 9 - 9 + 16a &= 21 - 9 \\ 16a &= 12 \\ \frac{16a}{16} &= \frac{12}{16} \\ a &= \frac{3}{4} \end{aligned}$$
The solution is $\frac{3}{4}$.

17. $$\begin{aligned} 6 - x &= -5 \\ 6 - 6 - x &= -5 - 6 \\ -x &= -11 \\ x &= 11 \end{aligned}$$
The solution is 11.

18. $$\begin{aligned} 2.8 - 3w &= 4.6 - w \\ 2.8 - 3w + 3w &= 4.6 - w + 3w \\ 2.8 &= 4.6 + 2w \\ 2.8 - 4.6 &= 4.6 - 4.6 + 2w \\ -1.8 &= 2w \\ \frac{-1.8}{2} &= \frac{2w}{2} \\ -0.9 &= w \end{aligned}$$
The solution is -0.9.

19. $$\begin{aligned} 2.9y + 1.7 &= 3.5 + 2.3y \\ 2.9y - 2.3y + 1.7 &= 3.5 + 2.3y - 2.3y \\ 0.6y + 1.7 &= 3.5 \\ 0.6y + 1.7 - 1.7 &= 3.5 - 1.7 \\ 0.6y &= 1.8 \\ \frac{0.6y}{0.6} &= \frac{1.8}{0.6} \\ y &= 3 \end{aligned}$$
The solution is 3.

20. $$\begin{aligned} 2.85a - 7 &= 12.85a - 2 \\ 2.85a - 2.85a - 7 &= 12.85a - 2.85a - 2 \\ -7 &= 10a - 2 \\ -7 + 2 &= 10a - 2 + 2 \\ -5 &= 10a \\ \frac{-5}{10} &= \frac{10a}{10} \\ -\frac{1}{2} &= a \end{aligned}$$
The solution is $-\frac{1}{2}$.

Page 740 Lesson 7-2

Exercises 1–10 For checks, see students' work.

1.
$$\begin{aligned} 6(m-2) &= 12 \\ 6m - 12 &= 12 \\ 6m - 12 + 12 &= 12 + 12 \\ 6m &= 24 \\ m &= 4 \end{aligned}$$
The solution is 4.

2.
$$\begin{aligned} 4(x-3) &= 4 \\ 4x - 12 &= 4 \\ 4x - 12 + 12 &= 4 + 12 \\ 4x &= 16 \\ x &= 4 \end{aligned}$$
The solution is 4.

3.
$$\begin{aligned} 5(2d+4) &= 35 \\ 10d + 20 &= 35 \\ 10d + 20 - 20 &= 35 - 20 \\ 10d &= 15 \\ \frac{10d}{10} &= \frac{15}{10} \\ d &= \frac{3}{2} \text{ or } 1.5 \end{aligned}$$
The solution is 1.5.

4.
$$\begin{aligned} w + 6 &= 2(w-6) \\ w + 6 &= 2w - 12 \\ w - w + 6 &= 2w - w - 12 \\ 6 &= w - 12 \\ 6 + 12 &= w - 12 + 12 \\ 18 &= w \end{aligned}$$
The solution is 18.

5.
$$\begin{aligned} 3(b+1) &= 4b - 1 \\ 3b + 3 &= 4b - 1 \\ 3b - 3b + 3 &= 4b - 3b - 1 \\ 3 &= b - 1 \\ 3 + 1 &= b - 1 + 1 \\ 4 &= b \end{aligned}$$
The solution is 4.

6.
$$\begin{aligned} 7w - 6 &= 3(w+6) \\ 7w - 6 &= 3w + 18 \\ 7w - 3w - 6 &= 3w - 3w + 18 \\ 4w - 6 &= 18 \\ 4w - 6 + 6 &= 18 + 6 \\ 4w &= 24 \\ w &= 6 \end{aligned}$$
The solution is 6.

7.
$$\begin{aligned} 4(k-6) &= 6(k+2) \\ 4k - 24 &= 6k + 12 \\ 4k - 4k - 24 &= 6k - 4k + 12 \\ -24 &= 2k + 12 \\ -24 - 12 &= 2k + 12 - 12 \\ -36 &= 2k \\ -18 &= k \end{aligned}$$
The solution is −18.

8.
$$\begin{aligned} 3x - 0.8 &= 3x + 4 \\ 3x - 3x - 0.8 &= 3x - 3x + 4 \\ -0.8 &= 4 \end{aligned}$$
The sentence $-0.8 = 4$ is never true, so the solution is $\varnothing$.

9.
$$\begin{aligned} \frac{5}{9}g + 8 &= \frac{1}{6}g + 1 \\ \frac{5}{9}g - \frac{1}{6}g + 8 &= \frac{1}{6}g - \frac{1}{6}g + 1 \\ \frac{10}{18}g - \frac{3}{18}g + 8 &= 1 \\ \frac{7}{18}g + 8 &= 1 \\ \frac{7}{18}g + 8 - 8 &= 1 - 8 \\ \frac{7}{18}g &= -7 \\ \frac{18}{7}\left(\frac{7}{18}g\right) &= \frac{18}{7}(-7) \\ g &= -18 \end{aligned}$$
The solution is −18.

10.
$$\begin{aligned} \frac{s-3}{7} &= \frac{s+5}{9} \\ 63\left(\frac{s-3}{7}\right) &= 63\left(\frac{s+5}{9}\right) \\ 9(s-3) &= 7(s+5) \\ 9s - 27 &= 7s + 35 \\ 9s - 7s - 27 &= 7s - 7s + 35 \\ 2s - 27 &= 35 \\ 2s - 27 + 27 &= 35 + 27 \\ 2s &= 62 \\ s &= 31 \end{aligned}$$
The solution is 31.

11.
$$\begin{aligned} 3(3x+4) - 2 &= 9x + 10 \\ 9x + 12 - 2 &= 9x + 10 \\ 9x + 10 &= 9x + 10 \\ 9x + 10 - 10 &= 9x + 10 - 10 \\ 9x &= 9x \\ x &= x \end{aligned}$$
The sentence $x = x$ is always true.
The solution set is all numbers.

12a. Let x represent the first integer. Let $x + 1$ represent the second consecutive integer. Let $x + 2$ represent the third consecutive integer.
$4[x + (x + 1) + (x + 2)] = 48$

12b.
$$\begin{aligned} 4[x + (x+1) + (x+2)] &= 48 \\ 4(3x + 3) &= 48 \\ 12x + 12 &= 48 \\ 12x + 12 - 12 &= 48 - 12 \\ 12x &= 36 \\ x &= 3 \end{aligned}$$
$x + 1 = 3 + 1$ or 4
$x + 2 = 3 + 2$ or 5
The three integers are 3, 4, and 5.

Page 740 Lesson 7-3

1. $5 \geq 2t - 12$
$5 \stackrel{?}{\geq} 2(11) - 12$
$5 \stackrel{?}{\geq} 22 - 12$
$5 \not\geq 10$
This sentence is false.

2. $7 + n < 25$
$7 + 4 \stackrel{?}{<} 25$
$11 < 25$ ✓
This sentence is true.

3. $6r - 18 > 0$
$6(3) - 18 \stackrel{?}{>} 0$
$18 - 18 \stackrel{?}{>} 0$
$0 \not> 0$
This sentence is false.

4. $3n + 2 < 26$
$3(3) + 2 \stackrel{?}{<} 26$
$9 + 2 \stackrel{?}{<} 26$
$11 < 26$ ✓
This sentence is true.

5. $h - 19 < 13$
$28 - 19 \stackrel{?}{<} 13$
$9 < 13$ ✓
This sentence is true.

6. $20m \geq 10$
$20(0) \stackrel{?}{\geq} 10$
$0 \not\geq 10$
This sentence is false.

7. $b \geq 4$

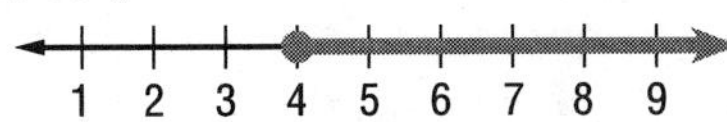

8. $x < -2$

9. $y > 2$

−3 −2 −1 0 1 2 3 4 5

10. $m \leq 0$

−3 −2 −1 0 1 2 3 4 5

11. $p > -1$

−4 −3 −2 −1 0 1 2 3 4

12. $q \geq -3$

−6 −5 −4 −3 −2 −1 0 1 2

13. Let s represent the number of students. $s \geq 295$
14. Let b represent the original bill. $b + 15 > 80$
15. Let n represent the number. $8n - 2 < 15$
16. Let a represent the age of voting citizen. $a \geq 18$
17. Let w represent the weight of one dozen jumbo eggs. $w \geq 30$
18. Let s represent the amount of grams of cereal. $s \leq 5$

Page 740 Lesson 7-4

Exercises 1–19 For checks, see students' work.

1. $m + 9 < 14$
$m + 9 - 9 < 14 - 9$
$m < 5$

2. $k + (-5) < -12$
$k + (-5) + 5 < -12 + 5$
$k < -7$

3. $-15 < v - 1$
$-15 + 1 < v - 1 + 1$
$-14 < v$ or $v > -14$

4. $-7 + f \geq 47$
$-7 + 7 + f \geq 47 + 7$
$f \geq 54$

5. $r > -15 - 8$
$r > -23$

6. $18 \geq s - (-4)$
$18 \geq s + 4$
$18 - 4 \geq s + 4 - 4$
$14 \geq s$ or $s \leq 14$

7. $38 < r - (-6)$
$38 < r + 6$
$38 - 6 < r + 6 - 6$
$32 < r$ or $r > 32$

8. $z - 9 \leq -11$
$z - 9 + 9 \leq -11 + 9$
$z \leq -2$

9. $-16 + c \geq 1$
$-16 + 16 + c \geq 1 + 16$
$c \geq 17$

10. $d + 1.4 < 6.8$
$d + 1.4 - 1.4 < 6.8 - 1.4$
$d < 5.4$

11. $-3 + x > 11.9$
$-3 + 3 + x > 11.9 + 3$
$x > 14.9$

12. $-0.2 \geq 0.3 + y$
$-0.2 - 0.3 \geq 0.3 - 0.3 + y$
$-0.5 \geq y$ or $y \leq -0.5$

13. $h + 5.7 > 21.3$
$h + 5.7 - 5.7 > 21.3 - 5.7$
$h > 15.6$

14. $t - 8.5 > -4.2$
$t - 8.5 + 8.5 > -4.2 + 8.5$
$t > -4.3$

15. $-13.2 > w - 4.87$
$-13.2 + 4.87 > w - 4.87 + 4.87$
$-8.33 > w$ or $w < -8.33$

16. $a + \frac{5}{12} \geq \frac{7}{18}$
$a + \frac{5}{12} - \frac{5}{12} \geq \frac{7}{18} - \frac{5}{12}$
$a \geq \frac{14}{36} - \frac{15}{36}$
$a \geq -\frac{1}{36}$

17. $7\frac{1}{2} < n - \left(-\frac{7}{8}\right)$
$\frac{15}{2} < n + \frac{7}{8}$
$\frac{15}{2} - \frac{7}{8} < n + \frac{7}{8} - \frac{7}{8}$
$\frac{60}{8} - \frac{7}{8} < n$
$\frac{53}{8} < n$
$6\frac{5}{8} < n$ or $n > 6\frac{5}{8}$

18. $\frac{2}{3} \leq a - \frac{5}{6}$

$\frac{2}{3} + \frac{5}{6} \leq a - \frac{5}{6} + \frac{5}{6}$

$\frac{4}{6} + \frac{5}{6} \leq a$

$\frac{9}{6} \leq a$

$1\frac{1}{2} \leq a$ or $a \geq 1\frac{1}{2}$

19. $-7.42 \leq d - 5.9$

$-7.42 + 5.9 \leq d - 5.9 + 5.9$

$-1.52 \leq d$ or $d \geq -1.52$

Page 741 Lesson 7-5

1. $6p < 78$

$\frac{6p}{6} < \frac{78}{6}$

$p < 13$

The solution is $p < 13$. You can check this solution by substituting a number less than 13 into the inequality.

2. $\frac{m}{-3} > 24$

$\frac{m}{-3}(-3) > 24(-3)$

$m > -72$

The solution is $m > -72$. You can check this solution by substituting a number greater than -72 into the inequality.

3. $-18 < 3b$

$\frac{-18}{3} < \frac{3b}{3}$

$-6 < b$ or $b > -6$

The solution is $-6 < b$. You can check this solution by substituting a number greater than -6 into the inequality.

4. $-5k \geq 125$

$\frac{-5k}{-5} \leq \frac{125}{-5}$

$k \leq -25$

The solution is $k \leq -25$. You can check this solution by substituting -25 or a number less than -25 into the inequality.

5. $-75 > \frac{a}{5}$

$-75(5) > \frac{a}{5}(5)$

$-375 > a$ or $a < -375$

The solution is $-375 > a$. You can check this solution by substituting a number less than -375 into the inequality.

6. $\frac{w}{6} < -5$

$\frac{w}{6}(6) < -5(6)$

$w < -30$

The solution is $w < -30$. You can check this solution by substituting a number less than -30 into the inequality.

7. $8 < \frac{2}{3}c$

$\frac{3}{2}(8) < \frac{3}{2}\left(\frac{2}{3}c\right)$

$12 < c$ or $c > 12$

The solution is $12 < c$. You can check this solution by substituting a number greater than 12 into the inequality.

8. $\frac{m}{1.3} \geq 0.5$

$\frac{m}{1.3}(1.3) \geq 0.5(1.3)$

$m \geq 0.65$

The solution is $m \geq 0.65$. You can check this solution by substituting 0.65 or a number greater than 0.65 into the inequality.

9. $0.4y > -2$

$\frac{0.4y}{0.4} > \frac{-2}{0.4}$

$y > -5$

The solution is $y > -5$. You can check this solution by substituting a number greater than -5 into the inequality.

10. $-\frac{1}{2}d \leq -5\frac{1}{2}$

$-\frac{1}{2}d \leq -\frac{11}{2}$

$-2\left(-\frac{1}{2}d\right) \geq -2\left(-\frac{11}{2}\right)$

$d \geq 11$

The solution is $d \geq 11$. You can check this solution by substituting 11 or a number greater than 11 into the inequality.

11. $\frac{2}{7}t < 4$

$\frac{7}{2}\left(\frac{2}{7}t\right) < \frac{7}{2}(4)$

$t < 14$

The solution is $t < 14$. You can check this solution by substituting a number less than 14 into the inequality.

12. $\frac{1}{5}m \geq 4\frac{3}{5}$

$\frac{1}{5}m \geq \frac{23}{5}$

$5\left(\frac{1}{5}m\right) \geq 5\left(\frac{23}{5}\right)$

$m \geq 23$

The solution is $m \geq 23$. You can check this solution by substituting 23 or a number greater than 23 into the inequality.

13. $\frac{y}{-13} > -20$

$\frac{y}{-13}(-13) < -20(-13)$

$y < 260$

The solution is $y < 260$. You can check this solution by substituting a number less than 260 into the inequality.

14. $14t < 266$

$\frac{14t}{14} < \frac{266}{14}$

$t < 19$

The solution is $t < 19$. You can check this solution by substituting a number less than 19 into the inequality.

15. $\frac{g}{-25} \geq 8$

$\frac{g}{-25}(-25) \geq 8(-25)$

$g \geq -200$

The solution is $g \geq -200$. You can check this solution by substituting -200 or a number greater than -200 into the inequality.

16. Let n represent the number.

$-4n \geq -20$

$\frac{-4n}{-4} \leq \frac{-20}{-4}$

$n \leq 5$

Any number less than or equal to 5 is a solution.

Page 741 Lesson 7-6

1. $2m + 1 < 9$

$2m + 1 - 1 < 9 - 1$

$2m < 8$

$\frac{2m}{2} < \frac{8}{2}$

$m < 4$

The solution is $m < 4$. You can check this solution by substituting a number less than 4 into the inequality.

2. $-3k - 4 \leq -22$

$-3k - 4 + 4 \leq -22 + 4$

$-3k \leq -18$

$\frac{-3k}{-3} \geq \frac{-18}{-3}$

$k \geq 6$

The solution is $k \geq 6$. You can check this solution by substituting 6 or a number greater than 6 into the inequality.

3. $-2 > 10 - 2x$

$-2 - 10 > 10 - 10 - 2x$

$-12 > -2x$

$\frac{-12}{-2} < \frac{-2x}{-2}$

$6 < x$ or $x > 6$

The solution is $6 < x$. You can check this solution by substituting a number greater than 6 into the inequality.

4. $-6a + 2 \geq 14$

$-6a + 2 - 2 \geq 14 - 2$

$-6a \geq 12$

$\frac{-6a}{-6} \leq \frac{12}{-6}$

$a \leq -2$

The solution is $a \leq -2$. You can check this solution by substituting -2 or a number less than -2 into the inequality.

5. $3y + 2 < -7$

$3y + 2 - 2 < -7 - 2$

$3y < -9$

$y < -3$

The solution is $y < -3$. You can check this solution by substituting a number less than -3 into the inequality.

6. $\frac{d}{4} + 3 \geq -11$

$\frac{d}{4} + 3 - 3 \geq -11 - 3$

$\frac{d}{4} \geq -14$

$\frac{d}{4}(4) \geq -14(4)$

$d \geq -56$

The solution is $d \geq -56$. You can check this solution by substituting -56 or a number greater than -56 into the inequality.

7. $\frac{x}{3} - 5 < 6$

$\frac{x}{3} - 5 + 5 < 6 + 5$

$\frac{x}{3} < 11$

$\frac{x}{3}(3) < 11(3)$

$x < 33$

The solution is $x < 33$. You can check this solution by substituting a number less than 33 into the inequality.

8. $-5g + 6 < 3g + 26$

$-5g - 3g + 6 < 3g - 3g + 26$

$-8g + 6 < 26$

$-8g + 6 - 6 < 26 - 6$

$-8g < 20$

$\frac{-8g}{-8} > \frac{20}{-8}$

$g > -2.5$

The solution is $g > -2.5$. You can check this solution by substituting a number greater than -2.5 into the inequality.

9. $-3(m - 2) > 12$

$-3m + 6 > 12$

$-3m + 6 - 6 > 12 - 6$

$-3m > 6$

$\frac{-3m}{-3} < \frac{6}{-3}$

$m < -2$

The solution is $m < -2$. You can check this solution by substituting a number less than -2 into the inequality.

10. $\frac{r}{5} - 6 \leq 3$

$\frac{r}{5} - 6 + 6 \leq 3 + 6$

$\frac{r}{5} \leq 9$

$\frac{r}{5}(5) \leq 9(5)$

$r \leq 45$

The solution is $r \leq 45$. You can check this solution by substituting 45 or a number less than 45 into the inequality.

11. $\frac{3(n + 1)}{7} \geq \frac{n + 4}{5}$

$35\left[\frac{3(n + 1)}{7}\right] \geq 35\left(\frac{n + 4}{5}\right)$

$5 \cdot 3(n + 1) \geq 7(n + 4)$

$15(n + 1) \geq 7n + 28$

$15n + 15 \geq 7n + 28$

$15n - 7n + 15 \geq 7n - 7n + 28$

$8n + 15 \geq 28$

$8n + 15 - 15 \geq 28 - 15$

$8n \geq 13$

$\frac{8n}{8} \geq \frac{13}{8}$

$n \geq 1.625$

The solution is $n \geq 1.625$. You can check this solution by substituting 1.625 or a number greater than 1.625 into the inequality.

12. $\frac{n+10}{-3} \le 6$

$$-3\left(\frac{n+10}{-3}\right) \le -3(6)$$
$$n + 10 \le -18$$
$$n + 10 - 10 \le -18 - 10$$
$$n \le -28$$

The solution is $n \le -28$. You can check this solution by substituting -28 or a number less than -28 into the inequality.

13. Let x represent the number.

$$3x + 5 < 2x - 4$$
$$3x - 2x + 5 < 2x - 2x - 4$$
$$x + 5 < -4$$
$$x + 5 - 5 < -4 - 5$$
$$x < -9$$

Any number less than -9 is a solution.

Page 741 Lesson 8-1

1. Yes; each x value is paired with only one y value.
2. Yes; each x value is paired with only one y value.
3. No; 2 is paired with 9 and 36.
4. Yes; each x value is paired with only one y value.
5. No; 1 is paired with 0, 9, and 18.
6. Yes; each x value is paired with only one y value.
7. Yes; each x value is paired with only one y value.
8. No; -2 is paired with 4 and 8, and -1 is paired with 5 and 7.
9. Yes; each x value is paired with only one y value.
10. Yes; each x value is paired with only one y value.

Page 742 Lesson 8-2

1. Choose four values for y and substitute them into $x = 4$. Sample answer:

y	$x = 4$	x	(x, y)
2	$x = 4$	4	(4, 2)
3	$x = 4$	4	(4, 3)
5	$x = 4$	4	(4, 5)
6	$x = 4$	4	(4, 6)

Four solutions are (4, 2), (4, 3), (4, 5), and (4, 6).

2. Choose four values for x and substitute them into $y = 0$. Sample answer:

x	$y = 0$	y	(x, y)
1	$y = 0$	0	(1, 0)
5	$y = 0$	0	(5, 0)
6	$y = 0$	0	(6, 0)
0	$y = 0$	0	(0, 0)

Four solutions are (1, 0), (5, 0), (6, 0), and (0, 0).

3. First, solve the equation for y.

$$x + y = 2$$
$$y = 2 - x$$

Choose four values for x and substitute them into $y = 2 - x$. Sample answer:

x	$y = 2 - x$	y	(x, y)
2	$y = 2 - 2$	0	(2, 0)
1	$y = 2 - 1$	1	(1, 1)
0	$y = 2 - 0$	2	(0, 2)
-1	$y = 2 - (-1)$	3	$(-1, 3)$

Four solutions are (2, 0), (1, 1), (0, 2), and $(-1, 3)$.

4. Choose four values for x and substitute them into $y = 2x - 6$. Sample answer:

x	$y = 2x - 6$	y	(x, y)
0	$y = 2(0) - 6$	-6	$(0,-6)$
1	$y = 2(1) - 6$	-4	$(1,-4)$
2	$y = 2(2) - 6$	-2	$(2,-2)$
3	$y = 2(3) - 6$	0	(3, 0)

Four solutions are $(0, -6)$, $(1, -4)$, $(2, -2)$, and (3, 0).

5. First, solve the equation for y.

$$x - y = 5$$
$$-y = 5 - x$$
$$y = -5 + x$$

Choose four values for x and substitute them into $y = -5 + x$. Sample answer:

x	$y = -5 + x$	y	(x, y)
8	$y = -5 + 8$	3	(8, 3)
7	$y = -5 + 7$	2	(7, 2)
6	$y = -5 + 6$	1	(6, 1)
5	$y = -5 + 5$	0	(5, 0)

Four solutions are (8, 3), (7, 2), (6, 1), and (5, 0).

6. First, solve the equation for y.

$$3x - y = 8$$
$$-y = 8 - 3x$$
$$y = -8 + 3x$$

Choose four values for x and substitute them into $y = -8 + 3x$. Sample answer:

x	$y = -8 + 3x$	y	(x, y)
3	$y = -8 + 3(3)$	1	(3, 1)
4	$y = -8 + 3(4)$	4	(4, 4)
0	$y = -8 + 3(0)$	-8	$(0,-8)$
2	$y = -8 + 3(2)$	-2	$(2,-2)$

Four solutions are (3, 1), (4, 4), $(0,-8)$, and $(2,-2)$.

7. Choose four values for x and substitute them into $y = \frac{1}{2}x - 3$. Sample answer:

x	$y = \frac{1}{2}x - 3$	y	(x, y)
0	$y = \frac{1}{2}(0) - 3$	−3	(0,−3)
2	$y = \frac{1}{2}(2) - 3$	−2	(2,−2)
4	$y = \frac{1}{2}(4) - 3$	−1	(4,−1)
6	$y = \frac{1}{2}(6) - 3$	0	(6, 0)

Four solutions are (0,−3), (2,−2), (4,−1), and (6, 0).

8. Choose four values for x and substitute them into $y = \frac{1}{3}x + 1$. Sample answer:

x	$y = \frac{1}{3}x + 1$	y	(x, y)
0	$y = \frac{1}{3}(0) + 1$	1	(0, 1)
3	$y = \frac{1}{3}(3) + 1$	2	(3, 2)
−3	$y = \frac{1}{3}(-3) + 1$	0	(−3, 0)
6	$y = \frac{1}{3}(6) + 1$	3	(6, 3)

Four solutions are (0, 1), (3, 2), (−3, 0), and (6, 3).

9. First, solve the equation for y.

$$2x + y = -2$$
$$y = -2 - 2x$$

Choose four values for x and substitute them into $y = -2 - 2x$. Sample answer:

x	$y = -2 - 2x$	y	(x, y)
1	$y = -2 - 2(1)$	−4	(1,−4)
0	$y = -2 - 2(0)$	−2	(0,−2)
−1	$y = -2 - 2(-1)$	0	(−1, 0)
−2	$y = -2 - 2(-2)$	2	(−2, 2)

Four solutions are (1,−4), (0,−2), (−1, 0), and (−2, 2).

10. First, solve the equation for y.

$$2x + 3y = 12$$
$$3y = 12 - 2x$$
$$y = 4 - \frac{2}{3}x$$

Choose four values for x and substitute them into $y = 4 - \frac{2}{3}x$. Sample answer:

x	$y = 4 - \frac{2}{3}x$	y	(x, y)
6	$y = 4 - \frac{2}{3}(6)$	0	(6, 0)
0	$y = 4 - \frac{2}{3}(0)$	4	(0, 4)
3	$y = 4 - \frac{2}{3}(3)$	2	(3, 2)
−3	$y = 4 - \frac{2}{3}(-3)$	6	(−3, 6)

Four solutions are (6, 0), (0, 4), (3, 2), and (−3, 6).

11. First, solve the equation for y.

$$x + 2y = -4$$
$$2y = -4 - x$$
$$y = -2 - \frac{1}{2}x$$

Choose four values for x and substitute them into $y = -2 - \frac{1}{2}x$. Sample answer:

x	$y = -2 - \frac{1}{2}x$	y	(x, y)
0	$y = -2 - \frac{1}{2}(0)$	−2	(0,−2)
2	$y = -2 - \frac{1}{2}(2)$	−3	(2,−3)
−2	$y = -2 - \frac{1}{2}(-2)$	−1	(−2,−1)
−4	$y = -2 - \frac{1}{2}(-4)$	0	(−4, 0)

Four solutions are (0,−2), (2,−3), (−2,−1), and (−4, 0).

12. First, solve the equation for y.

$$2x - 4y = 8$$
$$-4y = 8 - 2x$$
$$y = -2 + \frac{1}{2}x$$

Choose four values for x and substitute them into $y = -2 + \frac{1}{2}x$. Sample answer:

x	$y = -2 + \frac{1}{2}x$	y	(x, y)
4	$y = -2 + \frac{1}{2}(4)$	0	(4, 0)
0	$y = -2 + \frac{1}{2}(0)$	−2	(0,−2)
2	$y = -2 + \frac{1}{2}(2)$	−1	(2,−1)
−2	$y = -2 + \frac{1}{2}(-2)$	−3	(−2,−3)

Four solutions are (4, 0), (0,−2), (2,−1), and (−2,−3).

13. Sample answer:

x	$y = x + 4$	y	(x, y)
−3	$y = (-3) + 4$	1	(−3, 1)
−2	$y = (-2) + 4$	2	(−2, 2)
−1	$y = (-1) + 4$	3	(−1, 3)
0	$y = 0 + 4$	4	(0, 4)

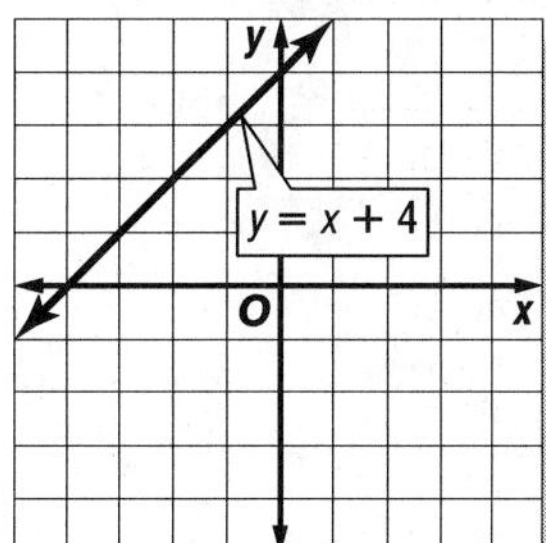

14. Sample answer:

x	$y = 4x$	y	(x, y)
-1	$y = 4(-1)$	-4	$(-1, -4)$
0	$y = 4(0)$	0	$(0, 0)$
1	$y = 4(1)$	4	$(1, 4)$
2	$y = 4(2)$	8	$(2, 8)$

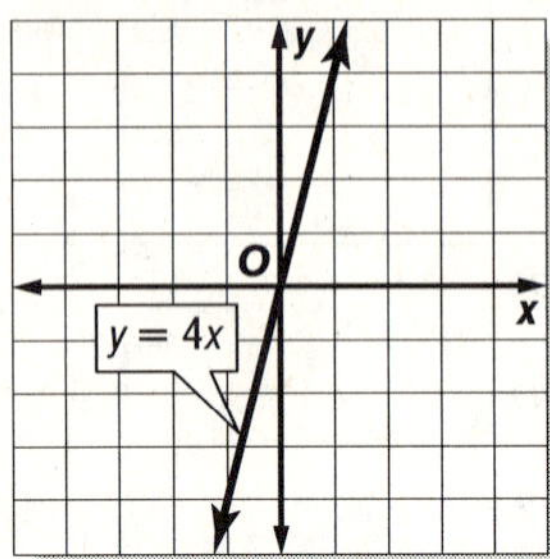

15. First, solve the equation for y.

$$x + y = 3$$
$$x - x + y = 3 - x$$
$$y = 3 - x$$

Sample answer:

x	$y = 3 - x$	y	(x, y)
-1	$y = 3 - (-1)$	4	$(-1, 4)$
0	$y = 3 - (0)$	3	$(0, 3)$
1	$y = 3 - (1)$	2	$(1, 2)$
2	$y = 3 - (2)$	1	$(2, 1)$

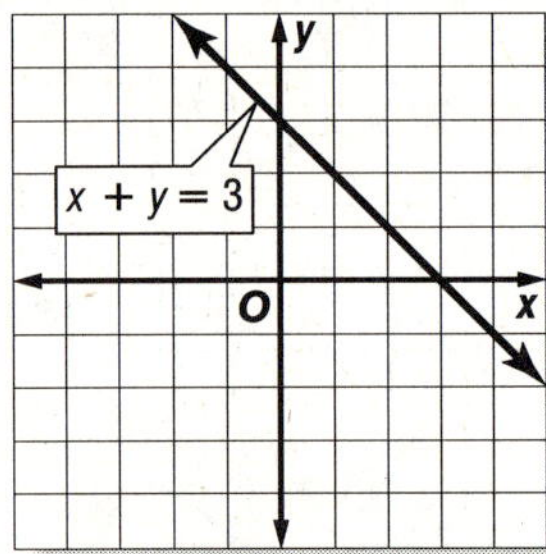

16. Sample answer:

x	$y = x - 3$	y	(x, y)
-1	$y = -1 - 3$	-4	$(-1, -4)$
0	$y = 0 - 3$	-3	$(0, -3)$
1	$y = 1 - 3$	-2	$(1, -2)$
2	$y = 2 - 3$	-1	$(2, -1)$

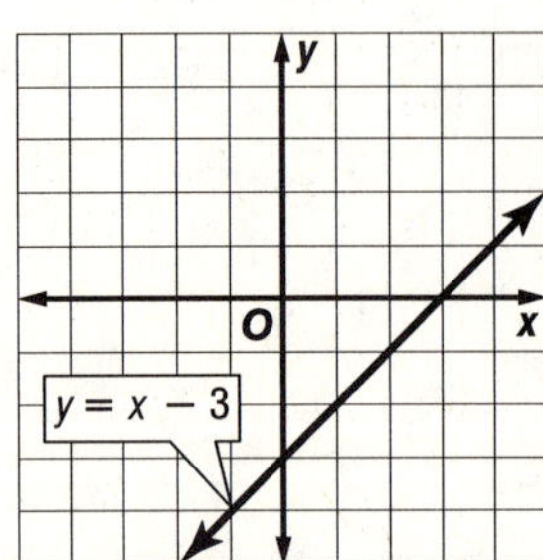

17. Sample answer:

x	$y = -2x + 5$	y	(x, y)
-1	$y = -2(-1) + 5$	7	$(-1, 7)$
0	$y = -2(0) + 5$	5	$(0, 5)$
1	$y = -2(1) + 5$	3	$(1, 3)$
2	$y = -2(2) + 5$	1	$(2, 1)$

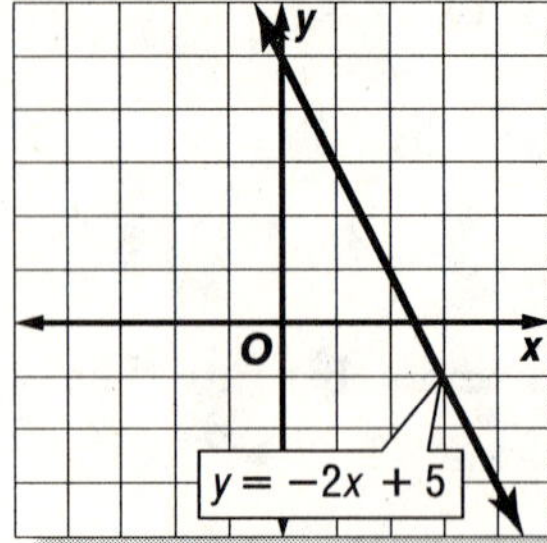

18. First, solve the equation for y.

$$2x + y = 6$$
$$2x - 2x + y = 6 - 2x$$
$$y = 6 - 2x$$

Sample answer:

x	$y = 6 - 2x$	y	(x, y)
-1	$y = 6 - 2(-1)$	8	$(-1, 8)$
0	$y = 6 - 2(0)$	6	$(0, 6)$
1	$y = 6 - 2(1)$	4	$(1, 4)$
2	$y = 6 - 2(2)$	2	$(2, 2)$

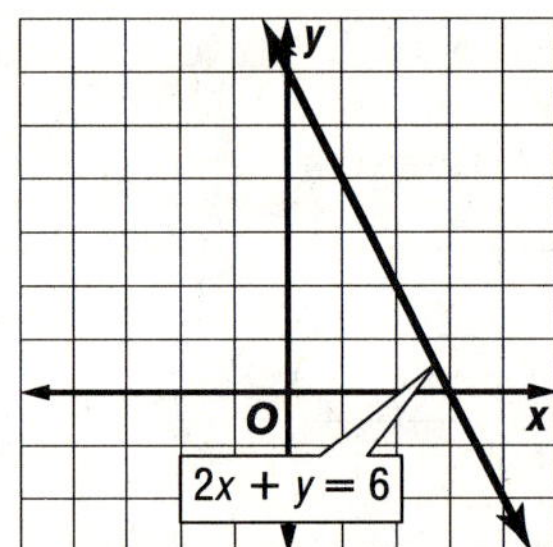

Page 742 Lesson 8-3

1. To find the x-intercept, let $y = 0$.

$$y = x + 7$$
$$0 = x + 7$$
$$-7 = x$$

The x-intercept is -7.

To find the y-intercept, let $x = 0$.

$$y = x + 7$$
$$y = 0 + 7$$
$$y = 7$$

The y-intercept is 7.

2. To find the x-intercept, let $y = 0$.

$$y = 3x + 12$$
$$0 = 3x + 12$$
$$0 - 12 = 3x + 12 - 12$$
$$-12 = 3x$$
$$-4 = x$$

The x-intercept is -4.

To find the y-intercept, let $x = 0$.

$$y = 3x + 12$$
$$y = 3(0) + 12$$
$$y = 12$$

The y-intercept is 12.

3. To find the x-intercept, let $y = 0$.

$$4x + 3y = 24$$
$$4x + 3(0) = 24$$
$$4x = 24$$
$$x = 6$$

The x-intercept is 6.

To find the y-intercept, let $x = 0$.

$$4x + 3y = 24$$
$$4(0) + 3y = 24$$
$$3y = 24$$
$$y = 8$$

The y-intercept is 8.

4. To find the x-intercept, let $y = 0$.

$$y = 8 - 2x$$
$$0 = 8 - 2x$$
$$0 - 8 = 8 - 8 - 2x$$
$$-8 = -2x$$
$$4 = x$$

The x-intercept is 4.

To find the y-intercept, let $x = 0$.

$$y = 8 - 2x$$
$$y = 8 - 2(0)$$
$$y = 8$$

The y-intercept is 8.

5. To find the x-intercept, let $y = 0$.

$$-5x + y = -10$$
$$-5x + 0 = -10$$
$$-5x = -10$$
$$x = 2$$

The x-intercept is 2.

To find the y-intercept, let $x = 0$.

$$-5x + y = -10$$
$$-5(0) + y = -10$$
$$y = -10$$

The y-intercept is -10.

6. To find the x-intercept, let $y = 0$.

$$y = \frac{2}{3}x - 7$$
$$0 = \frac{2}{3}x - 7$$
$$0 + 7 = \frac{2}{3}x - 7 + 7$$
$$7 = \frac{2}{3}x$$
$$\frac{3}{2}(7) = \frac{3}{2}\left(\frac{2}{3}x\right)$$
$$\frac{21}{2} = x$$

The x-intercept is $\frac{21}{2}$.

To find the y-intercept, let $x = 0$.

$$y = \frac{2}{3}x - 7$$
$$y = \frac{2}{3}(0) - 7$$
$$y = -7$$

The y-intercept is -7.

7. Find the x-intercept.

$$2x + y = 6$$
$$2x + 0 = 6$$
$$2x = 6$$
$$x = 3$$

Find the y-intercept.

$$2x + y = 6$$
$$2(0) + y = 6$$
$$y = 6$$

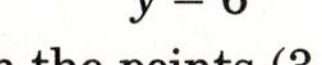
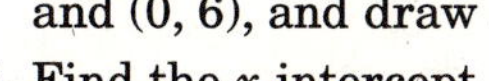
Graph the points (3, 0) and (0, 6), and draw a line through them.

8. Find the x-intercept.

$$-4x + y = 8$$
$$-4x + 0 = 8$$
$$-4x = 8$$
$$x = -2$$

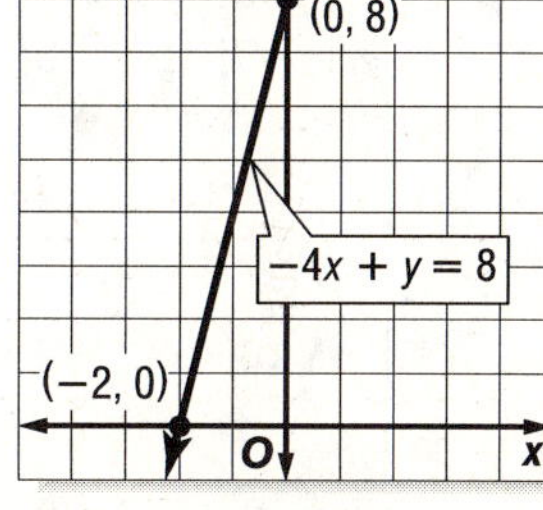

Find the y-intercept.

$$-4x + y = 8$$
$$-4(0) + y = 8$$
$$y = 8$$

Graph the points $(-2, 0)$ and (0, 8), and draw a line through them.

9. Find the x-intercept.

$$3x - 3y = -12$$
$$3x - 3(0) = -12$$
$$3x = -12$$
$$x = -4$$

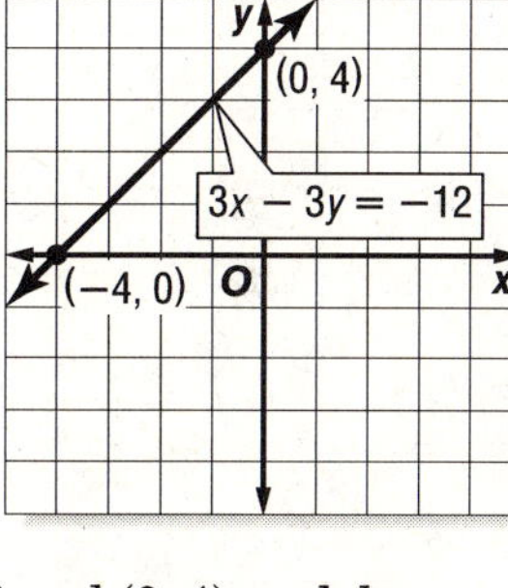

Find the y-intercept.

$$3x - 3y = -12$$
$$3(0) - 3y = -12$$
$$-3y = -12$$
$$y = 4$$

Graph the points $(-4, 0)$ and (0, 4), and draw a line through them.

10. Find the x-intercept.

$$x + 2y = -4$$
$$x + 2(0) = -4$$
$$x = -4$$

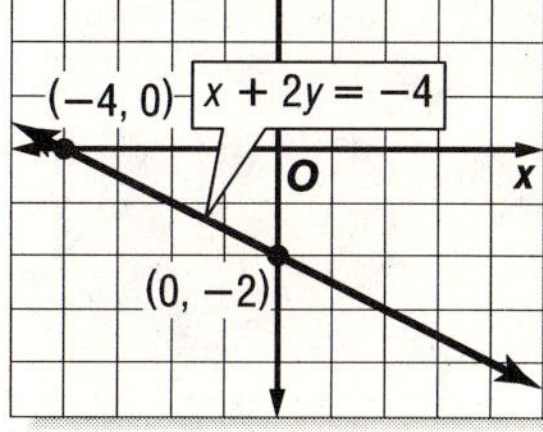

Find the y-intercept.

$$x + 2y = -4$$
$$0 + 2y = -4$$
$$2y = -4$$
$$y = -2$$

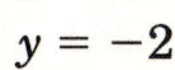

Graph the points $(-4, 0)$ and $(0, -2)$, and draw a line through them.

11. Find the x-intercept.

$y = -x - 6$
$0 = -x - 6$
$x = -6$

Find the y-intercept.

$y = -x - 6$
$y = -(0) - 6$
$y = -6$

Graph the points $(-6, 0)$ and $(0, -6)$, and draw a line through them.

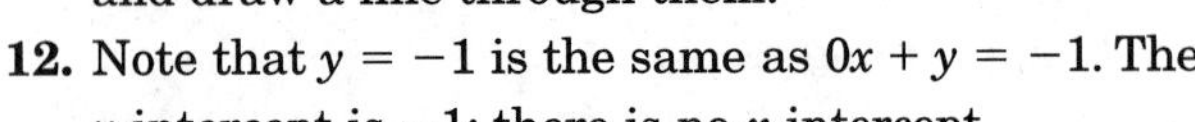

12. Note that $y = -1$ is the same as $0x + y = -1$. The y-intercept is -1; there is no x-intercept.

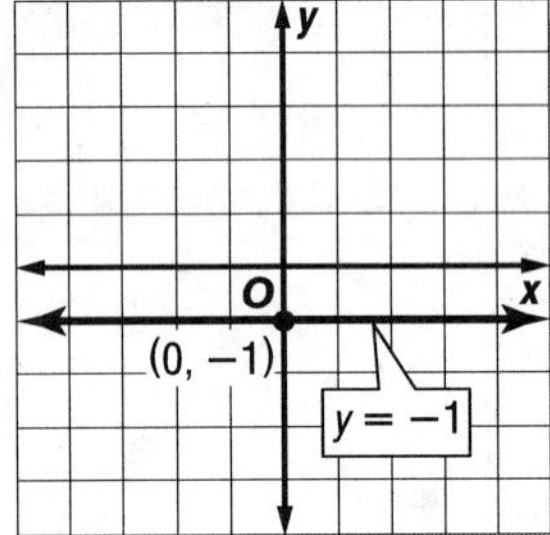

Page 742 Lesson 8-4

1. $m = \frac{y_2 - y_1}{x_2 - x_1}$
$m = \frac{4 - 1}{-1 - (-2)}$
$m = \frac{3}{1}$ or 3

2. $m = \frac{y_2 - y_1}{x_2 - x_1}$
$m = \frac{-2 - (-2)}{3 - (-2)}$
$m = \frac{-2 + 2}{3 + 2}$
$m = \frac{0}{5}$ or 0

3. $m = \frac{y_2 - y_1}{x_2 - x_1}$
$m = \frac{-3 - 8}{4 - 3}$
$m = \frac{-11}{1}$ or -11

4. $m = \frac{y_2 - y_1}{x_2 - x_1}$
$m = \frac{-9 - 5}{-3 - 4}$
$m = \frac{-14}{-7}$ or 2

5. $m = \frac{y_2 - y_1}{x_2 - x_1}$
$m = \frac{5 - 2}{0 - (-1)}$
$m = \frac{3}{1}$ or 3

6. $m = \frac{y_2 - y_1}{x_2 - x_1}$
$m = \frac{-4 - 2}{6 - 6}$
$m = \frac{-6}{0}$ or undefined

7. $m = \frac{y_2 - y_1}{x_2 - x_1}$
$m = \frac{1 - (-3)}{-4 - 8}$
$m = \frac{4}{-12}$ or $-\frac{1}{3}$

8. $m = \frac{y_2 - y_1}{x_2 - x_1}$
$m = \frac{10 - 5}{3 - 1}$
$m = \frac{5}{2}$

9. $m = \frac{y_2 - y_1}{x_2 - x_1}$
$m = \frac{-2 - 2}{-2 - 7}$
$m = \frac{-4}{-9}$ or $\frac{4}{9}$

10. $m = \frac{y_2 - y_1}{x_2 - x_1}$
$m = \frac{-19 - (-4)}{5 - 2}$
$m = \frac{-15}{3}$ or -5

11. $m = \frac{y_2 - y_1}{x_2 - x_1}$
$m = \frac{6 - 6}{7 - 5}$
$m = \frac{0}{2}$ or 0

12. $m = \frac{y_2 - y_1}{x_2 - x_1}$
$m = \frac{4 - (-3)}{-9 - (-6)}$
$m = \frac{7}{-3}$ or $-\frac{7}{3}$

13. $m = \frac{y_2 - y_1}{x_2 - x_1}$
$m = \frac{-10 - (-6)}{-5 - (-1)}$
$m = \frac{-4}{-4}$ or 1

14. $m = \frac{y_2 - y_1}{x_2 - x_1}$
$m = \frac{-5 - 9}{-4 - 5}$
$m = \frac{-14}{-9}$ or $\frac{14}{9}$

Page 743 Lesson 8-5

1. rate of change $= \frac{\text{change in } y}{\text{change in } x}$
$= \frac{40 - 20}{2 - 0}$
$= \frac{20}{2}$
$= 10$

There is an increase of 10 in./year.

2. rate of change $= \frac{\text{change in } y}{\text{change in } x}$
$= \frac{10 - 0}{20 - 0}$
$= \frac{10}{20}$ or $\frac{1}{2}$

There is an increase of $\frac{1}{2}$ ft/s.

3. rate of change $= \frac{\text{change in } y}{\text{change in } x}$
$= \frac{5.6 - 6}{1 - 0}$
$= \frac{-0.4}{1}$ or -0.4

There is a decrease of \$0.40/cookie.

4. Find k:
$y = kx$
$12 = k \cdot (-3)$
$-4 = k$

Write the equation:
$y = kx$
$y = -4x$

5. Find k:

$y = kx$

$45 = k \cdot (15)$

$3 = k$

Write the equation:

$y = kx$

$y = 3x$

6. Find k:

$y = kx$

$8 = k \cdot (18)$

$\frac{8}{18} = k$

$\frac{4}{9} = k$

Write the equation:

$y = kx$

$y = \frac{4}{9}x$

7. Find k:

$y = kx$

$7.6 = k \cdot (4)$

$\frac{7.6}{4} = k$

$1.9 = k$

Write the equation:

$y = kx$

$y = 1.9x$

Page 743 Lesson 8-6

1. slope $= 1$; y-intercept $= 9$

2. slope $= 2$; y-intercept $= -5$

3. slope $= -6$; y-intercept $= 0$

4. slope $= \frac{3}{2}$; y-intercept $= 0$

5. slope $= \frac{1}{3}$; y-intercept $= 8$

6. Write the equation in the form $y = mx + b$:

$$x + 2y = 12$$
$$x - x + 2y = 12 - x$$
$$2y = 12 - x$$
$$\frac{2y}{2} = \frac{12 - x}{2}$$
$$y = \frac{12}{2} - \frac{1}{2}x$$
$$y = -\frac{1}{2}x + 6$$

slope $= -\frac{1}{2}$; y-intercept $= -\frac{1}{2}$

7. The slope is 3 and the y-intercept is -2. Graph the point $(0, -2)$. Then go up 3 and right 1. Connect these points.

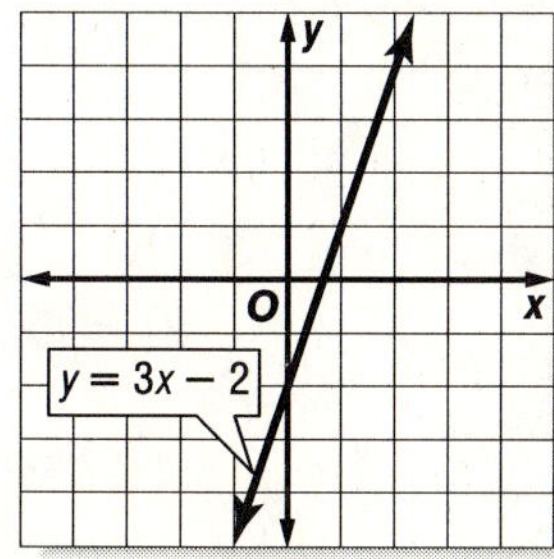

8. Write the equation in the form $y = mx + b$.

$$x - 3y = 9$$
$$x - x - 3y = 9 - x$$
$$-3y = 9 - x$$
$$\frac{-3y}{-3} = \frac{9 - x}{-3}$$
$$y = \frac{9}{-3} + \frac{1}{3}x$$
$$y = \frac{1}{3}x - 3$$

The slope is $\frac{1}{3}$ and the y-intercept is -3.

Graph the point $(0, -3)$. Then go up 1 and right 3. Connect these points.

9. The slope is $\frac{1}{2}$ and the y-intercept is 4. Graph the point $(0, 4)$. Then go up 1 and right 2. Connect these points.

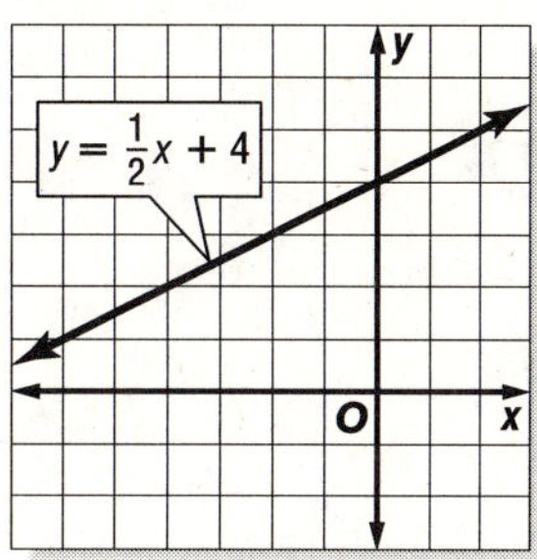

10. The slope is $-\frac{2}{3}$ and the y-intercept is -1. Graph the point $(0, -1)$. Then go down 2 and right 3. Connect these points.

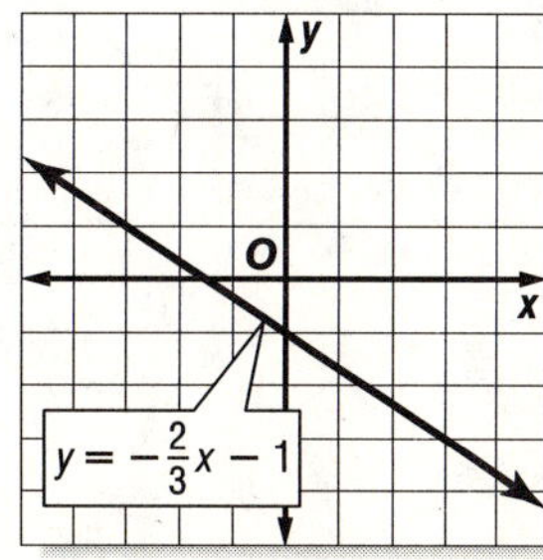

11. Write the equation in the form $y = mx + b$.

$$x - y = -4$$
$$x - x - y = -4 - x$$
$$-y = -4 - x$$
$$y = 4 + x$$
$$y = x + 4$$

The slope is 1 and the y-intercept is 4. Graph the point $(0, 4)$. Then go up 1 and right 1. Connect these points.

12. Write the equation in the form $y = mx + b$.

$$2x + 4y = -4$$
$$2x - 2x + 4y = -4 - 2x$$
$$4y = -4 - 2x$$
$$\frac{4y}{4} = \frac{-4 - 2x}{4}$$
$$y = -1 - \frac{1}{2}x$$
$$y = -\frac{1}{2}x - 1$$

The slope is $-\frac{1}{2}$ and the y-intercept is -1. Graph $(0, -1)$. Then go down 1 and right 2. Connect these points.

13. The slope is 1 and the y-intercept is 5. Graph the point (0, 5). Then go up 1 and right 1. Connect these points.

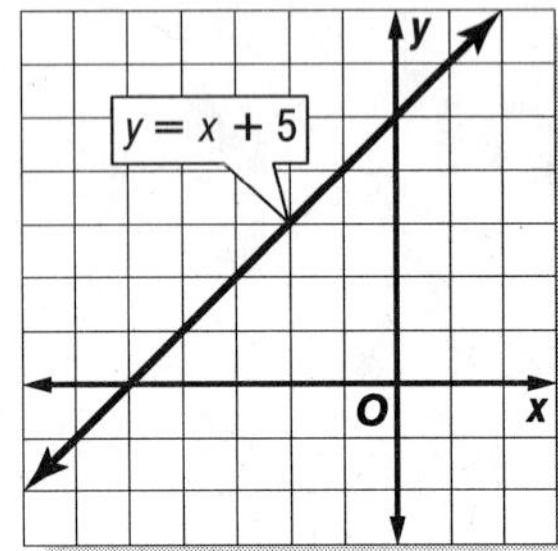

14. Write the equation in the form $y = mx + b$.

$$3x + y = 9$$
$$3x - 3x + y = 9 - 3x$$
$$y = -3x + 9$$

The slope is -3 and the y-intercept is 9. Graph the point (0, 9). Then go down 3 and right 1. Connect these points.

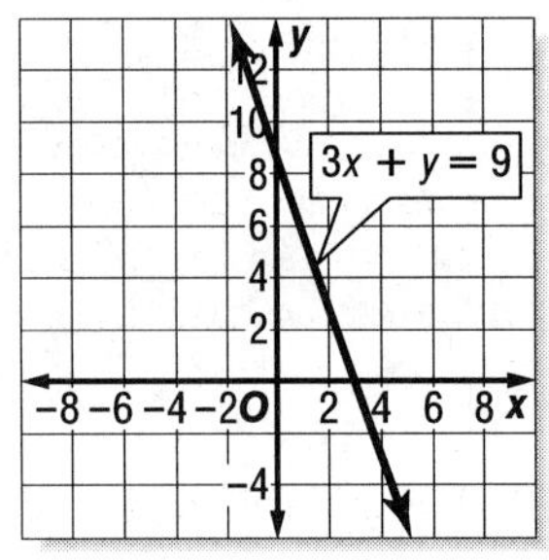

Page 743 Lesson 8-7

1. $y = mx + b$

$y = 3x + (-4)$

$y = 3x - 4$

2. $y = mx + b$

$y = \frac{3}{4}x + 1$

3. $y = mx + b$

$y = -7x + (-2)$

$y = -7x - 2$

4. $y = mx + b$

$y = \frac{5}{8}x + 9$

5. $y = mx + b$

$y = -\frac{1}{2}x + 0$

$y = -\frac{1}{2}x$

6. $y = mx + b$

$y = 0x + (-6)$

$y = -6$

7. Find the slope m:

$m = \frac{y_2 - y_1}{x_2 - x_1}$

$m = \frac{3 - 7}{0 - 4}$

$m = \frac{-4}{-4}$ or 1

Find the y-intercept b. Use the slope and the coordinates of either point.

$y = mx + b$

$7 = 1 \cdot (4) + b$

$3 = b$

Substitute the slope and y-intercept.

$y = mx + b$

$y = 1x + 3$

$y = x + 3$

8. Find the slope m:

$m = \frac{y_2 - y_1}{x_2 - x_1}$

$m = \frac{2 - (-6)}{-1 - 3}$

$m = \frac{8}{-4}$ or -2

Find the y-intercept b. Use the slope and either point.

$y = mx + b$

$-6 = -2(3) + b$

$-6 = -6 + b$

$0 = b$

Substitute the slope and y-intercept.

$y = mx + b$

$y = -2x + 0$

$y = -2x$

9. Find the slope m:

$m = \frac{y_2 - y_1}{x_2 - x_1}$

$m = \frac{0 - 7}{0 - 8}$

$m = \frac{-7}{-8}$ or $\frac{7}{8}$

Find the y-intercept b. One point given is (0, 0), which is the intercept point. So 0 is the y-intercept, b.

Substitute the slope and y-intercept.

$y = mx + b$

$y = \frac{7}{8}x + 0$

$y = \frac{7}{8}x$

10. Find the slope m:

$m = \frac{y_2 - y_1}{x_2 - x_1}$

$m = \frac{-6 - 4}{3 - 1}$

$m = \frac{-10}{2}$ or -5

Find the y-intercept b. Use the slope and either point.

$y = mx + b$

$4 = -5(1) + b$

$9 = b$

Substitute the slope and y-intercept.

$y = mx + b$

$y = -5x + 9$

11. Find the slope m:

$m = \frac{y_2 - y_1}{x_2 - x_1}$

$m = \frac{9 - 5}{3 - (-2)}$

$m = \frac{4}{5}$

Find the y-intercept b. Use the slope and either point.

$y = mx + b$

$9 = \frac{4}{5}(3) + b$

$9 = \frac{12}{5} + b$

$9 - \frac{12}{5} = \frac{12}{5} - \frac{12}{5} + b$

$\frac{45}{5} - \frac{12}{5} = b$

$\frac{33}{5} = b$

Substitute the slope and y-intercept.

$y = mx + b$

$y = \frac{4}{5}x + \frac{33}{5}$

12. Find the slope m:

$m = \frac{y_2 - y_1}{x_2 - x_1}$

$m = \frac{-1 - (-1)}{5 - 3}$

$m = \frac{0}{2}$ or 0

Find the y-intercept b. Use the slope and either point.

$y = mx + b$

$-1 = 0(5) + b$

$-1 = b$

Substitute the slope and y-intercept.

$y = mx + b$

$y = 0x + (-1)$

$y = -1$

Page 744 Lesson 8-8

1. Sample answer:

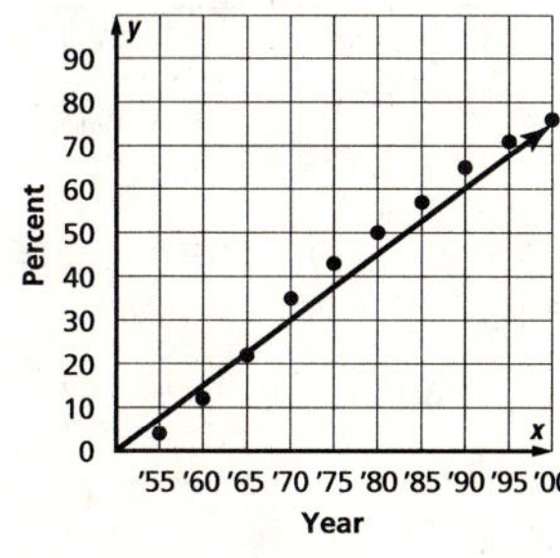

2. Sample answer: Choose any two points on the line, such as (1970, 30) and (1990, 60).

Find the slope m:

$m = \frac{y_2 - y_1}{x_2 - x_1}$

$m = \frac{30 - 60}{1970 - 1990}$

$m = \frac{-30}{-20} = \frac{3}{2}$ or 1.5

Find the y-intercept b. Use the slope and either point.

$y = mx + b$

$60 = 1.5(1990) + b$

$-2925 = b$

Substitute the slope and y-intercept.

$y = mx + b$

$y = 1.5x - 2925$

3. Sample answer:

$y = 1.5x - 2925$

$y = 1.5(2010) - 2925$

$y = 90\%$

Page 744 Lesson 8-9

1. $x + 2y = 6$

$y = -0.5x + 3$

Both equations have the same graph. Any ordered pair on the graph will satisfy both equations. Therefore, there are infinitely many solutions of this system of equations.

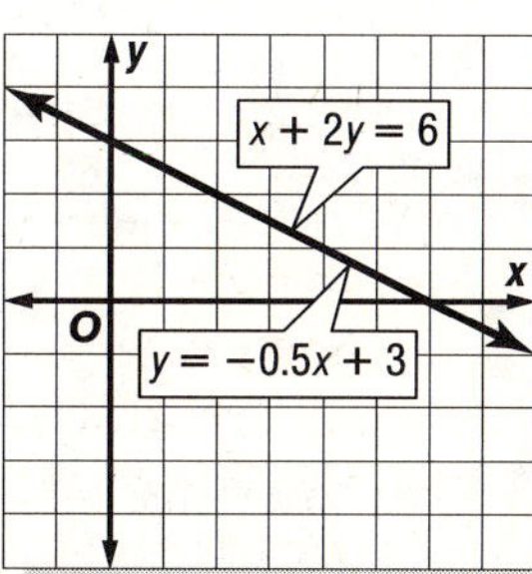

2. $y = -2$

$4x + 3y = 2$

The graphs appear to intersect at (2,−2).

Check:

$y = -2$	$4x + 3y = 2$
$-2 = -2$ ✓	$4(2) + 3(-2) \stackrel{?}{=} 2$
	$8 - 6 \stackrel{?}{=} 2$
	$2 = 2$ ✓

The solution is (2,−2).

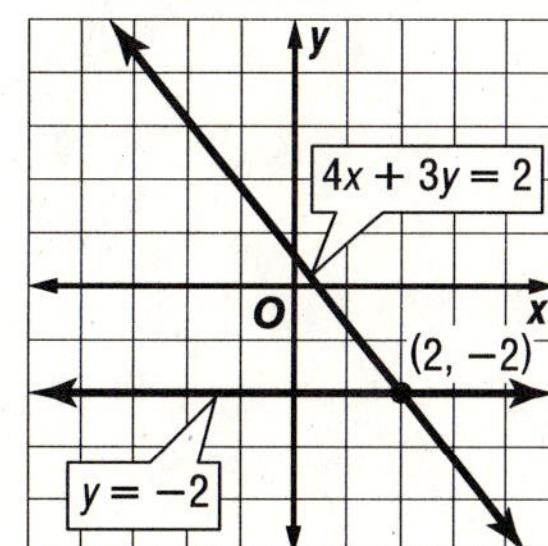

3. $y = x + 4$
$y = -2x + 4$
The graphs appear to intersect at (0, 4).
Check:

$y = x + 4$	$y = -2x + 4$
$4 \stackrel{?}{=} 0 + 4$	$4 \stackrel{?}{=} -2(0) + 4$
$4 = 4$ ✓	$4 = 4$ ✓

The solution is (0, 4).

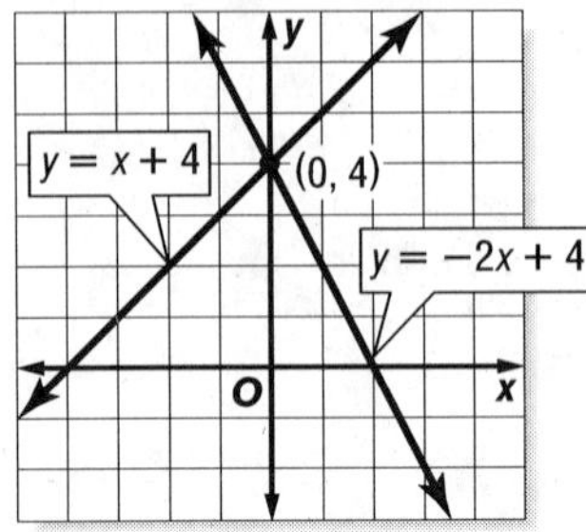

4. $y = 4 + \frac{2}{3}x$
$2x = 3y$
The graphs appear to be parallel lines. Since there is no coordinate pair that is a solution to both equations, there is no solution of this system of equations.

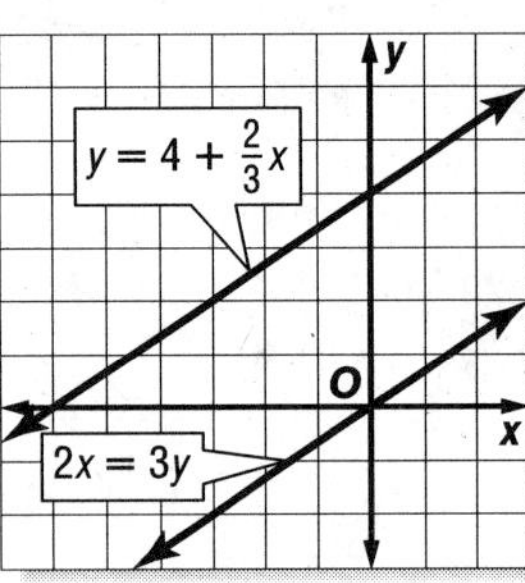

5. $y = x - 2$
$y = -\frac{1}{3}x + 2$
The graphs appear to intersect at (3, 1).
Check:

$y = x - 2$	$y = -\frac{1}{3}x + 2$
$1 \stackrel{?}{=} 3 - 2$	$1 \stackrel{?}{=} -\frac{1}{3}(3) + 2$
$1 = 1$ ✓	$1 \stackrel{?}{=} -1 + 2$
	$1 = 1$ ✓

The solution is (3, 1).

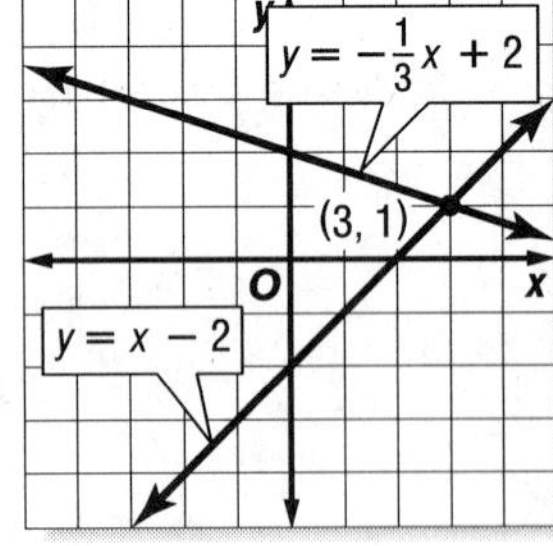

6. $y = \frac{1}{2}x + 6$
$2x + y = 1$
The graphs appear to intersect at (−2, 5).
Check:

$y = \frac{1}{2}x + 6$	$2x + y = 1$
$5 \stackrel{?}{=} \frac{1}{2}(-2) + 6$	$2(-2) + 5 \stackrel{?}{=} 1$
$5 \stackrel{?}{=} -1 + 6$	$-4 + 5 \stackrel{?}{=} 1$
$5 = 5$ ✓	$1 = 1$ ✓

The solution is (−2, 5).

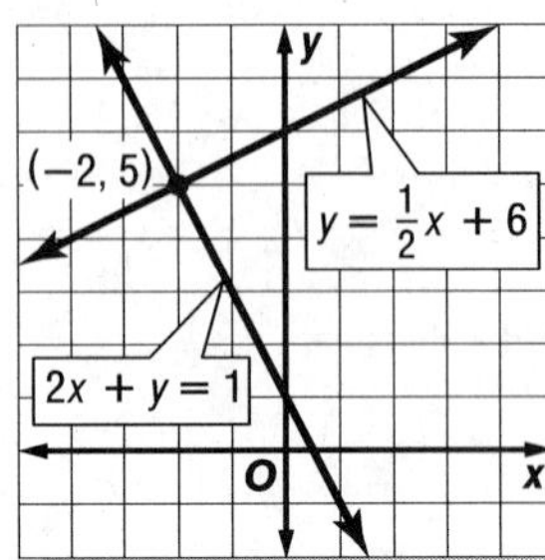

7. $x + y = 4$
$y = 2$
Replace y with 2 in the first equation.
$x + y = 4$
$x + 2 = 4$
$x = 2$
The solution is (2, 2). You can check the solution by graphing. The graphs appear to intersect at (2, 2), so the solution is correct.

8. $2x + y = 8$
$x = 2$
Replace x with 2 in the first equation.
$2x + y = 8$
$2(2) + y = 8$
$y = 4$
The solution is (2, 4). You can check the solution by graphing. The graphs appear to intersect at (2, 4), so the solution is correct.

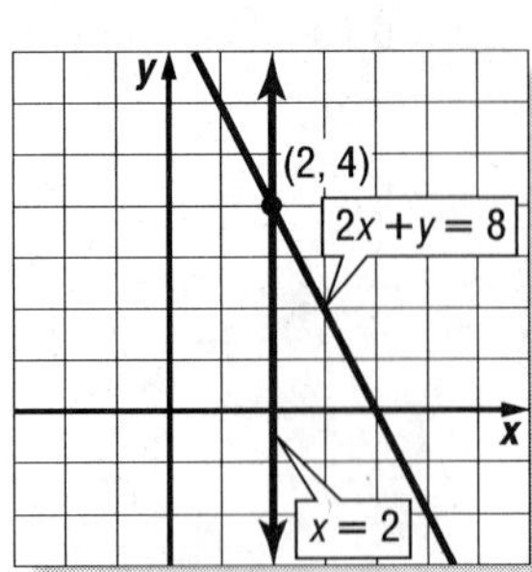

9. $y = x - 7$
$x = 7$
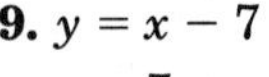
Replace x with 7 in the first equation.
$y = x - 7$
$y = 7 - 7$
$y = 0$
The solution is (7, 0). You can check the solution by graphing. The graphs appear to intersect at (7, 0), so the solution is correct.

10. $x = 3$
$y = 4$
The solution is (3, 4). You can check the solution by graphing. The graphs appear to intersect at (3, 4), so the solution is correct.

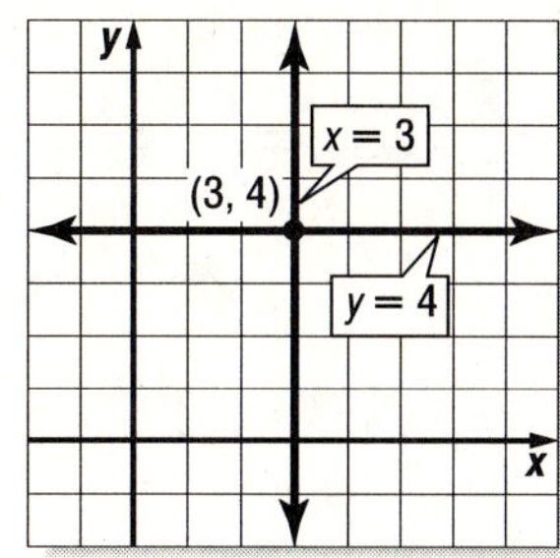

11. $y = 1$
$2y + x = 1$
Replace y with 1 in the second equation.

$2y + x = 1$
$2(1) + x = 1$
$x = -1$

The solution is $(-1, 1)$. You can check the solution by graphing.

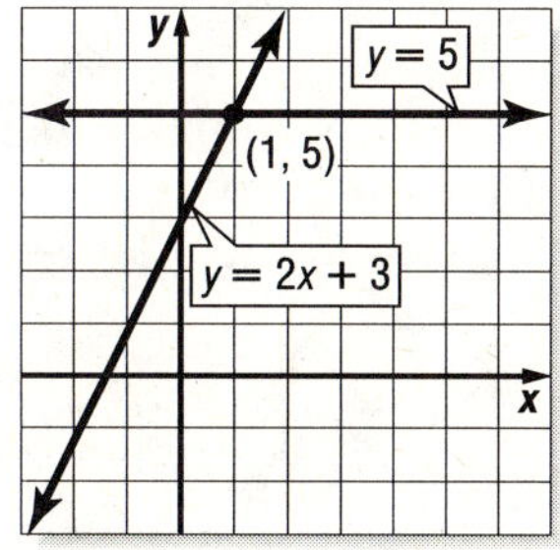

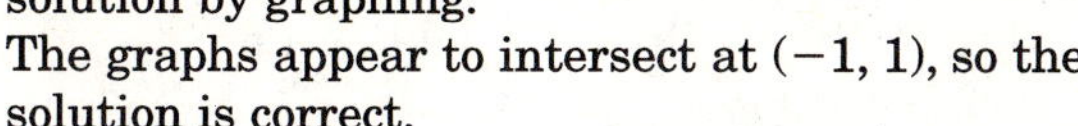
The graphs appear to intersect at $(-1, 1)$, so the solution is correct.

12. $y = 2x + 3$
$y = 5$
Replace y with 5 in the first equation.

$y = 2x + 3$
$5 = 2x + 3$
$2 = 2x$
$1 = x$

The solution is (1, 5). You can check the solution by graphing. The graphs appear to intersect at (1, 5), so the solution is correct.

Page 744 Lesson 8-10

1. $y > 2x - 2$
Graph $y = 2x - 2$. Draw a dashed line since the boundary is not part of the graph.
Test (0, 0): $y > 2x - 2$
$0 \overset{?}{>} 2(0) - 2$
$0 > -2$ ✓
Thus, the solution is all points in the region above the graph.

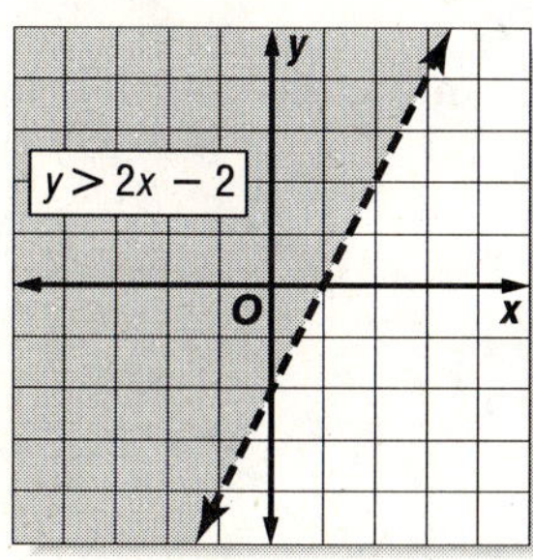

2. $y \geq x$
Graph $y = x$. Draw a solid line since the boundary is part of the graph.
Test $(-2, 3)$: $y \geq x$
$3 \geq -2$ ✓
Thus, the graph is all points in the region above the boundary.

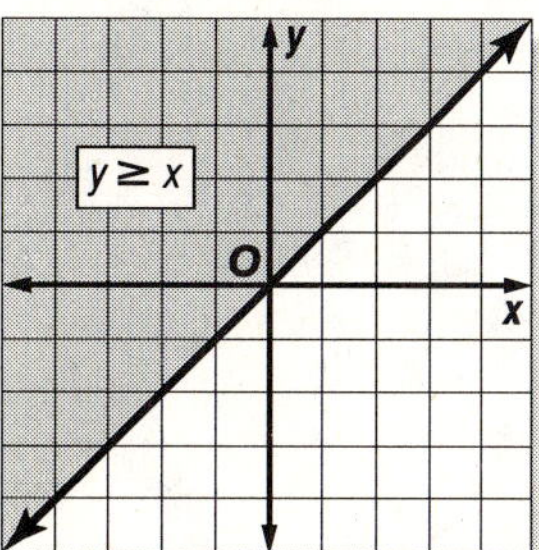

3. $y < 1$
Graph $y = 1$. Draw a dashed line since the boundary is not part of the graph.
Test (0, 0): $y < 1$
$0 < 1$ ✓
Thus, the graph is all points in the region below the boundary.

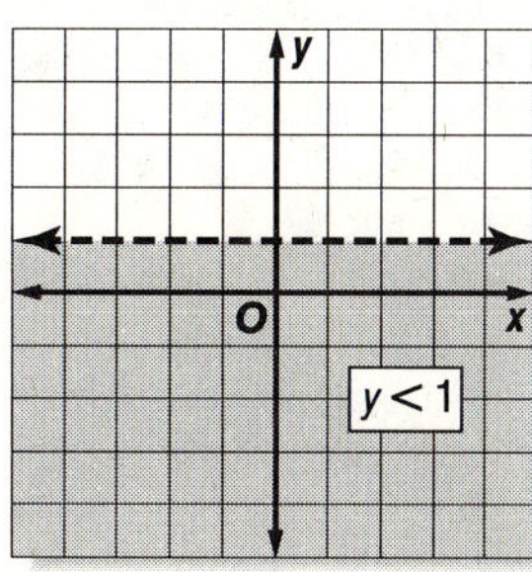

4. $x + y \leq -1$
Graph $x + y = 1$. Draw a solid line since the boundary is part of the graph.
Test (0, 0): $x + y \leq -1$
$0 + 0 \overset{?}{\leq} -1$
$0 \nleq -1$
(0, 0) is not a solution, so shade the other half plane.

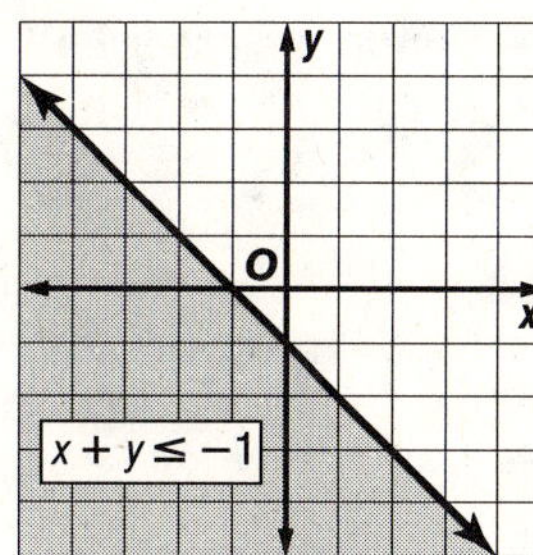

5. $y + 3x \leq 0$

Graph $y + 3x = 0$. Draw a solid line since the boundary is part of the graph.

Test $(-2, -1)$: $y + 3x \leq 0$

$$-1 + 3(-2) \overset{?}{\leq} 0$$
$$-1 - 6 \overset{?}{\leq} 0$$
$$-7 \leq 0 \checkmark$$

Thus, the graph is all points in the region below the boundary.

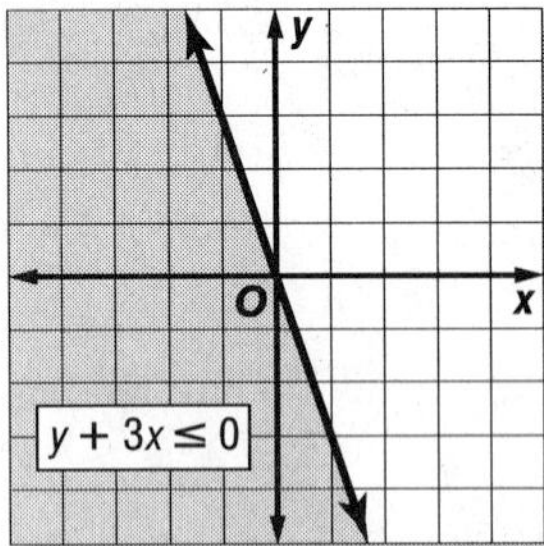

6. $x < -3$

Graph $x = -3$. Draw a dashed line since the boundary is not part of the graph.

Test $(0, 0)$: $x < -3$

$$0 \not< -3$$

$(0, 0)$ is not a solution, so shade the other half plane.

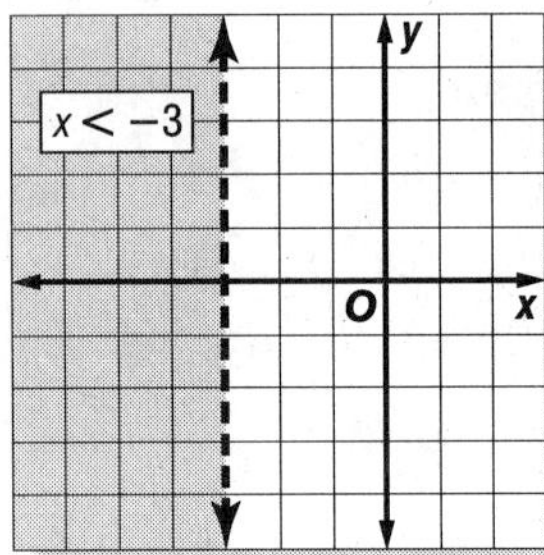

7. $2x + 3y \geq 12$

Graph $2x + 3y = 12$. Draw a solid line since the boundary is part of the graph.

Test $(0, 0)$: $2x + 3y \geq 12$

$$2(0) + 3(0) \overset{?}{\geq} 0$$
$$0 \not\geq 12$$

$(0, 0)$ is not a solution, so shade the other half plane.

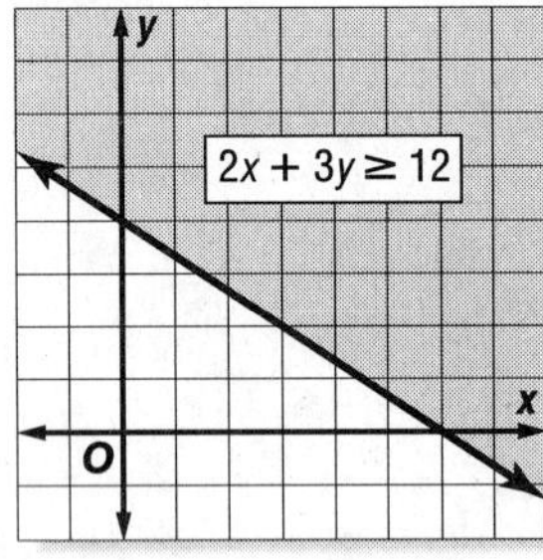

8. $-2x + y > -1$

Graph $-2x + y = -1$. Draw a dashed line since the boundary is not part of the graph.

Test $(0, 0)$: $-2x + y > -1$

$$-2(0) + 0 > -1$$
$$0 > -1 \checkmark$$

Thus, the graph is all points in the region below the boundary.

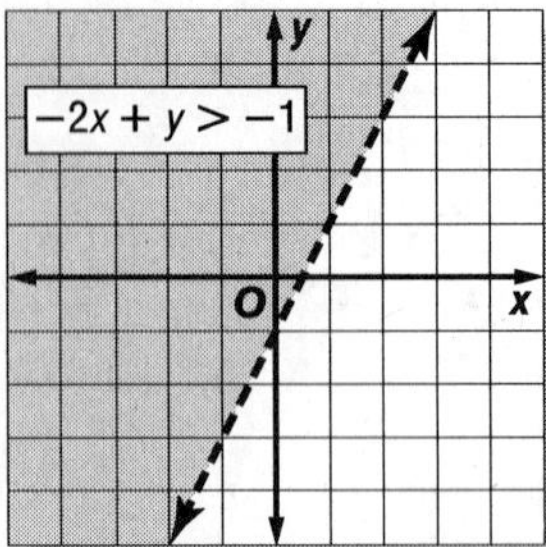

9. $y \geq -4$

Graph $y = -4$. Draw a solid line since the boundary is part of the graph.

Test $(0, 0)$: $y \geq -4$

$$0 \geq -4 \checkmark$$

Thus, the graph is all points in the region above the boundary.

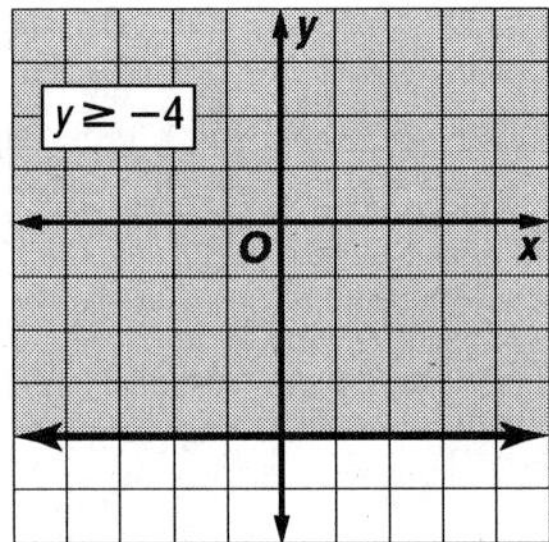

10. Let x represent the amount of time Seth can tutor, and let y represent the amount of time Seth can volunteer at a soup kitchen.

$$x + y \leq 9$$

11. $x + y \leq 9$

Graph $x + y = 9$. Draw a solid line since the boundary is part of the graph.

Test $(0, 0)$: $x + y \leq 9$

$$0 + 0 \overset{?}{\leq} 9$$
$$0 \leq 9 \checkmark$$

Thus, the graph is all points in the region below the boundary.

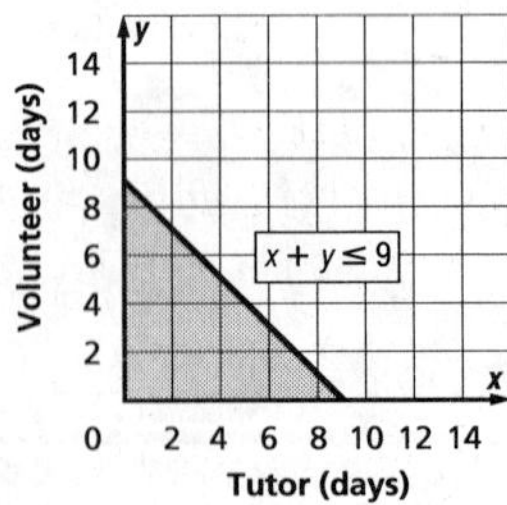

12. Sample answer: tutor 4 h, volunteer 5 h; tutor 6 h, volunteer 3 h

Page 745 Lesson 9-1

1. Since $6^2 = 36$, $\sqrt{36} = 6$.
2. Since $9^2 = 81$, $-\sqrt{81} = -9$.
3. Since $\left(\frac{1}{2}\right)^2 = \frac{1}{4}$, $\sqrt{\frac{1}{4}} = \frac{1}{2}$.
4. Since $12^2 = 144$, $-\sqrt{144} = -12$.
5. not possible
6. Since $(1.4)^2 = 1.96$, $\sqrt{1.96} = 1.4$.
7. not possible
8. Since $(0.7)^2 = 0.49$, $-\sqrt{0.49} = -0.7$.
9. Since $20^2 = 400$, $\sqrt{400} = 20$.
10. 4.6
11. 9.9
12. −7.7
13. 11.1
14. −18.7
15. 4.3
16. −6.5
17. −9.2
18. 13.5
19. $16 < 21 < 25$
$\sqrt{16} < \sqrt{21} < \sqrt{25}$
$4 < \sqrt{21} < 5$
$\sqrt{21} \approx 5$
20. $-100 < -85 < -81$
$-\sqrt{100} < -\sqrt{85} < -\sqrt{81}$
$-10 < -\sqrt{85} < -9$
$-\sqrt{85} \approx -9$
21. $4 < 7.3 < 9$
$\sqrt{4} < \sqrt{7.3} < \sqrt{9}$
$2 < \sqrt{7.3} < 3$
$\sqrt{7.3} \approx 3$
22. $1 < 1.99 < 4$
$\sqrt{1} < \sqrt{1.99} < \sqrt{4}$
$1 < \sqrt{1.99} < 2$
$1 \approx \sqrt{1.99}$
23. $-64 < -62 < -49$
$-\sqrt{64} < -\sqrt{62} < -\sqrt{49}$
$-8 < -\sqrt{62} < -7$
$-8 \approx -\sqrt{62}$
24. $64 < 74.1 < 81$
$\sqrt{64} < \sqrt{74.1} < \sqrt{81}$
$8 < \sqrt{74.1} < 9$
$\sqrt{74.1} \approx 9$
25. $784 < 810 < 841$
$\sqrt{784} < \sqrt{810} < \sqrt{841}$
$28 < \sqrt{810} < 29$
$28 \approx \sqrt{810}$
26. $-100 < -88.8 < -81$
$-\sqrt{100} < -\sqrt{88.8} < -\sqrt{81}$
$-10 < -\sqrt{88.8} < -9$
$-\sqrt{88.8} \approx -9$
27. $961 < 1000 < 1024$
$\sqrt{961} < \sqrt{1000} < \sqrt{1024}$
$31 < \sqrt{1000} < 32$
$\sqrt{1000} \approx 32$

Page 745 Lesson 9-2

1. N, W, Z, Q
2. W, Z, Q
3. Q
4. Q
5. Q
6. I
7. I
8. Z, Q
9. Q
10. $3\frac{3}{4} = 3.75$
$\sqrt{15} \approx 3.87$
Since $3.75 < 3.87$, $3\frac{3}{4} < \sqrt{15}$.
11. $-\sqrt{41} \approx -6.40$
Since $-6.40 > -6.8$, $-\sqrt{41} > -6.8$.
12. $\sqrt{27.04} = 5.2$
Since $5.2 = 5.2$, $5.2 = \sqrt{27.04}$.
13. $-\sqrt{110} \approx -10.49$
Since $-10.49 > -10.5$, $-\sqrt{110} > -10.5$.
14. $x^2 = 14$
$\sqrt{x^2} = \sqrt{14}$
$x = \sqrt{14}$ or $x = -\sqrt{14}$
$x = 3.7$ or $x = -3.7$
15. $y^2 = 25$
$\sqrt{y^2} = \sqrt{25}$
$y = \sqrt{25}$ or $y = -\sqrt{25}$
$y = 5$ or $y = -5$
16. $34 = p^2$
$\sqrt{34} = \sqrt{p^2}$
$\sqrt{34} = p$ or $-\sqrt{34} = p$
$5.8 = p$ or $-5.8 = p$
17. $55 = h^2$
$\sqrt{55} = \sqrt{h^2}$
$\sqrt{55} = h$ or $-\sqrt{55} = h$
$7.4 = h$ or $-7.4 = h$
18. $225 = k^2$
$\sqrt{225} = \sqrt{k^2}$
$\sqrt{225} = k$ or $-\sqrt{225} = k$
$15 = k$ or $-15 = k$
19. $324 = m^2$
$\sqrt{324} = \sqrt{m^2}$
$\sqrt{324} = m$ or $-\sqrt{324} = m$
$18 = m$ or $-18 = m$
20. $d^2 = 441$
$\sqrt{d^2} = \sqrt{441}$
$d = \sqrt{441}$ or $d = -\sqrt{441}$
$d = 21$ or $d = -21$
21. $r^2 = 25{,}000$
$\sqrt{r^2} = \sqrt{25{,}000}$
$r = \sqrt{25{,}000}$ or $r = -\sqrt{25{,}000}$
$r = 158.1$ or $r = -158.1$
22. $10{,}000 = x^2$
$\sqrt{10{,}000} = \sqrt{x^2}$
$\sqrt{10{,}000} = x$ or $-\sqrt{10{,}000} = x$
$100 = x$ or $-100 = x$

Page 745 Lesson 9-3

1. 120°; $m\angle PQW > 90°$, so $\angle PQW$ is obtuse.
2. 27°; $m\angle PQW < 90°$, so $\angle PQW$ is acute.
3. 90°; $m\angle TQW = 90°$, so $\angle TQW$ is right.
4. 140°; $m\angle SQW > 90°$, so $\angle SQW$ is obtuse.
5. 40°; $m\angle SQR < 90°$, so $\angle SQR$ is acute.
6. 153°; $m\angle VQR > 90°$, so $\angle VQR$ is obtuse.
7. 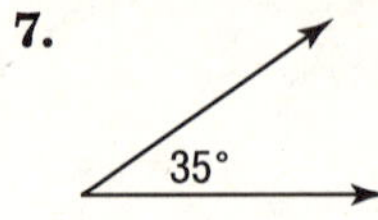

$35° < 90°$, so the angle is acute.
8. 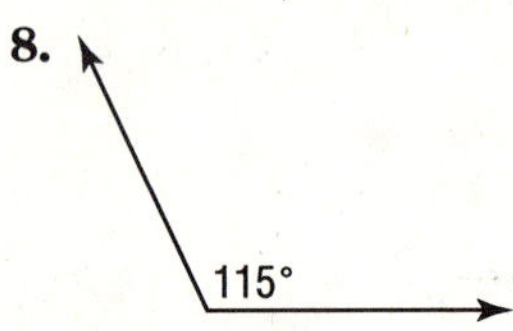

$115° > 90°$, so the angle is obtuse.
9. 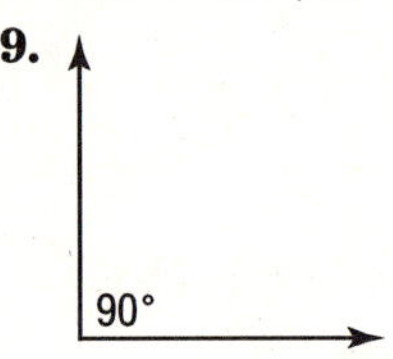

$90° = 90°$, so the angle is right.
10. 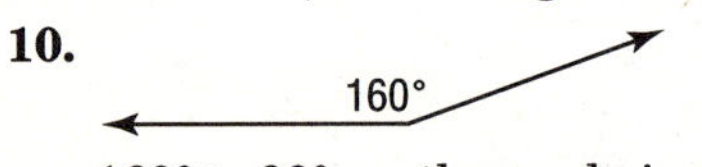

$160° > 90°$, so the angle is obtuse.
11. 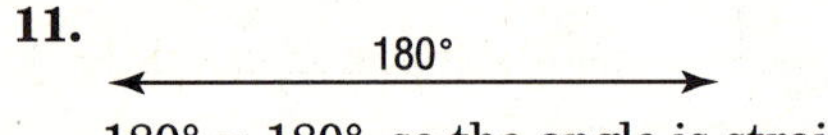

$180° = 180°$, so the angle is straight.
12. 18°
$18° < 90°$, so the angle is acute.

Page 746 Lesson 9-4

1. $x° + 56° + 90° = 180°$
$x + 146 = 180$
$x + 146 - 146 = 180 - 146$
$x = 34$
The value of x is 34. The triangle is a right triangle because it has a right angle.
2. $x° + 42° + 63° = 180°$
$x + 105 = 180$
$x + 105 - 105 = 180 - 105$
$x = 75$
The value of x is 75. The triangle is acute because all the angles are acute.
3. $x° + 16° + 18° = 180°$
$x + 34 = 180$
$x + 34 - 34 = 180 - 34$
$x = 146$
The value of x is 146. The triangle is obtuse because it has one obtuse angle.
4. $x° + 40° + 95° = 180°$
$x + 135 = 180$
$x + 135 - 135 = 180 - 135$
$x = 45$
The value of x is 45. The triangle is obtuse because it has one obtuse angle.
5. $x° + 31° + 90° = 180°$
$x + 121 = 180$
$x + 121 - 121 = 180 - 121$
$x = 59$
The value of x is 59. The triangle is right because it has a right angle.
6. $x° + 65° + 65° = 180°$
$x + 130 = 180$
$x + 130 - 130 = 180 - 130$
$x = 50$
The value of x is 50. The triangle is acute because all the angles are acute.
7. $x + 2x + 3x = 180$
$6x = 180$
$\frac{6x}{6} = \frac{180}{6}$
$x = 30$
Since $x = 30$, $2x = 2(30)$ or 60, and $3(x) = 3(30)$ or 90. The measures of the angles are 30°, 60°, and 90°.
8. $x + x + 2x = 180$
$4x = 180$
$\frac{4x}{4} = \frac{180}{4}$
$x = 45$
Since $x = 45$, $2x = 2(45)$ or 90. The measures of the angles are 45°, 45°, and 90°.
9. $x + 9x + 26x = 180$
$36x = 180$
$\frac{36x}{36} = \frac{180}{36}$
$x = 5$
Since $x = 5$, $9x = 9(5)$ or 45, and $26x = 26(5)$ or 130. The measures of the angles are 5°, 45°, and 130°.

Page 746 Lesson 9-5

1. $c^2 = a^2 + b^2$
$c^2 = 4^2 + 3^2$
$c^2 = 16 + 9$
$c^2 = 25$
$\sqrt{c^2} = \sqrt{25}$
$c = 5$
The length of the hypotenuse is 5 feet.
2. $c^2 = a^2 + b^2$
$c^2 = 6^2 + 8^2$
$c^2 = 36 + 64$
$c^2 = 100$
$\sqrt{c^2} = \sqrt{100}$
$c = 10$
The length of the hypotenuse is 10 inches.

3. $c^2 = a^2 + b^2$
$c^2 = 24^2 + 10^2$
$c^2 = 576 + 100$
$c^2 = 676$
$\sqrt{c^2} = \sqrt{676}$
$c = 26$
The length of the hypotenuse is 26 meters.

4. $c^2 = a^2 + b^2$
$c^2 = 7^2 + 24^2$
$c^2 = 49 + 576$
$c^2 = 625$
$\sqrt{c^2} = \sqrt{625}$
$c = 25$
The length of the hypotenuse is 25 meters.

5. $c^2 = a^2 + b^2$
$30^2 = 18^2 + b^2$
$900 = 324 + b^2$
$900 - 324 = 324 - 324 + b^2$
$576 = b^2$
$\sqrt{576} = \sqrt{b^2}$
$24 = b$
The length of the hypotenuse is 24 inches.

6. $c^2 = a^2 + b^2$
$20^2 = a^2 + 10^2$
$400 = a^2 + 100$
$400 - 100 = a^2 + 100 - 100$
$300 = a^2$
$\sqrt{300} = \sqrt{a^2}$
$17.3 \approx a$
The length of the hypotenuse is about 17.3 feet.

7. $c^2 = a^2 + b^2$
$9^2 = 3^2 + b^2$
$81 = 9 + b^2$
$81 - 9 = 9 - 9 + b^2$
$72 = b^2$
$\sqrt{72} = \sqrt{b^2}$
$8.5 \approx b$
The length of the hypotenuse is about 8.5 centimeters.

8. $c^2 = a^2 + b^2$
$32^2 = a^2 + 8^2$
$1024 = a^2 + 64$
$1024 - 64 = a^2 + 64 - 64$
$960 = a^2$
$\sqrt{960} = \sqrt{a^2}$
$31.0 \approx a$
The length of the hypotenuse is about 31.0 meters.

9. $c^2 = a^2 + b^2$
$65^2 = 32^2 + b^2$
$4225 = 1024 + b^2$
$4225 - 1024 = 1024 - 1024 + b^2$
$3201 = b^2$
$\sqrt{3201} = \sqrt{b^2}$
$56.6 \approx b$
The length of the hypotenuse is about 56.6 yards.

Page 746 Lesson 9-6

1. $d = \sqrt{(x_2 - x_1)^2 + (y_2 - y_1)^2}$
$AB = \sqrt{(-4 - 2)^2 + (2 - 6)^2}$
$AB = \sqrt{(-6)^2 + (-4)^2}$
$AB = \sqrt{36 + 16}$
$AB = \sqrt{52}$
$AB \approx 7.2$
The distance between points A and B is about 7.2 units.

2. $d = \sqrt{(x_2 - x_1)^2 + (y_2 - y_1)^2}$
$CD = \sqrt{[2 - (-3)]^2 + (4 - 9)^2}$
$CD = \sqrt{5^2 + (-5)^2}$
$CD = \sqrt{25 + 25}$
$CD = \sqrt{50}$
$CD \approx 7.1$
The distance between points C and D is about 7.1 units.

3. $d = \sqrt{(x_2 - x_1)^2 + (y_2 - y_1)^2}$
$EF = \sqrt{(1 - 6)^2 + [-6 - (-4)]^2}$
$EF = \sqrt{(-5)^2 + (-2)^2}$
$EF = \sqrt{25 + 4}$
$EF = \sqrt{29}$
$EF \approx 5.4$
The distance between points E and F is about 5.4 units.

4. $d = \sqrt{(x_2 - x_1)^2 + (y_2 - y_1)^2}$
$GH = \sqrt{(9 - 0)^2 + [-1 - (-1)]^2}$
$GH = \sqrt{9^2 + 0^2}$
$GH = \sqrt{81}$
$GH = 9$
The distance between points G and H is 9 units.

5. $d = \sqrt{(x_2 - x_1)^2 + (y_2 - y_1)^2}$
$IJ = \sqrt{[2 - (-8)]^2 + [2 - (-3)]^2}$
$IJ = \sqrt{10^2 + 5^2}$
$IJ = \sqrt{100 + 25}$
$IJ = \sqrt{125}$
$IJ \approx 11.2$
The distance between points I and J is about 11.2 units.

6. $d = \sqrt{(x_2 - x_1)^2 + (y_2 - y_1)^2}$
$KL = \sqrt{(-7 - 3)^2 + (-2 - 0)^2}$
$KL = \sqrt{(-10)^2 + (-2)^2}$
$KL = \sqrt{100 + 4}$
$KL = \sqrt{104}$
$KL \approx 10.2$
The distance between points K and L is about 10.2 units.

7. midpoint $= \left(\frac{x_1 + x_2}{2}, \frac{y_1 + y_2}{2}\right)$
midpoint $\overline{MN} = \left(\frac{3 + 7}{2}, \frac{5 + 1}{2}\right)$
$= (5, 3)$
The coordinates of the midpoint of $\overline{MN}$ are (5, 3).

8. midpoint $= \left(\frac{x_1 + x_2}{2}, \frac{y_1 + y_2}{2}\right)$
midpoint $\overline{OP} = \left(\frac{-6 + 0}{2}, \frac{2 + 8}{2}\right)$
$= (-3, 5)$
The coordinates of the midpoint of $\overline{OP}$ are (−3, 5).

9. midpoint $= \left(\frac{x_1 + x_2}{2}, \frac{y_1 + y_2}{2}\right)$

midpoint $\overline{QR} = \left(\frac{-2 + 4}{2}, \frac{7 + (-9)}{2}\right)$

$= (1, -1)$

The coordinates of the midpoint of $\overline{QR}$ are $(1, -1)$.

10. midpoint $= \left(\frac{x_1 + x_2}{2}, \frac{y_1 + y_2}{2}\right)$

midpoint $\overline{ST} = \left(\frac{-5 + 13}{2}, \frac{-3 + (-1)}{2}\right)$

$= (4, -2)$

The coordinates of the midpoint of $\overline{ST}$ are $(4, -2)$.

Page 747 Lesson 9-7

1. $\frac{12}{8} = \frac{x}{6}$

$6 \cdot 12 = 8 \cdot x$

$72 = 8x$

$\frac{72}{8} = \frac{8x}{8}$

$9 = x$

The value of x is 9 inches.

2. $\frac{20}{x} = \frac{8}{4}$

$20 \cdot 4 = 8 \cdot x$

$80 = 8x$

$\frac{80}{8} = \frac{8x}{8}$

$10 = x$

The value of x is 10 meters.

3. $\frac{AC}{AB} = \frac{DC}{EB}$

$\frac{9}{6} = \frac{x}{4}$

$4 \cdot 9 = 6 \cdot x$

$36 = 6x$

$\frac{36}{6} = \frac{6x}{6}$

$6 = x$

The value of x is 6 cm.

4. $\frac{LK}{LH} = \frac{LJ}{LG}$

$\frac{18}{10} = \frac{12}{x}$

$18 \cdot x = 12 \cdot 10$

$18x = 120$

$\frac{18x}{18} = \frac{120}{18}$

$x = 6\frac{2}{3}$

The value of x is $6\frac{2}{3}$ in.

Page 747 Lesson 9-8

1. $\sin A = \frac{6}{10}$

$\sin A = 0.6$

2. $\sin B = \frac{8}{10}$

$\sin B = 0.8$

3. $\tan F = \frac{5}{12}$

$\tan F \approx 0.4167$

4. $\sin E = \frac{12}{13}$

$\sin E \approx 0.9231$

5. $\cos A = \frac{8}{10}$

$\cos A = 0.8$

6. $\tan A = \frac{6}{8}$

$\tan A = 0.75$

7. $\sin 21° \approx 0.3584$

8. $\tan 83° \approx 8.1443$

9. $\cos 45° \approx 0.7071$

10. $\tan 10° \approx 0.1763$

11. $\sin 72° \approx 0.9511$

12. $\cos 3° \approx 0.9986$

Page 747 Lesson 10–1

1. Since $\angle 2$ and $\angle 1$ are supplementary, the sum of their measures is 180°. $180 - 38 = 142$. So, $m\angle 1 = 142°$.
2. Since $\angle 1$ and $\angle 4$ are vertical angles, they are congruent. So, $m\angle 4 = 142°$.
3. Since $\angle 2$ and $\angle 3$ are vertical angles, they are congruent. So, $m\angle 3 = 38°$.
4. Since $\angle 1$ and $\angle 6$ are corresponding angles, they are congruent. So, $m\angle 6 = 142°$.
5. Since $\angle 2$ and $\angle 5$ are corresponding angles, they are congruent. So, $m\angle 5 = 38°$.
6. Since $\angle 6$ and $\angle 8$ are vertical angles, they are congruent. So, $m\angle 8 = 142°$.
7. Since $x°$ and 32° are vertical angles, they are congruent. So, $x = 32°$.
8. Since $x°$ and 18° are supplementary angles, the sum of their measures is 180°.

$x + 18 = 180$

$x + 18 - 18 = 180 - 18$

$x = 162$

9. Since $x°$ and 52° are complementary angles, the sum of their measures is 90°.

$x + 52 = 90$

$x + 52 - 52 = 90 - 52$

$x = 38$

10. Since $x°$ and 94° are supplementary angles, the sum of their measures is 180°.

$x + 94 = 180$

$x + 94 - 94 = 180 - 94$

$x = 86$

Page 748 Lesson 10-2

1. $\angle A \cong \angle D$; $\angle B \cong \angle F$; $\angle C \cong \angle E$; $\overline{AB} \cong \overline{DF}$; $\overline{BC} \cong \overline{FE}$; $\overline{AC} \cong \overline{DE}$; $\triangle ABC \cong \triangle DFE$
2. $\angle G \cong \angle K$; $\angle H \cong \angle J$; $\angle I \cong \angle I$; $\overline{GH} \cong \overline{KJ}$; $\overline{HI} \cong \overline{JI}$; $\overline{GI} \cong \overline{KI}$; $\triangle GHI \cong \triangle KJI$
3. $\angle K$ corresponds to $\angle G$, so $\angle K \cong \angle G$.
4. W corresponds to L, and G corresponds to K, so $\overline{WG} \cong \overline{LK}$.
5. $\angle D$ corresponds to $\angle J$, so $\angle D \cong \angle J$.
6. K corresponds to G, and L corresponds to W, so $\overline{KL} \cong \overline{GW}$.
7. D corresponds to J, and G corresponds to K, so $\overline{DG} \cong \overline{JK}$.

8. $\angle W$ corresponds to $\angle L$, so $\angle W \cong \angle L$.

Page 748 Lesson 10-3

1.

vertex	2 right, 1 down	translation
$A(-1, 2)$	$+ (2, -1)$	$\rightarrow A'(1, 1)$
$B(3, 0)$	$+ (2, -1)$	$\rightarrow B'(5, -1)$
$C(-1, -2)$	$+ (2, -1)$	$\rightarrow C'(1, -3)$

The coordinates of the vertices are $A'(1, 1)$, $B'(5, -1)$, and $C'(1, -3)$.

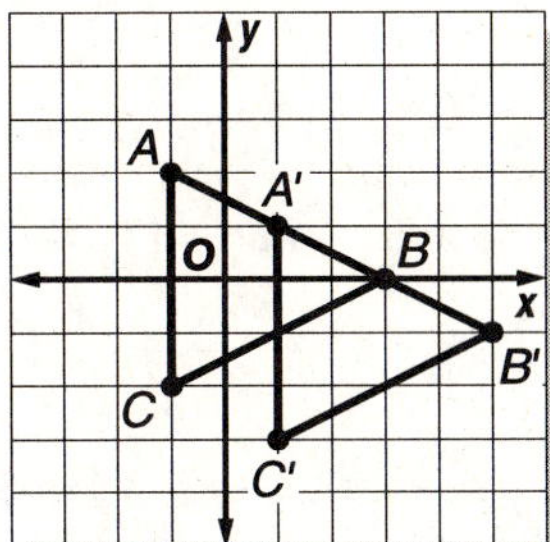

2.

vertex	3 left, 2 down	translation
$K(0, 5)$	$+ (-3, -2)$	$\rightarrow K'(-3, 3)$
$L(3, 0)$	$+ (-3, -2)$	$\rightarrow L'(0, -2)$
$M(0, -2)$	$+ (-3, -2)$	$\rightarrow M'(-3, -4)$
$N(-1, 0)$	$+ (-3, -2)$	$\rightarrow N'(-4, -2)$

The coordinates of the vertices are $K'(-3, 3)$, $L'(0, -2)$, $M'(-3, -4)$, and $N'(-4, -2)$.

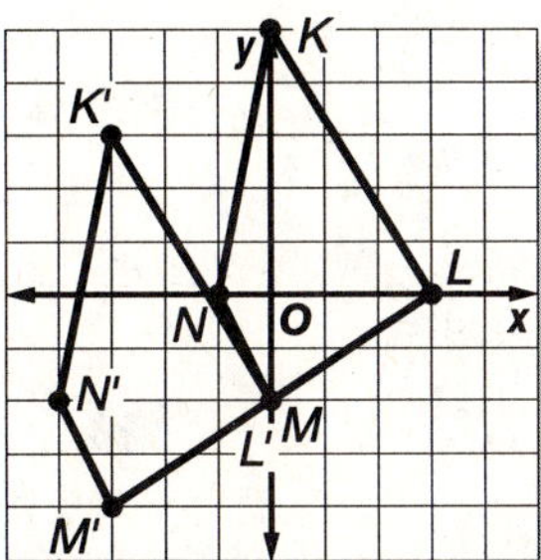

3. Use the same x-coordinate and multiply the y-coordinate by -1.

vertex			reflection
$I(3, 3)$	$\rightarrow$	$(3, -1 \cdot 3)$	$\rightarrow I'(3, -3)$
$J(3, -3)$	$\rightarrow$	$(3, -1 \cdot -3)$	$\rightarrow J'(3, 3)$
$K(1, -3)$	$\rightarrow$	$(1, -1 \cdot -3)$	$\rightarrow K'(1, 3)$

The coordinates of the vertices are $I'(3, -3)$, $J'(3, 3)$, and $K'(1, 3)$.

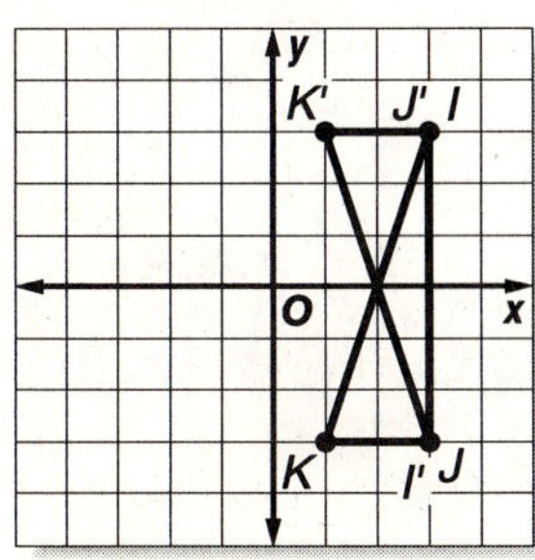

4. Use the same y-coordinate and multiply the x-coordinate by -1.

vertex			reflection
$D(0, 4)$	$\rightarrow$	$(-1 \cdot 0, 4)$	$\rightarrow D'(0, 4)$
$E(4, 2)$	$\rightarrow$	$(-1 \cdot 4, 2)$	$\rightarrow E'(-4, 2)$
$F(3, -4)$	$\rightarrow$	$(-1 \cdot 3, -4)$	$\rightarrow F'(-3, -4)$
$G(0, 1)$	$\rightarrow$	$(-1 \cdot 0, 1)$	$\rightarrow G'(0, 1)$

The coordinates of the vertices are $D'(0, 4)$, $E'(-4, 2)$, $F'(-3, -4)$, and $G'(0, 1)$.

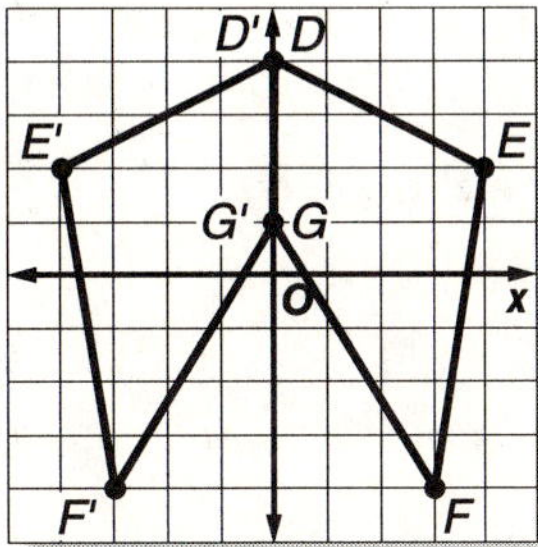

Page 748 Lesson 10-4

1. The sum of the measures of the angles is 360°.
$$x + 65 + 65 + 115 = 360$$
$$x + 245 = 360$$
$$x = 115$$
The value of x is 115. So, the missing angle is 115°.

2. The sum of the measures of the angles is 360°.
$$x + 85 + 40 + 125 = 360$$
$$x + 250 = 360$$
$$x = 110$$
The value of x is 110. So, the missing angle is 110°.

3. The sum of the measures of the angles is 360°.
$$2x + 82 + 134 + 96 = 360$$
$$2x + 312 = 360$$
$$2x = 48$$
$$x = 24$$
The value of x is 24. So, the missing angle is 2(24) or 48°.

4. The sum of the measures of the angles is 360°.
$$(x + 10) + (x - 5) + 93 + 86 = 360$$
$$x + x + 10 - 5 + 93 + 86 = 360$$
$$2x + 184 = 360$$
$$2x = 176$$
$$x = 88$$
The value of x is 88. So, the missing angle measures are 88 + 10 or 98°, and 88 − 5 or 83°.

Page 749 Lesson 10-5

1. The base is 10 in. The height is 3 in.
$$A = bh$$
$$A = 10 \cdot 3$$
$$A = 30$$
The area is 30 in^2.

2. The base is 14.2 cm. The height is 10.6 cm.

$A = \frac{1}{2}bh$

$A = \frac{1}{2}(14.2)(10.6)$

$A = 75.26$

The area is 75.26 cm^2.

3. The bases are 6 ft and 8.5 ft. The height is 2.5 ft.

$A = \frac{1}{2}h(a + b)$

$A = \frac{1}{2}(2.5)(8.5 + 6)$

$A = \frac{1}{2} \cdot 2.5 \cdot 14.5$

$A = 18.125$

The area is 18.125 ft^2.

4. The base is 6 m. The height is 9 m.

$A = \frac{1}{2}bh$

$A = \frac{1}{2} \cdot 6 \cdot 9$

$A = 27$

The area is 27 m^2.

5. $A = bh$

$32.3 = 3.4h$

$\frac{32.3}{3.4} = \frac{3.4h}{3.4}$

$9.5 = h$

The height is 9.5 in.

6. $A = \frac{1}{2}h(a + b)$

$70 = \frac{1}{2}h(8 + 12)$

$70 = \frac{1}{2}h(20)$

$70 = 10h$

$7 = h$

The height is 7 m.

Page 749 Lesson 10-6

1. The polygon has 6 sides. It is a hexagon. All sides and angles are congruent, so it is regular.
2. The polygon has 11 sides. It is an 11-gon. All angles are not congruent, so it is not regular.
3. A decagon has 10 sides. Therefore, $n = 10$.

$(n - 2)180 = (10 - 2)180$
$= 8(180)$
$= 1440$

The sum of the measures of the angles is 1440°.

4. A pentagon has 5 sides. Therefore, $n = 5$.

$(n - 2)180 = (5 - 2)180$
$= 3(180)$
$= 540$

The sum of the measures of the angles is 540°.

5. A nonagon has 9 sides. Therefore, $n = 9$.

$(n - 2)180 = (9 - 2)180$
$= 7(180)$
$= 1260$

The sum of the measures of the angles is 1260°.

6. A hexagon has 6 sides. Therefore, $n = 6$.

$(n - 2)180 = (6 - 2)180$
$= 4(180)$
$= 720$

The sum of the measures of the angles is 720°.

7. An octagon has 8 sides. Therefore, $n = 8$.

$(n - 2)180 = (8 - 2)180$
$= 6(180)$
$= 1080$

The sum of the measures of the angles is 1080°.

8. A 15-gon has 15 sides. Therefore, $n = 15$.

$(n - 2)180 = (15 - 2)180$
$= 13(180)$
$= 2340$

The sum of the measures of the angles is 2340°.

Page 749 Lesson 10-7

1. $C = 2\pi r$

$C = 2 \cdot \pi \cdot 5$

$C \approx 31.4$

The circumference is about 31.4 in.

$A = \pi r^2$

$A = \pi \cdot 5^2$

$A = \pi \cdot 25$

$A \approx 78.5$

The area is about 78.5 in^2.

2. $C = \pi d$

$C = \pi \cdot 9$

$C \approx 28.3$

The circumference is about 28.3 cm.

$A = \pi r^2$

$A = \pi \cdot 4.5^2$

$A = \pi \cdot 20.25$

$A \approx 63.6$

The area is about 63.6 cm^2.

3. $C = 2\pi r$

$C = 2 \cdot \pi \cdot 18$

$C \approx 113.1$

The circumference is about 113.1 ft.

$A = \pi r^2$

$A = \pi \cdot 18^2$

$A = \pi \cdot 324$

$A \approx 1017.9$

The area is about 1017.9 ft^2.

4. $C = \pi d$

$C = \pi \cdot 7.3$

$C \approx 22.9$

The circumference is about 22.9 cm.

$A = \pi r^2$

$A = \pi \cdot 3.65^2$

$A = \pi \cdot 13.3225$

$A \approx 41.9$

The area is about 41.9 cm^2.

5. $C = 2\pi r$

$C = 2 \cdot \pi \cdot 8.2$

$C \approx 51.5$

The circumference is about 51.5 ft.

$A = \pi r^2$

$A = \pi \cdot 8.2^2$

$A = \pi \cdot 67.24$

$A \approx 211.2$

The area is about 211.2 ft^2.

6. $C = \pi d$
$C = \pi \cdot 1.3$
$C \approx 4.1$
The circumference is about 4.1 yd.
$A = \pi r^2$
$A = \pi \cdot 0.65^2$
$A = \pi \cdot 0.4225$
$A \approx 1.3$
The area is about 1.3 yd^2.

7. $C = \pi d$
$C = \pi \cdot 5.2$
$C \approx 16.3$
The circumference is about 16.3 yd.
$A = \pi r^2$
$A = \pi \cdot 2.6^2$
$A = \pi \cdot 6.76$
$A \approx 21.2$
The area is about 21.2 yd^2.

8. $C = 2\pi r$
$C = 2 \cdot \pi \cdot 4.8$
$C \approx 30.2$
The circumference is about 30.2 cm.
$A = \pi r^2$
$A = \pi \cdot 4.8^2$
$A = \pi \cdot 23.04$
$A \approx 72.4$
The area is about 72.4 cm^2.

9. $C = \pi d$
$18.5 = \pi d$
$\frac{18.5}{\pi} = \frac{\pi d}{\pi}$
$5.9 \approx d$
The diameter is about 5.9 ft.

10. $A = \pi r^2$
$62.9 = \pi r^2$
$\frac{62.9}{\pi} = \frac{\pi r^2}{\pi}$
$20.0 = r^2$
$4.5 \approx r$
The radius is about 4.5 in.

Page 750 Lesson 10-8

1. Area of triangle:
$A = \frac{1}{2}bh$
$A = \frac{1}{2} \cdot 3 \cdot 5.5$
$A = 8.25$
Area of rectangle:
$A = bh$
$A = 8 \cdot 5.5$
$A = 44$
The area of the figure is 8.25 + 44 or about 52.3 ft^2.

2. Area of two semicircles:
$A = 2 \cdot \frac{1}{2}\pi r^2$
$A = 2 \cdot \frac{1}{2} \cdot \pi \cdot 3^2$
$A = \pi \cdot 9$
$A \approx 28.3$
Area of rectangle:
$A = bh$
$A = 6 \cdot 8$
$A = 48$
The area of the figure is 28.3 + 48 or about 76.3 cm^2.

3. First, separate the figure into two rectangles, 2.1 yd by 4.8 yd, and 3.2 yd by 6.4 yd.
Area of small rectangle:
$A = bh$
$A = (2.1)(4.8)$
$A = 10.08$
Area of large rectangle:
$A = bh$
$A = (3.2)(6.4)$
$A = 20.48$
The area of the figure is 10.08 + 20.48 or about 30.6 yd^2.

4. Area of triangle:
$A = \frac{1}{2}bh$
$A = \frac{1}{2} \cdot 12 \cdot 10$
$A = 60$
Area of semicircle:
$A = \frac{1}{2}\pi r^2$
$A = \frac{1}{2} \cdot \pi \cdot 5^2$
$A = \frac{1}{2} \cdot \pi \cdot 25$
$A \approx 39.3$
The area of the figure is 60 + 39.3 or about 99.3 in^2.

Page 750 Lesson 11-1

1. The figure has two parallel congruent bases that are rectangles, *ABCD* and *EFGH*, or *ADEH* and *BCGF*, or *ADFE* and *DCGH*, so the figure is a rectangular prism.
faces: *ABCD, ADHE, DHGC, BCGF, ABFE, EFGH*
edges: $\overline{AB}, \overline{BC}, \overline{CD}, \overline{DA}, \overline{EF}, \overline{FG}, \overline{GH}, \overline{HE}, \overline{AE}, \overline{DH}, \overline{CG}, \overline{BF}$
vertices: *A, B, C, D, E, F, G, H*

2. The figure has two parallel congruent bases that are triangles. *KHJ* and *NLM*, so it is a triangular prism.
faces: *HJK, LMN, HJML, KJMN, HLMK*
edges: $\overline{HJ}, \overline{JK}, \overline{KH}, \overline{LM}, \overline{MN}, \overline{NL}, \overline{JM}, \overline{HL}, \overline{KN}$
vertices: *I, J, K, L, M, N*

3. intersecting

4. $\overline{HK}$ or $\overline{LN}$

5. *HJK* and *LMN*

Page 750 Lesson 11-2

1. $V = \pi r^2 h$
$V = \pi(6)^2(15)$
$V \approx 1696.5$
The volume is about 1696.5 in^3.

2. $V = \ell \cdot w \cdot h$
$V = 10(3)(6)$
$V = 180$
The volume is 180 m^3.

3. $V = Bh$
$V = \left(\frac{1}{2} \cdot 4 \cdot 5\right)(8)$
$V = 80$
The volume is 80 ft^3.

4. $V = \ell \cdot w \cdot h$
$V = 25(4)(6)$
$V = 600$
The volume is 600 cm^3.

5. $V = \ell \cdot w \cdot h$
$V = 2.5(7)(12)$
$V = 210$
The volume is 210 yd^3.

6. $V = \pi r^2 h$
$V = \pi(4.6)^2(16)$
$V \approx 1063.6$
The volume is about 1063.6 mm^3.

7. $V = Bh$
$V = \frac{1}{2}(3.1)(1.7)(5)$
$V \approx 13.2$
The volume is about 13.2 cm^3.

8. $V = \ell \cdot w \cdot h$
$292.5 = 13(5)h$
$292.5 = 65h$
$4.5 = h$
The height is 4.5 in.

Page 751 Lesson 11-3

1. $V = \frac{1}{3}\pi r^2 h$
$V = \frac{1}{3}\pi(4)^2(10)$
$V \approx 167.6$
The volume is about 167.6 ft^3.

2. $V = \frac{1}{3}Bh$
$V = \frac{1}{3}(8)^2(12)$
$V = 256$
The volume is 256 cm^3.

3. $r = \frac{1}{2}(10)$
$r = 5$
$V = \frac{1}{3}\pi r^2 h$
$V = \frac{1}{3}\pi(5)^2(7)$
$V \approx 183.3$
The volume is about 183.3 yd^3.

4. $V = \frac{1}{3}Bh$
$V = \frac{1}{3} \cdot 6 \cdot 6 \cdot 9$
$V = 108$
The volume is 108 in^3.

5. $V = \frac{1}{3}Bh$
$V = \frac{1}{3}(3.25)^2(12)$
$V \approx 42.3$
The volume is about 42.3 ft^3.

6. $V = \frac{1}{3}\pi r^2 h$
$V = \frac{1}{3}\pi(3.6)^2(20)$
$V \approx 271.4$
The volume is about 271.4 cm^3.

7. $V = \frac{1}{3}Bh$
$V = \frac{1}{3}(185)(7)$
$V \approx 431.7$
The volume is about 431.7 m^3.

Page 751 Lesson 11-4

1. $S = 2\pi r^2 + 2\pi rh$
$S = 2\pi(6)^2 + 2\pi(6)(20)$
$S \approx 980.2$
The surface area is about 980.2 in^2.

2. $S = 2\ell w + 2\ell h + 2wh$
$S = 2(9)(5) + 2(9)(3) + 2(5)(3)$
$S = 174$
The surface area is 174 cm^2.

3. $S = 2\ell w + 2\ell h + 2wh$
$S = 2(6)(6) + 2(6)(6) + 2(6)(6)$
$S = 216$
The surface area is 216 ft^2.

4. $r = \frac{1}{2}(8)$
$r = 4$
$S = 2\pi r^2 + 2\pi rh$
$S = 2\pi \cdot 4^2 + 2\pi \cdot 4 \cdot 12$
$S \approx 402.1$
The surface area is about 402.1 m^2.

5. $S = 2\pi r^2 + 2\pi rh$
$S = 2\pi(2.5)^2 + 2\pi(2.5)(5)$
$S \approx 117.8$
The surface area is about 117.8 cm^2.

6. $S = 2\ell w + 2\ell h + 2wh$
$S = 2(4.9)(4.9) + 2(4.9)(4.9) + 2(4.9)(4.9)$
$S \approx 144.1$
The surface area is about 144.1 m^2.

7. $S = 2\ell w + 2\ell h + 2wh$
$S = 2(7.6)(8.4) + 2(7.6)(7) + 2(8.4)(7)$
$S \approx 351.7$
The surface area is 351.7 mm^2.

8. side $3 \times 10 = 30$
side $4 \times 10 = 40$
side $5 \times 10 = 50$
two bases $2\left(\frac{1}{2} \cdot 4 \cdot 3\right) = 12$
$S = 30 + 40 + 50 + 12$
$S = 132$
The surface area is 132 in^2.

Page 751 Lesson 11-5

1. Area of lateral face:

$A = \frac{1}{2} \cdot 6 \cdot 5.2$

$A = 15.6$

There are three faces, so the lateral area is 3(15.6) or 46.8 in^2.

Area of base:

$A = \frac{1}{2}b \cdot h$

$A = \frac{1}{2} \cdot 6 \cdot 5.2$

$A = 15.6$

$S = 46.8 + 15.6$

$S = 62.4$

The surface area is 62.4 in^2.

2. $S = \pi r\ell + \pi r^2$

$S = \pi(8)(15) + \pi(8)^2$

$S \approx 578.1$

The surface area is about 578.1 cm^2.

3. Area of lateral face:

$A = \frac{1}{2}b \cdot h$

$A = \frac{1}{2} \cdot 4 \cdot 9$

$A = 18$

There are four faces, so the lateral area is 4(18) or 72 ft^2.

Area of base:

$A = s^2$

$A = 4^2$

$A = 16$

$S = 72 + 16$

$S = 88$

The surface area is 88 ft^2.

4. $r = \frac{1}{2}(4.2)$

$r = 2.1$

$S = \pi r\ell + \pi r^2$

$S = \pi(2.1)(9.3) + \pi(2.1)^2$

$S \approx 75.2$

The surface area is about 75.2 m^2.

5. Area of lateral face:

$A = \frac{1}{2}b \cdot h$

$A = \frac{1}{2} \cdot 1.8 \cdot 3$

$A = 2.7$

There are four faces, so the lateral area is 4(2.7) or 10.8 mm^2.

Area of base:

$A = s^2$

$A = 1.8^2$

$A = 3.24$

$S = 10.8 + 3.24$

$S \approx 14.0$

The surface area is about 14.0 mm^2.

6. $S = \pi r\ell + \pi r^2$

$S = \pi(4)(7) + \pi(4)^2$

$S \approx 138.2$

The surface area is about 138.2 in^2.

7. $S = \pi r\ell + \pi r^2$

$S = \pi(7.6)(12.3) + \pi(7.6)^2$

$S \approx 475.1$

The surface area is about 475.1 cm^2.

Page 752 Lesson 11-6

1. $\frac{4}{3} \stackrel{?}{=} \frac{20}{10}$

$4(10) \stackrel{?}{=} 3(20)$

$40 \neq 60$

The corresponding measurements are not proportional, so the solids are not similar.

2. $\frac{8}{12} \stackrel{?}{=} \frac{4}{6}$

$8(6) \stackrel{?}{=} 12(4)$

$48 = 48$ ✓

The corresponding measurements are proportional, so the solids are similar.

3. $\frac{21}{x} = \frac{12}{4}$

$12x = 21(4)$

$x = 7$

The height is 7 feet.

4. $\frac{4}{10} = \frac{10}{x}$

$4x = 100$

$x = 25$

The slant height is 25 yd.

5. Sometimes; the radii and heights must be proportional.

6. Never; they are different shapes.

Page 752 Lesson 11-7

1. 3 significant digits
2. 1 significant digit
3. 2 significant digits
4. 4 significant digits
5. 2 significant digits
6. 2 significant digits
7. 4 significant digits
8. 3 significant digits
9. 3 significant digits

10. $32 + 16.9 = 48.9$

The least precise measurement, 32 yd, has 0 decimal places. So, round 48.9 to no decimal places, 49 yd.

11. $14.36 - 9.4 = 4.96$

The least precise measurement, 9.4 in., has 1 decimal place. So, round 4.96 to 1 decimal place, 5.0 in.

12. $20.86 - 0.375 = 20.485$

The least precise measurement, 20.86 cm, has 2 decimal places. So, round 20.485 to 2 decimal places, 20.49 cm.

13. $9.600 + 4.271 = 13.871$

The measurements all have 3 decimal places. So, round 13.871 to 3 decimal places, 13.871 m.

14. $\begin{array}{r} 8 \\ \times 6.2 \\ \hline 49.6 \end{array}$

The answer cannot have more significant digits than the measurement 8 ft. So, round 49.6 to 1 significant digit, 50 ft^2.

15. $\begin{array}{r} 7.50 \\ \times 3.01 \\ \hline 22.575 \end{array}$

The answer cannot have more significant digits than the measurements given. So, round 22.575 to 3 significant digits, 22.6 m^2.

16. $\begin{array}{r} 41.61 \\ 18.4 \\ \hline 60.01 \end{array}$ $\quad \begin{array}{r} 60.01 \\ -\ 3.65 \\ \hline 56.36 \end{array}$ $\quad \begin{array}{r} 56.36 \\ +\ 7.371 \\ \hline 63.731 \end{array}$

The least precise measurement, 18.4 in., has 1 decimal place. So, round 63.731 to 1 decimal place, 63.7 in.

Page 752 Lesson 12-1

1.

Stem	Leaf
3	2 7
4	4 9
5	3 9
6	1 9

5|3 = 53

2.

Stem	Leaf
0	3 5 8
1	
2	1 4 6
3	0 5 5 8 9

2|1 = 21

3.

Stem	Leaf
0	5 6
⋮	
7	3 4 9
⋮	
15	3 7

7|3 = 7.3

4.

Stem	Leaf
17	1 2 9
18	1 1 2 6
19	3 8

18|6 = 186

5.

Stem	Leaf
5	5
6	2 7 9
7	1 4 5
8	0 1

7|5 = 75

6.

Stem	Leaf
9	8
10	6 9
11	1 7
12	1 5 6
13	2
14	2

13|2 = 132

7.

Stem	Leaf
1	7
2	1 4
3	7
4	
5	4
6	9
7	7
8	6
9	2

8|6 = 86

8.

Stem	Leaf
5	4 9
6	1 7
7	3 5
8	2 3 9
9	3
10	2

6|7 = 6.7

9. The greatest value is 98.

10. The most values occur in the interval 80–89.

11. The median, or middle value, is $\frac{81 + 84}{2}$ or 82.5.

Page 753 Lesson 12-2

1. The greatest value is 96 and the least value is 23. So, the range is 96 − 23 or 73.

To find the interquartile range, first list the data from least to greatest. Then find the median and the upper and lower quartiles.

lower half | upper half

23, 35, 37, 44, 49, 61, 95, 96

$LQ = \frac{35 + 37}{2}$ or 36 | $UQ = \frac{61 + 95}{2}$ or 78

median

The interquartile range is 78 − 36 or 42.

2. The greatest value is 80 and the least value is 12. So, the range is 80 − 12 or 68.

To find the interquartile range, first list the data from least to greatest. Then find the median and the upper and lower quartiles.

lower half | upper half

12, 24, 30, 30, 35, 38, 39, 53, 62, 80

LQ (30) ↑ median ↑ UQ (53) ↑

The interquartile range is 53 − 30 or 23.

3. The greatest value is 30.1 and the least value is 3.2. So, the range is 30.1 − 3.2 or 26.9.

To find the interquartile range, first list the data from least to greatest. Then find the median and the upper and lower quartiles.

lower half / upper half

3.2, 4.7, 5.3, 6, 7.15, 9.19, 30.1

↑ LQ (4.7) ↑ median (6) ↑ UQ (9.19)

The interquartile range is 9.19 − 4.7 or 4.49.

4. The greatest value is 891 and the least value is 181. So, the range is 891 − 181 or 710.

To find the interquartile range, first list the data from least to greatest. Then find the median and the upper and lower quartiles.

lower half / upper half

181, 193, 271, 711, 791, 818, 861, 891

$LQ = \frac{193 + 271}{2}$ or 232 | $UQ = \frac{818 + 861}{2}$ or 839.5

median

The interquartile range is 839.5 − 232 or 607.5.

5. The greatest value is 59 and the least value is 20. So, the range is 59 − 20 or 39.

The median is 38. The lower quartile is $\frac{24 + 27}{2}$ or 25.5. The upper quartile is $\frac{45 + 47}{2}$ or 46. So, the interquartile range is 46 − 25.5 or 20.5.

6. The greatest value is 95 and the least value is 40. So, the range is 95 − 40 or 55.

The median is 67. The lower quartile is 47 and the upper quartile is 79. So, the interquartile range is 79 − 47 or 32.

Page 753 Lesson 12-3

1. The median is 43. The extremes are 8 and 103. The lower quartile is 26 and the upper quartile is 69.

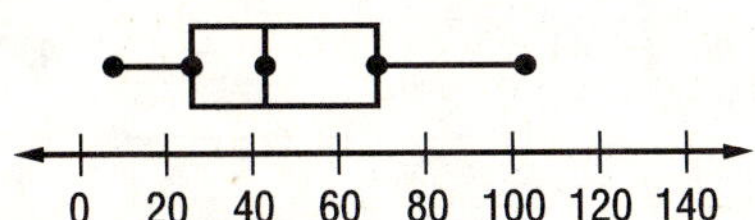

2. The median is 30.5. The extremes are 19 and 48. The lower quartile is 26 and the upper quartile is 37.

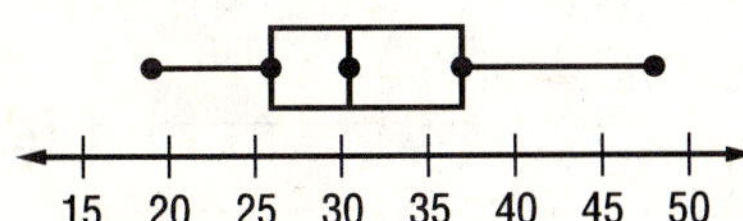

3. The median is 215. The extremes are 109 and 327. The lower quartile is 134 and the upper quartile is 278.

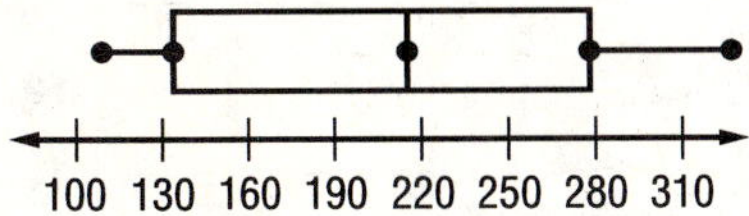

4. The median is 365. The extremes are 126 and 578. The lower quartile is 288.5 and the upper quartile is 434.5.

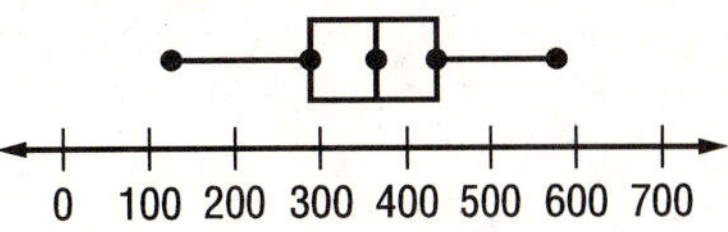

5. The tallest player is 68 inches tall.

6. The lower quartile is 56 and the upper extreme is 68. So, 75% of the players are between 56 and 68 inches tall.

7. How much the values are spread out.

Page 753 Lesson 12-4

1.

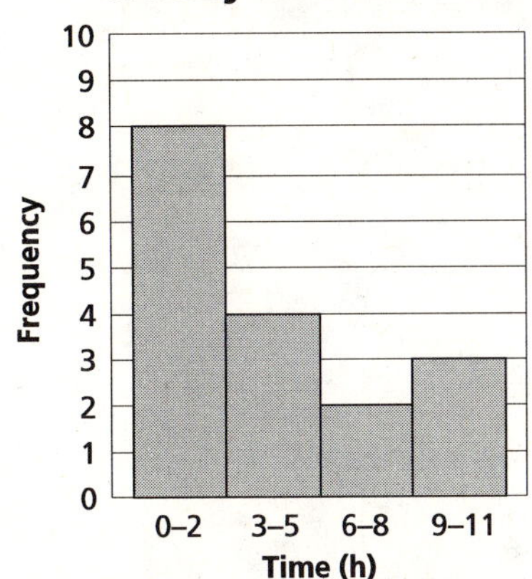

2.

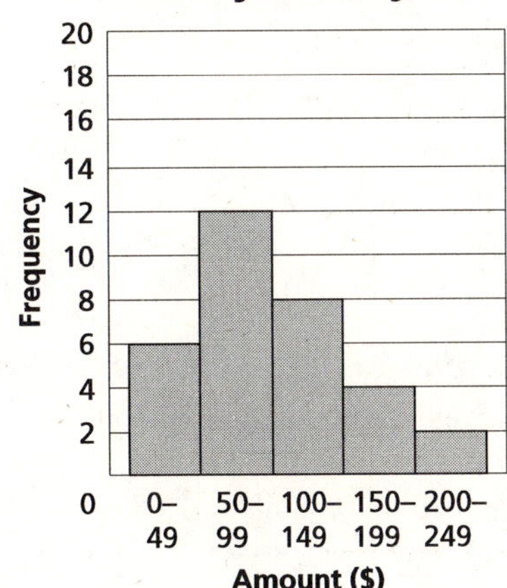

3.

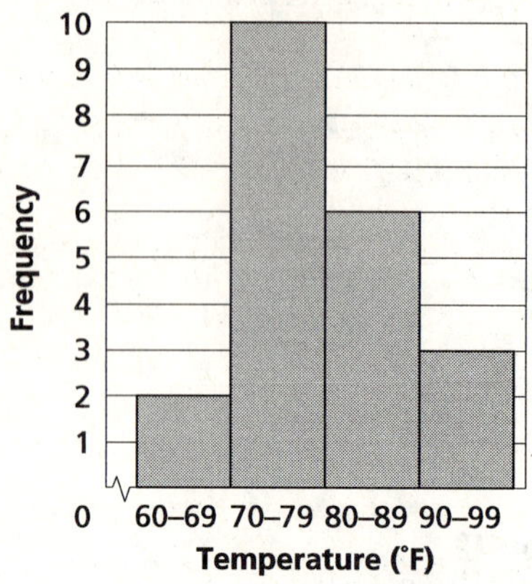

4.

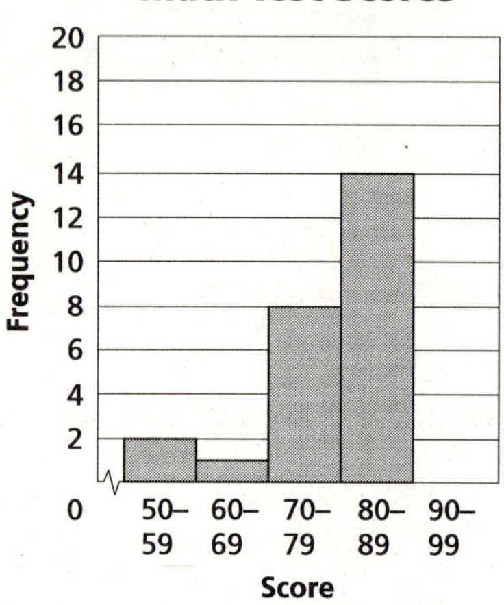

Page 754 Lesson 12-5

1. Different scales are used.
2. Graph A; it shows a less dramatic increase and decrease.

Page 754 Lesson 12-6

1. Use the Fundamental Counting Principle.

choices of metal		choices of weight		choices of shape		total possible rings
3	·	3	·	4	=	36

There are 36 possible engagement rings.

2. Use the Fundamental Counting Principle.

choices of length		choices in color		choices for shoulders		total possible dresses
4	·	2	·	2	=	16

There are 16 possible dresses.

3. Use the Fundamental Counting Principle. There is only one choice for the first digit and only one choice for the last digit. Each of the other digits have 10 choices, 0 to 9.

$1 \cdot 10 \cdot 10 \cdot 10 \cdot 10 \cdot 10 \cdot 1 = 100{,}000$

There are 100,000 possible phone numbers.

4. Use the Fundamental Counting Principle.

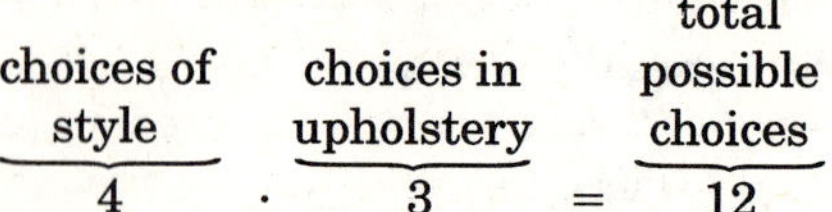

There are 12 possible chairs.

5. First find the number of outcomes.

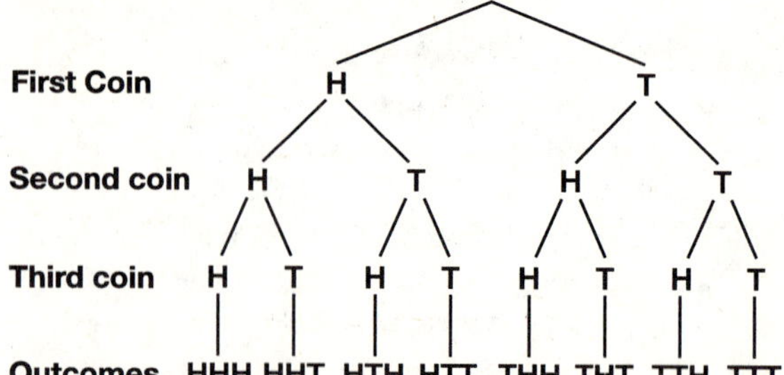

There are 8 possible outcomes. Look at the tree diagram. There is only 1 outcome that has three tails.

$P(\text{three tails}) = \frac{1}{8}$

The probability of three tails is $\frac{1}{8}$.

6. There are $6 \cdot 6$ or 36 possible outcomes from the roll of a number cube.

There are 18 odd sums. So, the probability of getting an odd sum is $\frac{18}{36}$ or $\frac{1}{2}$.

7. First find the number of outcomes.

Die	Coin	Outcomes
1	H	1, H
	T	1, T
2	H	2, H
	T	2, T
3	H	3, H
	T	3, T
4	H	4, H
	T	4, T
5	H	5, H
	T	5, T
6	H	6, H
	T	6, T
7	H	7, H
	T	7, T
8	H	8, H
	T	8, T
9	H	9, H
	T	9, T
10	H	10, H
	T	10, T

There are 20 possible outcomes. Look at the tree diagram. There are 7 outcomes in which the coin lands on tails and the die lands on a number greater than 3.

$P(\text{tails and number greater than 3}) = \frac{7}{20}$

The probability is $\frac{7}{20}$.

Page 754 Lesson 12-7

1. combination; $C(7, 4) = \frac{P(7, 4)}{4!} = \frac{7 \cdot 6 \cdot 5 \cdot 4}{4 \cdot 3 \cdot 2 \cdot 1}$ or 35 ways
2. permutation; $P(6, 6) = 6 \cdot 5 \cdot 4 \cdot 3 \cdot 2 \cdot 1$ or 720 ways
3. combination; $C(5, 3) = \frac{P(5, 3)}{3!} = \frac{5 \cdot 4 \cdot 3}{3 \cdot 2 \cdot 1}$ or 10 ways
4. permutation; $P(6, 6) = 6 \cdot 5 \cdot 4 \cdot 3 \cdot 2 \cdot 1$ or 720 ways
5. combination; $C(12, 3) = \frac{P(12, 3)}{3!} = \frac{12 \cdot 11 \cdot 10}{3 \cdot 2 \cdot 1}$ or 220 ways
6. There are 5 vertices on a pentagon. Since the order of the line segments is not important, this is a combination of 5 vertices taken 2 at a time.
   $C(5, 2) = \frac{P(5, 2)}{2!} = \frac{5 \cdot 4}{2 \cdot 1}$ or 10
   There are 10 lines segments that can be drawn between any two vertices of a pentagon.

Page 755 Lesson 12-8

1. There are 6 spins possible. There is 1 way to spin blue and there are 5 ways to spin a color that is not blue.
   Odds of spinning blue = 1:5
2. There are 6 spins possible. There are 2 colors whose names contain less than 5 letters, *red* and *blue*, so there are 6 − 2 or 4 colors with 5 or more letters.
   Odds of spinning a color whose name contains less than 5 letters = 2:4 or 1:2
3. There are 6 spins possible. There are 5 colors whose names begin with a consonant, and there is 1 color whose name does not begin with a consonant.
   Odds of spinning a color whose name begins with a consonant = 5:1 or 6:0 if *y* is considered a consonant.
4. There are 6 spins possible. There are 3 colors that are red, yellow, or blue, and there are 6 − 3 or 3 colors that are not red, yellow, or blue.
   Odds of spinning red, yellow, or blue = 3:3 or 1:1
5. There are 10 rolls possible. There are 6 rolls that are less than 7, so there are 10 − 6 or 4 rolls that are not less than 7.
   Odds of rolling a number less than 7 = 6:4 or 3:2
6. There are 10 rolls possible. There are 5 rolls that are odd numbers, and there are 5 rolls that are numbers that are not odd.
   Odds of rolling an odd number = 5:5 or 1:1
7. There are 10 rolls possible. There are 5 rolls that are composite numbers, 4, 6, 8, 9, and 10. There are 10 − 5 or 5 rolls that are numbers that are not composite.
   Odds of rolling a composite number = 5:5 or 1:1
8. There are 10 rolls possible. There are 3 rolls that are divisible by 3, and there are 10 − 3 or 7 rolls that are not divisible by 3.
   Odds of rolling a number divisible by 3 = 3:7

Page 755 Lesson 12-9

1. $P(\text{3 nines in a row}) = \frac{4}{24} \cdot \frac{3}{23} \cdot \frac{2}{22}$
   $= \frac{24}{12{,}144}$ or $\frac{1}{506}$
2. $P(\text{black jack, then red queen}) = \frac{2}{24} \cdot \frac{2}{23}$
   $= \frac{4}{552}$ or $\frac{1}{138}$
3. $P(\text{nine of clubs, black king, red ace})$
   $= \frac{1}{24} \cdot \frac{2}{23} \cdot \frac{2}{22}$
   $= \frac{4}{12{,}144}$ or $\frac{1}{3036}$
4. $P(\text{4 face cards in a row}) = \frac{12}{24} \cdot \frac{11}{23} \cdot \frac{10}{22} \cdot \frac{9}{21}$
   $= \frac{11{,}880}{255{,}024}$
   $= \frac{15}{322}$
5. There are 14 cards from 6 to 19. There are 7 cards that are even.
   $P(\text{13 or even}) = P(13) + P(\text{even})$
   $= \frac{1}{14} + \frac{7}{14}$
   $= \frac{8}{14}$ or $\frac{4}{7}$
6. There are 14 cards from 6 to 19.
   $P(\text{13 or less than 7}) = P(13) + P(\text{less than 7})$
   $= \frac{1}{14} + \frac{1}{14}$
   $= \frac{2}{14}$ or $\frac{1}{7}$
7. There are 14 cards from 6 to 19. There are 7 even cards and 7 odd cards.
   $P(\text{even or odd}) = P(\text{even}) + P(\text{odd})$
   $= \frac{7}{14} + \frac{7}{14}$
   $= \frac{14}{14}$ or 1
8. There are 14 cards from 6 to 19. There are no cards greater than 20.
   $P(\text{14 or greater than 20})$
   $= P(14) + P(\text{greater than 20})$
   $= \frac{1}{14} + \frac{0}{14}$
   $= \frac{1}{14}$
9. There are 14 cards from 6 to 19. The events are not mutually exclusive. There are 9 cards that are even or less than 10: 6, 7, 8, 9, 10, 12, 14, 16, and 18.
   $P(\text{even or less than 10}) = \frac{9}{14}$

10. There are 14 cards from 6 to 19. The events are not mutually exclusive. There are 11 cards that are odd or greater than 10: 7, 9, 11, 12, 13, 14, 15, 16, 17, 18, 19.

$P(\text{odd or greater than } 10) = \frac{11}{14}$

Page 755 Lesson 13-1

1. Yes; it has one term, so it is a monomial.
2. No; there is a variable in the denominator of $\frac{6}{x}$.
3. Yes; it has one term, so it is a monomial.
4. Yes; it has three terms, so it is a trinomial.
5. No; there is a variable under the radical sign of $\sqrt{w}$.
6. Yes; it has three terms, so it is a trinomial.
7. Yes; it has one term, so it is a monomial.
8. Yes; it has two terms, so it is a binomial.
9. No; there is a variable in the denominator of $\frac{x}{y}$.
10. The degree of a nonzero constant is zero, so the degree of 38 is 0.
11.

term	degree
$4b$	1
9	0

The greatest degree is 1. So the degree of $4b + 9$ is 1.

12. c has degree 1 and d has degree 1. The degree of cd is 1 + 1 or 2.
13. The variable x has degree 1, so the degree of $4x$ is 1.
14.

term	degree
a^2	2
-6	0

The greatest degree is 2. So the degree of $a^2 - 6$ is 2.

15.

term	degree
$11r$	1
$5s$	1

The greatest degree is 1. So the degree of $11r + 5s$ is 1.

16. x^2 has degree 2 and y has degree 1. The degree of x^2y is 2 + 1 or 3.
17.

term	degree
n^2	2
$-n$	1

The greatest degree is 2. So the degree of $n^2 - n$ is 2.

18. a^2 has degree 2 and b^2 has degree 2. The degree of $6a^2b^2$ is 2 + 2 or 4.
19.

term	degree
$3y^2$	2
-2	0

The greatest degree is 2. So the degree of $3y^2 - 2$ is 2.

20.

term	degree
$9cd^3$	4
-5	0

The greatest degree is 4. So the degree of $9cd^3 - 5$ is 4.

21.

term	degree
$-5p^3$	3
$8q^2$	2

The greatest degree is 3. So the degree of $-5p^3 + 8q^2$ is 3.

22.

term	degree
w^2	2
$2x$	1
$-3y^3$	3
$-7z$	1

The greatest degree is 3. So the degree of $w^2 + 2x - 3y^3 - 7z$ is 3.

23.

term	degree
$\frac{x^3}{6}$	3
$-x$	1

The greatest degree is 3. So the degree of $\frac{x^3}{6} - x$ is 3.

24.

term	degree
$-17n^2p$	3
$-11np^3$	4

The greatest degree is 4. So the degree of $-17n^2p - 11np^3$ is 4.

Page 756 Lesson 13-2

1. $$\begin{array}{rr} & -6m + 7 \\ (+) & \underline{9m - 2} \\ & 3m + 5 \end{array}$$

2. $$\begin{array}{rr} & 12y - 4 \\ (+) & \underline{-8y + 9} \\ & 4y + 5 \end{array}$$

3. $$\begin{array}{rr} & 5x + y \\ (+) & \underline{9x - 2y} \\ & 14x - y \end{array}$$

4. $$\begin{array}{rr} & 7c^2 - 10c + 5 \\ (+) & \underline{4c^2 - 4c - 8} \\ & 11c^2 - 14c - 3 \end{array}$$

5. $\begin{array}{rl} & 2a^2 + 5ab + 6b^2 \\ (+) & \underline{3a^2 \qquad\quad - b^2} \\ & 5a^2 + 5ab + 5b^2 \end{array}$

6. $\begin{array}{rl} & 3d^3 + 2d^2 + 6d - 4 \\ (+) & \underline{\qquad - 4d^2 \qquad - 3} \\ & 3d^3 - 2d^2 + 6d - 7 \end{array}$

7. $(3a + 4) + (a + 2) = (3a + a) + (4 + 2)$
$= 4a + 6$

8. $(8m - 3) + (4m + 1) = (8m + 4m) + (-3 + 1)$
$= 12m - 2$

9. $(5x - 3y) + (2x - y) = (5x + 2x) + (-3y - y)$
$= 7x - 4y$

10. $(8p^2 - 2p + 3) + (-3p^2 - 2)$
$= (8p^2 - 3p^2) - 2p + (3 - 2)$
$= 5p^2 - 2p + 1$

11. $(-11r^2 + 3s) + (5r^2 - s) = (-11r^2 + 5r^2) + (3s - s)$
$= -6r^2 + 2s$

12. $(3a^2 + 5a + 1) + (2a^2 - 3a - 6)$
$= (3a^2 + 2a^2) + (5a - 3a) + (1 - 6)$
$= 5a^2 + 2a - 5$

13. $(3m - 5n) + (-6m + 8n)$
$= (3m - 6m) + (-5n + 8n)$
$= -3m + 3n$
when $m = -2$ and $n = 4$,
$-3m + 3n = -3(-2) + 3(4)$
$= 6 + 12$
$= 18$

14. $(m^2 + 2p^2) + (-4m^2 - 6p^2)$
$= (m^2 - 4m^2) + (2p^2 - 6p^2)$
$= -3m^2 - 4p^2$
when $m = -2$ and $p = 3$,
$-3m^2 - 4p^2 = -3(-2)^2 - 4(3)^2$
$= -3 \cdot 4 - 4 \cdot 9$
$= -12 - 36$
$= -48$

15. $(-2m + 3n + 4p) + (5m - 6n - 8p)$
$= (-2m + 5m) + (3n - 6n) + (4p - 8p)$
$= 3m - 3n - 4p$
when $m = -2$, $n = 4$, and $p = 3$,
$3m - 3n - 4p = 3(-2) - 3(4) - 4(3)$
$= -6 - 12 - 12$
$= -30$

Page 756 Lesson 13-3

1. $\begin{array}{rl} & 2a + 7 \\ (-) & \underline{a + 3} \\ & a + 4 \end{array}$

2. $\begin{array}{rl} & -3k^2 + 6k \\ (-) & \underline{4k^2 + k} \\ & -7k^2 + 5k \end{array}$

3. The additive inverse of $5x^2 + 5x - 4$ is $-5x^2 - 5x + 4$.
$\begin{array}{rlcrl} & 6x^2 - 4x + 11 & & & 6x^2 - 4x + 11 \\ (-) & \underline{5x^2 + 5x - 4} & \rightarrow & (+) & \underline{-5x^2 - 5x + 4} \\ & & & & x^2 - 9x + 15 \end{array}$

4. The additive inverse of $2r^2 + 3r - 7$ is $-2r^2 - 3r + 7$.
$\begin{array}{rlcrl} & 9r^2 \qquad + 1 & & & 9r^2 \qquad + 1 \\ (-) & \underline{2r^2 + 3r - 7} & \rightarrow & (+) & \underline{-2r^2 - 3r + 7} \\ & & & & 7r^2 - 3r + 8 \end{array}$

5. The additive inverse of $4n^2 + 2mn$ is $-4n^2 - 2mn$.
$\begin{array}{rlcrl} & 8n^2 + 3mn - 9 & & & 8n^2 + 3mn - 9 \\ (-) & \underline{4n^2 + 2mn} & \rightarrow & (+) & \underline{-4n^2 - 2mn} \\ & & & & 4n^2 + mn - 9 \end{array}$

6. The additive inverse of $-10ab + 6a^2$ is $10ab - 6a^2$.
$\begin{array}{rlcrl} & -5b^2 - 2ab & & & -5b^2 - 2ab \\ (-) & \underline{-10ab + 6a^2} & \rightarrow & (+) & \underline{10ab - 6a^2} \\ & & & & -5b^2 + 8ab - 6a^2 \end{array}$

7. $(3n + 2) - (n + 1) = (3n + 2) + (-n - 1)$
$= (3n - n) + (2 - 1)$
$= 2n + 1$

8. $(-3c + 2d) - (7c - 6d) = (-3c + 2d) + (-7c + 6d)$
$= (-3c - 7c) + (2d + 6d)$
$= -10c + 8d$

9. $(4x^2 + 1) - (3x^2 - 4) = (4x^2 + 1) + (-3x^2 + 4)$
$= (4x^2 - 3x^2) + (1 + 4)$
$= x^2 + 5$

10. $(5a - 4b) - (-a + b) = (5a - 4b) + (a - b)$
$= (5a + a) + (-4b - b)$
$= 6a - 5b$

11. $(-12a + 9b) - (3a - 7b)$
$= (-12a + 9b) + (-3a + 7b)$
$= (-12a - 3a) + (9b + 7b)$
$= -15a + 16b$

12. $(3w^3 + 5w - 6) - (5w^3 - 2w + 5)$
$= (3w^3 + 5w - 6) + (-5w^3 + 2w - 5)$
$= (3w^3 - 5w^3) + (5w + 2w) + (-6 - 5)$
$= -2w^3 + 7w - 11$

13. $(2t^2 - 5) - (t + 8) = (2t^2 - 5) + (-t - 8)$
$= 2t^2 - t + (-5 - 8)$
$= 2t^2 - t - 13$

14. $(x^2 + xy - 9y^2) - (3x^2 - xy + 3y^2)$
$= (x^2 + xy - 9y^2) + (-3x^2 + xy - 3y^2)$
$= (x^2 - 3x^2) + (xy + xy) + (-9y^2 - 3y^2)$
$= -2x^2 + 2xy - 12y^2$

Page 756 Lesson 13-4

1. $2(3a - 7) = 2(3a) - 2(7)$
$= 6a - 14$

2. $(8c + 1)4 = 8c(4) + 1(4)$
$= 32c + 4$

3. $n(5n + 6) = n(5n) + n(6)$
$= 5n^2 + 6n$

4. $t(2 - t) = t(2) - t(t)$
$= 2t - t^2$

5. $(3k - 5)k = 3k(k) - 5(k)$
$= 3k^2 - 5k$

6. $(a + b)a = a(a) + a(b)$
$= a^2 + ab$

7. $4n(5n - 3) = 4n(5n) - 4n(3)$
$= 20n^2 - 12n$

8. $-3x(4 - x) = -3x(4) + (-3x)(-x)$
$= -12x + 3x^2$

9. $6m(-m^2 + 3) = -6m(m^2) + 6m(3)$
$= -6m^3 + 18m$

10. $5(3x - 2) = 5(3x) - 5(2)$
$= 15x - 10$

11. $(2p + 9)8 = 2p(8) + 9(8)$
$= 16p + 72$

12. $m(3m - 4) = m(3m) - m(4)$
$= 3m^2 - 4m$

13. $-2w(6 - w) = -2w(6) + (-2w)(-w)$
$= -12w + 2w^2$

14. $ab(a + b) = ab(a) + ab(b)$
$= a^2b + ab^2$

15. $7t(-3t + 4w) = 7t(-3t) + 7t(4w)$
$= -21t^2 + 28tw$

16. $4x(2x + y) = 4x(2x) + 4x(y)$
$= 8x^2 + 4xy$

17. $(c^2 - 3d)2c = c^2(2c) - 3d(2c)$
$= 2c^3 - 6cd$

18. $-5z(z^2 - 9z) = -5z(z^2) + (-5z)(-9z)$
$= -5z^3 + 45z^2$

19. $-5x(2x^2 - 3x + 1)$
$= -5x(2x^2) + (-5x)(-3x) - 5x(1)$
$= -10x^3 + 15x^2 - 5x$

20. $7r(r^2 - 3r + 7) = 7r(r^2) - 7r(3r) + 7r(7)$
$= 7r^3 - 21r^2 + 7r^2$

21. $-3az(2z^2 + 4az + a^2)$
$= -3az(2z^2) - 3az(4az) - 3az(a^2)$
$= -6az^3 - 12a^2z^2 - 3a^3z$

Page 757 Lesson 13-5

1. Nonlinear; the graph is a curve.
2. Linear; the graph is a straight line.
3. Nonlinear; the graph is a curve.
4. Linear; the graph is a straight line.
5. Linear; the equation can be written as $y = -3x + 0$.
6. Nonlinear; the equation cannot be written as $y = mx + b$.
7. Linear; the equation can be written as $y = \frac{2}{5}x + 2$.
8. Linear; the equation can be written as $y = \frac{1}{7}x + 0$.
9. Nonlinear; the equation cannot be written as $y = mx + b$.
10. Nonlinear; the equation cannot be written as $y = mx + b$.
11. Linear; as x increases by 2, y increases by 2.
12. Linear; as x increases by 5, y increases by 6.

Page 757 Lesson 13-6

1.

x	$y = 3x^2$	(x, y)
-2	$3(-2)^2 = 12$	$(-2, 12)$
-1	$3(-1)^2 = 3$	$(-1, 3)$
0	$3(0)^2 = 0$	$(0, 0)$
1	$3(1)^2 = 3$	$(1, 3)$
2	$3(2)^2 = 12$	$(2, 12)$

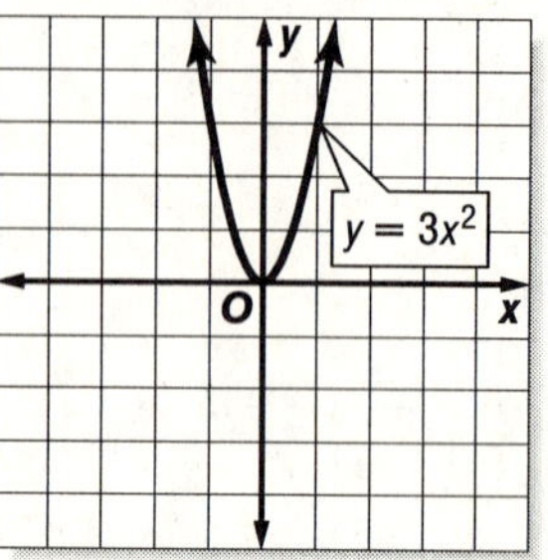

2.

x	$y = -2x^2$	(x, y)
-2	$-2(-2)^2 = -8$	$(-2, -8)$
-1	$-2(-1)^2 = -2$	$(-1, -2)$
0	$-2(0)^2 = 0$	$(0, 0)$
1	$-2(1)^2 = -2$	$(1, -2)$
2	$-2(2)^2 = -8$	$(2, -8)$

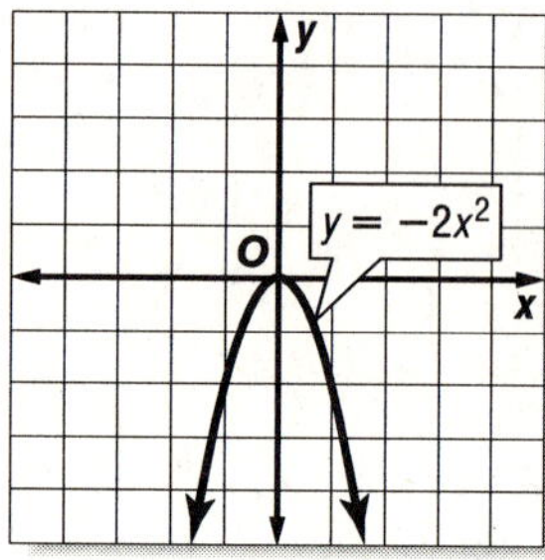

3.

x	$y = \frac{1}{2}x^2$	(x, y)
-2	$\frac{1}{2}(-2)^2 = 2$	$(-2, 2)$
-1	$\frac{1}{2}(-1)^2 = 0.5$	$(-1, 0.5)$
0	$\frac{1}{2}(0)^2 = 0$	$(0, 0)$
1	$\frac{1}{2}(1)^2 = 0.5$	$(1, 0.5)$
2	$\frac{1}{2}(2)^2 = 2$	$(2, 2)$

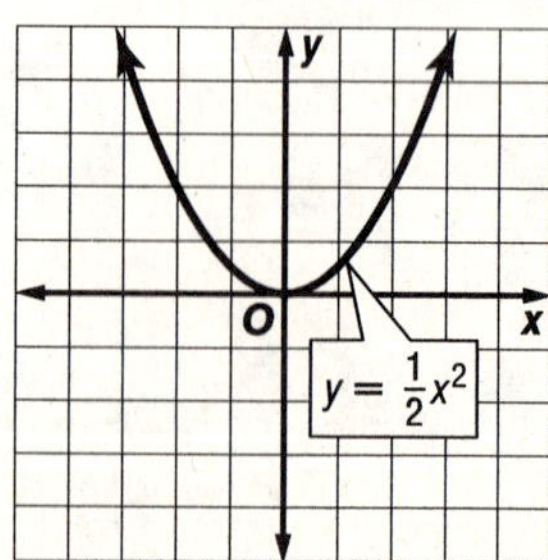

4.

x	$y = x^3$	(x, y)
−2	$(-2)^3 = -8$	(−2, −8)
−1	$(-1)^3 = -1$	(−1, −1)
0	$(0)^3 = 0$	(0, 0)
1	$(1)^3 = 1$	(1, 1)
2	$(2)^3 = 8$	(2, 8)

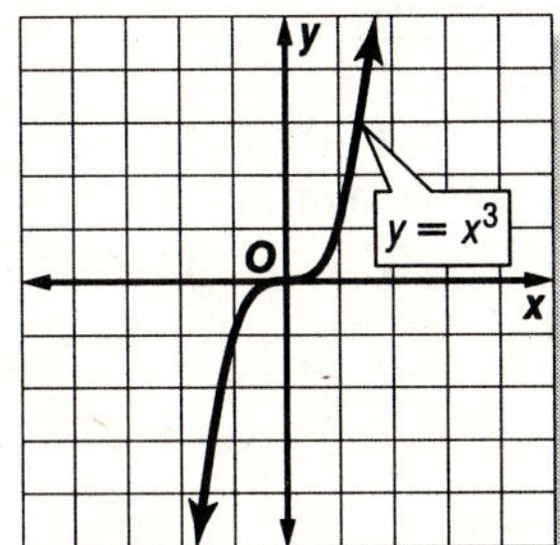

5.

x	$y = 0.3x^3$	(x, y)
−2	$0.3(-2)^3 = -2.4$	(−2, −2.4)
−1	$0.3(-1)^3 = -0.3$	(−1, −0.3)
0	$0.3(0)^3 = 0$	(0, 0)
1	$0.3(1)^3 = 0.3$	(1, 0.3)
2	$0.3(2)^3 = 2.4$	(2, 2.4)

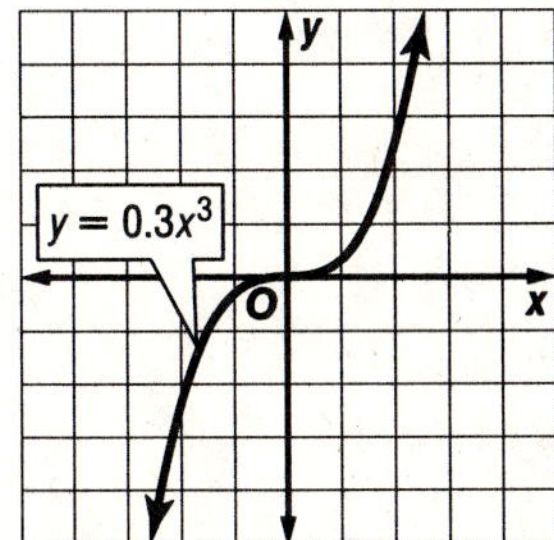

6.

x	$y = x^3 - 2$	(x, y)
−2	$(-2)^3 - 2 = -10$	(−2, −10)
−1	$(-1)^3 - 2 = -3$	(−1, −3)
0	$(0)^3 - 2 = -2$	(0, −2)
1	$(1)^3 - 2 = -1$	(1, −1)
2	$(2)^3 - 2 = 6$	(2, 6)

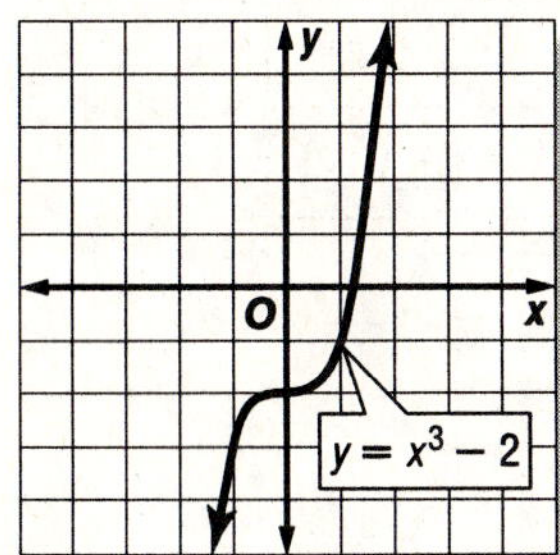

7.

x	$y = x^2 + 4$	(x, y)
−2	$(-2)^2 + 4 = 8$	(−2, 8)
−1	$(-1)^2 + 4 = 5$	(−1, 5)
0	$(0)^2 + 4 = 4$	(0, 4)
1	$(1)^2 + 4 = 5$	(1, 5)
2	$(2)^2 + 4 = 8$	(2, 8)

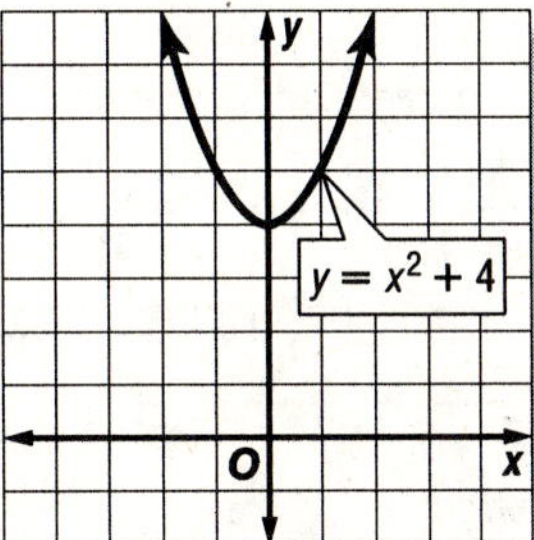

8.

x	$y = \frac{1}{3}x^2 - 3$	(x, y)
−2	$\frac{1}{3}(-2)^2 - 3 \approx -1.7$	(−2, −1.7)
−1	$\frac{1}{3}(-1)^2 - 3 \approx -2.7$	(−1, −2.7)
0	$\frac{1}{3}(0)^2 - 3 = -3$	(0, −3)
1	$\frac{1}{3}(1)^2 - 3 \approx -2.7$	(1, −2.7)
2	$\frac{1}{3}(2)^2 - 3 \approx -1.7$	(2, −1.7)

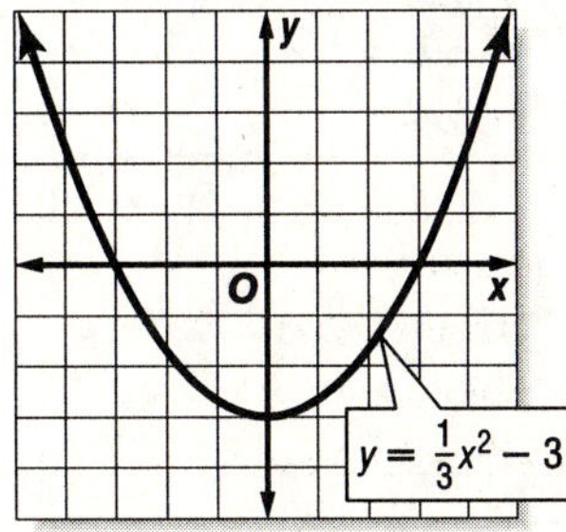

9.

x	$y = -0.5x^2 + 1$	(x, y)
−2	$-0.5(-2)^2 + 1 = -1$	(−2, −1)
−1	$-0.5(-1)^2 + 1 = 0.5$	(−1, 0.5)
0	$-0.5(0)^2 + 1 = 1$	(0, 1)
1	$-0.5(1)^2 + 1 = 0.5$	(1, 0.5)
2	$-0.5(2)^2 + 1 = -1$	(2, −1)

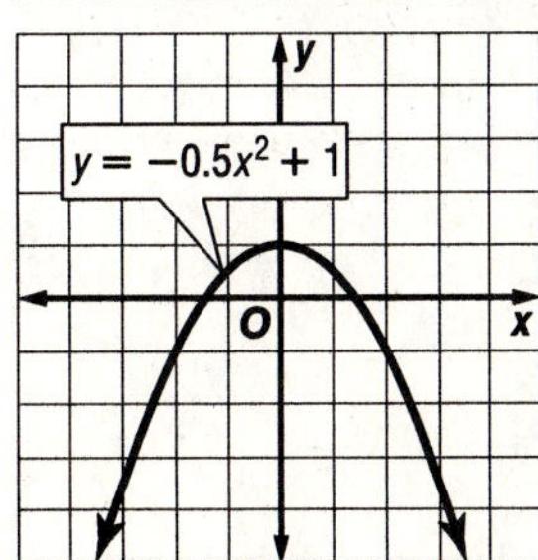

Mixed Problem Solving

Page 758 Chapter 1 The Tools of Algebra

1. Find the pattern.

Figure	1	2	3
Cubes	1	5	9

+4 +4

Each consecutive figure increases by 4 cubes. Assume the pattern continues.

Figure	3	4	5	6	7	8	9	10
Cubes	9	13	17	21	25	29	33	37

+4 +4 +4 +4 +4 +4 +4

There are 37 cubes in the tenth figure.

2. Estimate. $6.9 \text{ million} \div 262 \approx 7 \text{ million} \div 250$
$\approx 28{,}000$ mi

The shuttle traveled about 28,000 miles on each trip around the Earth.

3. \$35 = the cost of each sapling
\$7 = shipping and handling per order
Let x = the number of saplings.
So, $35(x) + 7$ = the total cost per order.
Substitute $x = 6$. $35(6) + 7 = 210 + 7$
$= 217$
The total cost for an order of six saplings is \$217.

4. 15 white chocolate hearts at \$4.25 each and 36 milk chocolate hearts at \$3.75 each and 22 milk chocolate assortments at \$7.45 each

$15 \times 4.25 \quad + \quad 36 \times 3.75 \quad + \quad 22 \times 7.45$

Expression:
$(15 \times 4.25) + (36 \times 3.75) + (22 \times 7.45)$
$= 63.75 + 135.00 + 163.90$
$= 362.65$
Sophia raised \$362.65.

5. Variable: Let x = the weight of an object on the moon.
Words: Six times more than on the moon
Expression: $6x$

6. Use the expression $6x$ to find the weight of the scientific instrument on Earth.
$6x = 6 \cdot 34$
$= 204$
The weight of the scientific instrument on Earth is 204 lbs.

7. volleyball net = 3 ft + 3 in.
distance from the floor to the bottom of the net = 4 ft + 8 in.
distance from the floor to the top of the net = (3 ft + 3 in.) + (4 ft + 8 in.)

8. (3 ft + 3 in.) + (4 ft + 8 in.)
= (3 ft + 4 ft) + (3 in. + 8 in.)
= 7 ft + 11 in.
The distance from the floor to the top of the net is 7 ft 11 in.

9. Let x = the number of papers Nick sold on Thursday.
Equation: $86 + 79 + 68 + x + 83 = 391$
$316 + x = 391$
$316 + 75 = 391$
$x = 75$
Nick sold 75 newspapers on Thursday.

10. Sample answer:
x = the number of blueberry bagels
$y = \frac{1}{2}x$ = the number of plain bagels

x	y
6	3
8	4
10	5
12	6

The relation is {(6, 3), (8, 4), (10, 5), (12, 6)}.

11. x = depth
$y = 35x + 20$ = degrees Celsius

x	$y = 35x + 20$
0	20
2	90
4	160

The ordered pairs are (0, 20), (2, 90), and (4, 160).

12.

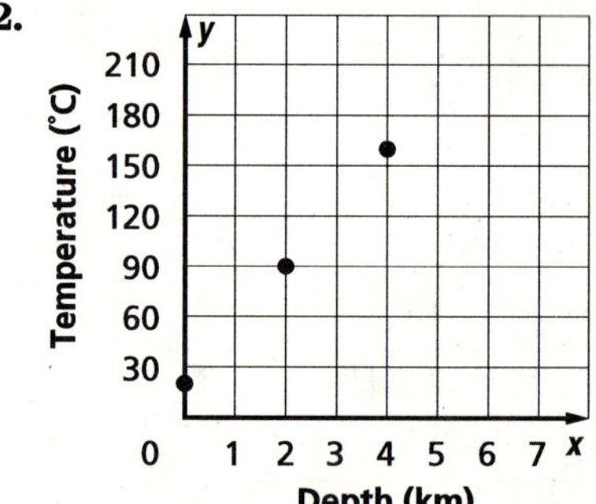

13. Positive; as experience increases, salaries increase.

14.

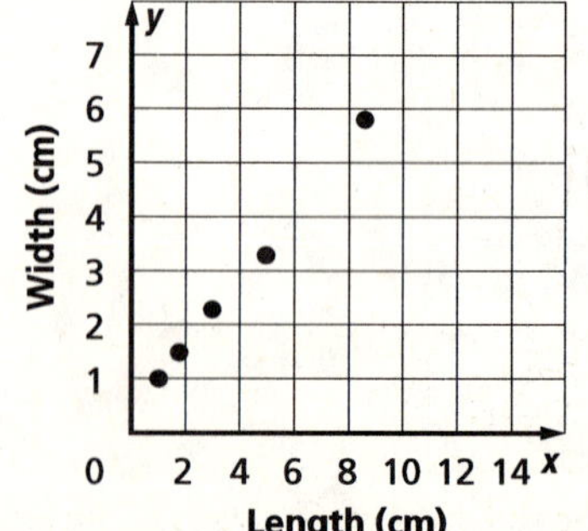

By looking at the pattern in the graph, we can predict that an egg 6 centimeters long will be about 4 centimeters wide.

Page 759 Chapter 2 Integers

1. $228 > 150$; $150 < 228$
2. $-200 > -500$; $-500 < -200$
3. $250 - 72 - 37 - 119 + 45$
   $= 250 + 45 - 72 - 37 - 119$
   $= (250 + 45) + (-72) + (-37) + (-119)$
   $= 295 + [-(72 + 37 + 119)]$
   $= 295 + [-228]$
   $= 67$
   Tino's balance was $67.
4. $112 - 252 = 112 + (-252)$
   $= -140$
   The average temperature of the surface of the moon is −140°C at night.
5. $-1500 + (-1250) = -2750$
   The ocean floor is 2750 meters below sea level.
6. $(-25) - (10) = -35$
   The difference is −35°F.
7. $5895 - (-155) = 5895 + 155$
   $= 6050$
   The difference in altitude is 6050 meters.
8. $-2 \cdot 28 = -56$
   The glacier will have advanced −56 feet.
9. Let y represent the amount of yardage the team was penalized each time.
   $-60 \div 4 = y$
   $-15 = y$
   The Wildcats were penalized 15 yards each time.
10. $-430 \div 5 = -86$
    The average distance per second is −86 ft.
11.

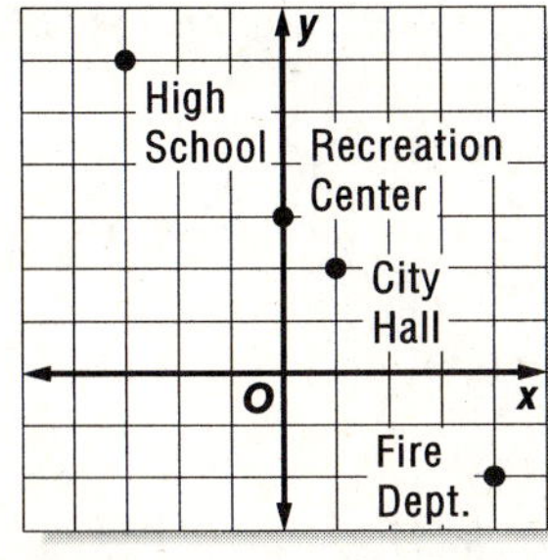

12. $A(2, 2)$, $B(-3, -1)$, $C(3, -3)$
13. $A'(2 + 2, 2) = A'(4, 2)$
    $B'(-3 + 2, -1) = B'(-1, -1)$
    $C'(3 + 2, -3) = C'(5, -3)$

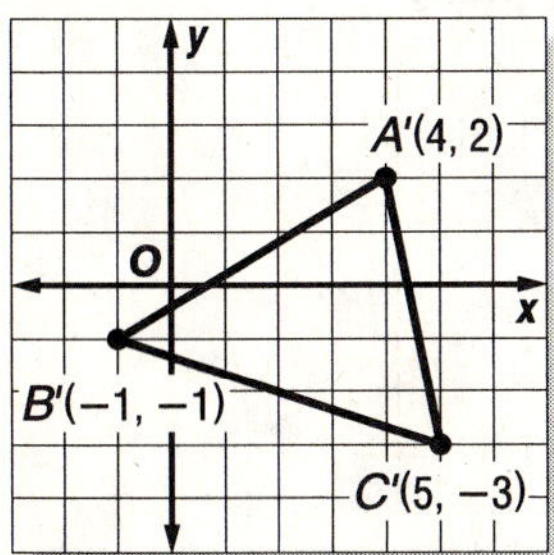

The triangle is moved two units right.

Page 760 Chapter 3 Equations

1. $23 \cdot (0.35) + 20(0.35) = 8.05 + 7.00$
   $= 15.05$
   The total is $15.05.
2. $3(23 + 15)$; $3 \cdot 23 + 3 \cdot 15$
3. $3 \cdot 23 + 3 \cdot 15 = 69 + 45$
   $= 114$
   The total cost is $114.
4. $3 \cdot x + 2 \cdot 10 + 14 = 3x + 20 + 14$
   $= 3x + 34$
   Kyung spent $3x + 34$ dollars.
5. $900 - 100 = 800$
   The maximum weight allowable for passengers is 800 lb.
6. $800 \div 5 = 160$
   The maximum average weight allowable is 160 lb.
7. $P = 4s$
   $72 = 4s$
   $\frac{72}{4} = \frac{4s}{4}$
   $18 = s$
   The length of each side is 18 cm.
8. Let p represent a single payment.
   $7p + 75 = 362$
   $7p + 75 - 75 = 362 - 75$
   $7p = 287$
   $p = 41$
   The amount of each payment is $41.00.
9. Let h represent the number of hours.
   $75 + 10h = 225$
   $75 - 75 + 10h = 225 - 75$
   $10h = 150$
   $\frac{10h}{10} = \frac{150}{10}$
   $h = 15$
   Marcie can play tennis for 15 h.
10.

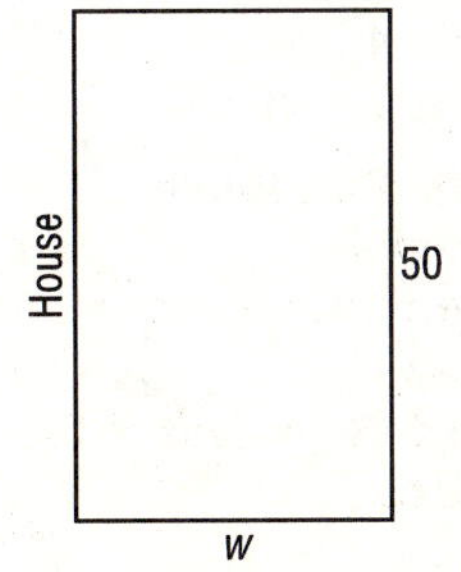

P = amount of fencing
$P = w + 50 + w$
$130 = 2w + 50$

11. $130 = 2w + 50$
    $130 - 50 = 2w + 50 - 50$
    $80 = 2w$
    $\frac{80}{2} = \frac{2w}{2}$
    $40 = w$
    The width of the garden is 40 ft.
12. Let x represent Kate's hourly pay.
    $40x + 50 = 410$

13. $40x + 50 = 410$

$40x + 50 - 50 = 410 - 50$

$40x = 360$

$\frac{40x}{40} = \frac{360}{40}$

$x = 9$

Katie was paid \$9 per hour.

14. Perimeter = sum of all three sides

$27 = x + (x + 2) + (x - 2)$

$27 = x + x + x + 2 - 2$

$27 = 3x + 0$

$27 = 3x$

$\frac{27}{3} = \frac{3x}{3}$

$9 = x$

$x - 2 = 9 - 2$ or 7, and $x + 2 = 9 + 2$ or 11.

The sides are 7 yd, 9 yd, and 11 yd.

15. $d = rt$

$600 = 55t$

$\frac{600}{55} = \frac{55t}{55}$

$10.9 = t$

The Flynn family will be driving for 10.9 h.

16. $a = \frac{f - s}{t}$

17. $a = \frac{f - s}{t}$

$= \frac{14 - 2}{6}$

$= \frac{12}{6}$

$= 2$

The acceleration of the motorcycle is 2 m/s^2.

Page 761 Chapter 4 Factors and Fractions

1.

Seats Per Table	Yes / No	Reason
6	No	136 is not divisible by 2 and 3.
8	Yes	136 is divisible by 8.
10	No	The ones digit of 136 is not 0.

The school should use tables that seat 8 students.

2. 3^x, when $x = 4$: $3^4 = 3 \cdot 3 \cdot 3 \cdot 3$

$= 81$

There are 81 tiles in row 4.

3^x, when $x = 5$: $3^5 = 3 \cdot 3 \cdot 3 \cdot 3 \cdot 3$

$= 243$

There are 243 tiles in row 5.

3^x, when $x = 6$: $3^6 = 3 \cdot 3 \cdot 3 \cdot 3 \cdot 3 \cdot 3$

$= 729$

There are 729 tiles in row 6.

3. $P = I^2R$

$P = (41)^2 \cdot 2$

$P = 41 \cdot 41 \cdot 2$

$P = 1681 \cdot 2$

$P = 3362$

The power lost from the wire powering the stove is 3362 watts.

4. $n = pq$

$1073 = pq$

Test 1073 for divisibility by primes. It is not divisible by 2, 3, 5, 7, 11, 13, 17, 19, or 23. It is divisible by 29.

$1073 \div 29 = 37$

So, $1073 = 29 \cdot 37$.

The values of p and q are 29 and 37.

5. Find the GCF of 48 and 60.

$48 = (2) \cdot (2) \cdot 2 \cdot 2 \cdot (3)$

$60 = (2) \cdot (2) \cdot (3) \cdot 5$

The GCF of 48 and 60 is $2 \cdot 2 \cdot 3$ or 12. So, each side of the pillows should be 12 in.

6. $\frac{\text{housing assistance}}{\text{total amount}} = \frac{6}{100} = \frac{\overset{1}{\cancel{2}} \cdot 3}{\underset{1}{\cancel{2}} \cdot 2 \cdot 5 \cdot 5} = \frac{3}{50}$

7. $\frac{\text{baskets}}{\text{free throw attempts}} = \frac{8}{14} = \frac{\overset{1}{\cancel{2}} \cdot 2 \cdot 2}{\underset{1}{\cancel{2}} \cdot 7} = \frac{4}{7}$

8. Since there are 60 minutes in an hour, there are $24 \cdot 60$ or 1440 minutes in a day.

$\frac{18 \text{ min}}{1 \text{ day}} = \frac{18 \text{ min}}{24 \text{ h}} = \frac{18 \text{ min}}{1440 \text{ min}} = \frac{1}{80}$

18 minutes is $\frac{1}{80}$ of the day.

9. $10^7 \div 10^3 = \frac{10^7}{10^3}$

$= 10^{7-3}$

$= 10^4$ or 10,000

Earthquake A was 10,000 times more intense than Earthquake B.

10. 6.67×10^{-11} Nm^2/kg^2 $= \frac{6.67}{10^{11}}$ Nm^2/kg^2

11. 6.67×10^{-11} Nm^2/kg^2 = 0.0000000000667 Nm^2/Kg^2

12. 10^{-7} meter $= \frac{1}{10^7}$ m

13. 10^{-7} meter $= \frac{1}{10^7} = 0.0000001$ m

14. two millionths $= \frac{2}{1{,}000{,}000} = 0.000002 = 2.0 \times 10^{-6}$

15. $\$4.72 \times 10^8 = 4.72 \times 100{,}000{,}000$

$= \$472{,}000{,}000$

Page 762 Chapter 5 Rational Numbers

1. $16\frac{5}{8}$ ● $16\frac{3}{4}$

16.625 ● 16.75

$16.625 < 16.75$

Yes, it will fit because $16\frac{5}{8} < 16\frac{3}{4}$.

2. $0.0025 = \frac{25}{1000}$

$= \frac{1}{40}$

It is $\frac{1}{40}$ in.

3. $\frac{1}{4} \times 15 = \frac{1}{4} \times \frac{15}{1}$

$= \frac{15}{4}$

$= 3\frac{3}{4}$

You can stay in the Sun for $3\frac{3}{4}$ hours.

4. $1\frac{1}{4} \times 22\frac{1}{2} = \frac{5}{4} \times \frac{45}{2}$
$= \frac{225}{8}$
$= 28\frac{1}{8}$

A coin is in circulation for about $28\frac{1}{8}$ years.

5. $12 \div \frac{2}{3} = \frac{12}{1} \cdot \frac{3}{2}$
$= \frac{\cancel{12}^{6}}{1} \cdot \frac{3}{\cancel{2}_{1}}$
$= 18$

18 guests would finish 12 pizzas.

6. $2 \times \frac{1}{4} = \frac{2}{1} \times \frac{1}{4}$
$= \frac{\cancel{2}^{1}}{1} \times \frac{1}{\cancel{4}_{2}}$
$= \frac{1}{2}$

$2 \times \frac{3}{8} = \frac{2}{1} \times \frac{3}{8}$
$= \frac{\cancel{2}^{1}}{1} \times \frac{3}{\cancel{8}_{4}}$
$= \frac{3}{4}$

$8 - \left(\frac{1}{2} + \frac{3}{4}\right) = 8 - \left(\frac{1}{2} \cdot \frac{2}{2} + \frac{3}{4}\right)$
$= 8 - \left(\frac{2}{4} + \frac{3}{4}\right)$
$= 8 - \frac{5}{4}$
$= \frac{8}{1} \cdot \frac{4}{4} - \frac{5}{4}$
$= \frac{32}{4} - \frac{5}{4}$
$= \frac{27}{4}$

$\frac{27}{4} \div 3 = \frac{27}{4} \cdot \frac{1}{3}$
$= \frac{\cancel{27}^{9}}{4} \cdot \frac{1}{\cancel{3}_{1}}$
$= \frac{9}{4}$ or $2\frac{1}{4}$

The columns should be set $2\frac{1}{4}$ inches wide.

7. $\frac{3}{8} + \frac{5}{8} = \frac{8}{8}$ or 1

The wall coverings are 1 inch thick.

8. $\frac{3}{8}$ ● $\frac{2}{5}$

0.375 ● 0.4

$0.375 < 0.4$

There are more residents from New York because $\frac{3}{8} < \frac{2}{5}$.

9. $\frac{1}{6} - \frac{1}{17} = \frac{1}{6} \cdot \frac{17}{17} - \frac{1}{17} \cdot \frac{6}{6}$
$= \frac{17}{102} - \frac{6}{102}$
$= \frac{11}{102}$

$\frac{11}{102}$ of the population bought bottled water in 2000 but did not in 1993.

10. Year 1

mean: $\frac{99.5 + 88.6 + 91.2 + 102.5 + 93.4}{5} = 95.04$

median: 88.6, 91.2, <u>93.4</u>, 99.5, 102.5
93.4

mode: none

Year 2

mean: $\frac{104.7 + 90.3 + 94.2 + 105.6 + 94.6}{5} = 97.88$

median: 90.3, 94.2, <u>94.6</u>, 104.7, 105.6
94.6

mode: none

11. $1372.2 - 684 = 688.2$

Texas produced 688.2 million barrels.

12. 9.95, 12.90, 15.85, 18.80
(+2.95, +2.95, +2.95)

It is an arithmetic sequence. The common difference is $2.95.

Page 763 Chapter 6 Ratio, Proportion, and Percent

1. $\frac{2.49 \text{ dollars}}{16 \text{ ounces}} \approx \frac{0.16 \text{ dollars}}{1 \text{ ounce}}$
$\approx \$0.16/\text{oz}$

$\frac{3.69 \text{ dollars}}{32 \text{ ounces}} \approx \frac{0.12 \text{ dollars}}{1 \text{ ounce}}$
$\approx \$0.12/\text{oz}$

The 32 oz bag is a better buy.

2. $\frac{4.5}{72} = \frac{x}{48}$
$4.5 \cdot 48 = 72 \cdot x$
$216 = 72x$
$\frac{216}{72} = \frac{72x}{72}$
$3 = x$

3 cups of flour would be needed.

3. $\frac{5 \text{ inches}}{55 \text{ feet}} = \frac{1 \text{ inch}}{11 \text{ feet}}$

The scale is 1 in. = 11 ft.

4. $80\% = \frac{80}{100}$
$= \frac{4}{5}$

$\frac{4}{5}$ of a marshmallow is air.

5. $\frac{3}{1000} = 0.003$
$= 0.3\%$

$\frac{3}{1000}$ is equal to 0.3%.

6. $\frac{760}{b} = \frac{32}{100}$
$760 \cdot 100 = b \cdot 32$
$76{,}000 = 32b$
$\frac{76{,}000}{32} = \frac{32b}{32}$
$2375 = b$

The total recommended daily value is 2375 mg or 2.375 g.

7. 288 is about 300.
560 is about 600.
300 is about $\frac{1}{2}$ of 600.
$\frac{1}{2}$ is 50%.
About 50% of the Calories are from fat.

8. $23.85 is about $24.
15% of $24 is $3.60.
The tip should be about $3.60.

9. $I = prt$
$I = 5500(0.0024)(1)$
$I = 13.20$
A person would pay $13.20 in interest.

10. $n = 2.5(50)$
$n = 125$
$50 + 125 = 175$
The number sold increased by 125, so she should expect to sell 50 + 125 or 175 hedgehogs.

11. 37% of ninth-graders consider brand names when buying jeans.
$\frac{a}{423} = \frac{31}{100}$
$a \cdot 100 = 423 \cdot 37$
$100a = 15{,}651$
$\frac{100a}{100} = \frac{15{,}651}{100}$
$a \approx 157$
You can expect 157 students.

12. There are 23 possible outcomes of red.
$P(\text{red}) = \frac{23}{56}$
The probability of selecting red is $\frac{23}{56}$.

Page 764 Chapter 7 Equations and Inequalities

1. $293{,}000 - 3500x$
2. $277{,}000 + 2000x$
3. $$293{,}000 - 3500x = 277{,}000 + 2000x$$
$$293{,}000 - 3500x + 3500x = 277{,}000 + 2000x + 3500x$$
$$293{,}000 = 277{,}000 + 5500x$$
$$293{,}000 - 277{,}000 = 277{,}000 - 277{,}000 + 5500x$$
$$16{,}000 = 5500x$$
$$\frac{16{,}000}{5500} = \frac{5500x}{5500}$$
$$2.\overline{90} = x$$
It will take about 3 years for the populations to be the same.

4. Let m represent the number of minutes.
Cost of first provider: $19.95 + 0.21m$
Cost of second provider: $24.95 + 0.16m$
$$19.95 + 0.21m = 24.95 + 0.16m$$
$$19.95 + 0.21m - 0.16m = 24.95 + 0.16m - 0.16m$$
$$19.95 + 0.05m = 24.95$$
$$19.95 - 19.95 + 0.05m = 24.95 - 19.95$$
$$0.05m = 5$$
$$\frac{0.05m}{0.05} = \frac{5}{0.05}$$
$$m = 100$$
The cost of the two Internet providers is the same when 100 minutes are used.

5. Let w represent the width.
Length $= 3(w - 2)$
$15 = 3(w - 2)$
$15 = 3w - 6$
$15 + 6 = 3w - 6 + 6$
$21 = 3w$
$\frac{21}{3} = \frac{3w}{3}$
$7 = w$
The width is 7 in.

6. Let f represent the number of fans.
$f > 100{,}000$

7. Let e represent the number of minutes she can spend on English homework.
$e + 35 \le 90$
$e + 35 - 35 \le 90 - 35$
$e \le 55$
She can spend *at most* 55 minutes on English.

8. Let x represent the amount he must still save.
$x + 285 \ge 375$
$x + 285 - 285 \ge 375 - 285$
$x \ge 90$
He must save *at least* \$90.

9. Let x represent the average number of spectators per mile.
$26x > 2{,}600{,}000$
$\frac{26x}{26} > \frac{2{,}600{,}000}{26}$
$x > 100{,}000$
On average there were *more than* 100,000 spectators per mile.

10. Let x represent her current spending.
$x \ge 2(54)$
$x \ge 108$
She now spends *at least* \$108.

11. $2x < 90$
$\frac{2x}{2} < \frac{90}{2}$
$x < 45$
Possible values of x are *less than* 45°.

12. Let x represent the amount he can spend on jeans.
$2(15.30) + x \le 85$
$30.60 + x \le 85$
$30.60 - 30.60 + x \le 85 - 30.60$
$x \le 54.40$
He can spend *at most* \$54.40 on jeans.

13. Let s represent the monthly sales.
$1000 + 0.03s > 2500$
$1000 - 1000 + 0.03s > 2500 - 1000$
$0.03s > 1500$
$\frac{0.03s}{0.03} > \frac{1500}{0.03}$
$s > 50{,}000$
The car salesperson must make *more than* \$50,000 in monthly sales.

14. Let x represent his minimum desired score on the fifth test.
$$\frac{73 + 85 + 91 + 82 + x}{5} \ge 82$$
$$\frac{331 + x}{5} \ge 82$$
$$5\left(\frac{331 + x}{5}\right) \ge 5(82)$$
$$331 + x \ge 410$$
$$331 - 331 + x \ge 410 - 331$$
$$x \ge 79$$
He must earn a score of *at least* 79.

Page 765 Chapter 8 Functions and Graphing

1. \$75 belongs to "\$70.01 and over," so the shipping cost is \$6.95.

2. \$30.01 – \$70.00

3. Yes, each price has a unique shipping cost.

4. Choose three values for x and substitute them into $y = 0.21x$.

   Sample answer:

x	$y = 0.21x$	y	(x, y)
1	$y = 0.21(1) = 0.21$	0.21	(1, 0.21)
2	$y = 0.21(2) = 0.42$	0.42	(2, 0.42)
3	$y = 0.21(1) = 0.63$	0.63	(3, 0.63)

5. $y = 0.21x$
   $y = 0.21(2.5)$
   $y = 0.525$
   The lightning is about 0.525 miles away.

6. To find the x-intercept, let $a = 0$.
   $a = 24{,}000 - 1500t$
   $0 = 24{,}000 - 1500t$
   $0 - 24{,}000 = 24{,}000 - 24{,}000 - 1500t$
   $\frac{-24{,}000}{-1500} = \frac{-1500t}{-1500}$
   $16 = t$
   The x-intercept is 16.

7. It represents the time when the altitude is zero and the jet has landed.

8. slope $= \frac{\text{vertical change}}{\text{horizontal change}}$
   slope $= \frac{70}{20}$
   slope $= \frac{7}{2}$

9. Let y represent the cost of gasoline, and let x represent the number of gallons bought.
   $y = kx$
   $27.80 = k(20)$
   $\frac{27.80}{20} = \frac{k(20)}{20}$
   $1.39 = k$
   $y = 1.39x$
   $y = 1.39(12)$
   $y = 16.68$
   It would cost \$16.68.

10. Write the equation in the form $y = mx + b$.
    $y = 1500 + 12x$
    $y = 12x + 1500$
    slope = 12; y-intercept = 1500;
    12 is the cost per item, and 1500 is the cost when no items are produced.

11. $y = 59 + 0.12x$

12. $y = 59 + 0.12x$
    $y = 59 + 0.12(30)$
    $y = 59 + 3.6$
    $y = 62.6$
    The daily rental cost is \$62.60.

13. Draw a line that best fits the data. Extend the line so that you can find the y value for the x value corresponding to October 2002.
    Sample answer: \$375 million

14. Let x = the number of student tickets, and let y = the number of nonstudent tickets.
    Cost: $3x + 5y = 590$
    Tickets: $x + y = 140$
    The graph of the system shows the solution of (55, 85).
    55 student tickets and 85 nonstudent tickets were sold.

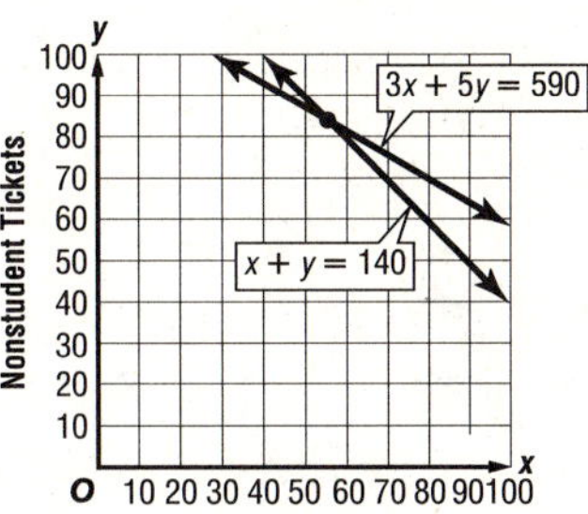

15. $3.5x + 5y \geq 35$

16.

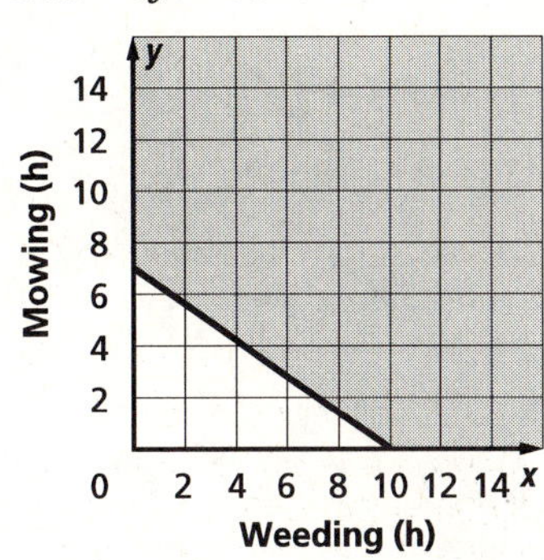

17. Sample answer: weed 4 hours, mow 6 hours; weed 8 hours, mow 2 hours

Page 766 Chapter 9 Real Numbers and Right Triangles

1. $4\text{ ft}^2 \times 100$ people $= 400\text{ ft}^2$
   Area of a square: $A = s^2$
   $400 = s^2$
   $\sqrt{400} = \sqrt{s^2}$
   $20 = s$
   Each side should be 20 ft long.

2. $d = 0.5gt^2$
   $55 = 0.5(32)t^2$
   $55 = 16t^2$
   $\frac{55}{16} = \frac{16t^2}{16}$
   $3.4375 = t^2$
   $\sqrt{3.4375} = \sqrt{t^2}$
   $1.85 \approx t$
   It takes about 1.85 seconds to hit the ground.

3. The angle is 10% of the circle.
   10% of 360 = $0.1 \times 360 = 36°$.
   36° is less than 90°, so the angle is acute.

4. The angle is 59% of the entire circle.
   59% of 360 = $0.59 \times 360 = 212.4$
   The angle labeled Fiction is 212.4°.

5. 36° west of north

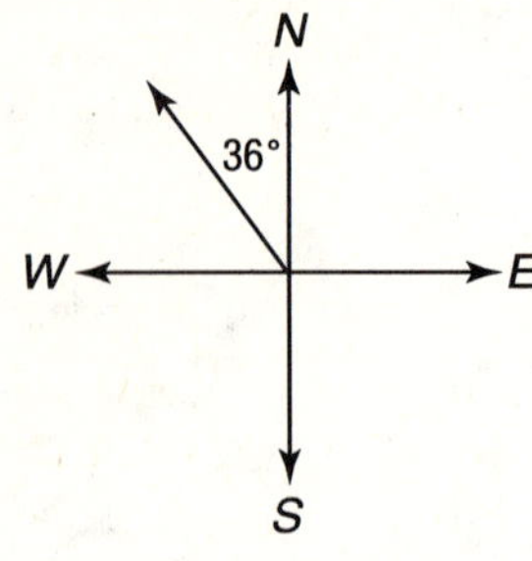

6. $x° + 65° + 90° = 180°$

$x + 155 = 180$

$x + 155 - 155 = 180 - 155$

$x = 25$

The value of x is 25°.

7. $c^2 = a^2 + b^2$

$c^2 = 90^2 + 90^2$

$c^2 = 8100 + 8100$

$c^2 = 16{,}200$

$\sqrt{c^2} = \sqrt{16{,}200}$

$c \approx 127.3$

The distance between home plate and second base is about 127.3 ft.

8. $c^2 = a^2 + b^2$

$24^2 = 6^2 + b^2$

$576 = 36 + b^2$

$576 - 36 = 36 - 36 + b^2$

$540 = b^2$

$\sqrt{540} = \sqrt{b^2}$

$23.2 \approx b$

The mast is about 23.2 feet high.

9. midpoint $= \left(\frac{x_1 + x_2}{2}, \frac{y_1 + y_2}{2}\right)$

$= \left(\frac{-4 + 6}{2}, \frac{9 + 3}{2}\right)$

$= (1, 6)$

The coordinates on the map where they should plan to meet are (1, 6).

10. $\frac{1.5}{3} = \frac{241}{x}$ (1 yd = 3 ft)

$1.5 \cdot x = 241 \cdot 3$

$1.5x = 723$

$\frac{1.5x}{1.5} = \frac{723}{1.5}$

$x = 482$

The height of the pyramid is 482 feet.

11. The triangles are similar.

$\frac{16}{15} = \frac{x}{30}$

$30 \cdot 16 = 15 \cdot x$

$480 = 15x$

$\frac{480}{15} = \frac{15x}{15}$

$32 = x$

The river is 32 m wide.

12. $\sin 45° = \frac{\text{opposite}}{\text{hypotenuse}}$

$\sin 45° = \frac{x}{75}$

$75(\sin 45) = 75 \cdot \frac{x}{75}$

$53 \approx x$

The kite is about 53 yards above the ground.

13. $\sin 60° = \frac{\text{opposite}}{\text{hypotenuse}}$

$\sin 60° = \frac{x}{15}$

$15(\sin 60°) = 15 \cdot \frac{x}{15}$

$13 \approx x$

The ladder reaches about 13 feet up the side of the house.

Page 767 Chapter 10 Two-Dimensional Figures

1. No; the corresponding angle has a measure of only 55°.

2. Sample answer: $\triangle CFB$

3. $\triangle ADE$

4. This translation can be written as (3, 2).

vertex		3 right, 2 up		translation
$A(-3, 5)$	+	(3, 2)	→	$A'(0, 7)$
$B(4, 5)$	+	(3, 2)	→	$B'(7, 7)$
$C(4, -3)$	+	(3, 2)	→	$C'(7, -1)$
$D(-3, -3)$	+	(3, 2)	→	$D'(0, -1)$

The coordinates are $A'(0, 7)$, $B'(7, 7)$, $C'(7, -1)$, and $D'(0, -1)$.

5. See students' work.

6. $A = \frac{1}{2}h(a + b)$

$A = \frac{1}{2} \cdot 140(280 + 200)$

$A = \frac{1}{2} \cdot 140(480)$

$A = 33{,}600$

The area of Indiana is about 33,600 mi^2.

7. octagon

8. square

9. trapezoid

10. The trays must fit together around a square table, so the angles must add to 90°.

The two angles at the bottom of the tray must be 90 ÷ 2 or 45°.

Since the angle sum of a quadrilateral is 360°, the top two angles are $\frac{360 - 45 - 45}{2}$ or 135°.

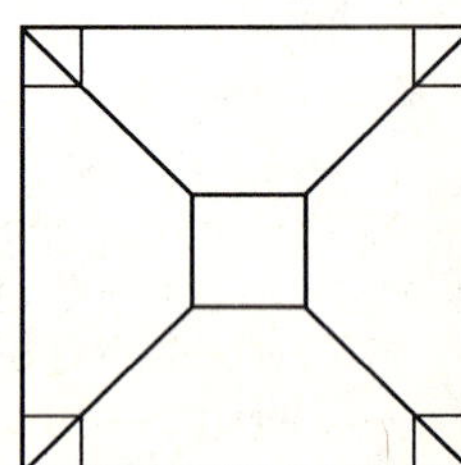

11. $A = \pi r^2$

$A = \pi \cdot 2^2$

$A = \pi \cdot 4$

$A \approx 13$

The area is about 13 mi^2.

12. $d = 250$, so $r = 125$.

$A = \pi r^2$

$A = \pi \cdot 125^2$

$A = \pi \cdot 15{,}625$

$A \approx 49{,}087.4$

The area is about 49,087.4 ft^2.

13. Area of rectangle:
$A = bh$
$A = 8 \cdot 5$ or 40
Area of semicircle:
$A = \frac{1}{2}\pi r^2$
$A = \frac{1}{2} \cdot \pi \cdot 3^2$
$A = \frac{1}{2} \cdot \pi \cdot 9$
$A \approx 14.14$
The area of the figure is 40 + 14.14 or about 54.14 ft^2.

Page 768 Chapter 11 Three-Dimensional Figures

1. A rectangular pyramid has 5 faces, 8 edges, and 5 vertices.
2. $r = \frac{1}{2}(2)$
$r = 1$
$V = \pi r^2 h$
$V = \pi(1)^2(6)$
$V \approx 19$
The volume is about 19 cubic feet.
3. $V = Bh$
$4.64 = 1.41h$
$3.29 \approx h$
The height is about 3.29 centimeters.
4. $r = \frac{1}{2}(21)$
$r = 10.5$

$V = \ell \cdot w \cdot h$	$V = \pi r^2 h$
$V = 30 \cdot 21 \cdot 5$	$V = \pi(10.5)^2 \cdot 4$
$V = 3150$ cm^3	$V \approx 1385.4$ cm^3

One rectangular pan holds 3150 cm^3, and 2 round pans hold about 2(1385.4) or 2770.8 cm^3.
The rectangular pan holds more batter.
5. $V = \frac{1}{3}Bh$
$V = \frac{1}{3}(34)^2(54)$
$V = 20{,}808$
The volume is 20,808 cubic feet.
6. The lateral area of one can is $A = 2\pi rh$.
The lateral area of 24 cans is
$24A = 24(2\pi rh)$
$= 24(\pi 7.6 \cdot 10.8)$
$= 1969.92\pi$
≈ 6188.7 cm^2
7. left side: $4.5(6) = 27$
right side: $4.5(6) = 27$
front side: $\frac{1}{2} \cdot 4(4) = 8$
back side: $\frac{1}{2} \cdot 4(4) = 8$
floor: $4(6) = 24$
$s = 27 + 27 + 8 + 8 + 24$
$= 94$
The surface area is 94 square feet.
8. Area of lateral face:
$A = \frac{1}{2}bh$
$A = \frac{1}{2} \cdot 30 \cdot 39.9$
$A = 598.5$
The lateral area is 4(598.5) or 2394.0 square meters.
9. $r = \frac{1}{2}(42)$
$r = 21$
$A = \pi r \ell$
$A = \pi(21)(47.9)$
$A = 3160.1$
The lateral area is about 3160.1 square feet.
10. $\frac{9}{4} \stackrel{?}{=} \frac{22.5}{10}$
$9(10) \stackrel{?}{=} 4(22.5)$
$90 = 90$ ✓
The measurements are proportional, so the tubes are similar.
11. $\frac{(1)^3}{(4)^3} = \frac{1}{64}$
$\frac{1}{64} = \frac{16}{x}$
$x = 1024$
The volume is 1024 cubic feet.
12.
$$\begin{array}{r} 8.25 \\ \times\ 23.7 \\ \hline 195.525 \end{array}$$
The least precise measurement, 23.7 ft, has one decimal place. So, round 195.525 to one decimal place, 195.5 ft^2.

Page 769 Chapter 12 More Statistics and Probability

1.

Stem	Leaf
2	3 5 8 8
3	1 2 3 3 6 9
4	2 5 7
5	1 3

4|2 = 42

2. **Participants**
The greatest value is 72 and the least value is 12. So, the range is 72 − 12 or 60.
To find the interquartile range, first list the data from least to greatest. Then find the median and the upper and lower quartiles.

lower half — upper half

12, 13, 13, 14, 18, 20, 23, 25, 35, 43, 56, 65, 67, 68, 72
(LQ = 14, median = 25, UQ = 65)

The interquartile range is 65 − 14 or 51.
Observers
The greatest value is 80 and the least value is 9. So, the range is 80 − 9 or 71.
To find the interquartile range, first list the data from least to greatest. Then find the median and the upper and lower quartiles.

lower half

9, 15, 16, 18, 23, 25, 39, 41, 42, 43, 43

↑ LQ

upper half

50, 51, 54, 55, 55, 60, 63, 70, 75, 75, 80

↑ UQ

Since there is an even number of observers, median $= \frac{43 + 50}{2} = 36.5$. The interquartile range is $60 - 25$ or 35.

3. The median is 1.12. The extremes are 0.92 and 1.21. The lower quartile is 0.95, and the upper quartile is 1.16.

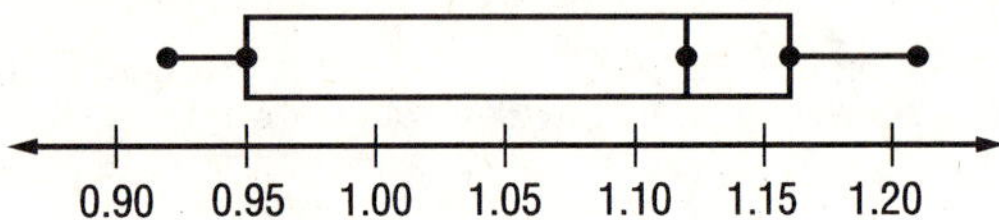

4.

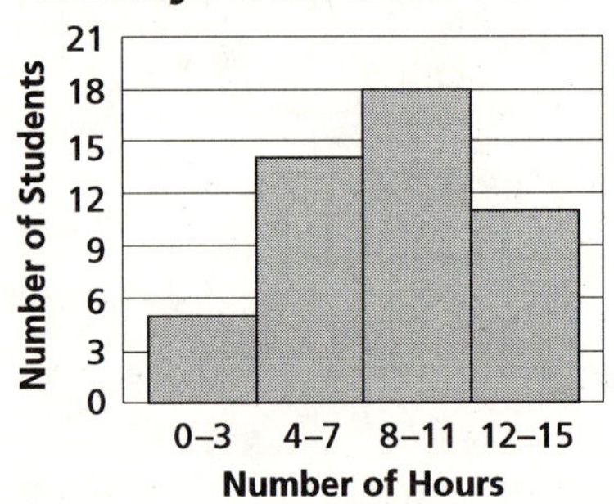

5. There are no title or labels on either scale, and the vertical axis does not include zero.

6. Use the Fundamental Counting Principle. Let n represent the number of choices of toppings.

choices of frozen yogurt		choices of serving size		choices of toppings		total possible sundaes
6	×	6	×	n	=	1512

$$36n = 1512$$
$$\frac{36n}{36} = \frac{1512}{36}$$
$$n = 42$$

They offer 42 different toppings.

7. Since order is not important, this is a combination of 15 players taken 6 at a time.

$C(15, 6) = \frac{P(15, 6)}{6!} = \frac{15 \cdot 14 \cdot 13 \cdot 12 \cdot 11 \cdot 10}{6 \cdot 5 \cdot 4 \cdot 3 \cdot 2 \cdot 1} = 5005$

There are 5005 squads possible.

8. The odds in favor of a student appearing on television are 1:3. So, for every 4 students, 1 student will appear on television and 3 students will not appear on television. The probability of appearing on television is $\frac{1}{4}$ or 25%.

So, $\frac{1}{4}$ of 32 students $= \frac{1}{4} \cdot 32 = 8$ students will appear on television.

9. The events are independent.

$P(\text{tune up and air filter}) = P(\text{tune up}) \cdot P(\text{air filter})$
$= (0.70) \cdot (0.50)$
$= 0.35$ or 35%

10. The events are mutually exclusive.

$P(\text{16 to 24 years old}) = P(\text{16 to 19}) + P(\text{20 to 24})$
$= (0.31) + (0.22)$
$= 0.53$ or 53%

Page 770 Chapter 13 Polynomials and Nonlinear Functions

1.

term	degree
$2xy$	2
$2y^2$	2
$2yz$	2

The greatest degree is 2, so the degree of $2xy + 2y^2 + 2yz$ is 2.

2. $A = \ell \cdot w$
$A = x(4y)$
$A = 4xy$
The expression $4xy$ has one term, so it is a monomial.

3. united inches = length + width
$= (3x - 5) + (x + 7)$
$= 3x + x - 5 + 7$
$= 4x + 2$
The size of the window is $4x + 2$ united inches.

4. Perimeter =
sum of lengths of legs + length of hypotenuse.
So, length of hypotenuse =
perimeter − sum of lengths of legs.
$= 4x + 4 - [(2x - 4) + (x + 3)]$
$= 4x + 4 - [(2x + x) + (-4 + 3)]$
$= 4x + 4 - (3x - 1)$
$= 4x + 4 + (-3x + 1)$
$= (4x - 3x) + (4 + 1)$
$= x + 5$
The length of the hypotenuse is $x + 5$ cm.

5. Area of large outer rectangle $= (2s + s) \cdot s$
$= 3s \cdot s$ or $3s^2$
Area of small unshaded rectangle $= s \cdot 3$ or $3s$
Area of shaded region
= area of large outer rectangle
− area of small unshaded rectangle.
$= 3s^2 - 3s$
The shaded area is $3s^2 - 3s$ units^2.

6. Area of outer rectangle:
$A = \ell \cdot w$
$A = (2x + 1) \cdot x$
$A = 2x^2 + x$
Area of inner rectangle:
$A = \ell \cdot w$
$A = (2x + 1) \cdot (2x + 1)$
$A = 4x^2 + 4x + 1$

Total surface area:

$A = 4(\text{area of outer rectangle}) +$
$(\text{area of inner rectangle})$
$= 4(2x^2 + x) + (4x^2 + 4x + 1)$
$= 8x^2 + 4x + 4x^2 + 4x + 1$
$= 12x^2 + 8x + 1$

The amount of cardboard needed is $12x^2 + 8x + 1$ units^2.

7. $12x^2 + 8x + 1 = 12(2.5)^2 + 8(2.5) + 1$
$= 12(6.25) + 20 + 1$
$= 75 + 21$
$= 96$

The surface area is 96 in^2.

8. linear; equation can be written as $y = 3.2x + 2500$

9. Nonlinear; the graph is a curve.

10.

x	$y = 2x^3$	(x, y)
−2	$2(-2)^3 = -16$	(−2,−16)
−1	$2(-1)^3 = -2$	(−1,−2)
0	$2(0)^3 = 0$	(0, 0)
1	$2(1)^3 = 2$	(1, 2)
2	$2(2)^3 = 16$	(2, 16)

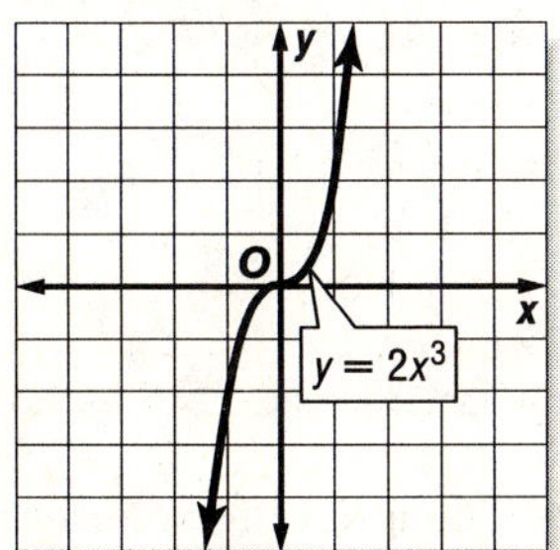